T0254801

CAMBRIDGE LIBRARY COLLECTION

Books of enduring scholarly value

Botany and Horticulture

Until the nineteenth century, the investigation of natural phenomena, plants and animals was considered either the preserve of elite scholars or a pastime for the leisured upper classes. As increasing academic rigour and systematisation was brought to the study of 'natural history', its subdisciplines were adopted into university curricula, and learned societies (such as the Royal Horticultural Society, founded in 1804) were established to support research in these areas. A related development was strong enthusiasm for exotic garden plants, which resulted in plant collecting expeditions to every corner of the globe, sometimes with tragic consequences. This series includes accounts of some of those expeditions, detailed reference works on the flora of different regions, and practical advice for amateur and professional gardeners.

A Dictionary of the Economic Products of India

A Scottish doctor and botanist, George Watt (1851–1930) had studied the flora of India for more than a decade before he took on the task of compiling this monumental work. Assisted by numerous contributors, he set about organising vast amounts of information on India's commercial plants and produce, including scientific and vernacular names, properties, domestic and medical uses, trade statistics, and published sources. Watt hoped that the dictionary, 'though not a strictly scientific publication', would be found 'sufficiently accurate in its scientific details for all practical and commercial purposes'. First published in six volumes between 1889 and 1893, with an index volume completed in 1896, the whole work is now reissued in nine separate parts. Volume 5 (1891) contains entries from *Linum* (the flax genus) to *oyster* (the subcontinent's best oyster beds were to be found 'on the coast near Karachi, Bombay and Madras').

Cambridge University Press has long been a pioneer in the reissuing of out-of-print titles from its own backlist, producing digital reprints of books that are still sought after by scholars and students but could not be reprinted economically using traditional technology. The Cambridge Library Collection extends this activity to a wider range of books which are still of importance to researchers and professionals, either for the source material they contain, or as landmarks in the history of their academic discipline.

Drawing from the world-renowned collections in the Cambridge University Library and other partner libraries, and guided by the advice of experts in each subject area, Cambridge University Press is using state-of-the-art scanning machines in its own Printing House to capture the content of each book selected for inclusion. The files are processed to give a consistently clear, crisp image, and the books finished to the high quality standard for which the Press is recognised around the world. The latest print-on-demand technology ensures that the books will remain available indefinitely, and that orders for single or multiple copies can quickly be supplied.

The Cambridge Library Collection brings back to life books of enduring scholarly value (including out-of-copyright works originally issued by other publishers) across a wide range of disciplines in the humanities and social sciences and in science and technology.

A Dictionary of the Economic Products of India

Volume 5: Linum to Oyster

George Watt

CAMBRIDGE
UNIVERSITY PRESS

University Printing House, Cambridge, CB2 8BS, United Kingdom

Published in the United States of America by Cambridge University Press, New York

Cambridge University Press is part of the University of Cambridge.
It furthers the University's mission by disseminating knowledge in the pursuit of education, learning and research at the highest international levels of excellence.

www.cambridge.org
Information on this title: www.cambridge.org/9781108068772

This edition first published 1891
This digitally printed version 2014

ISBN 978-1-108-06877-2 Paperback

ERRATUM.—VOLUMES III-V.

Attention has been drawn to the fact that an error runs through Volumes III, IV, and V of the Dictionary, wherein the consecutive numbers of letters G and L, in passing from Volume III-IV and from Volume IV-V respectively, have been partially duplicated. The numbers in those volumes, which had not been issued from the Press when the error was pointed out, have been corrected, and will, as corrected, form the future reference numbers. It is suggested, that those who possess uncorrected volumes should adopt a similar course, making the first reference number in Volume IV G 381 and in Volume V L 379.

EDGAR THURSTON,
Offg. Reporter on Economic Products
to the Government of India.

DICTIONARY

OF

THE ECONOMIC PRODUCTS OF INDIA.

BY

GEORGE WATT, M.B., C.M., C.I.E.,

REPORTER ON ECONOMIC PRODUCTS WITH THE GOVERNMENT OF INDIA.
OFFICIER D'ACADEMIE; FELLOW OF THE LINNEAN SOCIETY; CORRESPONDING MEMBER OF THE ROYAL HORTICULTURAL SOCIETY, &c., &c.

(ASSISTED BY NUMEROUS CONTRIBUTORS.)
IN SIX VOLUMES.

VOLUME V.,

Linum to Oyster.

Published under the Authority of the Government of India,
Department of Revenue and Agriculture.

LONDON:
W. H. ALLEN & Co., 13, Waterloo Place, S.W., Publishers to the India Office.

CALCUTTA:
Office of the Superintendent of Government Printing, India.
8, Hastings Street.

1891.

CALCUTTA:
GOVERNMENT OF INDIA CENTRAL PRINTING OFFICE,
8, HASTINGS STREET.

PREFACE TO VOL. V.

SINCE the publication of the Preface to the third Volume of this work very little has transpired which seems to call for any prefatory explanations. During the compilation of the present volume, however, the Government of India were enabled to procure the services of Dr. W. R. Clark, of the Indian Medical Service, as a third collaborateur. The articles written by Dr. Clark will be found distributed through the present as also in the subsequent volumes, and the same rule as pursued in the case of Mr. Duthie and Dr. Murray will be observed to have been followed with Dr. Clark, namely, to give the contributor's name in the top right hand corner of each page. The Editor thinks it would be quite superfluous in him to attempt to eulogise the high merits of the gentlemen who have been associated with him. Their labours will appeal more forcibly to the reader than would any words of commendation. It need only be added, therefore, that the continued courteous and considerate toleration of his collaborateurs, in all matters of detail, has reduced the Editor's supervision to one of mutual consultation.

GEORGE WATT,
Editor, Dictionary of the Economic Products of India.

SIMLA,
June 1891.

DICTIONARY

OF

THE ECONOMIC PRODUCTS OF INDIA.

LINUM, *Linn.; Gen. Pl., I., 242.* 1 (379)

Linum mysorense, *Heyne; Fl. Br. Ind., I., 411;* LINEÆ. 2 (380)

Syn.—L. HUMILE, *Heyne.*

References.—*Atkinson, Him. Dist., 306; Dalz. & Gibs., Bomb. Fl., 16; Gazetteers:—Mysore and Coorg, I., 56; Bombay (Kanara District), XV., 428.*

Habitat.—An annual herb with a corymbosely branched, slender, glabrous stem: found on the exposed hills of the Western Peninsula from the Konkan and the Deccan to the Nilghiri hills. Also on the Western Himálaya at Garhwál between 3,000 and 5,000 feet, distributed to Ceylon at altitudes of 4,000 to 5,000 feet.

L. strictum, *Linn.; Fl. Br. Ind., 411.* 3 (381)

Vern.—*Basant, bab-basant,* PB.

Habitat.—A small, herbaceous plant with yellow flowers, found on the Panjáb hills and in Tibet (at an altitude of 10,000 feet); cultivated in Afghánistán.

Oil.—Griffith says it is grown in Afghánistán on account of its oil-yielding SEED, not for flax. The oil very probably does not differ essentially from ordinary linseed oil. OIL. Seed. 4 (382)

L. perenne, *Linn.; Fl. Br. Ind., I., 411.* 5 (383)

Reference.—*Stewart, Pb. Pl., 20.*

Habitat.—A perennial herb, branching from the ground and growing to a height of from 1 to 3 feet; found in Western Tibet and Lahoul at altitudes of 9,000 to 13,000 feet.

Oil.—**Stewart** suggests that this may be the plant which afforded the wild linseed of Spiti which many years ago was sent to the Agri. Horticultural Society of India. **Aitchison**, in his *Lahoul, its Flora and Vegetable Products,* mentions this plant, but states that the people are ignorant of its fibre, and, when speaking of oils, he says none are used, except the oil prepared from species of **Impatiens**, and that even that oil is only employed to polish drinking cups. No plants, **Aitchison** adds, are cultivated in Lahoul as sources of oil. Ghí, or clarified butter, is the only substance used for lighting purposes. Thus these wild hill people appear to be alike ignorant of the oil and fibre of their indigenous **Linum**. OIL. Seed. 6 (384)

L. trigynum, *Roxb.;* see **Reinwardtia trigyna,** *Planch.;* Vol. VI.

(385) 7 **Linum usitatissimum,** *Linn.; Fl. Br. Ind., I., 410.*

LINUM, *Latin;* LIN, *Fr.;* LINO, *It. & Spanish;* LINHO, *Port.;* FLÜCHS, *Germ.;* VLAS, *Dutch.*

Syn.—L. TRINERVIUM, *Roth.*

Vern.—*Alsi, tísí,* HIND.; *Tísí, masiná* (=smooth), BENG.; *Tísí, chikna,* BEHAR; *Pesu,* URIYA; *Bijri* (Banda), N.-W. P.; *Tísí, alsí,* KUMAON; *Keun, alish,* KASHMIR; *Alísh, tísí, alsí,* PB.; *Zighir,* KASHGHAR; *Alási, javasa, javas,* BOMB.; *Alshi,* GUZ.; *Alshi, javas,* DEC.; *Alshi-virai,* TAM.; *Atasi, ullú súlú, madan-ginjalu,* TEL.; *Alshi, alashi,* KAN; *Cheru-chána-vittinte-vilta,* MALAY.; *Atasi, auma, malika, san, masrind* or *masína, masúna* (*kshauma,* according to Mason's *Burma*), *uma* (according to Sir W. Jones, SANS.; *Ziggar,* TURKI; *Kattán* (or *bazrut-kattán*), ARAB.; *Zaghú, zaghir, kutan* (or *tukhme-kaián*), PERS.

NOTE.—It has been found impossible to separate the names given to the plant, the fibre, the seed, or the oil: they are, therefore, all included together in the above enumeration.

References.—*Roxb., Fl. Ind., Ed. C.B.C., 277; Dalz. & Gibs., Bomb. Fl., Supp., 16; Stewart, Pb. Pl., 21; DC. Origin Cult. Pl., 119; Elliot, Flora Andhr., 17, 123; Mason, Burma and Its People, 517; Sir W. Jones, Treat. Pl. Ind., 107; Planchon, Hooker's Jour. of Botany, Vol. VII., 165, published 1848; Grierson, Peasant Life, 246; Carey, Asiatic Researches, X., 15; Pharm. Ind., 37; Dymock, Warden, & Hooper, Pharmacog. Ind., I., 239; Ainslie, Mat. Ind., I., 196, 612; O'Shaughnessy, Beng. Dispens., 213; U. C. Dutt, Mat. Med. Hind., 292; Dymock, Mat. Med. W. Ind., 2nd Ed., 116; Flück & Hanb. Pharm.; Fleming, Med. Pl. and Drugs, as in As. Res., Vol. XI., 170; Bent. & Trim., Med. Pl., 39; S. Arjun, Bomb. Drugs, 27; Murray, Pl. and Drugs, Sind, 93; Irvine, Mat. Med., Patna, 116; Butler, Med. Top. Oudh, 32; Rankine, Med. Top. of Sarem, 15; Year Book, Pharm., 1874, 268, 623; Moodeen Sheriff, Mat. Med. S. Ind. (proof seen by the writer), 69; Baden Powell, Pb. Pr., 420, 496-500; Atkinson, Him. Dist., 306, 740-771; Drury, U. Pl., 278; Duthie & Fuller, Field and Garden Crops, 40-42; Lisboa, U. Pl. Bomb., 215, 231; Birdwood, Bomb. Pr., 16, 280, 316; Royle, Ill. Him. Bot., 82; Christy. Com. Pl. and Drugs, VI., 10; Royle, Fib. Pl., 135-232; Liotard, Paper-making Mat., 15, 17, 24; Spons, Encyclop., 964-978, 1393-4, 2024; Balfour, Cyclop., I., 1127; II. 719; Smith, Dic., 178, 247; Ure, Dic., Indus., Arts, and Manu., II., 326, 375, also 876-879; Kew Off. Guide to the Mus. of Ec. Bot., 22; Simmonds, Trop. Agri., 399; Linschoten, Voyage to East Indies, I. 80; Gazetteers:—Mysore & Coorg, I., 58; Bombay, IV., 53; XII., 152; XVII., 270; C. P., 327; N.-W. P., I., 79; II., 159; III., 225; IV., lxix; Panjab, Hoshiarpur, 91; Gujrát, 78, 79; Gujranwalla, 53; Sialkot, 76; Hunter's Orissa, II., App., 15; Indian Forester, IX., 274; Settle. Repts.:—Chánda, 81, 96; Gujrat, vxxix.; Banda, 50; Azamgarh, 123; Bareilly, 82; Lahore, 9; Baitool, 77; Kángra, 24; Allahabad, 31; Hoshungabad, 276, 277, 288; Nagpur, 273; Nursingpore, 52; Nimár, 198; Kumaon, App., 34; Wardhá, 68; Madras, Man. of Admn., I., 288; Administration Repts.:—Bengal, 1882-83, 17; Quarterly Journals of Agriculture, V., 467; VI., 101, 449; X., 37; XI., 314.*

Habitat.—An annual herbaceous plant cultivated throughout the plains of India and up to altitudes of 6,000 feet above the sea.

HISTORY. (386) 8

HISTORY.

The history of both FLAX and LINSEED may be treated conjointly, since, to the India of the present day, linseed alone is of commercial importance. **Flückiger & Hanbury** give a brief but interesting sketch of this subject. "The history of flax, its textile fibre and seed," these authors write, "is intimately connected with that of human civilization. The whole process of converting the plant into fibre, fit for weaving into cloth, is frequently depicted on the wall paintings of the Egyptian tombs. The grave-clothes of the old Egyptians were made of flax, and the uses of the fibre in Egypt may be traced back, according to **Unger**, as far as the twenty-third

HISTORY.

century B.C. The old literature of the Hebrews and Greeks contains frequent reference to tissues of flax; and fabrics woven of flax have actually been discovered, together with fruits and seeds of the plant, in the remains of the ancient pile-dwellings bordering the lakes of Switzerland."

"The seed in ancient times played an important part in the alimentation of man. Among the Greeks, **Alcman** in the seventh century B.C., and the historian **Thucydides**, and among the Romans, **Pliny**, mention linseed as employed for human food. The roasted seed is still eaten by the Abyssinians." (Conf. with remarks below regarding its being eaten in India.)

"**Theophrastus** expressly alludes to the mucilaginous and oily properties of the seed. **Pliny** and **Dioscorides** were acquainted with its medicinal application, both external and internal."

"The propagation of flax in Northern Europe as of so many other useful plants, was promoted by **Charlemagne.** It seems to have reached Sweden and Norway before the 12th century."

The above passages have been taken from the *Pharmacographia*, a learned work which appeared subsequent to **DeCandolle's** historic sketch of **Linum** (*Geogr. Bot. Rais., 833*). These two works have brought together, in fact, all that can be said regarding the history of flax and linseed. It may not, therefore, be out of place to give here some of **DeCandolle's** more recent observations (*Origin. Cult. Pl.*): "The first important work on this subject," he writes, "was by **Planchon** in 1848. That botanist clearly showed the difference between **Linum usitatissimum, L. humile,** and **L. angustifolium,** which were little known. Afterwards **Heer,** when making profound researches into ancient cultivation, went again into the characters indicated, and by adding the study of two intermediate forms, as well as the comparison of a great number of specimens, he arrived at the conclusion that there was a single species, composed of several slightly different forms." Commenting on **Heer's** definitions of these forms **DeCandolle** adds:—"It may be seen how easily one form passes into another. The quality of annual, biennial, or perennial, which **Heer** suspected to be uncertain, is vague, especially for **angustifolium**; **Loret,** who has observed the flax in the neighbourhood of Montpellier, says:—'In very hot countries it is nearly always an annual, and this is the case in Sicily according to **Eussone**; with us it is annual, biennial, or perennial according to the nature of the soil in which it grows; and this may be ascertained by observing it on the shore, notably at Maguelone. Then it may be seen that along the borders of trodden paths it lasts longer than on sand, where the sun soon dries up the roots, and the acidity of soil prevents the plant from enduring more than a year.'" It may, in fact, be accepted that **Heer** established **L. usitatissimum** as a cultivated race derived from **L. angustifolium.** The annual flax (**L. usitatissimum**) of which there are two special forms, *viz.*, *α* **vulgare** and *β* **humile,** has not been found, with absolute certainty, in a wild state. It is the cultivated form met with in the greater part of India, while variety *β* extends into Persia. **Boissier** mentions a specimen collected by **Kotschy** at "Shiraz in Persia, at the foot of the mountain called Sabst Buchom." "This is perhaps a spot," adds **DeCandolle,** "far removed from cultivation; but I cannot give satisfactory information on this head."

"**Linum angustifolium,** which hardly differs from the preceding, has a well-defined and rather large area. It grows wild, especially on hills throughout the region of which the Mediterranean forms the centre, that is, in the Canaries, and Madeira, in Morocco, Algeria, and as far as the Cyrenaic; from the south of Europe as far as England, the Alps, and the Balkan Mountains, and lastly, in Asia from the south of the Caucasus to Lebanon and Palestine. I do not find it in the Crimea, nor beyond the Caspian Sea."

HISTORY.

Having thus reviewed the botanic evidence DeCandolle turns to the philologic : "The widely different commercial names indicate likewise an ancient cultivation or long use in different countries. The Keltic name *lin*, and Greco-Latin *linon* or *linum*, have no analogy with the Hebrew *pischta* nor with the Sanskrit names *ooma, atasi, utasi*. The varying etymology of the names, the antiquity of cultivation in Egypt, in Europe, and in the north of India, the circumstance that in the latter country flax is cultivated for the yield of oil alone, lead me to believe that two or three species of different origin, confounded by most authors under the name of **Linum usitatissimum**, were formerly cultivated in different countries, without imitation or communication the one with the other. I am very doubtful whether the species cultivated by the ancient Egyptians was the species indigenous in Russia and in Siberia." These were DeCandolle's remarks in his earlier work, and they were abundantly confirmed by subsequent research, for, in his *Origin of Cultivated Plants*, he continues : "My conjectures were confirmed ten years later by a very curious discovery made by Oswald Heer." DeCandolle then goes on to describe Heer's discovery, which has already been indicated by the above remarks regarding the discovery of the fruits of **L. angustifolium** in the lake-dwellings of Eastern Switzerland, denoting an acquaintance with **Linum** at a time when stone implements were being used. **L. angustifolium** is wild south of the Alps. The interest in Heer's discovery was confirmed by his finding also the seeds of **Silene cretica**, a plant foreign to Switzerland but abundant in Italy in flax fields. Hence Heer concluded that the Swiss lake-dwellers imported the seeds of their flax from Italy. The same form of flax has also been recognised in the peat-mosses of Lagozza in Lombardy. The prehistoric people of that region were, like the Swiss, lake-dwellers ignorant of hemp, and used stone implements, but possessed the same cereals and ate like them the acorns of **Qurcus robur**, *var.* **sessiliflora.** There was thus, DeCandolle adds, a civilization on both sides of the Alps in which **L. angustifolium** was used—a civilization probably anterior to the Aryan advent in Europe. This idea is confirmed, DeCandolle thinks, by philological considerations. The word *lin, llin, linu, linon, linum, lein, lan*, exists in all the Eurcpean languages of Aryan origin of the centre and south of Europe, but as it is not common to the Aryan languages of India, the cultivation of flax most probably took its origin with the Western Aryans, and before their arrival in Europe. "The name *flachs* or flax of the Teutonic languages comes from the old German *flahs*. There are also special names in the north-west of Europe—*pellawa, aiwina*, in Finnish ; *hor, härr, hor*, in Danish ; *hor* and *tone* in ancient Gothic. *Haar* exists in the German of Salzburg. This word may be in the ordinary sense of the German for thread or hair, as the name *li* may be connected with the same root as *ligare*, to bind, and as *hor*, in the plural *hörvar*, is connected by philologists with *harva*, the German root for *flachs ;* but it is, nevertheless, a fact that in Scandinavian countries and in Finland, terms have been used which differ from those employed throughout the south of Europe. This variety shows the antiquity of the cultivation, and agrees with the fact that the lake-dwellers of Switzerland and Italy cultivated a species of flax before the first invasion of the Aryans." "It is not known precisely at what epoch the cultivation of the annual flax in Italy took the place of that of the perennial, **Linum angustifolium**, but it must have been before the Christian era ; for Latin authors speak of a well-established cultivation, and Pliny says that the flax was sown in spring and rooted up in the summer." If it had been the perennial crop, the plant would have been cut so as to leave the roots in the ground. In a like manner the annual flax must have been grown in ancient Egypt, as it is at the present time since the old paintings shew it being uprooted. "Now it is

L. 8

HISTORY.

known," continues **DeCandolle,** "that the Egyptians of the first dynasties before Cheops belonged to a proto-semitic race, which came into Egypt by the Isthmus of Suez. Flax has been found in a tomb of ancient Chaldea prior to the existence of Babylon, and its use in this region is lost in the most remote antiquity Thus the first Egyptians, of white race, may have imported the cultivation of flax, or their immediate successors may have received it from Asia before the epoch of the Phœnician colonies in Greece, and before direct communication was established between Greece and Egypt under the fourteenth dynasty.

"A very early introduction of the plant into Egypt from Asia does not prevent us from admitting that it was at different times taken from the east to the west at a later epoch than that of the first Egyptian dynasties. Thus the western Aryans and the Phœnicians may have introduced into Europe a flax more advantageous than **L. angustifolium,** during the period from 2,500 to 1,200 years before our era."

"The sum of the facts and probabilities," concludes **DeCandolle,** "appear to me to lead to the following statements which may be accepted until they are modified by further discoveries :—

"1. **Linum angustifolium,** usually perennial, rarely biennial or annual, which is found wild from the Canary Isles to Palestine and the Caucasus, was cutivated in Switzerland and the north of Italy by peoples more ancient than the conquerors of the Aryan race. Its cultivation was replaced by that of the annual flax.

"2. The annual flax (**L. usitatissimum**), cultivated for at least four thousand or five thousand years in Mesopotamia, Assyria, and Egypt, was, and still is, wild in the districts included between the Persian Gulf, the Caspian Sea, and the Black Sea.

"3. This annual flax appears to have been introduced into the north of Europe by the Finns (of Turanian race), afterwards into the rest of Europe by the western Aryans, and perhaps here and there by the Phœnicians; lastly into Hindustan by the eastern Aryans, after their separation from the European Aryans.

"4. These two principal forms or conditions of flax exist in cultivation, and have probably been wild in their modern areas for the last five thousand years at least. It is not possible to guess at their previous condition. Their transitions and varieties are so numerous that they may be considered as one species comprising two or three hereditary varieties, which are each again divided into subvarieties."

Very little information of a historical character can be given regarding flax in India. The subject has not been carefully gone into, but from what has been written it would appear that the Muhammadans have always given more attention to it than the Hindus. According to most writers certain Sanskrit names, which occur in some of the early works, are assigned to it. If, for example, the *Kshaumá* garments, alluded to in the Rámáyana and Mahábharata, be accepted, as having been made, as many writers maintain, of *Kshuma* or *Ksumá, i.e.,* linen—synonyms *Umá, Haimavati, Atasí* and *Marsina*—then the fibre must have been well known to the Sanskrit speaking people from very ancient times. But there is some doubt on this point. *Kshauma,* the name which of all others would carry the Asiatic knowledge in linen furthest back, is by some writers supposed to denote silk: indeed, its resemblance to *Chumá,* the Chinese for grass-cloth, has been even suggested as worthy of consideration. In the Institutes of Manu, mantles of woven *Kshumá* are those directed to be worn by theological students of the military class. Commenting on this subject **Mr. Hem Chunder Kerr** (*Cultivation of, and Trade in, Jute, p. 11*) says "So recently as 300 years ago Kavi Kankana, in the *Bengali Chandi,* de-

L. 8
(386)

HISTORY.

scribed a female punished by being obliged to dress in *Kshauma* cloth and to tend goats." "In this instance," adds **Mr. Kerr**, "the cloth obviously was a coarse or sack cloth. In the Purânas references to the *Kshauma* cloth are frequent, but do not indicate its nature or character. In the *Ain-i-Akbarí* a linen cap is described as part of the dress of a Brahmachári (Vol. II., p. 483). Anyhow it is abundantly evident that, as in other parts of the world, so in India, the true flax was known and manufactured from very early times, but that within the last two hundred years it has entirely lost its ground; and in Bengal, in the present day, a man would be laughed at who would say to the cultivators that the stalks of their well-known *tisi* are rich in fibre, and could by proper management yield a valuable fibre." The writer has failed to find the passage, given above by **Mr. Kerr**, in **Gladwin's** or **Blochmann's** translations of the *Ain-i-Akbari*, nor has he come across, in that most valuable work, any other passage that alludes to flax or linen. Linseed is, however, mentioned in the *rabi* harvest crops of Agra, Allahabad, Oudh, Delhi, Lahore, Multan, and Malwa, so that one might be disposed to think the oil had been as well known in 1590 (date of the *Ain-i-Akbarí*) as at the present day. **Linschoten**, who visited India in the sixteenth century, mentions by name a fibre found in Ceylon which his translators have rendered flax, but he says nothing of flax or linseed in India. **Tavernier** is silent as to flax and linseed, though he describes the field-crops over a part of India that he certainly could not have crossed without noting the abundance of this crop, had it been as at the present day, of primary importance. **Linschoten** was one of the most remarkable travellers who ever visited India and the East. While speaking of China he says: "In China there is much Flax and Cotton, and so good and cheap that it is almost incredible." But may not the flax there mentioned have been China-grass, *chuma?* To the people of India edible oils only are of importance, and probably the non-edible ones were never more so than at present. Before the foreign demand, therefore, it seems highly probable that Linseed was of little or no value to them. Added to all this it is certainly significant that neither **Rheede** nor **Rumphius** describe any species of **Linum** as seen by them in India, and that even **Burmann** is silent on the subject, though his *Flora Indica* appeared so late as 1768.

The first detailed article, in fact, on the subject of flax which the writer has been able to consult, is that written by the late **Revd. W. Carey**, which appeared early in the present century. The correspondence of the Hon'ble the East India Company is of course dated a few years earlier, and **Roxburgh** wrote of it much about the same period as **Carey**. The flax or linseed plant, though very probably grown by the people of India long anterior to the date of **Carey's** paper, had apparently lost the importance once attached to it by the Muhammadan rulers. It would seem to owe its present value in Asiatic agriculture like tea, coffee, jute, wheat, cotton, &c., &c., to European influence and commerce. **Ainslie** in 1826 wrote of it: "There is a great deal of flax *now** cultivated in many parts of Upper India, and especially in Bengal, for making oil, and of *late years** it has also become an object in the lower provinces." Thus **Ainslie** viewed it as a modern agricultural crop. **U. C. Dutt** gives it its Sanskrit and Bengali names, but makes no mention of its properties having been known to the Sanskrit physicians. Even a stronger proof of this view of the Indian flax and linseed industry may be drawn from *Milburns' Oriental Commerce*. The first edition of that work appeared in 1813, but neither flax nor linseed are even mentioned in it by name. In the second edition, 1825 (issued by **Mr. T. Thornton**), it is stated,

* The italics have been given by the Editor of this work.

"Flax is very generally cultivated in Bengal and Behar for the oil which is obtained from the seeds, the stalks being rejected as useless." It may be here added that **Dr. Buchanan-Hamilton's** silence on the subject of flax or linseed is also very significant and strongly confirmatory of the suggestion here offered, that in India down to about the year 1830, neither of these substances were of commercial importance, and, in fact, could hardly be said to have been Indian agricultural products. **HISTORY.**

Buchanan-Hamilton, in his account of the Kingdom of Nepál, deals with the crops grown there in his time (1819), but makes no mention of Linseed. And what is even more remarkable he is silent on that subject, in his Statistical Account of Dinajpur, in Behar and also in his Journey through Mysore. The writer has repeatedly urged, in this work, that the mere existence of Sanskrit names, which by present universal admission allude to a certain plant, should not unreservedly be accepted, as **DeCandolle** very frequently does, to prove ancient cultivation in India. The names may be, as is doubtless frequently the case, comparatively modern adaptations, or they may denote, on the part of the early Sanskrit writers, a knowledge, or survival of a knowledge, that dates prior to their invasion of India.

Royle informs us that the first record of the exportation of Linseed from India occurs in 1832, when a **Mr. Hodgkinson** sent from Calcutta ten bushels. In 1860-61 the exports from India were 550,700 cwt., valued at R1,25,57,790; in 1880-81 5,997,172 cwt., valued at R3,69,81,265; and last year (1888-89) they were 8,461,374 cwt., valued at R5,05,79,221. Commenting on the Linseed trade of India, **Mr. J. E. O'Conor** points out that, in 1884-85, the figures of the export trade of Linseed during the five previous years showed an increase from 5,997,172 cwt. in 1880-81 to 8,716,596 cwt. in 1884-85, or nearly 46 per cent. In the succeeding year the exports reached their highest recorded extent, namely, 9,510,139 cwt. A diversion of the trade (**Mr. O'Conor** pointed out) from Calcutta to Bombay was striking. In 1880-81 Calcutta exported 4,065,341 cwt., and Bombay only 1,925,524 cwt.; whereas in 1884-85 they exported 3,757,018 cwt. and 4,989,578 cwt. respectively. This apparent migration does not, however, appear to have been maintained. Since last year (1888-89) Calcutta exported 5,659,492, and Bombay 2,797,246 cwt. For further particulars regarding the trade in linseed, the reader is referred to the section below which deals specially with that subject. (*Conf.* with the concluding paragraph of Trade.)

Properties and Uses.

Fibre.—The bast fibres of the stems yield FLAX which when bleached is known as LINEN. **FIBRE.** **Flax.** 9 (387) **Linen.** 10 (388)

Oil.—The OIL—LINSEED—is extensively used in the manufacture of paint, printing-ink, floor-cloth, artificial India rubber, oil varnish, and soft soap. The seed is nearly always adulterated, pure Linseed oil being in India almost unknown. In Russia it is adulterated with hemp seed, and in India being grown as a mixed crop with rape, it is rarely if ever pure. The OIL-CAKE is also an important article of trade and is both exported and sold in India as cattle food. **OIL.** **Linseed.** 11 (389) **Oil-cake.** 12 (390)

Medicine.—LINSEED is used for poultices. It is also taken internally in bronchial affections, diarrhœa, &c. LINSEED OIL is aperient, but is only rarely administered internally. It is chiefly employed in the preparation of liniment for burns. It has been thought unnecessary to deal with all the minor uses of so well known a substance as Linseed oil. The reader is referred to the Pharmacopœia and such other works for information of that nature. **MEDICINE.** **Linseed.** 13 (400) **Linseed oil.** 14 (401)

SPECIAL OPINIONS.—§ "The SEEDS are used internally for gonorrhœa and irritation of the genito-urinary system. The FLOWERS are considered a **Seeds.** 15 (402) **Flowers.** 16 (403)

MEDICINE.

cardiac tonic" (*Dr. Emerson*). "Decoction of it used extensively in chartable hospitals for gonorrhœa" (*Civil Surgeon S. M. Shircore, Moorshedabad*). "The seeds after being boiled are strained off, and the liquid used with great benefit in irritation of the genito-urinary passages" (*Honorary Surgeon E. A. Morris, Tranquebar*). "The seeds powdered and combined with sugar are useful as an aphrodisiac and in gonorrhœa. The oil is used as an article of diet in the country about Nagpur, its purgative properties not being perceptible" (*Narain Misser, Hoshangabad, Central Provinces*).

FOOD. Seeds. (404) 17

Food.—In many parts of India the SEEDS are eaten, especially those of the white form. In the passages quoted below frequent mention will be found of the various methods of preparing these seeds as an article of human diet so that the subject need not be here dealt with. The OIL is very

Oil. (405) 18

little used in India, hence the amount of cake available for cattle food is but small; the oil seed is exported. To a certain extent, however, LINSEED

Linseed Oil-cake. (406) 19

CAKE is used as a cattle food, and, in some parts of the country, it is purified and eaten as an article of human food also.

FIBRE. (407) 20

FLAX—THE FIBRE OF LINUM.

Messrs Cross, Bevan & King, in their Report on Indian Fibres and Fibrous substances shown at the Colonial and Indian Exhibition of 1886, say that the bast of **Linum usitatissimum** "forms a continuous ring, that sometimes two concentric fibrous zones are developed, the fibres in which have different features. In many cases, an aggregation into two groups is noticed. The fibres are in loose contact, the cortical parenchyma not largely developed." The average number of fibres in the bundle they found to be 5-10, and these were discovered to be easily sub-divisible; length 25-40 mm. They, however, noted that there were two types—*1st*, the normal, of small diameter, thick-walled, and polygonal; *2nd*, large, ovoid, with a considerable cavity resembling that in rhea. **Messrs. Cross, Bevan & King** seem, however, to have examined only English grown flax: they remark of India:—"The stems do not appear to develop the fibre, and in no part of India is the plant specially cultivated as a source of flax. Whether this is due to abundance of other fibres discouraging any effort to produce flax, or to the climate of India favouring the development of the oil at the expense of the fibre, or simply to the peculiar variety cultivated, being an oil and not a fibre-yielding form, does not appear to have been practically solved. We should be glad of an opportunity to thoroughly investigate this matter, and, with the approval and co-operation of the Government of India, would propose to procure a quantity of the best flax-yielding seed from two or three European countries. This we should simultaneously cultivate, half in India and half in Europe. If the Indian samples were cultivated and carefully reported on, according to a plan agreed upon, the produce would enable us to form some definite ideas. At the same time, it would be necessary to have selected samples of Indian linseed cultivated alongside of the European, both in India and in Europe. From such experiments we would be in a position to judge whether it was the effect of the climate, or defect of the form of linseed cultivated, that rendered the rejected stems of the Indian oil-yielding plant valueless as a source of flax.

"At the late Colonial and Indian Exhibition, numerous inquiries were made as to whether India could not be induced to cultivate flax. The large, bold linseed was much admired, especially the white form, and several flax spinners solicited and obtained samples of the Indian seed from a conviction that it was some defect in the cultivation that prevented the Indian plant from yielding a very superior flax. It was universally admitted that

FIBRE.

the time had now come for the English flax-spinners to seek a new source of supply, or some good substitute for flax

"While venturing to make the suggestion that experiments might be instituted on the subject of flax cultivation in India, we are fully aware that such experiments were performed from the beginning of the century down to about 1850. It is presumed, however, that the immense advances which have since taken place have prepared the way for the subject deserving renewed efforts. We are confident that, just as with hemp, so with flax, the remarkable progression of the jute trade hindered the establishment and growth of a flax industry. It is the opinion of most experts that a re-action has now set in, and that the demand for textiles of all kinds would justify a fresh effort being put forth. Moreover, it is presumed that the success of the wheat trade has abundantly demonstrated what may be done in India towards cultivating during the cold months crops peculiar to Northern Europe. The experiments made with flax some half a century ago were chiefly in Bengal, whereas the more natural regions would appear to be some parts of Behar, the Central Provinces, or Bombay and the Panjáb. That flax can be produced in India was demonstrated beyond all doubt (although it is probable that the fibre-yielding plant would be found to be of little or no value as a source of the oilseed); indeed, so successful were the early experimenters, that in a letter, dated November 22nd, 1841, the Government of India refused to afford any official assistance, because 'the cultivation of flax can no longer be considered a doubtful experiment since it appears from your (Agri-Horticultural Society's) report to have proved in many instances successful, and where successful, to be very fairly profitable. In spite of this apparently hopeful position, the cultivation of flax was discontinued in India shortly after; but, be it observed, that this was the very period at which jute cultivation was making rapid strides.'"

It will be seen (from the pages below) that to some extent the experiments proposed by **Messrs. Cross, Bevan & King** have already been performed. Much time and money have at least been expended in the effort to make India a flax-producing country, but while good results have here and there been obtained, as, for example, in the Panjáb, the much to be desired object has not been attained. It seems probable that the indifference of the cultivators to their own better interests when these are only to be secured by departures from their time-honoured methods of agriculture, combined with the difficulty of preparing the fibre in a mercantile form, are greater obstacles than physical peculiarities due to the Indian climate or soil. The proposals of **Messrs. Cross, Bevan & King** are, however, worthy of careful consideration, since, if carried out, they would be of great value in determining where these physical peculiarities amount to an obstacle against hopeful results. They would, in other words, indicate where the effort should, or should not, in the future be made, to conquer the obstructionist policy that existed, and exists now with almost greater power than at the beginning of the century, since the experiments then performed resulted in a disappointing record of failure. Before concluding these introductory remarks regarding flax, it may be as well to quote here a fuller account of the microscopal and chemical nature of the fibre than the above. In **Spons'** *Encyclopædia* it is stated that "of all the vegetable fibres, flax occurs in the greatest variety, as regards the length of the filaments, their colour, fineness, and strength; but the fibrous bundle always retains the character of being very readily divisible into its distinct filaments, by rubbing it between the fingers; it then becomes soft and extremely supple, while preserving a great tenacity." On account of repeated bending, the fibres are seen under the microscope to have X-like creases. Under the action of

FIBRE.

iodine they assume a transparent blue colour, the creases taking a darker shade. "The dimensions of the fibres are as follows: length, 0·157 inch to 2·598 inch; mean, about 1 inch; diameter, 0·0006 to 0·00148 inch; mean, about 0·001 inch. The chief characteristics of flax are its length, fineness, solidity, and suppleness. Its remarkable tenacity is due to the fibrous texture and thickness of the walls; its suppleness permits it to be bent sharply; its length is invaluable in spinning; and the nature of the surface prevents the fibres from slipping on each other, and contributes to the durability of fabrics made with them. Flax may be made lustrous like silk, by washing in warm water, slightly acidulated with sulphuric acid, then passing through bichromate of potash vapour, and gently washing in cold water. Samples of flax exposed for two hours to steam at two atmospheres, boiled in water for three hours, and again steamed for four hours, lost only 3·5 per cent. of their weight, while Manilla hemp lost 6·07; hemp, 6·18 to 8·44; jute, and 21·39."

Messrs. **Cross, Bevan & King** (in their report already quoted) give the results of their chemical analysis. In the article on **Marsdenia tenacissima** (see Vol. V, 189) the chemical nature of flax has been compared with four other fibres. The table there given may, therefore, be consulted, but it may briefly be said that flax was found by the above-mentioned chemists to possess 9·3 per cent. of moisture and 1·6 per cent. of ash. Under hydrolysis for five minutes it lost 14·6, after an hour 22·2 per cent. Of celluloses (one of the best tests for fibres) it was seen to have 81·9 per cent. By mercerising it lost 8·4 per cent. and by nitration gained, becoming 123·0 per cent. By acid purification it lost 4·5 per cent. and was found to yield 43·0 per cent. of carbon. The fibre experimented with was, as already stated, European flax, so that a comparison of the Indian fibre, with the above results, might be viewed as indicating the influence exercised by the Indian climate, soil, and method of cultivation, on the production of fibre, or might even demonstrate to what extent a plant cultivated for at least several centuries as a source of oil, was still capable of being utilized in the production of fibre.

CULTIVATION. Early experiments. (408) 21

CULTIVATION OF FLAX.

EARLY EXPERIMENTS IN INDIA.

The Government of India having procured on loan from the Board of Trade, Calcutta, the Proceedings of the Hon'ble the East India Company which contain the record of the early experiments in the cultivation of Flax, the writer found (after carefully reading through these curious manuscript volumes) that there was little more to be learned than had been published by **Roxburgh, Carey, Wisset, Royle,** and other authors whose works have long been in the hands of the public. The first experiments were performed between the years 1790 and 1799, and a further series of trials were again instituted about fifteen years later. The results were the same, though attended with a certain degree of success in the hands of the European expert cultivators, who at considerable cost were brought out for the purpose to India, the cultivation of flax did not extend from the Government farms to the fields of the agriculturists. In time the whole subject was thus allowed to lapse into absolute obscurity, the interest in both flax and hemp being extinguished through the commercial progression of India in other and more hopeful directions, *e.g.*, jute, which in time practically took the place of flax and hemp in popular favour.

There is thus little more to be said than is contained in **Royle's** work on Fibrous Plants, until the record be carried some forty or fifty years nearer the present date. Space will not admit of a reprint of **Royle's** article on Flax in its entirety, but since that work is not so accessible as its great value

CULTIVATION.

Early Experiments.

deserves, the following pages (up to page 24) which convey the main facts of practical interest, may be given :—

"REPORT ON THE CULTURE OF FLAX IN INDIA.

"India having, at least, for centuries grown the Flax plant, on account of the oil yielded by its seeds (Linseed), the country has very naturally been looked to as a source of Flax fibre, the supply of which is so greatly diminished by the war with Russia.* The Belfast Chamber observe that 'as India annually exports nearly 100,000 quarters † of seed to Great Britain and Ireland, it has been calculated that the plants which produced this quantity of seed would yield, annually, at least 12,000 tons of fibre—value, say £500,000, all of which now goes to waste.' Besides the above quantity of seed, much is also exported to North America and to other countries, and much is consumed in the country in the torm of oil, while the cake is in some places employed in feeding cattle. There can be no doubt, therefore, that the question is one of considerable importance, not only to this country, which requires such immense quantities of Flax fibre, but to India, which produces such enormous heaps of seeds, and is supposed to waste so much of valuable exportable material. But it does not follow that the production of fibre is in proportion to that of seed. Indeed, we have often to check vegetation in order to favour the production of flowers and fruit; while an undue growth of the parts of vegetation, that is, of the stem, branches, and leaves, is often obtained at the expense of the parts of fructification. The subject, however, has not escaped notice.

"The earliest attempt to produce Flax in India seems to have been made by Dr. Roxburgh about the beginning of this century, and as at that time the East India Company had established a Hemp farm in the neighbourhood of Calcutta, he was able to make many experiments on substitutes for Hemp and Flax. He also cultivated Hemp and Flax in the Company's farm at Reshera, in the neighbourhood of Calcutta.

"Of Flax, he says, it is very generally cultivated during the cold season in the interior parts of Bengal and Behar. 'Samples of the Flax have frequently been procured by the Board of Trade, and sent to England to the Honourable Court of Directors, so that it is from home we may expect to learn its properties. If the Flax has been found good, large quantities may be reared at a small expense, as the seed alone which the crop yields must be more than equal to the charges to render it profitable to the farmer' (*'Obs. on Subs. for Hemp and Flax,' p. 17*).

"The Author, as long since as the year 1834, stated in his *'Illusrations of Himalayan Botany*: 'In India the Flax is cultivated only on account of its seed, of which the mucilage is valued as a demulcent, in medicine, and the oil in the arts; but the plant, which in other countries is most valued, is there thrown away; and others, such as **Hibiscus cannabinus** and **Crotolaria juncea**, are cultivated almost in the same field for the very products which this would yield. It seems, therefore, worthy, of experiment whether a valuable product might not be added to the agriculturist's profits, without much additional expense.' And again, in the year 1840, the Author called attention to this subject in his *Essay on the Productive Resources of India.*.

"In the year 1839, moreover, a Company was established, by the influence of Mr. A. Rogers, at one time one of the Sheriffs of London, expressly for the growth of Flax in India. Money was subscribed; a Belgian cultivator and a Belgian preparer of Flax were sent out to Bengal, with both Riga and Dutch seed, and all the tools which are employed in the culture and preparation of Flax in Europe. A pamphlet moreover was published, in which full directions were given for the culture and preparation of Flax, and illustrated with figures of the various tools employed for this purpose. The subject was warmly taken up by the Agricultural Society of India, and a small committee appointed of members who took a special interest in the subject.

"The directions of the Irish Flax Society were printed in their *Proceedings* as well as those of Mr. Andrews from the *Northern Whig*. Translations of plain directions were made into the vernacular languages, which, as well as models of the tools, were distributed. The Gold Medal of the Society was offered for the production of a large quantity of Flax, and smaller prizes for the natives. Experiments were made by several members of the Society, in different parts of the Bengal Presidency, as well as by the Belgian farmers.

* The following formed the substance of a report prepared by Dr. Royle in August 1854, and is printed nearly as it was written.

† [See the present trade in Linseed, in the Chapter on that subject.— *Ed.*]

LINUM usitatissimum.

Early Experiments in

CULTIVATION. Early Experiments.

"Specimens of the Flax produced having been sent to Calcutta, comparisons were instituted between the samples produced by different individuals, and those from European and from indigenous seed. Mr. Deneef, the Belgian farmer, pronounced the samples worth from £44 to £60 a ton; and one, produced from country seed and heckled, was thought worth £66 a ton. Some of the specimens sent to Liverpool were valued at from £30 to £45; and those which were forwarded to me by the Secretary of the Agricultural Society were pronounced by Mr. Hutchinson, of Mark Lane, to be worth from £40 to £45 a ton. The experiments were made chiefly near Burdwan, Monghyr, and Shahabad; but the best native seed was obtained from the northern station of Saharunpore, and a white Linseed from the Saugur and Nerbuddah territories. A little Flax was also produced by Mr. Williams, at Jubbulpore, under the direction of Mr. Macleod.

"Mr. Leyburn gave as the result of his experiments near Shahabad, that the expenses of culture of a bigha of land, and the preparation of the fibre, amounted to R25-1-3; and the profits to R27-1-5, supposing the four maunds of Flax produced to be worth £35 a ton.

"In consequence of a communication which had been received from the Honourable Court of Directors, Sir T. H. Maddock, at that time Secretary to the Government of India, addressed a letter to the Agri.-Horticultural Society. In this the Society was requested, in order to assist the Government in determining on the measures proper to be adopted for improving the cultivation of Flax, to supply such accurate, detailed information as they may possess, or as they may be able to obtain.

"The Society accordingly prepared a report which contained everything that was known at that time on the subject of the cultivation of Flax in India. This was forwarded to the Government, and also published in their 'Proceedings' for November 1841. In this report, the Society took a very favourable view of the probabilities of the profitable culture of Flax in India.

"The Revenue Secretary to the Indian Government, on this, wrote (November 22nd, 1841) to the Agricultural Society, that—'The cultivation of Flax can no longer be considered a doubtful experiment, since it appears from your report to have proved in many instances successful; and where successful, to be very fairly profitable. His Lordship in Council is therefore much inclined to doubt whether any bounty or reward from Government is necessary, or would be justifiable.'

"Notwithstanding this favourable inference, the Flax Company did not go on with the cultivation; the various individuals who had taken up the culture did not proceed with their experiments; the several medals offered by the Agricultural Society seem never to have been claimed, and there are no appearances of the culture of Flax on account of its fibre in any of the places where the experiments were made. It is probable, therefore, that the success, which appeared sufficient when the experiment was of the nature of garden culture, was not realised when on a greater scale.

M. de Verinne, indeed, states that the experiment in the season of 1840-41 was a complete failure at Bullea, owing to too little seed having been sown, to the unusually dry weather at the late sowings, and to the improper time (the hot winds) in which the Flax was cleaned.

"Mr. Wallace, who had carried on the cultivation for three or four years at Monghyr, writes on the 8th July 1841: 'The crop has been in a great measure a failure this year About one-eighth of the produce that a favourable season would yield.' But in the year 1844, he again forwarded samples to the Agricultural Society, which were improved in cleanness and were also softer than the produce of former years, from the same cultivation. These were portions of several tons that had been grown at Monghyr, and which he intended shipping to Dundee, the port to which his last batch was sent. But Mr. Wallace, added, with regret, that after several years' labour, with a view to establish Flax cultivation at Monghyr, and after having taught the art of dressing the article to many parties, the speculation must be abandoned, unless the Government gave some encouragement. He therefore requested the assistance of the Society in bringing the subject to the notice of the authorities. It is stated at a subsequent meeting that the Committee of the Society, after being furnished with further details respecting the cultivation, did not feel inclined to refer the subject to the Government. But neither the details referred to nor the reasons of the Committee for their decision are given, and, therefore, we are unable to ascertain the real causes of failure after several years' trial.

"Mr. Henley, an intelligent merchant from Calcutta, having made some careful experiments on the culture of Flax, has favoured me with the following account:—

"'I have paid much attention to the fibres during my residence near Calcutta, and, not wishing to conclude from hearsay only, generally cultivated most things myself, having a large piece of ground available. I sent up to Baugulpore (an excel-

CULTIVATION.
Early Experiments.

lent Flax seed district), and obtained a considerable quantity of native grown Flax straw, after the removal of the seed. I had it collected from various fields, so as to obtain an average. This material was in every instance too *bushy* for the proper production of fibre, and the yield was very trifling, and in fact worthless for manufacturing purposes. The bushiness arose from the practice of the natives, who grow several plants, as you are aware, at once, in the same field. The Flax plants were consequently planted too far apart for fibre-yielding purposes.

"'Not yet fully satisfied on the question, I took a patch of land (three *cottahs*), the best I could pick out, fine, friable loam, fit for anything—it had been a cauliflower bed, and was therefore deeply spade-cultivated and highly manured—its last crop, cauliflowers, having nothing prejudicial to a Flax crop. I began very early in the season, had it turned up and laid for a fallow; two months after, again pulverized and weeded, and again—four times in all; with the addition of a large supply of fine old cow-dung. I had it now sown in the proper season, with the best Flax seed, very thickly planted, so as to draw it up as free as possible from lateral branches. Everything promised well. The field grew beautifully, and soon attained a height of three feet. I began to collect the crop, first, as soon as the flower had completed its growth and the seed-vessels began to form; secondly, as soon as the seed-vessels had fully formed and were filled with green but immature seeds; and lastly, after the seed was fully ripe. I took great pains in water-retting the samples—generally removing them from the water rather under-done, for fear of occasioning weakness in the fibre from over-retting.

"'In every instance, the quantity of fibre was small and weak, and very inferior to the samples of Flax deposited at the Agricultural Society's Museum, obtained from Jubbulpore, and other upper country districts. No Indian Flax, however, which I have seen, equals in nerve and general good qualities those of European growths.'

"As the above is no doubt a correct account of what occurs with the Flax plant in the moist climate of Bengal, of which the effects may perhaps have been aggravated by too great richness of soil, it might be inferred that a different result would take place in the drier climate of the upper or North-West Provinces of India. This is certainly the case, but though the product is different, it is not, from the shortness and brittle nature of the fibre, more suitable for the ordinary purposes of Flax.

Mr. Hamilton, of Mirzapore, one of the up-country stations alluded to, 'sent some bales of the stalks to Calcutta, for the inspection of the Belgians, and was told that the shortness of the stalks would prevent their manipulation.'

"It is evident, therefore, that there is some difficulty in producing good Flax in India. This difficulty is, no doubt, the climate; while the native methods of culture are the most unsuited to the production of good fibre. Mr. M'Adam, Secretary to the Royal Flax Improvement Society, has, in his Prize Essay on *The Cultivation of Flax*, well observed 'that a slow, steady growth is requisite for the quality and yield of fibre; also a temperate climate, that between the parallels of 48° and 55° being the best; and a continued supply of moisture from spring till autumn.' He also observes that 'the hot summers of Russia and of Egypt cause a dryness and brittleness of fibre, and prevent its retaining that elasticity, pliancy, and oiliness which characterise the Flaxes of Belgium, Holland, and Ireland.'

"But considered generally, it is not to be expected that a plant which attains perfection in Belgium, and is so successfully cultivated in the vicinity of Belfast, would succeed well in the hot and moist, but sometimes dry climate of Bengal. In fact, if the Flax was not one of those plants which, like the cereal grains and pulses, can be grown in the cold-weather months of India, it could not be cultivated there at all. But with this culture, we have the anomaly, of the seeds being sown in autumn, * when the climate is still hot and the ground moist, and the plant has to grow while the temperature is daily becoming lower and the soil drier—no irrigation being usually employed with these winter crops, though dew begins to fall as soon as the ground becomes cooled at night. In some places, the crop attains perfection in about ninety days, is collected in January, the coldest month; in others, not until February or March, when the rapid rise of temperature is favourable to the ripening of seeds, but not to the production of fibre.

"Of all parts of India there are none that appear to me better suited to the growth of Flax than the Saugur and Nerbuddah territories, where the soil is rich and prolific, and the climate a medium between the extreme moisture of Bengal and the dryness of the North-West Provinces. The Wheat of this district is considered superior to any seen in the English market, with the exception of what comes from Australia.

* In Egypt, also, the seeds are sown about the middle of November, in the plains which have been inundated by the Nile, and plucked in about 110 days.

CULTIVATION.

Early Experiments.

The Gram and the Linseed are also of finer quality than any produced elsewhere in India; while the suitableness of the climate for the production of good fibre is proved by the length and strength of the Jubbulpore Hemp, as grown by Mr. William as well as by the specimens of Flax which he has likewise grown.

"The Indian method of culture is certainly not suited to the production of fibre, but the seeds abound in oil. 'The yield of oil from a bushel of Indian seed is from 14¾lb to 16lb; of English or Irish, 10⅝lb to 12lb.' Therefore, it is evident that the Indan ryot succeeds in his object, as well as the Irish farmer, who grows the Flax plant for its fibre, but neglects to gather the seed: though this is not only a saleable product, but one which abounds in nutritious matter for his cattle, and would further afford the means of fertilising his fields. As it has been found difficult to persuade the Irish farmer to gather the double crop, I believe it would be hopeless to induce the Indian ryot to change a culture which is suitable for his purposes, without the aid of successful example in his neighbourhood. You might make him grow less seed, but I much doubt whether we should get him to produce any useful fibre; and without his co-operation it would be impossible to attain any considerable success. Indeed, the Agricultural Society of India have given it as their opinion, in one of their resolutions,—'That the culture and preparation of Flax in India, so as to be able to compete with the Flax of Belgium or Russia, can only be effected by practical European growers instructing native cultivators in the art; and, further, that an entire change in the mode of cultivation, as well as in the preparation of the plant, is necessary to produce the article in a proper state.'

"It has, indeed, been made a question, whether a good supply of fibre and of seed can be procured from the same crop. One gentleman, in reply to my inquiry, informed me (London, 4th July) that 'it has been found impossible to preserve both seed and fibre, *i.e.*, for the better qualities of each; and that the plan pursued is simply to gather before the seed ripens, when the delicacy and softness of fibre form the desideratum, but to leave the plant standing until the fibre is dried and greatly injured, in order to secure the superior seed fit for sowing;' and this is the result of information collected after a residence of many years in the interior of Russia. Another gentleman replies to the same inquiry, from Belfast, on the 8th July: 'It is not only quite practicable to have good seed and good fibre at the same time, but it is the universal rule in all countries except Ireland, where we have only been able to get the more intelligent farmer to abandon the wasteful practice of steeping the Flax stems without removing the seed. The finest Flax in the world is grown in Belgium, yet the seed is saved from it.'

"But as there is no doubt, from the experiments of the Indian Flax Company, and from other more recent facts, such as the production of Flax as far sonth as on the Shevaroy Hills, at Jubbulpore, and near Lahore, that Flax can be produced in India, it seems desirable to ascertain whether it cannot, by careful culture and improved processes, be produced as a profitable crop in some parts of the country; because, as I have before said, 'I cannot think that that which is done successfully in Egypt, is impossible in every part of India;' and there can be little doubt that, in some places, at least, coarse Flax could be produced, as well as some for the paper-makers.

"I am informed that Messrs. Hamilton, of Mirzapore, propose, this year, attempting the culture of Flax, in the tract of land of which they have a grant, in the Goruckpore district, and which I should consider a more favourable locality than any near Mirazpore. I would suggest that Mr. Williams, at Jubbulpore, should be requested to make an experiment. to ascertain the quality of the Flax which may be produced in that locality, as well as the quantity obtainable per bigha or acre, attempting at the same time to preserve the seed. Mr. Williams has already grown a little Flax; he is accustomed to the preparation of fibre, and the soil and climate are both, I conceive, more favourable than in most parts of India.

"I would also recommend that Dr. Jameson, the Superintendent of the Botanic Garden at Saharunpore, should be directed to make a small experiment, both in the plains and in the hills, in order to ascertain the same kind of facts respecting the Flax plant when grown according to European methods for the sake of the fibre. The Agricultural Society of the Panjáb are already attempting the culture, as the Secretary has addressed a letter to the Court, requesting an opinion respecting the quality of the Flax which they have already produced; but the specimens have not yet arrived. (Further information has, however, recently been received, and will be afterwards detailed.)

"Though I am well aware that Government experiments are not likely to prove profitable where those undertaken by individuals have failed, especially as these had

L. 21

CULTIVATION.

Early Experiments.

good scientific and practical advice, I am yet sanguine in thinking that experiments conducted in the localities I have indicated would give information which would be practically of great value for extensive tracts of country. The people are acquainted with the culture and preparation of *Sunn* fibre, and might easily be instructed by the European gentlemen to whom I have alluded, in applying the instructions for the culture of Flax in the *Proceedings of the Agricultural Society of India* for the years 1840 and 1841, including those prepared by Mr. Deneef, the Belgian farmer, after practical experience in India, published first in 1840, and then in 1842.

"I have not thought it necessary to refer to the opinions respecting the exhausting nature of Flax as a crop. By the methods of steeping the stalks in steam and hot water, it has been ascertained that the time required for the separation of fibre can be very greatly reduced; while the steep-water, where no fermentation has taken place, has been proved to be useful as manure water for the soil. Feeding cattle, moreover, upon a portion of the seed, produces manure which is invaluable in restoring much of what has been taken from the soil. But as these methods are not applicable to the present state of the culture in India, I will only allude to the probability of some of the mechanical methods of separating the fibre from the green flax, proving likely to be of useful application.

"Since, according to some accounts, considerable success attended the experimental culture of Flax in India, while others considered it a failure, it is desirable to ascertain the causes of this discrepancy, and to draw some conclusions which may be of use to other parts of India, if not to the places where the experiments were made. This we may probably effect, by analysing the statements of the different experimentalists.

"*Shahabad Experiments*.

"The cultivation of Flax in India in recent times seems to have begun at Shahabad, in 25° of north latitude, in the year 1837. In the *Proc. of the Agri.-Hortic. Society*, there is a communication from Mr. G. Leyburn, of Nunnoa Factory, giving an account of the sale in London, on 17th July 1838, of some Flax grown by him. 'The Flax, per *Windsor*, is landed sound. No. 1 sold for £28 per ton, and No. 2 for £14 per ton—nine months' credit. They are described as harsh, and without the softness characteristic of Russian Flax. Prices of the latter being lower than usual, P. T. R. selling here at this time at £40 per ton.'

"Mr. Leyburn states that he prepared his first sample of Flax in the common way, from plants which had borne seed. He sent them to Messrs. Truman and Cook who reported that any quantity of a similar article would find a ready sale, at £35 a ton. In the following year, Mr. Leyburn entered on the cultivation rather extensively, and succeeded in producing an article of lengthened staple, and of a quality vieing with the Flax of Russia. A portion of the cultivation was carried on in the bed of the Soane River, and part in the uplands of the district: some of it was prepared before the seeds were ripe. He calculates the probable profits of the culture to be:—

PER BIGHA.*	R	a.	p.	PRODUCE.	R	a.	p.
Rent of land	1	8	0	Linseed, 5½ mds. . . .	5	8	0
Ploughing	0	8	0	Flax, 4 mds., at (say) £35 per ton	46	10	8
Seed	1	8	0		52	2	8
Pulling, beating seed off, watering . . .	5	9	3	Deduct expense of cultivation, as per contra . . .	25	1	3
Packing, cartage, preparation (nearly all hand labour) . . .	16	0	0				
TOTAL .	25	1	3	Profit on one bigha of land .	27	1	5

"With a factory in full play and effective machinery, Mr. Leyburn considers that the manufacturing price of the article would be three to four rupees a maund, which is equal to about £9 or £12 a ton. But to give effect to the production of this article, the aid of European enterprise is necessary. (It is particularly deserving the attention of indigo-planters.) He failed in overcoming the deep-rooted prejudices

* The bigha of land in Shahabad is something more than the Bengal bigha, which is 1,600 square yards, or the third of an English acre.

CULTIVATION.

Early Experiments.

of the native cultivators, and could not induce them to enter on a cultivation which held out to them a prospect of more than ordinary profit for their labour.*

"*Experiments of the Flax Society.*

"We may now proceed to notice the efforts of the London Flax Experimental Society. Mr. Woollaston, in presenting, on their behalf, some specimens of Flax grown in Bengal, and prepared in Calcutta, observed that—

"'The object of the Society is not at this time to produce a large quantity, but to ascertain how good a quality can be readily obtained, the growth of India, and such as shall readily compete with the Russian and Belgian Flax in the Home market. This object has been already attained to a considerable extent. These samples far surpass the Russian Flax,' and he regretted that 'the Government of India have not responded to the recommendation of the Horticultural Society in granting a bonus to the Experimental Society of 10,000 rupees to further its objects.'

"'The seed received from England, Mr. Woollaston further remarks, has been distributed freely to all applicants who were desirous of trying the cultivation. The models of implements were sent out from Belgium, and facsimiles made for any person requiring them at the *boná fide* cost of the materials. Private profit or gain has never been allowed to interfere. Every kind of information, as far as possessed, has been freely imparted to all inquirers, and every endeavour made to excite an interest in the experiment.

"'Its importance in a national point of view is incalculable. Both as developing the resources of India, in enabling England to supply herself from her own possessions in a most important raw material, and in no longer making her dependent, for what may well be considered necessaries, upon a foreign and rival power. These observations, Mr. Woollaston considers, will apply in a great measure to Hemp also, in the cultivation and manufacture of which the Experimental Society are deeply interested. The successful introduction of these two staples into England, from this country, will not only prove a blessing of the largest degree to *India* generally, but be a severer blow to Russian aggrandisement and encroachment than the destruction of her fleets, or the annihilation of her armies.'

"The Agri-Horticultural Society having recommended that the bonus of R10,000 should be given for the furtherance of the objects of the Flax Society, Lord Auckland, who was at that time Governor General of India and was as warmly interested as any one in the improvement of its resources, was also a political economist: the Secretary to Government was directed to reply:

"'His Lordship cannot but regard with interest the public-spirited proceedings of the gentlemen who have come forward to promote the improvement of the cultivation of Flax in India, but it is only in very rare instances, and with the view of exciting a direct and general competition, that he would attempt by encouragement or bounty to influence the course of commercial and agricultural enterprise, and he does not feel that the case before him is one which would justify the special interference of the Government.'

"Mr. Deneef, the Belgian farmer, and Mr. Bernard, the preparer of Flax who had been sent to India by the Society, were of great use in examining the soil and giving directions on the mode of culture best adapted to the country, as well as in reporting on the different samples of Flax which were grown in the country. Mr. Deneef's directions for the cultivation of Flax, drawn up after he had had practical experience in the country, remain as a valuable document for the guidance of others. These we, therefore, reprint from the *Journal of the Agri.-Horticultural Society* for the year 1842, p. 393.

"PRACTICAL INFORMATION ON THE BEST MODE OF CULTIVATING FLAX IN BENGAL. BY MR. DENEEF, BELGIAN FARMER.

"'In accordance with my promise, I send you as follows, a detailed report of my observations since my arrival in India, on the cultivation of the Flax plant.

"'I will not enter on an explanation of the mode adopted in the cultivation of this plant in Europe, because nothing is easier than to do so theoretically, but will content myself with informing you, from my own practical experiments, of the means

* M. Bonnevie, indigo-planter at Rungpore, writes: "Having great difficulty to prevail on these ignorant cultivators to plant it—owing to a superstitious belief that the vengeance of an evil spirit will befall them for introducing the cultivation of a new article. Flax grows remarkably well here, and I have no doubt would succeed well in this district. The Zemindars now commence to show an inclination to improve agriculture in general."

CULTIVATION.

Early Experiments.

at our disposal in this country, which can readily be made available for the production of Flax and its seed.

"'1. Such portions of land as are annually renewed by the overflowing of the Ganges, or which are fresh and rich, are the best adapted for the cultivation of Flax.

"'2. After the earth has been turned up twice or thrice with the Indian plough, it must be rolled: because without the aid of the roller the large clods cannot be reduced, and the land rendered fine enough to receive the seed. The employment of the roller, both before and after sowing, hardens the surface of the earth, by which the moisture of the soil is better preserved, and more sheltered from the heat of the sun. About and near Calcutta, where manure can be obtained in great abundance for the trouble of collecting it, Flax may be produced of as good a quality as in any part of Europe. Manure is the mainspring of cultivation. It would certainly be the better, if the earth be well manured, to sow first of all, either *Sunn* (Indian Hemp), or Hemp, or Rice, or any other rainy-season crop; and when this has been reaped, then to sow the Flax. The tillage of the land, by means of the spade (*kodalee*) used by the natives (a method which is far preferable to the labour of the plough), with a little manure and watering at proper seasons, will yield double the produce obtainable from land tilled without manure and irrigation.

"'The mode of forming beds of six feet in width with intervening furrows, in use in Zealand and in Belgium, is very inconvenient in India, because great care must be taken to preserve the moisture of the soil; and on the other part, for the purpose of weeding, they are unnecessary. When proper Linseed, freed from mustard seed, is sown, I think that the Flax requires no weeding at all in India.

"'3. The proper time to sow the Flax in India is from the beginning of October until the 20th of November, according to the state of the soil. The culture must be performed, if possible, some time before the sowing. The Flax which I have sown in November was generally much finer and much longer than that sown in the former month, which I attributed to the greater fall of dew during the time it was growing. The quantity of country seed required to the Bengal *bigha* is twenty *seers*, but only fifteen *seers*, of the foreign seed, because it is much smaller and produces larger stalks. The latter should be preferred; it is not only more productive in Flax, but, owing to the tenderness of its stalks, it can be dressed much more easily.

"'4. The Flax must be pulled up by the roots before it is ripe, and while the outer bark is in a state of fusibility. This is easily known, by the lower part of the stalks becoming yellow; the fusion or disappearing of the outer bark is effected during the steeping, which may be fixed, according to the temperature; say, in December at six days, in January five, in February four days, and less time during the hot season. The steeping is made a day after the pulling, when the seed is separated, and then the stalks are loosely bound in small sheaves, in the same way as the *Sunn*. The Indians understand this business very well, but in taking the flax out of the water it should be handled softly and with great care, on account of the tenderness of its fibres. When it is newly taken out, it should be left on the side of the steeping pit for four hours, or until the draining of its water has ceased. It is then spread out with the root-ends even, turned once, and when dry it is fit for dressing or to be stapled.

"'5. To save the seed, the capsules, after they are separated from the stalks, should be put in heaps to ferment from twenty-four to thirty hours, and then dried slowly in the sun to acquire their ripeness.

"'6. When Flax is cultivated for the seed alone, the country Flax should be preferred. Six *seers* per *bigha* are sufficient for the sowing. It should be sown very early in October, and taken up a little before perfect ripeness, by its roots, separately, when it is mixed with mustard seed; the Flax seed being intended for the purpose of drying oil, is greatly injured by being mixed with mustard seed, by which mixture its drying qualities are much deteriorated. With regard to the dressing of the raw material, most of the coolies are now acquainted with the process, and I have not therefore alluded to it. Should you desire any further information on the subject, I am ready to afford it.'

"Mr. Deneef, in reply to some queries circulated by the Agri-Horticultural Society, observed that too dry or saline soils were injurious to the culture, but that his own had been a heavy clay soil; also, that the Bengal bigha contained 14,400 square feet, or one-third of an acre, and that he sowed of foreign seed 28℔; of American 36℔; but of plump Patna, or native seed, not less than 40℔, on account of its larger size. That the foreign seed cost R8 a maund of 82℔, while the native then cost R2-8.

"The acclimated American seed he found to succeed well in India. But on a previous occasion (Feb. 10, 1841), he had observed of some samples grown at Entally,

CULTIVATION. Early Experiments.

from acclimated English seed, from country seed, and some from Saharunpore seed, (from 30° of N lat.), that the sample from this last was very superior to the others. Of two samples from acclimated American seed, one grown in rather poor ground, the other in a rich soil—'The former,' he observed, 'is a most beautiful sample, containing great length of stalk with thinness; the other is of very little value, the goodness of the soil having caused the plant to become stunted and branchy.'

"Mr. Deneef further observed that he obtained the longest and finest fibres in sowing from the 25th of October to the 15th of November: this he ascribes to the plant being covered every morning with a heavy dew; while that which he had sown in the beginning of October, in the same soil and the same seed, was much shorter in stalks, but much more productive in seed,—'the rain being very scarce from the first days of October until the end of December, in this part of India.' (But the ground is still hot, and the temperature high at this period.) The bigha will yield 100lb of seed from foreign seed, and about 12 per cent. more from native seed.

"He concludes with an approximate account of the cost (amounting to £32) of raising a ton of Flax from Foreign seed, well dressed, and which would be worth £50 in the English market; stating that 80lb of Flax for a bigha of land is a very small product. 'When we shall be able to have the seed from our own product, R60 on that article will be saved.' The account is as follows:—

	R
30 bighas' rent for six months—the other six months for other crops	45
10 maunds of American or European seed, at R8	80
6 bighas of superior (spade) cultivation, at R5	30
24 bighas, four necessary ploughings, each R3	72
Sowing, malees, recolt, rippling, steeping, carrying, and petty expenses	48
28 maunds dressing, in a very clean way, at R3-8 . . .	98
Breaking of flax-tools	7
TOTAL .	380

Return:—1 maund Flax.
19 to 20 maunds Seed.
7 to 9 Codilla.

"'*Chittagong Flax.*—One of the most southern districts in the Bengal Presidency where Flax has been prepared, is that of Chittagong. A. Sconce, Esq., at that time Collector of Chittagong, forwarded, in March, 1843, some samples of Flax which he had grown there from acclimated Europe seed; that is, from seed re-produced for two or three years from imported seed, and sown there in the month of November. He suggested to the Society the awarding of small prizes to natives who cultivated the Flax on account of its fibre. His object being 'to interest chiefly those who are familiar with the cultivation of Linseed (which is common enough for the purpose of extracting oil) and the country *Sunn*.' He states that he had had an opportunity of observing in that season the very great difference between Flax grown from Europe and from country seed—the latter being softer and finer, but very much shorter, and very much weaker. If this is found to be the case by others, the length might probably be easily increased by cultivation. The samples examined by the Flax and Hemp Committee were reported on as follow:—

"'*Undressed Flax.*—This sample contains more Tow than Flax; it is badly prepared, dirty, and not adapted for the Home market; but the

"'*Dressed Flax*—Strong, clean, of very superior quality, but of short staple: if it were a little longer, say six inches, it would realise a very high price in the Home and Continental markets. Mr. Deneef said of it that the staple, though short, was most beautiful; but the mode of preparation (having been hackled) is 'too expensive to admit of its yielding a profitable return, even were it to sell at the value I affix to it, *viz.* £60 a ton.'

"'*Burdwan Flax.*—Four samples of Flax grown at Burdwan were presented to the Agri.-Horticultural Society, by Mr. J. Erskine, in July 1844. Of these, Nos. 1, 3, and 4 were the produce of acclimated and up-country seed mixed together—sown in October and November 1843, and reaped in February and March 1844. No. 2 was the produce of up-country seed, sown on 15th October, and reaped on the 27th February. Mr. J. Law having examined these samples considered them all as of a fair quality; and, judging from the prices of the different marks of the article then in Britain, valued Nos. 1 and 2 at £32, No. 3 at £34, and No. 4 at £30 per ton, landed in England.

"*Bullea Experiments.*

"Mr. de Verinne, Superintendent of Flax cultivation at Bullea (twelve miles below Benares), in reply to the queries, states, as already mentioned (September 20, 1841)

CULTIVATION.

Early Experiments.

that the experiment of the previous season had been a complete failure. He sowed 130 Duncanee bighas, each containing 28,336 square feet, double of the Bengal bigha. The soil was not manured, but ploughed seven and eight times. Somewhat sandy soils are the best. The hard soil which the natives select for growing Flax for the seed, remains in clods, and cannot be pulverized. He first sowed one maund of seed (from the Chupra district) per bigha. Mr. Bernard, one of the Belgian farmers, thought this too much; he reduced it to 20 seers, or 40lb., which proved too little, (and evidently so, as the same quantity is recommended by Mr. Deneef for the Bengal bigha, which is only half the size). He began sowing on the 16th of October, and concluded by the end of the month; the plant was ready for steeping on the 10th of February. He recommends sowing in the beginning of October, as there is moisture enough at the surface of the ground to sow broadcast. Early sowings, also, will in general do away with the necessity for irrigation, which is otherwise indispensable, and expensive. (But is not the greater heat both of the soil and of the sun more injurious than the greater dryness of the soil late in the season, when dew falls?) He states that from the 10th of September till the 20th of January, when the Flax was in seed, and had ceased growing, there was no rain. About fifty bighas were sown with drill-ploughs, because there was no moisture at the surface of the ground, but as Mr. Bernard disapproved of this mode, the rest was sown broadcast, when they were obliged to irrigate the land.

"The plants having been pulled by the 10th of February, and the seeds taken off, the stems were steeped in (indigo) vats. The first vat was steeped nine days; the second and third, ten days; and the fourth and fifth, eleven days, the weather having got cooler from the fall of a shower of rain. Range of thermometer, 60° to 70°. The plant for steeping was not perfectly ripe, but the small plants were left to ripen their seed.

"The crop was small, owing to the unfavourable season. Only 1 maund 25 seers of seed, and 70lb of Flax per bigha, while the Duncanee bigha ought to yield from 150 to 200lb of clean Flax.* The proportion of Flax to the Tow or Codilla varies according to the weather in which the Flax is cleaned; if prepared in the dry weather or hot winds, or from April to the end of June, the proportion is one-third Flax to two-thirds Tow; if prepared in damp weather, or from July to October, it is half to half. With regard to the cost, M. de Verinne says: 'Supposing the season to be an average one, and the produce of the bigha to be 150lb of clean Flax, 100 bighas would give 6¾ tons, and the cost, according to the annexed estimate, being R2,237, will show the cost per ton, landed in Calcutta, to be R331, or £33. In making up the estimate, I have calculated the expenses according to those of the experiment of last year. Only the produce has been valued at 150lb per bigha,' though 70lb only were obtained.

"Estimate of the probable expense for the cultivation of 100 bighas of plant, the manufacture, and the despatch of the produce to Calcutta—

	R	*a.*	*p.*
Land rent for 100 bighas, at R4-8	450	0	0
Irrigating the lands, if there is no moisture at the surface, at R1 per bigha	100	0	0
Six ploughings, at 4 annas each per bigah	150	0	0
Chikorage, or cleaning the fields before sowing, at 8 annas per bigha	50	0	0
150 maunds of seed, at R1-8 per maund	225	0	0
Plucking the plant, at R1 per bigha	100	0	0
Weeding, at 8 annas per bigha	50	0	0
Taking off the seed, at R1 per bigha	100	0	0
Filling the vats, taking the plant out, spreading and turning it, &c., at 12 annas per bigha	75	0	0
Breaking the flax for cleaning, at R1-4 per bigha	125	0	0
Cleaning the flax, at 2 annas per pound	468	12	0
Gunny bags, for bales	20	0	0
Making up the bales	14	0	0

* With these Indian returns of the produce per bigha (which is at Bullea two-thirds of an acre), we may contrast a statement by Dr. Hodges: "From the returns of the Royal Flax Society, and from my own inquiries, I would estimate the average produce of a statute acre in the North of Ireland of air-dried Flax straw, with bolls, at two tons, which, by the seeding machine, are usually reduced to 3,360 lb. By the various processes of the rural manufacturer, the amount of dressed Flax or fibre obtained averages from four to five cwt. per acre."

LINUM usitatissimum. **Early Experiments in**

CULTIVATION. Early Experiments.

	R	a.	p.
Boat-hire, at 8 annas per ton	50	0	0
Chaunder (person in charge of boat)	7	0	0
Carriage of the plant, at R1 per bigha	100	0	0
Four Zilledars, for six months, to look over the cultivation, at R3 each per month	72	0	0
	2,156	12	0
Exchange, at R3-12 per cent.	80	14	0
Co.'s R .	2,237	10	0

"*Monghyr Experiments.*

"'The culture of Flax was commenced near Monghyr, on the Ganges River, in the year 1839; and specimens were presented to the Agri.-Horticultural Society, in May 1840, and again in the month of September in the same year.

"'The strength of this Flax, as ascertained by Professor O'Shaughnessy of the Medical College, Calcutta, was as follows, and as compared with other kinds tried at the same time:—

Monghyr, undressed	40,000
Archangel	43,000
Baltic, dressed	42,033
Do. undressed	19,075
Irish, dressed	17,075

"Mr. Deneef considered it the best sample of India-grown Flax that he had seen.

"The sentiments of some of the members of the Flax Committee were as follow:—

"*Mr. Hodgkinson.*—'The samples of Flax are of middling quality; the fibre fine and strong, but deficient in cleanness and colour. The first defect arises from carelessness in scutching.'

"*Mr. Willis.*—'The Baltic *rough Flax,* which of all the specimens it is the most legitimate for us to compare with the Monghyr *undressed one,* is superior to it in colour, lustre, mellowness, and cleanliness.' The *Monghyr undressed Flax,* not having undergone the degree of cleansing, and preparative manipulation which has been given to the *Baltic rough Flax,* being more ligneous, &c., is not exhibited with all the comparative advantage it otherwise would have shown. The Monghyr *undressed* specimen seems to possess more *tow* in proportion than the rough Baltic one. Its length of fibre seems somewhat inferior to that of the Baltic one. Its strength of fibre seems good. But after all it seems so promising a production that I think the parties engaged in the experiment would do best to send home a good supply of it to the various markets of London, Liverpool, and Scotland, that they may derive the opinion of merchants, brokers, and manufacturers as their most true and unerring guide.

"Mr. Wallace again submitted, in August 1841, two samples of Monghyr-grown Flax, to show the improvement on last year's produce. He stated that they were average samples of thirty-four bales (nearly four tons) which had just been shipped by the *Mary Bannatyne,* for London.

"'*No. 1.*—Grown from country seed on a strong black soil, which had been inundated by the river, and retained its moisture through the season. The seed was sown on or about 10th November 1840. Twenty seers of clean seed to a bigha (the bigha is rather larger than that of Bengal); the plant was allowed to ripen fully, and the seed to come to full maturity. This was all saved by rippling combs, which separate it easily from the stalk. The soil received very little preparation; it was drilled with one plough, another plough following in the same track in which the seed was sown. The plant was pulled about the 20th March, and steeped for three days. The breaking or crushing of the plant was done by machinery; the scutching by hand. A man could clean of this quality of plant five seers a day. The outturn per bigha was 9 stone, such as the muster, and 3½ maunds of seed.

"'*No. 2.*—Grown from country seed on a light sandy soil, also inundated by the river; was sown about 8th October, with the same sort of seed; the land was well prepared, having had three or four ploughings; the seed was sown broadcast 25 seers per bigha; it was allowed to ripen fully; it was pulled 28th February, and steeped 4½ days. The outturn of this was about 1½ stone per bigha, and one maund of seed—no rain having fallen from the time of sowing, seven-eighths of the crop was lost; the dressing of this was similar to the other. A man could not scutch more than 2½ to 3 seers per day.'

L. 21

CULTIVATION.

Early Experiments.

"Mr. Wallace, in his reply to the queries of the Society, states that the bigha at Monghyr contains 3,600 square yards—three-fourths of an acre, and that the inundated land is to be preferred. Alluvial land will yield a crop if not too sandy, but the higher land is preferable if rain fall two months after sowing; that dry soil produces a good plant, but of very coarse quality. He sows at the latter end of September for *seed* only, but from 15th October and all November for fibre, to the extent of 60℔ to the bigha if sown broadcast, and 40℔ if sown in drills.

"The American and Europe seed, he states, produce double the quantity of fibre, half the quantity of tow, and of a finer quality, but only half as much seed as the native. Of this the price in the district was from R1-4 to R1-8. That he reaps from the 25th of February to the 10th of May. That the average crop is about 9 stone, or 126℔ per bigha; and that sixteen of these are required to produce a ton of Flax; with of tow, first quality 36℔, and second quality 14 ℔ per bigha; while there is an average crop of 3½ maunds of seed. The time of steeping varies from two to three days in the hot months, to four and five days in February, October, and November, and to seven and eight days, in the cold months of December and January; and that one day more is to be added for dry plants of the preceding season; and that the cost of Flax laid down in Calcutta, per ton of 20 cwt., is from £12 to £15, all expenses included.

"These expenses consist of rent of land, R1-8 to R2 per bigha; if the rayat uses his land in the other months, then four annas less than the above rates. Coolies obtainable at 1½ annas per day for rippling, steeping, or carrying to the factory; or one rupee for fourteen bundles of 3½ cubits in girth, tightly compressed. It is steeped in a vat, and taken out when the fibre will separate easily from the wood, and then dried in the sun. It is crushed by being passed under large iron rollers, and then placed on the edge of a board, where the fibre is separated from the wood by striking it with a wooden sword. The expense is R3 per bazár maund, and the conveyance to Calcutta, R20 per 100 maunds. Notwithstanding the small cost (£12 to £15) for a ton of Flax laid down at Calcutta, which was pronounced of good quality, and probably similar to the other Indian Flaxes, which were valued in England at various prices, varying from £35 to £45; also stating that the amount realised on the sale of the seed is calculated to have more than covered the expenses of cultivation, rippling, and steeping (*Report in Agric. Soc. Proc., Nov., 1841, pp, 38 and 95*): the speculation did not succeed. For we find in the *Proceedings* of the same Society for February, 1844, pp. 45 and 165, Mr. Wallace, intimating, in reference to his operations for several years past with a view to establish the cultivation of Flax at Monghyr, that he was afraid, from the serious drawbacks he had experienced, that he should be compelled to abandon the speculation unless some encouragement was afforded by Government. The Society having referred all the papers and samples of Flax received from Mr. Wallace to their Hemp and Flax Committee for report, it is stated that one of the members was in favour of an appeal to Government, while the other three were opposed to such a step; but as neither the facts nor reasons for either side are given, we are unable to draw any other general conclusions.'

"The foregoing rather detailed account of the experiments, culture, preparation, and cost of Flax grown in different parts of India, may appear to the reader to have occupied more attention than their importance entitles them to. But without going through this labour, it would be impossible to draw any satisfactory conclusions for the prosecution of any future experiments in the same or in other parts of the country, if such should be thought necessary. But it is first desirable to know what was thought of these fibres when sent to the markets of this country. This we are fortunately able to do, from some of the results having been published, and from some of the specimens having been sent to the India House. A detailed report is given from Liverpool of the first samples grown, and of which a report had also been made by Mr. Deneef. But he generally rated them higher than they were valued in this country, though we are unable to distinguish exactly the respective specimens in the two reports. The Secretary next submitted an extract of a letter which he had been favoured with by Mr. Hodgkinson on some samples of Flax (similar to those so favourably reported on by the Flax Committee of the Society) forwarded by him to Liverpool. (*Vide* 'Report' 1841, p. 41.)

"'The letter, which bears date January 30, 1841, is from Mr. Grey, and he says: 'From what I can judge, and having shown them to a friend here who has probably as much through his hands as any other in Liverpool, a partner of William Jackson, Son, & Co. Mr. Murray seemed far from sanguine about them, but I trust they indicate the capability of producing an article of great importance and extent.

"'*1st.*—The best is a lot (country Flax, native seed, 26th May 1840)—this may be worth £40 to £45 per ton here; it is finer, softer, and better than—

CULTIVATION. Early Experiments.

"'*2nd.*—The large parcel (country Flax, native seed, 27th May 1840)—which has a fine broad fibre, and not much inferior; it is worth £40, if in quantity equal to sample. The lengths are too unequal, which makes it fall upon the hackle, and is a disadvantage. Of these two samples the fibre is by no means weaker than of many other such Flaxes, and probably when this is the case it arises from the preparation.

"'*3rd.*—(Bengal, May, prepared by Belgians.) Dew-ripening weakens, I understand, the fibre, hurts the colour, and even prevents its bleaching as it ought to do, and for which such Flax would be used. Where water is obtainable for steeping this method should not be resorted to.

"'*4th.*—(Indian Flax, No. 1, grown in the neighbourhood of Calcutta, worth at least £30 per ton.) This is better, but seems, if I mistake not, also to be dew-ripened. It is worth £35 here, however.

"'*5th.*—These from imported seed don't seem equal to the produce of native.

"*6th.*—The heckled Bengal Flax does not show to advantage, being imperfectly dressed, and happens to be of a dry, hard nature.

"'*7th.*—(Country Flax, native seed, 26th May, 1840.) This mark is similar to the first, but you will distinguish it, being darker coloured and harsher. The first is the best, I think, decidedly. Weakness of fibre is an insuperable fault where it exists, and it may be perhaps avoided by better preparation. The Belgians' is very weak, and some of the others.

"'*Tows.*—One of these (Bengal Tow, native seed, 30th May 1840), seemingly the clearing or last tow, is a very good thing, worth in Dundee £30 to £33, I should say; the others from firmer tools before this, £20. Another of same mark as first worth perhaps £16. These are of great consumption, as recommended before to your attention Codillas from £12 @ 16 20 per ton would do well, and could be obtained from the waste in preparing the better Flax, observing always that the staple be good and the fibres strong, though they need not be of great length by any means. In Flax the longer the better, though not required beyond moderation, but the fibres should be *equal* and uniform, so as all to split and yield as much dressed as possible. These Flaxes on the whole resemble most the common Newry Flax, which costs £40 @ 50 per ton, wanting the natural sap, in which all these are deficient.

"'The quality I think will not be of the best for some time, but much that would sell largely in Dundee or even Belfast might be obtained; in Dundee everything is used, down to the coarsest; but Flax worth £40 to £60 per ton is most saleable, and to the most certain and best buyers. The Codillas and Tows there seems no doubt of, and Flax to bring from £30 to £45 per ton also.'

"The next report we have is on the samples of Flax grown in the following year (1840):—

"'Four specimens of country-grown Flax, prepared in Calcutta.—*Presented by Mr. H. Woollaston on behalf of the 'London Flax Experimental Society.'*

"'No. 1—Is a sample of Flax of last year's growth, from English seed, *not acclimated.* Six hundred pounds of this quality, Mr Woollaston mentions, were forwarded to London by the *Bucephalus;* and by the last mail Mr. Rogers advises that it was valued at £50 per ton.

"'No. 2—Is a sample from *acclimated* English seed, grown in Entally. The seed was sown last November, and the plant gathered in February, having been in the ground 85 or 90 days. Mr. Woollaston states that this sample is considered much *superior* to No. 1, and that Mr. Deneef attributes its superiority to the seed being acclimated, which renders the separation of the under-bark much easier, and leaves the Flax finer and softer. Mr. Deneef estimates its value compared with No. 1 at £56 per ton.

"'No. 3—Is a sample from the same seed and growth as No. 2—but consists of *picked* portions of plant, so as to furnish a specimen of the degree of fineness that it is possible to produce. This sample is superior even to No. 2, and nearly equal to the best produced in Belgium; much surpassing the £60 Belgian ordinary qualities. Mr. Deneef estimates its value at £60 the ton.

"'No. 4—Is a sample of Flax from *country* seed grown and prepared at Bowsing Factory, district of Burdwan; estimated in London at £40 to £45 the ton.'

"Some specimens of the Flax grown this year having been sent to the India House, and having, in March 1841, been examined by Mr. Hutchinson, of Mark Lane, he thought them very favourable specimens of so recent an experiment, as they seemed to be equivalent to Russian and Polish Flax, which was selling at that time for £40 a ton; and one of the specimens appeared of a quality which might sell for £45 a ton in the then state of the market.

"Mr. Enderby, then of the rope-manufactory at Greenwich, thought yarn made of it very good, and that nothing could prevent both the Flax and Yarn proving valuable

L. 21

CULTIVATION.

Early Experiments.

articles of commerce, if sent in quantities, and of uniform and sorted qualities. Messrs. Noble have recently informed the Author that they also had received specimens, and thought them equal to the middling and even better qualities of Russian Flax. There seems no doubt, from the concurrence of opinion, that the Flax produced in India was sufficiently good to stand favourable comparison with both Russian and Egyptian Flax, and was, therefore, of the kind which is much required, and which could be consumed here in the largest quantities. The question, therefore, is whether it can be produced at a cost so as to yield a profit to both planters and the merchants who would export from India to England.

"Observations on Experiments.

"On reviewing the accounts and the results of these experiments, it appears that though abandoned too soon in some situations, they were carried on for a sufficiently long period in others to allow of reliable deductions being drawn from them, if full information on all points had been supplied. The soil does not seem to have been complained of; but though drainage is essential in many parts of Europe, the power of irrigating will be found most useful in the East. The climate is obviously very different from what the plant meets with in Ireland, as not a drop of rain seems to fall from the time of sowing to that of reaping the crop; but heavy dews compensate for this deficiency during a part of the season. But as this dryness of climate prevails over a great part of India during the season of cultivation—that is, from the end of the rainy season to the beginning of the hot weather—it is evident that irrigation is necessary for such cultivation, and must not be too expensive. Perhaps the double monsoon of the Madras Presidency might in some localities afford a suitable climate if the temperature is not too high. A perfectly appropriate climate may no doubt be obtained at different elevations on the Himálayas, and in some of the mountainous ranges of the South of India; but European superintendence may not be available and the expenses of transit be too great. The Saugor and Nerbuddah territories appear favourable, from their more moderate climate; while some of the districts of the North-West of Sindh, and the Panjáb, may be found suitable, from the command of irrigation and the prevalence of a moderate temperature.

"The proportion of seed required having been ascertained by Mr. Deneef, we cannot but observe the unexpected results obtained from the sowing of some native seed—the Flax produced from Saharunpore seed having been considered to be of excellent quality, and, in Liverpool, to be the best of all the specimens sent. American seed was found to be suitable to the country; but it is remarked that when sown in a rich, it did not do so well as when sown in a poorer, soil. The whole question of what is the best seed for the untried soils and climates of new countries is one of considerable difficulty. It does not follow that seed from a rich soil and the most careful cultivation is necessarily the best for transference to a poorer soil and drier climate; indeed, the converse would, in many cases, appear to be the more suitable course. But even in the case of wheats, some from Australia and from the Nerbuddah, pronounced the finest in the English market, have never produced good crops in this country, notwithstanding the most careful attention. Indeed, the most advisable course appears to be to grow the best native seed, and as thickly sown as is found to be suitable, for the express purpose of changing the branching nature of the plant, and then making an exchange with the seed of other districts following the same course; in order to insure that interchange of seeds which is so beneficial for all kinds of crops, and is conspicuous in India in the indigo crops of Bengal being grown from the seeds of the North-West. In the directions for culture in Europe, we have seen that early sowing is necessary to produce good fibre, and late sowing for seed; but in India the reverse course is to be followed, for early sowing, in consequence of the high temperature, induces rapid but, from the dryness of the climate, stunted growth with an abundant production of seed; while later in the autumn the temperature is lower, the growth is slower, but the dews being heavy there is greater moisture, and this, with the slower growth, produces finer fibre. The early sowing is preferred by some, in order to save the expense of irrigation, but this must sometimes be with the sacrifice of the quality of fibre.

"With respect to the cost at which Flax can be produced in India, we find Mr. Leyburn stating that he gets four maunds of Flax, or 328℔, from about one-third of an acre of land, at a cost of £2-10s., or for about £17 a ton; and that this sold for £28 a ton in London; but we do not find that he proceeded with the cultivation. Mr. Deneef calculated the cost of Flax produced by him to be £32 a ton, with a profit of £8, which would afterwards amount to £14. Mr. de Verinne calculated the cost would be £31 a ton, with an average crop; though he did not get half the quantity. The cost of both is evidently too high, unless the finer qualities of Flax are produced. Mr. Wallace, who continued the longest, and produced Flax at the

CULTIVATION. Early Experiments.

cheapest rate—that is, from £12 to £15—ought to have succeeded, as he states that the expenses of cultivation were paid for by the seed. But we find him, as we have already stated, representing to the Agricultural Society that he should be unable to go on with the culture, from the discouragements he had met with, unless assisted by the Government. Though the difficulties are not specified, they must have been greater than appear from the published accounts, and therefore the facilities and the profit are not so great as they appeared to the *Agri.-Horticultural Society*,—who thus unintentionally induced the Government of Lord Auckland to consider that public aid was not necessary.

"It is very evident that such experiments can only be made under the superintendence of Europeans, when, if successful, they may be adopted by natives. In repeating the experiments in more favourable situations, it would seem very desirable at first to ascertain as accurately as possible the quantity of produce of ordinary quality obtainable per acre, with good cultivation in a favourable locality, on an average of years, and then to endeavour to improve the quality. The profits of the two kinds of cultivation and preparation are not very dissimilar in Europe. Though manufacturers may require more of the coarser qualities of Flax, planters will of course grow that for which they can get the best prices; though it will be safest at first to reckon only on getting the prices of Russian or Egyptian Flax."

As bringing **Royle's** account down to more recent times, the following special report, written by **Mr. W. H. D'Oyly**, Collector of Howrah (which appeared in the Proceedings of the Revenue and Agricultural Department for 1873), may be given. The report is of so great interest that it is feared it would be seriously injured by any attempt at abbreviation, though in some instances **Mr. D'Oyly** deals with some of the facts given above in the passage from **Royle's** *Fibrous Plants of India*—his report is, in fact, a review of **Royle's** statements and of all that had subsequently appeared. **Mr. D'Oyly**, in forwarding his report, accompanied it with a letter from which the following passages may be extracted, as showing the extensive research made by him which resulted in his report embracing chief facts continued in the Agri.-Horticultural Society's Journals as well as his own original enquiries:—

"While I was at home, or rather on my way there and back, I picked up all the information I could get from the journals of the Agri-Horticultural Society of India and several other books regarding the cultivation of flax in India; the enclosed report is the result.

"Dr. Forbes Watson very kindly took a great deal of trouble in hunting up all the records of experimental cultivation and manufacture in India, and supplied me with numerous extracts from the records of the India Office. I now send on the report to you, as it may be of some use to Government. It has been proved that there is nothing in the soil and climate of certain parts of India against the production of good fibre; in fact, very good fibre *has* been produced, but it has not yet been proved whether the cultivation and manufacture can be carried on *profitably*. Panjáb flax manufactured by Government was proved by sale at auction to be equal to Russian flax, which is better again than Egyptian. Now, if the Egyptians can make it pay, we ought to be able to make it pay also. If India could be made a flax-producing country it would be a grand thing not only for India, but for England.

"I believe I have cited chapter and verse for every statement I have made, so that there may be no mistakes.

"*Report on the cultivation of Flax in India.*

"OBJECT.—The object of this paper is to show, in as concise a form as possible, the results of the several experiments that have been made from time, to time both by Government and by private individuals, in the cultivation of the flax plant for fibre in various parts of India, and to urge the prosecution of further experiments in such a manner as to give them at least some chance of success. By success I mean something more than the success which has already attended many of the experiments which have been made; for, although it has been proved that fibre can be produced from the flax plant grown in this country when the cultivation is properly attended to, and that such fibre would find a ready sale in some of the home markets, still India has not become a flax-producing country.

"NATIONAL IMPORTANCE OF INDIA AS A FLAX-PRODUCING COUNTRY.—That England should be able to supply her manufactories with flax from India, instead of depending so much as she does on Russia, Prussia, Egypt, and other countries, is a matter of incalculable importance. I exclude Belgium, as there is at present no rea-

CULTIVATION.

Early Experiments.

son to anticipate any difficulties with regard to the supply from that country, and also because India, though perhaps she may not hope to equal Belgium in the *quality* of her flax, might still well compete with Egypt, Russia, and other countries. For the Dundee market the Indian flax that has already been produced in the Panjáb is admirably adapted. In the Dundee Chamber of Commerce, Mr. O. G. Miller, the Chairman, in 1858, speaking of the Panjáb flax, specimens of which were laid on the table, said that 'the fair value of the flax at the present time he considered to be fully £45 a ton overhead, and he would be glad to take a quantity of it at that price.' 'Indeed, a mixture of this flax with that of the Baltic *would much improve the yarn by imparting strength to it*. Such flax would be therefore admirably adapted for canvas yarn and other yarn where great strength was required. Every one then would agree that if a supply of *some thousand tons annually* of this flax could be obtained at a fair price, a very great boon would be conferred upon the linen manufactures of this country, and this district in particular would be benefited more than Belfast or any other town, for we have a hold of the coarser end of the linen trade, for which this flax is better adapted than for the finer spinning of Belfast and Leeds.' Dr. Royle, in his work on the *Fibrous Plants of India* (*page 141*), quotes from a letter written by Mr. J. McAdam as follows: 'Belgium, Holland, France, and Ireland can supply all the world with fine fibre, but Russia and Egypt cannot keep pace with the demand for coarse.'

"FAVOURABLENESS OF THE PRESENT TIME.—I believe that the time has now come when the cultivation and manufacture of flax can be carried on in India profitably—certainly with more hope of profit than it was reasonable to expect in former years—*first*, because higher prices can now be obtained in the home markets; *secondly*, because the means of communication both in India and also between India and England have been so much improved, and the rates of freight have lately been so much reduced;* and *thirdly*, because the great irrigation works lately undertaken by the Government of India in Behar and in other parts of the country, and which are being extended throughout immense tracts of dry and thirsty, and therefore hitherto comparatively unfruitful, lands, will supply the only thing wanting to enable farmers to turn these lands to good account,—lands now in many places lying waste to a great extent, and even where cultivated, obtainable in many parts at a very low rate of rent.

"THE FLAX PLANT OF INDIA.—The Indian plant called *ulsee* or *teesee* is a variety of the flax plant which is grown exclusively for the seed from which linseed oil is expressed. This variety has acquired certain characters from the peculiar method of culture adopted with a view to obtain as great a quantity of seed as possible; and this is done at the expense of the fibre, for the plant, in consequence of the peculiar method of culture, is always *short* and *bushy*. The same thing occurs with the plant known commonly in India as *ganja* or *bhang*, which produces an intoxicating drug; this is the true hemp plant (**Cannabis sativa**), and in India it is not cultivated for fibre, but for the 'resinous secretion of its leaves.' Dr. Royle in his work on the productive resources of India, says: 'In Europe it is well known that if it be wished to prevent a plant secreting the principles, bitter, acrid, or otherwise, which are natural to it, the practice is to exclude it from the influence of light and air, as in tying up lettuces and covering up celery; so to ensure a full secretion of the principles natural to a plant, an opposite treatment is necessary, and is practised with the hemp in India by openness of planting and consequent exposure to the full influence of light, heat, and air.' Hence the flax plant in India being valued on account of the oil of its seed, is sown in lines on the borders of fields, and not thickly, so that it becomes short and bushy, with many branches yielding a heavy crop of seed. This every one who has seen anything of the country where linseed is cultivated must have observed. To secure length and fineness of fibre, it is necessary to sow both hemp and flax very thickly in the same way that jute and sunn-hemp are sown.

"EARLY EXPERIMENTS.—It appears that the attention of Government was first directed to the cultivation of fibrous plants in the beginning of the present century.

* In the first part of the year 1855 the freights for jute varied from £5-5*s*. per ton to £6-5*s*. The quotations now (end of April 1873) are £2-12*s*. per ton. In July 1872 the steamer *City of Cambridge* took home 100 tons jute at £3 per ton, and in the same month the *City of Canton* took 300 tons jute at £2-10*s*. per ton. The *City of Lucknow* took 500 tons jute at £2-7*s*. 6*d*. per ton. In May 1872 the ship *Tythonus* took 625 tons jute at £2-5*s*. per ton. In both the years 1872 and 1873 the highest rates for jute did not exceed £4 per ton.

CULTIVATION.

Early Experiments.

The East India Company had hemp farms, and Dr. Roxburgh seems to have been the first to attempt the production of flax in India. Unfortunately no record appears to have been preserved of the result of his experiments. Dr. Royle who published a work on the fibrous resources of India, called attention to the subject in 1834, but little seems to have been done till 1839, when a company was formed in London called 'the London Experimental Flax Company.' By them a Belgian farmer and a Belgian preparer of flax were sent out to Bengal, and experiments were made with more or less success in several parts of Bengal, Behar, and the North-Western Provinces. Riga, Dutch, and country seed, were all tried. Generally, as might be expected, the Riga and Dutch seed proved better than the country seed, still some of the specimens of flax produced from the latter were most favourably reported on and considered by Mr. Deneef, the Belgian farmer, to be worth £66 a ton. This is probably a rather high estimate, for some specimens sent to Liverpool at the same time were valued at from £30 to £45 per ton; but it is right to add that some time after some Panjáb flax was actually sold at auction at £54-10s. per ton.* The Agricultural and Horticultural Society of India took the matter up warmly. Medals and prizes were offered; but although the results of experiments were encouraging, the cultivation did not till many years after extend beyond the few bighas in the several places where experiments were made.

"EXPERIMENTS IN SHAHABAD.—In the district of Shahabad, at a factory called Nonore, which is on the banks of the river Soane, Mr. Lyburn for several years (1840 and previously) persevered, and although his efforts were crowned with great success, still he wrote in a desponding way of the almost insurmountable difficulty he encountered in getting the natives to take to anything new. He must here have referred to the manufacture of the fibre, and not to the cultivation of the plant, for it is beyond doubt that the natives will always take to any cultivation that will pay well, and even to the preparation of the produce, when it is not attended with any great difficulty requiring more than ordinary care. Take, for instance, opium and jute, the cultivation of which has increased so enormously. But the preparation of the flax fibre is infinitely more difficult than that of jute or sunn-hemp, and requires European superintendence, without which the natives would never produce a marketable fibre. Even in the comparatively easy preparation of jute and sunn-hemp, a large proportion of the fibre is more or less damaged by carelessness in preparation. Mr. W. Stalkartt of Ghoosery, who owns an extensive rope manufactory, informs me that no less than one quarter of the sunn-fibre brought to him is damaged by carelessness in steeping; the stems are often left for too great a length of time in the water. I may add an extract from a report by Mr. Sturrock, Secretary, Chamber of Commerce, Dundee, dated 11th March 1860 (*Journal of the Agricultural and Horticultural Society of India, Part III., Vol. XII.*):—'As to the quality, it is not nearly so good as a lot sent home by Mr. McLeod, which I sold at £54-10s. per ton. It is not so good coloured, is altogether harsher, and has much less of the natural sap in it. This we consider must arise from the stock having been allowed to grow too long a time, or not having been pulled early enough, *and from over-steeping, or otherwise bad preparation.*

"Again, in the *Agricnltural and Horticultural Society's Journal for 1863, Part III, Vol. XIII.*, it will be seen that both Mr. McGavin and Mr. Stalkartt, in reporting on the weakness of samples sent to them for report, attribute the weakness to want of care in preparation. To return to Mr. Leyburn's experiments. It appears from the *Proceedings of the Agricultural and Horticultural Society iu November 1841* that 'Mr. Leyburn succeeded in producing an article of lengthened staple, and of a quality vieing with the flax of Russia and elsewhere. A poition of the cultivation was carried on in the sandy bed of the Soane river, and part in the uplands of the district." [Mr. Leyburn's results will be seen at page 15 above.—*Ed.*]

"In a foot-note the size of the bigha is not very clearly shown. The Shahabad bigha at present contains 27,225 square feet, and is therefore about two-thirds of an acre; but Mr. Leyburn puts down his bigha as 'something more than the Bengal bigha, which is 1,600 square yards, the third of an English acre.' Still, as he calculates the produce at four maunds of flax per bigha, which is equal to one ton from seven bighas, it would seem that the bigha referred to by him must be at least as large as the present Shahabad bigha. The factory maund is equal to $74\frac{11}{16}$lb, therefore four maunds would be equal to 2 cwt. 2 qrs. $18\frac{12}{16}$lb. At my request a search was made among the records of the Nonore factory by the present owner, Mr. Solano, for any papers that there might be relating to Mr. Leyburn's experiments, but unfortu-

* This, however, was a very small quantity, *viz.*, about two tons, which was the best that had been then prepared. See extract from Mr. Sturrock's report quoted in the next paragraph.

CULTIVATION.

Early Experiments.

nately the factory was burnt in the mutiny, and no papers on the subject have been found.

"MR. DENEEF'S ESTIMATE OF THE COST OF CULTIVATION AND PREPARATION.—Mr. Deneef, the Belgian farmer above alluded to, drew out an estimate of the cost of a ton of flax produced from English seed well dressed, and in his opinion worth £50 or more in the home market. It must be remembered that this showed the cost of cultivation, &c., near Calcutta, where it was, and always will be, higher than it is up-country. It will be seen that he calculates that thirty Bengal bighas* would be necessary to produce one ton of flax."

[Mr. Deneef's estimates will be found on page 18.—*Ed.*]

"EXPERIMENTS AT MONGHYR.—Mr. Wallace, at Monghyr, for years made experiments, and he perhaps may be said to have been the most successful producer as far as cost of production is concerned. He states in a letter, dated 21st August 1841, to Dr. Spry, Secretary to the Agricultural and Horticultural Society of India, that the expenses of *cultivation, rippling, and steeping, were more than covered by the proceeds of sale of seed*. The cost of preparation of three tons he gives as follows:—

	R
Scutching of three tons after being broken or crushed by the rollers	190
Gram and grass for horses and bullocks, syces' wages, &c.	94
Peons and carpenter	10
Mofussil expenses and carriage to Calcutta	60
Boat hire and travelling expenses	10
TOTAL	364

R364=£36, or £12 per ton.

"Mr. Wallace estimates the produce per bigha† as follows:—Flax, 'average crop 9 stone, 16 bighas for a ton; tow, 1st quality 36lb, and 2nd quality 14lb; seed, average crop 3½ maunds.' He considers that the most preferable land is that which is yearly inundated; chur lands, he says, will yield a good crop if not too sandy, but high lands are preferable if rain falls two months after sowing.‡

"EXPERIMENTS AT BULLEAH, IN GHAZEEPORE.—Mr. Deverinne's experiments at Bulleah, 1840-41, were the least satisfactory of all, but a reason is given for this, and it would appear that better results were obtained in other years from his estimate of average crops. In his letter to Messrs. Hamilton and Company, Calcutta, dated 20th September 1841, he estimates the cost of cultivation of 100 bighas, and of the preparation and despatch of the produce to Calcutta, at R2,237; the yield, he estimates, in an average season, at from 150lb to 200lb per bigha, or (say) 6⅔ tons per 100 bighas; the cost of one ton, therefore, landed in Calcutta, would be R331. The bigha referred to is the Duncanee bigha of 28,336 feet square, about double the size of the Bengal standard bigha. Mr. Deverinne adds that the proportion of flax to codilla is one-third flax to two thirds codilla, so that besides the 6⅔ tons clean flax, there would be 13⅓ tons codilla? The figures given above are taken from Mr. Deverinne's estimate of an average season; but his actual return in 1840-41 fell far short of these figures. This, however, is not to be wondered at when we see that the season 1840-41 was an unfavourable one. The plant was blighted in January; from the 10th September till the 20th January, when the flax was in seed, there was no rain; country seed was used; the sowings were too late, and, by the advice of Mr. Bernard, 20 seers of seed only were sown in each bigha, which was found to be much too small a quantity for the production of good flax; part was sown with drill ploughs and part broadcast; the flax was cleaned at an improper time, *viz.*, during the hot winds;‖ instead of get-

* The Bengal bigha here referred to contains 14,400 square feet (about one-third of an English acre, and about half of the Shahabad and Ghazeepore bighas). (See reports of experiments in those districts.)

† Of 3,600 square yards, ¾ths of an acre.

‡ Letter from Mr. J. Wallace to Dr. Spry, dated 2nd September 1841. Proceedings of Agricultural and Horticultural Society, November 1841.

§ I cannot help thinking that this must be a mistake for ⅔ flax to ⅓ codilla. Mr. Deneef puts down 1 ton flax to 9 maunds (or say 7 cwt.) codilla; and Mr. Wallace estimates 9 stone (126lb) flax to 36lb 1st quality tow, and 14lb 2nd quality tow.

‖ See *Balfour's Cyclopædia of India*, page 840; also letter from Mr. Deverinne to Messrs. Hamilton and Company, Calcutta, dated 20th September 1841, published in Agricultural and Horticultural Society's Journal for 1841, page 101.

CULTIVATION. Early Experiments.

ting a yield of 3 to 3½ maunds of seed per bigha, he harvested only 1 maund 25 seers per bigha.

"WITHDRAWAL OF GOVERNMENT AID.—Although Lord Auckland took a very great interest in the cultivation of flax, as indeed he did in all matters connected with the development of trade in the various fibrous productions of India, and though he requested the Agricultural and Horticultural Society 'to supply such accurate detailed information as they may possess, or as they may be able to obtain,' still nothing further was done by Government; for it appears that on receipt of the Society's report, the Secretary of the Government of India wrote a letter* to the Secretary of the Agricultural and Horticultural Society, from which the following passage is taken:—'The cultivation of flax can no longer be considered a doubtful experiment, since it appears from your report to have proved in many instances successful, and, where successful, to be very fairly prafitable. His Lordship in Council is therefore much inclined to doubt whether any bounty or reward from Government is necessary or would be justifiable for the support of this undertaking.' This was a rather unexpected blow; the Society seems to have almost dropped the question, and the flax company seems to have collapsed.

"GOVERNMENT EXPERIMENTS IN THE PANJAB.—Later on some very successful experiments were made by Government in the Panjáb, chiefly through the exertions of Mr. Cope. These were on a more extensive scale than those that had previously been made, and Lord Dalhousie took a personal interest in the matter. The plan proposd by Mr. Cope, and approved of by Government, was that premia should be given of R500, 350 and 200, for the successful cultivation of flax on areas to be respectively not less than 25, 20, and 15 acres; that Government should purchase the entire crop, seed, and fibre of the required length, paying at the market rate of the seed, with 25 per cent. added for the fibre; that simple instructions should be drawn out and made as widely known as possible.† These and some other resolutions were passed in 1854, and operations commenced during the autumn of the same year. On the pledge of Government to purchase the whole of the produce, large areas were sown with linseed. The cultivation increased from 3,453 acres in 1853-54 to 19,000 acres in 1854-55. These are the returns for eight districts only. The services of a German, Mr. Steiner, were secured; that part of the cultivation over which he was able to exercise immediate control yielded very satisfactory results; 110 maunds of respectable fibre were thus produced, and purchased by Messrs. Harton and Company of Calcutta.

"Unfortunately, however, the instructions given were not generally attended to. I will quote Mr. Cope's own words:—

"'But, as a general rule, it was found that the zemindars had so entirely neglected the tenor of the instructions communicated to them, that a great proportion of the plant was found, on examination, totally unsuited to the production of even a decent fibre, and *in no one instance* would the zemindar exert himself to the extent of *attempting even* the preparation of any fibre, although the difficulties in the way of doing so are by no means great.‡ This preparation of fibre is, however, an operation requiring labour, and labour is what the Panjábi cultivator dislikes of all other things if he can possibly avoid it, a fact that points to the *desirableness of introducing machinery*§ on an extensive scale, leaving the country-people to cultivate and increase the growth of whatever of a vegetable character may be applicable to the arts of Europe.'

"Again, Mr. Cope says: 'The Indian, or more correctly the Panjábi, in fact does not know what labour is. No wonder he prefers sitting smoking his *hukah*, now and then peeling his sunn stems by way of a change, to standing up like a man to break and scutch flax or any similar profitable, but distasteful, occupation. He may be brought to grow the plant, but I much fear that is all—for the present at least.' There is undoubtedly a good deal of truth in these remarks. All the flax

* From T. H. Maddock, Esq., Secretary to the Government of India, Revenue Department, dated Council Chamber, the 22nd November 1841, to H. H. Spry, Esq., Secretary to the Agricultural Society.

† See paper on the introduction of flax as a fibre-yielding plant into India, and especially into the Panjáb, by H. Cope, published in Vol. XI. of the Journal of the Agricultural and Horticultural Society of India, 1859.

‡ Here I think Mr. Cope is wrong, for the testimony of all other persons whose experience renders their opinions of great value, shows that the preparation of the flax fibre is more difficult than that of any other fibre—rhea perhaps excepted.

§ The italics are mine. In a paragraph below I have given my reasons for the belief that the use of machinery is necessary for preparation of this fibre in India.

L. 21 (408)

CULTIVATION. Early Experiments.

that was produced in the Panjáb was produced by hired labour, except in one isolated instance, namely, at Deenanuggur, where Mr. Cope found that a small quantity had been prepared by the cultivators.

"Owing to the inattention of the cultivators to the simple instructions given, the premia offered by Government were not *earned*, none of the prescribed conditions having been complied with. Mr. Cope, however, recommended that the premia should not be altogether withheld, and a sum of R1,491 was apportioned to the most deserving cultivators. Unfortunately, however, the payment *was delayed for two years*. It is therefore not to be wondered at that farmers and cultivators lost the small interest in the matter which had at first been excited by the offer of the premia. Lord (then Sir John) Lawrence, in the Panjáb report of 1855-56, says: 'In 1855 (the autumn of that year is here meant) about 25,000 acres of linseed (that is, half the superficial area sown in 1854-55) were sown, but the season, being dry, was unpropitious; no merchantable flax was obtained.' Now, irrigation would have prevented the loss caused by the dry season. Mr. Cope, in his paper above cited, says that Sir John Lawrence must have made a mistake, for 'the very fibre which has attracted so much attention at Belfast, Dundee, and Leeds, was prepared from flax grown in the districts of Gujranwalla and Lahore during the season of 1855-56.' The quantity sent home, however, was very small, being about two tons. Some 55 maunds, prepared by Mr. Steiner, were sent to Belfast, where the quality was tested. At a special meeting of the Chamber of Commerce at Belfast, the question of the cultivation of flax in India was discussed, and some yarn, spun from the Panjáb flax, was examined and pronounced to be of the value of from £35 to £45 per ton. Subsequently at Dundee a meeting of the Chamber of Commerce was held (September 1858), and three kinds of flax from the Panjáb were valued respectively at £48, £46, and £38 a ton. The Chairman, Mr. Miller, remarked on the strength of the Panjáb flax. In a letter, dated Lahore, 20th July 1859, written by Mr. T. H. Thornton, Personal Assistant to the Officiating Financial Commissioner, to the address of the Secretary to the Government of the Punjáb, the financial results of the experiment are given. In looking at these results we must remember, as Mr. Thornton says, that the object of this experiment 'was not to ascertain whether the exportation of flax could be remunerative, but to make investigations regarding the soil best adapted for its culture and to initiate the zemindars into the processes required for preparing the fibre for the European market.' 'It appears also that a considerable portion of the seed and inferior fibre was distributed gratuitously among the zemindars.' Attached to this letter is a statement of the sale proceeds of Gujranwalla flax, which I give below, as it shows what a large margin is left for the cost of cultivation and preparation:—

RECEIPTS.	£ s. d.	£ s. d.
Sale proceeds of 19 bales sold at Belfast, as per account	63 6 4	
Sale proceeds of 9 bales sold at Dundee, as per account	28 15 10	
		92 6 2
DISBURSEMENTS.		
From Wazírabad to Múltan, including cost of packing	4 10 1	
From Múltan to London	16 0 6	
„ London to Belfast per steamer, including warehousing in London, commission, &c., &c.	4 15 0	
From London to Belfast, including storing, commission, &c., at Belfast . . .	3 18 2	
For carriage of 9 bales to Dundee, &c., &c. . .	1 12 9	
		31 8 6
Net proceeds ...		60 14 8

"Now, it will be seen from the above that after paying all costs of carriage, packing, warehousing, and commission, there remained a balance of £60-14-8 to cover the expenses of cultivation and preparation of two tons (or rather less) of flax. Then again, over and above this there would be the proceeds from the sale of seed and codilla. Mr. Wallace of Monghyr found that the proceeds from the sale of seed alone covered the cost of cultivation, and he has shown that the expenses of preparation, packing, and carriage to port need not exceed £12 a ton—say £24 for the two tons; but even if we double this amount, there still remains a large balance for profit.

CULTIVATION.
Early Experiments.

"MR. GUBBINS' EXPERIMENTS AT ALLYGHUR.—In 1854-55 and 1855-56 Mr. Gubbins, then Judge of Allyghur, made some experiments, using both foreign and native seed, and sent some of the flax straw produced to the Agricultural and Horticultural Society in Calcutta, and also some specimens of fibre, prepared in a rough way, *without the usual process of retting*. The straw was forwarded to the Chamber of Commerce, Dundee. This was grown '*from native seed and after the native fashion solely for the sake of the seed*.' (See *Journal of the Agricultural and Horticultural Society, Vol. IX., 1854, p. 379.*) Mr. Robert Sturrock, Secretary to the Chamber of Commerce, Dundee, thus reported on it: 'From this it appears that if proper means as to sowing, growing, and cleaning are taken, India may be a flax-producing country; and in order that we may be provided with as much information as possible, could you state at about what price such fibre, prepared and ready as the sample, could be shipped from India? Unless it can be laid down here at a certain cost, it could not be used; and I have to mention that the Directors of the Chamber consider that if such flax could be laid down at a price not exceeding £35 a ton, it would be useful in the manufactures of this place. The price refers to this particular sample; *if better more could be got*.' These remarks referred to the following report:—'On examination you will find it, I think, to contain considerable promise, there being evidently nothing in the soil or climate of the district in which it is grown opposed to the production of a flax that could be used by spinners generally in this country. From the straw having been intrinsically a very poor article, such as the flax steepers in this country would scarcely purchase at all, you will find the quality of the finished sample also indifferent, but nevertheless it is a *marketable article*, and I think good results might follow if another experiment was made in which more pains was bestowed on the cultivation.' It must be remembered that this flax was raised from country seed sown thinly in the same way that natives sow for the sake of a good crop of linseed. Mr. Gubbins, to show the natives what use could be made of the enormous quantity of refuse *ulsee* straw which they throw away after the seed has been collected, made from some of this refuse straw a quantity of string which was purchased by a Calcutta firm. It does not appear whether his lessons yielded much fruit; but I suppose that there being no purchasers on the spot, the experiment was dropped, and now, as formerly, thousands of tons of *teesee* stalks are thrown away, from which at least a large quantity of string could be manufactured without expense and with very little trouble. In some parts of India the stalks are given as food to cattle; but cattle are equally fond of the refuse after the fibre has been extracted. In the steam retteries in England the water which runs through the flax stems is always used with the husks of flax and chopped straw to feed cattle, and forms a very nutritious food. (*See Mr. Blechynden's notes of a visit to Wishaw Flax Works, Journal of the Agri.-Horticultural Society for 1854, Vol. IX., p. 27.*)

"OTHER EXPERIMENTS ON A SMALLER SCALE.—Experiments on a small scale were made at Burdwan, and four samples were presented to the Agricultural Society by Mr. J. Erskine in July 1844; these were raised from acclimated and up-country seed, and were pronounced to be of a fair quality, and valued at from £30 to £34 per ton. In Chittagong Mr. Sconce, then Collector of that district (1843), produced some flax from acclimated Europe seed, and also some from country seed; the latter yielded a flax that was softer and finer, but much shorter and weaker than that raised from acclimated seed. At Jubbulpore Mr. Williams produced some flax of good quality, and Dr. Royle considered that this part of India is the best adapted for the growth of flax, the climate being a medium between the extreme dryness of the North-West Provinces and the moisture of Lower Bengal—an argument that would apply equally well to the province of Behar. The Jubbulpore hemp brought into notice by Mr. Williams, is well known, and Dr. Royle judged from its length and strength, as well as by the specimens of flax grown by Mr. Williams, that 'the climate of the Saugor and Nerbudda territories is eminently suitable to the production of flax.' The linseed of these territories is a white species, and seed collected therefrom was distributed for experiments to different parts of the country. Mr. Finch of Tirhoot found that three-fourths of the crop raised from this seed 'was destroyed by caterpillars, while the common linseed, grown in the vicinity of the white, was left untouched by them.' (*See Royle's Fibrous Plants of India*, p. 144.) Mr. Henley near Calcutta was unsuccessful. In the first year he procured from Bhaugulpore some flax straw, but the plant had been grown for seed in the native fashion, and the result was that it was too bushy for the production of fibre; the next year he took a patch of land—'fine friable loam fit for anything; it had been a cauliflower bed, and was therefore deeply spade-cultivated and highly manured.' (Letter to *Dr. Royle*, quoted above, p. 12.) He had the best seed sown very thickly. The plants came up well, and attained a height of three feet, but the result

CULTIVATION.

Early Experiments.

was that the fibre was small and weak, and inferior to the Jubbulpore and up-country flax.

"CONCLUSIONS.—From the results of the several experiments that have been made, I think it may be fairly inferred—*first*, that flax of a marketable quality, and of a quality for which there is a great and increasing demand, can be produced in several parts of this country, notably in Behar, the North-Western Provinces, the Panjáb, and the Central Provinces; *secondly*, that it is quite possible to manufacture flax fibre profitably where the area of cultivation is limited, and under the immediate supervision of a qualified European; *thirdly*, that the native cultivators generally will not prepare the flax, but that they can and will grow the plant in the way that is necessary for the production of good fibre *if it is made worth their while, i.e.*, if a price be paid for the produce equal to what the cultivators make by other crops on the same land at the same season; *fourthly*, that to firmly establish the cultivation of flax for fibre, it is absolutely necessary that Government should give every encouragement and help that it is in its power to give, and that whatever may be the nature of that encouragement and help, *it should not be withdrawn too soon*, as formerly it was, both in Lord Auckland's and subsequently in Lord Dalhousie's time.

"COLLECTION OF INFORMATION AS TO VALUE OF PRODUCE.—It remains to be seen whether flax can be cultivated *extensively* in this country in such a way as to yield a fair profit, *first*, to the cultivator, and *second*, to the purchaser of the plant and preparer of the fibre. To show this, it will first of all be necessary to calculate how much the latter would have to pay the former for the plant when ready for pulling, and for this purpose I have collected certain information from persons well qualified by long experience and intimate acquaintance with the agricultural classes to give accurate information on the subject. [The answers obtained were embraced in an appendix which it is not thought desirable to republish here, but the following remarks indicate the gentlemen from whom the facts shown in the appendix were obtained.—*Ed.*] I may mention here, says Mr. D'Oyly, that Mr. Mylne is a member of the firm of Burrows, Thomson and Mylne, who hold from Government a lease of the extensive Judgdeespore estates, which formerly belonged to the notorious rebel leader Koer Singh. This estate has been transformed by them from a jungle into one of the most valuable estates in that part of the country (Zillah Shahabad), and they bear the very highest character as zemindars. Mr. Charles Fox, with his elder brother, owns several indigo factories in Shahabad, and is the able manager of the Maharajah of Doomraon's extensive estates. Moolvie Syud Abdool Hye is the manager of the Court of Wards' estates in Shahabad, and is also a zemindar holding land in Zillahs Patna and Jaunpore. I am much indebted to all these gentlemen for the trouble they have taken to obtain for me as accurate information as possible.

"ESTIMATE OF COST OF FLAX STRAW AND OF THE MANUFACTURE THEREFROM OF FIBRE.—*Teesee* is sown by the natives for the seed crop in two ways—sometimes broadcast, and sometimes in lines with or round other crops. To make the cultivation of the same plant *for fibre* profitable to the raiyat, he must be paid for the produce of his field a sum equal to what he would get for the produce of the same field when cultivated in the ordinary way. When sown round the edges of barley, wheat, or gram fields the linseed yield per bigha would be about one maund, value R3, and the value of the produce of the other crops would be about R10, the value of fodder R2, and of the oil R3, or in all R18.* The value of seed saved for the yext year's sowings may be considered as a set-off against the value of seed sown. When sown broadcast, the yield of linseed in *a good year* might be from 5 to 7 maunds, or an average yield, taking good and bad seasons into account, would be about 3½ to 4 maunds. Syud Abdool Hye puts down the average yield at a somewhat smaller figure, but when we consider that Mr. Leyburn, Mr. Wallace, and others got 3½ maunds of seed per bigha from the plant that they cultivated for fibre, and remembering that when grown in the native fashion for seed the yield would be greater, I think we may look on Messrs. Mylne and Fox's figures as the more correct. Besides, in calculating what would be a fair sum to pay the raiyat, it is better, if we err at all, to err in his favour. Putting the average yield then down at 4 maunds, the value would be R12. In either case, therefore, in an *average season* the value of the produce of a bigha† of land from November till February or March would not exceed R16. From former experiments it appears that from 15 to 16 bigha are necessary to produce one ton of flax. If, therefore, raiyats were asked to grow the

* This is perhaps the outside value of a good crop. R16 would be rather above than under the average value of produce.

† The bigha of 27,225 square feet.

CULTIVATION. Early Experiments.

linseed plant in the way necessary to produce good fibre, and were paid at the rate of R16 per bigha for the produce, the purchaser would have to pay R250 for the produce of 16 bighas. The seed would be worth R168, calculating the yield at 3½ maunds per bigha, and the value at R3 per maund. Deducting, then, the value of the seed from the price paid to the raiyats, there would be a balance of R88 to be added to the cost of preparation, packing, and carriage to port, and if this cost does not exceed R120 (£12) per ton (this is what it cost Mr. Wallace at Monghyr), the flax could be shipped at Calcutta at £20-16*s.* per ton. The cost of freight, screwing, and insurance, should certainly never exceed £6 per ton (£4-10*s.* would be nearer the mark); so that a ton of flax might be landed in England, Scotland, or Ireland at £26-16*s.* a ton at the outside. Then, besides this ton of flax, there would be (say) 8 cwt. of codilla, which could be disposed of in Calcutta for from £3 to £3-10*s.** The ton of flax, if moderately well prepared, would fetch not less than £40, and probably more.† The expenses in England would be about £1-11*s.* per ton.‡ In the above calculation the value of the refuse, both for fodder and for manure, has not been taken into consideration. The following table will show in a more convenient form the figures above given:—

Expenditure.	R	a.	p.	Receipts.	R	a.	p.
Price of produce of 16 bighas	250	0	0	66 maunds seed, at R3 each .	168	0	0
Cost of preparation, packing, and carriage . . .	120	0	0	8 cwt. codilla (say) . .	32	0	0
Freight from Calcutta to England, insurance and screwing	60	0	0	1 ton flax	400	0	0
Expenses in England, warehousing, commission, &c.,	15	8	0				
Profit	154	8	0				
Total .	600	0	0	Total .	600	0	0

"If cost of preparation, &c., be raised to £20 per ton.—In the above calculation it will be observed that I have adopted Mr. Wallace's estimate of the cost of preparation, packing, and carriage, which is undeniably a low one, and probably would not cover the cost if machinery was employed with European supervision; but even if we raise these figures to R200 (£20) per ton, there would still be a profit of R74-8 (£7-9*s.*) per ton if the flax sold at £40 per ton, or a profit of R24-8 (£2-9*s.*) per ton if the flax realised only £35 per ton. Another R10 (£1) might safely be added to the profit, as the freight, insurance, and screwing would in all probability never exceed £5 per ton, but, on the contrary, be found generally to cost even less.

"Steam retteries.—The several processes which have to be gone through when only manual labour is employed, from the time that the plant is ready for pulling till the fibre is ready for scutching, require great care; and without constant and close supervision, such as it would be impossible to afford with any hope of ultimate profit, the natives would never turn out a marketable fibre. The pulling, stooking, stacking, rippling, water-retting, drying, &c., &c., require the greatest care and attention. By the introduction of steam retteries, fibre can be prepared immediately the plant is received from the cultivators, and within two days. An interesting account of a visit to the Wishaw Flax Works near Glasgow is given by Mr. Blechynden, and published at page 25, Part I., Volume IX. of the Journal of the Agricultural and Horticultural Society of India for 1854. In this Watt's patent process of preparing flax is explained: 'The flax straw is delivered at the works by the grower in a dry state with the seed on.' So that all that the cultivator has to do is to grow the plant, pull, stook, and deliver it at the works. One of the great advantages of the steam processes is that the seed and chaff is *all saved*, while it is generally lost, by the cold-steeping process. 'The shoves or refuse woody matters are also employed in these new works as fuel, whereby a great saving in coal is effected.' Another very great advantage that would be gained by the introduction of steam retteries into India would be that the risk of damage during the lengthy processes of preparation by hand during the hot season, when the hot winds prevail, would

* From what I have been able to ascertain, the owners of paper-mills would gladly give from R2-8 to R3 per maund for codilla.

† Turkish flax is now quoted at £40, and St. Petersburgh 12 head at £55.

‡ The total cost is less. I have added 10 shillings, as the commission on sale of flax would be double of what is put down for jute.

L. 21

(408)

CULTIVATION.

Early Experiments.

be almost altogether avoided. As regards the question whether machinery would pay in India, the best proof is that it pays in England, where a much higher price has to be paid for the flax straw. Mr. Blechynden, in his report of a visit to the Redford Flax Works, states that the proprietors contract with the farmers for the supply of flax straw at £4 per ton (*page 22, Part I, Vol. IX., Agri.-Horticultural Society's Journal for 1854*). A Committee of the Royal Flax Society of Ireland, in reporting* on Mr. Watt's system, state that 1 cwt. of flax straw weighed with the seed on yields 13½lb of fibre; so a ton of flax straw would yield 270lb of fibre. The necessary amount of flax straw, therefore, to produce 270lb of fibre would cost £4, and consequently the amount of flax straw necessary to produce one ton of fibre would cost a little over £33. If, therefore, it pays the manufacturers in England to pay £33 for an amount of flax straw which will produce a ton of fibre, it ought to pay manufacturers in India to pay the rayat £25 (R250) for the produce of 16 bighas; and I think I have shown that R250 would pay a rayat *well* for the produce of 16 bighas.

"PART OF BENGAL BEST ADAPTED TO CULTIVATION OF FLAX.—The part of Bengal which would be most suitable for many reasons for the experimental cultivation of the flax plant for fibre would be that comprised in the districts of Patna and Shahabad—*first*, because all the lands of these districts will be irrigable directly the Soane irrigation works have been completed; and *secondly*, because the soil, being a rich clay, is eminently suitable; and *thirdly*, because Dehree-on-the-Soane, where there is a large Government work-shop superintended by an officer of considerable experience and ability, would be an excellent site for the establishment of a steam rettery; and *fourthly*, because the means of communication by the several canals and branch canals will render carriage both cheap and safe. As regards the soil best adapted for flax, Mr. Deneef, the Belgian farmer, was of opinion that lands subject to inundation that are annually renewed by alluvial deposit are well suited. He writes to Dr. Spry, Secretary to the Agricultural Society, as follows:—'Les terrains qui sont renouvelés annuellement par les débordements du Ganges, ou qui sont frais et riches, conviennent le mieux pour la culture du lin.'

"Mr. Wallace thus reports† on the different kinds of lands;—

Alluvial	Will yield a good crop if not too sandy.
High land	Preferable, if rain falls two months after sowing.
Inundated	The most preferable of any.
Dosala (*i.e.*, two-year land)	Very good if not too low.
Loose and sandy . . .	Plant will not come to the same perfection.
Hard clay	Produces good plant, but of very coarse quality.

Mr. Deverinne was of opinion that 'the best lands are those in which there is a small mixture of sand; such lands are easily prepared, and can be well pulverised.' The large district of Shahabad is almost entirely surrounded by rivers from which spills annually inundate and renew the lands of the riparian estates. There are also in this district a great number of Government and Court of Wards' estates, so that almost every facility exists for experiments on a large scale.

"PROPRIETY OF GOVERNMENT ENGAGING IN TRADE.—There is a considerable difference of opinion as to the propriety or otherwise of Government engaging in trade. I would point out the case of Egypt. I think that there can be little doubt that it was owing entirely to Government operations that the flax trade became permanently established in that country. To this day flax is prepared by Government, and quotations in the home papers are given showing the prices of flax prepared by Government and of flax prepared by natives. The flax produced there is not quite equal to the *best* Russian, and it has been proved that flax can be produced in India which will command prices equal to those obtained for the best Russian marks. The only question, therefore, that remains to be solved is the question of cost; and I really do not see why, when the quality of Indian flax is equal to, if not better than, that of Egyptian flax, we should not be able to prepare the fibre here as cheaply as it is done in Egypt.

"FURTHER INFORMATION AS TO COST OF MACHINERY.—It was my intention, while at home on leave lately, to collect information as to the improvements in, and the

* In their 12th Annual Report.

† Letter to Dr. Spry, Secretary to the Agricultural and Horticultural Society, dated 2nd September 1841.

‡ Letter to Dr. Spry, dated 20th September 1841. These letters are all to be found in the Society's report to Government, submitted with a letter from Dr. Spry, the Secretary to the Society, dated 25th September 1841.

CULTIVATION.

Early Experiments.

cost of, machinery, the prices now paid for flax straw, the cost of preparation, and the demand for the coarser kinds of flax; but unfortunately my stay in England being of such short duration, I was unable to do so. I have, however, been making enquiries, and shall be glad hereafter to submit a supplementary report, if such information is likely to prove of any use to Government. I also propose to trace the growth of the trade in Egypt.

"CONCLUSION.—Dr. Forbes Watson very kindly supplied me with numerous extracts from the records of the India Office on the subject of flax cultivation in India. I only regret that I have not been able, from want of sufficient spare time, to enter more fully and completely into the subject than I have done. If, however, incomplete as it is, this report will in any way tend to bring the subject of flax cultivation in India into notice, and to the prosecution of further experiments, my object will have been gained. I trust that the result of such further experiments may be what is undoubtedly much to be desired, that India will some day become a flax-producing country for the benefit both of her agricultural classes and of the trade here and in the United Kingdom."

The Government of India considered **Mr. D'Oyly's** report of such interest that it was made the basis of an enquiry throughout India. It was accordingly printed and issued to Local Governments and Administrations, with a request for any further information which they might be able to furnish. The replies were in due course procured, and these, together with **Mr. D'Oyly's** report, were issued by the Government of India in 1874 in pamphlet form. Space cannot be afforded to republish all the replies, but the following jottings taken from them, together with passages derived from *Royle's Fibrous Plants* and the publications of the Agri-Horticultural Society of India, convey, it is believed, a fair conception of all that is now known of flax in the various provinces of India. The replies to the Government of India circular dealt unfortunately far more with linseed than flax, and even those that gave information regarding the fibre appear more often to be alluding to Hemp **(Cannabis sativa)** or San **(Crotalaria juncea)** than to Flax.

IN THE PANJAB. (409) 22

PRESENT CULTIVATION IN THE PROVINCES OF INDIA.

I.—PANJAB.

The following passages from *Royle's Fibrous Plants of India* give part of the information regarding this province up to the date of publication of that work in 1855:—

"Though Linseed is so extensively produced throughout India, we hear nowhere of the fibre being valued and separated. But when we get to the confines of cotton-producing districts—that is, into the Panjáb—we find that some Flax, prepared by the natives on their own account, is separated in the neighbourhood of Lahore. For we are told that the stalks of the linseed plant yield a fibre, which is made into twine, and used for the network of their *charpaes* or native beds. This information was elicited in consequence of inquiries originated by **Mr. Frere**, the distinguished Commissioner of Sind, from his desire to promote the culture of linseed in the province under his charge. The fibre, however, of the linseed plant is separated in still more northern parts, as some seed of Bokhara Flax sown in England was found to be that of the common Flax.

"The inquiries made respecting the growth of linseed in the Panjáb elicited the following facts:—

"'On the Cis-Sutlej it is stated that three seers to a bigha are sown broadcast when alone, or in drills; probably as an edging to other crops. Three maunds of linseed considered a good crop. The stalks and husks considered refuse and useless. The seed sells for 18 seers for the rupee to oilmen. A maund of seed yields of oil $10\frac{1}{2}$ seers; of oil-cake $29\frac{1}{2}$ seers. The oil sells for 5 and $5\frac{1}{2}$, and the oil-cake at 60 to 66 seers for the rupee.

"'In the Jalandhur Doáb, linseed is cultivated, especially in the *khadir* or inundated land of the Beás and Sutlej Rivers; but plants always small;

CULTIVATION. Panjab.

seed sells for 20 to 30 seers for a rupee. It is also cultivated in the Shírwul, or tract of country in which the soil is firm and covered by a deposit from the rains, with the sub-soil always moist—ripens before barley, and generally before wheat.

"'Linseed is also cultivated in the rich loamy soils of the Kángra district skirting the Himalaya, but chiefly in the eastern parts, on account of the seed, which sells for 20 to 100 seers for the rupee—flax itself burnt.

"'In the Panjáb, sown with barley and *mussúr* (or lentils) in *Kartic* (October and November), it ripens in *Chyet* (March and April); usually sown intermixed with the above crops, or in separate patches. Never irrigated, but grown along the Sutlej, in *khadir* land, or that which is inundated during the rainy season, but never manured.

"'In the Lahore division, it is grown chiefly about Sialkote and Dínanagar, and is the only part where the fibre seems to be made use of, as it is stated that the stalks yield a fibre which is made into twine, and used chiefly for the network of their *charpaes* or native beds. The price of the seed is about R2-8 per maund. Few localities are stated to be well suited to it, and the seed was in little demand. Twine made of the fibre was sent, but no notice seems to have been taken of its quality.'

"From the above details of cultivation, it is evident that though linseed is very generally known, it is nowhere extensively cultivated, as is evident from the price of the seed, which is dear in comparison with that of wheat. Some of the uses are well known; for instance, the oil is used as a drying oil, and the bruised seed, mixed with flour, is described by **Major Edwardes** as given as a strengthening food for cattle, and the oil-cake is no doubt employed for the same purpose; while the fibre is sufficiently valued to be separated in some, though burnt in most places. It would seem much in favour of the production of good fibre that the growth is much slower than in the southern provinces of Bengal."

In 1858, **Mr. Baden Powell** informs us, a quantity of native flax from Indaura, Kángra Valley, was sent to England and was considered to be the finest specimen sent from the Panjáb, it was valued at the high price of £55 to £60 a ton, and actually sold at £54-10*s*. "If," wrote Messrs. Kani & Co. of Dundee, "flax such as Colonel Burnett sent home (*i.e.*, the specimen alluded to above) could be put on board at Karáchi for £26 a ton, it would leave to both importer and exporter a handsome profit."

Inspired, apparently, by the above results, and in consequence also of the demand for coarse flax having then exceeded the supply, manufacturers in England became very anxious to obtain flax from India. The Belfast Indian Flax Company was created in 1861, with an agency at Sialkote, which had for its object the encouragement of the growth of flax in that district, with a view to its exportation to England. The sanguine hopes of the Company were, however, shortly afterwards seriously damped by the discovery that the whole of the seed sent out by their representatives in England had been so entirely damaged on the way, that no portion of it would germinate, an announcement which produced so depressing an effect on the members of the Company that it at one time appeared doubtful if they would have the resolution to persevere in the undertaking. On the spirited representation, however, of **Dr. Forbes Watson** of the India Office, who had very opportunely received a specimen of Sialkote flax, valued at between £60 and £70 per ton, the Home Government authorised a grant to the Association of £1,000 per annum for two years, on their engaging to carry on their operations for three years at all events. Under this stimulus the Company commenced its operations with renewed vigour. The result of the first experiment was highly satisfactory. The plants attained a good height, were rich in strong fibre, and altogether very superior to the Indian plant as ordi-

CULTIVATION.

Panjab.

narily reared, while the fibre shipped to England was pronounced by competent authority to be equal in quality to the best Irish flax. The prospects of the movement were thus considered "most cheering." Soon, however, serious difficulties arose. The Company could not grow the plants in sufficient quantity near the factory to make it remunerative to the zamindars. When grown at a distance in small quantities, the cost and delay involved in carrying the raw material to Sialkote were out of proportion to the sale-proceeds. The apathy of the peasantry was also a great difficulty. They could not be induced to persevere in the cultivation of the plant on the approved methods it was sought to introduce, the plants produced under the European method of cultivation being not near so rich in oil as the ordinary country seed. In these circumstances, the Company had to discontinue its operations, and ceased to exist in 1867. For some time afterwards the business was carried on by Messrs. Bertola, Cox & Co., but they too relinquished it in 1869.

It would appear (from the replies received by the Government of India) that linseed is now very little grown in this province, and that the fibre, flax, is only very occasionally utilised. In Karnal 17 *bighas* 10 *biswas* of land were devoted to the produce of the fibrous plant (during the year of report), which gave a produce of 52 maunds 20 seers. The cost of preparing one *bigha* of flax was ascertained to be R3; this produced 3 maunds at a price of R2-8 a maund. The report further states that in 1855-56 flax was sown, but the results were far from satisfactory. The zamindars do not seem to appreciate the culture of the plant. In the Jagraon tahsil of Ludhiána experiments were made in 1854, but the results were not favourable. In Gujarat flax was cultivated in 1863 on account of the Indian Flax Association of Belfast, but no notes of the produce were made. In Peshávar District flax is cultivated to a small extent and ropes manufactured from the fibre for domestic use. In Kángra it was stated that 4,432 acres were under **Linum usitatissimum.** In 1881-82 the area in Kángra under this plant was reported in the Gazetteer of the District to be 7,150 acres, being the largest figure for the Panjáb. The crop is grown in the valley, and is valued solely for the sake of the oil, no use being made of the fibre. Small care is bestowed upon its cultivation, the seed being simply thrown upon the ground between the stubbles of the newly-cut rice. The crop is very poor, but suffices to supply oil for local use.

The following passages in the Agri.-Horticultural Society of India allude to Panjáb Flax:—Journal (Old Series), IX., 139 (letter by **Mr. H. Cope**; (Proceedings, 1885, p. cxlv), Major Hollings' report on the suitability of the Panjáb for flax cultivation; X., 96 (report on samples); also Proceedings, 1858 (report on **Mr. Cope's** samples); XI., pp. 75-139—a detailed and important paper by **Mr. H. Cope** "On the introduction of flax, as a fibre-yielding plant into India, and especially into the Panjáb;" p. 188 "Report on the sale of Panjáb-grown flax in the markets of Dundee and Belfast; also (Proceedings, 1859) lvii., and (1860), VI., cv.

II.—NORTH-WEST PROVINCES.

IN THE N.-W. PROVINCES.

(410) 23

The Government of the North-Western Provinces forwarded, in reply to the Government of India's circular, copy of a letter from the Superintendent of the Botanical Gardens at Saharunpore, in which **Dr. Jameson** gives an account of the endeavours made up to 1861, in February of which year the letter was written, to extend the cultivation of flax in these provinces. **Dr. Jameson** relates how those trials proved fruitless, shows the causes of failure, and suggests the means which should, in his opinion, be adopted to ensure a lasting and profitable extension of the cultivation of the plant. "Before replying to your letter No. 1472A., dated 22nd October

CULTIVATION.

N.-W. Provinces.

last, with enclosures," Dr. Jameson wrote, "I deemed it necessary to examine what had been done regarding the cultivation of flax in India, and particularly in the Panjáb, preparatory to submitting a full and detailed report. To do this I found would only be going over a subject which had been fully investigated and exhausted by the late Dr. Royle in his work styled the *Fibrous Plants of India* and by Mr. Cope in a paper lately published in the *Journal of the Agricultural and Horticultural Society of India.*

"2. In these two publications full details will be found regarding all the experiments made in flax cultivation throughout the country.

"3. In the Panjáb the subject had been taken up with energy and activity, and good results had been gained. But in my late tour I ascertained that in almost every district where the plant had been cultivated, unless that of Sialkote, it had been discontinued; the experiment, therefore, so successfully begun was too prematurely abandoned.

"4. No doubt the question that flax fibre can be raised in the Panjáb fitted for the home market has been solved, and has thus passed from speculation to fact. But still, as far as the natives of the country are concerned, nothing has yet resulted. The cultivation, instead of extending, has diminished, and, had not the Deputy Commissioner of Sialkote taken up the subject with energy and activity, the experiment would have been fruitless, or, as remarked by Mr. Cope, would have died out in the Panjáb from sheer inanition, and that, too, originating in three causes—want of perseverance on the part of the local Government; second, want of enterprise on the part of British manufacturers; and third, want of activity, energy, and interest on the part of the native cultivators in the welfare and progress of the country.

"5. In a country like India, Government, when desirous of introducing a new product, or of rendering an old one, by a process of cultivation unpractised, valuable in the arts, must take the initiative, in order to overcome the prejudices of the ignorant, indolent, and slothful cultivators. There is no active and energetic middle class to direct and encourage the labours of the native farmers, and it is a well-known fact that even in Britain there is no class more difficult to persuade to adopt new and improved processes of cultivation and new ideas regarding farming than agriculturists; and, had not the policy lately introduced opened up the country to free trade, the old and routine system of cultivating the land would have by thousands been continued to this day in Britain.

"6. The British farmer is now compelled by the free importation of grain to resort to high and scientific cultivation and the best manures, in order to enable him to hold his own.

"7. Mr. Mechi and others, through means of their private experimental farms, have shown to their tenants and tenant farmers how to maintain their position, even though the British markets be extensively supplied with untaxed bread-stuff from abroad. In this country such spirited individuals are unknown, and therefore anything to be introduced for its improvement must, I respectfully beg to state, be initiated by Government. The system at present followed by native farmers in cultivating flax for its seed is miserable in the extreme; the *shove* or straw from whence the fibre is obtained being only used as fuel, or broken up and mixed with other substances and given to cattle. Let natives be shown that substantial advantages would occur to them by cultivating the flax properly, and that good marketable fibre can be obtained from it, for which there is always a ready and immense market, and I doubt not but they would soon take to the cultivation. But, though the cultivation in the Panjáb has in most places retrograded, yet still beneficial results have ensued from the experi-

CULTIVATION.

N.-W. Provinces.

ment instituted by Government, which may ultimately be of immense importance to the country.

"8. By the exertions of **Mr. D. Macleod** and others, the experiment was prominently brought to the notice of the flax manufacturers in Britain, where for years the supply of flax from home cultivation and foreign importation has been far short of the demand, and samples of the fibre laid before them, which were pronounced worth £55 per ton; and so satisfied were they, by the specimens exhibited, that the Panjáb was fitted to grow flax suited to the home market, that they formed a company, an 'Indian Flax Company, Limited, of the Panjáb,' in order to carry it on, and their agent, **Mr. Wightman**, has now settled in the Sialkote district, and has commenced operations. By him advances have been made to zemindars to cultivate flax, which they will repay him back in kind. He, too, has distributed acclimated seeds, and has applied to me for a large supply to extend his operations, which, however, I cannot meet. Land of his own he has none, and he is therefore entirely dependent on native cultivators. The system introduced is an admirable one, and will continue to be so as long as the company consider the interests of the native cultivator as well as their own, or, in other words, give him a fair remuneration for his labour. If this be done, and if the superintendence be confined to advice in the manner of cultivating the plant properly and preparing the fibre, and the distribution of acclimated seeds, and, above all, in a newly-acquired country like the Panjáb, where the inhabitants look to the district officers as their best advisers, if the countenance of Government through their district officers be continued, flax cultivation will rapidly spread, and the fibre become an important article of exportation.

"9. But the as yet small success gained in the Panjáb is not, in my humble opinion, sufficient to stimulate private enterprise to seek a field for operations in the North-Western Provinces.

"10. The services of **Mr. Cope** are not available, or if they were, they could only be procured at a rate which Government would not be prepared to meet. By him extensive mercantile transactions are carried on at Amritsar, and thus his time is fully occupied; nor is he acquainted with the methods of preparing flax.

"11. To grow the plant assistance is not required. This we can do. What are wanted are, *1st*, good scutchers and hecklers – men intimately acquainted with the processes of manufacture, and fitted to teach the natives of the country; *2nd*, the best kind of machinery used in preparing flax; *3rd*, a large supply of acclimated seeds; *4th*, the directions for the proper management of the flax crop, compiled by the Committee of the 'Royal Society for the Promotion and Improvement of the Growth of Flax in Ireland,' with a few alterations and modifications to suit the climate of the North-Western Provinces, might be translated into Hindí, printed, and distributed to native cultivators through district officers with much advantage.

"12. In the works alluded to all the information required on flax cultivation is to be found. The Agricultural and Horticultural Society of Calcutta, by publishing the most important information to be found in the reports and transactions of the 'Royal Society for the Promotion and Improvement of the Growth of Flax in Ireland,' have done all that is required to popularise the cultivation, so far as that can be done by the press, and in **Dr. Royle's** work on *The Fibrous Plants of India* the cultivator will find the same information condensed. To him, therefore, I would recommend this work as a text-book.

"13. Prizes have already been offered by the Panjáb Government for the best samples of prepared fibre, and for the largest quantity of land brought

(410)

CULTIVATION.

N.-W. Provinces.

under cultivation with flax, but with no beneficial results, as they remain unclaimed by any one. Such an inducement held out appears to be of doubtful utility, though it might, with much advantage, be done on a small scale by district officers.

"14. In the magnificent system of canal irrigation, the North-West Provinces has the means for flax cultivation far superior to that possessed by the Panjáb, and to encourage it, therefore, and meet the demands for acclimated seeds on an extensive scale, I would respectfully recommend that an experimental farm of from 50 to 60 acres be formed in the Saharunpore district, adjoining the garden, where irrigated land can be procured at a reasonable rate; that the incidental expenses, such as land, rent, water, &c., be met by the sale of the seeds, a certain quantity being reserved for district officers for distribution to zemindars.

"That if European instructors from regiments be available in this country two men be obtained from the ranks of any regiment for a short time to prepare the fibre, and teach natives how to scutch and heckle. That the fibre be sold, when prepared, to meet the wages of the parties who prepare it, a large sample being reserved for exportation and examination by British manufacturers. That acclimated seeds be given to district officers, particularly those whose districts are in part irrigated by canals, to distribute to zemindars. That zemindars repay in kind with the *shove* or straw the value of the seeds received, and that small rewards, such as those distributed by the Deputy Commissioner of Sialkote, be given to such cultivators as present the best samples of flax plants for scutching. These rewards were received with the greatest satisfaction by cultivators of Sialkote, as they were distributed publicly to the recipient by their own district officers.

"15. For the plants raised by zemindars, and fitted for preparing fibre, there might at first be difficulty in procuring a market. The finest samples, however, might with advantage be purchased by Government, and from them fibre prepared for the market as an encouragement to the best cultivators. This, of course could only be done to a limited extent. But when the field became extensive it would be high time for Government to discontinue the experiment and hand it over to private capitalists. But the seeds alone would ensure the zemindar against any loss. In fact, he would only be doing, though with better seeds, what he is now doing; flax of inferior quality and quite unfitted for preparing fibre being cultivated everywhere.

"16. In the Panjáb both **Mr. hWigtman** and **Mr. Cope** are prepared to purchase all fibre of good quality, and were it shown that flax capable of producing good fibre could be grown in the North-West Provinces, capital and funds to work it would, no doubt, be forthcoming to take it up.

"17. To do this it would be necessary to import some good seed from Courland and Livonia, and from which all the best Dutch seeds are forwarded. To the acre about 2 bushels, or 1⅓ maunds, are required. I would therefore recommend that a ton or 28 maunds be imported, which, with the acclimated seed which will be available this season, would enable me to cultivate 50 acres of land, an ample extent to spread the seed over the country.

"18. When the Honourable the Lieutenant Governor visited the garden last season, he saw two fields under cultivation with flax, the one with Russian seed, the other with indigenous seed, and the plants of the former from 3½ to 4½ feet, and the latter only from 2 to 2½ feet in height.

"19. From the former seed a supply was sent to **Mr. Cope**, then Secretary to the Agricultural and Horticultural Society of Lahore, and from it dates the commencement of the experiment of flax cultivation in the Panjáb."

LINUM usitatissimum.

Flax Cultivation

CULTIVATION.

N.-W. Provinces.

The Government of the North-Western Provinces also submitted a communication, received from the local Board of Revenue, which gave the results of the enquiries made by it as to the actual extent of the cultivation in these provinces. The substance of the papers forwarded may be here given, the broad results in regard to flax being as follows:—

"Flax (**Linum usitatissimum**) is not cultivated for fibre.

"The Commissioner of Allahabad has suggested that the richness of soil and large outlay required for the production of good flax fibre form an effectual bar to the successful cultivation of the plant for this purpose at anything like a moderate price. This statement traverses the conclusions drawn in **Mr. D'Oyly's** report, but is supported by the fact that the cultivation of flax for fibre has been abandoned. It appears to the Board that the exhausting effect of flax on the soil should not be overlooked, and that first experiments cannot be accepted as a crucial test.

"The Board recommended that a series of careful experiments should be instituted on a small area on the Bulandshahr Farm, both on ordinary land and on highly cultivated and manured land. The experiments, if carefully conducted, would prove whether flax could be profitably cultivated for fibre, and set the question at rest as regards the North-Western Provinces.

"The report furnished by **Mr. Ricketts**, the Commissioner of Allahabad, alluded to above, was as follows:—A little flax is grown in a few parts of the division for local consumption, for oil and cake. It is hardly ever grown as a field crop, but only in a row drilled in, every here and there, on the edge of a field, or in furrows 10 or 12 feet apart in wheat crops. The only soil suitable for flax in this division is a porous black soil, very like cotton soil, found only near the course of the Jumna.

"Regarding the cultivation of flax, I can say that a great effort was made during the Russian war to introduce the cultivation of flax into the Panjáb: good flax was grown with a long fibre fit for export to England for manufacturing purposes, but it was established that flax to be good requires a great deal of water, good soil, and plenty of manure, and that it is a very exhausting crop. These were obstacles to the extension of the cultivation which can only be got over (as in the case of cotton a few years ago) by great demand at high prices, and a certain market, not a remote sea-port, of which the cultivators have never heard, but in the markets in their neighbourhood, to which they are accustomed to resort.

"The native cultivator has no notion of growing flax for the sake of the fibre. It is only grown for seed. The same plant cannot produce both fibre and seed. If required for fibre, it must be forced in its growth with water and manure, be planted thick, and cut before the stalk gets hard,—that is, before the seed begins to ripen. When grown for seed, the plant is sown in drills far apart, and is preferred short in the stem,—so short that the same plant could not produce fibre of a sufficient length to be of any use."

Messrs. Duthie & Fuller take, however, a more hopeful view of a possible flax culture in these provinces than perhaps do any other writers. In their *Field and Garden Crops* they say: "It is improbable that flax culture could be extended on any other system than that followed by Indigo planters, under which the grower receives a cash advance at sowing time, together with a guarantee that his crop will be purchased at a fixed price. Flax fibre would be useless to a cultivator unless he was certain of gaining a sale for it. It does not seem that any energetic attempts have been made to extend flax culture on this system, and what efforts have been made to promote it have been confined to experiments which have indeed proved the possibility of successful flax-growing, but have given native cultivators no immediate incentive to undertaking it."

CULTIVATION. IN OUDH. 24 (412)

III.—OUDH.

The Chief Commissioner of Oudh submitted a copy of a memorandum drawn up by **Dr. McReddie**, Superintendent of the Oudh Central Jail. The practical information contained in that publication will render it valuable to those who may wish to undertake the cultivation of flax. It may therefore be quoted *in extenso* :—

"1. SOILS, AND PREPARATION OF SOIL.—The finest flax is grown on deep soils, argillaceous, sandy, and on the sandy clay commonly called loam. In strong, rich, damp soil flax will reach to a great height, but is always coarse. On light fertile soil it is shorter, but fine and silky. Deep ploughing is necessary, at least 11 inches. Fresh manure should not be used, but the manure, solid or liquid, added to previous crops made to answer. Newly broken-up pasture land does well for flax. All land must be perfectly pulverised and cleaned of all roots of every sort. Five different ploughings are necessary, with an interval of fourteen days between each, so that ploughing must begin at least two and a half months before sowing. After each ploughing all clods must be broken up, and the ground made quite even. In fact, as flax is a *rubbee* crop, the ground on which it is to be sown should be prepared in the same way as for wheat. If manure must be added, it should be done before, or at the beginning of, the rains. The manure should be spread evenly on the soil. Sheep manure should not be used.

"2. QUANTITY OF SEED PER ACRE, AND HOW SEED IS SOWN.—Flax is invariably sown broadcast; when sown for good fibre it is sown thick—230 or 240℔ (3 maunds) seed per acre. When sown for inferior fibre and seed 140 or 150℔ per acre will answer. It is never sown by drill.

"3. THE MONTHS FOR SOWING,—PRECAUTIONS.—From 15th or 20th October to the 20th November, according to season, *i.e.*, late continuance of rain, as flax will not stand much rain; hard rain soon after sowing would destroy the crop, hence the best time in Oudh would be the first week of November.

"For sowing, a calm day is chosen, if possible; if wind prevail, the sower must move with the wind on his side, not on his face or back. A uniform distribution of the seed would be insured by going over the ground twice and sowing half the amount of seed each time. The sowing is not to be done as for wheat, that is, not after the plough, as it were in drills, but as stated above, broadcast by sowers walking over the field. The progress of the sowing may be marked by a seed harrow going over the field in the line of the sower. When sown, the seed is harrowed into the soil, and a light hand-roller drawn by a man is passed over it; this last operation should not be omitted. Care must be taken not to bury the seed too deep.

"To get the seed to germinate quickly, it is well to sow immediately after a shower of rain.

"4. WHAT IS TO BE DONE AFTER SOWING.—When the plants are 2 inches high, the ground must be carefully weeded. This is done by a number of men kneeling amongst the flax and carefully removing by hand all the weeds; they must work facing the wind, so that the young plants may raise themselves with greater facility. If one weeding is not sufficient, the operation must be repeated. If not feasible earlier, the weeding must not be delayed beyond the time when the flax has obtained a height of 6 inches.

"Three or four waterings, not too copious, are necessary, according to season. In fact, water as often as you water wheat.

"5. HARVESTING.—Flax must be torn up by the roots, not cut down. The harvesting is to begin *before* the seed is quite ripe. Directly the bottom leaves of the stalk begin to fall off, and the last blossoms have disappeared, the time of pulling has come. A *good test* is to cut a capsule right across,

CULTIVATION.

Oudh.

horizontally. If the seeds have changed from a milky whiteness to a green colour, and are pretty firm, time to pull has come. By not allowing the seed to ripen on the ground, a better fibre is obtained, while, at the same time, the value of the seed is not lessened; for 'it is well known that most seeds, though not quite mature when gathered, ripen sufficiently after being plucked, provided that they be not detached until dry from the parent plant, all the sap which this contains contributing towards further nourishing and perfecting the seed.' Pulling is done by *catching the plants by the heads* and drawing rather obliquely; a handful is taken up each time, and the root-ends of the handful are kept as even as possible. Six handfuls are made into a sheaf. The sheaves are set up slightly, inclining to and leaning against each other with the root-ends downwards, and allowed to stand for five or six days. In pulling, the long stalks are laid by themselves, and the short separately, so as not to mix good and bad flax together. When plucked as directed above, the flax must be stored away under cover for three or four weeks, and then the seed is to be thrashed out.

"If, however, the grower thinks that he risks the quality of the seed by plucking before the seed has ripened, he may delay the process of plucking till the stalk is yellow up to the top, and all the leaves have fallen and the seed-vessels begin to open in the sun. The plants are then to be *pulled up by the roots* and laid on the ground. Clean grassy soil answers best in not too thick rows, in order to let the sun penetrate it thoroughly. At the end of four or five days it is turned, so that the plants underneath may be dried.

"Two plans of harvesting are given above; the latter probably is the one preferable in this country.

"The thrashing out the seed must be done as soon as possible after harvest. Thrashing is done by striking the plants with smooth round sticks. When the capsules have been removed from the plants, the seed is to be trodden out with bullocks, the same as is done with wheat. After removal of the seed, the flax is to be tied up in bundles of about 6℔ each; it is now ready for the process of retting or watering to be next described.

"A.—PREPARATION OF FLAX FOR SPINNING: *Steeping, retting, or watering.*—This may be done either while the flax is green immediately after the grain has been thrashed out, or when flax of superior quality is required, the plants must be perfectly dry and at least a year old before steeping. In this country, as the flax will be plucked by the end of March, it must be stacked until the rains—July or August—before steeping.

"Steep in ponds or tanks in which rain-water has collected, or ponds may be dug during the rains 10 feet wide, 4 feet deep, and any length desirable, and rain water allowed to fill in them. The flax tied in bundles, as previously described, is to be thrown into the ponds. The bundles or sheaves are not to have stones or weights on them, but every day each sheaf is turned with a pole to make the uppermost side change to the undermost. The water used must be pure, soft, and clear; well-water generally does not answer. The retting takes six or seven days, or only four if the weather be hot.

"To know whether the flax has been sufficiently steeped, the following test must be applied. Crack the stalk across the root, without breaking the fibre, and draw the latter towards the head of the stem, stripping it upwards. The fibre ought to come away easily, and should, moreover, hold together in ribbons; narrow and separate fibre would indicate the steeping to have gone too far. In short, when the skin peels easily along the stalk, the flax has been sufficiently steeped. The steeping should never be entirely completed in the water. It is best to finish by the agency of the dew. The process is to be carefully watched, and requires a little experience, so as not to under or overdo it. After removal from the water, each sheaf is set on end

CULTIVATION.

Oudh.

and allowed to stand for a day or two; the sheaves are then spread out on a field, pasture land, in very thin layers, to complete by the dew the retting process. This should be continued for about fifteen days. During the time the flax is spread out, it is frequently turned in hot damp weather at least once a day. If heavy rain threaten when the flax is spread, it is gathered into conical sheaves until the weather gets fine; rain soon causes it to rot. After steeping, flax should be kept five or ten months tied up in bundles, about 6℔ weight each; the best qualities may be kept two or three years. When dressed (the process to be next described) immediately after steeping it is heavy; it seems to get finer by keeping. It is stalked like a heap of grass or hay with the root-ends outwards in circular fashion, with a light thatch of grass and bamboos at the top; a *pucka* flooring, or simply a layer of *pucka* bricks, will ensure safety from white-ants; of course, high ground is to be selected for stacking.

"B.—DRESSING THE FLAX: *Breaking*.—The flax having been prepared for the last process (after steeping and drying) it is subjected to these either by hand or by machinery. It may be broken either with wooden mallets or it may be still better subjected to the action of Brasier and Hodgkins' (Liverpool) patent 'shamrock' hand-breaker. This machine will break—that is, extract—a very large portion of the woody matter, leaving the fibre uninjured and ready for the last or scutching process.

"Three men are required to work it; one to work the handles, one to feed, and one to receive the broken material. It is necessary to pass each handful of fibre twice through the rollers; one and a half maunds of fibre can be worked in one day. A machine costs at Liverpool £13. All charges included, it will cost £20 (R200) to any part of India.

"*Scutching*.—The flax is now ready to be *scutched*, or made ready for spinning by this final process. There is a scutcher attached to the breaker previously described, but I find that the scutching-blades get too brittle and soon get useless by working; this may be due either to the extreme dryness of the atmosphere during the cold season in this country, or to some fault inherent in the manufacture of the blades. At all events, I found that I could not scutch by the machine. Hand-scutching was then resorted to. A scutch is a wooden knife 2 inches long, 3½ inches deep, and half an inch thick, tapering to an edge. It is best made of *seesum* wood. A *scutching-board* is required on which to work the flax. This is an upright plank 51 inches high, 14 inches broad, 1¼ inch thick, firmly fixed in a solid block of wood. At the height of 37½ inches from the foot is a horizontal slit 1½ inch broad. With his left hand the scutcher introduces into the slit a handful or tuft of broken flax, so that it hangs down on the side of the scutching board; with his right hand he scrapes and chops at the flax with the scutch till all the remaining woody matter and all the broken pieces of fibre have been removed. The *long fibre* left in the hand of the scutcher is the marketable article; the shorter fibre is the tow fit for packing, &c, but not for spinning into thread. Each scutcher after he has learnt his work turns out 5 seers or 10℔ of flax per diem.

"NOTE.—This memorandum on flax cultivation was drawn up by me from, *first*, some papers given me for perusal by **Mr. E. C. Bradford**, who had paid £20 for the information contained in them at the British consulate in Brussels; *second*, from a work on 'Flax and Hemp,' by **Sebastian Delamer** (Publishers—London: Routledge, Warne, and Routledge). I acted strictly in accordance with the instructions given in my experimental cultivation of flax. I sowed imported (Riga) seed and country seed: of the former 300℔ were used; this quantity cost, inclusive of freight, inland carriage, R90. Of country seed 20½ maunds were sown. This quantity of seed, 20 maunds and 10 seers, sowed 8 acres of land. Four acres were

CULTIVATION.

Oudh.

rented land, which was sown, ploughed, watered, &c., at my own expense. Four acres were sown for me by a talukdar. I gave him the seed; all other charges were his, and he took the seed produced, giving me the straw (fibre). The seed produce of the 4 acres cultivated by myself was 29 maunds, equivalent in money value to about R57. The quantity of flax produced and exported as the produce of 8 acres was only 11 maunds and 28 seers, less than half a ton (the ton being 27 maunds and 10 seers); this was valued and sold in London, at £23 per ton, for £9-12*s*. The produce of Riga seed, if sufficient in quantity, would have fetched at the rate of £29 or £30 per ton; the produce of only country seed was not worth more than £23 per ton.

	R	a.	p.
The experiment cost	663	0	0
The return was	253	0	0
Loss . .	410	0	0

"The rail carriage from Hurdui to Bombay was R37-14, or about R70 per ton. Freight from Bombay to England is from £2-10*s*. to £3-10*s*. per ton, according to season. There are, besides charges for pressing into bales, inland charges on landing in England, &c. The produce of 8 acres, if all the seed had been imported, would have been 2 tons, which would have fetched a higher net value in England; but, as a set-off against this there would be the higher price paid for imported seeds (*i.e.*, country seed is from R2-4 to R2-8 per maund; 300℔ Riga seed cost R90, or at the rate of nearly R30 per maund), and of course increased carriage and freight on the larger quantity. The linseed as ordinarily sown and plucked by natives is *useless* for fibre, *i.e.*, from which to extract flax fit for spinning. It makes good *tow*, which, if it could by *machinery* be utilised for ropes and canvas, might, I think, prove of great value. Nor do I think pure country seed, even sown thickly in order to make the plant rise to a good height of much value. I sowed in good land, carefully attended to all the processes, but still the plants did not obtain a height of more than 19 or 20 inches; the plants from Riga seed were 2 feet 6 inches to 3 feet. The length of flax is of great importance, hence I believe that the flax from country seed will never be of any commercial value in England. What may be still worth trying is to import sufficient seed to sow about 50 or 100 acres of good land; part of the land to be sown for fibre, part only for seed; then in the next season to sow a larger area with acclimatised seed,—that is, seed the produce of the previous year's importation. The straw of two years' sowings may be worked into fibre, but not by manual labour. Machinery and steam-power are necessary, as a matter of economy, in the long run. The machinery is simple and not expensive, and I think it would be worth the while of a capitalist to try the experiment as indicated above."

The following passages in the *Journals of the Agri.-Horticultural Society of India* allude to flax in the North-West Provinces and Oudh:—IX., pp. 371-379 (Report by **Mr. C. Gubbins** on flax raised at Allyghur from foreign and native seed); XI., pp. 514-531 (correspondence regarding the cultivation of flax in the North-Western Provinces); continued pp. 593-613; Proceedings, 1860, cxi.; XIII., (Proceedings for 1864) report on samples prepared at Shajehanpore.

IN THE CENTRAL PROVINCES.

(413) 25

IV.—CENTRAL PROVINCES.

The Chief Commissioner of the Central Provinces reported, in reply to the Government of India's Circular letter, that flax was cultivated exclu-

CULTIVATION.

Central Provinces.

sively for its seed, and that it was always short and bushy, seldom exceeding one foot in height.

"**Captain MacDougall**, Honorary Secretary of the Nagpúr Agricultural and Horticultural Society, wrote that wherever the plant is raised in that locality, thick, indeed immoderately thick, sowing was the rule, which is the reverse of the practice in North India, and that the exception will only be met with at headlands of fields, where the individual plant, having air and room, tillers out and becomes bushy.

"'Again,' he continues, 'the inference to be gained from **Mr. D'Oyly's** wish to see thick sowing followed in order to gain fibre—namely, that the plant will run tall and thin—is refuted by the habit of the plant, as seen in our fields. Nowhere does the present system of thick sowing produce plants more than a foot in height, with perhaps five or six seed capsules.'

"**Captain MacDougall**, moreover, differs from all the opinions expressed as to the kind of soil which is most suitable to the growth of flax. He observes that it is plain that there is a great want in the plants grown in the Central Provinces of that woody tissue which constitutes the strength of the bark, the portion from which the flax would be scutched; and he attributes the want of this tissue to the plant being grown in the stiff black clay of the cotton soil, which, unless well stirred up and rendered friable, would prevent its roots permeating downwards into the subsoil and there gathering nourishment. He also gives another reason, *viz.*, that the plant is solely treated as a *rabí* crop instead of as a late *kharíf* one. 'Treated as a *rabí* crop,' he argues, 'the plant has not as good a chance of nourishment as a late *kharíf* crop would ensure. If during the cold weather there are heavy dews and occasional sharp showers of rain, such as sometimes fall about the Dewali and Christmas, the crop gains something; but still this is hardly sufficient, for the soil does not absorb the quantity of water thus given to any great depth, and the roots (unless it is a good free soil) have still to struggle hard to get downward to meet the moisture stored up in the subsoil. The result of the present system of cultivation is that the roots, from want of a nice, light, and free bed remain near the upper surface, and the life of the plant never exceeds four and a half months.'

"In order to gain a satisfactory result, this is what should, in **Captain MacDougall's** opinion, be done: 'If patient attempts are persistently made year by year to change the habit of the plant, and we could succeed in sowing it on well-ploughed and manured land in August with its life prolonged to February, I feel confident we should find a marked improvement both in flax and seed-bearing properties'

"**Captain MacDougall** cannot quite see the cause of the alarm as to the effects of thin sowing expressed by **Mr. D'Oyly**; and he says that a plant can tiller and yet give shoots. He compares the flax-producing plant to wheat, and observes that a single seed of the latter sown in good soil will give five and six (if not more) stalks, each of equal length, strength and productiveness.

"The Chief Commissioner, referring to the allusion made by **Mr. D'Oyly** in paragraph 13 of his report to certain experiments made by **Mr. J. B. Williams** in the Jubbulpore District, says that they have been discontinued, and practically they did not result in much. This gentleman, it seems, who for many years was in charge of the Thuggy School of Industry in that district, prepared the fibre, but had no idea of its commercial value. His fibre attracted notice in Calcutta, thanks to his having to send some valuable glass-ware to the care of Messrs. Colvin, Cowie & Co., which he packed with this fibre. It appears that **Mr. Williams** even received orders from a Calcutta firm to supply any quantity he could.

CULTIVATION.

Central Provinces.

But no account is given as to the particular manner in which these plants were cultivated, nor is it stated what the length of the flax was when prepared. **Mr. Williams'** business was subsequently purchased by Messrs. Maclean and Warwick, but it is not known whether this new firm still carries it on.

"An ex-partner of **Mr. Williams'** (a native) writes that the '*gundy sunn*' produced in Jubbulpore used to be purchased by them in co-partnership and forwarded to Calcutta, and that sometimes a little used to be shipped to England.

"In the publications of the Agri.-Horticultural Society no mention is made of Flax in the Central Provinces, though several papers deal with Linseed (see below).

IN BERAR. 26

V.—BERAR.

The Resident at Hyderabad transmitted to the Government of India the reports contributed by the local officers of the Assigned Districts.

The Officiating Commissioner of West Berar (**Major J. G. Bell**), for example, wrote :—

"In this division last year it was estimated that 162,693 acres of land were under oil-seeds, of which a large proportion may be taken to be linseed. An acre of good land in a good season is estimated to produce 250 seers of linseed, and the average price in 1869-70 was from R18 to R20 per candy of 540 seers. In 1872-73 the prices, it is said, rose to R65 per candy. Linseed is chiefly grown for the sake of oil, for the purpose of extracting which it is mixed with the *kurdee* seed (**Carthamus tinctorius**). The stalks of the linseed plant are thrown away or burnt, though occasionally they are used for thatching huts.

"The Assistant Cotton Commissioner, East Berar, reported that the rich and loamy soils in the Purna valley will be best adapted for cultivating flax, and also such lands as are subject to inundation by the waters of the Purna and Wurdha rivers. He adds that *ulsee* is now grown on the best and cleanest lands, but that he is unable to say whether flax cultivation would pay better than *ulsee* (linseed), and he is therefore desirous of growing some seed on an acre or two of land in the Purna valley.

"The Resident states that it appears that, although flax is cultivated to some extent in those districts, it is chiefly for the object of extracting oil rather than for the manufacture of fibre.

"This preference of the people for the extraction of oil is attributed to the shortness of the stems, as well as the greater expense and trouble of converting them into flax as compared with hemp [**Crotalaria ?**—*Ed.*]. The Resident states, moreover, that cotton, *jowari*, and wheat are the staple products, and that so long as the rayats find these more profitable than any others, there is but little hope of flax being taken up by them willingly for more extended cultivation than at present obtains. On the receipt of the circular a plot of land in the model farm at Akola was sown, however, with flax by way of experiment, and a communication on the result of this trial was promised in due course. But, as the farm has since been abolished, it may be inferred that the experiment came to nothing.

"The Resident expresses his readiness to give effect to any general measure which the Government of India may direct in view to the extension of the cultivation and manufacture of flax in Berar."

In Vol. X. of the *Journal of the Agri.-Horticultural Society of India* (p. 97), **Captain Ivor Campbell** states that he forwarded, in 1856, a sample of flax and coarse cloth made of it which he had prepared in North Berar. He then writes : "Although this district furnishes a large proportion of the linseed which is exported from Bombay, and last year there were more than 100,000 *bighas* of land under the cultivation, the people were not aware

CULTIVATION IN BOMBAY. 27 (415)

that the plant, which is sown wide apart, for seed only, produced any fibre."

VI.—BOMBAY.

The Government of Bombay submitted a summary of information which had been furnished by the Commissioner in Sind and the Revenue Commissioners of the Northern and Southern Divisions of the Presidency, in reply to the Government of India's Circular letter on the subject of Flax :—

"In Sind no flax is grown except in the Upper Sind frontier district, where flax alone is grown to a very limited extent for local use only, the zemindars making ropes from it. Two acres would cover the whole area grown annually for the whole district.

"The Collector of Karáchi, in 1869, obtained 11 maunds and 3 seers of linseed from Fleming & Co., and distributed it among fifty-three different zemindars in the Sihwan taluk. The result was not very successful : the average produce was threefold ; in some cases it was thirty-one times the amount of seed. The Collector attributes the failure of the crop to inattention and ignorance on the part of the zemindars, but considers that an experiment conducted at the Hola Government farm could not but have good results.

"In Belgaum and Kanara flax is grown to a very limited extent.

"It is reported by the Collectors of Poona, Dharwar, and Belgaum to be used in making cords and ropes ; oil is extracted from its seed, and the refuse is turned into cakes for cattle.

"*Northern Division.*—In this division flax is grown to a limited extent, oil being extracted from the seed for local use. But nowhere is it grown for the sake of the fibre.

"The Government of Bombay also forwarded two letters from **Mr. W. F. Sinclair** C.S., Assistant Collector, Kolaba, containing his observations on the possible development of the production of flax in that Presidency and the preparation of the fibre. In acknowledging the receipt of these letters, which are given below, the Government of India expressed a hope that His Excellency the Governor in Council might find it practicable to extend some encouragement to **Mr. Sinclair** in carrying out the experiments he proposed to undertake."

The following passages from **Mr. Sinclair's** letter are of considerable interest, and may, therefore, be reproduced here :—" The flax plant (**Linum usitatissimum**) is very largely cultivated in Khandesh, and also, I believe, in Násik and Nuggur. The linseed alone is used ; the fibre, which is useless as fodder, is thrown away. It is obvious that, since it pays to cultivate the crop for the sake of the seed alone, a very small success in utilising the fibre now wasted would be profitable. I may remark *aliter* that hemp also is much cultivated here without regard to the fibre ; but that of our Indian variety has been, I believe, proved by experiment to be of little value. The contrary, I think, is the case with the flax. The plant is short and branchy, unlike the long single stem of that raised in Europe from seed chiefly furnished by Russia and Belgium. (Ireland produces no linseed ; England little for agricultural purposes, the climate being unfavourable.) The Indian flax, therefore, is what cotton merchants call short in staple; but it is not deficient in tenacity, and would be worth, if properly prepared, at least £25 per ton in the Irish market. The present lowest price is £35 to £40 per ton, so I have allowed a good margin.

"It may be worth while to describe the Irish process of preparing flax. The plant is pulled up by the roots when the seed is nearly ripe, steeped in water for some time, and then rolled or beaten to break up the pith. It is next scutched, *i.e.*, submitted to an action like that of loosely-set

CULTIVATION.

Bombay.

scissor-blades, whereby the fibre is cleared of the broken refuse, called '*shoves*.' It is now 'flax,' and ready for the market, where it is purchased by the spinners. These, by processes which I need not describe, convert the finer sorts into 'line' for the manufacture of cloth, and the coarser into 'tow,' which is made up into cordage and sacking. To the latter class the fibre now producible in this Presidency would belong.

"The object which I have in view is the production of coarse 'flax,' *i.e.*, the article 'scutched,' and ready for the manufacture of tow. We require, first, water; secondly, the means of breaking and scutching. These operations are performed by steam, water, or hand-power. The very transformation which I wish to effect, *viz.*, that of the production of linseed alone into a combined trade in linseed and flax of inferior quality (for it has not been found possible as yet to save the seed, when the plant is plucked early to secure the finest fibre), is now going on in parts of England by means of small sets of machinery, which cost about £2,000 to set up there, and could hardly, I suppose, be brought into operation here under a cost of R40,000. But, after a careful enquiry into the history of the attempt made to set up a flax trade in the Panjáb, I am decidedly of opinion that the use of extensive machinery is in our case to be avoided.

"In the Panjáb, first-rate machinery, skilled labour, and Riga seed were imported. The article produced was at the top of the market, but the production was insufficient to keep the machinery going. The Belfast men got tired of waiting for profits, and the Government of advancing subsidies; the chief manager died, I think, and the whole concern rotted away, after a fashion not unfortunately outside of our own experience in industrial enterprise. The matter had been pushed too fast, and without regard to the caution of the rayat in venturing upon any new thing before he feels the ground firm under his feet.

"Upon understanding these things, I turned my attention during my leave, which was spent in the flax districts of Ulster, to the process of hand-beating and hand-scutching. These are chiefly performed by women; the former with the 'beetle' used in cookery (which would be efficiently replaced by its Indian representative, the *musal* or rice-pestle), and sometimes, where horse-power is available, the flax is rolled in a machine exactly like our Indian *chunamchakar* or lime-mill. The hand-scutching is done with tools that any village carpenter could make out of a two-dozen claret case, and so small that I brought the most important out with me in my portmanteau. The fibre, however, is inferior to that which is machine-scutched.

"I am certain that at a slight expense in tuition the rayats could be taught to use these tools and prepare an article saleable in Belfast and Dundee at remunerative prices; and that when at the end of a few years the trade had taken root, the time would come for the importation of seed and the erection of machinery, which could in many cases be set up in the cotton-ginning mills already in existence. No steam machinery is better fitted for use in India, because scutchers can 'fire up with *shoves*,' *i.e.*, depend chiefly upon their own refuse for fuel. I should add that they are more in danger from fire than most other trades, and can seldom effect an insurance.

"The process of production would be shortly as follows: After pulling, the flax would be rippled, *i.e.*, stripped of its seed by hand or by pulling it through a 'rippling comb,' which consists of a set of iron teeth set upright in a block. The seed would then be laid out to finish ripening, and the 'reed' or stalk stacked till the monsoon (flax is a *rabi* or cold-weather crop), when it would be steeped with proper precaution against the pollu-

CULTIVATION. Bombay.

tion of rivers by the poisonous 'flax broth,' broken and scutched by the cultivators' wives, and pressed or half-pressed, like cotton, for export to Belfast or Dundee, from both of which places I have valuable assurances of interest in the matter.

"At present the flax supply is diminishing in Ireland, from the high price of labour and the rapid conversion of arable land into pasture, which is the main feature in the phase of agricultural progress through which that country is now passing. I have the high authority of **M. Emile de Laveleye** for the fact that the former cause is producing a similar effect in Flanders. Of Holland I do not know. Russia produces much fair flax; Egypt some very bad and some very good. I am promised details with regard to the last-named country, and shall in the meantime be glad to answer any reference Government may please to make upon the subject, and to put my information and materials at the disposal of any officer who may feel inclined to take the matter up. I should propose the despatch of samples of this year's crop to England for examination, and the engagement of hands to teach the scutching for a couple of years."

Mr. Sinclair's further letter ran thus:—"The technical vocabulary of the flax trade is as follows:—

"Flax" means, firstly, the whole crop; secondly, the fibre after it has been scutched, in which state the spinners buy it.

"Linseed" or "seed,"—the seed.

"Reed,"—the stalk, of which the fibre is the outer part.

"Rippling,"—the separation of the seed from the reed.

"Scutching,"—the separation of the fibre from the pith.

"Breaking," "rolling," or "beating" is the operation of preparing the flax for scutching which is done either by mere pounding or by passing between rollers at a high pressure.

"'Shoves' are the broken refuse of pith and bad fibre which remains after scutching. In Ireland this refuse is generally used as fuel; but the Dundee men work it up into sacking, and it is also much used in making roofing-felt.

"'Retting' is the steeping of the flax in water to facilitate the separation of fibre from pith.

"'Beets' are the bundles or sheaves in which flax is tied up for the retting.

"'Scutching handle' is a wooden tool about 28 inches long, 4 inches wide, and 1℔ in weight. It is bevelled off at the edge.

"'Scutching-block,'—a plank set up on edge and so bevelled above as to correspond to the scutching-handle. It is nailed to a block, stool, or table, to keep it steady. The bundle of broken flax is held in the left hand over its edge, and turned and exposed to sharp drawing cuts of the scutching-handle, which separate shoves from the fibre. The bevelled edges of the block and handle are turned away from each other during this operation.

"So effective are these simple instruments that those worked by steam differ from them very little in shape or application.

"I would suggest that the model farms might, if they conveniently can, send in each six samples of 1℔ a piece, *viz.*, three of untouched straw and three of flax cleaned, the best way they can. A *chattiful* of any water (not containing lime in solution) would steep 3℔ of flax very easily. A few blows of a rice-pounder, or passing it under a *chunam*-mill or garden-roller, would break it, and some primitive attempt might be made at scutching so small a quantity. If the model farms think they cannot manage it, and will send me the flax, I will try it. The test of sufficient steeping is a green slime which comes off easily between finger and thumb;

CULTIVATION. Bombay.

that of sufficient dryness after steeping is the 'bowing' off the fibre, *i.e.*, its beginning to separate from the reed and form loops or bights along it. It is then ready to break and scutch.

"If we had these samples, they could be examined by proper authorities, and some estimate (though very rough) might be formed of the improvability of the native crop. Any particulars as to the average yield per acre of seed would be valuable, as tending to show what effect the importation of seed might have upon the oil trade.

"I can, if these views meet with concurrence, at any time furnish the addresses of firms who will be glad to examine samples, or answer any further reference that may occur to Government or to any officer concerned."

IN MADRAS. 28

VII.—MADRAS.

The Government of Madras submitted a copy of its proceedings, containing the information supplied by the Board of Revenue of that Presidency which had been compiled from the reports of the district officer, under its control. In these proceedings morespace is devoted to *gánja* or hemp, and *sunn* (**Crotalaria**), than to Flax. That portion of the communication which concerns flax is given below:—

"This plant is not cultivated in any of the following districts:—Vizagapatam, Nellore, Madras, Tanjore, Trichinopoly, Salem, South Kanara, and the Nilghiris, and is not mentioned in the reports from Coimbatore and Tinnevelly. It is grown most extensively in Bellary and the Kistna (total 10,500 acres). Linseed-oil is extracted from the seed and used by painters as well as medicinally; the seed, not the oil, is also exported to England. The plant is not used for its fibre, but the leaves when young are used in curries."

In Vol. IX. of the *Journal of the Agri.-Horticultural Society of India* (pp. 321 to 335) **Dr. John Mayer**, Professor of Chemistry in Madras, during 1856, contributed an analysis of the mineral constituents of the Flax plant, and of the soils on which the plant had been grown. The analysis had been worked out by himself and the late **Professor D. S. Brazier** of Aberdeen. As the results obtained by these chemists are frequently quoted, the reader is referred to the original article in the volume cited above.

IN MYSORE. 29

VIII.—MYSORE.

The Chief Commissioner of Mysore furnished the Government of India with a report prepared by **Colonel Boddam**, in whose opinion no part of the Mysore plateau was suited for growing flax, either for fibre or linseed-oil, because it required a colder climate and more copious dews to bring it on than they have in Mysore. He also stated that he had not seen *ulsí* grown anywhere south of Nagpúr;* that there it was sown at the beginning of the cold season (about November), in fields, at the same time as the wheat crop, on rich black cotton-soil; and that slight showers, which usually fall in that locality in November, start the crop, which is fostered by the heavy dews of that part of the country, while the cold is greater than in Mysore. He also observed that heavy rains ruin the crop, stripping off all the delicate leaves; he then continued,—"Knowing the value of linseed-oil for cattle fodder, I tried some two years ago in my garden; in the rains the leaves were stripped off, and the plants perished; in the cold weather it grew stunted, and, after giving a few flowers, died prematurely." **Colonel Boddam** concluded by suggesting that if an experiment were to be undertaken at the Lall Bagh, the seed should not be sown

* Nevertheless it is grown in Madras Districts, a good way south of Nagpur.

CULTIVATION

before the middle of November, to avoid the heavy rain usual in October and early in November.

IX.—BURMA.

IN BURMA. 30 (418)

"The Chief Commissioner of Burma reported that there was practically no flax cultivation in that province, but that it was grown to a very small extent here and there for the seed, and to a larger extent in Upper Burma. He asked for seed, so that he might be able to endeavour to introduce the cultivation of the plant into the province.

"Twelve maunds of good selected seed, purchased through Messrs. Moran & Co., were sent to him on the 6th instant, and it was suggested that he should try the experiment in four different localities, an acre being cultivated in each; that the ground should be selected at once, and properly prepared for the reception of the seed before the commencement of the sowing season, care being taken that the cultivation be conducted under such ordinarily favourable conditions as attend the growth of a crop by a cultivator interested in its success. He was also asked to note the cost of cultivation, and to submit a full report of the results of the experiment in due course.

"A pound of flax seed imported from Melbourne was obtained from the Agricultural and Horticultural Society of India, and was also sent to the Chief Commissioner for trial in the garden of the local Society or in some other suitable locality, the proceeds of the cultivation being compared with that of the indigenous seed." The results of the experiments carried out in accordance with the above arrangements, are noted below.

Two bags of the seed were sown at Zanganung in May 1875 by **Mr. Hernandez**, of the Burma Company, but resulted in a complete failure owing to the lateness of the sowing on a soil insufficiently prepared.

In Thayetmyo the experiment was entrusted to **Surgeon-Major Lamprey**, of Her Majesty's 67th Regiment. The seed, which was reported by him to be of a very fine quality, with a large and sound grain, reached him in September 1874; but owing to the unusually high rise of the river Irrawaddy in the month of October, and its late subsidence, it was not possible to sow it before the close of December, when the soil of the bed of the river was capable of being ploughed. The seed was sown over an area of one acre, and after six days had elapsed germination was apparent At first its growth was promising, but subsequently the growth seemed to be arrested, and the plant did not obtain a greater length than 14 inches, including the root fibres. It began to seed two months after sowing. The yield was not in proportion to the quantity sown, as it was barely in excess of that. **Dr. Lamprey** was of opinion that the seed would yield good oil, and stated that from a trial made in the hospital it was shown to be equally good, if not superior, to the linseed generally supplied by the Medical Stores. He was unable to obtain fibre from the plant, as the fibre was barely traceable and the inner portion of the stem was hard and woody in texture. In his opinion the conditions of the soil and climate did not seem suitable for the cultivation of the plant in the Thayetmyo district. One great obstacle to its cultivation was the growth of grass (*Kaing*), and on that account **Dr. Lamprey** strongly recommended the importation of English agricultural implements to ensure good results with experimental cultivations in the province.

In the Amherst district, no success attended the cultivation of the flax. The plant after growing to a certain height withered, but it was affirmed that the lateness of the season in which it was sown formed an obstacle to its growth. In Tavoy also the experiment was unsuccessful, the climate being considered unsuitable for the description of flax. Similar unfavourable

CULTIVATION.

Burma.

results attended the experiments at Meday. Kama, and Mindoon, and at the jail at Thayetmyo The attempts to cultivate the seed all failed.

In the Shwegyeen district the seed was apparently sown with a crop of wheat. The plants from this sowing were reported by the Agri.-Horticultural Society of Rangoon (to whom dried specimens were sent) to have been of a fair average height and to have seeded plentifully.

A more satisfactory result was obtained by **Captain Poole**, the Assistant Commissioner of Myanoung. "Of a bag containing 100lb, half the quantity was sown in the jail garden. there being no room for more. Four pieces of ground were selected, aggregating nearly one-twelfth of an acre, which had been previously well dug up and manured for the cultivation of vegetables. The four plots did not differ much in situation, except that one was a little lower than the others, and therefore more constantly under water during the rains. It was observed that the plants came up more quickly on this spot and that they looked stronger and better than the rest. The whole of the jail garden was rich clay; it was very fertile and an abundant supply of vegetables had been produced for many years. The site selected was apparently as favourable to the growth of flax as could have been chosen. The first sowing was on the 1st October 1874, the plants came above ground in ten days. A second sowing took place on the 16th, the plant appearing in five days. This was on the low ground before alluded to.

"In January the flax flowered, and in March 1875 the plants were gathered. About 220lb of seed were obtained and kept for subsequent sowing and for distribution to cultivators. There was unfortunately no one connected with the jail who had any experience in the preparation of the fibre, and consequently the Burmese method was adopted. The stalks were steeped in the river for five or six days and then taken out and dried, the fibre was then pulled off by the hand until nothing but the stalk remained. The gross quantity of the fibre obtained was about 30lb in weight: it was beaten out with a wooden "lutel" as in Ireland, and an attempt was made to spin some of it into thread, which failed. A small portion was made into rope and some also forwarded to the Rangoon Jail. **Captain Poole** strongly recommended more extensive cultivation of the fibre, as, in his opinion, the greater portion of the land in the province was suitable to its cultivation. He was also of opinion that there is nothing to prevent superseding the comparatively valueless production of crops known in the Burma revenue nomenclature as miscellaneous or *piné* cultivation. The enhanced value of the produce of the land would, he thought. amply justify an increase of the revenue on such lands." Whether **Captain Poole's** suggestion was ever carried out and was attended with the success he anticipated, the writer, from the imperfect data before him, is unable to say.

The pound of Melbourne seed, which was sent to the branch of the Agri.-Horticultural Society at Rangoon for experimental trial, was sown in November; the plants came up but sparsely, and had scarcely grown a foot in height before they came into blossom and began quickly to perish—being scorched up under a November's sun. In December five or six beds were planted with the Bengal seed which did better, and although not a success, since the plants did not grow above a foot in height, when they all flowered, the experiment was not altogether a failure, as a quantity of seed was gathered from the crop. **Mr. Hardinge**, the Honorary Secretary of the Society, mentioned that there was not the slightest difference between the plants raised from Bengal and Melbourne seed; that apparently the climate of Rangoon, when even at its coldest, was not favourable to the growth of the flax-plant with success, but that where greater cold is experienced, it would probably thrive better.

CULTIVATION IN BENGAL. 31 (49)

X.—BENGAL.

The proceedings of the Honourable the East India Company (to which reference has already been made) contain the financial and other results of the two experiments made at the close of the last and during the first two decades of the present century, with the object of introducing Flax cultivation into Bengal. These show that the attempt to introduce Bassorah Flax into Bengal and the North-Western Provinces failed. It appears from the correspondence that Arabian Flax cultivators were brought from Bassorah and flax seed was imported thence, but the final conclusion arrived at was that the soil of India was generally unfit for the cultivation of Flax and that the Arabian method of dressing the fibre by "beating" and "scratching" was so very tedious when compared with that used in England by "breaking" and "swingling" that if the cultivation were ever introduced in India on an extensive scale there was no doubt that the latter method of dressing should be adopted. The passages quoted above (pp. 11-24) from Royle's *Fibrous Plants of India*, exhibit the success or failure that attended the efforts (mostly of private individuals and mercantile firms), which were made to continue the efforts at establishing the new industry, down to about the year 1843. Some thirty years later Mr. D'Oyly's report once more reopened the enquiry, and the information collated on the basis of his essay has been briefly reviewed in the provincial notes above. Apparently Bengal contributed nothing further than Mr. D'Oyly's high expectations, and the subject may be said to have once more lapsed into obscurity.

The following references to the publications of the Agri.-Horticultural Society may be cited as having a bearing on Bengal:—Vol. I. (1842), p. 393. Mr. Deneef's "Practical information on the best mode of cultivating Flax in Bengal" (see p. 16 above); the details of the Chittagong experiments, Vol. II., pp. 275-281; the Monghyr experiments, Vol. III. (Proceedings), p 49; in Vol. VI. (New Series, 1878), p. vii. Mr. R. Macallister reported despatch of samples of Flax grown in Mr. Lethbridge's Indigo plantation at Otter in Tirhoot. High expectations were entertained by Mr. Macallister of the advantages of a combined Flax and Linseed cultivation if undertaken by the planters who have the means of sowing good seed thick, so as to encourage the formation of fibre as well as seed. The Secretary of the Society in commenting on Mr. Macallister's samples reviewed all the papers that had appeared on the subject, and then added: "None of these numerous experiments resulted successfully in a commercial point of view, though the produce was, in many instances, favourably reported on, as it could not apparently compete in the home market with the produce of Europe. Flax, for fibre, cannot, it is feared, be profitably grown in Lower Bengal; now, however, that a local demand has arisen for it, and the attention of indigo planters and others has been attracted to its cultivation and preparation, there is a fair prospect of success attending the renewed endeavours that are being made for the growth, for its fibre, in Upper Bengal, of this useful and valuable product."

Shortly after the date of the above correspondence in the Agri.-Horticultural Society's Journal Mr. Macallister published (1877) a note on Flax cultivation in India, from which the following passages may be extracted as conveying the new facts or suggestions which Mr. Macallister offers, and which, perhaps, have not been sufficiently urged in the above review of past literature of Indian Flax:—

"The common linseed stalk of this country, which is usually thrown aside as useless by cultivators after they have extracted the seeds, although short in fibre in consequence of thin planting and poor soil, can be made available for a fairly good class of fibre, if the stalk can be got out of the

CULTIVATION.

Bengal.

producer's hands before the exposure to the sun and hot winds has spoiled it. In the Mississippi and Missouri valleys of America, large quantities of linseed (flax seed) are annually planted with a view to utilise both seed and fibre, and both are utilised to the fullest extent with much profit to the planter. The quantity of seed put into the ground is about treble that in Bengal, consequently they not only get a good fibre, but obtain a better yield in seed. Machines patented by **F. A. Smith, Esq.**, of the Missouri Flax Works, High Point, Moniteem Co., Missouri, U. S. A., are used for threshing out the seed. Where labour is high, as in America, these machines are indispensable; here, with the cheap labour of the country, the usual method of threshing rice, *not the bullock tramp, because that spoils the fibre*, but the usual threshing by hand over a log of wood, or a stone, is the best. The fibre produced from native-grown linseed, although much inferior in quality to that grown, especially with reference to fibre, is nevertheless a fair substitute for Russian tow for all sorts of coarse goods, and if carefully worked out will fetch in the London market from £25 to £35 per ton according to quality.

"Good Russian Riga flax is now worth in the Calcutta market about R550 per ton, or R20 per Bengal maund. Indian flax can be produced nearly, if not quite, as good as Russian, and much cheaper; so that if, as I believe, numerous indigo planters are prepared to try a few acres, specially grown on prepared lands, and worked out with a view to better quality, as also to utilise their ryot's productions for the coarser quality, we are likely to get a long way on the road to success without much delay, and eventually to enrich the agricultural resources of the country to an important extent. Indigo planters especially would benefit by adopting flax raising as an auxiliary to indigo, as there need be no great outlay; they have engines, vats, and buildings; they have generally at their command the raw material, and can grow it for fine qualities; and what is of more advantage than all, the manufacture of this fibre serves at a season when they have little else to do.

"I have written the above rough remarks with a view to convey to others who are desirous to benefit by it such knowledge as I have gained from considerable experience, both in America and India. I have found in experiments, carried on at Dinapore during the past season, that Indian linseed straw grown by the native cultivators is rich in fibre, but, owing to the mode of growing that fibre, is harsh, and that when planted thickly and grown for fibre only, the fibre is almost equal to Russian. Cultivators who wish to make the most out of their labour would do well to plant double the usual quantity of seed to the acre, gather the stalk when the seed in the bolls is ripe or the bolls have turned brown, thresh out the seed after two days' sun, and immediately immerse the stalks in water and follow the process above stated. Thus, they will utilise the seed, and get a fibre useful to the mills, both here and in England, for making canvas for ships' sails, tarpaulins, &c., for which there is a large demand. The Rustomjee Twine and Canvas Factory at Ghoosery, near Calcutta, have this season manufactured from imported flax more than 300,000 yards of sail and paulin cloth, for which purpose this fibre would serve admirably. Producers will now find a market for their productions here in India."

No further reports have been published regarding the Tirhoot proposals of a combination of Indigo Flax cultivation—the most recently awakened interest in Flax—and it is thus feared, that, like all the other hopes and expectations raised during the past seventy, eighty, or hundred years, the subject has lapsed, once more, into the oblivion, from which, spasmodically, it has been rescued, through a consideration of the millions of pounds sterling which it is thought are annually lost owing to the people of India neglecting to utilise the stems of their linseed crops.

PAPER MATERIAL. 32

Paper Material.

It has recently been urged by the European Press that linseed cultivation has many advantages to the English and Scottish farmers. That although it is known to be an exhausting crop and hence is barred by many leases, an effort should be made to overthrow these objections through manuring. The reasons for this recommendation are, that the seed would give a fair return which, if combined with the proceeds of the sale of the uprooted stems (after being dried by stacking on the field like corn) would make the crop highly remunerative. These dried stems, it is urged, would, afford one of the best of paper materials. They would require no preparation, as fibre and wood could be pulped together. It is further pointed out that fowls fatten or lay marvellously if allowed to stray over a linseed field and they do little or no harm; while, according to the Russian peasant's experience, a certain percentage of linseed or linseed-cake given to cattle is the best food-material for causing a prolonged yield of very superior butter-yielding milk. Cattle so fed afford also the finest known manure, so that if this be applied to the field, the injury caused by the crop is more than compensated for by the numerous advantages to be obtained from a limited linseed cultivation on each farm.

The idea of utilising the Indian Flax stems as a paper material has more than once before been urged. With Esparto and other valuable grasses at their present cheap price, it is admitted, however, that the paper-maker cannot afford to pay for a fibre that requires to be prepared before being thrown into his vats. He can thus purchase the jute cuttings and ends, also old bags, ropes, and rags, but cannot afford to purchase new jute fibre, though jute is the cheapest of all fibres. But in the case of flax stems, as above recommended, no previous preparation is considered necessary. The objection urged by the native cultivators that a fibre-yielding linseed affords them an inferior oil goes for nothing. They are also not called upon to pluck the crop at a season more favourable to the fibre than the oil. They are not called upon to take measures to prevent the fibre being injured by exposure to the sun, nor are they required to learn a new industry, that of crudely separating the fibre from the flax stems. They are simply asked to dry these stems and sell them thus to the paper-maker rather than burn them as manure. Everything would thus seem to point to paper-making as the best channel of utilisation for the Indian linseed waste stems. But the linseed fields of India are, as a rule, far removed from the paper mills. The means of converting such stems into paper are peculiar, and a mill would probably have to be specially arranged before it could use linseed stems. And if this be so, most paper-makers would require some guarantee that the material would be forthcoming in sufficient quantity and at a sufficiently low rate. If it had to be carried by railway, for any distance, it would very likely be rendered thereby too expensive for the paper-maker and it would certainly never pay to export the Indian crude stems to Europe as a paper material.

Separation. 33

Separation and Preparation of Flax.

It seems unnecessary to deal with this subject in a separate section. Flax can hardly be said to be produced in India, and the best methods of cultivating the plant, the seasons of sowing and reaping, and every detail of the manipulation of separating and preparing the fibre for the market, have already been dealt with in the pages above (see pp. 16, 31-34, 40-44, and 47-50). Detailed information regarding the machinery employed in Europe, both in cleaning the fibre and in weaving it, will be found in *Spons' Encyclopædia.* The special report drawn up by the Royal

SEPARATION OF FLAX. Society for the Promotion and Improvement of the Growth of Flax in Ireland, gives details of cultivation and manufacture. This will be found in Vol. IX., *Jour. Agri.-Hort. Soc. India* (*Old Series*), *p. 60.* In that volume also the Secretary of the Society published a series of most valuable papers on flax culture and manufacture, thus bringing together a mass of material which should be carefully considered by persons who may desire to once more renew the enterprise of flax culture in India.

LINSEED. (422) 34

LINSEED.

The seed obtained from the cultivated plant—**Linum usitatissimum**—is known as Linseed or Flax-seed, the expressed oil as Linseed oil, and the cake as Linseed-cake. The history of the cultivation of Linseed in India has already been alluded to briefly, so that it but remains to bring together in this place such information as is available on the provincial cultivation and utilisation of the crop. The writer is, however, conscious that the material at his disposal is of so meagre a character that all he can hope to do is to indicate the plan upon which future facts might be collated by exhibiting the various headings into which the subject might be divided. The passages quoted (from a wide series of publications) manifest more than anything else the imperfect and disjointed state of the information that exists in India on this, one of our most important crops.

RACES. (423) 35

Races or Forms of Linseed.

To a certain extent the cultivation of Linseed has already been indicated by some of the remarks regarding Flax. The object being to promote flowering, not the production of long, straight, fibre-yielding twigs, the seed is sown much thinner than is generally the case in European flax cultivation. The result of this has been to develop several well-marked RACES, all of which possess one character in common, namely, the formation of a short, much-branched stem. Nothing definite, however, can be written regarding the forms of Linseed grown in India, since they have not been carefully worked out by botanists; but there are two important kinds, white-seeded and red-seeded, and within each of these divisions there appears to exist two forms, "bold" and "small," commercial terms that denote the character of the seed. The latter kind may, to some extent, be merely a consequence of imperfect cultivation, or of season of collection. The subject is too little understood, however, to allow of any definite inferences being drawn, but the white and red-seeded races seem perfectly distinct. It would, in fact, appear undeniable that a crop which has been cultivated for several centuries (if historic evidence be accepted), under the most diverse conditions of climate and soil, must have by now developed widely different races. And if this be so, it may safely be added that until the characters of these races have been established, it would be hopeless to expect the industry to divert at command from linseed to flax production. We must first establish the local influences and their effects upon the existing crop, and know, too, whether that crop manifests a favourable tendency to fibre formation, before the recommendation be offered that thick sowings should be made with the object of obtaining combined flax and linseed. This seems to have been the key-note of past failures. Foreign seed was forced on the cultivators—seed that often failed to grow at all, in certain of the districts in which it was arbitrarily sown, or which, where it did germinate, yielded a slightly better fibre but worse oil than the local stock. Disappointment, followed in time by antipathy against the new crop, was the not unnatural consequence. It would seem that to make flax cultivation gain a hold on the people of India, it must at first be a bye-product to seed cultivation. If the difficulties of locally cleaning the fibre, or of transit from the fields to cleaning mills, were once overcome, and it was

RACES.

found that the waste stems yielded a certain return, there would then be no difficulty in inducing the people to sow thicker in order to ensure both crops. In time they might be even educated to cultivate flax only, but at present failure is certain to follow any experiment that has professedly for its object the cultivation of flax and not of linseed. It would thus seem the natural course to thoroughly investigate the nature of the Indian races of **Linum** with the object of discovering those that manifest a tendency towards fibre production. With thick sowing and careful cultivation such forms might even be educated to produce both crops, and in this way a modified flax production might gradually be inculcated into Indian agriculture Spasmodic efforts to acclimatise highly prized European and Australian forms of flax are certain to have the same fate as the experiments already made. The path most likely to lead to success in the case of this staple of our trade, as with all the products of India, would seem to be careful study and selection of the indigenous stock.

The following jottings regarding the red and white-seeded forms may prove useful.

Red Linseed. 36 (424)

RED LINSEED.—There are several forms of this recognised in the trade, and these are apparently caused by soil or methods of cultivation. The season at which the seed is collected and the freedom of the plant from blight has much to say to its character, far more perhaps than the existence of distinct races, but, as already stated of this subject, we possess no definite knowledge. It is, however, admitted by most writers that the seed of one part of India is richer in oil than that of another, and it, therefore, seems probable that the drying property is equally variable.

White Linseed. 37 (425)

WHITE LINSEED.—The earliest detailed notices of this form occur in the *Journals of the Agri.-Horticultural Society of India*. Thus, in 1844, Colonel **J. B. Ouseley** sent a sample to the Society from the Nerbudda Valley. This was examined and reported on by **Henry Mornay** (*Vol. III., Selections, 98*) who wrote, "This beautiful seed is larger and plumper than the finest red seed, besides which the shell or outer cuticle is much thinner, which is very beneficial." It gave weight for weight fully 2 per cent. more oil and yielded its oil much more easily, while the cake was far softer and sweeter than that produced from the red seed. In a subsequent letter **Colonel Ouseley** returns to the subject (p. 249), and shows its relative value in 1835 to 1844 compared with wheat and gram. It then fetched the same price as the best qualities of wheat, and was one rupee more expensive than the red, per maní (about 5 maunds, 16 seers). **Colonel Ouseley** then concludes as follows: "White linseed is exported towards Bombay, and is to be found at Jubbulpore; but I understand none grows north of Rewah; from inhabitants of Oudh now here, I am informed it is unknown in that territory; it forms an article of trade south of the Nerbudda and is in great demand." **Mr. H. Cope** of Amritsar wrote in 1858 (see *Jour. Agri.-Hort. Soc., India, X., Proceedings (Old Series), lxxxviii*) of the white seeded flax:—"If I remember right it was you who told me that the Jubbulpore white linseed became brown again on sowing in other parts of India, showing that the white was a more local variety. In order to test the accuracy of what I believed was only a surmise on your part, I obtained a small quantity of mixed white and brown seed, carefully separated the white, sowed it myself, and have to inform you that so many of the flowers, which will of course produce brown seed, are blue, that I have no doubt these, which are white this year, will be blue next, and produce brown seeds, the whole returning to their primitive habit."

DISEASES. 38 (426)

Diseases of Linseed.

Various reports, here placed under contribution, will be found to allude to the precariousness of this crop. It is liable to injury from severe rain

DISEASES.

Rust. (427) 39

or excessive drought, but its greatest enemy is a parasitic fungus which causes great havoc, according to some writers often diminishing the crop by 50 per cent. The early literature of this disease is extremely interesting. Much was written of it in 1829 and the few succeeding years, but a gap of half a century has then to be passed over, until in modern official literature its ravages are deplored, but little or no attempt made to investigate its cause, or to discover the means by which to avert the ruin it periodically causes. In 1829, **Major Sleeman**, Agent in the Nerbudda territories, reported the calamity that had befallen the people in the form of a disease that had attacked the *alsí* crop first, then the wheat. He attributed this to a minute insect which penetrated the stem and absorbed the nourishment of the plant, thus ruining the crop. His report, accompanied with specimens, was submitted to **Dr. W. Carey**, who affirmed that the injury was really caused by a fungus like the wheat blight of Europe (then attracting attention and supposed to be caused by a fungus that passed from the barberry to the wheat). On obtaining this explanation **Major Sleeman** wrote: "The whole of the phenomena observed and described by me, seem now as reconcilable with the supposition of the growth of a parasitic plant as with that of an insect; for it is only a fungus instead of an insect feeding upon the sap, and when I discovered that the blight had first made its appearance last year in some fields of *alsí* (linseed), I began to believe, that the insects I had seen might be rather the effect than the cause of the calamity, and collected to feed upon these very minute but destructive mushrooms. I gave orders accordingly that the cultivators should in every village be recommended to watch their fields carefully and remove every blade of wheat or other plant the moment any red spots should be discovered upon it, as the only means in their power to preserve the rest.

"On arrival at my present ground I found that parts of every *alsí* field in the neighbourhood had begun to be affected, and that on some the plants were so lately covered with pale red spots as to appear to have altogether changed colour, as you will perceive by the samples forwarded with enclosed letter; and as the only means we have of preventing the dissemination of this disease over the wheat crops seems to be the removal of the *alsí* immediately, wherever it is discovered to be affected, I have thought it to be my duty to give orders to this effect. But orders would be unavailable, were I not to promise that the cultivators of these *alsí* fields should receive a remission of rent upon the lands from which the crops may be removed, since the blight injures but little the seeds already formed in the *alsí*, and does not appear till some few seeds upon each plant are formed, &c." "I have, therefore, promised that a remission shall be granted for all lands from which the *alsí* shall be removed in pursuance of this order."

Continuing his review of observations extending over two or three seasons, **Major Sleeman** wrote: "I now find that the *alsí* was last year affected by the disease as generally and still more early than the wheat, and that it commonly is so; and as it may be more subject to it, and is generally sown either among the wheat or around the borders of most wheat fields, it may be, if not altogether the source of the disease, the means of increasing and extending it. And as the cultivators can very easily substitute another crop for it, it may be worth while to consider whether it might not be expedient to prohibit the cultivation of this grain for a year or two." Later on he makes the observation: "At present not a leaf or a stalk of the wheat is affected, though close to patches of *alsí* entirely covered with the disease; no *alsí plant is affected till one or more of the seeds have been fully formed upon it*. The wind has for some days blown from the eastward, and since it began to do so, the blight has appeared" on the wheat.

At the time at which **Major Sleeman** wrote, the life history of wheat

DISEASES.

Rust.

blight was, even in Europe, a matter of speculation, so that it is not surprising that **Major Sleeman** made the mistake of supposing that in India it came to wheat from the linseed. We have long since known that this is impossible, and that Government and the people were alike deprived of revenue through the repressive measures against *alsí* briefly indicated by these passages. But so alarming had the blight become, that in March 1830 Mr. **Holt Mackenzie** addressed a letter to the Honourable **Sir Edward Ryan** desiring the co-operation of the Agri.-Horticultural Society. Referring to **Major Sleeman's** reports **Mr. H. Mackenzie** said:—"They relate to a blight, which appears to have occasioned much individual misery and has caused very large abatements to be necessary in the Government demand. If the Society can help us to remedy so great an evil, it will be entitled to no ordinary acknowledgments." It is not necessary in this place to allude to the famine that followed in 1831-32 through the almost total destruction of the wheat crops, nor to the calamity that overtook the starving population from trying to subsist on the pulse **Lathyrus sativus**, a grain that paralysed the lower extremities of a large number of the persons who ate it. Suffice it to say that it was then supposed that the blight not only went from the *alsí* to the wheat, but to the pulse named, and caused "the paralytic strokes" from which the people suffered. The reader is referred to **Lathyrus sativus** for further information on the subject of this remarkable property, and it need only be here added that there is nothing to show that the wheat blight was more connected with **Lathyrus sativus** than with **Linum usitatissimum.**

In concluding this brief review of the early controversy regarding *alsí* blight, it may be added that even **Major Sleeman** seems to have ultimately come to the opinion that it was quite unconnected with the wheat blight. This opinion has been recently confirmed by the investigations of **Surgeon-Major Barclay**, who has determined the nature of the fungi found on *alsí* and on wheat, and shown that they are of necessity utterly unconnected with each other. He has not, however, been able (from want of opportunity) to trace out their separate life-histories, but has made an important step in that direction by the discovery that in many parts of India the wheat blight is not the common European species (**Puccinia graminis**) which requires in its first stage to exist on the barberry, but is a species of **Puccinia** resembling **P. rubigo-vera**, which, in those countries in which its life-history has been traced (Europe and America), passes to a boraginaceous plant. In India the same life-history seems very improbable, since no borage has been found harbouring an **Æcidium.** This subject will be again reverted to under **Triticum** (which the reader should consult), but **Dr. Barclay's** account of the fungus found on **Linum** may be here given, since it is not merely of scientific value but is of economic interest also, as it establishes the independence of flax from wheat blight.

"I obtained," says **Dr. Barclay**, "excellent specimens of a species of **Melampsora**, gathered on 4th April 1890 at Dumraon (North-West Provinces). The leaves were very extensively attacked with orange red pustules, oval to round, but coalescing freely, and often involving most of the leaf surface. These pustules are mostly epiphyllous, and they are often surrounded by a wall of epidermis, giving them the appearance of the æcidial fructification of **Phragmidium.** In other parts, dark crests might be seen which were the teleutospore beds."

"The UREDOSPORES are pale orange red, and are accompanied by colourless capitate paraphyses, sometimes of very large size, the head exceeding the spores in diameter. They are round to oval, and the dried spores, when just immersed in water, measured $21-18 \times 18-16\mu$. But after lying 48 hours in water most spores become spherical, measuring $24-21\mu$ in diameter. The epispore is sparsely beset with spines. I could not ascertain

DISEASES.

Rust.

the number of germ pores. The paraphyses had heads measuring 30—28 × 20μ. The TELEUTOSPORES are long, cylindrical, or prismatic, single-celled bodies, very firmly adherent to one another laterally. They each exhibit a central nuclear space, and measured, after lying 48 hours in water, 54—56 × 10—9μ. They did not germinate after lying some days in water, and I conclude a period of rest is necessary before this can take place. This is no doubt **Melampsora Lini**, *Pers.*" The spore measurements show a slight difference from those given by European authorities, and the Indian form has large-headed paraphyses. These differences, however, are not important.

The above facts regarding Linseed blight have been obligingly furnished by **Surgeon-Major Barclay** from his forthcoming paper in the *Journal of Botany*.

Area under Linseed in India.

AREA. (428) 40

The various reports that exist on this subject admit the impossibility of arriving at definite information regarding the actual area under linseed, owing to the very general habit of raising it as a mixed crop. If intended for local consumption it is frequently grown along with mustard, both seeds being expressed at once, for their mixed oils. For this purpose also, pure linseed is very often mixed with mustard or other oil-seeds at the native oil-mills, and mixed oils of various recognised properties thereby produced. To the people of India the drying property of pure linseed is of no consequence, owing to their having no occasion to require an oil of that nature (see "**Pigments**"). Sometimes, however, linseed is grown with non-oil-yielding crops; for example, it may be seen forming lines through the gram or other pulse fields. The produce is collected separately, such pure seed being mostly intended for the foreign markets. But if grown as a mixed crop the greatest possible difficulty must necessarily be experienced in definitely determining the area under it. In the agricultural reports for 1885-86 and 1886-87 more of the provinces endeavoured to give the areas under linseed than in the succeeding years down to 1890. In Bengal, Mysore and Coorg, Ajmir, and Assam, no attempt has been made, but with the exception of Bengal the amount grown in these provinces is unimportant. It will be found that to produce the quantity shown as yielded by Bengal, the area cannot be on an average far short of 1,500,000 acres. Assuming this to be correct, the following were the areas under the crop during 1885-87:—

Provinces.	1885-86.	1886-87.
	Acres.	Acres.
I.—Bengal	1,500,000*	1,500,000
II.—North-West Provinces and Oudh	466,161†	499,673†
III.—Central Provinces	1,024,414	760,942†
IV.—Berar	621,893	386,374
V.—Bombay and Sind	435,912	411,197
VI.—Panjáb	32,281	24,454
VII.—Madras	30,677	16,558
VIII.—Burma (Lower only)	20,351	23,605
TOTAL	4,131,889	3,622,803

* It seems quite certain the actual area was over, rather than under, this figure since the estimate is framed on the assumption that it was entirely grown as a pure, not as a mixed, crop.

† This does not include mixed crops.

AREA.

In the subsequent reports to those from which the above figures have been compiled, the following provinces have reported their areas:—Bombay, including Sind, 309,786 acres in 1887-88, and 215,450 acres in 1888-89; Central Provinces, 633,928 acres in 1887-88, and 668,047 acres in 1888-89; Berar, 317,018 acres in 1887-88, and 304,620 acres in 1888-89; Panjáb, 36,158 acres in 1887-88, and 37,411 acres in 1888-89; and Madras, 12,985 acres in 1887-88, and 20,887 acres in 1888-89. On the whole the acreage under linseed seems to have remained stationary, the decline shown being in most cases a correction of the error for mixed crops. It seems thus safe to say that the total area in all India fluctuates about 4,000,000 acres, an acreage which, estimated on a mean yield of 2 cwt. to the acre, would at least produce the amount of seed annually shown to be exported. The Indian consumption of linseed is comparatively small, but to be quite safe the maximum area cannot exceed 5,000,000 acres, otherwise a lower average production must be assumed.

CULTIVATION OF LINSEED.

I.—BENGAL.

CULTIVATION IN BENGAL. 41 (429)

Although, as already stated, Bengal is the largest producing area for linseed, nothing can be learned regarding its cultivation. The Gazetteers are silent on the subject, and the Agricultural Department has only once briefly alluded to it. In the Administration Reports the traffic in the seed is sometimes dealt with; in the issue for 1888-89, for example, the following table occurs of the imports into Calcutta:—

		1887-88. Mds.	1888-89. Mds.
1.	Behar	50,82,873	41,99,502
	Bengal	11 48,696	12,19,966
2.	North-West Provinces and Oudh	16,21,659	20,33,292
3.	Assam	47,118	59,079
4.	Central Provinces and Rájputana	81,013	19,755
5.	Other places	8,922	10,694
		79,90,281	75,42,288

Area. 42 (430)

The Bengal area under linseed cannot be learned for certain (nor, indeed, that of any crop), but the above traffic returns afford the means of an approximate estimate. It would seem from the yield per acre, as given by all other provinces, that a minimum of 2 cwt. would be safe, but if this be under or over the mark of acreage production in Bengal, the statement here made could easily be corrected. On the assumption that the quantities, bracketed under No. 1, in the above statement of the Calcutta supply, were produced from a yield of 2 cwt. to the acre, Bengal and Behar must have had 2,225,560 acres in 1887-88, and 1,935,468 acres in 1888-89, under linseed. If the average yield proves by future enquiry to be more than here given, the necessary acreage would of course be lessened, but it seems safe to infer that the average annual acreage in Bengal under this crop cannot be below 1,500,000 acres, and is more probably nearer two million acres. In the passage which here follows it will be seen the Director of Land Records and Agriculture puts the yield at 2 maunds per acre (an estimate very probably considerably lower than the actual yield, for judging from the returns regarding the other provinces of India a yield of 3 cwt. to the acre would appear to be more nearly correct: *conf.* with remarks regarding Bombay), but if he be correct, the mean acreage in 1887-89 must have been 2,912,759 to produce the quantities shown as delivered at Calcutta from Bengal. The Director, in his Annual Report, 1886, fur-

CULTIVATION.

nished the following facts regarding Linseed, and it is to be regretted no other information on this very important crop is available:—

Bengal.

"Linseed is not a crop which is grown throughout the whole division of Burdwan. It is more largely cultivated on the *dearah* lands than anywhere else.

Soil.

43 "SOIL.—It is grown on all the different classes of soils comprised between the lighter class of clay and sandy loam. Linseed is supposed not to do as well in stiff clay as in light sandy soils. The land must be well drained, stagnant water being most injurious.

"In the cultivation of this crop one or other of the two following methods is adopted, according as it is grown in the interior on clay lands or on the *dearah* lands on the beds and banks of rivers.

"ON CLAY LANDS.—Here the plan adopted for growing linseed is the simplest imaginable. As soon as the paddy field has become sufficiently dry, linseed is broadcasted on the standing paddy at the rate of two seers per bigha. The paddy is harvested as usual, the linseed being left to be reaped in *Chaitra*.

"ON THE DEARAH LANDS.—Here linseed is sown either alone or mixed with wheat, gram, mustard, or *keshári*. Sometimes more than two of these crops are grown together. The land for linseed receives three or four ploughings and two or three harrowings. Linseed should not be buried deep, otherwise the seeds will not germinate properly. The seeds are, therefore, not ploughed in, but simply covered by passing the ladder over the field once or twice.

"When sown with such a crop as gram or wheat the plan adopted is this:—After wheat or gram has been sown, the land is ploughed. Linseed is now broadcasted, and the operation is finished by using the ladder twice.

Time of sowing	Aswin-Kartik=October.
Harvest time	Chaitra=March-April.
Quantity of seed for bigha	Three seers when sown alone, one and a half seers when mixed with other crops.

"The yield is very variable. About two maunds would be considered a good average crop when sown alone. Linseed straw is used as fuel."

Chief Linseed Tracts.

44 The principal oil-seed tracts in Bengal are along the banks of the Ganges, especially in Patna and Bhaugulpore Divisions. The damper districts are not well suited, but throughout Behar it is extensively grown. M. Deneef, in his practical information on the best mode of cultivating flax in Bengal (quoted above, p. 16) states that 6 seers per *bígha* is the amount of seed required for linseed. The plant should be pulled up by the root a little before perfect ripeness. In Vol. XII (p. 342) of the *Journals of the Agri.-Horticultural Society of India*, the following facts appear regarding linseed in Shahabad:—

"It is always grown in Shahabad as an auxiliary crop with wheat, *masoor*, barley, and other spring crops; and its bright blue flowers are a pleasing relief to the yellowish brown of the other cereals. I have never known it cultivated alone, so that I have no data as to its probable yield per *bigha* or acre. It would probably not be found to differ much from the Linseed crops of England in that respect were the cultivation equal, by which I mean, were the same labour bestowed upon its cultivation, which, however, is not the case in this district at least. Its favourite soil appears to be black clay (*kurile*), but it is sown largely in other soils, and the only difficulty to its indefinite extension appears to be the want of facilities for carrying the crop to market."

CULTIVATION IN N.-W. PROVINCES AND OUDH. 45

II.—NORTH-WEST PROVINCES AND OUDH.

Messrs. Duthie & Fuller (*Field and Garden Crops, II., 41*) give the following account of linseed cultivation in these provinces :—

Soil. 46

"*Soils.*—The distribution of linseed cultivation offers an interesting contrast to that of *tíl.* In both cases Bundelkhand is an important field of production, but for very different reasons. *Tíl* is grown on the light raviny lands which lie along courses of rivers and drainage lines, while linseed is grown on the heavy black *már* or cotton soil of which the level plains are formed. *Tíl,* in fact, prefers a light, and linseed a heavy, clay soil, and hence linseed is very largely grown in the eastern rice-growing districts, where *tíl* cultivation reaches its minimum. Linseed is also grown to a considerable extent in the sub-Himálayan districts. Like *til* it is hardly ever cultivated as a sole crop in the districts of the Ganges-Jumna Doáb, but, unlike *til,* its cultivation in this tract is confined to an occasional bordering to wheat or gram fields, and its production as a subordinate crop in a mixture is quite insignificant.

Area. 47

"*Area.*—Linseed cultivation thus is of insignificant importance in the Meerut Division and still more so in the Agra Division. In the Rohilkhand Division it is returned as occupying between 12,000 and 13,000 acres. In the Jhansi Division, which forms the western and least fertile portion of Bundelkhand, comprising the Hamirpur, Banda, and part of the Allahabad District, its area reaches 49,000 acres, or 4 per cent. of the total area under *rabi* crops. But its cultivation reaches its maximum in the Benares Division. The three districts of Azamgarh, Basti, and Gorakhpur return no less then 122,000 acres under linseed, which amounts to 6 per cent. on their total area cropped in the *rabi* season.

Method. 48

"*Method of Cultivation.*—Its method of cultivation varies very greatly in different localities. In the districts of the Ganges-Jumna Doáb it is as a rule merely sown in a line round the border of a wheat or barley field, or is grown in parallel lines across a field of gram. In Bundelkhand it is grown either alone or mixed in large quantities with gram, and in both cases the ground receives three or four ploughings during the rains preceding. The seed is sown broadcast at the rate of 8 to 12 seers to the acre. In the Benares Division it is largely grown on land which is under water during the rains, and in this case its cultivation is of the roughest possible description, no preparatory ploughings being given, but the seed simply scattered over the ground and ploughed in. It is very commonly grown in this fashion in rice fields, the rice stubble being left standing.

Irrigation. 49

"*Irrigation.*—Linseed is very rarely irrigated when grown by itself, except in the Basti and Gorakhpur Districts, where a quarter of the total linseed area is returned as receiving one or two waterings.

Harvesting. 50

"*Harvesting.*—The plants are cut down when ripe, and the seeds extracted from the capsules by beating.

Yield. 51

"*Average outturn.*—The average produce of linseed in Bundelkhand is from six to eight maunds per acre. In Basti and Gorakhpur it may be put as considerably more than this, ten maunds being probably not an excessive estimate."

It would seem probable that the above is a very high estimate of yield, accountable for, very probably, by the figures being those of a mixed crop with mustard. Allahabad, Benares, and Oudh, are the most important producing blocks. The following passages from the settlement reports afford additional information on the subject of linseed :—

Allahabad. 52

Of Allahabad it is stated (p. 31) that "In other districts this crop is usually grown mixed with gram, *masúr,* barley, and other *rabi* crops. In Allahabad, however, in the lowlands across the Ganges, and to a very great extent in the *márh* tracts south of the Jumna, linseed is grown alone. In the

CULTIVATION.

N.-W. Provinces and Oudh.

Doáb we have only some 40 acres, while in the trans-Ganges there were 3,000 acres and in the trans-Jumna tract 15,000 acres under linseed. The seed forms the export staple of Khairágarh, and is sent in great quantity down the Ganges to the eastern districts and Calcutta. The crop is a hardy one. It requires very few ploughings and no manure, weeding, or irrigation. From seven to eight seers per acre seed is required, and the produce averages three to four maunds per acre" (*Mr. F. W. Porter, C. S., in the Final Settlement Report of the Allahabad District, North-West Provinces*).

Azamgarh. 53

"Linseed (*tísí*) is sown alone throughout the field, or in separate rows in fields in which other crops are sown, or mixed with other crops. It is grown chiefly in clay soils, in fields from which an early rice crop has been taken, or like *latri*, in places where no better crop could be sown with as little tillage and without irrigation. For linseed the land gets little ploughings, sometimes none at all, the seed being simply scattered, like *latrí*, over the damp ground. It is also left unirrigated. Not very much linseed is raised in Azamgarh, but more perhaps than the settlement returns indicate. For, in some instances, land which produced both a rice and a linseed crop has been entered under the more important crop—rice. Ten or twelve seers of seed per acre should be sown for a full crop, but in Azamgarh, where little consequence is attached to the crop, so much is not always allowed. Ten maunds is a good return per acre. The plant suffers much from rust (*giruí*) in damp seasons. Linseed yields one fourth of its weight in oil. The oil-cake is given to cattle, but is also consumed by human beings, either mixed with *gúr* or alone. When eaten by human beings it is dignified with the name of *Pinné*. Linseed is also eaten by the people, being first pounded in an *okharí*, and then baked in dough" (*Mr. G. R. Reid, C.S., in the Settlement Report of the Azamgarh District*).

Banda. 54

"*Alsí* is, like *kodon*, not unknown elsewhere; but in this district (Banda) it is grown in a considerable area and linseed constitutes an important article of commerce. In this, as in other districts, *alsí* is frequently sown with other crops, but even alone it occupied at settlement nearly 2 per cent. of the cultivated area, nearly half of this being in the Banda tahsil. Some of this preponderance is no doubt due to the extent of *alsí* entered as mixed with gram in other pergunnahs, but the impression left on my recollection is that there is more and finer *alsí* in Banda than in other pergunnahs. If only a third area of the mixed crops of gram and *alsí* be assumed to be *alsí*, the crop covers over 16,000 acres, or 6 per cent of the area under cold weather crops" (*Mr. A. Cadell, C.S., in the Settlement Report for the Banda District, North-West Provinces*).

IN THE CENTRAL PROVINCES. 55

III.—CENTRAL PROVINCES.

Area. 56

In the report of the crop for 1888-89 it is shown that the provinces had 612,022 acres under Linseed. The principal districts are Wardha (118,036 acres), Nagpúr (104,789), Raipur (101,020), followed by Jubbulpore, Chanda, and Saugor, each with less than 100,000 acres. In the Agricultural and Revenue Reports for 1884-85 it is stated that linseed attains its greatest importance in Wardha and Chanda, where it covers 21·and 18 per cent. of the area under crop. In Nagpúr the percentage is 16, in Raipúr 11, and in Damoh 10. In every district it is grown more or less, and in some of the reports is said to be even meeting with greater favour than wheat. The total area under the crop has, however, contracted somewhat in recent years. The Director of Land Records and Agriculture reports that as it is sown later than wheat, it suffers less from failure of the October rain.

CULTIVATION. Central Provinces.

"The arch enemy," he adds, "of linseed is rust, and during this season it has, at all events, escaped damage from this source. Where a crop has germinated, the outturn will be fair. But over a considerable area germination has failed altogether, and the outturn will certainly fall very short of being satisfactory." In a forecast of the crop for 1885, the Director of Land Records and Agriculture wrote: "It is reported from two districts that there is a decrease in area owing to the ground having become too hard for sowing, but this is a great deal more than counterbalanced by a very general increase, resulting from linseed having encroached on land ordinarily sown with wheat and also from its having been sown on land which bore *kharif* crops, during the preceding monsoon months, but which, owing to the excessive rain, failed to give produce. There is also a steady annual increase in the area under linseed, owing to the active demand for it. The export of linseed is, next to that of wheat, the most important feature in the export traffic of these Provinces, and cultivation is steadily responding to the demand of the Bombay market. The increase in linseed cultivation is especially marked in districts such as Nagpúr, in which much of the soil is not of sufficient depth for the produce of wheat, and where the growth of linseed is increasing at the expense of the cotton and millet crops. It does not seem extravagant to assume an increase of at least 10 per cent. in the linseed area of the current year.

"The prospects of the standing crop are satisfactory: rain was much wanted for it a month ago, and in some places the plants were beginning to wither. But rain has now very generally fallen and the crop has immensely benefited. A little harm has resulted in the Raipúr district from the ten days of cloudy weather at the end of December, which have been prejudicial to the proper fertilisation of the flowers, but after making allowance for this a full crop may be expected in Raipur as well as in Biláspur, and the reports from other districts show that prospects are very nearly, if not quite, up to the average throughout the provinces."

Yield. 57 (445)

The following passages from the Nagpur Experimental Farm Reports afford additional information, especially as to yield of seed *per* acre:—"As in the case of wheat, two varieties were grown, one being the brown (or *katha*), and the other the white grained (or *haura*) kind. There was nothing experimental about the cultivation, which was conducted on precisely the same system as that followed by native cultivators. The land was prepared by frequent *bakharings* or bullock harrowings during the rains, and the seed was sown in October with the triple bamboo drill (or *tifan*). No manure or irrigation was used. The total area under linseed was 7·89 acres, comprising five fields which yielded a maximum outturn of 638 ℔, a minimum of 294℔, and an average of 428℔ to the acre. I have not, however, included in this a patch of remaining land on which linseed was sown as a speculation, but which produced only 96℔ to the acre. The white and red varieties were sown in two adjoining fields, the areas of which were respectively 1·46 and 1·05 acre. The white variety gave the largest return, 638℔ to the acre against 590℔. But this fact affords of course no safe basis for generalisation as to the relative productiveness of the two varieties. Both may be valued at the same figure, R16-4-0 per *khandi* of 382·5℔" (*1883-84, p. 7*). "Linseed was grown on 14·81 acres, the seed used being of the white or *haura* variety. The average outturn per acre was 215℔. The average outturn of last year was nearly double this, but was obtained on a much smaller area. The linseed crop was, however, much damaged this year throughout the country by the cloudy weather of December and January, which prevented the seed from setting properly, an injury to which linseed is very liable. Were it not indeed a rather precarious crop its cultivation would increase far more rapidly than it does, since

CULTIVATION. Central Provinces.

weight for weight it is twice as valuable as wheat, and will grow well on inferior classes of soil" (*1884-85 p. 6*).

The writer having failed to discover any detailed statement of the area under this crop, of the seasons and methods of sowing and of reaping, &c., &c., has republished the above passages from the forecasts and farm reports, since these deal with some of the results, and the difficulties and dangers of linseed cultivation. This confession of want of definite information has to be made in connection with almost every province in India, for, although the trade in linseed has steadily progressed, it has attracted only a fractional interest to that of wheat, although the latter crop is practically of only slightly greater value to India.

IN BERAR. (446) 58

IV.—BERAR.

In the Administration Report for 1883-84 (*p. 31*) of the Hyderabad Assigned Districts it is stated that the crop was much injured by excessive rain, that the area of cultivation declined by 83,000 acres, while at the same time the exports fell off considerably. The report continues:—

"The exports of linseed are almost exclusively of home produce: very little is imported. The exports and imports of linseed for the last five years have been as follows:—

Year.	Exports.	Imports.
	Mds.	Mds.
1879-80	1,13,000	3,700
1880-81	4,03,000	9,600
1881-82	5,89,000	19,000
1882-83	10,95,000	38,000
1883-84	8,20,000	60,000

"The growth of the trade during this period is almost as remarkable as that of wheat; and it is of more importance as being less dependent on imports from outside and consequently more stable in character, besides being the more valuable crop."

IN BOMBAY & SIND.

V.—BOMBAY AND SIND.

(447) 59 Districts.

(448) 60

The chief districts in the Western Presidency that produce linseed are Poona and Sholapur; Khandesh, Násik, and Ahmednagar; Gujarat and Kathiawar; and smaller quantities south of the Nerbudda and below the Ghâts. In the railway-borne trade report for 1883-84 it is stated that, as in the case of food-grains, the supply of linseed from external blocks is derived from the Central Provinces, holding the first rank, Berar next, with Hyderabad and Rájputána following as less important.

The following passages from the Gazetteers convey an idea of the methods of cultivation and other peculiarities in the Western Presidency. *Khandesh* (*Vol. XII., 152*).—"The average acre yield is from 250 to 280lb The cultivation is steadily spreading, owing to the Bombay demand. It forms one of the principal and most valuable exports. Deep loamy soils seem particularly well suited to the growth of the plant. The seed is bought wholesale by wealthy merchants from the cultivators." "The plant is too short and branchy to yield fibre of any value. It is never prepared, and many husbandmen are ignorant of the fact that the plant yields fibre. As nearly the whole of the seed is exported, little oil is pressed in the district." Of *Násik* (*Vol. XVI., 100*) it is simply stated to be sown in October and reaped in January. Of *Ahmadnagar* (*Vol. XVII., 270*) it is stated—"Sown in rich black soils, often with gram or wheat, in separate furrows or by itself as a separate crop, and, without water or manure, is harvested in February.

CULTIVATION. Bombay & Sind.

The seed is eaten in relishes or *chatnis*, and the oil, which is produced in the proportion of one pound of oil to four pounds of seed, is used in cookery. The fibre of the plant is not used." In *Sátára* (*Vol. XIX., 164*) it is said to be sown in November and harvested in February. In *Dharwar* (*XXII., 273*) it is known as *agashi* (KAN.) and *javas* (MAR.): in *Bijapur* (*XXIII., 319*) it is *alshi* (KAN.) and *javas* (MAR.). In *Kolhapur* (*Vol. XXIV., 171*) it grows with cotton, late *joári*, and wheat. It is harvested in about three and a half months. Pure linseed oil is expressed for painting purposes only. Most of the linseed grown in the State is sent to Bombay. Its average outturn is 375lb."

The following passages from the Experimental Farm Reports and the Special Reports of Crop Experiments to ascertain the Yield per Acre, may be here given:—

Method and Yield. 61 (449)

"The last crop selected for special enquiry was linseed. It is in this country grown as an oil-seed and never for flax. It is a most delicate crop, suffering from various diseases, of which a fungoid disease, probably rust, is the most common. Even where the attack is slight and in appearance the crop has not suffered, it is found that the capsules are only partly filled and the yield is thus often reduced by half without apparent injury. It is sown alone in Khándesh, but in the Deccan and Southern Marátha Country. chiefly as a row crop with wheat or gram and round the headlands. The framers of the formulæ credit the crop with a maximum yield of 500lb in every táluka in which it is grown. In Khándesh it is sometimes taken as a late crop following *bájri* of the same season. Linseed has seldom been selected by experimenters, and little is therefore on record as to its varying yield. In the season under report it was almost a complete failure in the Deccan and Southern Marátha Country from prevalence of the fungoid disease noticed above. An experiment was made in Dhárwár by Mr. H. R. Shirhatti which gave a yield on very poor soil (assessed at 14 annas 9 pies per acre, full of limestone nodules and very shallow) of 33lb per acre. It was here grown as a row crop. *Jowári* was sown with the 4-coulter drill and on every fourth row alternately, gram and linseed were deposited on the sown *jowári* by the seed tube. The yield of 33lb on every eighth row represents a yield of $9 \times 33 = 297$lb on the acre of linseed. It will be interesting if this crop could be selected specially again this year, care being taken to note whether (1) it is the sole crop of the year, or (2) the second crop after an early crop, or (3) a row crop. In the last case the number of rows of linseed and of the principal crop shown with it should be noticed" (*Crop Experiments, Bombay, 1885-86, p. 8*).

"Since the receipt of instructions of the Government of India to prepare forecasts of oil-seed along with those of wheat and cotton, this crop has attracted my particular notice, as it is grown in this presidency as an oil-seed and not for flax. This was consequently one of the crops selected last year for special enquiry, but only one experiment was made on it on a poor soil. So it was then desired that this crop should again be specially selected for experiment this year wherever available. Linseed is reported to have been grown in Násik this year to such an extent as to affect sensibly the area under wheat. In the field experimented on, linseed mixed with a little mustard was the sole crop of the year, and though the year's crop as a rule suffered from the fungoid disease, the plot under experiment escaped, and the yield (estimated at 14 annas) was 225lb of linseed with $4\frac{3}{4}$lb of mustard. The corresponding formulæ yield us 450lb of linseed. The formulæ estimate at indiscriminately 500lb the acre in every táluka without considering the soil and climate, &c., as was the case found in Dhárwár last year, appears rather high" (*Crop Experiments, Bombay, 1886-87, p. 7*).

LINUM usitatissimum.

Linseed Cultivation.

CULTIVATION. Bombay & Sind.

As the result of the crop experiments performed at Khándesh with linseed it was found, that while the yield worked out by the formula should have been 239℔ to the acre, the actual produce was 308℔. In another experiment the yield was 322lb. **Mr. Whitcombe** states that 360℔ would be a good estimate for an 11-anna crop.

IN THE PANJAB.

62

VI.—PANJAB.

Some twenty years ago **Mr. Baden Powell** wrote of the Panjáb that linseed was extensively cultivated on account of its oil-yielding seed. "In Kangra it is thrown in among the stubble after cutting the rice crop, and then springs up without any cultivation." In many of the settlement reports of the province mention is made of linseed. Thus in the Lahore report it is shown as sown in October and reaped in March, the land twice ploughed and five times watered with a yield of 20 seers per *kanál*. The total area under the crop is perhaps under 30,000 acres and the exports very insignificant.

The following passage regarding linseed cultivation in Kashghar may be here given:—"Extensively cultivated for its seed in all the western divisions. The seed is the chief source of the oil used in the country, and the cake is given to the stall-fed cattle. The oil-mill or press is worked by horses or oxen, and is similar to that used in the Panjáb. The seed is sown in April and May, and the crop is cut in October" (*Report of a Mission to Yarkand in 1873, 78*).

IN MADRAS & BURMA.

63

VII. & VIII.—MADRAS AND BURMA.

The amount of linseed grown in these provinces is too unimportant to necessitate separate notices in this work. The reports of the Saidapet farm should be consulted regarding the experiments that have been tried.

PRICES.

64

PRICES OF LINSEED IN INDIA.

In the *Transactions of the Agri.-Horticultural Society of India (Vol. VIII., pp. 146-204)*, a paper is given which deals with the market prices of staple articles of Indian trade in various districts during 1839. It is explained that the value of the rupee had been taken as 2 shillings and the seer as 2℔ 0 oz. 14 dr., avoirdupois. The prices are expressed at so many seers to the rupee, but reducing these to rupees per maund, so as to allow of comparison with more recent returns, the following may be given—

Localities.	Price per maund in the chief Marts during 1839.		Price per maund in a small neighbouring Village during 1839.	
	May.	December.	May.	December.
	R	R	R	R
Delhi	3	3	3-5	2-15
Benares	1-3 to 1-4	1-5 to 1-7	1-2-0	1-7-0
Dacca	1-5-0	1-2-0	1-1-0	1-1-0
Calcutta	1-4 to 1-7-0	...	...	...

The Department of Finance and Commerce annually publish a return of Prices and Wages in India: in that publication the wholesale rate of linseed in Calcutta is given since 1843. It would perhaps serve no good purpose to enumerate the prices in all these years, but the following may

PRICES.

be quoted :—January 1843, R2-5-0 a maund; July 1844, R1-15-0; January 1847, R2-3-0; July 1848, R2-3-0. Ten years later January 1857, R4-2-0 and July 1848, R4-2-0; January 1867, R4-12; July 1868, R4-10-0.

The following statement exhibits the prices during recent years from 1876:—

	PRICES DURING THE MONTH OF JANUARY.			
	CWTS.		MAUNDS.	
	In Calcutta.	In Bombay.	In Calcutta.	In Bombay.
	R a. p.	R a. p.	R a. p.	R a. p.
1876	5 7 5	5 9 6	4 0 0	4 1 6
1877	6 3 8	6 3 0	4 9 0	4 10 8
1878	6 7 9	6 8 0	4 12 0	4 12 1
1879	Not shown	5 2 0	Not shown	4 7 9
1880	Do.	7 2 0	Do.	5 3 2
1881	Do.	6 8 0	Do.	4 12 1
1882	5 8 9	6 4 0	4 1 0	4 9 2
1883	5 3 4	5 3 0	3 13 0	3 12 9
1884	5 14 3	5 11 0	4 5 0	4 2 7
1885	5 15 7	5 15 0	4 6 0	4 5 6
1886	6 2 4	5 15 0	4 8 0	4 5 6
1887	6 2 4	5 15 0	4 8 0	4 5 6
1888	5 12 11	Not shown	4 4 0	Not shown.
1889	6 13 3	Do.	5 0 0	Do.
1890	6 10 5	Do.	4 14 0	Do.

But perhaps a still more instructive table, obtained from the Reports of the Bengal Chamber of Commerce, would be the following:—

LINUM usitatissimum.

PRICES.

Prices of Linseed.

LINSEED.	1873. May.	1873. June.	1873. July.	1873. August.	1878. May.	1878. June.	1878. July.	1878. August.	1883. May.	1883. June.	1883. July.	1883. August.	1888. May.	1888. June.	1888. July.	1888. August.
Exchange D/Payment 6 M/S .	1 $11\frac{1}{16}$	1 11	1 $10\frac{7}{8}$	1 $10\frac{9}{16}$	1 9	1 $8\frac{15}{16}$	1 $8\frac{5}{8}$	1 $8\frac{9}{16}$	1 $7\frac{7}{8}$	1 $7\frac{7}{8}$	1 8	1 $8\frac{1}{16}$	1 $4\frac{3}{8}$	1 $4\frac{11}{32}$	1 $4\frac{7}{32}$	1 $4\frac{7}{16}$
Exchange D/Payment 3 M/S .	1 $10\frac{3}{8}$	1 $10\frac{11}{16}$	1 $10\frac{1}{2}$	1 $10\frac{3}{8}$	1 $8\frac{15}{16}$	1 $8\frac{7}{8}$	1 $8\frac{9}{16}$	1 $8\frac{7}{16}$	1 $7\frac{3}{4}$	1 $7\frac{23}{32}$	1 $7\frac{27}{32}$	1 $7\frac{29}{32}$	1 $4\frac{9}{32}$	1 $4\frac{1}{4}$	1 $4\frac{11}{32}$	1 $4\frac{3}{16}$
Price England, per 410 lb S.	65-66	64-65	64-65	61-9-61	52-53	52-53	52-53	52-53	43-43-6	41	41-6	44-44-6	36-6	35-9	36-3	37-6
Price Calcutta, per B. Md. R	4-10-6	4-10-0	4-11-0	4-8-6	4-8-6	4-11-0	4-14-0	4-15-0	3-11-6	3-12-0	3-14-0	3-15-0	4-1-0	4-3-6	4-5-0	4-5 0
Freight p. 100 Mds. Rail, Cawnpore, Calcutta, R	75	75	75	75	67	67	67	67	60	60	60	51	53	53	53	53
Freight p. Ton, Steamer, Calcutta, London £	3-5-0	3-7-6	3-5-0	3-17-6	1-10-0	1-10-0	1-7-6	1-7-6	2-2-6	2-0-0	2 0-0	2-0-0	1-12-6	1-6-3	1-8-9	1-13-9
Shipping Charges p. Ton, Calcutta	Average One Rupee per Ton.				Average One Rupee per Ton.				Average One Rupee per Ton.				Average annas 12 per Ton.			
Exports Bengal . .Cwt.	21,54,475				51,99,353				34,92,945				59,00,045			
Exports Bombay . . "	81,8·0				19,99,385				34 44,000				25,21,814			
Exports Sind . . . "	......				......				18				814			
Exports Madras, &c, . "	1,084				180				151				......			
TOTAL . "	22,37,439				71,98,918				67,37,114				84,22,703			

L. 64

(452)

Linseed Oil. (*G. Watt.*) **LINUM usitatissimum.**

PRICES.

The table on the page opposite shows the position of the Calcutta Linseed Trade during each fifth year since 1873; *viz.*, rate of exchange; price in Calcutta and in England; railway freight from Cawnpore to Calcutta; steamer from Calcutta to London; and shipping charges. It concludes by exhibiting the actual amount of linseed exported from each of the four principal ports. It will be seen that during these fifteen years exchange fell from 1*s.* 11$\frac{1}{16}$*d.* to 1*s.* 4$\frac{7}{16}$*d.* on bills at six months. Coincident with that depreciation in the value of silver, the English price of linseed declined from 66*s.* to 37*s.* 6*d.* per 410lb, but the Indian value fluctuated above and below an average of R4-5-8, reaching its maximum in 1879 at R5-2 and its minimum in 1876 at R3-9. Thus it may practically be said that the actual value of Indian linseed was attained about the time of the annexation in 1857, and that it has maintained an average price ever since, namely, R4-5-6 per maund. The most serious reductions that have taken place have been in the railway and marine freights and in the shipping charges. Taking Cawnpore as an example, and it is one of the most distant important centres of supply, the railway freight was reduced, during the past fifteen years, from R75 to R53 per 100 maunds, and the freight to England from £3-5-0 to £1-13-9.

LINSEED OIL.

Percentage of Oil obtained.

OIL.

Yield from seed.

65 (453)

In an official correspondence regarding the desirability of opening out Linseed Oil Mills in India, **Mr. Luchman Parshad Barmah**, Superintendent, Cawnpore Experimental Farm, gave the following particulars regarding the yield of oil:—"The amount of oil from a given weight of linseed, varies with the different varieties and also with the age of the seed. Fresh linseed, pressed just after it is gathered from the field, will give a larger quantity of oil than the same weight of old seed. *Bold* seed gives higher percentage of oil than *small* variety; and a white variety of Jalaun gives higher percentage than either of these. In the experiments tried last year at the Cawnpore Experimental Farm the following percentages were arrived at with the different varieties. The oil was pressed in a country *kolhu* and just after the harvesting of the seed:—

	*Percentage of oil to seed.	* Percentage of cake to seed.	Loss.
White seed	35·1	54·2	10·7
Bold red	31·2	64·8	4·0
Small	29·6	67·1	3·3

"In a letter to **Mr. (now Sir E.) Buck**, dated February 27th, 1880, **Sir James Caird**, of London, gave 130lb as the weight of oil in 410lb of seed, which amounts to 31·7 per cent. on the weight of seed pressed.

"The information collected from *télís*—men who have made oil-pressing their profession—gives the quantity of oil from 25 to 30 per cent. of the weight of seed pressed. I think we may fairly expect 30 per cent. of oil if good seed is employed."

In conversation with the Superintendent of a large Indian oil mill the writer learned that 30 per cent. would be an exceptionally low average yield of oil from good linseed. But it seems probable that with native

* The seed was weighed after it was thoroughly cleaned and was going to be pressed: oil and cake weighed a week after pressing when the dirt present in the oil had settled and the cake had lost its moisture.

OIL. Yield from seed.

appliances even less than 30 per cent. is obtained. The following passage alludes to native methods of extraction of the oil:—

"It is grown in South Shahabad principally for home consumption, and is used largely for lamps under the name of *Tease-ka-tél.* Under the native process, which is the same as previously described, it produces 15 per cent. of oil;—what it produces with the European method I am unable to say. The native method produces a tolerably clear-looking oil, but it smokes much in burning, showing that a large proportion of vegetable fibre finds its way through the rude press into the pure oil: considerable quantities are grown near the Ganges for export. The price varies much according to the season and the quantity in the market. I have known it sell for R40 or £4 sterling per ton, and I have known it fetch double the price" (*Jour. Agri.-Hort. Soc. India, Old Series, Vol. XII., 342*).

Quality. (454) 66

Quality of Indian Linseed Oil.

In 1855 the late **Dr. F. Royle** wrote on this subject:—"It may appear remarkable that Linseed oil should be imported into Calcutta, when so much linseed is exported for the express purpose of yielding its oil. This is in consequence of the linseed oil of India being considered as not possessed of the full drying properties of the oil prepared in Europe. But there is no doubt, this is owing entirely to the Indian Linseed being expressed before the mustard seed has been separated with which it is commonly mixed, in consequence of the two plants being often grown together. Mr. Bowen informed the author that, when connected with one of the light-houses in India, he had at one time under his charge some plate glass. This he made use of to separate the two seeds, by placing it on a slope: the round seeds of the Mustard rolled off, while the Linseed merely slipped down. These, when expressed, yielded as good drying oil as any he ever obtained from Europe. The same fact is confirmed by the following statements.

"In a report from **Mr. W. Ewin**, Branch Pilot, to **Captain W. Hope**, Master-Attendant at Calcutta, he acknowledges the receipt of five gallons of linseed oil, made at the Gloucester Mills, situated below Calcutta:—

"'I beg leave to say I painted my boat inside green with the above oil, without the assistance of turpentine, and it dried within the space of twenty, four hours. I do not hesitate to say therefore if the above oil, agreeable to the muster, be given, that it is equal to the linseed oil received from the Honourable Company's Marine Yard, said to be from Europe. Sandheads, H.C.P.V. *Sea Horse*, 2nd January 1837.'

"So **Mr. W. Clark**, commanding H.C.F.S.V. *Hope*, writes, 14th December 1836:—'I have to report, for the information of the Master-Attendant, in reply to his letter (No. 39) of the 7th ultimo, that I have painted the *Hope*, outside, with the Gloucester Mill oil on one side, and that supplied by the Naval Store-keeper on the other,—both laid on at the same time; and of the two I must give the preference to the former, in drying and bearing a better gloss.'"

Since the above was written the subject appears to have been spasmodically discussed, Indian writers maintaining that if carefully prepared (after having been freed from the rape seed with which it is purposely or accidentally mixed), the Indian linseed affords an oil by no means inferior to that obtained from European seed, while European reports, mostly, it is contended, from interested parties, affirm that it is considerably inferior. It will be seen from the remarks below that meanwhile the foreign exports of linseed have increased to a far greater extent than have those of almost any other single article of Indian agricultural production, so that there can be no doubt the oil has come into extensive use in Europe. It

OIL.

Quality.

is thus somewhat surprising that Indian capital should have, up to date, been almost entirely diverted to other channels instead of a fair proportion being devoted to a competition with European oil mills in the production of linseed oil. By this means unnecessarily heavy freight charges have had to be paid on one of India's most important crops, the cake and manure, which should have been left in the country, have enriched the cattle and fields of other parts of the world, leaving India to have a relatively smaller net gain in the linseed transactions than might be supposed from the magnitude of the figures of foreign export.

Samples of Linseed oil were sent (in 1887) from Madras to Her Majesty's Secretary of State for valuation. The reports of the brokers were not very favourable. The oil was pronounced "far below what it should be if proper appliances are used for crushing the seed." "The oil appears to have been expressed from Linseed which has been mixed with other sorts of seeds." "The result of this admixture is an unusual taste and odour." "The oil is deficient in the drying properties which characterise oil expressed from pure linseed." "The oil would be saleable here at about £1 per ton below the market price of merchantable linseed oil which to-day (21st April 1887) is quoted at £20-10 per ton." Similar opinions were expressed by the other brokers who examined the Madras samples. These opinions were duly published in India and called forth a protest from a well-known Calcutta firm, who maintained that opinions upon ordinary native-made linseed were likely to damage the trade in the pure article (such as that prepared by the Gourepore Company, Limited), since the oil expressed from Linseed and Mustard mixed, the ordinary native method, was certain to possess a much lower drying property than pure European linseed, or, indeed, than Indian linseed. In support of this statement the late **Dr. Waldie's** report on Gourepore Linseed oil was submitted for the information of Government. **Dr. Waldie** was, for the purpose of analysis, furnished with two samples of oil, one of the Gourepore ordinary trade oil, the other, English Linseed oil furnished for the purpose of comparison. **Dr. Waldie** wrote that the colour of the English oil was a little darker than the Gourepore; that the smell of both was nearly alike even when heated, neither oils giving evidence of the presence of any substance other than linseed. The specific gravity of the Indian oil at 60°Fh. was found to be 933 compared with water as 1,000, the English 934. The solubility of both oils in alcohol was practically the same (1·28 per cent Indian and 1·63 per cent. English). **Dr. Waldie** then tested the oils for purity and found the Indian to vary from the English no more than is known to be the case between most of the qualities met with in Europe (Russian, Dutch, English, &c.). With regard to their drying property **Dr. Waldie** wrote: "Both oils spread thin on small porcelain basins and exposed to a moderate or gentle heat, dried in about the same time, and the dried oils appeared equally firm." The English was browner in colour; but "there is no apparent difference by this experiment between the two in drying properties."

It would thus seem that either the Madras oils reported on by the London brokers was prepared indifferently, the seed not having been freed from the frequent adulterant—mustard seed—or that Madras Linseed yields a much inferior drying oil to that obtained from other parts of India. Mirzapore seed is reported to yield more oil and of a better quality than that of the Eastern Districts of Bengal. It would thus appear highly probable that not only is white linseed quite different from the red, but that the red seed of certain tracts of India affords better oil than that from other parts of the country. It would thus seem desirable, as one of the first steps towards the establishment of extensive oil mills in India, to have the exact properties of the oil of the various races of seed separately

OIL. and carefully prepared and tested. From the numerous reports that have appeared it would seem that the white linseed may expand into a trade of considerable magnitude, and that too even should it prove of less value as a drying oil.

Price of Linseed Oil and Linseed Cake.

Price. 67 The best oil sells (wholesale) in India from R1-8-0 to R1-12-0 per gallon for raw oil, and at R1-10-0 to R2 per gallon for boiled oil. The oil-cake is shipped to England for sale and is said to realise from £6 to £7 per ton.

For further particulars regarding the properties and uses of Linseed-cake see the Article OILS. Mills. 68

INDIAN LINSEED OIL MILLS.

The following passages from Mr. L. P. Barmah's report may be here given. He describes three methods of (and the machinery used in) expression of the oil:—

"(*a*) *Kolhu.*—This consists principally of a thick block of wood with a cavity in the centre, which receives the seeds, and a moveable rod which, with a few minor arrangements, is made to revolve in the cavity of the former piece and thus press against the seed; the oil which is thus expressed runs out through a spout. The cost of a medium-sized press of this description is about R6, and the bullock required to work it can be had for R10 or R12. It presses 7 seers linseed a day, working eight hours, or nearly 5 maunds a month. It is attended to by women, on whom it devolves as one of their household duties; but if worked by hired labour, a man on 2 annas per diem would be required for every three presses, which amounts to R1-4 per month per press; the man would look after the bullocks too. Taking all the year round, the cost of feeding a press bullock amounts to R0-1-3* per day, or R2-5-6 per month.

"The wear and tear of the press amounts to 5 annas per month. To these we may add interest at 6 per cent. on the capital invested in the manufacture of the *kolhu* and purchase of bullocks, amounting to a monthly expense of R0-1-4 nearly. Thus the total cost of working a *kolhu* per month, during which it will press 5 maunds seed, is R3-15-10:—

	R	*a.*	*p.*
Labour	1	4	0
Feed of bullocks	2	5	6
Wear and tear of the press	0	5	0
Interest on capital	0	1	4
	3	15	10

"Therefore, the cost of pressing 100 maunds seed by a native press amounts to R79-12-8."

"(*b*) *English Hand Press.*—This was once used by a firm in Cawnpore, but finding it difficult to dispose of the cake and oil, the attempt was abandoned. The press consisted of two strong screws and a number of iron plates. The seed was ground by an English grinding mill, placed in a piece of gunny cloth and then put in between each pair of plates; when all the plates were thus occupied, a fire was created by burning coal on the two sides and the screws tightened. The pressure assisted by heat expressed the oil, which ran down a channel into a reservoir where it collected. About $1\frac{1}{4}$ maund of linseed was thus pressed at a time, each pressing taking an hour and a half and done by four coolies, who received R0-2-6 for every pressing; two of them worked the press and the other two covered the seed in gunny cloth and took rest. About 10 maunds of seed was thus pressed

* Fodder and Cake.

OIL. Mills.

per day of 12 hours by one of these presses. The cost of pressing 10 maunds of seed by each press may therefore be taken at R4-9:—

	R	a.	p.
Labour employed, in grinding at 2 pice per maund of seed .	0	5	0
Ditto in pressing, at $2\frac{1}{2}$ annas per $1\frac{1}{4}$ maund .	1	4	0
Coal worth 4 annas for every pressing	2	0	0
Gunny cloth renewed after every 10 or 12 pressings . .	1	0	0
Total .	4	9	0

"From the reservoir the oil was carried to the boiling-pan, where water was added to it in the proportion of 1 to 40, boiled for about an hour and a half, and then removed to a strainer. When the oil sufficiently cooled down it was put in canisters. About 32 maunds of oil could thus be boiled in a day."

(*c*) *English Steam Press.*—**Mr. L. P. Barmah** gives particulars of an English steam press that was formerly established by Guru Prasad, a merchant in Cawnpore. About 100 maunds of seed could be pressed a day at a cost of R18-4-5, but wood, instead of coal, was employed, the cost of manufacture being (as afterwards shown in the correspondence on the subject) unnecessarily high.

Mr. J. E. O'Conor, in the Statistical Tables for British India, shows that there were in 1889-90 some seventy oil-mills and oil-wells in India. He does not mention, however, the Gourepore Company, Limited, the only Indian mill that is believed to be exclusively devoted to Linseed. The writer understands that the Gourepore Company produce about 700 gallons of very superior oil daily, which enters into competition with imported oil, and finds a ready and profitable sale. In the official correspondence, towards which **Mr. Lachman Parshad Barmah** contributed the report laid under frequent contribution here, it was contended that oil-mills should, if possible, be in the future established near the seaport towns. That the railway charges in India were more favourable to the carriage of seed than oil, hence the advantage of expressing the oil at the end instead of on the line of railway transport. The discussion that ensued seemed to proceed from the basis that the object in view should be the export of oil instead of oil-seed, thus leaving as much of the cake as possible in this country, to improve the cattle and soil of the regions on which so exhausting a crop, as linseed was systematically grown. The opponents of this idea of coastwise mills held that if once the seed was removed from the actual district of production, the cake would never find a market in India, as the cultivators could not afford to pay railway freight on its return. The oil hitherto produced in India has, however, found its best market in India itself, the cake alone being exported. It would thus appear that when (if ever) extensive oil-mills, for the purpose of exporting oil to the American, Australian, and European markets, are established in India, the cake also will have to be exported, so that, except in the sense of opening out a new industry, a new field of labour and of investment of Indian capital, this country will not benefit any more than it is now doing by the large market that it has established for its linseed. There would seem no good reason, however, why India should import any linseed oil. Mills should exist sufficient at least to meet the entire Indian market for the oil, for even if slightly inferior as a drying oil, this should be more than compensated for by a saving in price equivalent to the freight to Europe of the seed and the return charges in bringing the oil to the shores of India. Many years will doubtless pass by before the Indian cultivators will realise that it is in their best interests to keep a large proportion of the cake to feed their own cattle and manure the flax fields.

TRADE.

INDIAN TRADE IN LINSEED.

Foreign, (457) 69 **Dr. Royle**, some thirty years ago, wrote: "The large exports of Linseed from India have frequently been mentioned. It is desirable, therefore, to give some of the details. By these we may observe, that *though comparatively a recent trade,* the article is already known to other countries besides England. The first exports of Linseed were made from Calcutta by Mr. Hodgkinson and were—

In the year 1832,	to the extent of	10 bushels.	
" 1833	"	2,163	maunds.
" 1834	"	2,826	"
" 1837*	"	32,327	"
" 1839	"	1,67,601	"
" 1850	"	7,65,496	"

It has already been stated that no mention occurs in **Milburn's** *Oriental Commerce* of linseed, a fact that confirms **Royle's** statement that the article began to be exported from India only about 1832. In the year 1850-51, while Calcutta exported 765,496 maunds, valued at R15,30,902, Madras exported 801 cwt., valued at R2,271, and Bombay, 50,112 cwt., valued at R1,70,539. But in the following year Bombay is shown to have exported 114,309 cwt., and according to **McCulloch's** *Commercial Dictionary,* Great Britain imported from all sources, during 1851, 630,471 cwt. of linseed, of which Russia furnished 417,950 cwt., the "British Territories in the East Indies" standing next in importance, *viz.,* 93,814 cwt.

In the Annual Statement of the Trade and Navigation of British India with Foreign Countries for the year 1866-67, the value of the exports is shown to have been (for that year) R75,04,615. Along with gingelly, rape, and other oil-seeds, linseed was then subject to an export duty, and the amount of revenue thus realised on oil-seeds is shown to have been R5,57,808. Ten years later (1876-77), the exports of linseed amounted to 5,614,617 cwt., valued at R3,01,54,374, the duty having been some years before removed. In 1886-87 the exports amounted to 8,656,933 cwt., valued at R5,17,92,914; last year they were 8,461,374 cwt., valued at R5,05,79,221. Comparing the averages for the quinquennial periods from 1873-74 to the present date, it will be found that the trade shows an increase of 224 per cent. in weight and 278 per cent. in value. But these figures exhibit the foreign exports only, though it may be said they very nearly convey an idea of the total value of the Indian transactions in linseed. They show, at all events, that the trade has expanded from about 3 cwt. in 1832 to 8,461,374 cwt. in 1888-89. We may assume (for the purpose of demonstrating the full meaning of this expansion) that the Indian local demand has remained stationary during these years. To produce the extra amount demanded last year over that in 1832, the total area of linseed cultivation in India must, at the lowest estimate, have increased by 4,230,682 acres.†

If now we turn to the land or trans-frontier imports and exports, we find that, for the past three years, Nepál has exported to India (Bengal) on an
Trans-frontier. average some 175,000 cwt. of linseed, valued at R89,20,000. The trans-
(458) 70 frontier exports from India are unimportant, the major portion going to Kashmír.

* In the *Agri.-Horti. Soc. Jour.* for 1842, it is stated the exports from India in 1835-36 were, according to Mr. Bell of the Calcutta Custom House, 1,63,199 maunds or 6,044 tons.

† Estimated at 2 cwt. per acre, a very low yield. According to Bombay crop experiments the yield is variously stated from 190 to 239 and 360lb an acre or, say, 3 cwt.

TRADE.

Internal. 71 (459)

The total amounts of linseed shown as carried by rail and river, were in 1888-89 said to have been 1,14,08,389 maunds, valued at R4,35,19,349. The largest exporting province was Bengal with 54,05,056 maunds, the bulk of which went to Calcutta to meet the foreign trade from that port. Next in importance as exporting were the North-West Provinces and Oudh with 26,87,088 maunds, of which over 20,00,000 maunds also went to Calcutta. Bombay exported 7,55,052 maunds, of which nearly the whole went to its port town to meet the Bombay foreign exports. The Central Provinces, Berar, Rájputana, and Central India each exported nearly similar quantities and to the port town of Bombay. Thus it will be seen that the foreign exports from Calcutta are drawn from the Bengal Province and the North-West Provinces; the total imports into Calcutta from all sources amounted in 1888-89 to 75,22,764 maunds, while those into Bombay came to 36,51,935.

Coastwise. 72 (460)

The supplies drawn by Calcutta and Bombay (the chief marts in the foreign trade) are also slightly augmented by the coastwise transactions, particularly in the case of Bombay. For example, in the year 1888-89, Bombay received from Sind 1,794 cwt., valued at R11,154; from Goa 32,878 cwt., valued at R1,97,736; from Cambay 3,037 cwt., valued at R19,257; and from Kathiawar 129 cwt., valued at R783. It will thus be seen that the linseed exported from Bombay is drawn from Bombay Presidency, Sind, the Central Provinces, Berar, Rájputana, Goa, and Kathiawar. The Panjáb produces very little linseed, but doubtless what it does yield finds its way to Calcutta along with that from the North-West Provinces. From the above review of the internal trade in linseed it may be admitted to have been shown that the most extensive areas of production are Bengal and the North-West Provinces. This idea is confirmed by an inspection of the details of the total amount (8,461,374 cwt.) of the foreign exports, from which it will be seen that Bengal (*i.e.*, Calcutta) exported 5,659,492 cwt.; Bombay 2,797,246 cwt.; Sind 675 cwt.; and Madras 3,961 cwt. Of these foreign exports the United Kingdom took 5,295,175 cwt., valued at R3,06,36,195; France 1,375,689, valued at R88,39,664; the United States 712,042 cwt., valued at R44,94,350; and Holland 524,223 cwt., valued at R32,19,603. The balance went to other countries, of which Belgium figures as highest, having taken 218,193 cwt. It will thus be seen that not only has the Indian export trade in linseed increased during the past fifty years until it has assumed a gigantic form, but that the European demand has been the chief and only cause of this expansion. In 1851 (less than forty years ago) the total demand in Great Britain for linseed amounted to only 630,471 cwt., whereas last year India alone furnished the United Kingdom with 5,295,175 cwt.

Statistics cannot be obtained for the present or past Indian consumption of linseed, but it has certainly not materially increased during the past forty years, for no large and new industrial uses of linseed have been brought into existence—a statement abundantly proved by the fact that little or no linseed oil is expressed in this country by the Natives. As already remarked, there is practically only one linseed oil-mill in India, and the oil it produces is mainly, if not entirely, used up by exotic demands, since the people of India use only the small quantity of linseed necessary for the house paint employed by the well-to-do. The bulk of the people of India have not now, nor apparently did they ever have, any very important use for linseed or linseed oil—a statement confirmed by the want of definite historic evidence of cultivation between the development of present foreign trade in linseed and the classic records of the uses of flax, or what most writers accept as flax in Sanskrit literature.

(*J. Murray.*)

LIPPIA, *Linn.; Gen. Pl., II., 1142.*

[*1463;* VERBENACEÆ.

73 **Lippia nodiflora,** *Rich.; Fl. Br. Ind., IV., 563; Wight, Ic., t.*

Syn.—VERBENA NODIFLORA, *Linn.;* V. CAPITATA, *Forsk.;* BLAIRIA NODIFLORA, *Gærtn.;* ZAPANIA NODIFLORA, *Lamk.;* LANTANA SARMENTOSA *and* REPENS, *Spreng.;* PHYLA CHINENSIS, *Lour.*

Vern.—*Bhúi-okra,* HIND.; *Lúdra,* N.-W. P.; *Mokna, búkan, bakan, jalním,* dried plant=*gorakh múndí,* PB.; *Wakan,* SIND; *Tan,* DEC.; *Ratolia,* BOMB.; *Ratoliyá,* MAR. & GUZ.; *Podútalei,* TAM.; *Bokenakú,* TEL.; *Herimanadatta,* SING.; *Vashira,* SANS.

References.—*Dalz. & Gibs., Bomb. Fl., 198; Stewart, Pb. Pl., 166; Ainslie, Mat. Ind., II., 313; Dymock, Mat. Med. W. Ind., 2nd Ed., 599; Honigberger, Thirty-five Years in the East, II., 300; S. Arjun, Bomb. Drugs, 105; Murray, Pl. and Drugs, Sind, 175; Atkinson, Him. Dist., 315; Ind. Forester, XII., 19; Gazetteers:—N.-W. P., I., 83; IV., lxxvi.*

Habitat.—An evergreen undershrub, common in wet places throughout India and Ceylon, distributed to all tropical and warm-temperate regions.

MEDICINE.

Medicine.—**Ainslie** writes: "The tender STALKS and LEAVES of this low-growing plant, which last are in a slight degree bitter, the native practitioners prescribe, when toasted, in cases of children's indigestion, to the extent of two ounces in infusion, twice daily; it is also ordered as a drink for women after lying-in." **Stewart** states that it is considered cooling by the natives of the Panjáb, but in the time of **Honigberger** it appears to have been regarded as of very little medicinal value. He writes: "The natives know the plant, but very seldom use it." He himself considered it valuable in "Ischury, stoppage of the bowels, and pain in the knee-joint." **Dymock** states that it is used in Bombay as a demulcent in cases of gonorrhœa.

Stalks. 74
Leaves. 75

Liqueurs, see **Spirits,** Vol. VI.

LIQUIDAMBAR, *Linn.; Gen. Pl., I., 669.*

76 **Liquidambar Altingia,** *Bl.;* see **Altingia excelsa,** *Noronha;* Vol. I., 201.

77 **L. orientalis,** *Miller; DC., Prodr., XVI, 158;* HAMAMELIDEÆ.

LIQUID STORAX, LIQUIDAMBAR, ROSE MALLOES.

Syn.—LIQUIDAMBAR IMBERBE, *Aiton.*

Vern.—*Siláras, meih-síla, nágorigond,* HIND.; *Silha, silaras,* BENG.; *Siláras, salajet, usturuk,* BOMB.; *Silarasa,* MAR.; *Meih-síla, seláras,* GUZ.; *Neri-arishippál,* TAM.; *Shilá-rasam,* TEL.; *Rasamalla,* MALAY.; *Nantayu,* BURM.; *Rasamalla,* MALAYS.; *Silhaka,* SANS.; *Miah-sáyelah, usteruck, meati-lubani, salajet, meah, sillarus, cotter mija,* ARAB.; *Meih-síla, asle-lubní,* PERS.

References.—*Gamble, Man. Timb., 174; Ainslie, Mat. Ind., I., 405; O'Shaughnessy, Beng. Dispens., 255, 610; Moodeen Sheriff, Supp. Pharm. Ind., 169; U. C. Dutt, Mat. Med. Hind., 166, 318; Dymock, Mat Med. W. Ind., 2nd Ed., 313; D. Hanbury, in Pharm. Jour., XVI., 417, 461; XXII., 436; Flück. & Hanb., Pharmacog., 271; U. S. Dispens., 15th Ed., 1373, 1686; Bent. & Trim., Med. Pl., 107; S. Arjun, Bomb. Drugs, 130; Year Book Pharm., 1874, 293; Birdwood, Bomb. Pr., 81; Balfour, Cyclop., II., 721; Smith, Dic., 247; Ind. Forester, II., 181, 408; X, 435.*

Habitat.—"A handsome, umbraceous tree, resembling a plane, growing to the height of 30 to 40 feet or more, and forming forests in the extreme south-western part of Asia Minor. In this region the tree occurs in the district of Singhala, near Melasso, about Budrum (the ancient Halicarnassus), and Monghla, also near Giova and Ullá in the Gulf of Giova, and lastly near Marmorizza and Isgengak opposite Rhodes." "The tree is not known to grow in Cyprus, Candia, Rhodes, Kos, or indeed in any of the islands of the Mediterranean" (*Hanbury*).

MEDICINE. Resin. 78 (466)

Medicine.—LIQUID STORAX has long formed an important article of medicine in this country; indeed, **Flückiger & Hanbury** state that as early as the first century it was exported by the Red Sea to India. For many centuries it has been, and still continues to be, an important article of export from Bombay to China. **Hanbury's** investigations have proved that the solid storax of the ancients, which was derived from **Styrax officinale**, *Linn.*, and was always scarce and valuable, has, in modern times, entirely disappeared from commerce, and has been replaced by the balsam now under consideration.

The method of extraction and preparation is described by **Hanbury** as follows:—"The extraction of Liquid Storax is carried on in the forests of the south-west of Asia Minor, chiefly by a tribe of Turcomans called *Yuruks*. The process has been described on the authority of **Maltass** and **McCraith** of Smyrna, and of **Campbell**, British Consul at Rhodes. The outer bark is said to be first removed from the trunk of the tree and rejected: the inner is then scraped off with a peculiar iron knife or scraper, and thrown into pits until a sufficient quantity has been collected. It is then boiled with water in a large copper, by which process the resin is separated, so that it can be skimmed off. This seems to be performed with sea water; some chloride of sodium can therefore be extracted from the drug. The boiled bark is put into hair bags, and squeezed under a rude lever, hot water being added to assist in the separation of the resin, or, as it is termed, *yagh*, *i.e.*, 'oil.' **Maltass** states that the bark is pressed in the first instance *per se*, and afterwards treated with hot water. In either case the products obtained are the opaque, grey, semi-fluid resin known as Liquid Storax, and the fragrant cakes of foliaceous brown BARK, once common, but now rare in European Pharmacy, called "Cortex Thymiamatis."

Bark. 79 (467)

DESCRIPTION.—The balsam is a soft, viscid resin, usually of the consistence of honey, heavier than water, and greyish-brown in colour. It always contains water which by long standing rises to the surface. When heated it becomes dark brown and transparent; when spread out in a thin layer, it partially dries but does not lose its stickiness. When free from water it dissolves in alcohol, spirit of wine, chloroform, ether, glacial acetic acid, bisulphide of carbon, and most of the essential oils. It possesses a pleasant balsamic smell when old, but when recent, has an unpleasant odour of bitumen or naphthalin. Its taste is sharply pungent, burning, and aromatic.

CHEMICAL COMPOSITION.—For an account of the composition of this drug, which, not being of Indian origin, need not be fully discussed in this work, the reader is referred to the exhaustive description in **Flückiger & Hanbury's** standard work. It may be mentioned, however, that the principal constituents are *styrol*, a colourless, volatile, liquid hydrocarbon; *styracin*, a crystalline solid; *cinnamic acid;* a *resin*, and an *essential oil.*

THERAPEUTICS.—Liquid Storax is a stimulant expectorant like the Balsams of Peru and Tolu, and Benzoin, but except as a constituent of the Compound Tincture of Benzoin, is, at the present day, little prescribed in European medicine. Locally applied it is stimulant, antiseptic, and disinfectant, and has been much advocated of late years as an application for scabies and phthiriasis. For this purpose it is mixed with olive or linseed oil. It is interesting to find that this property of the drug appears to have been known to old Sanskrit writers, by whom it was considered "useful in affections of the throat, copious perspiration, and skin diseases" (*U. C. Dutt*). By Muhammadan physicians it is esteemed as a tonic, resolvent, suppura-

MEDICINE. tive, and astringent. Thus **Dymock** writes: "It is prescribed as a pectoral, and is thought to strengthen all the viscera; applied externally it is supposed to have a similar action upon the parts with which it comes in contact. It is a favourite application to swellings, and in Bombay is much used in orchitis, the inflamed part being smeared with it, and bound up tight in tobacco leaves." It is also largely used in perfuming medicinal oils. **Altingia excelsa** (*Vol. I., 201*) yields a very similar resin, which has been said to equal, in its medicinal properties, that now under consideration. **Waring**, however, wrote in the *Pharmacopœia of India*, that he had tried it as an expectorant without satisfactory results, and that, as far as his experience went, it appeared to be of little value medicinally.

TRADE. 80 **Trade**—As this drug is not returned separately in the statistical tables of imports, the total trade in the article cannot be specified. **Dymock** states, however, that the imports into Bombay in 1881-82 amounted to 363 cwt. from the Red Sea ports, and valued R16,154. It is probable that a considerable proportion was re-exported to China.

Liquorice, see **Glycyrrhiza glabra,** *Linn.*; Vol. III., 512.

Liquors, see **Malt Liquors,** p. 124.; also **Spirits,** Vol. VI.

Litharge, see **Lead,** Vol. IV., 602.

LITHOGRAPHIC STONES.

81 **Lithographic Stones,** *Geol. of Ind., III. (Economic, Ball), 556.*

References.—*Baden Powell, Pb. Prod., 45; Balfour, Cyclop. Ind., II., 728; in Sel. Rec., Mad. Govt (1865), II., 38; Adm. Rep., C. Prov., 80.*

DISTRIBUTION. **Distribution.**—Stones suitable for lithographic purposes are at present almost entirely imported from Europe and are sold by weight at high prices. An indigenous stone of this description is a great desideratum, since most native printing is by lithograph. This fact, combined with the high price of European stones, led to trials being made of Indian material soon after the introduction of lithography in 1822. Stones with the necessary qualities, however, are not of wide distribution; and since most of the limestones in India belonged to the more or less altered series of transition rocks, they were found to be hard, splintery, and difficult to dress and polish. They have accordingly proved only fitted for rough work, and, as **Ball** writes, "The fact that they are not used, at least to any appreciable extent, while high prices continue to be paid for European stones, is the strongest argument against their being of any substantial value."

The following are enumerated by **Ball** as the localities from which lithographic stones are said to have been obtained in India:—

Madras. 82 *Madras.*—Bellary, Karnul, Guntur, and Masulipatam. The best appear to have been derived from the metamorphosed limestone of the Karnul district, but even these are harder, more splintery, and less tractable than European stone.

Bengal. 83 *Bengal.*—The limestones of the Lower Vindhyan series early attracted attention, but are not sufficiently pure and homogeneous to be depended on.

Rewah. 84 *Rewah.*—In 1843 small samples from Búrwa were sent to the Asiatic Lithographic press and were favourably reported on, but further research appears to have proved them to be unsatisfactory.

Central Provinces. 85 *Central Provinces.*—Stones of a serviceable kind have been found in Raipur, and were at one time used in the jail press of that district.

Rajputana. 86 *Rájputana.*—**Ball** writes: "Of all the suggested substitutes of indigenous stone for that imported, a yellow limestone obtained in the Jesalmir State appears to have been of greatest promise" Attention was first drawn to it, about sixty years ago, by **Captain Boileau** in *Gleanings in Science, I.*, and the *Indian Review, III.* He stated that it was not suited for fine chalk drawings, but could be used for all other purposes with the ordinary materials. The chief point dwelt upon is the method of polishing which it was found necessary to adopt. The stone, being hard, brittle, and semi-crystalline, could not be polished with the ordinary sleek stone and pumice, but rubbers of lac and corundum, coarse, then medium, then fine, followed by polishing with calcined peroxide of tin, were found to produce good results. **DISTRIBUTION.**

Bombay.—The limestones of the Kaladgi and Bhima series have been tried, but have failed to give rise to a demand. **Bombay.** 87 (475)

Panjáb.—**Baden Powell** writes: "Some kinds of native stone are in use, but the European are always preferred for the finer kinds of work, and where great sharpness of delineation is required in the print." **Panjab.** 88 (476)

Ball, in summing up the remarks from which the above has been condensed, writes: "It would seem, judging from the geology, that in the Salt-range and in Cutch, there are better chances of finding a limestone suitable for the purpose than anywhere else in India." It may be mentioned that the best European stones are obtained in quarries in the oolitic rocks of Solenhofen near Munich, and in Pappenheim on the Danube. A perfect stone is compact and homogeneous, free from veins, flaws or spots, and of an even grey or drab colour.

LITHOSPERMUM, *Linn.; Gen. Pl., II., 860.*

[BORAGINEÆ.

Lithospermum officinale, *Linn.; Fl. Br. Ind., IV., 175;* 89 (477)

Syn.—MAGAROSPERMUM OFFICINALE, *Dcne.*

Vern.—Seeds=*Lubis firmun*, HIND.

References.—*Royle, Ill. Him. Bot., I., 304; Stewart, Bot. Tour in Hazara, &c., in Jour. Agri.-Hort. Soc. of Ind. (Old Series), XIV., 6; Aitch., Bot. Afgh. Del. Com., 90; O'Shaughnessy, Beng. Dispens., 497; Balfour, Cyclop., II., 728.*

Habitat.—An erect undershrub or herb, frequent in Kashmír at altitudes of 5,000 to 8,000 feet.

Medicine.—**O'Shaughnessy** writes: "The SEEDS are long, very white, and like small stones or pearls, on which account they have been popularly used as a remedy for stone." **MEDICINE. Seeds.** 90 (478)

Litmus.—A blue dye, prepared chiefly in Holland from the orchil-yielding lichen **Roccella tinctoria**; see **Lichens**, Vol. IV., 636.

LITSÆA, *Lam.; Gen. Pl., III., 161.*

A genus of evergreen, rarely deciduous, trees or shrubs, which comprises some 140 species, native of Tropical and Eastern Asia, Australasia, and the Pacific Islands; rare in Africa and America. Of these 65 are indigenous to India and Ceylon. The economic information relating to this genus has, as a rule, been detailed by Indian writers under the names of various species of **Tetranthera.** 91 (479)

Litsæa elongata, *Wall.; Fl. Br. Ind., V., 165;* LAURINEÆ. 92 (480)

Syn.—DAPHNIDIUM ? ELONGATUM, *Nees*; TETRANTHERA SIKKIMENSIS, *Meissn., in part.*

Vern.—*Paieli, púalay, phusri*, NEP.; *Phamlet*, LEPCHA.

References.—*Gamble, Man. Timb., 312; List of Trees, Shrubs, &c., of Darjiling, 65.*

Habitat.—A bush or small tree, native of the Subtropical and Temperate Himálaya, from Garhwál to Bhután, also of the Khásia mountains at altitudes from 5,000 to 6,000 feet.

TIMBER. (481) 93 **Structure of the Wood.**—Gamble describes **Daphnidium elongatum,** *Nees*, which has been reduced to this species in the *Flora of British India*, as a "large evergreen tree." It is possible, therefore, that the following description of the wood, and the vernacular names above enumerated, may in reality belong to another species. According to **Gamble** the timber is "yellow, turning olive-grey on exposure, moderately hard, even-grained, weight 34 to 41 lb per cubic foot. A very pretty wood, worthy of attention; used for building, chiefly as planking."

(482) 94 **Litsæa grandis,** *Wall.; Fl. Br. Ind., V., 162.*

Syn.—TETRANTHERA GRANDIS, *Meissn.*; POLYADENIA GRANDIS, *Nees.*

Reference.—*Kurz, For. Fl. Burm., II., 299.*

Habitat.—An evergreen tree, from 20 to 40 feet in height, not uncommon in tropical forests all over Pegu, Martaban, and Upper Tenasserim; found also in Penang, Malacca, and Singapore.

TIMBER. (483) 95 **Structure of the Wood.**—"Yellow with a beautiful lustre, rather heavy, close-grained, very soft, a fine fancy wood" (*Kurz*).

(484) 96 **L. polyantha,** *Juss.; Fl. Br. Ind., V., 162.*

Syn.—LITSÆA MONOPETALA, *Pers.*; TETRANTHERA MONOPETALA, *Roxb.*; T. MACROPHYLLA, *Wall.*; T. ALNOIDES, *Miq.*; T. FRUTICOSA and VERTICILLATA, *Ham.*; ? T. SEMECARPIFOLIA, *Wall.*; T. HEXANTHA, *Sieb.*; TOMEX PUBESCENS, *Willd.*

Vern.—*Meda, gwa, singraf, sangran, marda, kat marra, kakúri, kerauli, patoia, katmoria, papria, katmedh, kari, rand-kari*, HIND.; *Bara kúkúr chita*, BENG.; *Pojo*, SANTAL; *Sualu*, ASSAM; *Huara*, KACHAR; *Ratmanti, kadmero*, NEPAL; *Suphut*, LEPCHA; *Bút, mugasong*, MICHI; *Bolbek*, GARO; *Mendah, kari, kjera, toska, leja*, GOND.; *Leinja*, KURKU; *Randkorri, katmédh*, OUDH; *Rian, gwá, harein*, bark=*meda lakri*, PB.; *Ranamba*, MAR.; *Nara mamúdi, nara*, TEL.; *Ungdung*, BURM.

References.—*Roxb., Fl. Ind., Ed. C.B.C., 735; Brandis, For. Fl., 380., t. 45; Kurz, For. Fl. Burm., II., 299; Gamble, Man. Timb., 310; Stewart, Pb. Pl., 188; Campbell, Ec. Prod. Chutia Nagpur, No. 8447; Elliot, Fl. Andhr., 129; Ainslie, Mat. Ind., II., 227; Drury, U. Pl., 421; Cooke, Oils and Oilseeds, 80; Indian Forester, V., 212; VI., 239, 301, 316; VIII., 127; Gazetteer, Mysore and Coorg, I., 66.*

Habitat.—An evergreen bush or small tree, from 20 to 40 feet in height, met with from the Panjáb and the Salt Range along the foot of the Himálaya on which it ascends to 3,000 feet, eastwards to Assam and Burma, and southwards to the Satpura Range and Coromandel; distributed to Java and China.

OIL. Seed. (485) 97 **Oil.**—The SEEDS yield an oil, used in the manufacture of candles, also medicinally for ointments.

MEDICINE. Bark. (486) 98 **Medicine.**—Ainslie writes: "The BARK is mildly astringent, and has a considerable degree of balsamic sweetness." "It is used by the hill people in the cure of diarrhœa." **Stewart** writes: "The bark, with that of **Tetranthera Roxburghii,** *Nees* (**Litsæa sebifera,** *Pers., var.* **proper**), is officinal, being considered stimulant, and after being bruised, applied, fresh or dry, to contusions, and sometimes mixed with milk and made into a plaster." **Campbell** confirms the above, writing: "The powdered bark is applied to the body for pains arising from blows or bruises, or from hard work; it is also applied to fractures in animals." The SEEDS yield an OIL

Seeds. (487) 99 Oil. (488) 100 which is used medicinally. The medicinal properties above enumerated are very similar to those of the better known, and more largely employed, **L. sebifera,** *Pers.*, the vernacular names for which also strongly resemble—and, indeed, in certain dialects are identical with—those of this species.

TIMBER. (489) 101 **Structure of the Wood.**—Olive-grey, soft, not durable, soon attacked by insects, weight 38 lb per cubic foot, used for agricultural implements (*Gamble*).

DOMESTIC. Leaves. 102 (490)

Domestic.—The LEAVES are used in Assam to feed the *muga* silkworms (**Antheræa assama**, *Westw.*); they have a cinnamon-like smell when bruised" (*Gamble*).

103 (491)

Litsæa salicifolia, *Roxb.; Fl. Br. Ind., V., 167.*

Syn.—TETRANTHERA SALICIFOLIA, *Roxb.;* T. GLAUCA, *Wall.;* T. LAURIFOLIA, *Roxb.;* T. ATTENUATA, *Wall.;* T. LANCRÆFOLIA, *Roxb.* (LANCIFOLIA, *Kurz.*) T. SALICIFOLIA and SALIGNA, *Herb. Ind. Or. H.f. & T.*

Vern.—*Sampat*, NEP.; *Digilati*, MICHI; *Diglotti, súm*, ASSAM; *Chengphisol*, KACHAR.

References.—*Gamble, Man. Timb., 310; Trees, Shrubs, &c., of Darjiling, 65; Kurz, For. Fl. Burm., II., 300; Indian Forester, V., 212; Agri.-Hort. Soc. of Ind., Jour. (Old Series), VI., 29; XIII., 396, 399, 401, 411.*

Habitat.—An evergreen bush or small tree, of Northern and Eastern India, from Oudh and Nepál to Sikkim (ascending to 6,000 feet), met with also in Assam, Bengal, Chittagong, and Pegu.

The *Flora of British India* contains a description of six varieties, which, however, need not be enumerated here, as only one, **var. ellipsoidea**, is reported as possessing economic value.

DOMESTIC. Leaves. 104 (492)

Domestic.—The LEAVES are employed, like those of **L. polyantha**, to feed the *muga* silkworm in Assam.

105 (493)

L. sebifera, *Pers.*, **var. sebifera proper**; *Fl. Br. Ind., V., 158.*

Syn.—LITSÆA SEBIFERA and TETRANTHERA, *Pers.;* L. MULTIFLORA, *Blume;* L. CHINENSIS, *Lamk.;* TETRANTHERA LAURIFOLIA, MULTIFLORA, RACEMOSO-UMBELLATA, and ROXBURGHII, *Blume;* T. LAURIFOLIA, *? Jacq.;* T. ROXBURGHII, *Nees;* T. CAPITATA, *Herb. Roxb.;* T. APETALA, *Roxb.;* TOMEX TETRANTHERA and SEBIFERA, *Willd.;* SEBIFERA GLUTINOSA, *Lour.;* LAURUS INVOLUCRATA, *Kœnig.;* GAJA NIPELLI, *Jones.*

This species is divided into three varieties in the *Flora of British India*, *viz.*, 1, **sebifera proper**, 2, **glabraria**, and 3, **tomentosa**. Of these only the first, of which the synonyms have been above detailed, is of economic value.

Vern.—*Garbijaur, singrauf, medh, ménda*, bark=*maidá-lakri*, HIND.; *Kúhúr chita, ratún, garur*, bark =*maidá-lakri*, BENG.; *Suppotnyok*, LEPCHA; *Garbijaur, singrauf*, N.-W. P.; *Medh*, OUDH; *Medasak, chandna, gwá, rián, medachob*, bark=*méda-lakri, maidasak*, PB.; *Ménda*, C. P.; bark=*Maida-lakri*, leaves=*chickana*, BOMB.; bark=*Mirio*, GOA; *Maidá-lakadi*, MAR.; bark=*Maida-lakti, mushaippé-yetti, pishin-pattai*, TAM.; *Narra alagi, nara mamidi, meda*, TEL.; *Ong-tong, ung-dung, ungdungnet*, BURM.; bark=*Maghâse-hindi*, ARAB.; bark =*Kils*, PERS.

References.—*Roxb., Fl. Ind., Ed. C.B.C., 734, 735; Brandis, For. Fl., 319; Kurz, For. Fl. Burm., II., 298; Gamble, Man. Timb., 310; Thwaites, En. Ceylon Pl., 255; Stewart, Pb. Pl., 188; Elliot, Fl. Andhr., 114, 129; Campbell, Ec. Prod. Chutia Nagpur, No. 9821; O'Shaughnessy, Beng. Dispens., 548; Moodeen Sheriff, Supp. Pharm. Ind., 243; Dymock, Mat. Med. W. Ind., 2nd Ed., 671; Murray, Pl. and Drugs, Sind, 111; Irvine, Mat. Med., Patna, 64; Honigberger, Thirty-five Years in the East, II., 357; Baden Powell, Pb. Pr., 374, 600; Atkinson, Him. Dist., 316, 751; Birdwood, Bomb. Pr., 74; Cooke, Oils and Oilseeds, 79; Agri.-Hort. Soc. of Ind., Jour. (Old Series): —IV., Sel., 260; X., 33; XIII., 318, 319; (New Series):—I., 102; V., 71; Indian Forester:— III., 204; VI., 239, 301, 303, 304; X., 325.*

Habitat.—An evergreen tree of very variable habit, foliage, and inflorescence, widely distributed throughout the hotter parts of India.

FIBRE. Roots. 106 (494)

Fibre.—M. Dumaine states that the red ROOTS are employed as a paper material, and also for making a strong string in Hazaribagh (*Agri.-Hort. Soc. of Ind. Jour.* (*New Series*), *I., 71*).

Oil-yielding Plants.

OIL. 107 **Oil.**—An oil, obtained from the FRUIT, is employed in China and Java in the manufacture of candles. **Baden Powell** states that "it is used to make ointment and candles," but does not mention whether the latter observation refers to the Panjáb or not. No Indian author confirms his statement, and it appears probable that the greasy exudation is, in this country, utilised only medicinally.

MEDICINE. Bark. 108 **Medicine.**—The feebly balsamic, mucilaginous BARK is one of the best known and most popular of native drugs. **Dymock** states that it does not appear to have been mentioned by Sanskrit writers, and is only briefly noticed in Muhammadan works. He considers it probable that the drug has been adopted by Muhammadan physicians in India as a substitute for an Arabian drug, called *Maghath,* the botanical source of which is uncertain. At the present time it is largely employed as a demulcent and mild astringent in diarrhœa and dysentery. According to **Irvine** it is also esteemed as an aphrodisiac in Patna. Fresh ground, it is used either dry, or triturated in water or milk, as an emollient application to bruises, and as a styptic dressing for wounds. It is also supposed to be anodyne, and to act as a local antidote to the bites of venomous animals.

Oil. 109 The OIL from the berries is used in rheumatism; the LEAVES are mucilaginous and have a pleasant odour of cinnamon.

Leaves. 110

TIMBER. 111 **Structure of the Wood.**—"Greyish-brown, or olive-grey, moderately hard, shining, close and even-grained, seasons well, is durable and not attacked by insects. A fine wood, worth notice" (*Gamble*).

112 **Litsæa umbrosa,** *Nees; Fl. Br. Ind., V., 179.*

Syn.—LITSÆA CONSIMILIS, *Nees;* TETRADENIA UMBROSA and CONSIMILIS, *Nees;* TETRANTHERA UMBROSA, *Wall.;* T. PALLENS, *Don.*

Vern.—*Púteli,* NEP.; *Chira, chir-chira,* KUMAON; *Kanwal (a), tilbora, sara, jhatela, chírara, chírchira,* N.-W. P.; *Chírudi, chindi, chilotú rauli, shalanglú, charká,* PB.

References.—*Brandis, For. Fl., 382; Gamble, Man. Timb., 311; Trees, Shrubs, &c., of Darjiling, 65; Stewart, Pb. Pl., 188; Cooke, Oil and Oilseeds, 55.*

Habitat.—A small tree met with on the Temperate and Sub-tropical Himálaya from Kashmír to Sikkim, at altitudes of 3,000 to 7,000 feet, also on the Khásia mountains from 5,000 to 6,000 feet; distributed to Munnipur.

OIL. Fruit. 113 **Oil.**—A considerable amount of confusion appears to exist in the literature of Indian Economic Botany regarding an oil-yielding **Litsæa.** **Stewart** described an undetermined species in his *Panjáb Plants* as bearing an oil-yielding FRUIT. **Brandis** gives the same vernacular Panjáb names for the species now under consideration, as those enumerated by **Stewart** for his undetermined species, and states that the fruit yields an oil used for burning. He, however, unites **L. zeylanica,** *Nees,* with **L. consimilis,** *Nees,* and **L. umbrosa,** *Nees,* as one species, which appears to have led **Gamble** to make the mistake of stating that the first mentioned species also yields an oil used for burning. The writer can find no other reference to an oil obtained from **L. zeylanica,** and it appears probable that **Gamble** has inadvertently over-looked the fact that **Brandis,** though uniting the species, has been careful to describe the oil as obtained only from **L. consimilis.** Similarly **Gamble,** in his list of vernacular names for **L. zeylanica** (which he agrees with the *Flora of British India* in considering distinct), has enumerated the Panjábi and North-Western Provinces names of **L. umbrosa.** In this he is evidently mistaken, since **L. zeylanica** does not occur north of the Konkan.

TIMBER. 114 **Structure of the Wood.**—"Yellow, moderately hard, close-grained, weight 43lb per cubic foot; a good wood" (*Gamble*).

Litsæa zeylanica, *Nees; Fl. Br. Ind., V., 178; Wight, Ic., tt. 132, [1844, 1845.* 115 (503)

Syn.—LITSÆA OBLONGA, *Nees;* L. STRIOLATA, *Blume;* L. FOLIOSA, *Nees;* L. FURFURACEA, *Nees;* L. SCROBICULATA, *Meissn.;* L. TRINERVIA, *Juss.;* TETRADENIA CEYLANICA, FURFURACEA and FOLIOSA, *Nees;* TETRANTHERA FOLIOSA, PULCHERRIMA (*in part*) and FURFURACEA, *Wall.;* LAURUS CASSIA, *Linn.;* L. INVOLUCRATA, *Vahl.;* L. ZEYLANICA, *Herm.*

Vern.—*Belori,* NILGHIRIS; *Dawal kúrúndú,* SING.

References.—*Brandis, For. Fl., 382; Kurz, For. Fl. Burm., II, 306; Beddome, Fl. Sylv., t. 294; Gamble, Man. Timb., 311; Thwaites, En. Ceylon Pl., 257; Dalz. & Gibs., Bomb. Fl., 223; Lisboa, U. Pl. Bomb., 113; Gazetteer:—Bombay, XV., 441; Ind. Forester, II, 23; III., 204.*

Habitat.—A small tree with variable foliage, met with in Bhután, the Khásia mountains, Sylhet, Chittagong, Pegu, Tenasserim (ascending to 7,000 feet), and Martaban, also on the Western Coast from the Konkan southwards, and from Quilon to 7,000 feet in altitude on the Nilghiris; distributed to Ceylon, Penang, Malacca, Java, and Sumatra.

OIL. 116 (504)

Oil.—See remarks under the preceding species.

TIMBER. 117 (505)

Structure of the Wood.—"Reddish-white, with darker heartwood, moderately hard, growth slow, weight 36 to 38lb per cubic foot" (*Gamble*).

In Southern India it is used, according to Beddome, "for house-building purposes, planks, rafters, &c, being straight-grained and tough."

LIVERWORTS.

Liverworts, *Mitten, Indian Hepaticæ, in Jour. Linn. Soc., V., 89-108; Baillon, Bot. Med. Crypto., 51-55.* 118 (506)

The LIVERWORTS or HEPATICÆ constitute a group of the non-vascular Cryptogams, allied to the Mosses. They are cosmopolitan in range, but exist in greatest number in moist cool climates. Consequently, though they are to be found in nearly all moist parts of India, their number and relative importance in the vegetation increase from the plains to the temperate regions of the hills, and the Alpine Himalaya. But few are of any economic value, and, so far as the writer is aware, none are used in any way in India.

The following are the species that have been considered useful:—

MEDICINE. 119 (507)

Marchantia polymorpha, *Linn.; Mitten, in Jour. Linn. Soc., V., 125.*

One of the commonest of liverworts; found on the North-Western Himálaya from Jamu to Sikkim, between the altitudes of 6,000 and 14,000 feet. This species has been vaunted, more than any other, as a remedy for affections of the liver. It has also been described as a valuable medicine in certain chronic skin affections, phthisis, and anasarca.

Fegatella conica, *Corda; Mitten, 126.*

Occurs in the North-Western Himálaya and Western Tibet, between 7,000 and 10,000 feet. At one time a decoction of this liverwort was held in high repute as a remedy for calculous complaints, and as a diuretic. Certain species of **Jungermannia** have had antiscorbutic and antisyphilitic properties ascribed to them, due to the iodine contained in their fronds, and members of the genera **Riccia** and **Anthoceros** have also been considered of medicinal value.

From the gradual disuse into which these remedies have fallen, however, it would appear that the properties assigned to them were all more or less fanciful. The supposed action of many species in diseases of the liver may probably have arisen from a belief in the "theory of signatures." If so, it is all the more remarkable that they are not, and apparently never have been, employed in Indian medicine, in the more remote days of which this theory held such an important position.

LIVISTONA, *Br.; Gen. Pl., III., 929.*

(508) 120 [*a, b, c, d, App. xxiii.*; PALMÆ.

Livistona Jenkinsiana, *Griff., Palms of Brit India, 128, Pl. 226,*

Vern.—*Toko pat,* ASSAM; *Tailainyom, tulac-myom, porbong,* LEPCHA; *Htan-myouk-lu,* BURM.

References.—*Kurz, For. Fl. Burm., II., 525; Gamble, Man. Timb., 418; Trees, Shrubs, &c., of Darjiling, 86; Royle, Fibrous Pl., 97.*

Habitat.—A palm 20 to 30 feet in height, with a thick, round crown, commonly met with throughout Assam, but most plentiful in the Nowgong District.

DOMESTIC. **Domestic, &c.**—Griffith writes: "Major Jenkins tells me that 'this palm is an indispensable accompaniment of every native gentleman's house, but in some parts it is rare and the trees are then of great value.' I cannot call to my recollection having ever seen a *Toko* tree undoubtedly wild.

Leaves. (509) 121 The LEAVES are in universal use throughout Assam for covering the tops of *doolees* (palanqueens), and the roofs of boats, also for making the peculiar hats, or rather umbrella-hats (*jhapees*) of the Assamese. For all these purposes the leaves are admirably adapted by their lightness, toughness, and durability." The leaves are similarly employed by the Lepchas for thatching and umbrellas.

(510) 122 **L. speciosa,** *Kurz; For. Fl. Burm., II., 526.*

Vern.—*Thau,* MAGH.; *Tawtan,* BURM.

References.—*Kurz, Ind., Jour. As. Soc. Beng., xliii.; Fl. Burm., II., 204; Gamble, Man. Timb., 418.*

Habitat.—An evergreen lofty palm, frequent in the tropical forests of Chittagong, the eastern and southern slopes of the Pegu Yomah, and Upper Tenasserim.

DOMESTIC. Leaves. **Domestic.**—Gamble states that the LEAVES of this species are sometimes used for thatching in Chittagong instead of those of **Licuala.**

(511) 123 **Lizards,** see **Reptiles,** Vol. VI., Pt. I.

(512) 124

LOBELIA, *Linn.; Gen. Pl., II., 551.*

A genus of herbs or shrubs which comprises about 200 species, natives, for the most part, of America and South Africa. Fifteen are indigenous to India. One American species, **L. inflata,** *Linn.*, is of interest, since the dried flowering herb constitutes the Lobelia of medicine. This drug is much used in European practice as an expectorant, and to depress the respiratory centre and relax the bronchial muscles in asthma and whooping cough. In full doses it is diaphoretic, diuretic, and emetic, but is too powerful and dangerous an agent to be esteemed for these properties. The drug is officinal in the Pharmacopœia of India, and is imported from America, through Europe, in compressed, oblong, rectangular packages.

[*Ill., t. 135;* CAMPANULACEÆ.

(513) 125 **Lobelia nicotianæfolia,** *Heyne; Fl. Br. Ind., III., 427; Wight,*

Syn.—RAPUNTIUM NICOTIANÆFOLIUM, *Presl.*

Var., trichandra,—L. TRICHANDRA, *Wight.*

Vern.—*Deonal, bokenal, dhaval, dawal,* BOMB.; *Ras-ni,* SING.

References.—*Roxb., Fl., Ind., Ed. C.B.C., 170; Thwaites, En. Ceylon Pl., 170; Dalz. & Gibs., Bomb. Fl., 133; O'Shaugnessy, Beng. Dispens., 424; Dymock, Mat. Med., W. Ind., 2nd Ed., 468; S. Arjun, Bomb. Drugs, 81; Lisboa, U. Pl. Bomb., 265; Gazetteers:—Mysore & Coorg, I., 56; III., 17; Bombay, XV., 436.*

Habitat.—An erect herb met with in Malabar, on the Ghâts from Bombay to Travancore at altitudes of 3,000 to 7,000 feet, and in Ceylon.

MEDICINE. (514) 126 **Medicine.**—This species is said by Roxburgh to have been first described by Heyne, who found it near Bangalore. Little appears to be known regarding its medicinal properties, but the statement is made in the *Indian Pharmacopœia* that an infusion of the LEAVES is antispasmodic. The dry HERB and SEEDS are said to be extremely acrid, and according to Dymock the dust of the former irritates the throat and nostrils like tobacco. Lisboa states that the seeds contain an acro-narcotic poison, and that they are said to be preferred to Datura as a poison, when rapid effect is desired. No mention is made of the plant in native medical works, indeed it would appear to be more widely known as a poison, than esteemed as a drug. MEDICINE. Leaves. 127 Herb. 128 Seeds. 129

Domestic.—Graham states that the dried hollow STALKS are sold in the bazár at Mahableshwar, and are used as *Koluri* horns (called *pánwá* in the Konkan), for collecting herds of cattle and for scaring wolves. According to Gibson the Mahratta name *deonal* is probably given to the plant in allusion to the stems being similarly used as reeds for incantations (*Dymock*). DOMESTIC. Stalks. 130

Lobelia trigona, *Roxb.; Fl. Br. Ind., III., 423; Wight, Ic., t. 1170.* 131

Syn.—L. TRIANGULATA, *Roxb.*; L. STIPULARIS, *Roth.*; L. GRATIOLOIDES, *Roxb.*; L. SP., *Griff.*

Vern.—*Chauric arak*, SANTAL.

References.—*Roxb., Fl. Ind., Ed. C. B.C., 170; Dalz. & Gibs., Bomb. Fl., 133 (excl. syn.); Campbell, Ec. Prod., Chutia Nagpur, Nos. 7837, 9854, 9869; Gazetteers:—Bombay, XV., 436; N.-W. P., IV., lxxiii., X., 312.*

Habitat.—An annual herb, common in the Deccan and Ceylon from the sea level to 6,000 feet, found also in Assam, Bengal, and Burma.

Food.—Campbell states that the LEAVES are eaten as a pot-herb in Chutia Nagpur. FOOD. Leaves. 132

Locusts, see **Insects**, Vol. IV., 470; also **Pests**, Vol. VI.

Lodh Bark, see **Symplocos racemosa**, *Roxb.*; STYRACEÆ; Vol. VI.

LODOICEA, *Labill.; Gen. Pl., III., 939.*

[7, 8; PALMÆ.

Lodoicea Sechellarum, *Comm. & Labill., Bot. Mag., 2734, 5, 6,* 133

THE SEA COCOANUT; THE DOUBLE COCOANUT; COCOS DE MER, *Fr.*

Vern.—*Daryá-ká-náriyal*, HIND.; *Daryá-ká-nárél*, DEC.; *Jáhari-naral*, BOMB.; *Kadat-réngáy*, TAM.; *Samudrapu-tenkáya*, TEL.; *Katal-ténná*, MALAY.; *Daryá-nu-naríyal*, GUZ.; *Múdú-pol*, SING.; *Penle-on-si*, BURM.; *Ubdie narikaylum*, SANS.; *Nárjíle-bahrí*, ARAB.; *Nárgíle-bahrí*, PERS.

References.—*Brandis, For. Fl., 545; Ainslie, Mat. Ind., II., 126; Moodeen Sheriff, Supp. Pharm. Ind., 169; Dymock, Mat. Med. W. Ind., 2nd Ed., 804; S. Arjun, Bomb. Drugs, 147; Year Book Pharm., 1878, 288; Birdwood, Bomb. Pr., 93; Royle, Fib. Pl., 124; Balfour, Cyclop., II., 734; Treasury of Bot., II., 692; Ind. Forester, VI., 240.*

Habitat.—A tall palm, with a distinctly annulated stem and a crown of from 12 to 20 large leaves, met with in two or three rocky islands of the Seychelle group, north-east of Madagascar. It bears very large fruit, which takes several years to attain maturity, and often becomes from 40 to 50lb in weight.

Medicine.—Long prior to the discovery of the Seychelles the FRUIT which was found floating in the Indian Ocean, and washed ashore by the monsoons, was known and valued in India. Strange accounts of the origin of this mysterious fruit were naturally prevalent, such as that it was produced by some submarine plant. From this belief, and from the fact that it was only found when washed ashore, it obtained its many popular Asiatic and European names, all of which have the same meaning—sea-cocoanut. MEDICINE. Fruit. 134

Many marvellous medicinal properties were ascribed to these nuts by ancient physicians, both European and Asiatic, and they were consequently sold at high prices. Now-a-days, however, these properties have been

MEDICINE. recognised as fanciful and dependent solely on the rarity of the fruit. It is consequently no longer valued by Europeans, but **Dymock** informs us that it is still in great repute among the Arabs and natives of India as a tonic, preservative, and alexipharmic. **Ainslie** states that in his time the *Vytians* occasionally prescribed the KERNEL, given in woman's milk, in cases

Kernels. 135 of typhus fever, the dose being "a quarter of a pagoda weight twice daily," and adds, "It is also reputed antiscorbutic and antivenereal." **Dymock** mentions that in Bombay it is prescribed as a tonic and febrifuge in combination with LIGNUM COLUBRINUM (the small branches of **Strychnos colubrina,** *Linn.*). The special opinions below quoted show that it is also reputed to possess several other properties.

SPECIAL OPINIONS.—§ "*Daryai-naryal* is corrupted in Bombay into *Jehari-naryal*, which means 'poisonous cocoanut,' and is believed to be so by the common people. It is, however, non-poisonous, and is commonly given to children, mixed with the root of nux vomica, for colic. It seems to act mechanically, like Bismuth" (*Assistant Surgeon S. Arjun Ravat, L.M., Girgaum, Bombay*). "It is given by natives, rubbed up with water, to check diarrhœa and vomiting, especially in cholera" (*Assistant Surgeon Bhagwan Das, Rawal Pindi, Panjáb*). "The 'water' of the green fruit or its soft kernel is believed to be antibilious and antacid when taken after meals. The ripe fruit is also used similarly but is at the same time purgative" (*Assistant Surgeon Bolly Chund Sen, Calcutta*).

FOOD. Fruit. 136 **Food.**—When unripe the inside of the FRUIT containing a transparent jelly-like substance is eaten, but when ripe it becomes horny and useless for food. The crown of the palm is also said to be eaten in the Seychelles.

DOMESTIC. Shell. 137 **Domestic.**—The hard black SHELL is, in India, carved into ornaments and is also used for *fakirs'* drinking cups. The DOWN attached to the young

Down. 138 leaves serves for stuffing mattresses and pillows, the ribs of the CANES and

Canes. 139 FIBRES of the petiole are used in making baskets, and the young palms are

Fibres 140 also employed in basket-making.

TRADE 141 **Trade.**—**Dymock** writes:—"The nuts are now an article of export from the Seychelles, hundreds of them may be seen at Port Victoria, Mahé, whither they are brought from the island of Praslin. Value in Bombay R1¼ per ℔ for the dry kernel. Entire nuts fetch R1 to R2 each according to their size."

Logwood, see **Hæmatoxylon campechianum,** *Linn.;* Vol. IV., 198.

142 **LOLIUM,** *Linn.; Gen. Pl., III., 1202.*

A genus of grasses which, according to **Bentham,** comprises two or three species confined to temperate regions of the globe. The members may be readily distinguished from all others of the tribe HORDEÆ by the position of the spikelets on the rachis. They are so placed as to have their margins facing the rachis.

143 **Lolium perenne,** *Linn.; Duthie, Fodder Grasses of North India, 68.;*
PERENNIAL RAY OR RYE GRASS. [GRAMINEÆ.

References.—*Aitch. Bot. Afgh. Del. Com., 126; Von Müller, Select Plants, 226; Sutton, Permanent and Temporary Pastures, 40; Stebler and Schrotter, Best Forage Pl., 20; Rep. of Experimental Farm, Madras, 1871, 12; Jour. Agri.-Hort. Soc. of India (New Series), V. (Sel.), 59.*

Habitat.—A native of Europe, Northern Africa, and Western Asia, occurring in the Temperate and Alpine Himálaya up to an altitude of 15,000 feet. It is one of the best known of all grasses, is extensively cultivated as a pasture grass, and for hay, and is almost universally selected for lawn-culture. Several varieties occur in cultivation.

CULTIVATION.—This grass, though not perennial on all soils, nor under adverse conditions, is, according to **Sutton,** fully entitled to the name by

CULTIVATION.

which it has been known since 1611, the date of the earliest book on agriculture in which it is mentioned.

It thrives best in moist and rich loams and clay, but if the supply of moisture be adequate, can also be grown on loamy sands, as well as on calcareous and marly soils. On rich drained soil it is a success, but on heaths, dry sands, and parched soils, it does not grow well. As a general rule, it requires stiffness as well as moisture in the soil, and can be grown even on the heaviest clays, provided they are drained. One of the best features of the grass as an Indian fodder is the fact that if the soil be fairly compact, drought has little effect upon its vitality. Thus in a dry year in Europe or Australia it has been observed to persist while other grasses have succumbed.

The crop is more benefited than that of any other grass by the application of liquid manure and is also acted on very beneficially by irrigation, provided that the land be also well drained. Experiments in Europe have proved the maximum yield to be obtained in the year after sowing, the total of three cuttings from one acre being 9,300℔ of hay. The first cutting is the best. **Drs. Stebler & Schroter** state that one acre of ground requires 55℔ of seed, containing 71 per cent. (=38½℔) of pure and germinating grains. The same authors write: "For agricultural purposes perennial rye-grass is rarely sown alone. It may occupy about 80 per cent. of a mixture with white clover. Such a mixture forms the excellent pastures on the alluvial lands of North Germany. If for mowing much lower percentages should be used, especially when the soil is stiff. About 20 per cent. may be used for mixing with clovers to form 'clover grass.' For temporary grass meadows the amount should not exceed 10 per cent.; if permanence is important, 5 per cent. should rarely be exceeded. On a good soil, it should never be entirely absent from any mixture, because of its rapid and dense growth" (*Stebler and Schröter, Best Forage Plants*).

This grass has been introduced into Australia, has spread rapidly, and seems likely to become one of the most important of the pasture grasses in many parts of the country. Thus **Baron F. von Mueller** writes: "Rye grass stands the dry heat of Australian summers fairly well. It is likely to spread gradually over the whole of the Australian continent, and to play an important part in pasture, except in the hottest desert-tracts. It is one of the best grasses to endure traffic on roads or paths, particularly on soil not altogether light, and is also one of the few among important grasses which can be sown at any season in mild climes."

FODDER. 144 (532)

Fodder.—Though the productive and nutritive powers of the Rye grass are considerably less than those of other meadow hays, its valuable drought-resisting qualities, and its property of rapid growth, render it a most valuable fodder, and one that appears likely to be worthy of extended trial in India.

According to **Wolff**, 100 parts of the hay contain 79·2 of organic matter, consisting of albumen 10·2, fibre 30·2, non-nitrogenous extractives 36·1, and fat 2·7, giving a ratio of nitrogenous to non-nitrogenous nutriment of 1 : 7·3.

Though an excellent fodder and pasture grass, it has been said at times to produce disease in animals fed entirely on it. Thus **F. von Mueller** writes: "Sheep should not be continually kept on rye-grass pasture, as they may become subject to fits similar to those produced by **L. temulentum**, possibly due to the grass getting ergotised or otherwise diseased as many observers assert."

From available records it would appear that the introduction of this grass into India has only been attempted, on at all a large scale, in

FODDER. Madras. It is therefore disappointing to find the following report issuing from the Saidapet Farm: "The grass is not suited for this climate. It is possible that during a course of years it might be naturalised." Attempts made in the North-Western Provinces and Panjáb might, however, prove much more successful; indeed, a consideration of the nature of soil and climate required by the grass would seem to indicate that Madras, of all parts of India, may present the least favourable conditions to its growth.

(533) 145 **Lolium temulentum,** *Linn.; Duthie, Fodder Grasses of N. India,* [*68.*

THE DARNEL.

Syn.—LOLIUM ARVENSE, *With.*; L. ROBUSTUM, *Reich.*

Vern.—*Machní*, HIND.

References.—*Duthie, Indigenous Fodder Grasses of N.-W. Ind., 44; Bent. & Trim., Med. Pl., 295; U. S. Dispens., 15th Ed., 1689; Smith, Dic., 151; Gazetteers, N.-W. P., IV., lxxx; X., 321; Sel. Rec., N.-W. P., 1870, 179.*

Habitat.—An annual weed of cultivation, also occasionally met with as a wayside weed and in waste ground, common in the plains and hills of the North-West Provinces and the Panjáb. It extends throughout Europe and Western Asia, has been found in Madura and North Africa, and occurs as an introduced plant in the United States and Australia.

MEDICINE. Seeds. (534) 146 **Medicine.**—The SEEDS or grains were used medicinally by the ancient Greeks and Romans, but have never been officinal in India, nor in any modern Pharmacopœia. The plant is referred to in this place only because, occurring as a weed of cultivation, its seeds, which are generally supposed to be deleterious, may be found mixed with those of wheat or other cereals. Recent investigations have proved that the darnel grains are, if perfectly healthy and sound, quite inocuous. But they are particularly liable to the attacks of ergot, mildew, and other fungi, and in these conditions, or in certain of them, are poisonous. They then appear to act as a powerful gastro-intestinal irritant, and also to have a marked effect on the cerebro-spinal system, producing headache, giddiness, ringing in the ears, confusion of sight, delirium, convulsions, paralysis, and even death, with severe nervous symptoms. In the report of the Chemical Examiner in the North-West Provinces for 1869 several examples of acro-narcotic poisoning, following the use of wheat mixed with darnel, are cited. The writer of the report recommends that wheat likely to contain the grain should be purified by sifting it through a sieve, the holes of which are small enough to retain the wheat, but large enough to allow the darnel seed to pass through (*Sel. Rec. Govt. N.-W. P., 1870, 179*).

(535) 147 **LONICERA,** *Linn.; Gen. Pl., II., 5.*

A genus of erect or scandent shrubs, which comprises about 80 species, natives of the Temperate and Sub-Alpine regions of the Northern Hemisphere. Of these 23 are Indian, most of which are small shrubs found in the Himálaya. **L. periclymenum,** *Linn.*, is the woodbine or honey-suckle, a native of Europe, but cultivated in hill stations as an ornamental climber. Several other members of the genus are also cultivated for forming arbours, and as climbers on walls of houses, &c., for example, **L. sempervirens,** the Trumpet Honey-suckle.

(536) 148 **Lonicera alpigena,** *Linn.; Fl. Br. Ind., III., 15;* CAPRIFOLIACEÆ.

Syn.—L. WEBBIANA, *Wall.*; L. OXYPHYLLA, *Edgw.*

References.—*Boiss., Fl. Orient., III., 8; Brandis, For. Fl., 256; Gamble, Man. Timb., 217; Stewart, Jour. of a Bot. Tour in Hazára and Khagan, in Jour. Agri.-Hort. Soc. of Ind. (Old Series), XIV., 67; Atkinson, Him. Dist., 311.*

Habitat.—A large shrub of the Himálaya from Kashmír to Kumaon, at altitudes of 9,000 to 12,000 feet; distributed to the European Alps.

Structure of the Wood.—Moderately hard, used for firewood. TIMBER. 149 (537)

Lonicera angustifolia, *Wall.; Fl. Br. Ind., III., 13.* 150 (538)

Vern.—*Mithiga, jinjrú, pilrú, philkú, géang,* PB.

References.—*Brandis, For. Fl., 255; Gamble, Man. Timb., 217; Stewart, Pb. Pl., 113; Atkinson, Him. Dist., 311.*

Habitat.—A small shrub of the Temperate Himálaya from Kashmír and Kumáon to Sikkim, found between 6,000 and 12,000 feet.

Food.—It flowers in May to June, and produces a small, red, sweet FRUIT, which is eaten. FOOD. Fruit. 151 (539)

Structure of the Wood.—White, close-grained, hard, weight 60lb per cubic foot (*Gamble*). TIMBER. 152 (540)

L. glauca, *H.f. & T.; Fl. Br. Ind., III., 11.* 153 (541)

Vern.—*Shingtik, shea, shewa,* PB.

References.—*Stewart, Pb. Pl., 113.*

Habitat.—A dense, wiry undershrub found in the Temperate North-West Himálaya and Tibet, between the altitudes of 12,000 and 16,000 feet.

Medicine.—**Stewart** states that, in certain parts of Ladák, the SEEDS are given to horses for colic. MEDICINE. Seeds. 154 (542)

L. hypoleuca, *Done.; Fl. Br. Ind., III., 14.* 155 (543)

Syn.—LONICERA ELLIPTICA, *Royle.*

Vern.—*Kharmo, kodi, zhiko, rapesho,* PB.; *Sperái, gurásáh,* AFG.

References.—*Brandis, For. Fl., 256; Gamble, Man. Timb., 216; Stewart, Pb. Pl., 114; Atkinson, Him. Dist., 311; Royle, Ill., 236.*

Habitat.—A low shrub of the arid tracts of the North-Western Himálaya, at altitudes of 8,000 to 10,000 feet; lately found by **Mr. Lace**, in Ziárat, South Afghánistán, at an altitude of 8,000 feet.

Fodder.—The LEAVES are said by **Stewart** to be eaten by goats in the Panjáb, and by **Mr. Lace** to be considered, in South Afghánistán, a good fodder for camels, goats, and sheep. FODDER. Leaves. 156 (544)

L. quinquelocularis, *Hardwicke; Fl. Br. Ind., III., 14.* 157 (545)

THE HIMALAYAN HONEY-SUCKLE.

Syn.—LONICERA DIVERSIFOLIA, *Wall.;* L. ROYLEANA, *Wall.*

Vern.—*Bet kukri, bhat kúkra, cheraya, kurmali,* KUMAON; *Tita bateri, pákhur,* KASHMIR; *Phút, bakhrú, khúm, sái, dendrú, kliúnti, kraunti, takla, zbang, razbam, bijgái, jarlangei, adei,* PB.; *Jarlangai, gurásáh,* PUSHTU.

References.—*Roxb., Fl. Ind., Ed. C.B.C., 181; Brandis, For. Fl., 255; Gamble, Man. Timb., 216; Stewart, Pb. Pl., 114; Jour. Bot. Tour in Hazara, &c., Agri.-Hort. Soc. of India, Jour. (Old Series), XIV. 15, 47; Aitchison, Fl. Kuram Valley, 65; Baden Powell, Pb. Pr., 584; Atkinson, Him. Dist., 311; Gazetteers:—Bannu, 23; Dera Ismail Khan, 19; Ind. Forester, XIII., 68.*

Habitat.—A pubescent shrub of the Temperate Himálaya from Kashmír to Kumáon, between 4,000 and 12,000 feet, also found in the Súlimán range, the hills of the Trans-Indus Panjáb, and Southern Afghánistán.

Fibre.—**Aitchison** writes in his *Kuram Valley Flora,* "It sheds the external layers of its bark in long fibrous strips, resembling coarse hemp-fibre; this is collected and employed as rope, but has little or no strength, only suitable for stuffing mattresses and such purposes." FIBRE. 158 (546)

Structure of the Wood.—White, with a brown centre, very hard and close-grained, weight 52lb per cubic foot, used only for fire-wood (*Gamble*). **Mr. Lace** informs the writer that it is employed by the Patháns of Southern Afghánistán for making food utensils, handles to tools, ploughs, &c. TIMBER. 159 (547)

FODDER. Leaves. **Fodder.**—The LEAVES are used as fodder for cattle (*Stewart*).

160 **Lonicera rupicola,** *H. f. et T.; Fl. Br. Ind., III., 13.*

161 **References.**—*Duthie, Note on Trees and Shrubs of N.-E. Kumáon, in Indian Forester, XI., 3.*

Habitat.—A small rigid shrub met with in Tibet, the north of Sikkim and Kumáon.

DOMESTIC. 162 **Domestic, &c.**—Mr. Duthie writes, "Plentiful in Bijáns, forming near Kutti magnificent hedges between cultivated fields."

LOPHOPETALUM, *Wight; Gen. Pl., I., 362.*

163 **Lophopetalum Wallichii,** *Kurz; Fl. Br. Ind., I., 615;* CELASTRINEÆ.

Vern.—*Mondaing, múndain, konazo-ta-lú,* BURM.; *Toung-hmayo,* AND.

References.—*Kurz, For. Fl. Burm., I., 255; in Jour. As. Soc. Beng., 1872, pt. II., 299; Gamble, Man. Timb., 86; Indian Forester, VIII., 416.*

MEDICINE. **Habitat.**—A large glabrous tree, common in the open and more especially in the *Eng* forests all over Pegu and Martaban, down to Tenasserim

Bark. 164 (*Kurz*).

Root. **Medicine.**—According to Major Ford the BARK, ROOT, and FRUIT are

165 used in the Andaman Islands as a febrifuge.

Fruit. 166 **Structure of the Wood.**—"Pale, turning pale-brown, finely and rather loose-grained, hard, rather light, the annual rings very narrow, the heart-

TIMBER. 167 wood brown. Recommended for furniture" (*Kurz*). Major Ford states that it is used in the Andamans for making writing-boards.

168 **L. Wightianum,** *Arn.; Fl. Br. Ind., I., 615; Wight, Ic., t. 162; Ill., 178.*

Vern.—*Balpalé,* KAN.

References.—*Gamble, Man. Timb., 86; Bedd., Fl. Sylv., t. 145; Anal. Gen., lxv.; Dalz. & Gibs., Bomb. Fl., 48; Lisboa, Useful Pl. Bomb., 49; Gazetteer:—Bomb., XV., 430; Balfour, Cyclop., II., 740.*

Habitat.—A large evergreen tree of the Western Coast from the Konkan to Cape Comorin.

TIMBER. **Structure of the Wood.**—Reddish-grey, moderately hard, close-grained,

169 weight 28 to 29℔ per cubic foot (*Gamble*). It is much esteemed by the natives of South Kanara, where it is used for house-building.

Loquat.—The fruit of **Eriobotrya japonica,** *Lindl.; Vol. III., 257.*

170 LORANTHUS, *Linn.; Gen. Pl., III., 207.*

A genus of parasitic evergreen shrubs, which comprises about 350 species, of which 58 are known to be natives of India. Several attack and severely injure certain trees. The majority of the species are called *Pand* in Hindustani, *Ajerú* in Nepalese, *Badanike* in Telegu, and *Khyee-poung* in Burmese.

171 **Loranthus longiflorus,** *Desr.; Fl. Br. Ind., V., 214; Wight, Ic., t. 302;* LORANTHACEÆ.

Syn.—L. BICOLOR, *Roxb.;* L. KŒNIGIANUS, *Agardh.;* L. WIGHTIANUS, *Wall.;* L. IMBRICATUS and L. LINEATUS, *Edgew.*

Var.—falcata, *Kurz;* L. FALCATUS, *Linn.;* L. WIGHTIANUS, *Wall (in part).*

Var.—amplexifolia, *Thw.;* L. AMPLEXIFOLIUS, *DC.;* L. AMPLEXICAULIS, *Wall*

Var.—pubescens.

Vern.—*Bándá,* HIND.; *Bura-manda, ? bara-manda,* BENG.; *Banda,* SANTAL; *Ajeru,* NEPAL; *Prústi,* LEPCHA; *Patha,* BANDA; *Kaurak,* BHIL; *Panda, amút, parand, pand, banda,* PB.; *Bánda,* C. P.; *Vánda,* MAR.; *Vando,* GUZ.; *Yelinga, wadinika, velaga badanika, badanike, wajn, ippa wajna,* TEL.; *Khyee-paung,* BURM.; *Vánda, vanua, vrikshadani, vrikshabhaksha, vriksharúha,* SANS.

References.—*Roxb., Fl. Ind., Ed. C.B.C., 184, 301; Brandis, For. Fl., 397; Kurz, For. Fl. Burm., II, 321, 322; Gamble, Man. Timb., 320; Thwaites, En. Ceylon Pl., 134; Grah., Cat. Bomb. Pl., 86; Dalz. & Gibs., Bomb. Fl., 110; Stewart, Pb. Pl., 112; Elliot, Flora Andhrica, 19, 188, 190; DC. Prodr., IV. 304; Campbell, Ec. Prod., Chutia Nagpur, No. 9810; Atkinson, Him. Dist., 316; McCann, Dyes and Tans, Beng., 160, 161, 166; For. Ad. Rep. Ch. Nagpur, 1885, 33; Gazetteers:—Bombay, XV., 442; N.-W. P., IV., lxxvii.; Ind. Forester, I., 300, 301; VI., 238; VII., 180; VIII., 127, 370, 404; X., 309, 325; XII., App., 20*

Habitat.—A common shrub of the Temperate and Tropical Himálaya, at altitudes from 3,000 to 7,500 feet, extending from Jamu to Bhután, also found in the Gangetic Plains from Oudh eastwards to Assam, and southwards to Travancore, Ceylon, and Malacca. The varieties **falcata,** *Kurz.*, and **pubescens,** are confined to Western India and Ceylon.

The plant occurs as a parasite principally on trees of the following genera—**Acacia, Bassia, Bauhinia, Buchanania, Diospyros, Ficus, Mallotus, Mangifera, Melia, Morus, Prunus, Pyrus, Quercus,** and **Rottlera,** but may also be found on others. Like other members of the genus its seeds are surrounded by a viscid substance which passes uninjured through the intestines of birds, and enables it to obtain a firm attachment to a branch on which it may happen to be dropped. **Dr. Bidie,** commenting on the ravages caused by this parasite in the Nilghiri plantations, on the Apricot, Pear, Peach, and Australian Black-wood (**Acacia melanoxylon,** *R. Br.*), writes: "One or more large branches gets so covered with **Loranthus** that the whole, or nearly the whole, of the sap goes to the parasites, and thus the affected branches die of starvation down to the trunk. Branch after branch perishes in this way, and at length the tree, bereft of its foliage, and robbed of its sap, dies down to the root." It is to be regretted that there is apparently no means of getting rid of these destructive parasites;—excision is said to be useless; indeed, in the report from which the above passage is quoted, Dr. **Bidie** recommended discontinuing the plantations of **Acacia melanoxylon,** because the parasite rendered the timber worthless, save for firewood, and no means could be found of arresting its ravages.

TAN. Wood. 172 (560)

Tan.—The WOOD, prepared as follows, is largely used as a finishing tan-stuff in order to render leather soft:—The leaves are stripped from the shrub, and the sticks, laid together on a hide or mat, are bruised by two or three persons beating them with the *musal* or *gamal* (an iron rimmed wooden pestle used for pounding grain in an *úlkey*). After the skin has been subjected to the action of various tanning materials, and the process of preservation is complete, it is roughly cleaned, sewed up into the form of a bag, and hung upon a tripod. The bruised wood is then placed inside, and the skin is filled with water. It is allowed to remain in this state for a couple of days, and the process is then completed by rubbing one side or other with *khari* salt (*McCann*).

DOMESTIC. Bark. 173 (561)

Domestic, &c.—According to **Birdwood** the BARK of **var. falcata** is used in Kanara as a substitute for betel-nut.

Lotus, see **Nymphæa pubescens,** *Willd.*

Loxa bark, see **Cinchona officinalis,** *Hook.*; Vol. II., 300.

Lucerne, see **Medicago sativa,** *Linn.*; p.

LUCULIA, *Sweet; Gen. Pl., II., 43.*

174 (562)

Luculia gratissima, *Sweet; Fl. Br. Ind., III., 36;* RUBIACEÆ.

Syn.—CINCHONA GRATISSIMA, *Wall.*; MUSSÆNDA LUCULIA, *Ham.*

Vern.—*Dowari,* NEP.; *Simbrangrip,* LEPCHA.

References.—*Kurz, For. Fl. Burm., II., 71; Gamble, Man. Timb., 218; List of Trees &c., of Darjeeling, 47.*

Habitat.—A spreading shrub of the Temperate Himálaya from Nepál to Bhotán, at altitudes from 4,000 to 6,000 feet, distributed to the Ava Hills.

DYE. Leaves. (563) 175 **Dye.**—Gamble states that the LEAVES are used in dyeing, alone or mixed with those of **Hedyotis capitellata**, *Wall.*

DOMESTIC. Flowers. (564) 176 **Domestic.**—The handsome, pink, long-tubed flowers are much worn by the Paharias and Lepchas (*Gamble*).

(565) 177

LUFFA, *Linn.; Gen. Pl., I., 823.*

A genus of climbing plants, which comprises ten species, natives of the warmer regions of the Old World, and one indigenous in America. Of these four or five are truly wild in India.

(566) 178 **Luffa acutangula,** *Roxb.; Fl. Br. Ind., II., 615;* CUCURBITACEÆ.

Syn.—CUCUMIS ACUTANGULUS, *Wall. var.* amara may, for convenience of description, be considered separately.

Vern.—*Torai, jinga, turi,* HIND.; *Jhingá, jinga,* BENG.; *Janhi,* URIYA; *Paror jhinga,* SANTAL; *Ramtoroi,* NEPAL; *Puichenggah,* MAL. (S.P.); *Káli-taroi, satpatiya,* BUNDEL.; *Taroi, káh taroi, torai, satpatiya, jajinga,* N.-W. P.; *Torié,* KUMAON; *Gharúr gundoli,* KANGRA; *Káli tori, turái, jhinga,* PB.; *Túri,* SIND; *Dorka,* C. P.; *Turai, sirola, gonsali, jinga,* BOMB.; *Shirola,* MAR.; *Turin, ghisoda,* GUZ.; *Turai,* DEC.; *Pikunkai,* TAM.; *Burkai, bira-káya,* TEL.; *Hirékáyi,* KAN.; *Djinji,* MALAY.; *Tha-bwot-kha-wai, thapwot,* BURM.; *Jhingáka,* SANS.; *Khiyár,* PERS.

References.—*Roxb., Fl. Ind., Ed. C.B.C., 698; Voigt, Hort. Sub. Cal., 56; Stewart, Pb. Pl., 98; DC., Origin Cult. Pl., 271; Mason, Burma and its People, 471, 747; Elliot, Flora Andh., 27; O'Shaughnessy, Beng. Dispens., 351; U. C. Dutt, Mat. Med. Hind., 301; S. Arjun, Bomb. Drugs, 59, 203; Murray, Pl. and Drugs, Sind, 40; Baden Powell, Pb. Pr., 348; Atkinson, Him. Dist., 700; Econ. Prod. N.-W. P., V., 6; Drury, U. Pl., 382; Duthie & Fuller, Field and Garden Crops, Pt. II., 60, Pl. lxii.; Lisboa, U. Pl. Bomb., 158; Birdwood, Bomb. Pr., 158; Royle, Ill. Him. Bot., 219; Cooke, Oils and Oilseeds, 55; Stocks, Rep. on Sind; Agri.-Hort. Soc. of Ind., Trans., IV., 104; VII., 64, 66; Journ. (Old Series), IV., 202; Ind. Forester, IX., 201; Gazetteers:—Orissa, II., 180, App. VI.; Bombay, VIII., 183; N.-W. P., I., 81; IV., lxxii; Settle. Repts.:—Chánda, Kumáon, App. 33, 82; Kángra, 25, 28.*

Habitat.—Met with in North-West India, Sikkim, Assam, and Eastern Bengal; distributed to Ceylon and Malaya; it is also cultivated in most parts of India.

CULTIVATION.—The fruit, for which the plant is grown, is produced during the rainy season. Sowings should be made from March to the beginning of June. Rich soil should, if possible, be selected, and the seed sown in lines 5 feet apart. When the young plants are about 4 inches high, supports should be given for them to climb on. Until the rains begin the first sowings should be regularly watered. Two sowings—one early, the other late—will keep up a supply from July till October (*Ind. Forest., IX., 201*). No information of a trustworthy nature can be given regarding the area occupied by the crop.

OIL. Seeds. (567) 179 **Oil.**—The SEEDS of this species like those of most other Cucurbitaceous plants yield an oil. No definite information exists regarding the quantity obtainable, nature, or properties of this oil, but it is presumably similar to the more important commercial oils obtained from the melon, &c.

MEDICINE. Seeds. (568) 180 **Medicine.**—The SEEDS possess emetic and purgative properties, but to a much less marked degree than those of the variety **amara.** **Dr. Emerson**

Leaves. (569) 181 states that the LEAVES are used locally in splenitis, hæmorrhoids, and leprosy. **Aitchison** writes, "the ROOT is used in medicine."

Root. (570) 182 SPECIAL OPINION.—§ "The juice of the fresh leaves is dropped into the eyes of children in granular conjunctivitis, also to prevent the lids adhering at night from excessive meibomian secretion" (*Honorary Surgeon P. Kinslev, Chicacole, Ganjam, Madras*).

Food.—The FRUIT is highly esteemed by Natives, and is much eaten by them, either in curries or dressed with clarified butter. When half grown it is one of the best indigenous Indian vegetables, and when peeled, boiled, and dressed with butter, pepper, and salt is very palatable. When fully developed it is about a foot long, but if allowed to grow longer than 4 inches it rapidly deteriorates in quality, and becomes useless for the table. FOOD. Fruit. 183 (571)

Domestic.—Dr. **Stewart**, Cuttack, reports that the fibrous coat of the mature CAPSULE of this and the succeeding variety, forms a cheap and efficient flesh brush. DOMESTIC. Fruit. 184 (572)

Luffa acutangula, *Roxb.; var.* amara; *Fl. Br. Ind., II., 615.* 185 (573)

Syn.—L. AMARA, *Roxb.;* L. PLUKENETIANA, *DC.;* MOMORDICA TUBIFLORA, *Wall.*

Vern.—*Karvi-turi, karwitarui,* HIND.; *Ghoshá-latá, tito-jhingá, titotorai, tito-dhundul,* BENG.; *Kerula** (Atkins.), N.-W. P.; *Karvi-turái,* DEC.; *Kadú-sirola, kadú-dorka,* BOMB.; *Kadú-dodaká,* MAR.; *Pé-pirkkam,* TAM.; *Adavi-bíra, chédu-bíra, verri-bíra, sendu-bir-kai,* TEL.; *Koshátaki,* SANS.

References.—*Roxb., Fl. Ind., Ed. C.B.C., 699; Voigt, Hort. Sub. Cal., 57; Elliot, Flora Andhr., 10, 35, 37; U. C. Dutt, Mat. Med. Hind., 305; Dymock, Mat. Med. W. Ind., 2nd Ed., 342; S. Arjun, Bomb. Drugs, 59; Atkinson, Ec. Prod., N.-W. P., Pt. V., 6; Drury, U. Pl., 282; Lisboa, U. Pl. Bomb., 158; Balfour, Cyclop., II., 746; Agri.-Hort. Soc. of Ind., Trans., II., 110; VII., 64, 66; Jour. (Old Series), 309.*

Habitat.—Met with all over India, especially on the western side.

Medicine.—**Roxburgh** appears to have been the first European writer to notice the medicinal properties of this gourd. He writes: "Every part of the plant is remarkably bitter, the FRUIT is violently cathartic and emetic. The JUICE of the roasted young fruit is applied to the temples by Natives to cure headache. The ripe SEEDS, either in infusion or substance, are used by them to vomit and purge." In the *Pharmacopœia of India* the plant is described as bitter, tonic, and diuretic, but, from the vernacular names there given, *viz., kerúla* and *bindál*, it appears that the gourd meant may in reality be **Momordica Charantia** (*See note to vernacular terms*). **Dymock** writes: "The LEAVES are bitter, the fruit less so; the former in Bombay are used as an external application to sores in cattle. In dog-bite the pulp of the fruit is given with water; it causes vomiting and purging. The juice is applied in different kinds of bites, and the dried fruit is used as a snuff in jaundice. The root with equal parts of *jasúnd* root" (**Hibiscus rosa-sinensis,** *Linn.*) "and **Hemidesmus** is given with milk, cummin and sugar in gonorrhœa." From the following interesting note by **Moodeen Sheriff** it would appear that the seeds, if carefully prepared and administered, are of considerable value as a specific for dysentery. Should this prove to be correct, the drug would be a cheap, easily obtained, and most valuable substitute for Ipecacuanha. In any case, the method below recommended seems worthy of attention and careful trial. MEDICINE. Fruit. 186 (574) Juice. 187 (575) Seeds. 188 (576) Leaves. 189 (577)

SPECIAL OPINIONS.—§ "The mature and dry seeds of both **Luffa acutangula** and **L. amara** are emetics, but the action of the former is very irregular and uncertain. In some cases, they act in twenty to thirty-five grain doses pretty satisfactorily; but in others, they either do not act at all, or act violently and continue to produce vomiting for many hours.

* **Moodeen Sheriff** remarks: "*Karólá* or *Kerula* is the name assigned to this plant in some books, but is correctly the name of **Momordica Charantia,** both in Hindustani and Bengali."

MEDICINE.

On the other hand, the action of the seeds of L. **amara** is very sure, safe, and efficient in the same or somewhat smaller doses. The Hindu practitioners are aware that the fruit of **L. amara** is emetic, but they do not know what particular part of it possesses that property They use it, seeds and all, in infusion. One entire fruit, generally of middle size, is bruised and infused in some cold water at night, and the liquid is strained through and administered in the morning. The action of this draught is generally very irregular, uncertain, and often accompanied by griping, and is, therefore, very unsatisfactory. I have used the different parts of the fruit of this plant separately, and found the emetic property to reside in the kernel or cotyledons of the seeds. The seeds are dark-brown, oblong or oval, flat, rough with minute elevated dots, margin turned, and only distinct at the base. The length of the seeds varies from one-third to half an inch, and the breadth, from three to five lines; the kernel is albuminous, greenish-white, very bitter and oily. The kernel of the seeds is the best, and forms the only vegetable emetic in India which is equal to Ipecacuanha in the same quantity. In smaller doses, it is expectorant and also demulcent, owing to its containing albumen and oil. In addition to the above properties, it has a great control over dysentery. I have used this drug and also Ipecacuanha, separately, in several cases, in the same manner and doses, and found it to be at least quite equal to the latter. The dose of the kernel as an emetic is from twenty to thirty grains; as a nauseant, from eleven to fifteen grains; and as demulcent and expectorant from five to ten grains. When the kernel is rubbed and mixed with water it forms a greenish-white emulsion, which is the only form in which I have yet used it" (*Honorary Surgeon Moodeen Sheriff, Khan Bahadur, G. M. M. C., Triplicane, Madras*).

DOMESTIC. Rind. 190

Domestic.—The dried fibrous RIND is used by the natives in certain parts of the country as a brush for sizing paper (*Agri.-Hort. Soc. of India, Trans., II., 110*).

191 **Luffa ægyptiaca,** *Mill.; Fl. Br. Ind. II., 614; Wight, Ic., t. 499.*

Syn.—LUFFA PENTANDRA, *Roxb.;* L. RACEMOSA, *Roxb.;* L. CLAVATA, *Roxb.;* L. ACUTANGULA, *W. & A. (not of Roxb.);* L. CYLINDRICA, *Ræm.;* L. PETOLA *and* CATTU-PICCINA, *Seringe;* L. PARVALA, *Wall.;* L. GOSA, HEDERACEA, *and* SATPATIA, *Wall.;* BRYONIA CHEIROPHYLLA, *Wall.;* MOMORDICA LUFFA, *Linn.*

Vern.—*Ghiá-tarui, purula,* HIND.; *Dún-dúl, dhundul,* BENG.; *Bhol, bhatkerela, bhotkakrel,* ASSAM; *Palo,* NEPAL; *Ghiya taroi, ghiya tori,* N.-W. P.; *Tarod, ghiya taroi, turai, dhandal,* KUMAON; *Ghia tori, ghi turái, ghi gandolí,* PB.; *Turi, liasada,* SIND; *Dilpasand, tel doaka,* C. P.; *Ghosálí, parosi, parula, turi, gonsali,* BOMB.; *Turia,* GUZ.; *Gutti bira, néti bira, núne bira,* TEL.; *Tha-bwot, tha-pwot-kha,* BURM.; *Neyang-natta-colú,* SING.; *Rájakoshátaki, dirghapatolika,* SANS.; *Luff,* ARAB.; *Khujar,* PERS.

References.—*Roxb., Fl. Ind., Ed. C.B.C., 698, 699; Thwaites, En. Ceylon Pl., 126; Stewart, Pb. Pl., 98; DC., Origin Cult. Pl., 269; Elliot, Fl. Andhr., 67, 133, 138; Mason, Burma and Its People, 470, 747; U. C. Dutt, Mat. Med. Hind., 297, 315; Murray, Pl. and Drugs, Sind, 40; Med. Top. Ajmir, 149; Atkinson, Him. Dist., 310, 700; Ec. Prod. N.-W. P., V., 6; Drury, U. Pl., 282; Duthie & Fuller, Field and Garden Crops, II., 61, pl. lxiii.; Lisboa, U. Pl. Bomb., 158; Cooke, Oils and Oilseeds, 56; Balfour, Cyclop., II., 747; Kew Off. Guide to the Mus. of Ec. Bot., 71; Agri-Hort. Soc. of Ind., Trans., VII., 64; Jour. (Old Series), IV., 202; IX., Sel., 58; Indian Forester, IX., 201; Gazetteers:—Bombay, V., 26; N.-W. P., IV., lxxii.; Settlement Reports:—Chanda (C. P.), 82; N.-W. P., App., 33; Kangra (Pb.), 25, 28.*

Habitat.—A native of India, cultivated or naturalised in most hot countries of the world. In India it is common everywhere, and is often cultivated, especially in the plains.

CULTIVATION.—The seasons of sowing, of ripening of the fruit, and the method pursued in its cultivation are exactly similar to those of **L. acutangula**, and need not be again detailed. No returns are available of the area under the crop, with the exception of the incomplete figures given by **Messrs. Duthie & Fuller**. These writers state that during the rains of 1881, 256 acres was returned as the area in Allahabad, 199 in Meerut, 104 in Budaon, and smaller figures varying from 30 to 65 in Cawnpore, Bijnor, Pilibhit, Bulandshahr, Muttra, and Jalaun. The total area for all these districts was 849 acres. CULTIVATION.

Oil.—An oil is obtained from the SEEDS, probably similar to that of other CUCURBITACEÆ. OIL. Seeds. 192 (580)

Medicine.—The SEEDS are said to be emetic and cathartic like those of **L. acutangula.** MEDICINE. Seeds. 193 (581)

Food.—The FRUIT, which is smaller than that of **L. acutangula**, is also edible, and is similarly used in curries, &c., by the natives. FOOD. Fruit. 194 (582)

Domestic.—The DRY FRUIT, which is filled with an interwoven network of fibre, is used as a flesh-brush in Turkish Baths. DOMESTIC. Fruit. 195 (583)

Luffa echinata, *Roxb.; Fl. Br. Ind., II., 615.* 196 (584)

Syn.—L. BINDAAL, *Roxb.*

Vern.—*Jangthori*, SIND; *Kukar-wel*, BOMB.; seeds=*Wa-upla-bij*, GUZ.; seeds=*Deodágri*, MAR.; *Pani bira*, TEL.

References.—*Roxb., Fl. Ind., Ed. C.B.C., 699; Kurz, in Jour. As. Soc., 1877, II., 101; Dalz. & Gibs., Bomb. Fl., 102; S. Arjun, Bomb. Drugs, 59; Elliot, Fl. Andh., 140; Dymock, Mat. Med. W. Ind., 2nd Ed., 343; Murray, Pl. and Drugs of Sind, 40.*

Habitat.—Native of Gujarat, Sind, Purneah, and Dacca, distributed to Tropical Africa.

Medicine.—**O'Shaughnessy**, writing in 1841, mentions that the FRUIT is considered in North India to be a powerful remedy for dropsy. **Arjun** states that the fruit has purgative properties. **Dymock** writes: "I have not met with any notice of the medicinal use of this plant in European works on the Materia Medica of India." (The writer has evidently inadvertently overlooked the observations of **O'Shaughnessy** and **Arjun.**) "**Roxburgh** describes its botanical characters. In the Bombay Presidency it is found mostly in Gujarat, where it has a reputation among the Hindus on account of the bitter properties of the fruit, and is an ingredient in their compound decoctions. In the Konkan a few grains of the bitter fibrous contents of the fruit are given in infusion for snake-bite, and in cholera after each stool. In putrid fevers the infusion is applied to the whole body, and in jaundice it is applied to the head and also given internally; the infusion has also a reputation as a remedy for colic. The dried vine with the ripe fruit attached is brought to Bombay for sale along with other herbs from the province of Gujarat." MEDICINE. Fruit. 197 (585)

LUMNITZERA, *Willd.; Gen. Pl., I., 687.*

Lumnitzera racemosa, *Willd.; Fl. Br. Ind., II., 452;* COMBRETACEÆ. 198 (586)

Syn.—PYRRANTHUS ALBUS, *Wall.*; PETALOMA ALTERNIFOLIA, *Roxb.*; BRUGUIERA MADAGASCARENSIS, *DC.*

Vern.—*Kripa, kirpa*, BENG.; *Baireya*, SING.; *Yengyé, hmaing, yin-yé*, BURM.

References.—*Roxb., Fl. Ind., Ed. C.B.C., 361; Brand., For. Fl., 221; Kurz, For. Fl. Burm., I., 468; Beddome, Fl. Sylv., Anal. Gen., Pl., xxi.; Thwaites, En. Ceylon Pl., 103; Gamble, Man. Timb., 178; Mason, Burma and its People, 743; Lisboa, U. Pl. Bomb., 77.*

Habitat.—A small tree occurring on the coasts of India, Ceylon, and the Trans-Gangetic Peninsula, also on the coasts of the Andamans and Nicobars; distributed to Tropical Africa, Malaya, North Australia, and Polynesia.

TIMBER. (587) 199 **Structure of the Wood.**—**Roxburgh** writes: "The wood is remarkably strong and durable; it is much used for posts and other parts of the houses of the natives, but its chief consumption about Calcutta is for fuel."

LUPINUS, *Linn.; Gen. Pl., I., 480.*

(588) 200 **Lupinus albus,** *Linn.; DC. Prodr., II., 407;* LEGUMINOSÆ.

Syn.—L. SATIVUS, *Gater.*

Vern. *Turmás,* HIND.; *Túrmúz,* BENG.; *Túrmuz, zurmish,* PB.; *Turmus, bakila-i-misri,* PERS.

References.—*DC., Origin of Cultivated Pl., 325; Honigberger, Thirty-five Years in the East, II., 302; Dymock, Mat. Med. W. Ind., 2nd Ed., 292; Irvine, Mat. Med. Patna, 114; S. Arjun, Bomb. Drugs, 42; Year Book Pharm., 1874, 624; Baden Powell, Pb. Prod., 342; Church, Food Grains of Ind., 123; Smith, Dic., 234; Agri.-Hort. Soc. of Ind., Jour. (Old Series), II., (Sel.), 173, 174; Quarterly Jour. of Agric., IX., 317.*

Habitat.—An annual herb, native of the Levant, extensively cultivated in Southern Europe, Egypt, and Asia south of the Caucasus; said by **Church** to be also cultivated in some parts of India.

MEDICINE. Seeds. (589) 201 **Medicine.**—The SEEDS, called by the above vernacular names, are imported from Egypt and are used by the *Hakims* as a deobstruent, alterative, and anthelmintic. **Honigberger** states that in the Panjáb they were used in cases of "internal heat" and leprosy; **Irvine** describes them as carminative. The Persian name *turmus* is said by **Dymock** to be derived from the Greek θερμος, or the Coptic θαρμος, seeds used by the Greeks to counteract the effects of drink. In the south of Europe, at the present day, they are considered anthelmintic, and are also ground and boiled to form poultices. The active principle is an alkaloid called *Lupinin,* which has a very bitter taste and is poisonous to frogs, but apparently is not deleterious to man, even if given in rather large doses. It is dissipated by heat.

FOOD and FODDER. Seeds. (590) 202 **Food and Fodder.**—The lupin is largely cultivated in Southern Europe for the sake of its highly nutritious SEEDS, which are boiled and used as food. **Church** gives the following as the composition of 100 parts: water 12·5, albumenoids 31·7, starch 33·7, oil 5, fibre 13·5, and ash 3·6.

The same author states that the high amount of fibre renders a good deal of the albumenoids and oil unavailable for digestion. The seeds are also a good food for cattle.

DOMESTIC. Seeds. (591) 203 Plant. (592) 204 **Domestic, Agricultural, &c.**—A preparation of the SEEDS is employed in Tuscany as a cosmetic. The PLANT is a very valuable green manure and has been cultivated for this purpose from the earliest times in the south of Europe. The yellow lupin (**L. luteus**) is grown in Germany and some parts of England for the same purpose. The crop acts as a valuable leguminous rotation, improving the soil by the depth to which its roots penetrate.

Lupulus, see **Humulus Lupulus,** *Linn.*; Vol. IV., 302.

LYCHNIS, *Linn.; Gen. Pl., I., 147.*

(593) 205 **Lychnis indica,** *Benth.; Fl. Br. Ind., I., 225;* CARYOPHYLLEÆ.

Var.—indica proper, SILENE INDICA, *Roxb.*; MELANDRYUM INDICUM. *var.* GENUINUM, *Rohrb.*

Var.—fimbriata, *Wall.*; L. ERIOSTEMON, *Wall.*; MELANDRYUM INDICUM, *var.* FIMBRIATUM, *Rohrb.*

Habitat.—A tall, spreading, weak herb, of the Temperate Himálaya from Nepál to Marri, at altitudes of 5,500 to 10,000 feet.

Domestic.—Aitchison states that in Lahoul the ROOT and LEAVES are used for soap (*Stewart, Pb. Pl., 20*). DOMESTIC. Root. 206 (594) Leaves. 207 (595)

LYCIUM, *Linn.; Gen. Pl., II., 900.*

A genus of spinous shrubs, which comprises about 40 species, all natives of temperate and sub-tropical regions, especially of South Africa and South America; of these 3 are indigenous in India. 208 (596)

Lycium barbarum, *Linn.; Fl. Br. Ind., IV., 241;* SOLANACEÆ. 209 (597)

Syn.—L. EDGEWORTHII, *Dunal.;* L. DEPRESSUM *and* FOLIOSUM, *Stocks.*

Vern.—*Koh-tor,* BALUCH; *Barghauna, karghauna,* PUSHTU.

References.—*Aitchison, Bot. Afg. Del. Com., 91; Murray, Pl. and Drugs of Sind, 159; Smith, Dic., 57; Indian Forester, XIV., 362; Notes by Mr. Lace on Afgh. Pl.*

Habitat.—A spiny shrub of the Panjáb and Sind, distributed to Western Asia. The only point of interest connected with this plant, is, that during the Afghán Delimitation Commission many camels were supposed to have died at Omar-sha in consequence of eating the shrub when in fruit. This fact is the more curious, since, though belonging to the Nightshade order, neither this nor any other species of **Lycium** had been previously supposed to possess poisonous properties. Poison. 210 (598)

L. europæum, *Linn.; Fl. Br. Ind., IV., 240; Wight, Ic., t. 1403.* 211 (599)

Syn. LYCIUM INDICUM, *Wight.;* L. MEDITERRANEUM, *Dunal.;* L. SÆVUM, ORIENTALE, *and* PERSICUM, *Miers.;* L. INTRICATUM, *Boiss.;* L. ARABICUM, *Schweinf.*

Vern.—*Achmehudi,* MERWARA; *Gangro, ganger,* SIND; *Kangú, kúngú, kangi, ganger, mrál, chirchitta,* PB.

References.—*Brandis, For. Fl., 345; Gamble, Man. Timb., 273; Stewart, Pb. Pl., 156; Murray, Pl. & Drugs of Sind, 158; Baden Powell, Pb Prod., 584; Indian Forester, XII., App., 18; Gazetteers:—Bomb., V., 27; N.-W. P., IV., lxxv.*

Habitat.—A small, thorny shrub, not uncommon in the drier tracts of the Panjáb plains from Delhi west to the Súliman Range, found also in Sind, Gujerat, and the Deccan.

Medicine.—"The BERRIES are used medicinally as an aphrodisiac" (*Stewart*). MEDICINE. Berries. 212 (600)

Food and Fodder.—The BERRIES are eaten by Natives in certain parts of the Panjáb The PLANT is browsed by camels, goats, &c (*Stewart*). FOOD and FODDER. Berries. 213 (601) Plant. 214 (602)

Domestic.—The BRANCHES are used for fuel, and are made into wattled frames for the wall of huts (*Stewart*). DOMESTIC. Branches. 215 (603)

L. ruthenicum, *Murray; Fl. Br. Ind., IV., 241.* 216 (604)

Syn.—L. TATARICUM, *Pall.;* L. ARMATUM, *Griff.;* L. GLAUCUM, *Miers.*

Vern.—*Khichar, khitsar, kitserma,* LADAK.

References.—*Griff., Ic. Pl. Asiat., t. 415; Brandis, For. Fl., 346; Stewart, Pb. Pl., 157; Aitch., Bot. Afgh. Del. Comm., 91.*

Habitat.—A small, thorny shrub, of North Kashmír and Western Tibet; distributed to Afghánistán, Persia, and Central Asia.

Food.—The FRUIT, which is of the shape and size of a large pea, and deep reddish-purple when ripe, is eaten by Natives, but though sweet, is mawkish and flavourless. FOOD. Fruit. 217 (605)

Lycoperdon gemmatum, *Batsch;* THE PUFF BALL; *Kúnka,* PB.; FUNGI; see Vol. III., 455.

LYCOPERSICUM, *Mill.; Gen. Pl., II., 888.*

(606) 218 **Lycopersicum esculentum,** *Mill.; Fl. Br. Ind., IV., 237;* SOLANACEÆ.
THE TOMATO OR LOVE-APPLE.

Syn.—L. CERASIFORME, *Dunal.;* SOLANUM LYCOPERSICUM, *Linn.;* S. HUMBOLDTII, *Willd.;* S. PSEUDOLYCOPERSICUM, *Jacq.*

Vern.—*Gur-begun, tímoti, tamati,* HIND., BENG.; *Belati bengana,* ASSAM; *Wiláyati baigan,* N.-W. P.; *Bhatte,* PB.; *Wal-wangi,* BOMB.; *Vel-vángi, tamáte,* MAR.; *Viláyti vengan,* GUZ.; *Simie-takalie-palam,* TAM.; *Chhapar-badne,* KAN.; *Tamátie,* MALAY.; *Ka-yam-my-pong,* BURM.; *Maha-rata-tamátie,* SING.

References.—*Roxb., Fl. Ind., Ed. C.B.C., 190; DC., Origin Cult. Pl., 290; Mason, Burma and Its People, 471, 798; Year Book Pharm., 1873, 85; Atkinson, Him. Dist., 703; Econ. Prod., N.-W. P., V., 13, 19; Lisboa, U. Pl. Bomb., 167; Birdwood, Bomb. Pr., 170; Balfour, Cyclop., II., 752; Smith, Dic., 414; Agri.-Hort. Soc. of Ind.;—Transactions, III., 10, 197 (Pro.), 227, 239, 284; VI. (Pro.), 36; VII. (Pro.), 116; Journal (Old Series), IX. (Sel.), 55; X., 91; (New Series), IV., 38; V., 35, 44; Gazetteers:—Mysore and Coorg, I., 63; Bombay, VII., 40; Settle. Repts.:—Chanda, 82; Simla, 41.*

Habitat.—A trailing plant introduced from South America, and cultivated in many parts of India for its large red, or sometimes yellow, fruit. It is also found in many localities as an escape from cultivation. In the plains the seed is sown in autumn, and the fruit ripens during winter and spring. In the hills the plant grows more luxuriantly; and bears fruit throughout the summer and autumn months.

FOOD. Fruit. (607) 219 **Food.**—Natives are beginning to appreciate the FRUIT, but the plant is still chiefly cultivated for the European population. Bengalis and Burmans use it in their sour curries.

LYCOPODIUM, *Linn.; Baillon, Bot. Med. Crypt., 28.*

[LYCOPODIACEÆ.

(608) 220 **Lycopodium clavatum,** *Linn.; Bent. & Trim., Med. Pl., 299;*
THE CLUB-MOSS.

Syn.—L. INFLEXUM, *Sw.*

References.—*Baillon, Bot. Med. Crypt., 30; Flückiger and Hanb., Pharmacog., 731; Smith, Dic., 121; Balfour, Cyclop., II., 752.*

Habitat.—This plant, the common club-moss, is almost cosmopolitan. being found in the temperate and colder regions of both hemispheres. It occurs in hilly districts throughout Europe, Northern Asia, and North America, extending to within the Arctic circle and to the Himálaya. It also grows in Australia, the Cape of Good Hope, the Falkland Islands, Madagascar, Java, Japan, and tropical America.

MEDICINE. Spores. (609) 221 **Medicine.**—The minute SPORES, shaken out of the mature sporangia or capsules, form a light yellow powder which has been used in pharmacy since the seventeenth century. This powder is, however, not now regarded as possessing any medicinal virtue, and is employed externally only, for dusting excoriated surfaces and for placing in pill boxes to prevent the mutual adhesion of pills (*Pharmacographia*). In former times the spores and
Herb. (610) 222 the HERB were administered internally in retention of urine, plica polonica, and calculous complaints. Of late years it has been again coming into notice for similar diseases, and also as a remedy for dyspepsia, constipation with flatulence, hepatic congestion, and pustular skin eruptions.

CHEMICAL COMPOSITION.—The SPORES contain 47 per cent. of a bland, liquid, fixed oil, which does not solidify even at 15°C. By distilling **Lycopodium** with or without an alkali, **Stenhouse** obtained volatile bases in very small proportion (*Pharmacographia*). Certain allied species contain such a large amount of alumina in their ash, that they are employed as mordants in dyeing (*Baillon*).

Domestic, &c.—The chief use of **Lycopodium** is in the manufacture of fire-works and for producing artificial lightning at theatres. DOMESTIC. 223 (611)

LYCOPUS, *Tourn.; Gen. Pl., II., 1183.*

Lycopus europæus, *Linn.; Fl. Br. Ind., IV., 648;* LABIATÆ. 224 (612)

THE GIPSY-WORT.

Var.—exaltata,—L. EXALTATUS, *Linn.*

Vern.—*Gandamgúndú*, bazár plant=*jalním*, KASHMIR.

References.—*Stewart, Pb. Pl., 168; Gazetteer, N.-W. P., IV., lxxvi.*

Habitat.—A small, perennial herb found in marshy ground on the Western Himálaya, between altitudes of 1,000 and 6,000 feet; distributed to Europe, West, North, and Central Asia.

Medicine.—Part of the PLANT is sold in bazárs in Northern India, under the above name, as a cooling drug (*Stewart*). Surgeon-Major Calthrop reports that the LEAVES are used externally as a poultice to cleanse foul wounds. MEDICINE Plant. 225 (613) Leaves. 226 (614)

LYSIMACHIA, *Linn.; Gen. Pl., II., 635.*

Lysimachia candida, *Lindl.; Linn. Soc. Jour., XXV., 48;* PRIMU-[LACEÆ. 227 (615)

Syn.—L. SAMOLINA, *Hance.*

Habitat.—Found in the valley of Manipur in rice fields, between 2,000 and 3,000 feet in altitude (*Watt*).

Food.—Eaten by the Manipuris as a pot-herb along with fish (*G. Watt*). This fact is of considerable interest, since before Dr. Watt made the observation that this species was regularly used as a vegetable, no member of the PRIMULACEÆ was known to be edible. FOOD. Plant. 228 (616)

Lytta.—A genus of Coleopterous insects, certain species of which may be employed as substitutes for Cantharides (see Mylabris, p. 309).

(G. Watt.)

MABA, *J. R. & G. Forst.; Gen. Pl., II., 664.*

[*Fl. Br. Ind., III., 551;* EBENACEÆ.

1 **Maba andamanica,** *Kurz; Jour. As. Soc. Beng., 1876, Pt. II., 138;*

References.—*Kurz, For. Fl. Burm., II., 140.*

Habitat.—A shrub, 3 to 5 feet in height; frequently met with in the moister upper mixed forests of the Andaman and Nicobar islands.

TIMBER. 2 **Structure of the Wood.**—Bluish-grey. Weight 49lb per cubic foot.

3 **M. buxifolia,** *Pers.; Fl. Br. Ind., III., 551.*

Roxburgh translates the Tamil name into IRON WOOD.

Syn.—FERREOLA BUXIFOLIA, *Roxb.;* MABA EBENUS, *Wight, Ic., tt. 1228, 1229;* M. NEILGHERRENSIS, *Wight, Ill., t. 148.*

Vern.—*Gua koli, pisina,* URIYA; *Iramballi, eruvalli, humbilli,* TAM.; *Nella-madi, pishina, uti chettu, pisinika,* TEL.; *Mépyoung,* BURM.; *Kalu-habaraliya (Thuvarai, irumpalai,* TAMIL), SING.

References.—*Roxb., Fl. Ind., Ed. C.B.C., 724; Kurz, For. Fl. Burm., ii., 139; Pegu Prel. Rept., 86; Beddome, Fl. Sylv., cxlviii.; Gamble, Man. Timb., 247, xxv.; Trimen, Cat. Ceylon Pl., 51; Mason, Burma and Its People, 543, 782; Sir W. Elliot, Fl. Andh., 153, 187; Drury, U. Pl., 824; Indian Forester:—III., 237; VII., 127; X., 31; XII., 313; Man. Trichinopoly, 78; Gaz., Orissa, II., 181; also App. VI.*

Habitat.—A bush or small tree found in the Southern Deccan Peninsula, the Circar Mountains to Orissa, the Malay Peninsula, from Pegu to Malacca and distributed to Ceylon, North Australia, the Philippines, and Tropical Africa and Madagascar.

FOOD. Berries. 4 **Food.**—It flowers during the hot season, and produces BERRIES which, when ripe, are generally eaten by the natives, and are said to taste well. It is mentioned by **Shortt** in his list of famine foods of Madras.

TIMBER. 5 **Structure of the Wood**—**Roxburgh** says, "the wood is dark-coloured, remarkably hard, and durable It is employed for such uses, when its size will admit, as require the most durable heavy wood." This opinion is also given by **Moore** in his *Manual of Trichinopoly*, but **Gamble** remarks that the wood is "greyish white, moderately hard:" "the bush in Orissa," he adds, "is very common on poor soils" **Mason** speaks of it as a hard, tough, knotty wood which the Tavoyers select for anchors to their large boats. **Kurz** says, the "wood is dark-coloured, hard, and durable." In a report of the Forests of Ceylon by **F. D'A. Vincent**, Deputy Conservator of Forests, Madras, this tree is mentioned in a list of the undergrowth of the dry zone where it is known as *tuvere*. It is also alluded to as met with on similar soils in the Upper Godávari and Mudumalai forests.

6 **M. nigrescens,** *Dalz.; Fl. Br. Ind., III., 551.*

Vern.—*Kari,* KONKAN; *Kúla jhád,* MAR.; according to Dalzell & Gibson this is also known in Southern Bombay Forests as *Raktrúra*, a name, however, which Lisboa remarks is given to several other trees also.

References.—*Dalz. & Gibs., Bomb. Fl., 142; Bomb. Gaz. (Kanara), XV., Pt. I., 72, 437; Useful Pl., Bomb., 95 (incorrectly reduced to Maba buxifolia), 348.*

Habitat.—A small tree of the Western Gháts from the Konkan to Mysore. **Mr. Talbot** specially mentions it as occurring near the falls of Gersaffa and elsewhere in North Kánara.

FOOD. Berries. 7 **Food.**—**Mr. Lisboa** affirms that the BERRIES are eaten, but he is probably alluding to the more peculiarly South Indian species **M. buxifolia,** of which the berries are certainly eaten.

TIMBER 8 **Structure of the Wood.**—In the *Konkan Gazetteer* it is stated of this timber that it is small but remarkably hard and strong, and particularly prized for round rafters in native houses.

MACARANGA, *Thouars; Gen. Pl., III., 320.*

[EUPHORBIACEÆ.

Macaranga denticulata, *Müll. Arg.; Fl. Br. Ind., V., 446;* 9

Syn.—MACARANGA GUMMIFLORA, *Muell. Arg.;* MAPPA DENTICULATA, *Blume;* M. GUMMIFLORA, *Miq.;* M. TRUNCATA, *Muell. Arg.;* M. WALLICHII, *Baill;* and M. PANICULATA, *Wall.*

Vern.—*Burua,* CHITTAGONG; *Pawaing,* MAGH.; *Chakro,* GARO; *Jogi, mallata,* NEPAL; *Numro,* LEPCHA; *Toung-kpek-wan, taung-petwan,* BURM. Gamble distinguishes **M. gummiflora** as *Jogi mallata* in Nepalese and **M. denticulata** as simply *Mallata.*

References.—*Kurz, For. Fl. Burm., II., 287; Gamble, Man. Timb., 363.*

Habitat.—A small evergreen tree, often gregarious, found in the Sikkim hills, from 3,000 to 6,000 feet, but chiefly on old clearings. Also found in Assam, Khasia hills, Chittagong, and Burma.

Resin.—Mr. **J. N. Pickard** (of the Forest Department, Burma) reports that this yields a red resin, a fact also alluded to by **Kurz**; but no mention is made of the properties or uses of the substance. It is probable that the resins obtained from all the species of **Macaranga** are similar to that described below under **M. Roxburghii.** RESIN. 10

Structure of the Wood.—A fast growing tree, showing three rings per inch of radius, and attaining a height of 40 feet in ten years. Weight about 22 to 29lb a cubic foot. **Gamble** says that it is much used for fencing and temporary huts. **Kurz** remarks that the wood is red-brown, adapted for cabinet work. TIMBER. 11

M. indica, *Wight; Fl. Br. Ind., V., 446; Ic., tt. 1883 & 1949, f. 2.* 12

Syn.—MACARANGA FLEXUOSA, *Wight, Ic., t. 1909, f. iii.;* TREWIA HERNANDIFOLIA, *Roth.*

Vern.—*Boura,* BENG.; *Modula,* ASSAM; *Laikeza,* MICHI; *Lal mallata,* NEPAL; *Chánda,* MAR.

References.—*Kurz, For. Fl. Burm., II., 387; Beddome, Fl. Sylv., t., 287; Gamble, Man. Timb., 363; Indian Forester, I., 93; Drury, U. Pl., 284; Lisboa, U. Pl. Bomb., 124.*

Habitat.—An evergreen tree 50 to 60 feet high: found in Sikkim up to 3,000 feet, the Mishmi and Khasia hills, Western Gháts, and Andaman islands.

Resin.—Exudes a red resin similar to that of **M. denticulata** (*Kurz*). RESIN. 13

Structure of the Wood.—Grey, moderately hard: weight 33lb a cubic foot. TIMBER.

M. Roxburghii, *Wight; Fl Br. Ind., V., 448; Ic., t. 1949, f. 4.* 15

Syn.—MACARANGA WIGHTIANA, *Baill.;* M. TOMENTOSA, *Wight;* MAPPA PELTATA, *Wight (according to Sir W. Elliot),* and apparently also OSYRIS PELTATA, *Roxb.*

Vern.—*Chandkal,* KANARA; *Chándwar, chandora,* MAR.; *Vatte kanni,* TAM.; *Boddi chettu* (? Elliot), TEL.; *Upligi, upalkai, kanchupranthi,* KAN.; *Chenthakanni,* MYSORE; *Bukenda,* SING.

References.—*Roxb., Fl. Ind., Ed. C.B.C., 712; Gamble, Man. Timb., 362; Elliot, Fl. Andh., 29; Dymock, Mat. Med. W. Ind., 2nd Ed., 689; Jury Reports, Madras Exhibition; Drury, U. Pl., 284; Cooke, Gums and Gum-resins, 34-35; Indian Forester, X., 33; Gazetteers:—Bombay XV., Pt. I., 72; Mysore and Coorg, I., 52.*

Habitat.—A small resinous tree found in the Deccan Peninsula, from North Kánara to Travancore, and on the Circars. Distributed to Ceylon.

The young shoots and fruits are covered with an adhesive reddish secretion which has an odour of turpentine. This fact, mentioned by many writers on **Macaranga Roxburghii,** agrees admirably with **Roxburgh's** description of his somewhat doubtful plant **Osyris peltata.**

GUM. 16 **Gum.**—**Cooke**, in his account of the gums and resins formerly in the India Museum, quoting **Drury** apparently, says of this plant: "A gummy substance exudes from the cut branches and base of the petioles of these trees. It is of a light crimson colour, and has been used for taking impressions of leaves, coins, medallions, &c. When the gum is pure and carefully prepared, the impressions are as sharp as those of sulphur without its brittleness. Powdered and made into a paste it is reckoned a good external application for venereal sores." "The gum called *vutta thamary*, commented on in Madras jury reports, is the same substance: this is described as a simple pure gum of a crimson colour, exhibited from Travancore, and used for taking impressions of leaves, coins, medallions, &c." This same fact is alluded to in the *Mysore and Coorg Gazetteer*.

MEDICINE. Gum. 17 Plant. 18 **Medicine.**—In addition to the remedial properties assigned to the GUM in the above passage, the PLANT generally is used as a medicine. **Dr. Dymock**, for example, says: "The country people used the following in *jarandi* (*Angl.*, Liver):—One part of the young shoots, with three parts of the young shoots of *khoréti* (**Ficus asperrima**) are sprinkled with hot water and the juice extracted; in this is rubbed down two parts each of the barks of both trees. The preparation may be administered twice a day in doses of $\frac{1}{8}$th of a seer."

FOOD. Fruit. 19 **Food.**—Several writers allude to the FRUIT of this plant as having been eaten in times of famine.

MANURE. Leaves. 20 **Manure.**—**Drury** says: "The LEAVES afford a good manure for rice fields, and are much used for that purpose. Coffee-trees thrive well if planted under the shade of these trees, as the fallen leaves, which are large, enrich the soil."

Macaroni, see **Triticum** Vol. VI.; also **Vermecelli,** Vol. VI.

Macassar Oil, see **Carthamus tinctorius,** *Linn.*, COMPOSITÆ; Vol. II., 194.

Mace, see **Myristica fragrans,** *Houtt.*, MYRISTICEÆ; p. 311.

21 **MACHILUS,** *Nees; Gen. Pl., III., 156.*

The *Flora of British India* describes some ten species of Laurel as belonging to this genus. With the exception of M. **odoratissima,** none of the others appear to be of much importance. **M. edulis,** *King* (**Phœbe attenuata,** in **Gamble's** *Man. of Timbers, 308*) yields an edible fruit known in Dárjíling as *lepcha phal*, the tree being called *dudri* in Nepal and *phani* by the Lepchas. The wood of that species is much used in Dárjíling for tea chests. **M. macrantha,** *Nees,* is a fairly abundant tree from the Konkan southwards. It is known as *gumára* in the Konkan forests. It should thus be observed that **Phœbe attenuata,** *Nees,* as given in the *Flora of British India,* is a distinct plant from the *lepcha phal* described by **Gamble.**

22 **Machilus odoratissima,** *Nees; Fl Br. Ind., V., 139;* LAURINEÆ.

Syn.—MACHILUS INDICA, *Kurz;* M. RIMOSA, *Bl.;* LAURUS ODORATISSIMA, *Wall;* L. INDICA, *Lour.*

Vern.—*Dalchini, mith-patta, prora, badror, leddil, kálban, cháu, táura, chandna, shalanglú, múkrú, bajhol, shír,* PB; *Rare,* HAZARA; *Kawala;* HIND.; *Kawala, lali, jagrikat,* NEPAL; *Phamlet,* LEPCHA; *Súm,* ASS. *Dingpingwait,* KHASIA.

References.—*Brandis, For. Fl., 378; Kurz, For. Fl. Burm., II., 291; Gamble, Man. Timb., 308; Stewart, Pb. Pl., 188; Mueller, Select Extra-Tropical Plants, 7th Ed., 230; Atkinson, Him. Dist., 316; Indian Forester:—I., 95-99; III., 189; V., 35, 212; VI., 125; VIII., 404; IX., 359; XI., 355; XII., 286, 454; Gazetteer:—Gurdáspur Dist., 55; Kew Off. Guide to the Mus. of Ec. Bot., 112.*

Habitat.—A large tree of the outer sub-tropical and temperate Himálaya, from Marí eastwards; ascending to 8,000 feet. Also found in the

Khásia hills and Burma. In Assam it grows gregariously, in large forests, and is used for feeding the silkworms (**Antheræa assama**) which give the *muga* or *múnga* silk.

Fodder.—Goats even will not eat the LEAVES of this plant, but in Assam, as already remarked, they are extensively employed to rear the *muga* silk worms on. FODDER. Leaves. 23

Structure of the Wood.—Grey, darkening, and turning red on exposure, soft to moderately hard, even-grained. Weight 40lb per cubic foot. Used in Darjeeling, where it is very common, for building, chiefly for native houses. Also for tea-boxes (*Gamble*). TIMBER. 24

Machinery, Woods used for—see **Agricultural Implements**, Vol. I., 145.

MACLURA, *Nutt.; Gen. Pl., III., 363.*

Maclura tinctória, *D. Don;* URTICACEÆ. 25

THE FUSTIC.

References.—*Crookes, Hand-book of Dyeing, &c., 349, 405-406, 411, 672; Hummel, Dyeing of Textile Fabrics, 359, 364.*

Habitat.—A native of the West Indies, and of Central and South America; introduced into India.

Dye.—See the works quoted above. It can hardly be said to be as yet of any value in India. DYE. Wood. 26

Structure of the Wood.—Orange-yellow; hard. Weight 53lb per cubic foot. Used as a yellow dye and also in producing green shades. TIMBER. 27

MACROCHLOA, *Kunth; Gen. Pl., III., 1141.* 28

According to Bentham this, at most, constitutes a sub-genus of **Stipa** and includes **S. tenacissima,** *Linn.*, and **S arenaria,** *Boot.* (*Linn. Soc. Jour., Vol. XIX., 80*).

Macrochloa tenacissima, *Kunth;* GRAMINEÆ. 29

Syn.—STIPA TENACISSIMA, *Linn.*

References.—*Kew Report, 1877, 37; 1879, 33; 1880, 52.*

Habitat.—A rush-like grass, which grows plentifully on the sandy tracts of the Mediterranean coast, especially in Spain, Algeria, Morocco, and the Sahara. This is the true ESPARTO GRASS, which, from remotest times, has been used for making hats, mats, baskets, chairs, agricultural ropes, &c., and in which an immense trade has recently arisen for the manufacture of paper. **Saccharum ciliare,** *Anders.* (the *Munj*), has long been used for cordage, and forms a strong and useful rope, much employed by boatmen in the North-West Provinces. Perhaps the best Indian substitute for ESPARTO is, however, **Ischœmum angustifolium,** which see: but **Eriophorum comosum** (Vol. III., 266) and **Lygeum Spartum** are also so employed. It is admitted, however, by paper-makers that it will not pay to export Indian grasses nor to produce in India a bark fibre intended for competition with ESPARTO.

MACROPANAX, *Miq.; Gen. Pl., I., 945.*

Macropanax undulatum, *Seem.; Fl. Br. Ind., II., 738;* ARALIACEÆ. 30

Vern.—*Chinia,* NEPAL; *Prongsam,* LEPCHA.

Habitat.—A moderate-sized evergreen tree of the Eastern Himálaya, from 500 up to 5,000 feet: also common in Assam and Sylhet.

Structure of the Wood.—Soft, yellowish-white, even-grained. Weigh 30lb per cubic foot. TIMBER. 31

MACROTOMIA, *DC.; Gen. Pl., II., 862.*

32 **Macrotomia Benthami,** *DC.; Fl. Br. Ind., IV., 177;* BORAGINEÆ.

Vern.—It seems probable, as suggested by **Stewart**, that part of the *Gao-zabán* of the bazárs may be this plant. (See **Echium, sp.**, Vol. III., 200.)

Habitat.—Found at 10,000 to 13,000 feet on the Pir Panjál, and on the Western Himálaya frequent, from Kumáon to Kashmír.

MEDICINE. Plant. 33 **Medicine.**—The PLANT is considered useful in diseases of the tongue and throat (*Stewart*).

34 **M. perennis,** *Boiss.; Fl. Br. Ind., IV., 177.*

Habitat.—A herbaceous plant, met with in the Alpine Western Himálaya and Tibet, at altitudes of 10,000 to 14,000 feet; from the Karakoram and Kashmír to Kumáon.

MEDICINE. Root. 35 **Medicine.**—**Stewart**, speaking of this plant and of **Onosma echioides**, says: "The bruised ROOT of one or other or both is locally applied to eruptions, and is sent to the plains as the officinal *rattanjot* (see **Potentilla nepalensis**), which is also employed in dyeing wool. **Royle** assigned *ratanjot* to **Lithuspermum vestitum** (=**Arnebia hispidissima,** *DC.*). In Lahoul, Spiti, and Kanáwar, it is used by the Lamas to stain images, and as a red dye for cloth, being applied with *ghi* or the acid of apricots." **Stewart** then adds that the leaves of **Onosma** "appear to be most of the officinal *gao-zaban*" (see **Onosma,** Vol. V.). It may be remarked that the habitat of the two plants mentioned above should serve to prevent con-fusion. **Macrotomia perennis** commences at the altitudes at which **Onosma** disappears, as the latter ascends from 5,000 to about 9,000 feet above the sea. **Surgeon-Major J. E. T. Aitchison, C.I.E.**, informs the writer that in Afghánistán he found this and the next species used to relieve toothache and earache.

36 **M. speciosa,** *Aitch. et Hemsl.; Linn. Soc. Jour., XVIII., 81; XIX., 179.*

Habitat.—Hills above Kaiwás on exposed ridges from 9,000 to 12,000 feet, flowering in July.

MEDICINE. **Medicine.**—See the remark under **M. perennis.**

37 **Madar,** see **Calotropis gigantea,** *R. Br.;* ASCLEPIADEÆ; Vol. II., 34-49.

Madder, Indian, see **Rubia cordifolia,** *Linn.;* RUBIACEÆ; Vol. VI.

MÆSA, *Forsk.; Gen. Pl., II., 641.*

38 **Mæsa argentea,** *Wall.; Fl. Br. Ind., III., 510;* MYRSINEÆ.

Syn.—BÆOBOTRYS ARGENTEA, *Wall.*

Vern.—*Phusera, gogsa,* HIND.

References.—*Brandis, For. Fl., 283; Gamble, Man. Timb., 238; Atkinson, Econ. Prod., N.-W. P., Pt. V., 77; Himálayan Dist., 313.*

Habitat.—A large shrub of the outer Himálaya, found from Garhwal and Kumáon to East Nepal, at altitudes from 3,000 to 7,000 feet.

FOOD. Fruit. 39 **Food.**—Produces FRUIT which is larger than those of the other Indian species, and is eaten by the hill tribes.

40 **M. indica,** *Wall.; Fl. Br. Ind., III., 509; Wight, Ic., t. 1206.*

Syn.—M. MONTANA, *A. DC.* in part; BÆOBOTRYS INDICA, *Roxb.*; B. NEMORALIS, *Roxb.*

Vern.—*Ramjani,* BENG.; *Tamomban,* MAGH.; *Malmúriya,* SYLHET; *Bilauni,* NEPAL; *Purmo,* LEPCHA; *Phadupjoh,* MICHI; *Kalsís,* KUMAON; *Atki,* BOMB.; *Mata-bimbiya,* SING.

References.—*Roxb., Fl. Ind., Ed. C.B.C., 187, 188; Kurz, For. Fl. Burm., II., 99; Gamble, Man. Timb., 238, 239; Thwaites, En. Ceylon Pl., 172; Trimen, Cat. Ceylon Pl., 50; Atkinson, Him. Dist., 313; Lisboa, U. Pl. Bomb., 272; Bombay Gazetteer (Kánara), XV., Pt. I., 437.*

Habitat.—An evergreen, gregarious shrub or small tree, met with throughout India at altitudes up to 6,000 feet; common in the North-East Himálaya, Eastern Bengal, Manipur, Chittagong, and Burma. A specimen of this species has recently been sent to the writer by **Mr. A. L. McIntire**, of the Forest Department, collected in the Simla District. This is believed to be the most westerly Himálayan habitat hitherto recorded. The form known as **Perottetiana**, *A. DC.* (now reduced to a variety of **M. indica**) is found on the Nilghiri hills, and the variety **maxima** occurs in Assam. **Dalzell & Gibson** remark that **M indica** is very common along the Gháts. **Gamble** regards this as, next to **Artemisia**, the commonest woody plant in the Dárjíling district: in some places, especially abandoned cultivation, it forms almost alone a small coppice-like dense forest.

Food.—**Brandis** says the season of flowering varies from April to October, and that the BERRIES ripen three months later. These are eaten in Nepál. FOOD. Berries. 41

Structure of the Wood.—Growth fast (six rings to the inch); wood soft. Used only for fuel and rough house-posts. TIMBER. 42

Domestic, &c.—According to **Brandis** the LEAVES are used as a fish poison in Kánara. DOMESTIC. Leaves. 43

Mæsa macrophylla, *Wall.; Fl. Br. Ind., III., 510.* 44

Vern.—*Phusera*, KUMAON; *Bogoti*, NEPAL; *Tugom-kúng*, LEPCHA.

Habitat.—A large shrub or small tree met with in the Eastern Himálaya from Nepál to Bhútan, especially in second-growth forests.

Resin.—A resinous substance exudes on the bark being cut. RESIN. 45

Structure of the Wood. Light brown, moderately hard; the sapwood resinous. TIMBER. 46

M. rugosa, *Clarke; Fl. Br. Ind., III., 508.* 47

Habitat.—A small tree or stout shrub, with pretty white flowers; found in Sikkim between 5,000 to 7,000 feet, frequent in the upper valleys of the Teesta and Rutong.

Structure of the Wood.—Light brown, soft. The sapwood slightly resinous, annual rings marked by a dark line. TIMBER. 48

MAGNOLIA, *Linn.; Gen. Pl., I., 18.*

[MAGNOLIACEÆ.

Magnolia Campbellii; *H. f. & T.; Fl. Br. Ind, I., 41;* 49

RED MAGNOLIA.

Vern.—*Lal champ*, NEPAL; *Sigumgrip*, LEPCHA; *Pendder, patagari*, BHUTIA.

References.—*Gamble, Man. Timb., 5; List of Darjeeling Plants, 2; Indian Forester:—I., 88, 94, 98; V., 467; VIII., 404; XIII., 52.*

Habitat.—A lofty deciduous tree of Sikkim and Bhútan, from 7,000 to 10,000 feet; common also in Manipur (*Watt*); remarkable for its magnificent large pink or white flowers, which appear in April.

Structure of the Wood.—White, very soft. Weight 25℔ per cubic foot. Occasionally used for planking, but now scarce. TIMBER. 50

M. sphenocarpa, *Roxb.; Fl. Br. Ind., I., 41.* 51

Syn.—LIRIODENDRON GRANDIFLORUM, *Roxb.*; MICHELIA MACROPHYLLA, *Don.*

Vern.—*Dúli champa*, BENG. & SYLHET; *Burramtúri*, ASS.

References.—*Roxb., Fl. Ind., Ed. C.B.C., 452; Kurz, For. Fl. Burma, I., 24; Gamble, Man. Timb., 5; Agri.-Hort. Soc. Ind., Trans. V., 119; VII. (1840), 48; Journal, IV., 199; X. (Proc.), 32; (New Series) IV., 96; V., 67 & 68; VI., 35.*

Habitat.—An evergreen tree of the tropical Himálayan forests from Nepal to Assam, the Khásia hills, Chittagong, Burma: ascends to altitudes of 3,000 feet. **Masters,** in his account of the Angami Naga Hills, says, this tree is common on the lower undulations and plains immediately at the base of the mountains. Cultivated in gardens throughout the hotter parts of India—Calcutta to Mysore—on account of its sweetly-scented elegant flowers.

Mr. C. Nickels [*Jour. Agri.-Hort. Soc.* (*New Series*), *V.*, *67 and 68*] gives a useful note on the method of propagating this ornamental tree.

52 **MAGNESIA,** *Ball, in Man. Geology of India, III., 437.*

The principal sources of the salts of this metal, which are of economic value, are magnesite—the carbonate, and epsomite—the sulphate of magnesium. The metal also occurs as a silicate under three different forms—talc or steatite, meerschaum, and serpentine—which will be separately described.

53 **Magnesia,** *Mallet, Geology of India, IV. (Mineralogy), 145, 152.*

MAGNÉSIE, *Fr.*; BITTEVERDE, *Ger.*; MAGNESIA, *It.*

Vern.—Hydrated oxide = *Zahr mohra,* PERS., HIND., PB.; Sulphate = *gurm,* BHOTE.

References.—*Forbes Watson, Industrial Survey of India, II., 419, 422; Baden-Powell, Pb. Prod., 47, 99; Pharm. Ind., 339-342; Ainslie, Mat. Ind., I., 304, 629; Balfour, Cyclop., II., 769*

OCCURRENCE. 54 **Occurrence.**—According to **Ball,** epsomite, or natural sulphate of magnesia, is doubtless to be found as an efflorescence in many parts of India, since it is an occasional constituent of *reh,* but its occurrence appears to have been described only at Spiti. In that locality, **Mallet** found it as a plentiful efflorescence on black slate, in connection with considerable deposits of gypsum and arragonite. By collection and lixiviation of the fragments, he stated that a considerable supply might easily be obtained. **Mr. Mallet,** in his *Mineralogy,* (written seven years later than **Ball's** account) further states that epsomite is a frequent constituent of *reh,* and gives analyses of two samples from the Chánda district, which, by lixiviation and evaporation, afforded 16·02 and 11·86 of the salt respectively. He also mentions that the mineral occurs in considerable quantity in the Phurwalla salt mine of the Salt Range, impregnating, and efflorescing from, a bed of marl at least seven feet thick. It is also said to be found less plentifully in the other mines, in the thin seams of marl which sometimes separate the good salt layers from each other. **Dr. Warth** discovered an interesting hydrous sulphate, named *kieserite* mixed with sylvine (potassic chloride) in a lenticular deposit from two to three feet thick in the Mayo salt mines. Fifteen maunds of the mixture was collected, but no information exists of its having been met with since.

Epsomite has also been reported as occurring as an efflorescence on alluvial clay in the Nicobars Magnesite, or the carbonate of magnesia, is generally found in veins associated with other magnesian rocks such as serpentine, dolomite, &c. **Mr. Mallet** writes that "it forms innumerable veins in talcose, chloritic and hornblendic rocks over a large area in Salem, in the Madras Presidency. Associated with it are baltimorite, chalcedony, jasper, chromite, and talc. According to **Lieutenant Ochterlony,** it also occurs in Trichinopoli, Coimbatore, and Mysore, and **Newbold** says it is met with at Nellore.

Two analyses of the magnesite from Salem, made by **Prinsep,** and by **Sturneyer,** reveal that the mineral contains about 48 per cent. of magnesia. Veins of magnesite have been reported by **Blanford** as occurring in the serpentine of the Arakan range in Burma, and by **Oldham** under the same circumstances in Manipur. Impure magnesite has recently been discovered by **Captain Pogson** in the dolomitic limestones of the Happy Valley near Mussoree. This, examined by **Dr. Warth**, was found to yield 69·1 per cent. of the pure carbonate (*Mallet*).

Medicine.—The OXIDE (magnesia), the CARBONATE, and the SULPHATE are largely employed in European medicine, the two former as antacids and laxatives; the latter, as a purgative, refrigerant, and diuretic. MEDICINE. Oxide. 55 Carbonate. 56 Sulphate. 57

No record exists in the literature of India of the use of the compounds of magnesium as medicine excepting in the case of the impure hydrated oxide, *sahr-mohra*, which is employed in Northern India as a laxative, alterative and aphrodisiac. **Ainslie** talks of calcined magnesia as a useful remedy in sporadic cholera, and of the sulphate as a valuable purgative and alterative, but makes no mention of their being employed by Natives in the south of India, in parts of which the sulphate or natural epsom salts, as already stated, occurs plentifully. A recent note received from **Surgeon-Major Aitchison**, states that "a coarse natural product collected near Rúpshú, probably a sulphate of magnesia, called *gurm* by the Bhotes, is used by them as a purgative."

The salts above enumerated are all officinal in the *Pharmacopœia*, but are entirely imported for use. They are mostly artificially manufactured from sea-water, from magnesian limestones, or from magnesite, by treatment with sulphuric acid.

Industrial Use.—Besides the well known medicinal purposes to which Epsom salts are put, they are said to be also sometimes employed to give weight to cotton cloth. **Dr. McLeod**, and later **M. Sorel**, proposed to manufacture a water cement from the Salem magnesite, by mixing it, when powdered and calcined, with a solution of chloride of magnesia. INDUSTRIAL. Sulphate. 58

Mahogany, see **Swietenia Mahagoni**, MELIACEÆ; Vol. VI.

Mahonia nepalenis, *DC.*, see **Berberis nepalensis**, *Spreng.*; Vol. I., 440.

Mahwa or **Mahuá**, see **Bassia latifolia**, *Roxb.*, SAPOTACEÆ; Vol. I., 406

Maiden-hair Fern, see **Adiantum Capillus-veneris**, *Linn.*; FILICES; [Vol. I., 110.

Maize, see **Zea Mays**, *Linn.*; GRAMINEÆ; Vol. VI.

MÁJUN (vulgarly Májum). 59

Májun is an intoxicating sweetmeat prepared with *ghí* and sugar, the drug being *Bhang*, *Gánja* or *Charas* (the narcotics of **Cannabis sativa**, which see, *Vol. II., 113—118*); Opium or Poppy-seeds; or *Datura*. The preparation is flavoured with various spices, such as Cloves, Mastich, Cinnamon, Aniseed, Cummin, Cardamon, &c., &c. For further information, see the article **Narcotics**, pp. 324—326 and 328.

MALACHRA, *Linn.*; *Gen. Pl.*, *I.*, *205*.

Malachra capitata, *L.*; *Fl. Br. Ind.*, *I.*, *329*; MALVACEÆ. 60

Vern.—*Ran* or *ban-bhendi*, BOMB.

References.—*Voigt, Hort. Sub. Cal., 112; Dalz. & Gibs., Bomb. Fl., 9; Murray, Pl. and Drugs Sind, 60; Lisboa, U. Pl. Bomb., 227; Liotard, Paper-making Mat., 79, 80, 81; Report, Exp. Farm Khándesh, Bombay; Dec. 1880; Cross, Bevan & King, Report on Fibres of India, 54; Jour. Agri.-Hort. Soc. Ind., V., (new series), Pt. VI. (Proc. 1878), 25, 48; Indian Forester, VII., 179; Spons, Encyclop., 981; Balfour, Cyclop., 804; Official Corresp., R. and A. Dept. B. Proc., 1881.*

Habitat—An erect annual, with broad, heart-shaped leaves, covered with stiff hairs. The flowers are yellow or white, and form axillary or terminal heads. None of the earlier botanical writers allude to this plant, from which circumstance it is presumed to be a modern introduction. It is a native of west Tropical Africa (the Congo basin particularly), and of Tropical America. **Voigt** is the first Indian writer who makes mention of the plant, and in *his* day it would not appear to have been so plentiful as at the present time, since he simply remarks that it is domesticated about Serampore. **Graham**, and after him **Dalzell & Gibson**, affirm that the plant was introduced into Bombay from Brazil by the late **Mr. Nimmo**. It is now plentiful throughout the hotter damp tracts of India, more especially in the vicinity of Bombay.

FIBRE.

FIBRE. 61

Fibre.—It yields a fibre 8 to 9 feet long, with a silvery lustre, and almost as soft as silk. **Dr. King** reports that for paper-making this does not seem to promise much; it has, however, been experimented with in Bombay as a substitute for jute and was reported on favourably by the manager of a factory. The fibre was shown at the late Colonial and Indian Exhibition, an excellent sample having been presented by **Mr. A. B. Gupte.** At one time high hopes were entertained that Bombay would, in this fibre, possess the means of competing with the Bengal jute manufacturers. Just as the growth of the jute industry eclipsed all the other fibres that were, half a century ago, equally contesting the growing market for a cheap textile, so twenty years later, the immense popularity of the Bombay wheat trade seems to have extinguished the hopes once entertained of this fibre. The reports published regarding it are, however, unanimous, that it is little if at all inferior to jute, and that capital directed towards the cultivation and manufacture of the fibre is all that is needed to place the industry on a sound commercial basis. The efforts to introduce the cultivation of jute into Bombay may be pronounced a failure, and there would, therefore, seem every prospect of a new interest being taken in this neglected fibre. **Malachra** fibre was much admired at the Colonial and Indian Exhibition, and the experts, who examined it, were of opinion that Bombay would not much longer continue to import the sacking required for her grain trade, and that some enterprising manufacturer would soon be found willing to use this, or some other allied fibre in the preparation of gunnies for the local market. In the hope of usefully placing in the hands of the public existing information the following reprint may be here given.

Notes, &c., on the Experimental Cultivation of **Malachra capitata** *in Bombay.*

Mr. J. E. O'Conor, then Assistant Secretary to the Government of India in the Department of Revenue, Agriculture, and Commerce, on the 15th April 1878, wrote of this plant and its fibre:—

"Among the specimens of raw produce collected in Bombay for the Paris Exhibition were samples of fibre extracted from **Malachra rotundifolia** (so named in the list of Bombay contributions). Regarding that specimen, the following remarks were made in a memorandum attached to the list of contributions:

'From this plant, which grows abundantly during the rains in waste places in and near Bombay, **Dr. W. Gray, M.B.**, extracted a fine fibre and sent it to the proprietor of the Bombay Hemp and Jute Mill for examination. It was most favourably reported upon. At the beginning of the last monsoons the proprietor was, at his own request, taken to the Byculla flats in Bombay and the plant was shown to him. He remarked that the usefulness of this plant was not known, and reported that the discovery is 'a boon to Bombay.' He has subsequently taken steps in procuring fibre from this plant, and about forty day labourers were till now kept working by him at Chembive, a village near Coorla. A new industry has thus been started. This fibre perhaps deserves special attention.'

Struck with these remarks, I asked the officer who was in charge of the Bombay collections for Paris, and who had prepared the list (**Mr. F. F. Arbuthnot, C.S.,**

FIBRE.

Collector of Bombay), to be good enough to furnish me with some further information on the subject and samples of the fibre, as well as of the plant itself. Mr. Arbuthnot was not able to send me any of the fibre, the whole quantity prepared by the mill-owner referred to above after the monsoon having been already exhausted. He has mixed the fibre with jute and made gunny-bags of the mixture. Mr. Arbuthnot sent me samples of the plant, however, and the copy of a letter from the mill-owner to Dr. Gray on the subject of the fibre. The letter was to the following effect:—

'I have received through your friend, Mr. B., a sample of a nicely-cleaned new fibre taken from a jungle plant, and being new it gave me extra pleasure in testing it, and I can safely say that the new fibre is quite as good as jute. If this new fibre can be grown in quantity, it will be a great thing and a boon to this Presidency. The fibre is actually not the yarn of jute; nor do I know its true name. It seems like what the natives call *ranee bhendee*, or jungle *bhendee*; but even that is doubtful, as *bhendee* is scarcely so good as this fibre.

'The fibre is in length from 8 to 9 feet, thoroughly clear from gummy substance and dirt, has a nice silvery appearance with a peculiar lustre, and is almost as soft as silk. In passing the fibre through the machinery, damped with oil and water, as is commonly done with Bengal and Konkan jute, yarn was produced strong enough and nearly equal to that made from the second quality of Bengal jute. In the opinion of our European spinning-master, owing to the almost imperceptible difference between the yarn made from the new substance and Bengal jute, it is very suitable for weft; but if the plant is carefully grown and well looked after, the fibre would then, no doubt, rank fully equal to Bengal and Bombay Jute. Owing to the high prices ruling for jute in Bengal and elsewhere, the new fibre, if carefully prepared, would command a ready sale at R3-12 to R4 per Indian maund.' *

I submitted the specimens of the plant, and the papers received from Bombay, to Dr. King, Superintendent of the Botanical Gardens, Seebpore, who had kindly offered to identify and name the plant. He says that the plant is **Malachra capitata**, and not **M. rotundifolia**, which is found only in South America, and he adds—

'**Malachra capitata**, though probably originally a native of South America also, is now found everywhere within the tropics. It now grows sparingly about Calcutta; but, as it is not mentioned by Roxburgh in his *Flora Indica*, it appears probable that it did not occur at all here in his time. Wight and Arnott do not mention it in their *Prodromus Floræ Peninsulæ Indiæ Orientalis*, and it is, therefore, improbable that it was common in Southern India when that book was written. I should not anticipate any difficulty about growing the plant in Bengal; but whether it would yield as good a fibre or be as valuable as jute, I am quite unprepared to offer an opinion. The plant belongs to the new order MALVACEÆ. It is an erect annual (or occasionally perennial) shrub, covered everywhere with very stiff hairs. The leaves are broadly heart-shaped, almost rounded, and are borne on long stalks. The flower-heads are also carried on long cylindrical stalks which rise from the axils of the leaves. The flowers themselves are yellow or white in colour. There are about five or six of them on each head, and they are surrounded at their origin from the flower stalk by three or four half-kidney-shaped bracts. Each flower produces five seeds.'

The above is a popular description. Sir J. D. Hooker's description in the *Flora of British India* is to the same effect, though stated in botanical terminology. Sir Joseph Hooker writes of this plant (**M. capitata**) that it occurs throughout the hotter parts of India from the North-Western Provinces to the Carnatic, probably introduced.

The members of the family are usually found in marshy places within the tropics. It would seem, therefore, that this species would thrive well in Bengal; in fact, it is already common about Calcutta; and Voigt (*Hortus Suburbanus Calcuttensis*, page 112) refers to it as domesticated in his time (1841) about Serampore. Mr. Blechynden tells me that Graham, in his Catalogue of Plants in Bombay and its vicinity, alludes to **Malachra rotundifolia** as introduced by Mr. Nimmo about forty years ago.

The preparation of the fibre is the same as that of jute. When Dr. Gray operated on the fibre in Bombay, he steeped the stem in water for a week. It must be steeped when freshly cut; for, according to the experience of the mill-owner, who tried the fibre, if the stem is exposed to the sun and allowed to dry, great difficulty is felt in getting rid of the external bark, and the fibre obtained is coarse and inferior in quality.

* This letter was dated 20th July 1877.

MALACHRA capitata. **Malachra Fibre**

FIBRE.

The utilisation of this plant would be specially advantageous in the Bombay Presidency, where as yet attempts made to grow jute can hardly be said to have had any success. But the plant seems to merit attention in Bengal too. Growing as it does and flourishing without any attention in marshy soil, of which we have more than enough for all purposes in Lower Bengal, it would seem to offer the spinners and paper-makers an excellent substitute for, and addition to, jute. I believe the Bally Paper Mill Company are anxious to find some fibrous product capable of conversion into paper at low cost. Here is one to their hand, and I beg to recommend it to their attention, as well as to that of jute-spinner's and rope and twine makers.

In **Balfour's** *Class Book of Botany* (page 771) it is said that in Panama the leaves of this plant are used as an anthelmintic.

Reports obtained through the Agricultural and Horticultural Society of India on a sample of the fibre of " **Malachra capitata** *" prepared in Bombay by* **Surgeon-Major W. Gray.**

I have no doubt this fibre would prove a good substitute for jute for most purposes to which jute is applied; but it seems rather more harsh, and its spinning qualities should be tested in one of the jute-mills before giving a very decisive opinion. To ascertain whether it would prove economically a substitute for jute, we should require to know the yield of fibre per bigha or acre, and the cost of cultivation and manufacture.

(Signed) S. H. ROBINSON.

This sample is beautifully bright and clean, fair length, and good strength of staple, but somewhat harsh. I doubt if it would make a good warp yarn in itself, but mixed with good jute it would do so. The fibre for spinning is not so valuable as jute; it lacks the forked ends when broken, such as the latter possesses, and partakes of the character of the fibre known as *meshta*, which when broken looks as though it had been cut and left with square ends. Before its value as spinning fibre could be fairly assessed in competition with jute, it would be needful to show cost of production, outturn per *bigha*, &c.

(Signed) W. H. COGSWELL.

Fibre harsh.

(Signed) W. STALKART.

Extract, paragraph 3, of a letter from the Acting Collector of Tanna, No. 3298, dated 24th October 1878.

3. The seed of the **Malachra capitata**, which formed the subject of correspondence ending with Government Resolution under reference, was sown broadcast during the monsoon just over in *gurcharan* or waste land in Bandora taluka, Salsette, and in certain villages of Bassein and Kalyan talukas, and the result is reported to be a complete failure. This result is apparently attributable to the exceptionally heavy fall of rain, and I would recommend a further trial next season.

The following report was received from the Superintendent of the Victoria Gardens, Bombay, in 1879:—

'Not having any seed of the plant in stock, and none being procurable in the bazár, I gathered a quantity of young plants from the flats, as soon as they appeared after the commencement of the rains, and planted them in regular rows 3 inches apart, with a space of 1½ inches betwixt each plant. The soil was stony and loose, formed of a mixture of decayed *cutchra* and the natural soil of the garden. The plant grew very rapidly, and after three months' growth reached a height of 7 to 8 feet, each stalk perfectly straight, and measuring at the root from ¼ to ⅜ inch in diameter. After four months' growth the plants were cut down, all leaves, branches, and seeds removed, exposed to the sun for three days, and then steeped in water for five days; they were then removed stalk by stalk, and all the fleshy or pulpy part removed by pressing with the finger and thumb. The outcome is as follows: from a Bengal *bigha*, or an area of 1,600 square yards, fibre (the same as sample sent) can be produced weighing 560lb. When cultivated during the rains, the plant requires no attention whatever; but it must be grown very close together and not exposed to high winds. The cost of removing the fleshy matter from the fibre will amount to about R3 per *bigha*.'

The Superintendent, Khándesh Experimental Farm, Bombay, wrote (26th February 1879) to the Collector of Khándesh:—

'As directed in Government Resolution No. 2362 of 8th May 1878 (forwarded under your endorsement No. 2154 of 13th idem), I have the honour to report on the **Malachra capitata** (*Syn.* **rotundifolia**) as a fibre-producing plant.

2. This species of mallow was introduced into Khandesh presumably at a recent date, although there seems to be no very trustworthy information on the subject. It

FIBRE.

is known here by the name of *ran bhendee*, and the cultivators are more or less familiar with the quality of its fibres, which they sometimes extract.

3. It is found only in good soil, and makes its appearance soon after the first rain. Its leaves much resemble those of the cotton plants in its young state, and on this account often grows along with that crop for a month or more before being discovered.

4. I have conducted many experiments in the utilisation of this waste fibre as well as in the cultivation of the plant as an ordinary crop. I find that the best fibre is produced from green stems, and they are ready for cutting in the early *kharif* harvest, or in time to admit of a light *rabi* crop being sown afterwards, and that the crop does not exhaust the soil as in the case of *sunn* hemp.

5. The habit of the plant is very peculiar: it throws up in the first place a strong central stem, several feet high, furnished with a few straggling lateral shoots. The best fibre is got from an abundance of radical shoots, which remain procumbent and to a distance of 4 or 5 feet, where the ends become slightly bent upwards to admit of the thorough ripening of the cluster, or head of capsules which contain the seeds of the plant.

6. A bale of this fibre, along with seven other kinds, were forwarded to Government in accordance with the instructions contained in paragraph 8 of the Resolution No. 1881 of 7th July 1874. The whole was sent through the Secretary of State to the Dundee Chamber of Commerce, who merely recorded a general opinion on the whole batch without dealing with the merits of any particular sample.

7. The *ran bhendee* fibre is easily extracted, as the bark contains comparatively little of the gummy substance so common to many others of the fibre-producing plants.

8. Notwithstanding the large number of "miscellaneous fibre" shown at the late Mahiji Exhibition, two of the three prizes offered under this head were awarded to samples of *ran bhendee*.

9. I have had a parcel of the fibre prepared, and await your orders regarding its despatch.'"

The above compilation of available information was first published by the writer in the Selections from the Records of the Government of India, in the Revenue and Agricultural Department (1888-89), since which date no further particulars have been obtained. The fibre seems well worthy of careful investigation *1st*, as to whether it can be produced in Bombay at a price to compete with Jute; and *2nd*, whether it can be as cheaply and economically utilised.

Malabar Oil. 62

The ambiguous term 'Malabar Oil' is applied to a mixture of the oil obtained from the livers of several kinds of fish frequenting the Malabar Coast of India and the neighbourhood of Karachi. The species chiefly used belong to the genera **Ætobatis, Carcharias, Clupea, Pristis, Rhynchobatus, Silundia, Trygon,** and **Zygæna.**

See the article **Fish** in Vol. III., 363-397.

Male Fern, see the article **Ferns**, Vol. III., 323.

MALCOLMIA, *Br.; Gen. Pl., I., 77.*

[*146;* CRUCIFERÆ.

Malcolmia africana, *Br.;* and **M. strigosa**, *Boiss.; Fl. Br. Ind., I.,* 63

Vern.—*Khunserdia, patthra, páchan, chindka*, PB.

References:—*Stewart, Pb. Pl., 14; Murray, Pl. and Drugs Sind, 50; Stewart, Account of a Journey in Hazára and Khágán (Jour. Agri.-Hort. Soc. Ind., XII.); Aitchison, Rept. Botany, Afgh. Del. Com.; Lace, Report on the Vegetation of Quetta (in mss.).*

Habitat.—Small, herbaceous weeds found in the hotter parts of North India (Panjáb, Sind, Balúchistán, and Afghánistán), where, according to **Stewart**, the last mentioned species (in Hazára) is so abundant as to carpet the ground, and give it a heather-like purple hue. The first mentioned species ascends from the Panjáb plains to altitudes of 13,000 feet above the sea

FODDER. 64 **Fodder.**—These, along with a few other equally abundant cruciferous plants (principally **Lepidium Draba**), are greedily eaten by goats and sheep, and in Quetta may have much to say to the superiority of the mutton reared in that neighbourhood. **M. Bungei**, *Boiss.*, occurs in Peshín Valley (*Lace*). Of this species Aitchison reports that in the Hari-rud Valley it is in so great abundance, on gravelly soil, as to give a bright colour to the country. Mr. Lace states that in Peshín it is an important fodder plant, and doubtless this is the case also in Afghánistán.

65 MALLOTUS, *Lour.; Gen. Pl., III., 319.*

A genus of trees or shrubs, comprising some 70 species, natives of the tropical areas of the Old World. The arborescent forms yield soft, light, woods of no great economic value. In India there are 45 species: 2 of these occur throughout the hotter parts of the continent; 9 in Madras; 5 along the foot of the Himálaya; 6 in Assam; 15 in Burma; 4 in the Andaman islands; 7 in Ceylon; and 22 in the Malay Peninsula—Burma to Singapore. The following are the more important species.

[*1873;* EUPHORBIACEÆ.

66 **Mallotus albus,** *Muell. Arg.; Fl. Br. Ind., V., 429; Wight, Ic., t.*

Syn.—M. TETRACOCCUS, *Kurz;* ROTTLERA ALBA and TETRACOCCA, *Roxb.;* R. MAPPOIDES, *Dalz.;* R. PELTATA, *Wight;* R. PANICULATA, *Wall.*

Vern.—*Marleya,* SYLHET; *Jogi mallata,* NEPAL; *Numbong,* LEPCHA.

References.—*Roxb., Fl. Ind., Ed. C.B.C., 737; Brandis, For. Fl., 444; Kurz, For. Fl. Burm., II., 383; Beddome, For. Man., 208; Gamble, Man. Timb., 361; Dalz & Gibs., Bomb. Fl., 230; Bombay Gazetteer (Kánara), XV., Pt. I., 443.*

Habitat.—A small, evergreen tree, found in Sikkim, Eastern Bengal, Assam, Chittagong, the Western Ghâts, Mysore, and Ceylon. On the Himálaya it ascends to altitudes of 3,000 feet.

TIMBER. 67 **Structure of the Wood.**—Soft, white: weight 31℔ per cubic foot.

68 **M. muricatus,** *Muell. Arg., Fl. Br. Ind., V., 437.*

Under the above name, **Gamble, Kurz,** and other writers on Indian Economic Products have described a tree which yields a grey, moderately hard timber. The *Flora of British India* (*Vol. V., 436, 437, 439*) points out that the tree of Mysore and Travancore so designated is **M. muricatus,** *Beddome:* that of Ceylon **M. Walkeræ,** *Hook. f.:* and the Andaman plant is **M. andamanicus,** *Hook. f.*, the *Duk-mouk*, of the Burmese.

69 **M. nepalensis,** *Muell. Arg.; Fl. Br. Ind., V., 428.*

Syn.—M. OREOPHILUS, *Muell. Arg.; Gamble, Man. Timb., 362.*

Vern.—*Numbúngkor,* LEPCHA; *Safed mallata,* NEPAL.

Habitat.—A small tree of the Central and Eastern Himálaya: Nepál, Sikkim, altitude 5,000 to 7,000 feet, Khásia Hills 4,000 to 5,000 feet.

TIMBER. 70 **Structure of the Wood.**—White, soft; growth moderately fast, five rings to the inch of radius (*Gamble*).

71 **M. philippinensis,** *Muell. Arg.; Fl. Br. Ind., V., 442.*

THE MONKEY FACE TREE.

Syn.—ROTTLERA TINCTORIA, *Roxb.;* R. AURANTIACA, *Hook. & Arn.;* R. AFFINIS, *Hassk.;* R. MONTANA and MOLLIS, *Wall.;* CROTON PHILIPPENSIS, *Lamk.;* C. PUNCTATUS, *Retz.;* C. COCCINEUS, *Vahl.;* C. MONTANUS, *Willd.;* C. DISTANS, *Wall.;* C. CASCARILLOIDES, *Rauesch.*

Vern.—*Kambilá, kamúd, kamalá, kamilá, kameló, rúin, rúlú, kambhal, wussantha-ganda* (powder), HIND.; *Rori* (LOHARDUGGA); *Dhola sindur,* (BIRBHUM); *Sinduri* (DARJILING); *Kamilá, túng* (*kishur* or *késar*=saffron), *kamaláguri* (the dye powder), *kamalágundi,* BENG.; *Kumala, súndragundi, bosonto-gundi,* URIYA; *Rora,* SANTAL; *Gangai, puddum, jaggarú,* ASSAM; *Chinderpang, machugan,* GARO; *Sinduria, safed mallata,* NEPAL; *Puroa, tukla, numboongkor,* LEPCHA; *Baraiburi, sin-*

durpong, MICHI; *Koku*, GOND; *Reoni, roli, kamela* (BANDA); *Rúinia kamela*, (BIJNOUR); *Rori* (BUNDELKHUND); *Sindúria, puroahung*, N.-W. P.; *Rohni*, OUDH; *Rúen, riúna, roli, rauni, rerú*, KUMAON; *Kaimbil*, KASHMIR; *Kamela, kamal, kambal, kúmila, reini, reun, rúlyá*, PB.; *Kámbaila* (PESHAWAR), PUSHTU; *Rauni, rori, chamar gular, ningur*, C. P.; *Kúkú* (MELGHAT), BERARS; *Shendri, kapela, kamala*, BOMB.; *Shendri, shindur*, MAR.; *Kapilo*, GUZ.; *Kapli, kapila, kamela-mávu* (*? pod* =pollen), TAM.; *Kúnkúma, kápila, vassuntagunda* (powder), *chendrasinduri*, TEL.; *Kurku, rangamále, corunga-manje, sarnakasari, hulichellu, kunkuma*, KAN.; *Ponnagam* (? **Calophyllum inophyllum**), MALAY.; *Taw-tee-cteng, tan-thie-den*, BURM.; *Tawthadin*, SHAN; *Hamparandella*, SING.; *Kapila, kampilla* (the red mealy powder), *rechanaka* (*Punnaga* is incorrectly given in many books as Sanskrit for this plant,—see **Calophyllum inophyllum**), SANS.; *Kinbíl* (a word derived from the Sanskrit and now restricted in India to this plant), ARAB.; *Kanbélá*, PERS.

References.—*Roxb., Fl. Ind., Ed. C.B.C., 737; Brandis, For. Fl., 444; Kurz, For. Fl. Burm., II., 381; Beddome, Fl. Sylv., t. 289; Gamble Man. Timb., 361; Dalz. & Gibs., Bomb. Fl., 230; Stewart, Pb. Pl., 197; Grah., Cat. Bomb. Pl., 184; Mason, Burma and Its People, 512, 543, 761; Benth., Fl. Hongk., 307; Hooker, Him. Jour., I., 315; Miquel, Fl. Ind Bat. Suppl., 454; Rheede, Hort. Mal., V., t. 21, 24; Elliot, Flora Andhrica, 36, 86, 95, 96, 98, 103, 151, 168, 185, 189, 190; Pharm. Ind., 202; British Pharm., 167; Moodeen Sheriff, Supp. Pharm. Ind., 170; U. C. Dutt, Mat. Med. Hind., 232, 302; Dymock, Mat. Med. W. Ind., 2nd Ed., 709; Flück. & Hanb., Pharmacog., 572-576; U. S. Dispens., 15th Ed., 828; Bent. & Trim., Med. Pl., IV., t. 236; Murray, Pl. and Drugs, Sind, 34; Med. Top. Ajm., 142; Irvine, Mat. Med. Pat., 48; Buchanan, Journey through Mysore, Canara, &c., I., 168, 204, 211; II., 343; Baden Powell, Pb. Pr., 376; Atkinson, Him. Dist., 776; Drury, U. Pl., 285; Lisboa, U. Pl. Bomb., 122, 248, 258, 268, 275; Birdwood, Bomb. Pr., 78, 301, 336; Royle, Ill. Him. Bot., 329; McCann, Dyes and Tans, Beng., 18-20; Buck, Dyes and Tans, N.-W. P., 31; Liotard, Dyes, 56-58, 89, 90, 123; Wardle, Report on India Dyes, p. 5; Darrah, Note on Cotton in Assam, 33; Aplin, Report on the Shan States, 1887-88; Prof. Hummel, (Special Report in connection with Colonial and Indian Exhib.); Kew Reports, 81, 50; Kew Off. Guide to the Mus. of Ec. Bot., 119; For. Admn. Report Chutia-Nagpore, 1885, 34; Bomb. Gaz., XI., 25; XV., i., 75; Lamk, Encycl., II., 206; Indian Forester, Vol. III., 24, 204; Vol. IV., 230, 318, 323; VIII., 106, 119; IX., 413; X., 33, 325; XI., 367; XII., 188 (XXII.); XIII., 121; Gazetteers:—N. W. P., I., 34; II., 173; IV., 77; X., 317; Bombay, XV., 72; XVII., 26; Panjáb, Hoshiárpur District, 10; Ráwalpindi District, 15, 83; Peshâwar District, 2; Mysore and Coorg, I., 48, 436; II., 7; III., 22; Hunter, Orissa, II., 179, Ap. VI.; Manual of the Cuddapah District, 200; Settlement Report of the Upper Godávery District, 38; Central Provs. (Raipur Dist.), 76; (Chanda District), Ap. VI.; Agri.-Hort. Soc. (Old Series), IV., Pt. I., 210, 211; VI., Pt. I., 27; VIII. Sel., 178; XIII., 314, 353, 390, 391; XIV., 21, (New Series), I., 103; VI., Sel., 20; Honigberger, Thirty-five years in the East., 337.*

Habitat.—A small, evergreen tree, found throughout Tropical India; along the foot of the Himálaya from Kashmír eastwards (ascending to 5,000 feet); all over Bengal and Burma, Singapore, and the Andaman Islands; and from Sind southwards to Ceylon. Distributed to China, the Malay Islands, and Australia.

THE KAMÉLA DYE.

History of the Dye and Medicine.—Much has been written regarding the *Kaméla* dye (or as it is known in Europe *Kamala*). This is the powdery substance obtained as a glandular pubescence from the exterior of the fruits of **Mallotus philippinensis.** Even at the present day, however, *Kaméla* dye cannot be said to have obtained the position in European commerce, which its merits deserve. In India it is sometimes known by its Arabic name *Kinbíl* (a word derived from the Sanskrit name for the drug); but, according to certain Arabian writers, there were two kinds of HISTORY. 72

MALLOTUS philippinensis.

History of Kaméla.

HISTORY.

Kinbíl, the red Indian and the black or purple Abyssinian. It has now been ascertained that the darker coloured *Kinbíl* is the African *Wars* or *Waras* drug and dye, a substance obtained from a **Flemingia** (*see Vol. III., 400-403*). The Indian *Kinbíl* appears, in fact, to have at one time been chiefly used as an adulterant for *Wars*, and as such it first found its way to Europe. The history of the black *Kinbíl* has been more carefully worked out than the red, though what is apparently the latter is mentioned in the *Chakradatta-sangráha* and in the *Bhávaprakása* (A.D. 1535), and hence was probably known to the later Sanskrit medical writers. **Professor Wilson** was of opinion that the Arabians of the eighth century studied Hindu medicine before that of the Greeks, and that **Chakrapani Datta's** great work was translated into Arabic about A.D. 773. From that period the name *Kinbíl* probably dates. Arabian writers of an earlier period appear to allude exclusively to *Wars*, the synonym *Kinbíl* being given by commentators or translators. Thus **Paulus Aegineta** in the seventh century refers to *Wars* and **Ibn Khurdadbah**, who lived A.D. 869-885, states that from Yemen came stripe silks, ambergris, *wars*, and gum. *Kinbíl* is sometimes compared by the Arabian writers to saffron, hence in all probability the modern though inaccurate application of the vernacular name *késar* in some parts of India. This mistake was made by **Brandis, Gamble, Bentley & Trimen**, &c. In the tenth century, **Masundi** wrote of *Kinbíl* as a sandy substance of a red hue, which was useful as an anthelmintic and in the treatment of cutaneous diseases In the thirteenth century, **Kaz-wini** referred to *Wars* as a plant sown in Yemen which resembled Sesam. Most writers from that date carefully distinguish *Wars* from *Kinbíl*, and speak of the former as a dye and medicine used externally to remove freckles and taken internally to cure leprous eruptions. They seem to have understood that the red *Kinbíl* was an anthelmintic drug—a property not possessed by *Wars*. The *Makhzan el-Adwiya* makes the same distinction, but says the *Kinbíl* is the pulp of a fruit obtained from a tree growing on the mountains which has long stiff thorns. It is thus probable that **Mir Muhammad Husain** never saw either the tree or the fruits from which the *Kaméla* powder is obtained. His description of the fruit and of the powder obtained from the interior (instead of from the outer surface) is curiously enough almost identical with that incorrectly given by **Liotard, McCann**, and other popular modern writers on the dyes of India. Of *Wars*, **Mir Muhammad** remarks there is a black kind which comes from Ethiopia, called *Habshi* and a dull red kind known as Indian, the latter yielding an inferior dye. He then adds that the seeds of the *Wars* plant resemble those of *másh* (=**Phaseolus Mungo** *var.* **radiata**), a fact which recent investigation has confirmed, by the determination of *Wars* as the glandular pubescence from the pods of a **Flemingia.**

The vernacular names used in India to denote the *Kaméla* powder imply its colour, but they are the same names as are given to the disease jaundice. It is somewhat remarkable, however, that the earliest European writer who deals with **Mallotus philippinensis** should neither have heard of its dye, nor of its anthelmintic property. In 1684 **Rheede** (*Hortus Indicus Malabaricus*) described and figured the plant, but he speaks of the leaves and fruits as being made, along with honey, into a cataplasm employed in the treatment of itch and for application to the bites of snakes and other venomous animals. Similarly, the root, he adds, was made into a preparation found valuable in the treatment of contusions, since it possessed the property of dissolving coagulated blood. He does not appear to have heard of the powder, nor does he call the plant by a name in any way traceable to the names in modern use. He calls it *Ponnagam*, but, as with **Roxburgh's** *Poonag*, that name should in all probability be assigned to **Calophyllum ino-**

HISTORY.

phyllum. **Dr. Hove,** an indefatigable traveller and botanist, who visited Bombay in 1787, alludes to *Kaméla* apparently, but under another incorrect name—'*Kissury*,'—as being used to adulterate a permanent dye prepared from **Bixa Orellana** by combining the dye from the seeds with that prepared from the roots and leaves.

The reader will find (in the Selections from the Records of the Government of India, Revenue and Agricultural Department, 1888-89, pp. 53-58) an account of **Hove's** dye, and perhaps the facts there given may be found interesting, because of their bearing on the subject of *Kaméla*. Most writers speak of **Buchanan** as having been the first European writer who described *Kaméla* dye. The Diary of his Journey through Mysore, in which he gives particulars of the process then in use, is dated 1807, but **Roxburgh's** *Coromandel Plants* was published in 1798. In that work the subject is discussed in great detail, and the genus **Rottlera,** by **Roxburgh,** founded in order to isolate from **Croton** the *Kaméla* and one or two allied species. The specific name of **tinctoria** given by **Roxburgh** denotes its dye property as being then recognised, but in his *Flora Indica* he adds that " the root is said to dye red also."

The properties of both the dye and medicine are, at the present day, appreciated by the people in every province of India. It is mentioned in most of the Gazetteers and District Manuals, but the following special authors may be consulted regarding the substance :—

In Bengal, **Roxburgh, Dutt, Gamble, Liotard, McCann;** *in the North-Western Provinces,* **Brandis, Buck, Atkinson;** *in the Panjáb,* **Stewart, Baden-Powell;** *in Bombay and Sind,* **Birdwood, Murray, Dymock, Lisboa;** *in Madras,* **Elliot, Moodeen Sheriff, Bidie, Drury;** *and in Burma,* **Mason, Kurz,** &c.

The statement, made originally by **Roxburgh,** that the roots afford a red dye, has been confirmed by **Schlich** and by **Kurz.** All other writers, who allude to this subject, appear to have been compiling from **Roxburgh,** so that it would seem that the dye property of the roots is not generally known, or has, within the present century, been lost sight of, through the introduction of cheaper and more convenient dyes. A similar disappearance of knowledge has been incidentally alluded to above in connection with **Bixa Orellana**—a permanent dye having, according to **Dr. Hove,** been prepared in Bombay in 1787, through combination of the colour from the seeds with that prepared from the roots and leaves. *Kaméla,* he tells us, was employed to adulterate that permanent dye preparation from **Bixa.** But even at the present day certain writers speak of *Kaméla* as used to adulterate arnatto. It is thus probable that this practice was the outcome of the fact that the introduction of arnatto displaced *Kaméla,* the latter being continued as an adulterant only. Though much inferior to *Kaméla* in many respects, arnatto is a simpler and cheaper dye, eminently suited for temporary tinctorial purposes, such as the *Abír* of the *Holi* festival. The inference is thus probably admissible that *Kaméla* was the dye which **Buchanan** (*Statistical Account of Dinájpur*) informs us was displaced by the then recently introduced arnatto in the province of Behar. So far as the writer can discover, however, the dye property of *Kaméla* was at least not generally known in India much more than 200 years ago, and it seems highly probable that its medicinal virtues were not fully appreciated until a considerably later date. It is at least significant that painstaking investigators, such as **Ainslie, Elliot, Roxburgh, Honigberger,** and **O'Shaughnessy,** should have been ignorant of the fact (if fact it was, at the beginning and during the first half of the present century) that *Kaméla* was deemed by the people of India their most effectual anthelmintic drug. **Ainslie** does not so much as mention the plant by name, and **Elliot** and **Roxburgh** deal

History of Kaméla.

HISTORY.

with it only as a dye; the former author informs us that with the Telegu-speaking people it was in his time known as wild-arnatto (*Kondu* or *Karu-jáphara*). Throughout India there is a singular tendency to give to the dye the vernacular names for the disease jaundice (*i.e., Kápila, Kamila, Kaméla, Khamalai, &c., &c.*), a comparison doubtless suggested from the resemblance of the dye to the jaundice skin. But a comparative name, such as the above, stands every chance of being of modern origin, the more so since four or five other silk dyes are nearly always confused with the *Kaméla* of more precise modern writers. Of these may be mentioned, **Calophyllum inophyllum, Crocus sativus, Mesua ferrea, Ochrocarpus longifolius.** The first is apparently the true *Pannága* of Sanskrit writers, though **Roxburgh** gave that synonym to **Mallotus**; so also the **Crocus** is the true *Kesar*, though that name is sometimes given to *Kaméla*. While such confusions exist, we have the assurance of **Udoy Chand Dutt** that the *Kampilla* and *Rechanaka* of the more recent Sanskrit medical writers was **Mallotus philippinensis,** and since it was by them recommended as an anthelmintic, it seems probable that his determination is correct, and consequently that the knowledge of this drug may have migrated from India to the Arabian and Persian physicians, and returned again when an import trade was established in the Ethiopian, Arabian, and Abyssinian drugs with which it became confused by modern writers. **Flückiger & Hanbury** (*Pharmacographia*), **Bentley & Trimen** (*Medical Plants*), and several other authors state that **Mallotus philippinensis** is a native of Abyssinia, Southern Arabia, and India. This is probably a mistake, the genus **Mallotus** being sparsely distributed to Africa. There are, however, three or four Abyssinian species and several members of allied genera, some of which appear to possess the medicinal properties of the *Kaméla*. At one time, for example, considerable interest was taken in *Cortex musenæ*, the bark of **Croton macrostachys,** *A. Rich.* (**Rottlera Schimperi,** *Hochst*), a tree met with in Abyssinia, which, by the natives of that country, is held in esteem as an anthelmintic. The frequent reference, by the older writers, to Abyssinian *Kinbíl*, leaves it probable that we shall yet discover that an African or Abyssinian species of **Mallotus** may actually have produced the drug in question, though modern writers assume that the Abyssinian supply must have been drawn from India. If this suggestion proves correct, the Indian knowledge in *Kaméla* may have been the result of an Abyssinian imported drug being recognised as identical with an indigenous substance not previously utilized.

The earliest mention of the anthelmintic drug *Kaméla* by European writers on Indian Economics appears to be the brief allusion to it in *Royle's Illustrations of Himálayan Botany* published in 1839 (p. 329), but it is noteworthy that he does not give it the name *Kaméla*, nor any form of that word. This is followed, however (in point of date), by **Irvine** (*Materia Medica of Patná*), who gives an account of **Rottlera tinctoria** which he calls *Kupila*. He remarks: "The dust from the capsule of the fruit is used to dye silk yellow: considered as of a warm nature and given internally as an anthelmintic. Dose gr. ii to gr. v. Price per lb. R0-3-0." In another page he describes **Daphne Mezereon** as *Kaméla* and says "the seeds imported from Kábul are used as an irritant." He again reverts to MEZEREUM, but calls it *Mameera* (*see Dict. Econ. Prod., Vol. III., 25*). **Flückiger & Hanbury** (*Pharmacographia, 572*) followed by **Dymock** (*Mat. Med. West India, 2nd Ed., 709*) and several other writers appear to be in error when they state that this drug is mentioned by **Roxburgh** and by **Ainslie,** but they are probably correct in assigning to **Surgeon Mackinnon** of Bengal the honour of having first (in 1858) sent it to Europe. Shortly after that date it was placed in the British and Indian Pharmacopœias, but it cannot be said to have since made the same progress in Europe that it has done in India.

STATISTICS and PRICES. 73

Statistics and Prices of Kaméla.—With so extensive an amount of accessible literature as exists on the subject of *Kaméla*, it does not seem necessary that detailed information should, in this work, be given for each province. The tree is wild and apparently nowhere cultivated; the powder is obtainable in any local bazár and within easy reach of the chief seaports. If a demand were to arise, the supply might be almost indefinitely increased, without, for many years to come, necessitating cultivation. Though statistics cannot be given of the present Indian consumption or of the export, the writer, from personal knowledge of India, feels that he is within the mark when he affirms that considerably less than half the amount produced is at present utilised. It is quite customary to find, in the subtropical forests, miles of country, with here and there trees each bearing a mass of over-ripe powdery capsules, the *Kaméla* from which is simply being allowed to run to waste. Many writers allude to the extent of the local trade. Thus, for example, **Mr. Atkinson** informs us that 2,000 maunds are annually exported from Kumáon. The known extensive use of *Kaméla* as a silk dye is, however, a better criterion of the magnitude of the trade than the published statistics of any one district or province, since particulars of a limited number of districts only have been published, and the demand is for local rather than foreign markets.

The powder seems to vary greatly in price in the various districts of India: **Dr. Dymock** says that in Bombay it sells for R11 per maund of 41lb. **Lisboa** remarks: "If the berries be plucked too early, this dust (*Kaméla*) is mixed with another sort, of a greenish tint, which destroys the value of the article, and if not plucked at the right time, the dust will all disappear, being blown away by the wind, leaving the berries of a greenish-brown colour, and of no value. The article *Kaméla* finds a ready market, and is now worth one shilling and six pence a pound." **Buck** says that in the bazárs of the North-Western Provinces the *Kaméla* powder sells at R22 per cwt. The price in Bengal will be found, in the passage below, describing the process of dyeing which is pursued in the Lower Provinces. Regarding Nundydrug Division, Mysore, it is stated that one-eighth of the quantity produced is used locally, and that the average annual production is a little over 2 tons, and the average price R6 to R7 per maund.

CHEMISTRY. 74

Chemistry of Kaméla.—*Kaméla* powder was first examined by the late **Professor Anderson** of Glasgow University, and subsequently **E. G. Leube, Junior.** The opinions originally published by these chemists have been reproduced in all subsequent medical works which have appeared in Europe, America, and India, without apparently any additional information being brought to light. The powder is said to be aromatic: is but slowly wetted by water and yields but little colour even to boiling water, colouring it pale yellow. In the presence of alkaline carbonates and caustic alkalies, especially the latter, it forms deep red solutions. The extract prepared with soda imparts to silk a fine and durable fiery-orange colour, without further addition or the use of mordants: with cotton, on the other hand, it does not produce a good colour. The natural dye stuff contains 3·49 per cent. water, 78·19 resinous colouring-matters, 7·34 albuminous substances, 7·14 cellulose, and 3·84 ash, besides small quantities of volatile oil and a volatile colouring matter. The liquid distilled from the alcoholic extract has a yellow colour, and the odour of the original substance. The concentrated etherial extract of the colouring matter deposits a yellow crystalline substance called *Rottlerin.* The extract, prepared with boiling alcohol, deposits, on cooling, non-crystalline flecks of a substance having the composition of $C_{20} H_{34} O_4$. It may be obtained nearly

MALLOTUS philippinensis.

Process of Dyeing

CHEMISTRY.

colourless, by repeated solution and separation; it is sparingly soluble in ether and in cold alcohol, insoluble in water; not precipitated by lead or silver-salts. The alcoholic solution separated from these flecks leaves a dark-red resin, $C_{30} H_{30} O_7$, soluble in all proportions in alcohol and ether, insoluble in water, melting at 100°, and forming with acetate of lead, a deep orange-coloured precipitate of variable composition (*Anderson, as given in Watts, Chemistry*).

The authors of the Pharmacographia state, that so faras their experiments go, they have been able to confirm **Anderson's** results. They found that the resin was also soluble in glacial acetic acid or in bisulphide of carbon, but not in petroleum ether. The chemistry of the action of *Kaméla* as a purgative and anthelmintic does not appear to have been established: **Anderson** does not affirm that his *Rottlerin* is the active principle of the drug.

DYE. Kamela. 75

Process of Dyeing with Kaméla.—The brief account of the chemistry of this substance given above, expresses the rationale of its use as a dye. The ripe fruits are collected by the people, placed in a cloth or sac, and beaten until the glandular PUBESCENCE is removed from the exterior of the fruits. The powder thus obtained is then sifted to free it from the fruits and broken pieces, and in this condition it is ready for the market. The following account from **Roxburgh's** Coromandel Plants may be given in its original form. It conveys very nearly all we know on the subject, and has a historic interest as being the first account of the substance.

"The red powder, which covers the capsules, is a noted dyeing drug, especially among the Moors, and constitutes a considerable branch of commerce from the mountainous parts of the Circars. It is chiefly purchased by the merchants trading to Hyderabad and other interior parts of the Peninsula.

"When the capsules are ripe, or full grown, in February and March, they are gathered, the red powder is carefully brushed off, and collected for sale; no sort of preparation being necessary to preserve it.

"This substance like Annatto is difficultly acted on by water; it communicates no particular taste, either by infusion or decoction, and only a pale straw-colour, which acids scarcely alter, but alkales brighten and deepen. To spirits it very rapidly gives a rich, deep, flame-coloured orange, inclining to red. Alkaline salts enable water to extract a very deep blood-red, which, on agitation, produces an orange-coloured froth, and tinges the sides of the vial. Neither spirits nor alkaline solutions dissolve it, for the distinct minute grains of the powder are seen adhering, in their original state, to the sides of the vial, when shaken, but are now of a bright gold or orange-colour, about the size of very minute grains of sand: in this it differs widely from Annatto, which is soluble in both these menstruums. Alum, added to the alkalised infusion or decoction, renders the colour brighter and more permanent; tartar (to appearance) in a great measure destroys it, yet the mixture dyed white silk of a very beautiful colour, if possible, superior to any other I have tried.

"This red powder dyes silk a deep, bright, durable orange, or flame-colour of very great beauty. The Hindu silk-dyers use the following method:

"Four parts of *Wassunta-gunda*, one of powdered alum, two of salt of soda (native Barilla), which is sold in the bazárs, are rubbed well together with a very small proportion of oil of sesamum, so little as hardly to be perceptible; when well mixed, the whole is put into boiling water, proportionate to the silk to be dyed, and kept boiling smartly more or less time, according to the shade required, but turning the silk frequently, to render the colour uniform."

Sir E. O. Buck (*Dyes and Tans of the North-Western Provinces, 31*) says, "The dye obtained from the fruits is used in dyeing silk and wool, giving a rich flame colour of great beauty and permanence. It does not require a mordant, all that is necessary being to mix it with water containing about half its weight of carbonate of soda. The dye was very favourably reported on by the jurors of the Madras Exhibition as being especially valuable for silk. The bark of the tree is used in tanning leather. It is imported into these provinces from the Sub-Himálayan forests and Calcutta, and sells in the bazár at R22 per cwt. When used it is simply pounded and dissolved in water." **Colonel Beddome** (*Fl. Sylv., t. 289*) writes, "The red mealy powder of the capsules is a valuable product and might be a source of considerable revenue in many of our forest districts; it is used as an orange dye principally for silk. The ripe capsules are gathered in March and rubbed together or shaken in bags till the farina separates. It is known as *Kapli* or *Kaméla* powder." "The powder is much adulterated in our bazárs, but some collected carefully by the Forest Department realised a high price in the English markets." **McCann** gives (*Dyes and Tans of Bengal, 18*) the facts communicated, regarding the dye, to the Economic Museum by the various district officers of Bengal. The tree blossoms, he says, in the end of the autumn and the fruits ripen in January and February. From Bírbhúm the following report was received :—"Formerly *Kamaláguri* trees grew in abundance in Bhandibún and the adjacent villages, where the weavers were in the habit of paying a certain rent for them, but now they are not taken care of, and the weavers have given up dyeing *tasar* and silk." It would thus appear to have once upon a time been in a state of semi-cultivation in Bírbhúm. In Purí we are told by **McCann** the cost of collection of the powder is about R3 per maund, but in a table the selling price in six districts is shown as varying from R7 to R40 per maund. **McCann** then states, "the powder is only very sparingly soluble in either hot or cold water, but is completely dissolved in alkaline liquids, forming a dark red solution. The resinous yellow colouring matter may be separated from this red solution either by neutralising with an acid, or else by mere exposure to the air. In Bengal the red powder is dissolved by the addition of a solution of various alkaline ashes obtained by burning plants, and the development of the yellow colouring principle is in no case brought about by the addition of acids, but merely by allowing the cloth steeped in the red liquid to dry by exposure to the air. It is said not to require a mordant, but frequently alum is added for that purpose. The colour is sometimes heightened by the addition of turmeric."

SUBSTITUTES.—**Atkinson** (*Himálayan Districts, 776*) says the substances chiefly used to adulterate the powder are the pounded bark of **Casearia tomentosa** (the *chíla* of Garhwál) and a powder prepared from the red fruits of the banyan tree (**Ficus bengalensis**). On the other hand, he adds, *Kaméla* powder is itself used to adulterate annatto dye. In some districts *kaméla* is always used in combination with annatto, the one being supposed to strengthen the colour of the other. **Wardle**, in his recent Report on the Dyes of India, remarks that the colours produced by *Kaméla* are fast; it promises to be a very valuable dye-ware. (See **Mr. Wardle's** statement regarding the superiority of *kaméla* over that of *wars* as a silk dye in Vol. III., 402, of this work.)

TAN. Bark. 76

Tan.—The fact that the BARK is used as a tan has been incidentally mentioned above. This is referred to by **Atkinson** (*Himálayan Districts, l.c.*), **Buck** (*Dyes and Tans, North-Western Provinces, l.c.*), by **Kurz** (*Forest Flora Burma, l.c.*) and others. **Professor Hummel** of Leeds was, however, furnished with a sample from the Colonial and Indian Exhibition, and in his report on the same states that the bark was but a poor

Kaméla as a Medicine.

tanning material, possessing only 6·5 per cent. of tannic acid which imparts a deep reddish colour.

OIL. 77 Seeds. 78 **Oil.**—The SEEDS yield a bland oil which many writers recommend as worthy of investigation. It has not apparently been chemically examined, but it is used medicinally by the people of India.

MEDICINE. Kamela. 79 **Medicine.**—From the brief historic account given above, it will be seen that the early Arabian writers on *Kinbíl* spoke of it as being applied externally in certain skin affections, or as being taken internally to cure leprous eruptions. The distinction into *wars* and *kaméla*, the latter being defined as an anthelmintic, was not made by the Arabs until about the tenth century. In the seventeenth century, **Rheede**, the first European writer who in India described the drug, was apparently ignorant of any special merit possessed by the powder, but spoke of the PLANT (Plant. 80) as being employed in the preparation of a cataplasm which was found useful for external application. It was not, indeed, until well into the nineteenth century that the merits of the *Kaméla* powder were recognised.

From about 1780 to 1840 might be described as a period of earnest investigation into, and record of, the vegetable resources of India. In the field of research **Jones, Roxburgh, Ainslie, Buchanan, Elliot, Honigberger, O'Shaughnessy**, &c., worked and published their results. No mention is, however, made by any of these authors of the anthelmintic property of *Kaméla* powder, though they were all apparently familiar with the dye which it yielded. But, from the close of the period named, writers from every province, indeed almost from every district, speak of the anthelmintic property of the drug, as if this knowledge had been a matter of ancient history. It is impossible to believe that all the distinguished authors mentioned could have overlooked so important a subject, but it is equally difficult to credit so sudden an evolution on the part of the people of India. It seems much more likely that modern authors have compiled from one to two sources and that the anthelmintic drug, even at the present day, is more known to Government apothecaries than to the rural populations of India. The **Rev. A. Campbell**, who for many years has patiently devoted himself to the study of the Santals and made an admirable collection of all the plants, wild and cultivated, which are found in their country, says of **Mallotus philippinensis**:—"The fruit yields a red powder which is used as a dye and medicine, as also the bark. From the seeds a medicinal OIL (Oil. 81) is prepared." **Honigberger** mentions the plant as used in "anorexia, fever, giddiness, hemiplegia, hepatic and thoracic pains," but makes no mention of the specific property of the powder—the property which alone is dealt with by modern writers. Indeed, the use of the LEAVES (Leaves. 82), FRUIT (Fruit. 83), and BARK (Bark. 84) in the preparation of external applications, such as those described by **Rheede**, seems to have disappeared from modern Indian works, its place being taken by the undoubtedly more important property of an anthelmintic. Apparently, however, in Europe and America, *Kaméla* has gained some reputation as an external application in the treatment of itch, and the advocates of this use of the drug generally urge that it is extensively so employed in India. The writer can find no record of this being the case. **Dr. Dymock** (*Materia Medica, Western India*), for example, contents himself with giving a review of the drug (mostly compiled from **Flückiger & Hanbury**), and says practically nothing of its special applications in Western India except that it is viewed as an anthelmintic. **Waring** in his *Bazár Medicines* gives the chief facts regarding its value as an anthelmintic and describes the methods of usage:—"In medicine, the purplish-red powder has attained considerable repute as a remedy for Tænia or tape-worm. It has little or no effect on other forms of intestinal worms. The dose for an adult is from 2 to 3 drachms in honey, or a little aromatic water, no other medi-

MEDICINE.

cine being necessary before or after. In the above doses it acts freely on the bowels, causing, in many instances, considerable nausea and griping, though not generally more than is caused by other remedies of the same class; the worm is generally expelled in a lifeless state in the third or fourth stool. Should the first trial not prove successful, it may be repeated after the interval of a week, but should this be a failure also, it will be useless to continue its use further; other remedies may then be tried." **Garrod** found "its active purgative properties at times rather objectionable" (*Mat. Med.*, *313*). In most European and American works it is stated that the powder has been long used in India in the treatment of tape-worm. From what has been said it will be observed that the writer does not think the literature of the subject fully bears this out, and, indeed, with a large section of the people of India (the higher caste, vegetarian, Hindus) a drug which has its specific action on tape-worm was not likely to have been in much demand. The *United States Dispensatory* assigns to the RESIN extracted by ether, the active properties of the drug. **Royle** seems to have thought its action was mechanical, resembling that of cow itch (**Mucuna pruriens**). This opinion appears to have depended upon the microscopic structure of the grains. Their peculiarities have been thus described by **Flückiger & Hanbury**:—"The granules of *Kamala* are irregular spherical glands, 50 to 60 mkm. in diameter; they have a wavy surface, are somewhat flattened or depressed on one side, and enclose within their delicate yellowish membrane a structureless yellow mass in which are imbedded numerous, simple, club-shaped cells containing a homogeneous, transparent, red substance. These cells are grouped in a radiate manner around the centre of the flattened side, so that on the side next the observer, 10 to 30 of them may easily be counted, while the entire gland may contain 40 to 60."

Resin. 85

SPECIAL OPINIONS.—§ "There can be no doubt that it is *Kaméla* that has been generally employed in all Government Hospitals as it is supplied from the Saharanpore Botanical Gardens, and the **Mallotus philippinensis** is so common everywhere in the neighbouring Himálayan hills that there would be no object in adulterating it or substituting anything for it. There can be no doubt as to its efficacy as a purgative and anthelmintic in tænia" (*Brigade Surgeon G. A Watson, Allahabad*). "Purgative and anthelmintic in doses of half drachm to one drachm" (*Assistant Surgeon Shib Chunder Bhuttacharji, Chanda, Central Provinces*). "I have often found a two-drachm dose of *Kaméla* to be an efficient medicine in expelling tœnia, and it is not so nauseous to the taste as the extract of malefern" (*Surgeon-Major H. W. E. Catham, M.D., M.R.C.P., Lon, Bombay Army, Ahmednagar*). "*Waras* when applied externally, either mixed with water or oil, has a protective influence on the skin against dry and cold winds" (*Surgeon-Major Jayakar, Muskat*). "A commonly used anthelmintic. Uncertain in adults. Is also laxative" (*Assistant Surgeon Nehal Sing, Saharanpur*). "*Kaméla* powder is an efficacious anthelmintic, and its efficacy is increased when given along with purgatives in doses of from ʒiss. to ʒii" (*Assistant Surgeon Ram Chunder Gupte, Bankipore*). "Powder is anthelmintic, vermifuge, and cathartic, dose ʒi; externally used in scabies" (*Bolly Chand Sen, Teacher of Medicine*). "Used at the Civil Hospital, Umballa, Panjáb, as an anthelmintic. Can be obtained in the bazárs here" (*Brigade Surgeon R. Bateson, I.M.D., Umballa City*). "Ointment made of the powder (ʒi to ℥i) is used for skin diseases" (*Surgeon Anund Chunder Mukherji, Noakhally*). "Anthelmintic, purgative; doses ʒi to ʒii; used in tape-worm" (*First Class Hospital Assistant Choonna Lal, City Branch Dispensary, Jubbulpore*). "Is not to be compared with either santonin or male-fern" (*Surgeon-Major*

TIMBER. 86 E. *Sanders, Chittagong*). "A very inferior anthelmintic, not to be compared to santonin" (*Surgeon G. Price, Shahabad*).

Structure of the Wood.—Smooth, grey to light red, hard, close-grained, no heartwood. Annual rings indistinct. Pores small, uniformly distributed, scanty, often sub-divided. Faint indications of transverse bars. Weight 48℔ per cubic foot. The wood warps and shrinks and is not used except as fuel.

87 **Mallotus Roxburghianus,** *Muell. Arg.; Fl. Br. Ind., V., 428.*

Syn.—ROTTLERA PELTATA, *Roxb.; Wight, Ic., t. 1873.*

Vern.—*Kamli mallata, phusri mallata,* NEPAL; *Ním púteli,* BENG.; *Sirgúllum,* ASSAM.

References.—*Roxb. Fl. Ind., Ed. C.B.C., 737; Kurz, For. Fl. Burm., II., 383; Gamble, Man. Timb., 361.*

Habitat.—A small evergreen tree or shrub found in the Sikkim Himálaya, in Sylhet and Assam, in the Khásia Hills, Chittagong and the Martaban, ascending to 2,000 feet.

TIMBER. 88 **Structure of the Wood.**—White, moderately hard, close-grained. Weight 46℔ per cubic foot.

MALT LIQUORS.

89 **Malt Liquors.**—Under this head it is designed to give as concise an account as possible of the progress which the brewing of Malt Liquors—such as Ale, Beer, and Porter—has made in India and also to refer briefly to the indigenous processes by which exhilarating or intoxicating malt beverages are prepared by the people of this country. It may be explained that the portion of the article on the ancient history of the art of brewing, as here given, has been mainly taken from the pages of the *Encyclopædia Britannica,* and the Indian facts (on European Breweries) from a paper published by **Mr. H. Whymper** on 'Brewing in India' which appeared in *A. Boake & Co.'s Diary for the Brewing Room.*

HISTORY.

EARLY HISTORIC FACTS REGARDING MALT LIQUORS.

All countries, whether civilised or savage, have, in every age, prepared an intoxicating drink of some kind. The art was known and practised by the Egyptians many hundreds of years before the Christian Era, and afterwards by the Greeks, Romans, and ancient Gauls. In the second book of **Herodotus,** written about 450 B.C., we are told that the Egyptians, being without wine, made wine from corn. **Pliny** also informs us that they made wine from corn, and gave it the name Zythum. **Hellanicus,** speaking of the introduction of wine at Plinthium, a city of Egypt, states: "Hence the Egyptians are thought to derive their love and use of this liquor, which they thought so necessary for human bodies, that they invented a wine made from barley." The Greeks obtained their knowledge of artificial fermentation from the Egyptians, at a very early period. We find it mentioned, for example, in the writings of **Archilochus,** the Parian poet and satirist, who flourished about 700 B.C., that the Greeks were already acquainted with the art. Again, we learn from **Æschylus** (470 B.C.), from **Sophocles** (420 B.C.), and **Theophrastus** (300 B.C.), that the Greeks employed barley wine or beer (their Zythos) in daily life as well as at festive meetings. There is, in fact, little doubt that the discovery of beer and of its use as an exhilarating drink was nearly as early as that of wine. **Xenophon,** in his account of the retreat of the ten thousand, written 400 years B.C., mentions that the inhabitants of Armenia used a fermented drink made from barley. **Diodorus Siculus** states that the Galatians prepared a fermented beverage from barley, which they called *Zythos,* like the Egyptians. In the time of **Tacitus**

HISTORY.

(whose treatise on the manners and customs of the Germans was written in the first century of the Christian Era) beer was their usual beverage, and from his description, imperfect as it is, there can be no doubt that they understood the method of converting barley into malt. **Pliny** mentions its use in Spain under the name of *Celia* and *Ceria,* and in Gaul under that of *Cerevisia* or *Cervisia*. He says:—"The people of Spain in particular brew this liquor so well that it will keep good a long time. So exquisite is the cunning of mankind in gratifying their vicious appetites, that they have thus invented a method to make water itself produce intoxication."

The *Cervisia* of **Pliny** evidently takes its name from Ceres, the Goddess of corn. **Plautus** calls it *Cerealis liquor,* that is, liquor used at the solemn festival of that goddess. Beer and vinegar were the ordinary beverages of the soldiers under **Julius Cæsar**. From them the Britons are supposed to have learnt the art of brewing. Beer being so suitable to the climate of Britain, and so easily made by an agricultural people with plenty of corn, it was gladly welcomed and soon became the favourite beverage. After the departure of the Romans from Britain, the Saxons subdued the natives and learned from them the art of brewing.

Malt Liquors of Aboriginal Races.

Dr. H. Mann tells us that the Kaffre races of South Africa have made for ages, and still make, a fermented drink like beer from the seed of the millet (**Sorghum vulgare**), which is subjected to a process of malting in all essential particulars identical with our own. The natives of Nubia, Abyssinia, and other parts of Africa, also make an intoxicating drink of great power, called *bousa,* from the flour of the *teff* (**Poa abyssinica**) and from the *durrha* or millet, much esteemed by the natives, and preferred by many to palm or date wine, the common intoxicating drink in tropical countries. According to **Mungo Park**, the natives of Africa also make a beverage from the seed of the spiked or eared soft grass (**Pennisetum typhoideum**). The Russian drink *kwass* or *quass,* a thick sour beverage, not unlike *bousa,* is made of barley and rye flour, mixed with water and fermented. Formerly, the spruce-fir, birch, maple, and ash trees were tapped, and their juice used for beer-making in England, the first two, indeed, up to the last fifty years. *Koumiss,* the drink of the Tartar race, is fermented mares' milk. The Chinese malt beverage, *samshee,* is made from rice, and a similar liquor is prepared in Japan also from rice, known as *sake,* which is almost identical with the *Zu,* rice beer, of the Angami Nagas (*Conf.* with *Vol. II.,* 260) and the *pachwai* of India generally. The Kakhyans prepare their *sherú* from rice. The Lepchas, Lushais, as well as the Nagas, make a rice beer in which apparently the fermenting agent is spontaneously generated. The hill men of the Simla neighbourhood expose rice or maize water, flavoured with some bitter principle and spices till fermentation sets in. The Burman *congee* is a beer which the Khyens and Karens also use. In fact, throughout India a crude beer (*pachwai*) is prepared and was probably known from ancient times (*Conf.* with the account of Soma, *Vol. III.,* 246-251). The process of manufacture is of the most primitive kind. The ingredients are generally some fermentable substance such as malt from millet (*paisht*), from rice, barley, wheat, &c., or from the fruits or flowers of certain plants, particularly *Mahuá,* **Bassia latifolia** (*see Vol. I.,* 406-415); **Eugenia Jambolana** (*see Vol. III.,* 286); **Melia Azadirachta**—the *Nim* tree (*see p.* 211.); from dates, raisins, or other less important substances (see the article **Yeast,** *Vol. II,* 259-260). But by far the most prevalent Indian beverage is *tari* or toddy made from palm juice, *nim* or sugar-cane,—see **Borassus, Caryota, Cocos, Melia, Phœnix,** also the article **Narcotics,** &c.

M. 89

MALT Liquors.

HISTORY.

History of Brewing

Most of the Indian beverages are flavoured with drugs, often highly pernicious, and they are fermented by various substances (*see Vol. II., 259-260*). Harmless spices are also frequently added to flavour the liquor, or, as in the case of the *Zu* beer of the Nagas and the *Marwá* of Sikkim, a warm infusion from grain is consumed before fermentation has been established (*Vol. II., 260*). In the majority of cases the infusion is exposed in a warm place, for spontaneous fermentation; but the use of a special ferment, such as the common yeast of brewers and that too preserved in cakes (*see Vol. II., 257-260, also Vinegar, Vol. I., 72-77*), is not unknown.

History of European Brewing in India.

The history of the manufacture of malt liquors in India, is, to some extent, the history of a series of unsuccessful efforts at establishing an exotic industry, in a country then unfavourably placed for its prosperity as a remunerative enterprise It is only within recent years, as a consequence of the growth of large European communities and the existence of army contracts, given out by Government to the Indian brewers, that the industry has at last been able to firmly establish itself in this country.

The pioneer brewer in India appears, says **Mr. Whymper,** to have been a **Mr. Henry Bohle,** who commenced business at Meerut and Mussoorie in 1825. His attempts were, however, very disappointing, and in 1852 his business passed into the hands of his partner, **Mr. John Mackinnon,** the founder of the firm of that name now in Mussoorie. It was not, however, till about the year 1870 that success dawned upon the enterprise. In the meantime, between the years 1850 and 1860, several small breweries were opened in hill stations, most of which operated but for a short time and then failed. In fact, it may be said that one only, of the early breweries of Northern India has survived. It was started at Kussowlie by **Captain Bevan,** who, in 1854, finding it a fruitless enterprise, disposed of his interest to **Mr. Dyer.** The concern thereafter passed into the hands of a Company, and subsequently was bought by **Mr. Meakin,** who still retains an interest in it and has made it a success.

In 1860, a brewery on a more pretentious scale was started by **Messrs. Conill & Hay** in Simla. The lines on which it proposed to work may be said to have foreshadowed its failure. Even the bricks, which were employed in the construction of the buildings, were imported from England at an enormous cost. Expenditure on other branches of the concern were equally reckless, and the business closed and finally passed into the hands of **Mr. Meakin. Balfour** (*Cyclopædia of India*) says that in Southern India **Captain Ouchterlony** initiated the industry about 1850. He failed, and was followed by **Mr. Honeywell,** who may be said to have carried on the business ever since. A curious experiment, **Mr Whymper** tells us, was made at Bangalore not long after, *viz.*, to manufacture beer from imported concentrated wort, but it is probably needless to add that this venture also proved a failure. It would be beside the purpose of the present article to refer to the establishment of each and every brewery in India. Suffice it to say that there are now 25 breweries at work, of which 20 have been established since 1870, and of these 12 have sprang into existence within the past ten years (1879-89). This progress may be still further exemplified by the figures of outturn. In 1881 some 21 breweries were working and these produced 2,448,711 gallons, of which the Commissariat Departments purchased 1,764,927 gallons. During the succeeding eight years (1882-89) the production and Government purchases rose steadily until, in 1889, the figures stood at 5,165,138 made in India and 3,778,295 gallons purchased by Government. In the

HISTORY.

previous year the Government purchases of Indian beer amounted to 4,628,175 gallons.

Of the 25 breweries at work during 1889 the following were the more important:—

The **Murree Brewery Co., Limited,** at Murree (1,148,949 gallons), at Ráwalpindi, 205,632 gallons, at Ootacamund (336,558 gallons), at Bangalore (267,408 gallons), with smaller concerns at Quetta and Ceylon: **Meakin & Co.** at Poona (501,816 gallons), at Kasauli (450,000 gallons), with smaller breweries at Chakrata, Darjiling, Dalhousie, and Ranikhet. **Dyer & Co.** at Lucknow (340,038 gallons), at Mandalay (232,804 gallons), at Solon (133,272 gallons): **Mackinnon & Co.** at Mussoorie (183,591 gallons); also the **Crown Brewery Co.** carrying on business at Mussoorie, (411,183 gallons) and the **Naini Tal Brewery Co.,** at Naini Tal. The total outturn for the year was returned at 5,165,138 gallons.

Mr. Whymper, in concluding his historic sketch of Indian breweries, remarks:—

"There are few Indian, or Native, breweries in the Mysore State. They are of slight consequence. About 1875 a brewery was started at Bandora near Bombay. The peculiar feature of this establishment was that tidal water was used in brewing. This water was frequently quite salt and the beer was very nauseous; it however kept sound in a most remarkable manner. The beer was sold for some time in Bombay.

"The brewery, which works most satisfactorily, under the most trying conditions to be met with in India, is said to be that at Dapooree, near Bombay. This belongs to **Messrs. Meakin & Co.** The writer visited this brewery on the 22nd April 1886. The temperature of a well-shaded verandah at 8 that morning was 93°; at noon it was 106°; the brewery office at the same time was 100°. By using a five-ton ice machine as much as possible, the average pitch heats had been about 75° in that month. Nothing had been pitched under 72°. One gyle had to be pitched at 88°, it rose to 101°, at which the attemperators were able to hold it. Beers, brewed under nearly the same unfavourable conditions three months before, were examined and were perfectly sound to the palate. The writer is fully aware this will not receive ready credence in England. The owner, **Mr. H. G. Meakin,** is an elder brother of the Burton maltsters, and possesses an unusually venturesome spirit which has so far carried with it well-merited success.

"It must not be supposed that all brewers have anything like such unfavourable conditions to contend with as **Mr. Meakin** has had. The majority of Indian breweries are situated in the mountains of Northern India, or of the Madras Presidency. There is one brewery at Lucknow which has only a very short winter, but still it does have some cold weather, whereas the Dapooree one has none. The breweries in the Northern Hills (as the mountains are always called) have cold winters, some have as much as six months' good brewing weather, and **Messrs. Mackinnon** are so well situated that they can brew sound beers all the year round. The breweries in the Neilgherry Hills in Madras, and the brewery in the Ceylon Mountains, both being at an elevation of over 6,000 feet, can also brew every day in the year for export trade. The trade of the latter is principally with Lower Burma. **Sir Samuel Baker** was the pioneer brewer in Ceylon, but it is doubtful if he ever foresaw that Ceylon would eventually have an export beer business. The **Murree Brewery Co.** purchased the present brewery site from a German firm which did not succeed in brewing to meet the public taste.

"The brewery at Quetta has, perhaps, the most extraordinary climate of all Indian positions, the sun being so intensely hot, even in the winter months, that a brewer has to wear a sun helmet whilst at the same time he has to clothe himself in a fur-lined coat to protect himself from the biting cold which there is in the shade. Whilst prospecting for a brewery site, the servants of the Company suffered from both sun and from frost-bites. The cold which is occasionally experienced is too great to make it safe to employ much steam power, and although the Company, in the first instance, erected a steam plant, it had to be replaced by the open boiling system; pipes, pumps, and injectors, steam pressure gauges, and blow-off cocks were all frozen up, and burst in the most impartial manner."

MATERIALS.

MATERIALS AND METHODS OF BREWING.

Few industries can be said to have been more directly benefited by the chemist than that of BREWING. It is by no means the case, however, that all brewers adopt in their entirety the principles enjoined by chemical research. In many minor details they follow empirical laws and claim merit in certain processes, the knowledge of which they alone possess. The main facts of the chemistry of brewing will be found under Vinegar—**Acetum**—(*Vol. I., 72-77*), *viz.*, the transformation of insoluble starch, contained in grain, into soluble saccharine compounds. The methods by which this is accomplished in India and the subsequent fermentation, together with the details of brewing, that can be viewed as in any way peculiar to this country, will be found very briefly detailed below. This subject may be discussed under the headings of the chief materials used, *viz.*, Grain, Hops, Yeast, and Water.

Grain.
90

I.—THE GRAIN.

Barley is, of course, the grain employed in India. The following account of it is taken from **Mr. Whymper's** paper:—"The barleys used by the Indian brewer are entirely grown on the Peninsula. The breweries in the South imported Persian grain and English malt until very recently, but now northern barleys are carried down and malted at the breweries. The range of Indian barleys is considerable, and the quality varies to a great extent. Grain weighing nearly fifty-two pounds to the bushel is grown as far south as 24° north latitude. The best grain, however, is found about 28° north latitude, in the North-West Provinces, and extreme South Panjáb, where fifty-six pounds weight to the bushel is procurable. In Northern India a curious custom prevails of cutting the whole crop down to the ground when about to throw into ear. Cattle are fed on the green stuff so cut, and the barley is allowed to grow again, and, strange to say, it does not seem very much worse for this treatment. In the hills in Madras two crops a year are grown, but the grain is hardly ever allowed to ripen properly, and, consequently, occasions malting difficulties. Barley has been grown in Ceylon and used in brewing, but it is not likely to be permanently grown there, not being a sufficiently valuable crop. All Indian barleys require more warmth and moisture in malting, especially if grown on irrigated soil, than European barleys. Maltsters in England have complained of Indian barleys not germinating freely. It is open to question whether they have sufficiently allowed for the fact that they are dealing with grain grown under totally different conditions to that which they have been usually accustomed to malt. It should be remembered that Indian barley which would find its way to Europe is seed from an almost semi-tropical plant, and naturally requires much more warmth and coddling than English barley. It should be kept up to 60° in the cistern and kept thick on floors. It will not be injured by warmth when growing. The great drawback to its use is the large quantity of weevil found in some samples. The Indian crop is cut at varying dates, according to latitude, from March to May. The hot weather then sets in

MATERIALS. Grain.

and the grain undergoes a hot summer season and several months of monsoon weather before it is malted. The contrary holds good in England, where barley after harvesting is stored in cold weather until required for malting. The spread of weevil in Indian samples is thus very understandable. Natives believe that weevil will never be found in old buildings to the same extent as if stored in new granaries, and they attribute this entirely to the dampness of all new buildings. The writer believes there is something in this view, from facts which have fallen under his notice. There can be no doubt that the quantity of weevil can be minimised by shipments being made to England in June and July, and by great care being exercised in storing the grain in very dry places. Sun kills weevil."

Mr. Whymper has obligingly furnished for this work the following additional facts which in some respects supplement the passage quoted:—"Indian Barley, from growing in a warmer and drier climate than in England, requires more warmth in the malting process and less water than in England. Generally speaking, Indian barleys germinate more freely than English, French, or Belgian. I do not know of any other difference. There is no doubt Indian grain would be improved if zemindars could be induced not to cut the crop down for green fodder.

"Chevalier barley seed was given away freely by the **Murree Brewery Co.**, in the Hazara District about 1870 and in the Nilgiris in 1887-88, but with poor results. The following localities produce very good malting barleys:—**Hazára, Ludhiána, Delhi, Rewari, Fazilka** in the **Pánjab**; **Allahabad, Mirzapur** in the **North-West Provinces**; but ordinarily good barley can be procured almost anywhere in the **Panjáb** and **North-West Provinces**, and in parts of **Rájputána**. **Bombay** and **Madras Presidency** Breweries are supplied from the North. Indian Barley varies in weight from 46 to 56℔ per bushel."

The barleys of the Panjáb and North-West Provinces are fairly well adapted for brewing purposes, but it is generally found that the percentage of "extract," as compared with English grain, is below the mark. This has been traced to many causes:—Poor seed, unfavourable soil, the objectionable practice of preserving the grain in cow-dung, cutting down the crop for green fodder, and causing it to spring again and yield its grain, exposure of the grain to severe atmospheric changes, careless handling in packing and transit; these and many other defects tend to lower the value of Indian barley for the maltster. But perhaps the most pernicious practice of all is traceable to the middleman, *viz.*, the adulteration of new grain with old, the mixture being sold as fresh stock. When this is done, the brewer has no end of trouble and often heavy pecuniary loss owing to irregularity in germination.

In concluding these remarks regarding barley suitable for brewing, the following useful passage from *Spons' Encyclopœdia* may be quoted as giving the English experience:—"The selection of the barley used by the brewer calls for the exercise of much skill and judgment; unless the quality be of the very best, it is impossible to obtain good malt, and without good malt, it is useless to attempt to make good beer. A practised brewer can judge of the quality of his barley by its appearance. The heaviest, if in good condition, is always the best, the grains should be plump, and of a pale-yellow colour; they should have a thin skin, and a free, chalky fracture. That which has been grown in a light soil and harvested early, is also preferable. It is of much importance to the malster that barley be lodged in the stack for a few weeks before being thrashed, in order to allow the moisture from the soil to dry off before it comes into his hands. If this is done, the operation of drying in the kiln is avoided. In moist districts, however, where the grain never gets thoroughly dried, this process must invariably be had recourse to; the temperature of the kilns must never be

MALT Liquors.

Materials and Methods.

MATERIALS. Grain.

allowed to arise above 50° (120° F.). Care must be taken to avoid breaking or crushing the grains of malting barley, so as to minimise the chances of its becoming mouldy in the subsequent processes of malting, a contingency which should be avoided in every possible way. It should also be screened before steeping, in order that the grains may all be of equal size on the spiring floor. These remarks, of course, apply only to the brewer who is, as he ought always to be, his own maltster."

Hops. 91

II.—Hops.

Before fermentation hops are added to the wort, because beer made from barley alone has little or no flavour. Under **Humulus Lupulus** (Vol. IV., 302) will be found full particulars regarding the efforts which have been made in India to introduce the cultivation of this plant. It need, therefore, be only necessary to discuss here the qualities of imported hops that are in demand for the Indian breweries and the dangers to which stock is liable. There are known in trade two chief kinds of hops—red and green, the former being that most prized. The green is more easily grown, but does not possess the rich aroma of the red, and has, on the contrary, a more or less pronounced garlic-like smell. The distinction into red and green hops is not, however, always sharply defined, and, according to some writers, these properties are more due to care in culture and preparation, or to climate and soil, than to any racial distinction. Certain regions, for example, enjoy an exceptionally good reputation for cultivating hops, and, consequently, the name of the country of production often passes muster for the quality of the article.

The following are the principal kinds of hops known to Indian trade:—

Austrian. 92 *Austrian.*—Bohemia is the principal hop-cultivating province of the Austrian crown lands. The hops of the district of Saaz, which are of the red variety, are universally preferred.

German. 93 *German.*—Bavaria is the principal hop-cultivating country of the German Empire. The hops from Spalt (city and country) are most prized.

English. 94 *English.*—The hops of Kent, Sussex, Hereford, Hants, Worcester, and Surrey, are not an especially fine but very productive article.

American. 95 *American.*—The centre of the hop, cultivating district is Utica, in the State of New York. Both varieties are grown and are rising in favour.

Hops, as an article of commerce, occupy a very peculiar position. On the one hand, the yield varies very much from year to year; on the other, they cannot be stored like grain, for instance, without injury to their value, and therefore must be consumed as quickly as possible.

Hops are very liable to deterioration in the variable temperature of India. On this account special precautions have to be observed in packing and storing them. The influences which exert an injurious effect upon the quality of hops are moisture and atmospheric air. Several methods have been recommended to protect hops from these evils. Smoking with sulphur was once much in vogue and still is to some extent; but it is considered by experts to be of doubtful advantage by itself. Combined with pressure of the hops within an air-tight receptacle, it has been found to answer admirably. A double packing cloth is used, and in addition the bales are covered with varnished paper or enclosed in well-soldered tin or pitched wooden boxes.

Speaking of Hops **Mr. Whymper** (in the paper already alluded to) says:—"The hops used in India are nearly all imported, for, although there are hops growing in Kashmír and in Kúlu, the quantity as yet offered has been small. Some very fair quality hops have been grown by a **Major R. Rennick** in the Kulu Valley, and he is persevering with the

MATERIALS.

Hops.

growth. Australian hops are very serviceable, as they are picked about April and can be used in India early in June, thus supplying a new hop when most wanted." In reply to an enquiry as to the hops most generally used in India **Mr. Whymper** writes:—"I am not aware that any Indian brewers have a particular preference with regard to hops. All use various kinds or, in other words, "blend" various kinds in order to get regularity of flavour as is customary in England. The dry climate of India naturally tells against hops stored in this country, and the Indian brewers cannot thus safely hold very large stocks."

For further information regarding hops, the reader is referred to the article **Humulus Lupulus**, Vol. IV. 302. Also instructive papers in the Jour. Royal Agri. Soc., England, Vol. VII (1846) Chemistry of Hops, 210; IX (1848), Hop Culture, 532—582; XIV (1878), 723—736.

III.—Yeast and Fermentation.

Yeast and Fermentation 96

An interesting paper on the fermentation of brewing was read by **Mr. T. F. Garrett** at the Brewers' Congress which was held in London during 1887. The discussion which followed the reading of that paper also considerably amplified the facts laid before the Congress. Space cannot be afforded to quote very freely from **Mr. Garrett's** paper, which, as he claimed, was a popular *résumé* of the discoveries in connection with the fermentation of brewing, but the following passages are more especially interesting. **Mr. Garrett** traced the development of our knowledge of the subject from the discovery made by **Lavoisier** in 1788 that sugar was broken up, during fermentation, into alcohol and carbonic acid. He then dealt with the modifications accomplished in 1825, by **M. Caignard-Latour**, in the discovery of the yeast cells, and concluded with a sketch of the perfection of the theory and science of fermentation as brought about by **Pasteur**. **Mr. Garrett** alluded to **Pasteur's** discoveries in the following concise passages:—"The scientific genius, whose name is so familiar from his investigations in the matter of hydrophobia, started with **Caignard-Latour's** discovery of the growth of yeast in fermentation, and he, too, submitted yeast cells to close inspection under a powerful lens. He then subjected yeast to the addition of a solution of plain sugar and water, with the result that, although it began to grow at first, it soon left off and began to wither away—a kind of fermentation going on all the time; the yeast, therefore, must have been living upon itself. It was evident that the yeast contained in itself an element of its own nourishment that sugar did not, and this element he had no difficulty in deciding to be nitrogen. To the plain sugar solution of the previous experiment he then added ammonia, which contains a large proportion of nitrogen, and, with a satisfactory confirmation of this theory, the yeast grew and flourished amain, the sugar broke up into alcohol and carbonic acid gas, and the process of fermentation went on so long as any sugar remained to be broken up, or any nitrogen for the food of the yeast.

"He now turned his attention to the wort. That contained a nitrogenous substance in the shape of albumen, gluten from the malt, besides mineral salts and other not less important matters; and the discovery of these mineral salts, I may observe in passing, induced him to add some of them to the original ammoniated sugar and water solution, with the result that the yeast was decidedly more active and thorough in its work than it was without them. These mineral salts were the same as those usually found in hard or spring water, hence the very natural deduction that hard water was favourable for brewing—a discovery that brewers have not failed to adopt to their own uses where required, and to the improvement of their productions.

MALT Liquors

Materials and Methods

MATERIALS. Yeast and fermentation

"Thus far we have learned that, during a satisfactory fermentation, sugar is divided into alcohol and carbonic acid gas—the former remaining in the mass and the other being given off; and we also have learned that this action is in some way associated with the presence of yeast and its active growth, and that this plant requires a nitrogenous substance for its development."

"Let us now leave wort to itself, without adding yeast. Will it ferment? Certainly not. The juice of the grape and other fruits will, and so will certain other vegetable juices, but not wort. It will undergo a change, though a rapid change a change of decomposition, if you will, but not of *alcoholic* fermentation. The result will not give carbonic acid and alcohol, but other offensive gasses; it is *putrefaction*, and that you can tell by your nose. The early addition of yeast prevents this process of putrefaction, or leads the decomposition into another and pleasanter channel, but chemists have not failed to take advantage of this to declare that fermentation and putrefaction are in a measure the same. That is, however, an argument I shall not enter into here, for, unhappily, we are fast emerging from the *simplicity* of the process, into a wild, confusing and chaotic maze on contradictory discoveries, or shall we say flights of scientific imagination, that will require all our tact and discretion to avoid. We must pick out the grains of corn from the chaff for ourselves, in the absence of Pasteur, the great pioneer.

"There are many fermentations, all of which have some character in common, but which, nevertheless, differ so materially from each other that I must be excused if I do but give you their names. They are styled *amygdalous* fermentation, *butyrous* fermentation, *gallous* or *tannous, lactous, mucous, pectous, saccharous, sinapous, urinous, viscous;* and, of course, our ancient and respected *acetous.* If chemists are only as energetic and enthusiastic upon the subject of fermentations as they were a very few years back, the chances are that there are great many more fermentations invented by now, and so I shall be pardoned perhaps if I invent one—a name only, though—for our own particular use. Alcoholic fermentation is termed vinous, which applies to wine and beer alike. I should like to call that 'beerous' which applies to beer. However, as I do not wish to offend the ears of any term chemical friends that may be present, we perhaps had better adhere to the 'alcoholic.'

"As I said before, these numerous fermentations have some features in common, but in their actions, complications, and productions they differ widely from each other. It is my conviction that Pasteur intended to examine each individual fermentation by itself, but that as fast as he took away one head to examine it a hundred others cropped up; so that it is not to be wondered at if he has been scared, by this Hydra-fermentation, and taken to Hydrophobia as a relief. But he did much good service in the subject before he associated his great name with rabid rabbits and mad dogs, and we have every reason to be thankful to him for it.

"In the limit of this paper it would not be possible to treat individually of this vast assemblage of different fermentations. Our immediate interest is confined to the fermentation of beer, and even then we must admit of at least two *opposed* kinds, and they are *alcoholic* and *acetous* or vinegar fermentation.

"We go back to the yeast and the wort, and must take it for granted that the materials provided are good, that the wort is amply saccharine and the yeast the scion of a healthy stock; and I may mention here, that there are described by various chemists nearly a hundred varieties of yeast, due possibly more to circumstances than to breed. Two at least are known to you—the *high* and the *low*—and these may vary greatly according to the acciden-

M. 96

MATERIALS.

Yeas and fermentation.

tal conditions of their life and growth. Of the bitter wort we need take no further notice than to state that to its existence, purity, and the slight changes it undergoes are, in a measure, due the aroma and flavour of the beer.

"That the yeast should be healthy is very important, for the health of the beer depends upon it. Yeast is a very deteriorating stock, unless preserved in its pristine vigour by good management and good wort; should it be found to be weak or sickly, it should be discarded or sold to the bakers, who are not so particular as a rule, for it is better to part with a little yeast at once than spoil a whole series of brews. Once in possession of a goodly stock of yeast, take care that you keep it. Never mix inferior yeast with it, for you will not improve the bad, but will certainly weaken the good. Neither can inferior yeast be improved by working in strong or extraordinarily good wort. But above all things, keep your yeast from contact with unclean vessels, or anything, I need scarcely tell you, that has the slightest vinegar taint.

"We know that *alcoholic fermentation* in wort results from the addition of yeast. Without the yeast we have nothing better than putrefaction."

"Now what are the conditions most favourable to the growth of yeast? *Suitable food, warmth, and air.* Of the food and warmth we will speak presently. What about the air? Some chemists doubt if yeast does require air for its development; but other chemists have settled that point to their satisfaction. When the yeast is in the tun with the wort the carbonic acid gas that is given off lies above the wort and forcibly keeps the air from any further contact, but there comes along the brewer with his ranser, and he gives fresh energy to the yeast. But does he do this by mingling air with the wort? The processes that wort usually goes through before the yeast is added are all calculated to introduce air, and perhaps there may be some air with the yeast, at any rate we have confirmation, of its influence when admitted.

"If a bottle of freshly made wort be boiled so that all the air it contained is driven out, and then hermetically sealed in an atmosphere of carbonic acid gas, such as may be found in any fermenting tun at work, we have, to all intents and purposes, a bottle of wort with which atmospheric air is not in contact. Into this sealed bottle, by a very simple process, yeast can be introduced without air and then the opening be sealed up again. Fermentation is set up and goes on, but the yeast does not thrive; it rather withers away. Admit air *instead of yeast*, and a change slowly sets in that is not alcoholic fermentation. The liquid becomes cloudy, muddy, and eventually gummy and ropy, for a form of decomposition has set in, and a mould forms on the surface. But how did the germs get into the bottle? They must have been imported by the air, for had the air not been introduced into the hermetically-sealed bottle, no change would have taken place for generation after generation. The bottled wort would have kept as well as bottled fruit, or tinned milk, meat or vegetables. What did the yeast, then, introduce?

"**Pasteur** had the natural curiosity to ask himself that question, and moreover he endeavoured to answer it. Placing a drop or two of the discoloured fluid under the lens of his microscope, what did he see? I must not tell you now; but it was a sight that held him fascinated and spell-bound with wonder and admiration. It was the great secret revealed—the secret that generations of chemists had failed to discover.

"The experiment was repeated again and again, always with the same result—the yeast would not decompose without the air and the same astonishing changes took place; the introduction of yeast *after clouding had set in* was too late to save the beer, for instead of flourishing, as it should have

MALT Liquors.

Materials and Methods

MATERIALS.

Yeast and fermentation.

done, it yielded sway to the stronger power of putrefaction, or other mischievous fermentation that was now set up, and virtually gave up the ghost. How was this? The explanation is at hand.

"Next, air was passed into the bottle that had passed through asbestos, or a plug of cotton wool, and no change took place in the wort, or only a very little; but when air *unhindered* was admitted, the same decomposition set up in due course. Now, what had this asbestos or cotton wool arrested on the road? Evidently, the germs of decomposition. So **Pasteur** placed that under his lens, and then he saw, adhering to the wool, a number of little bodies. To these he added fresh wort, and, to his delight, they filled out and immediately became *plants and living creatures!* just the same, in fact, as those he had seen before in the decomposing wort.

"Here we have the great microbe discovery—the greatest discovery of the age, although it is microscopic.

To the presence of these germs **Pasteur** was quite entitled to ascribe the process of decomposition, fermentation, or putrefaction; the microbes of yeast, he ascertained, were quite distinct in character from those of putrefaction, and those of acetous fermentation, he showed again, were different from either of these, and all were different from each other.

"This microbe question is undoubtedly a great discovery, but how, or in what manner, they effect the various changes that they invariably attend requires much further investigation. That they are very materially concerned is shown by the fact that if wort be boiled, that is, raised to the boiling point, and then subjected only to air that has been heated to redness, so that all microbes are destroyed, no changes of ferment or putrefaction set in."

"The influence of air in the process of fermentation is interestingly shown if yeast be added to fresh wort in an open shallow vessel. This process is then excessively rapid. The rate of decomposition is intensified. The sugar is more speedily consumed by combustion, and then about 25 per cent. of the sugar is used by the yeast, leaving rather less than three-fourths of its quantity to be converted into alcohol and carbonic acid, and the product is proportionately weak, and, therefore, very liable to be overcome by acetous fermentation or putrefaction. Under ordinary circumstances only 5 per cent. of the sugar should be unconverted, leaving 95 per cent. to be converted into alcohol and carbonic acid—a great and fatal difference, because the presence of alcohol hinders putrefaction, and its absence permits it.

"Yeast, like every other plant or growing thing, requires warmth for its development; therefore it is arranged that, by the very process of breaking up the sugar, heat shall be evolved, some of which the yeast requires for the gentle nurture of its progeny or young cells, and the balance remains with the wort, and becomes sensible to the thermometer.

"If the sugar breaks up, or *burns* too fast, an excessive growth of sickly hothouse yeast follows as a matter of course; or if you prefer it, I will reverse the order of expression, and say that if the yeast grows so rapidly that the sugar is broken up by its agency too fast, then the sugar is extravagantly wasted in the process and less alcohol results."

"Again, yeastic fermentation must not be too slow or too long delayed in being set up, or other ferments which are not alcoholic will take the place of the yeast, having been imported by the air or utensils, and the two the brewer has most to fear are either vinegar germs or perhaps putrefaction. Therefore, the brewer has quite enough to do to steer and maintain his balance between the rocks and quicksands. Let him once disregard the one or the other, and damage to the beer ensues.

MATERIALS.

Yeast and fermentation.

"But the brewer regulates all this in his own way, and I doubt if we can teach him any better. He watches the indication of the thermometer, for full well he knows that anything unusual about that tells of something wrong with the brew. If the temperature be too high, the process is too quick and he fears the weakening result of excessive combustion; if too low the process is too slow and weak, and he has visions of vinegar fermentation. It is not my province to speak of practical brewing. You, gentlemen, are better able to cope with that subject than I; my task is to speak of the principles of fermentation as applied to brewing, and so we come back once more to the consideration of the yeast.

"The food for yeast is found in the nitrogenous substances provided by the malt, which are extracted by spurging, and it is important that this should exist in the wort in sufficient quantity to feed the yeast so long as there is sugar to be broken up. Should the nitrogenous matter be in excess, the yeast will go on devouring it at the expense of the alcoholic strength. Should the nitrogenous matter be exhausted before the sugar is consumed, the process is arrested too early, also at the expense of alcoholic strength, and either of these causes will affect the stability of the brew. Doubtless you have your remedies for these defects, based upon strictly scientific principles, confirmed by practical experience, so that I need not touch upon them. But how does the yeast effect its work? It is by actual contact with the sugar, for if a solution of ammoniated sugar-water in a bladder be suspended in working wort, it is absolutely unaffected. The mystery has not yet been quite satisfactorily explained, beyond that previously suggested by the law of affinity; the bonds of the sugar's combination of elements are loosened by the agency of the ferment, so that we must leave the further consideration of yeastic influence to the future, when something more shall be credited to the scientific labours of our genuine working chemists."

During the discussion which followed the reading of **Mr. Garrett's** paper on fermentation, the chairman at the meeting, **Mr. H. Whymper,** drew attention to the remark which had been made that malt wort if left to itself undergoes no alcoholic fermentation. As bearing on this subject, **Mr. Whymper** mentioned a discovery made by him, *viz.*, that barley dust would, however, establish fermentation in the wort, and hence in his opinion the outer coat of the grain bore the microbe necessary; in other words, it afforded, as he called it, "the seed of a wild yeast." This statement was confirmed by **Mr. L. Briant,** who on procuring barley under such circumstances as to remove any suspicion of yeast derived from a malt house obtained the ferment from its dust. Several gentlemen, present at the conference, stated that this was no new discovery, that it had been observed by several brewers before, but that no importance had been attached to the fact. The subject would, however, seem, as **Mr. Whymper** affirmed, to be of the very greatest interest, since it has been customary to suppose, grapes, apples, and one or two other fruits, peculiar in that they carry the mycrobes necessary for their own fermentation. **Mr. Whymper's** observation shows that this is not confined to such fruits, but that it exists with barley. And the writer would add, not with barley only, but very probably with many other grains, since the rice and millet beers made by the people of India are all known to become alcoholic if left to stand for a few days. Thus the Angami Naga rice beer which for the first three days, after the hot water has been poured over the grain, tastes like butter milk, in a day or two undergoes fermentation and becomes powerfully intoxicating. But, besides this fact, many plants are known in India to possess the direct power of establishing vinous fermentation in saccharine fluids. These, like the barley alluded to by **Mr. Whymper,** doubtless bear microbes, the na-

MALT Liquors.

Materials and Methods

MATERIALS.
Yeast and Fermentation

ture of which has not as yet been investigated. Thus, for example, **Mr. C. B. Clarke** pointed out that the Khásia Hill people were able to procure fermentation from the flowering spikes of **Rhyncospora aurea,** a cyperaceous plant found plentifully in that region. The writer, during a journey in Manipur, found that the powdered stems of a climber (probably a **Millettia**), mixed with rice flour and made into cakes, were similarly employed to establish fermentation in rice wort (*Conf.* with Vol. II., 259). It need only be added that other examples might be given, sufficient to justify the suspicion that the *tari* or palm juice so largely used in India as a source of yeast may become spontaneously charged with the germs of vinous fermentation.

The choice of yeast and its treatment for future use are of great importance. If the brewer desires a favourable course of fermentation he must first of all set the wort with good yeast. It may, indeed, be said that out of a hundred cases of defective fermentation ninety-nine can be traced to the bad quality of the yeast. Good brewing yeast is that which thrives best at a low temperature. The deterioration of yeast in this country is due to the high temperature at which fermentation has to be carried on, especially in the plains. A free use of ice would mitigate the evil; but at the same time it would seriously increase the selling price of the beer.

The Indian brewer has often on this account to propagate his exotic yeast in the cool climate of the hills and send to the plains supplies as required at the breweries. The result of this treatment is that the plant is in a constant state of deterioration and a continuous fresh supply from Europe becomes necessary.

Mr. Whymper has kindly furnished the author, with the following instructive reply to the enquiry regarding yeast:—"Indian brewers now very generally preserve yeast by drying it with charcoal powder. It only requires to be mixed with the charcoal and dried at a moderate heat. In this condition it will remain active for many months. This system has not been many years in use. Prior to my knowing of it, I used, whenever possible, yeast preserved in plasters of sorts. **Mr. Percy Adams** of Halstead, Essex, preserves good yeast made up in small marbles. Formerly I used dessicated yeast made up with powdered plaster-of-paris. This neyer gave such good results as the solid balls. But these methods of preservation are not absolutely necessary, as yeast can be produced in India. I can now undertake to start a true alcoholic fermentation in malt wort by the mere addition of the dust from the skin of barley. It may be said that it is now accepted that the dust or bloom of barley contains germs of several ferments."

Water.
97

IV.—WATER.

In the passage quoted above from **Mr. Garrett's** paper, which was read before the Brewers' Congress, reference has been made to **Pasteur's** discovery of the advantage of mineral salts in the water of the wort, fermentation being more vigorous and thorough in its operations. Commenting on this fact **Mr. Garrett** added that brewers had not failed to adapt to their own use the advantages thus demonstrated of a hard over a soft water, such as that to be obtained from wells or springs. The following passage from *Spons' Encyclopædia* gives the most prevalent ideas regarding the water suitable for brewing:—"A constant unfailing supply of good water is indispensable in brewing: though what really constitutes good water is a point upon which many brewers and chemists have long been at issue. Some rest their faith upon a soft water, others will use only the hardest water they can get, while others, again, are quite indifferent, and will use either. It is now, however, a generally accepted fact that water for brewing should not contain organic matter, but a considerable quantity of inor-

MATERIALS. Water.

ganic or saline constituents; these varying in nature and quantity, according as the beer to be made is required for keeping or for immediate consumption. English brewers are now agreed that the water should contain much carbonate and sulphate of lime. The former of these two ingredients is the most necessary, but they should both be present in the water from which ale is to be made; water used in brewing porter may contain the carbonate alone. For the best ales, the proportions seem to be from 10 to 20 grains a gallon of each. The excellence of the ales made by the Burton brewers is doubtless due to the quality of water used by them; it is very hard, and contains, as will be seen from the analyses given below, a large proportion of alkaline sulphates and carbonates, this is the best argument that can be brought forward in favour of the use of hard water. The supply is derived entirely from springs, and not, as some suppose, from the river Trent. It has also been urged, as an advantage, that hard water increases the quantity of saccharine matter held in the wort, thus heightening the flavour and preventing it from becoming acid." The following table represents two analyses of the water used by the Burton brewers:—

Two Analyses of Burton Water.

	(1)	(2)	
Chloride of sodium . . .	10·12	...	grains a gallon.
Sulphate of potash . . .	7·65	...	"
" lime . . .	18·96	54·40	"
" magnesia . . .	9·95	0·83	"
Carbonate of lime . . .	15·51	9·93	"
" magnesia . . .	1·70	...	"
" iron . . .	0·60	...	"
Silicic acid . . .	0·79	...	"
Cholride of calcium . . .	...	13·28	"
	65·28	78·44	

Mr. Whymper has obligingly furnished the writer with the analyses of some of the principal waters used by Indian breweries. These have been made by Government analysts and can, therefore, be depended upon as accurate. Space cannot, however, be afforded for the publication of these in this work, but in purity from organic and other injurious materials they are second to none, while they possess the necessary amount of salts to render them good brewing waters. Thus the water used by the Crown Brewery Company contains 46·62 grains of inorganic salts to the gallon. The proportion of these salts varies slightly; in one of the returns they amounted to only 41·8640 grains per gallon, the sulphate of magnesia being 14·06, carbonate of magnesia 7·57, sulphate of lime 8·78, and carbonate of lime 6·46.

Most of the Indian breweries use spring water filtered and boiled before being used.

V.—Malting.

Malting. 98

The first stage in brewing is that by which the grain is converted into malt. This technically is known as Malting and it consists of *steeping*, *couching*, *flooring*, and *kiln-drying*. In these various stages the grain is made to germinate and when it has reached the desired degree its further growth is arrested by the *kiln* heat. The chemistry of this operation has already been alluded to, namely, the conversion of the insoluble starch of the grain into a saccharine soluble materials. But it is not necessary to deal in this work with each of the stages of malting, since they are in India identical with those pursued in Europe.

VI.—Brewing.

Brewing. 99

This comprises the stages known as *grinding*, *mashing*, *boiling*, *hopping*, and *cooling* and afterwards of *fermenting*, *cleaning*, and *storing*.

MALT Liquors.

Materials and Methods of Brewing.

METHOD OF BREWING.

But as with MALTING it may be said that the Indian brewers follow closely on the lines of their European contemporaries and that they employ the same materials, machinery, and appliances.

The following information furnished by **Mr. H. Whymper** will, however, be read with special interest as giving facts of an Indian nature:—"The brewing season for nearly all Indian breweries is restricted by the short Winter. In Ootacamund the temperature allows brewing to be carried on all the year round, but elsewhere the season is from October to March. The worts are cooled in the ordinary manners first by exposure on shallow vessels termed coolers and thereafter by flowing over ordinary refrigerators through which cold water flows. Cellars are cooled by being left open in the coldest weather. No artificial means of cooling has yet been adopted, but the largest brewery (that recently erected at Rawalpindi) is now constructing powerful ice machinery for cellar cooling.

"The class of beers, &c., made in India is practically the same as in England, more light gravity beer is consumed, however, than in England. Wood is almost invariably employed as fuel except for drying malt on kilns when charcoal is used.

"Labour is much more expensive than in England (*a*) from the large number of men required to do the simplest job, (*b*) from the careless and indifferent manner in which work is done, and in which machinery is treated."

Pasteurisation.—"This process only applies to bottled beers. There is no known practical process of pasteurising beer in wood. Indian brewers generally are now adopting this process which consists merely in immersing beer when bottled in cold water and raising the temperature gradually by the injection of steam, until it reaches 135° to 140° Fahr., at which temperature it is maintained for about half an hour. By this process all germs which would ordinarily incite fermentation of sorts (vinous, lactic or hectic) are destroyed. As bottled beer ordinarily depends upon a fermentation in the bottle to obtain its briskness (otherwise its carbonic acid gas) and as no such fermentation takes place with pasteurised beer, it is necessary to bottle beer, which is to be pasteurised, highly charged with gas. Theoretically such beer should remain without change for ever, and practically it does remain sound for very much longer than it ordinarily would."

VII.—Barrels, Vats, Bottles, Corks, &c.

Barrels. 100

Vats. 101

Bottles. 102

Corks. 103

Before proceeding to discuss the trade in Beers and Ales, it is necessary to say something about the construction of beer-barrels, &c. The barrels or casks used in the breweries of this country are almost without exception constructed of oak, and are either made up in India from rough staves imported from the Baltic, or imported in shook, *i.e.*, bundles from London and remade in this country. Several attempts have been made to utilise the indigenous timbers of this country for barrel-making, but the extensive and various forests of India have failed up to the present moment in producing a wood good enough to replace the English or European oak. Sál (**Shorea robusta**) has been tried with some success in the construction of vats. White cedar from the Malabar Coast makes a good-looking vat, but its use is somewhat dangerous in consequence of the absorbent nature of the wood rendering it very liable to crust. The English oak would find a strong competition in the Indian Ash and Teak if the former could be got in larger quantity and the latter at cheaper rates. Deodar, and the wood of pines generally, impart their resinous properties to the beer. The brewers of India are very anxious to find a wood which would successfully compete with the expensive and indispensable English oak, but hitherto the efforts to find such a timber have been unsuccessful. The valuable characteristics of oak are its freedom from knots, its density,

METHOD OF BREWING.

durability, and lightness. It is also non-absorbent and thus not liable to impart its resin to the beer.

BOTTLES are purchased in the country. **Mr. H. Whymper** experimentally manufactured bottles at Jhelum from local materials for some time, but was unsuccessful and accordingly abandoned the enterprise. CORKS are all procured from England. BREWER'S GRAINS, that is, refuse malt, is usually sold by contract to zemindars or cattle owners, and in most cases is easily disposed of, but the price obtained fluctuates with the value of the fodder crop.

TRADE IN BEER AND OTHER MALT LIQUORS.

TRADE. 104

Mr. H. Whymper in his article on *Brewing in India* gives the following historic sketch of the trade in imported beer in this country :—"What was termed country-brewed beer was very generally made in India at the close of the last century and early in this. This is said to have consisted of about one-fifth of a bottle of porter (English), a wine-glassful of toddy or palm juice, some ginger and brown sugar, a squeeze of lime completing the ingredients. The toddy itself supplied the fermenting power and when this mixture had slightly fermented it was considered fit for use. This drink was in vogue when a London brewer, **Hodgson** by name, about the year 1816, began to ship a well-hopped and rather heavy beer to India. It quickly became known as India pale ale, and **Hodgson** speedily acquired a complete monopoly of the Indian trade. Until 1825 or 1826 he held the Indian market at his mercy, and his mercies were cruel. He kept out rivals at times by lowering his rates below cost price, and having stopped other brewers' shipments, up his prices went. It is reported that on more than one occasion he suddenly raised his price to £20 per hogshead. In 1824 he advanced to £24 per hogshead and refused all credit. The result was what he might have foreseen, a revolt of all interested in the beer trade taking place. This was about 1825. Very shortly after this we find the beers of **Bass, Allsopp** and '**Ind & Smith**' in the market against **Hodgson**, and by 1840 his beers were only a memory.

"There is no doubt in the writer's mind that had the early shippers only kept up the quality of their beers, the whole of the Indian trade might have been in their hands at this moment. Although plenty of good beer went out, the general quality almost invariably fell off or failed to meet the public taste after a few shipments. Occasionally such bad beers were in the market that it was inevitable people should say : 'Well, if we can't get better beer than that we might as well brew our own, as we couldn't possibly brew worse :' this certainly occasioned at least one essay in the early days of Indian brewing. The shipments of beer for the use of the army were not a very great, if any, exception to this rule. Plenty of good beer went out, but, every now and then, the Government was startled by thousands of hogsheads of beer proving so bad that they had to be run into the nearest sea or river. It was thus no peculiarly favourable local circumstances which caused the rise of the brewing industry in India. There are no such circumstances, there are difficulties which every English brewer who goes to India looks upon at first as insuperable; but beers which did not meet the public taste and were inferior and bad, coupled with high prices, gave the Indian brewer his chance. All that the Indian brewer has in his favour is being nearer to the markets he sells in. Against this he has endless difficulties ; he has to import and order, and often pay for ahead, his hops, casks, machinery ; he has to keep in reserve duplicates of everything likely to break down. He has to import all his supervising staff of servants ; if he gets an unsuitable man he has to put up with him. His most serious difficulty is, that owing to the above circumstances and from

MALT Liquors.

Trade in Malt Liquors.

TRADE.

having to do all his own malting, he has to employ three times the capital he would have to do in England for a similar trade to his own."

"The writer's experience," continues **Mr. Whymper,** "does not go back beyond 1866. In that year, and for several years after, the declared quantity was about 200,000 barrels; it is now about 60,000 barrels. The value was then about £600,000, and it now averages about £200,000. The quantity of beer brewed in India in 1866 was probably not more than 2,500 barrels, certainly not more than 3,000, whilst in the present year it will possibly reach to 170,000 barrels and will certainly be over 150,000 barrels. The limit of the whole trade to be done with the European population of India is probably 250,000 barrels."

"The trade will not likely expand beyond this until the Government relaxes certain rules, which, whilst they restrict the sale of beer in some districts, unquestionably foster the consumption of spirits. In Southern India, for instance, the brewer is not allowed to brew beer for native consumption above a certain alcoholic strength, and this strength is not sufficient for the native palate."

Turning to the official records of the BEER TRADE the following table may be given of the foreign beers brought into India. It will be observed that the imports fell off steadily from 1866 to 1878-79, but that from the latter date they have since steadily improved, until, at the present time, they are nearly as large as they were prior to the existence of Indian breweries:—

*Import of Malt Liquors from Foreign Countries.**

Years.	Gallons.	R
1866-67	Not given.	55,20,245
1867-68	2,268,298	43,57,701
1868-69	1,816,106	38,17,734
1869-70	1,898,762	41,35,199
1870-71	1,642,131	31,16,860
1871-72	1,499,877	30,53,186
1872-73	1,536,496	36,34,956
1873-74	1,435,345	33,79,155
1874-75	1,481,698	34,98,438
1875-76	1,143,157	26,81,065
1876-77	1,176,922	27,06,644
1877-78	1,328,077	31,30,700
1878-79	1,089,211	24,45,685
1879-80	1,065,347	25,42,620
1880-81	1,152,678	28,49,349
1881-82	1,199,395	28,46,121
1882-83	1,170,554	27,23,226
1883-84	1,261,444	30,32,236
1884-85	1,066,913	24,99,272
1885-86	1,299,408	30,06,098
1886-87	1,715,638	35,40,257
1887-88	2,138,518	39,71,534
1888-89	2,398,580	41,28,517

If to the imports of last year be added the amount of beer made in India during the year 1888-89 (*viz.*, 5,165,138 gallons), the total consumption in India must have been for that year 7,563,718 gallons, fully three times as much as in 1866. For further particulars regarding the trade in Malt Liquors the reader is referred to the article **Narcotics,** section MALT LIQUORS, pp. 328—330.

* Conf. with table on p. 329.

MALVA, *Linn.; Gen. Pl., I., 201.*

Malva parviflora, *Linn.; Fl. Br. Ind., I., 321;* MALVACEÆ. 105

Vern.—*Nárr, panirak, supra, sonchal, nanna, gogi ság,* PB.

References.—*Stewart, Pb. Pl., 23; Murray, Pl. and Drugs, Sind, 57; Pharmacogr. Indica, 228; Gazetteers:—N. W. Provs., IV., 68; X., 306; Journal, Agri.-Horti. Society XIV., 7; Indian Forester, VIII., 177.*

Habitat.—A small, spreading herb, found in the North-West Himálava (altitude 1,000 to 2,000 feet), also in Upper Bengal, Sind, and the Panjáb. Distributed through Europe, the Levant, Arabia, and Nubia.

Medicine.—The SEEDS are used as a demulcent in coughs and ulcers in the bladder. MEDICINE. Seeds. 106

Food.—Frequently eaten as a POT-HERB by the Natives, specially in times of scarcity. FOOD. Pot-herb. 107

Domestic.—In Kanáwar the ROOT is used by women to cleanse their hair; woollen cloth is also washed with it. **Bellew** states that the root is employed as *rísha khatmí* (*see* **Althæa rosea**). DOMESTIC. Root. 108

M. rotundifolia, *Linn.; Fl. Br. Ind., I., 320.* 109

Syn.—M. VULGARIS, *Fries.*

The *Flora of British India* describes two varieties under this species:—

Var. 1, borealis: Bengal: Mysore.

Syn.—M. ROTUNDIFOLIA, *Roxb.;* M. BOREALIS, *Wallm., et Boiss.;* M. ROTUNDIFOLIA, *var.* β in *W. & A. Prod.;* M. PARVIFLORA, *Huds.*

Var. 2, reticulata: Bengal and N.-W. Himálaya.

Syn.—M. ROTUNDIFOLIA, *var.* α in *W. & A. Prod.*

Vern.—*Sonchala, khubazi,* HIND.; *Kúkerai,* PUSHTU.; *Chandiri, khabazi* (seed and fruit), SIND; *Trikúla malle,* TEL.

References.—*Elliot, Flora Andh., 184; Aitchison, Botany Afghan Delimitation Commission; Murray, Pl. and Drugs, Sind, 58; Atkinson, Him. Dist., 306, 741; Gazetteers:—N.-W. Provs., IV., 68; Mysore and Coorg, I., 56.*

Habitat.—A much branched spreading herb found in the North-West Provinces, Kumáon, and Sind. Distributed through Europe and Western Asia.

Medicine.—The SEEDS possess demulcent properties. They are prescribed in bronchitis, cough, inflammation and ulceration of the bladder, and in hæmorrhoids; they are also externally applied in skin diseases. The LEAVES, being mucilaginous and emollient, are employed as an external application in scurvy; they are also reckoned useful in piles. MEDICINE. Seeds. 110 Leaves. 111

Food and Fodder.—In some parts of Sind, the LEAVES are eaten as a pot-herb. The SEEDS are also reported by **Mr. Lace** to be eaten by the people of Quetta, and the PLANT used as fodder for cattle. FOOD & FODDER. Leaves. 112 Seeds. 113 Plant. 114

M. sylvestris, *Linn.; Fl. Br. Ind., I., 320.* 115

THE COMMON MALLOW.

Vern.—*Viláyati-kangai, gúlkheir,* HIND.; *Khatmi* (PATNA), BENG.; *Kanji, tilchuni,* N.-W. P.; *Gul-i-khadmi,* AFG.; *Khabájhi,* SIND; *Khubazi,* BOMB.; *Vilayati-kangói,* DEC.; *Khubazi, khitmi,* ARAB.; *Khubází, towdrie,* (*nán-i-kulágh*=crow's-bread, *khitmi-i-kuchak*=small-khitmi), PERS. NOTE.—It will be seen from the remarks below that all the provincial names for this plant that have been derived from the Persian *Kangai,* or *Kangoi,* and hence probably refer to **Abutilon.**

References.—*Honigberger, Thirty-five years in the East, 304; Aitchison, Botany Afgh. Del. Com., 43; Ainslie, Mat. Ind., I, 205; O'Shaughnessy, Beng. Dispens., 214; Moodeen Sheriff, Supp. Pharm. Ind. 19, 170; Dymock, Mat. Med. W. Ind., 2nd Ed., 80; S. Arjun, Bomb. Drugs, 18; Murray, Pl. and Drugs, Sind, 58; Year Book of Pharm., 1874, 115, 623;*

The Common Mallow.

1878, 288; Irvine, Mat. Med. Patna, 49; Pharmacog., Ind., I., 204; Atkinson Him. Dist., 306, 741; Birdwood, Bomb. Pr., 10; Royle, Fib. Pl., 263; Gazetteers:—Mysore and Coorg, I., 58; Agri.-Hort. Soc. Ind. Journal, XIV., 15.

Habitat.—An erect, nearly glabrous herb, met with in the Western Temperate Himálaya, from Kumáon, at an altitude of 2,500 feet, to the Panjáb and Kashmír. Distributed to Europe, Northern Africa, and Siberia.

FIBRE. 116 **Fibre.**—The plant abounds in fibre (*Royle*).

MEDICINE. **Medicine.**—Most of the MALVAS contain much mucilage and have demulcent and emollient properties, but **M. sylvestris** perhaps enjoys the first rank as a medicinal species. To a large extent it takes the place, in Asiatic pharmacy, which **Althæa officinalis** (the Marsh Mallow) holds on the Continent of Europe. It appears to have been known to the early Muhammadan physicians, who probably derived their knowledge of it from the Greeks. The early Sanskrit medical writers do not seem to have placed much importance on malvaceous mucilaginous preparations. The modern Hindu doctors, however, following perhaps the Muhammadans, prescribe such preparations, but having adopted tropical substitutes for Malva, have caused a certain amount of ambiguity in their restricted application of vernacular names to the plants so employed. **Dr. Moodeen Sheriff** writes under **Abutilon indicum**—the COUNTRY MALLOW of popular writers:—"The word *kanghí* or *kangói* is not only used incorrectly in some books as synonymous with the Arabic and Persian words *khabbázi, khitmi, tódarí* (*the names of three different drugs*), but is also confounded with the word *kangóní* or '*Coongoonie*' as it is generally written. The latter is one of the Dukhni synonyms of the seeds of **Setaria italica**" (*Conf. with Dictionary Econ. Prod., I., 14—17*). **Dr. Moodeen Sheriff** recommends that when these words are used to denote **Malva sylvestris** *var.* **mauritiana** (the Bengal and Western India form of the species), the prefix *viláyatí* (=foreign) should be given, thus *viláyatí-kangói*. According to **Moodeen Sheriff**, therefore, **Abutilon** is the true *kanghí* of India, and the association of that name with **Malva sylvestris** would therefore be incorrect. This may be so, but it is more likely that *kanghí* has a generic and medicinal signification, denoting the mucilaginous **Malvaceæ** rather than any one species of **Abutilon** or **Malva**. At all events, the confusion that has arisen, is in the application of that name and its synonyms to closely allied plants which all possess the same or nearly the same properties. The *balá* of Sanskrit writers was a cooling preparation made, apparently, from the roots of various species of **Sida**. In the Pharmacopœia of India the mucilaginous substitute recommended for mallow is made from **Hibiscus esculentus**. This is described as a valuable emollient and demulcent, also diuretic. Of the same nature are the preparations, used by the people of India, made from **Corchorus** (see Vol. II., 540 and 543).

Birdwood suggests that **Malva sylvestris** is probably the μαλαχη χεροατα (Malakhí) of **Dioscorides**, which was known to the Egyptians as *Khokorteen*. But **Alpinus** figures and describes **Corchorus** as Melochia. **Dymock**, accepting the same rendering, remarks that "the Mahometans probably derived their knowledge of its medicinal properties from the Greeks. **Mauluna Nafí** describes three varieties, *viz.*:—

1st—A cultivated kind called *Malokia.*
2nd—A large wild kind called *Khitmí.*
3rd—A small wild kind called *Khubází.*

The author of the *Makhzan-el-Adwiya* pronounces the last mentioned to be the article now known as *Khubází.*"

Plan 117 **Dymock** adds that "all parts of the PLANT are commended in Muhammadan works on account of their mucilaginous and cooling properties, but the FRUIT is considered to be most efficient." **Irvine** speaks of the SEEDS as generally employed, the dose being in infusion ʒii to ʒ *ss.* **Honigberger** remarks "the seeds are used by the Hakims in cough, and ulceration of the bladder." Commenting on the nature of the *tódarí* seeds of the bazárs he says there are two kinds, the white (the seeds **of Polyanthes tuberosa**) and the coloured (the seeds of **Malva sylvestris**). From the passage quoted above it will be seen that **Moodeen Sheriff** dissents from the opinion that *Tódarí* is a Persian synonym for *Khubází*, and **Dr. Dymock** has determined the Persian seeds imported at the present day into Bombay under the name *Towdrí* to be those of **Lepidium iberis.** MEDICINE. Fruit. 118 Seeds. 119

Ainslie says that "the **Hindu** doctors prescribe the expressed JUICE, internally, in gonorrhœa, and give an infusion of the ROOT as a drink in fevers." From the fact, however, that **Ainslie** speaks of the plant, to which he refers, as being very common in South India, growing by road-sides and having yellow flowers, it seems probable that he refers to **Abutilon** and not to **Malva.** The LEAVES of **Malva sylvestris** are generally reported to be employed in the preparation of an emollient cataplasm. **Murray** writes that the plant is largely used by native drug-sellers in the formation of a decoction which contains in addition to **Malva,** rose petals and sugar-candy. This is said to be prescribed in strangury. The mucilaginous property of the plant is found beneficial as an external application in irritation of the skin, and a poultice of the leaves is sometimes employed in fomentation, very much after the same manner as Marsh Mallow is used on the Continent of Europe. Juice. 120 Root. 121 Leaves. 122

Trade.—The fruit is imported from Persia into Bombay under the name *Khubází*: it is worth about 4 annas per pound (*Dymock*). **Irvine** (*Mat. Med. Patna*) says that in his time it sold at 5 annas a pound. TRADE. 123

Food.—Was eaten by the Romans as a vegetable, and where it occurs in India it is also eaten by the people like most other species of Malva. FOOD. Plant. 124

Malva verticillata, *L.; Fl. Br. Ind., I., 320; Wight, Ic., t. 950.* 125

Syn.—M. NEILGHERRENSIS, *Wight.;* M. ALCHEMILLÆFOLIA, *Wall.*

Vern.—*Laffa,* ASSAM.

References.—*N.-W. P. Gazetteer, X., 306; Proceedings of the Rev. & Agri. Dept, Agri. File No. 6, Serial No. 28 of 1888 (condition of the People of Assam).*

Habitat.—An erect annual, or perennial herb, in the temperate Himálaya (ascending from 6,000 to 12,000 feet), from Assam and Sikkim to Kumáon and Lahoul. It is found in the cornfields of the Nilghiris. Distributed through Europe, Abyssinia, Egypt, Amoorland, and China.

Food.—This HERB is grown in patches on homestead land in Assam, where it is a very general custom among the natives to boil the leaves and tender shoots, and eat them as spinach with rice. FOOD. Herb. 126

Manalú Oil of Kanara is said to be used for lamps. The plant which yields this oil is not known. 127

Mandioca or **Manioc Meal,** see **Manihot Glaziovii,** *Müll.*, p. 157.

MANDRAGORA, *Juss.; Gen. Pl., II., 900.*

Mandragora microcarpa, *Bert.*, SOLANACEÆ. 128

THE OFFICINAL MANDRAKE.

Vern.—*Luckmuna, luckmunie, lufah,* HIND.; *Yebruj,* BENG.; *Kaat-júti,* TAM.; *Loofahat,* MALAY.; *Ustrung, serag-al-coshrob, ussul-ul-lufah* (root), *lufah,* (plant), *tufah-ul-shitan* (fruit), ARAB.; *Yabrooz, merdumgeeah,* PERS. The above vernacular names are given on the authority of **Sir George Birdwood's** *Bombay Products,* 61; also **Irvine's** *Mat. Med. Patna,* 61.

Habitat.—Indigenous in South Europe and Asia Minor. The indigenous species met with in India (M. **caulescens**, *Clarke, Fl. Br. Ind. IV.,* *242*) occurs in Alpine Sikkim at an altitude of 12,000 to 13,000 feet. It is not known whether the roots of that species possess the medicinal properties assigned by older writers to the Mandrake.

MEDICINE. Root. 129 **Medicine.**—This plant has been mentioned by **Birdwood** among his drugs, but without giving any information about its medicinal virtues. The *Pharmacographia* says that the ROOT, as also that of **M. officinarum,** and of **M. vernalis**, are very nearly allied, in appearance and structure, to the roots of **Atropa Belladonna,** *L.* (see Vol. I., 351—353). **O'Shaughnessy** (*Beng. Disp.*, *466*) says that the root was celebrated in the magic rites and the toxicology of the ancients and is known now in the bazárs of Central Asia and Northern India. Its properties are said to be identical with those of **Belladonna** although weaker. **Dr. Dymock** informs the writer that it is worn as a charm in India. **Irvine** mentions that the drug is used in Patna as a narcotic in doses of ⅛ to ½ grain.,

130 **MANGANESE,** *Ball, Geology of Ind., III., 326.*

The ores of Manganese are numerous and somewhat widely disseminated, though they rarely occur in any quantity in one place. **Ball** states that the commonest ores in India are Manganite, or the gray oxide; Wad, or the earthly protoxide; Pyrolusite, or the black peroxide; Psilomelane, a combination of the oxide with baryta; Hausmanite or peroxide occurring with other ores of the metal; and Braunite, or binoxide in combination with iron peroxide, silica and magnesia.

131 **Manganese.**

Syn.—BRAUNSTEIN, GLASSEISE, *Germ.*; SAVON DU VERRE, *Fr.*; BRUINSTEIN.

Vern.—Peroxide=*kolsa-ka-pathar, ingani, missi siyá*, HIND.; *Nijni, injani, ingani, jugni*, PB.; *Iddali kalu*, TEL.

References.—*Baden Powell, Pb. Pr., 100; Mason's Burma, 570, 587, 735; Balfour, Cyclop., II., 845; Ure, Dict Indus. Arts and Manu., III., 35; Madras Manual, of Admn., II., 36; Manual, Coimbatore District, 23; Settlement Reports Nagpore District, Sup. 276; Tropical Agril., Feb. 1889, p. 509; Forbes Watson's Ind. Surv., I., 406.*

OCCURRENCE. 132 **Occurrence.**—The following short account of the presence of the ores of manganese in India has been principally extracted from **Ball's** *Economic Geology* (*l. c.*), to which the reader is referred for fuller information:—

MANGANESE ores are chiefly found in the older crystalline or metamorphic rocks, but they occasionally occur in younger sedimentary and unaltered formatios. In India, indeed, a not unfrequent source is laterite, though in such a rock, as might be expected, the deposits are not constant over large areas. "It is possible," **Ball** writes, "that manganese is much more abundant in this association than is generally thought, since on the weathered surface it resembles ordinary laterite and might easily escape detection."

Madras. 133 Ores of manganese occur in MADRAS, in the Nilghiris, Mysore, Kadapah, Karnul, Bellary, Vizagapatam, and Hyderabad. Bengal. 134 In BENGAL, manganiferous limonite is found in some abundance in the neighbourhood of Chaibassa in Singbhum. Central Provinces. 135 In the CENTRAL PROVINCES, a deposit of manganese ore is met with in the neighbourhoodof Gosalpur in the Jabalpur district, and in the metamorphic rocks north of Nagpur a rich black oxide is said to be abundant; an impure ore, probably of little value if saleable at all, has also been found in the red clays of the Kamthi series around Malagarh in the Berar, Wun district, and an impure psilomelane in the South Rewah coalfield. Rajputana. 136 In RAJPUTANA a mixture of limonite, magnetite, and oxide of manganese is found in the iron mines of Bhangarh in Alwar State, and small veins of oxide of manganese occur in fault rock near Dabunda in Bundi State. Deposits are said to have been found near Wodoorti, a locality in the Dharwar district of BOMBAY, but no recent account of the ore is available. In LOWER BURMA, the occurrence of manganese ores has been described at three localities in the vicinity of the Great Tenasserim river; one on the bank of the Thugoo stream, one on the Therabuen and the third at an intermediate spot where the Great Tenasserim intersects an outcrop of the ore. No opinion could be formed as to the extent of the deposit owing to the scarcity of stream sections, and the thickness of the vegetation, but it was thought to be not improbable that a bed of ore, several square miles in extent, existed, which united the three localities. Even without this being necessarily the case, it was said that sufficient to pay working could be obtained at these points. The ores consisted of black and grey oxides and wad. **Mason** states that he has seen manganiferous iron from one of the islands south of the Mergui. Pyrolusite has been obtained from UPPER BURMA, but nothing is known as to the mode of its occurrence.

OCCURRENCE. Bombay. 137 Lower Burma. 138 Upper Burma. 139

CHEMICAL COMPOSITION.

ANALYSES OF INDIAN ORES.—Specimens of ore from Vizagapatam and Bimlipatam, were analysed by **Dr. A. J. Scott** in Edinburgh, who found them to contain 73·7 and 76·1 per cent. of red oxide of manganese respectively. The former, which from the analysis is considered by **Dr. Scott** to approximate most nearly in character to a variety called "marcellin" from **St. Marcellin** in Piedmont, is said to occur in large irregular masses of several tons' weight, probably included in laterite. Manganese oxide is reported to be obtainable in Bimlipatam for 2 annas a maund (*Vizagapatam Dist Man., 155*). Samples from Golaspur in the Jabalpur district were analysed by **Mr. Mallet** and were found to contain 75·86 per cent. of manganese calculated as the protosesquioxide and 9.96 of oxygen, giving 15·25 per cent. of available oxygen—an amount considerably above the average. This valuable ore was found to occur somewhat obscurely in laterite, but did not form either a regular lode or vein, nor was there any apparent connection between it and the underlying transition rocks. **Mr. Medlicott**, who examined the deposit, however, saw no reason for doubting, in spite of the irregular mode of occurrence, that a large supply of the ore might be obtainable.

An analysis of a sample of the ore from Ramtek in Nagpur was made by **Mr. Mallet**, who found that it contained 78·64 to 79·39 per cent. of the sesquioxide, giving 9·71 per cent of oxygen, a somewhat lower proportion than the ordinary commercial ores. The deposit, however, is near the surface, would be easily worked, and has been described as being ten feet thick and extending in a north-west to south-east direction for a quarter of a mile.

MEDICINE. Oxide. 140 Permanganate. 141

Medicine.—Several of the compounds of this metal are, owing to their oxidising properties, of considerable value as disinfectants. At one time a mixture of sulphuric acid with the black OXIDE was used for purposes of fumigation, but is little employed now-a-days. *Condy's-fluid* consists of an impure PERMANGANATE of potash, which salt is a valuable disinfectant, and mild astringent.

INDUSTRIAL USES. Glass-making. 142 Procelain painting. 143

Industrial Uses.—The uses to which the ores of manganese are put in the arts are somewhat varied. The peroxide is extensively employed in glass-making, to destroy the green colour of glass, which it does by converting the protoxide of iron into the peroxide; when added to excess it gives the glass a red or violet colour. The same oxide is used in porcelain painting for the fine brown colour which it yields. It is also employed for glazing pottery, and in the preparation of enamels. Its most valuable property, however, is the ease with which it gives off oxygen off the application of heat, a property which is largely taken advantage of in the arts. It is also used in the manufacture of chlorine and calcium chloride.

Of late years the ores of manganese have been extensively utilized in the manufacture of iron and of steel, by the Bessemer process, the latter especially. Manganese in the metallic state is said to deprive iron of its magnetism. Lately a process of application of this principle has been invented in England, by means of which the metal in the proportion of 27 per cent. is mixed with the steel used for ship-building; and it is contended that this mixture deprives the steel of its magnetic influence on the ship's compasses.

144 **MANGIFERA,** *Linn.; Gen. Pl., I., 420.*

A genus of trees which belongs to the Natural Order ANACARDIACEÆ and comprises some 30 species, which are found in tropical Asia, chiefly in the Malay Peninsula.

145 **Mangifera fœtida,** *Lour.; Fl. Br. Ind., II., 18;* ANACARDIACEÆ.

Syn.—M. HORSFIELDII and M. FŒTIDA, *Miq.*

Vern.—*La-móte,* BURM.; *Bachang,* MALAY.

References.—*Kurz, For. Fl. Burm., I., 305; Miq., Fl. Ind. Bat., I., Pt. 2, 632; Mason, Burma and Its People, 448, 774.*

Habitat.—A large tree, native of Malacca, Penang, and Singapore (cultivated in Southern Tenasserim) and distributed throughout the Malay Peninsula.

FOOD. Fruit. 146 **Food.**—Produces pink or dark red flowers, and a coarse-flavoured FRUIT which is eaten by the natives, and for which the tree is cultivated. **Mason** says: "This is a large Mango cultivated at Mergui, and is quite a favourite with the Natives. It has an odour resembling the dorian, and like that has been introduced from the Straits." **Rumphius** states that the fruit when eaten excites "cold exanthemata and fevers," and that it ought never to be used except when very mature.

147 **M. indica,** *Linn.; Fl. Br. Ind., II., 13.*

THE MANGO TREE.

Syn.—M. DOMESTICA, *Gærtn.*

Vern.—*Am, ámb, amchur* (unripe fruit), *am-ki-gúthlí* (seeds), HIND.; *Am, ambra,* BENG.; *Uli,* KOL.; *Ul,* SANTAL; *Jegachu, bocho,* GARO; *Ghariám, ám,* ASSAM; *Am,* URIYA; *Tsarat-pang,* MAGH.; *Ambe,* KURKU; *Ama,* BAIGAS; *Marka,* GOND; *Amb, ám, ánv,* N.-W. P.; *Am, amb, mawashi,* PB.; *Amb, amú,* SIND; *Amba, Am, Ambecha jhar,* DEC.; *Ambo, amba, am, amb,* BOMB.; *Amba,* MAR.; *Ambo,* GUZ.; *Am, ámb, ánv,* BUNDELKHAND; *Maá, mangas, mam-marum,* TAM.; *Elamávi, mámadi, mámid, mámidi, makandamu, guggu-mámidi, tiyya mámidi, racha mamidi, mávi, mamadichitú, tiya mamidi,* TEL.; *Mavina, mávu, amba,* KAN.; *Mava, mampalam, mánna.* MALAY.; *Thayet,* BURM.; *Makandamu, etamba* (wild), *amba* (cultivated), SING.; *Amra, chutu* (the juicy), *madha-dút* (messenger of spring), SANS.; *Amba, naghyak,* PERS.

References.—*Roxb., Fl. Ind., Ed. C.B.C., 215; Brandis, For. Fl., 125; Kurz, For. Fl. Burm., 304; Beddome, Fl. Sylv., t., 162; Gamble, Man. Timb., 107; Dalz. & Gibs., Bomb. Fl., 51; Stewart, Pb. Pl., 45; DC. Origin Cult. Pl., 200; Rheede, Hort. Mal., IV., t., 1, 2; Elliot, Fl., Andhr., 50, 64, 110, 113, 162, 183; Mason, Burma and Its People, 447, 774; U. C. Dutt. Mat. Med Hind., 140; Dymock, Mat. Med. W. Ind., 2nd ed., 196; Pharmacographia Indica, I., 381; S. Arjun, Bomb. Drugs, 32; Murray. Pl. and Drugs, Sind, 87; Year Book Pharm., 1880, 506; Irvine, Mat. Med. Patna, 120; Butler, Med. Top. Oudh and Sultanpore, 4; Macleod, Med. Top. Bishnath, 16; Buchanan, Statistics of Dinagepore, p. 159; Baden Powell, Pb. Pr., 338, 397; Atkinson, Him. Dist, 711, 741; Ec. Prod., N.-W. Provs., Pt. V., 57; Lisboa, U. Pl. Bomb., 53, 250, 257, 259, 279, 284, 289, 291; Birdwood, Bomb. Pr., 18, 146, 219, 261; Royle, Ill. Him. Bot., 53, 257; Atkinson, Gums and Gum-resins, I., 7; McCann, Dyes and Tans, Beng., 85, 139, 144, 160, 165, 168; Buck,*

Dyes and Tans, N.-W. P., 83; Liotard, Dyes, 33, 112, App. VIII.; Rep. Hort. Gar., Lucknow, March 1884, 2; 1885, 4; Rep. Bot. Gar. Ganesh Khind, Poona, 1882, 4; 1885, 4; 1885, 3; Watson, Rep., 4, 20, 33, 46, 50; Wardle, Rep. Dyes and Tans, Ind., 41, 43; Ayeen Akberi (Gladwin's), Trans., I., 84; II., 3, 41, 43; Blochmann's, 67; Wallace, India in 1887, 243; Smith, Dic., 263; Kew Off. Guide to the Mus. of Ec. Bot., 36; Kew Off. Guide to Bot. Gardens and Arboretum, 44, 71; Kew Bulletin, 1889, 23; Darrah, Note on Cotton, 31; Report, Forest Admn., Ch.-Nagpur, 1885, 6, 29; Linchoten, East Indies (1598), Ed. 1885, II., 23 to 26; Agri.-Hort. Soc., Ind:—Trans., I., 21; II., 13-16 (App.), 298, 300, 307; III., 58, 61, 65, 67, 68; IV., 104, 149, 179; VI. (Pro.), 60; VII., 104; Pro., 45, 186; VIII., 222; Journals, I., 262-273; II., Selc., 373; III., Pro., 182; IV., 54, 215; Selec., 89, 141; Pro., xci.; VI., 44; Sel., 107; Pro., 107, 108; VII., 71; Sel., 55; Pro., 31; VIII., Sel., 136, 138, 165, 178; IX., Sel., 56; X., 1, Pro., 40, 105; XI., Pro., 48; XIII., Sel., 59, 60, 61, 63; New series:—I., Sel., 59, Pro., 37; IV., Pro., 13, 14; V., 74; VI., 118-123, 141, 142 Pro., 2; VII., 325-327, Pro., (1883), 124, 136, 137, 148; VIII., 260-277, 292-305, Pro., 42, 43; Gazetteers:—Bombay, II., 39, 355; IV., 23, 24; V., 23, 24, 285, 360; VI., 12; VII., 38, 39, 40, 41; VIII., 94, 95; XI., 98; XIII., 23, 294; XV., 21, 72; XVIII., 41; N.-W. Provs.:—I., 80; III., 238; IV., lxx.; X., 308, 711; Oudh, II., 313; Mysore and Coorg, I., 53, 59; Settlement Reports:—Panjáb, Guzrát Dist., 134; Karnal Dist., 16; Hoshiárpur Dist., 12; Sialkot Dist., 11; Simla Dist., xliv., App. II. H.; Kohát Dist., 30; Rohták Dist., 78; Delhi Dist., 27; Hazára Dist., 94; N.-W. Provs., Shajehánpur Dist., IX.; Allahabad Dist., 38; Central Provs., Upper Godávery Dist., 38; Mundla Dist., 88; Chindwára Dist., III.; Nimar Dist., 201; Port Blair, 1870-71, 33; Hunter, Orissa, II., 5; App. I., 158, App. IV., 179; App. VI.; Aplin, Rep. Shan States, 1887-88; Manuals:—Bombay, Rev. Accts., p. 102; Madras, Trichinopoly Dist., 78; Cuddapah Dist., 56, 263; Indian Forester, I., 363; III., 201, 237; IV., 130; VI., 240, 298, 321, 338; VIII., 400; IX., 211; X., 31, 470, 543; XI., 18; XII., 73, 188; XXII., App., 27; XIII., 120; Indian Agriculturist, Augt. 14th, 1886; 15th June, and 10th Augt. 1889.

Habitat.—A large, glabrous tree, found on the Tropical Himálaya, at altitudes of 1,000 to 3,000 feet, from Kumáon to the Bhután Hills, the Khásia Mountains, Burma, Oudh, Lower Hills of Behar, and in the Western Peninsula from Khandesh southwards. Cultivated as far west as Muscat, in all Eastern Asia, and general in the tropics. **DeCandolle** writes: "It is impossible to doubt that it is a native of the south of Asia or of the Malay Archipelago, when we see the multitude of varieties cultivated in these countries, and the number of ancient common names."

History.—From its indigenous home in India, which, according to **De Candolle**, was the region at the base of the Himálaya, especially towards the east, and in Arracan, Pegu, and the Andaman Islands, the cultivation of the fruit must have spread at an early age over the Indian Peninsula. According to **Rumphius**, it has been introduced into certain islands of the Asiatic Archipelago within the memory of "living men," while in others it has existed from a remote date. HISTORY. 148

There is no doubt that it has been known and cultivated all over the peninsula of India from a very remote epoch. It is closely connected with Sanskrit mythology and is mentioned in many of the old tales and folk lore of the Hindus. **Linschoten**, in his *Voyage to the East Indies*, mentions several varieties of mango, and his description indicates a very wide spread cultivation at that date. **Abul Fazl**, in the *Aín-i-Akbari* written about the same time (300 years ago), describes a large number of cultivated races and states that "Mangoes are to be found everywhere in India, especially in Bengal, Gujrát, Málwa, Khándesh, and the Deccan," localities all famous for the fruit at the present day. Talking of Behar he notices an interesting race, produced by cultivation (which also exists at the present day), "not so high as the ordinary stature of a man, and producing very delicious fruit."

Cultivation of

CULTIVATION. **Cultivation.**—The editor is indebted for the following account of the chief races of cultivated Mango, and of the methods pursued in growing them, to Mr. Maries of Darbhangah, an expert on the subject:

Origin. 149 *Origin.*—"The cultivated mangoes of India have arrived at a great stage of perfection and consist of very numerous races, although these are unknown to most people, except as Bombays, Lungrahs (Lengras), and Maldas.

The many dozens of sorts sold in the bazárs under these three names, have given the idea that there are only three kinds of mangoes fit to eat. These three names really represent three distinct strains of cultivated fruits. It is interesting to note the changes that have taken place in these fruits The form or shape has continued almost the same as that of the wild varieties, but the flavour has developed from "tow and turpentine" to something too exquisite to express in words, each good variety having a flavour of its own. On examining, for example, the outline of the Kangra varieties, one notices the true shape of the Bombay "afooz," one of the finest mangoes; also in the Tirhoot mangoes, one sees a great similarity to a sort called "Kishunbogh." I have seen two types of wild mangoes; one very variable from Kangra, and one from Sikkim, but these may be viewed as manifesting the two great shapes of cultivated mangoes.

Selection. 150 *Improvement by selection.*—The latter wild sort is evidently the progenitor of the Malda cultivated varieties; the Kangra form might naturally be viewed as the ancestor of the Western Indian sorts, but these two wild varieties almost unaltered have been met with under cultivation in Tirhoot, and they produce all the different families of cultivated mangoes, as the result of accidental or artificial selection. Till recently mangoes were always planted from seedlings, and even now this is frequently the case. When the trees fruited, the good sweet ones were allowed to grow, while the sour and worthless were cut down and used as firewood. In this way, selection took place and is going on at the present time in Tirhoot and Northern Bengal, and I suppose in other districts. The intercrossing of the flowers of the primary races has produced innumerable subvarieties of fruit of all sorts, sizes, and quality, only the best of which have been grown and propagated to any extent.

Propagation. 151 *Propagation.*—Mangoes are propagated by inarching, that is grafting by approach. They can be grafted in other ways, but inarching is the simplest. They can be also grown from seeds. In fact if only the finest and best sorts be selected, the chances are that 50 per cent. will be as good as the fruit sown, a few better and the rest worse. I should advise planting seedling mangoes, where grafts are difficult to obtain, taking for the seed only such sorts as *Afooz Puary, Kishenbogh, Durbhungah, Bombay, Fuzlee* and good forms; and then only from well-formed. quite ripe fruits. The season of ripening too might be prolonged if such kinds as *Rhori Budaya, Mohur Thakoor*, and other *Budaya* sorts were used for seed. This was done on rather a large scale in Durbhungah A good mango seed should never be thrown away; always plant it if possible.

Soil. 152 *Soil and Cultivation.*—Mango trees grow everywhere in the plains of India. The home of the tree in the Himálaya is from 1,000 to 2,000 feet. It seems to grow as well in a swamp as on a bund, but the best fruits and finest trees in the plains are always produced on trees grown on raised ground. The soil does not seem to interfere much with the tree. In Bengal, it grows equally well in a rich deep river deposit, in clayey, or in sandy soil. In Gwalior we have fine trees in ***kankar***, with enormous crops of fruit. The best place to plant mangoes is on a raised, well-drained piece of land with a good depth of soil. When the trees are young, the land between them should be well cultivated every year, and round the young trees the ground should be dug up and stirred frequently. When the trees are about 10 feet high, the ground for a space of 10 or 12 feet all round should be dug up in January or February and manure well mixed with the soil. In Bengal, where irrigation is not generally necessary, this manuring would be best done after the fruit has been gathered. In Central India I do it in February. Where mango trees are irrigated, as they are in Gwalior, Allahabad, and other similar localities, no water should be given after the rains to fruiting trees; allow the plants to dry up and get well ripened. If this be done, a crop of fruit every year will be the result. Our trees are about 20 feet high. We do not irrigate after the rains. We dig up and manure the ground around the trees in February, and when the fruit is set we water from a well, every 8 or 10 days, till it becomes ripe, and the rains begin. This treatment has been carried on for two years; we had a good crop last year; this season we have an enormous one. If irrigation is carried on all the year round, the flowering season is brought on prematurely, the flowers are deformed, and become large masses of leafy flowers that hang on the trees for months and produce no fruit. CULTIVATION.

Planting.—The best time to plant mangoes is in the rains—July. The native method of growing plantain trees round them is very good, but instead of planting one or two for "luck," I grow four or five for shade and protection, taking them out when the mango tree is strong enough to stand the climate. In laying out a plantation the trees should be put at least 30 feet apart. The holes for planting should be prepared six months beforehand, dug up well, and a little very old manure mixed with the soil. Planting 153

Cultivated Races.—Of cultivated sorts of mangoes I have collected upwards of 500, and from these have selected 100 good ones. Mangoes may be obtained to fruit in succession, from May till November. Thus the Bombay *Afoos* fruits in May. The *Kuabogh* of Tirhoot also does so, while the *Budavas* and *Kalkees* fruit in September to November in Tirhoot and Malda. In 1885, in Durbhungah, I had mangoes every day for five months. In that year a list of fruits was selected and grafts were made of all. These were grown in model plantations. One plantation of about 125 trees was selected from stock derived from the Madras Horticultural Society, Calcutta Nursery men, Chanchal estate, Malda, and Bombay: the sorts planted were as follows:— Races. 154

Madras.—Peter mangoe, Goa, Mulgova, Komaine, Ameercola, Dilpusund, Wallajah pusund, Office pusund. Madras. 155

Malda.—Fuzlee bewa, Bura jalli bund, Chota jalli bund, Latcuspu, Mohunbogh, Lumba budaya, Dilshoj. Malda. 156

Durbhungah.—Kuabogh, Durbhungah-Bombay, Gopalbogh, Kakoria (cucumber-mango), Gobinpoor-ka-Sinduria, Khupurwa (camphor). The melon mango, peculiar to this district, is known as Naroi-ka-kerbuza, Mohidinugger kerbuza, Dhoola walla kerbuza also Nursinghbogh, Maharaj pusund, Derruna, Kishenbogh, Gowrays, Bhoopolie, Kurrelna, after the fruit Kurela and many others. Durbhungah. 157

Bombay.—Pieary, Afooz, Salem favourite.—In laying out a plantation of mangoes the trees should be so arranged that the season of ripening comes in order: the early mangoes planted to the east, medium mangoes in the middle, late magoes, west. In the large Durbhungah plantation of about 65 acres, there are three sections arranged as above. Bombay. 158

The following descriptive list of good sorts of mangoes may be found useful; it is referred to five sections:—

I. *Afooz.*—This is the celebrated Bombay mango, a lovely orange colour, with reddish flesh. It is really not a Bombay fruit at all, but probably Afooz. 159

CULTIVATION. Races.

came originally from Salem. Absurd prices are often paid for this fruit, as much as R60 per 100 being given by dealers. Like most mangoes this should never be eaten fresh, but should be gathered ripe from the tree and laid upon a shelf for a few days to fully mature. Weight, 8 to 12 ozs.

Kuabogh. 160 *Kuabogh.*—A Tirhoot mango, of which the quality is as good as the finest Afooz. It is a small green fruit and ripens early. The name signifies 'crow's food.' It weighs about 4 ozs. Season May.

Durbhungah-Bombay. 161 *Durbhungah-Bombay.*—This is the Bombay of the up-country gardens and about the best known mango. A very old plantation exists at Norgona, Durbhungah, and another called the Lakh Bagh, near Somaspur, had once upon a time a lakh of trees, said to be of this kind. Season May, June.

Safada. 162 *Safada.*—A whitish variety of the above, better in quality I think. Season May, June.

Gopalbogh. 163 *Gopalbogh.*—A Malda sort. This is the celebrated Malda kind said to be equal to the Afooz. It hangs till late in June, and is a superior fruit. It is very like Durbhungah-Bombay but smaller. Weight, 6 ozs.

Kakoria. 164 *Kakoria* (*cucumber-plantain*).—Very like a cucumber, often 7 to 10 inches long by 2½ inches wide; a most luscious, refreshing fruit; weighs from 10 ozs. to 1℔. It is a plentiful variety in Tirhoot, but is seldom gathered in good condition, and is often sour. When gathered ripe from the tree, and kept for a couple of days, it is a perfect fruit.

Kurrelna. 165 *Kurrelna.*—Named after Kurela (**Momordica**). This is a variety of the cucumber mango, smaller and covered over with greenish warts like a Kurela fruit. It weighs 8 to 12 ozs. Season July.

Banka. 166 *Bánka* that is, 'twisted.' This is a large green fruit totally unlike any mango I know. It is twisted, weighs 1℔ and has a strong flavour; it is a very rare sort. Season July.

Ameercola. 167 *Ameercola.*—Madras fruit; weighs 10 ozs., has a rough skin like an orange, a very peculiar shape, and very distinct. Season July.

Dilpusund. 168 *Dilpusund.*—Several fruits bear this name, and the one I received from Madras is like the Durbhungah *chupki* (flat) or *chupra*. It is a desirable fruit and very good looking. Season July.

Durma. 169 *Durma or Derrima*, from Lawanie Tirhoot. The true sort is one of the finest of mangoes. It varies in size from 8 ozs. to 1℔, and is a round yellowish fruit of most exquisite vanilla-like flavour; the flesh is rather hard, but melts in the mouth. There is another variety of this, a red fruit, which at first sight might be mistaken for a Blenhiem orange apple. Season June and July.

Kishenbogh. 170 *Kishenbogh Durbhungah*—A celebrated fruit, which, since the railway has been opened, is sold with Gowraya Malda by thousands in the Calcutta markets. It is a round fat mango, of first rate quality. Season July.

Kishenbogh.—This fruit often hangs on the tree till the seed germinates inside. I have had several examples of this, in which the young plant has grown completely out of the fruit. The flesh of the mango in these cases had become quite hard, and tasted like a carrot.

Lerrua. 171 *Lerrua or Lerrna*—(from *Laddu*, a sweetmeat). This is the most beautiful of all mangoes, the mixture of orange red and green, in stripes and blotches, resembles the colouring of a ripe apple. Season July.

Shah pusund. 172 *Shah pusund*—(generally called Malda). A fine large, irregular shaped fruit of fair quality, largely grown as it is hardy and a good cropper; some of the fruits weigh 2℔. Season June and July.

Gowraya. 173 *Gowraya Malda.*—A Tirhoot mango, also called "*Safada Malda*" and "*Tikari.*" A good specimen of this is one of the finest mangoes in India. It cannot be mistaken, as it is the type of the large class of raised stoned mangoes. It has an aroma and flavour distinct from those of any fruit I know. The skin is as thin as writing paper, and the stone so tender

CULTIVATION. Races.

that when cut, the knife often goes through it. There are many forms of this race; the best I have named after **Mr. Buckley**, as Buckley's Gowraya Malda.

Kumukht.—The skin of this is rough and leathery; it is a very irregular **Kumukht. 174**
shaped round fruit, often with the pistil scar or "Nak" developed in a most curious way; a fine flavoured and rare fruit. Season July. Weighs 8 ozs.

Buhpali.—A small ovoid mango, often perfectly crimson, vermillion, **Bhupall. 175**
and yellow in colour, perhaps the best of all mangoes. I have obtained it from several places in Tirhoot, but always of the same fine quality. Season July. Weighs 6 ozs.

Inerna—(meaning spontaneous). This is the largest mango, some speci- **Inerna. 176**
mens attaining a weight of 4℔. It is of good flavour, but is a rare fruit. Season July and August. It came up from seed in a native gentleman's garden in Durbhungah and only one tree was supposed to exist.

Nursinghbogh—A blue mango weighing 1½℔. It can be readily distin- **Nursinghbogh 177**
guished by its leaf which frequently is as much as 18 inches long. Season July and August. A good fruit.

Maharaj pusund—A Tirhoot fruit of fine quality though common and **Maharaj Pusund. 178**
well known. Weight 6 ozs. Season July.

II. *Kerbuza mangoes.*—We come now to a distinct class of mangoes **Kerbuza. 179**
called in Tirhoot kerbuzas or melons, from the musk scent they possess. There are three good kinds, all of which are of fine quality and ripen late in July, *Naroika kerbuza, Mohedenugger kerbuza, and Dhoola walla kerbuza,* all three should be in every collection.

III. *Budayas.*—The above mangoes are generally all over by the end of **Budayas. 180**
July, but sometimes hang till August. The class of mangoes called Budayas and Maldas (true) have all peculiar shaped fruits, and seldom ripen before the middle of July, and with care and protection will keep till October. These fruits may be seen hanging on the tree in October protected by little Bamboo baskets from wasps, birds, &c. In 1885, I had some fruits gathered fresh from the tree in excellent condition on the 30th October.

Khari Budaya—Ripens first. Season July-August. Weighs 8 ozs.

Terha Kellua—(The crooked plantain). Always a long, ugly fruit, with **Terha Kellua. 181**
the stalk on one side, hence the name. Weighs 1 to 1½℔. It comes from Chanchal in Malda.

Fuzlee Bewa.—The large mango one sees in Calcutta, weighing 1 to **Fuzlee Bewa. 182**
2℔, very common in the bazar there in August; these fruits sometimes fetch as much as 1 rupee each.

Julli bund (seed in a net), because after the skin has been taken off, **Jalli bund. 183**
the flesh appears to be in a yellow thread net; this is from Malda, and is an excellent fruit. Season August and September. Weighs 1 to 1½℔.

Durbhungah Budaya—or *souria budaya*—a very first class mango, flattish and good looking; it has no fibre, a very thin skin, and a small thin stone; ripens August. Weighs 10 ozs. to 1℔.

Nukkna Lungra.—So named because the pistil scar develops into a **Nukkna Lungra. 184**
prominent nose-like projection. This is a Durbhungah mango and is a very good sort. Season, August and September.

Mohunbogh—From Malda and Monghyr, a very large, round, irregular **Mohunbogh. 185**
shaped fruit, 1½℔ weight, of fair flavour.

Mohur-Thakoor.—One of the latest and best mangoes, very ugly, and **Mohur Thakoor. 186**
very irregularly shaped. They hang on the tree till October; weight 1 to 1½℔

Tars—The native name of the Borassus palm. This mango is just **Tars. 187**
like the fruit of palm of the same name; it weighs 1 to 1½℔, is good eating and ripens in September.

MANGIFERA indica.

The Mango Tree.

CULTIVATION. Races. IV. *Barramassia* (meaning twelve months). There are several varieties of the perpetually fruiting mango, none very good. They are grown more as curiosities than anything else.

Barramassia. V. *Luttea* —The creeping mango.

Luttea. 188 This is really not a creeping mango, but a decumbent variety, produced by grafting. There are several varieties of it. **Mr. Chatterjee**, the Calcutta Nurseryman, has one variety that grows along the ground and bears small roundish fruits. Another is trained on a *machan*, and bears large fine-shaped fruits. It is a true mango, and a cultivated sport. I have had samples of fruit of *Luttee am* from Tirhoot, and they prove to be "*Shah pusund*" and "*Dhoola walla kerbuza*" I am informed these two trees were originally staked and trained down to the ground, and bore fruit in this way. After a considerable space was thus covered, eventually the training ceased, and the trees at once grew up and formed straight stems."

[This form has probably given rise to the accounts of creeping and vine-like mangoes mentioned in the *Áin-i-Ákbari*, by **Wallace** in his "India in 1887," and by other writers. *Conf.* with **Willughbeia edulis** under INDIA-RUBBER, Vol. IV., 363.—*Ed.*]

GUM. 189 **Gum.**—The bark yields a gum, which, according to **Atkinson**, is frequently sold in the bazárs as gum arabic.

DYE & TAN. Bark. 190 Leaves. 191 **Dye, Tan, and Mordant.**—The BARK and LEAVES yield a yellow dye which is not much used. In Monghyr, the BARK is employed with that of **Bassia latifolia, Punica granatum**, and **Bauhinia variegata** for dyeing yellow, and in Lohardaga it is used in combination with the barks of several other trees in obtaining a permanent black. The Magistrate of Chittagong states that in his district "the juice of the bark obtained by simple beating and mixed with lime yields a fleeting green dye." The Forest Officer, Palanpur, Bombay, recently sent to the Editor a piece of cotton cloth dyed a bright rose-pink with mango bark, turmeric and lime—perhaps one of the best colours in the admirable collection of Palanpur dyes kindly furnished by that gentleman. Samples of the bark were sent by Government among other dye stuffs to **Mr. Wardle** for examination. It yielded, by his processes, a series of very beautiful, though generally light and more or less yellow, shades of brown, slate, and drab, when used with cotton, silk, and

Fruit. 192 wool. The pulp of the FRUIT, also experimented with by him, produced yellowish drab or grey shades, which were little affected by the different processes employed for silk or cotton.

The bark is employed for tanning in the Dacca district and Bankura; the leaves are similarly employed by the poorer classes in Oudh. The dry unripe fruit is largely used as a mordant specially in dyeing with safflower.

OIL. 103 Seed. 194 **Oil.**—**Dr. Cooke** states that the SEEDS contain a large percentage of oil, but no information apparently exists, either as to the method of preparing it, or as to its uses.

MEDICINE. Fruit. 195 **Medicine.**—The FRUIT has long been considered a valuable medicine both by Hindu and Muhammadan physicians, and has formed the subject of many articles by writers on the Materia Medica of the East Thus, in the *Bhavaprakasa*, a confection made of the juice of the ripe fruit, sugar, and aromatics is recommended as a restorative tonic. It is, however, unnecessary to enter into a detailed account of the opinions of older writers on what is after all an unimportant drug. The following extract from the recently published *Pharmacographia Indica*, together with the somewhat numerous list of Special Opinions below, may, therefore, suffice to indicate the principal medicinal properties supposed to be possessed by the fruit:—

"Shortly, we may say that the ripe fruit is considered to be invigorating and refreshing, fattening, and slightly laxative and diuretic; but the rind

MEDICINE.

and fibre, as well as the unripe fruit, to be astringent and acid. The latter when pickled is much used on account of its stomachic and appetising qualities. Unripe mangoes peeled and cut from the stone and dried in the sun form the well-known *A'mchúr* or *Ambosi* (*A'mrapesi*, SANS.), so largely used in India as an article of diet; as its acidity is chiefly due to the presence of citric acid, it is a valuable antiscorbutic; it is also called *A'm-ki-chhitta* and *A'm-khushk.* The BLOSSOM, KERNEL, and BARK are considered to be cold, dry, and astringent, and are used in diarrhœa, &c., &c. The smoke of the burning LEAVES is supposed to have a curative effect in some affections of the throat. According to the author of the *Makhzan*, the Hindus make a confection of the baked pulp of the unripe fruit mixed with sugar, which in time of plague or cholera they take internally and rub all over the body; it is also stated in the same work that the midribs of the leaves calcined are used to remove warts on the eyelids. Mangoes appear to have been known to the Arabs from an early date as a pickle; they were doubtless carried to Arabian ports by Indian mariners. Ibn Batuta, who visited India A.D. 1332, notices their use for this purpose. The powdered seed has been recommended by Dr. Kirkpatrick as an anthelmintic (for lumbrici) in doses of 20 to 30 grains, and also as an astringent in bleeding piles and menorrhagia. (*Phar. of India.* 59)." It may be here noted that this property of the seed is described by Paludanus in his Notes on Linschotan's Travels. He writes: "Being raw it is bitter of taste and is therefore good against worms, and looseness of the belly; against worms when it is eaten raw, and against looseness of the belly when it is roasted." "From the fruit just before ripening, a gummy and resinous substance exudes, which has the odour and consistence of turpentine, and from the bark a GUM is obtained which is partly soluble in cold water." Ainslie says that the gum-resin mixed with lime-juice or oil is used in scabies and cutaneous affections. The juice of the ripe fruit dried in the sun so as to form thin cakes (*Amras* or *Amaut*, HIND.; *Ampapoli*, MAR.; *Amravarta*, SANS.) is used as a relish and antiscorbutic. Mango bark and fruit have been lately introduced by Dr. Linguist to the notice of European physicians (*Practitioner, 1882, 220*); he recommends it for its extraordinary action in cases of hæmorrhage from the uterus, lungs, or intestines. The fluid extract of the bark or rind may be given in the following manner:—Ext. Fl. Mangif. Ind., 10 grains; water, 120 grains. Dose—One teaspoonful every hour or two, or the juice of the fresh bark may be administered with white of egg or mucilage and a little opium."

Flower. 196

Kernel. 107

Bark. 198

Leaves. 199

Gum. 200

In addition, it may be stated that in the Panjáb and Sind a gruel made of the kernels is administered in cases of obstinate diarrhœa and bleeding piles, and that the seeds are also considered useful in asthma.

Chemistry. 201

CHEMICAL COMPOSITION.—The following is extracted from the *Pharmacographia Indica*:—

"Professor Lyon (1882) examined the dried unripe peeled fruit, and found it to contain water 20·98, watery extract 61·40, cellulose 4·77, insoluble ash 1·43, soluble ash 1·91., alkalinity of soluble ash as potash ·41, tartaric acid, with a trace of citric acid 7·04, remaining free acid as malic acid 12·66, total free acid per 100 parts air dry substance 24·93.

The orange colouring matter of the ripe mango is a chlorophyll product, readily soluble in ether, bisulphide of carbon and benzol, but less readily soluble in alcohol. It yields with these solvents deep orange-coloured solutions which are bleached by solution of chlorinated soda, and turned green by hydrochloric or sulphuric acids, the orange colour being again restored by an alkali.

MANGIFERA indica.

The Mango as a Medicine.

MEDICINE.

"The bark and seeds contain a tannin. Fifty grams of the powdered seed exhausted with alcohol, 90 per cent., filtered, the alcohol evaporated off on the water bath, and the residue dried over sulphuric acid, left an extract weighing 3·16 grams. Of this extract ·3 gram was of a resinous nature, and insoluble in water. The portion soluble in water, equivalent to 5·72 per cent. of the seed, gave the usual reactions of a tannin. The aqueous solution of the tannin was precipitated with gelatine, filtered, and the filtrate shaken two or three times with ether. No appreciable residue was obtained by the evaporation of this ethereal extract showing the absence of gallic acid. (*J. G. Prebble*)."

SPECIAL OPINIONS.—§ "The smoke of the burning leaves is supposed to have a preventive effect in hiccough" (*Civil Surgeon J. Anderson, M.B., Bijnor, North-Western Provinces*). "The unripe fruit roasted, dissolved in water and made into *sherbet* with sugar is freely taken by the natives to prevent sunstroke, the pulp is also rubbed over the body for the same purpose" (*Assistant Surgeon N. R. Banerjee, Etawah*). "The kernel of the seed is used for dysentery" (*Surgeon-Major P. N. Mookerjee, 32nd Regiment, Madras Native Infantry, Cuttack, Orissa*). "The powder of the dried kernel is useful in diarrhœa and chronic dysentery, as an astringent" (*Assistant Surgeon Nehal Sing, Saharanpur*). "Unripe mangoes toasted and made into *sherbet* form a reputed remedy for heat apoplexy" (*Assistant Surgeon T. N. Ghose, Meerut*). "I was lately told by a very intelligent patient that he had found the mango decidedly anthelmintic" (*Surgeon Major Farquhar, M.D., Ootacamund*). "*Amchur* is the very best antiscorbutic that I know. I have found it stamps out scurvy when lime juice and all other available remedies had been tried in vain" (*Brigade Surgeon C. Joynt, M.D., Poona*). "The dried kernel of the ripe fruit is used as an astringent in diarrhœa" (*Civil Surgeon R. Gray, Lahore*). "The liquid extract is as efficacious as *bael* in dysentery" (*Civil Surgeon G. C. Ross, Delhi, Panjáb*) "Flour made from the kernel of ripe mango seeds when dried, is made into *chápatis*, and eaten by men of low caste, in the North Western Provinces" (*Surgeon A. C. Mukerji, Noakhally*). "The kernel is a constant and unfailing remedy for diarrhœa and dysentery amongst the hill tribes of the Sourah Mahlias. In my travels as Deputy Superintendent of Vaccine, I had frequent opportunities of noting the effects of the drug. When the Sourahs came down to the plains and remained for a week or more, they were very subject to diarrhœa or dysentery. They then eagerly sought for the seeds and used half a kernel in the morning and half in the evening. This treatment they continued for two or three days with marked effect and perfect cure resulted in five days at latest" (*Honorary Surgeon E. A. Morris, Tranquibar*). "I have never observed any laxative effects from eating the ripe fruit or heard of its being eaten with this object; if true it would not be safe to eat the fruit in large quantities when cholera is prevalent, as it often is in India during the mango season. The kernels of the seeds are sometimes roasted and eaten as food by the poorer classes in times of scarcity" (*Brigade Surgeon G. A. Watson, Allahabad*). "The dried flowers, either in the form of decoction or powder, are used as a useful astringent in looseness of the bowels, chronic dysentery, and gleet" (*Assistant Surgeon S. Arjun Ravat, L. M., Girgaum, Bombay*). "The gum of the mango tree is used for cracked feet with good effect" (*Surgeon-Major J. North, Bangalore*). "The green fruit is softened by roasting, mixed with water and used by the natives of Upper India in sunstroke and burning of the body. *Amchur* and pickles prepared from green fruit are issued to prisoners in jails as antiscorbutics. The kernels are dried and stored for medicinal use. In times of scarcity the flour of dried kernels is used by the poor as an article of

MEDICINE.

diet" (*Assistant Surgeon S. C. Bhattercharji, Chanda, Central Provinces*). "The kernel (powdered) with resin and kurchi is given in dysentery. I have seen several cases cured by this. Dose: equal quantities of each ingredient mixed and about 15 grains given twice or thrice a day to adults" (*Assistant Surgeon N. N. Bhattacharjee, Tirhoot State Railway, Somastipore*). "The kernels powdered when thoroughly dry are used as food in the North-West, being made into *chapátis*," (*Narain Misser, Kothe Bazar Dispensary, Hoshangabad, Central Provinces*). "The unripe fruit cut and dried is a valuable antiscorbutic. It is now in use in Bengal jails" (*Surgeon R. L. Dutt, M.D., Pubna*). "The ripe fruit is laxative. The kernel of the seed is used as an astringent in diarrhœa. This is one of the ingredients of Pogson's Bael powder. The baked green fruit is made into a *sherbet*, and the pulp applied also externally in sunstrokes" (*Bolly Chand Sen, Teacher of Medicine*). "If the small white kernel of the mango stone be steeped in a little water and reduced to the consistence of paste, it may be applied to any part of the skin which *burns*, and it will soon have a cooling effect" (*Surgeon W. Wilson, Bogra*). The unripe fruit is used by the natives in the form of sherbet as a refrigerant and diaphoretic. The juice is used in fissures of the feet and between the toes or fingers" (*Civil Surgeon J. H. Thornton, B.A., M.B., Monghyr*). "The kernel of the stone has been frequently used in diarrhœa of children with success, in 1 to 3 grain doses alone or with dried bael" (*Assistant Surgeon N. L. Ghose, Bankipore*).

FOOD. Fruit. 202

Food.—The mango is a favourite FRUIT among both Natives and Europeans, and is very largely eaten throughout the country. In many parts of India, it serves as an important addition to the resources of a large section of the native population who own the trees. Fine, luscious fruits, weighing ½lb each, were, a few years ago, produced on an old tree in the Kew Gardens, London.

Besides being eaten as a ripe fruit, the mango is used as follows:—

"*When green*, the stone is extracted, the fruit cut into halves or slices, and (*a*) put into curries; (*b*) made into a pickle, with salt, mustard oil, chillies, and other ingredients; (*c*) made into preserves and jellies by being boiled and cooked in syrup; (*d*) boiled, strained, and with milk and sugar made into a custard known as mango-fool; (*e*) dried and made into the native '*ambchúr*,' used for adding acidity to certain curries; (*f*) when very young cut into small pieces, mixed with a little salt, and sliced chillies and milk added, it forms a 'tasty' salad.

Kernels. 203

"*When ripe* (*a*) it is made into curry which has a sweet acid, not unpleasant, taste; (*b*) it is cut into small pieces and made into salad with vinegar and chillies (the sour fruit is sometimes so used); (*c*) the juice is squeezed out, spread on plates and allowed to dry; this forms the thin cakes known as *amb-sath* (*Mr. L. Liotard*). The KERNELS are eaten in times of famine, and by the poorest classes in many parts of India they are boiled and eaten with greens. They are also ground into meal and mixed with various other ingredients to form the relish known as *ám-khatai*. When stuffed with coriander, turmeric, and other spices, and boiled in mustard oil, they are esteemed a great delicacy.

Preserves, chatnies, and pickles are made from the mango fruit and largely exported to England and elsewhere. **Linschoten** and **Rumphius** both describe a method of eating the fruit now almost unknown in the country, and probably introduced by the Portugese as suited to their tastes. The former traveller writes: "This is ye best and ye most profitable fruit in al India, for it yieldeth a great quantity for food and sustenance of the country people, as olives do in Spaine and Portingale. They are gathered when they are greene and conserved, and for the most part salted in pots, and

FOOD. commonlie used to be eaten with rice, sodden in pure water, the huske being whole, and so eaten with salt mangas, which is the continuall food of their slaves and common people, or else salt dried fish instead of mangas." "These salted mangas are in cutting like the white Spanish olives, and almost of the same taste, but somewhat savorie, and not so bitter." "There are others that are salted and stuffed with small pieces of greene ginger, and garlike sodden; those they call Mangas Recheadas, or Machar." **Rumphius** states that salted mangoes were also much eaten in curries with fish.

There seems to be little truth in the charge frequently brought against the mango, that it is a fruitful cause of boils. The blue stain produced on the cutting knife results from the presence of gallic acid in the pulp, which likewise contains citric acid and gum.

TIMBER. 204 **Structure of the Wood.**—Grey, coarse-grained, soft, weight 41lb per cubic foot.

It is used for planking, door and window frames, in Calcutta for packing cases, and in Behar for indigo boxes; canoes and Masula boats are also made of it (*Gamble*). When employed for packing cases it should be previously well seasoned, otherwise the acid it contains is stated to corrode the lead lining. It is stated to be fairly durable, if not exposed to wet, but is liable to be worm eaten. Bareilly chairs are reported to be generally made of mango wood.

DOMESTIC & SACRED. **Domestic and Sacred.**—The mango is held sacred by the Hindus and is inextricably connected with many of their mythological legends and folklore. The following extract from the *Pharmacographia* indicates some of these ideas:—

Tree. 205 "The mango, in Sanskrit *Amra, Chúta* and *Sahakara,* is said to be a transformation of Prajápati (lord of creatures), an epithet in the Veda originally applied to Savitri, Soma, Tvashtri, Hiranga-garbha, Indra, and Agni, but afterwards the name of a separate god presiding over procreation (*Manu,* xii., 121). In more recent hymns and Bráhmanas Prajápati is identified with the universe.

Twigs. 206 "The tree provides one of the *pancha-pallava* or aggregate of five
SPRIGS used in Hindu ceremonial, and its FLOWERS are used in Shiva wor-
Flowers. 207 ship on the Shivarátri. It is also a favourite of the Indian poets. The flower is invoked in the sixth act of Sakuntala as one of the five arrows of Kámadeva. In the travels of the Buddhist pilgrims **Fah-hian** and **Sung-yun** (translated by **Beal**), a mango grove (Ámravana) is mentioned, which was presented by Ámradárika to Buddha in order that he might use it as a place of repose. This Ámradárika, a kind of Buddhic Magdalen, was the daughter of the mango tree. In the Indian story of Súrya Bai (*see Cox, Myth. of the Arian Nations*) the daughter of the sun is represented as persecuted by a sorceress, to escape from whom she became a golden Lotus. The king fell in love with the flower, which was then burnt by the sorceress. From its ashes grew a mango tree, and the king fell in love first with its flower, and then with its fruit; when ripe the fruit fell to the ground, and from it emerged the daughter of the sun (Súrya Bai), who was recognised by the prince as his lost wife."

Lisboa further informs us that "In *Smritisar Granth* the twigs of the
Leaves. 208 tree are ordered to be used as tooth-brushes, and its LEAVES as platters in *panch pallav,* and for pouring libations; and the flower in the worship of *Shiv* on the day of *Maha Shivráti* in the month of *Mágh.* The leaves are also employed in adorning *mandaps* and houses on occasions of various ceremonies." The twigs and leaves are largely used for cleaning the teeth, and the twigs as a substitute for *pan.*

[518

Mangifera oppositifolia, *Roxb.*; see **Bouea burmanica**, *Griff.*; Vol. I.,

M. sylvatica, *Roxb.; Fl. Br. Ind., II., 15.* 209

Syn.—M. INDICA, *Wall., Cat. 8487, I.*

Vern.—*Kosham*, HIND. & BENG.; *Lakshmi am*, SYLHET; *Bun am*, ASSAM; *Chuchi am*, NEPAL; *Katúr*, LEPCHA; *Bagnul*, MICHI; *Hseng neng thayet, Sinmirthayet*, BURM.; *Kosámra*, SANS.

References.—*Roxb., Fl. Ind., Ed. C.B.C., 216; Voigt, Hort. Sub. Cal., 272; Kurz, For. Fl. Burm., I., 304; Gamble, Man. Timb., 108; Mason, Burma and Its People, 448, 774; U. C. Dutt, Mat. Med. Hind., 305; Indian Forester, IX., 28.*

Habitat.—A large evergreen tree met with in tropical Nepál and the Sikkim Himálaya, in Sylhet and the Khásia Mountains. According to Kurz, it is found also in the Andaman Islands and rarely in the tropical forests of the Martaban hills.

Medicine.—The FRUIT is dried and kept for medicinal purposes (*Roxb.*). MEDICINE. Fruit. 210

Food.—The FRUIT is eaten by the natives, though by no means so palatable as even a bad domestic mango (*Roxb.*). The LEAVES are used in Assam to feed the silkworm **Cricula trifenstrata.** FOOD. Fruit. 211 Leaves. 212

Structure of the Wood.—Grey, moderately hard. Weight 34 to 41 lb per cubic foot. It has been recommended and tried for tea chests, but when used unseasoned, has been found to corrode the lead foil, thereby spoiling the tea. TIMBER. 213

Mango, see preceding article on **Mangifera.**

Mango Fish or **Polynemus indicus**, see **Fish**, Vol. III., 391.

Mango ginger, see **Curcuma Amada**, *Roxb.*; SCITAMINEÆ; Vol. II., 652.

Mangosteen, see **Garcinia Mangostana**, *Linn.*; GUTTIFERÆ; Vol., III, [470.

Mangosteen Oil, Brindonia-tallow, or **Kokum-butter**; see **Garcinia indica**, *Chois.*; GUTTIFERÆ; Vol., III., 466.

Mangrove Bark, a valuable tanning material. The following are the chief barks known commercially by this name—arranged alphabetically:— 214

Avicennia officinalis, *Linn.*; VOL. I., 360. (THE WHITE MANGROVE.)

Bruguiera gymnorhiza, *Lamk.*; VOL. I., 541.

B. parvifolia, *W. & A.*

Ceriops Candolleana, *Arnott*; VOL. II., 261. (THE BLACK MANGROVE.)

C. Roxburghiana, *Arnott*; VOL. II., 261.

Kandellia Rheedii, *W. & A.*; VOL. IV., 565.

Rhizophora mucronata, *Lamk.*; VOL. VI. (THE TRUE MANGROVE.)

MANIHOT, *Adans.; Gen. Pl., III., 306.*

Manihot Glaziovii, *Müll. Arg.*; EUPHORBIACEÆ. 215

THE CEARA RUBBER TREE: OR SCRAP-RUBBER TREE.

Full information will be found regarding this tree, in Vol. IV., 374, under **India-rubber**, to which article the reader is referred.

M. utilissima, *Pohl.; Fl. Br. Ind., V., 239;* also **M. Aipi**, *Pohl.* 216

CASSAVA, TAPIOCA, MANIOC.

Syn.—JANIPHA MANIHOT, *Kunth*; also of **Sir W. Hooker**, *Bot. Mag., Table 3071, Vol. 58;* JATROPHA MANIHOT, *Linn.*

MANIHOT utilissima.

Introduction and Cultivation

These two plants, by some botanists, are regarded as separate species; by others the latter is deemed a variety of the former, if not at most a cultivated race only. It differs from **Manihot utilissima**, chiefly in the fact that it is not acrid to the taste and is, in fact, devoid of the poisonous principle of that species; hence it may be eaten fresh. On this account it is known as the sweet cassava, while the other form is the bitter cassava, which can only be eaten after having been specially prepared. But under each of these varieties, there are races or qualities of good and bad cassava which exhibit the effects of climate and soil, or indicate ancient cultivation. Thus, of Brazilian Tapioca, a writer in the *Planters' Gazette* says: "Now Rio tapioca is as much superior to the common flake tapioca as rice grown in Carolina is to rice grown in Moulmein or Arracan. Rio tapioca is also as much superior to the East India tapioca as Bermuda arrowroot is to potato starch."

Vern.—*Maravuli*, TAM.; *Marachini*, MALAY.; *Pulu pinan myouk*, BURM.

References.—*Roxb., Hort. Beng., 69; Voigt, Hort. Sub. Cal., 158; Kurz, For. Fl. Burm., II., 402; Gamble, Man. Timb., 348; Dalz. & Gibs., Bomb. Fl. Supp., 77; DC., Origin Cult. Pl., 59; Grah, Cat. Bomb. Pl., 183; Mason, Burma & Its People, 507; Bidie, Cat. Raw. Prod., Paris Exh., 92; Ainslie, Mat. Ind., I., 428-430; O'Shaughnessy, Beng. Dispens., 559; Moodeen Sheriff, Supp. Pharm. Ind., 171; Flück. & Hanb., Pharmacog., 250; Bent. & Trim., Med. Pl., 235; Year Book Pharm., 1873, 185; Atkinson, Econ. Prod N.-W. P., Pt. V., 22-23; Drury, U. Pl., 265; Lisboa, U. Pl. Bomb., 270; Firminger, Man. Gard. in Ind, 124; Spons, Encyclop., 1828; Smith, Dic., 96; Treasury of Bot., 717; Simmonds, Trop. Agri., 349-352; Rep. Bot. Gar Ganesh Khand, Poona, 1882-83, 5; Madras Admin. Man., II., 135; Nellore Man., 114; Admin. Report, Andaman Islands, 1885-86, p. 53; Agri.-Hort. Soc. India (numerous papers & reports on Samples of Tapioca prepared, but the following are the more important): Trans. II. (1838), 215; Journals (Old Series) VII., 236-248; IX. (Sel.) 77; XII., 175; I (New Series) 184-191; VIII. (Proceedings, 1886), lxxxi.; Tropical Agriculturist I., (1881-82), 853; II., 189-191; V., 126; VII, 561.*

Habitat.—Reasoning on his usual triple lines, **DeCandolle** strongly supports the opinion that the Manioc is a native of America. All the wild species known in the world (42 in number) are found in America. Ancient historic evidence points to Manioc having been cultivated in America prior to the arrival of Europeans. The natives of America have several ancient names for the plant. Added to these facts the cultivation in Asia and Africa is unmistakably modern. But, on attempting to locate the habitat of the plant to narrower limits, **DeCandolle** admits that great difficulty exists. If we do not accept its origin as in eastern tropical Brazil, he says, we must have recourse to two hypotheses: either the cultivated Maniocs are obtained from one of the wild species, modified by cultivation, or they are varieties which exist only by the agency of man, after the disappearance of their fellows from modern wild vegetation.

CULTIVATION.

CULTIVATION. Assam. 217

INTRODUCTION INTO INDIA AND EXTENT TO WHICH CULTIVATED.

I. ASSAM.—It seems probable that the Manioc was actually introduced into India long before the earliest record of it in botanical books. It is not, however, mentioned by **Linschoten** in 1598, although that distinguished traveller and careful historian of Dutch and Portuguese influence in the East, goes into great detail regarding the yams which he saw in India. **Voigt** tells us that it "had never flowered in Serampore, nor had it done so in the Hon'ble Company's Garden, Calcutta, in 1814, though introduced in 1794 from South America." It was apparently first taken to Ceylon in 1786, from Mauritius, by Governor **Van der Graaf**. It is repeatedly men-

CULTIVATION. Assam.

tioned in the early numbers of the Transactions and Journals of the Agri.-Horticultural Society of India. The first detailed paper is that in Vol. VII. (Old Series) where the discoveries in Assam made by **Jenkins, Hannay,** and **Masters,** are reviewed. **Hooker's** account of the plant is also reprinted in the journal from the *Botanical Magazine*, and also **Mr. John Bell's** note, written in 1833, in which papers are detailed certain experiments in the cultivation of the plant in Calcutta, and the manufacture of Tapioca is discussed. **Major Jenkins** wrote: "The plant is very common, and **Mr. Masters** calls it the sweet 'Manihot'; that it does not contain the poisonous qualities of the South American plant, we may be sure, from the fact of the root being sold as a yam, and the Assamese eating it uncooked; they call it the *hemalú alú*, the leaves having some resemblance to those of the *simul* or *hímul* cotton-tree (**Bombax malabaricum**)." The Secretary of the Agri.-Horticultural Society, in replying to **Major Jenkins,** drew his attention to the fact that "**Mr. Masters** makes no mention of this plant in particular in his 'Memoir of some of the Natural Productions of the Angami-Naga Hills, and other parts of Upper Assam' (*Jour., VI., 34*), but that in his paper previously published (*Vol. IV., 197*) entitled 'Botanical Observations in Upper Assam,' he observes that **Janipha Manihot** ? the *Gash-alú*, is often used for hedges, and is not unfrequent in the jungles near the hills; it does not appear to have any poisonous qualities in a green state, as the Assamese eat the root eagerly when raw." This may be, the writer goes on to state, **Major Hannay's** *hemalú alú*. In a later communication **Major Jenkins** confirms that opinion, adding that *it grows all over Assam,* is constantly used for hedge rows, but after two or three years the root is dug up to be eaten as yam, and is certainly quite harmless, as it is eaten raw. "I fear," continues **Major Jenkins,** "I could not send you the flowers; it seldom or never flowers. I at least do not remember to have seen it in flower, though I have always had some of the plants in my garden hedges. An old Assamese gentleman tells me he never saw the flower nor heard of it I may add that I have seen the same plant in Arracan." **Major Jenkins** also refers to having got from **Dr Wallich** plants of the West Indies Cassava, and that these were not distinguishable from the Assam stock. In still another communication **Major Jenkins** writes: "There is no barren waste or hill land about us in which this plant does not thrive; the root increases in size according to the period it is allowed to grow, from one to three or four years; and I suppose an ordinary-sized root may weigh from 10 to 20 seers, perhaps still more. I have never seen it cultivated in fields or plots, but it appears to be just stuck in the hedges (for which, whilst it grows, it forms a useful post), and when wanted, or at maturity, it is dug out. I think it is probable, as stated in **Dr. Ainslie's** work, that the plant is not indigenous to India. I do not recollect, at least, ever seeing it in a true forest or jungle." **Dr. Falconer** examined two specimens collected by **Masters** and reported: "The Indian plant has the palmately 5-7-parted leaves, glaucous underneath, of **Jatropha Manihot** or **Janipha Manihot,** and I believe it to be merely a variety yielding the 'sweet cassava.' The variety yielding the 'bitter cassava' we do not appear to have in India. The people of Bengal call the plant *rotí alú* (bread-potato) and eat it raw without any bad effects. The plant has not flowered in the Botanic Gardens, at any rate there is no record of it."

Some fifteen years later, the Agri.-Horticultural Society published a paper communicated by **Dr. A. C. Maingay** from Malacca, in which, referring to **Major Jenkins'** statement that up to 1850 the Assam plant had not been seen to form flowers, he gave drawings and a detailed botanical description of the Malacca plant which he had fortunately found both in flower and fruit. He identifies it with "**Manihot utilissima,** *Pohl.*," and shows the

MANIHOT utilissima.

Introduction and Cultivation

CULTIVATION. **Assam.**

fruit as 3 celled, 3 seeded, and as having 6 wings. The tapioca prepared from the root is of excellent quality, 'and there is already in the Straits Settlements a very large and rapidly increasing area devoted to its cultivation." The cultivators are wealthy Chinese who purchase tracts of forest land, destroy the trees, and grow the Manioc for a few years making very large profits. But the crop is an exhausting one, the soil soon begins to fail to yield a remunerative crop, the Chinese dispose of their purchases and migrate to new tracts, leaving the land so severely injured that, **Dr. Maingay** estimated, it would take 80 years before the indigenous growths would again cover the land. While thus deprecating a too extensive destruction of forest in order to foster tapioca cultivation, **Dr Maingay** furnishes useful information as to the yield and methods of manufacture. Commenting on the subject of the difference between the sweet and the bitter Manioc, he says: "It is a fact adduced by all classes of cultivators with whom I have conversed on the subject, that in planting from cuttings, they must on no account be inverted. If by accident this occurs the resulting tubers do not belong to the variety described as the sweet, but to that to which in the West Indies I imagine the term 'bitter cassava' has been applied. The effects produced by eating these without prolonged steeping and washing are giddiness and vomiting, but as their taste is sufficiently bitter to act as a warning, such unpleasant results seldom occur."

"Though the bitter cassava, a plant considered botanically identical with the present species, is highly poisonous, unless exposed to heat, it may not be generally known in India that it forms the basis of the famous West Indian stew called 'pepper pot,' and that it also enters largely into the composition of several kinds of sauce."

Burma. **218**

II. BURMA.—Many writers, like **Major Jenkins** in the passage quoted above, incidentally allude to Manihot as existing in Burma, both under cultivation and in a state of naturalisation. **Mason**, for example, says: "I am not aware that either tapioca or cassava is manufactured in Burma, but Manihot, the plant which produces both, is frequently seen in culture. The natives boil the root and eat it like a yam, though severe sickness is often induced by the use of it. The Karen name signifies 'tree yam,' and in the Burmese it is called the 'Penang yam,' which shews whence it was imported. Malays have told me that much of the sago and arrowroot which comes from Penang and Singapore is made from this plant, though the former is usually supposed to be prepared from the sago-palm; it is said that an acre of ground planted with the cassava tree yields nourishment to more persons than six acres cultivated with wheat." **Kurz** remarks that the plant is "generally cultivated by Burmans and Karens, especially in toungyas."

Andaman Islands. **219**

III. ANDAMAN ISLANDS.—In the Administration Report for 1885-86 there occurs the following passage which would point to Manihot being a successful and important crop in these islands:—"Tapioca, as a vegetable, presents results more remarkable than even the Otaheiti potato; while the latter can be planted only at the beginning of the rains and ripens eight months later, after which the tubers will not keep for any length of time, the former can be planted out at any time of the year. A shoot, thrust into the ground, will, at the expiration of eight or ten months, produce an average weight of 5℔ of tubers, or no less than 8,000℔ per *bígha*, allowing a space of 3′ × 3′ per each plant."

Madras. **220**

IV. MADRAS PRESIDENCY.—The Manioc is now so widely cultivated in South India that it may be said to occur more or less in every district. **Ainslie** (one of the earliest and most trustworthy writers on Indian medical subjects) says: "Having found that the **Jatropha Manihot** grew in great abundance and luxuriance in many parts of Lower India, I, some months before

of Manioc. (G. Watt.) MANIHOT utilissima.

CULTIVATION.

leaving that country, in 1814, attempted to make tapioca from the root, and perfectly succeeded, the first, I believe, that ever was made in our Indian dominions." He then remarks: "The tapioca plant is called in Tamul *maravullie*, and, from the circumstance of its having no Sanskrit, Arabic, or Persian name, I am led to think that it is not a native of Hindustan, but was probably brought hither, many years ago, by the Portuguese." While giving the prepared article a Tamul name, it is somewhat significant that **Ainslie** should make no mention of its cultivation in Madras. As grown on the Coromandel coast, the plants are said to be "more fibrous, and, therefore, inferior to those raised in Malabar." In the Nellore District Manual the plant is said to be the *Manupendulam* of the Telegu people. It is, however, in Travancore that the cultivation of this plant has assumed the greatest proportions. In the Madras Manual of Administration (*II., 135*) it is stated of that district that, within the last few years, the cultivation of tapioca has so extended that it has become a staple article of food. The following passage from the *Tropical Agriculturist* (*April 1882*), and which appeared originally in the *Journal of Applied Science*, deals mainly with the subject of Travancore Tapioca:—"The bitter cassava or tapioca plant (**Manihot utilissima**), which is a native of South America, is now largely grown in Travancore, where the soil seems so well suited to its cultivation as to warrant a still more extended growth. It is stated that, as the price of rice has risen of late years, tapioca has become the more essential as an article of food. It will grow in any soil, and needs but little care, except to preserve it from the depredations of cattle. After the roots are dug, the stem is cut into pieces about 4 inches long and planted some 3 feet apart, with a little ash or other manure. The root requires occasional weeding and earthing, and arrives at maturity in nine or ten months. Well boiled it is eaten with fish curry. It is sometimes given to cattle. In a green state the root does not keep long, but it can be sliced and dried in the sun, or grated and made into farina. A field of this valuable and nutritious root is planted at but little cost; its yield is very large, and its cultivation highly profitable. The produce has been estimated in Ceylon at 10 tons of green roots per acre; this weighs one-fourth when dried, and if the dried roots gave half their weight of flour, it would amount to 2,800℔ per acre. With some care and attention any amount of the granulated flour might be prepared for home use and export, but though this plant grows almost wild, the people do not take the trouble to prepare it."

Bombay. 221

V. BOMBAY.—Graham (*Cat. Bomb. Pl. 183*) says this plant is "easily cultivated, growing equally well in any soil or situation. It is said to have been first introduced by the Portuguese at Goa, and is now pretty common in Bombay gardens, but simply as an ornamental shrub; the natives do not seem to be aware of the uses to which it can be applied, and if they were, could only be driven to them by a scarcity of their common and inferior articles of food, afforded by the Cucumber and Arum tribes." **Dalzell & Gibson** write: "About 22 years ago" (? 1849) "attempts were made by the Agri.-Horticultural Society to extend the growth of this plant as useful for food, but the experiment, as might have been expected in a great bread-corn country like this, failed, since the produce is by no means equal in nutritive property to that of our numerous cereals."

N.-W. P. & Oudh. 222

VI. NORTH-WEST PROVINCES & OUDH.—Mr. Atkinson (*Econ. Prod., Pt. V., 22-23*) remarks that the plant grows luxuriantly in these provinces. He then quotes from the Journals of the Agri.-Horticultural Society the process of manufacture as given in some of the papers here reviewed. No further information is available regarding these provinces, but Mr. Atkinson gives certain facts about Bengal.

CULTIVATION. Bengal. 223

VII. Bengal — The following passage from an interesting letter addressed to the Government of India (in September 1887) by **Mr. R. Mitchell**, Emigration Agent, Calcutta, gives much useful information regarding cassava and yam cultivation in Bengal:—"I am more anxious about the introduction of the sweet cassava, as a means of sustenance for the poorer classes, because it yields large returns under the most primitive cultivation, especially in light friable soils, and flourishes in the poorest land, where yams or sweet potatoes will not yield any return commensurate with the labour spent on them.

"The cassava stands drought well and thrives where every root crop languishes for want of moisture, but it cannot resist cold, and the leaves drop at once when touched by the frosty air of December, although the stems retain their vitality. In arenaceous soils, such as are to be found in the neighbourhood of Calcutta and most of the delta of the Ganges, the cassava stick merely requires to be cut into lengths of twelve inches, pointed, and thrust into the soil at an angle of about 60°, the ground occasionally cleared of weeds for the first two months, when the plant takes full possession of the soil and its vigorous growth destroys everything under it.

"The green tops are excellent food for cattle, the stems would make inferior firewood, being too slight and brittle to be used for any other purpose, while the roots provide the most delicious and wholesome food. Roasted they taste like chestnuts and, properly boiled, they are much to be preferred to indifferent potatoes. The only fault about the sweet cassava is that it will not keep. It must go from the garden to the pot, if possible the same morning; and the day after it has been dug, it becomes hard, woody, and unfit for food. It can be grated, however, and exposed to the sun, or dried on heated plates and made into flour, when it will keep an almost unlimited time, or it may be readily converted into starch.

"The bitter cassava is more hardy than the sweet, and yields a much larger return per acre, but the prussic acid contained in it must be got rid of before it becomes an article of diet. This is accomplished by pressing and exposure to fire on an iron plate.

"In January next, I shall be in a position to supply you with sufficient sweet cassava stems to plant an acre. I am also trying experiments in the Hills with this plant and will let you know the result at a later date."

The following, **Mr. Atkinson** says, is a mode of extracting tapioca from the roots pursued in the Lower Provinces:—"The roots are first washed and stripped of the rind, then ground to a pulp, which is thrown into a clean cloth, and the acrid poisonous juice well wrung out. The pulp thus partially deprived of its impurity is exposed for a few hours to the influence of the sun, by which any remaining juice is successfully taken up. The mass is next mixed with clear water, strained, and the pulp thrown away. The milky substance thus obtained is allowed to settle, when the clear water is carefully drawn off, and the subsidence again and again watered until it becomes perfectly firm and white; it is then put in the sun until quite dry, crushed, and passed through a muslin sieve." Much after the same fashion as in the above quotation, the information regarding Bengal tapioca is incidentally alluded to by the most unlikely authors. **Firminger**, for example, says: "The plant thrives well in Bengal, and a considerable plantation of it is raised annually in the garden of the Agri.-Horticultural Society, though the manufacture of the tapioca is rarely, I believe, resorted to in India. The season for taking up the roots is in January, at the same time that cuttings are put down for the crop of the following year."

Most suitable Method of Cultivation.

CULTIVATION.

Mr. J. P. Langlois (No. 5 of the Series of Gardener's Notes published in the Agri.-Horticultural Society's Journals), gives the following directions regarding this plant :—

"*Soil.*—The plant will thrive in any soil, although a sandy loam is the best. Soil. 224

"*Cultivation.*—It requires no cultivation whatever, and is occasionally met with in Arakan, growing wild in the jungle.

"*Propagation*—By cuttings. Care should be taken to use the stronger branches. The cutting must be from two to three feet long, to be placed in the ground in an upright position, and in rows, four feet apart. Propagation. 225

"*Preparation.*—Twelve months after planting, the roots are fit to be dug up. They must then be well washed, and put into a trough with water, in which they are allowed to remain six hours, when the outer bark will be easily removed by a pressure of the hand. The next process is to grate the roots, and then press out the milky juice, which is poured into a flat tub. This is now suffered to rest for eight hours, when all the flour will subside to the bottom. The water is then poured off and the meal laid upon wicker frames to dry in the sun, for two or three hours. The flour is then placed upon hot plates, and well stirred, to prevent it burning. Preparation. 226

"The heat will cause the amylaceous substance to coagulate into small irregular lumps of a transparent and gelatiniform colour. The tapioca is then ready for use. This is the best mode of preparing Tapioca and is that pursued in Mauritius." **Simmonds** writes :—"No less than 30 varieties of Mandioc are grown in Brazil, and of all the crops it is the one that gives the best return and the least trouble."

CASSAVA AND TAPIOCA.

Manufacture of Tapioca.—In addition to what has been said under the paragraph "Bengal," as well as in **Mr. J. P. Langlois'** account, the following facts may be given regarding the manufacture of tapioca :—"The tubers, each weighing from 10 to 25lb, to which they attain in from 18 to 20 months, are first scraped and then carefully washed by hand labour or by placing them in a rotatory drum exposed to a stream of water, by which all impurities are removed. After this they are reduced to a pulp by being passed through rollers. This is carefully washed and shaken up with abundance of water until the farina separates and passes through a very fine sieve into a tub of water placed beneath. The flour so obtained undergoes eight or nine washings, as upon the care with which these are conducted depend very much its whiteness and price in the market. It is now collected into large heaps, placed on mats, and bleached by exposure to the sun and air. It is finally converted into the pearl tapioca of commerce, by being placed in a cradle-shaped frame covered with canvas cloth, in small quantities at a time, slightly moistened, and subjected to a rotatory movement. The mass gradually forms into small globules, each about the size of a No. 6 shot. Whilst still soft, these are taken out and dried in the sun, and lastly, while constantly stirred, are fired in a large shallow iron pan, which is occasionally rubbed on the inside with vegetable tallow, after which they are packed in bags ready for exportation." We cannot afford space to deal with this subject more fully, but the reader is referred to the numerous works quoted in the paragraph of **References**, more particularly to an article on Malacca Tapioca (*Tropical Agriculturist, II., 189*), which greatly amplifies what **Dr. Maingay** wrote and brings his account of the process up to modern times. MANUFACTURE. Tapioca. 227

Cassava.—The meal known as Cassava is only a cruder preparation than Tapioca. It is obtained by subjecting the grated root to pressure, to Cassava. 228

MANUFACTURE.

express the juice, and then drying and pounding the residual cake. Of this meal cassava cakes are made. These are prepared by gently heating the moistened meal, forming cakes of it, and then drying them in the sun.

Tapioca Meal. 229

TAPIOCA MEAL.—Is the precipitated starch from the expressed juice (described above under Tapioca) dried in the air without being roasted. It is the heating of the damp starchy precipitate that gives to tapioca its peculiar character. By this process the starch granules are swollen, many of them being burst and then agglutinated into rounded masses. A change is thus effected by which the starch of tapioca is rendered partially soluble in cold water, and in boiling water it forms a jelly-like mass.

MEDICINE. Tapioca. 230

Medicine.—The effects and uses of TAPIOCA are similar to those of starch. Speaking of the poisonous property of the plant, **Sir W. Hooker** wrote :—" It yields an abundant flour, rendered innocent indeed by the art of man, and thus most extensively employed in lieu of bread." "Such is the poisonous nature of the expressed juice of the Manioc, that it has been known to occasion death in a few minutes. By means of it, the Indians destroyed many of their Spanish persecutors. **M. Fernier**, a physician at Surinam, administered a moderate dose to dogs and cats who died in a space of 25 minutes, passed in great torments." "Thirty-six drops were administered to a criminal. These had scarcely reached the stomach, when the man writhed and screamed with the agonies under which he suffered, and fell into convulsions, in which he expired in six minutes." The poison contained in these roots has long been known to be Hydrocyanic acid.

Manilla Hemp, see **Musa textilis,** *Nees;* SCITAMINEÆ; **p. 302.**

231

MANISURUS, *Linn.; Gen. Pl., III., 1130.*

A name which denotes the resemblance of the spikes to a lizard's tail.

[*Econ. Prod., III., 424;* GRAMINEÆ.

232 **Manisurus granularis,** *Swartz.; Duthie, Fodder Grasses, 29; Dict.*

Vern.—*Trinpali,* HIND.; *Kangni,* AJMIR; *Dhaturo ghas,* RAJ.; *Agimali-gadi,* CHANDA; *Ratop,* BERAR; *Palanggini,* SANS. (according to Ainslie).

References.—*Roxb., Fl. Ind., Ed. C.B.C., 118; Ainslie, Mat. Ind. II., 434; Drury, U. Pl., 287; Dalz. & Gibs. Bomb. Fl., 300; Grah., Cat. Bomb. Pl., 234; Coldstream, Grasses of S. Pb. Pl., 14; Trimen, Cat. Ceylon Pl., 107; Dymock, Mat. Med. W. Ind., 856.*

Habitat.—A hairy, annual grass, recognisable by the globular shape of the sessile fertile spikelet of each pair. According to **Duthie**, it is found on the plains of Northern India, ascending the Himálaya to altitudes of 5,000 feet. **Roxburgh** simply remarks that it grows amongst bushes: **Dalzell & Gibson** say it is very common on barren land.

MEDICINE. Plant. 233

Medicine.—The only author who deals with this subject is **Dr. Ainslie,** all other writers having repeated his words without either adding to, or even confirming, the accuracy of the original observations. Even **Ainslie** simply says that the PLANT was shewn to **Dr. F. Hamilton** while in Behar, as a grass prescribed internally in conjunction with a little sweet-oil in cases of enlarged spleen and liver. It may be added that it is somewhat significant, however, that this Behar drug, if it be such, is not mentioned by **Irvine** in his *Materia Medica of Patna.*

FODDER. Plant. 234

Fodder.—**Mr. Coldstream** states that "it is not much relished by cattle. It is both grazed and stacked, but opinions differ as to its qualities. It is supposed to last five or six years in stack." **Mr. Duthie** adds that in Ajmir it is considered a good fodder grass.

MANNA.

235

Manna.

Vern.—*Shirkhisht*, HIND.; *Shir-khisht*, BENG.; *Shirkhisht, bed-khist, shakar taghár*, PB.; *Gazanjbin*, BOMB.; *Ména*, TAM.; *Ména*, TEL.; *Kapurrimba, manná*, MALAY; *Terenjabiri, mun shir-khist, sukkarul-ghushar*, ARAB.; *Shir-khist*, PERS.

References.—*Roxb., Fl. Ind., Ed. C.B.C., 574; Brandis, For. Fl., 22, 145, 302, 512; Stewart, Pb. Pl., App. 93; Pharm. Ind., 136; Ainslie, Mat. Ind., I., 209, 613; O'Shaughnessy, Beng. Dispens., 278, 295, 434, 454; Moodeen Sheriff, Supp. Pharm. Ind., 37, 82, 171, 239; Dymock, Mat. Med. W. Ind., 2nd Ed., 77, 218, 516; Dymock, Warden, and Hooper, Pharmacog. Ind., I., 161, 419, 583; Fleming, Med. Pl. and Drugs, as in As. Res. Vol. XI., 188; Flück. & Hanb., Pharmacog., 413; Hooper, Chem. Notes on Mannas; U. S. Dis-pens., 15th Ed., 921, 1256; S. Arjun, Bomb. Drugs, 16; Kanny Lall Dey, Indigenous Drugs of India, 70; Waring, Pharm. Ind., 136; Linschoten, Voyage East Indies in 1598, II., 100; Irvine, Mat. Med. Patna, 101; Honigberger, Thirty-five years in the East, Vol. II., 305; Baden Powell, Pb. Pr., 320, 361; Royle, Ill. Him. Bot., 275; Smith, Dic. Ec. Pl., 265, 401, 402; Davies, Trade and Res. N.-W. Frontier, pp. cxx, cxxvii, ccxcvi; Balfour, Cyclop., II., 852-3; Encycl. Brit., XV., 493; Treasury of Bot., II., 718; Indian Forester, XIII., 93; Hanbury, Historical Notes on Manna (Jour. Pharm. Soc. XI, 1870), 326 (also in Science Papers, p. 355); Aitchison, Trans. Linn. Soc. (2nd Ser.), Vol. III., 3, 42, 64; Aitchison, Plants and Plant Products of Afghánistán, Pharm. Soc. Gr. Brit., 8th December 1886.*

This is a saccharine exudation obtained from several plants naturally, and from others on the bark or epidermis being incised. This subject has been dealt with to some extent under **Fraxinus ornus**, *Linn.*; Vol. III., 442-443.* The facts there given will not be repeated here, and as the MANNA chiefly used in India is imported, the subject has scarcely more than a scientific interest. The following are the plants reported to yield the substance:—

Alhagi camelorum, *Fisch.*; } Vol. I., 165; Vol. III., 443.
A. maurorum, *Desv.*; }
Astragulus, sp. in Persia.
Atraphaxis spinosa, *Linn.*; Vol. III., 443.
Calotropis gigantea, *R. Br.*; Vol. II., 37, 47; Vol. III., 443.
Cedrus Libani, *Barr.*; Vol. III., 443.
Cotoneaster acutifolia, *Linn.*; according to **Aitchison**.
C. nummularia, *F. et M*, Vol. III., 443.
Fraxinus ornus, *Linn.*; Vol. III., 442-444.
Musa superba, *Roxb.*; p......
Palmæ, various species.
Pinus excelsa, *Wall.*; Vol. III., 443; Vol. VI.
Quercus incana, *Roxb.*; Vol. VI.
Rhododendron arboreum, *Sm.*; Vol. III., 443.
Tamarix sp, Vol. VI.
Salix sp., according to **Stewart**.
Salsola fœtida, *Del.*; according to **Stewart** and **Aitchison**; Vol. VI.

One of the earliest, and at the same time most interesting, accounts of Indian Manna (written by a European) is that which occurs in the Journal of **John Huyghen van Linschoten's** Tour to the East Indies in 1598. "Manna," he says, "commeth out of Arabia and Persia, but most out of the Province of Usbeke, lying behind Persia in Tartaria: the manna yt is brought from thence in glasse kalles, is in peeces as bigge as preserved almonds, but of another fashion, and have no other speciall form, but like

* Please correct two misprints in Vol. III., page 443:—For **Araphaxis** read **Atraphaxis**. Cancel the words "samples have" in line 21 from the top.

broken peeces: it is whitish, and of taste almost like sugar, but somewhat fulsome sweetish like hony: the Persians cal it Xercast and Xerkest, that is to say, milke-of-trees, for it is the dew yt falleth upon the trees, and remayneth hanging upon the leaves, like water that is frozen and hangeth in drops at gutters and pentises. It is then gathered and kept in glasse kals and so brought into India and other countries, for in India they use it much in all sorts of purgations.

"There is another sort of Manna called Tiriamiabūn or Trumgibūn* which they gather from other leaves and hearbes: *that* commeth in small peeces as big as Hempe seed and somewhat bigger, which is red and of a reddish colour. Some thinke this manna groweth on the bodies of the trees as Gumme doth: it is much used in Ormus and Persia for purgations, but not in India so much as the first sort.

"There is yet another sorte, which commeth in great peeces, with the leaves among it; it is like the manna of Calabria; this is brought out of Persia into Bassora and so to Ormus and [from thence into] India, and is the dearest of all the rest. There commeth also a Manna [that is brought] in leather bags or flasks, which in Turkey and Persia they use to ride withall and is melted like Hony, but of a white colour and in taste like the other sortes of Manna, being altogether used for purgations, and other medicines."

The Manna known to Muhammadan writers as *Taranjabín* is obtained from **Alhagi maurorum. Mir Muhammad Husain** says of this, that is collected in Khorasán, Mawarunnahr, Kurjistan, and Hamadan, and that the plants are cut off, then shaken in a cloth, to separate the Manna. **Dr. Aitchison** informs us that the country around Rui-Khauf is famous for its *Taranjabín,* and that, in addition to that obtained by shaking the bushes, an inferior sort is prepared by washing the twigs and boiling down the fluid. **Aitchison** regards *Taranjabín* as more digestible than *Shirkhisht.* According to **Dymock** fine clean samples of *Taranjabín* are sometimes obtainable in Bombay during the season of import (November to January), but unless very carefully preserved, it soon spoils, running together and becoming a brown sticky mass.

Shirkhisht is probably, as stated by **Moodeen Sheriff**, a generic name in India for any form of Manna, being that by which the imported European article is sold in the bazárs. Its Persian usage has a specific meaning, however, being the Manna from **Cotoneaster nummularia. Mir Muhammad Husain** points out that *Shirkhisht* or *Shirkhushk* is not, as generally reported, a honey dew which falls upon the trees in Khorasán, but is an exudation from the tree called Kashira—a small tree with yellow mottled wood, much valued for making walking sticks.

Gasangabín is the name which very probably should be restricted to Tamarisk Manna, though it is generally used as synonymous with *Shirkhisht.* **Dr. Aitchison** found that **Tamarix gallica**, not **T. mannifera**, was the source of this substance. **Dr. Fleming** wrote in 1810 that Alhagi manna was then regarded as far inferior to the Calabrian. He gives *Shirkhisht* as the Persian and *Terenjabín* as the Arabic names for manna.

Honigberger describes *Turunjebín* and *Shirkesht,* but adds that a manna obtained in India is known as *Tíghul.* This, he says, is what the Sadus at Lahore import from Hindustan and sell by the name of *Shukurí Tíghal.* He concurs with **O'Shaughnessy** in thinking this may be obtained from **Calotropis gigantea** or some nearly allied plant.

CHEMISTRY 236

Chemistry.—Dr. Warden of Calcutta has kindly furnished the following note on the chemistry of this substance:—The chief constituent of manna is

* *Shirkhisht* and *Taranjubin* are Persian names for two kinds of Manna.

mannite or mannitol, a hexahydric alcohol discovered by **Proush** in 1806. Mannite is found in the sap of many plants and also in fungi. It is crystalline, and only slightly sweet to the taste. It does not reduce an alkaline cupric solution : it slowly ferments with yeast. A mixture of concentrated nitric and sulphuric acid converts mannite into a hexanitrate, which is explosive on percussion. Mannite also gives rise to several other derivatives when treated with certain other acids. Under the oxidizing influence of platinum, black mannite is converted into mannitic acid, and, according to **Hanbury & Flückiger,** also into mannitose, a sugar probably isomeric with glucose. By the action of nitric acid it is changed into saccharic acid. In the best specimens of manna from 70 to 80 per cent. of mannite occurs. Dextroglucose, water, a very small amount of reddish brown resin with offensive odour and a subacid taste, and a substance called fraxin are also among the constituents of manna. Solutions of certain samples of manna exhibit a fluorescence which was attributed by **Gmelin** to the presence of æsculin, the fluorescent glucoside contained in the bark of the horse chestnut, but **Flückiger & Hanbury** state that the fluorescence is due to fraxin, a body closely resembling æsculin, and occurring not only in the bark of the manna and common ash, but also, associated with æsculin, in that of the horse chestnut. **Stokes,** on the other hand, describes a second fluores, cent principle as being associated with æsculin in thehorse chestnut. which he has named panin. **Flückiger & Hanbury** describe fraxin as being faintly astringent and bitter, and soluble in water and alcohol, Dilute acids convert it into fraxetin and glucose. Madagascar manna, obtained from **Melampyrum nemorosum,** contains an isomeride of mannitol, melampyrite, dulcite or dulcitol. This principle is also contained in the sap of other plants. It may be artificially produced with mannitol, when a solution of milk and sugar which has previously been boiled with dilute sulphuric acid. is treated with sodium amalgam. It is crystalline and scarcely sweet. Nitric acid converts it into mucic acid, which is isomeric with saccharic acid It forms compounds with acids (*Graham*). Since the above was written an interesting note on the chemistry of mnnna by **Mr. David Hooper, F.O.S.,** has appeared, to which the reader is referred for further information. — CHEMISTRY.

(*W. R. Clark.*)

MANURES. 237

Manures and Manuring.

Vern.—*Khad, khau, eru, paus,* HIND.; *Khádar, khadaur, khaddhi, gondaura, goa, karsi, ghúr, gánaura,* BEHARI; *Páus, pásá, khát,* N.-W. P.; *Kurri khár, khát kúra, kallar, sarra,* PB.; *Páusu,* SANS.; *Zibl,* ARAB.

References.—*Liebig, Natural Laws of Husbandry, 131 et seq.; Anderson, Agricultural Chemistry, 152-265; Johnston, Agric. Chem. and Geol., 198 318; Baden Powell, Pb. Pr., I., 95, 204, 205, 214, 416; Benson Manual and Guide, Saidapet Farm, Madras, 19—34; Wallace, India in 1887, 79-84, 224, 285; Hove's Tour in Bombay, 118-119; Schrottky, The principles of Rational Agriculture applied to India, 106-156; Annual Report, Agric. Dept., Madras, 1878, 24-29, 87-99; 1882-83, 64-65; 1883-84, 37-46; Annual Reports, Madras Experimental Farms, 1872, 34-41; 1873, 26-28; 1875, 31-36; 1877, 32-37, 98-102; Annual Report, Dept. Land Records and Agric., 1887-88, 18-19; Settlement Reports:—Panjáb, Bannu Dist., 83, 84; Montgomery District, 101; N.-W. P., Azimgarh, 101; Banda District, 55; Central Provinces, Upper Godavery District 28; Annual Report of Director of Agric. Dept., Bengal, 1885-86, App. I., iii-v, ix, xxx-xxxiii; Gazetteer:—Bombay, VII., 79-91; VIII, 179; Agri.-Hort. Soc. Ind., Transactions, I., 20, 29, 42, 59; III., 185, 186; Journal I., 207, 295-296; V, Part I., 19, 101; Part II., 20-23, 44-60; X., Part I., 92-94; XIV, Part I., 133-135, 198; Spons, Encycl., II., 1256-1277; Encyclop. Brit., XV., 505-512; Balfour, Cyclop. Ind., II., 858-859.*

In India, systematic manuring has practically been neglected by the

Natives. They have been accustomed, from father to son, to carry on the same rough system of manuring, collecting for that purpose merely such substances as are inexpensive and easily obtainable. Not only this, but they have also, by the almost universal use of dried cow-dung as fuel, and, by the custom of allowing the urine of their cattle to run to waste through insufficient littering, neglected the most important source of manure and the one that lay readiest to their hand. Caste prejudices, too, have to a great extent forbidden the employment of the night soil of large towns, hence depriving the Natives of India of that important supply of valuable manure.

Various attempts have been and are being made by Government, and by the municipalities of the cities, to educate native cultivators to an intelligent conception of the value of manures, and large quantities, both of natural and of artificial manures, are used by the European planters in India; but neither the example of these, nor the precepts and efforts of the Government, have as yet done much to arouse the Native to a sense of the benefits arising from the systematic use of manures.

In dealing with the Manures available in this country, we shall divide the subject into three classes—Animal, Vegetable, and Mineral.

ANIMAL MANURES

I.—Animal Manures.

Farm-yard Manure.

238 (1) **Farm-yard Manure.**—Although this is the most valuable manure and most easily available to the agriculturist, yet no attention is given by the Natives generally to its collection and due preservation. Most of the Natives keep cattle, but they take no care that their droppings and urine are preserved. The droppings are in most cases sun-dried and used as fuel (*brattís* or cow-dung cakes); while the urine, through want of litter, is allowed to soak into the floors of the cowsheds and places where the cattle are picketed, and so is very generally lost. The practice of manuring is almost entirely confined to gardens, especially the large market gardens around cities, and some portions of irrigated lands, where particularly exhausting crops, such as sugarcane and tobacco, &c., are grown. No manure pits are formed, hence the manure usually applied to these grounds, except in the vicinity of towns where night-soil is procurable, and where the religious scruples of the native agriculturist do not prevent its use, are village ashes, weathered, sun-dried cattle dung, and the manure supplied by the folding of cattle and sheep on the land at night. During the day these feed anywhere on waste land where they can pick up a bite; and as they get no addition to the food thus obtained, it may be admitted that their excreta can hardly fail to be of poor quality. The native Indian agriculturist, although he appreciates the value of sheep folding, rarely thinks of raising green crops to be eaten on the land. The statement made by some agricultural authorities, that the use of cattle dung for fuel does not cause any loss, since the ashes made by the combustion of the dung are quite as efficacious a manure, seems to the writer to be but an attempt to revive the old "mineral theory" exploded forty years ago. Bulk and dilution are within certain limits essential for the utilization of a manure, and hence the superiority of farmyard manure over ashes. Further, when the solid excrements of cattle are collected for fuel, the urine is always lost, a large proportion of the ashes is wasted, and the fertilising properties of the portion actually used as manure are seriously lessened by the careless way in which the ashes are stored. From the Annual Report of the Superintendent of Government Farms (Madras) for the year ending 31st March 1875, the following interesting extract on the subject of farmyard manure may be quoted:—

"It is a mistake to suppose that the fertilising effects of farmyard manure can be measured, and determined by its composition as shown by analysis. Field experiments have shown over and over again that one ton of

ANIMAL MANURES.

Farm-Yard Manure.

farmyard manure when applied to the soil, will produce vastly greater effects than a dressing of mineral manure containing the ash equivalents of the ton of manure. While the effect of the ashes on the physical state of the soil would be almost imperceptible, that produced by the farmyard manure would be great and highly beneficial. In no country would the benefits that are conferred on a soil by an application of farmyard manure be greater than in India, where famines so frequently result from a drought. The power of a soil to absorb moisture from the air and to retain that moisture in a healthy condition depends almost entirely upon the quantity, and state of the organic matter that soil contains, but, with reference to this matter, I cannot do better than direct attention to the statements made by the late **Professor Voelcker** in a report upon plots of land which had for many years been manured by one kind of manure only. Speaking of the plot that had continuously been manured with farmyard manure, he says:—'**Dr. Gilbert** informs me that whilst the pipe drains from every one of the other plots (dressed with mineral manures) in the experimental wheat field run freely four or five or more times annually, the drain from the dunged plot seldom runs at all more than once a year, and in some seasons not at all. The fact is, the accumulation of decomposing organic matter in the plot lightens the soil, promotes the disintegration of the clayey portions, and altogether renders the surface soil more porous, and capable of retaining much more water.'"

The improved sanitation of India, which has to such an extent checked the outbreak and spread of the epidemics that were wont to decimate the people, and has thus been instrumental in increasing the population, loses its chief merit if coincident poverty takes place through deficient food supply. If the food produced in the country (through expansion of cultivation and improvement in yield) is to keep pace in its increase with the increasing population, all practices that tend to lessen the food-producing powers of the land should be checked, and none more necessarily so than the burning of cow-dung as fuel, a custom which deprives the agriculturist of his cheapest and most easily available source of manure.

Town Refuse. 239

(2) **Town Refuse.**—Night-soil and refuse from towns and villages is another valuable source of manure generally neglected by the Indian agriculturist. Although he is to some extent aware of the value of night-soil as a manure (witness the fact that "land close to the village site which is frequented by the villagers for purposes of nature is rented at three times the rate at which land near the boundary of the village is rented," **Mr.** (now **Sir**) **E. C. Buck** in a report on the employment of city refuse for agricultural purposes at Furrukhabad), still the Indian agriculturist allows an enormous loss of fertilising matters, especially throughout the villages and rural districts, by not collecting human excreta and the waste materials from habitations and utilising those fully as manure. The excreta deposited around human dwellings, both on cultivated and on uncultivated ground, dry up quickly and lose a great part of their fertilising powers, while the ashes and sweepings from the villages, which might be utilised to mix with and deodorise the excreta, without interfering with the manural properties, are simply heaped about, no effort being usually made towards their utilisation.

In large towns, of course, the excreta and other waste materials are always collected and removed from human habitations, but in such cases they are too often simply buried in the ground at some convenient locality and thus lost to the cultivators generally. This must be a serious loss to the agriculturist, since we may roughly calculate that the solid and liquid excreta of each individual, produce per annum about 6℔ of ammonia and 3℔ of phosphates, even after a due allowance has been made for the poverty of the food of the people in India.

ANIMAL MANURES.

Town Refuse.

The question of the disposal of night-soil and its utilisation for manural purposes is one which has been, of late years, carefully considered by the various Provincial Governments, and many attempts have been, and are being, made by the municipalities of the large towns to induce the rayats to collect, and remove, to their land, the excreta of town populations; but, even apart from their objections on the score of caste prejudices, they are not usually willing to do so on account of the unpleasantness of the work and the poverty of the stuff through its admixture with useless matters. The North-West Provinces and Oudh seem to have made most progress in utilising night-soil as a source of manure; but in the other provinces attempts are also being made in the same direction and with varying success. The Sanitary Commissioner of Bengal, in a recent note on the subject, commends the shallow temporary trenching on fields, pursued in Howrah, as a desirable system to be extended throughout the province. The fields thus manured may, he says, be cultivated within three months. But, in tracts of low-lying country subject to inundation during the rainy months, it seems questionable if that system could be followed during the rains.

In all the large cities, the night-soil is, of course, removed at the expense of the municipalities, and after removal the question arises as to how it can be most advantageously disposed of. In many places it is merely removed and buried in some convenient locality, but where attempts are made at utilising it, three courses are open (1) to trench it in, and lease the ground to cultivators, (2) to sell it after removal as manure, or (3) to convert it into poudrette and sell the product thus manufactured. As a result of numerous experiments on the subject of the utilisation of night-soil in the various provinces, the following conclusions have been arrived at (1) that where there is plenty of available land, a supply of cheap water ready at the right moment, and a sufficiency of cultivators who appreciate to the full the value of trenching land with night-soil, the first means of disposal is the most advantageous, since it not only repays the cost of removal, but also yields a clear income; (2) selling the night-soil and delivering it to cultivators at their fields at the largest price obtainable, or allowing sweepers to remove and dispose of it as best they can, both yielded the same result—neither profit nor loss to the municipality; (3) the poudrette system is the most costly of all since it requires a considerable outlay on the part of the municipality without yielding any corresponding income.

Animal Refuse. 240

(3) Animal Refuse.—This includes the blood, offal, and bones of the cattle and sheep which are slaughtered, the carcasses of domesticated and wild animals that die, and fish. As in the case of *Farm-yard Manure* and *Town Refuse* it may at once be admitted that the materials embraced under the present paragraph constitute another largely neglected means of preserving the fertility of the land. **Mr. W. A. Robertson**, in his report on some manural substances yet unutilised in this country, states that, in the Presidency of Madras alone, the amount of blood, offal, and bones of the sheep killed in that Presidency in the year 1871-72 would have been sufficient to manure highly 44,000 acres of land. Besides the diminution of stock by slaughter, there is annually an enormous loss from disease and other natural causes: the manural matters thus produced are, under existing circumstances, almost entirely wasted. Slaughter-house refuse is a highly nitrogenised manure, and would, therefore, be specially good for crops like sugarcane and maize. Fish manure also is scarcely ever utilised in this country, although large quantities of fish-oil are manufactured, the refuse from which, if properly conserved and utilised would form an invaluable substitute for guano, and not only prove a rich source of manure on our Eastern and Western Coasts, but also in time become a good marketable commodity and afford a livelihood to many poor people in these parts. Very occasionally in Bengal a manure of rotten fish is applied to the roots of trees.

ANIMAL MANURES.

Bones. 241

(4) **Bones.**—By far the most valuable source of manure which is lost to the country proceeds from the non-utilisation of the various bone manures. All the different preparations of bone, whether boiled bones, crushed bones, bone black, or superphosphate, are very valuable to the farmer, from the amount of phosphoric acid they yield; but, in spite of this, they are never used by the ordinary cultivator in any part of India. This, in addition to the very prevalent religious objection, is partly on account of the expense that would be entailed in crushing the bones and dissolving them with sulphuric acid—the most valuable method of using them—and partly also to the vegetarian habits of the agricultural people of India, not creating in their immediate vicinity a large supply of bones. As an example of the well known benefits that would arise from the extensive use of some of the various forms of bone manure with Indian crops, we may quote a letter in the *Indian Agriculturist* of November 2nd, 1889, by **Mr. B. C. Basu,** Assistant to the Director of Land Records and Agriculture, Bengal. He says:—"I have obtained this year, on the Seebpore Experimental Farm, an absolute increase of 26 seers of cleaned jute over the usual produce of half a *bígha* of land with an application of 1 maund of bone meal. The latter may be made in the country at 12 annas a maund, while the value of the increase in yield cannot be less than R2-8." In **Mr. C. Benson's** Saidapet Experimental Farm Manual, page 28, there is a notice of a series of experiments made on the comparative value of different manures. Amongst others, bone-dust was used, with the result that 37℔ of bone-dust valued at R1 caused an increase in a crop of **Sorghum** or Chinese sugar-cane of 48℔ of grain and 420℔ of straw, or a total increase per acre of 864℔ of grain and 7,560℔ of straw, valued at R45-8-3, giving a clear profit per acre of R27-8-3. In this series of experiments, saltpetre gave the best results; bone stood second on the list, but "probably," says the Manual, "if the continued effects of each were considered, bone-dust would stand first." In an interesting note on the use of and trade in bones in Bengal, prepared by **Mr. B. C. Basu,** an account is given of some experiments made, through rayats, to test the efficiency of bone-meal as a manure for paddy. The results of these were very encouraging as they showed an average increase of 570℔ of paddy per acre by manuring with 240℔ of bone-meal. "The money value of the increase in a year of ordinary prices may be taken as R9-8, while the price of 240℔ of bone-meal applied per acre is only R6 at its present high price in Calcutta." "Until, however, bone-meal can be offered to the public at cheaper rates than oil-cake, there is no hope for its general adoption in native agriculture."

A considerable and increasing export trade in bones exists in India, which is not a matter of congratulation to the Indian agriculturist. In the year 1884-85, 18,383 tons of bones were exported from India; while in 1888-89 this trade had nearly doubled itself, 35,393 tons, valued at R17½ lakhs, having been exported during the latter period. As very nearly half the weight of a quantity of mixed bones is made up of phosphate of lime, we thus see that in the year 1888-89 alone, about 17,000 tons of calcium phosphate were removed from the soil and taken out of the country altogether. If this continues for many years, the result may be anticipated of a serious diminution in the already small proportion of phosphates which exists in the Indian soil This export trade, however, is very far from representing the actual quantity of bones that must be available throughout India. According to **Mr. J. E. O'Conor's** Statistical Tables of the Foreign Trade of British India, in the year 1888-89, 6,606,142 undressed and 1,447,544 dressed hides, making a total of

ANIMAL MANURES.

Bones.

8,053,686 hides, *i.e.*, skins of cows, bullocks and buffaloes, were exported to other countries. From this may be deducted 48,449 hides which were imported into India from foreign countries, during the same period, and which may or may not have been again exported, thus leaving at least a net total export of 8,005,237 hides of cows, bullocks, and buffaloes that were slaughtered or died in India during the year 1888-89. Assuming that the bones of an average sized cow, bullock, or buffalo weigh 80lb (*Benson, Saidapet Experimental Farm Manual, p.* 25), we have a total of 285,901 tons of bones from the cattle from which these hides were obtained. Besides this, a large quantity of hides is annually used in India for manufacture into leather, and the bones from these should also be available. But there is still a further source of bones, for India exports large quantities of dressed and undressed skins chiefly from sheep and goats, the bones of which animals should be a valuable source of manure since in the year under notice these skins amounted to over twenty-four millions. When a due allowance has been made for the bones of the animals represented by the hides and skins of commerce, even this estimate would not include the bones of the very large number of cattle that annually die in the villages and whose bodies are thrown into the nearest nullah or water-course, without either their skins or their bones being utilised, nor would it provide for the vast numbers of wild animals that die annually throughout the country and whose carcasses rot in every jungle; but it will suffice to show that, although a considerable export trade in bones is springing up, yet the figures of that trade by no means represent the amount of bones that are annually available as manure. At present, as already remarked, bones are not directly used for manure, in any form, by the ordinary Indian agriculturist. Indirectly, as they decay, a portion of their nitrogen and phosphorus finds its way into the soil; but this natural process is a very wasteful one. In some instances, caste prejudices debar the rayat from the use of bones, but even where this is not so, the conservatism of the Indian cultivator and the absence of any simple cheap method by which bones may be crushed and reduced to the best form for application to the fields, have hitherto hindered the use of this manure. In the note (above mentioned) **Mr. Basu** gives details of a cheap method adopted by him for obtaining a supply of bone meal. He says:—" Raw bones may be bought in the Mofussil at prices not exceeding 8 annas a maund. They are then ground in a mortar with a heavy *dhenki* (or pestle such as is used for husking grain). With this appliance three men (two at the treadle and one at the mortar) made 20 seers of fine meal and 20 seers of roughly broken bones in five and a quarter hours. At this rate one maund of bone meal would be obtained in ten and a half hours. Taking a man's wages at 3 annas a day of eight hour, the cost of making one maund of bone-meal would be about 12 annas. To facilitate the work of grinding bones, they may be previously softened by some process of fermentation, for instance, by collecting them in a heap, moistening them with liquid manure and covering them over with stiff clay, from time to time moistening the mass with fresh liquid and allowing it to remain thus for 6 to 8 months, in the course of which time the bones will have become quite soft and fragile and can be easily reduced to powder."

Bone Mills. 242

Of late years, in Calcutta, Bombay, and Lahore, bone mills have been started, and thus both bone-meal and superphosphate are manufactured in India, but the present high price of these articles (R32 to R60 per ton) debars the native cultivator from employing them, and their use is as yet confined to experimental farms and the land occupied by English planters. A considerable quantity of the bone-meal manufactured in Calcutta and Bombay is exported to Europe.

ANIMAL MANURES.

Guano. 243

(5) **Guano.**—One of the most important animal manures used in European countries is guano, which owes its value to the ammonia, phosphates, potash, and soda it contains, as well as other constituents of plants in small amounts, but in a readily available condition. Insignificant quantities of this valuable manure have been imported to India, for experimental purposes. It has been ascertained that it is very efficacious as an application for preventing sugarcane cuttings from being attacked by white-ants, while at the same time it affords that crop the food needed during its growth. The large expense entailed by the importation into this country of South American guano will always preclude its use by the Indian agriculturist, but efforts might be made to utilise deposits of guano which have been found in various parts of India, such as the bat guano of the caves in the Kurnool District, in Moulmein, and near Vizagapatam, and the bird guano in the Nicobar and Andaman Islands (see **Callocalia nidifica**, Vol. II., 508). Of the former, an experimental analysis was made by the Chemical Examiner of Madras, which showed that bat guano was particularly rich in ammonia; but its value as a manure is little recognised locally, and no records of its experimental use are available. Samples of the latter (bird guano) were sent to the Superintendent of the Botanic Gardens, Calcutta, for practical trial, but no information has been received as to the result of this experiment.

Guano, both from **Callocalia nidifica** and from bats, has, however, been extensively employed by Chinese agriculturists in the Malay Peninsula in the cultivation of the cocoanut, nutmeg, clove and *sirih* vine (**Piper Betle**) with good success, and it seems strange that the extensive supplies of this substance in India, alluded to above, are not made use of.

II.—Vegetable Manures.

VEGETABLE MANURES. 244

Various vegetable substances, such as the boughs and leaves of bushes and trees, indigo refuse, wood ashes, weeds of every description, green, dry, or burnt, oil cakes of various kinds, the soft deposits of tank beds containing all sorts of vegetable substances, are, to some extent, used as manures by the Natives of India. They are usually, however, simply thrown upon the land and dug or ploughed, into the fresh state, into the soil where, during decomposition, they very frequently form a nidus for insects of various sorts, which prove injurious to the living plants. Except by the market gardeners around large cities, no pits or heaps for the collection and maturation of vegetable manures are formed by the ordinary Indian agriculturist, and it is only what comes to his hand readily at the time that the peasant uses, and he almost never tries to gather together the vegetable refuse that falls during the year for utilisation at the proper season.

Green Manures. 245

(1) **Green Manures.**—Plants that contain a milky juice, such as the *madar* (**Calotropis gigantea**) and the milk-hedge (**Euphorbia Tirucalli**) are specially preferred by the rayat; but besides these, various other plants are favourite manures. A complete list of all that are so used is almost impossible, since it would include nearly every plant that grows wild in India; but the following are the principal and those that are most generally employed or are supposed by the Natives to have some special manural value:—

Adhatoda Vasica, *Nees.*
Calotropis gigantea, *R. Br.*
Cassia auriculata, *Linn.*
Cedrela Toona, *Roxb.*
Datura, (species).
Dodonea viscosa, *Linn.*
Euphorbia Tirucalli, *Linn.*
Holarrhena antidysenterica, *Wall.*
Indigofera paucifolia, *Delile.*
Jatropha Curcas, *Linn.*
Melia Azadirachta, *Linn.*
Mirabilis Jalapa, *Linn.*
Ocimum sanctum, *Linn.*
Pongamia glabra, *Vent.*
Solanum, (species).

Vegetable Manures.

VEGETABLE MANURES.

Adhatoda. 246

In his Selections from the Records of the Revenue and Agricultural Department, Government of India, **Dr. Watt** notes that the leaves of **Adhatoda Vasica** (see Vol. I., 109) are much used in the Panjáb as a vegetable manure for paddy fields, and he states that in the Sutlej valley they are believed by the cultivators to possess the additional virtue of killing the aquatic weeds that spring up after the fields are inundated and which would, if left alone, materially injure the crop. An enquiry was instituted by him to ascertain whether a knowledge of this alleged property of the leaves of **Adhatoda Vasica** was universal, and he found that they were almost always prized as a manure wherever the plant occurred, and in some districts their powers of destroying low forms of vegetable life were well known to the natives. (*Conf.* with remarks under **Oryza sativa**, Vol. V.)

Burning the Soil. 247

(2) **Burning of the Soil—Jumming or Rábbing.**—Burning of the weeds in heaps is less practised in India than in Europe, but the aboriginal tribes are fond of cutting down patches of forests and burning the trees, bushes, and weeds on the surface soil, preparatory to tilling the land for a temporary cultivation, which lasts for but a few years, when these predatory cultivators migrate to other scenes, where they may renew their extravagant system of culture. In Bombay and South India, a civilised modification of this system is pursued, where seed beds or even fields are prepared by a method of manuring there known as *rub.* This consists of burning the surface soil by means of layers of dried manure, leaves and branches, &c., then ploughing in the ashes. For a full account of this process the reader is referred to the article **Oryza sativa.**

Jumming. 248

Rubbing. 249

Indigo refuse. 250

(3) **Set.**—Indigo vat refuse (*Conf.* Vol. IV., 401) is largely used in some places as a manure; but as it is also employed for fuel in indigo factories, it is not always easily obtainable. Sometimes it is neither used as a manure nor as fuel; but simply allowed to waste,—a striking example of the carelessness of the ordinary cultivator with regard to his conserving of manures. In some parts of Bengal the refuse from indigo vats is thought a particularly valuable manure for the cultivation of tobacco, and is much sought after as such.

Green Soiling. 251

(4) **Green-soiling.**—This method, although very suitable for dry sandy soils where the summer crops are precarious, a condition of things largely prevalent over the greater part of India, is almost unknown to the Natives. In the various experimental farms the production during the summer of a crop of horse gram (**Dolichos uniflorus**), indigo (**Indigofera tinctoria**), or *sun* hemp (**Crotalaria juncea**), and the ploughing of this into the soil before the sowing of the winter crop, has proved vastly successful. It increases the outturn of grain in some cases by more than 70 per cent., and is much to be recommended for fields where farmyard manure cannot be applied with profit, owing either to long distance or to scarcity of manure. Although this method of "green soiling" seems to be so productive, it appears to be very little known to the Natives of India. Instances are mentioned in the Report on the Nagpur Experimental Farm for the year 1883-84, p. 13, where in the Chindwara District a crop of **Crotalaria juncea** was ploughed in, as a manure for sugar-cane, and **Professor Wallace** (in his *India in 1887*), speaking of the same crop, remarks:—"A green manuring is sometimes given as in Guzarat by ploughing in a crop like *tag*—**Crotalaria juncea.**" **Baden Powell** also states that "in some parts of the Ambálah division a practice exists of occasionally growing a coarse kind of millet (**Panicum frumentaceum**) which is ploughed into the soil green as a manure." Green soiling appears to be used considerably among the Indian tea planters, as a means of increasing the yield of tea. Between the tea bushes they sow a crop of mustard in rows, and, when full grown and on the point of flowering, the mustard is dug into the ground between the tea

VEGETABLE MANURES.

plants, covered, and left to decompose (*Indian Agriculturist*, June 16, 1888). In spite of its suitability to the climate and soil of India, no very extensive use, however, appears to be made of this system of manuring. The agriculturist, either through greed or the innate conservatism of his character, usually prefers his meagre hot weather crop, to any prospective augmentation of the winter one, by methods which do not bear the stamp of ancestral use.

Oil-cake. 252

(5) **Oil-cake.**—The use of the various forms of oil-cake as manures will be dealt with in another portion of this work (see **Oils & Oil-cakes** Vol. V.)

III.—Mineral Manures.

MINERAL MANURES.

Mineral manures, although occurring in many parts of India, are not used by the Natives generally. They may be enumerated thus:—

Lime. 253

(1) **Lime.**—The use of lime as a manure is in India practically overlooked and perhaps fortunately so, for did the native agriculturist know its stimulating properties on a soil stinted of other manures, he would probably soon, by its help, liberate all the insoluble mineral substances in the soil, which in the natural state would have become slowly available to the plant, and thus anticipating the supplies of future years, leave the soil barren. The high price of fuel is a great obstacle to its use in India, but in the vicinity of large towns where accumulations of refuse exist, the lime might be burnt with these, and thus made available for agricultural purposes at a low cost. There is no doubt that with judicious management, many lands in India might be vastly improved by the use of lime; but it would be injudicious, in the present state of agriculture, to furnish the ordinary native agriculturist with so violent a stimulant for the land. As a manure for leguminous crops, lime, as well as the next manure of this series, gypsum, is very valuable. (*Conf.* with Vol. II., 151).

Gypsum. 254

(2) **Gypsum**—Sulphate of calcium, in the form of the refuse from the soda-water manufactories, has been used successfully as a manure on the Sydapet Experimental Farm in Madras. In this form it contains not only the pure sulphate of calcium, but also calcium carbonate, various chlorides, sulphates, and sulphurets of calcium and sodium, a mixture which is even more valuable as a manure, while in the neighbourhood of large towns it can be obtained at much less cost than pure gypsum.

The results of the application of gypsum in India have been very encouraging; the use of sulphate of lime in the Sydapet Experimental Farm so increased a crop that a clear profit of R14 was realised over what was gained on an unmanured crop, a result due entirely to the manure. Its use as a manure, however, is at present greatly checked by the high railway freight which is charged on it; but this would probably be at once remedied if it came into demand with cultivators. (*Conf.* Vol. IV, 195.)

Nitrates. 255

(3) **Nitrates.**—The nitrates of potash and soda are both employed as manures in Europe, principally on account of the nitrogen they contain, although nitrate of potash is valuable also from the potash that it can supply to the plants.

In Europe the high price of nitrate of potash has prevented its general application, and, consequently, nitrate of soda has been used; but in India where a coarse nitrate of potash could be manufactured at a very low cost, there is nothing to prevent the use of the more valuable salt. It is of special value as a top-dressing, applied when the plants are one or two inches high, and such are its powers of increasing the yield, that in some experiments with chemical manures made at the Madras Model Farm, it was found that an application of one cwt. per acre nearly doubled the crop and gave the farmer a clear profit of thirty rupees over what was gained from a crop not so manured. It is said that the nitrates should be used

MINERAL MANURES.

Nitrates.

in grain crops combined with common salt, as the latter checks the tendency to run to straw; and since the common nitrate of potash contains chloride of sodium in considerable amount as one of its impurities, this would be the form most valuable to the agriculturist. Almost the same remarks, however, apply to the use of the nitrates of potash and soda as to that of lime. Used alone, they only act for one season or at most two, and they are mainly effective when employed on soils already well manured with organic substances. In England, nitrates are always employed when the weather is showery, and as the same atmospheric state is not available in India, the application of saltpetre without irrigation will not pay. To get its full value as a manure it should be applied on irrigated lands immediately after irrigation (*Ind. Agric. Gazette, 1887, 564*). In spite of the abundance, however, with which this salt occurs in India, it is little used by the native farmers. Many of them have learnt its value through the experiments on the various model farms throughout the country, yet their natural aversion to purchased manures and the fact that the manufacture of crude saltpetre can only be conducted under license, for which an annual fee is charged, prevent their using this manure. Until the fiscal restriction is removed, at any rate in favour of the cultivators, we cannot expect that the use of the nitrates as manure will become general in India.

The protection of the salt revenue, requiring as it does, that saltpetre refineries should be worked under restriction, probably enhances the cost of production of nitrate of potash, and makes its competition with nitrate of soda difficult so far as the manure trade is concerned. Notwithstanding the largely increased demand in Europe for nitrates as manures, the export trade in nitrate of potash is practically no larger now than it was twenty years ago (*O'Conor, Trade Review, 1888-89*). In 1888-89 the export trade in saltpetre amounted to 420,503 cwt., while in 1884-85 it was 451,917 cwt. Nearly one-third of this amount went to China, Hongkong, and the Straits Settlements, over one-third to the United Kingdom, while the remainder was distributed in small quantities to various other European countries and to the United States. The price of nitrate of potash, however, renders it impossible that it can compete as a manure with the crude nitrate of soda from South America, in spite of the greater manural value of the potash salt.

Chloride of Sodium. 256

(4) **Chloride of Sodium.**—In Europe chloride of sodium or common salt has at various times been employed as a manure, but its effects are variable and uncertain, so that its use has of late years rather diminished than increased. Its employment in conjunction with nitrate of soda and potash has already been alluded to. It is said to enable the plant to absorb more silica from the soil; but this view is not supported by any definite experiment. It is generally used as an auxiliary with lime or the nitrates. Heavy dressings of salt are sometimes applied to pasture land to improve the herbage and destroy insect pests. Salt is also used to prevent the too rapid decomposition of manures, a purpose for which in India it might be employed with benefit, as farmyard manure is so rapidly decomposed that in a few months it becomes a fine mould which, although very valuable, does not answer the purposes of the ordinary article.

The high price of salt in India is, however, a serious drawback to its use by the ordinary agriculturist, and although it has been employed on the experimental farms, yet we cannot hope to see its general introduction among the native farmers until some means of denaturalisation is discovered so as to render it unfit for human consumption; while it remains fit for use by cattle and as manure. Various experiments have been tried and a large reward offered by Government to attain this end, but as yet all the methods tried have been unsuccessful, pure salt being easily

MINERAL MANURES.

recovered from the preparations. The methods used in Germany for the denaturalisation of salt, where salt for manure is issued duty free mixed with charcoal dust, ashes, lamp-black or ordinary soot in different proportions, although effective there, would not be so in India, where the salt duty is much heavier, and cheaper means of restoration exist. (*Spons' Encycl.; Encycl. Brit.*)

(5) **Coprolite.**—This name was originally applied by **Dr. Buckland** to substances found in many geological strata which he believed to be the dung of fossil animals. Its signification has, however, now been widened to include also other phosphatic concretions. Considerable interest has been aroused in agricultural circles in India of late years by the discovery of coprolite at Masúrí in the North-Western Provinces, and of fossil bones in the alluvium of the Jumna. If these contain, as was said on their discovery, more than 50 per cent. of tricalcic phosphate, and the supply is as abundant as is anticipated, the use of the superphosphate of lime as a manure in India may be hoped to become more general. With regard to the coprolites at Masúrí, some confusion was at first caused by apparently contradictory chemical analyses made by different authorities, but this was found to be due to the inequality of the fossiliferous strata, and it now appears indubitable that the greater part of the Masúrí rock is at any rate as valuable as the Cambridge coprolite strata. The desiderata of cheap carriage and a cheap supply of crude sulphuric acid to convert the tribasic into the soluble phosphate, at present alone curtail its use. Should these be met, there is no doubt that a considerable industry in this form of mineral manure would arise. Coprolite. 257

(6) **Ammoniacal Liquor.**—The ammoniacal liquor of gas works is another good manure which ought to be utilised wherever available. Of course in India this is only so in the neighbourhood of a few large towns; but there it is specially applicable, as it is invaluable to the market gardener for the cultivation of vegetables. Ammoniacal Liquor. 258

(7) **Other Mineral Manures.**—Various other mineral manures, such as the commercial ferrous sulphate, basic slag, the carbonates of soda and potash, have been used in Europe with good effect; but at present in India, they are either not available, or the cost at which they can be obtained precludes their use by the ordinary cultivator. Other Mineral Manures. 259

(*G. Watt.*)

MAOUTIA, *Wedd.; Gen. Pl., III., 391.*

Maoutia Puya, *Wedd.; Fl. Br. Ind., V., 592;* URTICACEÆ. 260

Sometimes called WILD-HEMP by early writers and also PUA-HEMP.

Syn.—BŒHMERIA PUYA, *Hook.;* B. FRUTESCENS, *Don.* (not of *Thunb.*); URTICA PUYA, *Ham., in Wall. Cat.*

Vern.—*Pói, púa,* HIND.; *Yenki* (Limbu), BENG.; *Puya,* NEPAL; *Kyinki, kienki,* LEPCHA; *Púya,* KUMAON; *Sat sha yuet,* BURM.

References.—*Brandis, For. Fl., 436; Kurz, For. Fl. Burm., II., 429; Gamble, Man. Timb., 323; Atkinson, Him. Dist., 317, 798; Royle, Fib. Pl., 368; Kew Off. Guide to the Mus. of Ec. Bot., 124; Hannay, in Jour. Agri.-Hort. Soc., VII (Old Series), 223; Memoir of Dehra Doon, 21; Madden, in Journ. As. Soc. Beng., XVIII., I., 622; Indian Forester, XIV., 269, 273, 275; Watt, in Col. & Ind. Ex. 1886, Commercial Reports.*

Habitat.—A native of the Tropical Himálaya (ascending to 4,000 feet in altitude), distributed from Kumaon and Garhwál eastward to Nepál, Sikkim, the Khásia hills, and the Assam Valley and thence to Burma, the Straits Settlements, and Japan.

Fibre.—Although many authors refer to this FIBRE and state that it closely resembles Rhea, and may be prepared and used in the same FIBRE. 261

FIBRE.

manner, yet no one seems to have performed independent experiments with the *poi* fibre, and the available information is, therefore, of an indefinite nature. **Dr. Campbell's** description of the fibre, published in **Royle's** *Fibrous Plants*, is the only complete account available up to date. The following extract will be found interesting :—

Description. 262

"*Description.*—The leaf is serrated, of a dark green colour above, silvery white below, not hairy or stinging; and has a reddish pedicle of about three inches long. The seed forms in small currant-like clusters along the top of the plant, and on alternate sides about an inch apart. Two small leaves spring from the stem at the centre of and above each cluster of seed.

"*Habitat.*—The *Pooah* is not cultivated but grows wild and abundantly in the valleys throughout the mountains of Eastern Nepál and Sikkim, at the foot of the hills skirting the Terai to the elevation of 1,000 to 2,000 feet, and within the mountains up to 3,000 feet. It is considered a hill plant, and not suited to the plains or found in them. It does not grow in the forests, but is chiefly found in open clear places, and in some situations overruns the abandoned fields of the hill people within the elevations which suit it. It sheds its leaves in the winter, throws them out in April and May, and flowers and seeds in August and September.

Cutting. 263

"*When Used.*—It is cut down for use when the seed is formed; this is the case with the common flax in Europe. At this time the bark is most easily removed and the produce is best. After the seed is ripe it is not fit for use, at least it is deteriorated.

Preparation. 264

"*How Prepared.*—As soon as the plant is cut, the bark or skin is removed. This is very easily done. It is then dried in the sun for a few days; when quite dry, it is boiled with wood-ashes for four or five hours; when cold, it is beaten with a mallet on a flat stone until it becomes rather pulpy, and all the woody portion of the bark has disappeared; then it is well washed in pure spring water and spread out to dry. After exposure for a day or two to a bright sun it is ready for use. When the finest description of fibre is wanted, the stuff, after being boiled and beaten, is daubed over with wet clay and spread out to dry. When thoroughly dry the clay is rubbed and beaten out; the fibre is then ready for spinning into thread, which is done with the common distaff.

Uses. 265

"*Uses.*—The *Pooah* is principally used for fishing nets, for which it is admirably adapted on account of its great strength of fibre and its extraordinary power of long resisting the effects of water. It is also used for making game-bags, twine and ropes. It is considered well adapted for making cloth, but is not much used in this way.

"**Dr. Falconer** recognised the *Pooah* as the **Bœhmeria frutescens,** *Don*,, of Botanists, common at lower elevations on the Himálaya from Garhwál to the Sikkim hills (Ganges to Burrampooter). In the outer hills of Garhwál and Kumáon it is called *Pooee*, and the tough fibre is used there for making nets. In Darjiling, **B. frutescens** goes by a similar name. *Pooah* and the fibre is used for similar purposes. It was first described by **Thunberg,** who distinguishes it from the textile species, **Bœhmeria (Urtica) nivea,** which grows there in abundance.

"**Captain Thompson,** to whom the specimens of the *Pooah* fibre were sent, says of it, that 'when properly dressed, it is quite equal to the best Europe flax, and will produce better sail-cloth than any other substance I have seen in India. I observe from **Dr. Campbell's** communication that mud is used in the preparation, which clogs it too much, &c. My Superintendent, **Mr. W. Rownee,** who understands the nature of these substances, tells me that if potash were used in the preparation (which is invariably done with Russian hemp and flax) instead of clay or mud, that the colour

FIBRE.

would be improved, the substance rendered easy to dress, and not liable to so much waste in manufacturing'" (*Royle, Fibrous Plants, 368—370*).

The above quotation regarding **Maoutia Puya** and the incidental allusion to it in the remarks regarding Rhea have very nearly recorded all that has to be said. The large bale of the fibre, shown at the Colonial and Indian Exhibition, and which was obtained from Assam, was either, as **Messrs. Cross & Bevan** suspect, a badly-prepared sample or the fibre is quite worthless. The former explanation would be in accordance with all previous reports, for, although admittedly inferior to Rhea, it would be hard to believe that a fibre so popular with the fishermen could be so utterly worthless as **Messrs. Cross & Bevan's** analysis would make it out to be. The following demonstrates the results of their chemical examination of the fibre:—

Moisture	11·2
Ash	8·2
Hydrolysis for 1 hour in 1 p. c. Na_2O	62·7
Cellulose	32·7

According, therefore, to **Messrs. Cross & Bevan's** observations, **Maoutia Puya** would be the least worthy fibre in India, since it has the next to the lowest amount of cellulose and loses more of its weight under hydrolysis than any other fibre examined by them. In their remarks regarding the fibre, however, these distinguished chemists affirm: "A large bale of this fibre was shown at the Colonial and Indian Exhibition, and by experts pronounced identical with Rhea. It now remains to be carefully ascertained in India whether there would be any advantage in cultivating this plant in place of **Bœhmeria nivea**. We can only repeat what we have said elsewhere, that every effort should be put forth in India to ascertain the peculiarities of every plant allied to Rhea. In some respects the true Rhea is too strong, and a Rhea-like fibre, a little inferior in point of quality, that could be more easily cultivated and more cheaply separated from the twigs, would in all probability prove a more profitable and more acceptable fibre than Rhea, which has occupied, and justly occupied, the minds of experts for the last few years.

"The specimen of this fibre exhibited was very inferior in many respects. It is introduced here, as presumably its rightful place, when normally prepared.

"Not only was the specimen inferior in point of preparation, but it was found in the microscopic examination impossible to isolate the ultimate fibre, by reason of its breaking up under the needles. Many of the fibres of the URTICACEÆ show this tendency to brittleness; but with special attention to cultivation and the conditions of growth, these defects can in all probability be removed."

It is needless to add anything further except to emphasise what has already been said—namely, that if **Messrs. Cross & Bevan's** analysis be confirmed as representing the fibre, an effort should be made to replace its cultivation with the true Rhea or China-grass, or at all events to see that consignments of *Poi* are not sent to Europe under the name of Rhea.

The above compilation appeared in the publication edited by the writer—Selections from the Records of the Government of India (*Vol. I., Pt. II. (1888-89), 312-315*); it has since transpired that the bale of supposed *Poi* fibre examined by **Messrs. Cross, Bevan, & King** was in all probability not *Poi*. In connection with the preparation of the fibres for the Imperial Institute an authentic sample of that fibre has been secured. On this subject **Mr. Gammie** of Mungpoo, Dárjíling, furnishes the following information:—"The whole sample has been prepared by the method pursued by the Nepálese and Lepchas.

FIBRE. "The bark is peeled off the stems in long strips; boiled in water thickened with common wood-ashes until it is pulpy; then as much as possible of the adhering bark is separated from the fibre by alternately beating with a wooden mallet and washing in cold water. After this the water is rinsed out, and each bundle of fibre is thickly covered with a paste of micaceous clay, and dried. When thoroughly dry, the clay and the remaining bark are easily shaken off, leaving the fibre in a state fit for use. If fibre is required free from dust, it is repeatedly rinsed until the water runs clear, and then re-dried.

"The white or bluish-white clay found here and there near streams is preferred, as it gives the fibre a good colour.

"This clay, by being fused with fire, is re-converted into common micaceous, schistose stone.

"If the appearance of the fibre is of no consequence, yellow clay is said to be as effective.

"I do not know whether the action of the clay is altogether mechanical or not. A few samples which were prepared by treatment with lime and chalk were coarse in appearance and rough to the touch; those treated by clay, on the other hand, were soft and silky. Although the *Pooa* is rather a common plant, it is seldom gregarious to any extent as far as I know; so that the collection of a large quantity entails an expenditure which must exceed the value of the fibre extracted. I obtained five maunds of stems, by contract, for three rupees per maund, but I question if I could obtain them at the same rate again, as the people had to search far and wide for even that quantity. At a moderate estimate the further cost to manufacture the fibre was five rupees, making a total of twenty rupees.

"The fresh stripped bark weighed 63 ℔ and yielded only 4 ℔ of fibre. The cost of producing one pound of fibre would, therefore, be five rupees.

"*Pooa* is chiefly used for fishing nets and lines. I am told that formerly the Lepchas made cloth from it, but the contraction and expansion readily caused in it by atmospheric changes made it uncomfortable and undesirable for wearing apparel."

It would thus seem that little hope need be entertained of obtaining this fibre from the wild stock. Should it prove of value (when Mr. Gammie's sample has been submitted to commercial and scientific tests) and to possess advantages over that of Rhea the plant would have to be cultivated.

266 **MARANTA,** *Linn.; Gen. Pl., III., 649.*

A genus of SCITAMINEÆ (the Ginger Family) named in honour of **Maranti,** a Venetian botanist and physician of the sixteenth century. It gives its name to the Tribe MARANTEÆ and is characterised by having a terminal few-flowered but branched panicle, the flowers borne on slender ebracteate peduncles with narrow deciduous sheathing bracts at the origin of the ramifications. Corolla tube cylindrical, often gibbous: staminal tube contained within the corolla, petaloid with a solitary anther on one side. Ovary by abortion 1-celled and 1-seeded with two small empty cells. Fruit ovoid, somewhat oblique with a fleshy green pericarp. There are some ten species in the genus, all natives of America, with one or two widely cultivated in the Old World on account of the starch contained in their tubers.

267 **Maranta arundinacea,** *Linn.; Fl. Br. Ind., VI., 198;* SCITAMINEÆ.

WEST INDIAN ARROWROOT—A name given to distinguish it from East Indian, the produce of **Curcuma angustifolia,** *Roxb.*;—see Vol. II., 652-655. [*Ger.* ARROW-ROOT, *Fr.*; AMERIKANISCHES STARKMEHL, ARROWMEHL,

Syn.—It seems probable that M. RAMOSISSIMA, *Wall.*, and M. INDICA, *Tussac*, are synonyms for a cultivated form of M. ARUNDINACEA, *Linn*, and are not specifically distinct, as some writers maintain.

Vern.—*Tikhor (tickhur)*, HIND.; *Tavkil*, MAR.; *Ararut*, GUZ.; *Kúaka neshasteh*, DEC.; *Kuva mavú* or *kuamau*, TAM.; *Tavaksha*, KAN.; *Kúa, Kughei*, MALAY; *Pen-bwa*, BURM. It seems probable that all the vernacular names here given, strictly speaking, refer to the starch obtained from East Indian arrowroot—**Curcuma angustifolia**—though they are doubtless now-a-days given to any form of that farinaceous substance.

References.—*Roxb., Fl. Ind., Ed. C. B. C., 10; Voigt, Hort. Sub. Cal., 575; DC., Orig. Cult. Pl., 81; Mason, Burma and Its People, 507, 806; Lindley and Moore, Treasury of Bot., 720; Flück. & Hanb., Pharmacog., 629; U. S. Dispens., 15th Ed., 1693; O'Shaughnessy, Beng. Dispens., 646, 647; Bent. & Trim., Med. Pl., 265; Smith, Ec. Dic., 25; Shortt, Manual of Indian Agric., 301; Atkinson, Him. Dist. (Vol. X., N.-W. P. Gaz.), 704; Simmonds, Tropical Agriculture, 347; Bombay, Man. Rev. Accts., 103; Gazetteers:—Mysore and Coorg, I., 70; Agri.-Hort. Soc. Ind.:—Trans., II., 79-81, 393, 418; Journal (Old Series), II., 215, 262, 266, 316, 365; III. (Proc.), 167, 243, 282; VIII. (Proc.), 24.*

Habitat, History, & Food.—Authors are agreed that the name arrowroot was derived from the alleged alexipharmic properties of the plant. The juice of the fresh ROOT was employed by the Mexican Indians as an external application against the action of the poison used on their arrows. The earliest authentic mention of the plant occurs in **Sloane's** *Catalogue of Jamaica Plants* (1696), where it is called *Canna India radice alba alexipharmace.* It was first discovered in Dominica, thence sent to Barbadoes and, subsequently, to Jamaica. **Patrick Browne** (in 1756) mentions it as cultivated in gardens in Jamaica, and, in addition to alluding to its alexipharmic virtues, informs us that "the root washed, pounded fine, and bleached makes a fine FLOUR and STARCH," which was sometimes used as food when provisions were scarce. In 1750, **Hughes** (*Natural History of Barbadoes, 221*) spoke of the starch made from the roots as far excelling that of wheat.

FOOD. Root. 268

Flour. 269

Starch. 270

According to some writers, arrowroot was introduced into England by **Houston** about 1732, but in **Renny's** *History of Jamaica* (232), 24 cases are said to have been the first consignment to England, and these were exported in 1799. Some confusion seems to have been made by writers both in Europe and India between the farina from this plant and the East Indian arrowroot, if not also with tapioca, the Brazilian arrowroot. There appears to be no doubt of the fact that all forms of the arrowroot plant were derived originally from America. The so-called arrowroot of India, referred to by **Bentley & Trimen** and others under the name of **Maranta indica**, a species which had narrower, sharper, and smoother leaves than the ordinary plant, was either **Curcuma angustifolia** or one of the earliest introduced forms of **Maranta**, which came to India before the first record of introduction. That the true arrowroot plant must have been introduced about the beginning of this century, there seems no doubt, but in Bengal at least it did not attract attention much before 1830. The earliest mention of a plant belonging to the genus **Maranta** is curiously enough associated with Sylhet. Thus **Wallich**, in 1833, recognised the plant, which the *Agri.-Horticultural Society of India* had been cultivating under the impression that it was the true arrowroot, as being "**Maranta ramosissima** introduced into the Botanic Gardens from Sylhet about twelve years ago." This, he said, was quite "different from the West India **M. arundinacea.**" The Sylhet plant was probably the **M. indica** of some writers, but, as remarked above, the more recent authors confused that with **Curcuma angustifolia**, the arrowroot plant of Southern and Western India. We have thus no record of the introduction of **M. ramosissima** into Sylhet, nor for that matter into India, nor indeed have we any distinct record for **M. arundinacea**, but that they are both American plants there seems no doubt. If any form of **Maranta** be a native of India, the singular silence of **Roxburgh** (for he alluded to

HABITAT, HISTORY & FOOD.

arrowroot as an American plant only which yielded a similar farinaceous substance to that obtained from **Curcuma angustifolia**) would be difficult to get over; indeed his silence practically proves that, in his time at least, the true arrowroot plant had not been even introduced. In the XIth Volume (1810) of the *Asiatick Researches*, for example, **Roxburgh** says: "I shall only add, on the subject of this nutritious powder, that it is very similar to the powder which is obtained in America from the roots of **Maranta arundinacea**, and which is known in Europe by the name of Indian arrow-root." **Ainslie**, who wrote some time after **Roxburgh** (1823), says that "**Maranta arundinacea** has lately been brought to Ceylon from the West Indies, and thrives well at the *Three Korles*, where arrowroot is now prepared from it and reckoned of the finest quality." It is remarkable that **Wallich**, if he meant that **M. ramosissima** was a native of Cachar, should not have said so, since he must have known that that opinion was opposed to **Roxburgh** and **Ainslie**. **Voigt** (*Hort. Suburb., Cal.*, 575) speaks of it in 1845 as cultivated in the West and East Indies; he adds, that it flowers in the rainy season, from which fact it may be inferred that it was, at the time he wrote, regularly cultivated in India. **Firminger**, who in 1863 brought out the first edition of his *Manual of Gardening in India*, while comparing the arrowroot of **Curcuma angustifolia** with the true article, says: "I cannot tell why any but the genuine kind should be produced at all in this country, or whether any difficulty is experienced in the cultivation of **M. arundinacea** on the Madras side; in Bengal the plant may be obtained in any abundance, and cultivated with the greatest ease. **Dr. Jameson** states too, that it thrives in the Saharanpur district and throughout the North-West Provinces."

The earliest direct mention of the introduction of **Maranta** to India is the passage to which reference has been made The *Agri.-Horticultural Society of India* obtained roots from the Cape of Good Hope. These were cultivated and gave a net profit of R2,307-10 from 4 *bíghas* of land, planted in 1831. A demand for roots thus arose, and the Society distributed all they had and were prevented from indenting for a further supply from the Cape by **Dr. Wallich**'s report, that the plant could be had in Sylhet. No mention is made of the true West Indian stock having been procured, but doubtless it is grown in India now, though it seems probable, from the facts here mentioned, that the form known as **M. ramosissima** must have been widely distributed, and is thus probably cultivated by many persons who are ignorant of its not being the best West Indian arrowroot. The available information on this subject is, however, so imperfect that little more can be done than to indicate, as already done, our imperfect knowledge.

CULTIVATION. Bengal. 271

METHOD OF CULTIVATION AND PREPARATION OF THE FARINA.

1st, Bengal.—The account given by **Firminger** may, perhaps, be accepted as the best statement, although, of course, he says nothing as to the extent of its cultivation, and his description is of a general nature, hence more or less applicable to all India:—"The root," he writes, "should be put in the ground in the month of May. Drills should be made about three or four inches deep and two feet apart, in which the roots should be laid at the distance of a foot and a half from one another, and the earth covered over them. As the plants grow, they should be earthed up in the same manner that potatoes are. They love a good rich soil, and plenty of water during the time of their growth, which latter, indeed, they get naturally, as their growing time is during the rains. They bear their small white flowers about August, and in January or February the crop may be taken up for use. A month or two previous, however, water should be entirely withholden, to

CULTIVATION.

Bengal.

allow the roots to ripen. They are of a pure ivory-white colour, and should be as large as moderate-sized carrots. The smaller ones should be reserved for a fresh planting, and the pointed ends also of the larger ones, at the extremities of which the eyes are situated, should be broken off, three inches in length, and kept for the same purpose.

"The mode of preparing the arrowroot is very simple. The roots after being well washed should be pounded to a pulp in a wooden mortar, which may be hired for the occasion from the bazár. The pulp should be thrown into a large vessel of water, which will become turbid and milky, a portion of the pulp remaining suspended in it as a fibrous mass. The fibrous part should be lifted up, rinsed, pounded again in the mortar, thrown again into the water, lifted up a second time, rinsed, and then thrown away. The milky-looking water should be then strained through a coarse cloth into another vessel, and when the sediment has settled, the water should be poured gently off and clean fresh water poured upon the sediment. This, after having been well stirred up, should be strained through a fine cloth, and on settling the water should again be carefully and gently drained away. The sediment, which is then fine pure arrowroot, should be dried on sheets of paper by exposure to the sun."

In the *Journals of the Agri-Horticultural Society of India* frequent mention is made of Chutia Nagpur arrowroot. Thus, Dr. **Mouat** reports on a sample prepared in 1843 at Purulia. In the same year Babu **Sumbu Chunder Ghose** forwarded to the Society twelve canisters of arrowroot grown and manufactured at Bírbhúm. The report on these samples was to the effect, that the farina was of excellent quality, well prepared and dried, without any disagreeable odour or flavour. "It forms a good jelly, and is well adapted for all the purposes to which this mild and nutritious substance is applied."

At the Colonial and Indian Exhibition excellent samples of arrowroot, prepared and tinned in the European method, were shown by Mr. **H. H. Abdoollah** of Colootolla near Calcutta. It is known that several other makers produce arrowroot which successfully competes, both in quality and price, with the imported powder, but definite particulars as to the extent of cultivation are not available.

N.-W. P & Oudh. 272

2nd, North-Western Provinces and Oudh.—**Atkinson** writes: "This root has been successfully cultivated in these provinces. The tubers are ready for digging in January, and should be scraped and well washed before being powdered, in order to remove the acrid poisonous juice. An ordinary piece of tin punched with holes makes a good grater. After being washed the powder should be dried in the sun before being stored. The arrowroot produced from these tubers at Haldwani, near Naini Tál, by Mr. **Fraser** has been pronounced by experts both here and in Europe to be equal to the best West Indian arrowroot." In a report of the *Lucknow Horticultural Society, 1843,* mention is made of the despatch to Calcutta of a sample of arrowroot made from the plants grown in the gardens.

Madras. 273

3rd, Madras.—The extent to which the true arrowroot is cultivated in South India cannot be discovered. Many writers allude to arrowroot, however, and speak of it as an important crop, but very probably the chief article of trade is the East Indian form **Curcuma angustifolia.** It is significant, however, that **Sir Walter Elliot** (*Fl. Andh.*, *142*) should state that **M. ramosissima,** *Wall.*, the *Pála ganda* of the Telegu people is "wild in all the hill forests." That name, according to **Moodeen Sheriff,** would appear to denote rather **Alpinia Galanga,** and **Sir Walter** does not refer to **Curcuma angustifolia,** although, while dealing with **Curcuma sp.** (the *Nakka pasupu*), he remarks that it is "wild Turmeric, **C. montana** vel **angustifolia**?" It seems thus probable that he made the mistake, by no means infrequent

MARANTA arundinacea.

West-Indian Arrow-root.

CULTIVATION.

Madras.

about the time he wrote (1859) of regarding the wild arrowroot of India as **M. ramosissima.** **Shortt**, in an essay on arrowroot cultivation in Madras, gives a somewhat incoherent account of the difference between **Maranta** and **Curcuma**. He, however, furnishes certain facts which must he accepted as applicable to both forms since he does not say of which he is dealing. Although the writer is therefore unable to give any information as to the relative extent to which these plants are cultivated, **Shortt's** essay may be here quoted :—"Arrowroot is largely grown in Travancore,* in Malabar and other districts of Southern India. The farina is manufactured in a rough and rude way; attention is not given to the cultivation of the plant, so as to increase the quantity and quality of starch. The mode of culture followed by the native producer is to roughly plough up sufficient land, level it, and plant out the rhizome or root stalks, at about one foot apart either way, before the commencement of the rains. Thereafter little or no attention is given to the crop. As a rule, manure is seldom used and the plants are not irrigated, except, perhaps, in exceptional instances, whilst crop after crop is taken off the same soil.

"For the proper and scientific culture of arrowroot, care should be taken in the selection of the soil. The crop grows best in a soft, loose sandy soil, or a somewhat porous loamy one, which admits of the rhizomes forming readily and enlarging to their fullest extent. The land should be well ploughed and freely manured with rotten farm-yard manure, which should be thoroughly mixed with the soil. It should then be levelled and formed into beds of a convenient size, so as to admit of the plants being irrigated when necessary, and the plants should be planted out 2 feet apart either way. On these hills† we prefer to place out the seed stalk in lines, 2 or 2½ feet apart either way, and when the plants have attained a height of 6 to 8 inches, to dig a trench, one foot wide, between every two lines, about half a foot deep; the soil thus obtained is used to ridge up the line of plants. This plan will also prove useful if it become necessary to irrigate the plants, which can be done by the trench between the lines, if care is taken to keep the ridges along the lines of plants well earthed up at all times. The plants should have plenty of room to enable them to form large clusters.

"For seed we generally prefer the root-stalks with their fibrous roots broken up, and the stalks cut short to about four inches in length. The first inch or two of the rhizomes will also answer the purpose of seed, in fact any portion of the rhizome of the size of an inch, containing a joint, will answer the purpose equally well. The rhizomes are fit to be taken up in about nine or ten months. They are planted out in the month of March and taken up in the following January or February. After the plants are put out, they are hand-watered twice a week, till they become established, but, where facilities exist, other modes of irrigation may be resorted to; and about once a month the field should be weeded, the earth loosened freely around the plants, and the stalks kept well earthed up."

Dr. Shortt furnishes an estimate of the expense of cultivation which amounts to R50 per acre. He then adds :—"The average crop of rhizomes produced on an acre of land is 2,500 pounds, yielding 400 pounds of farina, the average is one-fourth the quantity of the cormus, but for safety I have fixed it at one-sixth the quantity which at four annas the pound will realize R100; the retail price of arrowroot is from 12 annas to R1 the pound, and according to my estimate, deducting R50 for cost of cultivation, there is a balance of R50 as the net profit. With care and attention in the cultivation and preparation of the arrowroot, the profit will be found to exceed the estimate greatly."

* *Conf.* with Manihot, pp. 160 & 161. | † Shevaroys.

CULTIVATION.

Dr. **Shortt's** description of the process of manufacture does not materially differ from that given above by **Firminger**, and it need not, therefore, be reproduced here.

Bombay. 274

4th. Bombay.—**Graham** (*Cat. Bomb. Pl., 212*) alludes to **Maranta ramosissima,** *Wall.* (*Pl. As. Rar. 3 t., 286*) as a native of Sylhet which had been introduced into Bombay by **Mr. Nimmo,** from Bengal. "It has numerous long clavate tubers, and much resembles the West India arrowroot" (*Wall.*). The dry air, says **Graham** (quoting **Gibson**), of the Deccan in the cold season seems to affect its development. **Dalzell & Gibson** in their *Bombay Flora* repeat the same statements under the two names, **M. Zebrina** and **M. ramosissima,** two plants which **Graham,** and correctly so, kept distinct. In **Dr. W. Gray's** sketch of the botany of Bombay the remark occurs, "**M. arundinacea,** the West Indian arrowroot, exists in a few gardens, and, judging from its luxuriant growth, is capable of being profitably cultivated in Bombay."

Burma. 275

5th, Burma.—**Mason** says: "The true arrowroot plant was introduced several years ago by **Mr. O'Riley,** and is beginning to be largely cultivated. The arrowroot made is not inferior in quality to any imported; while it is sold for half the price, at a good profit. A gentleman at Tavoy has sold a considerable quantity for exportation this year, and has orders for more than a thousand pounds of the next crop" (*Burma and Its People. 507*). Mention is made of **Mr. O'Riley's** introduction of the root into Burma in the *Proceedings of the Agri-Horticultural Society of India (1844)*. He there states that he had been so successful with the few bulbs of **Maranta arundinacea** furnished to him by the Society that he intended to extend its cultivation considerably. **Mr. O'Riley** adds that, without exception, the persons to whom he had given tubers had pronounced it superior to **Speed's,** and that as the gardens of the natives possess a fine free soil of the richest description, he considers its introduction into Tenasserim Province of some importance. In the same year **Major D. Williams** sent to the Society samples of a bulb and farina prepared from it which he considered far superior to arrowroot. He then remarked "the arrowroot plant grows all over Arracan and is eaten as a vegetable." From this it would be difficult to know what either of the two plants alluded to may have been, but very probably neither of them were **Maranta.** In a later communication, however, **Major Williams** furnishes fuller details from which it would appear that **Dr. Wallich** determined his bulb that yielded a far superior arrowroot to the true plant to be **Tacca pinnatifida.** But the Arracan arrowroot, which was eaten as a vegetable, does not appear to have been determined. Could it have been the sweet cassava which in Assam, about the same time as the above appeared, was also stated to be plentiful and to be eaten as a vegetable?

Maranta dichotoma, *Wall.*, see **Phrynium dichotomum,** *Roxb.*; Vol. VI.

MARBLE, *Ball, in Man. Geol. of Ind., III, 455—471, 686.* 276

The term marble is geologically restricted to limestone or carbonate of lime capable of receiving a polish. It occurs in several different forms, two of which are unicoloured, *viz.*, white and black, while many varieties are streaked and parti-coloured. The veining or colouring is derived from the presence of accidental minerals, frequently metallic oxides, also in many cases from imbedded fossil shells, corals and other organisms.

Marble, *Mallet, Geol. of Ind., IV. (Mineralogy), 150.* 277

MARBRE *Fr.*; MARMOR, *Ger.*; MARMO, *Ital.*

Vern.—*Safaid-pattar, Kalai-ka-pattar, Shah-neaksadi,* HIND. & PB.; *Pulalani, marmar,* MALAY.; *Sung-i-marmar,* PERS.

MARBLE.

The subject of the occurrence of MARBLE in India has already been discussed, together with that of LIMESTONE, in the article CARBONATE OF LIME, Vol. II., 143—152. For more detailed information the reader is referred to *Ball, in Man. Geology of India, l. c.,* in which the subject has been very fully and elaborately discussed.

The following note has been kindly furnished by **Mr. Medlicott** on the extent to which marble is worked in India:—

Jabalpur. 278 Himalaya. 279 Rajputana. 280

"Marble of inferior quality is found in the metamorphic rocks of the Peninsula and the Himálaya, but the only place where it is worked is in Rájputána, where the marbles of *Jaipur* and *Jodhpur* are celebrated over the whole of India. The marble rocks at Jabalpur also yield marble (a dolomite) of good quality, a block sent to the Paris Exhibition having been declared to be equal to Italian marble for statuary purposes: as a rule, the beds are much jointed and crushed, so that it would be difficult to obtain large blocks in any quantity."

Mr. F. R. Mallet in his *Mineralogy* gives the following analyses of three well-known Indian dolomites:—

	I.	II.	III.
Calcium carbonate	55·48	59·7	60·5
Magnesium carbonate	43·55	37·8	38·7
Ferrous carbonate	·36	...	...
Oxide of iron and alumina	...	1·0	·3
Insoluble	·61	·8	
	100	99·3	99·5

I. The white saccharine marble of the Jabalpur rocks referred to above as equal to Italian stone for statuary purposes.

II. Light grey saccharoid, and III., white, almost crypto-crystalline marble from the Titi river, Western Duars.

In addition to the Rájputána marbles referred to by **Mr. Medlicott**, records exist of the stone being worked, to a small extent, in other parts of the country. Thus, it is stated that the shell-marble of Kerudamangalam is made into table-tops, paper weights, and small ornaments in Trichinopoly. This marble, when polished, is of a dark grey colour, and is marked like the well-known "Purbeck" stone with white sections of included shells (*Man. of Trichinopoly Dist.*, *69*).

Kerudamangalam. 281 White. 282 Black. 283 Yellow. 284 Grey. 285 Veined. 286

Mr. Baden Powell states that a yellow marble called *Shahmaksadi* obtained from Manairi, Yusufzai, is cut into charms and ornaments in the Peshâwar district (*Pb. Prod.*, *37*). He also mentions white, black and grey marble from Delhi; an inferior marble, *Kalai-ka-pattar*, from Karnaal in Hissar; grey marble from Bhunsi in Gurgáon; black marble from Kashmír; and a form of translucent marble, *safaid-pattar* from Sháhpúr. Most of these marbles, however, are used like limestone for making fine qualities of *chunam*. A veined marble (*abri*) found in the Kowagarh hills of Ráwalpindí is occasionally worked into cups and ornamental objects, but the cost is great on account of the hardness of the stone and the absence of skilled labour. The pillars of the pavilion in the garden of Bairam Khán at Attock are made of this beautiful stone (*Ráwalpindí Gaz.*, *11*). An inferior form of marble is frequent in the Sutlej Valley which does not appear to have been worked. The beautiful semi-transparent white marble obtained from the Toygun hills in Upper Burma is extensively used for carving the well-known sitting and recumbent figures of Gaudama, to be found in the pagodas, &c., of many parts of that country.

Sutlej. 287 Burma. 288

Margosa tree, see **Melia Azadirachta,** *Linn.*; MELIACEÆ; p. 211.

Marigold, see **Calendula officinalis,** *Linn.*; COMPOSITÆ; Vol. II., 24.

[SITÆ ; Vol. VI., Pt. II.
Marigold, African—the *genda* of India, see **Tagetes erecta,** ; COMPO-
[*Ind. III., 309.*
Marigold Burr, see **Bidens tripartita,** *Linn.;* COMPOSITÆ ; *Fl. Br.*
Marigold, French, see **Tagetes patula,** COMPOSITÆ ; Vol. VI., Pt. II.
[II., 50.
Marigold, Marsh, see **Caltha palustris,** *Linn.;* RANUNCULACEŒ ; Vol.
Marjoram, see **Origanum vulgare,** *Linn.;* LABIATÆ.
Marking Nut Tree, see **Semecarpus Anacardium,** *Linn.;* ANACAR-
[DIACEÆ ; Vol. VI., Pt. I.

MARLEA, *Roxb.; Gen. Pl, I., 949.* [CORNACEÆ.

Marlea begoniæfolia (begonifolia), *Roxb.; Fl. Br. Ind., II., 743;* 289

Syn.—M. AFFINIS, *Dcne.;* M. TOMENTOSA, *Endl. ex Hassk.;* DIACICARPIUM TOMENTOSUM, *Blume;* D. ROTUNDIFOLIUM, *Hassk.;* STYLIDIUM CHINENSE, *Lour.;* STYRAX JAVANICUM, *Blume.*

Var.—alpina, *H. f. & T.,* found in Sikkim at altitudes of from 6,000 feet to 9,000 feet.

Vern.—*Marlea, marlisa,* (SYLHET), ASSAM ; *Timil,* NEPAL ; *Palet,* LEPCHA ; *Garkum, budhal, túmbri,* N.-W. P. ; *Tumri,* KUMAON ; *Siálú,* (WARDWAN) ; *Prot,* KASHMIR ; *Mandrá, bodará* (BIAS) ; *Siálú,* (CHENAB) ; *Tilpattra, chit pattra, Kúrkni,* (JHELUM) ; *Padlú* (RAVI) ; *Budánár, memoká,* (KANGRA) PB. ; *Tapuya,* BURM.

References.—*Roxb., Fl. Ind., Ed. C.B.C., 326; Brandis, For. Fl., 251; Kurz, For. Fl. Burm., I., 544, 545; Gamble, Man. Timb., 211; Stewart, Pb. Pl., 93; Miq. Fl. Ind. Bat., I., pt. I., 744; Baden Powell, Pb. Pr., 585; Atkinson, Him. Dist., 311; Indian Forester, XIII., 57.*

Habitat.—A tree (often small, but sometimes attaining a height of 60 feet), found throughout Northern India, at altitudes from 1,000 feet to 6,000 feet; common from the Panjáb to Bengal and Burma. Distributed to China and Japan. The variety **alpina** is met with in Sikkim where it ascends to 9,000 feet.

Fodder.—The LEAVES are sometimes given as fodder to cattle and goats. (*Stewart*) (*Conf. with Vol. III., 430*). **FODDER. Leaves.** 290

Structure of the Wood.—Wood white, soft, even-grained. Weight 42lb per cubic foot (*Gamble*). **TIMBER.** 291

Domestic Uses.—"The WOOD is used for houses in Sylhet" (*Roxb.*). **DOMESTIC. Wood.** 292

Marmelos, see **Ægle Marmelos,** *Corr.;* RUTACEÆ ; Vol. I., 117.
Marmots, see "**Rats, Mice, Marmots**"; Vol. VI., Pt. I.
Marron, the French for **Castanea vulgaris,** *Lam.,* which see, Vol. II., 227.
Marron d'eau, the French for **Trapa bispinosa,** *Roxb.*; Vol. VI., Pt. II.
Marrow, Vegetable, see **Cucurbita Pepo,** *DC.;* CUCURBITACEÆ, Vol.
[II., 641.

MARRUBIUM, *Linn.; Gen. Pl., II., 1206.*

Marrubium Malcomii, *Dalz.,* see **Micromeria capitellata,** *Benth.;*
M. vulgare, *Linn.; Fl. Br. Ind., IV., 671.* [p. 244 ; LABIATÆ. 293

Habitat.—This plant occurs in the Western Temperate Himálaya: Kashmír between 5 and 8,000 feet. **Stewart** and one or two other authors allude to this plant, but apparently its medicinal properties are not known to the people of India.

Medicine.—In Europe it enjoyed the reputation of possessing bitter tonic properties for which it was useful in many complaints. It still holds a **MEDICINE. Plant.** 294

MEDICINE.

place in the American Pharmacopœia where it is described as laxative in large doses; is given to increase the secretion from the skin and occasionally from the kidneys. It was formerly regarded as deobstruent, and was recommended in chronic hepatitis, jaundice, amenorrhœa, phthisis, and various other chronic affections. By its mild tonic properties it may exercise a beneficial influence, but it has no specific property, and hence it is now mainly used as a domestic medicine (*U. S. Dispens., 15th Ed., 926*).

MARSDENIA, *R. Br.; Gen. Pl., II., 772.*

295 **Marsdenia Roylei,** *Wight; Fl. Br. Ind., IV., 35;* ASCLEPIADEÆ.

Vern.—*Murkúla* (Himálayan Districts), HIND.; *Murkila*, KUMAON; *Pathor* (Chenab); *Tar, verí* (Salt Range), *Kurang* (Simla), PB.

References.—*Brandis, For. Fl., 333; Gamble, Man. Timb., 266; Stewart, Pb. Pl., 145; Wight, Contr. Bot. Ind., 40; Atkinson, Him. Dist., 313, 794; Royle, Fibrous Pl. Ind., 305; Journ. Agri.-Horti. Soc. Ind., XIV., 44; (New Series) I., 94.*

Habitat.—A large, twining shrub, of the Eastern and Western Himálaya. At Simla it ascends to altitudes close on 7,000 feet, but is most plentiful at about 3,000 to 5,000 feet. In Sikkim it is mentioned as met with at 4,000 feet. It may, therefore, be described as a warm temperate plant.

GUM. Caoutchouc. 296

Gum.—The milky sap contains CAOUTCHOUC, but in so small quantity as to be of no value.

FIBRE. 297

Fibre.—It yields a FIBRE, of which fishing nets and strong ropes are manufactured. This has not, however, been scientifically examined, so that no opinion can be passed as to its relative value to the fibre of the next species. **Royle** mentions, however, that a sample of this fibre was sent to the great Exhibition of 1851 from Nepal. The plant might be extensively cultivated and probably more easily than that of **M. tenacissima,** as it is a free growing semi-succulent plant from which the fibre could probably be more readily separated than would be the case with the more tropical species. **Mr. W. Coldstream,** the Deputy Commissioner of Simla, saw an exceptionally large fish being hauled in from a stream in that district, the line being of so fine a quality as to excite his astonishment. He was shown the plant from which it had been prepared, and judging from **Mr. Coldstream's** description it was very probably the species here dealt with. The writer is not at any rate aware of any other Himálayan plant that would yield a fibre of the strength requisite for fishing lines.

MEDICINE. Fruit. 298

Medicine.—The unripe FRUIT is powdered and given as a cooling medicine (*Stewart*).

299 **M. tenacissima,** *Wight & Arn.; Fl. Br. Ind., IV., 35; Wight, Ic.,* RÁJMAHÁL HEMP. [*t. 590.*

Syn.—ASCLEPIAS TENACISSIMA, *Roxb.;* A. TOMENTOSA and A. ECHINATA, *Herb. Madr.;* GYMNEMA TENACISSIMA, *Spreng.*

Vern.—*Tongus*, HIND.; *Jítí* (Rájmahál Hills), *chití* (PALAMOW), BENG.; *Babal Ják*, CENTRAL INDIA; *Haba*, (?) BOMB.; *Jiti, chiti*, TAM.; *Muruvá-dúl*, SING.

References.—*Roxb., Fl. Ind., Ed. C.B.C., 258; Voigt, Hort. Sub. Cal., 537; Brandis, For. Fl., 333; Kurz, For. Fl. Burm., II., 201; Gamble, Man. Timb., 265; Trimen, Cat. Ceylon Pl., 55; Wight, Contr. Bot. Ind., 41; Atkinson, Him. Dist., 794; Royle, Fib. Pl., 304; Christy, Com. Pl. and Drugs, VI., 11, 37; Liotard, Paper-making Mat., 5, 37; Balfour, Cyclop., Vol. II., 886; Smith, Dic., 221; Kew Off. Guide to the Mus. of Ec. Bot., 97; Jour. As. Soc. P., II. (1867), 82; 1885, VII. (New Series) Proc., lxxi.; Gazetteers:—Mysore and Coorg, I., 56; N.-W. P., I., 82; IV., lxxiv.; Indian Forester, IX., 274; X., 195; XI., 369; XIV., 273; Journ. Agri.-Horti. Soc. Ind. (Old Series), III., 221-228; IX., 151; Pro., 128; (New Series) IV., Pro. (1873), 52; VII., Pro., 71.*

Jetí Fibre. (*G. Watt.*) **MARSDENIA tenacissima**

Royle calls this the Jetee Fibre. It is remarkable that a plant which yields so valuable a fibre and a caoutchouc should, practically, be unknown to the people of India. Indeed, by following the usual line of reasoning, based on vernacular names, the plant might be viewed as doubtfully a native of India, whereas it is nowhere cultivated, exists only in a wild state, and is, therefore, undoubtedly a native of this country.

Habitat.—A climbing plant, distributed throughout the lower Himálaya (ascending to 5,000 feet) from Kumáon to Assam and Burma: also found on the lower hills of Bengal—Rajmahal, Chittagong, &c. The plant is fond of dry barren localities, twining on the bushes and small trees. It might be extensively cultivated.

Gum.—A milky juice exudes from cuts on the stem, which thickens into an elastic substance, or CAOUTCHOUC which acts in the same way as India rubber in removing black lead marks (*Roxburgh*). GUM. Caoutchouc 300

Fibre.—Very little more can be said regarding this FIBRE than has already appeared in the volume of Selections from the Records of the Government of India published by the Editor of this work. In *Spons' Encyclopædia* (p. 982) it is stated: "The bark of the stems yields a valuable fibre, which is extracted by cutting the stems into sections, splitting them, drying them, steeping them in water for about an hour, and scraping them clean with the nails or with a stick. The hillmen simply dry the stems and altogether dispense with retting. About 6lb of clean fibre is a good day's work. The fibres are fine and silky, and of great strength, a line made of them breaking at 248lb dry and 343lb wet, as against hemp at 158 and 190lb. It is used locally for bow-strings and for netting." According to **Roxburgh** the plant was discovered in 1800. "During the rains," he says, "the natives of Rájmahál cut the shoots into lengths at the insertion of the leaves, peel off the bark, and with their nails, or a bit of stick on a board, remove the pulpy part." FIBRE. 301

The figures given above in **Spons'** account of the fibre are those first published by **Roxburgh**, and the facts appear to have been compiled from **Royle's** acccount of the fibre (*Fibrous Plants of India, 304*). **Royle** justly adds that "the plant is suited for better purposes than rope-making, besides not being eligible for this purpose, from its comparative rarity and mode of preparation. **Mr. Taylor** remarks that it might, however, be easily cultivated (*Jour. Agri.-Hort. Soc., 1844, V., 221*). One of the chief features of this fibre is its great elasticity, since it is, according to **Royle**, the second best of all the fibres in India.

Only a very small amount of it was shown at the Colonial and Indian Exhibition, but the sample was universally admired. Some of the experts indeed viewed it as a very superior quality of Rhea. According to **Messrs. Cross, Bevan & King's** chemico-microscopic examination of the fibre, it is very considerably superior to Rhea. The following table exhibits the results obtained by these chemists:—

	Moisture.	Ash.	Hydrolysis. (a) (Five mints.)	Hydrolysis. (b) (One hour).	Cellulose.	Mercerising.	Nitration.	Acid Purification.	Carbon percentage.
Marsdenia tenacissima	4·5	1·5	6·2	10·1	6·2	4·6	131·0	0·8	44·6
Bœhmeria nivea (Rhea)	9·0	2·9	13·0	24·0	80·3	11·0	125·0	6·5	...
Linum (Flax, European)	9·3	1·6	14·6	22·2	81·9	8·4	123·0	4·5	43·0
Calotropis gigantea .	7·3	2·5	13·0	17·6	76·5	...	153·5	8·5	44·0
Crotalaria (Sun-hemp) .	8·5	1·4	8·3	11·6	83·0	11·3	150·5	2·7	44·3

FIBRE.

To allow of comparison four other fibres have been shown alongside of **Marsdenia** (the Rájmahál-Bowstring). That fibre heads the list in percentage of cellulose and loses considerably less than any of the others, either under hydrolysis with caustic soda or in the acid purification, while it holds the third place in increased weight by nitration. These are facts the value of which cannot be over-estimated. They point the fibre out as being, from a scientific stand-point, far more worthy of experimental cultivation than **Rhea** or any of the other fibres with which in the above table it has been compared. The one point of uncertainty regarding it, which practical experiments alone can solve, is its yield of fibre per acre as compared to the cost of cultivation,—in other words, the price at which it could be put down in the textile markets. The ultimate fibres are 5 to 20 mm. in length, *e.g.*, nearly as long as those of flax, and two or three times as long as those of sunn-hemp or of jute, though of course very much shorter than the fibres of Rhea. But from this point of view Rhea stands by itself, as its ultimate fibres (40 to 200 mm.) are far in excess of any otherk nown fibre. **Messrs. Cross, Bevan & King** say of **Marsdenia**, "Next to Rhea it must rank in point of fineness and durability, and we cannot urge its claims to the attention of Government in too strong terms. If it can be shown that the fibre could be cultivated at all, it might then become a question whether the *haba* or Rhea could be produced the cheaper."

It seems probable that to arrive at good results the long young twigs of the plant had better be treated by some chemical decorticating process, such as that of **Favier**, instead of being cut into short lengths and decorticated mechanically. The shortness of the fibre-ribbons, as usually met with, would presumably be viewed as unfavourable, but since this is by no means a necessity it might be well to adopt some process of decortication that would produce ribbons the full length of the twigs.

The plant is too scarce and unimportant-looking for its merits to come by the usual "private enterprise" means to be recognised by the manufacturer. It must be cultivated, and that too perhaps for a good many years, before a final opinion can be pronounced. It is a climber and does not appear to grow either rapidly or profusely, but there is no knowing what it might do under careful management. Very likely the allied species **M. Roylei** might, as suggested above, be found a more suitable species for experimental cultivation, but of course in warm temperate regions only, such as Kúlu, Simla, Kumáon, Kashmír, the Nilghiris, &c. It might indeed be even possible to grow it in the warmer parts of Europe. **Marsdenia** is, however, too valuable a fibre to be longer ignored, and it would serve a public good were the various Botanic Gardens and Agri.-Horticultural Societies to take its experimental cultivation under their special charge. Were the cultivation of **M. tenaciassima** to prove remunerative, the plant might be reared in every hedgerow of India, but, being a climber, difficulties exist with which the Indian cultivator of fibre crops has not as yet attempted to deal. In order to avoid these difficulties—the expense and trouble of constructing supports for a climbing plant—it would be a good step to ascertain whether it could be induced to crawl over the ground instead of requiring support. Although, as stated, it might be grown in every hedgerow over the entire length and breadth of the plains of India, success could alone be ensured by the production of a stock that might be planted in the usual way over a limited area.

302 **Marsdenia tinctoria**, *Br.; Fl. Br. Ind., IV., 34; Wight, Ic., t. 589.*

Syn.—Marsdenia monostachya, *Wall.*; Asclepias tinctoria, *Roxb.*; Pergularia tinctoria, *Spreng.*; P. parviflora, *Blume.*; Cynanchum tingens, *Herb. Ham.*

Vern.—*Riyong*, (TEESTA VALLEY), BENG.; *Kali lara*, NEPAL; *Ryom*, LEPCHA; *Mai-nwai, mai-dee*, BURM.; *Tarúm-akkar*, SUMATRA.

References.—*Roxb., Fl. Ind., Ed. C. B. C., 225; Voigt, Hort. Sub. Cal., 537; Brandis, For. Fl., 332; Kurz, For. Fl. Burm., II., 201; Gamble, Man. Timb., 265; Wight, Contrib. Bot. Ind., 40; Grah., Cat. Bomb. Pl., 119; Mason, Burma & Its People, 510, 801; McCann, Dyes and Tans, Beng., 126; Liotard, Dyes, 109; Balfour, Cyclop., Vol. II., 886; Indian Forester, XI., 326; XII., App., 17; Gazetteers:—N.-W. Provs., I., 82; IV., lxxiv.; Agri.-Hort. Soc. Ind. Trans., VIII., 89; Jour., III., 231, 232; VI., 50, 51, 142, 148; X., 293-294.*

Habitat.—A tall, climbing shrub, of the North-Eastern Himálaya and Burma; occasionally cultivated in the Deccan and elsewhere in India, but only experimentally except in Burma where it assumes some importance; distributed through Sumatra, Java, and China.

Fibre.—Like the preceding, this species yields a FIBRE, but the plant is collected more on account of its dye than its fibre. FIBRE. 303

Dye.- The LEAVES of this climber yield INDIGO. This fact has been published repeatedly, but apparently never put to commercial test. On this subject **Roxburgh** wrote:—"The leaves of this plant yield Indigo, as mentioned by **Mr. Marsden**, and by **Mr. Blake**, in the first volume of the *Asiatick Researches*. I have also extracted it from them by hot water. The few experiments I have yet made do not enable me to say positively in what proportion they yield their colour, but it was of an excellent quality; and as the plant grows very readily from layers, slips, or cuttings, I think it very well worthy of being cultivated, particularly as it is permanent like the **Nerium**" (**Wrightia tinctoria**), "so that a plantation once formed will continue for a number of years; and if we are allowed to draw a comparison between the leaves of this plant and those of **Wrightia tinctoria**, the quantity of colour they may yield will be in a larger proportion than that from the common indigo plant." "Some more experiments I have made with the leaves confirm what is above related, not only respecting the quality of the Indigo, but also that the proportion is considerably greater than is obtained from **Indigofera tinctoria**. I have, therefore, warmly recommended an extensive cultivation thereof." DYE. Leaves. 304

Throughout the *Journals of the Agri.-Horticultural Society* the subject of **Marsdenia** Indigo is here and there referred to, but up to date no advancement seems to have been made towards utilizing the substance. For example, a sample was communicated in 1844 by **Mr. E. O'Riley** of Tenasserim to the Society. Commenting on that sample **Griffith** gave the extract from **Marsden's** *History of Sumatra* alluded to above by **Roxburgh**. "There is another kind of Indigo, **Marsden** says, called in Sumatra *taram akar*, which appears to be peculiar to that country, and was totally unknown to botanists to whom I shewed the leaves upon my return to England in the beginning of the year 1780. The common kind is known to have small pinnated leaves growing on stalks imperfectly ligneous. This, on the contrary, is a vine, or climbing plant, with leaves from three to five inches in length, thin, of a dark green, and in the dried state discoloured with blue stains. It yields the same dye as the former sort; they are prepared also in the same manner, and used indiscriminately, no preference being given to the one above the other, as the natives informed me; excepting inasmuch as the *Taramakar*, by reason of the largeness of the foliage, yields a greater proportion of sediment." Indigo. 305

Mason alludes to the fact that this indigo plant is, to a certain extent, cultivated by the Karens and sometimes the Burmans. This is said to be a creeper indigenous in some parts of the country and which yields a good indigo, "though not equal to the **Ruellia** (**Strobilanthes**) indigo." **Drury**, upon what authority is not known, says: "**M. tinctoria** is cultivated in

Northern India, being a native of Sylhet and Burma. The leaves yield more and superior indigo to the **Indigofera tinctoria**, on which account it has been recommended for more extensive cultivation" This would appear to be a too liberal reading of **Roxburgh's** statement. No record exists of its being cultivated in Northern India and Bengal except, perhaps, the plants raised by **Roxburgh** in the Botanic Garden, Calcutta. This subject has too long remained, however, in obscurity; it would seem well worthy the attention of planters. If any one of the three species of **Marsdenia** could be grown with the double object of affording dye and fibre, it seems probable the maceration to extract the indigo might prove an initial stage in the separation of the fibre, and thus render it possible to cheapen both products. The fibre of **Marsdenia** is of such extreme fineness and strength that, if produced commercially, Rhea and China grass would most probably be driven out of the market. It seems probable that, in India at least, the **Bœhmeria** fibres will never become important crops. The **Marsdenias** are natives of this country, could be readily cultivated by cuttings, and, being perennials, they might be grown at small cost. Every thing in fact points to the superior claims of these plants over almost any other of known economic value which has not as yet found a place in European commerce.

MARSILEA, *Linn.*

A genus of Cryptogams named in honour of **Count Marsigli,** the founder of the Academy of Science at Bologna.

306 **Marsilea quadrifolia,** *Linn.; Baillon, Traité de Botanique, Médicale Cryptogamique, 39;* MARSILEACEÆ.

Vern.—*Súsni-shak,* BENG.; *Chatom aruk,* SANTAL; *Paflú,* KASHMIR; *Tripattra, godhi,* PB.; *Mudugu támara, munugu támara, chick-lintakura, chitlinta kúra* (according to **Elliot**), TEL.; *Chitigina soppu,* KAN.

References.—*Roxb., Fl. Ind., Ed. C.B.C., 745; Voigt, Hort. Sub. Cal., 739; Thwaites, En. Ceylon Pl., 378; Dalz. & Gibs., Bomb. Fl., 309; Stewart, Pb Pl., 266; Burmann, Fl. Ind., 237; Grah. Cat. Bomb. Pl. 243; Elliot, Fl. Andhr., 117, 120; Stewart, Journ. Bot. Tour in Hazára (in Journ., Agri.-Hort. Soc. Ind.; Vol. XIV., 6); Rev. A. Campbell, Rept. Econ. Prod., Chutia-Nagpur, No. 7889; Atkinson, Him. Dist., 322; Jour. As. Soc., P. II., No. II., 1867, 81; Gazetteers:—N.-W. P., I., 86; IV., lxxx; Mysore and Coorg,; I., 71; Indian Forester, XIV., 390.*

Habitat.—This sub-aquatic plant (closely allied to the ferns) is found growing abundantly on the margins of tanks in Bengal and northwards to the Panjáb, also on the hills up to 5,000 feet.

FOOD. Pot-herb. 307 **Food.**—It is regularly eaten as a POT-HERB by the natives of Bengal, and probably in the Panjáb and other parts of India. The same species occurs in France where, according to **Baillon**, it is eaten in times of scarcity. Two other species, **M. hirsuta,** *R. Br.* and **M. Drummondii,** *R. Br.*, form the well-known *nardú,* of Australian writers, so often mentioned as furnishing food to travellers in that country. From these a sort of coarse bread and a gruel or broth are made. **Mr. J. H. Maiden** (*Native Plants of Australia, 135*), says that the *Nardú* "is much relished by stock. It is, however, better known as yielding an unsatisfactory human food in its spore-cases."

MARTYNIA, *Linn.; Gen. Pl., II., 1055.*

308 **Martynia diandra,** *Glox.; Fl. Br. Ind., IV., 386;* PEDALINEÆ.

TIGER CLAW OR DEVIL'S CLAW.

Vern.—*Háthajori, bichu,* HIND.; *Bagh noki,* BENG.; *Sher núi,* BEHAR; *Bag lucha,* SANTAL; *Sher núi (tiger-claws),* N.-W. P.; *Bichú?, hatha-*

jori (=fruit), PB.; *Vinchú* (Poison: scorpion), BOMB.; *Vinchú*, MAR.; *Garuda mukku* (hawk's beak), *télu kondi chettu* (Scorpion's tail), TEL. All the above vernacular names are clearly of modern origin and denote the hooked fruit.

References.—*Roxb., Fl. Ind., Ed. C.B.C., 496; Stewart, Pb. Pl., 149; Elliot, Fl. Andhr., 58, 180; Rev. A. Campbell, Rept. Econ. Prod., Chutia Nagpur, No. 8166; Dymock, Mat. Med. W. Ind., 2nd Ed., 555; Baden Powell, Pb. Pr., 364; Gazetteer:—Bombay, XV., 439; Indian Forester, XII., App. 18, 28; Indian Agriculturist, Jan. 1889.*

Habitat.—An American weed now common in the Gangetic plains, Chutia Nagpur, Bombay, and elsewhere in India. It is a rank coarse herb with capsules beaked by strong curved spines.

Oil.—The Rev. **A. Campbell** states that the Santals distil a medicinal OIL from the fruit; he does not mention the purpose for which the oil is, however, used. OIL. 309

Medicine.—The FRUIT is officinal in the Panjáb bazárs (*Stewart*). It is sold in the drug shops as an antidote to scorpion stings, hence the name *Bichu*, HIND., and *Vinchu*, MAHR. Its properties are very likely entirely imaginary, being suggested on the theory of signatures from the resemblance of the sharp hooks of the fruit to the sting of the scorpion, the claws of the tiger, &c. A writer in the *Indian Agriculturist* dwells on this property of antidote to venomous bites and stings. MEDICINE. Fruit. 310

SPECIAL OPINIONS.—§ "The fruit has received the name *Bichu* (scorpion), not from its use in scorpion bites, but its two curved hooks which resemble the tail of the scorpion. It is a useless substance" (*Assistant Surgeon S. Arjun Ravat, L. M., Girgaum, Bombay*).

Mastich or Mastache, see **Pistacia Lentiscus,** *Linn.* ANACARDIACEÆ: [Vol. VI., Part I

MASTIXIA, *Bl.; Gen. Pl., I., 950.* 311

Mastixia arborea, *C.B. Clarke, Fl. Br. Ind. II., 745;* CORNACEÆ.

Syn.—BURSINOPETALUM ARBOREUM, *Wight, Ic., t. 956.*

Vern.—*Diatalia*, CEYLON.

References.—*Thwaites, En. Ceylon. Pl., 42; Beddome, Fl. Sylv., t. 216; Dalz. & Gibs., Bomb. Fl., 28; Gamble, Man. Timb., 211; Lisboa, U. Pl. Bomb., 82; Indian Forester, X., 34.*

Habitat.—A large tree with dark green foliage found in Cachar, the Nilghiri Mountains, and in Ceylon at altitudes of from 4,000 to 7,000 feet.

Structure of the Wood.—Said to be of good quality, but no definite information exists regarding it. **Beddome** remarks that it is very abundant in the dense Western Ghát forests, from Kanara to Cape Comorin. It is thus doubtless of considerable importance to the people, though its properties do not appear to have been investigated. TIMBER. 312

M. tetrandra, *C. B. Clarke; Fl. Br. Ind., II., 745.* 313

Beddome, in reducing *Wight's* **Bursinopetalum** to **Mastixia,** referred this name to the above species, remarking that the "tetramerous form is certainly not a distinct species, as both forms occur on one and the same tree." If this opinion be confirmed, the name of the tree should be **M. arborea,** *Bedd.*

MATRICARIA, *Linn.; Gen. Pl., II., 427.* 314

Matricaria Chamomilla, *Linn.; Fl. Br. Ind., III., 315;* COMPOSITÆ.

GERMAN CHAMOMILE or PERSIAN CHAMOMILE, the true medicinal Chamomile being **Anthemis nobilis,** *Linn.;* see Vol. I., 264.

Syn.—M. SUAVEOLENS, *Linn.*

Vern.—*Bábunáh, babúna, suteigul* (Trans Indus), PB.; *Bábúná,* GUZ.; *Bábunaj ?,* ARAB.; *Bábunah,* PERS. It seems probable that the above names are more frequently given in India to the imported drug **Anthemis nobilis** (which see), than to this plant.

References.—*Roxb., Fl. Ind., Ed. C.B.C., 605; Stewart, Pb. Pl., 127; Dymock, Mat. Med. W. Ind., 2nd Ed., 448; Flück. & Hanb., Pharmacog., 386; U. S. Dispens., 15th Ed., 196 & 934; Bent. & Trim., Med. Pl., 155; S. Arjun, Bomb. Drugs, 80; Year Book Pharm., 1874, 626; Baden Powell, Pb. Pr., 357.*

Habitat.—A much branched herb found in the Upper Gangetic plain, and distributed to Northern Asia and westwards to the Atlantic.

OIL. 315 **Oil.**—An essential OIL is obtained by distillation, which, to a certain extent, possesses antispasmodic properties.

MEDICINE. **Medicine.**—It does not seem necessary to do more than indicate the literature of this drug. It is in India, as in Europe, only used as a substitute for true chamomile, and though it might be easily supplied at less price than the imported article, it is scarcely, if at all, used in India. The

Flower heads. 316 dried FLOWER HEADS are officinal, and are said to be stimulant, tonic, and carminative. They are employed in constitutional debility, hysteria, dyspepsia, and intermittent fevers. The warm and strong infusion of the flowers is emetic, while a weak infusion acts as a tonic and febrifuge.

Oil. 317 In flatulence and colic, chamomile OIL is generally regarded the most effectual of all remedies. The *Indian Pharmacopœia* says the *babunaka phul* forms a perfect substitute for the European Chamomile (*see* **Anthemis nobilis**). "In Persian works the flowers are described as stimulant, attenuant, and discutient. There is a popular opinion among the Persians that the odour of the flowers induces sleep and drives away noxious insects; they also say that the chamomile tea applied to the genitals has a powerfully stimulating effect" (*Dymock*).

MATT, COBALTIFEROUS.

318 **Matt, Cobaltiferous.**

Since the date on which the article *Cobalt* was written, an interesting correspondence has taken place regarding the "*cheep*" or cobaltiferous matt found in certain parts of Nepal. As this is of some interest, the opportunity has been taken to refer shortly to the subject in this place, leaving the reader, for a full detail of the correspondence, to consult the *Selections from the Records of the Government of India, Revenue and Agricultural Department, Vol. I., 61,* or the *Indian Agriculturist, November 1889, 663.* In January 1888 a parcel of "*cheep*" was forwarded by a **Mr. Ricketts** to the Resident in Nepál with the information that "there were several smelting mines of the stuff, and that it is only obtained close by the copper mines in Nepal." The name of the locality is "Kachipatar, Argah Zillah, Sowrobhar, about 80 miles north of Doolho. The price on the spot is R30 to R35 per maund, and they sell it here" (at Doolho) "from R40 to R50 per maund." A sample sent from the Resident in Nepál to the Government of India, Revenue and Agricultural Department, was subjected to an analysis and thus reported on by the Director, Geological Survey of India: "The sample of 'cheep,' sent to you by **Mr. Ricketts** in June last, and forwarded to this office, has been analysed by **Mr. E. J. Jones** with the following result:—

Loss at 100°C	0·40	Cobalt	13·97
Insoluble in acids	0·70	Iron	68·82
Sulphur	20·41		

Dr. W. King then stated that, from latest available information, cobal-oxide was worth £717 per ton; that 5 tons of "cheep" would yield 1 ton of

Cobaltiferous Matt. (*G. Watt.*) MATS & Matting.

the oxide, and at Mr. Ricketts' valuation could be obtained for from £270 to £455; and that, therefore, there appeared to be a fair margin to allow of the matt being profitably exported if obtainable in sufficient quantity. In answer to this report the Resident of Nepál stated that, according to Mr. Ricketts, the present supply is about 400 maunds a year, and that this might, in all probability, be increased by a greater demand.

Since the date of the last letter (October 1889), no record exists of any attempt having been made to export the "matt."

MATS AND MATTING.

319

Mats and Matting.

Syn.—MATTEN, *Dut., Ger.*; NATTES, *Fr.*: ESTEIRAS, *Port., Sp.*; STUOJE, STOJE, *It.*; PROGOSHKI, *Rus.*; HASSIR, *Turk.*

Vern.—*Chattai*, HIND.; *Motha*, KUMAON; *Chatái*, PB.; *Chattai*, GUZ.; *Tikar, bogor, galeran, klasa*, MALAY.

References.—*Stewart, Pb. Pl., App., 93; The Journal of Indian Art, III., 10, 14; Report of Commercial Conference, Colonial and Indian Exhibition, 3; Baden Powell, Pb. Pr., 517; Pb. Manufactures, 85, 303; Royle, Fib. Pl., 234; Birdwood, Indian Arts, II., 298; Mukharji, Art Manufactures of India, 310; Grierson, Behar Peasant Life, 150; Balfour, Cyclop., II., 896; Settlement Reports; Central Provs., Upper Godáveri Dist., 43, 44; Panjáb, Pesháwar Dist., 19; Muzaffargarh Dist., 103; Manuals:—Madras Administration, I., 361; Cuddapah Dist., 53; Admn. Reports, Bombay, 1871-72, 369, 398; Reports (official):—Kumáon, 280.*

It is not intended to deal in this work with the subject of **Mats and Matting** further than to afford a key by which the reader may be able to discover the chief materials employed in their construction. The references given above may, however, prove useful to persons desirous of discovering detailed accounts of the various methods pursued in manufacturing mats. **Sir George Birdwood** alludes very briefly to some of the more famous mats, such as those of Pálghát on the Malabar coast; Midnapur, "admired for their fineness and classical designs of the mosaic-like patterns of stained glass"; *Sitalpati* or Eastern Bengal cooling mats, used for sleeping on; *darmá* employed in Bengal in the construction of huts: *Sedge* or *Mádur* mats much prized for carpeting floors: and Sylhet ivory mats. But there are many others, such as palm mats, bamboo mats, reed mats, *dib* or **Typha** mats, **Aloe** and *Munj* mats, &c., &c.

The following are the more important plants used in mat-making:—

Agave americana, *Linn.*

Aloe fibre mats are now largely made in Madras, Hazáribágh, Bulandshahr, &c.; Vol. I., 140.

Arundhinaria falcata, *Nees*; Vol. I., 335.

A stunted Himálayan bamboo.

Bambusa, Dendrocalamus, and other forms of bamboo; Vol. I., 387.

Some of the finer coloured bamboo mats met with in India are prepared in Midnapur and Madras. In some parts of Bengal, bamboo matting is used in the construction of huts in place of *Darma*, and in Burma, Assam, and the North-Western Provinces this is nearly always the case.

Borassus flabelliformis, *Linn.*; Vol. I., 504.

Calamus, several species, but chiefly **C. Rotang,** *Linn.*

Cane mats, see Vol. II., 98-102 (*Conf.* with *Hoey, Trade and Manuf. N. Ind., 72-75*).

Cocos nucifera, *Linn.*

Two kinds of mats are made from this palm—one, and by far the most important, from the coir (see Vol. II., 432); the other, by plaiting the leaves to form what are known as *Cadjans* (see *Vol. II., 433*).

MATS & Matting.

Plants used for making mats.

Cyperus, various species of Sedge, but chiefly **C. tegetum,** *Roxb.*

From this the *mádur* grass mats of Calcutta are made. In Madras **C. corymbosus,** *Roxb*, takes its place. The species employed in making the *Masland* mats of Midnapur has not apparently been determined (see Vol. II., 682-689).

Gossypium—Cotton. See Carpets and Rugs; Vol. II, 176-182.

Hedychium spicatum, *Ham.;* Vol. IV., 207.

On the Himálaya the dry leaves of this plant are twisted and woven into the ordinary sleeping mats used by the hill people.

Ischœmum angustifolium, *Huck.;* Vol. IV., 527.

Bhábar mats are largely made in some parts of the North-West Provinces.

Licula peltata and Livistona Jenkinsiana; Vol. IV., 639.

Two palms met with in Assam and Burma, yield leaves which are largely used for mats, umbrellas, &c.

Nannorrhops Ritchieana, *Wendl.;* p

The palm used in making the Peshawar, Kohat, and other Panjáb mats, known as *Patta.* These are described by **Baden Powell** (*Panjáb Manufactures*).

Pandanus odoratissimus, *Willd.;* Vol. VI., Pt. I.

The leaves of the screw pine are largely used for making mats in the localities where the plant occurs, *viz.*, the Andaman Islands, South India, Bengal, &c.

Phragmites Roxburghii, *Trim.;* Vol. III., 27; also VI., Pt I.

The substance from which the *Darma* mats of Bengal are made,—not bamboo, as stated by **T. N. Mukharji** in *Art Manufactures.*

Phœnix dactylifera and P. sylvestris; Vol. VI., Pt. I.

The leaves of the various forms of date-palm are extensively platted into mats, and by the well-to-do people these are often spread on floors beneath other better-class mats or carpets (*Baden Powell, Panjáb Manufactures*).

Phrynium dichotomum, *Roxb;* Vol. VI., Pt. I.

This plant has, by most modern writers on Economic Products, been incorrectly referred to **Maranta.** Its real position is more probably in the genus **Clinogyne,** *Salisb.* From the stems of this plant the famous *Sitalpáti* mats of Eastern Bengal are prepared.

Saccharum ciliare, *Anders.;* Vol. VI., Pt I.

By the most recent botanical investigations, this species has now been made to include the two forms known in Indian works on Economic Botany as **S. Munja** and **S. Sara.** From the former and, to a small extent, from the latter, also, the famous *Munj* mats and carpets are made. The industry in these mats is mainly confined to the jails of Upper India, as for example Delhi, Allahabad, Lucknow (*Conf.* with *Hoey, Trade and Manuf. N. Ind. 65*).

Scirpus (Malacochæte pectinata, *Nees.*); Vol. VI.

The plant used in making Kashmír mats.

Typha angustifolia, *L.;* and **T. latifolia,** *Willd;* Vol. VI., Pt. II.

These aquatic plants afford leaves which are regularly employed in making mats. In Bengal these are known as *hogla* and in the Panjáb as *dib* mats.

Wool, Hair, &c.; Vol. VI., Pt. II.

Many of the finer qualities of mats are felted or woven from these materials, but, as they are nearly always highly ornamental, they are art manufactures and would not, therefore, be treated of in this work.

There are doubtless other materials used, especially among the hill tribes of Assam, but the above embraces the more important matting as well as basket materials.

MATTHIOLA, *Br.; Gen. Pl., I., 67.* 320

Matthiola incana, *R. Br.; Fl. Br. Ind., I., 131;* CRUCIFERÆ.

PURPLE GILLY FLOWER or COMMON STOCK.

Vern.—*Todri saféd, todri lila,* PB.; *Todri safed,* SIND.

References.—*Dals. & Gibs., Bomb. Fl., 4; Stewart, Pb. Pl., 14; Pharmacog. Ind., I., 120; Dymock, Mat. Med. W. Ind., 56; Murray, Pl. and Drugs, Sind, 48; Balfour, Cyclop. Ind., III., 907.*

Habitat.—Cultivated as a cold season garden annual throughout India, but on some parts of the Himálaya it becomes a perennial. **M. odoratissima,** *Br.*, is a common indigenous species on the higher ranges of the N.-W. Himálaya.

Medicine.—The SEEDS are of three kinds—yellow, red, and white; used in infusion in cancer; are expectorant; mixed with wine given as an antidote to poisonous bites (*Dr. Emerson*). According to **Stewart** these seeds constitute one of the kinds of *todrí* which are reckoned aphrodisiac. MEDICINE. Seeds. 321

MAYODENDRON, *Kurz, Prelim. Forest Rept. Pegu, App. D.,* [*p. i., ii.* 322

Mayodendron igneum, *Kurz; Fl. Br. Ind., IV., 382;* BIGNONIACEÆ.

Syn.—SPATHODEA IGNEA, *Kurz, Jour. As. Soc. Beng. (1871), Pt. II., 77.*

Vern.—*Mawkpyit,* SHAN; *Ekarit,* ? BURM.

Habitat.—A tall tree with a girth of about five feet, found in Martaban and distributed to Ava and Yunan. **Mr. Oliver,** Conservator of Forests, Upper Burma, has recently furnished a specimen of this tree, together with the above vernacular names. He remarks that it occurs in the moist forests of the Namyin valley.

Structure of the Wood.—Definite information does not exist regarding the timber of this tree, but presumably, like the other members of the order to which it belongs, it is soft and of inferior quality. TIMBER. 323

Meadow saffron, see **Colchicum autumnale,** *Linn.;* LILIACEÆ Vol. II., 501.

Mecca Balsam, see **Balsamodendron Opobalsamum,** *Kunth.;* BURSERACEÆ, Vol. I., 369.

MECONOPSIS, *Vig.; Gen. Pl., I., 52.*

Meconopsis aculeata, *Royle, Ill. Him. Bot. 67, t., 15 (colour of flowers wrongly shown as pink); Fl. Br. Ind.; I., 118;* PAPAVERACEÆ. 324

Vern.—*Guddi kúm* (JHELUM), *Gúdi* (RAVI), *Kanda*ʳ (SUTLEJ), *Kanta* (SIMLA), *Pb.*

References.—*Stewart, Pb. Pl., 69; Bot. Mag. t. 5456; O'Shaughnessy, Beng. Dispens., 184; Pharmacog. Ind., I., 112; Gazetteers:—N.-W. P., X., 304; Panjáb, Simla District. 12.*

Habitat.—A spiny herbaceous plant with pale blue flowers; found in the Western Himálaya from Kashmír to Kumáon, at altitudes of from 10,000 to 15,000 feet.

Medicine.—The ROOT is officinal in Kashmír as a narcotic, and is in Chumba regarded as poisonous. **O'Shaughnessy** mentions having given a drachm of an alcoholic extract to a dog without producing any perceptible effect. MEDICINE. Root. 325

326 **Meconopsis nipalensis,** *DC.; Fl., Br. Ind., I., 118.*

Syn.—PAPAVER PANICULATUM, *Don.*

References.—*Pharmacog. Ind., I., 112; Gazetteer, N.-W. P., X., 304; Honigberger, Thirty-five years in the East, II., 306, 352.*

Habitat.—Found in the temperate Himálaya at altitudes from 10,000 to 12,000 feet, in Sikkim and Nepál.

MEDICINE. Root. 327 **Medicine.**—According to **Honigberger** the ROOT is officinal in Kashmir, being regarded as a narcotic.

M. Wallichii, *Hook.; Fl. Br. Ind., I., 119.*

Habitat.—A slender, stellately pubescent and softly hairy plant, met with in the Temperate Himálaya at altitudes of 9,000 to 10,000 feet.

MEDICINE. Root. 328 **Medicine.**—It seems that this and the two preceding species possess similar properties. The authors of the *Indian Pharmacographia* appear to have examined the ROOT of the present species, and their analysis may, therefore, be here reproduced as expressing all that is known regarding these Indian drugs.

CHEMICAL COMPOSITION.—"The root dried by exposure to air, and reduced to a fine powder, lost 8 per cent. of moisture at 100°C. The ash amounted to 12·7 per cent., and contained a marked amount of manganese. The alkalinity calculated as KHO, after separation of lime, was equal to 8·6 per cent. Digested with light petroleum ether, ·48 per cent. of a pale yellow, viscid, transparent, odourless extract was obtained. With the exception of a few white flocks the extract was soluble in absolute alcohol. On spontaneous evaporation shining laminæ separated, which under the microscope consisted of rhombic plates and needles: oil globules were also visible. The alcoholic solution of the extract was strongly acid. The amount of crystalline matter was too small to admit of the nature of the fat acid being determined. After exhaustion with light petroleum ether, the powder was dried by exposure to air, and then digested with ether. On evaporating off the ether, ·41 per cent. of a fragrant, soft, indistinctly crystalline residue was left. The extract was heated with dilute hydrochloric acid, and the soft, yellow, insoluble residue separated by filtration. The acid solution was rendered alkaline with ammonia, and then agitated with ether. On separation of the ether only a minute trace of residue was left, which did not respond to alkaloidal reagents. The yellow residue insoluble in HCl was treated with ammonia, and the turbid mixture agitated with ether. The ether left on evaporation a yellow, soft, non-crystalline residue, without taste or odour, which had the properties of a neutral resin. The aqueous alkaline solution after the separation of the ether, yielded yellow flocks when treated with dilute acids, which were re-dissolved by alkalies: this principle had the properties of a resin acid. The fragrant odour of the etherial extract was probably due to a trace of benzoic acid.

"After treatment with ether the powder was again dried, and then digested with absolute alcohol. The alcoholic solution was of a pale greenish colour, and possessed a marked greenish-yellow fluorescence; examined spectroscopically no absorption bands were visible. On evaporation, the alcoholic solution yielded 1·07 per cent. of extractive, yellow in colour, and possessing a somewhat fragrant odour. The extract was partly soluble in water. The aqueous solution did not possess any particular taste; it yielded slight precipitates with alkaloidal reagents; with ferric chloride no coloration was produced. On evaporation and ignition a trace of ash was left, possessing an alkaline reaction. The portion of the alcoholic extracts insoluble in water, dissolved in alcohol, yielding a greenish solution, with acid reaction, and greenish-yellow fluorescence. The powder, after treatment with alcohol, yielded 12 6 per cent. of extractive to cold water. The aqueous solution was yellowish-brown in colour; alkaline in reaction it afforded no coloration with ferric chloride; it slightly reduced alkaline solution of copper on boiling." MEDICINE.

MEDICAGO, *Linn.; Gen. Pl., I., 487.* 329

Medicago denticulata, *Willd.; Fl. Br. Ind., II., 90;* LEGUMINOSÆ.

Syn.—M. CANESCENS, *Grah.;* M. POLYMORPHA, *Roxb.*

Vern.—*Mainá,* PB.

References.—*Roxb., Fl. Ind., Ed. C.B.C., 589; Stewart, Pb. Pl., 71; Murray, Pl. and Drugs, Sind, 114; Gaz. N.-W. P. (Bundelkhand), I., 80; (Agra) IV., lxx; Agri.-Hort. Soc. Jour., XIV., 9.*

Habitat.—A field weed in the plains and low hills of Bengal, North-West Provinces, Oudh, the Panjáb, and Sind. Distributed to Abyssinia, Europe, Japan, China, &c.

Fodder.—It is largely gathered for cattle-fodder, as it is considered good for milch cows. **Stewart** remarks that it is said to be cultivated. (See *Vol. III., 416.*) FODDER. Herb. 330

M. falcata, *Linn.; Fl. Br. Ind., II., 90.* 331

YELLOW LUCERNE.

Syn.—M. SATIVA, *Wall.; Cat. No. 5945 C.D.;* M. PROCUMBENS, *Besser.*

Vern.—*Rishka, hol,* AFG., LAHOUL.

References.—*DC. Origin Cult. Pl., 102; Best Forage Plants by Stebler and Schröter, Transl. by A. W. McAlpine, 147; Birdwood, Bomb. Pr., 126; Report, Agri. Dept., 1881-82, 236.*

Habitat.—A sub-erect perennial, met with in Kashmir, Ladak, and Kunáwar, at an altitude of from 5 to 13,000 feet. Distributed to Afghánistán, Persia, and Europe.

M. lupulina, *Linn.; Fl. Br. Ind., II., 99.* 332

THE TREFOIL.

References.—*Stewart, Pb. Pl., 71; Best Fodder Plants by Stebler and Schröter, Transl. by McAlpine, 153; Permanent & Temporary Pastures by Sutton, 71; Atkinson, Him. Dist., 308; Gaz. N.-W. P. (Bundelkhand), I., 80; (Agra) IV., lxx; Jour. Agri.-Hort. Soc. Ind., XIV., 9.*

Habitat.—A native of the tropical and temperate tracts of the North-West Himálaya, ascending from the Indus valley and Gangetic plain to 10,000 or 12,000 feet in altitude.

Fodder.—A common WEED, collected frequently for fodder. Its flowers resemble hop cones, hence its specific name. It mixes well with grasses and clovers for artificial pastures. FODDER. Herb. 333

M. sativa, *Linn.; Fl. Br. Ind., II., 90.* 334

LUCERNE OF PURPLE LUCERNE.

Syn.—Some difference of opinion prevails as to the Botanical position to be assigned to the cultivated forms of Lucerne. The *Flora of British India* suggests that **M. sativa** may be but a cultivated state of **M. falcata,** characterised by the pod forming a double spiral and by the flowers being usually purple. Many writers, however, regard **M. sativa, M. falcata** and **M. media** as forming but one species, while others depart so far from that position as to admit the forms indicated as varieties under one common species, and still others hold that all three are distinct. On the other hand, by still a further series of authors **M. media** is pronounced a hybrid between **M. sativa** and **M. falcata.** Whatever botanical view be taken, **Stebler** and **Schroter** very justly add that to the agriculturist the three forms of Lucerne are very distinct, both in yield and suitability to environment. The Indian literature of the subject is, however, too imperfect to allow of a critical account being written. An attempt has, therefore, been here

History of Lucerne.

made to refer all available information given under any of the species of **Medicago** indicated above, to its specific position, but to compile the general information into the concluding account of Lucerne irrespective of the plant or plants meant.

Vern.—*Wilayti-gawuth*, HIND.; *Hol*, LADAK; *Spastu*, PUSHTU; *Sebist, rishka, dureshta*, AFG.; *Yurushea* (green) and *beda* (dry), YARKAND; *Vilayti-ghas*, GUZ.; *Vilayti-hullu*, KAN.; *Alfafa, alfasafat, alfalfa, fisfisat*, ARAB.; *Isfist*, PERS.

References.—*Dalz. & Gibs., Bomb. Fl. Supp., 21; Stewart, Pb. Pl., 71; DC. Origin Cult. Pl., 102; Aitchison, Bot. Afgh. Del. Comm., 48; Saidapet Exp. Farm. Manual, 53; Murray, Pl. and Drugs, Sind, 113; Atkinson, Him. Dist., 308; Lisboa, U. Pl. Bomb., 277; Birdwood, Bomb. Pr., 126; Royle, Prod. Res., 220; Smith, Dic., 270; Bomb. Man. Rev. Acc., 102; Gaz. Bombay, V., 25; N.-W. P. III., 225; Mysore and Coorg, I., 59; Indian Agri. Sept. 7th, 1889; Indian Forester, X., 111; XIV., 367; Agri. Dept. Reports (Exp. Farms), Madras, 1877-78, 16 & 97; Hyderabad, Sind, 1885-86, 31; 1886-87, 7; 1887-88, 3; Agri.-Hort. Soc. of India, Transactions & Journals quoted below.*

LUCERNE.

Habitat.—**De Candolle** says of this plant: "It has been found wild, with every appearance of an indigenous plant, in several provinces of Anatolia to the south of the Caucasus, in several parts of Persia, in Afghánistán, in Baluchistán, and in Kashmír. In the south of Russia, a locality mentioned by some authors, it is, perhaps, the result of cultivation, as well as in the south of Europe. The Greeks may therefore have introduced the plant from Asia Minor, as well as from India which extended from the north of Persia. This origin of lucerne, which is well established, makes me note, as a singular fact, that no Sanskrit name is known. Clover and sainfoin have none either, which leads us to suppose that the Aryans had no artificial meadow." **Stebler** and **Schroter** state that lucerne is indigenous to the following countries "Asia, Anatolia, Southern Caucasus, Persia, Afghánistán, Baluchistán, and Kashmír."

HISTORY. 335 **History.**—Lucerne was known to the Greeks and Romans: they called it in Greek *medicai*, in Latin *medica* or *herba medica*, because it was brought from Media at the time of the Persian war, about 470 years before the Christian era. The name lucerne is sometimes supposed to be derived from the valley of Luzerne in Piedmont. **De Candolle**, however, suggests a more rational derivation. "The Spaniards," he says, "had an old name, *eruye*, mentioned by **J. Bauhin**, and the Catalans call it *userdas*, whence, perhaps, the *patois* name in the south of France, *laduzerdo*, nearly akin to *luzerne*." "It was so commonly cultivated in Spain that the Italians have sometimes called it *herba spagna*. The Spaniards have, besides the names already given, *mielga*, or *melga*, which appears to come from Medica, but they principally used names derived from the Arabic—*alfafa, alfasafat, alfalfa*." The botanical evidence favours the inference derivable from the names of the plant, namely, that its original habitat extended from the north-west frontier of India to the shores of the Mediterranean. The writer possesses in his private herbarium many sheets of **M. falcata,** collected from undoubtedly wild sources in Persia, Kashmír, Chamba, &c., and from cultivation in many localities on the plains of India, more especially Behar. Of **M. sativa,** one sample bears no remark as to its being wild or cultivated, namely, that collected by **Dr. Giles**, during the Gilgit expedition; all the others from Baluchistán (collected by **Mr. Lace**), from the plains of India, &c., are expressly stated to have been collected from fields. The *Flora of British India* states that **M. sativa** is grown for forage in Madras, Bengal, and the North-West Provinces. The *Transactions, and Journals of the Agri.-Horticultural Society of India*, in this case as in many others, throws the most direct light on the origin of the Indian forms of

HISTORY.

lucerne. We read, for example, in the *Transactions* (*Vol I, 72, 79*) that **Mr. W. Moorcroft** noticed the variability of the flowers in the wild plant. His remarks are so interesting as to justify our republishing them here. He wrote, speaking of the higher ranges of Lama Yooroo in Kashmír: "I witnessed so striking a difference between the condition of the yellow lucerne near the summits of the dry mountains of Lama Yooroo (Ladak) and of the same plants when skirting the water-courses of Drass as might almost have countenanced a suspicion that there was a greater difference than what arose from locality alone." In a foot-note to the above it is stated:—"Lucerne in its natural state bears a yellow flower of a rich scent and is of great longevity; under the influence of cultivation it runs through a diminished sulphur tint into whiteness, becomes green with a stain of red, and settles permanently in pink and purple; it also loses its fragrance and becomes short-lived." **Mr. Moorcroft** then alludes to the cultivation of lucerne in Pusa (1823), remarking, "I caused the Government to expend considerable sums in wells and other arrangements for the watering of lucerne grounds, of which the supply was hardly ever adequate in the dry season, and the plants of which died when their crowns were long submerged in the rains. The facts I have now seen in regard to the almost aquatic nature of this plant lead to a suspicion that, if a modification of the float system had been adopted on the edges of a river with a very slow current during the largest portion of the year, and which embraced a great portion of the grounds in a crescent, that an immense quantity of excellent forage might have been raised, and the expense of wells, the labour of cattle and of gardeners might have been saved." In a further paper on Prangos hay, **Mr. Moorcroft** wrote of Imbal or Droz that he found yellow lucerne, a spontaneous product; he wrote of it—"it is of a constitution more hardy than that of Europe, requires no other culture than that necessary for sowing it, and lasts in vigour for a long series of years." He adds, "it is submitted that as it naturally grows along with Prangos, it would be well to imitate this habitude; the joint yield is vastly greater than that of the richest meadow land, and is produced in this country on a surface of a most sterile nature, in regard to other herbage, hence, is respectfully suggested the propriety of furnishing a few pounds of this seed to the Cape of Good Hope to be sown along with the Prangos."

In a report of Karnal (1836) mention is made of the experimental cultivation of lucerne. It is there stated that, "the plant thrived well, but being so common no remark seemed necessary." In the Proceedings of the Society for 1838 mention is made of **Mr. Hodgson** having sent "from Nepal seed of lucerne grass." In the Proceedings for the succeeding year, **Lieutenant Nicolson** mentions that in the meadows near Kábul lucerne is very commonly grown as food for horses and cattle. This same fact is again alluded to in the Journal (*Vol. I, 105*) when **Sir Alexander Burnes** gave particulars of the artificial grasses of Kábul including lucerne. **Sir Alexander's** account of the method of cultivation pursued in Afghánistán will be found in a further paragraph. The subject of Kábul lucerne, however, seems to have attracted considerable interest since in Vol II. (*Selections*), p. 297, we read that "Kelat lucerne, of which a large quantity was furnished by Government six months ago (1843), has been partly sown and partly distributed. It has vegetated well, but I fail to detect the smallest difference between it and our Deccan species Its superior luxuriance at Kelat, &c., must be owing to climate rather than to species." The writer of the passage quoted was **Dr. A. Gibson**, Superintendent of the Government Botanical Gardens, Bombay, and the allusion to "our Deccan species" is, therefore, extremely interesting as indicating an early cultivation of lucerne in Western India.

Passing over a gap of nearly forty years, we next read, in the *Journals of the Agri-Horticultural Society*, of lucerne as being experimentally grown in the Saidapet Farm, Madras, from English seed. Of Benares, a writer reported the failure of a crop in 1878, and in 1884 lucerne is stated to have done well in Silos. Thus it will be seen that only occasional mention is made of lucerne in India, though enough to prove that its cultivation has been at least tried during the greater part of the past 100 years. In the recent reports of Government Experimental Farms, it is stated that the Australian and European forms were found not to succeed so well as "the country kind known as Púna Lucerne." Even the Púna Lucerne does not seem to have "an entire immunity from the attacks of insects or hurt from the heat, and so many of the plants die during the hot weather that no heavy cutting is ever afterwards got from the plots."

CULTIVATION OF LUCERNE.

CULTIVATION. 336

The following interesting account regarding Kábul may be here given. **Sir Alexander Burnes** wrote in 1841:—"There are three kinds of grasses cultivated in Kábul—*rishku*, or lucerne; *shuftul*, or a kind of trefoil; and *sibarga*. The first and the last continue to yield crops for some years, but the trefoil (*shuftul*) is an annual.

The lucerne (*rishku*) is sown in spring, generally about the vernal equinox; for each *jureeb* (or about half an English acre) two seers of Kabul (or about 28℔ English) are required as seed. In forty days it comes to perfection and is cut down, and will yield four full-grown crops ere winter sets in, but by early cutting six or eight crops may be drawn,—the last may sometimes be inferior from premature cold. One *jureeb* yields on an average ten camel-loads of grass at each cutting; as a camel carries about 500℔, this is a produce of 5,000℔ the *jureeb* or 10,000℔ the English acre; and for four or five crops 40,000℔ English. The third crop is considered the best, and from it the seed is preserved. Of this the half acre sown with two seers Kábul will yield 40 seers or about 560℔. This plant requires the best black soil, much manure, and is watered five times each crop, in fact whenever it droops. It is sometimes sown along with barley, but in that case the grain by exhausting the soil injures the crop. The seed is never exported, but the grass is so plentiful, though all the cattle are fed on it, as much to exceed the consumption; it is, therefore, dried, and that produced at any distance from a market is generally stored in this manner and sold during the winter. A camel load of it (or about 600℔ English), whether green or dry, sells for one Kábul rupee, a coinage of which 115¼ are equal to 100 Company's rupees. Lucerne generally lasts for six years, but it will yield for ten years, if manure be abundantly scattered over it."

In the report of the Mission to Yarkand in 1873, the following further particulars are given which show the importance of lucerne at the present time in Upper India or immediately beyond the frontier:—

"Sown in August and September: sprouts in March and April. Is cut three times in six months, and after each receives a top-dressing of manure, and free irrigation, one sowing lasts three years, after which the roots decay. When sown, the seed is mixed with an equal quantity of barley, otherwise the lucerne does not thrive. It is extensively grown as a fodder crop, and is stored in bundles for winter use."

A writer in the *Indian Forester* (*X., 111*) says that "in dealing with the important question of fodder-reserves, it is profitable to notice the great success which has attended the cultivation in this country of guinea-grass (**Panicum jumentorum**) and lucerne (**Medicago sativa**). Lucerne is grown in small quantities in most places where Europeans are to be found, but guinea-grass is not so generally known." The writer then adds: "Lucerne

CULTIVATION.

is cultivated with very little difficulty. It should be sown broadcast on ground well broken up and manured. According to **Pogson** and other authorities, the spring is the best time for sowing, and lime the best manure. The outturn of lucerne varies according to circumstances, and should not be less than that of guinea-grass. In 1883, 4 *bíghas* under lucerne at Dera Ghazi Khan produced 930 maunds of green fodder in six months (January to June)." In the *Bombay Gazetteer* for the district of Cutch it is said to be grown as food for horses and to thrive well. In the *Saidapet Manual and Guide* it is stated, " under irrigation this plant produces a large quantity of valuable fodder. A few pounds of English seed sown in September 1869 grew satisfactorily and yielded three cuttings of excellent fodder ; it did not appear to be injured by the heat of the sun, though the thermometer exposed in the sun part of the time registered 135°, and the crop then looked vigorous and healthy. A similar result was obtained with some seed sown in 1876."

In concluding this account of lucerne it may be said European experience has formulated the following facts regarding successful cultivation :—

1. Dry seasons, and a warm sunny exposure suit lucerne best.
2. The best soil is a warm calcareous one ; cold impervious clay being unsuitable.
3. The ground must be kept well cleaned of weeds ; grass is its greatest enemy.
4. Clean seed must be secured, that is, seed free from admixture with **Medicago denticulata, M. maculata,** &c.
5. The crop should be reaped before flowering.
6. Its cultivation is remunerative only where the crop can be allowed to grow for at least three years.

MEDICINES.

337

The reader is referred to the remarks under **Domestic & Sacred** (*Vol. III., 191*) for an explanation of this subject heading. Space will not permit of a collective article on Medicines being here given. The magnitude of such a review may be learned from the fact that in the writer's *Catalogue of the Economic Products*, shown at the Calcutta International Exhibition (Vol. V.), 1,248 indigenous drugs of India have been briefly described, making a volume of 503 pages. But doubtless, in the preparation of the material for the present work, that list has been increased to close on 2,500 substances (taking animal, vegetable, and mineral all into account) which have medicinal virtues, rightly or wrongly, assigned to them by the people of India. A bare list of the names of such substances would be comparatively valueless. What might be of value would be a careful classification under Therapeutic sections. This will be found, however, in many works, such as *O'Shaughnessy's Bengal Pharmacopœia*, 113 to 187, *Stewart's Panjáb Plants*, App. 77-106, &c. In the official correspondence conducted in 1880 by the Home Department, Government of India, regarding a proposed new edition of the *Pharmacopœia of India*, much valuable information was brought together regarding the indigenous drugs that might be used for the imported ones, of each therapeutic class. (See also *Lisboa, Useful Plants of Bombay*, being Vol. XXV. of the *Bombay Gazetteer*, pp. 254-263.)

Medlar, Indian, see **Pyrus Pashia,** *Ham.*; ROSACEÆ; Vol. VI. 338

MEERSCHAUM.

Meerschaum, *Ball, in Man. Geology Ind.*, 445-446. 339

This well known substance is a hydrous magnesium silicate. It is chiefly obtained in Asia Minor, Greece, Moravia, Spain, &c., but **Mr. Ball** remarks that it would be in no wise a surprise if the magnesite deposits of Salem and the adjoining districts, or the magnesium clays and serpentines of the Nicobar Islands, were found to contain Meerschaum

340 ## MELALEUCA, *Linn.; Gen. Pl., I., 705.*

Melaleuca Leucadendron, *Linn.; Fl. Br. Ind., II., 465;* MYRTACEÆ.

The remarks which here follow are believed to be mainly a compilation of information regarding Cajupat (var. β), but the habitat is that of the type form of the species. It has not been found possible to isolate the economic facts given by authors under the two varieties respectively.

341 **Var. α—Leucadendron;** *Roxb., Fl. Ind., Ed. C.B.C., 591.*

Syn.—MYRTUS LEUCADENDRON, *Linn.*

References.—*Arbor alba Cuju Puti, in Rumph. Amb. Herb., II., 72, t. XVI.*

The *Flora of British India* remarks that this form is cultivated in India, and **Roxburgh** informs us that it was introduced into the Royal Botanic Gardens, Calcutta, in 1811. It is a much larger tree than the next form—the true Cajuput-oil tree. It would, however, seem desirable to ascertain if the oil, even though of inferior quality, could be obtained from this tree since the uses of the oil in arts might be thereby greatly extended.

Var. β—minor; *Roxb., Fl. Ind., Ed. C.B.C., 590.*

342 The Cajuput-oil of commerce is apparently prepared from this form.

Syn.—M. MINOR, *Sm.*; M. CAJUPUTI, *Roxb.*; M. LEUCADENDRON, *Lam.*; M. VIRIDIFLORA, *Gœrtn.*; M. SALIGNA, *Blume.*; M. CUMINGIANA, and LANCIFOLIA, *Turez.*

References.—*Arbor alba minor Cuju Puti,* RUMPHIUS. *Amb. Herb., II., 76, t. XVII.; Roxb., Fl. Ind., Ed. C.B.C., 590. Roxburgh says this plant has been grown in the Royal Botanic Gardens, Calcutta, since 1797-98.*

Collective References, &c., to both forms.

Vern.—*Kayaputi,* HIND.; *Cajuputte, ilachie,* (PATNA) BENG.; *Káyákuti,* BOMB.; *Cajupútá,* MAR.; *Kijápúté, kayápúte,* TAM.; *Cajuputi, káyú pútia,* MALAY.; *Tram,* COCHIN-CHINA.

References.—*Roxb., Fl. Ind., Ed. C.B.C., 590-592; Roxb., Trans. London Med. Bot. Soc., 1829; Voigt, Hort. Sub. Cal., 45; Kurz, For. Fl. Burm., I., 472; Pegu Rept.; LX Gamble, Man. Timb., 188; Mason, Burma and Its People. 491,744; Laureiro, Flora Cochin-China, II., 468; Pharm. Ind., 90; British Pharm. (1885), 283, 380; Flück. & Hanb., Pharmacog., 277, 278; U. S. Dispens., 15th Ed., 1003; Fleming, Med. Pl. & Drugs (Asiatic Reser. XI) 185; Ainslie, Mat. Ind., I., 259; O'Shaughnessy, Beng. Dispens., 337; Irvine, Mat. Med. Patna, 24; Sakharam Arjun, Cat. Bomb. Drugs, 56; Bent. & Trim., Med. Pl., 108; Dymock, Mat. Med. W. Ind., 2nd Ed., 331; Year Book Pharm., 1874, 632; 1879, 466; Med. Topog., Ajm., 132; Watts', Dict. Chemistry, Vol. I., 710-713; VII, 231; VIII., 370; Birdwood, Bomb. Prod., 36; Crawford History of the Indian Archip., I., 513; Gazetteers:—Burma, I., 131; Mysore and Coorg, I., 60; Indian Forester, VI., 124; XI, 274, 275, 277.*

Habitat.—An evergreen tree, often of large size, found in Tenasserim, Mergui, and Malacca; distributed to the Malay Islands and Australia.

HISTORY. OIL. 343 **History.**—The account given by **Flückiger & Hanbury** on this subject (and indeed not of the history only but of every feature of Cajuput) has been practically reproduced by all subsequent writers. **Ainslie** (of what may be called modern authors) appears to have been the first Indian writer who described the drug, as **Roxburgh** was the first botanist who drew attention to the peculiarities of the plant. Both these writers quote freely from **Rumphius**, the Governor of Amboyna, who studied the plants found in the Dutch East Indies, between 1627-1702. **Rumphius** gave a detailed

HISTORY.

account of the uses of this plant and the preparation of its aromatic oil. The wood, he says, is fragile and not of much use for building purposes. The fruit, seed, and leaves have a strong aromatic odour and hence they are used as an aromatic and stomachic tonic. The people of Java, however, prepared from them many special medicines called *dju-djambu* which were employed for so many diverse diseases that it was difficult to say what their exact action might be. Some of these potions were, however, specially serviceable for checking debility and putting the stomach to rights. Others were valuable in the treatment of convalescent women after childbirth, as they tend to contract the internal organs. Men praise, **Rumphius** continues, preparations of this drug as valuable in the treatment of cephalalgia, but of this property, he remarks, definite proof has not as yet been adduced. The Amboyans also use the leaves, which they macerate with the flowers in new oil, and afterwards impregnate it with the smoke of benzoin and other aromatics. Of this they make their *Minjac mony,* that is, the perfumed oil with which they anoint their heads. Certain Javanese and Malays also fill their pillows with these leaves for the sake of the pleasant odour, but this to our notion, adds **Rumphius**, is far too strong. The leaves collected on a warm day and dried in the open air are also placed in their clothes' chests to drive away various insects through their powerful odour. To increase this action the leaves are often rubbed between the hands which causes a more liberal discharge of their cardamom-like odour. It is said also that the pillows filled with these leaves drive vermin out of bedsteads. The fruits are collected and sold separately in sweet-smelling baskets. The fruit is, in fact, the part of the plant mainly used in the preparation of the medicinal potions, but these preparations are always made up with other ingredients.

The leaves, if collected on an exceptionally warm day and placed within sacks, even though they be quite dry, burn with such a vehemence as to become moist, almost as if they had been macerated in water. If, however, they are treated differently and macerated in water, so that they ferment during night, and if they be then distilled, an oil is extracted from them which is thin, pelucid, and volatile, but in such small quantity that even from two bags of these leaves scarcely three drachms of the oil are obtained. The odour of the oil is like that of the strongest cardamoms. Two drops of it in ale or wine excite violent perspiration; in fact, India does not possess a more powerful sudorific.

The above may be accepted as the substance of what **Rumphius** actually wrote. One author has given one sentence, another a second, and it has thus transpired that his meaning has been somewhat distorted. The passage regarding distillation is as nearly as possible a literal translation from the original, and it will be seen that it is left doubtful whether the natives distilled the oil or only the Dutch did so, during **Rumphius'** time. **Flückiger & Hanbury** suggest the latter, and in this view they may be correct, since the information regarding the distillation forms a separate paragraph after the undoubtedly native uses have been detailed. The learned authors of the Pharmacographia then give the history of the introduction of Cajuput into Europe. It appears, they say, "to have been first noticed by **J. M. Lochner**, of Nürnberg, a physician to the German Emperor. About the same time (1717), a ship's surgeon, returning from the East, sold a provision of the oil to the distinguished apothecary **Johann Heinrich Link**, at Leipzig, who published a notice of it and sold the supply. It began then to be quoted in the tariffs of other German apothecaries, although it was still reputed a very rare article in 1726." "In France and England, it was, however, scarcely known till the commencement of the present century, though it had a place in the Edinburgh Pharmacopœia of 1788. In the

MELALEUCA Leucadendron. **Cajuput Oil.**

London Price Current, we do not find it quoted earlier than 1813, when the price given is 3*s.* to 3*s.* 6*d.* per ounce, with a duty of 2*s.* 4½*d.* per ounce."

TRADE. 344 **Manufacture and Trade.**—Early mention (1792) is made of its preparation in the Island of Bouro, and **Bickmore**, an American traveller who spent three months in that island in 1865, states that Bouro then produced 8,000 bottles annually. **Flückiger & Hanbury**, in continuing the account of the preparation, add: "The Trade Returns of the Straits Settlements published at Singapore show that the largest quantity is shipped from Celebes, the great island lying west of Bouro." "The oil is imported from Singapore and Batavia, packed in glass beer or wine bottles. From official statements it appears that the imports into Singapore during 1871 were as under:—

From Java	445 gallons.
,, Manilla.	200 ,,
,, Celebes	3,895 ,,
,, other places	350 ,,
TOTAL .	4,890

Of this large quantity, the greater portion was re-shipped to Bombay, Calcutta, and Cochin-China."

MEDICINE. Oil. 345 **Medicinal Properties.**—"Cajuput OIL is very fluid, transparent, of a fine green colour, has a lively and penetrating odour analogous to that of camphor and cardamom and a warm pungent taste. It is very volatile and inflammable, burning without any residue. The sp. gr. varies from 0·914 to 0·9274. Its composition, according to **Blanchet & Sell**, is $C_{10}H_{16}H_2O$, and by repeated distillation over phosphoric oxide the hydrocarbon $C_{10}H_{16}$ called *Cajuputene*, can be obtained. The oil is, therefore, said to contain *Cajuputene hydrate* or *Cajuputol*. It boils at 175°C (347° Fh.)" "The green colour has been ascribed to a salt of copper, derived from the vessels in which the distillation is performed; and **Guibourt** obtained two grains and a half of oxide of copper from a pound of the commercial oil. But neither **Brande** nor **Gœrtner** could detect copper in specimens examined by them; and **M. Lesson**, who witnessed the process for preparing the oil at Bouro, attributes its colour to chlorophyll, or some analogous principle, and states that it is rendered colourless by rectification. **Guibourt**, moreover, obtained a green oil by distilling the leaves of a **Melaleuca** cultivated in Paris. A fair inference is that the oil of Cajuput is naturally green, but that, as found in commerce, it sometimes contains copper, either accidentally present, or added with a view of imitating or maintaining the fine colour of the oil (*U. S. Dispensatory*)." The copper may be removed by distillation with water or agitation with a solution of ferrocyanide of potassium. The colour is thus destroyed, but it may be restored by exposure to copper filings (*Mr. Edward Hirted, Pharm. Jour. & Trans.* (*3*), *II.*, *804*). The high price of Cajuput oil has led to its adulteration, oil of rosemary, or that of turpentine, impregnated with camphor and coloured with the resin of milfoil is said to be the most common adulterant. The quantity of copper in Cajuput is, however, too small to render the oil unfit for medicinal use (*Watts' Dict. Chem.*).

The oil is highly stimulant, producing a sense of heat with increased fulness and frequency of the pulse, exciting in some instances profuse perspiration. **Ainslie** says, "Kijapútí oil is hitherto but little known to the native practitioners of India; it is in use, however, amongst the European medical men of that country, who recommend it, when mixed with an equal quantity of some mild oil, as an excellent external application in chronic rheumatism. The Malays are in the habit of prescribing it internally, and I understand with great success, in what they call *pítambúdí* and *lúmpú* (epi-

MEDICINE.

lepsy and palsy). It is, no doubt, a highly diffusible stimulant, antispasmodic, and diaphoretic, and may be efficaciously given in dropsy, chronic rheumatism, palsy, hysteria, and flatulent colic; the dose from two to six or even seven drops, on a lump of sugar." Ainslie adds that it dissolves caoutchouc or India rubber, by which means a good varnish may be made. It is officinal in the Indian and British Pharmacopœias: in the former it is said to be used with advantage in depression of the vital powers. "In cholera it has been lauded but on insufficient grounds. It proves useful also in flatulent colic, painful spasmodic affections of the stomach, hysteria, &c. Externally it forms a valuable embrocation in rheumatic, neuralgic and other painful affections, in paralysis, &c."

Chemistry. 346

CHEMISRTY.—The chemical nature of this substance has already been dealt with, but the account given in the Pharmacographia (reproduced in *Dymock's Mat. Med., W. India*) may be consulted. In the volumes of *Watts' Dictionary of Chemistry* (*l.c.*), the substance is dealt with in detail and the properties of its compounds with chlorine, bromine, and iodine investigated. The oil is then stated (but apparently incorrectly) to be "prepared in India," the green colour being accounted for as due to "a resinous colouring matter dissolved in it in very small quantity." "The colour of the crude oil is also partly due to copper, the presence of which may be accounted for, either by the use of a copper head in the distilling apparatus of the Hindus, or by intentional adulteration, resorted to for preserving the green colour of the oil."

SPECIAL OPINIONS.—§ "Very useful application in chronic rheumatism" (*Surgeon-Major and Civil Surgeon G. Y. Hunter, Karáchi*). "Stimulant, carminative, useful in flatulence and colic; rubefacient externally; applied to cold extremities in collapse of cholera and fever" (*Assistant Surgeon S. C Bhattacharji, Chánda, Central Provinces*). "Have used it as a stimulant and rubefacient for local application only in chronic rheumatism" (*Assistant Surgeon Nehal Sing, Saháranpur.*) "Cajuput oil is a powerful restorative in cholera" (*Civil Surgeon G. C. Ross, Delhi, Panjáb*).

TIMBER. 347

Structure of the Wood.—Reddish-brown, hard (*Gamble*).

(*J. Murray.*)

MELANOCENCHRIS, *Nees; Gen. Pl., III., 1169.*

348

Melanocenchris Royleana, *Nees; Duthie, Fodder Grasses of N. India, 54;* GRAMINEÆ. See "**Food and Fodder for Cattle,**" (Vol. III., 424).

Melanogaster durissimus, *Cooke;* FUNGI; see Vol. III., 455; also **Truffle,** Vol. VI., Pt. II.

MELANORRHŒA, *Wall.; Gen. Pl., I., 421.* 349

A genus of trees, the juice of which forms a varnish, and which comprises four species, all natives of India or the Malay Archipelago.

Melanorrhœa glabra, *Wall.; Fl. Br. Ind., II., 25;* ANACARDIACEÆ. 350

Vern.—*Thit-sae-yaing, thitsi, thitse,* BURM.

References.—*Kurz, For. Fl. Burm., I., 317; Mason, Burma and Its People, 514, 774; Liotard, Dyes, App. IX.*

Habitat.—Found in the forests of Tenasserim and Mergui.

RESIN. 351

Resin.—This tree yields a similar exudation to that of the next mentioned species, but there is no record of its having been applied to any use in the arts.

352 **Melanorrhœa usitata,** *Wall.; Fl. Br. Ind., II., 25.*

THE BLACK VARNISH TREE OF BURMA.

Vern.—*Kheu,* MANIPUR; *Thitsi* or *thit-tse, thitsibin,* BURM.; *Súthan* TALEING; *Kiahong,* KAREN.

References.—*Kurz, For. Fl. Burm., I., 318; Gamble, Man. Timb., 110; Mason, Burma and Its People, 514, 774; Pharm. Ind., 60; Gums and Resinous Prod. (P. W. Dept. Rep.), 32, 38, 62; Liotard, Dyes, App. IX.; Cooke, Gums and Resins, 120; Alpin, Report Shan States, 1887-88; Gazetteer:—Burma, I., 126, 134; Agri.-Hort. Soc.:—Ind. (Trans), VI., 127, (Pro.) 95; VII., (Pro.) 23-24, 25, 41; Journals (old series), IV., 215; VII., 73; IX., Sel., 45; XI., 446; Indian Forester, I., 362; II., 172, 181; VIII., 400, 412, 419; XIV., 394; Spons' Encycl., 1692; Balfour, Cyclop. Ind., II., 920; Smith, Dict., 426.*

Habitat.—A large deciduous tree, frequent in the open forests (especially the *In* and hill *In* forest—see **Dipterocarpus,** Vol. III., 160-171), rare in the dry forests, from Prome, Pegu, and Martaban down to Tenasserim; also found in Ava and Manipur. It ascends to an altitude of 3,000 feet.

OLEO-RESIN. 353 **Oleo-resin.**—Every part of the tree abounds in a thick, viscid, greyish, terebinthinate fluid, which soon assumes a black colour on exposure to the air. This is the famous black varnish or *thitsi* of the Burmese, by whom it is very extensively employed not only in the arts but in medicine.

The tree was first reported on by **Dr. Wallich,** who gave a description of its habitat, method of growth, and of the oleo-resin which it yields, in his *Plantæ Asiaticæ Rariores.* He writes: "In the neighbourhood of Prome, a considerable quantity of VARNISH is extracted from the tree, but very little is obtained at Martaban, owing, as I was told, to the poverty of the soil, and partly also to the circumstance of there being none of the people in that part whose business it is to perform the process. This latter is very simple:—Short points of a thin sort of bamboo sharpened at one end like a writing pen, and shut up at the other, are inserted in a slanting direction into wounds made through the bark of the trunks and principal boughs, and left there for twenty-four to forty-eight hours, after which they are removed, and their contents, which rarely exceed a quarter of an ounce, emptied into a basket made of bamboo or rattan previously varnished over. As many as a hundred bamboos are sometimes seen sticking into a single trunk during the collecting season, which lasts as long as the tree is destitute of leaves, namely, from January until April; and they are renewed as long as the juice will flow. A good tree is reckoned to produce from $1\frac{1}{2}$ to 2, 3, and even 4 viss annually, a viss being equal to about $3\frac{1}{2}$℔ avoirdupois. In its pure state it is sold at Prome at the rate of one tical, or 2*s.* 6*d.*, per viss. At Martaban, where everything was dear when I was there, the drug was retailed at 2 Madras rupees per viss; it was of inferior quality and mixed with sesamum oil, an adulteration which is often practised.

Varnish. 354

"The extensive use to which this varnish is applied indicates that it must be very cheap. Almost every article of household furniture destined to contain either solid or liquid food is lacquered by means of it. At a village close to Pagam on the Irrawaddy, called Gnauní, where this manufacture is carried on very extensively, I endeavoured to obtain some information relating to the precise mode of lacquering; but I could learn nothing further regarding this than that the article to be varnished must first be prepared with a coating of pounded calcined bones; after which the varnish is laid on thinly, either in its pure state, or variously coloured by means of red or other pigments. I was told that the most essential, as well as difficult, part of the operation consists in the process of drying, which must be effected in a very slow and gradual manner, for which purpose the articles are placed in damp and cool subterraneous vaults, where

OLEO-RESIN.

they are kept for several months, until the varnish has become perfectly dry.

"Another object for which the drug is extensively employed is as a size or glue in the process of gilding; nothing more being required than to besmear the surface thinly with the varnish, and then immediately to apply the gold leaf. If it is considered how very extensively the art is practised by the Burma nation, it being among their most frequent acts of devotion and piety to contribute to the gilding of their numerous religious edifices and idols, it will be evident that a great quantity of the drug must be consumed for this purpose alone. Finally the beautiful *Páli* writing of the religious order of the Burmans on ivory, palm-leaves, or metal is entirely done with this varnish in its native and pure state." Little can be added to the above exhaustive account, but it may be noted that according to **Mason** quoting **Major Berdmore**, the varnish mixed with bone-ashes is also used as a paste for sticking glass on boxes and images.

The following more detailed account by **Sir D. Brandis** of the method of collection is also of interest:—

"The trees which have been tapped are at once known by triangular scars about 9 inches long and 5 inches broad, the apex pointing downwards.

"On some trees we counted 40—50 of these scars, and some of them at a height of 30 feet. To work the higher scars the Shans use a most ingenious ladder which is permanently attached to the trees. It consists of a long upright bamboo with holes cut through at intervals of 2—3 feet. Through each hole are passed two flat bamboo sticks driven with their pointed ends into the bark. These form the spokes of the ladder and are about 12 inches long. The scars or notches to extract the varnish are made with a peculiarly shaped chisel about 15 inches long, the handle is of iron, of one piece with the chisel and about 9 inches long, the lower end thicker, hollow, and closed with a bamboo plug. The chisel is wedge-shaped, about 6 inches long (the edge half an inch broad), and forms an obtuse angle with the handle.

"With this instrument two slanting slits meeting at an acute angle, are made upwards through the bark, and the triangular piece of bark between the two slits is thus slightly lifted up, but not removed. A short bamboo tube about 6 inches long, with a slanting mouth and a sharpened edge, is then horizontally driven into the bark below the point where the two slits meet, and the black varnish which exudes from the inner bark near its contact with the wood runs down into the bamboo tube, which is emptied at the end of ten days, when it ceases to flow. A second cut is then made so as to shorten the triangular piece of bark which had been separated from the wood when the first cut was made. A shorter triangular piece of bark remains, ending in an angle less acute than before.

"The bamboo tube is then moved a little higher, and the edges of the original cut are cut afresh. The varnish then runs out for another ten days, after which the scar is abandoned. The trees vary in yield exceedingly; a crooked tree with scanty foliage which we examined was said to yield a good outturn, while some of the largest trees were said to yield very little. We saw trees tapped which had a diameter of only 9 inches. Moungmyat informed us that one man could make and look after 1,200 scars; that he could do 200 in a day, so that the whole number occupied six days, which left four days for rest. They only work in those parts of the forest where the tree is abundant and trees fit to tap stand close together. The tree yields nothing while it is leafless in the hot season, and the best season for working is from July to October. One man collects 40—50 viss (146 to 182℔) in one season: at Tyemyouk the viss sells for 12 annas and at Rangoon for one rupee."

OLEO-RESIN. In Manipur the tree attains very large dimensions and forms extensive forests from the top of the Kabo Valley for many miles in a northerly and north-easterly direction towards the Chinese frontier. The natural varnish is used, as in Burma, for many purposes, among which may be mentioned that of painting river-crafts, vessels destined to contain liquids, and scabbards. It is said to be conveyed to Sylhet for sale by the merchants who come down annually with horses.

Physical Characters. 355 PHYSICAL CHARACTERS.—The varnish is thick at ordinary temperatures, and of a dull leaden grey colour, but wherever it comes in contact with the air it assumes, in a very short time, a shining black surface. Alcohol, spirits of turpentine and benzole, combine with and dissolve it, rendering it more fluid. It may also be diluted with gold size, which tends to improve its drying properties, and intensify its colour, whilst the solvents above enumerated have a tendency to turn it brown. The varnish is very commonly adulterated with gingelly oil. It has peculiarly acrid properties, and hence has to be handled, when in the fresh state, with great care, for it frequently produces violent erysipelatous swelling, accompanied by pain and fever. These effects are said to be more marked in Europeans than in Natives accustomed to collecting the substance.

Trade. 356 TRADE.—The varnish is little known or appreciated outside the area of its production. Attempts have been made to introduce it into European commerce, but since it has no special application and is so long in drying, it is stated to have no value in the European market.

MEDICINE. Oleo-resin. 357 **Medicine.**—Black varnish is extensively employed by the Burmans as an anthelmintic in cases of Ascaris lumbricoides (round worm), as a remedy for which it is said to possess considerable power. It is administered as an electuary, prepared with an equal proportion of honey, the mixture having been subjected for some hours to the action of heat. The dose is one, two, or three table-spoonfuls of the electuary, according to the age of the patient, and is followed in a few hours by a dose of castor-oil, which causes the expulsion of the worms in a lifeless state, thus shewing that the remedy exercises a specific effect on the entozoa. The extremely nauseous taste of the drug and the largeness of the dose required, are great objections to its employment. It appears probable, however, that its activity resides in a volatile oil, which, if procurable in a pure state, would be well worthy of an extended trial.

The erysipelatous swellings caused by the fresh juice, in certain constitutions, are said to be effectually removed by the local application of an infusion of teak-wood—**Tectona grandis,** *Linn.* (*Pharm. Ind.*).

TIMBER. 358 **Structure of the Wood.**—Dark red with yellowish streaks, turning very dark after long exposure, very hard, close and fine grained, weight from 54 to 62℔ per cubic foot. When green it sinks, but when dry it floats in water. It is employed for making tool handles, and anchor stocks, and is said in the *Gazetteer of Burma* to be preferred by charcoal burners to the wood of any other tree. It has lately been recommended for buildings, railway sleepers, gun-stocks, sheaves, block-pulleys, and other purposes for which a strong but not very heavy wood is required. **Mason** states that in some Christian villages in Burma the posts of the chapels are made exclusively of this wood. The utilization of the timber in Munipur is said to be to some extent interfered with by the dread which the natives possess of the irritant effects of the oleo-resin.

359 **MELASTOMA,** *Linn.; Gen. Pl., I., 746.*

Melastoma malabathricum, *Linn.; Fl. Br. Ind., II., 523; Wight, Ill., t. 95;* MELASTOMACEÆ.

THE 'INDIAN RHODODENDRON.'

Syn.—? M. OBVOLUTUM, *Jack;* TREMBLEYA RHINANTHERA, *Griff.*

Var.—adpressum, *Wall.*; M. ANOPLANTHUM, *Naud.*

Vern.—*Choulisi,* NEPAL; *Tungbram,* LEPCHA; *Shapti, tunka,* MICHI; *Myetpyai, myetpyé,* BURM.; *Katakalúwa, mahabowittya, bowitteya,* SING.

References.—*Roxb., Fl. Ind., Ed. C.B.C., 372; Kurz, For. Fl. Burm., I., 503; Gamble, Man. Timb., 199; Thwaites, En. Ceylon Pl., 106; Dalz. & Gibs., Bomb. Fl., 92; Mason, Burma and Its People, 428, 744; Lisboa, U. Pl. Bomb., 156, 245; Gazetteer, Bombay, XV., 72; Ind. Forester, IV., 241.*

Habitat.—A spreading shrub, found growing very abundantly throughout India, from the sea level up to an altitude of 6,000 feet, except towards the Indian Desert. It is not found out of India, *i.e.*, the authors of the *Flora of British India* have narrowed the description of the species to the Indian typical plant which is not found in Malaya, &c. The variety **adpressum** occurs from Mergui to Singapore and in Penang.

Dye.—The FRUIT yields a purple dye used for cotton-cloths (*Lisboa*). DYE. Fruit. 360

Food.—The ovoid, truncate FRUIT has an edible pulp, which is said to strongly resemble the blackberry of temperate regions in taste and flavour. FOOD. Fruit. 361

Domestic, &c.—Gamble states that "this is probably the *lutki* bush on which the silkworm **Attacus atlas** is often found, and fed on which it gives a very fine silk." DOMESTIC. Leaves. 362

MELIA, *Linn.; Gen. Pl., I., 332.* 363

A genus of trees which belongs to the Natural Order MELIACEÆ, and comprises five species; natives of India and the Malay Archipelago.

Melia Azadirachta, *Linn.; Fl. Br. Ind., I., 544; Wight, Ic., t. 17;*

THE NEEM, OR MARGOSA TREE. [MELIACEÆ

Syn.—M. PARVIFLORA, *Moon.*; M. INDICA, *Brandis*; AZADIRACHTA INDICA, *Adr.*

Vern.—*Nim, bál-nimb, ninb, nimb,* HIND.; *Nim, nimgachh,* BENG.; *Nim,* KOL.; *Nim,* SANTAL; *Agas,* PALAMOW; *Betain,* KUMAON; *Nim, mahá-nim, bukhain, drekh, bakam,* fruit=*darkonah,* PB.; *Nimuri,* SIND.; *Limbo,* C. P.; *Nim, bál-nimb, baká-yan,* BOMB.; *Limba, kadu khajur, nimbay, limbácha-jháda,* MAR.; *Limba, libado, limbado, limb, dánu-jhada, kohumba,* GUZ.; *Nim,* DEC.; *Vémbu, véppam, véppa-maram,* TAM.; *Vépa, yapa, yeppa, taruka, nim-bamu,* TEL.; *Bévina-mara-kadbevina-mara, heb-bavu,* KAN.; *Véppa, ariya-véppa,* MALAY.; *Thin, bawtamaka, tamá-bin, thamáká, kamáká,* BURM.; *Kohumba, nimbu-nimba-gahá,* SING.; *Nimba, arishta, nimba-vrikshaha,* SANS.; *Níb, ázád-darakhte-hindí,* PERS.

References.—*Roxb., Fl., Ind., Ed., C. B. C., Brandis, For. Fl., 67; Kurz, For. Fl. Burm., I., 212; Beddome, Fl. Sylv., t. 13 (14 by mistake); Gamble, Man. Timb., 69; Dalz. & Gibs., Bomb. Fl., 36; Stewart, Pb. Pl., 32; Burmann, Fl. Ind., 1; Pharm. Ind., 53; Flück. & Hanb., Pharmacog., 154; Fleming, Med. Pl. & Drugs (Asiatic Reser. XI.), 171; Ainslie, Mat. Ind., I., 453; O'Shaughnessy, Beng. Dispens., 244; Irvine, Mat. Med. Patna, 77; Honigberger, Thirty-five years in the East, II., 307; Moodeen Sheriff, upp. Pharm. Ind., 63; Mat. Med. S. Ind. (in mss.), 98; U. C. Dutt, Mat. Med. Hindus, 136; Murray, Pl. & Drugs, Sind., 84; Bent. & Trim. Med. Pl., 62; Dymock, Mat. Med. W. Ind., 2nd Ed., 168; Dymock, Warden and Hooper, Pharmacog. Ind., I., 322; Year-Book Pharm., 1873, 41; 1878, 290; Birdwood, Bomb. Prod., 15, 260, 279; Baden-Powell, Pb. Pr., 335, 557; Drury, U. Pl. Ind., 59; Atkinson, Him. Dist. 741; Useful Pl. Bomb. (Vol. XXV., Bomb. Gaz.), 40, 196, 215, 241, 257, 258, 279, 285, 399; Econ. Prod. N.-W. Prov., Pt. I. (Gums and Resins), 11; Stocks, Report on Sind; Gums and Resinou Prod. (P. W. Dept. Rept.), 1, 21, 42, 44, 49, 50; Cooke, Oils and Oilseeds, 57; Gums and Resins, 9; Indian Fibres and Fibrous Substances, Cross Bevan, King, & Watt, 55; Buchanan, Journey through Mysore and Canara, &c., Vol. I., 9, 250; Statistics Dinajpur, 154; Moore, Man., Trichinopoly, 76; Settlement Reports:—Central Provinces, Chindwára,*

MELIA Azadirachta.

The Neem Tree.

110; Nimár. 306; Beláspore, 77; Gazetteers:—Bombay, IV., 23; V., 23; VI., 13; VII., 39, 40, 41; XIII., 26; XV., 72; Panjáb, Rohtak, 14; Sialkot, 11; Karnál, 16; N.-W. P., III., 33; Mysore and Coorg, I., 52, 58; II., 7; Oudh, III., 71; Agri.-Horti. Soc., Ind.:—Transactions, VI., 241, VIII., 22; Journals (Old Series), IV., 208; VIII. (Sel.), 136, 178; IX., Sel., 295, 410, Sel., 47; XI. (Pro.), 24; XII., 348; XIII., 309, 350; (New Series) II., 234; VII., 146, 147; Agri.-Hort. Soc., Panjáb, Proc., 1857; Indian Forester, II., 173; III., 201; V., 497; VI., 125; VII., 264; VIII., 403; IX., 357; XII., 188, App., 1, 27; XIII, 69, 120, 339; XIV., 391.

Habitat.—A large tree of 40 to 50 feet in height, common, wild or more often cultivated, throughout the greater part of India and Burma.

GUM. **Gum.**—The BARK exudes a clean, bright amber-coloured gum, which
364 is collected in small tears and fragments. It is said to form a portion
Bark. of the commercial gum gattie and of East India gum. It is considerably
365 esteemed medicinally as a stimulant. In the *Pharmacographia Indica* it is described as not bitter, fully soluble in cold water, and unaffected by neutral acetate of lead. "It gives a curdy white precipitate with basic acetate, a reddish gelatinous precipitate with ferric chloride, is unaffected by borax, is slightly reduced by boiling with Fehling's solution which it turns a dull red colour. Iodine does not affect it, but it precipitates with oxalate of ammonia. It makes a weak mucilage, and is of little value."

DYE. **Dye.**—Hove in the account of his Tour in Bombay (1787) mentions the
Gum. *ním*, and states that the tree yields a bitterish GUM in great abundance,
366 "which I understand the silk-dyers use in every preparation of their
Oil. colours." In the *Gazetteer of Mysore and Coorg* it is stated that the OIL is employed in dyeing cotton cloths; a statement repeated by **Lisboa**, who
367 adds that it imparts a deep yellow colour to the fabric. The writer can
Bark find no other mention of the dye properties of either gum or oil. **Stocks**
368 states (*Report on Sind*) that the BARK is used to dye red.

Fibre.—The BARK yields a fibre which is of little economic value, but is
FIBRE. commonly employed in the local manufacture of rope. **Dr. Watt**, in the
Bark. Report on Indian Fibres exhibited at the Colonial and Indian Exhibition,
369 writes: "It would never pay, however, to extract this fibre for commercial purposes, since the trees take years to grow, and would be killed by a wholesale process of decortication."

OIL. **Oil.**—A fixed, acrid, bitter oil, deep yellow, and of a strong disagreeable
370 flavour, is extracted from the SEED by boiling or pressure. It is already
Seed. manufactured to a considerable extent and forms an article of export from
371 Madras chiefly to Ceylon. It is employed medicinally as an anthelmintic and antiseptic, and is also considerably used by the poorer classes for burning in lamps, but is said to smoke offensively.

Chemical CHEMISTRY OF THE OIL.—The oil and other products of the *ním*
Composition. have recently been very carefully analysed by **Surgeon-Major Warden**,
372 who published his results originally in the *Pharmaceutical Journal*, and has reproduced them in the *Pharmacographia Indica*. As these are of great interest and may serve to decide the commercial utility of this cheaply prepared and abundant oil, they may be here quoted in entirety:—

"Margosa or Ním oil extracted from the seeds had a specific gravity of ·9235 at 15·5°C.; at about 10°–7° C. it congealed without losing its transparency. After standing for about 36 hours the recently expressed oil deposited a white sediment, which, examined microscopically, was found to be amorphous. The colour reactions of margosa oil were not characteristic. With concentrated sulphuric acid a rich brown colour was yielded, and a strong garlic odour evolved. By **Massie's** test with nitric acid the oil became almost immediately of a reddish colour; after standing about one hour and thirty minutes the colour was pale yellow. The elaidin

OIL
Chemical Composition.

reaction conducted according to Poutet's directions yielded a solid firm yellowish product after eighteen hours, the temperature in the laboratory varying between 89° and 93° F. Exposed in a thin layer on a glass plate to a temperature of 100°C. for some days the oil did not dry or become tacky. The oil was easily soluble in ether, chloroform, carbon bisulphide, benzole, &c. Absolute alcohol, agitated with it was coloured greenish; on separating the alcohol, and evaporating off the spirit, an extract was obtained which consisted of oil, from which a small residue, whitish in colour, separated on standing. The alcoholic extract was very bitter, and possessed in a marked degree the peculiar odour of the oil The whitish residue deposited from the oil separated by alcohol, and examined microscopically, did not appear crystalline. Margosa oil after repeated agitation with alcohol was found to have lost its bitterness and almost wholly its alliaceous odour.

"A known weight of the oil was saponified with alcoholic potash, the alcohol completely evaporated off, and the soap dissolved in water. On agitating the aqueous solution of the soap with ether, 1·60 per cent. of ether extract was obtained of an orange-yellow colour and bitter. This extract, treated with 60 per cent. alcohol, left a small amount of white residue, which had the character of a wax. The aqueous solution of the soap, after separation of the ether, was heated for some time to remove dissolved ether, the solution was then mixed with dilute sulphuric acid in excess, and the insoluble separated from the soluble fat acids in the manner recommended by Allen. The soluble fatty acids amounted to 3·519 per cent., the insoluble to 89·128 per cent. The volatile acids consisted of butyric and a trace of valeric acid. During the distillation to separate the fluid from the volatile fatty acids, a small amount of a snow—white fatty acid passed over; this acid had a melting point of 43·6° C., which corresponds with the fusing point of lauric acid. A weighed portion of the insoluble fatty acids, from which the lauric acid had not been separated, was dissolved in alcohol, and titrated with normal standard soda, using phenolphthalein as an indicator, ·288 gram of the acids required 1 c.c. of caustic soda for neutralization. No attempt at separating the fixed fatty acids was made; they probably consisted of a mixture of stearic and oleic acids, with a small amount of lauric acid.

"Examined by Reichert's distillation process, 2·5 grams of the oil gave a distillate which after separation of the lauric acid, which had distilled over, required 4·6 c.c. of decinormal soda for neutralization, phenolphthalein being used as an indicator.

"The saponification equivalent of the oil was determined by Koettstorfer's method, and was equal to 284, the percentage of caustic potash required to saponify the oil being 10·72.

"A preliminary examination of the oil having indicated the presence of sulphur, a quantitative estimation of the amount present was made and found equal to ·427 per cent The oil after repeated agitation with alcohol was found to contain only ·109 per cent. of sulphur.

"The extract obtained by agitating the oil with absolute alcohol has already been referred to; it was examined in the following manner:—The oily extract was treated with 60 per cent spirit, allowed to stand, and the clear yellow alcoholic solution decanted from the insoluble oil; the alcoholic solution thus obtained was evaporated to dryness, mixed with ammonia, and agitated with ether. The ether solution was marked *A*. The aqueous solution, after separation of the ether, was mixed with dilute hydrochloric acid, and again agitated with ether. The ether separated of a yellow colour, and below it some flocks of a dirty yellow hue, which refused to dissolve after prolonged agitation. The ether solution was marked *B*.

OIL. Chemical Composition.

From the aqueous solution the insoluble flocks were separated by filtration and marked *C.* The filtrate was not further examined.

"*Examination of ether solution A.*—The solution was agitated with dilute hydrochloric acid, to remove any principles of an alkaloidal nature. The ether was then separated and evaporated; the resulting extract was pale amber in colour, viscid at first, very bitter, and had a marked odour of the oil. It contained sulphur. It was easily soluble in 60 per cent. alcohol, ether, chloroform, &c., but insoluble in acids, or in caustic alkaline solutions. It had the properties of a neutral resin.

"The hydrochloric acid solution was of a yellow colour; it was mixed with ammonia, which occasioned a white precipitate, and agitated with ether. The etherial solution on evaporation left a yellow residue, not readily soluble in dilute acids. The dilute sulphuric acid solution was bitter, and yielded a precipitate with alkaline carbonates and hydrates, phosphomolybdic, and picric acids, potassio-mercuric iodide, chloride of gold and perchloride of platinum. This principle had therefore the properties of an alkaloid.

"*Ether solution B.*—On evaporating the ether solution *B*, a dark reddish bitter extract was obtained, soluble in alkaline solutions, and re-precipitated in yellowish flocks by dilute acids. It had the properties of an acid resin.

"*Precipitate C.*—The precipitate was well washed, and dissolved in alcohol; on evaporation a brittle darkish residue was obtained, soluble in alkaline solutions, re-precipitated in yellowish flocks by acids, soluble with very great difficulty in ether, easily soluble in chloroform. This principle thus also had the properties of an acid resin.

"In addition to the principles above described as being present in the oil, an examination of the cake left after expression of the oil, indicated the presence of another neutral principle, insoluble in ether or alkaline solutions, but dissolving in chloroform (*Pharm. Journ.*, *1888*).

"According to **Branet** the seeds contain from 40 to 45 per cent. of oil.

Oil-cake. 373

"Margosa cake is used as a manure in planting districts in Southern India. Two samples had the following composition:—

	1	2
Moisture	6·08	9·93
Organic matter	84·50	83·15
Ash	9·42	6·92
	100·00	100·00
Nitrogen	5·07	5·41
Phosphoric anhydride	1·40	1·33

"The powdered cake, like linseed meal, makes a very useful luting in chemical and physical laboratories, and is not liable to the attack of insects (*Pharmacog. Ind.*)."

MEDICINE.

Medicine.—Almost every product of this invaluable tree is largely employed medicinally in India. The parts used and their physiological actions have been arranged by **Moodeen Sheriff** as follows:—

Root-Bark. 374 Bark. 375 Fruit. 376

The ROOT-BARK, BARK, and YOUNG FRUIT—tonic and antiperiodic.

Oil. 377 Seeds. 378 Leaves. 379

The OIL, SEEDS, and LEAVES—local stimulant, insecticide, and antiseptic.

Flowers. 380

The FLOWERS—stimulant-tonic and stomachic.

Gum. 381

The GUM—demulcent-tonic.

Toddy. 382

The TODDY—refrigerant, nutrient, and alterative-tonic.

The bark, leaves, and fruit have been used in Hindu medicine from a very remote period, and are indeed mentioned in the earliest Sanskrit medical writings, *viz.*, those of **Susruta**. The very names of the tree seem to indicate a remote knowledge of its medicinal properties, *nimba*="the sprinkler,"

MEDICINE.

arishta=" relieving sickness," *pichumarda*=leprosy destroying." **U. C. Dutt**, in his account of the Sanskrit opinions of the plant, writes:—" The bark is regarded as bitter, tonic, astringent, and useful in fever, thirst, nausea, vomiting, and skin diseases. The bitter leaves are used as a pot-herb, being made into soup and curry with other vegetables. The slightly aromatic and bitter taste which they impart to curries thus prepared is much relished by some. The leaves are, moreover, an old and popular remedy for skin diseases. The fruits are described as purgative and emollient, and useful in intestinal worms, urinary diseases, piles, &c. The oil obtained from the seeds is employed in skin diseases and ulcers. The bark is used in fever in combination with other medicines." "The fresh juice of the leaves is given with salt in cases of intestinal worms, and with honey in skin diseases and jaundice. The juice of *ním* leaves and of emblic myrabolans, quarter of a tola each, are recommended to be given with the addition of clarified butter in prurigo, boils, and urticaria.

" As an external application to ulcers and skin diseases *ním* leaves are used in a variety of forms such as poultice, wash, ointment, and liniment. A poultice made of equal parts of *ním* leaves and sesamum seeds is recommended by **Chakradatta** for unhealthy ulcerations." As is customary in Sanskrit medicine, the *ním*, leaves, bark, &c., are seldom prescribed alone, but enter into the composition of numerous complex preparations,—for an account of which the reader is referred to **Dutt's** *Materia Medica of the Hindus*.

This useful tree with its multitudinous valuable properties naturally attracted the attention of the Muhammadans on their arrival in India, and was called by them *Azaddaracht-i-hindi* from the resemblance which it bore to their own *Azedarach*, the Persian lilac. Their knowledge of the medicinal properties of the tree having been derived from the Hindus, they naturally use its various products in the same way, and consider them cold and dry.

The above notice of the properties ascribed to the *ním* by ancient Sanskrit writers might almost exactly apply to the virtues which it is supposed to possess at the present day,—virtues many of which have been strongly confirmed by European practitioners and writers. The value of the bark in the treatment of periodic fevers is noticed by **Fra Bartholemo, Sonnerat, Garcia de Orta, Christoval Acosta**, and other old writers, but it was first prominently brought forward in 1803 by **Dr. D. White** of Bombay. Later **Dr. W. R. Cornish, Dr. Wyndour**, and others, carefully examined and experimentally tested it, and the result, as expressed by the former, is to the effect that margosa bark is nearly as effective in the treatment of intermittent fever as cinchona or arsenic. **Dr. Forbes** (*Madras Med. Reports, 1855*) arrived at a similar conclusion, a conclusion which has generally been corroborated by later investigators.

The following are the virtues ascribed to the various parts of the tree and the diseases for which they have been recommended by modern writers on *Indian Materia Medica* and *Therapeutics;* notably by **Moodeen Sheriff** in his forthcoming *Materia Medica of Madras.*

The bark, root-bark, and young fruit are useful in slight cases of intermittent fever and general debility. The root-bark is more active and speedy in its action than the bark and young fruit. The oil has proved a useful local stimulating application in some forms of skin disease, ulcers, rheumatism, sprains, &c., and is antiseptic It is also a useful adjunct to *chaulmugra* oil (see **Gynocardia odorata**) in cases of leprosy. Its antiseptic property might be taken advantage of for the manufacture of a medicated soap, since the oil readily saponifies. This soap might be very serviceable for the purpose of washing sores, &c., and for the general uses to

M. 382

MEDICINE.

which carbolic soap is now put. An interesting use of the oil is mentioned by **Buchanan Hamilton**, who states that in Madras about an ounce is given to every woman immediately after she is delivered of a child. The dry seeds possess almost the same properties as the oil when bruised and mixed with water or some other liquid, but do not make a cleanly preparation. A strong decoction of the fresh leaves is a slight antiseptic and may be used instead of a weak solution of carbolic acid. A hot infusion of the leaves is much used for fomenting swollen glands, bruises, and sprains, and appears to be anodyne. The flowers are useful in some cases of atonic dyspepsia and general debility. The gum is, from its medicinal properties, a better auxiliary to other remedies than Gum arabic and Feronia gum, in catarrhal and other affections accompanied by great weakness. The toddy or fermented sap of the tree appears to be of great service in some chronic and long standing cases of leprosy and other skin diseases, consumption, atonic dyspepsia, and general debility. The reader is referred to the numerous "Special opinions" quoted below for a more exhaustive account of the opinions of practitioners in India.

Moodeen Sheriff recommends the following preparations:—Of the root-bark, bark, and young fruit, a decoction, tincture, and powder; of the leaves, a decoction and paste or poultice; of the kernels, a solution or emulsion with water; the oil, alone, or with *chaulmugra;* of the flowers, an infusion; and the gum and toddy, as a mucilage and alone.

He suggests the following European drugs from which they might be efficiently substituted:—The root-bark, bark, and young fruit, for cinchona bark and gentian; the oil, nuts, and leaves, for carbolic acid; the flowers, for elder flowers, and coriander, aniseed and dill oils; the gum, for acacia and feronia gums; and the toddy for elm bark, diluted phosphoric acid, Jamaica sarsaparilla, and *chaulmugra* and cod-liver oils.

The fermented sap or *ním* toddy has specially powerful properties attributed to it. The sap is either yielded spontaneously, or is extracted artificially. In the former case, a clear and colourless liquid flows in a thin stream or continuous drops from two, three, or more parts of the plant, and continues to do so from three to four, six, or even seven weeks. Regarding sap obtained artificially. **Moodeen Sheriff** writes: "The *ním* trees, which yield the sap artificially, seem to be very rare, for I have heard only of three or four such plants. All these are said to have been pretty young and large, and were found near water or on the banks of nullahs or water-courses which were constantly wet. The sap was extracted in the following manner:—A moderate-sized and fresh-looking root being exposed by removing the earth, it was either cut through, or only to half of its circumference, from below, and then a vessel was placed beneath to receive the liquor. which began to dribble or flow in a very small and thin stream. The sap thus collected is supposed to be identical with that produced by the tree spontaneously, but is comparatively very small in quantity, amounting generally to only from 2 to 6 bottles in 24 hours. I think if many of the margosa plants growing near water be tried in the manner just explained, a much larger percentage of them will be found capable of yielding the liquor than is generally supposed." The same writer gives an account of one famous tree in Mylapore near Madras, which produced sap every third or fourth year for four occasions, after which it died. He writes: "On each occasion before the sap began to flow there was always, for three or four days, a distinct and peculiar rushing or pumping noise of a liquid within the trunk, which did not entirely cease till the discharge actually commenced from three or four parts of the plant." When this phenomenon occurred. the people of the neighbourhood flocked to the tree and bought the drug which they held in high esteem.

MEDICINE.

The *ním* tree is generally supposed by its presence to materially improve the health of a neighbourhood. Believed to be a prophyllactic against malarial fever, and even against cholera, it is frequently planted near buildings and villages. Even Europeans believe in this property to some considerable extent, especially in the North-West Provinces and Oudh, and villages surrounded with *ním* trees are frequently cited as proverbially free from fever, when neighbouring villages suffer severely. It is extremely doubtful, however, whether this tree exercises a beneficial effect to a greater extent than any other. A somewhat similar effect is supposed by the natives to be produced on syphilis, the air waved with a *ním* branch being considered a cure for that disease.

A plant with so many reputed properties has naturally been much used by European practitioners in India, and has obtained a place in the *Indian Pharmacopœia.* The officinal preparations are the powdered bark, the fresh leaves, a decoction and tincture of the former, and a poultice of the latter. The bark is said to be astringent, tonic and antiperiodic, the leaves to be a stimulant application to "indolent and ill-conditioned ulcers."

Chemistry. 383

CHEMISTRY OF THE BARK, LEAVES, &c.—The chemistry of the gum and oil has been already discussed. That of the bark and leaves is of particular interest from a medicinal point of view, and is described as follows by **Flückiger & Hanbury.**

"Margosa bark was chemically examined in India by **Cornish** (1856), who announced it as a source of a bitter alkaloid to which he gave the name of *margosine*, but which he obtained only in minute quantity as a double salt of margosine and soda, in long white needless. The small sample of bark at our disposal only enables us to add that an infusion produced with perchloride of iron a blackish precipitate and that an infusion is not altered by tannic acid or iodo-hydrargyrate of potassium. If the inner layers of the bark are alone exhausted with water, the liquid affords an abundant precipitate with tannic acid; but if the entire bark is boiled in water, the tannic matter which it contains will form an insoluble compound with the bitter principle, and prevent the latter being dissolved. It is thus evident that to isolate the bitter matter of the bark, it would be advisable to work on the liber or inner layers alone, which might readily be done as they separate easily."

According to the more recent researches of **Broughton** published in the *Madras Monthly Jour. of Med. Science* and quoted in the *Pharm. Jour., 1873*, and the *Year-Book of Pharmacy, 1873, p. 41*, the bitter principle is due to a resin, which it is very difficult to obtain in a state of purity. **Broughton** succeeded in obtaining a nitro-compound, which yielded a silver salt, not however crystalline, from which he ascribes to the resin the formula $C_{36}H_{50}O_{11}$, that of the nitro-compound being $C_{36}H_{46}(NO)_4O_{11}$. The resin is not, therefore, an alkaloid, since it contains no nitrogen. If required for medicinal purposes, the most suitable and convenient mode of administration would be an alcoholic solution of the resin.

The leaves also contain a bitter principle, more readily soluble in water than the resin above described, of which it is a hydrate. This substance also occurs in the bark, and closely resembles the resin in properties. The leaves contain no peculiar alkaloid, and the powerful smell of the tree was found not to be due to the presence of a sulphuretted oil as had been surmised.

SPECIAL OPINIONS.—§ "*Ním* oil is a valuable remedy in Veterinary Surgery for foul sores. It is stimulating and healing" (*Surgeon-General W. R. Cornish, F.R.C.S., C.I.E., Madras*). "The leaves made into a pulp may be applied externally over the mammæ as a lactifuge" (*Surgeon W. F.*

MELIA Azadirachta.

Medicinal Properties of

MEDICINE.

Thomas, 33rd M.N.I., Mangalore). "I have extracted the bitter principles from the bark and have found them a very efficient febrifuge. A strong decoction of the bark used every hour in remittent fever has had the desired effect when other febrifuge remedies had failed. In ulcers and skin diseases a poultice made of the leaves acts, I think, as an antiseptic, not as a topical stimulant" (*Surgeon K. D. Ghose, M.D., M.R.C.S., Khoolna*). "I frequently use the infusion of the bark as a tonic and antiperiodic with the best results" (*Honorary Surgeon E. A. Morris, Tranquibar*). "Used here as a tonic in convalescence from fevers. For this purpose a very cheap and useful mixture in dispensary practice consists of Quinetum grs. v., Nit. Hyd. dil. m. x., and infusion of Ním bark ℥i three times a day" (*Surgeon-Major L. C. Nanney, Trichinopoly*). "The leaves applied to the breasts arrest the secretion of milk" (*Surgeon-Major J. North, Bangalore*). "*Margosa* bark is used as tonic vermifuge (for Ascaris vermicularis)" (*Surgeon-Major H. D. Cook, Calicut, Malabar*). "The powdered bark is frequently used as an antiperiodic in dispensary practice, but is much inferior to cinchona and its preparations" (*Surgeon G. Price, Shahabad*). "The oil of the seeds is useful for the destruction of lice, and as an application in urticaria and eczema" (*Narain Misser, Kothe Bazar Dispenisary, Hoshangabad, Central Provinces*). "The bark in the form of decoction is a fairly efficient antiperiodic in mild agues. The leaves used as a poultice form a very useful application for foul ulcers often exciting a healthy action when other remedies have failed. The seeds are prized by the natives who express an oil from them. The latter is used for lighting purposes, and as an application in skin diseases" (*Surgeon S. H. Browne, M.D., Hoshangabad, Central Provinces*). "As a febrifuge, tonic and alterative, the decoction of the bark is efficacious in fevers. The leaves are applied to ulcers and in a variety of skin diseases. The fresh juice is used as an alterative in leprosy and skin diseases, but I found it of no use in a case of leprosy after a prolonged trial. A popular belief exists that a leper can be cured if he can live exposed under a *ním* tree for 12 years. The slender twigs are largely used as tooth-brushes (*datan*), the continued use of which is said to keep the system free from all complaints and certainly keeps the mouth and breath clean and sweet" (*Assistant Surgeon S. C. Bhuttacharji, Chanda, Central Provinces*). "An ointment made from the leaves fried in *ghí* and subsequently mixed with wax makes a good stimulating application for indolent ulcers, the infusion of the bark is useful as an antiperiodic and tonic, the dose of the latter from 1 to 2 ounces" (*Surgeon E. S. Brander, I.M.D., Rungpore*). "The oil of the seeds in doses of 30 minims to ℥i with *pán* is useful in asthma. The ashes of the bark are used as an application in eczema when there is much discharge" *Hospital Assistant Lal Mahomed, Hoshangabad, Central Provinces*). "The oil is used externally as a stimulant in convulsions and collapse from fevers and cholera" (*Honorary Surgeon T. R. Moodelliar, Chingleput, Madras Presidency*). "Astringent, tonic and febrifuge in form of decoction prepared from the bark thus:—Bruised *ním* bark 2 oz., water 1 pint, boil for half an hour: strain and again boil, this time down to 8 ounces. Soothing poultices and fomentations are made from the leaves" (*Civil Surgeon C. M. Russell, Sarun*). "A decoction of *ním* bark is useful as a tonic and antiperiodic in the milder forms of malarious fever. It is also efficacious in the treatment of boils and would seem to possess alterative properties" (*Surgeon R. D. Murray, M B., Burdwan*). "The oil of the seed is chiefly given internally for checking convulsions, fits, &c.; externally it is used as turpentine" (*V. Ummegudien, Mettapollium, Madras*). "In confluent smallpox I have seen *ním* oil act as a charm when applied over the whole

MEDICINE.

body; it decidedly promotes recuperative efforts of the skin and thus exerts a favourable influence in the course of this disease. The action is greatly superior to gingelly or cocoanut oil and carbolic acid. It forms a valuable remedy for fleas in dogs and for obstinate mange in the same animals. As an antiseptic stimulant applied to wounds, it deserves investigation" (*Surgeon W. G. King, M.B., Madras*). "A decoction of the leaves is used as a wash for unhealthy ulcers. The green leaves are considered antibilious and are used by the natives as an adjunct to some curries" (*Surgeon A. C. Mukerji, Noakhally*). "The bark is a good substitute for quinine in the chronic stage of fever when the latter drug disagrees with a delicate stomach. The oil is also used for scabies and superficial ulcers" (*Civil Medical Officer W. Forsyth, F.R.C.S., Edin., Dinájpore*). "Decoction of the bark given in cases of chronic fever complicated with diarrhœal Leaves most useful for fomentations and poultices" (*Civil Surgeon S. M. Shircore, Moorshedabad*). "The decoction of the bark is a valuable febrifuge and antiperiodic. The sprouts with black pepper form a good antiperiodic in chronic intermittent fever" (*Assistant Surgeon Nanda Lall Ghose, Bankipore*). "The leaves are used as a bitter tonic, also in the form of paste or poultice as a remedy for unhealthy ulcers and sores, and in the form of decoction as an antiseptic lotion. The bark is a substitute for cinchona bark, and a bitter principle prepared from it is used by native physicians as a substitute for quinine. The oil expressed from the seeds is a good application in old unhealthy sores and in leprosy. The tender leaves taken internally are used as a cure for leprosy" (*Civil Surgeon J. H. Thornton, B.A., M.B., Monghyr*). "I have used the decoction of the bark for its tonic and antiperiodic properties in chronic cases with good results. A decoction of the leaves has been used as a lotion for wounds and ulcers with some success. The young leaves are often used cooked as an article of food and fried with brinjal to which they impart a mild bitter taste. Also made into a curry with other vegetables. They are believed to have alterative properties" (*Civil Surgeon D. Basu, Faridpore, Bengal*). "A decoction of the bark is useful in fever and during convalescence; the leaves in fomentations for sprains and glandular swelling; the oil in rheumatism. All are highly antiseptic" (*Civil Surgeon G. C. Ross, Delhi, Panjáb*). "The leaves—fried, powdered, mixed with ghi, and made into an ointment—form a very efficacious application in sloughing sores and sinuses" (*Assistant Surgeon T. N. Ghose, Meerut*). "A milky fluid which runs from some of the old trees is considered an alterative tonic" (*Civil Surgeon R. Gray, Lahore*). "In 1872 (at Bhuj) I treated a great many cases of ague among sepoys with decoction of *ním* bark: they recovered and returned to duty; but the proportion of readmissions among them was very discouraging" (*Surgeon-Major H. DeTatham, M.D., M.R.C.P., Lond., Ahmednagar*). "The oil of the seeds is a good application in cases of ulcer and excoriation of the scalp—the result of want of cleanliness" (*Civil Surgeon J Anderson, M.B., Bijnor, N-W. Provinces*). "Extensively used in the form of poultices to abscesses, carbuncles, and boils. Taken by natives as an anti-scorbutic" (*Civil Surgeon J. McConaghey, M.D., Shajehanpore*). "I have used the bark as an antiperiodic, but not with satisfactory results. Poultices of the leaves seem to suit some indolent ulcers" (*Surgeon G. G. Ward, 5th N. L. I., Mhow*). "Native travellers, whenever possible, sleep under the *ním* at night-time, as the tree is said to ward off fever. The leaf ash mixed with ghi is useful as an external application to psoriasis" (*J. Parker, M.D., Dy. Sanitary Commissioner, Poona*). "(1) Decoction of *Ním* is a valuable tonic and alterative, and is very useful in chronic fevers, during convalescence from febrile complaints, and in cutaneous eruption. (2) As

MEDICINE. a wash for unhealthy sores, it acts as an alterative and antiseptic. (3) A mixture of the bruised leaves with turmeric is rubbed over the body during the desquamative stages of eruptive diseases, such as small-pox. It hastens the separations of the crusts, probably acting as an antiseptic, and promoting the healing of sores. (4) A poultice of the leaves is very useful to hasten suppuration or to disperse inflammation. (5) The leaves are occasionally used with brinjal or other vegetables, and act as an anthelmintic. (6) The expressed oil of the seed I tried in convict lepers for four years in the Bankura Jail. It was used in the form of a bath. In almost all the cases great relief was obtained. The results were as follows :—(*a*) In anæsthetic cases, great improvement followed in sensation and the eruption diminished. In some cases the improvement was so great that they could not be distinguished from healthy men. With hard work or any other debilitating cause, there was, however, a return of the original symptoms. (*b*) In tubercular cases, both thickening and ulceration became less, and in some cases entirely disappeared. (*c*) In the mixed form the effects were no less marked. I tried both this and *Garjan* oil, and I consider *ním* oil the better of the two. The results were noted in my annual medical reports of the Bankura Jail for the years 1879 to 1881" (*Surgeon R. L. Dutt, M D., Pubna*).

FOOD & FODDER. Leaves. 384 **Food and Fodder.**—The LEAVES are cooked with other vegetables in the form of a curry, or are simply parched and eaten. They impart a bitter taste to the food, but this seems to be liked by Natives. They are also used for cattle fodder. **Lisboa** states that the small sweet pulp of
Fruit. 385 the FRUIT is eaten in the Bombay Presidency, especially in times of scarcity.
Toddy. 386 The TODDY obtained from the tree has already been described

TIMBER. 387 **Structure of the Wood.**—Sapwood grey, heart-wood red, very hard, beautifully mottled, weight from 45 to 53℔ per cubic foot, transverse strength of bar 2'×1"×1" from 550 to 720℔ (*Gamble*). It is strong, clean-grained, and resists the attacks of worms. It is much used for the construction of carts, for making agricultural implements, and, in South India, for furniture. Owing to its sanctity and durability, it is largely employed by the Hindus to make idols. A contributor, forwarding specimens to the Agri.-Horticultural Society of India in 1861, wrote :—"I can almost confidently state that it resists the action of atmosphere infinitely better than '*toon*' or '*sísú*,' whilst in point of value the *ním* can be procured at one-third less the rate than the latter" He recommended it for carpentry work, and thought it admirably suited for door panels, rails, sash frames, and furniture. It has also been highly recommended for making trunks and chests, which withstand the attacks of white-ants, and are said also to render the contents proof against the ravages of other insects.

DOMESTIC. 388 **Domestic, Sacred, and Agricultural Uses.**—As already stated. the
Twigs. 389 TWIGS are largely used as tooth-cleaners, and the
Oil. 390 OIL is employed for burning. The
Seeds. 391 SEEDS and the oil obtained from them are used as applications to the hair by the women of Sind, both on account of their odour, and also to kill vermin. (*Conf.* with **Detergents,** Vol. III., 84-92.) The
Leaves. 392 LEAVES are largely used to protect clothes, books, papers, &c., from the ravages of insects, but are inferior to camphor for this purpose, and require to be frequently renewed. The *ním* is extensively planted as an avenue tree for which it is excellently adapted. The leaves and twigs are used for manure, and the
Oil-cake. 393 OIL-CAKE, as already stated, is considerably employed for the same purpose in the planting districts of Southern India.

The tree is held sacred by the Hindus and takes part in many of their ceremonies. It is believed that when nectar was being taken to heaven from the world below for the use of the gods, a few drops fell on the *ním*. Hence, on New Year's Day of *Shahalivan shak*, Hindus eat its leaves in

DOMESTIC.

the hope that they will thereby acquire freedom from disease (*Lisboa*). **Buchanan**, in his *Journey through Mysore*, relates that, "once in two or three years the Coramás of a village make a collection among themselves, and purchase a brass pot, in which they put five branches of **Melia Azadirachta** and a cocoanut. This is covered with flowers and sprinkled with sandal-wood water. It is kept in a small temporary shed for three days, during which time the people feast and drink, sacrificing lambs and fowls to Marima, the daughter of Siva; at the end of the three days they throw the pot into the water." In a passage in the *Statistics of Dinajpur*, the same author states that the leaves are much used by holy men to help them to resist the allurements of beauty. The authors of the *Pharmacographia Indica*, commenting on the former passage, write:—"This practice is known in other parts of India as *Ghatásthápan*, and is considered to avert ill-luck and disease." Amongst certain castes the leaves of the *ním* are placed in the mouth as an emblem of grief on returning from funerals. **Hove**, in his *Tour in Bombay*, writes, "The Gentoors worship this tree, and their barren women invoke and perform the same ceremonies round it every morning as they usually do in the other Pergunnahs about the **Ficus religiosa**."

Melia Azedarach, *Linn.; Fl. Br. Ind., I., 544; Wight, Ic., t. 160.*

THE PERSIAN LILAC; BEAD TREE.

Syn.—MELIA SEMPERVIRENS, *Sw.*; M. BUKAYUN, *Royle*.

Vern.—*Drek, bakáin, bakáyan, betain, deikna, bakarja, maha-ninb,* HIND.; *Ghoránim, mahá-ním,* BENG.; *Gara nim,* KOL.; *Thamaga,* ASSAM; *Bakainú,* NEPAL; *Bukain,* N.-W. P.; *Chein, kachen, bakáin, dhék, drek, jek,* seed=*habbulbán,* PB.; *Fakyána,* PUSHTU; *Bakayun. drek,* SIND; *Maha limbo, malla nim, muhli,* C. P.; *Gouri-nim, gouli-nim,* DEC.; *Nimb, maha-limbo, drek, bakayan, wilayati nim,* BOMB.; *Limbara bakána-nimb, wilayati-nimb,* MAR.; *Bakan limbodo,* GUZ.; *Malai, vembu, malai-veppam,* TAM.; *Turuka vépa, makánim, konda-vepa,* TEL.; *Bévu, chik bévu, kadbevina mara, bettada-bevina,* KAN; *Mullay vaempú,* MALAY.; *Ta-ma-ka, ka-ma ka,* BURM.; *Maha-nimba, lunu-midella,* SING.; *Mahánimba, himadruma, parvata-nimba-vrikshaha,* SANS.; *Hab-ul-bán,* ARAB.

References.—*Boiss., Fl. Orient., I., 954; Roxb., Fl. Ind., Ed. C.B.C., 369; Brandis, For. Fl., 68; Kurz, For. Fl. Burm., I., 212; Beddome, Fl. Sylv., t. 14; Gamble, Man. Timb., 70; Dalz. & Gibs., Bomb. Fl., Supp., 15; Stewart, Pb. Pl., 33; Mason, Burma and Its People, 411, 7-8; Sir W. Elliot, Fl. Andh., 184; Pharm. Ind., 55; Ainslie, Mat. Ind., I., 453; O'Shaughnessy, Beng. Dispens., 16; Moodeen Sheriff, Supp. Pharm. Ind., 172; Mat. Med. S. Ind. (in MSS.), 97; U. C. Dutt, Mat. Med. Hindus, 308; Sakharam Arjun, Cat. Bomb. Drugs, 25; Murray, Pl. & Drugs, Sind., 85; Bidie, Cat. Raw Pr., Paris Exh., 108; Dymock, Mat. Med. W. Ind., 2nd Ed., 171; Dymock, Warden and Hooper, Pharmacog. Ind., Vol. I., 330; Year-Book Pharm., 1875, 375; 1880, 204; Birdwood, Bomb. Prod., 279; Baden-Powell, Pb. Pr., 585; Atkinson, Him. Dist. (Vol. X., N.-W. P. Gaz.), 307, 741; Useful Pl. Bomb. (Vol. XXV., Bomb. Gaz.), 41, 286; Gums and Resinus Prod.; P. W. Dept. Rept. 4, 14; Cooke, Oils and Oilseeds, 57; Gums and Resins, 19, 20; Watson, Rept., 4; Settlement Report:—Panjáb, Jhang, 22; Hazára, 10; Peshâwar, 13; Gujrát, 3; Delhi, cciii; N.-W. P., Shahjehánpore, 9; Gazetteers:—Bengal, Orissa, II., 160; Bombay, XIII., 26; XV., 72; XVII., 21; Panjáb, Karnal, 16; Hoshiárpur, 10; Jhang, 17; Amritsar, 4; Sealkot, 11; Rohták, 14; Delhi, 18; Ludhiána, 10; Mysore and Coorg, I., 52, 58; Indian Forester, II., 171, 291, 292; III., 21, 200; V., 181; VI., 238; VII., 222, 250; VIII., 35; IX., 516; XI., 269, 272, 390; XII., 37, 453, 552, App. 27; XIII., 69, 120; XIV., 391.*

Habitat.—A tree which attains a height of 40 feet, and has a short erect trunk and broad crown; it is commonly cultivated in India, but is wild in the Sub-Himálayan tract at altitudes of from 2,000 to 3,000 feet. It was probably introduced into Southern India by the Muhammadans.

The Persian Lilac.

GUM. 394 **Gum.**—The Persian Lilac yields a brownish adhesive gum similar to that of the *ním*, and not unlike that of the wood-apple, **Feronia elephantum,** in appearance. It is apparently not used to any extent.

DYE. Leaves. 395 **Dye.**—**Bidie** includes the LEAVES of this plant amongst his dye materials, and states that they contain green colouring matter and an astringent principle. The writer can find no record, however, of these being taken advantage of either in dyeing or tanning in any other part of India.

OIL. 396 **Oil.**—A fixed OIL is extracted from the SEEDS, which, according to **Birdwood**, is similar to that of the *ním*. It appears to be little known or
Seeds. 397 used.

MEDICINE. **Medicine.**—Hindu writers on Materia Medica seem to have almost entirely neglected the Persian Lilac in favour of their own *ním*. It has, however, long been used by the **Arabs** and **Persians**, who brought a knowledge of its virtues with them into India. They consider the root-bark,
Flowers. 398 fruit, flowers, and leaves to be hot and dry, and to have deobstruent, resolvent, and alexipharmic properties. Thus, "the FLOWERS and LEAVES are
Leaves. 399 applied as a poultice to relieve nervous headaches. The JUICE of the
Juice. 400 leaves, administered internally, is said to be anthelmintic, antilithic, diuretic, and emmenagogue, and is thought to relieve cold swellings, and expel the humours which give rise to them" (*Dymock*). In America a decoction of the leaves has been employed in hysteria, and is believed to
Bark. 401 be astringent and stomachic. The leaves and BARK are used internally and externally in leprosy and scrofula, while a poultice of the flowers is believed to have vermicide properties and to be a valuable remedy in eruptive
Fruit. 402 skin diseases. The FRUIT has poisonous properties, but is used in leprosy and scrofula, and is worn as a necklace to avert contagion. In the Panjáb
Seeds. 403 the SEEDS are prescribed in rheumatism, and in Kángra they are pounded and mixed with apricots as an external application for the same disease.
Oil. 404 In Bombay strings of the seeds are suspended over doors and verandahs during the prevalence of epidemics to avert the disease. The OIL is said to possess similar properties to that of the *ním*, and according to **Ainslie** this
Gum. 405 species also yields a similar toddy. **Emerson** states that the GUM is used as a remedy for splenic enlargement.

Several parts of the Persian Lilac are considerably employed in America. Thus, "the root bark has obtained a place in the secondary list of the United States Pharmacopœia as an anthelmintic. It has a bitter nauseous taste, and yields its virtues to boiling water. It is administered in the form of decoction (4 ozs. of the fresh bark to two pints of water, boiled to one pint), of which the dose for a child is a table spoonful every third hour, until it sensibly affects the bowels or stomach, or a dose may be given every morning and evening for several days and then be followed by a cathartic" (*Pharmacop. Ind.*). **Moodeen Sheriff** states that, after a careful trial of the above preparation, he has arrived at the conclusion that "if the root-bark is vermifuge at all, it is very weakly so."

Other preparations have been used in America. The dried berries in whisky have been employed against ascarides, tape-worm, and verminous diseases, and the pulp of the berries stewed in lard has been used with success against scald head. A fluid extract and syrup prepared from the bark have been recommended, the latter containing vanilla which is said to wholly disguise the bitter and disagreeable taste of the drug (*Year-Book of Pharmacy* (*1875*), *375*). A recent writer on the subject, **Mr. Jacobs,** states that, when prepared in March or April while the sap is ascending, unpleasant effects have been observed, such as stupor, dilatation of the pupil, &c., which symptoms, however, pass off without perceptible injury to the system. There appears little doubt that if given in large doses the bark, leaves, and fruits are all toxic, producing narcotism followed by death. Dr. Burton Brown (*Panjáb Poisons*) records a case in which a European girl ate the berries, became insensible and died. Descourtilz says that six to eight seeds cause nausea, spasm, and choleraic symptoms, sometimes followed by death. MEDICINE.

CHEMICAL COMPOSITION.—Mr. Jacobs subjected the bark to a careful analysis and found the inner bark or liber to be extremely bitter, devoid of astringency, and not to give any indication of the presence of tannin. The outer bark, on the other hand, he describes as very astringent, and as giving abundant precipitates with gelatin and ferric chloride. The active principle was found to be a light yellow, non-crystalline bitter resinous substance without alkaloidal properties. He sums up as follows:— Chemistry. 406

(1) The activity of the bark resides in the liber (or inner bark), and this alone should be employed. (2) The active principle is a yellowish-white resin. (3) The drug is one of the best anthelmintics, and a fluid extract, prepared with diluted alcohol, or a tincture, would be a valuable preparation that would seem to deserve a place in the Pharmacopœia [*Year-Book of Pharm.* (*1880*), *206*].

From these results it would seem that the inner bark is a medicine worthy of more extended trial than it has yet received in India.

Fodder.—According to Aitchison the FRUIT (poisonous to man) is greedily eaten by goats and sheep. FODDER. Fruit. 407

Structure of the Wood.—Sapwood, yellowish-white; heartwood, soft, red, weight from 30 to 40lb per cubic foot. It is very handsomely marked and polishes well. According to Beddome, Brandis, and Kurz it warps and splits easily, but Gamble writes: "Our specimens split only very slightly, and we are inclined to think it is better than it has been supposed to be." It is said to be used in Bengal, like that of the *ním*, for making idols, but is generally employed for making furniture. TIMBER. 408

Domestic.—The STONE from the fruit is used all over India as a bead, being perforated and strung into necklaces, rosaries, and long strings. As already stated, it is supposed to act as a charm against disease (*see* **Beads**, Vol. I., 432). In America the LEAVES and FRUIT are employed, like those of the *ním* in India, as a preservative against the attacks of insects. Thus a decoction of the berries is sprinkled on plants to prevent the depredations of grubs, and the leaves and berries are used to preserve dried fruit and clothing from the attacks of insects. DOMESTIC. Stone. 409 Leaves. 410 Fruit. 411

Melia dubia, *Cav.; Fl. Br. Ind., I., 545.* 412

Syn.—M. COMPOSITA, *Willd.*; M. SUPERBA and M. ROBUSTA, *Roxb.*; M. AUSTRALASICA, *Adr. Juss.*; M. ARGENTEA, *Herb. Ham.*

Vern.—*Dingkurlong*, ASSAM; *Lapshi*, NEPAL; *Eisúr, limbarra, nimbarra*, the fruit=*kálá khajur, kurru khajur*, BOMB.; *Kariaput, nimbara*, MAR.; *Kadu khajur*, GUZ.; *Mallay vembu*, TAM.; *Bévu letta-bévu, kád-bévu, karibevin, ara-bevu*, KAN.; *Limimidella*, SING.

References.—*Roxb., Fl. Ind., Ed. C.B.C., 369; Brandis, For. Fl., 69; Beddome, Fl. Sylv., t. 12; Thwaites, En. Ceylon Pl., 59; Dalz. & Gibs., Bomb. Fl., 36; Dymock, Mat. Med. W. Ind., 2nd Ed., 173; Pharmacographia Indica, I., 332; Year-Book Pharm., 1878, 288; Drury, U. Pl., 292; Lisboa, U. Pl. Bomb., 42; Ind. Forester, VIII., 29; X., 3, 33, 38.*

Habitat.—A large, handsome, deciduous tree, met with in the Eastern Himálaya, South India, Ceylon, and Burma.

Medicine.—Dymock describes the FRUIT in his *Materia Medica of Western India* as follows:—"The pulp of the fruit has a bitter nauseous taste. It is a favourite remedy amongst the labouring classes for colic, half a fruit being the dose for an adult. It appears to have hardly any purgative properties, but is said to relieve the pain most effectually. In the Concan the juice of the green fruit with a third of its weight of sulphur and MEDICINE. Fruit. 413

MEDICINE.

an equal quantity of curds heated together in a copper pot is used as an application to scabies and to sores infested with maggots." The dried fruit is supposed to be the *Arangaka* of Sanskrit writers.

Chemistry. 414 CHEMICAL COMPOSITION.—" The bitter principle of the fruit is a white crystallizable glucoside soluble in ether, alcohol, and water; it is precipitable from its aqueous solution by tannin and alkaloidal reagents, but not by plumbic acetate, and it has a slight acid reaction. Sulphuric acid dissolves it with a deepening of colour discharged on the addition of water. Boiled with diluted hydrochloric acid, it is decomposed in less than half an hour into glucose and a colouring matter. Petroleum ether removes a fatty oil of nauseous property, and ether dissolves a tasteless wax of greenish colour soluble in boiling alcohol and only slightly in petroleum ether; besides these constituents, malic acid, glucose, mucilage, and pectin occur in the fruit." (*Phar. Ind.*)

TRADE. 415 TRADE.—**Dymock** states that the fruit is sold in the bazárs of South Western India for R1-4 per lb.

TIMBER. 416 **Structure of the Wood.**—Sapwood grey, heartwood reddish-white, soft, light, weighing only from 23 to 26lb per cubic foot, cedar-like, not easily attacked by white-ants, but not so strong and durable as that of the *ním.* It is often used by planters for building purposes, and in Ceylon is employed for ceilings, and out-riggers of boats. It has also been recommended for tea boxes and similar purposes.

DOMESTIC, 417 **Domestic, &c.**—**Beddome** recommends the tree highly for avenue purposes, on account of its handsome appearance and rapid growth from seedlings.

418 MELILOTUS, *Juss.; Gen. Pl., I., 487.*

A genus of annual or biennial herbs, which comprises about twelve species, spread through the temperate regions of the Old World. Of these, three are natives of India.

419 **Melilotus officinalis**, *Willd.; Fl. Br. Ind., II., 89;* LEGUMINOSÆ.

Syn.—M. MACRORHIZA, *Pers.*; M. ALTISSIMA, *Thuill. not of Wall.*; TRIFOLIUM OFFICINALE, *Willd.*

Vern.—*Aspurk*, HIND.; *Bau-piring*, BENG.; *Zirir*, PERS.

References.—*Roxb., Fl. Ind., Ed. C.B.C., 588; O'Shaughnessy, Beng. Dispens., 292; Dymock, Mat. Med. W. Ind., 2nd Ed., 211; Year-Book of Pharm., 1877, 295; Atkinson, Him. Dist., 308; Birdwood, Bomb. Prod., 29; Smith, Dic., 271; Gazetteer, N. W. P., IV., lxx.*

Habitat.—A tall, robust herb, wild in Nubra and Ladák, at altitudes of 10,000 to 13,000 feet; cultivated here and there in India.

MEDICINE. **Medicine.**—This is considered by some to be the *μελίλωτος* of **Dioscorides**, by others **M. hamosa**, *Link.*, a Persian species, is believed to have

Coumarin. 420 been that drug. Both possess a marked odour of *coumarin.* That substance may be extracted from either species, as well as from the Tonka bean, and from **Asperula odorata**, *Linn.*, **Liatris odoratissima**, *Willd.*, and **Galium triflorum**, *Mich.*, by exhaustion with ether, when the coumarin crystallises out. The substance is then purified by repeated crystallisation with alcohol It is also now manufactured synthetically from salicylol and salicylic aldehyde. Coumarin is almost insoluble in cold water, but readily soluble in hot dilute acids and alcohol, has an agreeable aromatic odour, a burning taste, and the vapour acts strongly on the brain. When administered internally to dogs in doses of 7 to 10 grains, it produces great and even fatal depression, and in man, doses of 30 to 60 grains occasion nausea, giddiness, depression, vomiting, and drowsiness. It is now chiefly employed in the proportion of 1 to 50 to disguise the odour of iodoform, but has been recommended as a styptic, and in the form of an oily preparation as an external remedy for bruises. By Arabian writers **M. hamosa** is held in high esteem as a remedy in a great variety of disorders; it is considered to be suppurative and slightly astringent, and is much used as a plaster to dispel humours and cold swellings. It is described in all Muhammadan works as *Iklíl-el-malik* (*Dymock*).

M. **officinalis** was considered by the Greeks, and is still supposed in parts of India, to possess similar properties, but it has little therapeutic value, and is not now employed in European medicine.

Fodder.—In Europe and Western Asia the PLANT is cultivated as a food for cattle. **FODDER. Plant. 421**

Melilotus parviflora, *Desf.; Fl. Br. Ind., II., 89.* 422

Syn.—M. INDICA, *All.*; M. MINIMA, *Roth.*; TRIFOLIUM INDICUM, *Linn.*

Vern.—*Banmethi*, HIND., BENG.; *Sinji*, PB.; *Zir*, SIND; *Vana methiká*, SANS.

References.—*Roxb., Fl. Ind., Ed. C.B.C., 588; Stewart, Pb. Pl., 72; U. C. Dutt, Mat. Med. Hind., 322; Murray, Pl. & Drugs of Sind, 112; Atkinson, Him. Dist., 308; Gazetteer, N.-W. P., IV., lxx; Indian Forester, XII., App., 11; XIII., 69.*

Habitat.—A slender herb met with in the Bombay Presidency, Bengal, North-West Provinces, and the Panjáb. Cultivated, or found as a cold season weed of cultivation.

Medicine.—The SEEDS like those of **M. officinalis** are supposed to be useful in bowel-complaints, especially infantile diarrhœa, given as gruel. Like most of the other species of this genus it contains coumarin. **MEDICINE. Seeds. 423**

Fodder.—See Vol. III., 416. **FODDER. Plant. 424**

MELOCANNA, *Trin.; Gen. Pl., III., 1214.* 425

A genus of Bamboo of the sub-tribe MELOCANNEÆ (see Vol. I., 370-389). [GRAMINEÆ.

Melocanna bambusoides, *Trin.; Kurz, For. Fl. Burm., II., 569;*

Syn.—BAMBUSA BACCIFERA, *Roxb.*

Vern.—*Múli, metunga, bish*, BENG.; *Kayoung-wa*, MAGH.

References.—*Roxb., Fl. Ind., Ed. C.B.C., 305; Gamble, Man. Ind. Timb., 429.*

Habitat.—A tall bamboo with stems from 50 to 70 feet long and from 12 to 13 inches in girth, found in Eastern Bengal, Chittagong, Arracan, and Tenasserim. It is the common gregarious bamboo of the Chittagong hills.

Food.—The FRUIT is large, pear-shaped, 3 to 5 inches long; edible. **FOOD. Fruit. 426**

Structure of the Wood.—Of good quality, durable, straight, with straight knots; very largely cut and exported for house-building, mat-making and other purposes (*Gamble*). For an account of the general uses of the bamboo, see Vol. I., 378-389. **TIMBER. 427**

Domestic.—"It yields more or less *tabashír* of a siliceous crystallisation; sometimes it is said the cavity between the joints is nearly filled with this, which the people call *chúna*, lime" (*Roxb.*). See Vol. I., 383. **DOMESTIC. Tabashir. 428**

MELOCHIA, *Linn.; Gen. Pl. I., 223.*

Melochia corchorifolia, *Linn.; Fl. Br. Ind., I., 374;* STERCULIACEÆ. 429

Syn.—M. TRUNCATA, *Willd.*; M. SUPINA, *Linn.*; M. AFFINIS, *Wall.*; M. PAUCIFLORA, *Wall.*; M. CONCATENATA, *Wall.*; RIEDLEIA CORCHORIFOLIA, *DC.*; R. TRUNCATA, *W. & A.*; R. SUPINA, *DC.*; R. CONCATENATA, *DC.*; SIDA CUNEIFOLIA, *Roxb.*; VISENIA CORCHORIFOLIA, *Spreng.*

Vern.—*Tiki-okra*, BENG.; *Thuiak' arak'*, SANTAL.

References.—*Roxb., Fl. Ind., Ed. C.B.C., 505; Thwaites, En. Ceylon Pl., 30; Dalz. & Gibs., Bomb. Fl., 24; Campbell, Ec. Prod., Chutia*

Nagpur, 7822; Atkinson, Him. Dist., 306; Gazetteers:—N.-W. P., IV., lxix.; Bomb. XV., 428.

Habitat.—An erect branching herb or under-shrub, generally distributed throughout the hotter parts of India, from Kumáon, where it reaches an altitude of 4,000 feet, to Sikkim, Malacca, and Ceylon.

FIBRE. **Fibre.**—The stems yield a FIBRE (*Campbell*).

430 FOODS Leaves. **Food.**—The Rev. **Mr. Campbell** states that the LEAVES are eaten as a vegetable by the Santals of Chutia Nágpur.

431 **Melochia velutina,** *Beddome; Fl. Br. Ind. I., 374; Wight, Ic., t. 509.*

Syn.—VISENIA UMBELLATA, *Wight;* V. TOMENTOSA, *Miq.;* RIEDLEIA TILIÆFOLIA, *DC.;* GLOSSOSPERMUM VELUTINUM, *Wall.*

Vern.—*Methúri,* BOMB.; *Al-abada,* AND.

References.—*Kurz, For. Fl. Burma, I., 149; Beddome, Fl. Sylv., Anal. Gen., t. 5, f. 3; Gamble, Man. Timb., 45; Dalz. & Gibs, Bomb. Fl., 24; Lisboa, U. Pl. Bomb., 25; Gazetteer, N.-W. P., IV., lxix.*

Habitat.—A shrub or small tree of the Andaman Islands, Burma, and the Malay Archipelago, widely distributed through the hotter parts of India from the North-West Provinces to the Konkan and Burma.

FIBRE. 432 Bark. 433 **Fibre.**—A strong fibre, called in the Andamans *betina-da,* is prepared from the BARK. From this a stout cord is manufactured which is woven into the turtle-net of the Andaman Islands, known as *yoto-tépinga-da.*

TIMBER. 434 **Structure of the Wood.**—Whitish, very light, even-grained, soft, silvery glossy; good for nothing except perhaps the construction of children's toys (*Kurz*).

MELODINUS, *Forst.; Gen. Pl., II., 694.*

435 **Melodinus monogynus,** *Roxb.; Fl. Br. Ind., III., 629; Wight, Ic., t. 242 (Excl. fig. of fruit) & 394;* APOCYNACEÆ.

Syn.—ECHALTIUM PISCIDIUM, *Wight (Excl. fig. of fruit);* NERIUM PISCIDIUM, *Roxb. (Excl. descr. of fruit).*

Vern.—*Sadul kou, echalat,* SYLHET and KHASIA.

References.—*Roxb., Fl. Ind., Ed. C.B.C., 244, 259; Royle, Fibr. Pl., 303,*

Habitat.—A tall, milky climber, of the Sikkim Himálaya, Assam, Sylhet, and the Khásia mountains, ascending to altitudes of 4,000 feet.

FIBRE. 436 Bark. 437 **Fibre.**—The BARK contains a quantity of fibrous matter, which the natives of Sylhet substitute for hemp. **Roxburgh,** while steeping some of the young shoots in a fish pond in order to accelerate the removal of the bark and to clean the fibre, found that many if not all the fish were killed, hence the name **Nerium piscidium.** Like the fibre of many other apocynaceous plants, this is long and tough.

FOOD. Fruit. 438 **Food.**—"The FRUIT is eaten by the natives of Sylhet; the taste of the firm pulp in which the seeds are immersed is sweet and agreeable" (*Roxb.*).

Meloe trianthema; COLEOPTERA; see Vol. IV., 471, also **Mylabris,** p. 309.

439
MEMECYLON, *Linn.; Gen. Pl., I., 773.*

Memecylon edule, *Roxb., Fl. Br. Ind., II., 563;* MELASTOMACEÆ.

THE IRON WOOD TREE.

The following twelve varieties are included, in the *Flora of British India,* under this species:—

Var. 1, typica,—M. EDULE, *Roxb.;* M. EDULE *var. α, Thwaites, Enum.;* M. UMBELLATUM, *Burm.;* M. TINCTORIUM, *Kœn., Wight, Ill., t. 93;* M. GLOBIFERUM, *Wall., Cat., 4108.*

Var. 2, ramiflora,—M. EDULE, *Lamk.*; M. SESSILE, *Wall.*, *not* M. RAMIFLORUM, *Griff.*
Var. 3, capitellata,—M. CAPITELLATUM, *Linn.*
Var. 4, ovata, —M. OVATUM, *Sm.*; M. EDULE *var.* γ, *Thwaites*; M. TINCTORIUM *var.* β, *W. & A.*
Var. 5, læta,—M. CAPITELLATUM, *Thwaites.*
Var. 6, rubro-cærulea, *Sp. in Thwaites, Enum. Cey. Pl.*
Var. 7, cuneata, *Sp. in Thwaites, Enum.*
Var. 8, leucantha, *Sp. in Thwaites, Enum.*
Var. 9, scutellata,—M. MYRTIFOLIUM, *Wall.*; M. OBTUSUM, *Wall.*; M. PUNCTATUM, *Presl.*; M. SCUTELLATUM, *Naud.*
Var. 10, Thwaitesii,—M. UMBELLATUM, *Thwaites, not of Burmann.*
Var. 11, Rottleriana.
Var. 12, molesta.

For a description of the botanical characters of these varieties, the reader is referred to the *Flora of British India.*

Vern.—*Anjan, anjana, yálki, karpa,* BOMB.; *Limba,* MAR.; *Cashamaram, kásá, casan-elai,* TAM.; *Alli, alli topalu, alli-aku,* TEL.; *Limbtoli,* KAN.; *Kanyavuh,* MALAY.; *Myen-phæ-te-nyet,* BURM.; *Kaian, dædi-kaha, welli-kaha,* SING.; *Pi-tanig,* AND.

References.—*Roxb., Fl. Ind., Ed. C B.C., 325; Kurz, For. Fl. Burm., 512-516; Beddome, Fl. Sylv., t. 206; Gamble, Man. Timb., 198; Thwaites, En. Ceyl. Pl., 110, 111, 112; Dalz. & Gibs., Bomb. Fl., 93; Mason, Burma and Its People, 41; Sir W. Elliot, Fl. Andh., 13; Thesaurus, Zey., t. 31; Bidie, Cat. Raw Pr., Paris Exh., 53, 112; Dymock, Mat. Med. W. Ind., 2nd Ed., 325; Birdwood, Bomb. Prod., 298, 331; Drury, U. Pl. Ind., 290; Useful Pl. Bomb. (Vol. XXV., Bomb. Gaz.), 156, 245; Liotard, Dyes, App. II.; Moore, Man., Trichinopoly, 78; Gribble, Man., Cuddapah, 263; Gazetteers:—Bombay, XV., 73, 434; XVII., 19, 25; Burma, I., 137; Mysore and Coorg, I., 437, 451; Indian Agriculturist, May 18th, 1889; Indian Forester, III., 202; X., 31; XII., 313.*

Habitat.—An exceedingly common plant, met with in the East and South of India and in Ceylon, Tenasserim, and the Andaman Islands. **Mr. C. B. Clarke** in the *Flora of British India* states that the first three varieties run completely together, while var. 4 recedes further from the type, and the remaining varieties are called species by most authors. The economic information given below has been chiefly recorded under the species corresponding to var. 1. typica.

Dye.—The LEAVES are employed in South India for dyeing a "delicate yellow lake." In conjunction with myrobolans and sappan wood, they produce a deep red tinge much used for dyeing grass mats, and also good for cloth. Samples experimented with by **Mr. Wardle** gave "a yellow colour to wool, *eri* silk, and bleached Indian *tasar* with a tin mordant, wool being dyed the deepest colour." **Mr. Wardle** writes: "They possess scarcely any tannin; on *tasar* silk without a mordant, they give a nice clear but light brown colour. There is a pretty yellow tinge on the *eri* silk." In another passage he adds: "This dye-stuff produces good light colours, but would not be of much use in the dye-house, owing to the very small amount of colouring matter it contains." **DYE. Leaves. 440**

The FLOWERS are employed by Native dyers as an adjunct to *chay*-root, for bringing out the colour, in preference to alum. By themselves they produce an evanescent yellow. **Flowers. 441**

Medicine.—The LEAVES are supposed by the Natives to be cooling and astringent, but though occasionally given internally in leucorrhœa and gonorrhœa, they are chiefly employed as a lotion in cases of conjunctivitis. "They should be bruised in a mortar and infused with boiling water. **Dr. Peters** found them in use in Belgaum, as a remedy of considerable **MEDICINE. Leaves. 442**

MEDICINE. Bark. 443 reputation, for gonorrhœa. In the Konkan the BARK with equal proportions of cocoanut kernel, *ajwan* seeds, yellow zedoary and black pepper, is powdered and tied up in a cloth for fomentation or applied as a *lép* to bruises" (*Dymock*).

FOOD. Berries. 444 **Food.**—The plant flowers in the beginning of the hot weather, and produces astringent, pulpy BERRIES, which, when ripe, are eaten by the Natives.

TIMBER. 445 **Structure of the Wood.**—Hard, close-grained, durable, and valuable for many purposes, but difficult to work. **Beddome** suggests that it might be employed as a substitute for box.

DOMESTIC. 446 **Domestic.**—The shrub is very handsome when covered with its dense bloom of blue flowers, and is well worth cultivation as an ornamental plant.

Memorialis, *Ham.*, see **Pouzolzia,** *Gaud.*; URTICACEÆ.; Vol. VI., Pt. I.

447

MENTHA, *Linn.; Gen. Pl., II., 1182.*

A genus of strongly scented, perennial herbs, which comprises about 25 species, natives of the North Temperate regions. Of these only two are natives of India, but several of the other mints occur in Indian gardens, and as escapes from cultivation are met with on roadsides near water.

Mentha arvensis, *Linn.; Fl. Br. Ind., IV., 648.* LABIATÆ.

THE MARSH MINT.

Var. javanica, — M. JAVANICA, *Blume;* M. SATIVA, *Roxb.;* M. ARVENSIS, *Thwaites*

Vern.—*Púdinah,* HIND.; *Pódína,* BENG.; *Pfudnah,* SIND; *Pudinah,* BOMB.; *Pudíná,* MAR.; *Pudina,* GUZ.; *Pudíná, i-ech-chak-kirai,* TAM.; *Pudíná, íga-engili-kúra,* TEL.; *Chetni-maragu,* KAN.; *Putiyina,* MALAY.; *Bhúdina,* BURM.; *Odú-talan,* SING.; *Naanaaul-hind, naanáæ-hindí, habaqulhind, fódanaje-hindí, fótanaje-hindí,* ARAB.; *Púdinah,* PERS.

References.—*Roxb., Fl. Ind., Ed. C.B.C., 460; Thwaites, En. Ceylon, Pl., 239; Ainslie, Mat. Ind., I., 241; O'Shaughnessy, Beng. Dispen., 489; Dymock, Mat. Med. W. Ind., 2nd Ed., 510; S. Arjun, Bomb. Drugs, 101; K. L. Dey, Indig. Drugs of Ind., 72; Atkinson, Him Dists., 315; Lisboa, U. Pl. Bomb., 168; Birdwood, Bomb. Pr., 64, 194, 224; Kew Off. Guide to the Mus. of Ec. Bot., 105.*

Habitat.—A herb of the Western Himálaya, found in Kashmír at altitudes of 5,000 to 10,000 feet. **Sir J. D. Hooker** in the *Flora of British India* remarks of *var.* **javanica**: "I suspect this is introduced and is **M. sativa,** *L.*, to which **Boissier** refers as a synonym to **M. arvensis,** *L.*" It is frequent in the gardens of Europeans in India, where it grows freely and easily.

OIL. 448 **Oil.**—Like most of the other species of this genus, **M. arvensis** and its variety yield an essential oil similar to that of peppermint, but inferior to it in aroma and quality.

Herb. 449 MEDICINE. Plant. 450 **Medicine.**—The dried PLANT is refrigerant, stomachic, diuretic, and stimulant. It is used by the natives as a remedy for jaundice, and is frequently given to stop vomiting. The scent of the fresh herb is said to relieve fainting. **Ainslie** states that this mint is placed by the Arabians and Persians amongst their *mulattifát* (attenuentia). **Fleming** observes that it fully possesses the aromatic flavour, as well as the stomachic, antispasmodic and emmenagogue virtues common to most species of the genus.

SPECIAL OPINIONS.—§ "Cooling and stomachic. Juice of fresh leaves also applied to relieve headache" (*Assistant Surgeon S. C. Bhuttacharji, Chanda, Central Provinces*). "Useful as a stimulant and carminative" (*Assistant Surgeon Nehal Sing, Saharanpore*). "A decoction of the leaves is used for stopping vomiting and nausea." (*Surgeon A. C. Mukerji,*

Noakhally) "The cold infusion of the plant is a good carminative for infants" (*Assistant Surgeon N. L Ghose, Bankipore*).

Food.—The LEAVES are eaten, and a *chatni* prepared from the fresh herb is in use all over Bengal. FOOD. Leaves. 451

Domestic.—The dried PLANT powdered is used as a dentifrice. DOMESTIC. Plant. 452

Mentha piperita *Linn.; D.C. Prodr., XII., 169.* 453

PEPPERMINT.

References.—*Ainslie, Mat. Ind., I., 242; Flück. & Hanb. Pharmacog., 481; S. Arjun, Bomb. Drugs, 204; Year Book Pharm., 1875, 217; 1879, 467; Lisboa, U. Pl. Bomb., 168; Birdwood, Bomb. Pr., 224; Piesse, Perfumery, 179; Smith, Dic., 319; Gaz., Mysore and Coorg, I., 64; Rep. on the Nilghiri Bot. Gardens, 1880-81, para. 55; 1881-82, 49 and 56; 1882 83, 54.*

Habitat.—An erect, usually glabrous perennial, found cultivated and spontaneous in most temperate regions of Europe, Asia, and North America. Its origin is very doubtful, indeed it is problematical whether it occurs truly wild any where at the present day. In the opinion of **Bentham** it is possibly a mere variety of **M. hirsuta,** *Linn.*, with which it is connected by numerous intermediate forms. It is cultivated on a large scale in England, France, Germany, and North America on account of its oil, and is grown to a small extent in India for culinary purposes. Experiments have been made with a certain amount of success in growing the plant at the Nilghiri Gardens for the purpose of obtaining the oil for the Medical Stores. In 1880-81 a good crop was produced which yielded a small quantity of oil of excellent quality. In 1881-82, the oil was extracted at the Gardens and forwarded to the Medical Stores, where, however, it was unfavourably reported on. In 1882-83 the crop was a failure.

Oil.—The HERB yields a volatile strongly aromatic oil, obtained by sun-drying cuttings of the plant, a process which increases the yield about 7 per cent., and thereafter distilling them. A wooden still is employed, into which the plants are packed and tramped down with the feet. This is then heated to about 212° *Fh.* by means of steam, the oil passes over mixed with watery vapour, and on cooling is skimmed off the top of the water. OIL 454 Herb. 455

CHARACTERS AND CHEMICAL COMPOSITION.—**Flückiger & Hanbury** write: "The oil is a colourless, pale yellow or greenish liquid, of sp. gr. varying from 0·84 to 0·92. We learn from information kindly supplied by **Messrs. Schimmel & Co.**, Leipzig, that the best peppermint grown in Germany carefully dried affords from 1 to 1·25 per cent. of oil. It has a strong and agreeable odour, with a powerful aromatic taste, followed by a sensation of cold when air is drawn into the mouth. We find that the Mitcham oil examined by polarised light in a column 50 mm. long, deviates from 14°·2 to 10°·7 to the left, American oil 4°·3." Characters. 456

The oil consists of two substances—one fluid, the other solid; the former has not yet been thoroughly investigated, but according to the learned authors of the *Pharmacographia*, appears to consist chiefly of the compound $C_{10}H_{18}O$; the latter, which is sometimes deposited from the oil when cooled to—4°C, in the form of colourless hexagonal crystals, is called *menthol* or peppermint camphor. The proportion of this varies much in different oils, is largely afforded by eastern mints, and is found in commerce in an almost pure state as "Chinese oil of peppermint." Menthol has the taste and odour of peppermint, melts at 42°C, and boils at 212°C. The Chinese oil has sometimes a bitterish after-taste, but by recrystallization acquires the pure flavour.

Dymock states that most of the oil of peppermint imported into Bombay is of the sort above described as Chinese oil of peppermint. He believes it to be derived from **Mentha canadensis,** *var.* **glabrata,** an opinion

OIL.

confirmed by the remark of **Mr. Holmes**, quoted in the *Pharmacographia, viz.*, that he found the mother plant coming nearest to **Mentha canadensis.** That plant is described in **DeCandolle's** *Prodomus* however as being purely a native of North America, and according to **Bentley** and **Trimen**, Chinese oil is derived from **Mentha javanica,** *Bl.* However this may be, there appears to be no doubt that the oil distilled in Canton forms a large proportion of the supply obtained by India.

Dymock describes it as follows: "The Chinese oil is high coloured and very pungent, and has not the delicate flavour of the best English oil, but for all practical purposes it may be considered a perfect substitute for it. It comes from China in quarter catty bottles, with a Chinese label. Four of these bottles are generally packed in a tin box; value R8 to 10 per ℔; value of menthol R16 per catty."

The oil is largely used in medicine, and is also employed as a flavouring and perfuming agent. Crystallized menthol is also considerably employed in medicine.

MEDICINE. Herb. 457 Oil. 458

Medicine.—Peppermint is the most agreeable and powerful of all the mints, possessing aromatic, carminative, stimulant, antispasmodic and stomachic properties. These qualities are especially possessed by the OIL, which, as the most convenient and elegant preparation, is generally prescribed. It is needless to enter into a detailed account of the numerous applications of this well-known medicine. Suffice it to say that the oil, with its preparations, a water and a spirit, are officinal in the Indian Pharmacopœia, and that the herb and oil of this and other species of mint are largely employed as carminative and flavouring agents by the vegetarian natives of India. In European medicine it has lately been recommended for internal administrations, as an antiseptic in cases of phthisis and diphtheria.

Menthol. 459

MENTHOL has lately acquired a considerable reputation in European medicine as an antiseptic and antineuralgic, and has been used with success as an external application in sciatica, neuralgia, toothache, and ringworm. In the *Lancet* for 1885, II., 128, it is stated that it may be employed with advantage as a local anæsthetic to mucous membranes in a 20 to 30 per cent. solution in alcohol or ether. The anodyne virtues of menthol have long been known to the Chinese and Japanese who employ the "Chinese oil of peppermint," which, as already stated, is very rich in menthol, as a local remedy for neuralgia.

FOOD Leaves. 460

Food.—The LEAVES are employed for culinary and confectionary purposes.

461

Mentha sylvestris, *Linn.; Fl. Br. Ind., IV., 647.*

Var. incana,—M. INCANA, *Willd.*

Var. Royleana,—M. ROYLEANA, *Benth.*

Vern.—*Podina*, HIND.; *Padína*, N.-W. P.; *Babúri, vien, yúra, púdna-kúshma, koshú, belanne*, leaves=*mushk tara*, PB.; *Shamshabái*, PUSHTU; *Púdina, vartalau*, BOMB.; *Boo-dee-na*, BURM.

References.—*Stewart, Pb. Pl., 169; Boiss., Fl. Orient., IV., 543; Aitchison, Bot. Afgh. Del. Com., 95; Mason, Burm., 497, 791; O'Shaughnessy, Beng. Dispens., 489, 490; Dymock, Mat. Med. W. Ind., 2nd Ed., 615; Baden Powell, Pb. Pr., 365; Atkinson, Him. Dist., X., 315; Ec. Prod., N.-W. P., V., 18; Gaz., N.-W. P., IV., lxxvi.*

Habitat.—A herb met with in the temperate Western Himálaya and Western Tibet, at altitudes of 4,000 to 12,000 feet; distributed to Afghánistán, Temperate Europe, and Western and Central Asia. It is frequently cultivated in gardens in the plains of India.

MEDICINE. Leaves. 462 Plant. 463

Medicine.—The LEAVES and a decoction of the PLANT are considerably used by the natives of the Himálaya as a carminative and for the other purposes to which peppermint is generally applied. The leaves of *var.*

incana are employed as an astringent, also for rheumatic pains. Trans-Indus, a decoction is said to be used in fevers and heat apoplexy and for no other purpose. **Dymock** states that the plant is very generally cultivated in gardens in the Konkan, where it is much used as a domestic remedy on account of its mildly stimulant and carminative properties. It is often made into a medicinal *chatni*, which is eaten to remove a bad taste in the mouth during fevers.

Food.—**Atkinson** includes this species in his list of cultivated vegetables. It is probably used, like the other mints, as a carminative adjunct to other green food. FOOD. 464

Mentha viridis, *Linn.; DC. Prodr., XII., 168.* 465

THE SPEARMINT.

Vern.—*Pahári pudína*, HIND.; *Púndia*, BENG.; *Pahári pudina*, N.-W. P.; *Púdina, púdina kúhi, pahári podina*, PB.; *Phúdino*, SIND; *Pudina, pahadi-pudina*, BOMB.; *Pudina*, MAR.; *Phudino*, GUZ.; *Pudiná*, TEL.; *Nagbó, shah-sufiam, púdneh*, PERS.

References.—*Stewart, Pb. Pl., 196; Elliot, Fl. Andh., 158; Mason, Burma and Its People, 497, 791; Pharm. Ind., 166; O'Shaughnessy, Beng. Dispens., 489; Fleming, Med. Pl. and Drugs, as in As. Res., Vol. XI., 172; Flück. & Hanb., Pharmacog., 479; U. S. Dispens., 15th Ed., 940; Bent. & Trim., Med. Pl., 202; S. Arjun, Bomb. Drugs, 101; Baroda Durbar, C. I. E.; Year-Book Pharm., 1879, 468; Baden Powell, Pb. Pr., 365; Atkinson, Him. Dist., 741; Ec. Prod., N.-W. P., V., 19; Lisboa, U. Pl. Bomb., 168; Birdwood, Bomb. Pr., 64, 224; Stocks, Rep. on Sind; Piesse, Perfumery, 155; Smith, Dic., 387; Kew Off. Guide to the Mus. of Ec. Bot., 105; Agri.-Hort. Soc., Ind., Transactions, VIII., Pro., 409; Journals (Old Series), X., 23; (New Series), IV., 32, 33.*

Habitat.—A coarse woolly herb of wide distribution, considered by **Mr. Baker** to be indigenous to the North of England. **Mr. Bentham**, on the other hand, regards it as not improbably a variety of **M. sylvestris**, *Linn.*, perpetuated through its ready propagation by suckers. It grows in kitchen gardens in most parts of Europe, Asia, and the Cape of Good Hope, North and South America, &c., and is cultivated for medicinal purposes in Mitcham, Surrey, and in the United States. It is commonly grown in native gardens all over the plains of India.

Oil.—"Spearmint yields an essential oil (*Oleum Menthæ Viridis*) in which reside the medicinal virtues of the plant. **Kane**, who examined it, gives its sp. gr. as 0·914, and its boiling point as 160°C. The oil yielded him a considerable amount of stearoptene. **Gladstone** found spearmint oil to contain a hydrocarbon almost identical with oil of turpentine in odour and other physical properties, mixed with an oxidised oil to which is due the peculiar smell of the plant" (*Flückiger & Hanbury*). This oil is said by **Stocks** to be extracted from the plant for medicinal use by the Natives of Sind. In the United States it is employed as a perfume by soap manufacturers, and as a flavouring agent by confectioners. OIL. 466

Medicine.—The PLANT and the OIL extracted from it are considered hot and dry, and are administered to prevent vomiting, and as a stimulant and carminative in indigestion and colic, also with purgatives to prevent griping. MEDICINE. Plant. 467

The oil is largely used as a domestic medicine in the complaints of childhood, as a stomachic, carminative, and expectorant. **Dr. Emerson** states that the SEEDS are mucilaginous, that the leaves are given in fever and bronchitis, and that a decoction is used as a lotion for aphthæ. In European medicine the oil is prescribed similarly to that of **Mentha piperita**, and a water made from the oil is used as a vehicle for other carminative medicines. Both are officinal in the Indian Pharmacopæia. Oil. 468 Seeds. 460

FOOD. Plant. 470 **Food.**—The PLANT is cultivated in native gardens to be used as a carminative in curries, and other forms of food. Its flavour is preferable to that of peppermint, and the plant is, therefore, more commonly used by Europeans for culinary purposes.

471 **MENYANTHES,** *Linn.; Gen. Pl., II., 819.*

A genus of perennial herbs, belonging to the Natural Order GENTIANACEÆ, and comprising three species, natives of the North Temperate and Sub-arctic Zones. Of these one is of economic interest, *viz.*, **M. trifoliata,** *Linn.*, found at high altitudes in Kashmír (*Fl. Br. Ind., IV., 130*). The leaves of this plant, the BUCKBEAN, or BOYBEAN, were formerly considered a valuable tonic, and are still, by certain writers, reckoned as one of the best of Gentians, though not officinal in any Pharmacopœiā and but little used except in domestic medicine. It is not mentioned by any writer on Indian Materia Medica, from which it may be assumed, as might naturally be expected from the rarity of the plant, that it is little, if at all, employed by the natives, even of Kashmír.

472 **MERCURY,** *Ball, in Man. Geology of Ind., III., 170.*

This metal occurs both native and in combination with other substances. It is distinguished from all other metals by being liquid at even the lowest temperatures to which moderate climates are subject. To this exceptional property it owes the synonyms of "quick silver" in English, and of hydrargyrum (from ὕδωρ, water, and ἄργυρος, silver) in Græco-Latin.

The most abundant ore of the metal is the sulphide, (HgS.) or cinnabar, which, when pure, constitutes the pigment vermillion. In nature it is as a rule accompanied by a greater or less proportion of the native metal, probably derived from the ore by some secondary reaction. It also occurs as a native amalgam of silver, or with chlorine, as in "horn mercury." The mercury of commerce, an important substance in medicine and the arts, is chiefly prepared from cinnabar, and is said to be produced for the use of the whole world to the extent of about 4,000 tons annually.

473 **Mercury,** *Mallet in Man. Geology of India, IV. (Mineralogy), 4, 21.*

Syn.—MERCURE, VIF-ARGENT, *Fr.*; QUECKSILBER, *Germ.*; MERCURIO, *Ital.*; AZOGUE, *Sp.*

Vern.—THE METAL=*Párá*, HIND., BENG., MAR.; *Páro*, GUZ.; *Irasham*, TAM.; *Rasam, páda-rasam*, TEL.; *Páda-rasá*, KAN.; *Rassam*, MALAY.; *Padá*, BURM.; *Rasadiyá*, SING.; *Párada, páradaha, rasa, rasam*, SANS.; *Zibaq*, ARAB.; *Simáb, jivah*, PERS.

CINNABAR=*Shangarf, shangraf, hingól*, HIND.; *Hingól, shangraf*, BENG.; *Lingam, jádi-lingam*, TAM.; *Ingili-gamu*, TEL.; *Ingaliká*, KAN.; *Cháyilyam, cháliyam, játi-lingam*, MALAY.; *Hen-tha-pada-yaing*, BURM.; *Lingam, játi-lingam*, SING.; *Inghúlam*, SANS.; *Shanjarf, zanjafr*, ARAB.; *Shangarf, hingól*, PERS. Mr. Moodeen Sheriff remarks: "*Zanjarf, dardár, sarúre-ahmar*, and some other names are found applied to Cinnabar in some Persian and other works; but they are neither restricted to it, nor in use at present."

MERCURIC CHLORIDE (Corrosive sublimate)=*Shavíram, shavír*, HIND.; *Víram, shav-víram*, TAM.; *Shavíramu, víramu*, TEL.; *Shavíram*, SING.; *Rasa karpura?* (U. C. Dutt—see remarks under mercurous chloride), SANS. Mr. Moodeen Sheriff states that the bazár drug with the above names is pure with the exception of a small admixture of the subchloride (calomel), and is quite fit to be used as a substitute for the imported drug. He adds: "In some books *sulaimání* is considered to be applicable to Corrosive Sublimate, but this is incorrect, because the *sulaimání* contains arsenic in its combination."

MERCUROUS CHLORIDE, OR CALOMEL=*Ras-kapúr*, HIND.; *Rasha-karup-púram, púram*, TAM.; *Rasa-karpúramu, púramu*, TEL.; *Rasa-karppú-ram*, MALAY.; *Rasa-karpúram*, SING. **Mr. Moodeen Sheriff** states that the medicine obtained by these names in the Indian bazárs is, according to chemical tests, an impure calomel, though it occurs in crystalline masses, and contains the perchloride as its chief impurity.

References.—*Mason, Burma and Its People, 562, 729; Pharm. Ind., 379-388; U. S. Dispens., 15th Ed., 747; Ainslie, Mat. Ind., I., 540-561; II., 348-356; O'Shaughnessy, Beng. Dispens., 697; Irvine, Mat. Med. Patna, 83; Moodeen Sheriff, Supp. Pharm. Ind., 156-158; U. C. Dutt, Mat. Med. Hindus, 27-38; Baden-Powell, Pb. Pr., 63, 103, 115, 360; Forbes Watson, Indian Prod., 406, 422; Econ. Prod. N.-W. Prov., Pt. III. (Dyes and Tans), 46; Crookes, Hand-book, Dyeing, &c., 152, 154; Liotard, Dyes, App. V.; Man. Madras Adm., Vol. II., App VI., 33; Bomb. Adm. Rep., 1871-72, 373; Gazetteer, Bengal, Hunter, Orissa, II., 27.*

Occurrence.—Although the discovery of this metal has been reported more than once, its occurrence in India is still open to doubt. Cinnabar, also, is not known to occur in the country, but, according to **Dr. R. Sanders**, "is found in Tibet," by which he probably means that portion of the country between Eastern Bhután and the Sangpo river where he travelled. The reports of mercury or mercurial ores having been found in India are given as follows by **Ball**:— OCCURRENCE. 474

MADRAS.—In 1858 **Brigadier Fitzgerald** reported, to the Madras Government, the existence of mercury in a bed of laterite at Cannancre. Excavations were made, but there is no record of the result. Possibly there was some mistake in the identification of the substance. Madras. 475

AFGHÁNISTÁN.—According to **Captain Hutton**, mercury is said to occur in Gurmsael (Garmsir) at Pir Kisri, where it is dug out of the ground and sold as a medicine at R2 to R3 per tola. The cinnabar used in Báluchistán is said to come from Persia, India, and Turkey. **Captain Drummond** mentions that a specimen of cinnabar was brought to him by a villager, who stated that he had found it close to Sultanpur, near Jellalabad, but after an examination of the ground it was concluded that it had been dropped there by accident. Afghanistan. 476

ANDAMANS.—Several reports exist in ancient and modern literature of mercury in the Andaman Islands, but no absolute confirmation of these reports exists. Since the British occupation nothing has been discovered to justify the belief in the existence of mercury in these localities, though enquiries have been made, and the natives from different parts shown the metal, when in Port Blair, in the hope that they might recognise it. **Ball** writes: "The rocks of the Andamans are of early tertiary or late cretaceous age with intruded volcanic rocks and serpentine. They apparently have a close resemblance to the rocks which in California now yield a large proportion of the mercury of commerce." One of the reports above alluded to, points to the probability of the deposit really existing in the Little Andaman. If it does occur, there, and there only, its discovery may still be far distant owing to the hostility and barbarity of the inhabitants of that island. Andamans. 477

Dye.—CINNABAR, ground and mixed with water, is used in dyeing, giving a fresh pink tint (*shingrafi*) to cloth. It is said by **Liotard** to be employed more by private persons than by professional dyers, and to cost R140 per cwt. It is also used as a pigment (vermillion) by artists and decorators in all parts of the country. It is said to be manufactured in Calcutta, by combining mercury with sulphur, and subliming the mass. **Ainslie** states that it was, in his time, an export from Surat to Madras, and a recent communication states that it is still manufactured in that place to a small extent, and exported through Bombay to China. In Europe VERMILLION, pure sulphide, is similarly employed in painting, decorating, DYE. Cinnabar. 478 Vermillion. 479

DYE. Iodides. 480 dyeing, and calico-printing. The BIN-IODIDE which yields a very rich scarlet is also similarly used, but is very fugitive. The YELLOW IODIDE is also occasionally utilized as a pigment, rarely as a dye.

MEDICINE. The metal. 481 **Medicine.**—The value of mercury in medicine has been long known in Arabia, Persia, and India. It is generally agreed that the nations of the East were the first to employ the metal, or its salts, in the cure of leprosy and other cutaneous diseases, but it may be questioned whether the natives of India were before or after the Arabs in discovering the valuable properties of this powerful drug. According to **Fallopius**, the first physicians in Europe who made use of mercury in venereal cases, lived towards the end of the fifteenth century, and were induced to try it from the writings of **Rhayes, Avicenna,** and **Mesne**, who lived and wrote from 500 to 200 years before. The first mention of mercury in Sanskrit medicine is rather doubtful. According to certain authors, it is referred to in *charaka*, the oldest Sanskrit work on medicine, under the name of *rasa*, but it appears very doubtful whether this term really did refer to mercury. According to **U. C. Dutt**, it is neither mentioned in *Charaka* nor *Susruta* (a work probably almost contemporary with the former), nor in the next work on Hindu medicine, the *Ashtánga-hridaya-sanhitá*. It is, however, certainly mentioned in the *Chakradatta-sangraha*, a work written previous to the eighth century A.D. In this book the preparations of the metal with sulphur and vegetable substances are mentioned, but those produced by sublimation and chemical combination with acids, salts, &c., appear to have been unknown to the writer. From this fact **U. C. Dutt** argues that in all probability mercury was just coming into use at the time of the publication of that work. In later works it occupied a gradually progressive, more and more important position, till in later days it came to be regarded as the most important medicine in the Hindu Pharmacopœia.

The valuable properties of the numerous modern preparations of the metal, found in all European Pharmacopœia, are too well known to warrant even a passing notice in this place. The officinal preparations in the Pharmacopœia of India are similar to those in the British Pharmacopœia. The methods of preparation, and the virtues ascribed to the compounds of the metal in Hindu and Muhammadan medicine, will, accordingly, be alone described. As already stated, mercury is, perhaps, the most important drug in Hindu medicine. Thus **U. C. Dutt** writes as follows :—"*Párada* literally means 'that which protects,' and mercury is so called because it protects mankind from all sorts of diseases. It is said that the physician who does not know how to use this merciful gift of God is an object of ridicule in society." Mercury is said by Sanskrit writers "to be bright like the mid-day sun externally and of a bluish tinge internally," and it is enjoined that if of a yellowish-white purple or variegated colour, it should not be used in medicine. If administered in an impure state, it is said to bring on a number of diseases, and is, accordingly, purified before use. The methods of purification generally followed by the *Kavirájas* are described as follows by **Dutt**:—"Mercury is first rubbed with brick-dust and garlic, then tied in four folds of cloth and boiled in water over a gentle fire for three hours in an apparatus called *Dolá yantra*. When cool, it is washed in cold water and dried in the sun. Some practitioners use betel-leaves instead of garlic for rubbing with the mercury. Mercury obtained by sublimation of cinnabar is considered pure and preferred for internal use. Cinnabar is first rubbed with lemon juice for three hours, and then sublimed in the apparatus called *Urddhapátana yantra*. The mercury is deposited within the upper pot of the apparatus in form of a blackish powder. This is scraped, rubbed with lemon juice and boiled in water, when it is fit for

MEDICINE.

use." The purified metal obtained by one of these processes is employed in the preparation of mercurial compounds, of which four are described, namely, black, white, yellow, and red, called respectively *krishna, sweta, pila*, and *rakta bhasmas*. For an account of the method of preparation of these, which are all evidently impure salts of the metal, the reader is referred to U. C. Dutt's exhaustive article on the subject in his *Materia Medica of the Hindus*.

Sulphides. 482
Perchloride. 483

The four preparations principally prescribed at the present time are the BLACK SULPHIDE, called *kajjali;* CINNABAR; the red preparation called *rakta bhasma* or *rasa sindura* (evidently an impure sulphide); and the *rasa karpura* or *sweta* (PERCHLORIDE of mercury). Mercury is supposed to be imbued with six tastes, and to be capable of removing derangements of all the humours, to be the first of alterative tonics, and when combined with other appropriate medicines to cure all diseases, acting at the same time as a powerful tonic, and improving the vision and complexion.

It is especially employed in fever of all descriptions, in affections of the lungs, as an alterative in skin diseases, and both externally and internally in leprosy and syphilis. As is usual in Sanskrit medicine, it is rarely given alone, but enters into a number of complicated prescriptions, each of which is supposed to possess some peculiar virtue. For a description of these the reader is referred to the long and interesting article in Dutt's *Hindu Mat. Med.* The modern practitioners of Hindu medicine are acquainted with the good results obtainable by inunction and fumigation which they occasionally employ.

Ainslie enters into an exhaustive consideration of the Tamil methods of preparing the drug, which appear to have been directly derived from the teaching of Sanskrit medicine, and remarks: "However much we may be inclined to smile at some of their strange mixtures, it must be confessed that the characterising principles are generally correct, and that everything considered there is in the present state of knowledge amongst the *Vytians* and *Hakîms*, more to call forth our wonder, than excite our contempt." These preparations, administered by the Tamil practitioners, were supposed to be valuable remedies in contraction of the sinews ?, venereal affections, skin diseases, ulcers, chronic and virulent boils and carbuncles, scrofulous affections, leprosy, fistulas, itch and hypochondriasis, mostly diseases for which the drug enjoys a reputation in European medicine.

By the practitioners of Arabian and Persian medicine also, the actions and virtues of mercury have long been appreciated, and, as already stated, the writers of these countries supplied the information from which western medicine first obtained its knowledge of the drug.

INDUSTRIAL USES. 484

Industrial Uses.—The uses of mercury in the arts are numerous and highly valuable, as in gilding, in making barometers, thermometers, mercurial air pumps, for silvering mirrors, &c., but are too well known to require any description in this place. In India the metal is employed chiefly in making looking-glasses, also, in certain localities, for making an amalgam of gold for gilding metal, and occasionally to aid in the separation of gold from its ores by the amalgam process.

MERIANDRA, *Benth.; Gen. Pl., II., 1194.* 485

Meriandra bengalensis, *Benth.; Fl. Br. Ind., IV., 653;* LABIATÆ.

BENGAL SAGE. [*Br.*

Syn.—SALVIA BENGALENSIS, *Roxb.;* S. DIANTHERA, *Roth.;* S. ABYSSINICA,

Vern.—LEAVES=*Káfúr-ká-pát,* HIND.; *Káfúr-ká-pátta,* DEC.; *Kafúr-ká-pattá, sesti,* BOMB.; *Shíma-karpúram-áku,* TAM.

The above names simply signify "leaf of the camphor plant," and owe their origin to the camphoraceous odour of the herb.

References.—*Roxb., Fl. Ind., Ed. C.B.C., 49; Gamble, Man. Timb., 301; Dalz. & Gibs., Bomb. Fl., Addend., 66; Elliot, Fl. Andh., 168; Pharm. Ind., 168; Ainslie, Mat. Ind., I., 359; O'Shaughnessy, Beng. Dispens., 487; Moodeen Sheriff, Supp. Pharm. Ind., 173; Lisboa, U. Pl. Bomb., 168.*

Habitat.—A camphoraceous, bitter herb or shrub, native of Abyssinia, cultivated in India, especially in Bengal.

MEDICINE. Leaves. 486 **Medicine.**—The LEAVES differ little in medicinal properties from those of the Sage, excepting that they have a strong smell of Camphor. **Ainslie** writes of this plant: "Sage is but little employed medicinally by the Hindus; the Mahometans cultivate it in their gardens and use it for the same purposes that we do; preparing with the leaves a sort of grateful tea, which they prescribe in certain stages of fever, and as a gentle tonic and stomachic. The leaves ought to be carefully dried in the shade; they then have an agreeable fragrant odour, with a warm, bitterish, aromatic and grateful taste, and are considered as tonic, carminative, and slightly astringent. The infusion alone, or mixed with honey and vinegar, makes an excellent gargle in cases of sore-throat." The leaves are much used in native practice, an infusion being employed like that of sage as a gargle for sore-throat, and a mouth wash in aphthæ. It is also said to diminish or arrest the secretion of milk (*Pharm. Ind.*).

FOOD. Leaves. 487 **Food.**—The LEAVES are much used in Bengal as a condiment with curries, &c.

488 **Meriandra strobilifera,** *Benth.; Fl. Br. Ind., IV., 652.*

A small shrub with grey bark, found on dry rocks, especially limestone, from Simla to Kumáon at altitudes of 5,000 to 6,000 feet. Its vernacular names and properties are similar to those of **M. bengalensis.**

489 **MESEMBRYANTHEMUM,** *Linn.; Gen. Pl., I., 853.*

A genus of herbs which belongs to the Natural Order FICOIDEÆ, and comprises some twelve species, natives of the hot regions of Asia, Africa, Australia, and the West Indies. None of these are natives of this country, but attempts have been made to introduce several species, especially **M. crystallinum,** *Linn.*, into Northern India, in the districts in which *reh* lands occur. This species, the "Ice plant," is believed to absorb an extraordinary amount of alkaline salts from the soil, and has been cultivated in America to remove excess from land on the sea-coast and in salty deserts.

The plant is also of economic value, large quantities being collected in the Canaries for the sake of the ash, which is exported to Spain for the use of glass-makers. Others have been recommended as sand binding plants, while one, the "Elephant Fodder" of the Cape of Good Hope, has been recommended as likely to prove very useful as a fodder plant in the dry plains of North-Western India.

MESUA, *Linn.; Gen. Pl., I., 176, 981.* [GUTTIFERÆ.

Mesua ferrea, *Linn.; Fl. Br. Ind., I., 277; Wight, Ic., t. 117—119;*

Syn.—M. SPECIOSA, *Chois.*; M. PEDUNCULATA, *Wight*; M. COROMANDELIANA *and* ROXBURGHII, *Wight*; M. SALICINA, WALKERIANA, *and* PULCHELLA, *Planch. & Trian.*; M. SCLEROPHYLLA, *Thwaites*; M. NAGANA, *Gard.*

Vern.—*Nágkesar, naghas,* HIND.; *Nágésar, nágkesar,* BENG.; *Nageshvoro, nágeswar,* URIYA; *Nagkeshur,* BEHAR; *Nahor,* ASSAM; *Nahshor,* MICHI; *Nogkesar,* PB.; *Nágchampa, thorlachampa,* BOMB.; *Nagachampa, nágchápha,* MAR.; *Nangal, mallay nangal, shiru-nágap-pú, nagasháp-pú,* TAM.; *Nang,* TINNEVELLY; *Nága késara, nága késaramu,* *geja-pushpam*, TEL.; *Naga sampigi, nagsampige*, KAN.; *Behetta-champagam, velutta-chenpakam*, MALAY; *Kaing-go*, MAGH.; *Ken-gau, gangau*, BURM.; *Ná, deya-ná, na-gaha*, SING; *Nágakesara, késaramu, kinjalkamu, nága-késaram*, SANS.

References.—*Roxb., Fl. Ind., Ed. C.B.C., 437; Kurz, For. Fl. Burm., I., 97; Beddome, Fl. Sylv., t. 64; Anal. Gen., xxiii.; Gamble, Man. Timb., 27; Thwaites, En. Ceyl. Pl., 407; Mason, Burma and Its People, 401, 402, 751; Sir W. Elliot, Fl. Andh., 57, 90, 121; Sir W. Jones, Treat. Pl. Ind., V., 139, No. 57; Pharm. Ind., 32; O'Shaughnessy, Beng. Dispens., 239; Irvine, Mat. Med. Patna, 74; Medical Topog., 147; Moodeen Sheriff, Supp. Pharm. Ind., 174; Mat. Med. S. Ind. (in MSS.), 44; U. C. Dutt, Mat. Med Hindus, 119, 310; Sakharam Arjun, Cat. Bomb. Drugs, 24; K. L. De, Indig. Drugs, Ind., 72; Murray, Pl. & Drugs, Sind., 69; Dymock, Mat. Med. W. Ind., 2nd Ed., 84; Dymock, Warden and Hooper, Pharmacog. Ind., Vol., I., 170; Birdwood, Bomb. Prod, 14; Baden-Powell, Pb. Pr., 333; Drury, U Pl. Ind., 291; Useful Pl. Bomb. (Vol. XXV., Bomb. Gaz.), 13, 214, 289; Cooke, Oils and Oilseeds, 58; Gums and Resins, 119; Darrah, Note on Cotton in Assam, 33; Aplin, Rep. on Shan States; Man. Madras Adm., Vol. II., 64; Gazetteers:—Bombay, XV., 73,—XVIII., 49; Orissa, II., 68, 160; App. II., & IV.; Mysore and Coorg, II., 7; III., 16; Agri.-Hort. Soc. Ind.:—Journals (Old Series), 118 120; (New Series), I., 323; VII., (Pro.), cxciv.; Indian Forester, IV., 292; VI., 125; VIII., 106, 112, 126, 371, 403; IX., 316, 426, 580, 606; X., 33, 532; XI., 199, 254, 288, 319, 320; XIV., 119, 339.*

Habitat.—An evergreen tree, wild in the mountains of Eastern Bengal, the Eastern Himálaya, Assam, Burma, South India, Ceylon, and the Andamans; also cultivated in other parts of India.

Oleo-resin.—A tenacious and glutinous oleo-resin, with a sharp aromatic odour, exudes round the base of the tender FRUIT. It has been recommended as a substitute for Canada Balsam, but does not appear to be utilised by the natives. It may also be obtained, by incision, from the BARK or ROOT, and is said, when diluted with oil of turpentine, to make a very good wood varnish (*Hannay*). (*See* paragraph "CHEMICAL COMPOSITION.")

OLEO-RESIN. 490 Fruit. 491 Bark. 492 Root. 493

Dye.—*Spons' Encyclopædia* says the FLOWER-BUDS of this plant are used in India for dyeing silk; "they were once introduced into the London market under the name of *nag-kassar*, apparently a corruption of the Hindustani and Bengali name *nagesar*." **Dr. Dymock**, however, informs the Editor that this is quite a mistake, and that the flower-buds referred to are those of **Ochrocarpus longifolius.** **Darrah** states that the FLOWERS of **Mesua** are said to yield a mordant.

DYE. Flower-buds. 494 Flowers. 495

Oil.—In Ceylon a thick, and dark coloured oil is obtained from the SEEDS. It is employed both for burning in lamps and as an external application to sores. It is also largely expressed by the inhabitants of North Kanara for use as an embrocation in rheumatism. The authors of the *Pharmacographia Indica* describe it as follows:—"The seeds yield to ether 31 5 per cent. of fixed oil, the kernels alone gave 72·9 per cent. The oil thus obtained had a deep yellow colour, formed orange-coloured mixtures with sulphuric and nitric acid, was partially soluble in alcohol, and had a specific gravity of 0·972 at 17°C., a temperature at which it began to set, on account of the crystallization of the more solid fats. The hard pericarp contained a considerable amount of tannin It is sometimes offered for sale and costs R4 per maund in Kanara." In the *Proceedings of the Agri.-Horticultural Society of India, IX, cxciv.* (New Series), it is stated that the seeds are so oleaginous that they are used in certain localities by natives instead of a lamp, for when merely shelled and dried they burn with a bright flame. An attar is also prepared from the flowers (see paragraph "CHEMICAL COMPOSITION").

OIL. 496 Seeds. 497

Medicine.—The DRIED FLOWERS are much used by Hindu physicians as a fragrant adjunct to decoctions and oils. They are regarded as

MEDICINE. Dried Flowers. 498

MESUA ferrea.

Nagkesar.

MEDICINE. astringent and stomachic, and useful in thirst, irritability of the stomach, excessive perspiration, &c. A paste of the flowers, with the addition of butter and sugar, is recommended by most of the late Hindu writers as an application to bleeding piles and burning feet (*U. C. Dutt*). In many localities they are used for cough, especially when attended with much expectoration.

Leaves. 499 **O'Shaughnessy** states that the flowers and LEAVES are employed in Bengal as remedies for snake-bite.

Bark. 500 The BARK is mildly astringent and feebly aromatic, but according to **Dymock** is not bitter as stated in the Pharmacopœia. **Rheede** states that it is given as a sudorific combined with ginger.

Oil. 501 The OIL of the seeds is used as an embrocation in cases of rheumatism, as a healing application to sores, and as a local remedy for itch and other skin diseases.

Chemistry. 502 CHEMICAL COMPOSITION.—The authors of the *Pharmacographia Indica* write: "The chief principle of **Mesua ferrea** appears to be an oleo-resin, which abounds in all parts of the tree, and is obtained pure from the young fruits. The fresh tears sink in water, melt between 50° and 60° C., and partially dissolve in rectified spirit, amylic alcohol, and ether, but wholly in benzol. Boiled with solutions of soda or ammonia, the resin forms a clear mixture, precipitable by acids, in a white curdy condition. The solution in spirit has an acid reaction, and is dextro-rotatory when examined by polarised light; the solution gives a precipitate with alcoholic plumbic acetate, soluble when heated. From the partial solubility there are probably two resins present Submitted to distillation, 0·6 per cent. of a fragrant essential oil was obtained: this was of a pale yellow colour, and possessed in a high degree the odour of the flowers, and resembled that of the exudation of the Chio Turpentine."

SPECIAL OPINIONS.—§ "A fixed oil, expressed from the dried kernel of the seeds, is used in the Dinajpore and Rungpore Districts of North Bengal for the treatment of unhealthy ulcers and skin diseases attended with a purulent discharge, as in rupia. It is not so useful in itch" (*C. T. Peters, M.B., Zandra, South Afghánistán*). "The expressed oil of the seeds is used as an embrocation in rheumatism, and a decoction of the bark and roots is used as a bitter tonic in convalescence after exhausting diseases" (*Surgeon-Major D. R. Thomson, M.D., C.I.E., Madras*).

FOOD. Fruit. 503 **Food.**—The tree flowers in April and May, and produces in June and July a reddish and wrinkled FRUIT, like a chesnut in size, shape, and taste, which is eaten by the Natives.

TIMBER. 504 **Structure of the Wood.**—Heartwood dark red, extremely hard, sometimes prettily veined; weight from 62 to 76℔ per cubic foot. It is highly prized for posts in buildings, bridges, gunstocks, and tool handles, but would be much more utilised than it is were it not for its great hardness, weight, and the difficulty of working it. **Captain Hannay** states that the heartwood when properly seasoned is perfectly proof against the attacks of white-ants, and that it is susceptible of a beautiful polish from simple hand-rubbing. It has been found to answer for sleepers equally well with *pynkado*, but the cost of cutting the hard wood, its weight, and the cost of freight from the Tenasserim forests to Calcutta, prevent its being much used, since the total cost is scarcely covered by the price of the sleeper (*Gamble*).

DOMESTIC. 505 **Domestic, &c.**—This beautiful tree with its large white flowers is much cultivated near houses and in avenues, as it affords an excellent shade.

Flowers. 506 The FLOWERS are much admired, both for their beauty and fragrance, and with bunches of the delicately coloured young leaves are used in Assam as hair decorations. The flowers are also worn by the Assam frontier people of both sexes stuck through the ear lobes (*Hannay*). When dried they are mixed with other aromatics, such as white sandal wood, and used for perfuming ointment (*Drury*). According to **Sir W. Jones** one of the five arrows of Kamadeva, the Indian cupid, is tipped with the wood of **Mesua**. [Vol. III., 506.

Methonica superba, *Hern.*, see **Gloriosa superba,** *Linn.*; LILIACEÆ,

METROXYLON, *Rottb.*; *Gen. Pl., III., 935.* 507

A genus of the Natural Order PALMÆ which comprises six species natives of the Malay Archipelago, New Guinea, and Fiji. From the pith of one of these, **M. Sagu,** *Rottb.* (a native of the Moluccas, Sumatra, and Borneo), part of the granulated SAGO of commerce is prepared.

Mezenkuri, see **Silk,** Vol. VI., Pt. II. [LÆACÆ, Vol. III., 24.

Mezereum or **Mezereon,** see **Daphne Mezereum,** *Linn.*; THYME-

MICA, *Ball in Man. Geology of India., III., 524.* 508

The group of minerals known collectively under this name have several characteristics in common, which vary with the combining ratios of the bases and the silicon of which they are composed. The light-coloured micas generally belong to the kind known as *muscovite,* and the black to *boitite;* other varieties are *lepidolite* and *lepidomelane.* Of these the only one having any extensive economic importance is muscovite (*Ball*). In India, however, the other forms are employed in medicine and will therefore be shortly noticed.

Mica, *Mallet in Man. Geology of India, IV., 96-98.* 509

TALC (common though erroneous commercial name); Mica, *Fr.*; GLIMMER, *Ger.*; TALCO, *Ital.*

Vern.—*Abrak,* HIND.; *Appracam,* TAM.; *Appracam,* TEL.; *Kokabá-larz, minirum,* SING; *Abhra,* black mica = *vajrábhra, krishnábhra, sheá-bhra,* SANS.; *Talk,* PERS.

References.—*Mason, Burma and Its People, 583, 734; Ainslie, Mat. Ind., I., 421-423; Irvine, Mat. Med. Patna, 8; U. C. Dutt, Mat. Med. Hindus, 76; Baden-Powell, Pb. Pr., 42, 43; Forbes Watson, Indian Prod., I., 411; Econ. Prod. N.-W.-Prov., Pt. III. (Dyes and Tans), 46; Man. Madras Adm., Vol. II., 40, 77; Settlement Rept.:—Central Provinces, Chánda, 106; Gazetteers:—Orissa, II., 75; Bombay, V., 360; Panjáb, Gurgáon 13; Sháhpur, 13; Tropical Agriculturist, 1889, 801; Balfour, Cyclop. Ind., II., 941; Ure, Dict. Indus. Arts. & Man., III., 82.*

OCCURRENCE. 510

Occurrence.—Although MICA is one of the most widely distributed minerals in India, its occurrence in plates of sufficient size to be of commercial value is limited to a few tracts. Plates of useful size are generally found in veins of coarsely crystalline granite. Thus at Kolur in the Vizagapatam district of Madras, mica of sufficient size to be worth collecting is obtainable; but as it sells at 24lb to the rupee, it cannot be of very good quality. The principal source of supply is from the Hazáribágh district of Bengal, where it is mined at Dhamvi, Quadrumma, Dhub, and Jamatra. In the adjoining districts of Gya and Monghyr, at Rajow, similar mica is obtained in plates 9 inches square. In the northern part of the Hazáribágh district, plates of muscovite up to a foot or more in diameter may be obtained; indeed **Mallet** states that he has seen plates measuring 20 × 17 inches and 22 × 15, and that he was informed that the miners had sometimes obtained considerably larger ones. The unaltered mica in this district has a smoke-brown or reddish-brown colour, occurs in plates of moderate thickness, and is highly transparent—the ruby mica of commerce. Occasionally it is pale, or olive green. In the same district **Mallet** states that "very large crystals of dark brown, apparently uniaxial mica" (biotite) "are sometimes met with," also a certain amount of "lead-

MICA. **Economic uses of Mica.**

OCCURRENCE. grey and violet-grey lepidolite." Plates of muscovite nearly a foot in length are found in some of the granite veins of Mysore and furnish mica for painting on. In some parts of the Western Ghâts and on the table lands to the east, mica is found in plates large enough for windows and lanterns. **Mr. Brough Smyth** mentions "large plates" as occurring in some of the quartz veins of the metamorphic rocks of the Wynaad. **Irvine** states that in parts of Rájputána "very large plates of talc" (mica) "can be extracted," and **Mallet** describes plates of fair size from the Chattarbhaj hills north-east of Tonk and from Jaipur which are, however, inferior in quality to fair Házaribágh mica. Large crystals are found in the granite veins at Wangtu bridge on the Sutlej, and **Baden Powell** records "a fine specimen in large plates from Gurgáon" as having been exhibited at the Lahore Exhibition of 1864 (*Ball, Mallet, &c.*).

DYE. 511 **Dye.**—The powdered mineral is occasionally employed in calico-printing, and frequently by washermen to give a sparkle to cloth, to which the particles adhere.

MEDICINE. 512 **Medicine.—U. C. Dutt** informs us that four varieties of talc (mica) are described by Sanskrit writers, namely, white, red, yellow, and black. Of these, the white is used for making lanterns and the black employed medicinally. Before use it is purified by being heated and washed in milk; the plates are then separated and soaked in the juice of **Amarantus polygamus,** *Linn.* (*tandulia*), and *kánjika* for eight days. It is then reduced to powder by being rubbed with paddy within a thick piece of cloth; the powder passes through the interstices of the fabric and is collected for use. In this form it is called *dhányábhraka*. It is further prepared for medicinal use by being mixed with cow's urine and exposed to a high degree of heat within a closed crucible for a hundred times. The process is said to be sometimes repeated one thousand times. When this is the case, the preparation is called *sahasra putita abhra* and is sold for as much as R8 per tola. Mica thus prepared is a powder of a brick-dust colour, and saline, earthy taste. It is considered tonic and aphrodisiac, and is used in combination with iron in anæmia, jaundice, chronic diarrhœa and dysentery, chronic fever, enlarged spleen, urinary diseases, impotence, &c. Its efficacy is said to be increased by combination with iron. Dose:—grains six to twelve (*Mat. Med. of Hindus*).

The only materials of therapeutic value present to any extent in this preparation must be a little iron and potash, with a mass of inert alumina and silica. As usual in Sanskrit medicine, the drug is rarely administered alone, but enters into the composition of many complicated preparations, the other ingredients of which probably are alone of value. For a description of these the reader is referred to **U. C. Dutt's** *Materia Medica of the Hindus*. **Ainslie** states that the "Vytians" consider mica to have virtues in pulmonic affections, and a dark sort to be of value in "flux cases." He further mentions that the Chinese imagine it to have the power of prolonging life.

DOMESTIC. 513 **Domestic, &c.**—The uses of mica all depend on its transparency, elasticity, the thinness of its foliæ, and the ease with which it can be cut; while for special purposes its property of withstanding great heat renders it invaluable. It may be substituted for glass in lanterns, doors of furnaces, windows, as a glazing material for pictures, for the backing of mirrors, &c. In India, it is very largely employed for ornamenting temples, palaces, and many of the banners, robes, &c., employed in ceremonies. It is also used in very fine fragments, or as powder for ornamenting pottery and ordinary clothes, and is a favourite substance with native artists for painting on.

MICHELIA, *Linn.; Gen. Pl., I., 19.* 514

A genus of lofty trees, which comprises about twelve species, natives of the temperate and tropical mountains of India.

Michelia Cathcartii, *Hook. f. & Th.; Fl. Br. Ind., I., 42; H. f. [Ill., Him. Pl., t. 7;* MAGNOLIACEÆ. 515

Vern.—*Kala champ,* NEPAL; *Atokdúng,* LEPCHA.

References.—*Gamble, Man. Timb., 6; Ind. Forester, I., 94; VIII., 405.*

Habitat.—A large tree which is found in the temperate forests of the Sikkim Himálaya, at altitudes of 5,000 to 6,000 feet.

TIMBER. 516

Structure of the Wood.—Sapwood large, white; heartwood dark olive brown, moderately hard, weight 41 lb. It is used for planking, and would do well for tea-boxes (*Gamble*).

M. Champaca, *Linn.; Fl. Br. Ind., I., 42; Wight, Ill., I., 13, 14, [t. 5, f. 6.* 517

Syn.—MICHELIA RUFINERVIS, *DC.*; M. DOLDSOPA, *Ham.*; M. AURANTIACA, *Wall.*; M. RHEEDII, *Wight.*

Vern.—*Champá, champac, champaca,* HIND.; *Champá, champaka,* BENG.; *Kanchanamu, chámpá,* URIYA; *Tita-sappa,* ASSAM; *Oulia champ,* NEPAL; *Champa,* N.-W. P.; *Chamúti, champa, chamba,* fruit=*chamakhri, chamoti,* PB.; *Champa, chápha,* BOMB.; *Pivalá-chápha, sonachápha, kud chámpa,* MAR.; *Ráe champo, champo, píto-champo,* GUZ.; *Champa,* DEC.; *Shampang, shembugha, shimbu, sempangam,* TAM.; *Shampangi-puvvu, champakamu, chámpéyamu, kánchanamu, gandhaphali, hémángamu,* TEL.; *Sampage-huvvu, sumpaghy, kola sampige, sampige,* KAN.; *Bongas jampacca, champakam,* MALAY.; *Saga,* BURM.; *Champakamu, chámpéyamu, kánchanamu, sappu, gand 'hap' halí, hémángamu, hémapushpakamu,* SING.; *Champaka,* SANS.

References.—*Roxb., Fl. Ind., Ed. C.B.C., 453; Voigt, Hort. Sub. Cal., 12; Brandis, For. Fl., 3; Kurz, For. Fl. Burm., I., 25; Gamble, Man. Timb., 6; Stewart, Pb. Pl., 5; Mason, Burma and Its People, 408, 740; Sir W. Elliot, Fl. Andh., 33, 57, 68, 69, 81, 166; Sir W. Jones, Treat. Pl. Ind., V., 128; Rheede, Hort. Mal., I. t., 19; Rumphius, Amb., II., 199, 202, tt., 67, 68; Pharm. Ind., 6; O'Shaughnessy, Beng. Dispens., 193; Taylor, Topog. of Dacca, 56; Moodeen Sheriff, Supp. Pharm. Ind., 174; Mat. Med. S. Ind. (in MSS.), 8; U. C. Dutt, Mat. Med. Hindus, 294; Sakharam Arjun, Cat. Bomb. Drugs, 5, 204; K. L. De, Indig. Drugs Ind., 73; Dymock, Mat. Med. W. Ind., 2nd Ed., 23; Dymock, Warden and Hooper, Pharmacog. Ind., Vol. I., 42; Cat. Baroda Darbar, Col. Ind. Exhib., No. 128; Baden-Powell, Pb. Pr., 326, 585; Drury, U. Pl. Ind., 292; Atkinson, Him. Dist. (Vol. X., N.-W. P. Gaz.), 304; Useful Pl. Bomb. (Vol. XXV., Bomb. Gaz.), 2, 289, 391; Cooke, Oils and Oilseeds, 59; McCann, Dyes and Tans, Beng., 90; Ain-i-Akbari Blochmann's Trans., Vol. I., 76, 82; Linschoten, Voyage to East Indies (Ed. Burnell, Tiele and Yule), Vol. II., 36; Statistics Dinajpur, 154; Darrah, Note on Assam, 31; Man. Madras Adm., Vol. II., 144; Gazetteers:—Orissa, II., 159, 179; Bombay, V., 23, 285; VII., 40; X., 40; XIII., 24; XV., 73; XVIII., 45; N.-W. P., I., 78; IV., lxvii; Mysore and Coorg, I., 48, 57; II., 7; Agri.-Hort. Soc., Ind.:—Trans., II., 167; V., 119; VII., 48; Journals (Old Series), IV., 123, 124, 199; IX., 399, (Sel.) 53; XIII., 345; Indian Forester, I., 88; III., 189, 200; IV., 47, 101, 229; VIII, 127; XII., (App.) 27; Spons' Encycl., 1422, 1694; Balfour, Cyclop. Ind., II., 942; Smith, Ec. Dic., 105.*

Habitat.—A large evergreen tree, with yellow sweetly-scented flowers, cultivated throughout India from the Ravi southwards, and up to 5,400 feet in the North-West Himálaya. Wild in Nepál, Bengal, Assam (ascending to 3,000 feet), Burma, in the Nilghiris, and the forests of the Western Gháts as far as Kanara.

DYE. Flowers. 518

Dye.—The FLOWERS when boiled yield a yellow dye which is sometimes used as a base for other colours. They communicate an agreeable perfume to the fabric.

MICHELIA Champaca.

Medicinal uses of the Champac.

OIL. 519 **Oil.**—The SEEDS are said to yield a fatty oil, but it appears doubtful if this be really the case. The FLOWERS yield a volatile oil, and the Seeds. 520 LEAVES when distilled produce a sweetly-scented water. The otto somewhat resembles *Ilang* (*Conf.* with Vol. II., 93), for which it is used as an Flowers. 521 adulterant. **Rumphius** describes an oil scented by the flowers, as used Leaves. to anoint the body and hair.

522 **Medicine.**—The FLOWERS and FRUIT are considered bitter and cool MEDICINE. Flowers. remedies, and are used in dyspepsia, nausea, and fever. The LEAVES, anointed with *ghí*, and sprinkled over with powder of cumin seeds, are said 523 in the *Baroda Durbar Catalogue, Col. Ind. Exhib.*, to be put round the head Fruit. in cases of puerperal mania, delirium, and maniacal excitement. **Taylor** 524 states (*Topography of Dacca*) that the flowers mixed with sesamum oil form Leaves. an external application which is often prescribed in vertigo. The flowers 525 beaten up with oil are also applied to fœtid discharges from the nostrils. According to **Rumphius**, the flowers are useful as a diuretic in renal diseases and in gonorrhœa. **Rheed** estates that the dried ROOT and ROOT-BARK, Root. mixed with curdled milk, is useful as an application to abscesses, clearing 526 away or maturing the inflammation, and that, prepared as an infusion, Root-Bark. it is a valuable emmenogogue. He also states that the perfumed OIL pre- 527 pared from the flowers is a useful application in cephalalgia, ophthalmia, Oil. and gout, and that the oil of the seeds is rubbed over the abdomen to 528 relieve flatulence. **Rumphius** mentions that the young leaves, contused and macerated in water, when instilled into the eyes are believed to clear the vision. In Dacca the juice of the leaves is given with honey in cases of colic (*Taylor*).

Bark. In more modern times the BARK has attracted notice. It is describ- 529 ed in the non-officinal list of the *Pharmacopœia of India* as having febrifuge properties, and **Dr. Kani Lall De** states that it is an excellent substitute for guaiacum. It is, however, little used by the Natives and is not as a rule to be met with in the drug shops. In the *Gazetteer of Orissa,* the bark is described as stimulant, expectorant, and astringent, the seeds and fruit are said to be useful for healing cracks in the feet, and the root is described as purgative. **Moodeen Sheriff**, in his forthcoming *Materia Medica of Southern India,* mentions the flowers as a very efficient stimulant, antispasmodic, tonic, stomachic and carminative, and describes an infusion, decoction, and tincture, particularly recommending the last.

SPECIAL OPINIONS.—§ "Juice of fresh leaves is said to act as a vermifuge" (*Surgeon-Major B. Gupta, M.B., Pooree*). "The bark is valuable as a tonic and febrifuge,—*vide Indian Medical Gazette* for February 1875, page 38, for further particulars" (*Civil Surgeon B. Evers, M.D., Wardha*). "Said to have diuretic properties" (*J. Parker, M.D., Deputy Sanitary Commissioner, Poona*). "Expressed juice of the leaves is used as febrifuge tonic, but it is inferior to other bitter tonics" (*Civil Surgeon S. M. Shircore, Moorshidabad*). "An oil prepared by macerating the fresh flowers in sweet oil is a useful application in subacute rheumatism" (*Civil Surgeon J. Anderson, M.B., Bijnor, North-Western Provinces*). "The fresh flower-buds are largely imported by post into Simla, packed between pieces of banana stem" (*Surgeon-Major J. E. T. Aitchison, Simla*). "I have lately found that the flowers of **Michelia Champaca** are not only the best, cheapest, and most convenient part of that plant, but also one of the cheapest, commonest, and most useful drugs in this country. Before adopting this drug in my practice, I made some trials of it on some healthy persons, including myself, and found its physiological actions to resemble those of Spirit Ammon. Arom., Tinct. Card. Co., Tinct. Cascarillæ and a few other such drugs. They may be used in infusion or tincture, prepared in the

MEDICINE.

ordinary way, the proportion of the flowers, in coarse powder, being two ounces and a half to one pint of boiling water in the former, and to the same quantity of rectified spirit in the latter. The tincture of the flowers is much more efficient than their decoction, and to ensure the best effects of the former in severe or special cases, it should be used repeatedly in the same way as other stimulants are under similar circumstances. The dose of the infusion is from 2 to 3 ozs. and of the tincture from 1 to 2 drachms, three or four times in the 24 hours. The bark of the plant is antiperiodic and is used in decoction, prepared by boiling it in two pints of water, till the liquid is reduced to 1 pint. The dose of the decoction is from 1½ to 3 ozs. Although **M. Champaca** is met with in many places of Southern India, it is very scarce in some localities. Madras is an example of the latter, and there are not more than two or three plants in the whole of this city. The dry flowers, however, are common and abundant in the bazár" (*Honorary Surgeon Moodeen Sheriff, Khan Bahadur, G.M.M.C., Triplicane, Madras*).

Food.—The little straw-coloured FRUIT, which is said to be eaten, ripens in the cold season. According to **Captain Hannay**, the BARK is eaten with *pán* by the natives of Assam. FOOD. Fruit. 530 Bark. 531

Structure of the Wood.—Soft, seasons, and polishes well; sapwood white, heartwood light olive brown; growth moderate; weight from 37 to 42℔ per cubic foot. It is very durable,—for example, a specimen cut by **Griffith** in 1836 is still in Calcutta and is as sound as if fresh cut. It is used for furniture, house-building, carriage work, and native drums. In Northern Bengal it is considered valuable for planking, door panels, and furniture; and in Assam for buildings and canoes (*Brandis; Gamble*). **Stewart** states that in the Panjáb it is one of the *padsháhí* trees, *i.e.*, reserved for *Rájas*, and **Baden Powell** adds that the tree is seldom used notwithstanding its fine timber, because it is considered sacred. TIMBER. 532

Domestic and Sacred.—The *múnga* silkworms, according to **Captain Jenkins**, are sometimes fed on this tree in Assam. He, however, probably alludes not to the *múnga*, but to some other closely allied silkworm, as yet not determined, since the *champa-pattea-mung* silk has properties that would separate it from the ordinary *munga*. The strongly scented yellow FLOWERS are worn in the turban by males and in the hair by females throughout the Panjáb and other parts of the country. **Rumphius** states that the natives of Malay, Java, and Macassar use the flwers similarly, place them amongst their clothes for the sake of their perfume, and employ them largely for decorating their nuptial beds. The flowers also form a frequent offering at Hindu shrines, and the tree is often planted in the vicinity of temples. DOMESTIC & SACRED. Leaves. 533 Flowers. 534

Michelia excelsa, *Blume; Fl. Br. Ind., I., 43; Wight, Ill., I., 14.* 535

WHITE MAGNOLIA.

Syn.—MAGNOLIA EXCELSA, *Wall.*

Vern.—*Bara champ, safed champ,* NEPAL; *Sigugrip, pendre,* LEPCHA; *Gók,* BHUTIA.

References.—*Griff., Ic., IV., 655; Gamble, Man. Timb., 6; Indian Forester, I., 94; VIII., 405; XI., 315, 355.*

Habitat.—A lofty, deciduous tree, found in the Temperate Himálaya from Nepál to Bhután, at elevations of 5,000 to 8,000 feet, and on the Khasia Hills.

Structure of the Wood.—Soft; sapwood small, white; heartwood olive-brown, glossy, yellow when fresh cut; growth rather slow; weight 33 to 34℔ per cubic foot. It is very durable and is used for building, chiefly for planking, door and window frames, and for furniture. It is the principal wood employed for these purposes in the Darjíling Hills (*Gamble*). TIMBER. 536

537 **Michelia nilagirica,** *Zenk.; Fl. Br. Ind., I., 44; Wight, Ic., t. 938.*

Var. α Wightii,—M. OVALIFOLIA, *Wight.*

Var. β Walkeri,—M. WALKERI *and* M. GLAUCA, *Wight.*

Vern.—*Pila champa,* HIND.; *Shempangan, sempagum, shembugha,* TAM.; *Walsappú, sapú,* SING.

References.—*Beddome, Fl. Sylv., t. 62; Dymock, Warden, and Hooper, Pharmacog. Ind., I., 43; Drury, U. Pl., 292; Balfour, Cyclop., II., 942; Gazetteer, Mysore and Coorg, I., 57; Ind. Forester, II, 22; X., 35, 552; Madras Man. Adm., II., 110.*

Habitat.—A tall tree, or shrub at high elevations, found in the higher forests of the Western Gháts and Ceylon at altitudes of 5,000 to 6,000 feet.

OIL. 538 **Oil.**—The reputed oil of this tree (*Pharm. Jour., Oct. 22, 1887, 334*) was in reality distilled from the bark of **Cinnamomum Wightii** (*Pharmcog. Ind.*).

MEDICINE. Bark. 539 **Medicine.**—Attention has recently been drawn to the febrifuge properties of the BARK. It is said to have been made into decoctions and infusions and used as a febrifuge, but there is no evidence of this property being known to the Natives of India. An examination of the bark made in Europe yielded the following results:—"The powdered bark gave 10·6 per cent of moisture, and left 9·7 per cent. ash. It contained a volatile and a fixed oil, acrid resins, tannin giving a greenish black colour with ferric salts, sugar, a bitter principle, mucilage, starch, calcium oxalate, &c. Search was made for alkaloids and mannite but with negative results. A decoction did not give the usual blue colour with iodine until a considerable quantity of the reagent had been added, a reaction peculiar to cinnamon and cassia barks" (*Pharm. Ind.*).

TIMBER. 540 **Structure of the Wood.**—"Strong, close, and fine-grained, but very hygrometrical; it is used for building purposes, beams, and rafters" (*Beddome*).

541 **M. oblonga,** *Wall.; Fl. Br. Ind., I., 43.*

Syn.—M. LACTEA, *Wall.*

Vern.—*Sappa, phulsappa,* Ass.

Reference.—*Gamble, Man. Timb., 7.*

Habitat.—A tree of the Khasia Hills and Assam.

TIMBER. 542 **Structure of the Wood.**—Sapwood white, heartwood dark grey, soft; weight 40lb per cubic foot. It is used in Assam for canoes and rough furniture (*Gamble*).

543 **MICROMERIA,** *Benth.; Gen. Pl., II., 1188.*

A genus of herbs or undershrubs, which belongs to the Natural Order LABIATÆ, and comprises about sixty species, of which three are natives of India. Of these **M. biflora,** *Benth.* (*Fl. Br. Ind., IV., 650*), is described by **Stewart** as having a weak odour of thyme; while **M. capitellata,** *Benth*, (*Fl Br Ind IV., 649*), a small plant inhabiting the Nilghiris and the Western Gháts, Parisnath, and the Western Himálaya, has the aromatic and carminative properties of **Mentha piperita.**

Mildew, see **Fungi and Fungoid Pests,** Vol. III., 455.

544 **MILIUSA,** *Lesch.; Gen. Pl., I., 28, 958.*

A genus of trees or shrubs which comprises seven species, all Indian. In addition to the species described below, it may be remarked that **M. Roxburghiana,** *Hook. f. & T.* (*Sungden,* LEPCHA; *Tusbi,* SYLHET), **M. sclerocarpa,** *Kurz*; and **M. macrocarpa,** *Hook f. & T.*, are small timber trees with hard, rather heavy wood.

Miliusa velutina, *H. f. & T.; Fl. Br. Ind., I., 87;* ANONACEÆ. 545

Syn.—UVARIA VELUTINA, *Dunal*; U. VILLOSA, *Roxb.*; GUATTERIA VELUTINA, *A. DC.*

Vern.—*Dom-sál, gidar-rúkh, kári,* HIND.; *Lakhari,* KOL.; *Ome,* SANTAL; *Bari kári, kajrauta, kharrei,* OUDH; *Gidar-rúkh, gwiya, goá-sál, dom-sál,* N.-W. P.; *Kári,* C. P.; *Peddachilka dúdúga, nalla dadúgá,* TEL.; *Thabutgyi,* BURM.

References.—*Roxb., Fl. Ind., Ed. C.B.C., 456; Brandis, For. Fl., 6; Kurz, For. Fl. Burm., I., 47; Beddome, Fl. Sylvt., 37; Gamble, Man. Timb., 9; Elliot, Fl. Andh., 124; Campbell, Ec. Prod., Chutia Nagpur, No. 9235; Gazetteers:—N.-W. P., IV., lxvii; Burma, I., 127; Ind. Forester, III., 200; IX., 437; X., 325.*

Habitat.—A large deciduous tree, met with in Garhwal, Behar, Malwa, Orissa, Malabar, and Pegu.

Food.—Flowers in March and April, produces BERRIES in June and July. They are very much like black cherries, and according to Campbell are used as food in Chutia Nagpur. FOOD. Berries. 546

Structure of the Wood.—Heart and sap-wood not distinct, sulphur-yellow when fresh, light brown when old; moderately hard; weight from 40 to 50lb per cubic foot. It is easily worked and durable, but liable to warp, and is used for small beams, cart-poles, yokes, agricultural implements, spear-shafts and bars (*Brandis*). TIMBER. 547

Milk Plants, see **Calatropis**, Vol. II., 33; also **Euphorbia**, Vol. III., 294; and the members of the ASCLEPIADEÆ, APOCYNACEÆ, EUPHORBIACEÆ, and URTICACEÆ generally. For further information, see **India-rubber**, Vol. IV., 337-382.

MILLETS. 548

In those localities of India, in which rice is not cultivated as the staple food, the millets are exceedingly important. This is markedly the case in wheat-producing districts, a fact which shows conclusively that the bulk of Indian wheat is grown essentially for foreign trade. With the exception of Burma, rice may almost be said to be grown for home consumption; but where it cannot be cultivated, the millets invariably take its place, as the staple food-crop of the mass of the people. The total area of land under millets in 1888-89 has been estimated as 35,154,468 acres, of which Bombay had 15 million, Madras 11½ million, the North-West Provinces 1¾ million, the Panjáb nearly 5 million, and Berár a little over 2 million acres. In Bengal millets are scarcely cultivated, excepting to a small extent by the hill tribes. The exports of millets to foreign countries during the past two years were in quantity and value as follows:—

	Cwt.	R
1887-88	640,705	18,47,620
1888-89	611,960	19,84,220

The following are the more important millets arranged alphabetically, for further information regarding which the reader is referred to the article on each in its respective position in this work.

Eleusine Coracana, *Gærtn.*; GRAMINEÆ. RAGÍ or MARÚA MILLET.

A procumbent grass, most probably a native of India, and widely cultivated during the rainy season, but chiefly in South and West India. It has been grown in Egypt during modern times, is mentioned by Sanskrit authors under the name of *Rájika*, but apparently was not known to the Arabs, Greeks, or Romans. This is looked upon as the staple grain of Mysore. A fermented liquor is made from it, also a kind of beer. Ragi in 1889 sold in Madras at from 28·42 to 31·36 seers per rupee.

MILLETS.

List of important Millets.

Panicum frumentaceum, *Roxb.;* SHAMULA OF SAWAN.

This is the most rapid growing of all the millets, affording, through its early ripening, a cheap grain which comes into market before the main autumn food-crop is harvested. It is subject to the danger of destruction through excessive rain and blight. The grain is wholesome and nourishing, and is a favourite food with the poorer class.

P. miliaceum, *Linn.;* CHENA OF THE COMMON MILLET.

This is viewed as a native of Egypt and of Arabia, and was evidently introduced at a very early date into India. In Europe, Egypt, and Asia its cultivation is pre-historic. In point of value as a food-stuff, it is supposed to be inferior to **Setaria italica,** *Beauv.*, and fetches accordingly a much lower price. It also grows very much more slowly than that species, but has one advantage of great importance to the poorer hill tribes, *viz.*, it may be successfully cultivated on indifferent soils up to an altitude of 10,000 feet. The straw is of no use as a fodder, and is accordingly thrown away.

Paspalum scrobiculatum, *Linn.;* KODA OR KHODON MILLET.

A native of India luxuriating in light, dry, loose soils, cultivated in the rainy season. It is far more extensively grown than any of the other minor millets, owing to the readiness with which it may be cultivated. It is a common and cheap grain, and an important article of food with the poorer classes, particularly those who inhabit the mountains and the more barren parts of the country; but it is considered unwholesome, as it tends to produce diarrhœa. It is said to have an intoxicating effect. The straw is given as fodder to cattle.

Pennisetum typhoideum, *Rich.;* THE SPIKED MILLET OR BAJRA.

A native of tropical Asia, Nubia, and Egypt. Cultivated to a large extent in Northern and Southern India, especially on the Coromandel Coast and in the North-West Provinces during the rainy season. The grain is used chiefly by the lower classes of natives, and is eaten in the cold season. It is considered heating, but more nutritious than rice. Bajra in 1889 sold in the North-West Provinces at from 15·73 to 19.7 seers per rupee. The fodder is much used.

Setaria italica, *Beauv.;* GERMAN OR ITALIAN MILLET, THE KANGNI OF THE NORTH-WEST PROVINCES.

This is supposed to be a native of China, Japan, and the Indian Archipelago. It is, however, extensively cultivated in India, both on the plains and on the hills ascending to 6,000 feet above the level of the sea. Two crops may be taken off the same field a year, but this millet is chiefly grown as a subsidiary crop; there are two varieties. The grain is much approved as an article of food. The flour is made into pastry and is valued as a food for invalids. The Brahmins specially esteem it. The grain is in demand as a food for cage-birds. The straw is not much valued.

Sorghum vulgare, *Pers.;* THE GREAT MILLET OR GUINEA CORN; JUAR CHOLUM, *Hind.*

This is perhaps the most important of the millets, and with *bajra* is regularly exported. According to some authors, this plant is indigenous to India and China. It is cultivated in Lower Egypt at the present day under the name *dourra*, and an analogous form is wild in equatorial Africa. DeCandolle inclines to the view that it is more probably a native of Africa than of Asia.

Juar in 1889 sold in the Panjáb at 23·46 to 29·65 seers per rupee.

MILLETTIA, *W. & A.; Gen. Pl., I., 498.* 549

A genus of trees or large climbing shrubs which belongs to the Natural Order LEGUMINOSÆ, and comprises some forty to fifty species, which are dispersed through the tropics of the Old World. Of these, about 24 are natives of India. Few are of much economic value. **M. pendula,** *Benth (Fl. Br. Ind., II., 105)—Thinwin,* BURM —found in the savannah, and dry lower hill forests of Burma up to 2,000 feet, has a very hard heavy wood, beautifully streaked and of a dark colour. It is used for cross pieces of harrows, but is worthy of attention owing to its beautiful grain and colour. **M. atropurpurea,** *Benth. (Fl. Br. Ind., II., 108)—Danyinnie,* BURM.—and **M. leiogyna,** *Kurz (Fl. Br. Ind., II., 109),* natives of the forests of Martaban, Tenasserim, Malacca, and Penang, are said by *Kurz* to yield red resins, of which no economic information is available. **M. auriculata,** *Baker (Fl. Br. Ind., II., 108)—Hehel,* SANTALI; *Hél,* KOL.; *Gurar,* KHARWAR; *Mandh,* OUDH; *Gonjha, ganj,* KUMAON; *Gúrúr,* GONDI; *Murari,* KURKU; *Gonjo,* NEPAL; *Brúrik,* LEPCHA—is a very common large climber of the Sub-Himálayan tract, from the Sutlej to Bhutan, ascending to 3,500 feet. Its roots are applied to sores on cattle to kill vermin; and are also used to poison fish *(Campbell, Ec. Prod., Chutia Nagpur, 7566).* Dr. Watt found a species of **Millettia** used in Manipur as a fermenting agent. See **Malt Liquors,** p. 136.

MILLINGTONIA, *Linn.; Gen. Pl., II, 1040.* 550

Millingtonia hortensis, *Linn. f.; Fl. Br. Ind., IV., 377;* BIGNONIACEÆ.

THE INDIAN CORK TREE.

Syn.—BIGNONIA SUBEROSA, *Roxb.*

Vern.—*Mini-chambeli, akas-nim,* HIND.; *Mach-mach,* URIYA; *Nimi-chambeli, akas-nim,* BOMB.; *Beratu mara,* KAN.; *Kát malli,* TAM.; *Aykayet,* BURM.

References.—*Roxb., Fl. Ind., Ed. C. B. C., 495; Brandis, For. Fl., 347; Kurz, For. Fl. Burm., II., 238; Beddome, Fl. Sylv., t. 249; Gamble, Man. Timb., 274; Dalz. & Gibs., Bomb. Fl. Supl., 55; Lisboa, Pl. Bomb., 104; Gazetteers:—N.-W.-P., I., 82; IV., lxxiv.; Burma, I., 139; Bombay, V., 27; Ind. Forester, XII., 139; XIII., 339; XIV., 142; Gribble, Man. of Cuddapah, 71; Settlement Rept., N.-W.-P., Shahjehanpur, IX.*

Habitat.—A large tree, cultivated in avenues and gardens in most parts of India; indigenous in Burma and the Malay Archipelago, perhaps also wild in Central India, on the Upper Godavery. **Kurz** says it is rather rare in the tropical forests from Martaban down to Tenasserim.

TIMBER. 551

Structure of the Wood.—Soft, yellowish white, weight 42lb per cubic foot, takes a fine polish, and is adapted for furniture and ornamental work.

DOMESTIC. Bark. 552

Domestic.—From the BARK an inferior kind of cork is made.

MILLSTONES. 553

"The usual conception of a millstone is that it should be a hard, tough, coarse, silicious sandstone or grit, and these characters have become perpetuated in the term millstone grit, which has been conferred upon a group of rocks underlying the coal measures. In the absence of sandstones or grits many other rocks, such as quartzites, gneiss, granite, and trachyte, are used, and this is particularly the case in India owing to the cost of carriage of the best stones to distant points. As a general rule, within the rocky tracts of India, the natives, if they happen to belong to these sections of the population who use ground meal, have shown considerable intelligence in selecting the material best suited for the purpose, and rude quarries which have been worked from time immemorial, are generally to be found in such regions. Where hard and tough rocks are not to be found, softer ones are used, with the natural result that the meal or flour contains a greater or less amount of grit and dust" (*Ball*).

554 **Millstones**, *Ball in Man. Geology of India, III., 559.*

References.—*Baden Powell, Pb. Prod., 58; Settlement Rept., Hazára Dist., 105; Bombay Gazetteer, V., 22.*

OCCURRENCE. 555 **Occurrence.**—Information regarding the actual sources of millstones, in the various districts of India, is very incomplete. **Ball** writes, "Excluding the inferior qualities of millstones made from gneiss, granite, &c., the oldest rock which is used for this purpose is a kind of arkose or grit, which occurs in the older transition rocks. The Vindhyan series affords a variety of materials of different degrees of density and texture. Among the Gondwána rocks the Barakar grits are perhaps the most largely employed, and in many of the coal fields they are quarried rudely at the surface for this purpose." A quarry has been worked for many years at Jutkuttia, in the Karakpur hills, Behar, in a bed of coarse arkose which presents some resemblance to the true millstone grits, but the origin of its component minerals from crystalline rocks is more directly and prominently apparent.

In Bombay, thick-bedded sandstones of the Kaladgi series, seen at Gudur, Parvate, Guldegudd, and elsewhere, have been recommended by **Mr. Foote** as likely to be well suited for the manufacture of large millstones. **Mr. Wynne** alludes to several rocks in Cutch which furnish tough millstones. These are silicious grits, which occur both in the jurassic and sub-nummulitic groups, and a very similar rock of nearly black colour is found in the tertiary beds at Karimori hill; they are also obtained near Chunduya, west of Anjar.

TRADE. 556 **Trade.**—Factories at the presidency towns, as a rule, import the millstones they require from Europe. **Ball**, commenting on this, writes, "It will be long before India will produce anything equal to the burrstones which have, to a great extent, superseded the ordinary millstones." It is somewhat interesting to remark, however, that millstones of superior quality were exhibited at the Colonial and Indian Exhibition (in the Indian Economic Section) and that these attracted considerable attention.

MIMOSA, *Linn.; Gen. Pl., I., 593.*

[Vol. VI., Pt. I.

Mimosa dulcis, *Roxb.*; see **Pithecolobium dulce**, *Benth.*; LEGUMINOSÆ;

M. pudica, *Linn.; Fl. Br. Ind., II., 291.*

THE SENSITIVE PLANT.

Vern.—*Lajálú, lajjávati*, HIND.; *Lájak*, BENG.; *Lájwánti*, KUMAON; *Lájwanti*, PB.; *Zhand*, PUSHTU; *Lájálu, lájri*, MAR.; *Lájálu, risámani*, GUZ.; *Total-vadi*, TAM.; *Pedda nidra kanti, atta patti*, TEL.; *Mudugudavare*, KAN.; *Hte-ka-yung, takayung*, BURM.; *Váráhakrántá, lajjálu*, SANS.

References.—*Roxb., Fl. Ind., Ed. C.B.C., 423; Elliot, Flora Andh., 187, 149; Ainslie, Mat. Ind., II., 432; U. C. Dutt, Mat. Med. Hind., 307, 322; Dymock, Mat. Med. W. Ind., 2nd Ed., 275; Pharmacographia Indica, I., 538; Catal. Baroda Darbar, Col. Ind. Ex., No. 130; Baden Powell, Pb. Pr., 345; Atkinson, Him. Dist., 309; Gazetteers:—Mysore and Coorg, I., 59; Bombay, XV., 433; Peshawar, 26; Ind. Forester, X., 35.*

Habitat.—A small shrub, naturalised over the greater part of Tropical and Sub-tropical India, probably introduced from Tropical America. In some parts of the lower provinces it occurs in such abundance, along roadsides, as to constitute the chief if not sole vegetation, and gives to such tracts a singular appearance as the traveller leaves behind him a field of drooping leaves.

MEDICINE. Root. 558 **Medicine.**—Ainslie informs us that "a decoction of the ROOT is considered on the Malabar Coast to be useful in gravellish complaints. The

Leaves. 559 Vytians of the Coromandel side of India prescribe the LEAVES and root in cases of piles and fistula: the first are given in powder, in a little milk of

MEDICINE.

the quantity of two pagodas weight or more, during the day." Baden Powell describes the same uses of the plant in the Panjáb. Dymock states that by Muhammadan writers it is considered resolvent, alterative, and useful in diseases arising from corrupted blood and bile. The JUICE is also applied externally to fistulous sores. It is gathered at special times, and its administration is accompanied by a certain ceremony. "In the first week, all bilious diseases and fevers are cured; in the second, piles, jaundice, &c., and in the third leprosy, scabs, and pox. In the Konkan the leaves are rubbed into a paste and applied to hydrocele; and their juice with an equal quantity of horse's urine is made into an *anjan*, used to remove films of the conjunctiva" (cornea?) "by setting up an artificial inflammation." The root is used as a charm for certain forms of cough, being tied round the neck (*Mat. Med. W. Ind.*). Surgeon-Major Oalthrop, Morar, writes that the juice of the leaves is used to impregnate cotton wool for a dressing in any form of sinus.

Juice. 560

CHEMICAL COMPOSITION.—The authors of the *Pharmacographia Indica* write, "The tapering thin roots of **M. pudica** contain 10 per cent. of tannin of such a nature as to form a good black ink with salts of iron. The ash of the root amounts to 5·5 per cent."

Chemistry. 561

Mimosa rubicaulis, *Linn.; Fl. Br. Ind., II., 291.* 562

Syn.—M. MUTABILIS, *Roxb.*; M. OCTANDRA, *Roxb.*; M. ROTTLERI, *Spreng.*

Vern.—*Shiah-kanta*, HIND.; *Shiah-kanta, kúchi-kanta*, BENG.; *Sega janum*, SANTAL; *Agla, kinglí, hingreí*, N.-W. P.; *Rál, riaul, didridr, arlú, allá, kikkri*, fruit=*deo khádir*, PB.; *Allá*, RAJ.; *Hujírú*, SIND; *Bída, chandra, undra, ventra*, TEL.

References.—*Roxb., Fl. Ind., Ed. C.B.C., 423; Dals. & Gibs., Bomb. Fl., 85; Stewart, Pb. Pl., 72; Elliot, Flora Andh., 28, 186, 191; Campbell, Ec. Prod., Chutia Nagpur, 8451; Murray Pl. and Drugs, Sind, 136; Baden Powell, Pb. Pr., 585; Atkinson, Him. Dist., 309, 741; Gazetteers:—Mysore and Coorg, I., 59; N.-W. P., I., 80; IV., lxxi; Rawal Pindi, 15; Agri.-Hort. Soc. Ind. Jour. (Old Series), IX., 420; XIII., 309; Ind. Forest., IV., 227; VIII., 101, 110, 417; IX., 13, 451; XII., App. 12.*

Habitat.—A large, straggling, prickly shrub, which is found throughout the greater part of India, and ascends to 5,000 feet in the Western Himálaya; distributed to Afghánistán.

Medicine.—Stewart states that in Chamba the bruised LEAVES are applied to burns, and that the FRUIT also is medicinal. Atkinson attributes to the SEEDS the same properties as those of **M. pudica**, and states that the leaves are prescribed in the form of an infusion for piles. Campbell writes that in Chutia Nagpur "the powdered ROOT is given when, from weakness, the patient vomits his food; the fruit and leaves are also used medicinally."

MEDICINE. Leaves. 563 Fruit. 564 Seeds. 565 Root. 566

Structure of the Wood.—Sapwood yellowish-white; heartwood red, hard; weight 41 to 52lb; used for making gun-powder charcoal.

TIMBER. 567

Domestic, &c.—It is a valuable hedge plant.

DOMESTIC. 568

M. scandens, *Linn.*, see **Entada scandens, *Bth.***; Vol. III., 245. 569

MIMUSOPS, *Linn.; Gen. Pl., II., 661.*

A genus of trees which comprises some 30 species, natives of the tropics of both hemispheres. Of these four or five are natives of India.

[*1586*; SAPOTACEÆ.

Mimusops Elengi, *Linn.; Fl. Br. Ind., III., 548; Wight, Ic., t.* 570

Vern.—*Mulsári, maulser, bakul mulsari, maulsarau*, HIND.; *Bakul-bakal, bohl*, BENG.; *Baulo, baul*, URIYA; *Maulsári*, N.-W. P.; *Maul siri, maulsari*, PB.; *Ghólsari, bhólsari*, C. P.; *Borsali*, BOMB.; *Ovalli, wowli, vavoli, bakúla*, MAR.; *Bolsari, borasalli*, GUZ.; *Barsoli*, MEYWAR;

Mogadam, magila-maram, TAM.; *Pogada, pogada-mánu*, TEL.; *Bokal-boklu, mugali, kanja, pogada*, KAN.; *Elengi*, MALAY.; *Khaya, kha-ya, gung*, BURM.; *Múnemal*, SING.; *Bakul, vakula*, SANS.

References.—*Roxb., Fl. Ind., Ed. C.B.C., 318; Voigt, Hort. Sub. Cal., 341; Brandis, For. Fl., 293; Kurz, For. Fl. Burm., II., 123; Beddome, Fl. Sylv., t. 40; Gamble, Man. Timb., 245; Dalz. & Gibs., Bomb. Fl., 140; Stewart, Pb Pl., 136; Mason, Burma and Its People, 404, 482; Sir W. Elliot, Fl. Andh., 154; Sir W. Jones. Treat. Pl. Ind., V., 110; Taylor, Topography of Dacca, 58; Moodeen Sheriff, Supp. Pharm. Ind., 175; U. C. Dutt, Mat. Med. Hindus, 188, 322; Sakharam Arjun, Cat. Bomb. Drugs, 83; Dymock, Mat. Med. W. Ind., 2nd Ed., 480; Cat. Baroda Durbar, Col Ind. Exhib.; Birdwood, Bomb. Prod., 269, 286; Baden Powell, Pb Pr., 585; Drury, U. Pl. Ind., 292; Useful Pl. Bomb. (Vol. XXV., Bomb Gaz.), 91, 164, 223, 392; Econ. Prod. N.-W. Prov., Pt. V. (Vegetables, Spices, and Fruits), 78; Gums and Resinous Prod. (P. W. Dept. Rept.), 15, 27; Gums and Resins, 20; McCann, Dyes and Tans, Beng., 136, 152, 159, 160, 165, 168; Selections, Records Govt. India (R. & A. Dept.), 1888-89, 93; Statistics Dinajpur, 152; Moore, Man. Trichinopoly, 79; Gribble, Man., Cuddapah, 262; Settlement Rept., N.-W. P., Shahjehanpore, IX; Gazetteers:—Bombay, V., 285; VII., 42; XI., 25; XIII., 23; XV., 40; XVII., 23; XVIII., 44; Panjáb, Hoshiár-pur, II; N.-W. P., I., 182; IV., lxxiii; Burma, I., 131; Mysore and Coorg, I., 62; II., 8; Rájputána, 27; Agri-Horti. Soc. Ind.:—Journals, (Old Series) IV., 220; VIII., Sel., 178; XIII., Sel. 61.*

Habitat.—A large evergreen tree, frequently cultivated in India, wild in the Deccan and Malay Peninsulas.

GUM. 571 **Gum.**—It yields the *Pogada* gum of Madras (*Birdwood*). No information is obtainable regarding the chemical composition or properties of this gum.

DYES. Bark. 572 **Dyes and Tans.**—The BARK is used in certain districts of Bengal, either by itself or in combination with that of **Terminalia tomentosa** (*ashna*), for dyeing shades of brown (*McCann*). Samples sent to **Mr. Wardle** were reported on as follows:—"This bark is scarcely, in the slightest degree, astringent; it only contains the small amount of brownish-red colouring matter usual with the majority of barks I have examined." With cotton a light grey colour was obtained, and with silk various shades of reddish-drab, drab, and fawn. The bark is also employed as a tan in various parts of the country. Examined by **Professor Hummel** of Leeds it was found to contain 4 per cent. of Tannic acid, and to yield a pale reddish, slightly turbid decoction. Its money value compared with Divi-divi at 12*s.* per cwt., was 1*s.* 1*d.* per cwt.; with Valonia cups at 16*s.* per cwt., 2*s.* 2¼*d.*; with Ground Myrabolans at 7*s.* 6*d.* per cwt., 1*s.* 13½*d.*; and with Ground Sumach at 13*s.* 3*d.* per cwt., 2*s.* 8½*d.* The tanning material is, therefore, of very little commercial value.

OIL. 573 **Oil.**—The FLOWERS contain a volatile oil from which a sweet-scented water is distilled. From the SEEDS a fixed oil is obtained by expression,
Flowers. 574 which is used for culinary purposes, for burning and for medicine. Ac-
Seeds. 575 cording to **Beddome**, it is also employed by painters.

MEDICINE. Fruit. 576 **Medicine.**—This highly ornamental tree has long been valued for its medicinal properties, and is mentioned by early Sanskrit writers. Thus the unripe FRUIT is recommended by **Chakradatta** as astringent, and a useful masticatory for fixing loose teeth. The same writer states
Bark. 577 that the BARK is astringent and that a decoction is useful as a gargle in diseases of the gums and teeth (*U. C. Dutt*). The same uses prevail to the present day in most parts of the country. Thus, in the *Catalogue of the Baroda Durbar Exhibits* at the Colonial-Indian Exhibition, it is stated that the unripe fruit is astringent and is chewed for fixing loose teeth; that the bark is also astringent and is used in discharges from the mucus membranes of the bladder and urethra, as a gargle in relaxation

of the gums, and as a febrifuge. **Dymock** states that a similar use is made of the unripe fruit in the Konkan, and that the fruit and FLOWERS along with other astringents are used to prepare a lotion for sores and wounds. He further writes that, according to **Mir Muhammud Hussain,** "a snuff made from the dried and powdered flowers is used in a disease called *ahwah,* which is common in Bengal. The symptoms of this disease are strong fever, headache and pain in the neck, shoulders, and other parts of the body. The powdered flowers induce a copious defluxion from the nose and relieve the pain in the head." **W. Coldstream, Esq., C.S.,** writes that the bark is much sought after in the Panjáb as a medicine for increasing fertility in women, the trees being frequently killed owing to the extent to which the bark is stripped. In Kánara the water distilled from the flowers is employed as a stimulant. **Taylor** mentions that the SEEDS, bruised and made into a paste, are used to form suppositories in cases of obstinate constipation. **MEDICINE. Flowers. 578**

SPECIAL OPINIONS.—§ "The green fruit is astringent. The bruised seeds after removal of kernel are used in constipation of children, but are highly irritant" (*Bolly Chand Sen, Teacher of Medicine*). "The seed, reduced to paste and mixed with old *ghí,* is used as a suppository in cases of constipation of children. I have frequently seen it employed in this way, and have found hard scybellæ passed within 15 minutes after the introduction of the suppository into the rectum" (*Civil Surgeon R. Macleod, M.D., Gya*). "Decoction of the bark is an excellent astringent gargle" (*Civil Surgeon S. M. Shircore, Moorshedabad*). "The pulp of the ripe fruit is sweetish and astringent and has been successfully used in curing chronic dysentery" (*Surgeon-Major B. Gupta, M.B., Purí*). "Applied to relieve headache" (*Assistant Surgeon S. C. Bhuttacharji, Chanda*).

Food.—The tree produces small fragrant flowers in abundance during the hot season. They fall in showers and are succeeded by small, oval BERRIES, which are yellowish when ripe, and have a small quantity of sweetish pulp, sometimes eaten by the poorer natives. **Lisboa** states that the natives in certain parts of Bombay make a preserve from the fruit. The OIL yielded by the seeds is employed for culinary purposes. **FOOD. Berries. 579 Oil. 580**

Structure of the Wood.—Sapwood large, whitish, very hard; heartwood red; weight about 60℔ per cubic foot (*Gamble*). **Beddome** writes, "It is close and even-grained, pinkish to reddish-brown in colour and takes a good polish." It is used in house-building, for cart shafts and cabinet work, and is said to last for fifty years. **TIMBER. 581**

Domestic and Sacred.—The tree is chiefly cultivated for its ornamental appearance, and its fragrant flowers. The latter are valued for making garlands, are sometimes used for stuffing pillows, and the attar distilled from them is esteemed as a perfume. According to **Drury,** the tree is chiefly grown in Southern India round the mausoleums of Muhammadans. **DOMESTIC. 582**

Mimusops hexandra, *Roxb.; Fl. Br. Ind., III., 549; Wight, Ic.,* [*t. 1587* **583**

Syn.—M. INDICA, *A. DC.;* M. KAUKI, *Wall., not of Linn.*

Vern.—*Kshiri, khír, khirni,* HIND.; *Khirkhejur,* BENG.; *Charkuli,* URIYA; *Raini,* GOND; *Khirni,* N.-W. P.; *Rain, khirni,* RAJ.; *Rájan, kerni, rayani,* BOMB.; *Rájana,* MAR.; *Rankokari, rayan, khirni, kaira,* GUZ.; *Palla, kannu palle,* TAM.; *Pala, palle panlo, palla pandu,* TEL.; *Palé, palú,* SING.; *Do-gota,* AND.; *Rájádani, kshirini,* SANS.

References.—*Brandis, For. Fl., 291; Beddome, For. Man., 142; Gamble, Man. Timb., 246; Grah. Cat. Bomb. Pl., 106; Dalz. & Gibs., Bomb. Fl. 140; Elliot, Flor. Andh., 142; U. C. Dutt, Mat. Med. Hind., 314; Dymock, Mat. Med. W. Ind., 2nd Ed., 482; S. Arjun, Bomb. Drugs, 83;*

Ec. Pd. N.-W. P., t. V., 78; Drury U. Pl., 293; Lisboa. U. Pl. Bomb., 164; Gums and Resinous Prod. (P. W. D. rept.), 15; Cooke, Gums and Gum-resins, 20; Oils and Oilseeds, 60; Gazetteers:—Orissa, II., 179; Bombay, III., 16, 198; IV., 23; V., 26, 285, 360; VI., 12; VII., 40, 42; XIII., 25; N.-W. P., I., 82; IV., lxxiii; Ind. Forester, IV., 230, 292; VIII., 29, 117, 208, 410; X., 31, 39, 547, 548; XI., 490; XII., (App.), 28; XIII., 25, 121; Madras, Man. Administration, I., 114; Gribble, Man. Cuddapah, 262.

Habitat.—A large tree found on the mountains of South India, extending to the sandstone hills of Pachmari, north of the Godavari; also wild in Ceylon, and cultivated in North-West India.

GUM. 584 **Gum.**—This species, like the preceding, yields a gum which appears to be of little value.

OIL. 585 **Oil.**—The SEEDS are said to yield an oil, regarding which no information is available beyond the statement that it is used in Baroda to adulterate *ghí*.

Seeds. 586

MEDICINE. Bark. 587 **Medicine.**—This species has much the same properties as the preceding; thus **Dymock** writes, "The BARK which is used medicinally is exactly similar to that of **M. Elengi.** In the Konkan the MILKY JUICE made into a paste with the leaves of **Cassia Fistula** and the seeds of **Calophyllum inophyllum** is applied to boils. The juice of a **Loranthus** which grows upon the tree is extracted by heat, and given with long pepper in cramp"

Milky Juice. 588

FOOD. Fruit. 589 **Food.**—It flowers in November-December and produces an olive-shaped, yellow BERRY, which is eaten. In the *Baroda Gazetteer* it is stated that the dried fruit is eaten by Hindus on fast days when cooked food is forbidden, and that in the hot weather the poorer classes eat it largely mixed with whey. It is, however, powerfully astringent and cannot be eaten with impunity by those unaccustomed to its use. **Lisboa** states that it is said to be the chief article of food of the poorer classes in Gujarát during the hot weather months. In other parts of the country it appears to be chiefly eaten during times of scarcity.

TIMBER. 590 **Structure of the Wood.**—Heartwood red, very hard, tough, close and even-grained; weight, 60 to 72lb per cubic foot. Used for sugar-mill beams, oil-presses, house-posts, &c., and recommended by **Brandis** as an excellent wood for turning.

Mimusops Kauki, *Linn.; Fl. Br. Ind., III., 549.*

THE OBTUSE LEAVED MIMUSOPS; POMA OT FRUCTA D'ADAO, *Port. at Goa.*

Syn.—M. BALOTA, *Blume;* M. DISSECTA, *Br.;* M. HOOKERI, *A. DC.;* M. BROWNIANA, *Benth.*

Vern.—*Khirni,* HIND.; *Khirni,* root=*khirni lodh,* PB.; *Khirni,* BOMB.; *Kauki,* MAR.; *Búa sau,* MALAY.

References.—*Roxb., Fl. Ind., Ed. C.B.C., 318; Voigt, Hort. Sub. Cal., 341; Brandis, For. Fl., 293; Dalz. & Gibs., Bomb. Fl., Suppl., 50; Stewart, Pb. Pl., 136; Mason, Burma and Its People, 463; S. Arjun, Bomb. Drugs, 204; Murray, Pl. and Drugs, Sind, 140; Baden Powell, Pb. Pr., 365, 585; Drury, U. Pl., 293; Lisboa, U. Pl. Bomb., 92, 164; Birdwood, Bomb. Pr., 167; Gazetteers:—N. W.-P., III., 33; Panjáb Hoshiárpur, 11, Gujrát, 11; Settlement Reports:—Central Provs., Chanda, App. VI.; Panjáb, Gujrát 135.*

Habitat.—A large tree, met with in Burma at Amherst, also in the Malay Peninsula, and tropical Australia; occasionally cultivated as far west as Hoshiárpur, Multán, Lahore, and Gujranwála.

GUM. 591 **Gum.**—A viscid gummy juice exudes on incision from the BARK which may probably be converted into an inferior kind of Guttapercha (*Lisboa*).

Bark. 592 It may be mentioned in connection with this statement that **M. Manilkara,** *Don.* (=**Achras Sapota,** *Linn.*, Vol. I., 80), a tree cultivated in Bengal for its fruit, also yields a substance resembling Guttapercha, which is known in Mexico as CHICLE-GUM. Several other American species of the genus yield similar gums, of which **M. globosa** may more particularly be mentioned as yielding GUM BALATA.

GUM. Chicle-gum. 593

Gum Balata. 594

Oil. - An oil is said to be expressed in Burma from the SEED.

OIL. 595 Seed. 596

Medicine.—The powdered SEEDS are used as an application for ophthalmia, and are also employed internally as a tonic and febrifuge. They are considered hot and moist, and are prescribed in leprosy, thirst, delirium, and disorders of many of the secretions. In the Panjáb they are also considered anthelmintic. The ROOT and BARK are believed to be astringent, and are given in infantile diarrhœa after being ground with water and mixed with honey. The LEAVES, boiled in gingelly oil, and added to the pulverised bark, are considered a good remedy in Beriberi (*Rheede*). The MILK of the tree is said by **Dr. Emerson** to be employed in inflammation of the ear, and in conjunctivitis. The leaves ground and mixed with the root of curcuma and ginger are used as a cataplasm for tumours.

MEDICINE. Seeds. 597 Root. 598 Bark. 599 Leaves. 600 Milk. 601

Food.—The FRUIT, known in Goa as *pome* or *fructa d' Adao* (Adam's apple or fruit), resembles an Ahmedabad *bhór* (**Zyzyphus jujuba**), is slightly acid, and is eaten (*Lisboa*). **Mason** states that the dried fruit is imported by the Chinese in Burma from Singapore. **Drury** states that the acid and esculent fruit is said to increase the appetite.

FOOD. Fruit. 602

Mimusops littoralis, *Kurz; Fl. Br. Ind., III., 549.* 603

ANDAMAN BULLET WOOD.

Syn.—M. INDICA, *Kurz, not of A. DC.*

Vern.—*Kappali*, BURM.; *Dogola*, AND.

References.—*Kurz, For. Fl. Burm., II., 123; And. Report, 42; Jour. As. Soc., 1871, Pt. II., 70; Brandis, For. Fl., 292; Gamble, Man. Timb., 240.*

Habitat.—A large evergreen tree, met with in the coast forests of the Andaman Islands and in Upper Tenasserim. In the former locality it forms nearly pure forests on the level lands behind the beach, and in the mangrove swamps.

Dye.—Major **Ford** states that the BARK is used in the Andamans to produce a red dye.

DYE. Bark. 604

Structure of the Wood.—Sapwood light-coloured, heartwood smooth, reddish or pinkish-brown, close-grained, durable, very hard, and heavy, handsome, but apt to split. Average weight, according to **Brandis**, 67·9lb, value of P. between 748 and 1091, average 895. It is used in the Andamans for bridges and house-posts, and has been tried for sleepers.

TIMBER. 605

MIRABILIS, *Linn.; Gen. Pl., III., 3.*

Mirabilis Jalapa, *Linn.; DC. Prodr., XIII., pt. II., 427;* 606

[NYCTAGINEÆ.

MARVEL OF PERU; sometimes called the "Four o'clock plant" from its flowers opening in the afternoon.

Vern.—*Gulabbás, gule-aabbás, gúla-básh*, HIND.; *Krishno-kéli, gúlá-bás*, BENG.; *Gúl-bánsa*, N.-W. P.; *Gulabbás, abási*, PB.; *Abhasie*, SIND; *Gulbhaji, gul-abbas*, BOMB.; *Gulá-bash*, DEC.; *Pattaráshu*, TAM.; *Bhadrákshi, chandra-malli, chandra-kánta*, TEL.; *Chandra-mallige, gulamaji, sanja-mallige*, KAN.; *Anti-montáram, anti-malari*, MALAY.; *Mizu-bin, myæ-zu*, BURM.; *Sindrika-gahá*, SING.; *Krisnakeli*, SANS.; *Zahr-ul-ajl*, ARAB.; *guli-aabbás*, PERS.

References.—*Gamble, Man. Timb., 302; Elliot, Flora Andh., 34; Mason, Burma and Its People, 433, 781; Rumphius Amb., V., 253, t. 89; Rheede, Hort. Mal., X., t. 75; Pharm. Ind., 184; Ainslie, Mat. Ind., II., 284; Moodeen Sheriff, Supp. Pharm. Ind., 175; U. C. Dutt, Mat. Med. Hind., 305; Dymock, Mat. Med. W. Ind., 2nd Ed., 657; Fleming, Med. Pl. and Drugs, as in As. Res., Vol. XI., 172; S. Arjun;*

Bomb. Drugs, 205; Med. Top. Ajm., 150; Baden Powell, Pb. Pr., 369; Atkinson, Him. Dist., 316, 741; Lisboa, U. Pl. Bomb., 203, 390; Birdwood, Bomb. Pr., 68; Smith, Dic., 269; Gazetteers:—Mysore & Coorg, I., 65; Bombay, XV., 441; N.-W. P., I., 84; IV., lxxvii.; Indian Forester, III., 237; X., 35; Settlement Rept., C. P., Chanda, App. VI.

Habitat.—Cultivated or spontaneous over the greater part of India, being equally plentiful in the hotter valleys of the North-West Himálaya (from the plains up to 7,000 feet in altitude) and in the far east in Bengal, Manipur, and Burma. The flowers are yellow, purple or magenta, and in cultivation variegated or double. The stems of the yellow-flowered forms are yellowish, of the others red. The plant is often so prevalent near village sites as to exclude all other vegetation.

MEDICINE. Root. 607 **Medicine.**—The tuberous ROOT of the Marvel of Peru was once supposed to be the jalap of the shops, but that has long since been known to be produced by an **Ipomœa** (*Conf.* with Vol. IV., 488). It is believed by native practitioners to be gently aperient, but when subjected to clinical trials by **Shoolbred & Hunter**, and later by **O'Shaughnessy**, its purgative powers were found to be feeble, uncertain, and unworthy of attention.

Leaves. 608 The bruised LEAVES form a favourite application amongst the natives for abscesses and boils, to hasten the suppurative process. **Waring** states that on one occasion he saw a greatly increased amount of pain and inflammation follow the use of this application, in the case of a lady who had been induced to apply it to a boil by her native attendant (*Pharm. Ind.*). **Dymock** notices the same use of the leaves, which was, indeed, first described by **Rumphius**. The former author also states that in the Konkan the dried root powdered and fried in *ghí* with spices, is given with milk as a *paushtik*, or strengthening medicine, and rubbed with water is applied as a *lep* in contusions.

SPECIAL OPINIONS.—§"The leaves are used as a stimulant poultice for boils and abscesses" (*Assistant Surgeon Nehal Singh, Sahárunpur*). "Four o'clock leaf is used as a poultice to boils, also mixed with oil as a stimulant to ulcers" (*Surgeon-Major L. Beech, Coconada*). "The heated leaves I have frequently used instead of poultices to promote suppuration" (*Honorary Surgeon E. A. Morris, Tranquebar*).

FOOD. Leaves. 609 **Food.**—The LEAVES are said to be largely used as a vegetable at Ooson in the Salem District.

DOMESTIC. Root. 610 **Domestic.**—**Rumphius** informs us that the Spanish ladies in Ternate, employed the powdered ROOT mixed with rice flour and sandal wood oil as a cosmetic.

Seeds. 611 The powdered SEEDS are still used for the same purpose in Japan.

Mochras, see **Bombax malabaricum,** *DC.;* Vol. I., 487.

612 MODECCA, *Lam.; Gen. Pl., I., 813.*

A genus of twining herbs or undershrubs, which belongs to the Natural Order PASSIFLOREÆ, and comprises comparatively few species, natives of the tropics of the Old World. Of the six Indian species, only one is of interest, *viz.*, **M. palmata,** *Lam.; Hondala,* SING.; a native of the Western Peninsula and Ceylon. According to **Thwaites** "the root is said to be poisonous, and is used by the Cinghalese as a medicine" (*En. Pl. Zey., 128; Fl. Br. Ind., II., 603*).

613 MOLLUGO, *Linn.; Gen. Pl., I., 857.*

Mollugo Cerviana, *Seringe; Fl. Br. Ind., II., 663;* FICOIDEÆ.

Syn.—M. UMBELLATA, *Seringe;* PHARNACEUM CERVIANA, *Linn.*

Vern.—*Ghimá sák,* BENG.; *Pada,* MAR.; ***Parpadagum,*** TAM.; *Parpataka,* TEL.; *Pat-paadagau,* SING.

References.—*DC., Prodr., I., 392; Boiss., Fl. Orient., I., 756; Thwaites, En. Pl. Zeyl., 24; Gazetteers, N.-W. P., IV., lxxii.; X., 310.*

Habitat.—A herb found from the Panjáb to Ceylon in the hotter and drier parts of India, not in Bengal.

Medicine.—Thwaites states that the PLANT is used as a medicine in fevers; Dr. Peters in a special note informs us that it has the reputation of promoting the flow of the lochial discharge. **MEDICINE. Plant. 614**

Mollugo hirta, *Thunb.; Fl. Br. Ind., II., 662.* 615

Syn.—GLINUS LOTOIDES, *Linn.*; G. PARVIFLORA, *Wall.*; PHARNACEUM PENTAGONUM, *Roxb.*; TRYPHERA PROSTRATA, *Blume.*

Vern.—*Dúsera-sag,* BENG.; *Kottruk,* SIND; *Porprang, gandi búti,* dried plant=*zakhmi haiyát,* PB.

References.—*Roxb., Fl. Ind., Ed. C.B.C., 275; Kurz, in Jour. As. Soc., 1877, pl. ii., 110; Dalz. & Gibs., Bomb. Fl., 16; Stewart, Pb. Pl., 101.*

Habitat.—A herb found commonly throughout India and Ceylon.

Medicine.—The dried PLANT is a common drug in the Panjáb and Sind, where it is believed to be a useful purgative in diseases of the abdomen. **MEDICINE. Plant. 616**

Food.—Roxburgh states that the tender SHOOTS are used by Natives in their curries. **FOOD. Shoots. 617**

M. stricta, *Linn.; Fl. Br. Ind., II., 663.*

Syn.—MOLLUGO TRIPHYLLA, *Lour.*; M. LINKII, *Seringe*; M. PENTAPHYLLA, *Linn.*; PHARNACEUM STRICTUM, TRIPHYLLUM, *and* PENTAPHYLLUM, *Spreng.*

Vern.—*Pita gohum,* URIYA; *Jhurusa,* MAR.; *Zharas,* BOMB.; *Verrichátarási,* TEL.

References.—*Rheede, Hort. Mal., X., t. 26; Kurz, in Jour. As. Soc., 1877, pt. II, 111.; Dalz. & Gibs., Bomb. Fl., 16; Elliot, Fl. Andhr. 191; Dymock, Mat. Med. W. Ind., 2nd Ed., 75; Lisboa, U. Pl. Bomb., 160; Gazetteers, N.-W. P., V., lxxii.; X., 310.*

Habitat.—A very common herb throughout India and Ceylon.

Medicine.—Believed to be stomachic, aperient and antiseptic, given to women to promote the menstrual discharge (*Dymock*). **MEDICINE. Herb. 618**

SPECIAL OPINION.—§ "The bitter leaves are antiperiodic" (*Surgeon-Major W. D. Stewart, Cuttack*).

Food.—Lisboa states that it is eaten as a pot-herb in all seasons. **FOOD. Herb. 619**

Mollusca, see **Oysters,** Vol. V.; **Pearls** Vol. VI.; and **Shells,** Vol. VI.

MOLYBDENUM, *Ball in Man. Geol. of India, III., 161.*

Molybdenum. This metal occurs generally as the sulphide, or molybdenite, and has never been found native; an oxide resulting from the alteration of the sulphide sometimes occurs. Molybdenite has occasionally been found in small quantities in the crystalline or metamorphic rocks of India, especially in those of Chutia Nágpur. Its appearance is so like graphite that it is often mistaken for it. Specimens have been found in the galena mines of Hazáribágh. 620

Uses.—Molybdenite is employed in the preparation of a blue pigment for pottery; otherwise the metal is not employed for industrial purposes (*Ball*). **PIGMENT. Molybdenite. 621**

Momea, see **Cannabis sativa,** *Linn.*; Vol. II., 115.

MOMORDICA, *Linn.; Gen. Pl., I., 825.*

A genus of climbers with simple tendrils which comprises some 26 species, natives of the tropics of both hemispheres, chiefly of Africa. Of these five or six are natives of India. 622

[BITACEÆ.

623 **Momordica Balsamina,** *Linn.; Fl. Br. Ind., II., 617;* CUCUR-

Vern —*Kurelo-jangro,* SIND.; *Mokha,* C. P.; *Mokah,* ARAB.

References.—*Lisboa, U. Pl. Bomb., 158; Atkinson, Ec. Prod., N.-W. P., Pt. V., 7; Ainslie, Mat. Ind., II., 275; Murray, Pl. and Drugs, Sind, 41; Birdwood, Bomb. Prod., 159; Chanda Settlement Rept., App. VI.; Gazetteers, N.-W. P., IV., lxxii.; Stocks, Rept. on Sind.*

Habitat.—A climber met with in Sind and the Panjáb.

MEDICINE. Fruit. 624 **Medicine.**—According to **Atkinson,** it is occasionally employed in native medicine. **Ainslie** writes, "The FRUIT, **Hasselquist** informs us in his *Iter Palestinum,* is famous in Syria for curing wounds; it is a fleshy ovate berry, ending in acute points. The natives cut it open and infuse it in sweet oil, which they expose to the sun for some days, until it becomes red, and then preserve it for use; dropped on cotton, and applied to a fresh wound, they consider it as a vulnerary, little inferior to the balsam of Mecca."

FOOD. Fruit. 625 **Food.**—The young green FRUIT is eaten as a pickle; when ripe it is from 1 to 3 inches long, rostrate and orange red, and is eaten as a vegetable in stews, &c.

626 **M. Charantia,** *Linn.; Fl. Br. Ind., II., 616; Wight, Ic., t. 504.*

Syn.—M. HUMILIS; M. MURICATA, *DC.*; M. SENEGALENSIS, *Lamk.*; CUCUMIS AFRICANUS, *Bot. Reg.*

Vern.—*Karélá, kareli, karólá,* HIND.; *Karalá, uchchhe, poti-kakar, jethuya, bárá masiya,* BENG.; *Karena,* URIYA; *Kakral, kakiral,* ASSAM; *Karela, karola,* N.-W. P.; *Kurela,* KUMAON; *Karélá, karílá,* PB.; *Karelo,* SIND; *Karlí,* C. P.; *Kárlá,* BOMB.; *Kárlí, karale,* MAR.; *Kárelá, karélo, karelu,* GUZ.; *Karélá,* DEC.; *Pávakká-chedi, pava-kai,* TAM.; *Kákara, úra kakara, tella kákara, metta kákara,* TEL.; *Kágala-káyi,* KAN.; *Kaippa-valli, pávakká-cheti, panti-pávél, kappakka,* MALAY.; *Ke-hin-gá-bin, kyet-hen-kha,* BURM.; *Karawila, battú-karawilla,* SING.; *Sushavi, kára-valli-latá, káravella,* SANS.; *Qisául-barri,* ARAB.; *Símá-hang, karelah,* PERS.

References.—*Roxb, Fl. Ind., Ed. C.B.C., 696; Kurz., in Jour. As. Soc., 1877, II., 102; Thwaites, En. Ceyl. Pl., 126; Dalz. & Gibs., Bomb. Fl., 102; Stewart, Pb. Pl., 98; Mason, Burma and Its People, 471, 747; Sir W. Elliot, Fl. Andh., 77, 115, 156, 177, 186; Rheede, Hort. Mal., VIII., t. 9; Rumphius, Amb., V., t. 151; O'Shaughnessy, Beng. Dispens., 351; Moodeen Sheriff, Supp. Pharm. Ind., 175; U. C. Dutt, Mat. Med. Hindus, 303, 319; Sakharam Arjun, Cat. Bomb. Drugs, 59, 206; Murray, Pl. & Drugs, Sind., 41; Dymock, Mat. Med. W. Ind., 2nd Ed., 340; Year-Book Pharm., 1879, 214; Birdwood, Bomb. Prod., 158; Baden Powell, Pb. Pr., 348; Drury, U. Pl. Ind., 205; Atkinson, Him. Dist. (Vol. X., N.-W. P. Gaz.), 700, 742; Duthie and Fuller, Field and Garden Crops, II., 62, t., lxiv.; Firminger, Man. Gard., 125; Useful Pl. Ind., Bomb. (Vol. XXV., Bomb. Gaz.), 158; Forbes Watson, Econ. Prod. N.-W. Prov., Pt. V. (Vegetables, Spices, and Fruits), 7; Royle, Ill. Him. Bot., 219; Stocks, Report on Sind; Darrah, Note on the Condition of the People of Assam; Settlement Reports:—Panjáb, Kángra, 28; N.-W. P., Allahabad, 36; Kumáon (app.), 33; Central Provinces, Chánda, 82; Gazetteers:—Orissa, II., 180; Bombay, V., 26; VIII., 183; XIII., 294; XV., 435; Panjáb, Gurgáon, 17; N.-W. P., I., 81; IV., lxxii.; Mysore and Coorg, I., 61; Agri.-Horti. Soc. Ind.:—Transactions, III., 61, 200; Pro. 228; VI., 104; VII., 64; Journals (Old Series), IV., 202; IX., Sel. 58; Indian Forester, IX., 162, 202; X., 162.*

Habitat.—Cultivated throughout India, distributed to Malaya, China, and Tropical Africa. There are several cultivated forms differing in shape and size of the fruit. The rainy season kind, called *kareli,* has rather smaller fruits and is more esteemed than the hot-weather variety, known in some districts under the name of *karela.* The fruit of the latter is longer and smoother; that of the former is more ovate, muricated and tubercled. Under these two primary classes there are numerous subordinate forms which have specific local names.

CULTIVATION.—As may be gathered from the above, this vegetable is cultivated at two distinct seasons of the year, and everywhere throughout the Peninsula. Regarding the best method to be followed in the cultivation of the hot-weather variety, **Mr. Gollan**, of Saharanpur, writes (in the *Indian Forester, IX., 162, 202*):—"It should be sown in the end of February and all through March in rich soil. The ground should be laid out in beds, and the seeds sown in lines 2 feet apart, and the same distance allowed between each seed. Water should be given twice a week until the ground is covered, afterwards once a week will be sufficient The first sowing will come into use about the middle of April, and successive sowings made in March will keep up the supply until the beginning of the rains." The rainy season variety must be sown in June, and supports for it to climb upon are necessary. CULTIVATION.

Medicine.—The FRUIT is considered tonic, stomachic, and cooling, and is used in rheumatism, gout, and diseases of the spleen and liver. **Rumphius** states that it was much esteemed in Amboina, where it was supposed to purify the blood, and to dissipate melancholy and gross humours. The fruit and LEAVES are both administered internally in leprosy, piles, jaundice, and as an anthelmintic. The latter are said by **Rumphius** to be used by Indian obstetricians to purify the blood and generate milk in the puerperal condition. He states also that a leaf was placed in the mouth of the newly-born infant to clear its breast and intestines of all mucus, excrement, &c. **Dymock** informs us that in the Konkan a third of a seer of the LEAF-JUICE is given in bilious affections as an emetic and purgative, alone or combined with aromatics; it is also applied externally for burning of the soles of the feet, and round the orbit as a cure for night-blindness. The fruit of the uncultivated form is said to act as a febrifuge. The ROOT is also used medicinally, being considered astringent and warm; and in the Panjáb is, according to **Honigberger**, applied externally to piles. The whole PLANT combined with cinnamon, long pepper, rice, and the oil of **Hydnocarpus Wightiana**, *Blume*, is employed by the Hindus as an external application in scabies and other cutaneous diseases.

MEDICINE. Fruit. 627

Leaves. 628

Leaf-Juice. 629

Root. 630

Plant. 631

SPECIAL OPINIONS.—§ "A case of death from violent vomiting and purging caused by the administration of this juice to a child, has fallen within my practice" (*Assistant Surgeon S. Arjun Ravat, L.M., Girgaum, Bombay*). "The fruit is used as a vegetable, cooked as curry. It is bitter and possesses mild laxative, antibilious, and tonic properties" (*Civil Surgeon D. Basu, Faridpore, Bengal*). "The expressed juice with chalk is used in apthæ, and is also an emmenagogue in dysmenorrhœa. It is applied externally to the scalp in pustular eruptions" (*Surgeon-Major D. R. Thomson, M.D., C.I.E., Madras*). "Useful as an application to burns, and allays the irritation of boils. Commonly prescribed as an anthelmintic, and as a purgative for children" (*Civil Surgeon J. Mc-Conaghey, M.D., Shajahanpore*). "The fruit is anthelmintic" (*V. Ummegudien, Mettapolian, Madras*).

Food.—The FRUIT, which is of a bright orange-yellow colour, and from 1 to 6 inches long, is eaten cooked in curries, or sliced and fried. Treatment in hot water or salt and water is necessary previous to cooking or frying to take away a portion of the bitterness. When sliced, dried, and kept in an airy place, it remains good for many months. The LEAVES were eaten in the Khandésh District, Bombay, during the famine of 1877-78.

FOOD. Fruit. 632

Leaves. 633

Momordica cochinchinensis, *Spreng.; Fl. Br. Ind., II., 618.* 634

Syn.—M. MIXTA, *Roxb.*; M. DIOICA, *Wall.*

Vern.—*Kákrol, gul-kakra,* HIND. & BENG.; *Adavi kákara,* TEL.; *Samong nway,* BURM.; *Karkataka,* SANS.

References.—*Roxb., Fl. Ind., Ed. C.B.C., 696; Kurz, in Jour. As. Soc., 1877, Pt. II., 102; Elliot, Fl. Andhr., 11; U. C. Dutt, Mat Med. Hindus, 303; O'Shaughnessy, Beng. Dispens., 349; Agri.-Hort. Soc. of Ind., Transactions, VII., 64; Journals (Old Series), IV., 202.*

Habitat.—Common in Bengal, Tenasserim, and the Deccan; distributed to Formosa and the Philippines.

MEDICINE. Fruit. 635 **Medicine.**—FRUIT about 4 inches in diameter, ovate, pointed, bright red, covered with conical pointed tubercles. **O'Shaughnessy** says that **Ainslie** has erroneously called this the *makal*, thereby causing confusion with colocynth, and that the *kakrol* is an edible but medicinally inert fruit.

FOOD. Fruit. 636 **Food.**—The FRUIT is occasionally used for food in Bengal (*O'Shaughnessy*).

637 **Momordica Cymbalaria,** *Fenzl.; Fl. Br. Ind., II., 618.*

Syn.—LUFFA TUBEROSA, *Roxb.*; L. AMARA, *Wall.*

Vern.—*Kadavanchi,* MAR.

References.—*Roxb., Fl. Ind., Ed. C.B.C., 699; Dymock, Mat. Med. W. Ind., 2nd Ed., 341; Agri.-Hort. Soc. of Ind., Transactions, VII., 64.*

Habitat.—A plant with a large tuberous root, found in the Deccan Peninsula, Mysore, and the Konkan.

MEDICINE. Tuber. 638 **Medicine.**—**Dymock** writes, "The whole plant is acrid; it is mentioned here as a number of the TUBERS were forwarded to the Chemical Analyser to Government from Satara as having been found in the possession of a person suspected of administering drugs to procure abortion. My specimen was grown from one of these tubers, which still retained vitality. **Dr. Lyon**, the Chemical Analyser, informs me that on reference to the records of his office, he finds that the *Kadavanchi* tubers have been three times sent to him within the last four years, as having been used to procure abortion."

639 **M. dioica,** *Roxb.; Fl. Br. Ind., II., 617; Wight, Ic., tt. 505, 506.*

Syn.—M. BALSAMINA, *Wall., not of others*; M. WALLICHII, *Rœm.*; M. RENIGERA, HAMILTONIANA, *and* HEYNEANA, *Wall.*; M. MISSIONIS, *Wall.*; M. SUBANGULATA, *Blume*; TRICHOSANTHES RUSSELIANA, *Wall.*; BRYONIA GRANDIS, *Wall.*

Vern.—Leaf=*kanchan arak,* fruit=*karla,* SANTAL; *Bat karíla,* ASSAM; *Gol-kándra, gol-kankra. ghosal-phal,* N.-W. P.; *Dhar karela, kirara,* PB.; *Katwal,* C. P.; *Kurtoli, karatola, karantoli, vantha-koratola,* BOMB.; *Kartoli,* MAR.; *Kuntola, kantolán,* GUZ.; *Kurtoli,* DEC.; *Palúpaghel-kalung,* TAM.; *Púágákara, ágákara, pótu ágákara, pótu kandulu,* TEL.; *Gid hágalu,* KAN.; *Erimapasel,* MALAY.; *Sapyit, sabyet,* BURM.; *Túmba-karawilla,* SING.; *Vahisí,* SANS.

References.—*Roxb., Fl. Ind., Ed. C.B.C., 696; Stewart, Pb. Pl., 98; Mason, Burma and Its People, 471, 747; Elliot, Fl. Andh., 12, 156, 157, 158; Campbell, Ec. Prod. Chutia Nagpur, No. 9496; Ainslie, Mat. Ind., II., 274; Dymock, Mat. Med. W. Ind., 2nd Ed. 339; Atkinson, Him. Dist., 701; Ec. Prod., N.-W. P., Pt. V., 8; Drury, U. Pl., 306; Lisboa, U. Pl. Bomb., 158; Birdwood, Bomb. Pr., 159; Gazetteers:—Mysore and Coorg, I., 55; Bombay, XV., 435; N.-W. P., I., 81; IV., lxxii.; Settlement Report, Chánda, Cent. Provs., App. VI.*

Habitat.—A diœcious climber with tuberous roots; met with throughout India from the Himálaya to Ceylon and Singapore, ascending to an altitude of 5,000 feet on the hills.

MEDICINE. Tuber. 640 **Medicine.**—The mucilaginous TUBER is used medicinally, especially that of the female plant, which is larger than that of the male. **Ainslie** notices

Root. 641 the use of the ROOT by Hindu doctors, writing, "They prescribe the mucilaginous tasted root in the form of electuary in cases of bleeding piles, and in certain bowel-affections connected with such complaints. the dose about two drachms or more twice daily." **Rheede** states that the PLANT

Plant. 642 is "truly cephalic;" for mixed with cocoanut, pepper, red sandal-wood, and other ingredients to form a liniment, it relieves all pains in the head. **Dymock** informs us that the JUICE of the root is a domestic remedy in the Konkan for the inflammation caused by contact with the urine of the House-Lizard (*pál*). MEDICINE. Juice. 643

SPECIAL OPINIONS.—§ "The powder of the dried fruit introduced into the nostrils is said to give rise to repeated sneezing" (*Surgeon-Major W. D. Stewart, Cuttack*). "The tuberous root of the female plant is used in Belgaum as an expectorant, and externally in ague cases as an absorbent. The root of the male creeper is used in ulcers, especially those caused by snake bites. The unripe fruit is used as a vegetable and given as a delicacy to patients recovering from fever" (*C. T. Peters, M.B., Zandra*).

Food.—Flowers during the wet and cold seasons, and produces a FRUIT which, when green and tender, is eaten in curries by the natives. The tuberous ROOTS of the female plant are also eaten. FOOD. Fruit. 644

Root. 645

Momordica echinata, *Muhl.;* see **Luffa acutangula,** *Roxb.;* p. 94.

M. muricata, *DC.;* see **Momordica Charantia,** *Linn.;* species above.

Monetia barlerioides, *L'Herit.;* see **Azima tetracantha,** *Lam.;* [Vol. I, 361.

Mongoose-plant, see **Ophiorrhiza Mungos,** *Linn.;* RUBIACEÆ.

MONKEYS, *Blanford, Fauna of Br. Ind., Mammalia, I., 3.* 646

Monkeys. The name usually given to most of the animals comprised in the two sub-orders **Anthropoidea** and **Lemuroidea** of the order **Primates**. The first of these sub-orders comprises four families of monkeys, *viz.*, the **Simiidæ**, or Anthropoid apes; the **Cercopithecidæ**, or Old World apes, monkeys (properly so called), and baboons, with the exception of the Anthropoid apes; and the **Cebidæ** and **Hapalidæ**, families confined to America. The second sub-order comprises only one family represented in South-Eastern Asia, namely, the **Lemuridæ**.

Few of the monkeys are of economic value, but certain yield furs of commercial interest—see List of Furs, Vol. III., 459; many are highly destructive to crops, and owing to their semi-sacred character in India, little is done to stop their depredations or diminish their numbers.

Monks' hood, see **Aconitum Napellus,** *Linn.;* Vol. I., 95.

MORCHELLA, *Will.; Cooke, Fungi, II., 655.*

Morchella esculenta, *Pers.; Baill., Bot. Med. Crypt., 121;* FUNGI. 647
THE MORELL.

Vern.—*Kana kach, girchhatra* (hills), *khumb* (plains), PB.

References.—*Stewart, Pb. Pl., 268; Baden Powell, Pb. Prod., 257, 384; Royle, Ill. Him. Bot., 440; Agri.-Hort. Soc. of Ind., Journals (Old Series), IV., Sel., 267.*

Habitat.—This fleshy fungus is found abundantly in Kashmír, Chamba, and many parts of Northern Panjáb.

Medicine.—It is believed to be aphrodisiac and narcotic. MEDICINE. 648

Food.—The Morell is commonly eaten by Natives in the regions where it is found, and is the only fungus used as an article of food by Muhammadans. It retains its flavour well when dry, and is largely exported from Kashmír and the hills of the Panjáb to Lahore, Amritsar, and other large stations. It is apparently very similar to the Morell, so much esteemed in European cooking, and is considered a great dainty by Natives. It imparts a rich mushroom-like flavour to soups, gravies, &c., and when dried is, according to Dr. Stewart, "a not unsatisfactory addition to a stew." FOOD. 649

650 **MORINA,** *Linn.; Gen. Pl., II., 158.*

A genus of perennial herbs which belongs to the Natural Order DIPSACEÆ, and comprises seven or eight species, natives of Western and Central Asia. Of these, six are indigenous to India, but are of little economic interest. **M. Coulteriana,** *Royle (Fl. Br. Ind., III., 216)*, a native of the Sub-alpine Himálaya from Kashmír to Garhwál, is stated by **Aitchison** to be thrown on the fire as an incense by the Buddhists in Lahoul, giving an agreeable perfume (*Jour. Linn. Soc., X., 78*). **M. persica,** *Linn. (Fl. Br. Ind., III., 216), Bekh-akwar,* HIND, common on the Western Himálaya from Kushmír to Kumáon, is mentioned by **Baden Powell** in his list of drugs. **Dr. Dymock**, in a letter to the editor, states that he suspects the same species may prove the source of the Red Bahen or Bahman of the Persians.

651 **MORINDA,** *Linn.; Gen. Pl., II., 117.*

A genus of erect or climbing shrubs or trees which comprises about 40 species, all tropical. Of these, seven are natives of India.

652 **Morinda angustifolia,** *Roxb.; Fl. Br. Ind., III., 156;* RUBIACEÆ.

Syn.—M. SQUARROSA, *Ham.*

Vern.—*Dáruharidrá,* BENG.; *Asugach, asukat, kchaitun, chenung, ntan,* ASSAM; *Ban hardi, hardi,* NEPAL & LEPCHA; *Yiyo,* BURM.

References.—*Roxb., Fl. Ind., Ed. C.B.C., 184; Brandis, For. Fl., 278; Kurz, For. Fl. Burm., II., 61; McCann, Dyes and Tans, Beng., 38, 145, 154; Darrah, Note on Cotton in Assam, 28; Wardle, Dye Report, 30, 81; Balfour, Cyclop, II., 984.*

Habitat.—An erect bush, or small tree, met with in the tropical Himálaya, wild and cultivated, from Nepál eastward to Assam, the Khásia and Nágá Hills, Chittagong, and Tenasserim. It attains an altitude of 6,000 feet on the Himálaya at Sikkim, and of 4,000 feet on the Khásia Mountains.

DYE. Bark. 653 Wood. 654

Dye.—The BARK and WOOD yield a good yellow dye, which is extensively employed by the natives of the Darjíling district and of Assam, also to a small extent in Burma. In the latter country it is said to be cultivated in Toungyas for its dye-yielding properties; in the two former regions **McCann** states that it is apparently never cultivated, the wild plant being utilised. **Darrah**, on the other hand, writes that it is occasionally cultivated around homesteads in Assam. The process followed in Darjíling and the Gáro Hills is described as follows by **McCann**:—"The dye is prepared by

Root. 655

pounding the bark of the ROOT and boiling it in water; it is then strained, and the water boiled over again to the required consistency. In the Gáro Hills the cotton thread which is dyed with *chenrung* undergoes a preliminary process of cleaning, resembling that usual elsewhere in *ál* or *ách* dyeing. The thread is rubbed well with pounded sesamum seed and the leaves of *bambi* or *daggal* (**Sarcochlamys pulcherrima**) and then left alone for two days after which it is well washed. The dye solution is made by pounding the roots, steeping them in water, and then boiling three or four times in succession. Into this decoction, when cool, the thread prepared as above is placed, and the whole heated gently, but not boiled, in a pot. When the liquor is cool, the thread is removed, and again washed. It is then steeped in the dye liquor, heated, taken out and washed once more, and this is repeated two or three times until the colour is sufficiently fixed. The Gáros do not dye their clothes with this, the thread is first dyed in the way described and is afterwards woven into cloth"

The account given by **Darrah** of the method of dyeing followed in Assam differs somewhat from the above, and is of sufficient interest to be worthy of quotation in full:—"The roots are cut into very small chips,

DYE.

and thrown either into cold or boiling water. The cold water process gives a very bright red, while the solution in boiling water gives a very pale red. The colour varies according to the greater or less care bestowed on the dyeing, and is said to be very permanent. Chips of the bark of *leteku* (**Baccaurea sapida,** *Muell. Arg.*), and leaves of the *bhoomrati* (**Symplocos spicata,** (*Roxb.*), are boiled with the *ásugach* as a mordant.

The following is the process adopted by the Phakials:—"The thread, having been steeped in mustard oil, or in oil obtained from a pig or elephant, is boiled for an hour or two, and then exposed in the sun to dry for twenty days. When thoroughly dry, it is washed and boiled with wood ashes in water and put out in the sun to bleach. Roots of the *ásugach* are then cut up and pounded. Water and wood ashes are added, and, the thread having been placed in the mixture, the whole is warmed over a fire. It is then allowed to stand for a day, after which the thread is taken out and exposed to the sun. The shade of red depends on the number of times the thread undergoes the last steeping process and subsequent exposure. The oftener this is repeated, the darker is the colour produced." Samples examined by Mr. Wardle were reported on as follows, "This root contains a moderate amount of yellow colouring matter, and by the application of various methods will produce rich golden dyes. It is said to produce a red dye, but the sample sent to me would not yield it. It is probably used in connection with *manjit*, or sometimes mistaken for it. The Madras specimen contains a fair amount of red colouring matter combined with yellow, but the Darjíling specimen and that from Dinájpore yield only yellow dyes." Similarly, in the cases of the other species of **Morinda,** Mr. Wardle admits his having failed to obtain the *ál* reds of India, affirming that all yield only yellow and orange shades. In a letter written subsequent to his visit to India, however, he writes, "But on visiting India, I found undoubted evidence in Calcutta, the North-West Provinces, Rájputana, &c., of the constant employment of both those species, yielding excellent reds on cotton, both in prints and dyes."

Morinda citrifolia, *Linn.; Fl. Br. Ind., III., 155.* 656

THE INDIAN MULBERRY; the TOGARI WOOD of Madras.

Considerable confusion has long existed in Indian botanical and economic literature regarding the dye-yielding species of **Morinda,** and especially the forms referable to this species. Many writers, with some degree of evidence in their favour, have considered **M. tinctoria** to be but a wild form, and probably the progenitor of the cultivated **M. citrifolia,** *Linn.* Others have supposed **M. bracteata** of Roxburgh to be the wild form, with apparently much less likelihood of truth. The *Flora of British India* has recently described the species as including the following varieties, which, in accordance with the rule uniformly pursued in writing this work, may be given:—

Var. *α*, citrifolia, PROPER,—M. CITRIFOLIA, *Linn.*

Var. *β*, bracteata,—M. BRACTEATA, *Roxb.*

Var. *γ*, elliptica.

At the same time, however, it may be remarked that var. **bracteata** of the *Flora* will probably by future writers be restored as a species, for the reduction of that plant to **citrifolia** appears quite untenable. **M. citrifolia** proper is always a cultivated plant in India, and is probably also so in Polynesia. But . **bracteata,** *Roxb.*, is a remarkably common sea-board plant all round the Indo-Malayo-Burma-Andaman coasts, and is apparently never cultivated anywhere, and besides this sharp distinction between the two, each possesses characters which seem sufficient to separate them

MORINDA citrifolia.

Ál Dye.

specifically. In any case **M. tinctoria**, which has always been more or less united with **M. citrifolia** in economic writings, would appear (if it also be not entitled to the rank of a species) to be much more nearly related to that plant and much more likely to be its sylvan form than **M. bracteata**. In this light, therefore, it must be held incorrect to assign **M. bracteata** the position of a variety under **M. citrifolia**.

Vern.—Var. α, citrifolia, proper—*Al, ách, ak, barra-al*, HIND.; *Ach aich, achhu*, BENG.; *Ál*, C. P.; *Aal, bártundi, abri, álan, ainshe, nága-kuda*, BOMB.; *Al, surangi, bartondi, nága kundá*, MAR.; *Al, saraoji*, GUZ.; *Al*, DEC.; *Mina-maram, munja-pavattay*, TAM.; *Núna, torugu, molagha, maddi*, TEL.; *Sira njikadi, maddi*, KAN.; *Ka-da-pilva*, MALAY.; *Nyah-gyi, nie-pa-hsæ*, BURM.; *Ahu-gaha*, SING.; **Var. β, bracteata**—*Hurdi, huldi-kunj, rouch*, BENG.; *Mhan-bin, yaiyæ*, BURM.; *Yahú-gaha*, SING.

References.—*Roxb., Fl. Ind., Ed. C.B.C., 182, 183; Voight, Hort. Sub. Cal., 385; Brandis, For. Fl., 277; Kurz, For. Fl. Burm., II., 60; Beddome, Fl. Sylv., t. 220; Thwaites, En. Ceyl. Pl., 144; Dalz. & Gibs., Bomb. Fl., 114; Stewart, Pb. Pl., 115; Mason, Burma and Its People, 463, 512, 786; Rheede, Hort Mal., I., t. 52; Ainslie, Mat. Ind., II., 254; Irvine, Mat. Med. Patna, 7; Medical Topog., Ajm., 125; Sakharam Arjun, Cat. Bomb. Drugs, 206; Dymock, Mat. Med. W. Ind., 2nd Ed., 401; Birdwood, Bomb. Prod., 332; Drury, U. Pl. Ind., 296; Useful Pl. Bomb (Vol. XXV., Bomb. Gaz.), 88, 162, 200, 246, 392, 400; Atkinson, Econ. Prod., N.-W. Prov., Pt. III. (Dyes and Tans), 15-20; Crookes, Hand-book, Dyeing, &c., 391; Liotard, Dyes, 49, 53, 66, 69, 134; McCann, Dyes and Tans, Beng., 20-38; Wardle, Dye Rept., 32, 33, 39; Com. Pl. & Drugs, VIII., 45; Report, Agricultural Dept. Bomb., 1888-89, App xii.; Settlement Report:—Central Provinces, Chánda, App. vi.; Nimar, 198; Mundlah, 88; Narsinghpur, 53; Banda, 49; Chindwára, 23; Gazetteers:—Bombay, VII., 37; VIII., 183; XII., 24; XIII., 23; XV., 436; XVIII., 44; N.-W. P., I., 57, 81; Mysore and Coorg, III., 28; Agri.-Hort. Soc. Ind.:—Journals (Old Series), VII. (Pro.), 24; IX., (Sel.) 44, 54, 58; XIII., 359; Indian Forester, III., 203, 237; VIII., 378; X., 33.*

Habitat.—A small tree or bush, cultivated throughout India. **Var. citrifolia** proper is much the most important form economically, since it yields by far the greater portion of the *ál* dye of Indian commerce. It is accordingly the most largely cultivated, being met with in all parts of the country. **M. bracteata** is regarded by **Roxburgh** as a native of Ganjam in Orissa, while **Thwaites** considers it to be wild in Ceylon. It is not infrequent in the forests of the Andaman Islands (*Kurz*), and occurs sparsely in those of the Terai near the Tista (*Schlich*). **Var. elliptica** (a form intermediate between **M. angustifolia** and **M. citrifolia** in character of foliage) is found in Tavoy, the Konkan, and Malacca.

DYE. Root. 657

Dye.—The *ál* dye, yielded by the above and other species of **Morinda**, has been frequently described, but considerable confusion exists regarding the species from which it is mostly obtained. **M tinctoria**, *Roxb.*, has been considered by several authors as the chief source of *ál*, and, indeed, it is, even to this day, frequently confused with **M citrifolia** in official returns, &c. **M. angustifolia**, as already described, yields a dye which is apparently similar in every respect to the *ál* of the greater part of India. The two former species are the more likely to be confused owing to the fact that they bear the same, or very similar, vernacular names, but there appears to be no doubt that, while both yield a red dye, **M. citrifolia, var. citrifolia** proper, is the plant of which the wood, root, or root bark is most frequently met with under the name of *ál* or *ách*, and is the chief source of the dye of commerce. Experiments by **Wardle** with the root barks of **M. citrifolia, M. tinctoria**, and **M. angustifolia**, yielded almost identical results (see remarks under **M. angustifolia**). In certain localities the dyers make a distinction between *ál* and *ách*, but it is believed that they allude merely to different qualities of the dye stuff, and not to the products of different species. Accepting this assumption as correct, a résumé of available information regarding the cultivation of the tree, the outturn and price of dye stuff, and the methods employed in dyeing with it, as practised in the different provinces, may be here given. DYE.

ÁL DYE.

I. North-West Provinces.—The cultivation of *ál* is chiefly confined to the Bundelkhand districts of these provinces (Bánda, Hamírpur, Jhánsi, and Jaláun), although it is also grown in the southern parts of the Futehpur and Cawnpore districts bordering on the Jumna. It does best on the peculiar black soil of Bundelkhand (*Buck*). In the Revenue Administration Report for 1887-88, 4,158 acres are returned as having been under the crop in Hamírpur, 2,191 in Jhánsi [1,450 (*Buck*)], and 512 in Jaláun. N.-W. PROVINCES. 658

The method of cultivation, production of the dye, and process followed in using it, have been fully described by **Mr. Wright, Sir E. C. Buck**, and more lately by **Colonel Ternan**; from the accounts given by these writers the following facts have been compiled :— Statistics. 659

STATISTICS PER ACRE.

Ploughings.	Time of sowing.	Seed.	Weeding.	Cutting.	Digging.	Chopping.	OUTTURN. Root.
5	Sawan (July).	2 mds.	8 times. Ploughed twice.	20 men a day.	10 men a month.	4 men a day.	10 mds.

The crop is grown in black (*már*), friable soil which has been previously sown with a spring crop for two or three years, but which does not receive manure. On the first fall of rain it is ploughed with the *bakhar* plough not less than five times, oftener if possible. This implement is a sort of hoe plough, used in the Bundelkhand districts in place of the ordinary short conical share; it consists of a hoe-shaped piece, about 20 inches broad and 5 deep, which enters into the ground about 8 inches, and is much more powerful in breaking up the soil and eradicating weeds than the ordinary plough. Towards the end of July, the seed is sown broadcast, about two maunds to the acre, and thoroughly mixed in the ground with the *bakhar*. Early rain after sowing is necessary to the welfare of the plant, which sprouts in about twenty days. It is weeded four times, and has to be protected from injury by cattle. In the following July, the soil is turned up round the young plants, and the land is weeded twice. In the rains of the third and fourth years the field is ploughed between the plants, to allow the rain to reach the roots, and at the end of 3½ years from sowing the roots have matured and are fit to be collected. Sowing and Ploughing. 660

During the second, third, and fourth years seed is collected, about 20 seers, according to **Colonel Ternan**, being obtained per *bigha* the second year, and about 10 seers in the two following. Towards the end of December, 3½ years after sowing, the trees are cut down, the roots are dug up with pickaxes, collected, sorted, and cut into lengths, each root being divided into three according to thickness. The cut root is then carefully dried and packed close in gunny bags. The three classes of root have very different commercial values, and are known by distinct names. The thinnest or best is known as *hárgharka, bhára*, or *bár*, and fetches about R8 per maund; the middle part of the root, *lari, jharan, pachmer*, is worth R4 Collection. 661

MORINDA citrifolia.

Ál Dye.

N.-W. ROVINCES.

per maund, and the thick end, or most inferior part, *pachhkat ghatiya, lari*, fetches only R2 per maund. Very thick roots, called *katerdo*, are valueless for purposes of dyeing, but are peeled and mixed with the others as an adulterant. The price of the dye has fallen off greatly during the past few years, and in former times was worth as much as R20, R10, and R9, according to its class. According to **Colone. Ternan**, the three sorts are generally mixed before being sold in the proportion of 1½ seers of the thin, 2 seers of the medium, and 3 seers of the thick quality. An acre of ground generally produces from 10 to 14 maunds of root, one-third of each class, and 6 maunds of seed.

Yield. 662

Profit and cost of Cultivation. 663

In former times, when the dye was of greater value, the cultivation was profitable, but now, as a rule, it is said to be generally attended with loss. Though the actual cultivation does not cost much, digging the roots is an operation requiring much care, and consequently much time and labour. The total outlay per acre is said by **Mr. Wright** to amount to over R89, of which the chief items are R37-8 for digging, R16 for watching, and the same sum for rent. The last item appears to be high, indeed in Jalaun the acre rate is said to be only R2-8 for the best *már*. The value of the whole produce (at 10 maunds of the root per acre) comes to only R46, so the loss, if the figures given be correct, must, as a rule, be considerable. It is noteworthy, however, that the cost of cultivation as returned from other Provinces is much less than that calculated by **Mr. Wright**, and that in all cases the profit obtained is said to be large.

Method of Dyeing. 664

The method of using the dye is described as follows by **Sir E. C. Buck**:—"In *ál* dyeing the previous preparation of the cloth is the most noticeable thing. The colouring matter is, as with safflower, not extracted from the plant till in the actual operation of dyeing. The roots are mixed with a little sweet-oil and ground to powder in a handmill. Cloth is dyed by being, as a rule, boiled with this powder. The kind of cloth most frequently dyed with *ál* is the coarse fabric known as *khárua*. The following is a description of the process of dyeing which is for the most part applicable to every other kind of cotton cloth:—

"The cloth is well washed and soaked in water with which some powdered sheep's dung has been mixed; the next step is the bleaching (ver. *merái*). For this a mixture is made of 4℔ of castor-oil, 5℔ of the alkaline earth known as *rassi*, and 1¼℔ of sheep's dung in about 30 gallons of water. The cloth is steeped in this mixture for twelve days, when it is washed in clear water. It is then soaked in a decoction of myrobalan (**Terminalia Chebula**) in water, and after that in alum water. It is then ready for the application of the dye For each piece of cloth measuring 8 yards × 1 yard, 30℔ of powdered *ál* are mixed with water and boiled. Into the mixture, while boiling, the cloth is thrown and boiled with it till it has become dyed to the shade required. It is then cleaned and washed. It is sized by being dipped in a solution of gum and water and beaten smooth with wooden clubs.

"The term *khárua* is properly applied to this cloth when it has been dyed a dull-red colour in this manner; before dyeing it is generally known as *ikri*. The manufacture and dyeing of this cloth is almost peculiar to the neighbourhood of Mau Ránipur, in the Jhánsi district, and the following mention of it is made in the settlement report of that district:—'The town of Mau Ránipur is chiefly peopled by *dhobis* (washermen) and *chipis* (the caste who dye in *ál*), besides the *baniyas* who are engaged in the *khárua* trade, which is an important branch of industry and one of great consequence to the prosperity of the place.'" Dyeing a bale of sixty pieces of cloth is said to cost R28, a bale of undyed *ikri* sells for R56, and after dyeing for R90, so that the profit to the dyer amounts to R6. The annual

N.-W. PROVINCES.

value of the export from Mau Ránipur in *khárua*, and other cotton cloths and *ál* dye is stated to be R6,80,000 Besides *khárua*, the well known red *sálu* is dyed with *ál*, either by steeping it in a tincture of the dye-stuff or by the process above described. In the former case the colour is fleeting, in the latter permanent. The colour produced is a dull red, not so brilliant, but far more lasting than that of safflower dye. It is little employed in the production of compound colours or intermediate shades. Sir E. C. Buck, in a letter to the editor, says that one of the most remarkable properties of *Khárua* cloth (*i.e.*. cloth dyed with *ál*) is the fact that it is not attacked by white ants. On this account it is universally used to wrap round the account-books of bakers and shop-keepers.

II. Central Provinces.—The reports from these provinces include under *ál* both **M. tinctoria** and **M. citrifolia**, but it is probable that, in the greater number of cases, the latter is referred to. In the Urya district of Sambalpur and the Telegu district of Upper Godaveri, a wild or spontaneous species (*anchi*) is said to be the only one known or used, and has generally been spoken of as **M. tinctoria**, but may possibly be simply a spontaneous form of **M. citrifolia**. This supposition is supported by the fact that in other districts the cultivated *ál* only is known, and is spoken of by reporting officers as **M. citrifolia**. Both of these forms, whatever they may be, are taken collectively by the Chief Commissioner on a note on the subject, republished in a condensed form, as follows, in Liotard's *Dyes*:— CENTRAL PROVINCES. 665

Statistics. 666

DISTRICTS.	Area in acres.	Cost of cultivation per acre.	Value of produce per acre.	Total quantity of produce in maunds.
		R a. p.	R a. p.	
Narsinghpur	1,225	39 12 0	55 0 0	6,135
Damoh	1,556	36 8 0	56 0 0	8,272
Balaghat	125	15 0 0	40 0 0	625
Seoni	25	33 0 0	60 0 0	150
Saugor	979	82 2 0	92 0 0	3,108
Nagpur	2,171	21 0 0	62 0 0	5,880
	6,081	...	...	24,170

These figures give an average production of four maunds per acre. For the Jubbulpore, Damoh, Sambalpur, and Upper Godaveri districts figures are not furnished—in the two latter the plant is said to be found growing wild.

The method of cultivation, pursued throughout the province, may be summarised by saying that the seed is sown broadcast in July; that the plants are allowed to stand for three years; that they grow two feet high in that period; and that the roots are dug up in October, dried in the sun, cut to pieces, and then stored for sale. Almost all the *ál* produced is used within the district in which it is grown, Damoh alone sending 5,000 maunds to out districts. The wild variety is said to be collected from the forests and exported in large quantities, but to what markets is not stated, except in the case of Sambalpur, which is said to send 1,000 maunds to Madras and to Nagpur, about half each way. Sowing and gathering. 667

The average prices of merchantable *ál* vary from 3 seers of the thin, to 10 seers of the thick quality per rupee. Price. 668

An interesting account of the *ál* cultivating class is to be found in the Settlement Report of the Nimár district, "It is wholly in the hands of a Cultivating Class. 669

MORINDA citrifolia.

Ál Dye.

CENTRAL PROVINCES.

caste called *Alees*, who, to maintain their monopoly, excommunicate any of their members found to have sold or given away any of the seed. The *Alees* seldom own cattle, and usually rent a ready-prepared field for three years, as their crop cannot be grown successively on the same land, and they practise no other cultivation.

"The *ál* is sown in the first year along with *jowarí*, both for profit and to shade it from the sun. It stands alone the second year and is not gathered till the third. To collect the roots intact, the soil is carefully dug out to the extreme depth it penetrates, three or four feet. The land is thus returned to the owner greatly improved by the deep working it has received, and he immediately reaps a magnificent crop of *toda* or cotton from it. The cultivators are, therefore, by no means averse to rent their land to the *Alees*, who also pay very high rent for suitable fields ready-prepared. The writer has known R600 paid for 17 acres for the three years' lease, that is, R11-12 per acre per annum, and this for land worth a rent, for ordinary purposes, of not more than R1 per acre. The profit of the cultivation must therefore be very high, though it is quite impossible to ascertain what it really is, from the suspicious *Alees*."

BERAR. Area and Yield. 670

III. **Berar.—Liotard** writes :—"The average extent of land under *ál* (the produce of **M. citrifolia** and **M. tinctoria**) and the total average produce are estimated to be yearly thus :—

District.	Acres.	Produce in lb.
Amraoti	1,602	3,775,936
Akola	1,432	4,924,340
Elichpur	1,173	964,976
Buldana	30	15
Wun	...	...
Basim	...	...

Cost & Profit. 671 Method of Cultivation. 672

The average cost of cultivation is R25 per acre, the average price of the produce is 2 annas per lb., and the average profit is R30 per acre. *Ál* is sown between the 15th June and 15th July, immediately after the first fall of rain: 216lb of seed are used for each acre, and is sown in rows in ground twice ploughed. It germinates in about a month. In the first year the plot sown is weeded three times, in the second year it is weeded once, and in the third year the plant attains the thickness of a man's thumb or finger, the thicker plants are called *jambora*, and the thinner are the *ál* proper. The root is cut into small pieces, and thus sold."

MADRAS. 673

IV. **Madras.—M. citrifolia** is grown almost everywhere in this Presidency, and is largely used, most of the red cotton turbans and other red cloths so much employed in South India being dyed by the root-bark. The red colouring matter, called *serinjewe* by the dyers, is generally fixed by means of alum. The colouring matter of the wood is apparently not utilised, probably owing to the superiority and comparative cheapness of the root-bark. **Dr. Balfour** states that the latter could be obtained in large quantities and at cheap rates, the smallest and best roots fetching only from R8 to R10 a maund of 82lb. Large quantities are, at the present time, exported from Malabar to Guzerát and Northern India. The *ál* dye is gradually supplanting the more expensive red obtained from *chay* root, once famous as yielding the fine reds of Madras handkerchiefs and turbans.

BOMBAY. Area. 674

V. **Bombay.**—In 1888-89, 1,155 acres were returned as under the crop in the whole Presidency, of which 875 were in Khándesh, 273 in Sholápur,

Al Dye. (*J. Murray.*)

BOMBAY.

6 in Surat, and 1 in Bijápur. **M. citrifolia** also grows spontaneously in the Khándesh jungles where it is collected by the Bheels. **M. tinctoria** is said to be cultivated in the northern part of Gujerát. The price of the root-bark is said to be 12 seers for a rupee, or a little under R7 a maund. The only detailed description of the cultivation and utilisation of *ál* in this Presidency is by **Mr. Narayan Daji**, in his work on the *Art of Dyeing in Western India*, published in 1873, from which the following has been compiled. Price. 675

The method of cultivation is similar to that described elsewhere, and need not be again described. The roots obtained are known collectively as *suranji*, and the three commercial kinds, as *gujerati*, the thin sort; *gháli*, the thick; and *malabári*, the mixed or medium kind. **Mr. Daji** expresses the opinion that the best sort may be the root of **M. tinctoria**, but it is more probable that the dye-stuff is, in Bombay as elsewhere, simply selected artificially for commercial convenience. Method of Cultivation. 676

Ál is described as being used in the same way as madder, for which indeed it is employed as a substitute. The reds obtained are not inferior to those of madder in quality, but are said to be less permanent, and to be capable of being produced in fewer shades. The mordant used is alum, and the process in no way differs from that of madder dyeing, except in the fact that "galling" so necessary in the latter is dispensed with. **Mr. Daji** describes three different processes at length. The first, for the production of "morinda red," is a "cold" process, requiring no boiling, and is generally used for twist. The thread or fabric is first washed in plain water and dried, then steeped in a mixture of castor oil, carbonate of soda (*kháro*) and water, called *khárani*. After soaking in this fluid for a night it is dried, exposed to the sun for seven hours, steeped in a little fresh water, well beaten, and kept tied in a moist state for one night. Next morning it is again exposed to the sun, and these operations are repeated for eight days, at the end of which time it is thoroughly washed and carefully dried. The twist by this process acquires a soft silk-like texture. It is dyed by steeping it for four days in a mixture of solution of the dye-stuff with saturated solution of alum, during which time it is frequently worked up. At the end of this operation it is taken out, washed, dried, and finally steeped in a solution of carbonate of soda, which renders the colour brighter. Dyeing. 677

A "morinda purple" is produced by steeping this morinda red twist in a "peculiar black dye," which **Mr. Daji** does not describe. Cloth is generally dyed by a "hot process," after previous treatment with filtered myrobalan liquor, fresh goat's dung, and a solution of alum and turmeric.

VI Bengal.—The following description of the cultivation and utilisation of the *ál* tree in the Lower Province is given by **McCann** under the name of **M. tinctoria**, *Roxb.*:—"This small tree seems to grow in most parts of Bengal in sufficient quantity to meet the demand for the dye, which is almost entirely local, and it is consequently cultivated to only a slight extent. The only estimates received of the area under cultivation are the following:— BENGAL. Area. 678

Nadiyá . . .	1 bigha.	Sárun . . .	18 acres.
24-pergunnáhs . .	50 bighas.	Darbhangah . .	4 kathas.
* Dinájpur . .	450 bighas.	Balasor . .	10 acres.

* (Ganj . . . 42 bigahs.
Thákurgáon . . . 200 bigahs.)

It is, however, cultivated in many other districts, but only to a slight extent, and in such a manner that the area under cultivation cannot be ascertained. It is occasionally grown for home use in gardens, sometimes in

MORINDA citrifolia.

Al Dye.

BENGAL.

small patches in homesteads or betel-nut plantations (Kuch Behar). In Orissa it is everywhere allowed to grow as a weed in the weavers' homesteads for use in dyeing, the weavers being in general the dyers in that part of the country. In far the greater part of Bengal the plant is not cultivated at all, being found in sufficient quantity in the jungles.

Propagation by cuttings. 679

"In Bengal the plant is everywhere, except in Orissa, propagated by slips or cuttings, and not from seed, as would seem to be the case in other parts of India: in Orissa, although generally grown from cuttings, it is occasionally raised from seed. It grows in all soils, but best in rich soils: the soil best suited to its growth is ordinary black soil with a slight admixture of sand. In the Darbhangah district the best soil is that locally called *baugar*.

"The method of cultivation as followed in the Patna Division, from which fullest particulars have been received, is as follows:—The slips or cuttings from the previous crop, about a foot and a quarter long, are laid down from January to April (*Paush* to *Chaitra*) in a bed, so as to form a stock, and are watered regularly every third day until the rains set in. At periods varying from middle of June (*Darbhangah*) to September (*Sáran*), these slips are transplanted into a field previously carefully prepared for the purpose by ploughing and weeding. About a fortnight after transplanting the field is weeded; the plants are watered regularly until new leaves shoot out, after which nothing in the way of cultivation is done for over three years, but the plant is left to mature of itself. A moderate quantity of rain is necessary for the perfection of the plant, both excess and deficiency being injurious. In the cold weather following the third year, by which time the plants have attained a height varying from 2 feet to 8 feet, according to the quality of the soil, whilst the straight, spindle-shaped roots extend into the ground to a depth of 3 or 4 feet, the roots are dug up,

Collection. 680

generally in January and February. At the same time the portion of the plant above ground, except the upper part, is cut into slips of about a foot and a quarter in length, which form the cuttings from which the plant is propagated as above and the next crop secured.

Trenching. 681

"In other parts of Bengal where the plant is cultivated, after the fields have been carefully ploughed and weeded, deep trenches are dug and the soil from these formed into parallel ridges, generally a little over 3 feet apart, but in Bákarganj only 18 inches. On these ridges the cuttings or slips are planted at a distance of about 18 inches apart. In Bákarganj apparently the trenches between the ridges are filled with fine earth, and the cuttings planted in these furrows, instead of on the ridges as elsewhere. In the Birganj jurisdiction of the Dinájpur district the plant is only cultivated on the strips of land around fields devoted to other purposes. In this district the plant is propagated by cuttings from the roots of mature plants, 3 to 4 inches long, instead of slips from the stem. The period of planting out the cuttings, the amount of care devoted to the growing plant, and the number of years allowed to elapse before the roots are dug up, vary considerably in different districts. In all districts except those of the Patna Division, and sometimes in Kuch Behar, the cuttings are at once planted, on the ridges, without being first grown in a bed and subsequently transplanted to the fields. This planting of slips takes place about the beginning of March in Bákarganj, from April to June in Dinájpur, in the beginning of July in Orissa, in June in Nadiyá, and in August in the 24-Pergunnahs. There is probably some error in the returns from these last two districts, as the period of planting can scarcely differ so widely in two districts so close together.

"In the Nadiyá district considerably more attention seems to be devoted to the growing plant than elsewhere. The slips being planted in June, the

BENGAL.

soil about their roots is dug up and loosened in September and February, and kept carefully free from weeds in the intervals between these diggings. This process of digging is, apparently, repeated every half-year until the plant is mature.

"The period required for the maturing of the roots seems to vary considerably in different districts, depending, no doubt, on local peculiarities of soil and climate. In the Patna districts and in Bákarganj, the roots are ready for digging up at the beginning of the fourth year after the planting of the cuttings; in Dinájpur and Kuch Behar, the roots attain maturity in either three or four years; in 24-Pergunnahs in about five years; and in Nadiyá the plant is said to be fit to yield the dye in ten years. If this latter statement be correct, taken along with the more elaborate digging and weeding required in the cultivation of *ach* in this district, it would seem to imply that the soil of Nadiyá is less adapted than that of any other part of Bengal for this particular plant. Where the plant grows wild, there seems to be generally no special season for digging up the roots: the soil about them is turned up and a few of the roots cut off, the plants being left standing.

"It must be understood that the plant is still young when the roots become best suited for dyeing purposes, and that if allowed to grow it would ultimately attain the height of a large mango-tree, but the roots after the first three or four years no longer yield the dye in any quantity. The height attained by the tree, when the roots are best suited for dyeing purposes, is given variously as 2 to 3 feet (Patna), 6 to 8 feet (Darbhanga), 6 to 8 cubits (Dinájpur): the circumference of the trunk is stated to be half a cubit (Dinájpur).

"The Collector of Balasor states that *achhu* is considered to be a most exhausting crop to land. After a crop of *achhu*, land has to lie fallow for from three to five years. As a consequence, a very high rent (R10 per acre) is charged.

Yield. 682

"As regards the average quantity of root obtained from each tree, in the Dinájpur district we are told that each tree produces on an average 1 seer of root. The Collector of Balasor states that an acre grows from 2,000 to 4,000 plants, producing from 5 to 10 maunds of *achhu*. This would give about 1 seer of root for 10 plants, which would seem to imply that the plant is only one-tenth as productive in Orissa as in Dinájpur. But the Orissa returns are probably inaccurate."

McCann then gives a table embodying all the figures regarding the cost and profit of cultivation, the average outturn, &c. This table is full of obvious discrepancies which are noted and commented on by Dr. McCann himself who observes:—"Omitting returns which differ so widely from the others as to throw doubt upon their correctness, as, for example, those from Bákarganj and Balasor, we may sum up these figures as follows:—

Cost of cultivation varies from R50 to R100 per bigha (distributed, of course, over several years).

Average outturn . . .	10 maunds to 65 maunds per bigha.
Selling price	R1 to R18* per maund.
Profit	R85 to R100 per bigha.

It will thus be seen that the cultivation of *ál* is fairly remunerative. The Collector of Balasor remarks that 'a weaver by cultivating an acre of *achhu* is able to recoup himself for all charges of cultivation as well as to supply all his home requirements.'"

* **Dr. Schlich**, late Conservator of Forests, Bengal, reported that for Bengal in general the price varies from R6 to R20 per maund. But as R20 is higher than the price mentioned in any one district, we have omitted it from this summary.

MORINDA citrifolia.

Ál Dye.

BENGAL. Different qualities of root. 683 The selling price depends on the fineness of the root, the thin roots being by far the most valuable. In some places the bark stripped from the stem of the plants, as also the twigs, are used for dyeing purposes, the dye they yield being much inferior to that given by the roots. In Maimansingh the juice of the leaves seems to be occasionally used as a dye for cotton to which it imparts a reddish brown colour. In Sáran the price of the root is said to vary from R18 per maund for the thinnest. to R14 for that ½ inch in diameter. Roots over that size are thrown away as worthless. In Darbhangah the roots fetch R6, the bark, R4 and the twigs, R3.

Trade between districts. 684 Regarding the trade in *ál* McCann writes, "As stated above, the plant is generally grown solely for local use, and there is in consequence very little trade in it between one district and another. Of the total annual produce of the Dinájpur district, 30,000 maunds, ⅜ths are exported to Purniah, Jalpáigurí, and Rangpur, adjoining districts; ⅞ths of the produce of the Sáran district (540 maunds) is exported into Tirhut; part of the produce of the Darbhangah district is exported to Benáres and Patná: all the dye produced in the Nadiyá district is sent to Calcutta. A certain quantity also finds its way from the Patná Division into the North-West Provinces. The Collector of Bákarganj reports that in that district the *rayats* sell the roots to *mahajans,* who export it to the North-West Provinces and the Panjáb for its dye, and that the annual export to the North-West Provinces is from 12,000 to 15,000 maunds. There is no other mention of trade in this dye between one district and another. No particulars have been received relating to the competition of aniline or other foreign dyes with *ách,* except from Bákarganj, the Collector of which reports that in his district it was formerly cultivated over 10,000 to 12,000 bighas of land, but that owing to the import of foreign dye-stuff (? aniline) its cultivation has now almost ceased."

Describing the preparation of the dye and the processes of dyeing, the same writer continues:—

Use in dyeing. 685 "*Ál* is used principally for dyeing the thread or yarn out of which the coloured borders of the cotton garments worn by the lower classes are woven. In the northern part of the Rájsháhi Division (Dinájpur, Kuch Behar) it is used also for dyeing silk thread to form the borders of the coarse silk fabrics known as *erendi* or *endi* cloth. This is the chief use in dyeing to which *ál* is put in all districts of Bengal; but besides this it is used in many places to dye pieces of coarse cotton cloth. or thread to be afterwards woven into such cloth. The cloth most frequently dyed with *ál* is the coarse cotton fabric known as *khárua,* as is also the case in the N.-W. Provinces. Frequently the natives dye the cotton thread or yarn for their own garments with *ál* in their own houses; the dyed thread is then given to the weavers to be woven. (Lohárdagá.)

"The colours given by *ál* vary from a *reddish-yellow* through *pink* and various shades of *red* up to a dark *brownish-red.* The dye contained in the root-bark seems to be the best red, whereas that contained in the woody part of the roots is more yellow than red, and consequently where the wood preponderates over the bark the resulting dye is reddish-yellow (Bográ). Very full reports of the methods adopted for *ál*-dyeing have been received from all parts of Bengal, from which it would seem that the methods vary very considerably in different parts of the province.

Preparation of cloth. 686 "Before being dyed with *ál,* the material is almost universally submitted to a preliminary process of preparation, more or less elaborate. The most usual process is to steep the cloth for three or four days in a mixture of powdered castor-oil seeds and cow-dung or sheep-dung; it is

BENGAL. Method of Dyeing. 687

then removed, well washed in soft water, and dried Sometimes the cloth is *boiled* in this decoction, instead of being merely *steeped* in it (Santál Parganás). The dung is occasionally omitted. Thus, in the Santál Parganás the thread for dyeing is sometimes prepared by being steeped for five or six days in water containing powdered castor-seeds alone, and in Lohárdaga, thread is sometimes simply well rubbed with charred castor-oil seed and then dried. Some other kind of oil-seeds may be used instead of, or in addition to, the castor-seed: and plantain ashes, the alkaline earth called *khar* (ক্ষার), or some other kind of alkaline earth, may be employed instead of, or in addition to, the sheep or cow-dung. In Balasor, the thread is prepared for dyeing by being steeped in the juice of thoroughly ripe pumpkins or water-melons (*boital* or *kudoo*, and *panikakhdru*), in the proportion of 1 seer of thread in ½ seer or 16 ounces of pumpkin juice. The following processes of preparing the cloth are employed in the districts mentioned:—

"Thread well rubbed with a mixture of charred *castor-oil seed* reduced to a paste with water and *cow dung*, and kept in this paste for two days. It is next steeped and rubbed daily for 21 days in water made from the ashes of the bhoosa of *sirgooja* (the oil-seed **Guizotia abyssynica**, *Cass.*), boiled in this ash water, and finally rinsed in cold water. Proportions: one seer thread, one seer castor-oil seed, two chittacks cow-dung. (Lohárdaga.)

"Cloth first cleaned and dried; then mix the *ashes of plantain* bark and leaves with a quantity of water sufficient to wet the cloth to be dyed; to this add *castor-oil seed* fried and pounded, and mix well, and afterwards mix with this decoction *mustard-oil*. Steep the cloth in this for one night; in the morning wash it in this same decoction, and dry it. Repeat this process of steeping and washing the cloth in the same water, and subsequently drying, every day for fifteen days. Then wash the cloth in fresh water. On drying, it is ready to receive the dye. Proportions: one seer of castor-seed for every 60 yards of cloth, one powa of mustard-oil for every seer in weight of cloth. (Dinájpur.)

"Cloth steeped in a mixture of *mustard-oil* and *cow-dung* for one night. Next day it is dried in the sun, and the next night is steeped in water containing the *ashes of plantain leaves*. Next day it is again dried, and on the third night is again steeped in water containing plantain ashes. Next day it is washed in soft water and dried in the sun. Proportions: cloth one seer, mustard-oil one powa, and cow-dung one powa. (Bákarganj.)

"Cloth or yarn is steeped three or four times in a mixture of *oil* (castor-oil?) and *country salt* (the ashes of plants, ক্ষার), dried and washed as often in soft water. (Kuch Behar.)

"Cotton thread is first mixed with *oil* and *alkali*, and kept in this state for a month. It is then well washed and dried. (Jalpáiguri.)

"Cloth or thread is steeped in a mixture of cold water, *castor-oil*, a little *cow-dung*, and any of the *alkaline ashes* (the ashes of plants) used by washermen. At intervals the material is rubbed vigorously about in this mixture, then taken out and dried in the sun, and again replaced. This is continued for four or five days, when it is washed in cold water, and finally dried in the sun. (Cuttack Tributary Mehals.)

Preparation of dye. 688

"The dyeing solution is prepared by cutting the roots, or sometimes only the bark of the roots, into small pieces, pounding them well, or grinding them in a *dhenki*, and steeping or boiling in water. Generally the material to be dyed, prepared as above, is boiled with the roots in water, but sometimes, after the roots have been treated as above, they are taken out and thrown away, and the cloth or thread is simply boiled in the water contain-

MORINDA citrifolia.

Ál Dye.

BENGAL.

Auxiliaries employed. 689 ing the dye extracted by the above process from the roots (Dinájpur). A great variety of substances are used as mordants or auxiliaries: these are *lodh* bark or leaves, **Symplocos racemosa** (Lohárdagá, Monghyr, Balasor) *bhauria* or *bhauli* leaves, **Symplocos theæfolia** (Dinájpur, Kuch Behar); *latkan* bark, **Bixa Orellana** (Kuch Behar), *sood* tree bark (?) (Santál Parganás), myrobalans, **Terminalia Chebula** (Patná Division, Monghyr), turmeric, **Curcuma longa** (Sáran, Darbhangah, Monghyr), alum (Bírbhúm, Dinájpur Nadiyá, Patná Division), *dhao* flowers, **Woodfordia floribunda** (Darbhangah, Monghyr), *babul*, **Acacia arabica,** gum arabic, and other gums Darbhangah, Monghyr), lime (Bográ), saltpetre (Monghyr), *bohari*, **Cordia Myxa** (Darjíling), *geru, i e.*, red-ochre (Darbhangah), *thalthelang*), bark and leaves, **Acacia Farnesiana** (?) (Jalpáiguri). Different combinations of these substances are employed in various districts. They may be applied to the cloth after the cleaning process described above before the application of the dye; they may be mixed with the dye before the cloth is steeped or boiled in it; they may be applied to the cloth along with the materials used for cleaning and bleaching the cloth; or, again, the materials required for preparing the cloth, these mordants and accessories may all be made into one decoction along with the dye-stuff, and the cloth at once steeped or boiled in this complex liquid. Most of these materials seem to act as mordants in fixing the colour, generally modifying the intensity of the red dye produced: the gums are of course employed to give tenacity to the colours."

A detailed account of the various processes is then given, to which the reader is referred for further information.

McCann states that in the three districts—Purí, Huglí, and Midnapur—the plant yielding *ál* is **M. citrifolia**, not **M. tinctoria**, but that the methods of cultivation and utilization are exactly similar. From the figures given by the Collector of Huglí, however, it would appear that the cultivation of **M. citrifolia** is less expensive and considered more profitable than that of the so-called **M. tinctoria**; also that while the root in general sells at pretty much the same rate, the finer varieties fetch a much higher price. A peculiar form of *ál* called *hurdi* is obtained from **M. bracteata** in the Darjíling district.

RAJPUTANA and CENTRAL INDIA. 690

VII. Rajputana and Central India.—M. citrifolia and **M. tinctoria** are both said by **Dr. Balfour** to be commonly cultivated in Bundi, Kotah, and Mewar, and **Irvine** states that the root of the latter is largely exported from Ajmere for use as a dye. It is doubtful to what extent these remarks apply to **M. citrifolia**, or to **M. tinctoria**.

NIZAM'S DOMINIONS. 691

VIII. The Nizam's dominions.—Liotard writes: "Reports received tell us that it is known as *serinji* in these territories. It is not produced in the Tandur district, but sold by merchants from Hanamkonda and Nirmal at (*hali-sicca*) R2 for 12 seers, and is used to a small extent in dyeing the *khadi* cloth. One of the officers of His Highness thinks that the soil of the Edlabad Taluk is well adapted for the cultivation of the *serinji* plant." These remarks appear to apply to **M. citrifolia**.

BURMA. 692

IX. Burma.—Mason states that the Burmans prepare their red dyes most usually from the roots of two or three species of **Morinda**, but that the species most generally seen in cultivation is **M. citrifolia**, *Roxb.* (**var. citrifolia** proper). **Kurz** states that this species is generally cultivated in native gardens throughout the province, and District officers have reported it as growing apparently wild, and very plentifully, in the Tavoy and Toungoo districts. **Balfour** mentions that **M. bracteata** is common throughout the Province of Pegu, and is frequently cultivated near *Phoungyee* homes. This is the only reference the writer can find to the cultivation of **M. bracteata**, and is probably a mistake for **M. citrifolia** or **M.**

tinctoria. **J. W. Oliver, Esq.**, Conservator of Forests, Burma, in a note kindly furnished to the editor, states that, as far as he can make out, there are three common forms of **Morinda** in Pegu—1, *Nibase taunge*, said to be the best, cultivated by the Karens; 2, *Nibase saukse*, wild in swamp forests; and 3, *Nibase segyi*, wild in **Dipterocarpus** forests. The writer is unable, from want of material, to determine what these forms are. BURMA.

In Bassein 88 acres are said to be cultivated with the *nyaw-gyi*, the sowing being done in June or July. Information as to the quantities and value of the dye produced is unfortunately wanting, nor is it known to what extent the cultivated plant is **M. citrifolia** or **M. tinctoria**, some reports mentioning one, some the other.

X. Assam.—The chief dye-yielding species of Assam is apparently **M. angustifolia**, for an account of which the reader is referred to the description of that species. ASSAM. 693

FOREIGN TRADE. 694

FOREIGN TRADE IN ÁL DYE.

Of late years *ál* root has been occasionally sent to Europe, but no regular foreign trade in the dye-stuff appears to exist. The colour yielded is a fast but dull red, and appears to have little chance of competition with those obtained from madder. It is possible, however, that certain of the orange and golden yellows obtained by **Mr. Wardle** may find favour and thus open out a field for foreign trade in the dye.

CHEMISTRY. 695

CHEMICAL COMPOSITION.

The root bark has been carefully examined by several chemists with somewhat varying results. According to **Anderson** it was found to yield, by exhaustion with boiling alcohol, a crystalline principle, *morindin*, $C_{28} H_{30} O_{15}$, to the presence of which the dyeing properties of the plant are due. This substance is deposited on cooling, as an impure flocculent material, mixed with a red precipitate. After repeated crystallization from alcohol acidulated with hydrochloric acid, it forms satin-like, needle-shaped, yellow crystals. It is sparingly soluble in cold water, more readily in boiling, but is deposited as a gelatinous mass from the latter on cooling. It is sparingly soluble in alcohol when cold, more readily when boiling, but dissolves best in weak spirit. It is insoluble in ether. In alkaline solutions it dissolves readily forming an orange-red liquid; lime and baryta water yield red precipitates, while acetate of lead forms a crimson precipitate. Morindin dissolves in concentrated sulphuric acid, forming a deep purple-red solution. Heated in a close vessel it melts and boils, emitting beautiful orange vapours, which condense to red, needle-shaped crystals, insoluble in water, but soluble in alkaline liquids, giving a violet colour. This substance, which **Dr. Anderson** calls *morindone*,* seems to be the same as that formed under the influence of sulphuric acid. The ammoniacal solution of morindone yields, with alum, a red lake, and gives a blue precipitate upon addition of baryta water.

Professor Rochleder considers morindin to be identical with the ruberythric acid which he obtained from madder; but later investigations have proved that it differs in being insoluble in ether, and in its behaviour with alkalies. The same chemist regarded morindone as identical with alizarin, but **Dr. Anderson's** analysis shews the composition of the former to be C., 65·81, H., 4·18; figures which do not agree with the composition of alizarin, but agree very nearly with those which **Dr. Schützenberger** found for purpurin, *viz.*, C., 65·6, H., 3·2 (*Crookes*).

MEDICINE. Fruit. 696 Leaves. 697

Medicine.—**Ainslie** states that the Cochin-Chinese place **M. citrifolia** amongst their medicinal plants, believing the FRUIT to be deobstruent and emmenagogue. **Dymock** writes that th LEAVES are used in Bom-

MEDICINE. bay as a healing application to wounds and ulcers, and are administered internally as a tonic and febrifuge. **Mr. Brownlow**, in a note on the timber trees of Cachar, states that the young leaves of **var. bracteata** are intensely bitter and are used by the Natives for the sequelæ of fevers (*Agri.-Hort. Soc. Ind, Jour.* (*Old Series*), *XIII.*, *359*). According to **Irvine** the

Root. 698 ROOT of **citrifolia** proper is employed in the Panjáb as a cathartic.

SPECIAL OPINION.—§ "The charred leaves made into a decoction with mustard are a favourite domestic remedy for infantile diarrhœa. The unripe berries charred and mixed with salt are applied successfully to spongy gums" (*Native Surgeon T. Ruthman Moodelliar, Chingleput, Madras Presidency*).

FOOD. Fruit. 699 **Food.**—The FRUIT when green is frequently eaten in curries; when ripe it is also used as an article of food, and is said by **Mason** to be a great favourite with the Burmans. In the *Agri.-Hort. Soc. of Ind. Jour.* (*Old Series*), *IX.*, *Sel. 58*, it is stated that the albuminous SEEDS of **var. bracteata**

Seeds. 700 when roasted are used in Pegu as articles of food.

TIMBER. 701 **Structure of the Wood.**—Yellow or yellowish-brown, weight from 30 to 40lb per cubic foot, durable, used for making plates, dishes, and gun-stocks, and, according to **Lisboa**, suitable for turning. The above weight is given by **Gamble** for the wood of **M. exserta,** probably referable to this species.

702 **Morinda exserta,** *Roxb.; Fl. Br. Ind., III., 156.*

This is reduced by the *Flora of British India* to a mere sexual form of either **M. citrifolia** or **M. tinctoria.**

703 **M. persicæfolia,** *Ham.; Fl. Br. Ind., III., 157.*

Syn.—M. LANCEOLATA, *Wall.*

This shrub is a native of Burma, Chittagong, and Singapore, and is distributed to Sumatra and Siam. **Gamble** describes it as yielding a good dye. Several reports exist from the Chittagong division regarding a **Morinda** dye-stuff in extensive use there, which may be the product of this plant, but unfortunately sufficient information is not forthcoming to lead to the definite identification of the species.

704 **M. tinctoria,** *Roxb.; Fl. Br. Ind., III., 156.*

This is considered by many Indian botanists, as has already been explained, to be merely a wild form of **M. citrifolia,** *Linn.* It is possible, if not probable, this may be so, and that **M. bracteata,** regarded by others as the sylvan form of the cultivated *ál*, is a distinct and almost entirely wild species. The following are the varieties recognised by the *Flora of British India*:—

Var. α, tinctoria proper—M. TINCTORIA, *Roxb.;* M. ASPERA, *W. & A.;* M. COREIA *and* NODOSA, *Ham.;* M. LEIANTHA, *Kurz;* M. CITRIFOLIA, *Bedd., Fl. Sylv., t. 220.*

Var. β, tomentosa—M. TOMENTOSA, *Heyne;* M. NAUDIA *and* CHACUCA, *Ham.;* M. STENOPHYLLA, *Spr.;* M. ANGUSTIFOLIA, *Roth., not of Roxb.*

Var. γ, multiflora—M. MULTIFLORA, *Roxb.*

Var. δ, aspera—M. ASPERA, *W. & A.*

Forma exserta, see **M. exserta,** *Roxb.*

Vern.—*Al*, HIND.; *Ach, auch, dárnharidrá*, BENG.; *Achhu*, URIYA; *Chailí, bankatari*, SANTHAL; *Larnong, asú khat*, ASSAM; *Ach, áich*, C. P.; *Manjishta*, BOMB.; *Maddi chettu, múlaga chettu*, TEL.; *Seing laing*, SINGPHUS; *Achchhuka*, SANS. The same terms as those already given for **M. citrifolia** are indifferently applied to that species or to **M. tinctoria,** but those above enumerated appear to be most commonly used for the latter.

References.—*Roxb., Fl. Ind., Ed. C.B.C., 182, 183; Brandis, For. Fl., 276, 277; Kurz, For. Fl. Burm., II., 59; Beddome, Fl. Sylv., t. 220; Thwaites,*

En. Ceylon Pl., 145; Dalz. & Gibs., Bomb. Fl., 114; Elliot, Flora Andh., 109, 119, 183; Campbell, Ec. Prod., Chutia Nagpur, No. 8427; U. C. Dutt, Mat. Med. Hind., 289; Irvine, Mat. Med. Patna, 7; Baden Powell, Pb. Pr., 44; Darrah, Note on Cotton in Assam, 34; Lisboa, U. Pl. Bomb., 88, 246; McCann, Dyes and Tans Beng., 20-38; Liotard, Dyes, 49, 51-53, 54; Wardle, Dye Rept., 13; Gazetteers:—Bombay, XVIII., 44; Orissa, II., 179.

Habitat.—Found throughout India from the Sutlej eastward, and southward to Ceylon and Malacca; generally wild.

Dye.—The WOOD, BARK, and ROOT-BARK yield a red dye, apparently identical with that of **M. citrifolia** (see *al* dye, p. 263). DYE. Wood. 705 Bark. 706 Root-bark. 707

Medicine.—Irvine states that the ROOT, like that of **M. citrifolia**, is used internally as an astringent. MEDICINE. Root. 708

Food.—"The green FRUITS are picked by the Hindús and eaten with their curries" (*Roxb.*). FOOD. Fruit. 709

Structure of the Wood.—Similar to that of **M. citrifolia**, described under **M. exserta**, *Roxb.*, by Brandis, Gamble, &c. According to Roxburgh it is "hard and very durable, variegated with red and white," and "employed for gun-stocks in preference to all other kinds." TIMBER. 710

Morinda umbellata, *Linn.; Fl. Br. Ind., III., 157.* 711

Syn.—M. SCANDENS, *Roxb.*; M. TETRANDRA, *Jack*; M. PADAVARA, *Juss.*

Vern.—*Al*, BOMB.; *Núna, kai, núna marum*, TAM.; *Múlúghúdú*, TEL.; *Maddi chekhe*, KAN.; *Kirri-walla*, SING.

References.—*Roxb., Fl. Ind., Ed. C.B.C., 184; Kurz, For. Fl. Burm., II., 62; Thwaites, En. Ceylon Pl., 145; Ainslie, Mat. Ind., II., 253; Dymock, Mat. Med. W. Ind., 2nd Ed., 402; Drury, U. Pl., 297; Lisboa, U. Pl. Bomb., 162, 200; Gazetteers, Mysore and Coorg, I., 436; Ind. Forester, III., 237; Moore, Man. Trichinopoly, 79.*

Habitat.—A diffuse shrub met with in the hilly regions of Eastern Bengal, ascending to 4 000 feet on the Khásia mountains, also in the Malay Peninsula, and in South-Western India, in the Southern Konkan, the Nilghiri Hills, the mountains of Travancore, and in Ceylon.

Dye.—Like all other members of this genus, the ROOT yields a dye, but of a brilliant yellow, not a red, colour. DYE. Root. 712

Fibre.—Thwaites states that the Singalese employ the tough stems instead of ropes for binding fences. FIBRE. 713 Stem. 714

Medicine.—Ainslie informs us that the LEAVES, made into a decoction in conjunction with certain aromatics, were employed in his time by the "Tamool doctors" in cases of diarrhœa and lientery. MEDICINE. Leaves. 715

Food.—"The green FRUITS are used in curries and the ripe ones eaten" (*Lisboa*). FOOD. Fruit. 716

MORINGA, *Juss.; Gen. Pl., I., 430.* 717

Moringa concanensis, *Nimmo; Fl. Br. Ind., II., 45;* MORINGEÆ.

Vern.—*Sainjna, súnjna, segora, hegu, segu*, RAJ.; *Múah*, SIND; *Sainjna*, BOMB.

References.—*Grah., Cat. Bomb. Pl., 43; Dalz. & Gibs., Bomb. Fl., 311; Brandis, For. Fl., 311; Gamble, Man. Timb., 114; Murray, Pl. and Drugs, Sind, 44; Lisboa, U. Pl. Bomb., 57, 101; Indian Forester, XII., App. 10.*

Habitat.—A small tree met with on the dry hills of Rájputana, Sind, and the Konkan.

Gum.—E. A. Fraser, Esq., Assistant Agent to the Governor General, Rájputana, states, in a note furnished to the editor, that it yields a white gum of little value. GUM. 718

MORINGA pterygosperma. The Horse-radish Tree.

FOOD. Fruit. 719 Flowers. 720 Roots. 721 **Food.**—The unripe FRUIT and FLOWERS are eaten as a pot-herb. Murray states that the ROOTS, like those of the next species, have a pungent flavour, and are said to be used as a substitute for horse-radish.

Moringa pterygosperma, *Gærtn.; Fl. Br. Ind., II., 45; Wight, Ill., t. 77.*

THE HORSE RADISH TREE.

Syn.—MORINGA OLEIFERA, *Lam.*; M. ZEYLANICA, *Pers.*; M. POLYGONA, *DC.*; HYPERANTHERA MORINGA, *Vahl.*; H. DECANDRA, *Willd.*; GUILANDINA MORINGA, *Linn.*

Vern.—*Shajnah, shajná, ségvá, soanjna,* HIND.; *Sojná, sujuna, sajina,* BENG.; *Munigha, sajiná,* URIYA; *Mulgia,* KÓL.; *Munga arak,* SANTAL; *Sahajna, senjna, sújna,* N.-W. P.; *Soanjna, sánjna, senjna,* PB.; *Swanjera,* SIND; *Mungé-ká-jhár,* DEC.; *Saragvo, sekto, sujna, mangai, segat, sanga, shegva,* BOMB.; *Munagácha-jháda, badadí-shingá, hajháda, shevaga, shevgi,* MAR.; *Saragavo,* GUZ.; *Músing,* GOA; *Murungai, morunga,* TAM.; *Munaga, adavi-munaga, káru munaga, múraga,* TEL.; *Nugge-giḍá,* KAN.; *Murinna,* MALAY.; *Dándalonbin, danthalone, daintha,* BURM.; *Murungá,* SING.; *Shóbhánjana-vrikshaha, sóbhánjana, sigru,* SANS.

References.—*Roxb., Fl. Ind., Ed. C.B.C., 360; Voigt, Hort. Sub. Cal., 78; Brandis, For. Fl., 129; Kurz, For. Fl. Burm., I., 68; Beddome, Fl. Sylv., t. 80; Gamble, Man. Timb., 114; Dalz. & Gibs., Bomb. Fl., 314; Stewart, Pb. Pl., 19; Rev. A. Campbell, Rept. Econ. Pl., Chutia Nagpur, No. 7820; Graham, Cat. Bomb. Pl., 43; Mason, Burma and Its People, 516, 749; Sir W. Elliot, Fl. Andh., 11, 53, 87, 119; Sir W. Jones, Treat. Pl. Ind., V., 115; Pharm. Ind., 61; Flück. & Hanb., Pharmacog., 73; U. S. Dispens., 15th Ed., 23; Fleming, Med. Pl. & Drugs (Asiatic Reser., XI.), 168; Ainslie, Mat. Ind., I., 175; O'Shaughnessy, Beng. Dispens., 269; Irvine, Mat. Med. Patna, 68; Taylor, Topog. of Dacca, 58; Honigberger, Thirty-five years in the East, II., 311; Moodeen Sheriff, Supp. Pharm. Ind., 176; U. C. Dutt, Mat. Med. Hindus, 117, 318; Sakharam Arjun, Cat. Bomb. Drugs, 50; K. L. De, Indig. Drugs Ind., 75; Murray, Pl. & Drugs, Sind., 44; Dymock, Mat. Med. W. Ind., 2nd Ed., 206; Dymock, Warden and Hooper, Pharmacog. Ind., Vol. I., 390; Official Corresp. on Proposed New Pharm. Ind., 225, 228, 234, 310; Birdwood, Bomb. Prod., 149, 220, 267, 282; Baden-Powell, Pb. Pr., 397, 424; Drury, U. Pl. Ind., 297; Atkinson, Him. Dist. (Vol. X., N.-W. P. Gaz.), 308; Stocks, Rep. on Sind; Useful Pl. Bomb. (Vol. XXV., Bomb. Gaz.), 57, 151, 218, 278, 397; Forbes Watson, Econ. Prod. N.-W. Prov., Pt. I. (Gums and Resins), 3, 5; Pt. V. (Vegetables, Spices, and Fruits), 44, 59; Royle, Gums and Resinous Prod. (P. W. Dept. Rept.), 4, 19, 26, 36, 50, 51, 53, 68; Crookes, Hand-book, Dyeing, &c., 517; Gums and Resins, 33; Christy, New Com. Pl., V., 45; Piesse, Perfumery, 390; Ain-i Akbari, Blochmann's Trans., Vol. I., 64; Statistics Dinajpur, 158; Nicholson, Man. Coimbatore, 41; Moore, Man., Trichinopoly, 79; Settlement Reports · Panjáb, Jhang, 21; Central Provinces, Mundlah 88; Chánda, 82; For. Ad. Rept., Chutia-Nagpur, 1885, p 29; Gazetteers:—Bengal, Orissa, II., 180; Bombay, V., 24, 285; VII., 42; X., 404; XV., 73; XVII., 26; XVIII, 52; Panjáb, Karnal, 16; Muzaffargarh, 23; Amritsar, 4; Hoshiárpur, 10; Jhang, 16; N.-W. P., I., 80; IV., lxx.; Sind, 559; Mysore and Coorg, I., 52, 68; Indian Forester, II. 292; III., 201; VI., 238; XII. (App.), 10; Spons' Encycl., 1378, 1621, 1674; Balfour, Cyclop. Ind., II., 986.*

Habitat.—A tree, wild in the sub-Himálayan tract from the Chenáb to Oudh; commonly cultivated in India and Burma on account of its leaves, flowers, and pods, all of which are eaten.

GUM. 722 **Gum.**—It yields a gum which is white when it exudes, but gradually turns to a mahogany colour on the surface; used in native medicine. This exudation belongs to the tragacanth or hog-gum series, but, owing to its commonly dark colour, is of no European commercial value. Samples vary in colour from dark mahogany, red, or pink to almost white, and in shape from stalactitic pieces to tears. Like other gums of the tragacanth series, it is insoluble in water. About 8·3 per cent. is soluble in alcohol, 7·85 of the remainder in ether, while the residue is almost completely dissolved by alkalies.| This gum is one oft hose frequently called MOCHARAS or MOCHRAS.

Dye.—Ainslie states that the WOOD is used in Jamaica for dyeing a blue colour, "for which purpose I cannot learn that it is employed in India." The gum is used in calico-printing.

Mr. Christy, in his *New Commercial Plants*, includes this amongst DYE
East Indian tans, the bark, according to him, being used and known under Wood.
the vernacular name of *subanjuna*. As far as India is concerned, this is a 723
mistake; the plant is far too valuable as a vegetable producer to be used
as a tan, even should it possess the necessary properties.

Fibre.—The BARK yields a coarse fibre from which mats, paper, or FIBRE.
cordage might be prepared. 724

Oil.—The SEEDS yield a clear, limpid, almost colourless oil (according Bark.
to Cloez 36·2 per cent.), rather thick at ordinary temperatures, easily ex- 725
tracted by simple pressure. It has a sp. gr. of 0·912–0·915 at 60° Fh., is
fluid at 77° Fh., thick at 59° Fh., and solid at lower temperatures. It is OIL.
almost devoid of odour and flavour, saponifies slowly, and does not turn 726
rancid. Composed essentially of oleine, margarine, and stearine, after Seeds.
separation of the, solidifiable portion by cooling, it is one of the best 727
lubricants for fine machinery. The oil from this species and that from **M. aptera**, *Juss.*, are commercially termed BEN OIL and are highly valued as lubricants by watch-makers. It is, however, seldom made in India and does not form an article of export, a fact which is the more remarkable when one remembers the great extent to which the tree is cultivated. India might easily, and apparently profitably, supply the whole world with BEN or MORINGA OIL, and it is to be hoped that attention may be directed to the subject. In addition to its value as a lubricant, it is highly esteemed by perfumers, owing to its great power of absorbing and retaining even the most fugitive odours In the West Indies it is said to be used as a Salad oil. [For Benne Oil see **Sesamum**.]

Medicine.—The medicinal virtues of this plant have long been known MEDICINE.
and appreciated in India. It is frequently mentioned by Chakradatta,
also in the *Bhávaprakása*, and other works on Sanskrit medicine. By
ancient writers the ROOT is described as acrid, pungent, stimulant, and
diuretic when given internally, rubefacient when applied externally. The Root.
SEEDS have similar properties ascribed to them and are called *sveta* 728
maricha, or white pepper. In the *Bhávaprakása* two varieties of *Sobhán-* Seeds.
jana or **Moringa** root are described, namely, a white and a red. The root 729
of the former is said to be a stronger rubefacient, while that of the latter
is preferred for internal use. A decoction of the ROOT-BARK is recommended
by Chakradatta, and in the *Bhávaprakása*, for ascites, enlarged spleen or Root-Bark.
liver, internal and deep-seated inflammation, and calculous affections. It 730
is also directed to be used externally as a plaster or decoction over inflamed parts. The fresh juice of the root-bark is recommended for much
the same diseases as the decoction, and is also said to relieve otalgia
when poured into the ears, while the GUM is said to be used for the same
purpose (*U. C. Dutt*). Gum.

Dymock informs us that Muhammadan writers describe the FLOWERS as 731
hot and dry, and consider that they expel cold humours, disperse swellings, Flowers.
act as a tonic and diuretic and increase the flow of bile. The JUICE of the 732 Juice.
root is prescribed by them with milk, as diuretic, antilithic, digestive, and 733
useful in asthma. A poultice made from the root is supposed to reduce
swelling, but is said to be very irritating and painful to the skin. The Pods.
PODS are esteemed as a vegetable, and act as a preventive against intes- 734
tinal worms (*Dymock, quoting the Makhzan*). Oil.

The root and OIL were early noticed by English writers on Indian 735

MEDICINE.

Materia Medica. Thus **Fleming** wrote: "In medicine the root of the young tree completely supplies the place of the horse-radish, whether employed externally as a rubefacient, or used internally in cases of palsy, chronic rheumatism and dropsy as a stimulant. The expressed oil of the seeds is used externally for relieving the pain of the joints in gout and acute rheumatism." **Ainslie** also notices the green root, giving the same information, and adds that it was prescribed by native doctors in doses of about one scruple in intermittent fever, was also used in epilepsy and hysteria, and was considered a valuable rubefacient in palsy and chronic rheumatism. He quotes **Fleming's** account of the oil. Later, **Taylor** writes that the fresh root, mixed with mustard seed and green ginger, is used as an external application in rheumatism and is also administered internally in enlargement of the spleen and dyspepsia. **Honigberger** states that "the FRUIT is

Fruit. 736

administered by the *Hakims* in affections of the liver and spleen, articular pains, tetanus, debility of nerves, paralysis, pustules, patches, Indian leprosy, &c.," and recommends the root for soreness of the mouth and throat, and the gum for dental caries.

Later writers have, as a rule, simply repeated certain of the above enumerated opinions in their accounts of the drug; and in the *Pharmacopœia of India* it has a place on the secondary list, and is briefly described. **Stewart** adds to the above information that the gum is used in the Panjáb for rheumatism and as an astringent. **Dymock** also informs us that in Bombay a decoction of the root-bark is employed to relieve spasm, and gives the following interesting account of the medicinal use of the plant in the

Bark. 737

Konkan:—"The BARK of the wild tree is ground with Plumbago root, pigeon's dung, and chicken's dung, and applied to destroy guinea-worm.

Leaves. 738

Four tolás of the JUICE OF THE LEAVES of the cultivated plant are given as an emetic. The gum is said to be used to produce abortion, but it is difficult to obtain any reliable evidence on a point of this nature; it would be quite possible to use it as a tent to dilate the os uteri, as it is very tough, and swells rapidly when moistened." In Bengal half-ounce doses of the bark are employed internally for the same purpose.

The root is in all probability the only part of real medicinal value, and would appear to be a perfect substitute for the true horse-radish, **Cochlearia Armoracia**. It has been examined by **Broughton** and was found to yield by distillation with water an essential oil of disgusting odour and much pungency. It differs from the oil of mustard and garlic, the odour being distinct and more offensive (*Pharmacographia*).

SPECIAL OPINIONS.—§ "The root is used as a blistering agent in cases of enlarged liver in young children, and is also said to be used as an abortive" (*W. Forsyth, F.R.C.S., Civil Medical Officer, Dinajpore*). "An infusion of the fresh root is used as a stimulant and diuretic in dropsical affections, also as a gargle in hoarseness and relaxed sore throat" (*Civil Surgeon J. H. Thornton, M.B., Monghyr*). "The gum is frequently applied to disperse glandular swellings" (*Assistant Surgeon S. C. Bhattacharji, Chanda, Central Provinces*). "The young leaves, the flowers, and the young pods are used as food in West Bengal. The flowers are hot, dry, tonic, and useful in catarrhal affections" (*Civil Surgeon D. Basu, Farídpore, Bengal*). "The leaves, ground into a paste with a few pods of garlic, a bit of turmeric and a little salt and pepper, are given internally in cases of dog-bite and applied externally over the bite. In five or six days the wound heals and the inflammatory and febrile symptoms subside" (*Surgeon-Major D. R. Thomson, M.D., C.I.E., Madras*). "The flowers, boiled in milk, are said to be aphrodisiac" (*Native Surgeon T. R. Moodelliar, Chingleput, Madras Presidency*). "The root can be used as a rubefacient and counter-irritant in rheumatic cases" (*Surgeon R. L*

MEDICINE.

Dutt, M.D., Pubna). "The juice of the bark is used for mange in horses" (*Surgeon-Major P. N. Mookerjee, 32nd M. N. I., Cuttack, Orissa*). "The bark of the root made into a paste with water can be used as a substitute for cantharides" (*Assistant Surgeon S. C. Bose, Bankipore*). "The root-bark, which is used as a substitute for horse-radish by Europeans, makes an excellent rubefacient plaster. The outer bark of the tree is also similarly used. I have found it very useful as a carminative stimulant given internally in flatulent colic, in combination with asafœtida and ginger. These ingredients are pounded together and given in the form of a bolus (30 grains) twice or three times daily. The leaves cooked with the *lye* obtained from the ashes of the plantain tree are made into a sort of soup and used as an article of food in North Bengal, and considered a good antiscorbutic, but both the leaves and pods which are also used in curries are usually supposed to be heating" (*C. T. Peters, M.B., Zandra, South Afghánistán*). Lye. 739

Food.—The tree flowers in February and produces a long whip-like bean in March and April. The LEAVES, FLOWERS, and PODS and even the TWIGS are used with various condiments, and cooked in many ways as pot-herbs, and the pungent root is employed as a garnish instead of mustard. The fruit is also made into a pickle, which is described as "most nauseous to Europeans," but which appears to have been long known and appreciated by the natives of Northern India, being mentioned (300 years ago) in the list of pickles given in the *Áín-i-Akbari*. The ROOT has long been known to Europeans in India as an efficient substitute for horse-radish, and is thus described by Fleming: "The root of the young tree, when scraped, so exactly resembles horse-radish as scarcely to be distinguished from it by the nicest palate." It is now universally used like true horse-radish, as a condiment with roast beef, &c. FOOD. Leaves. 740 Flowers. 741 Pods. 742 Twigs. 743 Root. 744

Fodder.—In many parts of the country the TWIGS and LEAVES are largely lopped for fodder. FODDER. Twigs. 745 Leaves. 746

Morphia, see **Papaver somniferum**, *Linn.*; Vol. VI., Pt. I.

Morrhua or **Cod-liver-oil**, see **Fish**, Vol. III, 368; also **Gadus** [**morrhua**, *Linn.*; Vol III., 462.

Mortar, see **Carbonate of Lime**, Vol. II., 142; also **Cements**, Vol. II., [245; and **Cocos nucifera**, Vol. II., 455.

MORUS, *Linn.*; *Gen. Pl., III., 364.* 747

A genus of trees or shrubs which, according to the *Genera Plantarum*, comprises from ten to twelve species, which have been reduced to five by Bureau. In the *Flora of British India* three are described as natives of India, while two, **M. alba**, *Linn.*, and **M. atropurpurea**, *Roxb.*, are stated to be extensively cultivated. All the Indian species are united by Bureau under **M. alba**, *Linn.*, and following him many Indian writers have described all the Indian forms under one name. The economic information regarding each species cannot, therefore, as a rule be separated, but it appears probable that all possess very similar features, and may, therefore, all be used similarly to **M. indica**, which, as the commonest of the Indian White Mulberries, will be most fully described.

Morus alba, *Linn*; *Fl. Br. Ind., V.. 492*; URTICACEÆ.

THE WHITE MULBERRY.

Syn.—M. ALBA (*in part*), *Bureau*; M. SERRATA, *Wall.*; M. TATARICA, [*Linn.*

Vern.—*Tút, túl, túlklu, chinni, chún*, HIND.; *Tút*, PB.; *Tut, chinni, satur, tutla, shah-tut*, BOMB.; *Uppu nute*, KAN.; *Tula*, SANS.; *Túth, tút*, ARAB. & PERS.

The White Mulberry.

References.—*Roxb., Fl. Ind., Ed. C.B.C., 658; Voigt, Hort. Sub. Cal., 283; Brandis, For. Fl., 407, t. 47; Gamble, Man. Timb., 327; Dalz. & Gibs., Bomb. Fl., Suppl., 80; Stewart, Pb. Pl., 218; Aitchison, Bot. Afgh. Del. Com., 109; DC., Origin Cult. Pl., 149; Baden Powell, Pb. Pr., 585; Atkinson, Him. Dist., 317; Lisboa, U. Pl. Bomb., 126, 172, 235; Ayeen Akbery, Gladwin's Trans., I., 84; II., 135; Gazetteers:—Mysore and Coorg, I., 70; N.-W. P., IV., lxxvii.; Karnal, 16; Ind. Forester, IV., 230; V., 478; VI., 3, 4, 10; XII., 37, App. 28; Rep., Agri. Dept. Bengal, 1886, App., lviii.*

Habitat.—A deciduous, monœcious tree, cultivated in the Panjáb, the North-West Himálaya, and Western Tibet, ascending to 11,000 feet; distributed either wild or cultivated to Afghánistán, and North and West Asia. The original wild area of the species is involved in great obscurity, an obscurity which has been increased by the fact that botanists have at different times given the species conflicting characteristics. Thus **Bureau** in *DC. Prodromus* reduced all the Indian species (now again recognised as distinct in the *Flora of British India*) to **Morus alba,** while other writers have described mere cultivated forms as distinct species. **DeCandolle** writing of **Morus alba,** *Linn.*, as defined by **Bureau,** remarks: "The antiquity of its culture in China and in Japan, and the number of different varieties grown there, lead us to believe that its original area extended eastward as far as Japan; but the indigenous flora of Southern China is little known, and the most trustworthy authors do not affirm that the plant is indigenous in Japan." "It is also worthy of note that the white mulberry appears to thrive especially in mountainous and temperate countries, whence it may be argued that it was formerly introduced from the North of China into the plains of the South." **Sir J. D. Hooker** writes, "wild or cultivated in Afghánistán, North and West Asia," but **DeCandolle** states that he believes the tree not to be truly wild, but to have been naturalised in Persia, Armenia, and Asia Minor at a very early epoch. It is interesting to note that **Sir J. D. Hooker's** opinion is confirmed by that of **Aitchison,** who, in his *Botany of the Afghán Delimitation Commission,* writes: "An indigenous tree near water at Badghis, at an altitude of above 3,000 feet." If not indigenous to Northern India the tree appears to have been cultivated from remote times in the North-West Himálaya. In more recent years, it has extended to the plains of the Panjáb, Bombay, and even Bengal, where it is grown partly for feeding the silkworm with its leaves, partly for the sake of its fruit. It is cultivated in the same way as the more commonly grown species **M. indica,** *Linn.*, to which the reader is referred for an account of the methods employed.

FIBRE. 748 **Fibre.**—see **M. indica,** *Linn.*

MEDICINE. Juice. 749 **Medicine.**—**Baden Powell,** writing of **M. nigra** (by which he probably meant the black fruited form of **M. alba**), remarks that the sweet, deep red JUICE of the fruit is used for sore throat, and acts as a pleasant refrigerant in cases of fever. Fruit. 750 The FRUIT is employed by *hakíms,* as a remedy for sorethroat, dyspepsia, and melancholia, Bark. 751 while the BARK is considered purgative and anthelmintic.

FOOD & FODDER. Fruit. 752 **Food and Fodder.**—The Indian white mulberry flowers in February, and the fruit ripens in May and June The FRUIT varies greatly in different localities, and it is difficult to determine whether the descriptions of many authors in reality refer to the fruit of this species or to that of the **M. nigra.** **Brandis** writes: "There are many varieties, sweet and acid, and of all shades of colour, from white to a deep blackish-purple. The large white kind of the Peshawar valley (*shah tút*) is one of the best." The following kinds are, according to **Stocks,** cultivated in Baluchistán:—*Siah,* colour black and white mixed; *bedana* (seedless); *pedwani* (grafted), with delicious small pearly, white fruit; *shah tút* (royal mulberry); *khar tút* (jackass mulberry). So many are the gradations in quality and colour, that **Brandis** considers it not impossible that, in the Panjáb and Afghánistán at least, it may be found impossible to maintain the distinction between the two species, **M. alba** and **M. nigra.** FOOD & FODDER.

The fruits of all the better kinds are considered great delicacies by the natives whether fresh, dried, or preserved, and in Kashmír and Afghánistán constitute a considerable portion of the food of the inhabitants during autumn. **Baden Powell** states that in Afghánistán the fruit is dried and made into flour, the bread from it being nutritious and fattening.

The LEAVES are extensively used for feeding silkworms, especially in Kashmír. They are also considered excellent fodder and are given to cattle when they can be spared. They have been recently very highly recommended as a powerful lactagogue, a contributor to the *Indian Agriculturist* (August 10th, 1889) stating that with a seer of leaf morning and evening the yield of a cow increased from three to five measures of milk. Leaves. 753

Structure of the Wood.—Yellow or reddish-brown, hard, even-grained, weight 38 to 56lb per cubic foot. It seasons and polishes well, and is used for building, boats, furniture, and agricultural implements. TIMBER. 754

Industrial Uses.—For the value of this species in sericulture, see **Silk.** INDUSTRIAL USES. 755

Morus atropurpurea, *Roxb.; Fl. Ind., Ed. C.B.C., 658.* 756

Vern.—*Chin-ki-tút.*

References.—*Fl. Br. Ind., V., 491; ? Stewart, Pb. Pl. (under M. sinensis), 218; Baden Powell, Pb. Prod., 585.*

Habitat.—A Chinese species closely allied to, if not a variety of, **M. alba**, with long, cylindric, dark purple fruit, cultivated in India. **Stewart** (under the name above cited) describes a Chinese species as follows: "The cultivation of this species has been largely extended in the Panjáb, chiefly from the Saharanpur Botanic Gardens, with a view to sericulture, and now there are thousands of them in the Bári Doáb, &c., ready for use, when the proper method of managing the worm has been found."

M. indica, *Linn.; Fl. Br. Ind., V., 492; Wight, Ic., t. 674.*

Syn.—MORUS ALBA, *var.* INDICA, *Bureau*; M PARVIFOLIA, *Royle*; ? M. ACIDOSA, *Griff.*; M. CUSPIDATA, *Wall.*; M. ALBA, *var.* CUSPIDATA, *Bureau.*

Vern.—*Tút, tutri,* HIND.; *Tút,* BENG.; *Nuni, bola,* ASSAM; *Chhota kimbu, kimbu,* NEPAL; *Mekrap, nambyong,* LEPCHA; *Singtok,* BHUTIA; *Tút, tútri, sháh-tút,* N.-W. P.; *Túl, tútri, karan,* PB.; *Tút,* bark=*tú t. jo-kul,* SIND; *Tut, tutri, ambor. setur, tula ambor,* BOMB.; *Tut,* MAR.; *Shetur,* GUZ.; *Tút,* DEC.; *Kambili-púch, mushu hattai,* TAM.; *Kambali, kambali-búchi,* TEL.; *Hippal-nerali,* KAN.; *Posa,* BURM.; *Shálmalivrikshaha, tula, tuda,* SANS.; *Tút,* ARAB.; *Shihatuta, tú',* PERS.

References.—*DC., Prod., XVII., 243; Roxb., Fl. Ind., Ed. C.B.C., 658; Brandis, For. Fl., 408; Kurz, For. Fl. Burm., II., 468; Gamble, Man. Timb., 328; Stewart, Pb. Pl., 218; Mason, Burma and Its People, 456, 775; Rumphius, Amb., VII., t. 5; Moodeen Sheriff, Supp. Pharm. Ind., 117; U. C. Dutt, Mat Med. Hindus, 321; Murray, Pl. & Drugs, Sind., 31; Birdwood, Bomb. Prod., 177; Baden Powell, Pb. Pr., 266, 271, 511, 585; Atkinson, Him. Dist. (Vol. X., N.-W. P. Gaz.), 317, 742; Stocks, Rep on Sind.; Useful Pl. Bomb. (Vol. XXV., Bomb. Gaz.), 126, 172, 235, 390; Forbes Watson, Econ. Prod. N.-W. Prov., Pt. V. (Vegetables, Spices, and Fruits), 44, 82; Royle, Fibrous Pl., 343; Gums and Resinous Prod. (P. W. Dept. Rept.), 15; Cooke, Gums and Resins, 21; Rep. of Agricultural Dept. Bengal, 1886, App., lviii; Settlement Report:—Panjáb, Lahore, 15; N.-W. P., Sháhjehánpur, ix.; Gazetteers:—Panjáb, Simla, 12; Siálkot, 11; Sháhpur, 70; Lahore, 11; Hazára, 13; N.-W P., III., 33, 248; Mysore and Coorg, I., 70; Indian Forester, IV., 292; V., 37; VI., 3.*

Habitat.—A moderate-sized, deciduous tree or shrub, found in the Temperate and Sub-tropical Himálaya from Kashmír to Sikkim, ascend-

MORUS indica.

The True Indian Mulberry.

ing to 7,000 feet; also met with wild and cultivated for silkworm feeding, in Bengal, Assam, and Burma; distributed to China and Japan. In North India the tree is leafless during the cold season, the new leaves appearing from the middle of February to March or even to April. It flowers in March to April and the fruit ripens in May, June, or later at great elevations.

CULTIVATION.

CULTIVATION.—As already stated, the Indian white mulberry is largely cultivated in many parts of India for purposes of sericulture. In Assam the silk of the *Pat* worm (**Bombyx textor** and **B. crœsi**) is produced on this tree (see **Silk**). The method of cultivation varies slightly in different localities, but it may be sufficient to give a detailed account of the course followed in the Burdwán district of Bengal, extracted from a Report of the Agricultural Department of that province:—

Soil.

757 "*Soil.*—In the opinion of the cultivators, mulberry can be grown on any description of high and well-drained land, success depending more on the mode of the cultivation and the quantity of manure used than on the nature of the soil. It seems, however, that it does better in clay loam than either in stiff clay or sandy soils.

Preparation.

758 "*Preparation of the field.*—Deep cultivation is an absolute necessity, in the cultivation of the mulberry. For this purpose the ordinary country plough is altogether inefficient. As soon as the rains are over and the earth is still soft, the land is dug to a depth of about 18 inches by the *kodali*. This is generally done in *Aswin* to *Kartik* (15th September to 15th November). This depth is attained at the cost of much labour and expense. The large clods are broken by hammering. The field is then ploughed twice and levelled with the ladder. If there be no rain and the ground be dry, it will be necessary to irrigate the field before ploughing; the land should be well weathered.

Planting.

759 "*Planting.*—When the land has been well prepared, holes are made over the field 18 inches apart, each hole being 18 inches by 18 inches and 6 inches deep. Mulberry is propagated by cuttings. When an old field is cut down, the plants are divided into pieces about 9 inches long, the tops, as well as very thin and dry stems, being rejected. The cutting is done by placing the stalks on a hard substance and striking them with a sharp knife, care being taken at the same time not to split the ends. The cuttings are then made into bundles tied with grass close to the top and bottom. A hole is now made near a tank, and the bottom of it is worked into mud. The bundles are sunk into this mud to a depth of a few inches, and are then covered with more of the same substance. The cuttings are kept in this way for about a month, and are from time to time watered. When the buds have put forth shoots about two inches long they are taken out for planting.

"In the holes made in the fields, as above described, these cuttings are placed horizontally from one to three in each hole. They are then covered with earth and watered from a *kulsi*. The cuttings are well pressed into the earth, taking particular care not to break the shoots. As long as the cuttings do not take root they should be watered once a week. When the plants become about a cubit high the whole field is flooded, and after a week it is hoed with the *kodali*, the earth that was raised in making the holes being thus spread around the plant. After the plants have attained a height of two to three cubits, it is not necessary to irrigate the field more frequently than once $1\frac{1}{2}$ to 2 months.

Plucking.

760 "*Plucking the leaves.*—The cuttings are planted in *Kartick* to *Agrahayan* (15th September to 15th December). In *Falgun* (15th February to 15th March) the plants will be in a fit condition for plucking the leaves.

At first only the leaves should be plucked, but when the plants have grown sufficiently, the twigs can be removed without injury. **CULTIVATION.**

"*After-treatment.*—The field should be hoed once in May, and in the end of June and beginning of July, weeds should be smothered by turning up the soil. In *Chaitra* (15th March to 15th April) of the 2nd year, before the commencement of the leaf-plucking season, the field is manured with pond mud, the quantity of mud used being often so large as to alter the whole nature of the soil. About 400 maunds of mud are applied per bigha. The mud receives no previous treatment before being brought to the field, but before being mixed with the soil, it is first placed in heaps between the rows of plants, and exposed to sun and air, and is mixed with the soil at the hoeing time in May. **After-treatment.**

"The fields are thus manured with pond mud once in every three years. The plants are kept up for ten or twelve years, when they are ratooned, *i.e.*, they are cut down at the bottom, when the field is hoed and manured, New plants come up, and these are kept up for five years more, when a new field ought to be made. Sometimes a second ratooning is given, and the plants are kept up for another five years" (*First Annual Report of the Director of the Agricultural Department, Bengal, 1886, App., p. lviii.*).

Attempts have been made to cultivate **M. alba** in the Panjáb on a large scale for purposes of sericulture, but without success (see **Silk**).

Gum.—Mulberry gum was amongst the samples exhibited by **Lieutenant Hawkes** at the Madras Exhibition of 1855. It is mentioned by **Watson** and **Cooke**, but no account of its properties or uses appears to exist in the literature of Indian products. **GUM. 761**

Fibre.—The BARK of a mulberry (probably **M. alba**, *Linn.*) seems to have been employed in China from very early times for paper-making. Thus **Marco Polo** informs us that "the Grand Kán causes the bark to be stripped from these mulberry trees, the leaves of which are used for feeding silkworms, and takes from it that thin rind which lies between the coarse bark and the wood of the tree. This being steeped and afterwards pounded in a mortar until reduced to pulp, is made into a paper, resembling that which is made from cotton." **Royle** commenting on this passage remarks that the fibre from the countless number of shoots pruned from the trees in their cultivation to feed silkworms, and now thrown away as useless, might yield good half-stuff for the paper-maker. "**Mr. Henley**," he adds, "informs us that after leaving India, he produced very good specimens of half-stuff from the bark of these rejected stems." The bark separates when the cut stems are steeped in water, and when pounded up leaves a mass having much of the good qualities of linen rag "half-stuff." No economic information exists regarding a fibre obtainable from **M. indica**, but there is little doubt that this species would yield a fibre of similar roperties. The TWIGS are employed in many localities, on account of their toughness and strength, for binding bundles of fuel, loads, &c. **FIBRE. 762 Bark. 763 Twigs. 764**

Medicine.—The FRUIT has an agreeable aromatic and acid flavour, is cooling and laxative, allays thirst, and is grateful in fevers. The BARK is supposed to be vermifuge and purgative. The ROOT is considered anthelmintic and astringent. **MEDICINE. Fruit. 765 Bark. 766 Root. 767**

SPECIAL OPINIONS.—§ "The syrup of the fruit is used as a gargle in inflammatory affections of the throat, and for relaxation of the uvula" (*Civil Surgeon R Gray, Lahore*). "A decoction of the leaves is used as a gargle in inflammation and thickening of the vocal cords" (*Civil Surgeon J. Anderson, M.B., Bijnor, N.-W. P.*).

Food and Fodder.—The FRUIT is inferior to that of **M. alba** and is little used as an article of food except in times of scarcity. The LEAVES are used for feeding silkworms, and are also valuable as fodder. **FOOD & FODDER. Fruit. 768 Leaves. 769**

TIMBER. 770 **Structure of the Wood.**—Yellow with darker streaks of various colours; hard, very similar to that of **M. alba**; weight 42 to 47lb per cubic foot. Used in Assam for boat oars and furniture; it does not seem to be employed in making tea boxes, but would be very suitable for that purpose.

INDUSTRIAL USES. 771 **Industrial Uses.**—As already stated, it is a highly important cultivated shrub in Indian sericulture. For an account of the worms reared on it, &c., see **Silk**, Vol. VI.

772 **Morus lævigata**, *Nall.; Fl. Br. Ind., V., 492.*

Syn.—MORUS ALBA, *var.* LÆVIGATA, *Bureau;* M. GLABRATA, *Wall.*

Var. viridis, *Bureau;* M. VIRIDIS, *Ham.*

Vern.—*Tút,* HIND.; *Tút, sháh-tút, siyáh-tút,* KUMAON; *Tút,* PB.; *Malaing,* [BURM.

References.—*Brandis, For. Fl., 409; Baden Powell, Pb. Pr., 585; Atkinson, Him. Dist., 317; Gazetteers:—Amritsar, 4; Ráwalpindi, 15; Ind. Forester, IV., 47; XI., 355; XII., 64; Agri.-Hort. Soc. Ind., Journ. XIV., 22.*

Habitat.—A medium-sized tree, wild and cultivated in the Tropical and Sub-tropical Himálaya from the Indus to Assam up to 4,000 feet, also cultivated in Bengal; wild in Martaban and Tenasserim.

FOOD. Fruit. 773 **Food.**—The flowers appear in the cold weather, and the long cylindrical, yellowish-white or pale purple FRUIT ripens in March to May, and is eaten, though insipidly sweet and of little value.

TIMBER. 774 **Structure of the Wood.**—Similar to that of **M. indica,** said in the Rawalpindi district to be an excellent timber.

775 **M. serrata,** *Roxb.; Fl. Br. Ind., V., 492.*

Syn.—MORUS ALBA, *L.? Wall. Cat., 4648 A;* M. ALBA, *var.* SERRATA, *Bureau;* M. PABULARIA, *Dcne.;* M. VICORUM, *Jacquem.*

Vern.—*Kimu, himu,* HIND.; *Karún, tút, káura, túlúkúl, soá, án, shta, chimu, kimu, krúm, shahtút, tulklú, kárt-tút, chún,* PB.

References.—*Roxb., Fl. Ind., Ed. C.B.C., 658; Brandis, For. Fl., 409; Gamble, Man. Timb., 328; Stewart, Pb. Pl., 219; Bureau in DC. Prodr., XVII., 242; Baden Powell, Pb. Pr., 586; Indian Forester, VIII., 272; Settlement Report, Simla Dist. (app.), xlii.; Gazetteers:—Panjáb, Simla, 9; Gurdáspur, 55.*

Habitat.—A large, deciduous tree, found in the North-West Himálaya between 4,000 and 9,000 feet. It is often of very large size; **Dr. Stewart** noted several trees of 20 feet girth, and one at the Hindu temple at Barmaor, Chamba, of 28 feet girth.

FOOD and FODDER. Fruit. 776 **Food and Fodder.**—The FRUIT is little valued, but the tree is extensively lopped for cattle fodder.

Leaves. 777

TIMBER. 778 **Structure of the Wood.**—Sapwood small, white; heartwood yellow or brown, moderately hard; weight 35 to 36lb per cubic foot. It works well, does not warp, and takes a beautiful polish, showing a golden lustre. It is used for troughs, agricultural implements and for cabinet-work, and is much esteemed by the Simla wood-carvers (*Gamble*).

Moschus moschiferus, *Linn.;* THE MUSK DEER, see Vol. III., 58.

MUCILAGE.

779 **Mucilage;** PLANTS YIELDING—

Abutilon asiaticum, *G. Don;* } seeds and leaves; MALVACEÆ.
A. indicum, *G. Don;* }
Acacia arabica, *Willd.;* the gum; LEGUMINOSÆ.
Adansonia digitata, *Linn.;* pulp of the fruit; MALVACEÆ.
Althæa officinalis, *Linn.;* the whole plant; MALVACEÆ.
Cetraria islandica, *Ach.;* the whole plant; LICHENES.

Corchorus Antichorus, *Rœusch.;* } the whole plant; TILIACEÆ.
C. capsularis, *Linn.;*
C. fascicularis, *Lam.;*
Gmelina asiatica, *Linn.;* the leaves and twigs; VERBENACEÆ.
Hibiscus esculentus, *Linn.;* the fruit; MALVACEÆ.
Ipomæa biloba, *Forsk.;* } the stems and branches; CONVOLVULACEÆ.
I. Turpethum, *Br.;*
Kydia calycina. *Roxb.;* the bark; MALVACEÆ.
Lallemantia Royleana, *Benth.;* LABIATÆ.
Lepidium Iberis, *Linn.;* the seeds; CRUCIFERÆ.
L. sativum, *Linn.;* the seeds.
Malva sylvestris, *Linn.;* the whole plant; MALVACEÆ.
M. rotundifolia, *Linn.;* the seeds; MALVACEÆ.
Mentha viridis, *Linn.;* the seeds; LABIATÆ.
Ocimum Basilicum, *Linn.;* } the seeds; LABIATÆ.
O. gratissimum, *Linn.;*
Pithecolobium dulce, *Benth.;* the gum; LEGUMINOSÆ.
Plantago ovata, *Forsk.;* the seeds; PLANTAGINEÆ.
Zizyphus Jujuba, *Lamk.;* the fruit; RHAMNEÆ.

For further information regarding these the reader is referred to the article regarding each in its respective alphabetical position.

MUCUNA, *Adans.; Gen. Pl., I., 533.* 780

A genus of twining annuals or perennials, which comprises about twenty species, natives of the Tropics. Of these ten are met with in India.

Mucuna monosperma, *DC.; Fl. Br. Ind., II., 185;* LEGUMINOSÆ. 781

THE NEGRO BEAN.

Syn.—CARPOPOGON MONOSPERMUM, *Roxb.;* MUCUNA CRISTATA, *Ham.;* M. CORYMBOSA, *Grah.;* M. ANGUINA, *Wall.;* CARPOPOGON ANGUINEUM, *Roxb.*

Vern.—*Sonagáravi, mothi-kuhili,* BOMB.

References.—*Roxb., Fl. Ind., Ed. C.B.C., 553; Dalz. & Gibs., Bomb. Fl., 70; Dymock, Mat. Med. W. Ind., 230; Gaz., Bomb., XV., 432.*

Habitat.—A woody climber of the East Himálaya and Khásia Hills; also met with in Assam, Chittagong, Pegu, Tavoy, and the Hills of the Western Peninsula and Ceylon, up to 3,000 feet.

MEDICINE. Seed. 782

Medicine.—The large flat nearly circular SEED is used as an expectorant in cough and asthma, and, applied externally, as a sedative.

SPECIAL OPINION.—§ "The seeds are used as an expectorant in cough and asthma, also in affections of the tongue. Externally they are said to be sedative. The pods are roundish and hairy. Each pod contains a solitary seed which is about an inch in diameter earshaped, flat, the entire convex margin is occupied by the hilum" (*C. T. Peters, M.B., Zandra, South Afghánistán*).

FOOD. Seed. 783

Food.—The SEED is a favourite vegetable, and is said by Dymock to be eaten by Brahmins as a restorative when fasting.

M. nivea, *DC.; Fl. Br. Ind., II., 188.* 784

Syn.—CARPOPOGON NIVEUM, *Roxb.;* M. NIGRA, *Ham.*

Vern.—*Khamach, alkushi,* BENG.

Habitat.—Met with in Burma and Bengal, perhaps only a cultivated variety of M. pruriens.

FOOD. Fruit. 785

Food.—Cultivated during the cold season for the sake of its abundant and useful FRUIT. The large, fleshy, tender legumes have long been known and valued as a vegetable by the Hindus, and, according to Roxburgh, are, when dressed, like French beans, a most excellent vegetable for European tables. The exterior velvet-like skin should be removed before cooking.

786 **Mucuna pruriens,** *DC.; Fl. Br. Ind., II., 187; Wight, Ic., t. 280.*

THE COWHAGE PLANT.

Syn.—CARPOPOGON PRURIENS, *Roxb.*; DOLICHOS PRURIENS, *Linn.*; M. PRURITA, *Hook.*; M. UTILIS, *Wall.*

Vern.—*Kivánchh, kiváchh, goncha,* HIND.; *Akolshi, kámách, bichhoti,* BENG.; *Kaincho,* URIYA; *Karyani,* MONGHYR; *Alkusi,* KOL.; *Etka,* SANTAL; *Kouatch,* NEPAL; *Goncha,* N.-W. P.; *Kawanch, kúnch, kanaucha, konch kari, gúnchgaji,* PB.; *Kách-kúri, kánch-kúri,* DEC.; *Kuhili,* BOMB.; *Kavacha, kuhili,* MAR.; *Kivánch, kancha,* GUZ.; *Punaik-káli,* TAM.; *Pilli-adugu, dúla-gondi, pedda-dúlagondi,* TEL.; *Nasagunigidá, turachi-gidá,* KAN.; *Náyik-korana,* MALAY.; *Khwele, khuele,* BURM.; *Achátiyapalbe,* SING.; *Kapikachchhu, átmaguptá, vánari,* SANS.

References.—*Roxb., Fl. Ind., Ed. C.B.C., 553; Dalz. & Gibs., Bomb. Fl., 70; Rev. A. Campbell, Rept. Econ. Pl., Chutia Nagpur, No. 8482; Mason, Burma and Its People, 490, 768; Pharm. Ind., 73; Flück & Hanb., Pharmacog., 189; U. S. Dispens., 15th Ed., 1702; Fleming, Med. Pl. & Drugs (Asiatic Reser., XI.), 166; Ainslie, Mat. Ind., I., 93, 596; O'Shaughnessy, Beng. Dispens., 297; Irvine, Mat. Med. Patna, 46; Medical Topog., 141; Moodeen Sheriff, Supp. Pharm. Ind., 177; U. C. Dutt, Mat. Med. Hindus, 147, 293, 303; Sakharam Arjun, Cat. Bomb. Drugs, 42; K. L. De, Indig. Drugs Ind., 75; Murray, Pl. and Drugs, Sind., 124; Bent. & Trim., Med. Pl., t. 78; Dymock, Mat. Med. W. Ind., 2nd. Ed., 229; Dymock, Warden and Hooper, Pharmacog. Ind., I., 447; Birdwood, Bomb. Prod., 29; Baden-Powell, Pb. Pr., 320, 341; Atkinson, Him. Dist. (Vol. X, N.-W. P. Gaz.), 309, 742; Useful Pl. Bomb. (Vol. XXV., Bomb. Gaz.), 257; Gazetteers:—Bombay, XV., 432; Panjáb, Peshάwar, 27; N.-W. P., IV., lxxi.; Mysore and Coorg, I., 59; Indian Forester, XII., App. 11.*

Habitat.—A lofty annual climber, with large, dark purple papilionaceous flowers, and golden-brown, velvety legumes not unlike those of a sweet pea. It occurs commonly throughout the tropical regions of America, Africa, and India, in the latter of which it is to be found all over the plains. **M. utilis,** *Wall.*, is a cultivated form.

MEDICINE.

Medicine.—The English names Cowhage or Cowitch are derived from the Hindi *Kiwach*, the earliest European writer who notices the plant having stated that it was called "in Zurrate (Surat) where it groweth, "Couhage." It has long been known and valued in Indian medicine being mentioned by **Susruta** and in the *Bhávaprakása*. The former describes the SEED as a powerful aphrodisiac, and states that it should be prescribed with the fruits of **Tribulus terrestris** in sugar and tepid milk. In the latter work a preparation called *vánari vati* is described which is said to be the best of aphrodisiacs. It is made by boiling the seeds in cow's milk, then pounding and decorticating them, and making a confection with the mass. The ROOT had also medicinal virtues ascribed to it in Sanskrit medicine, being considered tonic, and useful in diseases of the nervous system, such as facial paralysis, hemiplegia, &c. (*U. C. Dutt*). **Dymock** states that similar properties are ascribed to the seeds (*hab-el-kulai*) in Persian works, and informs us that a *paushtik* for spermatorrhœa is made in the Konkan from the seeds with several other drugs. According to **Ainslie** "a strong infusion of the root, sweetened with honey, is given by the Tamool doctors in cases of Cholera Morbus." **Campbell** states that the root is prescribed as a remedy for delirium in fever in Chutia Nágpur, and that when powdered and made into a paste it is applied to the body in dropsy, a piece of the root being also tied to the wrist and ankle. The seed is believed to absorb scorpion poison when applied to the part stung.

Seed. 787

Root. 788

Cowhage has been known and used in European medicine since 1640, chiefly as a vermifuge. The employment of the HAIRS of the LEGUME for

Hairs of the Legume, 789

this purpose originated in the West Indies and except of recent years has been quite unknown in India. **Flückiger & Hanbury** state that it began to attract considerable attention in England in the latter part of last century, when it was strongly recommended by **Bancroft** and by **Chamberlaine.** It was introduced into the Edinburgh Pharmacopœia in 1783 and into the London Pharmacopœia of 1809. Noticed by **Rheede, Rumphius,** and **Ainslie** it soon became considerably employed as a vermifuge by European physicians in India, and was made officinal in the Pharmacopœia of 1868, though now almost entirely discarded from European medicine. It is prescribed as an electuary or confection with honey or treacle. The effect of the hairs is purely mechanical, and though efficient in cases of round and thread-worms, it is coarse, bulky, and inferior to many other anthelmintics. MEDICINE.

An ointment prepared with the hairs acts externally as a local stimulant and mild vesicant.

SPECIAL OPINION.—§ "The seeds powdered in doses of $\frac{1}{3}$ drachm to 40 grains are useful in leucorrhœa and as an aphrodisiac, also in cases of spermatorrhœa" (*Lal Mahomed, 1st Class Hospital Assistant, Main Dispensary, Hoshangabad, Central Provinces*).

Food.—The young tender PODS are cooked and eaten as a vegetable. FOOD. Pods. 790

Mudar, or **Madar,** see **Calotropis gigantea,** *R. Br.*, and **C. procera,** *R. Br.*; Vol. II., 34 and 49.

Múdúga Oil, see **Butea frondosa,** *Roxb.*; Vol. I., 552.

MUKIA, *Arn.; Gen. Pl., I., 829.*

[CUCURBITACEÆ.

Mukia scabrella, *Arn.; Fl. Br. Ind., II., 623; Wight, Ic., t. 501;* 791

Syn.—M. MADERASPATANA, *Kurz*; BRYONIA SCABRELLA, *Linn. f.*; B. WIGHTIANA, *Wall.*; B. MADERASPATANA *and* ALTHÆOIDES, *DC.*; CUCUMIS MADERASPATANUS, *Linn.*; KARIVIA JAVANICA, *Miq.*; TRICHOSANTHES DIOICA, *Wall.*

Var. gracilis, BRYONIA GRACILIS, *Wall.*

Vern.—*Bilári, agumaki,* HIND.; *Gwála-kakri,* N.-W. P.; *Bilári, agumarki, gwála-kakri,* KUMAON; *Chiráti, bellari,* SIND; *Musu-musukkai,* TAM.; *Kutaru-budama, putribudinga,* TEL.; *Sa-tha-khiva, thabwot-kha,* BURM.

References.—*Roxb., Fl. Ind., Ed. C.B.C., 702; Dalz. & Gibs., Bomb. Fl., 100; Stewart, Pb. Pl., 98; Mason, Burma and Its People, 746; Bidie, Cat. Raw Pr., Paris Exh., 53; Atkinson, Him. Dist., 301, 702; Ec. Prod., N.-W. P., V., 12; Drury, U. Pl., 87; Gazetteer:—N.-W. P., I., 81; IV., lxxii.; Bombay, V., 25; XV., 435.*

Habitat.—A climber common throughout India in the plains, ascending the hills as far as subtropical warmth extends.

Medicine.—The SEEDS in decoction are sudorific. The ROOT similarly prepared is useful in flatulence, and when masticated relieves toothache. MEDICINE. Seeds. 792 Root. 793

SPECIAL OPINION.—§ "The tender SHOOTS and bitter LEAVES are used as a gentle aperient and recommended in vertigo and biliousness" (*C. T. Peters, M.B., Zandra, South Afghánistán*). Shoots. 794 Leaves. 795

Mulberry, see **Morus alba** and **M. indica,** pp. 279, 281.

Mulberry-paper Tree, see **Broussonetia papyrifera,** *Vent.*, Vol. I., 538

Mules, see **Horses,** Vol. IV., p. 297.

Múltáni mitti, see **Clay,** Vol. II., 361; see also **Iron oxides,** Vol. IV. p. 520; and **Pigments,** Vol. VI., Pt. I.

Mumiai, or **Momea,** see **Cannabis sativa,** *Linn.;* Vol. II., 115, 116.

796 **MUNDULEA,** *DC.; Gen. Pl., I., 497.*

A small genus of shrubs which belongs to the Natural Order LEGUMINOSÆ, and comprises three species, of which one is a native of India. This plant, **M. suberosa,** *Benth.* (*Fl. Br. Ind., II., 110*), is a native of the Hill Valleys of the Western Peninsula and Ceylon. It is known in the former locality as *surti*, or *supti*, and its seeds are used for poisoning fish.

Munga, see **Silk,** Vol. VI., Pt. II.

MURRAYA, *Linn.; Gen. Pl., I., 304, 992.*

[RUTACEÆ

797 **Murraya exotica,** *Linn.; Fl. Br. Ind., I., 502; Wight, Ic., t. 96;*

Syn.—CHALCAS INTERMEDIA *and* C. PANICULATA, *Roem.*

Var. α.—Corymbs many flowered, ovary 2-celled,—M. EXOTICA, *Auct.;* M. EXOTICA, *and* BREVIFOLIA, *Thwaites.*

Var. β.—Arboreous, corymbs few flowered, ovary 2-celled,—M. PANICULATA, *Jack.;* M. SUMATRANA, *Roxb.;* CHALCAS PANICULATA, *Linn.*

Var. γ.—Ovary, 4-5-celled,—M. GLENIEII, *Thwaites.*

Vern.—*Marchula, juti, bibsar,* HIND.; *Kamini,* BENG.; *Simali,* NEPAL; *Shitzem,* LEPCHA; *Raket-berár,* GOND; *Murchob,* KUMAON; *Chulajuti, marchula juti,* BOMB.; *Kunti,* MAR.; *Naga golunga, naga-golugu,* TEL.; *Makay, tha-nat-kha,* BURM.; *Attaireya,* SING.; *Machalla,* AND.

References.—*Roxb., Fl. Ind., Ed. C.B.C., 362; Brandis, For. Fl., 48; Kurz, For. Fl. Burm., I., 190; Beddome, Fl. Sylv., Anal. Gen., xliv., t. 7, f. 2; Gamble, Man. Timb., 61; Thwaites, En. Ceylon Pl., 45; Dalz. & Gibs., Bomb. Fl., Supl., 12; Aitchison, Cat. Pb. and Sind Pl., 28; Elliot, Fl. Andhr., 121; Mason, Burma and Its People, 500, 534, 760; Pharmacog., Indica, I., 265; Atkinson, Him. Dist., 307; Lisboa, U. Pl. Bomb., 32; Balfour, Cyclop., I., 820; II., 1013; Gazetteers:—Mysore & Coorg, I., 69; N.-W. P., IV., lxix.; Indian Forester, IV., 345; VIII., 412; XIV., 373; For. Admin. Rep. Chutia Nagpur, 1885, 28; Balfour, Cyclop. Ind., II., 1013.*

Habitat.—A shrub, or small tree, met with in the outer Himálaya from the Jumna to Assam, ascending to 4,500 feet; also in Behar, South India, Burma, and the Andaman Islands. It is often planted for ornament, and is sometimes called "Satinwood" at Port Blair, and "Cosmetic Bark Tree" in Burma. Variety α occurs only in Northern India; variety β is chiefly found in the Western Peninsula and Ceylon; while variety γ is reported only from Ceylon.

MEDICINE. Flowers. 798 **Medicine.**—The authors of the *Pharmacographia Indica* state that DeVrij has separated a glucoside from the FLOWERS which he has named *Murrayin;* its composition is $C_{18}H_{22}O_{10}$. They, however, give no account of the plant being used medicinally, nor of the properties of the glucoside.

TIMBER. 799 **Structure of the Wood.**—Light yellow, close-grained, very hard, apt to crack, weight 62℔ per cubic foot. It resembles boxwood, and has been tried for wood-engraving, for which it seems suitable if well seasoned; it is also used for handles of implements.

DOMESTIC. Bark. 800 **Domestic.**—Mason states that in Burma the fragrant BARK is more universally used as a cosmetic than Sandalwood.

M. Kœnigii, *Spr.; Fl. Br. Ind., I., 503; Wight, Ic., t. 13.*

THE CURRY-LEAF TREE.

Syn.—BERGERA KŒNIGII, *Linn.*

Vern.—*Harri, katnim, bursunga,* HIND.; *Barsúngá, karia-phulli,* BENG.; *Bhursunga, básang,* URIYA; *Húmwah,* MICHI; *Gandla, gánt, gani,* *bowála*, N.-W. P.; *Gandla, gani*, KUMAON; *Gandla, gándalú, gandi, bowala, ganda-nim*, PB.; *Karé-pák, karyá-pák, karyá-pát*, DEC.; *Karri-nim, karria-pat, gora-nimb*, BOMB.; *Karipat, karé-pákácha, karhi-nimb, jhirang, jirani*, MAR.; *Kadhi, limbdo, gora-nimb*, GUZ.; *Karu-véppilai, karu-vémbu, kamwepila*, TAM.; *Chunangi*, HYDERABAD; *Kari-vépa, karepak*, TEL.; *Kari-bevu, kari-bévana, yelé*, KAN.; *Karea-pela, kariya-pála*, MALAY.; *Pido-sin, pindo-sin*, BURM.; *Karri-pincha*, SING.; *Watu-karapincha*, CEY.; *Surabhi-nimbu, kristna-nimba, nimba-patram, surabhi*, SANS.

References.—*Roxb., Fl. Ind., Ed. C.B.C., 362; Brandis, For. Fl., 48; Kurz, For. Fl. Burm., I., 191; Beddome, Fl. Sylv., Anal. Gen., xliv; Gamble, Man. Timb., 61; Trimen, Sys. Cat. Cey. Pl., 15; Dalz. & Gibs., Bomb. Fl., 29; Stewart, Pb. Pl., 29; Aitchison, Cat. Pb. and Sind Pl., 29; Sir W. Elliot, Fl. Andh., 83; Pharm. Ind., 49; Ainslie, Mat. Ind., II., 139; O'Shaughnessy, Beng. Dispens., 232; Moodeen Sheriff, Supp. Pharm. Ind., 178; Sakharam Arjun, Cat. Bomb. Drugs, 21; Waring, Pharm. Ind., 49; Dymock, Mat. Med. W. Ind., 2nd Ed., 130; Dymock, Warden & Hooper, Pharmacog. Ind., Vol. I., 262; Official Corresp. on Proposed New Pharm. Ind., 238; Birdwood, Bomb. Prod., 141, 218; Baden-Powell, Pb. Pr., 570; Drury, U. Pl. Ind., 78; Atkinson, Him. Dist. (Vol. X., N.-W. P. Gaz.), 307, 705, 742; Useful Pl. Bomb. (Vol. XXV., Bomb. Gaz.), 149, 222, 397; Forbes Watson, Econ. Prod. N.-W. Prov., Pt. V. (Vegetables, Spices, and Fruits), 25, 28; Cooke, Oils and Oilseeds, 62; Gazetteers:—Bombay, V., 24; XV., 429; N.-W. P., IV., lxix.; Orissa, II., 158, 180; Mysore and Coorg, I., 69; Indian Forester, IV., 323; VIII., 101; X., 325.*

Habitat.—A small tree of the outer Himálaya from the Rávi to Sikkim, ascending to 5,000 feet, also met with in Assam, Bengal, Burma, South India, and Ceylon.

Oil.—**Birdwood**, and quoting him, **Cooke**, state that the SEEDS yield an oil, known as Simbolee or Limbolee oil. This is, however, probably a mistake, since Limbolee oil is obtained from **Melia Azadirachta**. The LEAVES of **M. Kœnigii** contain a strongly odoriferous essential oil. OIL. 801 Seeds. 802 Leaves. 803

Medicine.—The LEAVES, BARK, and ROOT are used in native medicine as tonic and stomachic. They were noticed by **Roxburgh**, who remarks, "The bark and root are used as stimulants by the native physicians. Externally they are also used to cure eruptions and the bites of poisonous animals. The green leaves are described (*sic*) to be eaten raw for the cure of dysentery; they are also bruised and applied externally to cure eruptions." **Ainslie** states that an infusion of the washed leaves stops vomiting. The plant has obtained a place in the secondary list of the Indian Pharmacopœia, where it is remarked that **Drs. E. C. Ross** and **Shortt** had reported favourably on the leaves, bark, and roots as tonic and stomachic. **Moodeen Sheriff** in his forthcoming work describes the leaves as slightly stimulant and carminative, and regards them as "a weak substitute for peppermint." **Dymock** states that a decoction of the leaves is sometimes given with bitters as a febrifuge." MEDICINE. Leaves. 804 Bark. 805 Root. 806

CHEMICAL COMPOSITION.—The authors of the *Pharmacographia Indica*, quoting **Mr. J. G. Prebble**, give a very complete account of the chemical composition of the leaves. They yield to distillation a small quantity of oil resembling that obtained from the leaves of **Ægle Marmelos**. Treated with ether, and dried, a greenish black resin was obtained equivalent to 7½ per cent. of the leaves. This resin extracted with water and evaporated yielded a small residue equivalent to 3 per cent. of the resin. The aqueous extract was slightly acid, and reduced Fehling's solution. On acidifying a similarly prepared alcoholic extract with sulphuric acid, a few granular crystals were obtained, which from their reactions are supposed to have been a glucoside, and were provisionally named *Kœnigin*. Chemistry.

SPECIAL OPINIONS.—§"Stomachic, given with **Mentha arvensis** in the form of *chutney* to check vomiting" (*Surgeon-Major H. D. Cook, Calicut,*

Malabar). "Antibilious" (*Surgeon-Major D R. Thomson, M.D., C I.E., Madras*). "Cultivated in gardens for the leaves, which are used in making green chutnee" (*W. Dymock, Bombay*). "Used as an antiperiodic. *Baids* prescribe the powdered root mixed with honey and juice of betel-nut" (*Civil Surgeon John McConaghey, M.D., Shajahanpur*).

FOOD. Leaves. 807 **Food.**—It is largely cultivated on the plains on account of its LEAVES, which are used, either fresh or dry, to flavour curries.

TIMBER. 808 **Structure of the Wood.**—Greyish-white, hard, and durable, weight 43℔ per cubic foot. It is used for agricultural implements.

Mushroom, see **Agaricus,** Vol. I., 129; **Fungi,** Vol. III., 455; **Morchella esculenta,** p. 259; and **Truffle,** Vol. VI., Pt. II.

809 **MUSA,** *Linn.; Gen. Pl., III., 655.*

A genus of the Natural Order SCITAMINEÆ, which comprises some twenty species, several of which are regarded by certain writers as mere cultivated varieties. **Kurz** in two articles on the Plantains (*Jour. Agri.-Hort. Soc. of Ind. (Old Series), XIV., 299 (New Series); V., 112-168*) describes eighteen Indian species, and refers the cultivated forms to nine, namely, **M. troglodytarum, M. cormeulata, M. Rumphiana, M. zebrina, M. rhinozerotis, M. nana, M. Basjoo, M. sapientum** (in which he includes **M. paradisiaca,** *Linn.*), and **M. textilis.** His later paper in which the synonymy was altered, unfortunately never reached completion, so with the exception of a short analytical table, no botanical definition of the species as defined by him exists. In the earlier paper he referred most of the plantains of Burma and the Archipelago to **M. simiarum,** *Rumph.*, and nearly if not quite all the Indian forms to **M. sapientum.** No later literature on the subject appears to exist, and in want of material by which to arrive at some satisfactory conclusion, it may suffice, for the purposes of the present article, to describe the several cultivated forms of banana and plantain under the name of **M. sapientum,** and the Manilla hemp under **M. textilis.** Most of the other species are at best very unimportant economically, and possess no peculiar characters, and, since many of them are doubtfully botanically distinct, they may for the present be omitted.

[*200;* SCITAMINEÆ.

Musa ornata, *Roxb.; Fl. Ind., Ed. C.B.C., 224; Fl. Br. Ind. VI.,*

Vern.—*Ramanigi-kula,* CHITTAGONG; *Chavaya, rankela,* BOMB.

References.—*Dalz. and Gibs., Bomb. Fl., 272; Lisboa, U. Pl. Bomb., 204.*

Habitat.—Common on sides of precipitous crags at Mátherán, Rám Ghát, and Khandála in Bombay, also met with in Chittagong.

FOOD Scape. 810 **Food.**—**Lisboa** writes, "The SCAPE and the convolute leaf sheath which immediately surround it, are cut into pieces, boiled, and made into a dish with spices, or they are dried and pounded into a kind of flour out of which cakes are made. They are resorted to especially in times of scarcity."

[*Hort. Soc. (New Series), V., 112.*

811 **M. sapientum,** *Linn.; Kurz, Notes on the Banana, in Jour. Agri.-*

THE BANANA, or PLANTAIN.

Syn.—MUSA PARADISIACA, *Linn.*

Vern.—*Kéla, kach-kula, maoz-kula,* HIND.; *Kala, kach-kula,* BENG.; *Kéla, khéla, múz,* PB.; *Kewiro,* SIND; *Kéla, kél,* BOMB.; *Kél, kadali,* MAR.; *Kéla,* GUZ.; *Mouz, máoz, kél,* DEC.; *Vazhaip pazham, valei,* TAM.; *Anati, amti, ariti, kommu-ariti, nalla-ariti, chakrakéli-ariti, bonta-ariti, kadali,* TEL.; *Bálé, bálénaru,* KAN.; *Vasha, vazhap-paghan, vellacoi, pizang,* MALAY.; *Ya khaing, ya-thi-lan, napiyá-si, huga-pyau, ngetpyau, nga-pyi-sthi,* BURM; *Wal-kaihil, kehol,* SING.; *Kadali, rambhá,* SANS.; *Tulhtula, mouz, shajrátul-talh, shajratul-mouz,* ARAB.; *Tulhtula, mous,* PERS. In addition to the above, many distinct names exist for different cultivated races. For the more important of these the reader is referred to the text.

References.—*Roxb., Fl. Ind., Ed. C.B.C., 222, 223, 225; Voigt, Hort. Sub. Cal., 598; Thwaites, En. Ceyl. Pl., 321; Trimen, Sys. Cat. Cey, Pl., 92; Dalz. & Gibs., Bomb. Fl., Supp., 88; Stewart, Pb. Pl., 229; Aitchison, Cat. Pb. and Sind Pl., 147; DC., Orig. Cult. Pl., 304; Graham, Cat. Bomb. Pl., 212; Mason, Burma and Its People, 449, 450, 806; Sir W. Elliot, Fl. Andh., 14, 16, 30, 33, 93, 123; Ainslie, Mat. Ind., I., 316; Moodeen Sheriff, Supp. Pharm. Ind., 178, 179; U. C. Dutt, Mat. Med. Hind., 257, 301; Waring, Bazar Med., 119; Dymock, Mat. Med. W Ind., 2nd Ed., 777; Watts, Dict. Chemistry, Vol. III., 1063; Johnston (Church Ed.), Chemistry of Common Life, 81, 85, 87; Birdwood, Bomb. Prod., 179; Baden Powell, Pb. Pr., 379, 512; Drury, U. Pl. Ind., 300; Atkinson, Him. Dist. (Vol. X., N.-W. P. Gaz.), 318, 742; Useful Pl. Bomb. (Vol. XXV., Bomb. Gaz.), 174, 239, 279, 284; Forbes Watson, Econ. Prod. N.-W. Prov., Pt. V. (Vegetables, Spices, and Fruits), 44, 90; Royle, Fibrous Pl., 78; Liotard, Mem. Paper-making Mat., 5, 7, 10, 14, 15, 18, 23, 24, 28, 30, 31, 48, 49, 54; Indian Fibres and Fibrous Substances, Cross, Bevan, King, & Watt, 9, 45; Kew Bulletin, 1889, 27; Kew Reports, 1881, 61; Simmonds, Tropical Agriculture, 455; Christy, New Com Pl., VI., 48; Darrah, Note on Condition of People of Assam, App. D; Rep. Bot. Gardens, Poona, 1885, 4; Linschoten, Voyage to East Indies (Ed. Burnell Tiele, & Yule), Vol. II., 37, 39, 40, 41, 42; Man. Madras Adm., Vol. I., 360; II., 85; Morris, Account Godavery, 12; Moore, Man., Trichinopoly, 72, 79; Man. Rev. Accts., Bombay, 102; Settlement Reports:—Central Provinces, Nimar, 200; Nagpur, 274; Chánda, 82; Madras, Godaveri Dist., 151; Gazetteers:—Bombay, II., 277, 280, 284, 288, 291, 295; VII., 42; XV., 445, Pt. II., 21; XVI., 103; XVIII., 48; N.-W. P., I., 80; III., 33; IV., lxxviii; Orissa, II., 179; Mysore, and Coorg, I., 53, 55, 70; II, 7; III., 45; Agri.-Horti. Soc. Ind.:—Trans., I., 52-53; II. (App.), 300, 314; III., 56, 61, 67-70 (pro.), 229; IV., 145, 149; VIII., 58-64; Journals (Old Series), IV., 230; VI., Sel., 1-3; Pro., 32; VII., 69; IX., Sel., 56; X., 61, 343, 344; XIII., Sel., 54, 59, 61; XIV., 296, 299; (New Series), I., Sel., 10, 19, 20; III., Pro., 83; V., 112-133, 147, 154; VI., Pro., 36; VII., 3, 4; VIII., 118, 120, 124; Indian Forester, VIII., 414; IX., 274; Ind. Agri. Gazette, 1887, 624; Rep. Horti. Gard., Lucknow, March 1884, 3; Balfour, Cyclop. Ind., II., 1015.*

Habitat.—A perennial herb, of 8 to 15 feet in height, extensively cultivated throughout India, nearer the coast tracts than inland, chiefly for its fruit. **Roxburgh** describes it as wild in Chittagong. **Rumphius** speaks of a wild variety with small fruits in the Philippine islands. **Sir Joseph Hooker** and **Dr. Thomson** found it wild in the Khásia Hills, and **Thwaites** met with the plant in the rocky forests of the centre of Ceylon. Other writers mention the wild plantain as occurring in Bombay, the North-West Provinces and other parts of India, but it is doubtful how far these are only spontaneous. There is little doubt that the plantain is a native of India and Southern Asia, and from thence has spread under cultivation in all directions, reaching even South America. **DeCandolle** considers it probable that it was early introduced by the Spanish and Portuguese into San Domingo and Brazil; but adds, "If, however, later research should prove that the banana existed in some parts of America before the advent of the Europeans, I should be inclined to attribute it to a chance introduction, not very ancient, the effect of some unknown communication with the islands of the Pacific or with the coast of Guinea, rather than to believe in the primitive and simultaneous existence of the species in both hemispheres."

RACES.—Much confusion exists in Indian economic and botanical literature in the nomenclature of the species of **Musa**. As already indicated in the generic note, recent literature on the subject is scanty; indeed, no botanist with sufficient material to enable him to accurately determine **RACES. 812** the limits and characters of the Indian species appears to have written on the subject. **Roxburgh** describes four species as indigenous to India, most of which are retained by modern botanists, but only one of which is of much economic interest. That species, **M. sapientum,** he believed to include the plantain or vegetable-like fruit, and the banana, or edible fruit, both of which he considered to be cultivated forms, or at most varieties of the same plant. Other botanists have separated these two under the names of **M. sapientum** and **M. paradisiaca,** but later writers, including **DeCandolle, Kurz,** and others have, as a rule, accepted the species **M. sapientum** as embracing both forms.

But the number of cultivated races, bearing fruits which differ widely in appearance and quality, is very large. **Roxburgh** states that he obtained in India three kinds of "plantain" and about thirty of "banana"; **Rheede** enumerates eight; writers in Madras mention from fifteen to twenty; Bombay authors describe about ten, &c, &c. **Moore** in his *Catalogue of Ceylon Plants* enumerates thirty-nine kinds of what he considers to be **M. sapientum** and **M. paradisiaca,** and eight others under the names of doubtful species, **M. rosacea** and **M. troglodytarum,** possibly also forms of **M. sapientum. Mason** states that in Burma he collected the names of twenty-five different kinds, **Captain Ripley** enumerates nineteen, and **Dr. Helfer** remarks that twenty kinds are cultivated in Tenasserim. The most important of these will be again referred to in the chapter on "CULTIVATION," and all will be included under the species **M. sapientum.** Future investigation may demonstrate the necessity of the establishment of more than one species for the cultivated plantains and bananas, but available evidence at present points to the advisability of considering them as one.

CULTIVATION.

CULTIVATION. 813 The value of the fruit as an easily cultivated and nutritious article of food, added to the utility of almost every other part of the plant, causes the cultivation of the plantain to be very nearly general over all but the extreme north-west of India, from the sea level to 5,000 or 6,000 feet. Indeed, there is scarcely a cottage in India that has not its grove of plantains. In certain localities large gardens are cultivated, the fruit being in this case grown for sale, in others it is merely cultivated in small patches in homestead lands for the immediate use of the cultivator.

Madras. 814 **I. Madras.**—The method of cultivation followed in this Presidency has been fully described by **Dr. Shortt,** and by the authors of the *Manuals of the Godáveri* and *of the Coimbatore Districts.* It is chiefly grown on wet lands, also in gardens in certain villages. The land is ploughed thoroughly, pits one foot cube are dug 10 to 12 feet apart, in which, during December and June, or January and April, young shoots are planted. A little manure is occasionally given, and drainage is always established to carry off superfluous water, plantains requiring a deep, porous, well-drained soil. Irrigation is effected by flooding the soil, and stopping up the drains at the outlet after the whole ground is covered. The water is allowed to soak in for a day and the outlets are then opened; the land is hoed once a month, and, three months after planting, a manure of wild indigo and dung is hoed in. Hoeing is stopped as soon as the flowers appear, but begins again after gathering the crop at the end of the first year. The plants last from three to four years, the latter being the case only in the best soils. Each stem bears only once, it is then cut down, and its place supplied by a shoot from the stock, which in its turn bears, is cut down, and replaced. It is estimated that from each group of plantains seven or eight bunches worth R2 are obtained, together with small additions from sale of the leaves and stems. In round figures R1 per plant

CULTIVATION.

Madras.

pays all the expenses including assessment, and as there are from 300 to 400 trees per acre, a net profit is realised in about four years of, say, R350. In **Nicholson's** *Manual of the Coimbatore District* it is stated that there are two modes of growing plantains. These are known locally as *pakkavâlai* in which the plants are set 6 or 7 feet apart, and *thúrúválai*, in which they are planted 10 or 12 feet apart. The former is practised on high and *kalár*, and the latter on rich deep wet lands. In *pakkavâlai* the mother or main shoot only is allowed to stand, in the latter all the shoots remain except the middle one (*nadukan*), which is removed. *Pakkavâlai* gives 8 annas per stem in the three years of its life as net profit, and since there are from 600 to 800 stems in one acre, the total profit may be estimated at from R300 to R400.

In garden lands the shoots are planted at from 6 to 8 feet apart at any time during the year, but usually in January; the land is then well manured by sheep, and afterwards thoroughly ploughed. The garden is irrigated twice a week and hoed once a month during the whole three years of the life of the plant. In the *Settlement Report of the Western Delta Taluqs of the Godavery District*, the total expenditure per acre during the four to six years of the average life of the plantain, in that locality, is estimated at R93. The returns, which commence from the second year, comprise the profits accruing from the fruit and leaves, to which is added profits derived in the fifth and sixth years from the sale of young trees. The produce of each pit in leaves and fruit is estimated at 4 annas annually—a considerably smaller sum than that allowed in Coimbatore. But even this low estimate gives an annual yield of the value of R62-8 per acre, or in six years R229.

Experiments conducted in the Saidapet Experimental Farm showed an expenditure in three years of R172 per acre, and a total profit after only one year and nine months' existence, of R60 per acre, including the estimated value of the standing trees. The yield of fruit by weight per acre, during the months when the plantation was bearing, amounted to between 2,500 and 2,600 pounds. It was found that the yield of food-stuff from the bunches amounted to about 49 per cent. of their weight, or a rate per acre of 1,250 pounds.

Nearly all writers on the economic botany of Madras make frequent reference to the importance of the cultivation of plantains in the province. No statistics are, however, available of the total area under the fruit, nor of the approximate total production. A considerable Local trade is carried on between outlying gardens and the large towns, and in addition to the fruit, the minor products derived from the plant appear to be utilised to a considerable extent. Dried fruits from the more southern districts are said to be exported, in some considerable quantity, to Ceylon.

Dr. Shortt describes thirteen different cultivated forms of the plantain, all of which have distinct vernacular names, but it is unnecessary in such a work as the present to enter into a detailed description of these. It may, however, be mentioned that the kind popularly known as the "guindy" plantain (*puvaly*, TAM.) is considered the best, and is described as round, small sized, with a very thin rind, luscious, sweet, and of a most delicate flavour. A good bunch may contain over a thousand fruits. This kind is used entirely as a table fruit, being considered too valuable for cooking purposes. The *rustali* is, however, the sort generally sold as table plantains, though not of such good quality as the former. A large plantain, known in Tamil as *monthen*, is one of the commonest cooking fruits of the Presidency. The *kathali*, cultivated in the Tinnevelly and Malabar Districts, is considered sacred and used chiefly as offerings to the gods. **Shortt** states that the fruit is gathered, as a rule, long before it reaches maturity

CULTIVATION.

and is artificially ripened by being smeared with *chunam* (fresh lime) and hung up. In certain districts, it is placed in a large pot filled with straw, artificially heated, and smoked for three or four days, after which the fruit is found quite ripe. More rarely the fruit is packed in straw and buried two or three feet deep for a few days. All these methods ensure the fruit being obtained not overripe, but destroy its flavour to a great extent.

Bengal. 815

II. Bengal.—The plantain is extensively cultivated throughout Bengal, more especially along the banks of rivers. As in Madras, many cultivated races exist, which in the *Bengal Agricultural Department Report* are classed under two heads. First, those grown for their sweet and fine flavoured ripe fruit; and, second, those grown for their half ripe fruit, which is used as a green vegetable. Of these two classes the latter, known under the general name of *kachkala*, can alone be regarded as a farm crop. Plantains, which belong to the other class, require to be grown in gardens, with the usual care and attention devoted to fruit cultivation.

Kachkala plantains will grow in almost any soil, except stiff clay and barren sand. The crop, however, as well as the finer class of banana, does best on the newly raised earth on the embankments of canals and tanks. Ordinarily, it is planted on standing *aus* paddy, *kachu*, *begun*, or turmeric, and is said to succeed best in a *begun* field. The usual time for planting is the end of *Jaistya* and the whole of the month of *Ashar* (from the beginning of June to the end of July). Two small shoots are allowed to grow from each plant, all the rest being dug out and used for planting if necessary. The shoots are set from 12 to 15 feet apart. The crop requires no watering, as it is planted in the rainy season. When plantain is put in among standing *aus* paddy, a crop of *kalai* (**Phaseolus Mungo** *var.* **radiatus**) may also be taken off the field after the rice has been harvested. But no third crop can be grown when the plantain has been planted among *kachu* (**Colocasia antiquorum,** *Schott.*) or *begun* (**Solanum Melongena,** *Linn.*). When these secondary crops are off the ground in the months of *Baisakh* and *Jaistya* (15th April to 15th June), the field is ploughed two or three times. The plants begin to bear fruit one year after the time of setting, and after the first year the ground is devoted to plantains alone, no other crop being sown. During the rainy season the weeds are destroyed by turning over the soil with a *kodali*, and in the month of *Aswin* or *Kartik* (15th September to 15th November) the land should either be hoed once, or given two or three ploughings. With care and cultivation, a plantain field may be kept up for more than ten years.

The foregoing account is condensed from a description returned from the District of Burdwan, but is doubtless equally applicable to other localities in Bengal. In the same district the plantain is said to be subject to the attacks of a large black insect known as the *antopoka*, which takes up its abode at the crown of the root-stalk just below the mid-leaf. As a consequence of its attack the leaf-stalks soon begin to wither and the plant dies. No remedy for this pest is known to the rayats. All the trees in the clump of which one has been thus attacked should be uprooted and removed from the field. Injury is also sometimes done by the earth-worm, and the plants are said to be liable to destruction by storms.

The cost of cultivation of 100 trees is estimated at R12-12.

Liotard describes five chief cultivated forms of plantain in Bengal, which he says are known commonly as the "table plantain," the *champa*, the *Dhakkai*, the *kantali*, and the *kanch kolla*. The first of these is the best, and is grown entirely for the consumption of Europeans and well-to-do natives. It is in much best condition during the rains, indeed the few procurable in the cold season are very inferior. The *champa* is the next best and like the preceding is of finest quality during the rains. The *Dhakkai*, a long fruit with light pink, soft flesh, is only found in abundance in the east of the province, while the *kantali* is much inferior, and only eaten as fruit by the poorer classes. As already stated, the term *kanch kolla* is employed generically to embrace all field cultivated, coarse plantains The fruit of this class is stringy, astringent, and unpalatable; it is hardly ever allowed to ripen and is mostly used when unripe as a vegetable in curries, &c.

CULTIVATION.

III. **Bombay.**—The plantain is cultivated to a large extent as a garden crop in Bombay, but would appear to be of less importance from an agricultural point of view than in Madras and Bengal. **Bombay. 816**

In the Poona District it is stated to be commoner than any fruit except the mango. Young shoots are planted in gardens at any time of the year. They require a rich soil, and are watered once in ten or twelve days. The plants are generally removed after they have borne fruit once and fresh shoots substituted. In Nasik the same system of allowing only one shoot from a root-stock to bear fruit is stated to be general. This shoot is styled the daughter or *kár*, and, when it has borne fruit, the plantation is generally destroyed. But occasionally a grand-daughter or *nát* is allowed to grow. After the plantation is abandoned, the ground is generally used for chillies, ground-nuts, and other similar crops.

No statistics are available for the area under the plantain in Bombay, owing to the fact that the Agricultural Department returns include it under the general heading of "fruit trees." It may, however, be stated that the *Surat Gazetteer* gives an area for that district of 198 acres. **Mr. Woodrow** of Poona has described eight cultivated forms of the fruit, as commonly met with in Western India. The best of these are known locally as:—*Bajapúri*, a long-pointed, three-cornered fruit with thick skin, yellow, and fine flavoured; *sonekale*, considered the best of all, small, cylindrical, yellow, thin-skinned, of very superior flavour; *raikále* or *rajkále*, a large fruit with thick red skin and of delicious flavour; and *kúli*, similar to the last, but yellow, also of excellent flavour.

IV. **Sind.**—No records exist of the extent of cultivation of the plantain in this province. Experiments were conducted at the Government Farm at Hyderabad between 1872 and 1876. The expenditure amounted to R464, the produce realised R375, and valuing the plants on the ground at the close of the experiment at R100, the total value of the produce would amount to R475. The return would have been much larger, had not two severe storms committed extensive damage. The cost of cultivation in most parts of Sind must, however, always be great. Irrigation represents, even in the Experimental Farm, a charge of about R28 per annum per acre cropped. In later years the severe frosts, occasionally experienced in winter, were also found to be most detrimental. Though the soil was very sandy it produced, with compost manure, a very heavy crop. **Sind. 817**

V. **North-West Provinces.**—According to **Atkinson**, several cultivated races are found in these provinces, but none attain the perfection that characterises those produced in Bengal. The nearer the Bengal frontier the better they are. The best kind is known as *champa*, next comes a small fruit, *chíni-champa*, about the size of a thumb, and occurring in thick compact clusters. A large reddish variety, known as the *rám-kela*, is inferior to the others. The *dhakka* or *dhakkai* variety already mentioned in the account of Bengal is occasionally, though rarely, grown in these provinces, but is frequently imported. **N.-W. Provinces 818**

VI. **The Panjáb.**—The larger, coarse vegetable-like plantain is that usually found in the Panjáb, and much inferior as a fruit to the well-flavoured plantain of Bengal. **Baden Powell** states that the art of making **Panjab. 819**

CULTIVATION. flour from the plantain, or of preserving the fruit is unknown in this province. **Stewart** remarks that it is largely grown in places towards the east of the Panjáb plains, and in the Siwálik tracts and outer hills, while on the Sutlej it may be seen up to 4,000 feet. It becomes very rare towards the North-West.

Assam. 820 **VII. Assam.**—Returning to the Eastern warmer and more moist regions of India we find the plantain again a very common and important plant. Cultivation is carried on much in the same way as already described in the account of Bengal. **Darrah** mentions the following three as the best kinds: *malbhag*, *pura*, and *ban tulsikol*.

Burma. 821 **VIII. Burma.**—The plantain is largely cultivated in this province. **Mason** states that it holds the same place in Burma, as the apple does in England and the United States, and is largely used both as a fruit and as a vegetable, the greater proportion being eaten with rice and meat in the place of potatoes. He further remarks that he knew the Burmese names of twenty-five distinct "varieties," but does not give these. **Kurz** states that three wild plantains occur in the country, *viz.*, **M. rubra** (*tau-hnek-pyau*), **M. glauca** (*nat-hnek-pyau*), and **M. sapientum**, *tau-knek-pyau*, *saip-cho*, or *ya-khaing*. **Mr. E. Oates, C.E.**, obtained fourteen kinds of cultivated fruit for him, and mentioned two others, while **Captain Ripley** described nineteen met with in Arracan, and **Dr. Helfer**, twenty in Tenasserim. Many of the Arracan plantains are described as of excellent flavour, those which appear to be the best being known as *rakoing-hnset-pyau-bhí* (**Musa arakanensis**, *Ripley*), *may-dau-letthé*, *nathabú*, *byat-taus*, *moung-bya*, *pím-we*, *wet-tsway*, *hpí-gyan*, and *moung-net*.

DYE & TAN. Ashes. 822 Leaves. 823 Bark. 824 Fruit-Rind. 825 Sap. 826 **Dye and Tan.**—The ASHES of the LEAVES, the BARK (?) and the FRUIT-RIND, are employed in many of the dyeing processes practised in Bengal (*McCann*). The latter is also used as a tan, and for blackening leather. The SAP contains a considerable amount of tannin, and stains cloth a dark, almost black colour, which is fairly permanent, is very difficult to wash out, and may be employed as a substitute for marking ink.

FIBRE. 827 **Fibre.**—The fibre of the plantain has long been used by the natives of India for cordage purposes, for mats, and to a smaller extent for making coarse paper. It attracted early attention from writers on economic subjects, from the fact that it so much resembles Manilla hemp, the product of **M. textilis**. As far back as 1822 we find that it had been compared with the fibre of the latter, and allowed to be inferior, the specimens examined having been in each case extracted from plants grown experimentally in Calcutta. In 1846 a **Mr. May** showed **Dr. Royle** some beautiful specimens of note and letter-paper made from plantain fibre. He was at that time anxious to establish a manufactory for plantain paper in Calcutta, but for some reason abandoned the project. **Dr. Hunter** of Madras sent the fibre and tow of Indian plantain, in a well-cleaned state, to the exhibition of 1851, along with specimens of thick rope, fine cord, and paper of all degrees of fineness from thick packing paper to some "almost as thin as silver paper." A sample of the tow was also sent to the Ordnance Department for trial, and was reported on as "undoubtedly of a very superior description and admirably adapted for packing." The remark was also made that from its soft elastic character it appeared to be a desirable substitute for coir in stuffing hospital beddings, &c. **Dr. Hunter's** mode of extracting the fibre was as follows:—"Strip off the different layers and clean them in the shade, if possible, soon after the plant has been cut down. Lay a leaf-stalk on a long flat board with the inner surface uppermost; scrape the pulp off with a blunt piece of hoop-iron fixed in a groove in a long piece of wood. When the inner side, which has the thickest layer of pulp, has been cleaned, turn over the leaf and scrape the back of it. When a bundle of fibres has

FIBRE.

been thus partially cleaned, it ought to be washed briskly in a large quantity of water, so as to get rid, as quickly as possible, of all the pulpy matter, which may still adhere to the fibres. It may be readily separated by boiling the fibres in an alkaline ley, or in alkaline soaps, but not in the Indian soaps made with quick-lime, as these are too corrosive. When the fibres have been thoroughly washed, they should be spread out in thin layers or hung up in the wind to dry. If exposed to the sun when in a damp state, a brownish-yellow tinge is communicated, which cannot be easily removed by bleaching. Exposure during the night to the dew bleaches them, but it is at the expense of part of their strength."

Dr. Royle (*Fibrous Plants of India*) in 1855 devoted a considerable amount of attention to the fibre. Experiments made by him with fibre prepared in Madras, shewed that it bore a weight of 190lb, but some from Singapore bore not less than 390lb, while a "salvage" of Petersburgh hemp of the same length and weight broke at 160lb. A twelve-thread rope of plantain fibre made in India broke with 864lb, while a similar rope of pine-apple fibre broke with 924lb. **Dr. Royle** concluded :—"Even from these experiments it is evident that plantain fibre possesses sufficient tenacity to be applicable to many, at least, of the ordinary purposes of cordage. The outer fibres may also be converted into a useful kind of coarse canvas, as has been done by **Dr. Hunter**; and the more delicate inner fibres most probably into finer fabrics, as is the case with those of **M. textilis** when equal care has been taken in the preparation and separation of the fibres, and there is some experience in weaving them." Of late years attention appears to have been more directed towards introducing the superior **Musa textilis** than towards utilizing the valuable cordage and paper fibre of which such a large quantity exists in India. At the same time it must be remembered that it is utilized to a certain extent by the natives, especially in Madras, and that though the cost of preparation is very small, the cost of collecting in sufficient quantity to furnish a large rope or paper-making concern with sufficient raw material would be large. A consideration of the facts detailed in the accounts of cultivation shows that the plant is only in a very few places cultivated over any extended area. As a rule, plantains occur in small groups scattered about villages, and in gardens; consequently a not inconsiderable amount of trouble and expense would necessarily be involved in collecting the cut down stems. It may, however, be mentioned that **Dr. Royle** suggests that the plant might be cultivated in fields for the purposes of fibre near towns, where the fruit might be readily saleable or that the extraction of the fibre might be combined with a concern for preserving the fruit or converting it into flour either of which "ought to pay all the expenses of, and afford some profit on, the culture" He recommends that a rough crushing mill, such as the rollers of a sugar mill or an enlarged *churka*, should be used in the vicinity of these fields to extract the fibre roughly on the spot and thus save cost of carriage. **Mr. Henley** basing his calculations on **Dr. Royle's** statements stated, in a lecture before the Soceity of Arts (*Jour. of Soc. Arts, II., 486*) that he was of opinion that contracts could be made to obtain the rough fibre at from R1-8 to R2-8 a maund, deliverable at a central depôt within a radius of twenty miles.

This estimate, however, must be considerably under the mark. Experiments conducted with the wild plantain fibre in the Nicobar Islands under the orders of Government, shewed that thoroughly cleaned fibre could not be produced there, at a cost of much under 6 annas per pound.

Recent investigations by **Messrs. Cross & Bevan** shew the fibre to contain 13·4 per cent. of moisture, and 64·6 of cellulose. It loses by

MUSA sapientum. **Medicinal Properties of**

the process of Hydrolysis, 11 per cent. when boiled for five minutes in a 1 per cent. solution of caustic soda, and 33 when boiled for one hour.

MEDICINE. Fruit. 828 **Medicine.**—The unripe FRUIT, called *mochaka* in Sanskrit, is considered cooling and astringent; it is much used in diabetes in the form of a *ghrita*, composed of plantain flowers, root-stock, and unripe fruits, *ghi*, cloves, cardamoms, and several other drugs. This medicine is generally prescribed in doses of two *tolas* along with some preparation of tin or other metallic drug (*U. C. Dutt*). Leaves. 829 Young plantain LEAVES are used as a cool dressing for blisters, burns, &c., and to retain the moisture of water dressings. They may also be used as a green shade in ophthalmia and other eye diseases. Root. 830 Stem. 831 The ROOT and STEM are considered tonic, antiscorbutic, and useful in "disorders of the blood" and venereal disease. Sap. 832 **Emerson** states that the SAP forms a valuable drink and mouth-wash to allay thirst in cholera. According to **Dymock Mir Muhammad Husain** states in the *Mazkhan*, that the kind of plantain called *málbhok* is used as a poultice to burns, while that called *bolkad* is boiled and employed as an ointment for the syphilitic eruptions of children. Ashes. 833 He also notices the use of the ASHES on account of their alkaline properties, and of the root as an anthelmintic. **Ainslie** writes, "The plantain is one of the most delicious of all the Indian fruits, and one of the safest for such as have delicate stomachs, being entirely free from acidity; it is, moreover, very nourishing, and is always prescribed as food by the Hindu practitioners for such as suffer from bile and heat of habit."

The fruit has long been known and commented on by European writers. Perhaps the first authentic description is by **Pliny** who quotes the name *pala*, a term which still exists in Malabar. He states that the Greeks of **Alexander's** expedition saw it in India, and that sages reposed beneath its shade and ate its fruit (hence the name "**sapientum**"). In the middle ages it had some reputation as a medicine. **Avicenna** wrote that it engendered choer and phlegm, and that it spoiled the stomach, but that it was good for heat in the stomach, lungs, and kidneys and provoked urine. **Rhasis** stated that the fruit was hurtful to the "maw," **Serapio** that it was in the end of the first degree warming, diuretic and aphrodisiac. **Paludanus**, the commentator and friend of **Linschoten**, confirms these statements, and from personal observation, supports the remark that the fruit breeds "a heaviness in the mawe." In modern times it is employed medicinally by Europeans as an antiscorbutic only, and as a mild, demulcent astringent diet in cases of dysentery, but several other less well known properties are attributed to different parts of the plant in the following special opinions:—

SPECIAL OPINIONS.—§ "The ripe fruit of the finer varieties of the plantain is useful in chronic dysentery and diarrhœa. The dried fruit of the larger varieties is a valuable antiscorbutic. In North Bengal the dried leaves and in fact the entire plant is burnt, and the ashes dissolved in water and strained yield an alkaline solution containing chiefly potash salts, which is used in curries, especially as a cure for acidity, and antiscorbutic, and where common salt is scarce this is used by the people for seasoning their curries" (*C. T. Peters, M.B., Zandra, South Afghánistán*). "I have known a diet of green plantain well boiled, and curds (*dohi*), sweetened with sugar or seasoned with salt according to taste, to be of singular benefit, in cases of dysentery and diarrhœa. (2) Ripe plaintain, well beaten up with pulp of old tamarind and sweetened with old treacle or sugar-candy is a household remedy among the natives of Bengal for dysentery, at the commencement of the attack. (3) Flour made out of green plaintain dried in the sun is used in the form of *chappatis* in certain parts of Tirhoot in cases of dyspepsia with troublesome flatulence and

MEDICINE.

acidity. I have known one case in which it agreed remarkably well when even a diet of plain sago and water brought on a severe attack of colic. The *chappatis* are taken dry with a little salt" (*Assistant Surgeon N. C. Dutt, Durbhanga*). "A combination of ripe plaintain, tamarind, and common salt is most efficacious in dysentery. I have used it in many cases both of the acute and chronic forms of the disease, and seldom failed to effect a cure. It may, in fact, be said to be a specific, and I can confidently recommend it to the profession as well as to the public. It is simple, easily procurable, and may safely be administered to a child. It is not disagreeable to take, has no bad effects, and is on the whole preferable to **Ipecacuanha.** In simple cases a single dose is sufficient, as a rule three or four doses are required to effect a cure. The patients should be kept quiet and placed on low diet. The dose for an adult is,—ripe plantain one ounce, the pulp of ripe tamarind half an ounce, common salt quarter of an ounce; well mixed and administered immediately. It may be given two or three times a day" (*Civil Surgeon R. A. Parker, M.D.*). "The juice of the tender roots contains a large quantity of tannin and is used with mucilage for checking hœmorrhages from the genital and air passages. The ashes produced by burning the plant contain a large amount of potash salts, and are used as an antacid in acidity, heartburn, and colic. The tender fruit is used as a diet for patients suffering from hæmoptysis and diabetes" (*Civil Surgeon J. H. Thornton, B.A., M.B., Monghyr*). "The fruit is used with salt in dysentery. The powdered root is antibilious, also used in anæmia and cachexia" (*Surgeon-Major D. R. Thomson, M.D., C.I.E., Madras*)." "The juice of the plant is reported to possess styptic properties" (*Surgeon R. L. Dutt, M.D., Pubna*). "The juice of the bark and leaf is frequently given to children suffering from an overdose of opium. The juice of an ounce of bark mixed with an ounce of *ghí* acts as a brisk purgative" (*Surgeon J. McCloghey, Poona*). "I have often seen natives use the young unripe fruit curried, when suffering from dysentery or chronic diarrhœa" (*Honorary Surgeon E. A. Morris, Tranquebar*). "The root juice in which burnt borax and nitre are dissolved is given with success in ordinary cases of retention of urine. The juice of the flowers mixed with *curds* is used in dysentery and menorrhagia" (*Native Surgeon T. R. Moodelliar, Chingleput, Madras*). "The unripe fruit dried and powdered is used as an astringent in the diarrhœa of infancy; it is useful in most forms of diarrhœa, also in gonorrhœa" (*Hospital Assistant Lal Mahomed, Hoshangabad, Central Provinces*). "The juice of the root is used as an antidote to arsenical poisoning in the lower animals. Mixed with *ghí* and sugar and administered internally it is said to be useful in gonorrhœa" (*Deputy Sanitary Commissioner J. Parker, M.D., Poona*). "The unripe fruit, well cooked as a curry, makes excellent food for natives recovering from chronic diarrhœa. Leaves largely used in charitable dispensaries as a substitute for gutta percha tissue in surgical dressings" (*Brigade Surgeon S. M. Shircore, Moorshedabad*). "Plantain leaf is the cleanest and nicest dressing for a blistered surface that I know and is also useful in covering other dressings. A piece of plantain leaf introduced into the helmet on a hot day forms an effectual protection from the sun's rays, without appreciably adding to the weight of the head-dress" (*Surgeon-Major H. DeTatham, M.D., M.R.C.P., Lond., Ahmednagar*).

FOOD & FODDER. Fruit. 834

Food and Fodder.—The FRUIT of the cultivated forms of this species are sometimes popularly distinguished by the names of banana and plantain, according to whether they are eaten raw or cooked. These names are, however, very loosely applied, some calling any round and plump fruit "banana," others making a distinction in size only, the small being "banana," the large plantain. It is, therefore, advisable to reject the arbi-

FOOD & FODDER.

trary distinction which has arisen between the names, and to call all alike by the commoner name, plantain.

The finer varieties, grown as garden fruits, are, after the mango, the commonest and most prized of Indian fruits, while the coarser forms, cultivated for use as vegetables, form one of the staple articles of food in many parts of India and the Malay Peninsula. Thus **Rumphius** relates how in the latter region man begins life with plantains, the roasted fruit being used as pap for new born infants. In India it supplies in many localities the place of bread and potatoes, and is generally roasted, stewed, eaten in curries, or boiled in cow's milk. By Europeans it is sliced and cooked in the form of fritters.

The finer varieties, cultivated to be eaten raw, vary much in taste and flavour; some are acidulous, others acid-sweet, others pure sweet like sugar, while the inferior kinds are mawkish. The names of those considered the finest will be found in the account of the method of cultivation in each Presidency or Province.

Humboldt was perhaps the first to draw attention to the fact of the very large yield of plantains obtainable from one acre of ground, stating that the produce would support a much greater number of people than a similar area under wheat or any other crop. **Boussingalt**, following up this line of enquiry, gave the following as the produce per imperial acre of the raw fruit in three places, according to (1) **Humboldt's**, (2) **Gondot's**, and (3) his own observations :—

	Temp.	Produce per acre.	Of dry food per acre.
	Fh.	Tons.	Tons.
(1) In warm regions	81·5	72	19½
(2) At Cauca	78·8	59	16
(3) At Hague	71·4	25	6¾

The last column is calculated on the assumption that the fruit contains 27 per cent. of nutritive matter. For purposes of comparison with this table **Royle** showed that it would require a crop of twenty-seven tons of potatoes per imperial acre to yield the smallest of the quantities (of food-stuff) above mentioned as the yield of an acre of plantains.

It is interesting to notice, in connection with the above, that the large crop of food thus produced may be preserved for an indefinite period either by drying the fruit, or by preparing meal from it. Both of these processes, which have long been known and carried out in the West Indies and South America, are also carried on in India, though to a much smaller extent. **Linschoten** notices the practice as common in the sixteenth century, writing,—" These grow much in Cananor, in the coast of Malabar, and are by the Portingales called figges of Cananor: and by reason of the greater quantities thereof are dried, the shells being taken off, and so being dried are carried over all India to be sold." When the nearly ripe fruit is cut into slices and dried in the sun, a certain part of the sugar contained in the fruit crystallizes on the surface, and acts as a preservative. The slices thus prepared, if made from the finer varieties, make an excellent dessert preserve, and if from the coarser, may be used for cooking in the ordinary way. They keep well if carefully packed when dry, and ought to form a valuable antiscorbutic for long voyages. The fruit may also be similarly preserved whole by stripping off the skin and drying it in the sun.

FOOD & FOODER.

Plantain meal is prepared by stripping off the husk, slicing the core, drying it in the sun, and when thoroughly dry, reducing it to a powder, and finally sifting. It is calculated that the tresh core will yield 40 per cent. of this meal, and that an acre of average quality will yield over a ton. In the West Indies it is chiefly employed as a food for infants, children, and convalescents, and is undoubtedly of considerable value as a mild and fairly nutritious diet.

According to **Corenwinder**, the ripe fruit grown in Bengal contains 73·9 per cent. water, 4·82 albumin, 0·2 cellulose, 0·63 fat, 19·66 cane sugar and invert-sugar (together with organic acids, pectose and traces of starch), 0·06 phosphoric anhydride, and 0·73 lime, alkalies, iron, chlorine, &c. The ash of the ripe fruit was found to contain 47·98 per cent. of carbonate of potassium, 6·58 carbonate of sodium, 25·18 chloride of potassium, 5·66 alkaline phosphates, 7·50 charcoal, and 7·10 lime, silica, earthy phosphates, &c. (*Watts, Dic. Chem*). **Dr. Warden** has kindly communicated the following later and more complete analysis:—"The composition of the banana at different stages of maturity has been investigated by **L. Ricciardi.** The green fruit contains over 12 per cent. of starch, which disappears as the fruit ripens. It contains 6·53 of tannin and the ripe only ·34 per cent., so that as the fruit ripens this principle disappears, and this is also the case with the other organic acids which are present. The sugar in the fruit which ripens on the tree is almost entirely cane sugar, but in the fruit cut and ripened by exposure to air, the invert-sugar reaches about 80 per cent. while the cane sugar is reduced to about 20 per cent., calculated upon the sugar present. Proteid substances are present in the green fruit to 3·04 per cent. and in the ripe to 4·92 per cent. The green fruit yields 1·04 and the ripe ·95 per cent. of ash, which contains 23·18 per cent. of phosphoric anhydride and 45·23 per cent. of potash."

From these analyses it will be seen that, notwithstanding many extravagant statements regarding the nutritive qualities of the plantain, it is, by itself, by no means a perfect food, and requires the addition of some nitrogenous material. Nevertheless, with the addition of a little pulse, or lean meat, it is a most valuable food for tropical countries. About 6½℔ of the fruit, or 2℔ of the dry meal with ¼℔ of salt meat or fish is said by **Johnston** and **Church** to form the daily food allowance for a labourer in tropical America.

Besides the fruit several other parts of the plant are used as food. The **Flower-heads.**
FLOWER-HEADS (*mocha*, HIND.; *djantong*, MALAYS) of many kinds, are 835
cooked and eaten, generally in curries. The inner portion of the STEM **Stem.**
(the scape) is also eaten, and is said by **Kurz** to be brought into the 836
Calcutta bazars to the amount of half a ton daily. It is called *thor*, and
is generally prepared for food by boiling. A solution of the ash is fre- **Shoots.**
quently employed instead of salt in cooking vegetable curries, &c. The 837
SHOOTS and tops of young plants are also occasionally eaten as a vege- **Outer**
table, and are given as fodder to sheep and cattle. The OUTER SHEATHS **Sheaths.**
form a valuable fodder for elephants. The central portion of the stem, 838
and the ROOT-STOCK are said to be given to cattle to increase the quantity **Foot-Stock.**
of milk. 839

The fermented JUICE is made at Cayenne and the Antilles into a palat- **Juice.**
able wine, called "Vino di banana." A similar liquor is prepared in the 840
Congo region, where it has the reputation of being a preventative of malaria.

Domestic and Sacred.—Many parts of this useful plant are largely **DOMESTIC**
employed for domestic purposes. The DRIED PETIOLE is used without **& SACRED.**
further preparation for tying fences, training the betel vine to its support, **Dried petiole.**
and for numerous other purposes as a rough kind of twine, and the larger 841

MUSA textilis.

The Manilla Hemp.

DOMESTIC & SACRED. Trunk. 842 Leaves. 843 parts are made into little square boxes for holding snuff, drugs, &c. In the Archipelago, the TRUNK is cut into several pieces, which serve as hearths during festivities in the open air, and in Siam it is used for clarifying sugar. The LEAVES are much used for packing all sorts of small goods in the bazars, and are also employed as plates, being sold for this purpose for from 1 to 3 pies each. When dry they are employed by shop-keepers much as brown paper is in Europe. They are also used for making mats, and as thatch for temporary huts. In the Malay peninsula, the fresh leaves are employed as a water-proofing covering for the earthen pots or bamboo, in which rice is steamed. The ASH of the leaf and leaf stalk, rich in alkaline salts, is used instead of country soap or fuller's earth in washing clothes.

Ash. 844

Lisboa states that it is ordered in the *Vratráj* that females should worship the tree on the 4th of *Kartik shudh*, whereby their husbands are said to survive them, and their life is lengthened. It is also worshipped on the 3rd of *Shrávan*. The bunches of fruit are much used in certain festivities and ceremonials, and are generally placed by Hindus at the entrance of their houses on such occasions (especially at marriage), as appropriate emblems of plenty and fertility. The plantain called *kathali* is considered sacred in Madras, and is reserved as an offering to the gods.

Musa superba, *Roxb., Fl. Ind., Ed. C.B.C., 224.*

Vern.—*Chavai*, NASIK; *Chavaicha kanda*, POONA; *Chavlya kand*, KHANDESH, BOMB.

References.—*Dalz. & Gibs., Bomb. Fl., 272; Kurz, Notes on the Banana, 165; Lisboa, U. Pl. Bomb., 204, 235; Dymock, Mat. Med. W. Ind., 887, 889.*

Habitat.—A handsome plant, said by **Roxburgh** to be a native of Pegu, and by **Dalzell & Gibson** and other Bombay writers to be common at Mátherán Rám Ghát, and Khandála. **Dymock** states it in his Famine Plant List that it is found in Násik, Poona, and Khándesh.

FIBRE. 845 **Fibre.**—Like the other members of the genus this species is said to yield a fibre, probably similar to that of **M. sapientum**.

MEDICINE. Manna. 846 Dr. **Dymock** has recently found a sweet, translucent, jelly-like manna exuding from the plant, which, when dried at a low temperature, yielded 82·3 per cent. of fermentable sugar (*Hooper, Chem. Notes on Mannas, 1891*).

FOOD. Scape. 847 Leaf-sheaths. 848 Root. 849 **Food.**—The SCAPE and convolute LEAF-SHEATHS are, according to **Lisboa**, used as food similarly to those of **M. ornata**, especially in times of scarcity. The ROOT was also eaten in the districts of Poona, Násik, and Khándesh during the Deccan Famine of 1877-78.

849 **M. textilis,** *Nees.; Kurz, Notes on the Banana, 165.*

MANILLA HEMP.

References.—*Voigt, Hort. Sub. Cal., 579; Kurz, For. Fl. Burm., II., 504; Drury, U. Pl. Ind., 302; Royle, Fibrous Pl., 64; Liotard, Mem. Paper-making Mat., 48, 49, 51, 54-58; Note on Musa Fibre, 1881; Balfour, Note on Musa Fibre, 1880; McCann, Dyes and Tans, Beng., 155; Tropical Agriculture, 466; Christy, New Com. Pl., VI., 13, 50; Bot. Garden Ganesh-khind, Poona, 1882-83, 5; 1883-84, 12; 1885, 5; Dir., Agri. Bomb. Presidency, 1883-84, 9, 12; Report, Bot. Gard. Calc., 1884-85; Sahâranpur, 1885; Proceedings of Govt. of India regarding Manilla Hemp, 1861-1863, a correspondence with Madras; 1877, 5, p. 35, October 1-5; 1879, July, 17-29; 1880-81, 28-35; 1881, July, 39-48; October, 52, November, 6-11, 12-16; 1882, March, 16-17; April, 85, 86; June, 27-33; July, 8, 9; September, 57-90, December, 17-21; 1883, February, 1, 2; April, 1; 1884, February, 75; April, 17-18; June, 1; July, 32, 33; 1885, September, 1, 2; 1886, March, file No. 8, 1-12; 1887, June, 15-20; Watt, Manilla Hemp in Sel. from Rec., Govt. of Ind., Rev. and Agri.-Dept. I., 1—6; Man. Madras Adam, Vol. I., 360; Adm. Rep. Andaman Islands, 6, 53; Madras, 1882-83, 95; Gazetteers:—Mysore and Coorg, III., 46; Agri.-*

The Manilla Hemp. (*J. Murray.*)

Horti. Soc. Ind.: — VII., Pt. III., Pro. c. iii; Indian Forester, XI., 39; Balfour Cyclop. Ind., II., 1016.

Habitat. — A native of the Philippine Islands, which, like other species of the genus, has a marked tendency to the insular type, luxuriating in hot, moist climates, within certain limitations of the sea influence. The Andaman Islands and the West Coast of India might accordingly have been indicated as the regions where every effort should be directed towards its acclimatisation, and, as will be seen, in these regions alone has the plant undoubtedly taken hold. In the Philippine Islands the plant is stated by **Mr. Honey**, British Consul at Manilla, to thrive best in soil largely impregnated with decayed vegetable matters, and is therefore particularly suited to reclaimed forest lands. It is most healthy on high lands subject to considerable rainfall, and on soil of volcanic origin, and is said to suffer severely during long periods of excessive heat and drought. It can be grown from seed, but is generally reared from shoots like the ordinary plantain. It takes about four years to reach maturity, and the first crop is available two years after planting, but a much heavier production of fibre is obtained in the third and fourth.

FIBRE. 850

Fibre.—Manilla Hemp has been described as follows by **Dr. G. Watt**:—"It is a fibre of great strength and extreme lightness, hence eminently suited for rope-making. Only about 50,000 tons are, however, annually required, so that the world's demand for this substance is not great, and it would seem that there is little prospect of this quantity being much increased. The waste materials and worn-out ropes afford, however, the much-prized strong manilla-paper, and a new industry has recently been started in dyeing the fibre with logwood and copper so as to allow of its being used as a substitute for horse-hair. The quality varies according to the part of the plant from which it has been extracted — an unfavourable property, since, unless great care be exercised, mixed consignments are sent into the market which fetch much lower prices than need otherwise have been obtained. The collective name for the fibre in the Philippine Islands is *abacá*; *bandala* is the harder and stronger outer fibre, the inner layer being known as *aúpoz*, while the delicate fibres, obtained from the edge of the petiole, are known as *lupis*. This last-mentioned form was formerly to a small extent imported into Paris, being used in the preparation of a special kind of under-clothing. This form of the fibre is, however, locally manufactured into fine cloth and shawls, and the textile either from this form or from the *aúpoz* is to some extent exported to Europe. The mean length of the ultimate fibriles is 0·236 inches and the breadth 0·00096 inches. The central cavity is well developed; the walls are of a uniform thickness and the extremities taper gradually and uniformly. These are qualities which eminently fit the fibre for the rope-making industry but are ill adapted for textile purposes, although it must here be added that the *abacá* affords the textile fabrics worn by the entire population of the Philippine Islands."

The fibre has long attracted attention from writers on Indian economic subjects, and many endeavours have been made to introduce the plant into India. **Mr. Liotard** gives an exhaustive account of these attempts in his work on *Indian Paper-making Materials*, from which the following information has been mainly obtained. The first record of **M. textilis** being grown in India dates from 1822, when it was introduced into Calcutta. No account exists of how the plant then introduced thrived, and no further information on the subject exists till 1859, when fresh specimens were introduced into Madras from Manilla. In 1862 these were said to have thriven well, "in every respect like common plantains," and in January of that year **Dr. Hunter** exhibited specimens of fibre obtained from them before the Committee of the Agri.-Horticultural Society. In 1870, however, we find the

MUSA textilis.

History and Cultivation of

FIBRE.

Board of Revenue of that Presidency reporting that "the experiment with the Manilla Hemp had been attended with little or no success," from which it may be presumed that the species had been allowed to die out.

In 1873 the plant was introduced into the Andaman Islands, and in 1876 forty-eight young plants existed, all sprung from one original shoot. In 1879 some of the largest plants were cut down, and 43℔ of fibre extracted was sent to the Government of India, by whom it was submitted to the Agri.-Horticultural Society of India for report. The reply received from the Secretary of that Society was favourable, and the remark was made that, if procurable at a reasonable rate, it would soon become an article of commerce. The more recent history may be quoted in entirety from **Dr. Watt's** recent note in the *Selections from the Records of the Govt. of India, Revenue and Agri. Dept., I., 1888-89*:—

"It may be briefly explained that **Surgeon-General Balfour**, in a note dated 15th October 1880, urged the claims of **Musa textilis** as a future Indian Fibre. His note on this subject was communicated to the Government of India, by Her Majesty's Secretary of State, and ultimately issued to all Local Governments, Botanic Gardens, Chambers of Commerce, &c., for information or for whatever action might seem necessary. **Mr. Liotard** greatly enlarged the information furnished by **Dr. Balfour**, by giving a history of the attempts, hitherto made, to cultivate the plant in India, from its introduction in the Madras Presidency in 1858, down to the date of his note, January 22nd, 1881. As a result of the renewed interest thus awakened, various reports have appeared of the efforts put forth by the local authorities to realise the high expectations entertained both by **Dr. Balfour** and **Mr. Liotard**. The most satisfactory appears to have been that obtained in the Andaman Islands. The instructive memorandum recently furnished by **Mr. R. Blechynden** of the Agri.-Horticultural Society of India summarises the Andaman experiments, and furnishes a report on the value of the fibre produced. It will be seen that while the fibre is stated to have been good, the price at which it could be produced was practically fatal. It is possible this defect, with improved means of separating the fibre, may be overcome."

The following is the note by **Mr. Blechynden** referred to above, a contribution which is full of interest and giving particulars of the most approved methods of obtaining the fibre:—

"I have the honour, by direction of the Council of this Society, to acknowledge the receipt of your No. 144—8-20 F & S., dated the 9th June, forwarding two samples of **Musa textilis** fibre for examination and report, and enclosing copy of letter from **Colonel Cadell, V.C.**, Superintendent, Port Blair and Nicobars, and extract of a letter from the officer in charge, Northern District, to the Superintendent, Port Blair and Nicobars.

The two samples of **Musa textilis** fibre have been carefully examined and tried. The results obtained at the first trial were contrary to those expected, consequently further trials were made; they, however, confirmed the results first obtained, but this caused some delay in the preparation of the report and in replying to your communication.

As stated by the officer in charge, Northern District (Andamans), in his letter above referred to, the two samples of fibre were prepared by different processes—the larger sample weighing 13℔ having been prepared simply by scraping with a blunt knife, while the other sample, weighing 32℔, was steeped in running water and then beaten out.

The two samples closely resemble each other in appearance and are both of good merchantable quality; they are of fair colour and strength and compare favourably with the "*medium*" quality fibre from the

FIBRE.

Philippine Islands; they are not, however, equal to "*best.*" Hemp of the same quality could be sold at Calcutta now for, say, R20 per cwt., or say from £30 to £32 per ton in the London market; the price, however, is very fluctuating.

In the comparative trials of the two samples, the fibre prepared by the retting process proved itself the stronger, breaking with a weight of 266lb, whereas the fibre prepared by the scraping process broke at 224lb; this was contrary to the expectations which had been formed, as it was thought that the process of retting would have weakened the fibre.

In reference to the methods used for cleaning the fibre. By the information furnished it does not appear from what weight of raw material the samples were prepared, or whether there was a greater waste by one process than by the other. The retting process is unusual as respects the **Musa textilis**; and though in this instance the results have been favourable, it would be as well, before adopting it on a future occasion, to have careful trials made of the waste, cost of transporting the raw material to the water, handling, &c. Such experiments would also prove valuable records; were the results favourable it would probably prove a method which Indian labourers would more readily adopt from their acquaintance with it when dealing with jute; were such experiments determined on, it would be as well also to try the effect of retting in still water, to ascertain if it would injure the fibre.

The other method adopted, *viz.*, scraping with a blunt knife, is substantially the system in use in the Philippines. The Society is indebted to Mr. Wilkinson, British Consul at Manilla, for an account of the process actually in use there, published in the Society's Proceedings for July 1883:—

'The following is a description of the apparatus in use in the province of Albay, Island of Luzon, for extracting the fibres from the stalks of the wild plantain (**Musa textilis**), locally known as Abacá or Manilla hemp:—

'Two strong uprights are firmly fixed in the ground and connected by a cross bar, in the centre of which a large broad-bladed knife is fixed edge downwards on a block of wood fastened lengthwise on the bar; the knife has a strong handle, which is connected by a cord to a long bamboo made to act as a spring, by being tied in the middle and the butt parallel and above the bar; the free end thus forms a supple and powerful spring and holds the edge of the knife firmly against the block; below the bar, there is a treadle attached by a cord to the handle of the knife; the mode of operation is for the worker to stand opposite the knife, placing either foot on the treadle, which he depresses, thus forcing the knife handle down and the blade up; he then places a strip of stalk (called locally *Sifa*) between the blade and the block, leaving only enough to wrap round a stock on the near side; he then releases the treadle, and the knife by the action of the bamboo spring, nips the strip firmly against the block, and on the workmen drawing the strip through, the pulp is left behind. The apparatus is extremely simple and inexpensive.'

'In the *Bulletin* for April 1887 (No. 4) published by the authorities of Kew, there is a great deal of interesting information regarding the Manilla hemp. It is there stated that the whole supply comes from the Philippine Islands; the imports to Great Britain 'amounts to about 170,000 bales and to the United States about 160,000 bales, equal to about 50,000 tons per annum.' The imports to Calcutta are comparatively insignificant, being probably less than 300 tons per annum. It is stated in the Kew report that a labourer working under pressure 'can clean nearly 20lb of hemp per diem; but as a rule the quantity cleaned by one man working

FIBRE.

steadily day by day averages about 12℔; usually two men work together, one cutting down the stems and splitting them, while the other cleans the fibre. At the current rate of wages in 1879 one labourer's earnings were $7\frac{1}{2}d$. to $8d$. per diem.'

'The foregoing extract is given, as it may be a guide in future experiments. The amount of fibre cleaned per man in the experiment at the Andamans having been about 7 oz. and 5 oz. per head in the scraping and retting process respectively. With more experience, and an appliance such as has been described, a model of which I am instructed to say can be furnished to the Chief Commissioner if required, there is no reason why the results may not more closely approximate to that obtained in the Philippines. I am to add the following extract from the *Kew Bulletin* already quoted from :—

'After a systematic series of trials made by the Glenrock Company at Madras in 1885, it is stated that plants put out in 1864 grew well and yielded numerous shoots; 179 stems, weighing about 60 pounds each, were cut down for experimental purposes and passed through Death and Ellwood Machines. These produced 159℔ of clean fibre, or 1·49 per cent. of green stem. The cost of cleaning the fibre was at the rate of £6 per ton, while the fibre itself, described as 'poor, weak, and flaggy, with some clean fibre of good colour,' was valued in London at £10 per ton; the best alone was valued at £25 per ton. The minute upon this of the Government of Madras is that 'unless much improvement both in the method and cost of production of this fibre can be made, the cultivation cannot be made remunerative.'

'The fibre prepared in the Andamans being, as above stated, nearly equal to medium hemp imported from the Philippines, and being valued at R20 per cwt., and the data above given showing it is possible to produce the fibre at less than 6 pie per pound by hand labour, it would appear that there is every prospect of the cultivation of **Musa textilis** being made a productive one in the Andamans'" (*Blechynden*).

In addition it may be mentioned that **Mr. Liotard** insists strongly on the necessity of cutting down the plants before they bear fruit, of extracting the fibre as soon as possible after the stems are cut down, and of avoiding all necessary exposure to moisture or to the direct rays of the sun. **Dr. Watt**, commenting on **Mr. Blechynden's** memorandum and on the facts previously given by him, remarks, "The above abstract of the published facts regarding Manilla-fibre may be of some value in suggesting the course which should be followed in further efforts to acclimatise the plant and produce the fibre. The efforts which have been made to introduce the plant into Bengal have failed, the low temperature, as **Dr. King** reports, of the cold weather having proved too much for the plant. **Mr. Duthie**, in his reports of the experimental cultivation in Saharanpur, simply mentions the plant as growing in the gardens. In Madras and in Bombay the attempts to introduce the plant have proved more hopeful, especially in the districts possessing the necessary conditions briefly indicated. The Commissioner reports, for example, of the Dharwar District, that 'the result of planting **Musa textilis** is a great success, and I am of opinion, with regard to the experiment, that it might, with good treatment and without much difficulty and expense, succeed well.' The Director of Agriculture, Madras, remarks that 'There is sufficient evidence that the plant will grow in most parts of the Presidency, and I think the future operations should be confined to the Government farms in order to ascertain whether—

(1) the plant can compete with the ordinary plantain—**Musa sapientum**;

FIBRE.

(2) the difficulty of extracting the fibre can be overcome so as to make its production cheaper than the purchase of imported Manilla hemp;
(3) the fruit can be utilised;
(4) a hybrid variety can be raised between the **Musa textilis** and the **M. paradisiaca,** useful both as a fruit and a fibre.'

Mr. Robertson, of the Saidapet Experimental Farm, reports that the plants will not grow well unless planted in a deep well-manured soil, under regular irrigation. On this account the cost of growing the plants, until they reach the stage of cutting, is great. He adds: 'There seems little or no probability of **Musa textilis** becoming an established crop on this side of India. On the western coast, in some localities where irrigation and manuring are less urgently needed, the cost of producing the plants will be much less; and there some experiments should be tried.' He further states that 'there is no probability of really clean fibre being turned out by the ordinary process, at a cost less than 6 annas per pound, a cost far in excess of the value of the fibre.'

The cultivation of **Musa textilis** in the Andaman Islands would appear, from the reports furnished by the local authorities and by the Agri.-Horticultural Society of India, to have been eminently successful, the methods of separating the fibre being alone defective, since the article could only be produced at a price far above its mercantile value. How far this will prove insuperable must be left for further experiments, with improved machinery, to reveal, but it would seem desirable to guard against too high expectations as to any part of India being ever likely to become a formidable rival to the Philippine Islands in meeting the present comparatively limited demand for the fibre."

In July 1888 another letter was received by the Government of India from the Superintendent, Port Blair and Nicobars, stating that 1,415lb of **Musa textilis** fibre had been manufactured, and was ready for export, from which it may be presumed that the plant continues to thrive, and holds out good prospects of commercial success.

MUSK. 851

True musk, the dry, inspissated secretion of the preputial follicles of the Musk Deer, has already been fully described in the article on **Moschus mochiferus** under **Deer**, in Vol. III., 58. In continuation of that article a brief account of the medicinal and chemical properties of the substance may be given in this place, together with a few remarks regarding musk substitutes.

Musk and Musk Substitutes. 852

The term "musk" is, in common usage, applied in compound names to a number of products of both animals and vegetables characterised by the peculiar scent of the true perfume. Amongst these the chief Indian musk-scented animal is the so-called musk-rat, in reality a shrew (see Rats, Mice, Marmots, &c.), but its odourous secretion is not utilized.

Amongst vegetable musks may be mentioned the musk-plant proper (**Mimulus moschatus**) common in window-culture; **Ferula Sumbul,** see Vol. III., 339; and **Hibiscus Abelmoschus,** see Vol. IV., 229. The last mentioned is the only musk plant of commercial value. The reader is referred to the account in Vol. IV. for a description of the utilization of its seeds in perfumery. Despite the large number of products capable of affording more or less of a musk-like odour, the musk deer remains the only important commercial source of the perfume.

Perfume.—Musk is remarkable for the power, permanency, and subtility of its odour, everything in its vicinity soon becoming affected by it and retaining the scent for a long time. It has long been highly valued

PERFUME 853

PERFUME. in perfumery, and though now little used alone, is very largely employed to give permanence and strength to other odours. **Piesse** writes:—"It is a fashion of the present day for people to say 'that they do not like musk,' but, nevertheless, from great experience in one of the largest manufacturing perfumatories in Europe, I am of opinion that the public taste for musk is as great as any perfumer desires. Those substances containing it always take the preference in ready sale—so long as the vendor takes care to assure his customer 'that there is no musk in it.'" (*Perfumery, 258*). Perfumers use the scent principally for imparting an odour to soap, sachet-powder, and in mixing liquid perfumery. The alkaline reaction of soap is said to be favourable to the development of its odoriferous principle. Its fragrance is much affected and even completely destroyed by some other bodies, such as camphor, valerian, bitter almonds, and powdered ergot.

MEDICINE. 854 **Medicine.**—Musk has long been known and valued in Hindu medicine. In the *Bhávaprákasa* three varieties are described, named *rámrupa, nepála*, and *káshmíra*. The first is described as black and superior to the others, and probably consisted of China or Tibet musk imported *viá* Kamrup. That of Nepal is described as of bluish colour and intermediate quality, while the Kashmír musk was inferior. The drug was regarded by Sanskrit physicians as stimulant and aphrodisiac, and was employed in low fevers, chronic cough, general debility and impotence (*U. C. Dutt, Mat. Med. Hind., 279*).

In European medicine, musk is regarded as a powerful stimulant and antispasmodic, and is chiefly prescribed in the advanced stages of typhus, typhoid, and other diseases of an asthenic type. It has also been found useful in spasmodic asthma, laryngismus stridulus, whooping cough, epilepsy, and chorea, &c.

MUSSÆNDA, *Linn.; Gen. Pl., II., 64.*

[*t. 124*; RUBIACEÆ.

855 **Mussænda frondosa,** *Linn.; Fl. Br. Ind., III., 89; Wight, Ill.,*

Syn.—M. FLAVESCENS, *and* M. DORINIA, *Ham.* M. FORMOSA, *Linn.* M. VILLOSA, *Wall.*; M. CORYMBOSA, *Roxb.*; M. ZEYLANICA, *Burm.*

Var. α, zeylanica.

Var. β, ingrata, M. INGRATA, *Wall.* ?

Var. γ, laxa.

Vern.—*Bedina*, HIND.; *Asari*, NEPAL; *Tumberh*, LEPCHA; *Babina*, N.-W. P.; *Bhúta-kesa, lándachúta, bebana*, BOMB.; *Sarwadh, bhurtkasi, churtkasi, shivardole*, MAR.; *Vella-ellay*, TAM.; *Belíla*, MALAY.; *Maa-senda*, SING.

References.—*Kurz, For. Fl. Burm., II., 58; Dalz. & Gibs., Bomb. Fl., 111; Dymock, Mat. Med. W. Ind., 2nd Ed., 413; Atkinson, Ec. Prod., N.-W.-P., 91, 95; Lisboa, U. Pl. Bomb., 162; Birdwood, Bomb. Pr., 45, 162; Gazetteers:—Bombay, XV., Pt. I., 73; Mysore and Coorg, I., 70; Ind. Forester, XIV., 298.*

Habitat.—A handsome shrub, with yellow flowers and a large white calycine leaf, found in the North-East Himálaya, Bengal, South India, and Burma; often cultivated in gardens.

MEDICINE. Root. 856 **Medicine.**—The ROOT is used medicinally. **Dymock** states that in the Konkan half a *tolá* is given with cow's urine in white leprosy (*pándú-rog*).

Leaves. 857 In jaundice 2 *tolás* of the white calycine LEAVES are given in milk.

FOOD. Leaves. 858 **Food.**—**Atkinson** writes that the LEAVES are eaten as a pot-herb in the North-West Provinces, and **Lisboa** remarks that the white leaf of the calyx is eaten as a vegetable in Bombay.

DOMESTIC. Leaves. 859 **Domestic.**—In Southern India the LEAVES are considered a charm against demons.

860 **Mustard.**—Three species of **Brassica** yield Mustard and Mustard Oil, and are also grown for the well-known condiment which is prepared from

the seed. See **Brassica alba, B. juncea,** and **B. nigra,** Vol. I., 521, 528, and 530.

MYLABRIS.

Mylabris cichorii, *Fabr.; Stephenson, Med. Zool., t. 26, f. 5;* 861
M. indica, *Fressl.* [INSECTA. 862
M. pustulata. 863
M. punctum. 864

THE TELINI FLY, SUBSTITUTES FOR CANTHARIDES.

Vern.—*Télni, télni-makkhi,* HIND.; *Bad-bó ki-yirangi, zirangi,* DEC.; *Pinsttarin-i,* TAM.; *Blishtering-igelu,* TEL.

References.—*Pharm. Ind., 277; Moodeen Sheriff, Supp. Pharm. Ind., 179; Corresp. on New Pharmacop., 225, 230.*

These flies, together with **M. melaneura, M. humeralis, M. proxima, M. orientalis, Lytta assamensis, L. gigas, L. violacea, Epicauta nipalensis,** and **Meloe trianthema,** have all been described at different times as substitutes for cantharides, but since the best of all, **M. cichorii,** exists plentifully in most parts of India, the necessity of increasing the number of these insect vesicants is lessened. **M. cichorii,** common throughout India, is also to be found in Southern Europe, Egypt, and China.

MEDICINE. 865 **Medicine.**—This and the other species enumerated above have been highly extolled as substitutes for cantharides: **M. cichorii** was first brought to notice in 1809, and has since been highly recommended by **Drs. Burt, Hunter, Fleming,** and **Bidie.** The action of these insects is the same as that of the well-known cantharides **(Cantharis vesicatoria)** for which they afford a very efficient substitute as a vesicant. According to the *Pharmacopœia of India,* "They should not be substituted as an internal remedy for the Tincture of Cantharides, since the strength and operation of the latter is well ascertained, which is not the case with our article." **M. cichorii** is regarded as a more powerful blistering agent than European Cantharides, and is regularly employed as a substitute in the Government Medical Store Depôts.

SPECIAL OPINIONS.—§ "A very good substitute for cantharides" (*Surgeon-Major C. R. G. Parker, Pallaveram*). "Is quite equal to cantharides" (*Surgeon-Major E. Sanders, Chittagong*). "As good as cantharides, perhaps stronger" (*Bolly Chund Sen, Teacher of Medicine*). "Regularly used in the Government Medical Depôts as a substitute for cantharides and found very efficient" (*Brigade-Surgeon G. A. Watson, Allahabad*). "Very good vesicant" (*Assistant Surgeon Nehal Sing, Saharunpore*).

MYRIACTIS, *Less.; Gen. Pl., II., 262.*

866 A small genus of erect annuals, which belongs to the Natural Order COMPOSITÆ, and comprises three Indian species. Only one of these, **M. Wallichii,** *Less.* (*Fl. Br. Ind., III., 247*), is of economic interest. It is known as *baberi* in Assam, where it is grown in homestead lands for the sake of the LEAVES and tender SHOOTS which are used as food, either boiled in *khár* water, or fried in oil.

FOOD. Leaves. 867 Shoots. 868

MYRICA, *Linn.; Gen. Pl., III., 400.*

Myrica Nagi, *Thunb.; Fl. Br. Ind., V., 597; Wight, Ic., t. 764,* 869

THE BOX-MYRTLE, YANGMÆ OF CHINA. [*765;* MYRICACEÆ.

Syn.—M. SAPIDA, *Wall.;* M. INTEGRIFOLIA, *Roxb.;* M. MISSIONIS, *Wall.;* M. FARQUHARIANA, *Wall.;* M. RUBRA, *Sieb. & Zucc.;* NAGEIA JAPONICA, *Gærtn.;* ? M. JAVANICA, *Blume;* ? M. LONGIFOLIA, *and* ? M. LOBII, *Teysm. & Binnend.*

Vern.—*Káiphal, káephal,* HIND., DEC., SIND; *Káiphal, káyophul, satsarila,* BENG.; *Dingsolir,* KHASIA; *Kobusi,* NEPAL; *Kaphal, káiphal, karphal,* N.-W. P.; *Káiphal, kahi kahela, kaphal,* PB.; *Káyaphala kaiphal,* BOMB.; *Káya-phala,* MAR.; *Kariphal,* GUZ.; *Marudampattai,*

TAM.; *Kaidaryamu*, TEL.; *Marutamtoli*, MALAY.; *Katphala, kaidaryama*, SANS.; *Azúri, audul, quandól*, ARAB.; *Dárshíshaán*, PERS.

References.—*Brandis, For. Fl., 495; Kurz, For. Fl. Burm., II., 475; Gamble, Man. Timb., 391; Stewart, Pb. Pl., 202; Sir W. Elliot, Fl. Andhr., 76; Pharm. Ind., 217; O'Shaughnessy, Beng. Dispens., 611; Irvine, Mat. Med. Patna, 44, 98; Moodeen Sheriff, Supp. Pharm. Ind., 179; U. C. Dutt, Mat. Med. Hindus, 234, 304; Dymock, Mat. Med. W. Ind., 2nd Ed., 749; Baden-Powell, Pb. Pr., 385, 586; Atkinson, Him. Dist. (Vol. X., N.-W. P. Gaz.), 317, 742, 779; Forbes Watson, Econ. Prod. N.-W. Prov., Pt. III. (Dyes and Tans), 82; Pt. V. (Vegetables, Spices, and Fruits), 44, 88; Gazetteers, Panjáb, Simla, 12; Indian Forester, III., 163; IV., 241; VI., 25; VIII., 127, 405.*

Habitat.—An evergreen diœcious tree with an aromatic odour, met with in the Subtropical Himálaya from the Ravi eastwards, at altitudes from 3,000 to 6,000 feet; also found in the Khásia mountains, Sylhet, and southwards to Singapore; distributed to the Malay Islands, China, and Japan.

DYE & TAN. Bark. 870 **Dye and Tan.**—The BARK is occasionally used as a tanning agent for fancy leather work. In 1875-76, 304 cwt. of bark was exported from the North-West Provinces for tanning and medicinal purposes, of which 235 cwt. came from Kumáon. In more recent years about 50 tons is said to be the average export. **W. Coldstream, Esq., C.S.**, informs the editor in a private note that the bark is used in Sirmur (Simla District) for dyeing pink—*piazi*.

MEDICINE. Bark. 871 **Medicine.**—The BARK is described by writers on Sanskrit medicine as heating, stimulant, and useful in diseases supposed to be caused by deranged phlegm, such as catarrhal fever, cough, and affections of the throat. It enters into the composition of numerous formulæ for these diseases, in which it is combined with other stimulants and alteratives. The powdered bark is occasionally used as a snuff in catarrh with headache (*U. C. Dutt*). It is also used by Hindús at the present day, mixed with ginger, as a rubefacient application in cholera, &c., and according to **Irvine**, *kaiphal* and ginger mixed, is the best substance that can be employed for this purpose. **Dymock** writes, "Muhammadan writers tell us that the bark is resolvent, astringent, carminative and tonic; that it cures catarrh and headaches; with cinnamon they prescribe it for chronic cough, fever, piles, &c. Compounded with vinegar, it strengthens the gums and cures toothache; an oil prepared from it is dropped into the ears in earache. A decoction is a valuable remedy in asthma, diarrhœa, and diuresis; powdered or in the form of lotion the bark is applied to putrid sores: pessaries

Oil. 872 made of it promote uterine action. The usual dose for internal administration is about 60 grains. *Dahu-el-kandúl*, an OIL prepared from the

Flowers. 873 FLOWERS, is said to have much the same properties as the bark."

No critical chemical examination of the bark has been made, but, according to **Dymock**, the watery extract evaporated to dryness yields a brittle, highly astringent kino-like substance. In Bombay *kaiphal* fetches from one to two rupees per maund of 41℔, and is entirely imported from Northern India.

FOOD. Fruit. 874 **Food.**—The FRUIT, which is about the size of a cherry, has not much flesh, but is pleasantly sour-sweet and is eaten by Natives.

TIMBER. 875 **Structure of the Wood.**—Purplish-grey, hard, close-grained, apt to warp, weight 48℔ per cubic foot.

DOMESTIC. Bark. 876 **Domestic, &c.**—The BARK is said to be employed in the Khásia Hills to poison fish. Twigs. 877 The TWIGS are used in the Panjáb as tooth-brushes by persons suffering from overdoses of mercury.

MYRICARIA, *Desv.; Gen. Pl., I., 161.*

878 **Myricaria elegans,** *Royle; Fl. Br Ind., I., 250;* TAMARISCINEÆ.

Syn.—M. GERMANICA, *var.* LONGIFOLIA, *H. f. & T.*

Vern.—*Hámbú, úmbú*, PB.

References.—*Stewart, Pb. Pl., 91; Royle, Ill. Him. Bot., 214; Gazetteer, N.-W. P., X., 306.*

Habitat.—A bush found on the Western Himálaya from Garhwál to Ladak and in Western Tibet, at altitudes from 6,000 to 15,000 feet.

Medicine.—Aitchison states that the LEAVES are used as an application to bruises, &c., in Lahoul. MEDICINE. Leaves. 879

Fodder.—The TWIGS are browsed by sheep and goats in Ladak (*Stewart*). FODDER. Twigs. 880

Myricaria germanica, *Desv.; Fl. Br. Ind., I., 250.* 881

Syn.—M. BRACTEATA, *Royle*; M. HOFFMEISTERI, *Klotz.*; TAMARIX GERMANICA, *Linn.*

Var. prostrata, *Benth. & Hook, f.* (Sp.).

Vern.—*Bis, shálakát, hambúkh, kathí, humbú, úmbú, joaraktse,* PB.

References.—*Stewart, Pb. Pl., 91; Brandis, For. Fl., 23; Gamble, Man. Timb., 20; Royle, Ill. Him. Bot., 214, t. 44; Gazetteers:—N.-W. P., IV., lxviii.; X., 306; Indian Forester, IV., 233; XI., 2.*

Habitat.—A shrub of the Temperate and Alpine Himálaya, from Sikkim to Kumaon, at altitudes of 10,000 to 14,000 feet.

Fodder.—The BRANCHES are employed as fodder for sheep and goats. FODDER. Branches. 882

Structure of the Wood.—Hard, white, used for fuel. TIMBER. 883

MYRISTICA, *Linn.; Gen. Pl., III., 136.* 884

A genus of evergreen aromatic trees which comprises about 80 species, chiefly natives of Tropical East Asia, Malaya, and America; a few are African and one Australian. Of these 30 are met with in India and the Malay Peninsula.

Myristica fragrans, *Houtt.; Fl. Br. Ind., V., 102;* MYRISTICEÆ. 885

THE NUTMEG, MACE.

Syn.—M. OFFICINALIS, *Linn. f.*; M. MOSCHATA, *Thunb.*; M. AROMATICA, *Lamk.*

Vern.—*Jayphal* or *jáephal,* the nut=*jáé-phal,* the aril or mace=*jápatr,* HIND.; *Jayphal,* the nut=*jáe-phul,* the aril=*jótri,* BENG.; the nut=*Jáiphal,* the aril=*jauntari,* PB.; the nut=*Jáphal,* the aril=*joutri,* DEC.; the nut=*Jáiphal,* the aril=*jawantrí, jápatrí,* BOMB.; the nut=*Jáiphala,* the aril=*jáyapatri,* MAR.; the nut=*Jáyephal,* the aril=*jávantari, jápatri,* GUZ.; the nut=*Jádikkáy,* the aril=*jádi-pattiri,* TAM.; *Zevangam, jáji kayu,* the nut=*jájikaya, játi-phalamu,* the aril=*jápatri,* TEL.; the nut=*Jájikáyi,* the aril=*jápatri,* KAN.; the nut=*Játikká,* the aril=*játipattiri,* MALAY.; the nut=*Zádiphu,* the aril=*zádiphu-apóén,* BURM.; the nut=*Jádi-ká, sádika,* the aril=*vasávasi, vaduváshu,* SING.; *Játiphala,* the nut=*jáji-phalam,* the aril=*jajipatri,* SANS.; the nut=*Jouzbuvá, jouzuttib, jauz-ut-trib,* the aril=*basbásah, basbás,* ARAB.; the nut=*Jouzbóyah,* the aril=*bazbáz,* PERS.

References.—*Roxb., Fl. Ind., Ed. C.B.C., 742; Kurz, For. Fl. Burm., II., 282; Gamble, Man. Timb., 313; Stewart, Bot. Tour in Hazara, &c., 9; DC., Orig. Cult. Pl., 419; Mason, Burma and Its People, 499, 740; Sir W. Elliot, Fl. Andhr., 72, 73, 106; Pharm. Ind., 189; Flück & Hanb., Pharmacog., 502; Fleming, Med. Pl. & Drugs (Asiatic Reser., XI.), 188; Ainslie, Mat. Ind., I., 201, 259, 622; O'Shaughnessy, Beng. Dispens., 534; Medical Topog., 139; Moodeen Sheriff, Supp. Pharm. Ind., 180; U. C. Dutt, Mat. Med. Hindus, 224, 300; Murray, Pl. & Drugs, Sind., 37; Bent. & Trim. Med. Pl., 218; Dymock, Mat. Med. W. Ind., 2nd Ed., 661; Year-Book Pharm., 1879, 467; Birdwood, Bomb. Prod., 228; Baden Powell, Pb. Pr., 302, 373; Lisboa, Useful Pl. Bomb. (Vol. XXV., Bomb. Gaz.), 170; Royle, Prod. Res., 73, 74, 200; Hammel, Piesse, Perfumery, 153, 164; Simmonds, Tropical Agriculture, 484; Christy, New Com. Pl., VIII., 26; Linschoten, Voyage to East Indies (Ed. Burnell, Tiele & Yule), Vol. II., 52, 84, 86; Andaman Islands, Port Blair, 1870-71, 45; Indian Forester, VIII., 187; Smith, Ec. Dic., 289.*

Habitat.—A handsome, bushy evergreen tree, with dark shining leaves, which grows in its native islands to a height of 40 to 50 feet. It is found wild in the Moluccas principally in the small volcanic group of Banda.

History.—Towards the end of last century, the English introduced the tree into Bencoolen and Prince Edward's Island, and later into Malacca, HISTORY. 886

MYRISTICA fragrans.

Nutmeg and Mace Oils.

HISTORY.

the Islands of Singapore and Penang, as well as Brazil and the West Indies. It was also cultivated with some success by **Dr. Roxburgh** at the Botanic Gardens, Calcutta. In Bencoolen, nutmeg cultivation is still carried on, and the greater part of the trade supply is now derived from that place and Banda. Extensive plantations were formed in Penang and Singapore, in the early part of this century, and with so much success that for some years the exports from the latter island exceeded those from Banda. But in 1860 the trees were visited by a destructive blight, which the cultivators were powerless to arrest, and which ultimately led to the extermination of the trees and the ruin of the planters. Of late years the Chinese have been making hopeful efforts to introduce the plant, and to compete with the Spice Islands. It has also been established in India on the Nilghiri Hills, but with little commercial success, and though cultivated to a small extent in many parts of India and Ceylon, the future field for British nutmeg plantations seems to be in Jamaica. Notwithstanding this fact, mention is made of small amounts of nutmegs of Indian produce being exported to Europe.

OIL. 887 Nuts. 888

Oil.—The NUT yields an essential and a fixed oil. The former is white, acrid, pungent, and smells powerfully of nutmeg; the latter, "nutmeg butter," is yellowish in colour, and solid. It is extracted from refuse nuts by reducing them to powder, heating them in a water bath, and, while hot, expressing the oil. This solidifies on cooling into the mottled, orange-brown butter. It has a pleasant odour, and a fatty, aromatic flavour.

Mace. 889

Both MACE and the Nutmeg yield an otto or essential oil upon aqueous distillation. That from the former is yellow, with a strong odour of the mace and an aromatic flavour; that from the latter is nearly colourless, or white, with a strong odour and taste of the nut. Both are extensively used for perfuming soaps. The extent to which they are so utilized is at once seen by the enormous consumption of Nutmegs in Great Britain. The actual consumption is variously stated. **Piesse** (*Art of Perfumery*) writes that the "produce of Nutmegs in the Moluccas has been reckoned at from 600,000 to 700,000℔ per annum, of which half goes to Europe, and about one-fourth that quantity of Mace. The annual consumption of Nutmegs in Britain is said to be 140,000℔." **Simmonds**, on the other hand, writes that, during the five years ending 1870, the average was 592,736℔, valued at £37,756.

The Otto of Nutmegs enters largely into the composition of many articles of English perfumery, but especially into that of Frangipani. When used sparingly it combines pleasantly with lavender, santal, and bergamot. Formerly soap, known as Banda soap, was prepared from the fatty oil or Butter of Nutmegs. The trade in this article has died out, being replaced by ordinary soap perfumed with the otto of the Nutmeg.

Chemistry. 890

CHEMICAL COMPOSITION.—The fixed oil makes up about one-fourth of the weight of Nutmeg, the essential oil from three to eight per cent. The latter consists almost entirely of a hydrocarbon, having, according to **Cloez**, the composition $C_{10} H_{16}$. **Gladstone**, in 1872, confirmed the analysis of **Cloez**, and assigned to this hydrocarbon the name *myristicene*. The same chemist found in the crude oil, an oxygenated oil, *myristicol*, of very difficult purification, and possibly subject to change during the process of rectifying. It possesses the characteristic odour of the nut, and is isomeric with *carvol*. From the facts recorded by **Gmelin** it would appear that the essential oil sometimes deposits a stearoptene called *myristicin*. The authors of the *Pharmacographia* state that they have never obtained such a deposit, and imply that possibly this substance may have been merely *myristic acid*, a substance plentiful in the fixed oil.

The fixed oil or Nutmeg Butter contains the volatile bodies already described, to the extent of about six per cent., besides several fatty bodies, the most important of which is *myristin*, $C_3 H_5 (OC_{14} H_{27} O)_3$. This substance by saponification yields *myristic acid*, $C_{14} H_{28} O_2$, and glycerine.

HISTORY.

Another fat has also been observed, readily soluble in spirit, and accompanied by a reddish colouring matter, but its composition has not as yet been investigated.

Mace contains about 8 per cent. of a volatile oil, consisting for the most part of *macene* $C_{10} H_{16}$, similar to oil of turpentine, but possessing the smell and taste of nutmegs. In addition it contains a thick aromatic balsam (*Flückiger & Hanbury, Pharmacographia*).

MEDICINE. Nutmeg. 891

Medicine.—The NUTMEG was known at least to the later Sanskrit writers by the name of *játiphala*, but it is doubtful at what time it first became known in India. In the sixth century A.D. records exist of the nutmeg having formed an article of Arabian import from the East, and export to Europe, and the medicinal properties of the nut were probably then known to Arab and Persian physicians. **Ainslie** informs us that "it is considered by the natives of India as one of their most valuable medicines in dyspeptic complaints, and in all cases requiring cardiacs and corroborants; they likewise prescribe it to such puny children as appear to suffer much in weaning." Similarly MACE "is a favourite medicine of the Hindú doctors, who prescribe it in the low stages of fever, in consumptive complaints, and humoral asthma; and, also, when mixed with aromatics, in wasting and long-continued bowel-complaints, in doses of from grs. 8 to grs. 12, and sometimes to as much as dr. $\frac{1}{2}$; but they generally administer it cautiously from having ascertained that an overdose is apt to produce a dangerous stupor and intoxication; the same effect is ascribed to the Nutmeg by **Bontius**." According to the same author, the Arabians place Nutmeg amongst their "hepatica," and "tonica," and mace amongst their "aphrodisiaca," and "carminativa." **Dymock** writes that the Muhammadans consider Nutmegs and mace to be stimulating, intoxicating, digestive, tonic, and aphrodisiac; useful, especially when roasted, in choleraic diarrhœa, also in obstructions of the liver and spleen. According to **Rumphius**, the JUICE of the GREEN FRUIT, mixed with water, is used in Amboyna as a wash in apthous affections. The VOLATILE OIL is occasionally employed as an external stimulant, and the FIXED OIL, or Nutmeg Butter, is also considered stimulant and is considerably used by native practitioners in preparing rubefacient liniments. **Dymock** states that a paste made with the nuts is used as an external application in headache, palsy, &c., and is applied round the eyes to "strengthen the sight."

Mace. 892

Juice. 893

Green-Fruit. 894

Volatile Oil. 895

Fixed Oil. 896

The utilization of Nutmeg in European medicine is too well known to require any description; suffice it to say that it is regarded as stimulant, carminative, and in large doses narcotic, and is used in doses of ten to twenty grains in atonic diarrhœa, flatulence, colic, and some forms of dyspepsia. An infusion has been recommended as of great service in quenching the thirst of cholera patients. It is, however, chiefly employed as an adjunct and as a condiment. It is officinal in the Pharmacopœia of India, three preparations being described, *viz.*, the volatile oil, the spirit, and the expressed oil. The last mentioned is recommended as a useful application in cases of rheumatism, paralysis, and sprains.

SPECIAL OPINIONS.—"Is useful in cases of dysentery and is thus given by native doctors. A small excavation by means of a scoop is made in the kernel, a quantity of opium suited to the age of the patient is inserted into the pit and covered with the scrapings. The kernel is now thinly covered with some flour paste, and the whole roasted in hot ashes. It is then ground into a paste and the mass is given in divided doses at intervals. It is also tonic and anti-rheumatic. The powdered kernel with **Casuarina** fruit is used in toothache" (*Surgeon-Major D. R. Thomson, M.D., C.I.E., Madras*). "Nutmegs rubbed up with water form a useful application for parotitis and other glandular swellings" (*W. Forsyth, F.R.C.S., Edin. and U. C. Mukerji, M.B., C.M., Dinajpore*). "When made into an electuary with sugar and *ghí*, it is a popular domestic remedy for dysentery among children" (*Native Surgeon T. R.*

Moodelli ar, Chingleput, Madras). "Nutmeg, rubbed with oil on a piece of stone, is applied with friction in cases of cholera, and in cases when the internal organs appear to be congested" (*Civil Surgeon J. H. Thornton, B.A., M.B., Monghyr*).

FOOD. Nutmeg. 897 Mace. 898 **Food.**—Both NUTMEG and MACE are largely used by the Natives as a condiment. The widespread utilization of both in European cookery is too well known to require comment.

899 **Myristica Irya,** *Gærtn.; Fl. Br. Ind., V., 109.*

Syn.—M. JAVANICA, *Blume.*; M. SPHÆROCARPA, *Wall.*; M. EXALTATA, *Wall.*

Vern.—*Maloh*, BURM.; *Ereya*, SING.; *Mutwindá, choglum*, AND.

References.—*Kurz, For. Fl. Burm., II., 282; Gamble, Man. Timb., 314; Beddome, For. Man., 176; Gazetteer, Bombay, xviii., 46.*

Habitat.—A moderate-sized evergreen tree, native of Burma, the Andaman Islands, Malacca, and Ceylon.

TIMBER. 900 **Structure of the Wood.**—Dark olive-grey, hard, close-grained; weight 52lb per cubic foot. A handsome wood, which seasons well, takes a good polish, and is worthy of attention (*Gamble*).

901 **M. longifolia,** *Wall.; Fl. Br. Ind., V., 110.*

Syn.—M. LONGIFOLIA, *in part, Hook. f. & Thoms.*; ? M. LINIFOLIA, *Roxb*, **Var. erratica,** M. ERRATICA, *Hook. f. & Thoms.*; M. CORTICOSA, *Hook. f. & Thoms.*

Vern.—*Zadeitbo*, var.=*thit-thon*, BURM.

References.—*Roxb., Fl. Ind., Ed. C.B.C., 744; Kurz, For. Fl. Burm., II., 283; Gazetteer, Burma, I., 135.*

Habitat.—An evergreen tree of the Sikkim Himálaya, Assam, Sylhet, the Khásia Hills, Chittagong, Pegu, Martaban, and Tenasserim.

RESIN. 902 **Resin.**—The tree exudes a red resin (*Kurz*).

TIMBER. 903 **Structure of the Wood.**—Whitish, turning pale brown, rather heavy, fibrous, soon attacked by xylophages.

904 **M. malabarica,** *Lamk.; Fl. Br. Ind., V., 103.*

Syn.—?MYRISTICA TOMENTOSA, *Grah.*; M. DACTYLOIDES, *Wall.*; M. NOTHA, *Wall.*

Vern.—The nut=*Jangli jaiphal, ranjaiphal, kaiphal*, the aril=*rámpatri* BOMB.; *Kanagi*, the nut=*pindi-kai*, KAN.

References.—*Beddome, Fl. Sylv., t. 269; Gamble, Man. Timb., 314; Dalz. & Gibs., Bomb. Fl., 4; Dymock, Mat. Med. W. Ind., 547; Lisboa, U. Pl. Bomb., 110, 170, 214; Kew Off. Guide to the Mus. of Ec. Bot., 109; Gaz., Bomb., XV., 441.*

Habitat.—A tall, evergreen tree of the Konkan, Kánara, and North Malabar.

OIL. Seed. 905 **Oil.**—The SEED, when bruised and subjected to boiling, yields a quantity of yellowish concrete oil, which is used medicinally, and for purposes of illumination.

MEDICINE. Fruit. 906 **Medicine.**—As early as the time of **Rheede**, the FRUIT of this species appears to have been used for the purpose of adulterating the nutmegs and mace of **M. fragrans.** Of late years this practice has been revived, and according to **Dymock** *rámpatri*, once very cheap, is now worth R10 per maund, while the nuts fetch R2 per maund of $37\frac{1}{2}$lb. The former is exported to Europe (chiefly to Germany) as an adulterant for true mace.

Oil. 907 The plant occupies a place in the secondary list of the Pharmacopœia, where it is stated that the OIL has been represented as a most efficacious application to indolent and ill-conditioned ulcers, allaying pain, cleansing the surface, and establishing healthy action. For this purpose it requires to be melted down with a small quantity of any bland oil. **Waring** suggests that it may be found serviceable as an embrocation in rheumatism.

Seeds. 908 According to **Dymock**, the SEEDS in the form of a *lep* are used as an external application in Bombay.

SPECIAL OPINIONS.—§ "The seed, roasted and ground into powder, is given three times a day as an astringent for diarrhœa and dysentery. It has also narcotic properties" (*Surgeon-Major L. Beech, Cocanada*). "The expressed oil is used as a stimulating application to indolent ulcers; the pounded seeds are also used as a stimulating plaster" (*W. Dymock, Bombay*). "The expressed oil is stimulant to indolent and ill-conditioned ulcers, detergent and anti-rheumatic" (*Surgeon-Major D. R. Thomson, M.D., C.I.E., Madras*). "The fruit is much larger than the true nutmeg, but has little aroma. A concrete oil is prepared from it and used in rheumatism. The *arillus rámpatú* is considered to be a nervine tonic, and is used in stopping vomiting, also as a substitute for the true mace from which it differs in being destitute of aroma" (*C. T. Peters, M.B., Zandra, South Afghánistán*). MEDICINE.

Structure of the Wood.—Reddish-grey, moderately hard, weight 32 lb per cubic foot. Used for building. TIMBER. 909

Myrobalan, Chebulic, see **Terminalia Chebula,** *Rets.*; Vol. VI., Pt. II

Myrobalan, Emblic, see **Phyllanthus Emblica,** *Linn.*; Vol. VI., Pt. I.

Myrrh, see **Balsamodendron Myrrha,** *Nees.*; Vol. I., 367.

MYRSINE, *Linn.*; *Gen. Pl., II., 642.*

Myrsine africana, *Linn.*; *Fl. Br. Ind., III., 511*; MYRSINEÆ. 910

Syn.—M. GLABRA *and* SCABRA, *Gærtn.*; M. ROTUNDIFOLIA, *Lumk.*; M. BIFARIA, *Wall.*; M. POTAMA, *Don.*

Vern.—*Guvaini, pahari cha, chúpra,* N.-W. P.; *Bebrang, kakhum, kokkuri, karuk, gúgul, jutru, chachri, prátshú, branchu, bránti, khúshin, pápri, bandáru, binsin, atúlgán, shamshád, vávarang, khukan,* fruit—*bebrang,* PB.; *Baibarang, baring,* ARAB.

References.—*Brandis, For. Fl., 286; Gamble, Man. Timb., 239; Stewart Pb. Pl., 135; Baden Powell, Pb. Pr., 369; Atkinson, Him. Dist., 313, 743; Gazetteers:—Panjáb, Hazára Dist., 133; N.-W. P., IV., lxxiii.; Settlement Rep., Hazára Dist., 94.*

Habitat.—A small evergreen shrub found in Afghánistán, the Salt Range, and the outer Himálaya from Kashmír to Nepál, at altitudes from 1,000 to 8,500 feet.

Medicine.—The FRUIT, known as *bebrang* (a name also applied to that of **Samara Ribes**), is sold for medicine in all the bazárs of the Panjáb. It is said to be a powerful cathartic vermifuge, and is also used in dropsy and colic. The GUM is considered a "warm remedy" and is considerably employed as a remedy for dysmenorrhœa. MEDICINE. Fruit. 911 Gum. 912

SPECIAL OPINION.—§ "Is said to be alterative, and is used with Liquorice for strengthening the body and preventing the effects of age" (*Civil Surgeon J. Anderson, M.B., Bijnor, N.-W.P.*).

Food.—In the Settlement Report of the Hazára District the FRUIT is said to be commonly eaten by the poorer *zemindars.* FOOD. Fruit. 913

Structure of the Wood.—White, moderately hard, weight 49 lb per cubic foot. TIMBER. 914

Agricultural.—The PLANT has been recommended as suitable for hedges. AGRICULTURAL. Plant. 915

M. capitellata, *Wall.*; *Fl. Br. Ind., III., 512*; *Wight, Ic., t. 1211.* 916

Syn.—M. EXCELSA, *Don*; M. LUCIDA, *Wall.*

Var. lanceolata, *Wall.* (*sp.*),—M. WIGHTIANA, *Wall.*

Var. lepidocarpon, *Wight* (*sp.*).

Var. avenis, *A. DC.* (*sp.*),—M. PORTERIANA, *Wall.*; M. UMBELLULATA, *A. DC.*

References.—*Brandis, For. Fl., 286; Kurz in Jour. As. Soc., 1877, pt. ii, 221; Beddome, Fl. Sylv., t. 234; Gamble, Man. Timb., 239.*

Habitat.—A shrub or small tree of Nepál, Bhután, Assam, and the Khásia Mountains; **var. lanceolata** is also common on the hills of South India and Ceylon.

FOOD. Fruit. 917 **Food.**—The FRUIT is eaten in Southern India (*Beddome*).

TIMBER. 918 **Structure of the Wood.—Var. lanceolata** has a moderately hard, pink-coloured wood, said by Beddome to be durable and used by the natives.

919 **Myrsine semiserrata,** *Wall.; Fl. Br. Ind., III., 511.*

Syn.—M. ACUMINATA, *Royle*; M. SESSILIS, *Don.*

Vern.—*Parwana, kúngkúng, gogsa, bamora, gaunta,* HIND.; *Bilsi, beresi, kalikatha, bilauni,* NEPAL; *Tungcheong,* LEPCHA; *Chupra,* KUMAON.

References.—*Brandis, For. Fl., 285; Kurz, For. Fl. Burm., II., 105; Gamble, Man. Timb., 239; Atkinson, Him. Dist., 313; Royle, Ill. Him. Bot., 265; Ind. Forester, III., 183; XI., 273, 367, 369; XII., 551; XIV., 394.*

Habitat.—A shrub, small or middling sized tree, met with in the outer Himálaya from the Beás to Bhután, at altitudes from 3,000 to 9,000 feet, also on the Nattoung Hills of Martaban.

TIMBER. 920 **Structure of the Wood.**—Hard, red, weight 51lb per cubic foot. **Wallich** states that it is chocolate coloured, heavy, hard and handsome, and is used in Nepál for carpenter's work. It is, however, apt to split and is usually too small for anything but firewood (*Gamble*).

MYRTUS, *Linn.; Gen. Pl., I., 714, 1006.*

921 **Myrtus communis,** *Linn.; DC. Prod., III., 239;* MYRTACEÆ.

THE COMMON MYRTLE.

Vern.—*Viláyati mehndi, múrad,* HIND.; *Sutr-sowa. fruit=hab-úl-ás,* BENG.; *Viláyáti mehndi, múrad,* leaves=*múrad,* fruit=*hab-úl-ás, habhúl,* PB.; Fruit=*abhúlas,* SIND; *As, ásbiri, maurid, ismar, isferem,* fruit=*hab-úl-ás,* PERS.

References.—*Stewart, Pb. Pl., 93; O'Shaughnessy, Beng. Dispens., 333; Irvine, Mat. Med. Patna, 101; Medical Topog., 152; Sakharam Arjun, Cat. Bomb. Drugs, 56; Murray, Pl. & Drugs, Sind, 192; Dymock, Mat. Med. W. Ind., 2nd Ed., 335; Year-Book Pharm., 1874, 625; 1876, 13; 1879, 467; Birdwood, Bomb. Prod., 36; Baden-Powell, Pb. Pr., 349; Gazetteers:—N.-W. P., I., 81; Mysore and Coorg, I., 60; Agri.-Horti. Soc., Ind., IV., Sec. I., 119; Indian Forester, XI., 55; XII., 59.*

Habitat.—A shrub, indigenous in the area extending from the Mediterranean region to Afghánistán and Baluchistán, extensively cultivated in India.

DYE. Bark. 922 Leaves. 923 Berries. 924 **Dye.**—The BARK and LEAVES are used in tanning, the BERRIES for dyeing.

OIL. 925 Berries. 926 Leaves. 927 **Oil.**—A fixed oil is said to be obtained from the BERRIES, and according to **Baden Powell**, is believed to strengthen and promote the growth of the hair. In Europe an essential oil, distilled from the LEAVES, is largely employed in perfumery, under the name of *Eau d'Ange.* In the preparation of this oil the leaves, flowers, and fruit are all distilled together, about 5 oz. being obtained from 1 cwt. It is yellowish, or greenish yellow, and highly fragrant. No record exists of this essential oil being known to, or prepared by, the Natives, but **Irvine** states that the leaves and berries are employed for making scent in Patna.

MEDICINE. Leaves. 928 **Medicine.**—The LEAVES of the myrtle have long been known and valued in European medicine, aud have also been considered of importance by Muhammadan writers. Their virtues are too well known to require description, but it may be remarked that in Upper India they are considered useful in cerebral affections, especially epilepsy; also in dyspepsia, and diseases of the stomach and liver. A decoction is employed as a mouthwash in cases of aphthæ.
Fruit. 929 The FRUIT is carminative and is given in diarrhœa, dysentery, hæmorrhage, internal ulceration, and rheumatism.
Seeds. 930 The SEEDS, ground and mixed with antimony, are used to colour the eyelids.

(J. F. Duthie.)

NANNORHOPS, *H. Wendl.; Gen. Pl., III., 923.*

Nannorhops Ritchieana, *H Wendl.*; PALMÆ. 1

Syn.—CHAMÆROPS RITCHIEANA, *Griff.*

Vern.—*Masri*, HIND.; *Patha* (the fibre), *kilu, kaliun*, (Salt Range), PB.; *Msarái*, TRANS-INDUS; *Maisurrye*, PUSHTU; *Pfis, pesh, pease, fease, pfarra, pharra*, SIND & BALUCH.

References.—*Brandis, For. Fl., 547; Gamble, Man. Timb., 418; Stewart, Pb. Pl., 242; Aitchison, Cat. Pb. and Sind Pl., 142; Murray, Pl. and Drugs, Sind, 18; Royle, Fib. Pl., 92, 95; Balfour, Cyclop., 645.*

Habitat.—Usually a stemless, gregarious shrub, common on rocky ground in Sind, and in the Trans-Indus tracts of the Panjáb up to about 3,000 feet; also in Afghánistán and Báluchistán. Under favourable circumstances it develops a trunk which reaches to 14 feet in height. In the Saharanpur garden there are three fine specimens grown from seed obtained by **Dr. Jameson** from Kohat nearly 40 years ago.

Fibre.—The LEAVES are used for matting, fans, baskets, hats, sandals, and other articles. Rope is also made from the leaves and leaf-stalks. **Stewart** mentions that the rope, used for the construction of a bridge of boats across the Jhelum, during one season, when *munj* (**Saccharum**) was scarce, snapped at once under a strain which *munj* ropes could have resisted. FIBRE. Leaves. 2

Medicine.—**Bellew** states that the delicate young LEAVES, which have a sweet astringent taste, are in great repute for the treatment of diarrhœa and dysentery. They are also used as a purgative in veterinary practice. MEDICINE. Leaves. 3

Food.—The LEAF-BUDS are eaten as a vegetable. FOOD. Leaf-buds. 4

Domestic Uses.—The SEEDS are made into ROSARIES and exported from Báluchistán to Mecca. The stems, leaves, and petioles serve as fuel. The reddish-brown moss-like wool of the petioles, impregnated with saltpetre (or steeped in the juice of mulberry leaves according to **Bellew**), is used as TINDER for matchlocks. A rude kind of DRINKING CUP is made of the entire leaf-blade by tying together the tops of the segments (*Brandis*). DOMESTIC. Seeds. Rosaries. 5 Tinder. 6 Drinking Cup. 7

Naptha, see **Petroleum**, Vol. V.

NARAVELIA, *DC.; Gen. Pl., I., 4.*

Naravelia zeylanica, *DC.; Fl. Br. Ind., I., 7*; RANUNCULACEÆ. 8

Vern.—*Chagal-bati*, BENG.; *Sát jo yit*, BURM.; *Dayúpalú* or *narawella*, SING.

Habitat.—A scandent bush, found plentifully in the tropical Himálaya from East Nepál eastward to Bengal, Assam, &c., distributed to Ceylon. About Calcutta it is one of the most abundant of plants.

Fibre.—As with most of the CLEMATIDEÆ the STEMS are twisted into rough but useful ropes. FIBRE. Stems. 9

NARCISSUS, *Linn.; Gen. Pl., III., 718.*

Narcissus Tazetta, *Linn.*; AMARYLLIDACEÆ. 10

Vern.—*Nargis, irisa*, PB.

References.—*Stewart, Pb. Pl., 235; Aitchison, Cat. Pb. and Sind Pl., 148; O'Shaughnessy, Beng. Dispens., 657; Dymock, Mat. Med. W. Ind., 837; Royle, Ill. Him., 373.*

Habitat.—Indigenous in South Europe, North Africa, and Western Asia, extending to Persia. **Griffith** records it from Afghánistán, and **Mr. J. H. Lace** informs the writer that large quantities of the flowers are annually brought to Quetta from Kandahar for ornamental purposes. The flowers are gathered when in bud, and expand after being placed in water. This species, **Dr. Watt** informs the writer, is abundant on some of the hillsides below Simla where it flowers in November to February according to altitude, and has all the appearance of being wild. **Dr. Royle** mentions having found it in some apparently wild situations, which after enquiry he found to be the sites of old habitations and deserted gardens.

MEDICINE. Root. 11 **Medicine.**—The ROOT possesses emetic properties. It is also absorbent. As a perfume it is used to relieve headache.

SPECIAL OPINIONS.—§ "Imported into Bombay, dried and sliced, commonly sold in the bazár as a substitute for bitter hermodactyls" (*W. Dymock, Bombay*). "Purgative. It is poisonous" (*Surgeon-Major D. R. Thomson, M.D., C.I.E., Madras*).

(*G. Watt.*)

NARCOTICS.

12
Narcotics and Drugs.

HISTORY 13 HISTORY.—This term is applied medically to substances which are either anodyne in their action (that is, relieve pain) or soporific (produce sleep). The word narcotic is derived from *ναρκωτικός*, benumbing, or from *νάρκη*, numbness or torpor. Soporifics generally act also as anodynes, and various anodynes are antispasmodic. The list here given includes, however, not only all the narcotics proper, such as Opium, Indian-Hemp, &c., but also those of less pronounced properties such as Hops, Tea, Tobacco, &c. It has thus more a popular than a strictly therapeutic character. In fact, the term narcotics is here employed in its Indian fiscal acceptation as synonymous with the equally general term "Drugs." The enumeration below has thus been made to embrace mild stimulating preparations, adjuncts to fermentation, flavouring ingredients to beverages, and poisons,—substances which perhaps could scarcely be relegated to a position more likely to prove convenient to the enquirer after such information than the present.

The following tabular statement exhibits some of the more striking statistical records of the trade in narcotics. While it would (to arrive at a conception of the total trade) be incorrect to add together the value of the imports and exports from and to foreign countries, to those from and to the provinces of India coastwise, or to those carried by internal routes, still these are to a large extent independent transactions and afford employment to separate agencies. But, even were such an addition made, the total would very probably not exceed the actual production and consumption, since the exports are of necessity drawn from the surplus. The bulk of the amounts carried by rail is towards the seaports and is intended to meet the foreign exports, while the intra- and inter-provincial adjustments of supply and demand, as represented by coastwise trade and by a minor portion of road, rail, and river traffic, does not probably exceed £3,000,000. The production and consumption of Hemp, Opium, Spirits, and other excisable narcotics is to a large extent local, and regulated to meet but not exceed indigenous demands, so that with the exception of specially selected and registered tracts, which are concerned in the production of Ganja and Opium, for external markets, very little of the local traffic in such narcotics as Alcohol and Hemp (Bhang) would appear in trade returns.

TRADE IN NARCOTICS. 14 **Trade.**—From the table given on the next page it will be seen that last year the railways (mainly) carried close on £20,000,000 worth of narcotics, that the total value of the foreign transactions by sea and across the land frontier very considerably exceeded that amount. The coastwise traffic was less important, being only valued at about £2,000,000. But in these statements of pounds sterling the conventional rupee of two shillings has been accepted; doubtless the error thereby involved is more than compensated for by defects in road and river returns and by the intentional omission from the table of all unimportant narcotics.

TRADE IN NARCOTICS.

Value of the Chief Narcotics and Spirits, as recorded in Trade Statistics, Marine and Rail Returns, &c., for the year 1889-90.

NAMES OF PRODUCTS.	FOREIGN TRADE				INTERNAL TRADE		
	BY SEA.		BY TRANS-FRONTIER ROUTES.		BY COASTWISE.		BY ROAD, RAIL, AND RIVER.
	I. Imports.	II. Exports.	III. Imports.	IV. Exports.	V. Imports.	VI. Exports.	VII. Total Provincial Transactions.
	R	R	R	R	R	R	R
Areca Catechu—Betel-nut	c. 38,20,142	a. 42,675 b. 13,194	c. 946	a. 7,65,756	a. 60,80,241	a. 54,36,509	1,24,26,358
Bassia latifolia—Mahua	Nil	a. 195 b. Nil	Nil	Nil	Nil	Nil	Unimportant.
Camellia theifera—Tea	c. 31,79,373	a. 5,26,73,149 b. 20,58,226	(Seed) c. 11,550 (Tea) c. 58,106	(Seed) a. 80 (Tea) a. 13,55,806	a. 10,87,676 c. 1,61,395	a. 10,66,747 c. 2,01,317	4,94,55,642
Cannabis sativa—Indian Hemp, Ganja, &c.	Nil	a. } Unim- b. } portant.	c. 2,34,631	a. 22,475	Nil	Nil	d. 14,92,065
Coffea arabica—Coffee	c. 10,24,835	a. 1,88,42,430 b. 1,02,237	Nil	Nil	a. 32,52,206	a. 37,87,839	Unimportant.
Humulus Lupulus—Hops	c. 5,94,734	a. } Nil b. }	Nil	Nil	Nil	Nil	Ditto.
Malt or Fermented Liquors—Beer, Ale, Cider	c. 41,50,685	a. Nil b. 10,691	Nil	Nil	a. 52,947 c. 1,70,251	a. 51,299 c. 2,29,062	80,55,640
Nicotiana Tabacum—Tobacco	c. 13,92,573	a. 10,37,495 b. 18,027	c. 5,07,039	a. 9,13,509	a. 54,22,946	a. 43,37,896	1,60,67,408
Papaver Somniferum—Opium	c. 4,116	a. 10,50,80,808 b. 15	c. 4,20,088	a. 2,102	a. 72,956	a. 2,04,061	8,89,53,078
Wines and Spirits	c. 1,07,01,402	a. 556 b. 60,818	c. 565	a. 9,80,885	a. 32,93,350 c. 5,32,486	a. 34,49,000 c. 6,36,630	1,82,58,533
TOTAL	2,48,67,360	17,99,40,516	12,32,925	40,40,613	2,01,26,454	1,94,00,360	19,47,08,724

a.=Indian produce.
b.=Re-exports, Foreign produce.
c.=Foreign produce.
d.=Drugs other than opium.

TRADE IN NARCOTICS.

The above table will enable the enquirer after such information to discover the net exports or imports by striking the balance between the figures given in columns I., II., III., and IV. Thus, for example, the net import of foreign tea was last year valued at R11,79,253, and the net export of Indian grown tea at R5,40,28,955.

CHIEF INDIAN NARCOTICS.
15

ALPHABETICAL ENUMERATION OF THE CHIEF INDIAN NARCOTICS

For further information regarding the narcotics of India, the reader is referred to the list below, and from the brief notices there given to the detailed articles in their respective alphabetical positions in this work.

1. **Acacia arabica,** *Willd.;* LEGUMINOSÆ.

THE BABUL.

Bark of the root is used to flavour native spirits. See Vol. I., 25.

2. **A. ferruginea,** *DC.*

The bark is employed in the distillation of arrack from jaggery (*Beddome*).

3. **A. leucophlœa,** *Willd.*

The bark is used to facilitate the fermentation of spirits prepared from sugar and palm juice; and at the same time is supposed to increase the amount of the alcohol. See Vol. I., 53.

4. **Aconitum ferox,** *Wall.;* RANUNCULACEÆ.

INDIAN ACONITE.

The root of this and other Himálayan species is used medicinally as a powerful narcotic, sedative, and also criminally as a poison. See Vol. I., 84-92.

5. **Anacardium occidentale,** *Linn.;* ANACARDIACEÆ.

CASHEW-NUT TREE.

A spirit is distilled from the succulent fruit-stalk, by the people at Goa. See Vol. I., 233.

6. **Anamirta Cocculus,** *W. & A.;* MENISPERMACEÆ.

COCCULUS INDICUS.

The seeds contain a poisonous principle called *picrotoxin*. They are intensely bitter and have been employed as a substitute for hops in the manufacture of beer. They are also said to be used in Bombay to increase the intoxicating effects of country spirits, sold in retail. See Vol. I., 236; also FISH POISONS below, p. 327.

7. **Areca Catechu,** *Linn.;* PALMÆ.

ARECA, OR BETEL-NUT PALM.

This palm is viewed as a native of Cochin China, the Malayan Peninsula and Islands. It is cultivated throughout Tropical India; in Bengal, Assam, Sylhet, but will not grow in Manipur, and only indifferently in Cachar, Burma, and Siam. In Western India, below and above the ghats, it flourishes.

The following particulars regarding the narcotic properties of betel-nut have been taken from a paper read by Dr. W. Dymock before the Bombay Natural History Society:—

"The areca or betel-nut palm is supposed to be a native of the Malayan Peninsula and Islands, but is now met with only in a cultivated state. Some idea of the consumption of betel-nut in India may be formed

BETEL-NUT.

from the fact that in addition to her own produce India imports about 30,400,000 pounds of the nut from Ceylon, the Straits Settlements, and Sumatra. The exports are under 500,000lb, which go to Eastern countries frequented by Indians, such as Zanzibar, Mauritius, Aden, China, &c. Bombay is the chief centre of the export trade. It has long been known in the East that the fresh nuts have intoxicating properties and produce giddiness, and that the nuts from certain trees possess these properties to an unusual extent, and even retain them when dry, the produce of such trees being known as *Marjari supari* or intoxicating betel-nut. Ordinary betel-nuts have undoubtedly a stimulant and exhilarant effect upon the system, and are supposed to be aphrodisiacal. *Marjari supari* are produced by a small number of trees in most betel plantations. These trees cannot be distinguished from the others until they bear fruit, so that not unfrequently accidents happen, from the nuts becoming mixed with the produce of the plantation before their presence has been detected. The intoxicating properties of the betel-nut are greatly diminished by heat, and consequently many people only use the red nuts of commerce, which have undergone a process of cooking. The only account of the *Marjari supari* in European works appears to be that of Rumphius, which agrees in every respect with the particulars related by betel farmers in the neighbourhood of Bombay. He says:—'Many of the fresh nuts have the property of intoxicating and making giddy those who eat them, affecting them much as tobacco does those who are not used to it. Some of the old nuts also cause, in those not addicted to their use, great oppression on the chest and a sense of strangulation. These are called *Pinanga-Mabok*, 'intoxicating pinanga,' and are chiefly produced by the black variety of areca, which some consider a distinct species. Intoxicating nuts may be known by the central portion being of a red colour when cut open. I have already observed that I do not consider this black areca to be a distinct species, but a variety of the two species described by me, and found here and there amongst other trees—although some trees certainly occur, all the nuts of which are intoxicating, especially among those belonging to this third variety.' Rumphius adds that when these nuts have been eaten by mistake, either lime-juice or acid pickles are the best remedies. The above facts seem to indicate the return of a few plants to an original wild form now extinct, especially as the unripe nuts of the best trees produce similar effects in a less degree. The betel-nut, in Sanskrit *guvaka, puga,* and *kramuka,* in the vernaculars *supari,* when wrapped in the leaves of the piper-betel or *pan,* along with lime and spices forms the *bira* or *vira,* which is so much used by the natives of all parts of India, and is commonly presented by one to another in token of civility or affection. It is also given in confirmation of a pledge, promise, or betrothal, and among the Rajpoots is sometimes exchanged as a challenge: thus the expression *bira uthana* signifies 'to take up the gauntlet,' or take upon oneself any enterprise; *bira dalna,* 'to propose a premium' for the performance of a task: the phrase originated in a custom that prevailed of throwing a *bira* into the midst of an assembly, in token of an invitation to undertake some difficult affair, for instance, in the first story of the *Vetalapanchavinshati,* the king, when he sends the courtesan to seduce the penitent who was suspended from a tree nourishing himself with smoke, gives her a *bira. Bira dena* signifies 'to dismiss' either in a courteous sense or otherwise. A *bira* is sometimes the cover of a bribe, and a *bira* of seven leaves (*sat pan ka bira*) is sent by the father of the bride to the bridegroom as a sign of betrothal. At marriages the bride or bridegroom places a *viri* or cigarette-shaped *vira* between the teeth, for the other party to partake of by biting off the projecting half; one of the tricks played

BETEL-NUT.

on such occasions is to conceal a small piece of stick in this *viri*, so that the biting it in two is not an easy matter. The nut is also a constant offering to the gods at Hindú temples, and on grand occasions the *bira* is covered with gold or silver leaf.

"According to the *Hitopadesa* the betel-leaf so constantly used with the nut has thirteen properties: it is sour, bitter, heating, sweet, salt, astringent; it expels flatulence (*vataghna*), phlegm (*kaphanasana*), worms (*krimihara*); it removes bad odours, beautifies the mouth, and excites desire. Betel-nuts and leaves were known to the Greeks, the former as Hestiatoris or 'the convivial nut,' which appears to be a rendering of the Sanskrit names in Greek: the latter was doubtless the Malabathron or 'Indian leaf,' sometimes called simply *φυλλον* (*pan*), and sold in rolls in a dried state. **Dioscorides** speaks of their being threaded on strings to dry,—a practice still common in Bombay among the Indian traders, who send the leaves to their friends in Arabia, Persia, and elsewhere. The passage in **Dioscorides** *έν τω μελανίζείν τε ἄθραυστον καὶ ὁλόκληρον* is probably corrupt, and should be as suggested by **M. Vergilius** *έν τω μαλακίζειν τε αθραυστον καὶ ὁλόκληρον*, a reading which he found in one manuscript. As regards the fabulous growth of Malabathron recorded by **Dioscorides**, it was probably the tale of some traveller who had seen the practice of burning the jungle after the monsoon on the west coast of India. That Malabathron was not a cinnamon leaf is, I think, proved by **Dioscorides** in his chapter on **Cassia**, describing its leaves as like those of the pepper plant. Until very recently the betel-nut was considered by European medical writers to be simply astringent, and the intoxicating properties of the *bira* were supposed to be due to the spices and leaf; but the rapid progress of organic chemistry and physiology during the last few years has led to the discovery of intoxicating properties in the nut, while **Dr. Kleinstuck** has shown that the essential oils of betel leaves are of much use in catarrhal affections, inflammations of the throat, larynx, and bronchi, exerting an antiseptic action; and has also used them with advantage in diphtheria. The juice of four fresh leaves diluted may be given as a dose when the oils are not readily obtainable.

"In 1886, **Herr E. Bombelon** announced that the betel-nut contained a liquid volatile alkaloid, but did not describe its composition and properties. As it seemed probable that the physiologically active constituent was to be looked for in this alkaloid, **Herr Jahns** was induced to investigate the subject more closely, and has reported the results recently to the Berlin Chemical Society. He found in the nut three alkaloids—Arecoline $C^8 H^{13} NO^2$; Arecaine $C^7 H^{11} NO^2 + H^2O$; and a third alkaloid which could not be closely examined, as the quantity obtained was very small.

"Of these alkaloids arecoline is undoubtedly the active principle of the betel-nut. It was found that full-grown rabbits died within a few minutes after the subcutaneous injection of twenty-five to five milligrams of the hydrobromide and hydrochloride; cats succumbed after the injection of ten to twenty milligrams. The most dangerous action of arecoline consists in the slowing of the heart's action by small doses, or even its stoppage, just as takes place with muscarine. Simultaneously with the heart's action the respiration is also affected, causing a feeling of suffocation; and purging may take place when it is given in poisonous doses; a strong contraction of the pupil of the eye was observed. Atropine was found to counteract the poisonous effects of the alkaloid, so that the addition of a seed or two of *dhatura* to the *vira*, as sometimes practised in India, is really antidotal. It was also found during the experiments on animals that the organism may become gradually tolerant to the poison of the areca-nut, as in the case of smoking and chewing tobacco." See Vol. I., 298-301.

8. Balanites Roxburghii, *Planch.*; SIMARUBEÆ.

The bark yields a juice used as a fish poison. See Vol. I., 363, also Vol. III., 366.

9. Bassia latifolia, *Roxb.*; SAPOTACEÆ.

MAHUA OR MOHWA.

MAHUA FLOWERS.

The spirit resulting from the fermentation of the flowers is largely consumed by the natives inhabiting the area of country occupied by this tree. *Mahua* is, in fact, one of the chief sources of alcohol throughout the table-land of the central and southern portions of India. In official correspondence on this subject local authorities have urged the facility which exists for illicit distillation and the practical inability of the existing police force to check such a trade over the sparsely populated wild hilly tracts where the tree grows. The Chief Commissioner of the Central Provinces alluding to this subject wrote:—"This tree occurs abundantly all over these provinces and the process by which spirit is distilled from the *Mahua* flower in the wilder parts of the country is of the simplest character—a couple of earthen pots and a piece of hollow bamboo, to form a tube, constituting the distiller's apparatus. There is not a district in some portion of which spirit cannot, under these circumstances, be distilled illicitly without much fear of detection, and experience has proved most convincingly that unless the inherited taste of these people for this stimulant is satisfied by the establishment of shops within their reach, where they can buy taxed spirit, they will resort to illicit distillation, and render themselves liable to the penalties of the criminal law."

Some few years ago a considerable trade was done in exporting *Mahua* flowers, chiefly to France, but recent legislation which prohibits the import, has extinguished the traffic. *Mahua* flowers have, therefore, been shown in the table given at page 319, not because of the present value of the export traffic, but because of the immense importance of the flowers, in the internal traffic, in spirits and spirituous materials. For further information see Vol. I., 409.

10. B. longifolia, *Willd.*

MAHUA OF SOUTH INDIA.

A spirit is also prepared from the flowers of this species. See Vol. I., 416.

11. Bhang, the least injurious form of **Cannabis sativa** (which see).

12. Bojah, a kind of beer made from millets. See **Eleusine** (Vol. III., 241) and **Sorghum** (Vol. VI).

13. Borassus flabelliformis, *Linn.*; PALMÆ.

PALMYRA PALM.

The toddy prepared from the fermented juice, or *ras*, is largely consumed. The distillation of the toddy yields palm wine or *arak*. See Vol. I., 497.

14. Calotropis gigantea, *R. Br.*; ASCLEPIADACEÆ.

An intoxicating liquor called *Bar* is said to be prepared from this plant by the tribes of the Western Ghâts. See further in Vol. II., 47.

15. Camellia theifera, *Griff.*, TERNSTRŒMIACEÆ.

TEA.

See the article **Camellia,** Vol. II., 73-85, also **Tea,** Vol. VI.

16. Canavalia ensiformis, *DC.*, and var. **virosa** (the wild form); LEGUMINOSÆ.

INDIAN HEMP.

Birdwood says:—"This is a common narcotic in the Konkan. The pods are shred with French beans, boiled and eaten, when intoxication follows." See Vol. II., 97.

17. Cannabis sativa, *Linn.;* URTICACEÆ.

INDIAN HEMP.

Ganjá, Charas, Bhang, Majun or *Májum*, see Vol. II., 109-110, 113-118.

The following brief account appeared in the Colonial and Indian Exhibition Catalogue and may be here reproduced, since in some respects it amplifies the facts already given in this work:—

"Hemp is in India almost exclusively cultivated on account of its narcotic property. There are three distinct forms of the drug, *viz.*, *ganjá*, *charas*, and *bhang*. These are used in the various provinces of India as follows:—

"(*a*) In Eastern Bengal the plant is cultivated for the agglutinated female flower-tops known as *ganjá*. This is sold in two forms—round and flat—and the drug is always smoked, a small quantity being mixed with tobacco. In 1883-84 2,493 acres were under *ganjá*. This gave employment to 1,972 persons and yielded 8,982 maunds. The cultivation, manufacture, and sale is regulated by law. Permission is granted on license, and the produce is compulsorily placed in Government stores. There are two charges made by Government—a license to trade and sale, and a direct duty per maund on actual amounts removed from store. During the year 1883-84 the Bengal Government realised a net revenue of R19,73,713, being a charge of R335 a maund on amount consumed. The consumer paid from R16 to R20 a seer, or R640 to R800 a maund (*e.g.*, £64 to £80 for 84℔).

"(*b*) In the North-West Provinces the cultivation of *ganjá* is prohibited, but *bhang* or the young leaves and twigs are largely collected from the semi-wild plant. These are made into a greenish intoxicant liquor known as *hashish*. The supply of *ganjá* consumed in these provinces is imported from the Central Provinces, Bombay, and Bengal. Including *bhang*, the consumption in 1883-84 amounted to 6,690 maunds, which gave a total revenue of R5,53,356. There is no direct duty levied on amount consumed, and the revenue is raised entirely by farming the retail shops. The Government obtained in this way only R82-11 per maund, while in some parts at least of these provinces the drug was retailed at the same price as in Bengal.

"(*c*) In the Panjáb *ganjá* is very little used, but the consumption of hill-grown *bhang* is very extensive. From Kashmír, Ladak, and Afghánistán *charas* is largely imported into the Panjáb. This is a resinous-like substance found on the flower-tops and twigs, collected by rubbing the flower-tops between the hands or by causing men to run violently through the fields. The resin or *charas* adheres to their naked bodies and is scraped off. It is commonly reported that from Nepál a fine quality of this substance, known as *momea*, is obtained. **Dr. Gimlette**, Residency Surgeon, Nepál, reports, however, that as far as Katmandu and its neighbourhood is concerned the name *momea* is unknown except as applied to an extraordinary medicinal preparation in which human fat forms an important ingredient. (For further information see Vol. II., 115-116.) **Dr. Gimlette** further states that *charas* is, in Nepál, prepared by rubbing the flower-tops between the hands. A small amount of *charas* is produced in Sind by causing men clad in skins to run through the fields. The chief supply of *charas* comes to India, however, from across the Panjáb frontier, and is conveyed under a permit system (but free of duty) nearly all over India. On reaching the frontier of Bengal it has, however, to pay the heavy import duty of R8 a seer. About 600 maunds of *charas*

INDIAN HEMP.

are annually consumed in the Panjáb, but the imports are over 5,000 maunds. As this large amount enters and passes over India almost quite free of duty, it naturally affects most materially the Indian-grown *ganjá* and *bhang*. *Charas* is stronger than *ganjá* and fetches a slightly higher price in Bengal; but in the Panjáb it is sold at R4 to R13 a seer (2lb); and in Bombay for about R2 a seer. The total revenue derived in the Panjáb from hemp narcotics was in 1883-84 only R1,44,640, or a revenue of R31 a maund on consumption.

"(*d*) In Bombay *ganjá* is very extensively cultivated, the consumption being over 9,000 maunds, and the revenue only R1,61,599, or about R17 a maund. There is no fixed duty, but the revenue is realised, as in the North-West Provinces and the Panjáb, by farming the retail shops. *Ganjá* is sold in Bombay at R2 a seer.

"(*e*) In the Central Provinces cultivation of *ganjá* on a large scale takes place according to a system very much similar to what prevails in Bengal. The shops are farmed, but over and above, a fixed duty per maund is levied on amounts removed from the stores. The production in 1883-84 amounted to 6,356 maunds, and the local consumption to only 768 maunds The total revenue realised from all sources amounted to R1,29,207, or R168-7 per maund on amounts consumed locally. Removals from the stores to other provinces appear to pay no duty, so that exports from the Central Provinces may be put down in Allahabad free of duty, while from Bengal the same article pays R5 a seer to Government.

"(*f*) Assam consumes annually a very considerable amount of Bengal *ganjá*, and the provincial treasury is credited with a large revenue therefrom; this in 1883-84 amounted to R363 a maund, the total revenue being R9,31,691.

"(*g*) Madras cultivates *ganjá* and *bhang* to a certain extent, and doubtless it imports these substances from neighbouring provinces also; but a separate account of the revenue derived therefrom is not published, so that the condition of the Southern Presidency cannot be compared with other parts of India. Hemp narcotics are apparently not consumed in Burma.

"A very great inequality thus exists in the revenue derived from hemp narcotics in the various provinces of India. The substance, in one or other of its forms, is more or less used all over India,—either smoked as *ganjá* and *charas*, or consumed as *hashish* liquor, or eaten in the form of a special sweetmeat, known as *majun*, the last two forms being preparations from *bhang*. The total consumption in 1884 for all India amounted to only 16,378 maunds, leaving, as shown by the published figures of production, 11,264 to be either in stock or consumed without paying duty.

"Intimately associated with the religious systems of Hinduism, hemp intoxication has become an established luxury and has been inherited by the corresponding classes of Muhammadans. It should be recollected that habitual indulgence in hemp is nowhere so prevalent as habitual alcoholic intemperance. Hemp-narcotism does not establish the same irresistible craving as alcohol. The indulgence is rather accidental or occasional, than habitual, so that it is more widely diffused than might at first sight be inferred. The agricultural classes, who, of course, constitute the bulk of the population of India, rarely if ever indulge in hemp narcotics. It is the artizans, mendicants, and domestic servants who are the chief consumers; the middle and upper classes partake of hemp only at certain religious observances, and even then but to a small extent. These facts necessarily narrow the community who partake in this indulgence, and it is probable that a percentage of two persons in every 1,000 of population would express

INDIAN HEMP.

very nearly the number of consumers. The utmost that a man could smoke of *ganjá* per annum has been estimated at one and a half seers."

In a long and detailed note on Excise Administration (***published in the Supplement to the Gazette of India, March 1st, 1890***), the Government of India traces the Bengal trade in *ganjá* from 1868-69 to 1887-88. The main facts there shown may be briefly summarised:—At the commencement of the period there were 4,073 shops licensed for the sale of the drug, at the close of the period 2,949. The revenue realised was in 1868-69 R8,95,240, in 1887-88 R22,44,970. The amount consumed declined from 8,442 maunds to 6,550 maunds, while the incidence of taxation per seer increased from R2·6 to R8·5. The report concludes:—"Notwithstanding the large increase of population which must have taken place between 1868-69 and 1887-88, the number of shops for sale of *ganjá* has decreased by more than 27 per cent., and the consumption by more than 22 per cent., while simultaneously the revenue has increased by more than 150 per cent. and the incidence of taxation per seer has increased by more than 226 per cent."

18. **Caryota urens,** ***Linn.;*** PALMÆ.

A toddy is prepared from the fermented sap. See Vol. II., 208.

19. **Chandu.**

Opium or ***kafa*** boiled down, distilled and prepared in the Chinese fashion.

20. **Charas.**

A peculiar form of the resinous narcotic of **Cannabis sativa.** It is chiefly imported into India across the Panjáb frontier. See **Cannabis sativa.** Charas-smoking is described by **Honigberger** (Thirty-five years in the East, pp. 152—156).

21. **Cissampelos Pareira,** ***Linn.;*** MENISPERMACEÆ.

An ardent spirit is said to be distilled from the root of this plant in Garhwal. See Vol. II., 327.

22. **Cleistanthus collinus,** ***Benth.;*** EUPHORBIACEÆ.

In Chutia Nágpur the fruit and bark of this tree are used as a fish poison. See **Lebidieropsis orbicularis,** ***Muell;*** Vol. IV., 616.

23. **Clerodendron serratum,** ***Spreng.;*** VERBENACEÆ. See Vol. II., 375.

24. **Cocos nucifera,** ***Linn.;*** PALMÆ.

COCOA-NUT PALM.

The toddy and the fermented juice are largely consumed; as also the spirit distilled from the latter. See Vol. II., 454-455.

25. **Coffea arabica,** ***Linn.;*** RUBIACEÆ.

COFFEE.

The reader should consult the article in Vol. II., 460-491.

26. **Datura fastuosa,** ***Linn.;*** and other species; SOLANACEÆ.

DATURA. See Vol. III., 29-43.

27. **Eleusine Coracana,** ***Gærtn.;*** GRAMINEÆ.

Maruá, or *rágí*—A beer and spirit is made from this millet which is known as *bojah*. See Vol. III., 241.

28. **Erythroxylon Coca,** ***Lamk.;*** LINEÆ.

COCA. See Vol. III., 270-277.

29. **Eugenia Jambolana,** ***Lamk.;*** MYRTACEÆ.

JAMAN.

A spirit is distilled from the juice of the ripe fruit, and in Goa a wine is also prepared from the same. See Vol. II., 286.

FISH POISONS.

30. **Euphorbia Tirucalli,** *Linn.;* EUPHORBIACEÆ.
The milk is used to poison fish. See Vol. III., 302, also 366.

31. **Fish Poisons.** See Vol. III., 366.
Mr. Hooper has kindly furnished the Editor with an extract from the *Druggists' Bulletin for November 1890,* in which he published an account of the chief Fish Poisons used in India. One of the best known poisons of this nature, says Mr. Hooper, is **Cocculus indicus** (No. 6 of above list). The berries are made into a paste with boiled rice, and very small pieces of the mass are sufficient to make fish, birds, and small animals insensible. The poisonous principle of these berries has been ascertained to be *picrotoxine.* In Jamaica the bark of **Piscidia Erythrina** is employed for poisoning fish: the bark is macerated with the residue of rum distillation, placed in a basket and dragged up and down the river till the fish are intoxicated. The active principle is similar to *picrotoxine.* In the South Sea Islands quite a different drug is used for this purpose, *viz.,* **Lepidium piscidium,** a kind of cress which possesses an extremely pungent taste. The effects of this drug are to cause the fish to float on the surface of the water in a helpless state of insensibility. But besides **Cocculus indicus** already mentioned, there are other fruits, barks, and roots, Mr. Hooper tells us, that are employed in India as fish-poisons. The seeds and bark of **Mundalea suberosa** are so used in Southern and Western India. The bark of **Walsara piscidia** also acts effectually.

Randia dumetorum is an old-fashioned fish-poison which is used from the Himálayan Tarai southwards. For this purpose the pulp from the inside of the fruit is utilized. **Crotalaria paniculata** has recently been brought to light as a fish-poison used in Tanjore. So also the fruits of **Diospyros montana** have been stated to be employed for that purpose in Travancore. Mr. Hooper adds, however, that he is disposed to think there must have been some mistake, as no poisonous principle could be detected in them. The seeds of the **Barringtonias,** Mr. Hooper informs us, are used in South India for killing fish, as also the following:—the bark of **Flueggea Leucopyrus;** the bark of **Securingea obovata;** the bark of **Berberis aristata;** the fruit-pulp of **Gynocardia odorata;** the root of **Millettia pachycarpa;** the leaves of **Macaranga,** various species; and the flower-heads of **Spilanthus Acmell.**

It may be added that the peculiar principle *Saponin* is contained in several of the above drugs though their special use has been discovered in remote parts of the globe and they belong to widely different natural orders of the Vegetable Kingdom. Fish captured through the effects of that poison would not be unwholesome as human food.

32. **Flueggea microcarpa,** *Bl.;* EUPHORBIACEÆ.
Syn.—PHYLLANTHUS VIROSUS, *Roxb.*
Included in O'Shaughnessy's list of narcotics. The bark is said to intoxicate fish.

33. **Ganja.**
The chief narcotic of **Cannabis sativa** (which see); chiefly prepared in Bengal, the Central Provinces, and Bombay. It is smoked like opium.

34. **Grewia asiatica,** *Linn.;* TILIACEÆ.
PHALSA.
The fermented juice of the fruit is drunk as a beverage. See Vol. IV., 178.

35. **Hashish.**
A green narcotic beverage prepared from *bhang.* See **Cannabis sativa.**

36. **Hordeum vulgare,** *Linn.;* GRAMINEÆ.
BARLEY.

This grain is largely employed in the preparation of beer and spirits. See **Malt Liquors,** Vol. V., 128.

37. **Humulus Lupulus,** *Linn.;* URTICACEÆ.

HOPS.

Contains a narcotic and also a tonic principle which give to beer its peculiar odour and flavour. See **Humulus,** Vol. IV., 302; also **Malt Liquors,** Vol. V., 130.

38. **Hydnocarpus venenata,** *Gærtn.* also **H. Wightiana,** *Bl.;* BIXINEÆ.

This fruit acts as a narcotic poison, and is used to intoxicate fish. See Vol. III., 366, also Vol. IV., 308.

39. **Hyoscyamus niger,** *Linn.;* SOLANACEÆ.

HENBANE.

Sedative, anodyne, and antispasmodic. See Vol. IV., 319.

40. **Ilex paraguayensis,** *St. Helaire;* ILICINEÆ.

MATE OR PARAGUAY TEA.

See Vol. IV., 328.

41. **Kapha** or **Kafa**; see **Madak.**

42. **Lactuca Scariola,** *Linn.,* **var. sativa,** *Linn.;* COMPOSITÆ.

LETTUCE.

The inspissated juice, known as lactucarmin, acts as a sedative, and has been recommended as a substitute for opium. See Vol. IV, 579.

43. **Lasiosiphon eriocephalus,** *Dcne.;* THYMELACEÆ.

Syn.—L. SPECIOSUS, *Dcne.*

Leaves and bark are poisonous and are used to kill fish. See Vol. III., 366, also Vol. IV., 589.

44. **Lebidieropsis orbicularis,** *Müll-Arg.;* see **Cleistanthus collinus,** *Benth.*

45. **Ligustrum Roxburghii,** *Clarke;* OLEACEÆ.

In South India the bark of this tree is put into the toddy of **Caryota urens** (*birly-már*) to accelerate fermentation. See Vol. IV., 646.

46. **Lolium temulentum,** *Linn.;* GRAMINEÆ.

DARNEL.

The grain of this grass has from very early times been regarded as an acro-narcotic poison. Recent experiments tend to show that this is due to the grains becoming ergotized, and that healthy darnel seed is innocuous. See Vol. V., 90.

47. **Loranthus longiflorus,** *Desr.,* **var. falcata,** LORANTHACEÆ.

Bark of wood used in Kánara instead of betel-nut. See Vol. V., 92.

48. **Madak.**

A narcotic preparation made by boiling down and inspissating the juice of opium or *kaphá.* *Kapha* or *kafa* is the juice of opium collected on rags. It is largely smuggled. See **Papaver somniferum.** (Vol. VI.)

49. **Majun** or **Majum.**

A sweetmeat rendered narcotic by the addition of Bhang. See **Cannabis sativa.** Honigberger says this sweetmeat in his time was largely used in Turkey, Arabia, India, and Egypt.

50. **Malt** or **Fermented Liquors.**

BEER, ALE, CIDER AND THE INDIAN EQUIVALENTS.

The reader should consult the article **Malt Liquors** (Vol. V., 124—140), as also the remarks in this enumeration regarding **Eleusine, Hordeum, Oryza, Sorghum,** &c. The following table and comments on the Beer trade of India is taken from the Memorandum on Excise Administration (*Published in the Gazette, March 1st, 1890*); it may be read in amplification of the article on **Malt Liquors** more especially of the table at page 140:—

BEER AND ALE.

The Table below furnishes information regarding the Consumption of Beer in India since 1877.

YEAR.	Beer manufactured in India.	IMPORTED BEER.		TOTAL.	CONSUMPTION		
		* For general consumption.	* For consumption by troops.		By Troops.		By others than troops.
					Indian beer.	* Imported beer.	
	Gallons.	*Gallons.*	*Gallons.*	*Gallons.*	*Gallons.*	*Gallons.*	*Gallons.*
1877	2,164,048	1,328,077	3,123,128	6,615,253	954,933	3,123,128	2,537,192
1878	1,522,769	1,089,211	1,965,222	4,577,202	869,270	1,965,222	1,742,710
1879	1,569,026	1,065,347	2,156,325	4,790,698	872,296	2,156,325	1,762,077
1880	1,974,578	1,152,978	1,695,959	4,823,215	1,298,773	1,695,959	1,828,483
1881	2,448,711	1,199,395	1,708,596	5,356,702	1,764,927	1,708,596	1,883,179
1882	2,594,667	1,170,554	1,486,234	5,251,455	1,699,914	1,486,234	2,065,307
1883	2,597,298	1,261,444	1,906,520	5,765,262	2,027,169	1,906,520	1,831,593
1884	2,778,680	1,066,913	1,505,062	5,350,655	2,030,499	1,505,062	1,815,094
1885	3,150,342	1,299,408	375,396	4,825,146	2,266,801	375,396	2,182,949
1886	4,403,638	1,715,638	152,064	6,271,340	3,339,361	152,064	2,779,915
1887	5,085,030	2,138,518	387,788	7,611,336	4,178,658	387,788	3,044,890
1888	5,352,191	2,398,580	415,816	8,166,587	4,628,175	415,816	3,122,596

* Official years.

MALT LIQUORS.

The Memorandum commenting on these figures states:—"These figures do not bear out the assertion that the increased consumption of beer 'is Native and not English.' The consumption of beer by the British troops in India has increased. This is not necessarily inconsistent with **Mr. Caine's** assertion that the British soldier is more temperate than formerly, as the quantity of beer supplied to the British soldier by the canteens only represents a portion of the total consumption of liquor by the army, and there has been an increase in the number of the troops from 60,000 to 70,000, dating from 1885; there has also been a marked decrease in the consumption of spirits by British soldiers. Natives of India as a rule do not drink beer made after the European method, but there is a considerable and increasing European and Eurasian population in India, and increase in the consumption of beer by the general population is almost entirely confined to these classes" (*Memorandum on Excise Administration*).

51. **Meconopsis aculeata,** *Royle ;* PAPAVERACEÆ.

The roots are reputed to be exceedingly narcotic (*O'Shaughnessy*). See Vol. V., 197.

52. **Melodinus monogynus,** *Roxb. ;* APOCYNACEÆ.

Syn.—NERIUM PISCIDIUM, *Roxb.*

The leaves, wood, and root contain a narcotic poison. See *Roxburgh Fl. Ind., 244,* regarding its action as a fish poison. See also Vol. V., 226.

53. **Momea** or **Mumai,** see **Cannabis sativa.**

54. **Mundulea suberosa,** *Benth. ;* LEGUMINOSÆ.

The leaves are used as a fish-poison. See Vol. III., 365, also Vol. V., 288.

55. **Nerium odorum,** *Soland. ;* APOCYNACEÆ.

The root, bark, and leaves contain a powerful narcotic poison. See p. 348.

56. **Nicotiana Tabacum,** *Linn.,* and **rustica,** *Linn. ;* SOLANACEÆ.

TOBACCO.

See Vol. V., pp. 341 to 428.

57. **Osyris arborea,** *Wall. ;* SANTALACEÆ.

THE HIMÁLAYAN SUBSTITUTE FOR TEA.

58. **Oryza sativa,** *Linn. ;* GRAMINEÆ.

RICE.

It is perhaps unnecessary to refer in any detail to the fact that rice is one of the chief grains employed in India in the preparation of Malt or Fermented Liquors (*Pachwai*) (see Vol. V., 124—140) or in the distillation of spirits. (See p. 332.) In addition to the passages of this work just cited, the reader should, however, also consult the article **Oryza sativa.** The following quotation will exemplify the ease with which both a sort of beer and a spirituous liquor may be prepared, and will thus convey some tangible idea of the importance of rice as one of the Indian sources of alcohol:—

"The preparation of rice beer presents no more difficulty than the infusion of a pot of tea, and has this additional facility that, whereas tea, under the circumstances in which we most know it, is an imported product, which has to be purchased, rice is the staple food of the country, and being cultivated and stored in large quantities, every Kachari and every Miri has it always at hand. The mode of preparation is to steep rice in water for two or three days, then boil it with certain herbs, which act as a yeast, and the result is beer. It is rarely sold, each person manufacturing his own supply, and borrowing from his neighbours if he falls short."

RICE-SPIRITS AND BEER.

"*Pachwai* is manufactured so simply that the process is only a little more complicated than the preparation of rice for ordinary meals or the brewing of a pot of tea." "It should be noted that the rice-beer is sometimes made to undergo a further process of distillation, and the distillate is called *phatika.* It is almost as cheaply and simply made as the rice-beer itself: take two earthen-ware pots; in one of them put the rice-beer, and on the top of it stand the other pot, first knocking a hole in the bottom of it. Stand a saucer on the bottom of the higher pot, and stop the mouth of this pot, with a vessel made in the shape of an inverted cone and filled with cold water. Now set the whole thing on a fire. The fumes rise, through the hole, into the upper jar, are condensed against the cold inverted cone, and drop into the saucer in the form of spirits." "If people want to obtain spirits, they can do so either by getting smuggled spirit from the hills, or by brewing and distilling for themselves by methods with which they are quite familiar, and which are very cheaply and easily available" (*Extract from a Memorandum by the Chief Commissioner of Assam*). For further particulars consult the paragraph below on Spirits.

59. Pachwai, a kind of beer made from rice; see **Oryza sativa** above.

60. Papaver somniferum, *Linn.;* PAPAVERACEÆ.

OPIUM.

See Vol. VI.

61. Peganum Harmala, *Linn.;* RUTACEÆ.

Recommended as a substitute for opium. See Vol. VI.

62. Phœnix sylvestris, *Roxb.;* PALMÆ.

WILD DATE-PALM.

The sap is fermented into toddy. See Vol. VI.

63. Picrasma quassioides, *Benn.;* SIMARUBEÆ.

INDIAN QUASSIA.

Bark used like the officinal quassia to kill insects. See Vol. VI.

64. Piper Betle, *Linn.;* PIPERACEÆ.

The leaves act as a mild stimulant when chewed, together with lime and Areca nut. See Vol. VI.

65. Pistacia integerrima, *J. L. Stewart;* ANACARDIACEÆ.

According to O'Shaughnessy (*Beng. Pharm., 184*) the leaves and seeds of this plant are narcotic. See Vol. VI., Pt. I.

66. Prunus Amygdalus, *Baill.;* ROSACEÆ.

ALMOND.

The kernels of this and of the apricot (**P. armeniaca**), also the leaves and kernels of the Bird Cherry (**P. Padus**); the Himálayan Cherry (**P. Puddum**); and the leaves, flowers, and kernels of the peach (**P. persica**), contain varying quantities of prussic or hydrocyanic acid. See Vol. VI.

67. Rhododendron arboreum, *Sm.;* ERICACEÆ.

The leaves and flowers of this and other Himálayan species contain a narcotic poison, which proves fatal to sheep and goats. The honey of bees visiting Rhododendrons is also often poisonous. See Vol. VI., Pt. I.

68. Ruellia suffruticosa, *Roxb.;* ACANTHACEÆ.

See Vol. II., 259, also Vol. VI., Pt. I.

69. Saccharum officinarum, *Linn.;* GRAMINEÆ.

The spirit known as rum is obtained chiefly by the distillation of the uncrystallisable portion of the expressed juice of this plant.

SPIRITS.

70. Sapium indicum, *Willd.;* EUPHORBIACEÆ.

The juice acts as a narcotic poison, and the seeds are used for intoxicating fish.

71. Solanum nigrum, *Linn.;* SOLANACEÆ.

Employed as a narcotic by the hakims. Contains *solanine* united with malic acid. The fruits are very dangerous, and act in the same manner as those of belladonna. The extract of the whole plant acts like lactucarium (*O'Shaughnessy*).

72. Sorghum vulgare, *Pers.;* GRAMINEÆ.

JOWÁR.

In Poona a native beer, called *boja*, is brewed from *jowári* grain malted, and *bhang* is added as a substitute for hops (*Dalz. & Gibs.*). See **Eleusine Coracana,** *Gærtn.* (Vol. III., 241), and **Sorghum** (Vol. VI.).

73. Spirits or Alcohol.

Space cannot be afforded here to deal with this subject in detail, and the reader is, therefore, referred to **Spirits** in Vol. VI., and the article **Alcohol,** Vol. I., 161-162, where fuller particulars will be found. It may be explained that any study of the traffic in spirits must be referred to at least two main headings, *viz.*, locally prepared spirits and imported spirits. The former is distilled from many substances, but chiefly from the following:—**Bassia latifolia** flowers, *mahua;* **Borassus flabelliformis,** *tari* juice; **Caryota urens,** toddy; **Cocos nucifera,** toddy; **Eleusine Coracana,** *marua*, millet; **Hordium vulgare,** barley; **Oryza sativa,** rice; **Phœnix sylvestris,** date-palm toddy; **Saccharum officinarum,** sugar-cane, rum; and **Sorghum vulgare,** *jowár,* millet. For details of the various methods of distillation from these products, the reader is referred to the remarks in their respective places in this review of narcotics, as well as to their separate places in the various volumes of this work. The following passage may, however, be here given as exemplifying the ease with which alcoholic beverages can be, and doubtless are, prepared outside the limits of excise control:—

Extract from a Minute forwarded by the Government of Bombay, to the Government of India.

"When illicit markets for toddy were opened in all directions, and tappers were allowed to take or send their toddy to any shop or distillery they pleased, the transport regulations broke down, and any one who wanted toddy for illicit distillation was able to carry it where he pleased on the plea that he was conveying it to some shop or distillery. How greatly the facilities for illicit distillation were thus increased will be understood when it is remembered that any pot of fermented toddy can be converted into a ready charged still, and that distillation can be set going anywhere within the space of less than five minutes. All the apparatus necessary, besides the pot of toddy, is an earthen saucer, and a little wet earth wherewith to close tightly the mouth of the pot and a small bowl to be placed floating on the surface of the toddy in the pot. If a pot of fermented toddy thus treated is set to boil, and the saucer closing its mouth is kept cool by pouring water on it, the spirit given off from the boiling toddy in the shape of steam is condensed on the underside of the saucer and drips from the saucer into the bowl floating on the toddy ready to receive it. Two or three bottles of strong spirit can thus be made in a couple of hours from an ordinary sized pot of toddy. Distillation of this kind can be carried on anywhere, in the houses or in the fields, or in the jungles; wood and water are plentiful in all these coast talukas. It was the custom of the whole countryside to make toddy spirit in this primitive fashion before reforms were taken in hand in 1877-78. In those days such stills were in almost daily use in every village and hamlet of the toddy-producing tracts."

SPIRITS.

The following table gives a complete statement of the liquor and drug traffic in Bengal—the largest and most densely populated province of India :—

"The number of shops for the sale of all kinds of liquor and drugs in Bengal is shown in the following table for each year since 1870 :—

Year.	Distilled Liquor.	Drugs.	Tari.	Pachwaii.
1870-71	8,193	15,565	21,670	1,687
1871-72	8,937	16,615	21,689	1,815
1872-73	7,271	17,119	22,351	1,844
1873-74	6,812	10,975	22,873	2,006
1874-75	6,152	9,240	19,424	1,863
1875-76	5,294	7,424	19,265	1,661
1876-77	5,183	7,092	18,844	1,677
1877-78	5,267	7,248	19,077	1,717
1878-79	6,751	6,999	19,048	1,739
1879-80	6,878	6,877	21,579	2,066
1880-81	7,369	7,145	25,563	2,034
1881-82	6,874	6,796	30,268	2,135
1882-83	5,634	6,499	30,311	2,159
1883-84	5,740	6,513	20,138	2,259
1884-85	5,502	6,291	19,577	2,168
1885-86	5,298	6,005	19,555	2,162
1886-87	5,310	5,922	19,471	2,195
1887-88	5,112	6,059	19,051	2,157
1888-89	4,539	6,092	18,467	2,203

"It will be observed that since 1870-71 the number of shops for the sale of distilled liquor has been reduced by 3,654, or more than 44 per cent.; the number of shops for the sale of drugs has been reduced by 9,473, or more than 60 per cent.; the number of shops for the sale of *tari* has been reduced by 3,203, or more than 10 per cent.; the number of shops for the sale of *pachwai*, or rice-beer, has been increased by 516, or more than 23 per cent." (*Supplement to the Gazette of India, March 1st, 1890.*)

The figures for the other provinces show a similar result, *viz.*, increased taxation and diminished facilities for obtaining alcoholic beverages and drugs. By *tari* is meant the fermented juice of the palms enumerated above, and by *pachwai* a kind of beer prepared from rice.

The main facts regarding the traffic in imported spirituous liquors is shown by the remarks which may now be given from the official report which has furnished the chief data for the account here given of narcotics :—

Importation of Spirits into British India by Sea from Foreign Countries.

During	Quantity.	Amount of customs duty realised.	Rate of duty per gallon of London proof.
	Gallons.	Rx.	R
1870-71	461,323	184,309	} 3
1871-72	671,626	167,417	
1872-73	723,609	184,068	
1873-74	608,824	193,261	
1874-75	674,987	215,240	

SPIRITS.

Importation of Spirits into British India.—continued.

During	Quantity.	Amount of customs duty realised.	Rate of duty per gallon of London proof.
	Gallons.	Rx.	R
1875-76	704,874	225,667	
1876-77	654,527	255,128	
1877-78	737,714	275,983	
1878-79	692,384	286,847	
1879-80	814,334	321,846	
1880-81	848,238	307,905	4
1881-82	842,739	337,497	
1882-83	949,169	353,624	
1883-84	894,420	355,210	
1884-85	857,970	344,993	
1885-86	936,984	363,075	
1886-87	1,064,386	429,708	
1887-88	1,084,487	462,935	5
1888-89	1,119,367	482,854	

Rx. = 10 rupees or the conventional pound sterling.

"It will be seen that in 18 years the rate of duty has been increased by 66 per cent., and that, nevertheless, the quantity consumed has increased by 142 per cent. and the revenue by 161 per cent. The fact that the rate of increase of revenue is not proportionate to the increase in quantity and in duty is an indication that the spirits now imported contain on an average less alcohol per gallon than in former years. In quantity of pure alcohol imported the increase is about 50 per cent. The increase of the Excise Revenue proper, *i.e.*, excluding the duty on imported wines, spirits, and beer, between 1870-71 and 1888-89, has been from Rx. 2,374,465 to Rx. 4.705,346, or an increase of 98 per cent., against one of 161 per cent. in the case of imported spirit.

"The figures we have just given show clearly that an increase in the revenue derived from excise, or even in the quantity of liquor consumed, does not necessarily indicate any relaxation of the restriction on the liquor trade. We have adopted what is admitted to be the most efficacious means of restricting the consumption of imported spirits by imposing a high duty and raising the rate from time to time. The imports have, nevertheless, increased; and the increase has been greater, proportionately, than that of the Excise Revenue proper. We believe that a portion of this increase is due to the higher rates of duty gradually imposed on spirit manufactured in the country which have made it possible for imported spirit to compete in some places with country spirit. If we are correct in this conclusion, there cannot have been any stimulus of the sale of country spirit by reducing the cost at which it may be procured. But apart from that, the point we wish to bring out is that the fact of an increase having taken place in a similar branch of revenue where we have admittedly adopted the best possible means of restricting consumption is sufficient to show that a mere increase in the Excise Revenue is not a ground for condemning our excise administration" (*Memorandum on Excise Administration published by the Government of India in the Gazette, March 1st, 1890*).

74. **Strychnos Nux-vomica,** *Linn.*; LOGANIACEÆ.

NUX-VOMICA.

The bark and seeds contain strychnine and brucine. See Vol. VI.

75. **Tárí.**

The juice of certain palms which is consumed as a kind of beer or distilled in the preparation of spirits. See **Borassus, Caryota, Cocos, Phœnix,** &c.

N. 15

WINE.

76. **Theobroma Cacao,** *Linn.;* STERCULIACEÆ.
CHOCOLATE.
See Vol. VI., Pt. II.

77. **Tephrosia suberosa,** *DC.*, see under Fish-poisons, Vol. III., 366; also **Mundulea suberosa,** *Benth.* p. 288.

78. **Viscum monoicum,** *Roxb.;* LORANTHACEÆ.
Specimens taken from Nux-Vomica trees in Cuttack were found to contain strychnine and brucine (*O'Shaughnessy, Beng. Dispens., 375*).

79. **Vitis vinifera,** *Linn.;* AMPELIDEÆ.
The Gape and Wine, Brandy, &c. See Vol. VI., Pt. II.

80. **Walsura piscidia,** *Roxb.;* MELIACEÆ.
The bark is used in the Circars for poisoning fish. See Vol. VI., Pt. II.

81. **Withania somnifera,** *Dun.;* SOLANACEÆ.
The leaves and stems are supposed to be narcotic and diuretic, and by some are considered to be the **Strychnos hypnoticus** of **Dioscorides.** See Vol. VI., Pt. I.

EXCISE SYSTEM IN AND REVENUE FROM NARCOTICS.

REVENUE. 16

This subject is perhaps referable to three main headings: Liquors, Opium, and Drugs, *e.g.*, *Bhang, Ganjá,* &c. In the present brief sketch of the Indian Excise System it is unnecessary to do more than deal with the main features of Excise as affecting Spirituous Liquors. Information regarding Opium will be found in the article **Papaver somniferum** and of Indian Hemp in the remarks above under **Cannabis sativa,** also in Vol. II., 103-126. This isolation of Opium and Indian Hemp is the more necessary since the major portion of the Opium revenue is not returned as Excise. The items of opium which do appear in Excise are the duty and the license fees obtained from the right to sale and from the amount consumed in India.

The Government of India in its Memorandum on Excise Administration (to which repeated reference has been made) points out that there are numerous difficulties which prevent the adoption of a uniform system with indigenous liquors, but that these difficulties do not exist with imported beverages. The system that would be preferred, were uniformity possible, would be the imposition of a fixed duty per gallon of spirits proportionate to the alcoholic strength. "This is known in its simplest form as the Central Distillery system, because under it all liquor is distilled at a public distillery, centrally situated with reference to the tract to be supplied, and watched by a preventive establishment appointed by Government. No liquor is allowed to leave the distillery until the duty has been paid. The history of Excise Administration in India shows that the line of progression has been from the Farming system, inherited from the Native States which preceded British rule, towards a system under which each gallon of spirits pays a fixed duty. The earliest system, that of uncontrolled farming, the farmer paid a lump sum for the right to manufacture and sell liquor in a special tract of country; there was no limit as to number of shops and he made what profit he could out of his farm. The next step was the limitation of the number of shops. A further step was the establishment of the outstill, under which only the right to manufacture and sell at a specified shop is granted. The chief objection both to Farming and Outstill systems is that there is no control over the rate of duty per gallon, and, consequently, it may be to the interest of the farmer or licensee to steadily lower prices. It then becomes necessary to consider whether an attempt can be successfully made to establish a system under

REVENUE.

which each gallon of spirits shall pay a fixed duty. This may be done roughly and imperfectly by limiting the capacity of the outstill and fixing a minimum rate of duty for the right to work the outstill. If the outstill can only produce a certain number of gallons of spirit in the month and must pay a certain tax every month, we know that each gallon of spirit will bear, at least, a certain rate of duty and cannot be sold below a certain price. The practical difficulty in the way of securing this result is the facility which the licensee possesses for distilling in a still other than that which has been licensed. This difficulty is in some cases insuperable. A perfect remedy lies in the establishment of a Central Distillery system, but unfortunately the practical experience of years, and the results of many experiments, have shown that it is impossible to maintain this system in all parts of India.

"Under the Farming and Outstill systems the interest of the monopolist is enlisted for the suppression of illicit distillation, and sources of information and means of detection are available to him which are not accessible to officials. This is not the case under the ordinary Central Distillery system. Under that system it is to the interest of the distiller both to distil illicitly outside the distillery and to smuggle liquor out of the distillery without payment of duty. The shop-keeper and the subordinate officials share the profit. Thus it was proved in the course of a judicial enquiry in the Gya District of Bengal that out of the spirit made at one outlying distillery only one gallon out of three paid duty to Government, the duty on the second being retained by the preventive officer, who allowed the distiller to take out every third gallon free.

"The key to the numerous and complicated systems of Excise which prevail in different parts of British India lies in the fact that they are attempts to combine the monopoly and fixed duty systems with the object of securing that every gallon of spirit shall bear a certain rate of duty in places where it is not possible to work the fixed duty system in its simplest form" (*Supplement to the Gazette of India, March 1st, 1890*).

The Government of India in its Resolution on the Memorandum on Excise Administration exhibits the reasons why a Central Distillery system has not been universally adopted. These arguments may be here summarised:—

1st—Where the quantity of liquor consumed is small, the cost of a central distillery, and of an establishment to guard it and to prevent outstill distillation, is prohibitive.

2nd—It is found impossible on the scale of pay which can be given to obtain for those establishments men who are thoroughly trustworthy.

3rd—The means of communication are too defective, thus prohibiting the transport of liquor from central distilleries to remote and rural populations. Besides, the less harmless beverages cannot stand the effects of climate and handling on being carried from one place to another. Any attempt, therefore, to enforce a central distilling supply would be to force the people to use stronger alcoholic drinks instead of the locally prepared beers or mild liquors which take the place of beer among the people.

4th—The difficulty of preventing illicit distillation is very great. The Government on this point remarks (taking Assam as an example):—"In nearly the whole of Assam we must be content with enforcing direct restrictions on the *sale* of spirits, sale to the public being a transaction which cannot be altogether concealed, while distillation may be carried on in secret without fear of detection." The difficulty of preventing Native States from sending liquor into British territory is also very great, and illicit transactions in this direction are of no mean importance.

Even in provinces such as Bengal where the Central Distilling system

REVENUE.

has for some time been in full operation, smuggling prevails to a large extent. This is especially the case in districts where the *mahua* tree abounds, in thinly-peopled tracts or in centres inhabited by the lower classes. In cities it occurs but to a small extent, the excise administration being there able to control the traffic.

The extent to which the Central Distillery system has been enforced in the various provinces of India may be learned from the following paragraphs:—

"It was between 1860 and 1864 that the Central Distillery system was generally introduced into Bengal and the Upper Provinces of India.

"In the Province of Bengal it was found necessary to abandon the Central Distillery system in a few places very shortly after it was introduced. It was also found that the high uniform rate of duty could not be enforced in all places where the Central Distillery system was maintained. The choice, therefore, lay between a low uniform rate of duty and a scale varying from place to place. The latter alternative was adopted. The subsequent substitution on an extensive scale of the Outstill for the Central Distillery system after 1877 was not made without cause. The objections to the Central Distillery system were obvious, and it was hoped that certain advantages, such as the suppression of fraud, of illicit manufacture and smuggling, and a consequent increase of revenue would be obtained. These advantages were actually secured, but the change was accompanied by the removal of the restriction on the capacity of outstills, and an increase of drinking followed which has rendered it expedient to return to the Central Distillery system where practicable, and to impose various restrictions on the outstill system where it must be retained.

"In the North-Western Provinces certain tracts had to be removed from the operation of the Central Distillery system, and though the uniform rate has been retained, this has only been rendered possible by fixing the rate at a comparatively low figure, and increasing the numbers of shops so as to reduce the temptation to illicit practices.

"In the Panjáb, the Central Distillery system has been introduced, and with two trifling exceptions, maintained throughout the Province, the rate of duty is high, and it is practically uniform; on the other hand, illicit distillation and smuggling are more common than in any other Province, and the number of shops for the sale of liquor has been increased, though there is still only one shop for the sale of liquors of all kinds to nearly 10,000 of the population.

"In 1878 the Central Distillery system was generally introduced in Bombay, but it was found necessary that it should be accompanied with the grant of monopolies and with a 'minimum guarantee,' the amount of the guarantee being fixed by competition. The system of a 'minimum guarantee' fixed by competition is now being altered, as it was found to be open to objection in some respects.

"In 1869 the Central Distillery system began to be introduced in Madras. It was accompanied with various conditions and special provisions, and has since been extended to nearly all the districts of the Presidency, having now devolved into a system which is almost the Central Distillery system in its simplest form and with a high rate of duty. The success of the system adopted in Madras is, however, accompanied by a large number of shops" (*Memorandum on Excise Administration*).

Concluding its Resolution on the information collected from all parts of India (from which the *Memorandum on Excise Administration* was compiled), the Government of India expresses the policy it has pursued and intends to continue in these words:—

"The practical measures which we propose to adopt in future in

REVENUE.

furtherance of our declared policy comprise—(1) the abolition of the Farming or Outstill systems in places where it is found practicable to do so; (2) the gradual introduction of the Central Distillery system in its least complex form; (3) the imposition of as high a rate of duty on country liquor as it will bear, subject to the limitation that such duty shall not exceed the tax levied on imported liquor; and (4) the restriction of the number of shops. Where the Outstill-system is retained, we shall, as far as possible, enforce the limitation of the capacity of the still, and in some instances a minimum selling price.

"We do not anticipate that the carrying out of this policy in a rational manner and with reasonable regard to the circumstances of the country will lead to any loss of revenue. On the contrary, we believe it will be as successful from the financial as from every other point of view."

The gross receipts for spirits, wines, malt liquors, opium, and drugs, &c., returned under the Excise Department and for the corresponding imported articles under Customs may be seen from the following table which gives the figures for the past five years:—

	Customs duty.				
	Malt Liquor.	Spirits.	Wines and Liqueurs.	Opium.	Total.
	R	R	R	R	R
1884-85 .	6,609	3,44,993	49,658	286	4,01,546
1885-86 .	8,128	3,63,075	46,961	279	4,18,443
1886-87 .	10,712	4,29,708	47,776	428	4,88,624
1887-88 .	13,178	4,62,935	51,678	254	5,28,045
1888-89 .	15,205	4,82,854	50,224	305	1,14,019

Excise.					Total revenue derived from customs duty and excise.
Spirits.	Drugs.	Miscellaneous.	Opium.	Total.	
R	R	R	R	R	R
25,48,289	3,88,511	35,732	8,44,654	38,17,186	42,18,732
26,99,200	3,92,444	44,829	8,44,177	39,80,660	43,99,103
28,37,880	4,18,037	40,345	8,63,073	41,59,335	46,47,959
29,72,077	4,34,725	34,156	8,77,377	43,18,335	48,46,380

Details not available.

(*J. F. Duthie.*)

NARDOSTACHYS, *DC.; Gen. Pl., II., 153.*

17 **Nardostachys Jatamansi,** *DC.; Fl. Br. Ind., III., 211;*

SPIKENARD. [VALERIANEÆ.

Syn.—N. GRANDIFLORA, *DC.*; PATRINIA JATAMANSI, *Don*; VALERIANA JATAMANSI, *Wall.*

Vern.—*Jatamánsi, bálu-char, bál-chhar, bál-chir,* HIND.; *Jatamánsi,* BENG.; *Bekh-kurphus,* BEHAR; *Haswa, naswa, jatámángsi,* NEPAL; *Jatamansi, pampe, paumpe,* BHUTAN; *Mási,* GARHWAL; *Bhút-jatt, kukil-i-pot,* KASHMIR; *Jhata-mánsi,* DEC.; *Balacharea, sumbul,* BOMB.; *Jatamasi, kalichhad,* GUJ.; *Jatamáshi,* TAM.; *Jatamámshi,* TEL; *Jeta-mávashi* KAN.; *Jeta-mánchi,* MALAY; *Jaramánsi,* SING.; *Jatámánsi,* SANS.; *Sumbulu'l-hind, sunbuluttibe-hindi, sumbulul-aasáffir,* ARAB.; *Sunbuluttib,* PERS. As is the case with Valerian, cats are fond of the smell of the root of this plant, hence it is, sometimes, called *Billi-lotan* in the Deccan.

References.—*Roxb., As. Res., IV., 433; Voigt, Hort. Sub. Cal., 435; Stewart, Pb. Pl., 118; Pharm. Ind., 120; Ainslie, Mat. Ind., II., 367; O'Shaughnessy, Beng. Dispens., 403; Moodeen Sheriff, Supp. Pharm. Ind., 253; U. C. Dutt, Mat. Med. Hind., 180; Dymock, Mat. Med. W. Ind., 417-419; Fleming, Med. Pl. and Drugs, As. Res., Vol. XI, 183; Flück. & Hanb., Pharmacog., 312; U. S. Dispens., 15th Ed., 1706; S. Arjun, Bomb. Drugs, 72; Murray, Pl. and Drugs, Sind, 180; Bidie, Cat. Raw Pr., Paris Exh., 32; Waring, Bazar Med., 78; Year-Book Pharm., 1879, p. 214; Irvine, Mat. Med. Patna, 11; Med. Top. Ajm., 128; Linschoten, Voyage to E. Indies, II., 126 (Edition 1885); Atkinson, Him. Dist., 743; Birdwood, Bomb. Pr., 46; Royle, Ill. Him. Bot., 242-244, t. 54; McCann, Dyes and Tans, Beng., 143; Piesse, Art Perfumery, 203-4, 305; Balfour, Cyclop., II., 1062; Smith, Dic., 387; Treasury of Bot., 777; Kew Off. Guide to the Mus. of Ec. Bot., 83; Ind. For. X. (1); Home Dept. Cor., regarding Pharm. Ind., 222-239; Sir W. Jones, Asiat. Res., II., 405; & IV., 109; LeMaout & Decaisne, Descrip. & Analyt. Bd. (Eng. Ed.), 491; Madden, Ann. & Mag. Nat. Hist., Ser. 2., Vol. XVIII., 449.*

Habitat.—A perennial herb of the Alpine Himálaya, which extends eastwards from Garhwál, and ascends to 17,000 feet in Sikkim.

History.—For the identification of the well-known fragrant root of this plant with the true spikenard of the ancients we are indebted to **Sir William Jones,** whose classical treatise, published in Vol. II. of the *Asiatic Society's Researches,* deals exhaustively with all previous investigations regarding its supposed origin. The uncertainty which had so long obscured its botanical origin may to a great extent be accounted for by the inaccessible habitats of the plant; and, indeed, it may be observed that both **Sir W. Jones** and **Dr. Roxburgh** (see *As. Res.,* IV., 433) failed to obtain specimens of the true **Nardostachys.** In both instances the plates accompanying their papers represent **Valeriana Wallichii,** a totally different plant, though of the same Natural Order, and found at much lower elevations on the Himálaya. In spite of this error as to the exact botanical determination of *Jatamansi,* the arguments brought forward by **Sir W. Jones** removed all doubts as to the identification of the drug called *Jatamansi* with the *Sumbulu'l Hind.* of the Arabs and the Indian spikenard of **Dioscorides.** This conclusion was in reality based upon the examination of the bazár drug which was easily obtainable, together with the results of philological enquiries made of the best informed Muhammadan physicians. By **Linnæus** and other writers it was supposed that the true *nard* was a grass, many scented kinds of which, chiefly of the Genus **Andropogon,** occur in India. **Sir W. Jones,** who at one time, during the course of his investigations, was tempted to trace a gramineous origin for *Jatamansi,* failed to discover a grass having any resemblance to *Jatamansi.* "I am not, indeed," he remarks, "of opinion that the *nardum* of the Romans was merely the essential oil of the plant, from which it was denominated, but am strongly inclined to believe that it was a generic word, meaning what we now call *átar* and either the *átar* of roses from Kashmír and Persia, that of *Cetaca* or **Pandanus** from the western coast of India, or that of *Agaru*—aloe-wood—from Assam or Cochin China, or the mixed perfume, called *abir,* of which the principal ingredients were yellow sandal, violets, orange-flowers, wood of aloes, rose-water, musk, and true spikenard: all those essences were costly; and, most of them being sold by the Indians to the Persians and Arabs, from whom, in the time of **Octavius,** they were received by the Syrians and Romans, they must have been extremely dear at Jerusalem and at Rome." Further he says:—"The *nard* of Arabia was, probably, the **Andropogon Schœnanthus,** which is a native of that country; but, even if we suppose that the spikenard of India was a reed or a grass, we shall never be able to distinguish it among the many Indian species of **Cyperus, Andropogon, Schœnus, Carex,** and other genera of those

HISTORY. 18

HISTORY.

natural orders, which here form a wilderness of sweets and some of which have not only fragrant roots, but even spikes in the ancient and modern senses of that word." "My own enquiries having convinced me that the Indian spikenard of **Dioscorides** is the *Sumbulu'l Hind,* and that the *Sumbulu'l Hind* is the *Jatamánsi* of **Amarsingh,** I am persuaded that the true *nard* is a species of Valerian, produced in the most remote and hilly parts of India, such as Nepál, Morang, and Bhután, near which **Ptolemy** fixes its native soil: the commercial agents of the Devarája call it also *Pampi,* and, by their account, the dried specimens which look like the tails of ermines, rise from the ground, resembling ears of green wheat both in form and colour,—a fact which perfectly accounts for the names *Stachys, Spica, Sumbul,* and *Kushah,* which Greeks, Romans, Arabs, and Persians have given to the drug, though it is not properly a spike, and not merely a root, but the whole plant, which the natives gather for sale, before the radical leaves have unfolded themselves from the base of the stem."

It is most probable that the wrong plant was purposely sent to **Sir W. Jones,** in place of the true *Jatamansi,* which appears to have been so highly valued by the Bhután authorities that strict orders had been given forbidding the exportation of living plants from that country. **Dr. Wallich** and **Dr. Royle** independently detected the imposture, and the true plant was sent home and described by **Don** in his *Prodromus Flor. Nep.* under the name of **Patrinia Jatamansi.** It was afterwards removed by **DeCandolle** to the Genus **Nardostachys. Dr. Royle,** in his *Illustrations of Himálayan Botany,* p. 242, thus writes, in confirmation of the conclusions arrived at by **Sir W. Jones**:—

"Notwithstanding the proofs, adduced by **Sir W. Jones,** it has been said that the grounds are insufficient on which the *Jatamansi* of the Hindús has been considered to be the spikenard of the ancients. Having followed the course pointed out by that eminent orientalist, without taking exactly the same steps, it is not uninteresting to state that I arrived at precisely the same results. **Dioscorides** describes three kinds of *nard,* of the first and principal of which there are two varieties, Syrian and Indian; the latter is also called *Gangites,* from the River Ganges, near which, flowing by a mountain, it is produced. The second kind is called 'Celtic,' and the third 'mountain nard.' On consulting **Avicenna,** we are referred from *Narden* to *Sunbul,* pronounced *Sumbul,* and in the Latin translation from *Nardum* to *Spica,* under which the 'Roman,' the 'mountain,' the 'Indian' and 'Syrian' kinds are mentioned; and *Senbel* misprinted *Seubel,'* is given as the synonymous Arabic name. This proves, as stated by **Sir W. Jones,** that *Sambul,* in Persian dictionaries, translated, 'the Hyacinth—the spikenard, to which the hair of a mistress is compared, an ear of corn, &c.'—was always considered by Arabian authors as synonymous with the *Nardos* of the Greeks. On consulting the Persian works on Materia Medica in use in India, and especially the *Mukhzun-ool-Udwieh,* we are referred from *Narden,* in the Index, to *Sumbul,* in the body of the work. Under this name, however, four separate articles are described—1st, *Sumbul-hindí;* 2nd, *Sumbul-roomí,* called also *Sumbul-ukletí,* and *Narden ukletí.* But the first alone is that with which we have at present any concern. The synonyms of it, given by Persian authors, are Arabic, *sumbul-ool-teeb* or fragrant nard; Greek, *narden;* Latin, *nardoom;* Hindee, *balchur* and *jatamasí.* The last is the Sanscrit name and that which was given to **Sir W. Jones,** as the equivalent of *sumbul-hindí,* and which he informs us, like other Sanskrit names applied to the same article, has reference to its resemblance to locks of hair."

HISTORY.

Madden, referring to a second species of **Nardostachys** (described by **DeCandolle** under the name of **N. grandiflora** and reduced in the *Fl. Br. Ind.*, to **N. Jatamansi**), remarks that it flourishes at similar elevations in Kumáon (13,000 to 14,000 feet), but is a larger plant, and with a more agreeable smell. "The perfume and properties of the genus are," says **Madden**, "very nearly those of **Valeriana celtica** and **V. phu**; and it is curious enough that the radical leaves of the last two species (the roots of which are substituted in Western Asia for the spikenard) are simple and bear a considerable resemblance to those of **Nardostachys**.

Garcia d'Orta gave two localities in Central India, *viz.*, Chitor and Mandu, for the true spikenard. The same statement was made by **Linschoten** in the account of his Voyage to the East Indies, though his description was evidently intended to be that of the Himálayan plant. This error, which, strange to say, **DeCandolle** was also led into, may be accounted for by the fact of the abundant growth, at those very localities in Central India, of a sweet-scented **Andropogon** (**A. Schœnanthus**), a grass which at one time was supposed might be the source of Indian spikenard.

From a general view of the early history of *Jatamansi*, there is much reason to believe that the ointment of spikenard, alluded to by St. John, also the alabaster box of ointment mentioned by St. Mark continued as the principal ingredient the essence yielded by this plant. Spikenard is also mentioned more than once in the "Song of Solomon." As the ointment was usually described as having been "poured" when used, its consistency must have been rather of the nature of an oil, in which condition *Jatamansi* mixed with a variety of other perfumes is in common use as a hair-wash among Indian women of the present day.

DYE. Root-stock. 19

Dye.—**McCann** states that the ROOT-STOCK of this plant is used in Lohárdaga (Bengal), as an auxiliary in dyeing along with *Kamalágundi* (**Mallotus philippinensis**).

OIL. 20

Oil.—Fifty-six pounds of *Jatamánsi* distilled by **Kemp** yielded 3 fluid ounces of a brown-coloured essential oil, with a molecular rotation of 19·5 in 100 mm., and a sp. gr. (at 82° F.) of 0·9748 (*Dymock*).

MEDICINE. Root. 21

Medicine.—The ROOTS of this plant are aromatic and bitter in taste. They are supposed to possess tonic, stimulant, and antispasmodic properties, and are often employed in the treatment of epilepsy, hysteria, and convulsive affections. They are also considered deobstruent, diuretic, and emmenagogue, and are recommended in various diseases of the digestive and respiratory organs. **O'Shaughnessy** found *Jatamánsi* a good substitute for Valerian. It is said to be useful in jaundice, affections of the throat, and as an antidote for poisons. It is popularly believed to have the power of promoting the growth and blackness of hair. It has been noticed by **Ainslie** that in Lower India the people prepare a fragrant and cooling liniment from this drug, which is applied to the head; *Jatamánsi* is also used internally as a blood-purifier. According to **U. C. Dutt** it is much employed as an aromatic adjunct in the preparation of medicinal oils; but except as an ingredient of complex prescriptions, it does not appear to have been used internally.

SPECIAL OPINIONS.—§ "Used in palpitation of heart" (*Surgeon-Major D. R. Thomson, M.D., C.I.E., Madras*). "Antispasmodic, aromatic, doses 10 to 20 grains, used in hysterical affection, palpitation of the heart, chorea, flatulence" (*1st Class Hospital Assistant Choonna Lall, in charge of City Branch Dispensary, Jubbulpore*).

TRADE. 22

Trade.—**Atkinson** (*Gaz. Him. Dist.*, 743) mentions that its roots, together with those of certain species of Valerian, are sent down to the plains to the extent of about 20 maunds a year.

TRADE. The bazaár price of *Jatamansi* is now about eight annas per pound. There is very little demand for it in these days, and practically no export out of India. In ancient times when it formed the chief ingredient of the most precious ointments, it is said that a pound of genuine Indian spikenard would realise one hundred denarii.

NAREGAMIA, *W. & A.; Gen. Pl., I., 331.*

3 **Naregamia alata,** *W. & A., Prod., I., 117; Fl. Br. Ind., I., 542; Wight, Ic., t. 90;* MELIACEÆ.

GOANESE IPECACUANHA.

Syn.—TURRÆA ALATA, *Wight, ex W. & A., l.c.*

Vern.—*Pittvel, pittpápra, pittmári, tinpáni,* MAR.; *Nela-naringu, nala kanu-gida,* KAN.; *Nela-naregam,* MALAY.; *Trifolio,* GOA.

References.—*Gamble, Man. Timb., 69; Dalz. & Gibs., Bomb. Fl., 36; Rheede, Hort. Mal., X., t. 22; Dymock, Mat. Med. W. Ind., 174; Pharmacog. Ind., 333; Hooper, in Pharm. Journ., XVIII., 317; Drury, U. Pl., 307.*

Habitat.—A small, glabrous, undershrub, with trifoliolate leaves, found in Western and Southern India.

HISTORY. 24 **History.**—Garcia d'Orta, for many years, during the sixteenth century, physician to the Viceroy's Court at Goa, was, according to **Dymock**, acquainted with the drug, which he called *Avacari* (meaning emetic), but he had not seen the plant. He mentions a successful cure of a case of dysentery treated with a decoction of the bark in rice water. **Dymock** supposes the Goanese name *Trifolio* to be a translation of the Marathi name *Tínpáni.*

MEDICINE. Roots. 25 Stems. 26 **Medicine.**—The drug consists of the creeping ROOTS and the slender STEMS attached to them. It is mentioned by **Rheede** as a remedy in South India for rheumatism and itch. **Dymock** states that in the Konkan the Hindus use the stems and leaves in decoction with bitters and aromatics as a remedy for biliousness. According to the *Pharmacographia Indica* it has recently been tried in Madras in acute dysentery and also as an emetic and expectorant with results similar to those obtained from Ipecacuanha, given in equal doses. The root has a pungent aromatic odour.

CHEMISTRY. 27 CHEMICAL COMPOSITION.—The ether extract was found by **Hooper** to contain an alkaloid, an oxidisable fixed oil, and a wax.

Narthex Asafœtida, *Falc.,* see **Ferula Narthex,** *Boiss.,* Vol. III., 339; [UMBELLIFERÆ.

NASTURTIUM, *Br.; Gen. Pl., I., 68.*

28 **Nasturtium officinale,** *Br.; Fl. Br. Ind., I., 133;* CRUCIFERÆ.

WATER CRESS.

Vern.—*Piriya-hálim,* N.-W. HIM.; *Lút-putiah* (according to **O'Shaughnessy**), DEC.

References.—*Voigt, Hort. Sub. Cal., 67; Stewart, Pb. Pl., 14; Aitchison, Cat. Pb. and Sind Pl., 4; DC., Origin Cult. Pl., 438; O'Shaughnessy, Beng. Dispens., 186; Pharmacog. Ind., I., 130; U. S. Dispens., 15th Ed., 1706; Murray, Pl. and Drugs, Sind, 49; Atkinson, Him. Dist., 708; Foods of N.-W. Prov., Pt., 34; Royle, Ill. Him. Bot., 70; Balfour, Cyclop., II., 1064; Smith, Dic., 142; Treasury of Bot., 778; Journ. Agri.-Hort. Soc., 1871-74, IV., 23; XIV., 5.*

Habitat.—An aquatic herb, with creeping or floating stems, cultivated in many parts of India. On the Himálaya it occurs in a semi-wild state near hill stations.

MEDICINE. Plant. 29 **Medicine.**—The PLANT is widely known for its antiscorbutic and stimulant properties, and is also largely eaten to increase appetite.

Food.—To a small extent the PLANT is eaten by the natives, but it is more frequently collected and sold at the market places frequented by Europeans. FOOD. Plant. 30

NAUCLEA, *Linn.; Gen. Pl., II., 31.*

[266; RUBIACEÆ. 31

Nauclea Cadamba, *Roxb.*, see **Anthocephalus Cadamba,** *Miq.*, Vol. I.,

N. cordifolia, *Roxb.*, see **Adina cordifolia,** *Hook. f., & Benth.*, Vol. I.,

N. ovalifolia, *Roxb.; Fl. Br. Ind., III., 27.* [144. 32

Vern.—*Shal*, SYLHET.

References.—*Roxb., Fl. Ind., Ed. C.B.C., 173; Voigt, Hort. Sub. Cal., 375; Pharm. Ind., 117; Journ. Agri.-Hort. Soc. Ind., IX. (1857), App., ccxxiv. and ccxlvi.*

Habitat.—A tree, said to be found in the forests of Sylhet. In the *Flora of British India* it is placed amongst the doubtful species, with the remark—"possibly **Adina sessilifolia.** Don refers it to **Uncaria elliptica,** which is not a Khasian plant."

Medicine.—In a communication to the Agri.-Horticultural Society of India, Lieutenant **Stewart** mentions that "the BARK of this tree is used extensively among the tribes of the Cachar frontier as a cure for fever and bowel complaints, and that it possesses a bitterness equal almost to the Peruvian bark." MEDICINE. Bark. 33

N. parviflora, *Pers.*, and **N. parvifolia,** *Willd.*, see **Stephegyne parvifolia,** *Korth.*, Vol. VI.

N. rotundifolia, *Roxb.*, see **Stephegyne diversifolia,** *Hook. f.*, Vol. VI.

Nectandra Rodiæi, *Schomb.* (LAURINEÆ), is the greenheart, or *Bibiri* tree of British Guiana. The BARK of this tree, which is officinal, contains an alkaloid used as an antiperiodic in ague. The TIMBER is much valued for ship-building. 34 MEDICINE. Bark. 35 TIMBER. 36

Neeradimootoo Oil. 37

Vern.—*Jungli badam-ka-tel*, HIND.; *Mootoo, yennai*, TAM.

This OIL was sent to the Madras Exhibition under several names. It is generally prescribed by native practitioners as a valuable medicine. The exact source of this oil has not as yet been determined. MEDICINE. Oil. 38

NELUMBIUM, *Juss.; Gen. Pl., I., 47.*

Nelumbium speciosum, *Willd.; Fl. Br. Ind., I., 116;* NYMPHÆACEÆ. 39

THE SACRED LOTUS—PYTHAGOREAN OF EGYPTIAN BEAN.

Syn.—NYMPHÆA NELUMBO, *Linn.;* N. ASIATICUM, *Rich.;* NELUMBO INDICA, *Poir.;* CYAMUS NELUMBO, *Smith.*

Vern.—*Kanwal, kanval*, HIND.; *Padma, padama*, BENG.; *Padam*, URIYA; *Besenda* (Bijnor), *pabbin* (Muzafarnaggar), N.-W. P.; *Pamposh, kánwal kakri*, and *bhe* or *phe* (root), *gatte* (seed), PB.; *Pabban* (plant), *beh* (root), *paduro* (seeds), *nilofir* (drug), SIND.; *Kungwelka-gudda*, DEC.; *Kamala, kankadi*, BOMB.; *Tavarigadde, tavaribija*, KAN.; *Dudha-malida-kand*, KHAND.; *Pand-kanda*, POONA; *Shivappu-támara-ver, ambal*, TAM.; *Erra-támara-veru*, TEL.; *Tamara* (Rheede), MALAY.; *Pa-dung-ma*, BURM.; *Nelum*, SING.; *Padma, nalina, aravinda, mabotpala, camala, cuseshaya, sabafrapatra, sarasa, panceruba, támarasa, sarasiruba, rajiva, visaprasuna, pushcara, ambhoruna, satapatra* (all the preceding from Sir W. Jones' works), *padmachári* (Ainslie), SANS.; *Nilufer, ussulneelufir*, ARAB.; *Nilufer, nilufu, beykhneelufir*, PERS.

NELUMBIUM speciosum.

The Sacred Lotus.

References.—*Roxb., Fl. Ind., Ed. C.B.C., 450; Voigt, Hort. Sub. Cal., 9; Thwaites, En. Ceylon Pl., 14; Grah., Cat. Bomb. Pl., 5; Dalz. & Gibs., Bomb. Fl., 7; Stewart, Pb. Pl., 9; Aitchison, Cat. Pb. and Sindh Pl., 3; W. & A., Prod., I., 16; Elliot, Fl. Andhr., 141, 160, 173, 179; Taylor, Topogr. of Dacca, 47; Ainslie, Mat. Ind., II., 410; Moodeen Sheriff, Supp. Pharm. Ind., 17, 18; U. C. Dutt, Mat. Med. Hind., 110; Dymock, Mat. Med. W. Ind., 2nd. Ed, 37; Pharmacog. Ind., 70; S. Arjun, Bomb. Drugs, 8; Murray, Pl. and Drugs, Sind, 71; Bidie, Cat. Raw Pr., Paris Exh., 122; Year-Book, Pharm., 1879, 290; Baden Powell, Pb. Pr., 329; Atkinson, Foods, N.-W. P., 23; Drury, U. Pl., 309; Lisboa, U. Pl. Bomb., 143; Birdwood, Bomb. Pr., 5, 136; Royle, Ill. Him. Bot., 65; Balfour, Cyclop., II., 1080; Smith, Dic., 359; Treasury of Bot., 781; Kew Off. Guide to the Mus. of Ec. Bot., 10; Kew Off. Guide to Bot. Gardens and Arboretum, 25; Jour. As. Soc., 1867, 79, 82; LeMaout and Decaisne, Descript. and Analyt. Bot. (Eng. Ed.), 212; Baron v. Mueller, Select. Extratrop. Pl., 255.*

Habitat.—A large aquatic herb with peltate leaves, and handsome rosy, red or white flowers, found all over India, and extending as far north as Kashmir. The flowers appear during the hot season, and the seeds ripen towards the end of the rains.

FIBRE. Stalks. 40

Fibre.—The long STALKS of the Lotus yield a sort of yellowish white fibre, which is used principally for the wicks of sacred lamps in Hindú temples; and the Hindú doctors are of opinion that the cloth prepared from this fibre acts medicinally as a febrifuge (*Baden Powell*).

MEDICINE Filaments. 41 Seeds. 42 Leaves. 43 Juice. 44

Medicine.—The FILAMENTS, known under the name of *kinjalka*, are astringent and cooling, and prescribed in the burning of the body, bleeding from piles, and menorrhagia. The SEEDS are considered medicinal, and used to check vomiting. They are also given to children as diuretic and refrigerant. The large LEAVES form cool bed-sheets, useful in fever accompanied by much heat and burning of the skin. It is said that the milky viscid JUICE of the leaf and FLOWER-STALKS is a remedy in diarrhœa, and that the PETALS are slightly astringent.

Root. 45

A sherbet of this plant is used as refrigerant in small-pox, and is said to stop eruption; it is given also in all eruptive fevers. The ROOT is employed as a paste in ringworm and other cutaneous affections (*Dr. Emerson*).

SPECIAL OPINIONS.—§ "The dry petals of **Nelumbium speciosum** are from 2½ to 3½ inches long, elliptical, pink, crimson, reddish-brown or pale-white, and possess no distinct smell or taste The seeds or nuts are hard and dark-brown, round, oval or oblong, about the size of the seeds of a soap-nut tree, with a white, albuminous and slightly sweetish kernel. The root occurs in the bazár in small and circular pieces, varying in diameter from 3 or 4 lines to 1 inch with several holes arranged in circular form with a solitary and generally smaller one in the centre; odourless and slightly mucilaginous in taste. The holes are the result of the cut ends of the spiral tubes. The physiological actions, therapeutic uses, preparations and doses of the petal, seeds, and root of this plant are precisely the same as those of the corresponding parts of **Nymphæa Lotus**" (*Honorary Surgeon Moodeen Sheriff, Khan Bahadur, G.M.M.C., Triplicane, Madras*). "The flowers and seeds are eaten by natives" (*Brigade Surgeon G. A. Watson, Allahabad*). "Is also used as a nervine tonic" (*Civil Surgeon J. Anderson, M.B., Bijnor, N.-W. P.*). "Refrigerant in fever, ½ drachm with sugar and water" (*Assistant Surgeon Nehal Sing, Saharunpur*). "Found useful as an infusion in cases of fever acting as a diaphoretic, and also said to act as powerful diuretic" (*Civil Surgeon F. F. Perry, Julunder City, Panjáb*). "The root, flower, stalks, and leaves are used as a refrigerant in fevers, including solar fever" (*First Class*

Hospital Assistant Lal Mahomed, Main Dispensary, Hoshangabad, Central Provinces). "Bees fed upon the *padma* flowers give what is called *Padma-madhu* much used in eye diseases in Bengal" (*Surgeon R. L. Dutt, M.D., Pubna*). "The tuber boiled in gingelly oil is rubbed on the head to cool the head and eyes. The expressed juice is also sometimes employed instead of sliced pieces of the tuber" (*Native Surgeon T. R. Moodelliar, Chingleput, Madras Presidency*). "The pistils are used with black pepper externally and internally as an antidote in snake poisoning" (*Civil Surgeon J. H. Thornton, B.A., M.B., Monghyr*).

FOOD. Roots. 46 Stalks. 47 Seeds. 48 Flowers. 49

Food.—The tender farinaceous ROOTS or rhizomes are eaten by the natives. In Kashmír, and parts of the Panjáb, **Dr. Stewart** says, the roots are dug out in October when the leaves dry up, are then sliced, and used cooked or pickled. The STALKS are eaten as a vegetable. The oblong nut-like SEEDS twice the size of peas, and, when ripe, so hard as to require a hammer to break them, are eaten by the natives either raw, roasted, or boiled. In parts of the Bombay Presidency and elsewhere in India, the roots are much sought after as food in times of famine. **Dr. Stocks**, in his list of wild edible plants of Sind, states that the FLOWERS and LEAF-STALKS are eaten, and that the roots are much esteemed by the natives, and are sold in every bazár in Sind. In China the roots are served up in summer with ice, and are also kept in salt and vinegar for the winter (*Loudon*).

SACRED. Flowers. 50

Sacred and Domestic Uses.—**U. C. Dutt**, in his *Hindu Mat. Med.*, referring to this and the allied species of **Nymphæa**, says:—"These beautiful plants have attracted the attention of the ancient Hindus from a very remote period, and have obtained a place in their religious ceremonies and mythological fables; hence they are described in great detail by Sanskrit writers. The FLOWERS of **N. speciosum** are sacred to Lakshmi, the goddess of wealth and prosperity." **Herodotus** describes the plant with tolerable accuracy, comparing the receptacle of the flower to a wasp's nest. **Strabo** and **Theophrastus** likewise mention the plant as a native of Egypt. Sculptured representations of it abound among the ruins of Egyptian temples, and many other circumstances prove the veneration paid to this plant by the votaries of Isis. In a manuscript of **Dioscorides** supposed to be of the twelfth century, formerly in the Rinuccini library at Florence, there is a figure of the **Nelumbium** under the name *kuamos*, while under the name *lotus*, a tolerably good representation of **Celtis australis** is given (*Treasury of Botany*). The wicks made from the spiral FIBRES of the leaf-stalks are burned by the Hindús before their idols, and the LEAVES are used as plates on which offerings are placed (*Smith, Econ. Dict.*). The NUTS are strung as beads. **Dr. Royle** alludes to the Egyptian mode of sowing this plant, by enclosing it in a ball of clay before throwing it into the water, as being still practised in India.

Fibre. 51 Leaves. 52 Nuts. 53

NEPENTHES, *Linn.; Gen. Pl., III., 166.*

Nepenthes.—A genus constituting the Natural Order NEPENTHACEÆ, and containing about thirty species, nine of which occur in British India. They are climbing or prostrate evergreen undershrubs, and are chiefly remarkable on account of the curious pitcher-like apendages which terminate the leaves. The inner surface of these pitchers is covered below the middle with glands which secrete water. The long tough STEMS (of **N. distillatoria** called *Bándúrawel* in Ceylon) are used for tying fences, and for other purposes, by the Sinhalese. 54

DOMESTIC. Stem. 55

NEPETA, *Linn.; Gen. Pl., II., 1199.*

Nepeta ciliaris, *Benth.; Fl. Br. Ind., IV., 661;* LABIATÆ. 56

Vern.—*Zúfa yábis*, PB.; *Júfa*, SIND.

References.—*Stewart, Pb. Pl., 170; Dymock, Mat. Med. W. Ind., 2nd Ed., 612; Atkinson, Him. Dist., 315.*

Habitat.—Western Himálaya, between 6,000 and 8,000 feet.

MEDICINE. Plant. 57 **Medicine.**—**Stewart** says that the PLANT is given in sherbet for fever and cough; the correct identification of the species alluded to by him is, however, open to question.

58 **Nepeta elliptica,** *Royle; Fl. Br. Ind., IV., 658.*

Reference.—*Stewart, Pb. Pl., 170.*

Habitat.—Temperate Western Himálaya from Kashmír to Kumáon.

MEDICINE. Seeds. 59 **Medicine.**—The SEEDS are said to be medicinal.

SPECIAL OPINION.—§ "One drachm of seeds infused in cold water in dysentery" *Assistant Surgeon Nehal Sing, Saharunpur*).

60 **N. floccosa,** *Benth.; Fl. Br. Ind., IV., 662.*

Vern.—*Chongmongo,* LADAK.

Reference.—*Stewart, Pb. Pl., 170.*

Habitat.—A woolly herb which occurs in Western Tibet, between 7,000 to 11,000 feet, and, according to **Stewart**, at from 10,000 to 16,500 feet in Ladak.

FODDER. 61 **Fodder.**—Said to be browsed by goats and sheep (*Stewart*).

62 **N. glomerulosa,** *Boiss., Fl. Or., IV., 651.*

Vern.—*Chingan butai,* PUSHTU.

Habitat.—Baluchistán hills up to 8,000 feet.

MEDICINE. Plant. 63 **Medicine.**—**Mr. J. H. Lace** mentions that it is used for indigestion.

N. malabarica, *Linn.*; see **Anisomeles malabarica,** *R. Br.*; Vol. I., 254.

64 **N. ruderalis,** *Ham.; Fl. Br. Ind., IV., 661.*

Syn.—GLECHOMA ERECTA, *Buch.*

Vern.—*Billi lotan, badranj boya, bebrang khatai,* PB.; *Niasbo,* NEPAL.

References.—*Roxb., Fl. Ind., Ed. C.B.C., 460; Voigt, Hort. Sub. Cal., 458.*

Habitat.—An annual which occurs in the hilly parts of Northern and Central India, and ascends to 8,000 feet on the Himálaya.

MEDICINE. Plant. 65 **Medicine.**—Supposed to be a cardiac tonic. According to **Dr. Buchanan** it is used by the Nepálese internally as a remedy for gonorrhœa. The decoction is used as a gargle in sore-throat; and it is largely employed in fevers (*Dr. Emerson*). **Stewart** says:—"It seems to be part at least of the officinal *billi lotan* which has been assigned to various Labiate plants, and is probably obtained from several."

NEPHELIUM, *Linn.; Gen. Pl., I., 405.*

66 **Nephelium lappaceum,** *Linn.; Fl. Br. Ind., I., 687;* SAPINDACEÆ.

Syn.—SCYTALIA RAMBOUTAN, *Roxb.*; EUPHORIA NEPHELIUM, *DC.*; DIMOCARPUS CRINITA, *Lour.*; NEPHELIUM ECHINATUM, *Nororh.*

References.—*Roxb., Fl. Ind., Ed. C.B.C., 329; Voigt, Hort. Sub. Cal., 95; Gamble, Man. Timb., 97; DC., Origin Cult. Pl., 315; Balfour Cyclop., II., 1085; Smith, Dic., 248; Treasury of Bot., 784.*

Habitat.—A tall tree of the Malay Peninsula, where it is known under the names of *ramboutan* or *rambosteen.* There are several varieties in cultivation.

FOOD. Aril. 67 **Food.**—Cultivated for its edible pulpy ARIL.

68 **N. Lit-chi,** *Camb.; Fl. Br. Ind., I., 687; Wight, Ic., t. 43.*

THE LITCHI.

Syn.—EUPHORIA PUNICEA, *Lamk.*; E. LIT-CHI, *Juss*; SCYTALIA LITCHI, *Roxb.*; N. DIMOCARPUS, *H. f. & T.*

The Litchi. (*J. F. Duthie.*) **NEPHELIUM Lit-chi.**

Vern.—*Litchi*, HIND. (originally Chinese); *Kyetmauk*, BURM.; *Lichi*, BOMB.

References.—*Roxb., Fl. Ind., Ed. C.B.C., 328; Voigt, Hort. Sub. Cal., 95; Kurz, For. Fl. Burm., I., 293; Gamble, Man. Timb., 97; Grah., Cat. Bomb. Pl., 29; Dalz. & Gibs., Bomb. Fl., Suppl., 13; DC. Origin Cult. Pl., 314; Trimen, Hort. Zeyl., 19; Atkinson, Econ. Prod., N.-W. P., Part V., 57; Lisboa, U. Pl. Bomb., 52, 150; Birdwood, Bomb. Pr., 143; Liotard, Dyes, 33; Spons', Encyclop., 1668; Balfour, Cyclop., II., 1086; Smith, Dic., 248; Treasury of Bot., 784; Kew Off. Guide to Bot. Gardens and Arboretum, 42, 70; Rep. Agri. Dept., 1881-82, 220; Ind. Gardener, 274-275; Linschoten, Voyage to E. Indies (Eng. Ed. 1885), I., 131.*

Habitat.—A handsome evergreen tree, introduced from South China, and now cultivated largely in India for its delicious fruit. **DeCandolle** says that "Chinese authors living at Pekin only knew the Litchi late in the third century of our era. Its introduction into Bengal took place at the end of the eighteenth century."

Cultivation.—The only successful mode of propagation known at present is by layering. The climate of Bengal has produced the best results hitherto, in regard to the quality of the fruit. Very fine fruit is, however, obtainable in the Lucknow Horticultural Garden and in the Saharanpur Government Garden, from a few trees, originally introduced from Calcutta. This shows the tree to be sufficiently hardy to merit greater attention towards extending its cultivation in those parts of India, where the climate is found to be suitable. **Dr. Bonavia**, in an article on the Lichi contributed to the *Pioneer*, remarks:—"Here then is a fruit tree, which resists the heaviest rains, and stands the hottest winds, and also the frosts of these provinces (North-West Provinces). Moreover, it bears annually an abundant crop of fine, well-flavoured and aromatic fruit, which can readily be sent to distant markets without injury. Instead of being planted by the one or two, it should be planted by the thousand. From all I know of the hardiness and fruitfulness of this remarkable tree, I feel confident that if any individual (or company), possessing the necessary capital, were to plant an extensive orchard of litchi trees, say where canal water would be easily obtained, or where well water is within easy reach, he would very probably make a good life-long business of it." [As remarked, however, this result has been abundantly attained in Bengal, and although statistics of the extent of the trade cannot be given, it may be said that in the Lower Provinces the litchi tree is almost co-extensively cultivated with the mango. It comes into season a little before that fruit, and in the larger cities such as Calcutta is sold in every fruit-dealer's shop, the streets for a month or six weeks being literally bestrewn with the rind and large seeds, rejected by the way-side consumers. The fruit to be enjoyed must, however, be eaten as soon after being plucked as possible. When fresh, the great bunches look like bright pinkish strawberries, but they rapidly lose their bloom and assume a dirty brownish colour. The dried fruit, as sold in Europe, bears no possible resemblance to the deliciously bitter sweet pulp of the fresh litchi. *Ed.*] **CULTIVATION. 69**

Medicine.—In China the LEAVES are stated to be officinal as a remedy for the bites of animals. **MEDICINE. Leaves. 70**

Food.—The FRUIT is nearly round, and about an inch and a half in diameter. The edible portion is the sweet semi-transparent jelly-like pulp, or aril, which covers the seed, and the whole is enclosed in a thin reddish or brownish brittle shell, which is rough with warty protuberances. The Chinese dry the fruit which then becomes blackish, and in this state it may often be seen in London fruit shops. The fresh fruit has a very pleasant acid flavour, and is much liked both by Natives and Europeans in this country. **FOOD. Fruit. 71**

72 **Nephelium Longana,** *Camb.; Fl. Br. Ind., I., 688.*

THE LONGAN.

Syn.—EUPHORIA LONGANA, *Lamk.*; SCYTALIA LONGAN, *Roxb.*

Vern.—*Ashphal*, BENG.; *Wumb, wumb-ashphal*, BOMB.; *Vomb*, MAR.; *Púvati*, TAM.; *Puna*, COURTALLUM; *Malahcota*, KAN.; *Kyetmauk*, BURM.; *Mora, murale*, SING.

References.—*Roxb., Fl. Ind, Ed. C.B.C., 329; Voigt, Hort. Sub. Cal., 95; Kurz, For. Fl. Burm., I., 294; Beddome, Fl. Sylv., t. 156; Gamble, Man. Timb., 97; Thwaites, En. Ceylon Pl., 58; Grah., Bomb. Pl., 29; Dalz. & Gibs., Bomb. Fl., 35; DC. Origin Cult. Pl., 315; W. & A., Prod., 113; Smith, Mat. Med. of China, 155; Lisboa, U. Pl. Bomb., 52, 150; Balfour, Cyclop., II., 1086; Treasury of Bot., 784; Kew Off. Guide to the Mus. of Ec. Bot., 34; Bomb. Gaz., XV., 68.*

Habitat.—A moderate-sized evergreen tree, found in Mysore, the Western Ghâts, Eastern Bengal, and Burma; also in Ceylon. It is called *Longan* in China, from which country, according to **DeCandolle**, it was introduced into the Malay Peninsula some centuries ago.

MEDICINE. Fruit. 73

Medicine.—In China the FRUIT is reputed to be nutrient, stomachic, and anthelmintic.

FOOD. Fruit. 74

Food.—The FRUIT is smaller than that of the *lichi*, and contains an acid pulpy aril resembling it in flavour.

TIMBER. 75

Structure of the Wood.—Red, and moderately hard. **Kurz** says that it is good for furniture and takes a fine polish. It is used for building purposes in Ceylon (*Thwaites*).

NEPTUNIA, *Lour.; Gen. Pl., I., 592.*

76 **Neptunia oleracea,** *Lour.; Fl. Br. Ind., II., 285;* LEGUMINOSÆ.

Syn.—MIMOSA NATANS, *Roxb.*; DESMANTHUS NATANS, *Willd.*; D. LACUSTRIS and STOLONIFER, *DC.*

Vern.—*Páni-najak, páni-lájak*, BENG.; *Laj-alú*, PATNA; *Páni-lájak*, BOMB.; *Sunday-kiray*, TAM.; *Niru-talvapu, nidra-yung*, TEL.; *Nitti-todda-vaddi* (**Rheede**), MALAY.

References.—*Roxb., Fl. Ind., Ed. C.B.C., 420; Dalz. & Gibs., Bomb. Fl., 84; Irvine, Mat. Med. Patna, 60; Lisboa, U. Pl. Bomb., 199; Journ., Agri.-Hort. Soc. of Ind., IX., 420; Ind. For., III., 236.*

Habitat.—A floating aquatic herb common in tanks throughout the greater part of India.

Medicine. Plant. 77

Medicine.—Used as refrigerant and astringent (*Irvine*).

FOOD. Plant. 78 Pods. 79

Food.—The PLANT is eaten as a pot-herb, and the PODS, sometimes as a vegetable.

NERIUM, *Linn.; Gen. Pl., II., 713.*

80 **Nerium odorum,** *Soland.; Fl. Br. Ind., III., 655;* APOCYNACEÆ.

SWEET-SCENTED OLEANDER.

Syn.—N. ODORATUM, *Lamk.*; N. LATIFOLIUM and INDICUM, *Mill.*

Vern.—*Kanér, kanél, karber, kuruvira*, HIND.; *Karabi*, BENG.; *Rajbaka*, SANTAL; *Alari*, MAL. (S.P.); *Kanyúr*, KUMAON; *Kanira, kaner, ganhira*, PB.; *Ganderái*, PUSHTU; *Jaur* (according to **Aitchison**), BALUCH.; *Kanirkejur*, DEC.; *Kanhera, ganira, kanir*, BOMB.; *Kaneri*, MAR.; *Kanera*, GUZ.; *Alari, aralivayr*, TAM.; *Gannéru, gheneru, kasturi-patte*, TEL.; *Kanagale, levagani-galu*, KAN.; *Alari*, MALAY.; *Karavira, asvamáraca, pratihása, sataprása, hayamáraca, chandáta*, SANS.; *Difli, sum-el-himar*, ARAB.; *Khar-zahrah*, PERS.

References.—*Roxb., Fl. Ind., Ed. C.B.C., 242; Voigt, Hort. Sub. Cal., 524; Brandis, For. Fl., 328; Kurz, For. Fl. Burm., II., 194; Gamble, Man. Timb., 264; Grah., Cat. Bomb. Pl., 114; Stewart, Pb. Pl., 142; Aitchison, Cat. Pb. and Sind Pl., 89; Boiss., Fl. Or., IV., 48; Rheede, Hort. Mal., IX., tt. 1 & 2; Elliot, Fl. Andhr., 57, 88; Pharm. Ind., 139;*

Ainslie, Mat. Ind., II., 23; O'Shaughnessy, Beng. Dispens., 445; Dymock, Mat. Med. W. Ind., 2nd Ed., 500; U. S. Dispens., 15th Ed., 1707; S. Arjun, Bomb. Drugs, 87; Murray, Pl. and Drugs, Sind, 150; Bidie, Cat. Raw Pr., Paris, Exh., 33; Year-Book, Pharm., 1881, 54; Irvine, Mat. Med. Patna, 57; Mason, Burma, 417, 799; Baden-Powell, Pb. Pr., 360, 586; Atkinson, Him. Dist., 743; Drury, U. Pl., 310; Lisboa, U. Pl. Bomb., 266; Birdwood, Bomb. Pr., 54; Royle, Ill. Him. Bot., 269; Balfour, Cyclop., II., 1086; III., 20; Smith, Dic., 296; Raj. Gaz., 27; Home Dept. Corr., regarding Pharm. Ind., 239; Encyc. Brit., XVII., 759; Firminger, Man. of Gardening, 496; Aín-i-Akbari (Blochmann's trans.), 85; Sir W. Jones, Works, V., 99.

Habitat.—An evergreen shrub, with milky juice, indigenous on the lower slopes of the Western Himálaya, extending westward from Nepál to Sind and Afghánistán; also in the hilly parts of Central and South India. It is cultivated in gardens throughout India, and occurs with red or white flowers, which may be single or double. Probably a variety only of the common Oleander, or Bay Laurel (**N. Oleander**, *Linn*), which extends from Europe eastward as far as Persia. The flowers of the latter are not scented

Medicine.—The ROOTS are highly poisonous when taken internally, but a paste is reputed to be useful in skin diseases. The fresh JUICE is described by the Sanskrit physicians as useful when dropped into the eye in ophthalmia, producing copious lachrymation. The drug is regarded by Muhammadan physicians as a very powerful resolvent and attenuant, but serviceable only for external application. "A decoction of the LEAVES is recommended to reduce swellings, and an OIL prepared from the ROOT-BARK in skin diseases of a scaly nature, and in leprosy" (*Dymock*). The use of this drug in leprosy is alluded to by nearly every writer, and it has also been spoken of as useful in itch. Several cases are on record of fatal results from the internal administration of the drug, the fatal symptoms being manifested by a depression of the nervous functions, and particularly of the heart. This plant has several Sanskrit names, such as *Asvamaraka* (which signifies "horse-killer"), *karavira, &c.* The Persian and Arabic name *Difli*, and others, *viz., Sum-el-himar* and *kharzahrah* all signify "Asses'-bane,"—a curious fact, since the Italian *Ammazza l'Asino*" (applied to the allied European plant) has the same meaning. The root is so commonly resorted to for the purpose of self-destruction by the women of India when jealous, that it is a proverbial taunt among females to say—"Go and eat of the *kaner* root." The leaves boiled in lard or oil yield a medicated ointment, which, when rubbed on the skin, is said to keep away insects which infest the person.

MEDICINE. Roots. 81 Juice. 82 Leaves. 83 Oil. 84 Root-bark. 85

CHEMICAL PROPERTIES.—**Mr. H. G. Greenish** has extracted from the bark two bitter principles—one soluble in chloroform, but little soluble in water, and to this substance he has given the name *neriodorin*. The second principle he has named *neriodorein*, a substance insoluble in chloroform, but very soluble in water. Both these are powerful heart poisons. Of the latter substance he found that 0·0016 grains injected hypodermically into a large healthy frog caused in 14 minutes diminution of the heart's pulsation from 70 to 12 beats per minute, followed by a temporary rise to 60. After the lapse of five minutes the heart ceased to act. **M. Latour** made a careful chemical examination of Oleander from which he obtained the following results:—(1) the poisonous principle resides in the leaves, bark, and flowers, but most largely in the bark; (2) this principle is of a resinous nature, and not volatile, and is found more largely in the wild than in the cultivated plant; (3) the solution of this resin in water is much facilitated by alkaline salts, and hence it exists in the watery extracts; (4) the distilled water of the bark and leaves possesses some activity which it owes to a small portion of the resin carried over with the stream. **Prof.**

Chemical Properties. 86

MEDICINE. E. Pelikan confirmed these results, and suggested that the drug, owing to its depressing influence on the heart, might be given as a substitute for **Digitalis.** The two principles isolated by **Greenish** are probably identical with those extracted by **Lenkowsky**, and named *pseudo-curarine* and *oleandrine*. Most Indian writers speak of a yellow resinous substance as the active principle.

Special Opinions.—§ "The milky juice of the leaves is applied to ringworm" (*Civil Surgeon, Aligarh*). "The roots are sometimes used to procure abortion" (*Civil Surgeon J. H. Thornton, B.A., M.B., Monghyr*). "It is used largely to procure abortion, and a poultice of the leaves fried in oil is often applied to wounds in man or beast, in which maggots have formed" (*Surgeon-Major C W. Calthrop, M.D., Morar*). "Over doses of the root cause tetanic symptoms" (*Surgeon A. Crombie, Dacca*). "In case of poisoning from the root, pulse was as low as 36, though of good volume" (*Surgeon-Major P. N. Mukerji, Cuttack, Orissa*). "Its root made into paste and then boiled with mustard oil is given in skin diseases" (*Civil Surgeon W. A. Galligan, Durbhanga*). "There is also a yellow variety, it is not common and I have only seen it grow near Seringapatam in the Mysore province. Further than knowing that it is used for suicidal purposes, I am unable to speak of its medicinal properties. When first I brought it to the notice of **Dr. J. Shortt**, this gentleman supposed that I mistook the '**exile Thevetia neriifolia**' for the yellow variety and was only satisfied after seeing the flowers I sent him" (*Honorary Surgeon E. A. Morris, Tranquebar*). "Oleander is a powerful heart poison" (*W. Dymock, Bombay*).

FODDER. Foliage. 87 **Fodder.**—The goat appears to be able to feed on the FOLIAGE of this plant with impunity, but it proves fatal to camels and all other animals.

TIMBER. 88 **Structure of the Wood.**—Greyish white, soft. The charcoal is sometimes used in making gun-powder.

DOMESTIC & SACRED. Flowers. 89 Stalks. 90 Wood. 91 Bark. 92 **Domestic and Sacred Uses.**—Amongst the Hindus the FLOWERS are collected as sacred offerings to Siva. **Stewart** says that the STALKS are in some places used as *hookah* tubes, the wood probably imparting a narcotic principle to the tobacco smoke. The rasped WOOD is used in parts of India for killing vermin. The powdered BARK of the common oleander is said to be employed by the peasantry in the South of France for a similar purpose. Some correspondence has recently taken place regarding the supposed property of this plant in checking the malarious tendencies of certain districts. This idea originated from the statement that the peasantry of France regarded the Oleander as not only ornamental but also as possessing beneficial properties when grown around their villages.

Nettle, see **Urtica.** Vol. VI., Pt. II.

Ngai Camphor, see **Blumea densiflora,** *DC.*; Compositæ, Vol. I., 458.

NICANDRA, *Adans.*; *Gen. Pl., II., 897.*

Nicandra indica, *Rœm. & Sch.*, see **Physalis minima,** *Linn.*, **var. indica**; Solanaceæ, Vol. I., Pt. I.

93 **N. physaloides,** *Gœrtn.*; *Fl. Br. Ind., IV., 240.*

Syn.—Atropa physaloides, *Linn.*; Physalis daturæfolia, *Lamk.*

References.—*Voigt, Hort. Sub. Cal., 514; Grah., Cat. Bomb. Pl., 140; O'Shaughnessy, Beng. Dispens., 466; Gazetteers:—Mysore and Coorg, I., 63; N.-W. Provs., IV., 75; X., 314.*

Habitat.—An annual, introduced from Peru, but now found as a weed on rich soils in many parts of India; it ascends to 7,000 feet on the Himálaya, and in Simla, for example, is threatening to contest the ground with Datura.

Medicine.—Nicandra is said to be diuretic (*O'Shaughnessy*). MEDICINE. Plant. 94

NICKEL.

Nickel; *Ball, in Man. Geol. of India, III.*, *326*. 95

A hard, tough, silver-white metal, fusible with difficulty and susceptible of magnetism. It forms a constituent of many ores, occurring in combination with sulphur, arsenic, antimony, carbon, and silica. It is an essential component of certain ores of cobalt, and is often found in pyrrhotite or magnetic pyrites. **Ball** states that in India traces of nickel have been found in several ores, especially in those in which cobalt also occurs in the mines of Rájputana. Specimens of pyrrhotite from the Khetri mines were found by **Mallet** to contain both cobalt and nickel, and the iron ores from Bhangarh were also found to possess the latter metal. Traces of nickel, in association with copper, have been observed by **Mallet** to occur in the veinstone in which the Kandahar gold is obtained.

Uses.—Owing to the scarcity of nickel ores, the metal is not known nor worked by the natives of India. In Europe it is principally used in the manufacture of German silver, and of late years the demand for it for that purpose has increased. **Ball** states that "at present (1880) it costs only three shillings per pound," and that one of the chief sources of supply is a silicate called "garnierite" which is worked to some extent in New California. None of the Indian ores appear to be sufficiently rich in the metal to render future working probable. USES. 96

NICOTIANA, *Linn.*; *Gen. Pl.*, *II.*, *906*. 97

To this genus belong the various kinds of tobacco, wild and cultivated. Over fifty species have been described, all of which are natives of the New World, with the exception of one found in Australia, and another in New Caledonia. The most important species, as far as this country is concerned, are **N. Tabacum**, and **N. rustica**; in fact, very few species out of the total number have been utilised to any extent as tobacco. As usually happens in the case of plants which have been under cultivation for many years, a large number of more or less distinct varieties has been produced in different localities, according to the nature of the climate, the composition of the soil, or the mode of cultivation. For the most part the names by which these different varieties are commercially recognised indicate the particular country or tract where they were produced. Thus, Virginian, Maryland, Kentucky, and Latakia (Loadicœa) tobaccos are all yielded by distinct varieties of **N. Tabacum**. In regard to the botanical source of Cuban and Havanna tobacco, **Senator Vidal** asserts that no other species than **N. Tabacum** (*var.* **macrophylla**) is grown in that island; **N. repanda**, which **Hanbury** and others believed to be the source of the finer kinds of Havanna tobacco, has not been found in that island, either wild or cultivated. "Shiraz" or "Persian" tobacco, formerly known botanically under the name of **N persica**, is, according to **DeCandolle**, the product of **N. alata**; it is, however, stated in the Report on the Royal Gardens, Kew, for 1877, p. 40, that some seed of the finest Shiraz tobacco, procured from Persia, produced plants of **N. Tabacum**. What is known as "East Indian" or "Turkish" tobacco is derived from **N. rustica**; and it has generally been supposed that Manilla tobacco was manufactured from this latter species. In all probability, however, the bulk of the tobacco now grown in the Philippines is **N. Tabacum**. Again, **N. multivalvis** is the native tobacco of the regions of the Columbia river, whilst **N. quadrivalvis** is the species said to be used by the people inhabiting the region of the Missouri. A very strong tobacco used in Chili is prepared from **N. angustifolia**.

The genus **Nicotiana** was named in honour of **Jean Nicot** of Nismes, who introduced the plant into France.

98 **Nicotiana rustica,** *Linn.; Fl. Br. Ind., IV., 245;* SOLANACEÆ.

TURKISH, or EAST INDIAN TOBACCO.

Vern.—*Biláeti* (European) or *kalkatiya* (Calcutta) *tamáku* (Behár), BENG.; *Kalkattia-tamáku*, N.-W. P.; *Chilassi-tamáku, kakkar-tamáku, kandahári-tamáku, kandahár-kakkar*, PB.

References.—*Voigt, Hort. Sub. Cal., 516; Dalz. & Gibs., Bomb. Fl. Suppl., 62; Stewart, Pb. Pl., 157; DC., Origin Cult. Pl., 141; Flück. & Hanb., Pharmacog., 469; U. S. Dispens., 15th Ed., 1416; Bidie, Cat. Raw Pr., Paris Exh., 14, 82, 83; Baden Powell, Pb. Pr., 290; Duthie & Fuller, Field and Garden Crops, 69; Royle, Ill. Him. Bot., 283; O'Conor, Rep. on Production of Tobacco in India (1873); Spons' Encyclop., 1325; Balfour, Cyclop., II., 1095; Treasury of Bot., 787; Morton, Cyclop. of Agri., II., 450.*

Habitat.—Believed to be a native of Mexico, and, according to **DeCandolle,** it is probably indigenous in California. Botanically it differs from the ordinary tobacco plant (**N. Tabacum**) in many important characters. The leaves are distinctly stalked, cordate, ovate, obtuse, rather leathery and somewhat crumpled. The flowers, which are in close panicles, have a short campanulate calyx with bluntish teeth, and a greenish corolla with the limb not much developed. It is called English tobacco on account of (as some say) its having been the first kind that was introduced into England; and, according to **Parkinson,** the tobacco prepared from it was preferred by **Sir W. Raleigh.** It is a hardier species than **N. Tabacum,** and requires a much shorter time to come to maturity; on this account it is considered to be better suited for cultivation in Europe. It is largely grown in West Africa and also in Egypt, and until quite recent times, was supposed to be the kind yielding the famous Latakia tobacco. The late **Dr. Stewart,** Conservator of Forests in the Panjáb, was the first who drew attention to the existence of this species in Upper India. In the neighbourhood of Lahore he found it (1865) being cultivated almost as extensively as the ordinary kind; and, under various names, it was met with in many other parts of the Panjáb, as at Multán, Hoshiarpur, and Delhi. **Dr. Stewart** says: "I have also seen it cultivated in some quantity in Pángi on the Upper Chenáb, from 7,500 to 9,400 feet; in Khágán and on the Kishenganga in the Jhelam basin from 3,300 to 4,500 feet; and in Ladák at 10,500 feet."

Tobacco. 99 This species is cultivated also in Kúch Behar, Rungpur, Sylhet, Cachar, and other parts of Eastern Bengal, Assam, and Manipur. The writer is informed by **Dr. Watt,** that in the latter country he saw no other kind in cultivation, though **N. Tabacum** is constantly met with as an escape. The Godaveri *lunka* tobacco is, to some extent, prepared from **N. rustica.** The tobacco made from this species is much stronger than that of **N. Tabacum,** with which it is usually mixed for smoking. It is said to be a more profitable crop, as more of it can be grown per acre than of the ordinary sort, and the prepared tobacco is said to fetch a higher price.

As sent to the market (in the Panjáb) the leaves are tied into bundles never twisted into ropes. It is sometimes made into snuff but never chewed. Molasses is not mixed with it, so its sweet taste is probably due, as **Mr. Baden Powell** remarks, to the addition of honey. In many places the entire plant, including the flowers, is made up for smoking. The vernacular names given to it would indicate its introduction into Northern India from opposite directions.

MEDICINE. 100 **Medicine.**—The medicinal properties are similar to those of **N. Tabacum.**

Nicotiana Tabacum, *Linn.; Fl. Br. Ind., IV., 245; Dunal in DC. Prod., XIII., 557.* 101

AMERICAN OT VIRGINIAN TOBACCO.

Vern.—*Tamáku, tumák, támbáka, támbáku, bujjerbhang,* HIND.; *Tamák,* BENG.; *Tamáku,* SIND; *Tambakhu,* BOMB.; *Tamáku,* MAR.; *Tamáku,* GUZ.; *Pugai-ilai, poga-yellei* (lit. smoke-leaf), *poghei, pogháko,* TAM.; *Tamáku, tamáqu,* DEC.; *Pogáku, dhumrapatramu,* TEL.; *Hogesappu,* KAN.; *Puka yila, pokala,* MALAY.; *Se, sac, sacpin,* BURM.; *Dungazha, dimkola, dungkola,* SING.; *Tanbák,* ARAB.; *Tanbáku,* PERS.

References.—*Voigt, Hort. Sub. Cal., 516; Dalz. & Gibs., Bomb. Fl. Suppl., 62; Stewart, Pb. Pl., 158; Rept. Pl. Coll. Afgh. Del. Com., 92; DC., Orig. Cult. Pl., 139; Graham, Cat. Bomb. Pl., 140; Sir W. Elliot, Fl. Andhr., 154; Le Maout & Decaisne, Descript. & Analyt. Bot. (Eng. Ed.), 581; Pharm. Ind., 178; British Pharm., 403; Flück. & Hanb., Pharmacog., 466; U. S. Dispens., 15th Ed., 1416; Fleming, Med. Pl. & Drugs (Asiatic Reser., XI.), 173; Ainslie, Mat. Ind., I., 447; O'Shaughnessy, Beng. Dispens., 471; Moodeen Sheriff, Supp. Pharm. Ind., 182; U. C. Dutt, Mat. Med. Hindus, 212; Sakharam Arjun, Cat. Bomb. Drugs, 97; Murray, Pl. & Drugs, Sind., 154; Bent. & Trim., Med. Pl., 191; Dymock, Mat. Med. W. Ind., 2nd Ed., 632; Year-Book Pharm., 1874, 170; 1877, 93; 1878, 72, 200; 1879, 118; 1880, 37, 40; 1881, 31, 453; 1882, 45; 1883, 119; 1884, 69, 118, 208; Watts, Dict. Chemistry, Vol. IV., 44; V., 849; VII., 851, 1161; VIII., 1391, 1981; Johnson, How Crops Grow, 378; Birdwood, Bomb. Prod., 209; Baden Powell, Pb. Pr., 288, 364; Drury, U. Pl. Ind., 311; Atkinson, Him. Dist. (Vol. X., N.-W. P. Gaz.), 755; Duthie und Fuller, Field and Garden Crops, 69; Royle, Ill. Him. Bot., 283; Prod. Res., 188, 249; Mandis, Manual of Cultivation and Preparation of Tobacco in Hungary, Vienna, 1886 (Eng. Transl.); Forbes Watson, Report on the Cultivation and Preparation of Tobacco in India (1871); J. E. O'Conor, Report on Production of Tobacco in India; Fairholt, Tobacco, Its History and Associations (1876); Note on Indian Tobacco (Govt. of India, Rev. & Agri. Dept., 11th Aug. 1886); Watt & McCarthy on Tobacco (Col. and Ind. Exhib., 1886); Buchanan, Journey through Mysore and Canara, &c., Vol. I., 52; II., 256, 315; III., 441; Rep. Agri. Dept., Beng., 1886, Append. LXVI.; Rep. Agri. Dept., N.-W. P., 1877-78, 7, 19; 1881, 11; 1882, 23, 31; Rep. Agri. Dept., Bombay, 1883-84, 13; 1884-85, 18, 20; 1886-87, 16, 24; Rep. Agri. Dept., Madras, 1885-86, 32; Rep. Agri. Dept., Burma, 1887-88, 27; Man. Madras Adm., Vol. I., 292, 344; Nicholson, Man. Coimbatore, 226; Robertson, Man. Neilgh. & Coimbatore, 27, 123; Man. of Kurnool, 168; Boswell, Man. Nellore, 402; Moore, Man. Trichinopoly, 73; Mackenzie, Man. Kistna, 336; Nelson, Man. Madura, pt. II, 106; Shortt, Man. of Ind. Agri., 137; Gazetteers:—India, VI., 499; Bombay, II., 408; III., 47; VII., 89, 97; XII., 165; Panjáb (Karnál), 182; (Jhang), 116; (Hoshiarpur), 90; (Dera I. Khan), 130; (Gujrat), 80; (Kángra), II., 59; (Montgomery), 119; (Simla), 58; Central Provinces, 503; Burma, I., 428; Mysore & Coorg, I., 97; Rajputana, I., 257; Agri.-Horti. Soc., Ind.:—Trans., II., 70, 171, 203; III., 33, 39, 41, 71, 97; VI., 243; VII., 79, 90; VIII., 163, 248; Journ., I., 231, 279; III., 219; (Sel.), 250; V., 101; VII., 65; VIII., 46, 64; IX., 160; XI., 551; XIII. (Sel.), 64; Ind. Agriculturist, Oct. 26th, 1889; Linn. Soc. Journ., XV., 246; Rep. R. Gardens, Kew (1877), 40; Wallace, India in 1887, 234; Aitchison, Prod. of W. Afgh., 140; Yule & Burnell Gloss. 144; Spons' Encycl., 1325, 1351; Encyclop. Brit., XXIII., 423; Balfour, Cyclop. Ind., II., 1095; Morton, Cycl. Agri., II., 450; Ure, Dict. Ind., Arts & Man., III., 927; Smith, Dict., 413; Treasury of Bot., 787; Mueller, Sel. Extra-Trop. Pl., 257; Bragge, Bibliotheca Nicotiana (1880).*

Habitat.—An erect, viscidly pubescent herb, with large sessile amplexicaul lanceolate acuminate leaves; corolla tube greenish, inflated above, and spreading into five broadly triangular pointed lobes of a dull pink or rose colour. Its origin as an indigenous species cannot be exactly determined, though there is no doubt of its being a native of some part of Central or South America. **DeCandolle,** in his *Origin of Cultivated Plants,* after reviewing all the evidence at his disposal, gives "Ecuador and the neighbouring countries" as the region where it most probably had its origin. It is now grown in nearly all the temperate and warm parts of the world, and has probably the widest range of any economic plant.

In many parts of India this species is met with as an escape from cultivation, and is found to be capable of maintaining itself in a semi-wild state; **N. rustica** is never found naturalized. In more temperate countries the reverse takes place. **N. rustica** is the species which wanders as an escape. On page 237 of the Report on Tobacco at the Colonial and Indian Exhibition, 1886, **Dr. Watt** wrote as follows:—"**N. Tabacum** has become an abundant weed in many parts of India; around Calcutta, for example, in every dark and damp lane through the villages and neighbouring bamboo jungles, and on every wall and roadside, a stunted form of **N. Tabacum** is found to be one of the commonest weeds, and, indeed, on the sandy islands of the Hooghly and Ganges, this plant has practically exterminated the indigenous vegetation, and may be seen covering miles of these newly formed tracts of country. The plants spring up at the close of the rains and flower in early summer." **Mr. C. B. Clarke** is of opinion that the above mentioned plant is not **N. Tabacum**, but **N. plumbaginifolia**, a native of Mexico and the West Indies. It is described in the *Flora of British India, Vol. IV., p. 246*, and is mentioned as being "the only species of **Nicotiana** which has established itself in India." **Dr. Watt**, however, informs the writer that he is still inclined to adhere to his own opinion in favour of the acclimatised tobacco of Bengal being mainly referable to **N. Tabacum**, or at least that the plant which is found as the almost exclusive vegetation on many of the *chars* (islands) of the Hooghly and Ganges rivers is so. The form found on roadsides in Bengal generally, he thinks, may be the narrow-leaved **N. plumbaginifolia**, as recognised by botanists, nevertheless **Dr. Watt** thinks that he has practically demonstrated by cultivation the effacement of the distinguishing characters of these two plants.

OIL. **Oil.**—"When distilled at a temperature above that of boiling water, tobacco affords an empyreumatic oil, which **Mr. Brodie** proved to be a most virulent poison." "This oil is of a dark brown colour, and an acrid taste, and has a very disagreeable odour, exactly resembling that of tobacco pipes which have been much used. It has been stated to contain

Seed. 102 nicotine" (*U. S. Dispens., 15th Ed., 1419*). "Tobacco SEEDS are said to yield, by pressure, about 36 per cent. of a greenish-yellow, mild, inodorous

Oil. 103 OIL, of specific gravity 0·923 at 15°, solidifying at 25°, and quickly drying when exposed to the air" (*Watts, Dict., Chem., V., 851*).

MEDICINE. Leaves. 104 **Medicine.**—The dried LEAVES of this plant are officinal. They are powerfully sedative and antispasmodic, and in overdoses an acro-narcotic poison; useful in "tetanus, dropsical affections, spasmodic affections of the abdomen, retention of urine, and as a means of inducing muscular relaxation, and thus aiding in the reduction of strangulated hernia and dislocations. As a general rule it is unfitted for internal administration, on account of the great nervous depression it produces. As a local application, it has been used for relieving pain and irritation in rheumatic swellings, syphilitic nodes, and skin diseases. Tobacco-smoking is sometimes resorted to in asthma, spasmodic coughs, nervous irritation, and sleeplessness" (*Pharm. Ind.*). "Native physicians consider the smoke to be disinfectant, and recommend it for fumigating cholera patients." "The WATER from the *hookah* is diuretic, and the black OIL which collects in the pipe stem is used on tents to heal up sinuses, and is dropped into the eye to cure night blindness and purulent conjunctivitis." "In the Konkan a paste made with snuff, lime, and the powdered bark of **Calophyllum inophllyum** (*undi*) is applied in orchitis. **Dr. Leith** of Bombay was in the habit of applying a poultice of tobacco leaves to the spine in tetanus with good results" (*Dymock*). Regarding the effects resulting from the immoderate use of tobacco, **Dr.**

MEDICINE.

Richardson, in a paper read by him at the Bath meeting of the British Association, pointed out that excessive snuffing is liable to occasion unmanageable forms of indigestion; that chewing and smoking tend to weaken the energy of the nervous system, impair digestion and the action of the liver, and in extreme cases are apt to produce an affection of the muscular system resembling 'paralysis agitans.' Notwithstanding the many evils resulting from the abuse of tobacco, there is no evidence, however, that moderate smoking is injurious to health.

SPECIAL OPINIONS.—§"The strong decoction of the leaves is used as a fomentation to relieve muscular tension and spasm as in tetanus" (*Assistant Surgeon N. L. Ghose, Bankipore*). "It is also used as a local application to relieve joint swelling in lymph scrotum" (*Surgeon-Major A. S. G. Jayakar, Muskat*). "The rind ribs of the leaves are used as a laxative among children by introducing a piece in the anus" (*Surgeon-Major H. D. Cook, Calicut, Malabar*). "In acute hydrocele application of moistened leaves proved efficacious in relieving pain and reducing swelling" (*Assistant Surgeon S. C. Bhattacharji, Chanda, Central Provinces*). "It is a common practice here with dames in families to introduce the smoothed stalk of the tobacco leaf into the rectum of infants to relieve constipation" (*Assistant Surgeon S. Arjun Ravat., L.M., Girgaum, Bombay*). "The green leaves are a favourite local application for orchitis. The stem of the leaf is introduced into the anus as a laxative with children" (*Surgeon-Major J. North, Bangalore*). "Tobacco leaves bruised or the residue deposited in the tube of a *huqqa* is used by natives for the expulsion of leech from the nostrils or throat" (*Assistant Surgeon Bhagwan Das (2nd), Civil Hospital, Rawal Pindi, Panjáb*). "Leaves dry are applied over the testicles in orchitis, to relieve pain and lessen the swelling; but such applications are almost invariably followed by nausea and vomiting" (*Surgeon A. C. Mukerji, Noakhally*). "The stalk or midrib of the leaf dipped in castor oil and introduced into the rectum acts as a purgative. It is, however, said to be dangerous. The infusion is used as an antidote in strychnine poisoning. The leaf applied to boils is anodyne. Tobacco smoking is used in anæmia and tabes mesenterica" (*Surgeon-Major D. R. Thomson, M.D., C.I.E., Madras*). "The leaves applied to gum boils prevent suppuration and relieve pain. The water of the *hookah* is used as a wash for unhealthy ulcers, especially when maggots are present" (*Civil Surgeon J. H. Thornton, B.A., M.B., Monghyr*). "Tobacco used in excess causes amaurosis. I have seen several cases in England of Tobacco amaurosis cured by giving up smoking, other conditions being the same. Tobacco leaves are applied over inflamed testes. Tobacco smoking is found useful in some forms of dyspepsia" (*Surgeon-Major Robb, Civil Surgeon, Ahmedabad*). "The powder of the dry leaves mixed with lime juice is used as an external application in cases of enlargement of the spleen. Fresh leaves applied in orchitis" (*J. McConaghey, M.D., Civil Surgeon, Shajahanpore*). "The infusion is a very efficient antidote for strychnia poisoning" (*R. Gray, Civil Surgeon, Lahore*). "It is salivative and sometimes chewed to relieve toothache" (*D. Basu, Civil Surgeon, Faridpore, Bengal*).

CHEMISTRY. 105

Chemistry.—Posselt and Reimann, who (in 1828) appear to have been the first to publish any reliable analysis of tobacco leaves, found in 1,000 parts:—

Nicotine	0·6
Nicotianine	0·1
Bitter extractive	28·7
Resin	2·7
Albumin and glutin	13·1
Gum	17·4

CHEMISTRY OF TOBACCO.

Malic acid, malates, and nitrates	9·7
Other potassium salts	1·2
Phosphate of calcium	1·7
Woody fibre, water, &c.	924·8

More recent investigations show, that the quantities of the organic and inorganic constituents vary greatly in different kinds of tobacco; and that the amount of nicotine as stated in the analysis given above is far below the average.

The organic components of tobacco leaves, in 100 parts, according to analysis made in the Paris factory laboratories, are as follows:—

Nicotine ($C_{10}H_{14}N_2$) 1·5—9 per cent.

Empyreumatic, or essential oil, quantity exceedingly small.

Nicotianine, a very small quantity.

Malic and citric acids, calculated as anhydrides,—10-14 per cent.

Acetic acid—small amount in fresh leaves, increasing during fermentation, and may amount to 3 per cent. in snuff.

Oxalic acid —1-2 per cent.

Pectic acid—about 5 per cent.

Resins, fats, and other bodies extractable by ether—4-5 per cent.

Sugar—a little in the leaves, more in the stems, disappears in fermentation.

Cellulose—7-8 per cent.

Albuminoids calculated from the nitrogen not present as nicotine, nitrates, or ammonia about 25 per cent.

"Nicotine is a colourless oily liquid of sp. gr. 1·027 at 15°C., deviating the pole of polarization to the left; it boils at 247° C., and does not concrete even at 10° C. It has a strongly alkaline reaction, an unpleasant odour, and a burning taste. It quickly assumes a brown colour on exposure to air and light; and appears even to undergo an alteration by repeated distillation in an atmosphere deprived of oxygen. Nicotine dissolves in water, but separates on addition of caustic potash". (*Flück. & Hanb., Pharmacogr., 467*). Its chemical formula is expressed by $C_{10}H_{14}N_2$. It was first obtained, in an impure state, by **Vanquelin** in 1809. "It is easily extracted from tobacco by means of alcohol or water, as a malate, from which the alkaloid can be separated by shaking it with caustic lye and ether. The ether is then expelled by warming the liquid, which finally has to be mixed with slaked lime and distilled in a stream of hydrogen, when the nicotine begins to come over at about 200°C" (*Flück. & Hanb., Pharmacog.*). The amount of nicotine is very variable in different kinds of tobacco, and depends to a great extent on the climate where the plant is grown, on the manner of its cultivation, and on the mode of curing adopted. It gives strength to tobacco, but not its own flavour or aroma.

Mr. Broughton, Quinologist to the Government of Madras, in his report on analyses made by him of several samples of tobacco, cultivated in **South India**, remarks as follows:—"What is usually called the strength of tobacco in smoking depends immediately on the amount of nicotine contained. A tobacco that contains over 4 per cent. of this powerful alkaloid is a strong intoxicating tobacco, while that which contains less than 3 per cent. is called mild. It has been found, as an invariable result of experiment, that the finest tobacco, as the **Havanna, Manilla, Cuban**, and others, do not contain more than 2 to 3 per cent of nicotine. To this result there is no exception that I am aware of, though of course by the constant custom of smoking strong tobacco, a few persons may even come to prefer it to the finer kinds." "It is remarkable that among the numerous **Indian** tobaccos, I have found but two instances in which as much nicotine is found as in **Virginian** or **French** tobaccos, where it amounts to

CHEMISTRY OF TOBACCO.

nearly 7 per cent." The amount of incotine increases with the age of the leaf, and thick leaves contain more nicotine than thin membraneous ones.

To estimate nicotine, weigh out 15 grains of tobacco, digest for twenty-four hours with alcohol of 85 per cent., acidified with 15 drops of sulphuric acid, so as to make 150 cubic centimetres. Evaporate 50 cubic centimetres of the filtered liquid and add iodo-hydrargyrate of potassium to the residue. The number of cubic centimetres employed multiplied by 0·00405 (0·0001 of the equivalent of nicotine), gives the quantity of alkaloid contained in five grains of tobacco (*Year-Book Pharm., 1874, 170*).

By distilling tobacco leaves with water a concrete volatile oil is obtained, to which the name nicotianine, or tobacco camphor, has been given. It is solid, has a strong odour of tobacco and a bitter taste. From one pound of leaves only two grains of this substance are obtainable. **Posselt & Reimann** found one part of nicotianine in 1,0000 parts of green tobacco.

Another active chemical ingredient in tobacco is an empyreumatic oil which is produced in the destructive distillation of tobacco. It resembles both in taste and smell an oil, obtained by a similar process, form the leaf of foxglove (**Digitalis purpurea**), and may be regarded as consisting of a harmless oil combined with a very poisonous alkaline substance.

"The percentage of nitrogen is greater than in any other cultivated crop; part of it exists as nitrates. The composition of tobacco is very variable; like all green crops, its constituents are much influenced by the nature of the soluble matters in the soil" (*Watts, Chem. Dict., VIII., part II., 1981*).

It is highly probable that ammonia is the volatilising agent of many odours, and especially of those of tobacco and musk. If a fresh green leaf of tobacco be crushed between the fingers, it emits merely the herbaceous smell common to many plants; but if it be triturated in a mortar along with a little quicklime or caustic potash, it will immediately exhale the peculiar odour of snuff. Now, analysis shows the presence of muriate of ammonia in this plant, and fermentation serves further to generate free ammonia in it; whence by means of this process and lime, the odoriferous vehicle is abundantly developed. If, on the other hand, the excess of alkaline matter in the tobacco of the shops be saturated by a mild dry acid, as the tartaric, its peculiar aroma will entirely disappear (*Ure, Dict. Indus., Arts and Man., III., 927*).

Ash.

THE ASH CONSTITUENTS.—The amount of mineral matter taken up by the tobacco plant from the soil is very considerable, four pounds of dry leaf containing on an average one pound of ash.

The greater part of the ash consists of insoluble salts, principally carbonate of lime; and the soluble part largely of potash salts, which may amount from 5 to 35 per cent.

The following table taken from **Johnson's** *How Crops Grow, p. 378*, gives the average of all trustworthy analyses of the ash constituents published up to 1865:—

Percentage of ash.	Potash.	Soda.	Magnesia.	Lime.
24·08	27·4	3·7	10·5	37·0
Phosph. acid.	Sulph. acid.	Silica.	Chlorine.	
3·6	3·9	9·6	4·5	

The amount of ash yielded by various samples of tobacco leaf analysed by **Nessler** varied between 19 and 27 per cent.; according to **Broughton** the ash yielded by Indian tobaccos was found to vary between 16 and 28 per cent. The total amount of ash, as well as the relative quantities of each constituent, appear to vary according to the composition of the soil, with the exception apparently of silica, which, in the opinion of Pro-

CHEMISTRY OF TOBACCO.

fessor **Johnston**, varies in quantity according to the age and rate of growth of the leaf. There appears to be no relation between the total quantity of ash and the quality of tobacco leaf. In addition to mineral salts proper, tobacco contains salts of ammonia and nitrates. In the leaf the proportion of nitrates is greater in the ribs of the leaves than in the blades; in the former it may be as much as 10 per cent. According to **Schlossing** the proportion of nitric acid does not affect combustibility, which depends on potash combined with organic acids. Cultivation experiments show that chloride of potassium used as manure does not add to the organic potash salts in the leaves, but the sulphate, carbonate, and nitrate do give up their potash for the formation of organic salts. (*Encyclop. Brit., XXIII., 425.*)

"**Schlossing** was the first who pointed out (1860) that the good burning qualities of a tobacco depend on the presence of potash in combination with a vegetable acid in it, and that a soil deficient in potash is unfit to produce tobacco of good quality. All the numerous analyses made since that time have tended not only to corroborate the assertion made by **Schlossing**, but to demonstrate also that it is not the total amount of potash, but the potash found as carbonate of potash, which existed in the plant in combination with a vegetable acid, that is the constituent chiefly affecting the combustibility of a tobacco. The complete analyses of **Nessler** have shown that, although a tobacco may contain a great amount of potash, it does not necessarily follow that the tobacco burns well. He found that some German tobaccos contained more potash than Havanna, although the latter burned much better than the former, and that a great amount of potash did not always indicate a great amount of carbonate of potash" (*K. Schiffmayer*).

The ash of Havanna cigars according to analyses made by **A. Percy Smith** (*Chem. News, XXVIII., 261, 324*) contains:—

Potassium	Sulphate	7·401
,,	Carbonate	9·012
Sodium	Sulphate	5·764
,,	Chloride	3·272
,,	Carbonate	1·039
Calcium	Sulphate	4·180
,,	Carbonate	45·400
Ferric oxide and Phosphoric acid		0·460
Calcium and Magnesium phosphates		9·210
Silica		9·641
Charcoal		3·162
Aluminium lithium carbonate		1·459
		100·000

An experiment undertaken by **Nessler** to ascertain the action of different kinds of manure on the combustibility of tobacco clearly shows that the quality of tobacco is improved in relation to the amount of potash salts in the soil capable of producing potassium carbonate in the ash. "The whiteness and permanency of the ash of a cigar depend," as **Mr. Broughton** says, "entirely on the amount of potassic carbonate it contains, the presence of potash salts certainly modifies the burning of tobacco in a peculiar way to the improvement of its flavour, and also positively facilitates its burning." According to **E. R. Durrwell** (*Bull. Soc. Chem- XXIV., 450*) the whiteness of the ash of good tobacco is due also to the salts of sodium as well as those of potassium, which swell up as the tobacco burns, and tear the fibres, thereby inducing complete combustion.

Products of Combustion.—Dr. **Richardson** in a paper read before the British Association at Bath in September 1864 showed that the products

CHEMISTRY OF TOBACCO.

of the combustion on smoking tobacco were:—water, free carbon, ammonia, carbonic acid, nicotine, an empyreumatic oil, and a resinous bitter extract. The water is in the form of vapour; the carbon in minute particles giving the blue colour to the smoke; the ammonia is in the form of gas combined with carbonic acid, a large portion of the latter remaining free; the nicotine, being a non-volatile body remains in the pipe; the empyreumatic substance is volatile, and gives to the smoke its peculiar odour; the resinous substance is black and has an intensely bitter taste. Tobacco that has not undergone fermentation yields very little free carbon, but it gives off much carbonic acid and ammonia, little or no water, and only the smallest possible trace of nicotine, a very small amount of empyreumatic vapour, and an equally small quantity of the bitter extract (the above description applies to Latakia tobacco). Turkish tobacco yields a large quantity of ammonia. Shag yields all the products in abundance, and the same may almost be said of Havanna cigars. Swiss cigars yield enormous quantities of ammonia, and Manillas yield very little.

Vohl & Eulenberg, who have carefully studied the physiological action of tobacco as a narcotic, failed to detect the presence of nicotine in the smoke of tobacco. "The fact," they say, "that very strong tobacco, which can scarcely be smoked in pipes, may be used for making cigars, is explained by the abundant occurrence of the highly volatile and intoxicating pyridine in pipe-smoking, whereas cigar-smoking produces only a small quantity of pyridine, but a large quantity of collidine. In general, pipe smoking produces a larger proportion of the more volatile bases." They are also of opinion that "the disagreeable symptoms felt by persons beginning to smoke, and the chronic affections which occur in those who smoke to excess, as well as the cases of poisoning from swallowing tobacco juice are due, not to nicotine, but to the pyridine and picoline bases. The idea that they were due to nicotine originated in the fact that picoline bases having a high boiling point, such as parvoline, resemble that alkaloid greatly both in smell and in physiological action" (*Watts, Dict. Chem., VII., 1162*). The conclusions arrived at by **Vohl & Eulenberg** regarding the absence of nicotine in tobacco smoke are not in accordance with recent investigation. **Kissling** (*Dingl. Polyt. Journ., ccxliv., 234-246*) states that when experimenting on cigars he found that a large proportion of the nicotine passes unaltered into the smoke; that the composition of tobacco smoke is highly complex, but beyond nicotine the only substances found in appreciable quantities are the lower members of the picoline series (*Encycl. Brit., xxiii., 425*). **G. le Bon. & G. Noel** (*Compt. Rend., June 28, 1880*) detected in tobacco-smoke, hydrocyanic acid, various aromatic principles, and a highly toxic alkaloid of a pleasant odour and which they believe to be collidine (*Year-Book Pharm. 1880, 40*).

TOBACCO.

HISTORY.

HISTORY. 106

The practice of tobacco-smoking was made known to Europeans about the year 1492, having been first observed by followers of Columbus when visiting the West Indian Islands. Accounts differ as to the particular locality where they first became acquainted with the practice. By some it is affirmed that Columbus and his followers first saw tobacco smoked in Cuba. According to another account it is said that the messengers sent by Columbus to explore the New World related that, on reaching the island of Guanahani (afterwards named by Columbus, San Salvador), they met natives holding in their hands bunches of burning herbs, the smoke of which they inhaled. Lobel in his *History of Plants* (1576) describes

NICOTIANA Tabacum.

Tobacco.

HISTORY.

these rolls of tobacco which were first seen in San Salvador as consisting of funnels of palm leaf with a filling of tobacco leaves. The natives called the plant *cohiba,* and the burning brand *tabaco.* The Spanish monk, Romano Pano, who accompanied Columbus (on his second expedition, 1494-96), is said to have reported that the natives of San Domingo rolled up the leaves of a plant called *guioja* or *cohoba,* and smoked them in pipes called *tobaco*; and he also described the habit of snuff-taking. Gonzalo Fernandez de Oviedo, Governor of San Domingo, in his *Historia General de las Indias* (1535), describes the *tabaco,* or pipe, as a branched tube shaped like the letter Y; and it appears to have been used by the natives, not as a receptacle for the tobacco, but as an instrument for collecting the smoke of the tobacco leaves which they kindled on a fire, the two upper branches were inserted in the nostrils, whilst the lower end was held in the smoke of burning tobacco. The last named author also alludes to the esteem with which the natives of San Domingo regarded the plant on account of its medicinal properties. The practice of chewing tobacco was first seen by the Spaniards on the coast of South America in 1502.

As the exploration of the New Continent proceeded, it became evident that the habit of using tobacco by the aborigines was universal, and must have been adopted from very early times. **DeCandolle**, in his *Origin of Cultivated Plants,* p. 139, says:—"At the time of the discovery of America, the custom of smoking, of snuff-taking, or of chewing tobacco, was diffused over the greater part of this vast continent. The accounts of the earliest travellers, of which the famous anatomist **Tiedemann** has made a very complete collection, show that the inhabitants of South America did not smoke, but chewed tobacco or took snuff, except in the districts of La Plata, Uruguay, and Paraguay, where no form of tobacco was used. In North America, from the Isthmus of Panama and the West Indies as far as Canada and California, the custom of smoking was universal, and circumstances show that it was also very ancient. Pipes in great numbers and of wonderful workmanship have been discovered in the tombs of the Aztecs in Mexico and in the mounds of the United States; some of them represent animals "foreign to North America."

The native names for the plant were found also to differ according to locality. In Mexico, where the growing herb was first recognised by Europeans, the word given for it was *petum* or *petun,* from which is derived **Petunia**, the name given by **Jussieu** to another genus of Solanaceæ. *Yetl* is another Mexican name for the tobacco plant, and in Peru it is called *sayri.*

The first tobacco plants were brought to Europe about the year 1560 by the naturalist **Francisco Hernandez** who was sent by Philip II. of Spain to explore Mexico. The plants did not attract much notice in Spain, although the dried leaves had been introduced and smoking had for several years become a practice in that country. It was from Portugal that the plant first became known to other countries in Europe. **Jean Nicot** (in whose honour the genus **Nicotiana** was named), when residing at the Portuguese Court as French Ambassador, obtained some tobacco seed from a Dutchman, and cultivated the plants in his garden at Lisbon. He is said to have effected some remarkable cures on some people belonging to his establishment, and from this circumstance he was induced (in 1561) to send some seed to the French Court. The fame of this herb, on account of the extraordinary medicinal virtues attributed to it, also from the fact of its having been patronised by the Queen, quickly spread throughout France, and received various flattering names, such as:—"Herba sancta," "herba panicea," "herbe de la reine," "herbe de l'Ambassadeur," &c. In Italy the plant was called "Erba Santa Croce" after the Cardinal

N. 106

of that name, who is said to have been the means of its introduction from Portugal. From Italy the knowledge and use of the plant spread gradually throughout Northern and Eastern Europe. HISTORY.

Tobacco was introduced into England in 1586 by **Captain Ralph Lane** on his return from Virginia with **Sir Francis Drake**, and the smoking of it was made fashionable by the example of **Sir Walter Raleigh** and others. **Sir Walter** had two years previously founded a colony in Virginia under **Lane**, and the colonists grew tobacco there to a large extent. The original import duty on Virginian tobacco in England was *2d.* per pound. In 1603, James I. raised the tax to *6s. 10d.* His famous *Counterblaste to Tobacco* was written about that time.

For several years after its introduction into Europe the plant was regarded chiefly in view of its supposed power to cure almost every imaginable malady. Practical experience at length demonstrated the fallacy of this belief, and every effort was made by emperors, kings, and popes to discourage the use of the plant, by means of severe and often cruel punishments. In Turkey persons convicted of smoking had their lips cut off, and snuff-takers were deprived of their noses, and in some cases were put to death. Nevertheless the use of tobacco rapidly and steadily spread until it became the most extensively used article of luxury in the world. The high tax imposed on foreign tobaccos led to the cultivation of the plant in the United Kingdom. This, however, was prohibited in 1660; the law not extending to Ireland until about 1830. In 1886 this law was repealed, and tobacco can now be grown as a field crop in Great Britain under certain restrictions.

Tobacco in India.

In India. 107

The tobacco plant was introduced into India by the Portuguese about the year 1605 during the latter part of the reign of Akbar. It has been asserted by some authors that the habit of smoking must have been practised in Asia long before the discovery of the New World, but no proof can be adduced in support of this theory; in fact, the entire absence of any allusion to the plant in the works of early travellers, and in the latest Sanskrit writings, and the universal adoption of a foreign name for the plant, are strong arguments in favour of the conclusion that the use of tobacco was unknown to oriental nations before the beginning of the seventeenth century.

At this period the influence of the Portuguese in the East was at its highest, and by all accounts it was through their means that the use of tobacco was first made known in Persia, Arabia, India, and China. The plant was known to be largely grown at their settlements in Ormuz, and at other places in the Persian Gulf. The plant is said to have reached Java in 1801. The following extract taken from the *Bahar-i-Ajan* (quoted by Blochmann in *Ind. Antiq., I., 164*) indicates the direction from which tobacco and its uses spread throughout India. "It is known from the Maásir-i-Rapimi that tobacco came from Europe to the Dakhin, and from the Dakhin to Upper India during the reign of Akbar Sháh (1556—1605), since which time it has been in general use."

Some extracts from ancient literature, quoted in **Yule & Burnell**, *Glossary of Anglo-Indian Words*, are of much interest in connection with the early history of tobacco in India. The first account bears the date 1604 or 1605:—

"In Bijápur I had found some tobacco. Never having seen the like in India, I brought some with me, and prepared a handsome pipe of jewel work." "His Majesty (Akbar) was enjoying himself after receiving my presents and asking me how I had collected so many strange things in so

HISTORY.

In India.

short a time, when his eye fell upon the tray with the pipe and its appurtenances: he expressed great surprise and examined the tobacco, which was made up in pipefuls; he enquired what it was, and where I had got it. The Nawab Khán-i-Azam replied: 'This is tobacco which is well known in Mecca and Madina and this doctor has brought it as a medicine for Your Majesty.' His Majesty looked at it, and ordered me to prepare and take him a pipeful. He began to smoke it, when his physician approached and forbade his doing so." "As I had brought a large supply of tobacco and pipes, I sent some to several of the nobles, while others sent to ask for some; indeed all, without exception, wanted some, and the practice was introduced. After that the merchants began to sell it, so the custom of smoking spread rapidly. His Majesty, however, did not adopt it" (*Asad Beg, in Elliot, VI., 165-167*).

The use of tobacco by the people of the East appears to have met with as much opposition on the part of those in power as was the case in European countries. The following passage from the Memoirs of Jahangir (Akbár's successor) is quoted by **Elliot** in *Ind. Antiq., VI., 851*:—

"As the smoking of tobacco (*tambáku*) had taken very bad effect upon the health and mind of many young persons," ordered (1617) "that no one should practise the habit. My brother Shah' Abbas, also being aware of its evil effects, had issued a command against the use of it in Iran." It is stated that in Lahore, during the reign of Jehangir, guilty persons were punished by having their lips cut. Another punishment, known by the name of *Tashír*, was also inflicted, by which the accused person was forced to ride on a donkey, face to tail, and with his face blackened.

Sikhs, Wahábis, and certain Hindu sects are still forbidden the use of tobacco by their religious guides, though allowed to indulge in hemp and opium to any extent. Mussalmáns now regard smoking as "an act indifferent," having come into practice since the death of the prophet. In spite of all such obstacles, the cultivation of the plant must have spread very rapidly throughout India; for in almost every district tobacco is grown as a crop, and there is hardly an out-of-the-way village where the familiar patch of tobacco, grown carefully as a garden crop, may not be seen. The following verse, which **Mr. Grierson** observes has passed into a proverb in Bihár, applies with equal force over the greater part of India:—

"Khaine kháe, na tamáku pie,
Se nar batáwa kaise jie."

i.e., "Show me the man who can live without either chewing or smoking tobacco."

The earliest attempts on the part of the British Government to improve the quality of Indian tobacco with a view to its export to Europe were made towards the end of the last century, about the time when it was proposed to establish the Calcutta Botanical Gardens. In 1829, experiments were undertaken in the three Presidencies with some Maryland and Virginian tobacco seed, according to instructions received from Captain Basil Hall regarding the mode of cultivation in Virginia. Successful results were obtained by the Agri.-Horticultural Society of India, and samples of the produce were sent by the Court of Directors of the East India Company to dealers and manufacturers in London, who pronounced them to be the best they had seen from India. The tobacco was valued at from 6*d*. to 8*d*. a pound, thus rivalling in quality some of the better preparations from America and the West Indies. In 1831, a bale of tobacco, grown in Guzerát, was sold in London at the rate of 6*d*. a pound. These samples were, however, exceptional ones; for, as a rule, the verdict on Indian tobacco exported to England has been anything but favourable.

HISTORY. In India.

The opinions, however, obtained of late years show very clearly that the chief defects of Indian-grown tobacco, as landed in England, are due to causes which can, for the most part, be remedied. For instance, in 1870, some samples raised in the Kaira and Ahmedabad districts from Havannah seed were sent to two brokers in London for their opinion. The tobacco was condemned because of the shortness and brittleness of the leaves, attributed to over-dryness; the colour was uneven, and the central portions of the packages were found to be rotten, which showed the packing to have been faulty.

Another defect, frequently complained of by professional experts as diminishing the value of tobacco exported from India, is the large amount of sand and dust found adhering to the leaves. Besides constituting an impurity it all goes to increase the weight upon which duty has to be paid. In those localities where dry dusty weather prevails during the final period of cultivation it is almost impossible to prevent a certain amount of dust from accumulating on the sticky glandular surfaces of the leaves, but it is evident that the bulk of the impurities mixed up with the leaves prepared according to native methods could with sufficient care be dispensed with.

CULTIVATION AND PREPARATION.

CULTIVATION. Area. 108

Area of Cultivation.—The total area in British India under tobacco in 1888-89 was estimated approximately as 800,000 acres, and the outturn as over 800 million pounds. No statistics are available for Bengal; and, with the exception of Mysore, there are none for the Native States. But **Mr. O'Conor**, in his report on "Tobacco in India" (1873), gives a much greater area as under this crop. He writes:—"Taking the average rate of produce per acre at 500lb, and the average daily consumption at ½ chittack per adult smoker, and allowing for exports, I calculate that there cannot be less than 2,000,000 acres altogether under tobacco in India, and I think that the facts, if it were possible to ascertain them, would prove this estimate to be far below the truth."

The most important tobacco-growing districts in India are:—Godáveri, Kistna, and Coimbatore in Madras; Rungpur and Tirhut in Bengal; and Kaira in Bombay. The famous *Lanka* tobacco is grown in the delta tracts of Godáveri and Kistna. The raw material for Trichinopoly cheroots is supplied chiefly from the Coimbatore and Madura districts.

Conditions necessary for the production of good tobacco leaf.—The essential conditions which determine the successful growth of tobacco are primarily dependent on climate and soil. A suitable climate is absolutely necessary; and generally speaking, it may be said that the finest and best flavoured tobaccos are produced in those countries where the climate is moist and warm—conditions which prevail in the regions where the species of **Nicotiana** are indigenous. As, however, tobacco is a quick-growing plant, it can be cultivated more or less successfully all over the warmer and temperate parts of the world, if care is taken to select the most suitable season of the year. In India, for instance, the plant is usually treated as a cold-weather crop; whilst in temperate countries, such as Europe, it must be grown during the summer months.

The kind of soil most suitable for the tobacco plant is found to be a friable well-drained sandy loam, not too rich in organic matter. The above description applies chiefly to its mechanical properties. Chemically the required condition may be inferred from a knowledge of the ash constituents of tobacco leaf of good quality.

The bulk of the tobacco grown in India is consumed by the native population, to suit whose taste and mode of using it, all that is required is the production of a leaf possessing sufficient flavour and strength. The Eu-

CULTIVATION.

Conditions necessary for.

ropean smoker, however, requires a different article, that is, a better prepared leaf, full flavoured, easily combustible, and with a minimum amount of nicotine. The shape and texture of the leaf is also of great importance, especially for the manufacture of cigars. In order to produce a leaf of this description care must be taken to select the kind of seed which will yield such a leaf. With suitable climate and soil, and the proper kind of seed, the cultivator will have secured the preliminary conditions towards the production of a well-shaped and well-flavoured leaf. Much, however, will depend on the mode of cultivation adopted, and still more on the after-preparation or curing of the leaf. The various operations concerned with the actual cultivation of the plant, *viz.*, *the preparation of the seed-beds, the sowing of the seed, the transplanting of the seedlings* into properly prepared ground, *weeding*, *watering*, *topping* the flowering stems, and the *removal of superfluous shoots*, guarding against *insect attacks* and *diseases*, and finally the judicious *plucking of the mature leaves*, will be alluded to in greater detail further on. The amount and quality of the manure required for any particular soil must be determined either by direct analysis of the soil, or by inference drawn from observation of the defective properties of the product. An excess of nicotine, for example, indicates a badly prepared or an improperly drained soil, or that the manure given was too highly nitrogenous. Again, if a tobacco burns badly, and with a dark ash we may conclude that the soil was deficient in carbonate of potash, that very important substance, which, combining with a vegetable acid, gives to tobacco leaf its good burning properties. In fact, a soil deficient in potassium carbonate will never grow good tobacco. There do exist virgin soils which contain naturally all the constituents necessary for the production of good tobacco, but even they after continuous cultivation must from time to time be replenished with suitable manure. In certain parts of Java it is said that tobacco requires no artificial manure, as the requisite mineral constituents are continually being supplied by volcanic action.

A valuable report on the "Cultivation and Preparation of Tobacco in India" by **Dr. Forbes Watson** was published in 1871, appended to which is a translation of a "Manual of Practical Operations connected with the Cultivation and Preparation of Tobacco in Hungary" by **Mr. J. Mandis**, one of the Government Inspectors for superintending tobacco cultivation in that country. The following extracts have been selected as possessing special interest in connection with tobacco operations in India at the present day. After commenting on the adverse opinions pronounced by the London brokers on the Kaira samples of tobacco forwarded by the Bombay Government in 1870, he remarks:—"These faults are either accidental, as in the case of the wrong mode of packing, or they are preventible by a careful method of curing the tobacco. The question of the acclimatisation of Cuban and American seed in India, intimately connected as it is with the question how far tobacco can be made one of the staples of Indian export, is far too important to be decided on such insufficient data as a few preliminary experiments can give. The example of the continental countries, France, Germany, and Austria, where the acclimatisation of the American varieties has been carried out successfully on a large scale, clearly proves that, provided the trials be made in the same systematic and persevering manner, a similar process will succeed in India also, as more favourable conditions of climate frequently prevail there." "The conditions and circumstances attending the industry in Austria present some points of resemblance to those occurring in India. The districts devoted to the growth of tobacco in Austria are situated principally in Hungary and Galicia, where the soil is similar in many respects to that of large districts in India, and the climate very continental and dry. Although the mode of

CULTIVATION.

Conditions necessary for.

cultivation and preparation is still to a great extent in a very backward state, the quality and quantity of the produce satisfy the internal wants; there, as in India, the improved and extended cultivation is principally insisted on with a view to an export into foreign countries." "According to the unanimous statement of the reports from India, the chief desire of the native cultivators is to obtain strength, which implies a high percentage of nicotine in their tobacco. But the Havanna tobacco is exactly that which contains the smallest known amount of nicotine, and the most esteemed varieties of tobacco generally appear to contain less of this principle than the common tobaccos, whilst, on the other hand, they are distinguished by richness of aroma. Exact scientific observations and experiments have established beyond doubt that complete acclimatisation, that is, a complete reproduction of the plant with all its distinguishing characteristics, is not obtained by merely using the seed of the desired variety. The acclimatisation depends on a concurrence of several conditions, and in the exact measure in which these conditions are fulfilled, the acclimatised variety will approach to, or recede from, its original character. The observations which bear on the whole question may be considered from a twofold point of view—1, How far are the botanical characters of the plant, the number, size, and shape of leaves, the position of the ribs, &c., preserved? 2, What are the changes in chemical composition ensuing on acclimatisation, as compared with the original plant? It may be remarked that if the scientific investigation were complete, the two points just mentioned would embrace the whole question. As matters stand, however, the data bearing on these points afford only means for sifting and ascertaining the meaning of the various practical observations made on the smoking qualities of the different kinds of tobacco, on their combustibility, strength, and aroma." "The individual and distinctive characters of the various tobaccos, as articles of consumption, depend mainly on the proportion of four elements:—*1st*, on those of the mineral constituents, among which the potassium salts are most important; *2nd*, on the amount of albuminous compounds; *3rd*, on the amount of nicotine; *4th*, on the amount of nicotianine and of the essential oil. The influence of the mineral constituents on the properties of the tobacco leaf, as at present established, seem to be twofold. In the first place, the ash seems to act by its great amount as a preserving and antiseptic principle, preventing and stopping the fermentation, thus facilitating the operation during the curing of the tobacco, and rendering the final commodity stable and unalterable. In the second place, the presence of a large quantity of potassium salts, and especially of the nitrate and carbonate, seems essential to assure a proper burning of the prepared leaf a most important point in the estimation of various kinds of tobacco. In so far then as these two properties are concerned, the value of the products will not depend on the kind of seed taken, but only on the soil on which the plant has grown and the kind of manure which has been employed." "The albuminous substances amount to 10 and more per cent. of the dry leaf before curing. Their presence in the prepared tobacco prevents the proper burning, and besides gives rise in burning to a disagreeable smell which overpowers the aroma of the leaf. It is the chief end of curing and fermentation to get rid of these substances." "The conditions which favour the formation of the compounds which give to the tobacco its aroma are sunshine, warmth, and a light airy soil, sandy or calcareous. Thus it comes to pass that a given variety of tobacco may either be very strong, but containing fewer of the aromatic principles, or rich in aroma and less strong, the latter kind being the more esteemed of the two. The Havanna is the example *par excellence* of a tobacco of this kind." "From this cursory view of the principal constituents of the prepared

N. 108

NICOTIANA Tabacum.

Tobacco.

CULTIVATION.

Conditions necessary for.

tobacco leaf, it appears that its outward shape and size, the strength and disposition of the ribs, &c., together with the character of the aroma, depend to a large extent on the kind of seed used, and will probably remain substantially unchanged in the acclimatised plant; whereas the amount and relative proportion of the mineral constituents depend exclusively on the soil and its cultivation; and the amount of nicotine and the quantity of the aromatic substances partly on the soil, and partly on the joint influences of temperature, sunlight, and moisture. The conclusion to which the above paragraph more especially points is the importance of the influence exercised by the soil." "Numerous experiments and observations bear out this view." "In the numerous notices about Indian tobacco published in the Proceedings of the Agri.-Horticultural Society of India there is repeated mention of a valuable variety of tobacco restricted in its growth within narrow limits, and different from the usual kinds grown in the villages around. The action of the soil on the plant is twofold. It acts by the chemical constituents contained in it, and by its state of aggregation and its physical properties. As the tobacco is a very exhaustive plant it wants an ample and rapid supply of its ash constituents and of ammonia. The want will be best supplied when the soil contains a great proportion of vegetable mould, as this will present a large proportion of the ash constituents in a soluble form. The physical properties of the soil which most influence the cultivation of tobacco are, its state of cohesion, its power of retaining water, and its power of absorbing heat." "For the cultivation of aromatic varieties of tobacco, a light loose soil, readily absorbing heat, is required, such as a sandy or calcareous soil. This kind of soil will never have a high retentive power for moisture, and this is of considerable importance, as stagnant moisture must be carefully avoided. On the other hand, the soil should always remain slightly humid." "The next important conclusion is, that in order to obtain the same combination of strength and aroma in the acclimatised plant, it must be placed under equally favourable conditions of temperature and moisture." "In this respect India is situated more favourably than the Europe States, where the foreign varieties of tobacco have been acclimatised, and where the principal advantage consists in the better shape of the plant, and in the position, number, and disposition of the leaves. In Europe the climate is such as not to allow of the full attainment of the original aroma; although, even as regards aroma, the plant raised from Cuban seed in Austria is favourably distinguished. The diversity of climate in India ought to enable us to put the acclimatised plant under conditions similar to those existing in the country where it is indigenous." "The cultivation and preparation of the acclimatized as well as of the native varieties of tobacco (intended for export) can only be carried out successfully when certain fundamental principles are acted on. The most important of these refer to the following points:—1, choice of seed; 2, proper system of manure and of rotation of crops; 3, proper system of cutting and gathering the ripe leaves; 4, proper mode of curing the leaves; 5, commercial assortment of the produce. First, as regards the seed, it may be observed that in France the greatest attention is now paid to its selection. Formerly, nearly every cultivator of tobacco provided his own seed; now, the Government administration has taken this matter into its hands, and grows its own seed, selecting with extreme care only the finest plants for it, and this seed it supplies to the farmers, who are prohibited from using any other. There are several advantages in this system. The seed proceeds exclusively from the very best varieties, and each variety is kept distinct in cultivation, so that the seed sown by the farmer is uniform in kind. This is a very important matter." "A proper system of manure and of

CULTIVATION.

Conditions necessary for.

rotation of crops is of great importance, because tobacco is a very exhaustive plant. The mineral substances essential for the growth of tobacco are chiefly bases; potash, lime, and magnesia,—whereas the amounts of phosphoric, sulphuric, and silicic acids are less important. Now as these latter substances are exactly those most important in the cultivation of grains, the position of tobacco in the rotation of crops is fixed. Of the store of available mineral substances in the soil, made up partly by the ever progressing decomposition of its constituents and partly by manure, the tobacco will principally exhaust the potash, lime, and magnesia only, whilst phosphoric, sulphuric, and silicic acids will go on increasing in amount. If now a crop of grain be taken from the same soil, these acids will be used up, whilst potash and lime will be accumulated, thus restoring to the soil the conditions for growing tobacco. A proper manure is of essential importance. Thus, on soils poor in lime the use of burnt lime or gypsum is recommended. Cow, sheep and goat's dung is most usually employed, besides sewage, which being rich in ammonia and potash does very good service."

"Next in importance is a proper system of cutting and gathering in the ripe plant. Carelessness in this respect is one of the greatest defects in the cultivation of tobacco as practised in India. Throughout Europe, in the United States, and in Cuba, the gathering does not take place at once, but extends over a long period, the leaves being taken one by one as each gets ripe." "The same care is taken in Turkey to collect only the ripe leaves. In Albania and in the district of Salonichi, where the finest Turkish tobacco is grown, the gathering extends over from three to four weeks, and takes place at five different periods." "If the gathering of all the leaves is done at the same time, then one cannot fail to gather one portion of them while yet unripe, and another portion when over-ripe. The consequences are almost equally fatal." "The proper moments for the gathering of the leaves are just the latter stages of ripening, when the mineral substances are rapidly increasing. Then the mineral substances begin to diminish again, thus reducing the combustibility of the final product; the gummose substances diminish equally, which renders the dried leaves less elastic and more crisp and brittle, and subject to being reduced into dust; and finally the proportion of nicotine is rapidly increasing, although in all finer qualities of tobacco an excess of it ought to be carefully avoided." "The curing of the leaves is perhaps the most important operation. It can be done properly, only when the previous operations have been executed with all the care insisted on in this report, for even the best material can be entirely spoilt by bad curing. Curing consists of a series of operations. The cut leaves are first allowed to wilt, in the next stage they acquire the proper colour, then they are dried, made into 'hands,' and finally undergo a fermentation. In the first three stages most of the usual methods of manipulation are defective. It is here to be noticed that in the course of his experiments in Dharwar, Mr. E. P. Robertson hit on the right principles. For an account of these experiments see paragraphs under Bombay Cultivation." "The leaf is subjected to a remarkable transformation during the curing. The organic substance undergoes the process of decomposition, water and carbonic acid are given off, and compounds are formed containing a higher percentage of carbon distinguished by a brown colour, and probably analogous to the brown compounds composing the mould produced by decay of vegetable matter. At the same time the albuminous substances are being partially destroyed during the whole process of curing, and especially during the final stage of fermentation. This chemical transformation is a gradual process, and requires time—weeks and months even—to develop itself fully, and during

Tobacco.

CULTIVATION.

Conditions necessary for.

this time there are other influences at work which may become injurious to the product if the utmost care be not taken. *1st,* If the leaves are allowed to dry too soon, and especially if they are exposed to the sun, the process of the decay of the substance of the leaf, and the decomposition of albuminous matter is left incomplete; some of the shaded portions of the leaves remain green, and the portions exposed to the sun get yellow, not in consequence of the internal decomposition, but solely because the chlorophyl, or green colouring matter, becomes bleached by the sun. Such a leaf will finally present an uneven colour, a chequered appearance, and, especially the portions which got dry in the sun, will be very brittle and crisp. *2nd,* If, in consequence of careless manipulation in the handling, the leaves are allowed to rub one against the other, or if moisture in drops collects on their surface, either by rain or by artificial moistening, or even by too violent a sweating, then those places begin to rot, become very deep brown or even black, the fibres of the leaves become injured, and the leaves altogether become brittle after drying. *3rd,* The fleshy midribs are a great difficulty in complete drying, and unless they get completely dry they will entail mouldiness, which may communicate itself to a great portion of the leaf. *4th,* The operations during the curing require a repeated handling of the leaves, and only a very methodic way of manipulation will prevent mechanical lesions, holes, and fissures, all which cause a serious diminution in the value of the produce, because such leaves cannot be applied to the manufacture of cigars. It will be seen that most of these points are taken into account in **Mr Robertson's** directions. (See under Bombay Cultivation.) The proper manner of sorting, packing, and arrangement for the market generally is of very great importance in ensuring commercial success. The sorting must have reference to three different objects: *1st,* it must refer to the ultimate destination of the tobacco, so that tobacco suitable for the production of cigars shall be separated from that which is to be used for cutting up and from that used for the manufacture of snuff; *2nd,* it must look to quality, that is, to the more or less successful curing, so that all the kinds of leaves are again sub divided into three or four portions of different quality, by which means a much better price is obtained for the whole quantity of tobacco, because, if a small percentage of good leaves be interspersed among a large number of second-rate quality, the manufacturer will disregard the good leaves altogether, and fix the price as if the whole bale were uniformly second-rate; *3rd,* it must take size into consideration. Good sorting is a most tedious and difficult operation, and can only be carried out successfully when the precautions here insisted on have been observed during the whole cultivation of the plant, that is, when the same kind of seed has been employed and the plants consequently all belong to the same variety, and when the gathering of the leaves has taken place in such a manner that the three to six different kinds found on the same plant have all been collected at different periods and kept separately. Besides, it needs a very good judgment to recognise the precise quality of every leaf and its special suitability for some particular destination, a destination often dependent not only on the inherent qualities of the leaf, but also on arbitrary trade customs." "The great difficulty in curing tobacco is the disposal of the midrib, which persists in either drying stiff or not drying at all. But why should it not be removed? Not entirely, for then the leaf would be split into two, but only on the back of the leaf, where the convex and greater part of the midrib projects. The operation, though a delicate one, would become easy to any child after an half hour's practice. The operator would take a leaf in his left hand, holding it between his finger and thumb at the stalk end. About half an inch from the end (the stalk end), he would

CULTIVATION. Conditions necessary for.

make an incision in the midrib with the thumb-nail of the right hand and turn up an end. He would then take hold of this with the finger and thumb of the right hand, and with an equable force pull off the midrib downwards towards the point of the leaf. As soon as it became very fine, and there was a danger of the leaf being torn, he would nip the midrib off with his finger and thumb. By this the concave or nearly flat surface of the midrib would be left on the upper side of the leaf, while on the back of the leaf the only sign of the midrib would be a narrow depression running down the centre of the leaf where the troublesome midrib had been.

"The sun-flower (**Helianthus**) might be advantageously grown among the tobacco, *1st*, for the shade it would give to the larger and coarser tobaccos required; *2nd*, for the admirable stringing rods (if string itself is not used) which their stems supply; *3rd*, as they would (if their leaves were ploughed into the ground) give almost the exact vegetable mould which is required by tobacco. From the exposition presented in this report, it is manifest that great and systematic care must be given to the acclimatisation and preparation of tobacco to ensure a good result. From the choice of the district and soil where the plant is to be grown, through all the stages of cultivation and preparation, every stage is of decisive influence on the final produce, and the neglect of any one of the manifold precautions will at once tell upon the marketable value of the leaf, and render the production unremunerative." "The very wide scope of the question is evident. It involves nothing less than the reform of a considerable branch of agriculture." "This reform can only be brought about gradually and by the force of example, which is the only manner in which agricultural reforms have ever been introduced. The possibility of growing superior varieties of tobacco must be demonstrated practically, and this not only in a few garden experiments, but on the same scale and in the same manner as it is to be conducted by the producers. The proposed measures are, therefore, not merely tentative, and promising only a problematical success; on the contrary, precedents applying very closely to the point show that it is only necessary to follow a known tract and apply principles which have already succeeded in the case of other commodities. And the similarity between the proposed cultivation of tobacco and that which has been successfully carried out in the case of tea is very great. In both instances the object aimed at is the production of a leaf containing certain active principles, and combining certain conditions as regards strength and aroma; and even the processes in the final preparation of the leaves are in some measure analogous and certainly require as much nicety of manipulation in the one case as in the other. More than this, the conditions under which the experiments on the tobacco will have to be conducted are far more favourable to their success and to their economical importance than was the case in the parallel instance of tea. Tea was an entirely new culture; the popular interest in it had to be created; whereas thousands of acres are already devoted to the cultivation of tobacco, and a vast mass of people are already interested in everything which affects the commerce in one of their own chief articles of production. It remains only to open to them the prospect of an advantageous export trade by showing the preliminary conditions which must be satisfied in view of this prospect."

Mr. O'Conor, in the concluding chapter of his exhaustive report, gives a useful summary of the conditions and prospects of the tobacco industry in India in 1873. There is no reason to suppose that since that date the quality of the bulk of Indian tobacco has in any way improved; and although comparatively little has been done to increase materially the

CULTIVATION Conditions necessary for.

export trade to Europe, still the results of private enterprise are continually showing to what extent improvement can be effected with capital aided by Western experience.

As to the possibility of improving the quality of Indian tobacco up to the standard required for European consumption, **Mr. O'Conor** remarks as follows :—"Undoubtedly a considerable time may be expected to elapse even after Indian tobacco generally has been improved up to the point where it will be fit for export to the European market, before any such very extensive demand will arise as to make a development of the area of cultivation necessary; but the actual extent of cultivation is very great, and it is desirable to consider whether efforts might not with great advantage be made to improve the quality of the tobacco now produced in this country up to the standard required by European taste, as well as to increase the yield of the cultivated area. In the first place, any well-directed effort to improve the methods of cultivation now pursued, must, *pro tanto*, re-act upon the general agriculture of the country. Tobacco is essentially a crop which requires high cultivation, and the example given by the successful growth of tobacco in accordance with sound principle will, of necessity, have a most beneficial effect upon native agriculturists. In the plains of this country all lands yield two crops in the year, and land which has been properly prepared for tobacco will require comparatively but little attention for the second crop, while this at the same time, in consequence of the high manuring and deep cultivation required for tobacco, may be expected to give much more than an average yield. The crops taken off tobacco land in the second year in America are always much heavier than those given by ordinary lands not previously treated for tobacco." "In the second place the importation of tobacco into India, caused as it is entirely by the presence of the European population of the country, would almost altogether cease coincidently with the improvement of the quality of the local produce up to the European standard. And in the third place, I conceive it to be an object of importance to improve as much as may be possible the quality of a crop which already occupies an enormous space in the cultivated land of India." "We are acquainted with many places in India which now produce abundant crops of good and tolerably well-flavoured tobacco, and it is to these places that the efforts of tobacco planters should be directed. It is useless to undertake the cultivation of this plant in tracts like the Berars and the Central Provinces, where the climate and soil are alike unfavourable to tobacco, and where consequently the yield is poor and the quality bad. The tobacco-growing tracts of India are comprised in Burma, many parts of Bengal, some small scattered tracts in the North-West Provinces, a large area in the Bombay and Madras Presidencies, and the Native State of Travancore. In these provinces the cultivation might be spread and the quality of the produce very greatly improved." "It has been grown in India for two hundred and fifty years; its cultivation is widely diffused, and in some places the produce enjoys good repute among Europeans as well as Natives. The improvement of the quality, and the development of the industry, so as to make tobacco a profitable article of export to the European market, now seem to lie with the private capitalist rather than with the Government. The question for consideration seems to be, how far the Government can and should encourage and sustain the well-directed efforts of individual or associated capital and energy in working out and extending an industry which has taken deep root in the country."

After the publication of **Mr. O'Conor**'s report, which clearly indicated the conditions and possibilities of the tobacco industry in India at that time, the question arose as to whether the Government, by importing

CULTIVATION Conditions necessary for.

skilled curers from America and other countries, and posting them in the best tobacco-growing tracts of India, could effect such a general improvement in the quality of Indian tobacco, as would enable it to compete with the bulk of the leaf imported into Europe from America. The opinion, however, of competent officers was, on the whole, unfavourable to the proposal to import foreign curers to instruct the people, as it was thought that more substantial good would arise by encouraging private enterprise. The progress which has taken place, especially in the Madras Presidency, affords a better indication of future success, than could be obtained by any number of experiments carried out under the direction of Government. Some extracts from a letter on this subject by **Mr. Broughton** are here given:—"From my experience of the natives of the Madras Presidency, I do not at all think they would follow any improved method of dealing with their tobacco, unless indeed a pecuniary advantage in the change could be made immediately apparent to them in their narrow experience. Substantial proof would, I think, be necessary of the money advantage to be obtained before a ryot would voluntarily alter the current method of manufacture which has so long been pursued in his district." "This could only, under any circumstances, occur in a few districts, as most tobacco cultivators are quite content with a local sale of their crop, and, furthermore, there are but few districts in this Presidency where tobacco, fit for European sale, can readily be grown. This latter is, I think, rendered evident in my various reports on tobacco. Furthermore, I do not believe there is any great art in the proper curing of tobacco. This, I think, is evident from the fact that tobacco is successfully cured in the Philippines by a people that are certainly not more intelligent than a native of South India. Even now the tobacco of the Godavery lunkas and Dindigul is also successfully cured by natives, though it requires a certain experience before their cheroots are thoroughly liked by Europeans. At the present time, there are European firms in South India, as **Messrs. Campbell & Co.**, Dindigul, and **Messrs. Roberts**, Coconada, who are attempting not unsuccessfully to produce Indian cheroots for European smokers. The judicious care necessary for the curing of tobacco, which is more necessary to success than special art, will, I think, be produced in the districts, as the trade of these firms extends. As a personal opinion, therefore, I cannot see that the introduction of a teacher from Manilla would be productive of much good to South India. My previous reports have abundantly shewn that it is in the more jungly districts of India that the best tobacco can be grown. Travancore, Burma, the wilder parts of Coimbatore, the Lower Pulneys, &c., have produced the best, with the exception of the centres cited above. The ordinary practice of placing good tobacco seed in the hands of *tahsildars* for cultivation in unfavourable districts can never result in success for many reasons. But I would respectfully recommend that small quantities of good seed should be issued to planters who would take a more intelligent care and interest in the growth and curing of the crop than can be expected from the ryot. The planter would also possess a better discrimination in the quality of his product than the latter."

Regarding the measures which should be adopted for improving the quality of Indian tobacco, **Messrs. Campbell & Co.** of Dindigul, who have had several years' experience in the preparation of tobacco, especially for the manufacture of cigars, thus express their opinion:—"In South India the climate is much against the methodical curing of tobacco, which ripens just as the atmosphere gets dry and parched; the consequence is the curing has to be hurried to prevent the leaf becoming brittle, and

CULTIVATION Conditions necessary for.

what, no doubt, the ordinary cultivator fears more to prevent, loss, of weight by the rapid evaporation of the moisture in the leaf. The curse of tobacco-growers in these districts is this hot parching atmosphere in February, April, and May; but for it Southern India might compete successfully with Cuba; hasty slurred-over curing ruins our leaf. We ourselves have in a small way remedied this by bringing in half-cured leaf from native gardens and slowly curing it in vaults built at considerable cost below ground where we could produce an artificial atmosphere. As this artificial atmosphere is absolutely necessary for the successful curing of tobacco in Southern India, we would suggest the erection of curing sheds in suitable localities where the produce of native gardens might be brought and slowly and methodically cured. The difficulty would be, that native gardeners would not bring their leaf unless it was purchased. Government might not care probably to make such large purchases. Moreover, intimate knowledge of the localities is necessary, as the leaf grown in the vast proportion of native gardens is of no use whatever for burning; the soil in which it was raised and the water in the well by which it was irrigated being deficient in the necessary 'salts.' This leaf is used for chewing, and there is a great demand for it. Coffee planters seldom cure their own produce. The berry from perhaps a hundred different estates goes to the same curer on the coast. Tobacco should be treated in the same way. The cultivator has not sufficient time to devote to the curing of his produce. Curing should be quite a different and separate industry. It would then be slowly and carefully undertaken and attended to. In South India cultivators of gardens irrigated by wells get usually three crops off their soil in one year; the consequence is, that no sooner has the tobacco been cut than the roots are torn up, the ground ploughed and grain sown; until this is done and finished, the tobacco is forsaken, and at its most critical moment the curing is neglected. Often have we heard the native gardener grudge the time that tobacco curing requires; again, where is there a native gardener who will go to the expense of building *pucka* to enable him to produce an artificial atmosphere? Now, as the tobacco gardeners in Southern India cultivate remarkably well, but fail in the curing, we have long recognised the fact that, as in indigo planting, so with tobacco, the produce should be raised by natives, and the curing should be done by Europeans. Indigo planters build factories in the native centres of cultivation, watch the culture, and carry the produce to the factories where European supervision has raised Indian indigo to its present high standing. So must it be with tobacco, and with assistance to the pioneers of the new industry from Government. we fear not that South India may some day see the factories of the European tobacco planters."

Mr. Tucker, in his note accompanying the report on the Statistics of the Inland Trade for 1888-89, makes the following remarks in regard to the condition and prospects of the tobacco industry in India:—"The principal barrier against a larger trade lies in the ignorance of the natives of the art of curing the leaf. A sufficiently profitable market might be found for ordinary Indian tobacco if it were properly cured; but the manner in which the narcotic is usually consumed by the natives themselves does not encourage a knowledge of curing. It is either chewed as a dry powder mixed with lime, or smoked in the form of a conserve or paste mixed with treacle and other ingredients. In either case, pungency rather than aroma or delicacy of flavour is the chief desideratum. The best prospect of improvement in Indian tobacco manufactures lies in the steadily increasing consumption among the richer classes of natives as well as the European population of the country cheroot and pipe tobacco."

MANUFACTURE: Chief Operations in. 109

Chief Operations.—Mr. Schiffmayer's paper on "Tobacco and its Culture," published in the Annual Report of the Madras Agricultural Department, 1877-78, contains much useful information on the cultivation and curing of tobacco, and more especially with reference to the methods best adapted to the climatic conditions of South India. Omitting the detailed directions concerning the cultural operations some extracts may be here given describing the various stages in the preparation of the leaf from the time of its ripening until it is ready for the market :—" The plants commence to ripen about three months after being planted; this is indicated by the leaves assuming a marbled appearance and a yellowish green colour. The leaves also generally become gummy and the tips bend downwards."

Harvesting. 110

" HARVESTING.—The leaf being matured, it should be harvested only after the dew is off the plants and not on a rainy day. There are two modes of harvesting—gathering the leaves singly, and cutting down the whole plant. Gathering single leaves admits of removing them from the plant as they ripen; the bottom leaves are removed first and the top leaves left some time longer until they have attained full maturity. The cultivator is thereby enabled to gather his crop when it possesses the greatest value. This plan necessitates, however, a great amount of labour, and in a hot climate the single leaves are apt to dry so rapidly as not to attain a proper colour, unless stacked early in heaps. Stacking in heaps, however, involves great risk of the leaves heating too much and developing a bad flavour, whereby the tobacco loses more or less in value. For Indian circumstances generally, I believe that cutting the whole plants is better adapted than gathering the leaves singly. For cutting down the plants, a long knife or chopper should be used. The man taking the plant with his left hand about 9 inches from the ground, and with the knife in his right hand, cuts through the stem of the plant just above the ground. If the plants are sufficiently wilted, he may lay the plant on the ground and proceed to cut down others; if, however, the plants are so brittle as to cause the leaves to be injured by laying them down, he should give the plants to another person to carry them at once under shade." After describing in some detail the construction of a framework on which the plants are to be hung in the shade until they are sufficiently wilted, **Mr. Schiffmayer** proceeds to give directions regarding the drying shed :—

"When cultivating tobacco on a small scale, any shed not in use will answer, provided that it contains a sufficient number of doors and windows to admit of regulating the circulation of air. A roof made of straw seems to answer very well. The shed should be high enough to admit of hanging three rows of tobacco in it, one above the other. The bottom tier for the first row should be about 3 to 5 feet from the ground according to the size of the plants, which should not touch the ground; the second tier should be 3 to 5 feet higher than the first, and the third 3 to 5 feet higher than the second, the whole being from 10 to 17 feet high from the bottom of the shed to the highest tier." "When cultivating tobacco on a large scale the same arrangements should be made, but the building may be made higher and be provided with a cellar in which to place the tobacco for the purpose of stripping, &c. The tobacco-shed being ready, the plants immediately after they have reached the shed are transferred from the waggon or cart, on the bamboos, to the lowest tier. No rule can be given as to the distance the bamboos should be placed from each other; the distance to be given varies according to the species of the plant, the degree of ripeness, and especially the state of the weather. The purpose of hanging the plant here on the lower tier is to cause the leaves to dry gradually to assume a good yellow colour, to cause a slight

MANUFACTURE: Chief Operations in.

fermentation in them, but to allow such circulation of the air between the plants as will facilitate the gradual escape of the moisture from the plants and to prevent the injurious development of ammonia and other combinations that give rise to the bad flavour in the tobacco. How to attain this must be left to the judgment of the cultivator; he, by frequent examination of the plants, and by careful observation of the changes going on in the leaves, will soon find out the right way to accomplish it." "The leaves must be examined carefully every day; there may be some plants which dry quickly, others more slowly; one plant may progress very well, whereas another hanging close by decomposes too rapidly, and again another too slowly." "The plants should remain on the lower tier until the leaves have turned yellow, which will take place within six to ten days according to circumstances; after this they should be hung on the upper tiers. There they should be hung more apart, each plant hanging free." "The light yellow colour of the leaves should change into a dark yellow golden colour or light brown. After hanging on the upper tier for about a week, the veins of the leaves will be nearly dry leaving only the midribs pliant. The drying of the leaf and the changing of its colour proceed gradually, commencing from the margin and proceeding to the midrib. At this time the plants may be hung closer together, the evaporation from the leaves being little." "All the windows may be kept open from this time; the tobacco may also be brought into an open shed or even hung outside exposed to the sun. In about a week more the midribs will be entirely dried up, and the tobacco fit for stripping.

Stripping. 111

"STRIPPING may be performed at any time, provided the leaves after being once properly dried have again become pliable. For stripping, such a number of plants as will furnish work for several days should be taken down on a morning when the plants have absorbed some moisture and have become elastic; they should be put in a heap and properly covered to check evaporation. If, however, the night-air should be so very dry that the leaves cannot absorb sufficient moisture to become pliable, then a moist atmosphere can be created either by steam, or by pouring water on the floor, or by keeping chatties with water in the shed. If this cannot be done, the tobacco must remain hanging until there is damp weather. Under no condition should the tobacco be stripped when not pliant, that is, if the leaves are so brittle that they would break when bent or rolled.

Sorting. 112

"SORTING.—Tobacco intended for smoking should be carefully sorted when stripped. There should be four sorts, *viz.*, *1st*, large equally good coloured untorn leaves; *2nd*, leaves of good size and colour but torn; *3rd*, leaves of inferior colour and bottom leaves, and *4th* the refuse containing shrivelled-up leaves, &c., to which may be added the suckers. Leaves under No. 1 when thin, elastic, and of good species, are mostly valued as wrappers (outside covers) for cigars. No. 2 may also be used as wrappers, but are less valued than No. 1; they are adapted for fillars and cut tobacco." "The leaves should then be made into hands, that is, 10 to 20 leaves should be tied together by twisting a leaf round the end of the stalks; each sort should be attended by a special man to avoid mixing." "When making the hands of the two first sorts, the man should take each leaf separately, smoothen the same on a flat board, leave it there and take another leaf, treating it in the same way and continue thus until a sufficient number is ready to make a hand. When the hand is ready, it should be laid aside, and a weight placed upon it to keep the leaves smooth.

Bulking. 113

BULKING.—Bulking means placing the tobacco leaves in heaps for the purpose of heating it, in order to attain colour and flavour; this is carried out in various ways, nearly all of which involving great labour and risk. In most instances tobacco loses more or less in value during the process called

MANUFACTURE: Chief Operations in.

curing. It must here be mentioned that the more care is taken in raising the crop, the less attention the tobacco requires in the shed. With a good species of tobacco grown on light friable soil, treated as laid down in this paper and the leaves dried as mentioned above, little care will be needed, after the leaves are dried and stripped. By the drying process described, the leaves will have undergone a slow fermentation which makes it unnecessary to watch or guide a regular fermentation afterwards, hence bulking, and fermenting as generally understood, are not required." "If the colour of the leaves is not uniform, or if it is desired to give them a browner colour, then the heaps must be made large and a somewhat moist atmosphere is required in the storing-room. This will cause fermentation to set in after a short time, and the heat to rise after some days so much that rebulking is required, which is done by putting the top leaves of the old heap at the bottom of the new one. Under such circumstances, the heap must be frequently examined during the few first weeks to prevent overheating. It is advisable to rebulk the tobacco also, even when not much heated after the first fourteen days, and again a month later to ascertain the exact state in which it is. Sometimes the tobacco becomes mouldy; this may occur especially with tobacco which has been manured with chlorides, which cause the tobacco to become more hygroscopic than when manured otherwise. If this occurs, the mould must be brushed off, and, if necessary, the tobacco dried. The tobacco may now remain heaped in the store-room until there is a chance for sale." "It is sometimes the custom to subject the tobacco leaves to some sort of improvement. There is no doubt that by proper application of ingredients the value of tobacco may be much enhanced. The most costly tobacco often commands a high price, not so much on account of its inherent flavour as on account of that given to it artificially. In most instances, the best course to be adopted is to leave the improvement of the leaves to the manufacturer. Many ingredients are employed to improve smoking tobacco. They tend—

(1) to make the tobacco more elastic and flexible;
(2) to remove the coarse flavour;
(3) to add a particular flavour;
(4) to improve the burning quality;
(5) to improve the colour.

To make the tobacco more flexible and pliant, the leaves are macerated in, or sprinkled with, a solution of sugar. In hot countries this is often necessary to give tobacco such an elasticity as to become fit for handling, especially when intended for wrappers, and may be done by an intelligent cultivator. To remove the coarse flavour, tobacco is often macerated in water or in a solution of hydrochloric acid; the more coarse the flavour of the tobacco, the stronger is the solution used. Sometimes tobacco is steeped in a mixture of sugar solution and diluted hydrochloric acid. To extract the fatty matter, tobacco is macerated in alcohol or spirit of wine." "In the maceration of tobacco for the purpose of influencing the flavour, the following ingredients are mostly in use:—Cognac, vanilla, sugar, rose-wood, cassia, clove, benzoin, citron oil, rose-wood oil, thyme, lavender, raisins, sassafras-wood, orange, and many others. The burning quality is improved by macerating tobacco in, or sprinkling it with, a solution of carbonate of potash, acetate of potash, acetate of lime, saltpetre, &c. Badly burning cigars inserted for a moment in such solutions are much improved. Tobacco treated with acetate of lime yields a very white ash. The colour of tobacco is sometimes improved by fumigating the leaves with sulphur and by the application of ochre and curcuma. Although it may be said that fine tobaccos generally do not require any impregnation with foreign matter for the sake of flavour, yet the manufacturer resorts frequently to it

NICOTIANA Tabacum.

Tobacco.

MANUFACTURE Chief Operations in.

to give the leaf a particular aroma. An inferior tobacco, however, which often would not find a market, is sometimes so much improved by artificial means, as to compete successfully with the genuine fine article. A special preparation of tobacco for snuff is seldom attepmted by the cultivator. With reference to the preparation of tobacco for export, the sorting of the leaf is of the utmost importance. Only first and second sorts should be exported. It would be well to remove the midribs whereby the cost of transport and customs duty would be greatly reduced. Finally, it must be mentioned that the value of a cigar depends, not only on the intrinsic value of the leaf, but to a great extent on the mode of manufacturing the article. Thus, the raw material may be of good quality, but if the maker does not classify the leaves properly, or, if he rolls his cigars too hard which must vary according to the qualities of the leaves, the cigar will burn badly. The best burning leaves must always be used for wrappers. If this should be neglected, the inside of the cigar burns faster than the covering, the air has no access to the burning parts, and the empyreumatical substances are volatilised without being decomposed. Such cigars, therefore, make much smoke and smell badly."

DISEASES. 114

Diseases of the Tobacco Plant.—The various diseases to which the tobacco plant is liable in India are, with the exception of those brought about by unseasonable weather, mostly preventible. No animals, except insects, will touch the tobacco plant in the field. Caterpillars and grubs are more or less troublesome in certain localities. The only effectual remedy is to employ coolies to collect and destroy the worms morning and evening. In America, where caterpillars are a more troublesome pest, turkeys are taken into the tobacco fields to feed on them. Guinea fowls have also been recommended for this purpose. In the Jhang district of the Panjáb the tobacco plant is said to suffer greatly from the attacks of the *mula*, a whitish brown woolly caterpillar, which devours the plant at the base of the stem just below the surface of the ground. These caterpillars are said to be more troublesome during rainy weather.

Diseases of the nature of blight frequently occur during unseasonable weather, or when an unfavourable wind is blowing, during critical stages of the plant's growth. In the Panjáb there is a disease known under the names of *tela* and *saresa* (glue), which attacks tobacco and many other crops; it appears as an oily substance deposited by an aphis, and rain is said to be the only remedy. **Mr. Nicholson** enumerates various diseases which are more or less prevalent in the Coimbatore district—*Sámbal* appears when the weather is cloudy and misty, especially at the time of topping or if the east wind then prevails; the leaves become ash-coloured and brittle, and quite unfit for use. *Poriyán* is also due to the east wind, especially if the ground is water-logged; it shows itself in minute black spots over the whole leaf and stem. *Murugan* or *murulie* is very destructive, and is also due to the east wind; the symptoms are a sickly look, roughness of the leaves, and brownish yellow spots. *Pachchei puluvu* is an insect pest (*Man. Coimbatore District*).

In North-West India *Sáwani* tobacco, *i.e.*, that which is cut in February, occasionally suffers from frost in districts west of Benares; the *asaárhi*, or later crop, is often injured by hail, especially in districts close under the hills. The leaves sometimes suffer from the attacks of a kind of grey mildew, known as *kápti* in the Azamgarh district.

Grierson mentions a disease called *Kachoha*, which attacks the tobacco plant in Behár; also a worm called *chhiri*, or *kanátha* which attacks tobacco and other crops. Another worm called *kenuán* is stated by him to be destructive by burrowing into the ribs of the leaves.

The most serious injury from which the tobacco plant occasionally suf-

DISEASES.

fers is caused by the parasitic growth of a fleshy leafless plant, known in Southern India under the name of *bodu,* and scientifically as **Orobanche nicotiana.** It is a kind of broom-rape, and belongs to a family of parasites. Being annuals, they are propagated by seed which, accumulating in the soil, and coming into contact with the particular kind of plant it requires to feed upon, attaches itself to it by its roots, and thus weakens the growth of its host, often destroying it altogether. In 1883 the tobacco crops on the Godáveri lankas were very seriously injured by the *bodu* parasite, and measures were taken by the Madras Government to ascertain the cause of these ravages with a view to discovering the best remedy for their prevention. The presence of the parasite to any great extent has been attributed by the natives to soil exhaustion, as well as to too much moisture in the soil, and they say it always increases on land not sufficiently silted. According to **Mr. Caine**, the spread of *bodu* is chiefly due to soils being worked when too wet, and all soils insufficiently or carelessly cultivated are rendered liable to the ravages of this parasite where tobacco is grown. The lanka soil, though rich, very soon cakes, and, unless kept carefully tilled, is apt to become damp and sour. A similar outbreak took place at the Poosa tobacco farm, which **Mr. Caine** thinks was caused by heaping as manure on the land enormous quantities of tobacco stems, which slowly decomposing formed hotbeds for the development of the disease. In any locality where the *bodu* parasite is detected, arrangements should at once be made to pull up the plants, which are easily detached, and burn them; and if possible the crop should be changed, as the land would no doubt contain a considerable quantity of the dormant seeds of the *bodu* plant.

PROPERTIES. 115

Properties of the Tobacco suitable for European Trade.—REPORT ON TOBACCO EXHIBITS AT THE COLONIAL AND INDIAN EXHIBITION, 1886.—A Conference was held in the Commercial Room of the Imperial Indian Court of the Colonial Indian Exhibition on the 8th July 1886. There were present **Sir Edward Buck, Dr. G. Watt, C.I.E.,** and **Dr. Forbes Watson;** also representatives of most of the leading tobacco firms in London and other important commercial centres.

The general opinion as regards the exhibits of leaf tobacco was that, while there was still room for improvement, some of the samples shown would find a distinct place in Europe. **Mr. Chambers (of Messrs. Grant & Chambers)** in the words of the report pointed out that "leaf tobacco had two distinct uses—*1st*, for cutting; *2nd*, for cigars. He gave it as his opinion that Indian tobacco was not suitable for cutting by itself, but that the better qualities might be taken up for certain cigars, or as substitutes for American tobaccos. In the cigar industry two kinds of leaf were required, one for the 'covers,' and the other for the 'fillars.' The former must be thin, silky, nut brown, with the lateral veins neither spreading at a right angle nor at too acute an angle from the midrib. The veins must be equidistant, and not too thick. The leaf should be about 1½ to 2 inches at the base on either side of the midrib, a tapering leaf not being suitable for covers, although useful for fillars. What was wanted for fillars was good quality, flavour, and burning power; and the smaller upper leaves were often found the best for this purpose. Most of the samples of leaf exhibited were pronounced almost unsaleable from being packed too hard, having a bad smell, and being often perforated, and even broken, and indeed in many cases almost rotten, due to imperfect curing. The process of curing adopted in India was pronounced imperfect, but even if this had not been the case the leaf was in most cases completely ruined from too hard packing. A few samples were, however, much admired, and one or two brokers were authorised by the manufacturers to take steps to pro-

NICOTIANA Tabacum.

Tobacco.

PROPERTIES.

cure large quantities if the price was found suitable. It was strongly urged that the leaf-stalks should be cut off at the base of the blade, and never exported to England. It was pointed out that about 75 per cent. of the tobacco cut for the pipe is sold, in the shape of "shag," at 3*d.* an oz. retail. The manufacturer sells this at 3*s.* to 3*s.* 2*d.* a pound, so that taking the average cost of tobacco at 6*d.*, and the duty at 3*s.* 6*d.*, the article is being sold at considerably below cost price. This is due to the fact that the raw tobacco comes into the manufacturer's hands in a dry state, and the profit is made on the increased weight due to moisture absorbed. A tobacco must, to use the technical expression, 'drink well,' to be profitable, but if the importer has to pay duty upon a heavy section of the parent stem dangling at the end of a long and heavy leaf-stalk, it is impossible that he can make a profit. Indian native leaf is also very injuriously coated with sand and dirt, which, apart from the trouble of cleansing which this necessitates, greatly increases the duty by raising the weight. England is, one might almost say, well enough supplied with tobacco without the aid of India, and with numerous disadvantages it is scarcely likely that India can take an important place in the supply of tobacco unless radical improvements are effected. Tobacco on importation should contain not more than 10 per cent. of water, for although the duty is raised when there is less than 10 per cent. of water, it is not lowered when there is more. This is a most important consideration, and one that cannot be too forcibly urged, for, as has been stated, the profit in the tobacco trade is due to water, but it is fatal if the importer has to pay duty on the latter." "Several of the gentlemen present showed British-made cigars, which in the wholesale trade were as cheap as the Indian cigars. It was pointed out that India, if it came into the British market, would have to compete against home-made cigars. It was stated that there were only one or two tobacco-producing countries that could manufacture good cigars, and these fetched an exceptionally high price, but in point of quality India could never compete with them. It was with the average quality of British cigars that India came into competition, and it had to be recollected that the immense difference between the duty on leaf and that imposed on cigars rendered it possible for the home manufacturer to undersell the Indian, in spite of cheap labour and all other considerations. It was, on the other hand, urged by some of the gentlemen present that through the large number of Anglo-Indians now resident in England a demand had been created for Indian cigars; that just as Indian tea had a distinct flavour of its own, there was something about the Indian-made cigars which commended itself to persons who had smoked a good 'Trichy,' or 'Burma.' These cigars smoked well, the ash remained firm, and they had the reputation of not affecting the head half so much as many of the 'smokes' sold in the London market." "A selection of the cigars shown were cut up and examined, and a number smoked during the Conference. It was pointed out that, in addition to being too green and damp (which thereby increases the duty unnecessarily), the cigars were too hard filled. Havanna and Manilla cigars were cut up and shown alongside of the Indian, and it was found that, instead of being packed in a bunch-like manner, the filling of the Indian cigars went throughout the entire length. This allowed of their being hard packed, a defect which made them too heavy, and difficult to smoke. At the same time, of course, the method of packing was admitted to have something to do with the firm ash produced, which in this respect was superior to that of most of the cigars sold in London. It was, however, urged that every consideration should be subordinated to the effect on duty, since it is the heavy import duty that kills the Indian cigar trade, and therefore light

PROPERTIES.

weight was of primary importance. The improvement, however, which, within recent years, has been effected in the Indian cigars, was admitted by nearly every one present. The approach to the size, form, and colour of the Manilla was a vast improvement on the immensely large cigars of former years, which, from their size and dense filling, required a straw before it was at all possible to smoke them. It was recommended that a selection of the Indian cigars and tobacco should be submitted to a chemical examination, for it did not follow that a light-coloured cigar was, as popularly supposed, a mild smoke. What was required in good tobacco was a minimum of nicotine."

Actual Cultivation in India.

CULTIVATION in Madras. 116

I. TOBACCO OPERATIONS IN THE MADRAS PRESIDENCY.—The history of the development of the tobacco industry in this Presidency, which is certainly foremost in regard to the quality of the tobacco it produces, is an interesting record of cultural experiments, undertaken on sound scientific principles, and followed up by an exhaustive series of chemical analyses. The facts brought to light by **Mr. Robertson,** Superintendent of the Madras Government Farm, and others concerned with cultivation and preparation, together with other facts, clearly indicated in the numerous analyses made by **Mr. Broughton,** the Government Quinologist, added immensely to our knowledge regarding the capabilities, not only of this Presidency, but indirectly of many other parts of India, for the production of good tobacco.

The cultivation of tobacco has rapidly extended in Madras within the last few years. In 1888-89 the estimated area was 87,860 acres, which shows an increase of over 9,000 acres, since 1882-83. The districts comprising the largest areas were Kistna (21,959 acres), Coimbatore (16,054 acres), and Godáveri (11,957 acres).

The following extract, taken from the Madras Manual of Administration, Vol. I., 292 (1885), contains a brief description of the various modes adopted in the cultivation and preparation of tobacco:—"Tobacco is grown more or less throughout the Presidency, with the exception of Malabar and the Hill Ranges, but the chief localities of production are the alluvial lands of the Godáveri district, where is grown the well known 'Lunka' tobacco (so named from the lunkas or river islands on which it is cultivated), and parts of the Coimbatore and Madura districts, from which the Trichinoploy cheroot manufacturers draw their supplies of raw material. The plant is grown on almost every description of soil, from black loam to sand, and from irrigated land to high arid sites. Alluvial lands are preferred, then high ground, and such places as deserted village sites, and backyards of houses, the latter on account of the salts impregnating the soil, and also probably for convenience of position as regards manuring and watching and curing the produce. Of the more esteemed tobaccos used for European consumption, the best of the Godáveri produce is grown on these alluvial lands which receive rich deposits of silt in the river floods and are out of the influence of the sea-freshes; while the Dindigul tobacco is produced on a carefully cultivated red loam to which an alluvial character has been artificially imparted. Some of the highest priced tobacco is grown on rich dry land under irrigation, but this, while suited for chewing, is too coarse in texture of leaf and too pungent in flavour for smoking. In some parts irrigation is practised and in others it is dispensed with, only a small quantity of water is supplied to the plant, and as a rule not by gravitation, but by mechanical appliances, and preferably from wells of brackish water containing potassic salts. Excessive damp is prejudicial, and the seed beds and soil gene-

CULTIVATION in Madras.

rally are superficially drained or stand high. The crop while young is gently watered by hand, and heavy rains detract from the quality of the tobacco, the tobacco grown on ordinary irrigable lands being generally inferior. The manures used are the droppings of sheep and goats penned on the land previously to cultivation, cattle-dung and urine, ashes and sweepings. In Nellore salt-earth is used. The manures are very plentifully applied to all soils except alluvial lands. The seed is invariably sown in seed-beds. The seasons of cultivation vary according to local climatic considerations. As a rule, sowing commences after the local rains from July to October, though tobacco is sometimes grown as a second crop commencing in January. The site of cultivation is thoroughly ploughed and manured, the seed germinates in some eight days after sowing, and the seedlings are transplanted in the course of some six weeks on attaining a height of five or six inches, into holes a foot to a yard apart, sometimes in ridges, sometimes on the flat surface of the field. In some localities the seed-beds and young plants are protected from the extreme heat of the sun by means of mats, &c., and all leaves except ten or twelve are nipped off to strengthen those left; the flowers are also promptly nipped off with the exception of those purposely left for seed. The leaves begin to ripen in the course of some two months from transplantation, and as soon as one or two turn colour, the whole crop is collected. This is effected generally by cutting the stem with a knife, though in Ganjam and the alluvial lands of Vizagapatam, the leaves are nipped off separately, and in part of Tanjore some leaves are first plucked in January and the stem and remaining leaves cut down in May or June. As a rule, no second crop is gathered, and where the after-sprouts are collected at all, they are of very inferior quality. The process of gradually drying and fermenting is effected by modes slightly differing in detail. In Nellore, for instance, the cut leaves are hung in the sun for two days, put in heaps, turned every two days, and ranged in layers for twenty days, during which time they are frequently turned. They are then tied in bundles, dipped in water, sweetened with date jaggery, and are then ready for sale. In other parts, as in the Salem district, the plants are left a day or two in the field, then exposed to the sun and dew alternately for a week, then wrapped in straw and buried in the ground for a week, after which the leaves are stripped from the stems, made into bundles, placed in straw, and put under heavy weights, with their ends exposed for six weeks, the piles in which they are laid being opened and turned every other day. In other localities the leaves, after drying in the fields for a day or two, are hung over poles or ropes, preferably in the shade, in regular drying sheds or in the cultivators' houses, and subsequently stacked in heaps, which are opened out and pressed together again at stated intervals until the requisite curing is effected. Occasionally the leaves are sprinkled with jaggery water or an infusion of the **Cassia auriculata** while drying, and in Coimbatore the festoons of leaves after being strung are hung up on the milk-hedge (**Euphorbia Tirucalli**) to acquire thence a flavour. State interference has been suggested in the case of this industry, but the Government have decided otherwise. Such interference has not been found necessary with indigo or coffee, and it was relinquished in the case of tea when that industry had made much less progress than tobacco has at the present time. The tobacco grown in this Presidency is at present inferior, but it seems clear that this is mainly due to the fact that there is a great demand for the coarse article, and that it is found to pay better to grow a large quantity of inferior leaf than to grow a smaller quantity of superior leaf. European capital would, however, doubtless improve the curing process."

CULTIVATION in Madras.

Kistna and Godaveri Deltas:—**Mr. C. Benson**, in an interesting note on the tobacco cultivation of these tracts, which produce the famous *Lanka* tobacco, writes:—"The land on which tobacco is grown in the Godáveri consists, as is well known, for the most part of alluvial islands lying within the banks of the Godáveri river and generally liable to be flooded annually, although some portions are so high that they are seldom overflowed. The soil of the '*Lankas*,' as the islands are usually called, is of course very variable; some parts lying low are covered with deep layers of coarse sand, and in other parts, both high and low, the soil varies from a light friable to a stiff loam. The best soil is a fine friable loam composed of the finer parts of the silt brought down by the river. As might be expected, the lankas are continually being altered in contour and size as well as being overlaid with deposits or washed down to lower levels. The soil, therefore, of lankas high up the river may be spread over those lower down during any season. Tobacco seems to be grown on any part of the lankas almost indifferently. It is grown even on coarse sand, provided it is not too deep, and that there is a layer of good soil not more than a foot or so below the surface. What is liked best is a new deposit of fine alluvium, nine inches to a foot deep, lying on a sand bank. Amongst such a diversity of soils and in such a changeable land there are fields which have grown tobacco uninterruptedly for many years, and others again which have only recently assumed their present form." "The greater part of the tobacco-growing area is annually fertilised by the deposit of silt which the river leaves." "The tobacco seed is sown about September or October in seed-beds which are very carefully prepared and cleaned, and generally situated near the villages where the rayats reside. These seed-beds are heavily manured, by folding cattle and sheep on the land and also with village sweepings, and the land is frequently stirred with the native plough until a good depth of loose mould is formed." "The preparation of tobacco fields begins after the last freshes have passed down the river." "After the weeds have been ploughed up, small holes are made in the soil, the depth varying with its nature, for, in cases where the good soil is overlaid with sand, an effort is made to reach the former, and also in some cases a little good soil is brought and poured into the hole around the plant when the sand is very deep. The plants are put in from two to three feet apart and are watered by hand from pots daily for a month or more. About three weeks after planting the land will usually be weeded, one weeding only being general." "The tobacco crop, a very important one in the Godáveri district, is, according to the agricultural statistics, still more so in the Kistna. In the last named district it is grown not only in the *lankas* of the river but around most of the villages. The system of growth in the Kistna is practically the same as in the Godáveri, except that the land not being naturally fertilised by river silt needs heavy manuring, and in fact, the land devoted to tobacco and chillies receives nearly all the rayat's manure. The tobacco land, however, is made to furnish manure to other fields in some places, for the soil is dug out and carried off to them as manure; this soil is doubtless very rich, not only from direct manuring, but also from being situated close to the village site. Folding with cattle is the favourite means of manuring combined with the use of village ashes and sweepings, the latter being especially used for the seed-beds." "When the tips of the leaves begin to dry, the leaves are cut in some places separately from the stem, and in others with a portion of it adherent. They are then generally laid out on straw in the shade for a couple of days and then strung up on strings close together for a fortnight or three weeks in a shed; after this they are placed in heaps and turned daily for about three weeks, then moistened, piled in larger heaps and turned once in two or

CULTIVATION in Madras.

three days for another three weeks, by which time they are ready for sale. The *Lanka* tobacco is said to be finer than that grown on the village sites, and therefore not so much appreciated by the rayats who prefer a coarse strong article."

In the Godáveri district, the *Lanka* tobacco, which is used for making cheroots intended for European consumption, is washed in cocoanut water in order to diminish the strength and pungent flavour of the tobacco. It is also considered to have a better flavour after it has been kept for a year.

Mr. H. Caine, the tobacco expert, was engaged by the Madras Government during the summer of 1888 to make a tour of inspection of the Godáveri *lankas*. He commenced work with the Malluka *lankas*, eight miles from Rajamundry, where he found the soil mostly silt and but little sand, and the tobacco much cleaner in texture, than on the other *lankas*. He purchased fifty green plants of the first growth, and managed to cure the leaf well enough to perceive that great improvement could be obtained by introducing the same method of shed-curing as pursued by the Government experiment in the Madura and Dindigul districts. In his opinion the mistake in the method of curing adopted here lies in keeping the tobacco, while in its green stage, too much exposed to sun and light, hence too rapid evaporation of the sap resulting in the colours being mottled green, yellow, and light red. The rayats were much taken with the appearance of the tobacco he had cured, and wished to know how it was done. Judging from what he had seen of the crop, **Mr. Caine** felt persuaded that the only method likely to create a demand for this tobacco in the European market would be in the manufacture of mahogany-coloured "cavendish" wrappers, and also yellow tobacco (bright smokers), as these two sorts would find a sale in England. In "cavendish" wrappers a long and large leaf would be requisite, and for pipe tobacco ("bright smokers") a medium sized leaf is most suitable. To obtain "cavendish" wrappers of the paper colour, however, a moist temperature is preferable, and in the town of Coconada itself on the canal bank the leaf would cure a much darker and more valuable colour than if the tobacco were cured on the *lankas*. The pipe tobacco would cure better in a dry quick heat, obtainable on the *lankas*. **Mr. Caine** thinks that a single year's trial would suffice to convince the brokers in England that good pipe tobacco can be grown in India in a cheaper way, and of a quality good enough to compete with Virginian tobacco on its own merits. The "bright smokers" tobacco, he believes, would be valued at not less than 8*d.* per pound in England. First grade "bright" would fetch 8*d.* to 1*s.* per pound and good "cavendish" wrappers from 8*d.* upwards. The quality of the present *lanka* cigars, he says, is execrable, being strong, rank, hot and saltish to the taste, besides gritty and full of sand (*Madras Times*).

COIMBATORE.—Tobacco is largely grown in all parts of this district as a cold weather crop on land dependent on irrigation from wells. The climate and soil appear to be all that are needed for the production of good leaf, but these advantages are almost entirely lost owing to the treatment the leaf receives after being cut. For, as **Mr. Robertson** says in his report on this district, "Nearly all the tobacco raised in this district is cured under the verandahs of the rayats' dwellings, or inside their dwellings. Sheds are never erected purposely in which to cure tobacco. The first stages of the curing are performed chiefly out of doors, the leaves being suspended in various ways while drying. I saw in some places rows of castor oil plants supporting lines over which tobacco leaves were suspended."

CULTIVATION in Madras.

The following table gives the results of analyses made by Mr. Broughton on certain samples of tobacco grown in this district :—

No.	Place of growth.	Per cent. of ash.	Per cent. of carbonate of potash in ash.	Per cent. of nicotine.
1	Coimbatore	22·85	2·94	3·32
2	Ditto	22·60	5·64	4·94
3	Ditto	19·92	7·92	4·90
4	Polláchi	24·98	4·39	1·95
5	Mukásiputhúr	25·73	7·67	2·24
6	Kurichi	26·39	7·93	1·17
7	Púndurei Sémúr	23·34	2·65	1·29
8	Aval Púndurei	23·49	19·97	1·46
9	Mánikampálaiyam	26·65	2·61	2·95
10	Mádhalli	23·45	5·33	2·32
11	Satyamangalam (A)	17·09	14·74	2·88
12	Ditto (B)	21·34	8·87	3·74
13	Ditto (C)	25·14	9·14	3·02

The most striking feature to be noticed in the above analyses is the varying proportion of carbonate of potash; another important fact is the small amount of nicotine in most of the samples. The best tobaccos are those from Erode, Coimbatore, and Satyamangalam, that of the southern taluks is coarser and stronger, the red soils are also preferred to the black. The three specimens from Satyamangalam were pronounced by Mr. Broughton to be of good quality, and the best received from Coimbatore, and he considered that that taluq would suit foreign tobaccos. In 1872 he stated that the foreign Havanna and Manilla tobaccos grown in Perandurei (Erode) taluk "were decidedly of good quality" and the best yet received. As usual, carbonate of potash was high (6 to 10 per cent.) and nicotine low ($2\frac{1}{3}$ per cent). Mr. Broughton was of opinion that these specimens showed that foreign tobaccos could be produced in South India of their foreign quality, and that good curing would have given the Manilla tobacco the high qualities of foreign Manilla Tobacco. The soil was apparently red loam; the manuring ordinary.

The following account of the cultivation and curing of tobacco in Coimbatore is taken from Mr. Nicholson's Manual :—

"The nurseries are prepared as follows :—

"The soil is liberally manured by folding sheep on it; it is then ploughed, formed into beds, and ashes and decayed vegetable matter are spread over it; it is then watered and allowed to stand. After a few days the ground is weeded, sown with seed which has been steeped and moistened for four or five days, and for the first week water is given daily, every other day for the second week, and after that twice a week till the seedlings are ready for transplanting. At six weeks or two months' old (November-December) they are transplanted into well-prepared soil, formed into beds by ridges, into the sides of which the seedlings are dibbled, and thereafter watered twice a week. They are hoed three or four times in the first three months after sowing; they are topped at the end of the third, suckered four times during the fourth and fifth, and cut down in the sixth month. It is usual to leave them on the field the first night after cutting; they are then heaped for two days, and on the third are hung up to dry for twelve or fifteen days. After that they are heaped to sweat for some days, the heaps being shifted occasionally, and weights placed upon them.

NICOTIANA Tabacum.

Tobacco.

CULTIVATION in Madras.

Subsequently they are bundled, each bundle being a convenient handful, again heaped, and the bundles constantly shifted thereafter till sold.

"Much of the tobacco is exported to the West Coast, as in the days of the monopoly. Malabar and Travancore, where it is almost a necessity, are mainly supplied with this tobacco. Some is sent to Trichinopoly for the manufacture of cheroots.

"The area cultivated in 1291 was 19,810 acres, and the outturn may be put at 15,000,000℔ of dry leaf. It is sold in the weekly markets in vast quantities at one to two annas per pound; a factory, say at Erode, Dhárápuram, Udamalpet, or Coimbatore, might do a good business."

Dr. Buchanan, who visited Coimbatore in 1801 gives a minute account of the cultivation of tobacco in that district. After describing in detail the preparation of the nurseries and the treatment of the seedlings after transplanting, he goes on to say that "the plants are cut down close by the ground, and left on the field until next morning, when they are tied by the root-end to a rope, and hung up all round the hedges. If it be clear weather, the leaves dry in ten days; but when the sun is obscured by clouds fifteen are required. When dry, the tobacco is placed in a heap under a roof, is covered with bushes, and pressed with stones for five days. The leaves are then removed from the stems, and tied up in bunches, which are again heaped up, and pressed for four days. After this they are made up into bundles, each containing some small and some large leaves." "These are heaped up again, and pressed for twice five days, having at the end of the fifth day been opened out, and new-heaped. The tobacco is then ready for sale. A good crop, from a Vullam-land, is 1,000 bundles, or 566⅓℔ from an acre. During the busiest part of this cultivation, eight oxen and ten men are required daily for one Vullam-land."

MADURA.—"Tobacco of a very superior description is grown in large quantities in some parts of the district, particularly at Védasandúr, and in the sub-division generally. Most classes of natives in Madura chew, snuff, and smoke in moderation, and large quantities of leaf are exported to Trichinopoly, Madras, and other places Tobacco is usually sown in the month of October or November, transplanted in December or January, and gathered in February or March. It requires watering once in four days" (*Nelson, Man. Madura, Part II., 106*). **Mr. H. W. Bliss**, in a letter dated June 1874, remarks that the natives of this district have little to learn in the way of cultivating the plant, and that a native tobacco garden presents as good a specimen of careful culture as can be seen anywhere; the only improvement which might be suggested would be as to the quality of the manure which is deficient in potash.

NELLORE.—"Tobacco is invariably cultivated in the rich soils adjoining the villages, or in land permanently improved by a long course of high farming. The land is always extensively manured and well ploughed up. The seed is first sown in some favourable spot, situated high and carefully prepared; and when the young plants are of sufficient size to plant out, the land is again ploughed up and divided at once into beds, formed with the *danti*, or large wooden hand-rake. The beds are laid out with great regularity and then watered, and whilst the soil is still wet, the young tobacco plants are transplanted at equal and regular distances. About the second, and again on the fifth day, the plants are watered, and by that time are taking root. Irrigation is continued at intervals of a few days; and when the plants are about six inches high, the soil is dug up two or three times during the course of the following fifteen days, and the plants are banked up in rows and watered, between the rows thus formed, every second or third day. Should the soil not partake of saltpetre earth, a little is applied to the plants, or mixed in the water let on to the land. As

CULTIVATION in Madras.

the plant blossoms, the flower is removed, and also all young shoots until the leaves have matured, when the whole plant is cut down and cured. Besides, in the garden land, tobacco in small patches is planted in each rayat's holding throughout the northern taluqs of the district. The young plants are watered with a *chatty* two or three times when first put out, and afterwards are left to mature with the rain. These patches always occur in the midst of dry cultivation, the young plants being put out during November" (*Boswell, Man. Nellore, 402*).

NILGIRI HILLS.—Tobacco, according to **Mr. Robertson**, used to be much more extensively grown on these hills than at present. The cultivators now find that the potato meets with a readier market and they cannot afford sufficient manure for both crops. "From my own knowledge," **Mr. Robertson** remarks, "of what constitutes a good district for tobacco, I believe that there are, on the Nilgiris, many localities admirably suited, not only for producing a really good tobacco, but, what is of far greater importance, for providing the climatical conditions required for curing and preparing the leaf for the market."

Mr. Broughton analysed two samples of Nilgiri tobacco with the following results:—

No.	Tobacco specimen from.	Per cent. of ash.	Per cent. of carbonate of potash in ash.	Per cent. of nicotine.
35	Nilgiris, Todanad	20·596	29·26	1·43
36	,, Koondahs	17·786	6·37	2·95

He alludes to those analyses as being important examples to show the possibility of modifying the constituents of a tobacco ash by special culture. "It is well-known," he says, "that the soil of the Nilgiris is exceedingly poor in salts of lime to a very exceptional degree, while from the decomposing felspar it is comparatively rich in potassic salts. The analyses show that the tobacco grown on these hills yields an ash in which the ordinarily occurring calcic carbonate is nearly entirely replaced by potassic carbonate, so that in one specimen the latter substance occurs in the ash to the large amount of 29·26 per cent. Though the cultivation on these hills is of the roughest possible kind, and produces consequently a most inferior product, I cannot but consider the result now mentioned as most important and suggestive." In a letter to **Mr. Robertson**, Superintendent of the Madras Government Farm, **Mr. Broughton** states that he has seen on these hills many tracts of land on which tobacco might be expected to thrive; that the soils are rich in potash, and the climate favourable for curing operations.

In Chingleput the tobacco plants after being cut are stacked on a layer of palm leaves and straw, over them is placed another layer of leaves and straw, and the whole is weighted with stones. After five or six days the plants are taken out and hung up by their stalks for a few days, till the stalks are dry, when they are taken down and placed in a close room and covered as before with palm leaves and straw pressed down with stones. Should the plants have become too dry and brittle, a few stalks are cut out and boiled with a sufficient quantity of water to which palm sugar or jaggery is added, and this decoction is sprinkled on the tobacco before stacking it the second time. The stack is turned once in three or four days, and when this has been done several times, the leaves are

CULTIVATION in Madras.

stripped off the stalks, and tied into bundles. These are again stacked as before and weighted. The bundles are re-arranged every three or four days for two or three weeks, when the tobacco is considered cured and fit for use (*Dr. Shortt*).

Tobacco, though not cultivated in Malabar, owing to the unsuitable climate, is largely used by the people of that district, and is believed by them to be highly beneficial as a febrifuge and a preventive of chills. Previous to the year 1807 there were no restrictions on the trade, which was chiefly with Coimbatore; but in that year tobacco in Malabar was made a Government monopoly, and a few years afterwards the monopoly was extended to Coimbatore, the whole crop being taken up by the Collector at certain rates. The balance, after despatching the required quantity to Malabar, was sold by Government agents to retailers for local consumption. This arrangement resulted in so many abuses that the Board of Revenue resolved in 1852 to abolish the monopoly (*Madras Times*).

In January 1886, the Government of India engaged the services of Mr. Caine, for a period of two years, to undertake the cultivation and curing of tobacco in the Andaman Islands; the result, however, proved a complete failure.

Commercial Value of Madras Tobacco.—The most recent published information the writer has seen regarding the commercial value of Madras tobacco is contained in a "Memorandum," dated India Office, 21st April 1890. Samples of three kinds known locally as "Bright smokers," "Oosikappal," and "Warralkappal," were despatched by the Madras Government to the India Office in order to obtain the opinions of English experts. The tobacco was cured by Mr. Caine. The sample of "Bright smokers" was found on opening to have heated badly, the other two kinds arrived in good condition. Equal portions were sent to fourteen of the leading brokers and manufacturers in the United Kingdom for their opinions.

After tabulating the detailed remarks given by each, the writer of the Memorandum draws up the following summary and conclusions :—

"'*Bright smokers.*'—Not well cured; heated and mildewed. Leaf of fair colour and size, but does not approach that of 'Bright Virginian' tobacco. Stalk very large, and veins coarse. Burns and smokes fairly well, flavour not good. Moisture, 12 per cent. The valuations range from 1*d.* to 5*d.* per pound.

"'*Osikappal.*'—Fairly cured, and in dry condition. Leaf thin and narrow, and blistered. Colour fair, but very unequal. Stalk very large, being 23·8 of entire leaf. Burns and smokes well, good flavour, poor aroma. Moisture 12·6 per cent. Absorbed 21 per cent. of water. Valuations in present condition from ½*d.* to 3*d.* per pound. If in 'strips' with (midrib excised), 4*d.* to 6*d.*, or if 'butted' (stalk docked), 3*d.* to 5*d.* per pound.

"'*Warralkappal.*'—The most perfect in curing and colour. Leaf of fair size, but rather narrow and papery. Veins coarse, and stalk large, being 24·2 of entire leaf. Burns well, fair flavour, but with black ash. Moisture, 13·4 per cent. Absorbed 18 per cent. of water. Value in present state 1*d.* to 4*d.* per pound. If in 'strips,' 5*d.* to 6½*d.* per pound."

"The general opinion of the manufacturers interviewed by me was that the present samples of tobacco showed a very decided improvement in growth and curing over previous samples sent from India. The excessive amount of stalk and the large veins of these Madras tobaccos were much objected to; and before they could successfully compete with American leaves it would be necessary to greatly reduce the present undue proportion of stalk to lamina. Most of the American tobaccos are now imported

CULTIVATION in Madras.

into this country in the form of 'stript' leaf, *i.e.*, with the midrib or stem entirely removed to within 2 inches or 3 inches of the tip of the leaf, and the leafy portion only packed in cases of 2 or 3 cwt., or else in the form of 'butted' leaves, *i.e.*, with 4 or 5 inches of the bottom chopped off, removing the coarsest part of the stem. The leaf is then tied in small bundles and packed in bales or cases. Manufacturers cannot afford to buy a leaf with nearly 25 per cent. of stalk, and pay a duty of 3*s.* 2*d.* per pound upon it. The only use of the stalks is to be ground up for snuff, or otherwise returned into bond as offal at a considerable loss on the duty paid. It must also be borne in mind that the 'stript' or 'butted' leaf saves a considerable amount in freight alone.

"For cutting for the pipe the Madras tobacco was declared to have too thin and brittle a lamina, breaking away from the knife with a large percentage of dust and waste. Also the dark colour of the ash is disliked by the English smoker, who insists on a white ash.

"For cigar-making, the excessive size of the midrib and prominence and white colour of the veins is much against the Madras tobacco. The cigar-maker likes a broad-leafed tobacco, with the veins small, and as nearly as possible at right angles to the midrib, especially if it is to be used for 'wrappers.' The veins in the varieties now under report were at a very acute angle with the midrib, and one manufacturer stated that with the 'Oosikappal' and 'Warralkappal' varieties it was only possible to make 80 to 100 wrappers out of one pound leaves, whereas from 400 to 450 wrappers can be made from the same quantity of ordinary American tobacco. The suggestion made by **Messrs. Wills & Co.** of Bristol and by **Messrs. Thomson & Porteous**, of Edinburgh, that a few bales of each variety of these tobaccos, 'stript' and 'butted,' should be put on the London market, appears to be practical and well worthy of attention. If this is done, the tobacco should be cased or baled as closely as possible, in the same manner as American tobacco is sent to the European markets. It was pointed out that the leaves of the present samples did not appear to have been sorted. It is of the utmost importance that the leaves should be properly classified for the purpose to which it is intended that they should be put, *viz.*, cutting for the pipe, in the form of roll or cake tobacco, or for making into cigars. It is suggested that, with any future consignment, information on these points should be supplied; in the present instance, no indication whatever was given as to special manufacture for which the tobacco had been grown and cured. Brokers and manufacturers are most conservative, and view with much suspicion any product, however familiar, presented for sale in a make-up unknown to the trade. Some manufacturers are inclined to think that it might be advisable to import Indian leaves, if better cured and packed, for the manufacture of cigars for Anglo-Indian smokers in this country, who, having acquired a taste for the peculiar flavour of Indian tobacco, would become willing buyers of it, particularly when made into cigars of the more perfect finish of the English manufacturers. There is no doubt a fashion and taste in smoking as in most other things. For instance, Indian tea has a marked and distinct flavour of its own, which at one time was much disliked, but is now looked upon as one of its best characteristics, whether used by itself or blended with China tea, until now it is actually driving the latter out of the English market."

In Burma. 117

II. CULTIVATION AND MANUFACTURE IN LOWER BURMA.—Tobacco is grown almost universally; and the area under cultivation is steadily increasing. In 1869-70, the area was stated to be 10,318 acres, whereas in 1888-89 it amounted to only 23,525 acres. The more important tobacco-growing districts are: Henzada, Thayetmyo, Tharra-

NICOTIANA Tabacum. — Tobacco.

CULTIVATION in Burma.

waddy, Prome, Sandoway, and the Arrakan hill tracts. The average yield per acre is estimated at 930lb, so that the total outturn in 1888-89 must have been not far short of 22 million pounds. Nearly the whole of this is consumed locally, in addition to large quantities imported from Bengal and Madras. The people of Burma all smoke—men, women, and children.

The cultivation is mostly confined to river-banks, sand-banks, and the dry beds of streams. The soil in such places being naturally enriched by alluvial deposits, a minimum of labour is required in the cultivation of the crop. The good quality of some of the Burma tobacco attracted attention many years ago. **Dr. Royle**, in his *Productive Resources of India, p. 187 (1840)*, thus wrote regarding the famous Arrakan tobacco:—"One of the results of **Dr. Wallich's** investigations was the bringing to notice some very superior tobacco, which obtained the name of Martaban Tobacco. This the author has already noticed in another work (*Ill. Him. Bot., 285*), in the following words: '**Dr. Wallich** states, that 'the sort is from Arrakan and not from Martaban;' and describes it as having 'a fine silky leaf: tried by many people, it had been pronounced the very best they had ever tasted, equal to, nay surpassing, the finest imported from Turkey and Persia.' An extensive tobacconist says, 'a finer and better-flavoured tobacco he never saw or tasted in his life.' One of the first brokers in the city says, 'the sample of leaf tobacco is certainly of a very fine quality, and appears to have been produced from some peculiar seed and a greatly improved cultivation and cure.' By many manufacturers 'it was supposed to be from the seed of Havanna or St. Domingo Tobacco.' For smoking, it is compared with Maryland Tobacco, having the same qualities, 'except the flavour, which is better, and more like Havanna.' The colour and leaf are, moreover, pronounced excellent for cigar-making; 'but if anything is against it for that purpose, it is the largeness of the principal stalk, and coarseness of the small fibres in the leaf.' The commercial gentleman by whom the tobacco was transmitted to the brokers pronounces it very superior, and the leaf as very fine, adding, that the price of 6*d*. or 8*d*. per pound might readily be obtained, perhaps more, with the improvements suggested.' As we purpose resuming this subject in the subsequent pages, it is at present only necessary to call attention to the remarkable fact of this tobacco, cultivated by the Burmese, being so excellent in quality, while the only other tobacco of those regions which has a European reputation is that of Shiraz, where the climate, it is important to remark, is not at all like that of Arrakan. But in one point both places have enjoyed the same advantage, that is, they have had Europeans settled in their neighbourhood, by whom it is more than probable that the improved culture of tobacco was taught. Shiraz, we know, is not far from Ormuz, so long occupied by the Portuguese, and Arrakan was the seat of a Roman Catholic Mission in the beginning of the seventeenth century."

The following account of tobacco cultivation in Northern Arrakan is given by **Mr. O'Conor** on p. 87 of his report:—"In Northern Arrakan, tobacco is cultivated by all the villagers on the banks of the Kaladan. It is said to be remarkably good. Whence it is derived is not known, but it seems likely that it was first raised from Manilla and other imported seed distributed by **Sir A. Bogle. Mr. St. John** says it is sown broadcast on the alluvial deposits along the banks after the fall of the river in November, the long elephant-grass having first been cut and burnt. The young plants are not transplanted, but well weeded and thinned out; a good deal, however, depends upon the season, as the plants require a little rain, though the heavy fogs no doubt do a great deal for them. When the

CULTIVATION in Burma.

plants are about two feet high, the shoots and lower leaves are broken off to make the good leaves grow longer. In April and May the leaves are picked and strung through the stalk on a thin bamboo skewer about one cubit in length, from twenty to thirty leaves on a skewer, and hung up in the house-roof to dry; after five or six days they are taken down and shaken about to prevent the leaves from adhering to one another; they are then re-hung, and after six or seven days, when quite dry, thrown into a large basket in which they undergo heavy pressure; after about a month and a half, when the rain has well set in, they are taken out and sorted into bundles. The tobacco is never exposed to the sun, and is kept till the rains for sorting, to secure pliancy in the leaf."

The good quality of the tobacco grown on the banks of the Sandoway river in Arrakan was brought to the notice of the Agri.-Horticultural Society in 1844 by **Captain Phayre.** The soil producing the best kinds is on old *char* land, and on the alluvial soil deposited by the numerous mountain streams of the Yoma-toung range. These streams begin to subside early in October, and as soon as the soil becomes sufficiently dry, it is most thoroughly ploughed and weeded. The best description sells in the Sandoway market for R10 per maund, second quality for R6, and inferior sorts at R3-8 to R4. Sandoway tobacco as imported into Calcutta is generally mixed with inferior sorts grown elsewhere (*Report by General Fytche, 1852*).

A very interesting description of the Kyoukkyee tobacco tracts is given by **Mr. Hough** in his report dated February 1882, from which the following extracts may be taken :—

"I have just finished a long projected trip to the Kyoukkyee tobacco tracts. I am not aware that any European has ever visited these regions with the deliberate object of making local enquiry about the tobacco which is grown there. What is commonly called 'Kyoukkyee tobacco' is cultivated chiefly by the Karens, who inhabit the hilly parts of the Kyoukkyee township of the Shwaygyn district. As it was impossible for me to visit all or even half-a-dozen of these places, I selected Htopeeden hill for local inquiry, that place being within the very heart of what may be called the tobacco-growing country." "At Htopeeden I visited one particular tobacco '*ya*' and saw at a distance many other *yas* on the hills around me. The *toungya* to which I more particularly refer was about three acres in extent: it had been sown with chillies, cotton, and tobacco. The tobacco was not growing uniformly over the *ya*, but was in patches, and at first sight it looked very much as if the plants had sprung up by chance. Close inspection showed that the patches had been more or less weeded, but, as in almost everything else connected with these Karens, a want of care and of cleanliness was apparent throughout the tobacco *yas*." "The plucking of the tobacco leaves was still going on. The tobacco plants were about 18 inches high, and they were at distances of two, three, and four feet from each other: many were growing in close contact with each other. There had been no atttempt at careful systematic planting. I brought away the most massive plant I could find. The largest leaf on this specimen measures 16 inches in length and 7 inches in width at its broadest part. All the plants were crowded with leaves, large and small, but, as a fact, only about nine leaves are taken from each plant, that is to say, the topmost whorl of about four leaves, and the whorl next below of about five leaves. The leaves are considered fit for plucking when they assume a yellow tinge. The topmost whorl of four leaves is made into 'extra fine' tobacco: the next whorl of about five leaves is used for inferior tobacco. The two kinds of leaves are of course kept apart. All the other leaves are abandoned.

CULTIVATION in Burma.

The *toungya* which I visited had been cleared of bamboos and small trees in March last, the felled jungle had been burnt in the usual way, and then the various crops were sown. The tobacco-seed is thus sown: Bunches of seed-capsules are fixed in the clefts of bamboos or tree-stumps at a height of about two feet from the ground. The capsules burst and the wind blows the seeds over the surface of the ground. Vacant spots are filled up subsequently by transplanting. In about a month after the sowing the young plants appear, and by this time the rains have well set in Transplanting is done by means of a crooked knife, which is also used for weeding. The blade is about eight inches long and the handle about 10 inches long. The tops of the plants are twisted off in *tasoungmong*, when the paddy crop is usually reaped, and until the leaves intended for manufacture have been plucked, the plant is not allowed to flower; in fact the main stem is not allowed to flower at all if leaves are reserved and nourished for plucking. The tops had been removed from the vast majority of the plants which I saw, and only the seed plants had been permitted to flower. Leaves are plucked in January and February. Having been plucked, six or eight leaves are placed flat against each other, stem to stem and point to point; they are then wrapped loosely with fresh plantain-leaf, and the bundles are placed upright round the sides of the interior of a '*teh*' or hut. The bundles are placed with the tips of the leaves upwards. A fire is kept burning in the hut throughout the period of seasoning of the leaves, which period lasts for three days, at the end of which time the leaves are sliced. The slicing is done with a small sharp knife held in the hand. A bunch of leaves is forced through a small hole in an upright post, and as the ends of the bunch emerge from the outer aperture they are sliced off with the knife. The sliced tobacco is then spread out on bamboo trays five feet long and two feet wide, and is exposed to the sun and the dew for three days and three nights continuously, the upper surface having been seasoned; the whole is deftly turned over and the lower surface is similarly exposed for three days and nights. The tobacco is then ready for the market. The best tobacco is of a lighter colour than the inferior tobacco and is less acrid when smoked. Much of the tobacco which goes to market has a greenish tinge showing faulty curing, but when extra pains have been taken with the curing, the tobacco is of uniform colour. The Karens told me that they cultivated tobacco for only one season in the same *ya*, and that the second season's growth in the same *ya* would be a failure. They also told me that next year paddy was to be sown in the very *ya* which I saw, thus indicating that the soil was still fit for another crop—of paddy at any rate. There is no attempt at terracing, nor is seed ever exchanged with neighbouring hills, nor is any attempt made to select good seed. Tobacco is cultivated on the lower spurs or slopes of the hills. The ground is of decomposed granite underlying a thick layer of rich vegetable soil." "The tobacco plants are sometimes attacked by insects, but not seriously. Burman traders from Kyoukkyee go into the hills and buy the tobacco on the spot at R1 per viss for the best tobacco and eight annas per viss for the inferior tobacco. This is all re-sold at double the original prices at Kyoukkyee. The Karen vendor carries the tobacco from the toungya to the Kyoukkyee market." "Tobacco in leaf is never prepared for the market by these Karens. How tobacco first came to be cultivated in these regions I could not discover; no one could tell me." "Individual *ya* cutters do not cultivate more than two or three viss weight of tobacco each." "The tobacco plant flourishes in the Karen localities which I have visited, and the Karens, as we know, can manufacture good tobacco. The 'Kyoukkyee tobacco' is certainly the best tobacco produced in Burma. I know a few

CULTIVATION in Burma.

Europeans who almost habitually smoke this tobacco. But the tobacco is produced in insufficient quantities and it is not sufficiently well cured."

There is a very good kind of tobacco in Burma called Kama tobacco, of which **Colonel Horace Browne**, Deputy Commissioner of Thayetmyo, thus writes:—"Kama tobacco never finds its way to Rangoon, but it bears a high repute all along the Irrawaddy as far as Mandalay, the best kind always fetching about 25 or 30 per cent. higher prices than any other. If other tobacco is selling at R20, first class Kama fetches R25 or R27 per 100 viss. It is not generally known, however, that this superior kind is the produce not of any indigenous but of acclimatised American seed. Some few years ago, some foreign seed (Virginia, I believe,) was supplied through the Deputy Commissioner to the cultivators who grow tobacco on the islands in the Irrawaddy just above Kama. The experiment has been a very successful one. The tobacco proved superior to the ordinary indigenous kind, which it is now rapidly supplanting in that locality. There are probably not less than 500 acres planted this year with the acclimatised American plants. It appears to grow just as well as the indigenous kind, from which, however, it is readily distinguishable, having long and pointed leaves, whilst the leaves of the Burman kind are broad and blunt. The only objection I have heard urged against it by the cultivators is, that it is more brittle than the Burman kind, and therefore takes more trouble to gather it; whilst the Burman leaf can be torn roughly from the stalk without injuring it. One person can gather twice as much of the Burman kind in one day as he can of the foreign. This is a serious defect in the eyes of a lazy cultivator. The method of cultivation pursued is the same for both kinds. The islands on which the plant is grown are submerged during the rains, and receive every year, therefore, a new and more or less rich alluvial deposit. If there be too much sand, the season of course is an unfavourable one. As soon as the islands appear above water, the highest spots are selected for nurseries." "The cultivator here, as elsewhere in Burma, does his best to spoil his produce by spreading it out to be dried by the sun upon the sand. He does this, now, not through ignorance, for every one of them is well aware of the superiority of house-dried tobacco, but because he does not believe that the extra price he would obtain would compensate him for the trouble and expense of building large drying-sheds." "One old cultivator, who appeared to be a connoisseur in tobacco, informed me that he generally dried a small quantity of tobacco in his house for home consumption, and the taste was something very fine, very different from that of the sun-dried, and far better than that of any tobacco brought from foreign countries" (*From Mr. O'Conor's Report, 1873*).

Early in 1883 **Mr. Cabaniss** was appointed an Assistant to the Director of Agriculture. Originally a Virginian planter, he was engaged by **Messrs. Begg, Dunlop & Co.**, in connection with their tobacco operations both at Ghazipur and Poosa. He afterwards undertook the management of a tobacco farm and factory of his own at Purneah. His appointment to the Agricultural Department was mainly for the purpose of developing the tobacco industry in Burma by practically demonstrating to the people how to improve the quality of their tobacco by better methods of cultivation and curing.

The following extracts taken from the General Administration Reports on Lower Burma for 1883-84 and 1884-85 show what progress has been made in this direction: "**Mr. Cabaniss**, the Assistant Director of Agriculture, has himself grown and cured tobacco both in Virginia and in India. He has journeyed twice through the parts of Burma which now grow tobacco: he has distributed Virginia seed to cultivators, and has instruct-

CULTIVATION in Burma.

ed them in the American system of drying the leaf. Briefly, the difference between the two systems is this. The Burman cultivator dries his tobacco leaf in the sun, and the process takes a few days only: under the American system the leaf is dried under cover, and the process takes as many weeks as the Burman process takes days. But the difference in price between sun-dried and shade-dried tobacco is that the latter fetches in Burma £7-10 per 100 viss (365lb), while the former fetches only £2-10 per 100 viss. A limited number of old tobacco-growers on the banks of the Ngawan and the Irrawaddy have, under **Mr. Cabaniss'** instruction, taken to the American system of drying. But the greatest extension of tobacco culture has radiated from the Kyauktan farm, to which Burmese cultivators have come for seed, for seedlings, and for instruction in shade-drying and in curing, from villages distant one, two, and three days' journey. Last season, within a small section of country round the Kyauktan farm, about ninety cultivators grew smaller or larger patches of tobacco, who had never grown tobacco in their lives before. Most of them made a good thing of it and got good prices for their leaf. Moreover, they found that tobacco employed them just at a time when there was not much work in the rice-fields. It is expected that this season the breadth of tobacco in the country near Kyauktan will be doubled or trebled. In order to show how much better a price tobacco dried and cured on the American system fetches than Burman tobacco, a public sale of all the tobacco produced on the Kyauktan and Ma-ubin Government farms has been held in Rangoon. Forty-eight parcels of about 50lb each were sold, realising an average rate of 1*s.* 8½*d.* per viss (3·65lb) as against 6*d.* per viss., the prevailing rate for ordinary Burman tobacco. The Kyauktan parcels ranged from 10½*d.* to 2*s.* 6*d.* per viss, and the Ma-ubin parcels from 1*s.* to 2*s.* It is hoped that the results of these sales will satisfy the Burman cultivators of the superior value of shade-dried leaf. Hereafter, when the improved system of drying and curing becomes more general, it will be possible for the Agricultural Department to arrange periodical sales of well-prepared tobacco in the Rangoon market. But the Burman cultivator at present is not quite satisfied about the new system of preparing tobacco, which occupies and keeps him out of his money some months longer than the indigenous system."

"During the coming season **Mr. Cabaniss** will supervise tobacco experiments on the Irrawaddy islands near Myanaung. This cultivation will be undertaken in presence of Burman cultivators, and, if possible, with their co-operation. The bulk of the tobacco cultivation of the province is on these islands: a large number of Burmans will therefore be able to profit by **Mr. Cabaniss'** teaching, and it is hoped that the system of shade-drying, which is now practised only on the Ngawan, will also be adopted on the Irrawaddy. In order to give a greater impetus to tobacco cultivation, the rate on all lands growing tobacco has lately been reduced to two shillings per acre where it exceeded that rate, and notices have been widely circulated amongst the cultivators informing them of this change" (*Administration Report for 1884-85, page 24*).

Mr. Cabaniss' report for 1885-86, though not as full as it might have been, owing to the disturbed state of the Province, is sufficient to show that the cultivators are beginning to appreciate the better results obtained by using good American seed and by adopting a more rational method of curing their tobacoo.

The higher prices they are now able to get is leading to extended cultivation, and the annually increasing outturn should, after a few years, cause a material diminution of the hitherto enormous quantity of imported leaf.

The principal localities in Upper Burma where tobacco is grown are:

CULTIVATION in Burma.

Ava, Mingyan, Pokokoo, and the neighbourhood of Mandalay. It is also grown in the Bhamo district for local consumption. and in the Shan States, between the Kachyen hills and China; some of this latter is exported across the Chinese border. **Captain Spearman**, formerly Assistant Political Agent at Bhamo, writing to the Political Agent at Mandalay (May 1872), says that in that district the leaves are dried in two different ways, though the price and flavour of each are the same. One method is to cut each leaf into three pieces, and to lay these out flat on the ground, or on a bamboo trellis work, in the sun; the other is to string 30 or 40 leaves together on a piece of split cane or bamboo running through holes made in the stalks, and hang them up in bunches.

In a letter from **Messrs. Sutherland, McKenzie & Co.** of Mandalay to the Political Agent (dated March 1872), it is stated, with reference to the tobacco of the Kachyeen hills, that when the crop is ready, the entire plants are pulled up by the roots, and hung up inside their houses. The leaves are then plucked as they are required. They say that by this mode of storing the tobacco, the leaves retain their sap longer, the tobacco is stronger than the ordinary Burmese leaf, and that, with proper treatment, would become very fragrant.

In Bengal. 118

III. CULTIVATION AND MANUFACTURE IN BENGAL.— Tobacco is very largely grown in this Presidency. As, however, no statistical records for crop areas are available for Bengal, it is impossible to state what is the total acreage. **Mr. O'Conor** was of opinion, in 1873, that the area under tobacco in Bengal could not be less than half a million acres. The chief tobacco-growing districts are Rungpur, Tirhut, Purneah, and Cooch Behar. The fine quality of certain samples of tobacco produced naturally in these districts, and in other parts of Bengal, very soon attracted attention as to the suitability of the climate and soil for attempting the acclimatisation of the better varieties, from America and elsewhere. The results of the numerous experiments, undertaken chiefly at the instigation of the Agri.-Horticultural Society, are fully recorded in the Journals and Transactions of that Society.

The following extracts relating to tobacco cultivation in Bengal have been taken from the General Administration Reports of that province:

"Tobacco is grown more or less extensively in every district in Bengal, but with the exception of Rungpur and the Dooars, Durbhanga, Purneah, 24-Purgannahs, the Chittagong Hill Tracts, and Nuddea, there is hardly a place in which it is sown for trade and export. It is, however, everywhere grown for local consumption. The cultivator ordinarily takes up a small plot of land at his homestead, near his cowhouse, for the convenience of manuring the land, as he always, if possible, manures his tobacco crop. In Baraset and elsewhere, where indigo cultivation has mostly died out, tobacco has been found to thrive well on the old indigo lands, and may be seen planted up to the very edge of the ruined vats. Tobacco is reared in a nursery in August, September, and October, is transplanted a month later, and the leaves are ready for gathering from January to March. The Rungpur or Kachar tobacco, as it is called, is distributed all over Eastern Bengal, and a not inconsiderable quantity leaves the country and goes to British Burma and elsewhere. The climate and soil of Rungpur are remarkably suited for the cultivation of tobacco, and it is possible that in a few years the growth of this plant may become of far greater importance than it now is. What is most wanted at present is to introduce improvements in the curing process. As evidence of the excellence of the Rungpur tobacco, it may be noted that a medal was obtained by a native of the district for a specimen which he exhibited at the Paris Exhibition of 1867. The quality of the tobacco grown at Rungpur is much liked by the

CULTIVATION in Bengal.

natives, and tobacco has been a staple of the district for at least a century past. The trade is chiefly in the hands of Mughs, forty or fifty of whom come annually to the district and export the tobacco to Calcutta and to Naraingunge and Chittagong for export to Burma. This tobacco is eventually for the most part made up into Burma cheroots, and is manufactured in Calcutta as well as in Burma. The produce varies from 6 to 12 maunds an acre, while the price varies from R3 to R7 per maund. The Burmese Mughs, who import tobacco into their own country for the purpose of making cheroots, select the very broad and thick-leaved plant, neither too mild nor too strong, and pay as much as R7 a maund for it. The variety most prized by them is called by the people *Háthi Kahn* from its resemblance to the ears of the elephant. In Dinagepur, Cooch Behar, and Julpaiguri also tobacco is widely cultivated, and the produce of Cooch Behar especially is highly esteemed." "In the districts of the 24-Pergunnahs and Nuddea, tobacco is generally grown merely for domestic use, but in a tract of country including the northern part of the Baraset sub-division of the 24-Pergunnahs and the southern portions of Ranaghat and Bongong, in Nuddea, the cultivation is of more importance and the crop is exported. The quality and price vary considerably. The best tobacco is called the *Hinglí* tobacco, from *Hinglí*, a village on the left bank of the Jomoona river, three or four miles west of the Gaighatta Thanna in Nuddea. Tobacco going by the name of *Hinglí* tobacco is grown generally over this area, and sells for from R5 to R7 or R8 per maund. The exports are mostly to Calcutta."

In Behár, tobacco is grown chiefly on the northern side of the Ganges. Three varieties are distinguished *viz., desi* or *barki, biláeti* or *kalkaltiya* **(N. rustica)**, and *jethua*. The latter kind is sown January to February, and cut May to June. Tobacco is most extensively grown in the Durbhanga district, and the best kind is produced in the Tájpur sub-division in Pergunnah Saraysa, which has become famous for the good quality of the leaf. "The average yield per acre is said to be about 18 to 20 maunds, and the price is about R5 per maund. The cultivators generally derive a very handsome profit, generally averaging R60 per bigha, and sometimes as much as R80 or R90. Tobacco leaves the district in various ways. Some of the rayats themselves export it in carts to Nepál or take it to Hajeepur and other river marts in the district, and there dispose of it, while many thousand maunds are bought up by the travelling merchants and transported by river and rail to the Upper Provinces and Bengal. The cultivators, however, generally dispose of it to purchasers from Gorakhpur, who sometimes buy it as it stands on the land."

The following account of tobacco cultivation in Behár is given by **Dr. Shortt** in his *Manual of Indian Agriculture* pp. 149 and 174:—The ground is prepared by free ploughing, and the clods are well crushed and opened out frequently to permit the dew to act upon the soil. The land is often allowed to lie fallow for a whole year, being kept free of weeds and manured with the usual dunghill rubbish consisting for the most part of cattle manure and wood ash. The sowing season commences in the beginning of June. The leaves attain maturity about the month of September. The first crop is cut and thrown aside for curing, and the second crop allowed to succeed. The stalks are then allowed to run to seed. The produce of a bigha ($\frac{1}{3}$ acre) is equivalent to twenty maunds from the first crop and fifteen from the second. Sometimes a third and fourth crop are taken off, the third yielding ten, and the fourth five maunds. The leaves are plucked and laid on the fields night and day, so as to have the benefit of the dew and sun, and they are then gathered and packed in straw or hill grass. Before they are packed, the stalks are cut away

CULTIVATION in Bengal.

and macerated in water for several days, and the infusion thus obtained is sprinkled freely over the leaves with the view of improving their aroma. This is repeated as often as is considered desirable, and after each application, the leaves are repacked. When they have continued thus sufficiently long, the packages are undone and some soil impregnated with alkalis is sprinkled upon the leaves till they are quite dry."

POOSA TOBACCO FARM.—This farm was established in 1877, and has since that time been worked under the direction of **Messrs. Begg, Dunlop & Co.** in connection with the Gházipur Farm (see under North-West Provinces). The climate of Poosa is much more favourable for tobacco cultivation on a large scale than at Gházipur; for this reason, and in the hope of being able to induce indigo planters of the neighbourhood to take up the industry, **Sir Edward Buck** strongly urged the advisability of handing over the Poosa Stud lands to the Calcutta firm. The results of their operations have hitherto been altogether successful.

In Assam. 119

IV. CULTIVATION AND MANUFACTURE IN ASSAM.—In Assam very little tobacco is grown. It is, however, largely imported from Bengal, and is consumed principally by the immigrant labourers, and by most of the hill tribes of Central and Northern Assam. The Mishmis and Abors of the extreme North-Eastern frontier are said to be the greatest smokers in the world, both men and women, and they are hardly ever to be seen without a pipe, generally made of brass, in which they smoke their strong home-grown tobacco. The Kukis, who inhabit the hill tracts of Central Assam, are fond of refreshing themselves with the oil which is collected in a reservoir attached to their bamboo pipes.

CULTIVATION in N.-W. Provinces. 120

V. CULTIVATION AND MANUFACTURE IN THE NORTH-WEST PROVINCES.—In 1888-89 an area of 41,288 acres was returned as occupied by tobacco. This shows an increase of over 2,000 acres as compared with the previous year's area. Taking the average outturn as 800lb per area, the total would be over 10½ million pounds. The largest tobacco-growing districts at present are Farukhabad and Bulandshahr.

The following extracts from *Field and Garden Crops of the North-West Provinces, I.,* 70-72, will give a good general idea of the native methods of cultivating and preparing tobacco in those Provinces:—

"The dryness of the soil and climate of these Provinces renders them unsuitable for the production of good tobacco, and the area under tobacco would be much smaller than it is were it not for the common occurrence of wells, the water of which is impregnated with nitrates (known as *khári*) and which is therefore especially suited for the production of the coarse pungent weed appreciated by the natives of the country. Possibly, too, on account of its comparatively late introduction, none of the higher castes of cultivators will grow it, and its cultivation is almost entirely restricted to the market gardener caste, known as the Káchi, Murao, or Sani. Tobacco cultivation may, therefore, be held to follow the distribution of *káchis* and of *khári* wells, and it is in consequence restricted within what would otherwise be considered very capricious limits." "The months for sowing and cutting tobacco vary considerably in different parts of the Provinces, but the seasons may be conveniently grouped into two. In one case the seed is sown in July and August, the seedlings planted out in October, and cut in February, while in the other case the seed is sown in November, seedlings planted out in February, and cut in April-May. Tobacco grown in the former season is known as *sáwani,* and that grown in the latter season as *asárhi.* Occasionally, after cutting a *sáwani* crop in February the roots are allowed to yield a *ratoon* crop in the following May, which is, however, always of very inferior quality. No particular

NICOTIANA Tabacum.

Tobacco.

CULTIVATION in N-W. Provinces.

rotation is used. The land is always heavily manured, so that the tobacco is occasionally grown after a crop of maize, and the field does not receive a fallow even in the kharif preceding. And in cases where strong manure is available, such as in the environs of large cities, tobacco commonly forms one of three crops which are regularly taken off the land each year, potatoes in the months November to February following after maize, and being succeeded by a crop of *asárhi* tobacco. The manure which is given to the land is so heavy as to make the natural character of the soil a secondary consideration. A loam is undoubtedly the most suitable soil, but if there is the inducement of a *khari* well, shift will be made with any kind of land, and since *khari* wells chiefly occur near old village sites, tobacco fields are often met with, the natural soil of which would, if uncorrected by manure, be little else than a collection of brickbats. Fields growing tobacco receive at least 200 maunds of the richest manure available to the acre, but since the land generally bears two crops within the year, the tobacco does not appropriate the whole benefit to itself. In some districts (Fatehpur, Allahabad, and Jaunpur) it is the practice to manure tobacco land by herding cattle on it at night, and in the Bijnor District almost the whole of the tobacco is grown on clearings along the forest border which are used to herd cattle in during the rains. The cultivators usually live in villages at some distance, but in the cold and hot weather months migrate to the jungles and establish their tobacco fields, returning as soon as they have cut their crop." "The soil must be very finely pulverised, and often owes its preparation more to the mattock than to the plough. If the plough is used it is driven through the land at least eight times, the log clod-crusher being dragged over the ground between every two or three ploughings, so as to reduce the soil to a condition as nearly resembling powder as possible. The seed is sown in nurseries, and planted out when about six inches high." "The soil must be thoroughly moist, and the seed is covered by brushing the earth over it by hand or by a twig brush. The soil round the seedlings must be kept always moist, and this when the seed has been sown after the end of the rains necessitates a light watering every third or fourth day. The seedlings are thinned out from time to time, and when six inches high, are transplanted by hand into the field, being placed in lines at a distance of six to eight inches apart. The thin planting practised in America finds no favour in this country. Transplanting is invariably carried out in the afternoon or evening, and the seedlings are often protected by screens from the heat of the sun for the first few days." "The cutting and curing has but little resemblance to the parallel operations in American tobacco culture. In districts west of Allahabad the practice is to cut the plants down whole close to the ground, but in the Eastern districts the leaves are often picked separately as they ripen. The plants or leaves are then allowed to lie on the ground, and wilt for a period which seems to vary greatly in different districts, and which is much longer when the plant is cut in February than when it is cut in April or May." "Tobacco intended for chewing is left out on the ground nearly twice as long as that intended for smoking; in the latter case, the leaves are carried in when of a black colour, and in the former case not until they have been burnt reddish brown (Etah). The leaves are carried in when damp with dew in the early morning, so as to run as little risk as possible of breakage." "If the plants were cut down whole they are now stripped, and the leaves are then heaped in a mass for fermentation, being arranged with their apices pointing towards the centre of the heap and their stalks outwards. Occasionally the heaping is carried out in a hole or trench in the ground. They are allowed to remain in this condition for a period varying from three days to a month,

CULTIVATION in N.-W. Provinces.

fermentation being occasionally assisted by a sprinkling of water, which should be brackish if possible. The temperature is carefully watched, and immediately it rises too high the heap is opened out, the leaves turned over and made up again. When sufficiently fermented the leaves are pliable and can easily be made into 'hands' or coils, which when finally dried are ready for sale. If no immediate market for them can be obtained, they are 'bulked,' *i.e.*, heaped in a corner of the cultivator's house, or occasionally hung from the roof, until they find a purchaser."

GHAZIPUR TOBACCO FARM.—The following extracts are taken from a "Note on Tobacco Culture and Curing" written in 1878 by **Sir Edward Buck**, under whose direction and influence the success of the undertaking was secured :—

"A Tobacco Farm was projected in 1875 on the stud lands at Ghâzipur, which consists of about 1,200 acres of culturable land. In that year little more was done than to divide the land into blocks, improve its irrigating capabilities, settle cultivators upon it, and collect manure. Meanwhile an arrangement was made with **Messrs. Begg, Sutherland & Co.**, of Calcutta and Cawnpore, under which they were to take over the farm on lease, on condition of their bringing over a skilled curer from America and continuing the cultivation of tobacco.

"The curer arrived in January 1876, and proceeded to cure, but under unfavourable conditions, a small quantity of leaf which had been raised at Cawnpore and Ghâzipur. The plant had been badly cultivated, and no proper curing-houses were ready. Notwithstanding these difficulties, sufficiently good results were obtained with the leaf sent to England to give reasonable expectations to every one concerned that Indian tobacco can be profitably raised and cured for the European market. The area under cultivation in the year 1876-77 has been about 50 acres.

"The outturn in good leaf from a properly cultivated acre of land ought not to be less than 800℔ an acre, and in America is said to average 1,000℔ an acre. Now, as the leaf can undoubtedly be cured up to a quality which will not fetch less than 5*d.* a pound in the English market, the gross outturn of cured leaf ought not to be worth less than 5*d.* This has been proved by the facts obtained in the Ghâzipur experiment.

"Eight hundred pounds an acre may, according to Indian notions, be a low estimate, but it must be remembered that a coarse-heavy leaf will not fetch such a good price, or be received so well in the European market, as a more delicate and better-flavoured leaf of less weight; 5*d.* a pound (although a very much higher price than that usually fetched by East Indian tobacco, which generally sells for 1*d.* or 2*d.*) is not a high standard for the bulk of cured leaf imported to England from other countries, and it is not at all unreasonable to expect that an average price of 6*d.* or 7*d.* may be altained. In fact, samples sent in considerable quantity last year from Ghâzipur to England and Australia were valued at prices which on the average were nearer 6*d.* than 5*d.* a pound (in bond).

"I may here point out that 1*d.* a pound profit is equivalent to a profit of from £150 to £200 on 50 acres; and that therefore for every extra 1*d.* which tobacco fetches above 5*d.*, or for every 1*d.* which the cost is reduced below 5*d.*, an income of about R1,500 to R2,000 per 50 acres may be realised.

"The climate of the North-Western Provinces (at any rate above Ghâzipur) is not favourable for tobacco operations upon a large scale. In the first place, the plant would be exposed to much danger from frost and hail; and in the second place, the climate is so dry, that while a great deal of irrigation is required for cultivation, a great deal of trouble must be incurred in regulating temperature during curing processes. There is little doubt that the climate of Upper Bengal is much more favourable.

CULTIVATION in N.-W. Provinces.

"The Gh//azipur experiment confirms the idea, which has often been mooted, that indigo planters or other gentlemen who have land in Upper Bengal should attempt the cultivation and curing of tobacco. Operations required for tobacco curing will apparently dovetail in well with those required for indigo manufacture. In the first place, the indigo godown can without great difficulty be converted into a curing-house; in the second place, it is clear of indigo at the season when it would be necessary to fill it with tobacco. The same argument applies to labour and supervision. The busy time with tobacco will be from November to February, when indigo has been already sent to market, and even if it is necessary to keep tobacco leaf in the curing-house till the rains, it ought to be packed before indigo is cut.

"In Virginia a planter never confines his farming operations to tobacco, which, on the contrary, only occupies a comparatively small portion of his farm area. One planter, indeed, seldom cures more than 50 acres of tobacco in a year, and hardly ever attempts to grow so much as 100 acres. The produce of 50 acres is usually as much as one curing-house and establishment can there undertake, and although it may be found that where labour is as plentiful as it is in this country, a large area may be undertaken, it must be left to time and experience to prove this."

"Tobacco importers in England declare that there is no chance for Indian leaf of good quality establishing itself firmly in the market unless a regular supply of large quantities is annually despatched. It appears, therefore, desirable that as many estates as possible should export properly cured leaf, if there is to be any chance of founding a regular trade between Bengal and Europe in the better class of leaf at remunerative price."

"The Gházipur experiments have succeeded in showing that Indian tobacco can approach American tobacco in quality, and is so valued in the English market; and also show that if an export trade can be established the profits will be sufficiently good to prosecute the industry. That the firm who has taken up tobacco enterprise at Gházipur believes in its capability is proved by the fact that it has imported a second curer from America, and has established a new centre of cultivation and curing in Upper Bengal, which country I have for some time recommended as likely to be the most favourable to success."

The results of operations at the two farms at Gházipur and Poosa, during 1882-83, are thus given in the North-Western Provinces and Oudh Administration Report for that year, p. 124:—

"Of an area of 523 acres under cultivation, 129 were set apart for home cultivation and the remainder leased out to independent cultivators, whose crops were purchased by the lessees. The total produce from both farms was 362,000lb or 36,000lb over and above that of last year. The rate of produce per acre at Gházipur and Poosa was 570lb and 793lb respectively."

Mr. **Caine**, in a Memorandum published by the Revenue and Agricultural Department of the Government of India in 1882, thus describes the system of cultivation and curing adopted in the Ghazipur Farm:—

"The tobacco fields are usually sheltered from the hot west winds by a high crop on that side, or, in the absence of this, by sowing a line of castor-oil plants or any other fast and high growing crop. The cultivation commences in July.

"As to rotation of crops we follow the American system, and tobacco is grown on the same land only once, or on rich land twice, in three years. The land usually lies fallow the third year, or should do, and in America a crop of oats is often sown, which crop is ploughed into the land just as the ears commence to form.

CULTIVATION in N.-W. Provinces.

"The manure at Gházipur and Poosa consists principally of cow-dung and vegetable manure, such as leaves, indigo *seet*, &c.; at Gházipur a good deal of night-soil and poppy trash. The land is manured yearly.

"Lands suitable for sugarcane and poppy are selected as being the richest. The land is ploughed from commencement of rains to time of planting, or earlier if feasible. A piece of good high land is selected for a seed bed, well ploughed, cleaned, and manured with good old manure (low ground would swamp). The ground, when soil has become well pulverised, is now marked off into beds 4 feet broad and running the whole length of the ground. The bed is slightly raised in the centre as a protection against heavy rain. A small ditch is cut between the beds to drain off the rain. Tatties made of straw or *arhar* twigs are put over the beds, and are raised 3 feet from the ground. The seed is sown in July, and a second sowing is made in August in case of accidents.

"Two table-spoonfuls of seed are sown over 100 square feet of seed beds. It is sown mixed with ashes.

"It is sown by a man who stands in the ditch running between the seed beds. After the seed is sown, the bed is beaten down gently with a plank or the naked feet of coolies. The seed germinates in eight days. The land is kept clean from weeds. The tatties are kept on for at least a fortnight. They are left off gradually, that is to say, they are first taken off for a few hours daily, in the morning and evening and at night, till the young plants get accustomed to the sun. They are a protection to the plants from the sun and also from heavy rain, which often washes out the earth from the roots of unprotected seedlings.

"The tatties must not be left on until the young plants are transplanted, or else the plants will be weak and unable to bear the sun.

"The land having been well ploughed and cleaned from the middle of June to the middle of August, is smoothed over with a *henga* (harrow), and the young plants being now large enough, they are transplanted when the leaves are not quite the size of a rupee. A cloudy or rainy afternoon is selected for the planting (the afternoon is better than the morning, as it gives the plants the whole night in which to take hold).

"The field is either marked out beforehand, by means of a long rope laid on the field, along and on which a few coolies are made to walk, and which leaves a clearly-defined line marked on the field; these lines are made first down the field and then across, each line being the same distance apart, or else a lighter rope marked with knots is thus laid on the field at the time of planting, and a plant is put in opposite each knot. It is very necessary for facilitating the after-working of the tobacco that the plants should be equidistant from each other. In rich land the plants are put 3 feet apart. In poorer soils they are only two feet and two and a half feet apart. No plants whose stems have become at all hard should be planted; they will certainly be stunted. Grubs should be looked for in the roots and stems, and all affected plants thrown away. If the ground is hard and clayey, it is desirable to stir the earth with a *khurpee* a little round the young plants three or four days after the planting.

"The land is usually hoed about ten days after planting. When the plants are from a foot to $1\frac{1}{2}$ feet high, the earth is thrown up round the roots of each plant.

"Irrigation is carried on whenever, from the appearance of the plants, it is required. The ground is hoed and the plants earthed up after each watering until the plants become too big to allow of men working in the field.

"When the plants are about 3 feet high, or if weakly looking, 2 feet, the top shoot is plucked off (this shoot is plucked off directly it makes its

CULTIVATION in N.-W. Provinces.

appearance in small or sickly plants), also the lower leaves which are dirty and draggled, and from 7 to 14 leaves are left, according to the strength and growth of the plant, the principal object being to get a few large and well developed leaves in preference to a quantity of small ones. The side shoots or suckers are plucked off the instant they appear, and are left on the field for manure.

"If a grub be found in a large plant, it should be cut out with all the affected part and all the portion of the plant above it; a side shoot may be allowed to grow which will give a fairly good plant.

"Tobacco ripens in about three months' time. It is cut during the months of November, December, January, February, to the middle of March. A ripe leaf has yellow spots on it. It has a crumply look, and if bent between the finger and thumb will break.

"The cutting commences directly there are sufficient plants ripe in a field to fill the curing barn. The plants are cut off bodily at the stem just below the lowest leaves of the plant. The plants when cut are left lying with their butts towards the sun in the field to wilt. The time a plant takes to wilt depends on the heat of the sun. Usually half an hour is sufficient. When wilted the plants are either carried or carted to the curing barn. There they are spiked on split bamboos.

"Each cooly is provided with an iron spike which he fixes like a spear-head on to the bamboo stick, he then takes a plant of tobacco in his hand, fixing first the other end of the stick into a hole in a block of wood provided for the purpose which he holds between his toes. The plant is placed with the butt on the spike about 5 or 6 inches from the end and the plant forced down over the spike on to the stick. From six to ten plants according to size are hung on one stick which is 4 feet long. These sticks are then hung in the barn; the stick should be hung so that the leaves may touch each other slightly, but should not press against each other. The barn is fitted up with a scaffolding of bamboos. The bamboos are 3 feet 6 inches apart and 4 feet above each other, the lowest tier of bamboos being 6 feet at least from the ground (where the tobacco is intended to be cured by fires). The barn is provided with as many doors as possible, those on the west side being made as airtight as possible. Ventilators in the roof made to open and shut are advantageous. Rooms can be made any size. A room from 35 to 40 feet high and 30 yards long by 15 yards broad is preferable, as it can be filled rapidly and will hold sufficient tobacco to cure well.

"When the barn is full (it should be filled as rapidly as possible in order to prevent the tobacco drying out in hanging) all the doors are closed and also the ventilators if any. It is left for two or three days. The planter can now tell whether the tobacco is drying up too rapidly or not fast enough. If the tails of the leaves curl up and break when handled, it shows that the tobacco is going up too fast; on the other hand, if there is a sour smell in the room and the plants sweat, the tobacco requires air and perhaps fires. In the first case the doors and ventilators are still kept closed and fires are lighted in different parts of the room, or if the house is filled with flues (which are preferable to open fires) hot air is carried through the room in the flues. The temperature will probably be raised to 80° Fahrenheit, but this can only be told by experience. The tobacco must be carefully watched, and if drying too fast, the temperature lowered and water sprinkled on the floor. Raising the temperature causes the tobacco to sweat, and the moisture thus created in the house makes the colour run in the leaves. The leaves should turn gradually yellow and then brown. If dried too rapidly it retains its original green colour. If it is intended to cure golden leaf, the temperature is raised to 140° Fahrenheit or higher

CULTIVATION in N.-W. Provinces.

at the stage in which the tobacco has changed to a yellow, but this curing cannot be attempted in a hot climate, except by an experienced curer. Golden leaf realises double the price that dark leaf does. The plants should originally all be in the same stages of ripeness to ensure success in bright or golden leaf curing. In the above, curing in a hot dry climate like Ghazipure is referred to; in Tirhoot in mild weather tobacco can be cured without any fires. This process will now be described. If, as in the second case stated above, after two or three days hanging in the barn the tobacco feels soft and there is a sour smell in the room and the plants may or may not sweat, then all the doors and ventilators should be opened and kept so until the sour smell is gone and the sweating has stopped; if that is not effectual then fires must be lighted. The curer must now be guided by the weather and must carefully watch the tobacco. If the tips of the leaves begin to curl, it is going too fast and the doors must be shut during the day and opened only at night to allow the cold air to circulate through the room, the main object being to make the tobacco dry up gradually to yellow, and the greater part of it will turn reddish yellow called medium bright. The temperature must be regulated by the doors and ventilators. This air-curing makes a lighter brown than the firing process, and can only be adopted in a climate in which there is a certain amount of moisture in the air. If the tobacco sweats badly, doors and ventilators must be opened and fires lighted and the heat raised till it stops. Green tobacco is preferable to sweated. There is another process followed in some parts of America called sun-curing. In this process a scaffolding is erected under the shade of a tree and after the plants are hung up, the whole is covered around and on the top with straw. The straw is opened out when it is found necessary to quicken the drying. This style of curing is hardly adaptable to the plains of India.

"The higher the barn is the better it is for curing purposes. The highest tobacco in a room is usually the best colour, if you have a thick roof, otherwise the centre is the best.

"The tobacco is generally cured, so far as its colour goes, in a fortnight or three weeks. It is left to hang through the hot weather in the barns, as the heat makes it too dry to handle. Early tobacco may be ready to bulk down in the Christmas rains. No tobacco should be bulked until the sap is entirely dried out. This can be seen by breaking the stem of the leaf. If bulked with sap in it, it will rot.

"Directly the rains commence in June and the tobacco has become soft and pliable, it is bulked down in heaps, in the curing room in which it is hung. The heaps are raised some 8 inches off the ground by a small scaffolding made up of bamboos and sticks, so that air can circulate underneath, and are covered over with straw or matting. The tobacco should not be bulked down in too moist a condition. The best order for bulking is when the tobacco is just soft enough to handle without breaking. If too soft it must be fired and allowed to come in order again. When all the tobacco is bulked down, the bulks must be opened and the leaves stripped from the stem and tied in 'bands' or 'bundles' with about 50 leaves in each band. The band is tied round with a leaf of tobacco tied round the upper ends of the leaves and tucked in at the centre of the bundle; these bands are now carried to the head barn or sorting room. They are re-bulked here in the same way as before. When all the tobacco is in the storing room, the bulks are again opened and the bands being untied, the leaves must be sorted. They should be sorted into, 1, long leaf dark, 2, short leaf dark, 3, long leaf bright, 4, short leaf dark, 5, lugs, that is, all torn or dirty and very small leaves, red and bright, being band-

CULTIVATION in N.-W. Provinces.

ed separately, and 6, green—six varieties in all. The sorting is most important and requires strict supervision. Care must be taken that the coolies do not make unnecessary breakage in handling the leaves. They should be tied in bands of from 15 to 20 leaves. These bands are again bulked and left in bulk till packed. The bright is divided into four varieties, should there be any golden leaf. Golden leaf is pure yellow. In this case you have 1, bright long leaf (that is, golden leaf), 2, bright short leaf, 3, medium bright long leaf, 4, medium bright short leaf. Lugs are often made into strips by taking out the thickest portion of the stem midrib; $\frac{2}{3}$ is taken out, $\frac{1}{3}$ of the way from the tail of the leaf. It sometimes sells best in this form.

"The tobacco is packed in hogsheads made of thin staves. The hogsheads are made 4 feet in height and about 3 feet in diameter for despatch to Europe, or else, after the native custom, in bales. The tobacco should be packed as dry as it can possibly be packed without breaking it. It is generally necessary to hang it again in a barn, the bands just slung across the stick, and fire it till sufficiently dry. If too dry the doors may be left open at night, when it will probably be found in the right order on the following morning. The bands are packed with the butts outwards and tails inwards. There are three lines in each row, two with their butts at the edge of the hogshead and tails meeting in the centre of hogshead, and one in centre of hogshead. The next row is commenced from the other side of the hogshead. When the hogshead is filled, it is pressed down with powerful screws and refilled till it can hold no more. It should contain 900℔ of leaf as nearly as possible."

CULTIVATION in Oudh. 21

VI. IN OUDH, tobacco cultivation, according to the statistical returns of 1888-89, occupied 13,374 acres, showing a decrease of over 2,000 acres as compared with the area of the preceding year. Of the district areas Sitapur and Kheri appear to be the most extensive.

CULTIVATION in the Punjab. 122

VII. CULTIVATION AND MANUFACTURE IN THE PANJAB.—The area under tobacco in 1888-89 was returned as 60,566 acres. In 1883-84 as many as 85,400 acres were said to have been occupied by this crop; in 1886-87, however, the area was reduced to 46,437 acres, whilst in the following year over 19,000 acres appear to have been added. The districts for which the largest areas were shown in the returns for 1888-89 were: Jullundur, Sialkot, and Lahore. A considerable portion of the Panjáb tobacco crop, especially in the Lahore district, consists of the *Kandahári* kind or that obtained from **N. rustica.** Other varieties, mentioned by **Mr. Baden Powell,** are—*Baghdádi*:—The seed of this is very much sought after by cultivators, on account of the abundance of the produce; it is not imported from the place whose name it takes, but probably came originally from thence. *Noki,* so called from its pointed lanceolate leaves. *Sambli,* this is grown in the Lahore, Amritsar, and Siálkot districts. *Purbi,* an expensive kind introduced from Bengal and grown near Lahore; it is chewed with *pán* and betel nut, and is sometimes smoked. *Baingani,* with leaves shaped like those of the *baingan* (**Solanum Melongena**). *Suráti,* a strong and bitter kind, introduced from Surat; the two last-named are rare. Other local names are given for some of the above. All these kinds are prepared alike, by burying in a pit after drying for a couple of days or so, and left to ferment. The best leaves are tied into bundles (*gaddi*), the rest are twisted into ropes, the latter selling for one-sixth to one-eighth less than the former (*Panjáb Prod.,* 288). The following information regarding the more usual methods of cultivation and curing adopted in the Panjáb is taken from a report submitted to the Government of India in November 1873:—"It seems to be immaterial whether the land selected for tobacco cultivation is high or low, provided

CULTIVATION in the Panjab.

ample means for irrigation exist. In three cases only,—parts of the Siálkot, Ludhiána, and Ráwalpindi districts,—are alluvial tracts selected. The general opinion is that alluvial lands are not suitable. Garden lands and '*goera*' or manured lands near the village sites, are almost invariably preferred. Irrigation, both from wells and canals, is practised, the plots being watered about once in every four days. Extensive manuring is requisite to success. Night soil, sheep and goat dung, stable litter and cow-dung, are used; the older the better. An admixture of saltpetre in the shape of '*kalar*' or '*reh*' efflorescence is found beneficial. Upon proper irrigation and manuring the success of the crop depends." "The sowing season commences in September. In most instances the seed is put into the ground during October and November, but in some places sowing continues till December, and in the Sháhpur, Multán, and Derah Gházi Khán districts till January. In the Pesháwar, Amritsar, and Ráwalpindi districts, seed is sown till February. The transplanting takes place during January, February, and March, and the cutting during May and June and sometimes in July. The plant is almost invariably cut down to the roots, and no second crop taken. The experiment of leaving the stems for a second crop has been tried in the Lahore, Gujránwálah, Karnál, and Ambála districts, but the outturn was exceedingly meagre, and it is not considered worth while to keep the land out of cultivation for the purpose. The usual mode of curing is as follows:—The crop when ripe, *i.e.*, when it turns red and yellow, is cut down and left on the field for two or three days. It is then stored in a place protected from the sun, either in a room or under ground, with layers of *dhák* leaves (**Butea frondosa**) or *Ák*, *madár* leaves (**Calotropis procera**), and the whole covered over with quilts, in which condition it is left for a period varying from eight days to a month. The stems of the *kakar* tobacco are utilised, those of other kinds are rejected, and the remainder tied up into bundles and exposed to the sun for a short time. In Derah Gházi Khán the tobacco during the curing process is sprinkled with a solution of impure soda (*sajji*), *gúr*, and water. The object of this is apparently to give it colour."

The following descriptions of local methods of cultivating and preparing tobacco are, for the most part, derived from certain of the District Gazetteers:—" In the Karnál district tobacco is very generally grown in the villages, but mostly for private consumption only, except where local peculiarities are especially favourable. The *desi* variety is almost exclusively cultivated, of which *bugdi*, *surnáli*, and *khajúri* are forms distinguished by the shape of the leaf. The plant grows best in a nice loamy soil, neither too stiff nor too open. A slight saline impregnation rather improves the plant, and the water of bitter wells, or of the dirty village ponds, is best. Canal water is too pure. There is a well in the village of Phúrlak, the tobacco of which is celebrated throughout the district. The seed is scarcely ever sown by the villagers, who obtain the young seedlings from the market gardeners of the towns, paying R1-4 for enough to plant an acre. The land is ploughed eight or ten times, dressed most carefully, and laid out in ridges some two inches high and eight inches apart, the seedlings being planted half way up the ridge on either side alternately and about eight inches apart, for if water lies about the stem, it injures the plant. This is done in *mágh* or *phágan*. They are then hand-watered with manure dissolved in water. Soiled manure is generally used as a top dressing, as less is thus required. The dung of goats and sheep is the best, and old dry cow-dung mixed with ashes. The field is watered every ten days or so, and the hoe is then freely used, so as to keep the earth about the roots open and the weeds removed. As the leaves grow, they are sprinkled with *reh* or ashes to keep off insects and improve the

CULTIVATION in the Panjab.

flavour; and the flowers are nipped off as fast as they appear. The plant is ready to cut in *jet* or *asárh*. The wholeplant is cut in the morning, and left in the fields for twenty-four hours to dry. Next day they are piled up and left to dry further. A hole is then dug and the plants are packed into it, covered up with *dhák* **(Butea frondosa)** or *ák* **(Calotropis procera)** leaves, and left to ferment for five to ten days. The leaves (*pát*) are then stripped and either tied up into hands (*júti*) or twisted into a thick rope. They are, if necessary, further fermented; and are finally dried and kept for use. When tobacco is wanted, the leaves are cut up and powdered with an equal weight of *gúr* in a mortar. After the plant has been cut, leaves sprout from the stump, and are picked and used by the poorer classes" (*Karnál Gazetteer*).

In the Hoshiarpur district tobacco is grown in small well-manured plots, and requires constant irrigation. If following maize, in rotation, it can be sown as early as November, but it is often grown after wheat and barley (cut green), and it is then sown in February or March. The earlier sowings ripen in April, and the later ones in May. After being cut the tobacco is left lying in the field for two days; it is then buried in a hole in the ground for a week, and after being beaten with sticks is made into twists and sold. Three varieties are grown in this district, *viz.*, *desi*, *gobi*, and *dhatura*; the last-named is the strongest and most liked, and is supposed to be good for goitre, and on this account is exported towards the hills. Land suitable for tobacco fetches about R20 per acre, and if well manured yields three crops,—maize in autumn, wheat (cut green), and lastly, tobacco.

In the Jhang district tobacco is looked upon as the most profitable of all crops, its selling price, per *kanál*, being nearly equal to that of sugarcane, while it occupies the soil for only three months. It does not require any more manure or more frequent waterings, nor does it exhaust the soil to the same extent. To ensure a good crop of an acrid and pungent leaf the soil must be heavily manured, but another crop can always be grown after it, either *jowár* or turnips, or even both. In preparing soil for tobacco, four ploughings ought to be given, accompanied by four rollings in order to break the clods. The manure is then spread. Sheep and goats' droppings are considered best. On the wells near Jhang, where tobacco is an important crop, about 400 maunds of this manure are given to the acre. After being spread and well mixed into the soil with two or three ploughings, the land is then rolled until all clods are broken. The seedbed having been carefully prepared, the seed is sown broadcast, and covered with an inch thickness of fine manure, and then watered; during the frosty months it is covered with grass. Transplanting commences in the middle of *phagan*. The transplants are obtained at the rate of four annas per square cubit, or are raised by the zamindar himself. They are watered at first every three or four days, and afterwards only once a week. The first weeding and hoeing is given as soon as the plants have taken good root. Two or three hoeings are given afterwards, and top dressings three or four times. The roots are seldom manured. The breaking off of the young shoots from the stem (*kallí bhanna*) involves much labour. As the plants ripen they are cut, small quantities at a time, and spread on the ground for three days. The leaves are not then stripped off, but the plants are heaped on the floor of a dark room and covered with blankets or *razáis*, and remain thus for a week. At the end of the week the leaves are stripped off the stalks, and twisted into ropes and carried off by the purchaser (*Jhang Gazetteer*).

In the district of Dera Ismail Khan, two sorts of tobacco are grown, *viz.*, *síndhar* and *garoba*; the latter is a coarse variety, requiring but little care in cultivation, and selling at about R2 or R3 a maund. *Sindhar* has an even, well-shaped leaf. It gives much more trouble to cultivate, and sells at R3 to R5 a maund, or for nearly twice as much as the other. About a third of the cis-Indus tobacco is exported. The Derah tobacco is mostly consumed in the tahsil. The people are in the habit of mixing the *garoba* tobacco with *Kandahári* tobacco (**N. rustica**), which is brought down by the *Pawindahs* and costs R8 or R10 a maund. The Daman people for the most part smoke this mixed tobacco. The *garoba* tobacco by itself is somewhat tasteless (*Gaz., Derah Ghazi Khán*).

CULTIVATION in the Panjab.

Captain Harcourt, in his account of Kulu, Lahoul, and Spiti, gives the following information regarding the preparation of tobacco:—A seed-bed having been carefully prepared, the sowing is performed by simply blowing the seed from the hand. This is done in April or May, or later on, in the rainy season, when no artificial irrigation is required. The irrigated crop ripens in August, and the other in September or October. When the ripe plant is cut down the leaves are twisted round it and left to dry. An acre, well prepared, ought to return 10 to 12 maunds of leaf. The selling price varies from 6 to 8 seers per rupee. The dry leaf has a very mild and pleasant perfume: tobacco forms one of the chief imports in Lahoul and Spiti; it comes from Kulu. The women in Spiti do not smoke.

VIII. CULTIVATION AND MANUFACTURE IN BOMBAY.—The total area in this Presidency, utilised for tobacco cultivation in 1888-89, was given as 57,187 acres, which shows a decrease of over 30,000 acres as compared with the returns of the previous year. In his report for 1888-89 the Director of Land Records and Agriculture attributes this remarkable diminution of area, affecting every district, to scanty and unseasonable rainfall, as also to the large stocks of previous years. The importation of foreign tobacco is also supposed to have contributed to some extent to the decrease. The best tobacco is grown in the Kaira and Khandesh districts. In the following year (1889-90) the total area was increased by 21,000 acres. Seasonable rainfall and the high prices which prevailed during the previous year are said to be the causes of this increase. The Director also states that only in Kaira and Belgaum is tobacco grown extensively as a field crop.

CULTIVATION in Bombay. 123

Mr. O'Conor in his report, p. 28, enumerates the more important experiments that have been undertaken to improve the quality of Bombay tobacco. In 1831 some American seed was tried for the first time according to the instructions taken from **Captain Basil Hall's** account of tobacco cultivation in Virginia. In 1831 the local Government received from the Resident in the Persian Gulf a maund of seed of the finest tobacco grown in Persia, together with full instructions for its cultivation. In 1839 **Mr. Elphinstone**, Collector of Ratnagiri, obtained successful results with some Shiraz and Kazeroon seed received through the Agri.-Horticultural Society. A crop was obtained of each kind during the monsoon, second crops were grown in the cold weather, the latter thriving the best. During the same year the Society distributed some Syrian seed. Numerous trials were made for some years with this seed and its produce, as well as with Persian seed, under the supervision of **Dr. Gordon** in the Hewrah Botanical Gardens. The Persian tobacco proved to be too mild to suit the tastes of the mass of the people. The Syrian tobacco, on account of its greater strength and flavour, was much appreciated by native cultivators, and was readily bought up in the bazár. It was, however, condemned by London brokers.

In 1887, some genuine Shiraz tobacco was sent by **Dr. Balfour**, late Surgeon-General in Madras, for trial in the Bombay Presidency. The Conservator of Forests appears to have been the only successful grower, and his was pronounced by the Secretary of the Agri. Horticultural Society, as very superior.

NICOTIANA Tabacum.

Tobacco.

CULTIVATION in Bombay.

In 1869, the Bombay Government distributed small supplies of seed of the Shiraz, Havanna, and other varieties to the Superintendents of Cotton Experiments, and to the Collectors of Kaira, Khandesh, Dharwar, and Karáchi for experimental cultivation. In most cases the trials were successful as far as the cultivation was concerned, but the produce was entirely spoilt owing to want of proper knowledge of the curing processes. The Shiraz, Havanna, and Maryland promised the best results. In Dharwar, the Superintendent of Cotton Experiments, although having plenty of excellent seed for distribution, could not induce the rayats to cultivate it on a large scale, as they feared the exotic plant would be more liable to injury in the stormy weather which prevails about the time when the leaf begins to mature. In Khandesh the experiments were successful, and the Superintendent was of opinion that these exotic varieties could be grown in Khandesh successfully and with little trouble provided the rainfall was up to the average. The most important results of trials undertaken in this Presidency were those obtained by **Mr. E. P. Robertson**, Collector of Dharwar. In a letter addressed to the Agri.-Horticultural Society of India in December 1870, **Mr. Robertson** wrote as follows:—"I have for two years cultivated exotic tobaccos, and have been making experiments in the hope of discovering some method adapted to the climate of this country for curing tobacco. I am convinced that the system of tobacco curing ordinarily in use in America and Cuba, is of no use whatever in this country. Here we have the dry air to contend against. The difficulty is to get the midrib to dry without making the tobacco quite brittle. If the tobacco is packed in hands to ferment, with the midrib not quite dry, no care in the world will prevent the tobacco becoming mouldy and bad. If the midrib is allowed to become quite dry, the tobacco becomes brittle and is condemned as useless. It is not useless however. Tobacco cannot be properly cured in this country till the midrib is quite dry." In the following month **Mr. Robertson** sent to the Revenue Commissioner of the Southern Division, Bombay, an account of several experiments made by him in curing tobacco. From this letter the following extracts are given:—"In all well-cured Havanna tobacco, the leaf, when ready, is soft, pliant, and kidlike; of course, it will become hard and brittle with exposure to the air, and especially to the hot air of this country, but if properly taken care of, it is fit, with but little previous preparation, for the manufacture of cigars. The native-cured tobacco has the defect of being very brittle, and further, by the system of curing pursued, it has lost most of its fragrance and good qualities. In 1865, I took a quantity of the best Neriad tobacco to England, and it was pronounced from its curing to be worthless. The point, therefore, is to discover some method by which tobacco can be cured in this country so as to be fit for the European market. The tobacco that I reared last year was cured by me according to directions given by **Colonel DeCoin**. At first, while there was some moisture in the air, the tobacco began to cure very well in the tobacco-house, but soon the weather changed, and the dry air rapidly dried up the leaves, and I was unable to proceed. To enable me, therefore, to pack the leaves into hands to ferment, I exposed some of the tobacco to the night dews, and in the morning had it made into hands about twelve leaves to each hand. The dews, however, failed last year, and then I had to cover the leaves with damp towels, and when moist I packed them into hands. This tobacco during fermentation became mouldy owing to two causes:—*1st*, that tobacco thus artificially damped is very difficult to deal with, and requires great and constant care; *2nd*, that the centre stem of the leaves was not sufficiently dried up. I cured almost all the tobacco in this manner, but being disheartened at its turning mouldy, notwithstanding great

CULTIVATION in Bombay.

care, I threw the balance away in the corner of a verandah room just as it was. This year I again planted Havanna, Shiraz, Virginian, Ohio, and Himálayan seed." "The seed all came up well, especially the Havanna and the Shiraz, the plants being very large and vigorous, many of the leaves being fully three times the size of the common country tobacco of this district." "As I had planted my tobacco early in the season, May, it ripened early, and when ready for cutting, the weather was still moist and favourable. The first batch was hung for three days rather closely packed on bamboos in the open air to obtain colour. It was then, while still soft, taken into the tobacco-house and hung closely packed on bamboos. At the end of about the eighth day after being placed in the tobacco house, the tobacco was of good color, and had gone through a slight fermentation. Some was then made into hands of twelve leaves each, and placed in heaps to ferment. All proceeded well for a few days, but on examining it again I found that the tobacco had commenced to become mouldy all along the stem. Just about this time I happened to examine the tobacco I had last year, when dry, thrown aside in a heap as useless. To my astonishment I found this tobacco to be most perfectly cured, and as soft and pliant as possible. None of it was mouldy, and it was in exquisite condition and flavour. I therefore determined to leave the tobacco I had in the tobacco-house to dry. This I did. A few days, however, after it had quite dried, we had a heavy shower of rain, when the moisture in the atmosphere rendered this tobacco soft and capable of handling. I at once made it into hands and placed it to ferment. The tobacco has been thus made up for some time, and not a single leaf is mouldy, and the tobacco promises to cure well and to be of good flavour. From the above I deduce the following important points with regard to the curing of tobacco in this country:—*1st*, If tobacco is to be cured properly, there must be no haste, and no attempt whatever at artificial moistening; *2nd*, tobacco should never be made into hands to undergo fermentation till the centre stem of each leaf is perfectly dried up. This is a most important point, as if there is any moisture in the stem, the tobacco is certain to become mouldy, and a very little mould taints the whole leaf; *3rd*, that tobacco when cut and hung and dry should be carefully kept till the next rains. It should then be exposed to the atmosphere. It will soon become soft and pliant. As soon as pliant enough to handle without breaking, it should be made into hands of about twelve leaves each; these should then be put into heaps of about 1½ feet to 2 feet high, and left quietly to ferment for about two months, at the end of which time the tobacco will be found to be perfectly cured. The heap should be occasionally looked at to take care it does not get hot inside, and it would be well to cover it over with plantain leaves, or even to wrap it up in bundles of paper to keep out the air. There is another point I would draw attention to, and that is, when the weather is dry, the tobacco when cut rapidly becomes dry, and it is difficult to obtain colour. After several experiments I found the following the most successful method in the very dry weather. When the plants are ready for cutting, cut them early in the day and place them on the ground, all close to one another, and leave them there till the next morning. Early in the morning turn them over and leave them thus exposed for a second day. The next morning early turn them again, and at about 10 to 11 o'clock, after they have had some sun, bring them into the tobacco-house and place them in heaps about nine inches to one foot in height. Leave them for a day to ferment and obtain colour, turn the heap over next day, and as soon as the leaves are of a good yellow colour, hang them from bamboos in the tobacco-house. It must be borne most carefully in mind that the operation of thus placing them in heaps in the tobacco-house to obtain colour is danger-

CULTIVATION in Bombay.

ous, and that the heaps require the greatest attention, for the leaves being full of moisture rapidly ferment, and become hot and mouldy; therefore, constant watching is necessary, and no plant should be allowed to remain after it has obtained colour, but should be at once hung up. It will, while hanging, continue to obtain a fuller and better colour. Hang till dry and proceed as detailed above. The great points are to obtain a colour before allowing to dry; to use no artificial method of moistening, to have patience and wait for the moist atmosphere of the next rains; then to make into hands, and leave to ferment for about two months. I venture to give the above results of my own experiments, hoping they may aid others and induce them to make further experiments, and at length we may, by this means, arrive at a simple and certain method of curing tobacco in this country fit for the European market. The present country method is of no use, and it is quite immaterial to the natives if the tobacco is mouldy, so long as it is strong and pungent. The method adopted in Cuba is also one that does not suit this country, as there they have constant showers, and all the instructions, say—'Avoid excess of moisture.' In this country, when tobacco is ready, we generally have a fine hot wind and no showers, and, therefore, we must wait till the monsoon. It is possible, however, that in Guzerat this delay might be avoided, and that an artificially damp moisture might be obtained in tobacco-rooms or houses by the use of wet *kuskus* or *jowaser* tatties."

In November 1871, **Mr. Robertson** sent various samples of tobacco, prepared by him, to **Mr. Broughton**, Quinologist to the Madras Government, and to **Dr. Lyon**, Chemical Analyst to the Bombay Government, for analysis. Regarding a sample of Virginian tobacco prepared in 1869, **Mr. Broughton** remarks:—"I find it to contain three per cent. of nicotine most certainly. It yields 22·9 per cent of ash, of which 19·67 per cent. consists of potassic carbonate. It thus appears that the specimen fulfils the conditions necessary to tobacco of good quality to European smokers. On actual smoking the tobacco is good, though not of special excellence; considered as an Indian production for the European market, it is a success. Its only fault appears to reside in the curing. Three per cent. is a somewhat small percentage of nicotine for Virginian tobacco, which is said to contain ordinarily a considerably larger amount. The soil sent contains but little calcic carbonate, and is suitable for tobacco cultivation." Seven samples were examined by **Dr. Lyon**, who reported as follows:—"The first point to be noticed is the great general similarity in chemical composition which exist between the mineral constituents of the various samples of Shiraz and Havanna tobacco. This resemblance is especially noticeable as regards the proportion between the basic constituents, *viz.*, the potash, lime, and magnesia. The resemblance tends to confirm the inference terminating the 12th paragraph of **Dr. Forbes Watson's** report (1871) on the cultivation of tobacco in India, *viz.*, 'that tobacco plants grown from different seeds on the same soil will possess substantially an identical composition as regards their mineral constituents.'" "I would next draw attention to the relatively large amount of chlorides in the tobacco (both Shiraz and Havanna) of 1871. It is possible that this is due to the presence of chlorides in the manure, &c., used. If so, it affords a proof of the facility with which the tobacco plant alters its mineral constituents according to the nature of the soluble matters present in the soil on which it is grown, and (showing its susceptibility to treatment) holds out hopes of our being able to improve the mineral constituents of the plant in any desired direction, by properly apportioning the soluble matters of the manure employed. I should think that the presence of chlorides in any large quantity in the plant must tend to lower its quality: the presence of

CULTIVATION in Bombay.

alkaline chlorides. in all probability, confers saltness, not only on the tobacco itself, but also on the smoke, from their volatilisation. It would, therefore, appear advisable to avoid as much as possible the presence of chlorides in the manure to be applied to the soil. It is curious that the percentage of nicotine is smaller in the Shiraz tobacco of each year than in the Havanna tobacco of the same year; and remarkable also that the nicotine is very much less in quantity in the tobacco (both Shiraz and Havanna) of 1871 than in the tobacco of any other year. This diminished percentage of nicotine in the tobacco of 1871 may be due to an alteration in the method of curing adopted. It is remarkable, however, that coincident with it we have a decided alteration, as shown by the increased proportion of chlorides present, in the mineral constituents of the plant. It would be worth while, I think, to conduct a few experiments something in this way. Samples of the same variety of tobacco might be grown on the same soil, under the influence of different manures subjected to the same process of curing, and then submitted to analysis. This appears to me to be a likely method of ascertaining the best way of improving the quality of the tobacco. The curing process also might be varied with a view of ascertaining the effect of alterations in it on the percentage of nicotine." **Mr. Robertson,** commenting on these two reports, says:—"Both **Dr. Lyon's** and **Dr. Broughton's** letters are in my opinion extremely interesting and instructive to those interested in tobacco cultivation. Their analysis proves that the opinion given in Calcutta that the samples sent by me to the Agricultural Society contained no nicotine, was erroneous. I have, in subsequent experiments made in 1871, on tobacco grown in the rains of 1871, closely followed as to curing the directions given by **Dr. Forbes Watson** of the India Office. But I purposely continued from year to year to plant on the same soil, manuring it highly and using a very considerable amount of ashes. **Dr. Lyon's** analysis, as remarked by him, proves very clearly the truth of **Dr. Forbes Watson's** remark in paragraph 12, page 6, of his report, as to the influence of acclimatisation on the composition of the ash. It will also be noticed how, year by year, nothwithstanding full manuring, the quantities both of potash and of nicotine have steadily gone on decreasing, till in 1871 none of the potash remained, and but little nicotine. These experiments are, I think, also a proof that as yet we do not fully understand the cultivation of tobacco in India, for they have been made entirely on red soil, a soil not generally considered as suitable for tobacco, and as irrigation was not available, the seeds were planted in May, and the crop was commenced to be gathered in September and October. That the red soil is well-suited for tobacco is clear from **Dr. Broughton's** remarks, *viz.*, that the 'specimen fulfils the conditions necessary to tobacco of good quality to European smokers.' It appears evident, therefore, that tobacco of good quality can be grown on common red soil if only the necessary condition of rotation of crop is atttended to, and if the seeds are planted about the middle of May. This is somewhat important, as tobacco is a very valuable and remunerative crop." "I would, however, venture to remark that the experience I have gained after four seasons' careful attention of the subject is that, as regards curing tobacco for the European market, it is next to impossible to hope ever to cure it in this country at a profit for export to the European markets. To cure tobacco well, is, even if following the most simple methods, exceedingly troublesome, and requires more care than natives will ever give to it, especially as by the native method with less trouble and care comparatively (when the trouble and care is considered as so much expenditure), as favourable a price can be obtained for their own tobacco in the native market as can. be obtained for Indian tobacco grown for the European

NICOTIANA **Tabacum.**

Tobacco.

CULTIVATION in Bombay.

market." In a report on some samples of tobacco grown and prepared at the Government Model Farm, Khandesh, Dr. Lyon writes:—"Before commenting upon these analyses, I have, in the first place, to point out that very little is known as to the influence which the chemical composition of a tobacco has on its quality and commercial value, and that it is futile, therefore, to attempt to draw any but very general conclusions as to the quality or value of tobaccos from the results of their analyses. Further, we know exceedingly little as to the influence which varying conditions of climate and soil have on the chemical composition of tobacco. It is therefore, much better, I think, for the present at any rate, to regard analyses such as those under report in the light of experiments undertaken with the view of collecting data from which ultimately conclusions may be drawn on these and other similar points. We are now, as it were, only at the first stage of the enquiry, only collecting data, and ought to content ourselves with recording facts, leaving their interpretation until such time as enough analyses have been made, and data collected, to give reasonable ground for supposing that conclusions grounded upon them will be fairly correct."

In 1885 some samples of tobacco grown at Nadiad, in the Kaira district of Bombay, were forwarded to London by the Bombay Government for valuation, and a few extracts are here given from the reports of various well-known tobacco manufacturers in Great Britain:—

"In determining the value of this article, we require to take into account the present value and supply of North American tobacco, as that we think would be its principal competitor for cutting or for roll. There is a plentiful supply of American for fillars, and the prices are low this season, and when such is the case other growths are at a discount. It is to be borne in mind, also, that the bulk of American is in the form of 'stripts' which is much more suitable for our purpose than leaf. To compete successfully, therefore, with that growth it would require to be fresh, clean, well cured, and dried, and to possess these two qualities in addition, it must smoke pleasantly; and take a good drink of water; in other words, it must possess the quality of absorbing a sufficient supply of moisture to overcome the cost and the duty, and yield a profit when sold at a low price two or three pence under the duty. We have tested the Indian growth on both these points, and find it deficient as compared with the American. There is another purpose for which it might suit, that is, as a fillar for cigars, but of that cigar manufacturers will be the best judges" (*Extract from the Report of Messrs. Thomson & Porteous, Edinburgh*).

"The stalk is too coarse and heavy for the narrow-shaped leaf, too much like Arrakan growth to find a market here, but might possibly do so on the Continent. The cause of this is the use of indigenous seed, and the only remedy the use of seed from Japan or Java, which carries a broad leaf on a thin stalk, and in good condition is worth from 5*d*. to 8*d*. per pound." (*Extract from Report of Messrs. Richard Lloyd and Sons, London*).

"The condition in which unmanufactured tobacco should be shipped from India is a very important point, particularly for tobacco suitable for spinning and cutting, and, unless the tobacco is carefully dried before packing in bale or hogshead, the results are very serious indeed. The shipper suffers in value and disappointment, and the tobacco has a stigma attached to it. We have seen what would have been good tobacco, had the drying been attended to, sold for export at 1*d*. per pound, quite unfit for the duty. The classification into size of leaf is very important, also colour and texture. There is another important matter, and that is the stemming, or withdrawal of the midrib after the tobacco is cured, before the process of

CULTIVATION in Bombay.

classification or after it. We think we are correct in stating that 75 per cent. of the importation from the United States is stemmed tobacco. Manufacturers who produce roll or cut prefer stemmed tobacco, unless when birds' eye is required, or the making of snuff. In curing tobacco for spinning or cutting there must be no fermentation; with tobacco for cigar-making it is somewhat different." "From the flavour and appearance of the sound leaves, we are of opinion that the tobacco had been somewhat fermented after curing. The sound leaves had very good colour for spinning, but quite unsuitable for cutting or cigars. The selection with regard to texture and size of leaf had not been attended to. Some of the leaves were suitable for cigars on account of texture." "Looking at the sample as a whole, we would describe it as 'nondescript.' The burning properties were very defective, the stem or midrib too fleshy or thick in comparison to size of leaf. We worked up the samples into roll, but the results were very unfavourable, the taste and flavour being very objectionable. In growing tobacco it is desirable that the planter or cultivator should have a specific object in view, *i.e.*, if his intention is to export his product, or to manufacture in India. Should he decide on exporting to the United Kingdom, it will be requisite for him to fix on producing tobacco suitable for either spinning, cutting, or the making of cigars. This arranged, select the most suitable seed, adapting the soil so as to produce a leaf with all the necessary qualifications, when the plant is matured, cutting and curing, or drying, after the American manner, if for spinning and cutting, and after the Cuban method for cigars. The importations of tobacco from the United States (for spinning and cutting) into the United Kingdom, and for cigars from Cuba, Mexico, United States of Columbia, Germany, and Holland, have to be feared by the Indian grower, he has to face all this competition. Still we think he will obtain a share of the business, provided he complies with the necessary conditions and is prepared to adapt himself to circumstances. The questions of proper drying before shipment, the classification of length and texture, and colour, freedom from dirt, sand, and fermented flavour, and the withdrawal of the stem after curing must all be carefully attended to in order to success." "We have since learned that a very large proportion of wrappers, or outside covers for cigars, are imported into the United Kingdom with the stem taken out or stemmed. To make tobacco-growing in India a success in ten years the Goverment should engage thoroughly practical Americans to instruct the cultivators, teach them how to grow, and how to cure, and see that their orders are strictly carried out; this applies to tobacco for spinning and cutting; the Cubans for tobacco for cigar-making otherwise the position will not be much better fifty years hence" (*Extract from the Report of Messrs. J. & F. Bell, Glasgow*).

"We believe that nothing can be done with the Indian tobacco in this country so long as the native seed is used; we should suggest that seed should be used and obtained from the 'Fostenlander' district of Java, or from British Honduras, which, in our judgment, are likely to produce good results in Indian soil. There is a large consumption of tobacco grown in the Dutch colony of Java, which might, we believe, be partly supplied from India, if the same seed be used, and the same care in cultivation be exercised" (*Extract from the Report of Messrs. Sales, Pollard & Co., London*).

Khándesh.—Tobacco was grown in Khándesh as early as 1660, and spread thence to Gujrát. Its area in 1888-89 was given as only 2,322 acres, and its export is now insignificant. So much has the local tobacco fallen off by careless tillage, that it is now almost set aside for the exotic Virginian and Shiraz hybrid. This was introduced as an experiment in

N. 123

CULTIVATION in Bombay.

1869-70 by **Mr. Fretwell,** Superintendent of the Model Farm. The two seeds were accidentally mixed together and the present crop is a cross between them. The Havanna seed was also tried, but was found too delicate for the climate and was given up. The local tobacco is considered very inferior to the mixed variety both in strength and flavour. Next to the alluvial soils, which are very limited, the grey soil on the sites of deserted villages is the best for tobacco. Failing this, black soil is chosen though light red is in some respects more suitable. In the grey soils of village sites very little manure is wanted. After more than one crop has been grown, an occasional dressing of old farm-yard manure is used.

Irrigation, though objected to by some, is, in **Mr. Fretwell's** opinion, especially in the dry east, necessary, not so much for the growth of the plant, as to bring the curing season before the middle of November when the air is still moist. Tobacco is generally grown in small plots of not more than one-eighth of an acre. The sowing season lasts from June to August, but is sometimes delayed till October. Twice during growth, the suckers are removed, but this is usually very carelessly done. In a native field, nearly all full-grown plants have suckers, rivalling the parent stem in luxuriance, and they flower both on the stems and suckers. The cultivators seldom show any signs of an attempt to limit the number of leaves. For this reason the leaves are not properly developed, and their strength and flavour never come to perfection. The cutting lasts from November to February, or about five and a half months from the time of sowing. At the time of cutting, the lower leaves are usually faded and yellow, the central ones in prime condition, and the upper ones unripe. Generally, the whole plant is cut and the flower-buds nipped, leaving a few inches of stem which again throws out fresh leaves. But these are worth very little and are seldom gathered. The Kunbis, from a feeling against destroying vegetable life, do not cut the plants themselves, but employ Bhils and others to do it for them. A few of the better husbandmen, especially among the Musalmans and those near the Government farm, pluck the leaves singly.

According to the common way of curing them, the plants, as they are cut, are laid in rows on the ground until the leaves lose their brittleness and become limp and flaccid. They are made into bundles, *erkás,* of four or five plants each and brought to some convenient place for drying, very often to the roof of the owner's house, and laid in close overlapping rows. When the colour of the leaves has begun to change, the rows are turned over, and this is done several times with many sprinklings of water till all are of nearly the same shade. At this stage, about twenty-five small bundles are made into large bundles, *judis,* tied together with a few fibres of the root of the *palas* (**Butea frondosa**) tree, sprinkled with water, stacked and covered with gunny cloth or *rosha* (**Andropogon Schœnathus**) grass, and loaded with heavy stones. To equalise the fermentation, every third day the bundles are turned, watered, and rebuilt. This water sprinkling, made necessary by the dryness of the climate, destroys the finer qualities of the tobacco. The process of curing is entirely performed in the open air and takes from five weeks to three months. The part near the stem is always mouldy, and the rest varies from the proper fawn colour to deep black. Much is absolutely rotten.

Blackened tobacco, though useless for any other purpose, is generally preferred by the natives. The present method of curing must continue, unless, by the help of irrigation, tobacco is sown in June and cut in November, and a drying house is made under ground and covered with thick thatch. Even with this care, the wind will probably be too strong to allow leaves to grow perfect enough to be made into cigars. The only improvement in curing, introduced on the Model Farm, is the cutting out of the

CULTIVATION in Bombay.

stem. The average acre cost of tillage (in 1880) varied from R30 to R35, and the outturn from about R144 to R250 (*Bombay Gaz.. XII., 165*).

In Nasik no well-to-do Kunbi will grow tobacco, because, as it is said, his aversion to destroy life will not permit him to nip off the flowering shoots. The cultivation is generally entrusted to Bhils or Kolis, who are paid half the produce for their labour. When the leaves are ripe, they are nipped off and three or four of them are laid one over the other in the sun to dry. They are turned from time to time, and after a fortnight sprinkled with water, sometimes mixed with the sap of mango-tree bark or the juice of a coarse grass called *surad*, and packed in underground pits, or, if the quantity be large, stacked closely in the open air for eight days. This heightens the colour of the leaf and improves its flavour (*Bombay Gaz., XVI., 101*).

Hove, in the account of his tour in Bombay (1787), page 113, thus describes the mode of cultivation and preparation of tobacco in Cambay at that date:—"The tobacco plant is of the same species as that cultivated in the West Indies. The soil is partly a light white clay, but the greatest part consists of a soft sand. The seed is sown under cover in baskets as soon as the first rain has fallen, from whence they transplant it about 3 feet from each other. Two months after these plants were set, the tops, except those which were intended for seeds, are cut off so as to give to the leaf a more vigorous growth. The leaves begin to turn yellow about October and are plucked off as they ripen and are suspended in shady places to dry. Those that are plucked at this time of the year (April) are the lateral shoots which are taken off as they arrive to a tolerable size, and are laid into barley straw where they acquire the yellow colour, when they are dried like the former. They besprinkle them with a kind of syrup not unlike treacle, which is extracted from the juice of the **Borassus** and **Corypha**, which gives them a particular flavour, and by the moistness they are enabled to spread them in small packages in their natural form, and press them lastly down between two boards by the help of stones, in which state they are exported to Surat and Bombay markets." "The price of tobacco is monopolised by the Company's brokers; therefore it is not well known to any persons but themselves. The Nabob is entitled to the half of every produce, but commonly leaves the poor farmer only the fourth part."

CULTIVATION in Sind. 124

IX. TOBACCO IN SIND.—"This plant is grown in the *kharif* season. The first crop is called *Néri*, the second is gathered about a month later, and is inferior; its name is *Bauti*, or *Bajara*. There are several varieties: The chief are, 1st, *Shikarpuri*, of which there are two kinds, *Talkh* (bitter) and *Mittho* (sweet); 2nd, *Sindi*, a dark and inferior article. All the Sind tobacco is sweated in cocks, and covered with mats, preventing the access of air. Hence its inferiority, when compared with American" (*Dr. Stocks*).

CULTIVATION in Central India. 125

X. CULTIVATION AND MANUFACTURE IN CENTRAL INDIA.—Dr. R. H. Irvine, late Residency Surgeon, Gwalior, gives the following description of the famous Bhilsa tobacco and the mode of cultivating it:—"The field has been celebrated for above one hundred years for producing this superior tobacco. It is situated eastward of, and adjoining to, the wall of the town of Bhilsa, and is completely enclosed by a wall of stones, containing an extent of six or seven bighas (say, about three acres). In this space, tobacco is alone cultivated, and no other crop. The ground is copiously manured, nearly the whole of the town manure being ploughed into the ground previous to planting out the tobacco plants. The manure is a compost of cow-dung, horse and sheep-dung, straw, fire-ashes, and general refuse, amongst which must not be overlooked the baked clay of broken earthen pots. This field alone produces the superior tobacco, though that plant is extensively cultivated of the ordinary quality round

CULTIVATION in Central India.

the town. The fine produce of this field is not to be bought. The soil is not alluvial, being far removed from any *nullah* or river. Besides the immense quantity of manure supplied, the plants are carefully weeded and thinned out, the strongest only left; the flower-buds are nipped off, saving the few allowed to remain for seed; and five or six leaves are the utmost extent allowed to remain on the plants. The seed is never changed. The soil appears to consist of débris of various kinds, and stiff loam, in nearly equal proportions. The débris is chiefly the remains of substances in the manure, small pieces of red and black earthenware and carbon; also of small pieces of *kankar* (calco-siliceous concrete, effervescing slightly with acids), and pieces of compact felspar and black schorl. The other half, or elutriated portion, is a stiff dark coloured loam." "The presence of the carbonate of potash is doubtless owing to the large quantity of wood-ashes contained in the manure applied. The chief, if not the only cause, of the goodness of the tobacco of this single field is the very careful and high cultivation applied." **Mr. O'Conor** says that in Central India, tobacco is grown on high, well-ploughed lands. The leaves in the process of curing are sprinkled with earth containing *reh* salts, and that they are also sometimes steeped in tobacco-water. A second crop is frequently taken from the stems left after the first crop has been gathered, and the yield is said to be abundant. In Malwa, *rusa* grass (**Andropogon Schœnanthus**) and *madar* leaves (**Calotropis gigantea**) are used in the fermentation of tobacco, and are supposed to improve its flavour. The grass contains an aromatic volatile oil, and the *madar* leaves an acrid principle, both of which are approved by native taste.

CULTIVATION in Central Provinces. 126

XI. CULTIVATION AND MANUFACTURE IN THE CENTRAL PROVINCES.—The area under tobacco in these Provinces (exclusive of the Feudatory States and Zamindaris) was stated to be 19,649 acres in 1888-89. The plant is cultivated in small patches in the vicinity of nearly all the villages, and the whole of the produce is consumed locally. The seasons for sowing and cutting vary in different districts. In Narsingpur the crop is harvested exceptionally early, *viz.*, in October. In other districts the stems are cut usually in February and March. The curing is done in a very rough and primitive fashion. The average yield in these Provinces is stated in **Mr. O'Conor's** report to be about 200lb to the acre, and the price from four to twenty pounds per rupee according to quality. The well-to-do classes mostly use imported tobacco.

CULTIVATION in Berar. 127

XII. CULTIVATION AND MANUFACTURE IN BERAR.—According to statistical returns tobacco occupied 15,760 acres in 1888-89. This shows a great falling off compared with the extent of its cultivation during the preceding year. The fluctuations observed in the returns for previous years are mainly to be accounted for by the varying character of the seasons. At present Amraoti is the largest tobacco-growing district. The quality of the tobacco is poor; and most of it is consumed locally. **Dr. Lyon** analysed a sample of tobacco from the Buldana district with the following results:—

Potash	1·73
Chloride of potassium	15·82
Oxide of iron and alumina	13·31
Lime	30·65
Magnesia	5·89
Carbonic acid	2·08
Sulphuric acid	3·68
Silica	26·84
Phosporic acid	traces.
	100·00
Percentage of ash	21·28

CULTIVATION in Berar.

Commenting on this analysis, **Dr. Lyon** says :—" The market-value of tobacco depends very much on delicacy of flavour and aroma, points not appreciable by chemical analysis, and on which the influence of the methods of curing or drying adopted is not yet made out. It is hence not only very difficult from an analysis alone to give an opinion as to the quality of a tobacco, but also even more difficult to suggest the proper remedial measures for inferiority of quality. I may, however, state that except in showing a slight deficiency of soluble extractive, the proximate analysis of this tobacco gives no indication of inferior quality. The analysis of the mineral constituents or ash shews, however, that the tobacco, like most Indian samples, contains less potash and more lime than the finer American tobacco." It would, therefore, appear that manures containing potash salts might be advantageously employed in the cultivation of this tobacco. The following information is taken from **O'Conor's** report :—" In Berar, as in the Central Provinces, the average return per acre is very poor indeed. In West Berar a rayat is satisfied if he gets 200℔ of cured leaf to the acre. In East Berar the yield is better, but still very poor. The highest recorded yield amounts to 500℔, but the average is little more than half this, *viz.*, 258℔. The yield might be greatly increased by liberal irrigation and more care generally. Prices vary greatly according to quality and other circumstances." " The average price is stated at from 3 to 5 seers per rupee, or from one anna and seven pie to two annas and eight pie per pound. There is hardly any competition in the trade in tobacco."

CULTIVATION in Mysore. 128

XIII. CULTIVATION AND MANUFACTURE IN MYSORE AND COORG.—The following information regarding tobacco operations in Mysore is taken from **Mr. O'Conor's** report, and from the Gazetteer. Tobacco is grown on land where *ragi* and other similar grains are cultivated. Arid sandy soil is preferred, though black soil is sometimes utilised. The seed is sown in June or July, *i.e.*, after the early rains have commenced. The seed-beds are prepared in the vicinity of wells or other sources of water supply. The seeds are sown mixed with dung, and after being pressed down smooth with the hand and watered, they are covered with mats or with the leaves of the date-palm. When the plants come up the coverings are removed. If the plants are wanted quickly more dung is given, and all weeds are removed. The seedlings are transplanted into fields previously prepared by frequent ploughings, and manured by cattle and sheep. The plants are placed in holes 18 inches apart and are filled in with a mixture of red earth, sand, and dung. As this is the rainy season artificial watering is not usually required. The plants are topped about a month and a half after having been set out, six or seven leaves only being allowed to remain on each stem, and the suckers are removed. The plants, which ripen between November and January, are cut down to within four or five inches of the ground; the stems are then split lengthwise, each portion carrying three or four leaves. These half stems are then strung upon a line passed through their lower ends, and for twenty days they are exposed to sun and air. Every third day they are turned, and should there be rain they must be covered with mats. The tobacco is then taken into a house and arranged in a heap. After being turned four or five times, three days intervening between each operation, the tobacco is considered fit for sale, and is made up by the merchants into bundles which include the stems. " The Battadpur tobacco has the reputation of being the best in this Province, the red soil on which it is grown is much superior to that round Bangalore, and the sample of the tobacco sent to **Mr. Broughton** was pronounced by him to be richer in potassic carbonate than many of the native specimens which he had tested. A plentiful supply of

CULTIVATION in Mysore.

the ashes of plants and of wood, stated by him to be required for manure, can also be easily procured from the jungles in the immediate neighbourhood" (*Letter from Officiating Commissioner, Nundydroog District, dated February 1874*). A tobacco produced in the Periyapetna taluk of the Ashtagram division is also said to be of very superior quality. The climate, however, of Mysore is on the whole considered to be unfavourable, owing to the dryness of the atmosphere during the critical time of curing.

The following interesting description of the cultivation and preparation of tobacco in the neighbourhood of Seringapatam is taken from **Buchanan's** *Journey through Mysore and Canara*, Vol III., 441 (1801) :—

"Tobacco is cultivated not only in gardens, but also in rice-land and dry fields. In the first and last cases, the cultivator pays the usual rent. When it is cultivated on rice-land, the State gets one-half of the produce. In the month preceding the summer solstice, the field is ploughed fourteen or fifteen times. In the month following, furrows at the distance of two cubits are drawn throughout the field, and are filled with water. In these, young tobacco plants from the seed-bed are placed, at nine inches distance, and a little dung is put at their roots. The young plants are then covered with broad leaves, and for four times are watered once a day. These leaves having been removed, the plants for three times get water once in four days; and even again on the twentieth day, should the rainy season not have then commenced. At the end of the month the whole field is hoed, and the earth is thrown toward the plants in ridges. At the end of the second month this is repeated, and at the same time all the leaves, except from six to nine, are pinched from every plant; and all new leaves, that afterwards shoot from the centre, are once in eight or ten days removed. When it begins to whiten, the tobacco is fit for cutting. After having been cut to the ground, the stems are allowed to lie on the field until next day, when they are spread on a dry place, and exposed to the sun. Here the tobacco remains nine days and nine nights. On the tenth morning some grass is spread on the ground; on this heaps of the tobacco are placed, and the roots are turned towards the circumference. The heap is covered with straw, and pressed down with a large stone. In these heaps the tobacco remains for nine days. The stems are then removed from the leaves, of which from six to ten, according to their size, are made up into a small bundle. These bundles are again placed in a heap, covered with straw, and pressed with a large stone. Every evening the heap is taken down; and, each bundle having been squeezed with the hand, to make it soft, the whole is again replaced as before. On the fifth evening the tobacco is spread out all night to receive the dew. Next day the heap is rebuilt, and this process of heaping, squeezing, and spreading out to the dew, must be in all performed three times; the tobacco is then fit for sale. The larger leaves of this tobacco seem to me to be well cured for the European market, being not so dry as usual with that cured in India, but moist and flexible. For three successive years, three crops of tobacco may be taken from the same field: but before a fourth crop some other article must intervene for at least one year; and after this plant, even in gardens, no second crop is admitted."

PREPARATIONS; Used by Natives. 129

NATIVE AND EUROPEAN PREPARATIONS OF TOBACCO.

Very little has been written on the subject of the various preparations of tobacco intended for consumption by the natives of India; the following information extracted from a few well known works will, however, give a general idea as to the composition of the tobacco mixtures in common use. **Grierson**, in his *Bihar Peasant Life*, thus describes the preparation of tobacco in that part of Bengal :—The tobacco is crushed by a beam acting as a lever and supported on pillars. At the lower portion of the beam is fixed a peg which, when the beam is lowered, crushes the tobacco placed in a hollow below. The fragments of tobacco are collected by a broom, and sprinkled with water; and afterwards made up into balls of different sizes. Tobacco intended for chewing is called *khaini, surti,* or *dokhta* in Patna); tobacco for smoking is known as *piáni* or *pini*; and snuff is called *nás* or *nas.* Tobacco which is spiced or scented is known as *khambira* or *khamwira ; sáda* is plain tobacco; and a mixture of the two is called *doras* or *dorusa.* The principal spices used in Bihar for mixing with tobacco are :—*jatamansi, chharila, sugand-wala,* and *sugand-kokila.*

PREPARATIONS: Used by Natives. Chewing-tobacco. 130

Mr. Baden Powell, in his *Panjáb Products,* says :—"Every kind of tobacco is either "*halka*" or "*phika,*" that is, weak or mild; and *kaura,* which is superior, strong, and pungent. Tobacco-buyers test the strength of tobacco by placing a bit on the tongue, and seeing whether it produces any irritation. Strong tobacco is heavier than mild." According to Dr. Dymock the *gúráku* which is used in the *hukah* is a mixture of tobacco and *gur* in equal proportions. This substance has the appearance of an extract; when required for use it is broken into fragments and packed in the *chilam,* and ignited by placing on the top a layer of burning cinders or of rice balls specially prepared for the purpose.* The following ingredients are mentioned by Mr. Baden Powell as being used in the preparation of the *khamira* smoked by the wealthier classes in the Panjáb :— Preserved apple, *gulkhand* (conserve of roses), *panri* (the dried clippings of *pán* leaves) *muskhbála* (a scented wood), sandalwood, *ilachi* (cardamoms), *khesra* (the arak or essence of the flowers of **Pandanus**), *kokanber* (wild jujubes), and occasionally the pulp of the pod of *amaltás,* (**Cassia Fistula**). A cheaper kind is made merely with sandalwood, wild *ber* fruit, and *gugal* gum (**Balsamodendron**), and sells as cheap as 7 seers per rupee. The real *khamira* is sold by the jarful, not by weight. The snuff used in the Panjáb is made of *noki* tobacco pounded; the best comes from Kabul, Kandahar, and Peshawar.

Smoking-tobacco. 131

Special tobaccos. 132

The tobacco used in the North-West Provinces for chewing and eating is all imported from Oudh, and is sold in the leaf at from ten to fifteen rupees a maund; it is bought in the raw state and prepared by the purchaser; that eaten with *pán* is a decoction prepared by boiling the leaves in water after carefully separating all the stalks and fibres and cleaning the tobacco from all impurities. After being twice boiled and allowed to stand for twelve hours, the water is drawn off, and, after being thickened with perfumes, is evaporated until it assumes the consistency of paste. The second, or inferior kind, which is eaten alone or with *pán,* is merely the dry leaf well saturated with water in which the stalks and fibres have been boiled. In the Moradabad district tobacco for smoking is usually prepared in three ways. The best sort consists of half tobacco and half molasses. With cheaper sorts a considerable quantity of *rák,* an impure carbonate of soda, is mixed, and the proportion of molasses is diminished. The flavour of inferior tobacco is said to be much improved by the addition of *rák.*

INDIAN CIGARS.—During the last twelve or fifteen years there has been a steady improvement in cigar manufacture, both as regards make and quality. This is more especially the case in regard to those prepared in the Madras Presidency. It must have taken a considerable time to acquire a taste for the old style of 'trichy'—a coarse, black, briny cheroot with a hollow reed inserted down the centre to enable it to draw properly. The present representative, with its carefully selected wrapper of thin glossy

Indian Cigars. 133

* In Bengal specially-prepared cakes of charcoal are used for this purpose. The trade in making these is very extensive.

PREPARATIONS.

Indian Cigars.

leaf, is so far superior as not only to have replaced the Manilla, which used to be the fashionable smoke in India, but is gaining, as statistics will show, a reputation in Europe. There are at present nine factories for the manufacture of cigars in Trichinopoly, and one in Malabar.

Dr. Shortt thus describes the native mode of manufacturing cigars in Southern India :—" The Indian cigar-maker, like all other Indian artisans, works with very few tools. To turn out the most approved cigars, he needs only a small board, a pair of scissors, and a smooth stone, weighing 5 or 6 pounds, to serve as a pounder. He first tears off the midrib, out of a parcel of leaves, which he sorts at the same time in two heaps. One of these contains the best leaves reserved for the covers, and those unfit for that purpose are put in the second heap, and are used to fill the interior of the cigars. Then a roll is made of the leaves intended for the covering. By this means the wrinkled leaves are stretched and made even. The roll is then beaten with the pounding stone to flatten the veins and further to stretch and stiffen the leaves, which, after this process, come out of the roll in bands as neat as fresh paper. The cigar-maker then takes up one of these bands and cuts out a crescent about 8 inches long and 2 inches broad at the centre : meanwhile his assistant takes up a few of the leaves, unserviceable for the covering, and gathers them into a rough cigar, which he covers with a coarse wrapper of tobacco. This is the core. The cigar-maker takes up one of these cores and by rolling it between the palm of his hand and the board, gives it a neat appearance. Much of the drawing qualities of the cigar depends on this stage of the process. If the core be very irregular, a little extra pressure will make it appear even, but the air passages will, at the same time, be choked. If the core be properly made, very little rolling is enough to give the cigar a neat appearance. When this is done, the outer covering called the robe is put on and the end is pasted. In cheroots both ends are then trimmed with the scissors, but in the Havanna-shaped cigars one end being conical does not require further manipulation. A cigar-maker and an assistant can roll from 500 to 1,000 cigars in a day of eight hours, and they are paid from 12 annas to R3-8 per thousand, according to the finish and the form of the cigars. Except for some qualities of Pondicherry cigars, no substance whatever is used in Southern India to improve the quality of tobacco. But for Pondicherry cigars the water contained in tender cocoanuts is said to form a valuable sauce, which is the technical name of the stuff used in preparing tobacco for cigar-making."

Dr. Shortt states that snuff is largely prepared at Masulipatam in the Kistna district, and is exported for sale all over India, chiefly to Calcutta, where it is readily bought up by Jewish, Armenian, and Arabian merchants for export to Persia and the Arabian Gulf. It is made of the best tobacco produced in the district; the mode of preparing it is as follows :—The leaves are cut into halves and the stem entirely removed. One half is dried in the sun and pounded into a coarse powder and the second half is boiled twice in salt water, and the juice extracted is used again in the place of salt water with a fresh supply of tobacco. When the juice becomes rather thick and gummy, it is poured into a big pot and left to cool. The tobacco powder is now put into large chatties and the juice, with which a little brandy has been mixed, is poured over the powder, and the preparation is allowed to stand for about a week. The snuff is then taken out and put into English wine bottles and it is ready for exportation.

"In both the upper and lower Carnatics, taking snuff is much more common than in Bengal: indeed, I have never been in a country where the custom was more prevalent. Smoking, on the contrary, is in great disrepute. The *hooka* is totally unknown, except among Mussulmans.

PREPARATIONS.

Indian Cigars.

The lower classes smoke cheroots, or tobacco rolled up in a leaf; but a Bráhman would lose caste by such a practice, and it is not considered as becoming, even among the richer part of the Súdra tribe" (*Buchanan, Journey through Mysore and Canara, I., 52*).

An interesting account of tobacco is given in a report published in connection with the Colonial and Indian Exhibition of 1886. The following extract regarding the manufacture of cigars and other preparation, of tobacco is here quoted:—

Colonial & other Foreign Cigars. 134

"In the Colonies cigars are made almost entirely by hand, and in England the same rule prevails in regard to the best qualities; but the poorer sorts are formed in moulds. The chief advantage of the better mode of manufacture is that comparatively little skill is required in forming the cigar into the necessary shape, and it can be done by young girls possessed of but little experience. The drawback is irregularity in the compression of the cigar, and consequently uncertainty in the burning These cigars are generally made of the more common leaf, *viz.*, German fillars with Japan or Java covers or wrappers, and they enter into competition rather with smoking or pipe tobacco than with foreign cigars. In the manufacture of the latter the West Indian and other Colonies possess the advantage of being the owners of good raw materials, and if the quality can be improved, or even maintained, there is every reason to expect a prosperous future. It is hardly creditable to Britons over the sea that they should be so largely dependent on Manilla and Sumatra for supplies of cigar wrappers, and that the best cigars in Britain should be of foreign origin. As soon as the cigars are formed they go to the sorter, by whom they are classified according to colour. They are then tied in bundles, forced into boxes, and set to dry in a warm room. The English cigar manufacturer has to lose the interest on the duty from the time the leaf is taken out of warehouse until the cigars go into consumption. He has also to go to the expense of grinding his stalks, which are practically worthless as an article of commerce, before he can obtain a return of the duty paid upon them. There are few articles of daily consumption so free from adulteration as the cigar. Attempts have been made to imitate the well known Havanna flavour by artificial means, but the result has not been very satisfactory.

Java Leaf. 135

Manilla. 136

Wrappers. 137

Havanas. 138

British Cigars. 139

"The manufacture of pipe tobacco for English consumption is, with the exception of a few fancy sorts, entirely in the hands of British manufacturers. Very little skill is required in this trade, but great ingenuity has been displayed in the various machinery for saving labour, both in cutting and spinning. It may be remarked that Australia and the Cape are both quite abreast of England in the adoption of machinery. In the United Kingdom the law does not permit the use of any ingredient except water and acetic acid in cut tobacco; and water, acetic acid, olive oil, and essential oil for flavouring, in roll and cavendish. A few years ago a strong demand set in in favour of highly flavoured tobacco, and smoking mixtures composed of flavoured cut-cavendish and ordinary cut tobaccos were largely used. This demand has now become considerably modified.

"Twenty or thirty years ago the working classes in this country for the most part smoked cut tobacco. The fashion may now be said to have changed in favour of roll; this is due largely to a somewhat unfortunate competition which set in a few years ago. For a considerable time the retail price had not varied and the public had become accustomed to it. When circumstances absolutely required an advance in price, certain traders endeavoured to meet the difficulty by causing the leaf to absorb more water. The competition then really became one as to how much water tobacco leaf could be made to absorb, and inasmuch as a rich leaf

NICOTIANA Tabacum.

Tobacco.

PREPARATIONS.

British Cigars.

was more liable to decompose in the presence of water than a poor one, the latter came largely into use In the meantıme a few large manufacturers continued to maintain their quality, and raised their prices, and to protect themselves, sent out their cut tobacco in packages bearing their own labels. The result was that this class of tobacco became a favourite amongst the middle and lower midde classes, and cut tobacco, apart from that sold in manufacturers' packages, has been largely superseded by roll. Formerly cut tobacco was confined almost exclusively to three classes, *viz.*: (1) Bird's eye, which was generally understood to represent a light leaf, and included the stem; (2) Returns generally prepared from the same leaf with the stem removed; and (3) Shag, generally prepared from a darker leaf, and subjected to a slight amount of sweating or fermentation. These three names are still continued in the trade, but manufacturers have for the most part adopted fancy ones. Public taste changes in the most unaccountable manner in regard to these brands; and when unequivocal success has been achieved for any one, it appears to be the chief aim of every other manufacturer to imitate as nearly as possible its name, price, and quality. A considerable demand prevails for flake tobacco, which only differs from ordinary cut in being pressed into a cake before being placed in the cutting-machine. This causes it to appear in uniform lengths and it is frequently packed in boxes and used for cigarettes.

Roll Tobacco. 140 Cavendish. 141

"The manufacture of roll and cavendish has undergone a complete change within the past twenty years. The old mode of manufacture consisted in placing the cake or roll in an open press at ordinary temperature for a month or two, when it assumed the well-known dark colour which distinguishes it. Now the roll or cake is subjected to heat in the press, and two or three days are sufficient for the process.

Pipe Tobacco. 142

"American leaf is for the most part used in the preparation of pipe tobacco, and it is remarkable that in every case it is the produce of a milder climate than that from which we obtain our cigar sorts. This would lead us to look to Australia and the Cape Colonies for the future supplies of this class of tobacco. At present the cultivation in these Colonies may be said to have hardly got beyond the experimental stage, but there are not wanting many hopeful signs in regard to the future. The consumption of cigarettes as yet gives no indication of abatement. Originally confined to Turkish and light Virginian, they may now be said to embrace nearly every description of tobacco; and such is the ingenuity displayed in the machinery by which they are manufactured, that about two hundred and fifty cigarettes per minute can be turned out completely finished by one machine.

Snuff. 143 Snuff boxes. 144

"The demand for *SNUFF in Europe has declined of late years, and this has no doubt discouraged the working out of new ideas. The stems or midrib of smoking or pipe tobacco-leaf are utilised by the snuff manufacturers, but the stems of cigar leaf are regarded as almost worthless for their purposes" (*Watt & McCarthy*). [Certain races of the inhabitants of India use snuff but to no great extent. Their snuff boxes are often very interesting. See the account under **Entada scandens**, Vol. III., 246.—*Ed.*]

Pipes. 145

PIPES AND OTHER KINDS OF TOBACCO HOLDERS.—The oldest known pipes, *viz.*, those discovered in the tombs of the Aztecs in Mexico, and in mounds in various parts of the United States, were made of clay or stone, often elaborately carved into the form of animals; and they had a single stem. As already mentioned, the first pipes seen in use were those curious bifurcated tubes belonging to the inhabitants of the West Indian islands, and through which the smoke was drawn into the nostrils. The rolls or bunches of burning herbs which are said to have been first seen

* Conf. with remark at p. 423 under Export Trade.

by Columbus and his followers in the island of San Salvador were practically cigars, the wrappers of which were constructed from the spathes of Indian corn. The practice of using a wrapper other than that of tobacco leaf is common in many other countries at the present day. In India, both in the plains and on the hills, the leaves of various trees are constantly used for this purpose. In Western India these wrappers, known by the name of *bidis*, are the leaves of **Bauhinia racemosa.** This tree is worshipped by Hindús during the Dasera feast in October, and its leaves are distributed amongst friends and relatives. The Santals twist up the leaves of *Sál* (**Shorea robusta**) into tobacco holders. The ordinary cigar or cheroot of Europeans, with its wrapper or leaf specially selected as an outer covering, is obviously an outcome of the primitive method of using a covering other than that of tobacco. Regarding the Santal wrappers, which are often shaped like pipes, Dr. Watt remarks:—"It might be assumed that the substitution of a bamboo joint or reed, in place of the wrapper that was consumed, and which flavoured the tobacco, would be a very natural and simple improvement. To affix to a bamboo joint a reed mouthpiece is to produce the form of pipe which is generally claimed as of European origin—the clay pipe. Bamboo pipes with reed mouthpieces are in use, as a matter of fact, among many of the hill tribes of India, Africa, and other parts of the world. It is also just possible that the suggestion to use a pipe emanated from the East, being borrowed from the ancient practice of smoking *ganja*, or the narcotic derived from hemp. Tobacco pipes closely resembling the *ganja* and opium pipes are in common use in many parts of India. Clay pipes are not made in India of the bamboo joint pattern; but pipes in that shape are made both in stone and brass by the inhabitants of the higher mountains, and these pipes are very similar to the metal pipes used by the Laplanders." In Spiti, according to Captain Harcourt, the men smoke from a long steel pipe which every zamindar carries in his waistcloth. A very primitive form of smoking, and one which is even now practised by some of the aboriginal tribes of India, as by Hindú inhabitants of the higher Himálayan ranges, is to make two connecting holes on the face of a sloping bank, into one is inserted the ignited tobacco, and from the other the smoke is drawn into the mouth. The awkward and uncomfortable position of the smoker naturally led to an improvement effected by the insertion of a bamboo pipe into the smoke hole.

PREPARATIONS.

Pipes.

Cigarette wrappers. 146

The ordinary pipe used in India is the *huka*, the primitive form of which is the *naryel*, a hollow cocoanut shell half filled with water. The tube, called *bota* or *gatta*, supporting the *chilam* or tobacco-holder, is passed through a hole at the top of the cocoanut until the end dips below the surface of the water; and the smoke is inhaled through a pipe inserted into another opening near the top of the shell. If a pipe is not handy the mouth is then applied directly over the opening. This form of *huka* is that which is used chiefly by Hindús. The Muhammadan pipe, or hubble-bubble, although the same in principle, offers a greater variety of forms, some of which are elaborately and elegantly fashioned. The water-vessel, if of the cocoanut shape, rests in a stand, or is shaped, usually in the form of a bell, so that it can be made to stand on the ground. These pipes, of which there are many varieties, differ chiefly as to mode of attachment, and shape of a smoking stem or *necha*. In some the *necha* issues from the top, in others from the side of the bowl. They may be straight, curved, bent at an angle, or jointed; or they may be in the form of a long flexible tube; these latter are called *pechwán* or *penchdár*, and are said to have been invented in Akbar's time. The flexible tube consists of a long coil of wire covered with cloth and more or less ornamented.

Chilam. 147

Huka. 148

Necha. 149

Pechwan. 150

NICOTIANA Tabacum.

Tobacco.

The *necha* or drawing tube is usually a reed or a piece of *sarkanda* (**Saccharum ciliare**) stem covered often with silk and gold thread.

TRADE. 151

TRADE STATISTICS.

In order to give some general idea as to the share which India takes in the tobacco trade of the world, the following extract from a report prepared by **Dr. G. Watt & Mr. McCarthy** (Colonial and Indian Exhibition Reports, 1886) is here given:—

"It is difficult to obtain trustworthy information regarding the world's trade in tobacco. The French Statistical Society estimates the annual consumption at 4,480 million pounds, and it is probable that the area under tobacco is not far short of six million acres. In the United States alone there are annually close upon 600,000 acres, and in India 817,000 acres. Viewing the trade of the United Kingdom alone, the following tabular statement of the consumption of tobacco and revenue therefrom is instructive as showing the magnitude of the trade and the steady way it has developed:—

Year.	Consumption.	Revenue.
	℔	£
1790	10,700,316	696,804
1800	19,490,199	1,388,193
1810	21,133,083	2,093,495
1820	16,519,515	2,904,251
1830	18,899,137	2,828,968
1840	22,971,406	3,431,908
1850	27,538,104	3,337,258
1860	35,106,641	5,529,400
1870	40,485,253	6,433,147
1880	48,191,555	8,433,538

"Last year (1885-86) the imports of tobacco were 83,227,297℔, and the consumption 53,055.984℔, and the exports were 8,084,574℔, while the total amount in bond was declared to be 93,767,377℔. The value of the tobacco imported last year into Great Britain was £3,900,599, and the revenue obtained from tobacco amounted to £8,975,140. Thus it would appear that in ninety-five years the consumption of tobacco in Great Britain and Ireland has increased from 10,700,316℔ to 53,055,984℔, and the revenue derived from tobacco by Government during the same period has increased from £696,804 to £8,084,574. It is noteworthy that of the imports only 4,307,789℔ were returned as manufactured tobaccos and snuff, and of this only 1,730,924℔ are stated to have been consumed. At the same time the United Kingdom exported 1,368,780℔ of manufactured tobaccos. These figures, when considered alongside of the quotation given as the total of the imports of the manufactured tobacco, convey some idea of the magnitude of the British or home trade in manufacturing or preparing tobacco for the consumer. The Havanna, Manilla, and other foreign-made cigars, must be viewed as but luxuries compared with the infinitely larger trade in British manufactured tobacco and cigars.

"Speaking of the cigar trade, our Colonies and India have to compete mainly with the home-made, and not with the finer qualities of foreign cigars. This fact has caused some authorities to advance the idea that the present system of imposing one uniform duty upon all manufactured tobaccos, whatever may be their value, is unjustly severe upon all our Colonial and Indian manufacturers. It is admitted that they cannot compete with the Havanna and Manilla cigars, and the low duty upon unmanufactured tobaccos, as compared with manufactured, allows the British manufacturer

TRADE.

to produce a cheaper article at home than can be sent from the Colonies and India to England. It is, however, believed that many of the Colonial and Indian tobaccos and cigars have properties of their own that would commend them to a large number of consumers, were it possible that they could be sold in England at a price that would enable them to compete with the British manufacturer.

"Since the 5th of April 1878, the duty on manufactured tobaccos has been 3*s.* 6*d.* a pound, provided it contains 10 per cent. moisture, and 3*s.* 10*d.* if less than that amount. Thus, while the duty is not lessened when the tobacco contains more than 10 per cent. moisture, it is raised when it contains less. The duty on cigars has, since the 4th of April 1879, been fixed at 5*s.* 6*d.* a pound, regardless of their intrinsic value, or the amount of moisture they contain. The question of moisture is one of the very greatest importance, for it may not inaptly be said that the profits made in the tobacco trade at the present day are entirely due to water. We are indebted to **Mr. J. Chambers** (of **Messrs. Grant, Chambers & Co.**) for much valuable information regarding the tobacco trade, and may here quote a passage from a private letter of his on the subject of profit: 'About 75 per cent. of the tobacco cut for the pipe is sold in the shape of 'Shag' at 3*d.* an ounce retail; the manufacturer sells this at 3*s.* to 3*s.* 2*d.* a pound, so that taking the average cost of tobacco at 6*d.*, and the duty at 3*s.* 6*d.*, the article is sold at considerably below cost price.' This is due to the fact that the article comes into the hands of the manufacturer in a dry state, and the profit is due to the increased weight, on the addition of moisture, in the process of manufacture. A tobacco, to use the technical term, must drink well to be profitable."

TRADE IN INDIAN TOBACCOS. 152

Trade in Indian Tobacco.*

[In dealing with this subject the following may be given as the more important sections:—1st, Exports; 2nd, Imports; 3rd, Internal Trade, *i.e.*, the movement from Province to Province, whether by sea or by road, rail, and canal; and 4th, Trans-frontier Trade. Each of these sections will be found briefly reviewed in the remarks below; the returns for each fourth year during the past twelve being more fully analysed, in order to show the growth or decline of trade in each class of Indian tobacco.

Export. 153

1st.—Export Trade in Indian Tobacco. One of the earliest references to the export of tobacco from India occurs in **Milburn's** *Oriental Commerce.* No mention is made of tobacco in the edition of 1813, but in the later issue (1825) it is stated that tobacco of a very superior quality was cultivated in the vicinity of Masulipatam, and that snuff was occasionally sent from India to England as presents. It would thus appear that Indian tobacco was practically a curiosity at the beginning of the century, but fifty years later it had assumed a distinct position in the foreign markets. Thus, in the first Annual Statement of the Trade and Navigation of British India, published in 1866-67, the Government of India show that the exports of raw tobacco were valued at R5,61,836; of that amount Aden received R3,37,743; Mauritius R94,517; the Straits Settlements R52,266, and the United Kingdom R7,088, smaller amounts going to France, America, &c. No mention is made of cigars being exported from India, though a large amount was annually imported. Of the exporting Provinces, Bombay in 1866-67 headed the list with R3,62,721, followed by Madras R1,15,079, and by Bengal with R76,914. Ten years later the exports were valued at R8,91,398. The following tabular statement exhibits the exports of Indian tobacco during the past

* The review of the Indian Tobacco Trade here given has been furnished by the Editor.

NICOTIANA Tabacum.

Tobacco.

TRADE in Indian Tobaccos.

Export.

twelve years under the headings Cigars and Other Sorts for Manufactured Tobacco, Unmanufactured Tobacco being exhibited separately:—

Exports of Indian Tobacco to Foreign Countries.

Years.	Manufactured.				Unmanufactured.	
	Cigars.		Other Sorts.			
	℔	R	℔	R	℔	R
1876-77 . .	190,136	1,17,445	205,033	22,578	10,508,720	7,51,375
1877-78 . .	189,742	1,43,946	317,887	38,750	10,594,604	7,47,675
1878-79 . .	196,759	1,21,786	247,743	30,176	13,279,158	11,11,260
1879-80 . .	130,324	96,633	407,148	33,439	10,874,623	11,67,025
1880-81 . .	207,005	1,59,953	198,811	26,504	13,267,325	12,21,853
1881-82 . .	223,470	1,48,136	515,463	46,581	9,791,392	9,55,659
1882-83 . .	220,019	1,40,788	228,228	25,287	10,653,549	9,86,358
1883-84 . .	251,926	1,62,618	256,465	25,324	18,577,276	14,88,773
1884-85 . .	238,109	1,69,049	166,328	21,064	15,620,864	13,09,789
1885-86 . .	230,924	1,58,892	205,456	23,735	10,752,397	10,65,835
1886-87 . .	273,209	2,11,391	193,996	27,036	9,868,834	9,57,156
1887-88 . .	296,829	2,42,344	254,819	29,415	10,888,807	10,22,760
1888-89 . .	370,075	3,47,913	175,269	21,917	6,799,880	6,67,668

A review of the returns for each fourth year from 1876-77 reveals the fact that not only have the exports increased during the past twelve years—94·6 per cent.—but that the consumption of Indian cigars in Great Britain is the chief cause of the increase; thus in 1876-77 out of a total valued at R1,17,445, the United Kingdom received only R36,224 worth, whilst the Straits Settlements took R46,044. Another striking fact of the year under reference may be mentioned, *viz.*, that of these exports the share taken by Madras was R59,168, British Burma contributing R40,836. Four years later (1880-81) out of a total of R1,59,953 worth, the United Kingdom took R54,948, the Straits Settlements R64,720. The Madras manufacture of cigars is, however, seen to have made a start, since of the total exports that Presidency contributed R79,940 worth, British Burma remaining stationary. In 1884-85, of the total exports—R1,69,049—the United Kingdom took R69,444 worth, the Straits R52,490. The Madras share of these exports was practically stationary, that of British Burma amounting in the year under notice to R36,892. In the last of the series of years selected above, the Madras exports and the British consumption are shown to have greatly increased. Of the total exports (R3,47,913), the United Kingdom took R1,62,414, and the Straits Settlements R86,965, but of these exports Madras furnished R2,20,405, Burma R58,359; the other Provinces, *viz.*, Bengal, Bombay, Sind, in that year, as in all previous years, contributed much smaller amounts.

The tobacco returned as "Other Sorts" will be observed, from the table given, to be much inferior in quality to the cigars. It consists chiefly of native preparations of tobacco which find their largest markets in the Maldives, the Straits Settlements, Ceylon, and Arabia, the chief exporting Province being Bengal; for example, out of the total exports of "other sorts" of tobacco shown in the table, Bengal contributed the following amounts, in 1876-77, R20,972 worth; in 1880-81, R20,071; in 1884-85, R16,019, and in 1888-89, R19,415.

Reviewing the trade in Unmanufactured Tobacco for the same periods as above, it is seen that in 1876-77, out of a total valued at R7,51,375, the United Kingdom took R2,53,613, Austria, R1,22,474, Mauritius, R1,20,888,

TRADE in Indian Tobaccos.

Export.

and Aden, R36,236. The shares contributed by each Province in these export were, from Bengal R6,09,058, Bombay R57,165, Madras R79,781. In 1880-81 out of the total valued at R12,21,853 the United Kingdom took R2,08,701, Mauritius, R1,23,775, and Aden, R5,83,589. The immense increase in the Aden trade is noteworthy, as it accounts for the high exports from Bombay; thus of the exports Bengal contributed R4,31,340, Bombay R6,53,187, and Madras R1,23,707. In 1884-85 the Bombay-Aden trade continued to expand, while the amounts consigned to the United Kingdom and the exports from Bengal showed indications of declining. In 1888-89 out of the total exports (R6,67,678), the United Kingdom does not appear to have received any, while Aden took R3,38,038, Straits Settlements, R1,61,255, and Mauritius, R90,243. The shares contributed by the Provinces were, Bengal R94,980, Bombay R3,47,273, Madras R2,04,747. From the facts given above it will thus be seen that the most striking feature of the Indian export trade in tobacco is the progression made by the Madras Presidency, not only in producing cigars, which are meeting the Indian market and checking the import of foreign cigars, but year by year in these finding a larger market in Europe.

Import. 154

2nd.—Import Trade.—In the Annual Statement of the Indian Trade for 1866-67 the total value of the tobacco imported into India is shown to have been R10,13,093, which gave an import duty of R91,667. The analysis of that traffic shows that America contributed R2,30,040, the Straits Settlements R2,60,284, Ceylon R1,72,180, other Asiatic Countries R1,62,315, while the receiving Provinces were in order of importance as follows:—Bengal R5,97,604, Bombay R1,92,287, Madras R1,87,674. It will be thus seen that twenty years ago the Madras cultivation of tobacco was not sufficient to meet the consumption in the Presidency, and that, like the other Provinces, it received large quantities from foreign countries.

At the same time while the value of the total import traffic in tobacco has steadily increased, the share taken by cigars has declined, and that too almost coincidentallywith the growth of the export trade in Indian-made cigars, already dealt with. The following table gives the figures of the Indian imports of tobacco during the past twelve years:—

YEARS.	MANUFACTURED.				UNMANUFACTURED.	
	Cigars.		Other Sorts.			
	lb	R	lb	R	lb	R
1876-77 . .	118,761	2,89,941	1,015,100	5,84,312	813,853	82,627
1877-78 . .	115,982	3,26,410	757,343	4,24,103	636,562	72,329
1878-79 . .	94,658	2,35,499	1,049,921	5,20,427	290,655	37,163
1879-80 . .	88,082	2,73,388	621,243	3,53,562	277,109	40,117
1880-81 . .	65,782	1,95,358	897,797	4,80,548	537,552	54,581
1881-82 . .	78,140	2,32,536	550,834	3,13,553	780,716	85,683
1882-83 . .	92,286	2,69,327	924,439	5,14,826	467,708	51,934
1883-84 . .	109,206	3,50,384	536,854	3,37,405	517,985	72,776
1884-85 . .	97,610	2,67,636	764,544	4,88,511	522,201	85,934
1885-86 . .	74,248	2,62,372	704,375	4,81,744	519,920	77,821
1886-87 . .	62,400	2,41,141	1,087,579	7,88,224	604,436	75,454
1887-88 . .	60,949	2,12,377	804,876	6,88,142	700,909	1,19,202
1888-89 . .	61,008	1,94,944	1,257,934	10,88,064	481,693	1,12,565

In the early trade returns no distinction was made into cigars and other manufactured tobaccos; this first appeared in the year 1870-71,

TRADE in Indian Tobaccos.

from which date the highest recorded imports occurred in 1883-84, *viz.*, R3,50,384, but in two other years, *viz.*, 1875-76 and 1877-78, the imports exceeded 3 lakhs. Allowing for fluctuations, such as those indicated, the imports seem to have steadily declined from 118,761℔ to 61,008℔.

Import.

One of the most noteworthy features of the Indian import trade is the prominence of the Straits and China in the supply to India of Manilla cigars. During the year 1876-77 this traffic amounted to R2,46,929, or 85·1 per cent. of the total value, and 88·2 per cent. of the total weight. Four years later (1880-81), these countries contributed cigars valued at R1,25,949, or 64·5 per cent. of the total value and 71·6 per cent. of the total weight of cigars imported. In 1884-85 the imports of these Manillas were valued at R90,690 or 33·9 of the total value and 60·5 of the total weight of cigars imported. In 1888-89 a similar result is noticeable; the Straits supplied cigars to the value of only R44,103 or 22·6 per cent. of the total value and 36·6 per cent. of the total weight of imported Manilla cigars from the Straits and China. Thus it will be observed that a marked decline took place in these Manillas during the period under notice, a decline which was much more pronounced in the value than in the weight of the article furnished. In fact, many writers hold that the degeneration in quality of these Manilla cigars proved one of the most powerful factors in stimulating the manufacture of Indian cigars. It may be pointed out also that, while the total weight of all cigars imported into India declined 50 per cent. during the period from 1876 to 1889, the value of these cigars was considerably enhanced, a fact which would seem to point to the growth of our import trade in high class cigars coincident with the decline already shown of the cheaper Manillas from the Straits and China. The most noticeable countries which appear to have grown into importance in the supply of these more costly cigars seem to be the United Kingdom, Malta, Egypt, and the United States.

Of the tobaccos returned in trade statistics as "Other Sorts," a prominent feature is the increase in the supply from the United States, which, during the twelve years under notice, expanded from 143,377℔, valued at R84,683 to 284,285℔, valued at R1,67,309. The tobacco, however, which is drawn from the United Kingdom is by far the most important. During the twelve years from 1876-1889 it has fluctuated between 400,000 and 800,000℔, the major portion being received by Bengal.

The supply in unmanufactured tobacco seems if anything to be declining. In 1876-77 India imported 813,853℔, valued at R82,627, all but 10,000℔ being obtained from Ceylon and imported by Madras. In 1880-81 the imports were 537,552℔, valued at R54,581, of which all but 30,000℔ were received by Madras from Ceylon. In 1884-85 522,201℔, valued R85,934, were imported, of which Ceylon furnished 431,508℔, Madras taking a little more than that amount of the total. In 1888-89 the imports were 481,693℔, valued at R1,12,565, of which Ceylon contributed 385,504℔, Madras taking, as before, a little more than that amount. It may be added that during the past ten years, the United Kingdom, Persia, Aden, and Turkey in Asia have steadily increased their supplies in unmanufactured tobacco.

The re-exports of foreign tobacco, since the year 1866-67, seem to have considerably improved. The transactions in foreign cigars, for example, amounted on an average to R8,000; of other sorts of tobacco to about R6,000.

Internal. 155

3rd.—Internal Trade.—Coastwise.—The total imports and exports of tobacco carried from port to port along the coast of India average in value from 45 to 55 lakhs of rupees. Of unmanufactured tobacco Burma is the chief receiving country, taking from Bengal and Madras, in nearly equal proportions about two-thirds of the tobacco carried interprovincially by sea. Of

TRADE in Indian Tobaccos. Internal.

the traffic in cigars Madras naturally takes the lead as an exporting Province; it sent to other Provinces, in 1888-89, R3,08,451, out of a total of R3,69,437 worth. Of that amount Bengal received cigars to the value of R2,14,679, and Burma R1,24,609, the other Provinces having taken considerably smaller amounts. The traffic in tobacco of other sorts is of much less importance; it was valued in 1888-89 at R85,012. The chief receiving Provinces were Burma, Bombay, and Bengal; the first mentioned took very nearly half the amount exported, and drew its supplies chiefly from Bengal.

Rail & River. 156

By Rail and River.—The total value of the exports in unmanufactured tobacco from Province to Province by land routes was returned in 1888-89 as R1,36,18,047, of which Bombay exported 52 lakhs, and Bengal 56 lakhs. Of these exports of unmanufactured tobacco from the Provinces Calcutta (Port Town) received R33,20,141, Bombay R13,58,536, and Madras R4,40,932 worth. Of the Provinces which do not appear to meet their own demands, the North-West Provinces stand first, and received in 1888-89 R11,33,907, mainly from Bengal. The Central Provinces come next with R7,86,529, drawn in equal proportions from Bengal and Bombay. But the Native States appear to produce considerably less tobacco than the Provinces of British India, for the returns of rail and river traffic show Rájputana and Central India to have received last year tobacco to the value of R36,71,163, the bulk of which was obtained from Bombay; thus these Native States may be said to constitute the chief market for the surplus Bombay production. The rail and river traffic in cigars is comparatively unimportant. The total value of the amounts carried were returned at R6,60,519, of which Madras furnished of that amount to other Provinces R4,14,594 worth.

Trans-frontier. 157

4th.—Trans-frontier.—The imports across the frontier of India are not very important and are drawn mainly from Nepál. In 1888-89 they were valued at R4,97,243, of which Nepál supplied R3,99,375 worth. The exports from British India across the frontier are slightly larger than the imports; they were for the past three years as follows:—

	Cwt.	R
1887-88	63,803	6,16,286
1888-89	59,038	5,72,669
1889-90	61,572	5,47,669

Of these exports it is somewhat curious that Nepál takes from British India about the same amount as it has already been stated to give, *viz.*, about four lakhs of rupees worth. Kashmír is next in importance, receiving on an average about R50,000 worth of Indian tobacco.

In concluding this brief review of the Indian trade in tobacco it may be remarked that the figures analysed by no means convey an absolute conception of the total trade. For example, the exports from the Provinces to the Port Towns doubtless to a large extent appear again as exports coastwise from Port Town to Port Town, or to Foreign Countries. An error would thus be involved by an attempt at totalising all the returns, if the actual production and consumption were the object in view. But as financial transactions they are probably quite independent, and indeed any error involved by totalizing them would be more than compensated for by the local consumption which is of necessity nowhere accounted for in trade

NIGELLA sativa. **Black Cummin.**

TRADE in Indian Tobaccos. returns. The total value of tobacco transactions in India may thus be said to have been in 1888-89 as follows:—

		R
Foreign	Exports	10,39,498
	Imports	13,95,573
Coastwise	Exports	43,37,896
	Imports	54,22,946
Carried by Rail and River		1,42,78,566
Trans-frontier	Exports	5,72,669
	Imports	4,97,243
	GRAND TOTAL	R2,75,44,391

The most serious error in the above statement of Indian transactions in tobacco lies in the coastwise returns, the exports from one Province being to a large extent the imports by another; but allowing for this it is probably safe to say that the annual value of the tobacco sales and of the local consumption in India is not far short of £25,000,000. It need only be added here that the whole trade in tobacco is entirely free of duty or other fiscal restriction.—*Ed.*]

NIGELLA, *Linn.; Gen. Pl., I., 8.*

158 **Nigella sativa,** *Linn.; Boiss., Fl. Or., I., 68;* RANUNCULACEÆ.

SMALL FENNEL, OR BLACK CUMMIN.

Syn.—N. INDICA, *Roxb.*

Vern.—*Kalónji, kálájíra,* HIND.; *Mugrela, kála-jira, kál-zira,* BENG.; *Túkm-i-gandna,* KASHMIR; *Shewa dáru, siyah-dáru* (Stewart), AFG.; *Kalonji, kálenjire,* BOMB.; *Karun-shirogam,* TAM.; *Nalla-jilakra,* TEL.; *Karijirigi, kare-jirage, karimsiragam,* KAN.; *Karun-chirakam,* MALAY.; *Samon-né,* BURM.; *Kaluduru,* SING.; *Krishna-jiraka, kárave, sushave,* SANS.; *Sh-ouniz, kamúne-asvad, habbatoussouda,* ARAB.; *Siyáh-dánah, siyah-biranj,* PERS.

Care must be taken to distinguish between the names *kálá-zirá* and *káli-ziri,* the latter being applied to the seeds of **Vernonia anthelmintica. Moodeen Sheriff** also points out that the literal meaning of the Arabic and Persian names *Habbatoussouda,* and *Siyá-dánah* in Hindustani is *Kála-dánah* (black seeds), which is the name in the latter language for the seeds of **Ipomœa (Pharbitis) Nil.**

References.—*Roxb., Fl. Ind., Ed. C.B.C., 450; Voigt, Hort. Sub. Cal., 4; Stewart, Pb. Pl., 4; Elliot, Fl. Andhr., 125; Pharm. Ind., 5; Ainslie, Mat. Ind., I., 128; O'Shaughnessy, Beng. Dispens., 164; Moodeen Sheriff, Mat. Med. of Madras, 5; Supp. Pharm. Ind., 183; U. C. Dutt, Mat. Med. Hind., 102; Dymock, Mat. Med. W. Ind., 2nd Ed., 15; Pharmacogr. Ind., I., 28; Fleming, Med. Pl. and Drugs, as in As. Res., Vol. XI., 173; U. S. Dispens., 15th Ed., 1707; S. Arjun, Bomb. Drugs, 4; Med. Top. Ajmir, 140, 145; Year-Book Pharm., 1873, 22; 1880, 224; Irvine, Mat. Med. Patna, 121; Baden Powell, Pb. Pr., 323; Atkinson, Econ. Prod., N.-W. P., V., 27; Him. Dist., 384; Birdwood, Bomb. Pr., 3, 217, 275; Royle, Ill. Him. Bot., 55; Cooke, Oils and Oilseeds, 63; Balfour, Cyclop. II., 1096; Kew Off. Guide to the Mus. of Ec. Bot., 8; Home Dept. Cor., 223.*

Habitat.—A native of Southern Europe. Extensively cultivated in many parts of India for its seeds. **Roxburgh** believed it to be indigenous in India. Its Sanskrit names indicate, however, its introduction at a very early period. It is supposed to be the "Black Cummin" of Scripture, the "μελανθιον" of **Hippocrates** and **Dioscorides,** and the "*Gith*" of **Pliny**; and by some authors this, or an allied species, is considered to be the "*Fitches*" mentioned in Isaiah (xxviii., 25, 27).

OIL. Seeds. 159 **Oil.**—The SEEDS contain two kinds of oil, the one dark-coloured, fragrant, and volatile, the other clear, nearly colourless, and of about the consistence of castor-oil. The former, according to **Dr. Dymock**, has a fine blue fluorescence. OIL.

Medicine.—The SEEDS are regarded as aromatic, carminative, stomachic, and digestive, and are used in indigestion, loss of appetite, fever, diarrhœa, &c. They are also said to increase the secretion of milk (*U. C. Dutt*). The Hakeems describe it as heating, attenuant, suppurative, detergent, and diuretic. They give it, too, as a stimulant in a variety of disorders which are ascribed to cold humours, and credit it with anthelmintic properties (*Pharmacogr. Indica*). **M. Canolle** observed that after doses of 10 to 40 grains of the powdered seed the temperature of the body is raised, the pulse accelerated, and all the secretions stimulated, especially those of the kidneys and skin; in doses of 10 to 20 grains they possess a well-marked emmenagogue action in dysmenorrhœa, and in larger doses cause abortion (*De l'avortement criminel à Karikal. Thèse de Paris, 1881*). MEDICINE. Seeds. 160

SPECIAL OPINIONS.—§ "Nigella seeds have been long used in medicine. As a tonic condiment they are praised by **Hippocrates.** In India they are chiefly used as aromatic adjuncts to purgative or bitter remedies. They are given to nurses mixed with curry. The officinal preparations are the tincture and powder. The former is an useful warm stimulant often added to other remedies" (*Notes added from O'Shaughnessy's Bengal Dispensatory and Bengal Pharmacopœia*). "Carminative and lactagogue in doses of ʒi in fever" (*Assistant Surgeon Nehal Sing, Saharanpur*). "The seeds fried, bruised, and tied in muslin rag, and smelt, relieve cold and catarrh of the nose by constant inhalation" (*E. W. Savinge*). "Carminative and stomachic. Used also to preserve woollen cloths from insects" (*Assistant Surgeon S. C. Bhattacherji, Chanda, Central Provinces*). "It is a powerful diuretic in complicated cases of general dropsy. The seeds are antibilious and when given internally arrest vomiting" (*Surgeon-Major D. R. Thomson, M.D., C.I.E., Madras*). "Seeds are used as condiment and to protect woollen clothes from being worm-eaten; they are mixed with powdered camphor and put within their folds" (*Surgeon A. C. Mukerji, Noakhally*). "Has been employed locally combined with other ingredients in cocoanut-oil in cases of eczema. ptyriasis capitis, &c., and found useful. Its action is mostly, I fancy, a stimulant, and probably a destroyer of pediculæ. The preparation for local application is as follows:— **Nigella sativa** bruised, two ounces; Psoralea corylifolia bruised, two ounces; Bdellium bruised, two ounces; Coscini radix, two ounces; Sulphur (Nillika gendagum), one ounce; Oleum cocoa, two bottles. All to be put in a bottle and placed in the sun for one week. The bottle to be well shaken occasionally" (*J. G. Ashworth, Apothecary, Kumbakonam, Madras*).

Food.—The SEEDS, which have a strong, pungent, aromatic taste, are much used by natives in curries and other dishes. They are also frequently sprinkled over the surface of bread along with sesamum seed. "French cooks employ the seeds of this plant under the name of *Quatre épices* or *toute épices*, and they were formerly used as a substitute for pepper" (*Treasury of Botany*). FOOD. Seeds. 161

Domestic Uses.—It appears to be a common practice in this country to scatter the SEEDS of this plant between the folds of linen or woollen cloths in order to prevent their being worm-eaten. DOMESTIC. Seeds. 162

[COMPOSITÆ.

Niger-seed oil, see **Guizotia abyssynica,** *Cass.;* Vol. IV., 186;

[SOLANACEÆ.

Nightshade, Deadly, see **Atropa Belladonna,** *Linn.;* Vol. I., 351;

NIPA, *Webb; Gen. Pl., I., 484.*

163 **Nipa fruticans**, *Wurmb.*; PALMÆ.

Vern.—*Gúlgá, gabna, golbhal* (fruits), BENG.; *Da-ne*, BURM.; *Poothadah*, AND.; *Gim-pól*, SING.

References.—*Roxb., Fl. Ind., Ed. C.B.C., 677; Voigt, Hort. Sub. Cal., 684; Kurz, For. Fl. Burm., II., 541; Gamble, Man. Timb., 425; Le Maout & Decaisne, Descrip. & Analyt Bot. (Eng. Ed.), 822; Spons' Encyclop., 985; Balfour, Cyclop., II., 1101; Smith, Dic., 287; Treasury of Bot., 790; Kew Off. Guide to Bot. Gardens and Arboretum, 57; Ind. For., I., 222; Mason, Burma, 524.*

FIBRE. Leaves. 164 FOOD. Fruit. 165 Spirit. 166 DOMESTIC. Fronds. 167 Seeds. 168 Foliage. 169

Habitat.—A soboliferous palm of the river estuaries and tidal forests of the Sundarbans, Chittagong, Burma, and the Andamans. The nuts of this, or of an allied species, are abundant in the tertiary formations of the island of Sheppey, at the mouth of the Thames.

Fibre.—The LEAVES are used for thatching houses, also for matting.

Food.—The inside of the large FRUIT is, when young, edible; a TODDY is obtained from the spathe (*Gamble*). This palm was known to the Dutch traveller, **Linschoten**, who, nearly 300 years ago, in his "Account of a Voyage to the East Indies," mentioned that it yielded an excellent wine.

Domestic Use.—Hats and cigar cases are made from the FRONDS. The SEEDS might be used for vegetable ivory (*Kurz*). The FOLIAGE when burnt yields a supply of salt which is used for various purposes.

NOTHOPEGIA, *Blume; Gen. Pl., I., 425.*

170 **Nothopegia Colebrookiana**, *Blume; Fl. Br. Ind., II., 40; Wight, Ic., t. 236*; ANACARDIACEÆ.

Syn.—GLYCOCARPUS RACEMOSUS, *Dalz.*

Vern.—*Amberi*, KAN.; *Bála*, SING.

References.—*Beddome, Fl. Sylv., t., 164; Thwaites, En. Ceylon Pl., 441; Dalz. & Gibs., Bomb. Fl., 51 (excl. syn).*

Habitat.—A small tree of Central and Southern India, extending to Ceylon.

FOOD. Fruit. 171

Food.—The sweet oily pulp of the FRUIT is eaten.

NOTHOSÆRUA, *Wight; Gen. Pl., III., 34.*

172 **Nothosærua brachiata**, *Wight; Fl. Br. Ind., IV., 726; Wight, Ic., t. 1776*; AMARANTACEÆ.

Syn.—PSEUDANTHUS BRACHIATUS, *Wight*; ACHYRANTHES BRACHIATA, *Linn.*; ÆRUA BRACHIATA, *Mart.*

Vern.—*Dhaula phindawri*, RAJ.

References.—*Dalz. & Gibs., Bomb. Fl., 217; Ind. For., XII.. App. 20.*

Habitat.—An annual, with minute woolly flowers, found in most parts of India; also in Burma and Ceylon.

FOOD. Plant. 173

Food.—The PLANT is used in Merwára as a vegetable.

NOTONIA, *DC.; Gen. Pl., II., 446.*

174 **Notonia grandiflora**, *DC.; Fl. Br. Ind., III., 337; Wight, Ic., t. 484*; COMPOSITÆ.

Syn.—CACALIA GRANDIFLORA, *Wall.*; C. KLEINIA, *Herb., Madr.*; N. CORYMBOSA, *DC.*

Vern.—*Wánder-roti* (Monkey's bread), MAR.

References.—*Voigt, Hort. Sub. Cal., 422; Thwaites, En. Ceylon Pl., 168; Dalz. & Gibs., Bomb. Fl., 132; Pharm. Ind., 126; O'Shaughnessy, Beng. Dispens., 46; Dymock, Mat. Med. W. Ind., 2nd Ed., 463; Lisboa, U. Pl. Bomb., 274; Balfour, Cyclop., II., 1108.*

Habitat.—A small shrub with fleshy leaves found in the hilly districts of the Deccan and Western Ghâts; also in Ceylon.

Medicine.—"This plant was brought forward in 1860, by **Dr. A. Gibson**, as a preventive of hydrophobia. The mode of administration is as follows:—About four ounces of the freshly-gathered STEMS, infused in a pint of cold water for a night, yield in the morning, when subjected to pressure, a quantity of viscid greenish JUICE, which, being mixed with the water, is taken at a draught. In the evenings a further quantity of the juice, made up into boluses with flour, is taken" (*Pharm. Ind.*). Experiments were tried at the European Hospital in Bombay in 1864, by **Dr. Dymock**, who says:—"In one-drachm doses it had a feebly aperient action; no other effect was observed. The dried plant was for a time issued to medical officers in Government employ, but no further information as to its properties would appear to have been obtained."

MEDICINE. Stems. 175

Juice. 176

SPECIAL OPINION.—§ "Further experiments with this plant have shown that it has hardly any physiological action upon healthy animals or men beyond a mild aperient effect in large doses" (*W. Dymock, Bombay*).

Nutmegs, see **Myristica fragrans,** *Houtt.*, p. 311; MYRISTICEÆ.

NUTS. 177

The word NUT, used in the popular or commercial sense, does not coincide with its botanical definition any more than in the case of the word FRUIT. In the botanical sense the term nut is applied to a particular class of indehiscent fruits which may or may not be edible, or indeed be of any economic value; whilst popularly it is rather loosely applied to various nut-like fruits, seeds, or tubers, usually possessing more or less value either for food, or ornament. In this latter sense the following list has been compiled of plants which yield the more important kinds of nuts used for economic purposes. In drawing up this enumeration the writer has practically only amplified the list given by **Dr. Watt** in the Colonial and Indian Exhibition Catalogue. Nuts used merely as ornaments are, however, omitted, as they have already been mentioned under the article "**Beads**" in Vol. I., 426. The non-indigenous species are marked with an *.

USEFUL NUTS. 178

1. **Æsculus indica,** *Colebr.;* Vol. I., 126; SAPINDACEÆ.
THE HIMALAYAN HORSE-CHESTNUT.

2. * **Aleurites moluccana,** *Willd.;* Vol. I., 163; EUPHORBIACEÆ.
BELGAUM WALNUT OR CANDLE-NUT.
A handsome tree introduced from the Malay Archipelago. The edible nut, which tastes like a walnut, contains a large quantity of useful oil.

3. * **Anacardium occidentale,** *Linn.;* Vol. I., 232; ANACARDIACEÆ.
THE CASHEW-NUT.
A tree indigenous in the West Indies, and now naturalised in many of the coast forests of Chittagong, Burma, and the Andaman Islands.

4. * **Arachis hypogœa,** *Linn.;* Vol. I., 282; LEGUMINOSÆ.
GROUND-NUT, EARTH-NUT, OR PEA-NUT.
An annual clover-like plant, originally of South America, and now extensively cultivated in Southern India and in most warm countries. The edible portion is the seed or pea, two or three of which are enclosed in a pod which ripens under ground. The seeds are universally offered for sale by fruit-sellers in India, and are roasted and eaten as nuts.

5. * **Areca Catechu,** *Linn.;* Vol. I., 291; PALMÆ.
THE BETEL-NUT.

USEFUL NUTS.

This palm is a native of Cochin China, and of the Malay Peninsula and Islands. Cultivated throughout tropical India. The nut is universally chewed in India as an ingredient of *pán*.

6. Artocarpus integrifolia, *Linn.;* Vol. I., 330; URTICACEÆ.

JACK FRUIT.

The nuts (true fruits in the strictly botanical sense) are roasted and eaten.

7. * Borassus flabelliformis, *Linn.;* Vol. I., 495; PALMÆ.

PALMYRA PALM.

A native of Africa, which has now become naturalised in, and is largely cultivated, all through the tropical parts of India, and beyond the tropics in Bengal and the southern portion of the North-West Provinces. The soft albuminous seeds of the unripe fruits, known as *tálsans,* are much eaten by natives.

8. Buchanania latifolia, *Roxb.;* Vol. I., 544; ANACARDIACEÆ.

CHIRAULI-NUT.

A tree of the Sub-Himálayan tract, and found generally throughout the hotter parts of India. The kernel is commonly eaten as a substitute for almonds.

9. Cæsalpinia Bonducella, *Fleming;* Vol. II., 3; LEGUMINOSÆ.

THE FEVER-NUT, PHYSIC-NUT, or NICKAR.

A shrub found all over India and throughout the tropics generally. The nuts, or seeds, powdered, are used medicinally as a powerful antiperiodic and tonic. They are also made into bracelets, necklaces, and rosaries.

10. * Castanea vulgaris, *Lam.;* Vol. II., 227; CUPULIFERÆ.

THE SWEET OR SPANISH CHEST-NUT.

Has been cultivated with partial success at various places on the Himálaya. A variety with smaller fruit was established in Dehra Dun by the late Dr. Jameson from seed forwarded from North China by Mr. Fortune.

11. * Cocos nucifera, *Linn.;* Vol. II., 415; PALMÆ.

COCOA-NUT.

This valuable tree, supposed by DeCandolle to have originated in the Indian Archipelago, is now extensively cultivated along the sea coasts of Tropical India and Burma. The edible portion is the tough albuminous substance which lines the hard shell or endocarp. For information concerning the various products of the cocoa-nut, and the uses to which they are applied, see the article "**Cocos.**" Vol. II., 515.

12. * Cola acuminata, *R. Br.;* Vol. II., 500; STERCULIACEÆ.

THE COLA-NUT.

A tree of Western Tropical Africa. The nuts or seeds have lately attracted attention as being likely to be used as a substitute for cocoa and chocolate. Cola or *goora* nuts are held in high estimation amongst the African negroes on account of properties similar to those possessed by the Coca leaf of South America.

13. Corylus Colurna, *Linn.;* Vol. II., 575; CUPULIFERÆ.

THE INDIAN HAZEL-NUT.

A moderate-sized tree of the North-West Himálaya. The nut is smaller than the European hazel-nut, but nearly as good in point of flavour. The true hazel-nut is occasionally met with in the bazárs of India, and is probably mported by Kabulis.

of the useful Nuts. (*J. F. Duthie.*) NUTS.

USEFUL NUTS.

14. **Cucumis,** *spp.;* Vol. II., 632; CUCURBITACEÆ.
The seeds of various forms of melon are roasted and eaten as nuts.

15. **Cyperus esculentus,** *Linn.;* Vol. II., 684; CYPERACEÆ.
The underground tubers, called *Kaseru* in the North-West Provinces, are used as food, as also are those of **C. jeminicus.** The tubers of other species are employed in perfumery, dyeing, and in medicine.

16. **Euryale ferox,** *Salisb.;* Vol. III., 303; NYMPHÆACEÆ.
THE GORGON-NUT.
The black seeds are roasted in hot sand and eaten as nuts.

17. **Jatropha Curcas,** *Linn.;* Vol. IV., 545; EUPHORBIACEÆ.
THE PHYSIC or PURGING NUT.

18. **Juglans regia,** *Linn.;* Vol. IV., 549; JUGLANDEÆ.
THE WALNUT.
The tree is wild and cultivated on the Himálaya. The nuts form an important article of trade over the greater part of India, a large proportion of the supply being brought down to the plains annually by Kabuli and other traders.

19. **Nelumbium speciosum,** *Willd.;* Vol. V., 345; NYMPHÆACEÆ.
The hard nut-like seeds of the Sacred or Egyptian Lotus are eaten either boiled or roasted; and are also sometimes strung as beads.

20. **Pinus Gerardiana,** *Wall.;* Vol. VI., Pt. I.; CONIFERÆ.
The seeds of the Neosa or Edible pine are collected and stored for winter use. They form an important article of food where the tree exists, and are brought into India in large quantities by Afghán traders.

21. * **Pistacia vera,** *Linn.;* Vol. VI., Pt. I.; ANACARDIACEÆ.
THE PISTACHIO NUT.
This nut also is imported into India in large quantities by Afghán fruit-sellers, the main portion of the supply coming probably from Persia where large orchards of this tree are maintained.

22. **Prunus Amygdalus,** *Baillon;* Vol. VI., Pt. I.; ROSACEÆ.
THE ALMOND.
Large supplies of almonds are annually brought into India from beyond the North-West frontier along with pistachio nuts.

23. **Semecarpus Anacardium,** *Linn.;* Vol. VI., Pt. I.; ANACARDIACEÆ.
THE MARKING-NUT.
The tree is abundant in Northern and Central India. The nut produced on the apex of the edible hypocarp, or enlarged fruit-stalk, is eaten. The pericarp, or outer shell of the nut, yields a black resinous juice, which, when mixed with lime, constitutes a good marking ink for cotton.

24. **Strychnos Nux-vomica,** *Linn.;* Vol. VI., Pt. II.; LOGANIACEÆ.
VOMIT-NUT.
A tree of Central and Southern India. The seeds or nuts yield a valuable medicine, which in overdoses acts as a powerful poison.

25. **S. potatorum,** *Linn. f.;* Vol. VI., Pt. II.
CLEARING-NUT.
The seed, which contains no poisonous matter, is used for clearing muddy water. It is also employed in native medicine as an emetic.

26. **Terminalia Catappa,** *Linn.;* Vol. VI., Pt. II.; COMBRETACEÆ.
THE COUNTRY ALMOND.
The coiled-up embryo contained in the nut is much eaten by natives. The kernels of **T. belerica** are also eaten, but are said to be intoxicating.

USEFUL NUTS.

27. * **Theobroma Cacao,** *Linn.;* Vol. VI., Pt. II; STERCULIACEÆ.
THE CACAO OR CHOCOLATE BEAN.
A tree of Tropical America. Cultivated in Southern India and Ceylon.

28. **Trapa bispinosa,** *Roxb.;* Vol. VI., Pt. II.; ONAGRACEÆ.
SINGARA-NUT.
A common aquatic plant found floating on tanks and lakes. The starchy nut is universally eaten, and in many parts of the country forms an important article of food.

LOGANIACEÆ.

Nux vomica, see **Strychnos Nux-vomica,** *Linn.;* Vol. VI., Pt. II;

NYCTANTHES, *Linn.; Gen. Pl., II., 675.*

[OLEACEÆ.

179 **Nyctanthes Arbor-tristis,** *Linn.; Fl. Br. Ind., III., 603;*

Syn.—SCABRITA SCABRA, *Linn.;* S. TRIFLORA, *Linn., Mant.*

Vern.—*Hár, siháru, harsinghár, saherwa, seoli, nibari,* HIND.; *Singhár, harsinghár, sephalika,* BENG.; *Saparung, kokra,* KOL.; *Saparom,* SANTAL; *Samsihar,* KHAWAR; *Gongo, seoli, gang-siuli,* URIYA; *Kher-sári,* GOND.; *Pakura, laduri, harsingar, kuri, sháili,* PB.; *Karassi,* BHIL; *Kirsahár,* BAIGAS; *Siralu,* (Hoshangabad), *Shiralli,* (Nimár) C. P.; *Harásingara, párijátaka,* BOMB.; *Khurasli,* MAR.; *Jayaparvati,* GUJ.; *Manja-pu,* TAM.; *Paghada, karchia, káru-chiya, pári-játamu, párujátamu,* TEL.; *Harsing,* KAN.; *Tsaybeeloo, hseik-ba-lu,* BURM.; *Párijatáka, sephalica, rájanikasa,* SANS.

References.—*Roxb., Fl. Ind., Ed. C.B.C., 29; Voigt, Hort. Sub. Cal., 552; Brandis, For. Fl., 314; Kurz, For. Fl. Burm., II., 155; Beddome, Fl. Sylv., t. 240; Gamble, Man. Timb., 254; Thwaites, En. Ceylon Pl., 190; Grah., Cat. Bomb. Pl., 111; Stewart, Pb. Pl., 141; Jones, As. Res., IV., 244; Elliot, Fl. Andhr., 86, 145; Mason, Burma, 410; O'Shaughnessy, Beng. Dispens., 57; U. C. Dutt, Mat. Med. Hind., 189; Dymock, Mat. Med. W. Ind., 2nd Ed., 473; S. Arjun, Bomb. Drugs, 207; Murray, Pl. and Drugs, Sind, 169; Year Book Pharm., 1880, 248; Irvine, Mat. Med. Patna, 48; Med. Top. Ajm., 143; Taylor, Top. Dacca, 57: Baden Powell, Pb. Pr., 359; Atkinson, Him. Dist., 743, 778; Lisboa, U. Pl. Bomb., 97, 223, 247, 290; Birdwood, Bomb. Pr., 300; Royle, Ill. Him. Bot., 268; McCann, Dyes and Tans, Beng., 41; Buck, Dyes and Tans, N.-W. P., 25; Liotard, Dyes, 26, 58, 59, 92; Piesse, Perfumery, 63, 64; Balfour, Cyclop., II., 1116; Treasury of Bot., 796; Kew Off. Guide to the Mus. of Ec. Bot., 94; For. Ad. Rep. Ch.-Nagpur, 1885, 32; Buchanan, Statistics of Dinajpore, 151; Linschoten, Voyage to E. Indies (Eng. Ed. 1885), II., 58-62, 68.*

Habitat.—A large shrub, with rough leaves and sweet-scented flowers, occurring in the Sub-Himálayan and Tarai tracts; also in Central India, Burma, and Ceylon. Common in gardens. The flowers open towards evening, and fall to the ground on the following morning.

DYE & TAN. Bark. 180 **Dye and Tan.**—The BARK is said to be employed for tanning.

Tubes. 181 The corolla TUBES are orange-coloured, and when severed from the limbs give a beautiful but fleeting orange or golden dye, which is mostly used for silk, sometimes in combination with turmeric. Half a seer of the dried corolla tubes will dye 60 yards of silk cloth. **Mr. Wardle** says:—"These

Flowers. 182 FLOWERS are fairly rich in yellow colouring matter, which is readily soluble in water. The infusion produces very bright yellow colours on mulberry silk when properly mordanted; and when the right processes are used this dye-stuff appears to be eminently suited for dyeing tussur silk yellow. It has not such a great affinity for wool." The author of the *Makhzan-ul-Adwiya* states that the white portion of the flowers yields a purple dye known in India as *Gulkama.* He says that directions for its preparation will be found in *Karabadien-i-kabir* (*Dr. Dymock*).

PREPARATION AND PROCESS OF DYEING.—The flowers, which fall in the early morning, are usually collected by women and children. The white limb, or upper portion of the corolla, is picked off, and the yellow tubes which remain are put out to dry in the sun, and afterwards stored for future use. The colour is extracted by steeping the dye-stuff in boiling water, and straining through a cloth. The fabric to be dyed, generally silk, is first drenched with water and then steeped in the liquid, either when hot, or after it has cooled. It is dried in the shade. The result is a bright orange colour, the shade of which can be deepened by further steeping. The colour is very fleeting, and no proper means of fixing it appear to be known. In Murshidabad, however, a little lime juice is added to the liquid, and this seems to render the colour less transitory. In other parts of Bengal alum is employed for the same purpose. It is said to be sometimes used in combination with *kusum* or *turmeric*, occasionally with *indigo* and *kath* (*McCann, Dyes and Tans of Bengal*). **DYE & TAN. Preparation. 183**

Buchanan (*Statistics of Dinajpore*) wrote that the inner BARK affords a red dye when beaten with a little lime; and that the flowers produce a beautiful, though perishable, purple dye. This latter colour is, as stated above, according to **Dr. Dymock**, mentioned by the author of the *Makhzan* as the product of the white portion of the flowers, and is called *gulkamah*. **Bark. 184**

Oil.—The fragrant FLOWERS of this plant contain an essential oil similar to that of jasmine. **OIL. Flowers. 185**

Medicine.—"The LEAVES are useful in fever and rheumatism. The fresh JUICE of the leaves is given with honey in chronic fever. A decoction of the leaves, prepared over a gentle fire, is recommended by several writers as a specific for obstinate sciatica" (*U. C. Dutt*). "Six or seven of the young leaves are rubbed up with water and a little fresh ginger, and administered in obstinate fevers of the intermittent type. The powdered SEEDS are used to cure scurfy affections of the scalp" (*Dymock*). It is antibilious and expectorant, and useful in bilious fevers (*Rai Kani Lal Dé, Bahadur*). **MEDICINE. Leaves. 186 Juice. 187 Seeds. 188**

SPECIAL OPINIONS.—§ "Juice of leaves commonly used in fevers" (*Assistant Surgeon S. C. Bhattacharji, Chanda, Central Provinces*). "Expressed juice of the leaves used as anthelmintic mixed with common salt. The use is very general" (*Surgeon A. C. Mukerji, Noakhalli*). "The expressed juice of the leaves acts as a cholagogue, laxative, and mild bitter tonic" (*Civil Surgeon J. H. Thornton, B.A., M.B., Monghyr*). "Used in fever in children as a diaphoretic and diuretic ℥ii in infusion" (*Assistant Surgeon Nehal Sing, Saharunpur*). "The fresh leaves are eaten in vegetable curries and *dal* as a mild tonic" (*Surgeon-Major B. Gupta, M.B., Pooree*).

Structure of the Wood.—Brown, close-grained, useful as fuel. **TIMBER. 189**

Sacred and Domestic Uses.—The FLOWERS are much used in Hindu worship, and as votive offerings. The plant is supposed to have been brought from heaven by the God Krishna for his wife Satyábháma. A tale is given in the *Vishnu Purána* to explain the origin of the name *Párijatáka*. "This shrub was a King's daughter, named Párijatáka. She fell in love with the sun, who soon deserted her, on which she killed herself, and was burnt. This shrub arose from her ashes. Hence it casts its flowers in the morning, as it cannot bear the sight of the sun" (*See* editor's note of *Eng. Ed.* (*1885*) *of Linschoten's Voyage to E Ind., Vol II., p. 62*). The name *párijátaka*, or *párijátamu*, is applied to a variety of plants with jasmine scent, *e.g.*, **Mimusops Elengi**, &c. The Coral-tree (**Eyrthrina indica**), and **Cochlospermum Gossypium** have also had the same names applied to them (*See Elliot, Fl. Andhr. l.c.*). **SACRED & DOMESTIC. Flowers. 190**

The rough LEAVES are sometimes used for polishing wood. **Leaves. 191**

NYMPHÆA, *Linn.; Gen. Pl., I., 46.*

192 **Nymphæa alba,** *Linn.; Fl. Br. Ind., I., 114;* NYMPHÆACEÆ.

THE WHITE WATER-LILY.

Vern.—*Brimposh, nilofár, kamúd,* KASHMIR.

According to **Sir George Birdwood**, this plant is the γυμΦαία of **Dioscorides**, σίδη of **Theophrastus**, and the *Nymphæ* of **Pliny.**

References.—*Stewart, Pb. Pl., 8; U. S. Dispens., 15th Ed., 711; Year-Book Pharm., 1874, 622; Birdwood, Bomb. Pr., 135; Smith, Dic., 435; Treasury of Bot., 797.*

Habitat.—Found in Kashmír Lake, altitude 5,300 feet. It is common in Europe, in ponds and streams, and fresh-water lakes.

DYE. Root-stocks. 193 **Dye.**—The ROOT-STOCKS are said to be used for dyeing purposes, as they contain gallic acid (*Treasury of Botany*).

MEDICINE. Root-stock. 194 **Medicine.**—Its mucilaginous and somewhat acrid ROOT-STOCK is administered in some countries for dysentery. According to **O'Shaughnessy** it is astringent and slightly narcotic. Flowers. 195 Its FLOWERS are reputed to be anti-aphrodisiac. Fruit. 196 An infusion of the flower and FRUIT is given in diarrhœa and as a diaphoretic (*Stewart*).

FOOD. Root. 197 **Food.**—The ROOT and Seeds. 198 SEEDS are eaten in times of scarcity.

Root-stock. 199 The ROOT-STOCKS contain a large quantity of starch, and are used in France in the preparation of a kind of beer.

200 **N. Lotus,** *Linn.; Fl. Br. Ind., I., 114.*

Syn.—N. RUBRA, *Roxb.*; N. EDULIS, *DC.*; N. ESCULENTA, *Roxb.*; N. PUBESCENS, *Willd.* **Sir George Birdwood**, in his *Bombay Products.* gives the following further identifications as probably correct :—N. LOTUS =LOTUS ÆGYPTIA *of* **Pliny**, *Bk. XIII., Chap. 17;* and of PROSPER ALPINUS, *de lant Ægypt, 2, page 49;* the N. SEU NEUPHAR ÆGYPTUM of **Vesling**; and CASTALIA MYSTICA of **Salisbury**; N. RUBRA *Roxb.*= CASTALIA MAGNIFICA of **Salisbury.** N. PUBESCENS, *Willd.*=CASTALIA SACRA, *Salisb.*; N. LOTUS, *Burm.*; N. INDICA MINOR, *Rumph.*; and AMBEL of **Rheede.**

Vern.—*Kanval, chota kanval,* HIND.; *Shάluk, sáluk, nal, koi* (parched seeds), *rakta kambal* (red var.), *chota sundi* (**N. edulis**), BENG.; *Dhabala-kain, rangkain,* ORISSA; *Kuni, puni, lorhi* (root), *napo* (seeds), SIND; *Alli-phul,* DEC.; *Kanval, nilophal,* GUJ.; *Alli-támarai, ambul,* TAM.; *Alli-támara, tella-kalava, koteka, erra kaluva* (red var.), *kalhárarmu,* TEL.; *Nyadale-huvu,* KAN.; *Ampala,* MALAY.; *Kyah-phyú kiya-ni,* BURM.; *Olu,* SING.; *Kamala, kumuda, kalháramu, hallaka, sandhyaka,* SANS.; *Nilufar,* ARAB.; *Nilufar,* PERS.

References.—*Roxb., Fl. Ind., Ed. C.B.C., 428; Voigt, Hort. Sub. Cal., 8; Thwaites, En. Ceylon Pl., 14; Dalz. & Gibs., Bomb. Fl., 6; Aitchison, Cat. Pb. and Sind Pl., 3; W. & A., Prod., 17; Drury, Handbook of Ind. Fl., 30s; Rheede, Hort. Mal., XI., t. 26; Elliot, Fl. Andhr., 52, 79, 103; Moodeen Sheriff, Supp. Pharm. Ind., 183; U. C. Dutt, Mat. Med. Hind., 109; S. Arjun, Bomb. Drugs, 7; Murray, Pl. and Drugs, Sind, 70; Irvine, Mat. Med. Patna, 119; Med. Top. Ajm., 143; Med. Top. Dacca, 58; Buchanan, Stat. Dinajpur, 169; Baden Powell, Pb. Pr., 329; Atkinson, Him. Dist., 304; Econ. Prod. N.-W. Prov., Part V., 91, 92; Drury, U. Pl. 314, 315; Lisboa, U. Pl. Bomb., 143; Birdwood, Bomb. Pr., 134; Balfour, Cyclop., II., 1116; Smith, Dic., 251-253; Treasury of Bot., 798; Kew Off. Guide to Bot. Gardens and Arboretum, 24; Bomb. Gaz VI., 15; Jour. As. Soc., Pt. II., No. II., 1867-82; Ind. For., III., 237; As. Res., IV., 285; Trans. Linn. Soc., XX., 29; Gazetteers:—Mysore and Coorg, I., 57; Gujrát Dist., 12; Mason, Burma, 739.*

Habitat.—A large aquatic herb, common throughout the warmer parts of India. Flowers white, pink, or crimson. A variety (**N. pubescens**) with pubescent leaves and smaller flowers also occurs.

Medicine.—The FLOWERS are used as a dry and cold astringent in diarrhœa, cholera, fever, and diseases of the liver, and are also recommended as a cardiac tonic. The powdered ROOT is prescribed for piles as a demulcent, also for dysentery and dyspepsia. The SEEDS form a cooling medicine for cutaneous diseases and leprosy, and are considered an antidote for poisons. **Taylor**, in his *Topography of Dacca*, mentions that "The flowers and STALKS of this species are reduced to powder, which is administered in cases of discharge of blood from the stomach and bowels."

MEDICINE. Flowers. 201 Root. 202 Seeds. 203 Stalks. 204

Food.—The ROOTS, FLOWERING STEMS, the young FRUIT, and the SEEDS are eaten in various parts of India The roots, which contain a large quantity of starch, are usually boiled, though sometimes eaten raw; the stems are cooked in curries; the unripe fruit is eaten as a vegetable; and the seeds are parched. **Taylor** says that the bulbous root, known by the name of *Shalúk*, and the seeds and stalks by that of *Sampala*, are sold in the bazárs as articles of diet, and the parched seeds are sold as *Koi* (*Topography of Dacca*)."

FOOD. Roots. 205 Flowering stems. 206 Fruit. 207 Seeds. 208

Dr. Dymock, in a recently published paper entitled "The Flowers of the Hindu Poets," gives the following account of the mythology of the Indian Lotus:—

"The queen of Indian flowers is the Lotus; the Hindús compare the newly-created world to a Lotus flower floating upon the waters, and it thus becomes symbolical of spontaneous generation. It is the *Padmamani* of the Buddhists. The golden lotus of Brahminic and Buddhistic mythology is the sun, which floats in the waters which are above the firmament, like an earthly lotus in the deep blue stream below; from it distils the Amrita, the first manifestation of Vishnu. Brahma and Buddha (the supreme intelligence) were born of this heavenly lotus. Lakshmi, the wife of Vishnu, the Indian Venus, is called Padmavati, because she is represented sitting on this flower. The Hindus see in the form of the lotus the mysterious symbol Svastika, which signifies the Bán and Sháluka combined. In Goa the common name for the plant is *Savak*, which I suppose to be a corruption of *Svastika*. The allusions to this flower by Indian poets are innumerable; no praise is too extravagant for it; it is the chaste flower, and its various synonyms are bestowed as names upon women. In the Koka Shastra the best class of women is called Padmini. The lotus is supposed to calm the pangs of love, but in the case of Sagarika in the Ratnavali, so violent was her malady that even this remedy was applied in vain. 'Take it all away' she says to Susamyata, 'I shall never obtain the object of my desire.' The red lotus is said by the poets to be dyed with the blood of Siva, that flowed from the wound made by the arrow of Kama, the Indian Cupid."

"There are three varieties of Lotus or Water Lily: The white, called Pundarika; the red, Kokanada; and the blue, Indivara. The entire plant is called Padmini; the fruit, Karmikara; and the honey formed in the flowers, Makaranda. The stalks of the leaves are called Mrinala. The face of a beautiful woman is compared by the poets to a lotus blossom, the eyes to lotus buds, and the arms to its filaments, in Sanskrit Kinjalka or Padmakesara. The bee is represented as enamoured of the lotus. Hari likens the eyes of Radha to a blue lotus, and when sorrowful, to a red one; the face of Hari is to his Radhika like a full blown lotus."

The allusions to the sacred lotus in ancient Hindú literature appear to refer equally to different kinds of **Nymphæa** as to **Nelumbium speciosum**, which, by many authors, has hitherto been regarded as the particular plant referred to. In the above extract the description of the "flower floating upon the waters," and the mention of its various colours—white, red, and blue—suggest **Nymphæa** rather than **Nelumbium**.

209 **Nymphæa stellata,** *Willd.; Fl. Br. Ind., I., 114.*

Vern.—*Nil-padma,* HIND.; *Nilsáphala, nilsápla,* BENG.; *Subdikaim,* ORISSA; *Bambher* (Bijnor), N.-W. P.; *Uplia-kamal,* BOMB.; *Nalla-kalava,* TEL.; *Cit-ambel* (Rheede), MALAY.; *Kya-nyu,* BURM.; *Nilotpala, utpala, indivara,* SANS.

References.—*Roxb., Fl. Ind., Ed. C.B.C., 428; Voigt, Hort. Sub. Cal., 9; Thwaites, En. Ceylon Pl., 14; Aitchison, Cat. Pb. and Sind Pl., 3; Drury, Hand-book of Ind. Fl., 30; W. & A., Prod., I., 17; Mason, Burma, 740; Elliot, Fl. Andhr., 70, 135; Rheede, Mal., XI., t. 27; Med. Top. Ajm., 144; Irvine, Mat. Med. Patna, 78; Taylor, Top. Dacca, 58; Atkinson, Econ. Prod., N.-W. P., Part V., 92; Lisboa, U. Pl. Bomb., 143; Birdwood, Bomb. Pr., 135; Balfour, Cyclop., II., 1116; Kew Off. Guide to Bot. Gardens and Arboretum, 24; Gaz. Mysore and Coorg, I., 56.*

Habitat.—Common throughout the warmer parts of India. It varies in the colour and size of the flowers. In the *Flora of British India* the following varieties are described:—

(1) **cyanea**; flowers medium-sized, blue, scarcely odorous. *Syn.*—N. CYANEA, *Roxb.*

(2) **parviflora**; flowers usually smaller, blue. *Syn.*—N. STELLATA, *Willd.*

(3) **versicolor**; flowers larger, white, blue, purple, or flesh-coloured; stamens very many. *Syn.*—N. VERSICOLOR, *Roxb.*

FOOD. Roots. 210 Seeds. 211 **Food.**—The ROOTS and SEEDS are frequently eaten, especially in times of scarcity.

NYSSA, *Linn.; Gen. Pl., I., 952.*

212 **Nyssa sessiliflora,** *Hook. f.; Fl. Br. Ind., II., 747;* CORNACEÆ.

Vern.—*Kalay, chilauni,* NEPAL; *Tumbrúng,* LEPCHA.

References.—*Kurz, For. Fl. Burm., I., 240; Gamble, Man. Timb., 81; Ind. For., IX., 377.*

Habitat.—A large tree found in the forests of the Sikkim Himálaya above 5,000 feet; also in Martaban between 4,000 and 6,000 feet.

TIMBER. 213 **Structure of the Wood.**—Grey, soft, even-grained.

DOMESTIC. 214 **Domestic Uses.**—Used for house-building and other purposes about Darjíling (*Gamble*).

OAK.

Oak, see **Quercus**, Vol. VI., Pt. I.

Oats, see **Avena sativa**, *Linn.*; GRAMINEÆ; Vol. I., 356.

OCHNA, *Schreb.*; *Gen. Pl.*, *I.*, *317*.

[*69*; OCHNACEÆ.

Ochna squarrosa, *Linn.*; *Fl. Br. Ind.*, *I.*, *523*; *Wight*, *Ill.*, *I.*, *t.* 1

Syn.—O. LUCIDA, *Lamk.*; O. RUFESCENS (?), *Thunb.*; O. OBTUSATA, *DC.*; O NITIDA, *Thunb.*

Vern.—*Koniari*, URIYA; *Champa baha*, SANTAL; *Kanak-champa*, BOMB.; *Sunari, yerra-juvi, sunúru, tammichetta*, TEL.; *Narole, mudah*, KAN.; *Hsen-way*, BURM.; *Kunuk-champa*, SANS.

References.—*Roxb.*, *Fl. Ind.*, *Ed. C.B.C.*, *449*; *Brandis*, *For. Fl.*, *60*; *Gamble, Man. Timb.*, *65*; *Dalz. & Gibs.*, *Bomb. Fl. Suppl.*, *17*; *Mason, Burma & Its People*, *764*; *Elliot, Fl. Andhr.*, *170*; *Graham, Cat. Fl. Bomb.*, *37*; *Revd. A. Campbell, Rept. Econ. Prod.*, *Chutia Nagpur, No. 8409*; *Lisboa, U. Pl. Bomb.*, *37*; *Gazetteers:—Mysore & Coorg, I.*, *58*; *Bombay, XV.*, *429*; *Ind. Forester, IX.*, *255*.

Habitat.—A shrub or small tree, found in Bengal, Assam, Burma, South India, and Ceylon.

Medicine.—The ROOT, which is long, twisted, and tuberous, is used by the Santals as an antidote in snake-bite; a decoction of the root is given in certain menstrual complaints, also for consumption and asthma (*Revd. A. Campbell*). MEDICINE. Root. 2

Structure of the Wood.—Reddish-brown, moderately hard, and close-grained; weight 51lb per cubic foot (*Gamble*). TIMBER. 3

O. Wallichii, *Planch.*; *Fl. Br. Ind.*, *I.*, *524*. 4

Syn.—O. ANDAMANICA, *Kurz*; O. NITIDA, *Wall.* (*not of Thunb.*); O. SQUARROSA, *Kurz* (*not of Linn.*).

Vern.—*Yodayah*, BURM.

References.—*Kurz, For. Fl. Burma, I.*, *205*, *and in Journ. Asiat. Soc. Beng.*, *XLI.*, *pt. ii.*, *295*; *Gamble, Man. Timb.*, *66*.

Habitat.—A small, deciduous, glabrous tree found in Burma. It is very frequent in the tropical forests of Martaban and Tenasserim, less so along the eastern and southern slopes of the Pegu Yomah (*Kurz l.c.*). The **O. andamanica** of **Kurz**, found by him in the Andaman Islands, is, in the *Flora of British India*, reduced to this species.

Structure of the Wood.—Light-brown, hard, close grained, and brittle; weight 54lb per cubic foot (*Gamble*). TIMBER. 5

Ochres, see **Pigments**, Vol. VI., Part I., and **Iron Oxides**, Vol. IV., 520.

OCHROCARPUS, *Thou.*; *Gen. Pl.*, *I.*, *175*.

[*Wight*, *Ic.*, *t. 1999*; GUTTIFERÆ.

Ochrocarpus longifolius, *Benth. & Hook. f.*; *Fl. Br. Ind.*, *I.*, *270*; 6

Syn.—CALYSACCION LONGIFOLIUM, *Wight*; CALOPHYLLUM LONGIFOLIUM, *Wall.*; MAMMEA LONGIFOLIA, *Planch. & Trian.*

Vern.—*Nág-késar-ké-phul* (the flowers), HIND.; *Nágésar-ke-phul* (the flowers), BENG.; *Suringi, támbra-nágkesar*, BOMB.; *Ran undi*, KONKAN; *Punnág, suringi*, MAR.; *Ráti-nág-kesar*, GUZ.; *Nágap-pú, nágasháp-pú, nágésar-pu* (the flowers), TAM.; *Sura-ponna*, TEL.; *Wanai, laringi* (male), *púne* (female), *suringi, gardundi*, KAN.; *Seráya*, MALAY.; *Nága-késaram-pushpam*, SANS.

References.—*Beddome, Fl. Sylv.*, *t. 89*; *Gamble, Man. Timb.*, *21*, *27*; *Dymock, Mat. Med. W. Ind.*, *2nd Ed.*, *85*; *S. Arjun, Bomb. Drugs*, *23*; *Pharm. Ind.*, *I.*, *172*, *173*; *Moodeen Sheriff, Mat. Med. of Madras*, *44*; *Lisboa, U. Pl. Bomb.*, *11*, *241*; *Birdwood, Bomb. Pr.*, *13*; *Liotard, Dyes*,

32; Balfour, Cyclop., I., 552; Kew Off. Guide to the Mus. of Ec. Bot., 16; Gazetteers:—Mysore & Coorg, I., 52, 68, 436; Bombay, XIII., 26; XV., 73; Ind. Forester, XII., 188 (xxii.).

Habitat.—A large, deciduous tree, found in the forests of the Western Peninsula from Kanara to the Konkans. It is said to be met with also in some parts of Burma.

GUM. Seed. 7 **Gum**—The SEED, which is as large as an acorn, when cut exudes a viscid gummy fluid.

DYE. Flower-buds. 8 **Dye**—The dried FLOWER-BUDS, known in commerce as *tambra nagkesar* (red *nagkesar*), are used for dyeing silk. They are about the size of cloves and of a red colour (*see Balfour, Cyclop., under* **Calophyllum longifolium**).

PERFUME. Flower-buds. 9 **Perfume.**—In some parts of India a perfume resembling that of violets is said to be extracted from the dried FLOWER-BUDS (*Gazetteer, Mysore & Coorg*).

MEDICINE. Flower-buds. 10 **Medicine.**—The dried FLOWER-BUDS possess mild, stimulant, carminative, and astringent properties, and are reported to be useful in some forms of dyspepsia and in hæmorrhoids (*Dymock*). They are used in the form of an infusion prepared in the ordinary way, the proportion of flowers to boiling water being one to eight (*Moodeen Sheriff*).

FOOD. Fruit. 11 **Food.**—"The FRUIT is eaten by children who call it *gori undi* or sweet *undi*" (*Dymock*).

TIMBER. 12 **Structure of the Wood.**—Hard, red, close, and even-grained; weight 55℔ per cubic foot (*Lisboa*).

DOMESTIC. 13 **Domestic Uses.**—The timber is used in Burma for planking and for the masts and yards of boats.

14 **Ochrocarpus siamensis,** *T. Anders.; Fl. Br. Ind., I., 270.*

Syn.—CALYSACCION SIAMENSE, *Miq.*; MAMMEA SIAMENSIS, *T. Anders.*

Vern.—*Tharapi, ta-lapi,* BURM.

References.—*Kurz, For. Fl. Burma, I., 94; Gamble, Man. Timb., 21.*

Habitat.—An evergreen tree, found in the forests on the dry hills of Pegu and Arrakan, and distributed also to Siam.

DOMESTIC. Wood. 15 **Domestic Uses.**—The WOOD is said to be good for cabinet-making (*M'Clelland*).

OCHROSIA, *Juss.; Gen. Pl., II., 700.*

16 **Ochrosia borbonica,** *Gmel.; Fl. Br. Ind., III., 638;* APOCYNACEÆ.

Syn—OCHROSIA SALUBRIS, *Bl.*; CERBERA PARVIFLORA, *Wall.*; C. SALUTARIS, *Lour.*

References.—*Kurz, For. Fl. Burma, II., 172; Gamble, Man. Timb., 62; Ind. Forester, VI., 238; VII., 244.*

Habitat.—A small evergreen tree, found in the tidal forests of the Andaman Islands, and distributed also to similar localities in Ceylon, the Straits Settlements, and Java.

GUM. 17 **Gum.**—Some species of this genus are said to yield gutta (*Baron von Müeller*), but no information can be obtained as to whether the Indian one does so or not.

OCIMUM, *Linn.; Gen. Pl., II., 1171.*

18 **Ocimum Basilicum,** *Linn.; Fl. Br. Ind., IV., 608; Wight, Ic., t. [868 (var. thyrsiflorum);* LABIATÆ.

COMMON SWEET BASIL; SWEET BASIL; COMMON BASIL.

Syn.—O. PILOSUM, *Willd.*; O. ALBUM, *Linn. Mant. not of Roxb.*; O. MINIMUM, *Burm. not of Linn.*; O. HISPIDUM, *Lamk.*; O. MENTHÆFOLIUM, *Benth.*; O. CARYOPHYLLATUM, *Roxb.*; PLECTRANTHUS BARRELIERI, *Spreng.*

Various varieties of **Ocimum Basilicum** have been described by **Bentham**, but **Sir J. D. Hooker**, in the *Flora of British India*, says he cannot distinguish these by any constant characters. "The plant," he observes, "varies much in robustness and hairiness of its parts. *Var.* **thyrsiflora** is a luxuriant state; *var.* **difforme** has abnormally deeply cut leaves; *var.* **purpurascens** is a purple-coloured state, imported from Persia; *var.* **glabratum** has very large fruiting calyces, sometimes one-fourth inch diameter, with a rather elongate upper lobe."

Vern.—*Bábul, kali-tulsi, babui-tulsi, sabzah,* HIND.; *Babui tulsi, khúb kalam, debunsha, pashanabheddie,* BENG.; *Dhála tulasi,* URIYA; *Dimbu baha, mali buha, bharbari,* SANTAL; *Furrunj-mushk, nigand, babri, tulsi, babúri, rehan, panr, niyazbo,* PB.; *Nazbo, sabajhi,* SIND; *Sabza,* MAR.; *Subjah, subze, salzat, tirunitru,* DEC.; *Tirnutpatchie, tirunitru,* TAM.; *Rudra jada, bhú-tulasi, vépudu pachcha, vibudi-patri,* TEL.; *Kam-kasturi, sajjebiya,* KAN.; *Tiru-nitru, pach-chá,* MALAY.; *Sawandatala,* SING.; *Varvara, munjariki,* SANS.; *Buklut-ul-zub, rihan, sháhasfaram, hebak, badrúj, asaba-ul-feteyat,* ARAB.; *Firanj-mushk, tureh-korasani, daban-shah, nazbu, ungúsht-kuni-zuckan,* PERS.

References.—*Roxb., Fl. Ind., Ed. C.B.C., 464; Voigt, Hort. Sub. Cal., 447; Stewart, Pb. Pl., 170; Mason, Burma & Its People, 501, 790; Elliot, Fl. Andhr., 26, 165, 191; Pharm. Ind., 167; Ainslie, Mat. Ind., II., 423; O'Shaughnessy, Beng. Dispens., 492; Moodeen Sheriff, Supp. Pharm. Ind., 184; U. C. Dutt, Mat. Med. Hind., 220, 323; Dymock, Mat. Med. W. Ind., 2nd Ed., 606; Fleming, Med. Pl. and Drugs, as in As. Res., Vol. XI., 174; S. Arjun, Bomb. Drugs, 102; Murray, Pl. & Drugs, Sind, 173; Year-Book Pharm., 1878, 288; Med. Top. Ajmere, 149; Irvine, Mat. Med. Pat., 49, 76; Atkinson, Him. Dist., 708; Lisboa, U. Pl. Bomb., 224; Birdwood, Bomb. Pr., 64, 224; Butler, Top. & Statis., Oudh & Sultanpore, 44; Aitchison, Afgh. Del. Com., 95; Balfour, Cyclop., II., 5; Smith, Dic., 399; Treasury of Bot., 801; Home Dept. Cor., 224; Gazetteers:—Mysore & Coorg, I., 64; N.-W. P., I., 83; IV., lxxvi.; Ind. Forester, XII., App. 19.*

Habitat.—An erect, herbaceous, glabrous or pubescent annual, indigenous in Persia and Sind; growing throughout India. It is cultivated throughout Tropical India from the Panjáb to Burma, and is said to be indigenous on the low hills in the Panjáb (*Aitchison*).

Fibre.—*Spons' Encyclopædia* mentions that it is cultivated to a small extent in the western portion of the Hughli district, on account of the strong fibre that it yields, which is employed for rope-making. [This must be a mistake.—*Ed.*] FIBRE. 19

Oil.—It is included in a list of vegetable fixed oils in *Spons' Encyclopædia.* The Natives of India, in their perfumery, distil ottos from some of the different species of **Ocimum**, but exact information on this subject is not available. "On distillation the PLANT yields a yellowish green volatile oil lighter than water, which, on being kept, solidifies into a crystalline camphor, isomeric with turpentine camphor" (*Gmelin's Handbook, XIV., 359*). OIL. Plant. 20

Medicine.—"The WHOLE PLANT has an aromatic odour, which is improved by drying. Its taste is aromatic and somewhat cooling and saline." (*Pharmacog. Ind.*). The SEEDS, which are much used medicinally in some parts of India, are small, black, oblong (one-sixteenth inch long), slightly arched on one side and flattened on the other, blunt pointed. They have no odour, but an oily, slightly pungent taste. When placed in water they become coated with a semi-transparent mucilage. Steeped in water, they form a mucilaginous jelly (*U. C. Dutt, Murray, Dymock, &c.*). Their properties are said to be demulcent, stimulant, diuretic, and diaphoretic. They are often used in infusion in India, especially by Native practitioners in the treatment of gonorrhœa and nephritic affections, and are regarded also as useful for dysentery and diarrhœa, especially in children for the MEDICINE. Plant. 21 Seeds. 22

MEDICINE. diarrhœa of dentition. A cold infusion of the seeds is said to relieve the after-pains of parturition (*Fleming, Asiat. Res., XI., 174; Baden Powell*). Dr. Irvine in his *Materia Medica of Patna* remarks that the seeds in doses of from ʒi to ʒiii are used as an aphrodisiac. The expressed juice of the

Leaves. 23 LEAVES forms the basis of a celebrated nostrum for the cure of ring-worm, and the bruised leaves are applied to parts stung by scorpions to lessen the pain. The powdered dried leaves are said to be an effectual means of dislodging maggots (*Dr. Newton*). The plant is also credited with diaphoretic properties. "The juice of the leaves mixed with ginger and black pepper is given in the cold stage of intermittent fever" (*Taylor, Topography of Dacca*).

SPECIAL OPINIONS.—§ "*Char-tukhm*—a well-known mixture of four seeds employed medicinally by all Native druggists in the Panjáb—usually consists of two seeds of **Ocimum pilosum, Plantago sp., Salvia sp.**, and another, altered and added to according to the taste or idea of the druggist" (*Surgeon-Major J. E. T. Aitchison, Simla*). "The seeds are useful in diarrhœa and dysentery, given in doses of 1 to 2 drachms to adults, and 4 to 5 grains to infants administered in syrup" (*Lal Mahomed, 1st Class Hospital Assistant, Hoshangabad, C. P.*) "The seeds washed and pounded are used in poultices for unhealthy sores and sinuses. They are also given internally with sherbet in cases of habitual constipation and in internal piles" (*Civil Surgeon J. H. Thornton, B.A., M.B., Monghyr*). "I have seen a teaspoonful of the seeds infused in a tumbler of water to which a little sugar is added, and the whole drunk daily as a demulcent in cases of genito-urinary diseases" (*Honorary Surgeon A. E. Morris, Tranquebar*). "The seeds are made into a sherbet and administered in fevers; they have diuretic properties" (*Surgeon-Major A. S. G. Jayakar, Muskat*). "The juice is dropped into the ears for the cure of earache and dulness of hearing. Roots are used for the bowel complaints of children" (*Surgeon-Major D. R. Thomson, M.D., C.I.E., Madras*).

FOOD. Plant. 24 Seeds. 25 **Food.**—The PLANT has a strong aromatic flavour like that of cloves and is often used for culinary purposes as a seasoning. The SEEDS are sometimes steeped in water and eaten. They are said to be cooling and very nourishing. In Kanawar they are sometimes eaten mixed in ordinary bread (*Stewart, Pb. Pl.*). They are largely employed, especially by the Muhammadans of Eastern Bengal, infused in water, to form a refreshing and cooling drink.

26 **Ocimum canum,** *Sims.; Fl. Br. Ind., IV., 607.*

Syn.—O. AMERICANUM, *Linn.*; O. ALBUM, *Roxb.* (*not of Linn.*).

Vern.—*Bharbhari*, SANT.; *Kukka tulasi*, TEL.; *Hin-talla*, SING.

References.—*Roxb., Fl. Ind., Ed. C.B.C., 463; Dalz. & Gibs., Bomb. Fl., 203; Grah., Cat. Bomb. Pl., 147; Sir W. Elliot, Fl. Andhr., 102; Rev. A. Campbell, Rept. Econ. Pl., Chutia Nagpur, No. 7876; Gazetteers:—Bombay, V., 28; XV., 441; N.-W. P. (Him. Dists.), IX., 315; Mysore & Coorg, I., 64.*

Habitat.—An erect, pubescent, herbaceous plant found on the plains and lower hills of India, from Assam, Bengal, Behar, and Central India to the South Deccan and Ceylon. It is usually abundant in a wild or semi-wild state about Native gardens. **Dalzell & Gibson** mention that Africa is supposed to be its native country.

MEDICINE. **Medicine.**—The **Rev. A. Campbell**, in his *Econ. Prod. of Chutia Nágpur*, says of this plant:—"During fever, when the extremities are cold, the

Leaves. 27 LEAVES, made into a paste, are applied to the finger and toe nails. The same preparation is used as a cure for parasitical diseases of the skin."

Ocimum gratissimum, *Linn.; Fl. Br. Ind., IV., 608.* 28

THE SHRUBBY BASIL.

Syn.—O. CITRONATUM, *Ham* ; O. ROBUSTUM, *Heyne.*

Vern.—*Ram-tulsi*, HIND.; *Ram-tulsi, ram-tulshi*, BENG.; *Banjere*, PB.; *Ramatulasa, tulsi, rán-tulsi*, BOMB.; *Ranatulasu*, MAR.; *Avachibávachi*, GUZ.; *Elumich-cham-tolashi*, TAM.; *Ram-tulsi*, DEC.; *Nimma tulási, ráma tulási*, TEL.; *Káttu-tuttuvá*, MALAY.; *Furanjmishk*, ARAB.; *Palangmishk*, PERS.

References.—*Roxb., Fl. Ind., Ed. C.B.C., 464; Dalz. & Gibs., Bomb. Fl., 203; Elliot, Fl. Andhr., 135, 163; Grah., Cat. Bomb. Pl., 147; Rheede, Hort. Mal., X., t. 86; Dymock, Mat. Med. W. Ind., 2nd Ed., 607; S. Arjun, Bomb. Drugs, 102; Lisboa, U. Pl. Bomb., 224.*

Habitat.—A larger plant than either of the preceding and more shrubby in character, cultivated in India in gardens throughout Bengal, Chittagong, East Nepál, and the Deccan Peninsula, and extending into Ceylon. In Western India it is said to be a common wild plant; but it is very doubtful if it is really indigenous.

Medicine.—The LEAVES diffuse a stronger fragrance than do those of any other member of the genus. Their juice is used as a remedy for gonorrhœa. In Mauritius baths and fumigations prepared with them are employed in the treatment of rheumatism and paralysis (*Pharm. Ind.*). **S. Arjun**, in his *Catalogue of Bombay Drugs*, states that the SEEDS of this plant are not used medicinally, but are sometimes eaten as a nutritive substance, and that a decoction of the leaves is of value in cases of seminal weakness. **MEDICINE. Leaves. 29** **Seeds. 30**

SPECIAL OPINION.—§ "Used in mercurial rheumatism and mercurial salivation. The seeds are given in headaches and neuralgia" (*Surgeon-Major D. R. Thomson, M.D., C.I.E., Madras*).

O. sanctum, *Linn.; Fl. Br. Ind., IV., 609.* 31

THE SACRED BASIL.

Syn.—O. MONACHORUM, *Linn., Mant.*; O. TENUIFLORUM, *Linn., Sp. Pl.*; O. INODORUM, *Burm. (not of Koen.).*

Two forms of this plant, which, however, scarcely deserve to be considered as varieties, are met with in cultivation. They are distinguished, by Sanskrit writers, on account of some difference in the colour of their leaves, the one being known as the *white* and the other the *black* (*U. C. Dutt*).

Vern.—*Kala-tulsi, tulsi, barandá, varandá*, HIND.; *Kala-tulsi, kural, tulsi, tulshi*, BENG.; *Bantulsi, tulsi*, PB.; *Tulas, tulasa*, BOMB.; *Tulasa,* [illegible] TAM.; *Tulasi, krushna-tulasi, gaggera-chettu*, TEL.; *Tulashi-gidá*, KAN.; *Niella tirtua, krishna-tulsi, nallu tirtta*, MALAY.; *Lun*, BURM.; *Muduru-tulla*, SING.; *Parnása, sorasaw, ajaka, tulasi, manjarika, tulashi*, SANS.; *Ulsi-badruge.*

References.—*Roxb., Fl. Ind., Ed. C.B.C., 463; Voigt, Hort. Sub. Cal., 448; Dalz. & Gibs., Bomb. Fl., 204; Stewart, Pb. Pl., 171; Aitchison, Cat. Pb. & Sind Pl., 114; Mason, Burma & Its People, 438, 790; Sir W. Jones, Treat., Pl. Ind., V., 130; Rumph., Herb Amb., V., t. 92, f. 2; Graham, Cat. Bomb. Pl., 147; Trimen, Cat. Ceylon Pl., 69; Sir W. Elliot, Fl. Andhr., 101, 124, 184; Pharm. Ind., 167; Ainslie, Mat. Ind., II., 426; O'Shaughnessy, Beng. Dispens., 493; Moodeen Sheriff, Supp. Pharm. Ind., 185; U. C. Dutt, Mat. Med. Hind., 219, 321; Dymock, Mat. Med. W. Ind., 2nd Ed., 605; S. Arjun, Bomb. Drugs, 207; Irvine, Mat. Med. Pat., 105; Baden Powell, Pb. Pr., 365; Atkinson, Econ. Prod., N.-W. P., IV., 41; Drury, U. Pl., 327; Lisboa, U. Pl. Bomb., 224, 279, 284, 286, 287; Birdwood, Bomb. Pr., 64, 225; W. W. Hunter, Orissa, II., 181, App. VI.; Kew Off. Guide to the Mus. of Ec. Bot., 60; Home Dept. Cor., 240; Gazetteers:—Mysore & Coorg, I., 56, 64; N.-W. P., I., 83; II., 506; IV., lxxvi.; Settlement Report, Chanda Dist., App. VI.*

OCIMUM sanctum.

The Sacred or Holy Basil.

Habitat.—A somewhat shrubby, herbaceous plant, found throughout India, Burma, and Ceylon, and distributed to the Malay Archipelago, Australia, and the islands of the Pacific. It is also met with in Western Asia and Arabia. In India it is found on the Himálaya, up to an altitude of 6,000 feet. Although common in waste places, it is more frequently cultivated, and it is thus doubtful whether this species is really indigenous to India.

MEDICINE. Leaves. 32 **Medicine.**—The LEAVES have expectorant properties, and their juice is used by native physicians in catarrh and bronchitis. This preparation also is applied to the skin in ring-worm and other cutaneous diseases. An infusion of the leaves is used as a stomachic in the gastric disorders of children, and in hepatic affections. The dried leaves are powdered and employed as a snuff in ozæna. They are also "an effectual means of dislodging maggots" (*Dr. J. Newton*). The ROOT is given in decoction as a diaphoretic in malarial fevers. The SEEDS are mucilaginous and demulcent, and are given in disorders of the genito-urinary system.

Root. 33

Seeds. 34

SPECIAL OPINIONS.—§ "Juice of the leaves combined with ginger given for colic in children" (*Civil Surgeon S. M. Shircore, Moorshidabad*). "The juice of the leaves poured into the ear, first-rate remedy for earache" (*V. Ummegudien, Mettapolian, Madras*). "The infusion of the leaves is extensively used as a carminative and laxative for children" (*Surgeon W. F. Thomas, 33rd M. N. I., Mangalore*). Its FLOWERS are used with honey, ginger, and onion juice in cough as an expectorant" (*Assistant Surgeon N. N. Bhuttacharji, Tirhoot State Railway, Somastipore*). "The juice of the fresh leaves sweetened with honey is given with benefit to children suffering from bronchitis" (*Assistant Surgeon R. C. Gupta, Bankipore*). "The juice of the leaves and flower-tops and the slender roots are used as an antidote in snake poisoning. The juice of the leaves is carminative, refrigerant, and febrifuge; also cholagogue and laxative. It is used as an expectorant in the bronchitis of children, externally it is used in ring-worm" (*Civil Surgeon J. H. Thornton, B.A., M.B., Monghyr*). "The juice is given internally in cases of febricula and catarrh of the nose and lungs" (*Native Surgeon T. R. Moodelliar, Chingleput, Madras*). "Leaves are said to possess diaphoretic properties" (*J. Parker, M.D., Deputy Sanitary Commissioner, Poona*).

Flowers. 35

FOOD. 36 **Food**—Occasionally grown as a pot-herb by Europeans, for which purpose it is very useful (*Atkinson*).

DOMESTIC & SACRED. 37 **Domestic & Sacred.**—The *Tulsí* is the most sacred plant in the Hindu religion; it is consequently found in or near almost every Hindu house throughout India. Hindu poets say that it protects from misfortune and sanctifies and guides to heaven all who cultivate it. The Brahmins hold it sacred to the gods Krishna and Vishnu. The story goes that this plant is the transformed nymph Tulasi, beloved of Krishna, and for this reason near every Hindu house it is cultivated in pots or on brick or earthen pillars with hollows at the top in which earth is deposited; it is daily watered and worshipped by all the members of the family. Under favourable circumstances, it grows to a considerable size, and furnishes a woody stem large enough to make beads from for the rosaries used by Hindus on which they count the number of recitations of their deity's name.

In the *Vrat Kaumudi*, one of the sacred books of the Hindus, a ceremony, called the *tulashi laksha vrat*, is ordered to be performed, when a vow is made, which consists in offering a *lac* of the leaves one by one to Krishna, the performer fasting till the ceremony is complete (*Lisboa*). *Nayavad*, another ceremonial sacrifice among the Hindus, consists in taking a brass dish containing some cooked food, and placing it before the god in a square previously marked out on the ground with the fingers

DOMESTIC & SACRED.

dipped in water. The worshipper then squats on a low stool, and taking two leaves of the *tulsí* in his right hand, he closes his eyes with his left, dips the leaves in water, and throws one upon the food, and the other, after five peculiar motions of the hand, on the god (*Lisboa*). The leaves are also used in the funeral ceremonies of the Hindus (*O'Shaughnessy*).

The leaves of the variety **villosum**, **Balfour** says, are employed in Courts of Justice in India. "Hindus are sworn by them. They are placed in the palm of the hand by a Brahman, who repeats the prescribed oath, and at the termination they are masticated and swallowed."

Ocotea lanceolata, *Nees,* see **Phœbe lanceolata,** *Nees*; LAURINEÆ; Vol. VI., Pt. I.

ODINA, *Roxb.; Gen. Pl., I., 423.*

[ANACARDIACEÆ.

Odina Wodier, *Roxb.; Fl. Br. Ind., II., 29; Wight, Ic., t. 60;* 38

Vern.—*Jingan, kiamil, kaimil, kimúl, kamlái, kashmala, ginyan, jhingan, mowen, mohin, moyen,* HIND.; *Jiol, bohar, jiyal, lohar, bhadi, jir, jial, jival, ghadi,* BENG.; *Dowka, dhoka,* KOL; *Doka,* S NTAL.; *Mooi, indrámai,* URIYA; *Hneingpyoing,* MAGH.; *Bara, dabdabbi, halloray,* NEPAL; *Kekeda,* KURKU; *Kaikra, gumpri, gharri,* GOND; *Jhingan, jibán, sindan harallú,* N.-W. P.; *Kiamil, kambal, batrín, kimlú, kemball, dhauntika, dila, kemal, koamla, sulambra, pichka, lidra, kamlai,* PB.; *Gob,* RAJ.; *Simati, moya,* SIND.; *Gúnja, mouni, mageer, moyeen,* C. P.; *Shimti, ginyan, moya, kimul, moina, moi, simati, moja, shembat, molarda, gajel,* BOMB.; *Moi, moja, moye, shimat, munldi, shimti,* MAR.; *Wodier, wude, odiyamaram, otiyam,* TAM.; *Gumpini, gumpna, dumpini, dumpri, dumper, dampara, udai-mdnu, gumpena chettu,* TEL.; *Shimti, púnil, gojal, udi, suggipatte, shimli,* KAN.; *Nabé, hnanbai, nabhai, nabhay, hnabé,* BURM.; *Jingini,* SANS.

References.—*Roxb., Fl. Ind., Ed. C.B.C., 336; Voigt, Hort. Sub. Cal., 275; Brandis, For. Fl., 123; Kurz, For. Fl. Burm., I., 321; Gamble, Man. Timb., 110; Dalz. & Gibs., Bomb. Fl., 51; Stewart, Pb. Pl., 46; La Maout & Decaisne, 303; Mason, Burma and Its People, 540, 774; Elliot, Fl. Andhr., 45, 65; Pharm. Ind., 60; Ainslie, Mat. Ind., 486; O'Shaughnessy, Beng. Dispens., 22; U. C. Dutt, Mat. Med. Hind., 301; Dymock, Mat. Med. W. Ind., 2nd Ed., 202; S. Arjun, Bomb. Drugs, 36, 207; Baden Powell, Pb. Prod., 396, 397; Atkinson, Him. Dist., 308, 744; Econ. Prod., N.-W. P., Pt. I., 5; Lisboa, U. Pl. Bomb., 54, 242, 250, 278; Watson, Rep., 4, 22, 53, 55; Balfour, Cyclop., III., 8; Kew Off. Guide to Bot. Gardens & Arboretum, 37; Home Dept. Cor., 239; Burm. Gaz., Vol. I., 134, 137; Jour. Agri.-Hort. Soc., 1875, V., 75; For. Ad. Rept, Chutia Nagpur, 1885, 29; Gazetteers:—Mysore & Coorg, I., 52, 59; III., 28; Bombay, VIII., 11; XIII., 26; XV., 73; N.-W. P., I., 80; IV., lxx; Panjáb:—Rawalpindi, 15; Hoshiárpur 11; Gurdáspur, 53; Sháhpur, 69; Settle Repts., Seonee, 10; Bhandára, 19; Nimár, 306; Baitool, 127; Chhindwárá, 107; Manuals of Administration, Trichinopoly, 79; Madras, I., 362.*

Habitat.—A large, deciduous tree, 40 to 50 feet in height, met with throughout the hotter parts of India from the Indus eastwards. It ascends in the Sub-Himálayan tract to an altitude of 4,000 feet; is found also in Assam, Madras (chiefly in a cultivated state), and in Burma, the Andaman Islands, and Ceylon.

GUM. 39

Gum.—From wounds and cracks in the bark of this tree, at some seasons of the year, there issues a GUM, at first yellowish white in colour (*kanne-ki-gond*), which takes on a brownish tinge, and afterwards, if it falls to the ground, becomes blackish (*jingan-ki-gond*). It usually exudes in October and "occurs partly in tears of a yellowish tinge, partly in colourless angular fragments which are full of fissures like those of gum-arabic. It has a disagreeable taste, is not astringent; about one half is completely soluble in water; the remaining portion forms a slimy mucilage, but is

GUM. not gelatinous" (*Dymock*). It is much used along with the gum of **Anogeissus latifolia** in calico-printing, and in Nepál as a paper-size. In Kumáon **Captain Campbell** states that it is used mixed with lime in white-washing. In Burma, **Kurz** says, it is employed as the basis of an inferior varnish. The Brahmins of Bengal use it to stiffen their Brahminical strings.

DYE & TAN. **Dye & Tan.**—The BARK of the tree furnishes a small amount of
40 brownish-red colouring matter, which produces on tasar-silk a golden or pale brown tint similar to that obtained from *lodh* bark (**Symplocos racemosa**) (*Wardle*). It is also used in tanning.

FIBRE. Bark. **Fibre.**—The BARK yields a good although coarse cordage fibre.
Medicine.—The BARK is very astringent, and, although not officinal, is
41 described in the Pharmacopœia of India. A decoction is said to be useful
MEDICINE. Bark. as a local application in cutaneous eruptions and obstinate ulceration, and to form an excellent astringent gargle. **Ainslie**, who gives a similar account
42 of its properties, remarks:—"The bark powdered in combination with *margosa* oil is considered by the *Vytians* a valuable application for old and obstinate ulcers." The GUM, beaten up with cocoanut milk, is applied to
Gum. 43 sprains and bruises, and the LEAVES boiled in oil are used for a similar
Leaves. purpose. It is given internally in asthma, and as a cordial to women.
44 Externally it forms the basis of many of the plasters employed for rheumatism. In **Taylor's** *Topography of Dacca* mention is made of a medi-
Juice. cinal use of this tree not referred to elsewhere. He says that the JUICE of
45 the green branches, in a dose of four ounces mixed with two ounces of tamarinds, is given as an emetic in cases of coma or insensibility produced by opium or other narcotics.

FODDER. **Fodder.**—The LEAVES and young SHOOTS afford fodder for cattle. In
Leaves. some places (Madras, Oudh, &c.) it is pollarded to supply fodder for
46 elephants (*Gamble*).
Shoots. **Structure of the Wood.**—The sapwood is large in amount and very
47 subject to worms, the heartwood when freshly cut is light red, but becomes
TIMBER. reddish brown on exposure. It is moderately hard, close-grained, seasons
48 well, and does not warp, but is not very durable (*Gamble*). It is, however, not liable to be attacked by white ants (*Gazetteer, Mysore & Coorg*). Average weight about 58 lb the cubic foot.

DOMESTIC. **Domestic Uses.**—The WOOD is used for a variety of purposes. Spear
Wood. shafts, scabbards, wheel spokes, cattle yokes, oil presses, and rice pounders
49 are made of it. It was tried for sleepers on the Madras and on the Oudh & Rohilkhand Railways, but did not succeed. **Kurz** recommends it as suitable for cabinet making; but no trial of the wood for that purpose has apparently as yet been made.

(*George Watt.*)

50 OILS, FATS, & OIL-SEEDS.

References.—*Agri.-Horti. Soc., Ind., General (India):—Journ., II., Pro., 464, 465, 539, 560, III., Pro., 47, 171, 183, 184, Sel., 248; IV., Sel., 211-214; V., Sel., 81; Pro., 27; VI., 219-222, Pro., 69, 114, 137, 148; VIII., 220-223, Pro., 24; IX., 47-51, 360-370 Pro., 140, 159; (Countries other than India):—Journ., IV., 113, Sel., 151, 153, Pro. 8; VIII., 164-167; IX., Pro., 6; XI., 551; New Series, VII., Pro., 32; Bengal:—Trans., IV., 102, 132; V., 64; VI., 239; VII., 45, 60, 61, 75, 80, 82; VIII., 165-168, 182-185; Journ., I., 102, 394-399; II., Sel., 503, 539-541; III., Pro., 47; IV., Sel., 210-215 Pro., 53, 71; V., Pro., 27; VI., 241, Pro., 3; VII., 18, 19, 40, 41; VIII., Pro., 8, 120; IX., 404; X., 2, 7, 17, 20, 223, 224, 351; XII., 109, 333-346 Pro., 27; New Series, II., 231, 232; V., Pro., 55, 56; Madras, Trans., VII., 86, 101; Bombay:—Journ II., Sel., 300, 301; III., Sel., 248; IX., 290, 295, 296; N.-W. P. and Oudh:—Trans I., 20; VIII., 165-168, 182, 184, 186, 188, 190; Journs., II. 52, 89; Sel., 207, 261; X., 65-69, Pro., 37; XIII., 296,*

298, 317 ; New Series, IV., Pro., 32 ; Panjáb :—Trans., VIII., 165-168, 182, 184, 188, 190 ; Journs., VIII., 143 ; XIII., 390, Sel., 57, 58 ; New Series, I., 75 ; Central Provinces :—Trans., VIII., 124-144, 183 ; Journs., III., Sel., 98, 99, Pro., 171, 183, 286 ; IV., Pro., 23, 74, 90 ; V., Pro., 61 ; New Series, V., Pro., 57, 58 ; Central India :—Trans. VIII., 183, 185, 187, 191 ; Journ., III., Sel., 249 ; Burma :—Trans., VI., 127, 128 ; VII., Pro., 188 ; VIII., 345 ; Journs., IV., Sel., 14, 15 ; Pro., 9 ; IX., Sel., 52, 54, 58, 59 ; X., Pro., 51 ; XI., 445-447 ; New Series, V., Pro. (1876), 38, 39 ; Assam :—Trans., VIII., 166, 168 ; Journs., IV., 120 ; VIII., Pro. 27 ; IX., 247 ; X., 223, 224, 342; Sel., 1, 6, 7 ; Mysore :—Journs., New Series, IV., Pro., 27 ; V., Pro. 16 ; Hawkes, Report upon the Oils of Southern India ; Cooke, Oils and Oil-seeds ; Balfour, III., 9-18 ; Spons' Encyclop., II., 1360-1484.

Oils suitable for Candle-making.

The Oils and Fats of India may be classified according to many systems as, for example, Chemical, Physical, Industrial, &c., &c. Thus some oils are said to be fixed, others essential : some are edible, others medicinal; and still others suited to various arts and industries according as they are drying or non-drying oils. Fats, tallows, and waxes, are chemically of the same nature as the substances popularly designated oils. In the present article it is intended to deal mainly with the fatty oils (except those exclusively of medicinal interest) and to assign the essential oils a place with perfumes; that is to say, to deal with the oils and fats that are of interest in the arts and industries, or to the people of India. It may be said that these belong to two great sections—I. Fatty Oils: II. Mineral Oils.

I.—THE PRINCIPAL FATS & FATTY OILS OF INDIA.

FATS & FATTY OILS. 51

Very little can be learned for certain as to the extent Indian Fats and Oils are employed by the Indian soap and candle makers. Writers on Indian economic subjects make mention of a limited number of vegetable oils in connection with these industries, but they do not definitely say whether they are actually used or are only suitable for those purposes. The following out of a possible list of some 300 oil-yielding plants, &c., may be mentioned as the more important :—

1. **Aleurites moluccana.**
2. **Anacardium occidentale.**
3. **Anamirta Cocculus.**
4. **Arachis hypogœa.**
5. **Bassia butyracea.**
6. **B. latifolia.**
7. **B. longifolia.**
8. **Brassica campestris.**
9. **B. juncea.**
10. **Buchanania latifolia.**
11. **Butter or Ghí.**
12. **Callophyllum inophyllum.**
13. **Camellia theifera.**
14. **Canarium commune.**
15. **Cannabis sativa.**
16. **Carapa moluccensis.**
17. **Carthamus tinctorius.**
18. **Cleome viscosa.**
19. **Celastrus paniculatus.**
20. **Cochlospermum Gossypium.**
21. **Cocos nucifera.**
22. **Croton Tiglium.**
23. **Dalbergia latifolia.**
24. **Dipterocarpus tuberculatus.**
25. **D. turbinatus.**
26. **Eriodendron anfractuosum.**
27. **Eruca sativa.**
28. **Euphorbia dracunculoides.**
29. **Fish Oil.**
30. **Garcinia indica.**
31. **G. Morella.**
32. **Gossypium, various species, &c.**
33. **Guizotia abyssynica.**
34. **Gynocardia odorata.**
35. **Helianthus annuus.**
36. **Jatropha Curcas.**
37. **Juglans regia.**
38. **Linum usitatissimum.**
39. **Melaleuca Leucadendron.**
40. **Melia Azadirachta.**
41. **Mesua ferrea.**
42. **Moringa pterygosperma.**
43. **Murraya Kœnigii.**
44. **Olea cuspidata.**
45. **Papaver somniferum.**
46. **Pongamia glabra.**

FATS & FATTY OILS.

47. **Prunus Amygdalus.**
48. **Rhus semialata.**
49. **R. succedanea.**
50. **R. Wallichii.**
51. **Ricinus communis.**
52. **Salvadora oleoides.**
53. **Sapindus Mukorossi.**
54. **S. trifoliatus.**
55. **Sapium sebifera.**
56. **Sesamum indicum.**
57. **Sterculia fœtida.**
58. **Tamarindus indica.**
59. **Tectona grandis.**
60. **Terminalia Catappa.**
61. **Tetranthera monopetala.**
62. **Vateria indica.**

Oils suitable for Candle-making.

To what extent any one of the above have become important oils in soap- and candle-making cannot at present be definitely ascertained. They may collectively, however, be designated the Indian vegetable fatty oils, most of which, at normal temperatures, show a tendency to become solid. This is certainly one point in their favour, but many of them are "non-drying oils," and hence doubtless contain for candle-making an injurious amount of oleic acid.

Classification of Oils.

52 It may be said that oils are primarily classified into—I., Fixed, and II., Volatile. Each of these may be referred to *1st*, Animal, *2nd*, Vegetable, and *3rd*, Mineral. Popularly (and in nearly every language in the world) fats are treated as different from oils. It, however, serves no good purpose to keep up this distinction, except to indicate the particular fatty oils which at ordinary temperatures remain in a solid or semi-solid condition. The volatile and the mineral oils agree with each other in consisting of carbon and hydrogen or in having, in addition, smaller proportions of oxygen, nitrogen, and sulphur than occur in the fatty oils. They cannot, therefore, in any respect be spoken of as fatty oils.

1st. ANIMAL.

53 *1st*—INDIAN ANIMAL FATS AND OILS.

Very little more can be said regarding these than will be found under Honey and Bees'-wax (*Vol. IV., 263-271*), and Tallow, the last mentioned substance will be found described in Vol. VI. Pt. II. It is probable that a larger amount of fish oil might be manufactured, along the entire coast, than appears to be the case (*Vol. III., 368*). Materials that might be utilised for this purpose are either allowed to rot on the beach or are used at most only as manures. The oil from the Indian porpoise (**Platanista indi,** ***Blyth***) is highly extolled for luminous purposes. The Indian whale (**Balœnopterus indica,** ***Blyth***) is regularly caught by the fishermen of the west coast, but definite information as to the nature of the oil obtained from that animal cannot be learned. Crocodile oil (**Crocodilus palustris,** ***Less.***, Vol. II., 592) enjoys the reputation of affording a larger amount of solidifiable fat than either neat's-foot or any fish oil. Of the latter the following are the fish specially mentioned as being employed in the manufacture of oil: **Labeo rohita** (the *rohu* or *rui*), **Pristis cuspidatus** (the saw-fish), **Rhynchobatus djeddensis** (the *ulavi*), **Barbus chola** (the bitter carp), and **Clupea longiceps,** which yields the Malabar oil-sardine. For these and many other oil-yielding fish see the article **Fish,** Vol. III., 363-397.

Porpoise. 54 Whale. 55 Crocodile. 56 Fish. 57

2nd VEGETABLE.

58 *2nd*—THE INDIAN VEGETABLE WAXES, FATS, AND OILS.

These may be classified as follows—

(A) Waxes.
(B) Solid or Semi-solid Fats.
(C) Liquid Oils—
  (*a*) Drying.
  (*b*) Non-drying or greasy.

Waxes.

59 (A) **Vegetable Waxes.**—There are several vegetable waxes known to European commerce, but only a very few of these are found in India.

The Japan wax tree is **Rhus succedanea**, a small tree which is common in the Temperate Himálaya. Another form of Japan wax is said to be produced through the action of an insect on one or two species of **Ligustrum** (several members of that genus also occur on the Himálaya). The white wax of China is the result of the parasitic action of **Coccus Pe-la** on **Fraxinus** or on **Ligustrum.** The Fig-tree wax of Java and Sumatra (largely used in candle-making) is obtained from **Ficus umbellata.** Perhaps the best vegetable waxes are, however, **Myrica** wax (**M. cerifera,** the candle berry myrtle) and Palm wax (**Ceroxylon andicola**). As a curiosity it may be added that the flush on the white gourd of India (**Benincasa cerifera**) is a form of vegetable wax.

VEGETABLE OIL. Waxes. Japan Wax. 60

China White Wax. 61

Java Wax. 62

Myrica Wax. 63

Palm Wax. 64

It will be observed that one at least of the vegetable wax plants (**Rhus succedanea**) occurs in India, but although that plant is reported to yield the substance as freely in this country as in Japan, it is apparently never collected. The Natives of most of the hilly tracts of India, where the tree abounds, appear to be either ignorant of the useful product which the berries afford, or are unacquainted with the method of collecting and utilising it which is pursued in Japan. It is thus probable that with a little encouragement and education they might be induced to plant the tree around their homesteads, and in this way, India might soon come to take a distinct place in meeting the demand for vegetable wax. Of the other plants alluded to in the preceding remarks some might easily enough be introduced into India, such, for example, as the Java wax-yielding fig. The white wax of China, like the white wax of India, is in reality an animal product closely allied to Lac.

Fats. 65

(B.) Vegetable Fats.—The list of 62 oils given above is practically an enumeration of the oils which should appear in this place. It may be of some value to add, however, that the following are those which would seem to offer the greatest inducement for practical experiments being made to test their utility for the candle trade:—

1. **Aleurites** (*the Candle Nut*). The oil from this fruit is generally known in India as *kekuna.*

5, 6, and 7. The species of **Bassia** (the Indian butter trees). The solid oil of **Bassia latifolia** was viewed at the late Colonial & Indian Exhibition as one of the most hopeful of Indian oils for the candle-maker.

14. **Canarium.** This is said to be very similar to cocoa-nut oil in its properties.

16. **Carapa.** A very solid oil even in India, which melts only at high temperatures.

21. **Cocos nucifera.** The cocoa-nut. It is believed this oil is largely used by the European soap- and candle- makers.

30 and 31. **Garcinia.** Several species of trees, common on the mountain ranges of the west coast, yield *kokam butter.*

48, 49, and 50. **Rhus.** As already urged, at least one species of the Japan wax tree occurs in India, and apparently trade might easily enough be opened out in its wax. The Natives of Nepál actually do extract that substance from **R. semialata,** but there does not seem to be any export trade in the article from that State, and information is wanting as to the purpose for which it is prepared.

55. **Sapium.** The tallow tree of China. This is said to be largely used for candle-making in China, and it is believed the oil is even exported to Europe for that purpose. India could supply a very large amount, if a demand were to arise for it.

62. Lastly, in the *Asiatic Researches, Vol. xx., 1825,* there occurs a short paper in which the merits of the fat from **Vateria indica** is highly commended for the purpose of candle-making.

VEGETABLE. Liquid Oils. 66

(C) Liquid Oils.—These may be conveniently referred to two useful sections—(*a*) drying and (*b*) non-drying or greasy oils. The latter owe their liquid character to the presence of a considerable proportion of oleic acid. As this subject will have to be dealt with in the paragraph on the Chemistry of Oils, it will suffice for our present purpose to enumerate a few of the leading oils that fall under (*a*) and (*b*):—

Drying. 67 Non-drying. 68

(*a*) DRYING OILS.	(*b*) NON-DRYING OILS.
Linum usitatissimum.	**Brassica campestris & juncea.**
Juglans regia.	**Sesamum indicum.**
Cannabis sativa.	**Arachis hypogœa.**
Papaver somniferum.	**Cocos nucifera.**
Cucurbita Pepo.	**Canarium commune.**
Helianthus annuus.	**Vateria indica.**
Nicotiana Tabacum.	**Sterculia fœtida.**
Euphorbia dracunculoides.	**Camellia theifera.**
Ricinus communis.	**Pongamia glabra.**
Gossypium herbaceum, arboreum, neglectum & barbadense, &c.	**Croton Tiglium.**

In the above lists the scientific names of the plants which afford the oils have been approximately arranged in the order of importance. Castor-oil and Cotton-seed oil have been placed at the bottom of the list of drying oils, because some writers seem to deny their drying property. Were the above oils arranged purely in point of commercial importance, Castor and Cotton oils would hold much more honourable positions.

MINERAL OILS. Candles. 69

II.—INDIAN MINERAL OILS.

The Mineral Oils of India have, in recent years, been rapidly rising to a very important position in the list of economic products of this country. Petroleum, of many different kinds, exists in various parts of the country, varying from the limpid oil of the Sheráni hills, to the thick, or in some cases nearly solid, petroleums of certain parts of Balúchistán and Assam. From these oils excellent material for illuminating purposes may be obtained, as well as solid paraffines for candle-making, and other solid hydrocarbons for patent fuel. For full information on this subject the reader is referred to the article **Petroleum, Vol. VI., Pt. I.**

CHEMISTRY. 70

CHEMISTRY OF TALLOW AND FATTY OILS.

Popularly, it might be said that all oils or fats (met with in commerce) consist of mixtures of certain definite oils known to the chemist. Each of these when isolated has properties peculiar to itself. Such an oil may be unsuited for certain industrial purposes, but it admits of being acted on by re-agents, or of being changed by chemical and physical agencies, and may be thus so altered as to be made highly serviceable for purposes for which it was previously unsuited. It has, in fact, been shown that theoretically the so-called oils are, chemically speaking, salts of a base with a fatty acid. Tallow, according to this explanation, is a mixture of the salts of glycerine, with the fatty acids—stearic, palmic, and oleic. Of these stearic and oleic may be accepted as representing the two great groups of fatty acids known as the *Adipic* and the *Acrylic*. Of the former Stearic is one of the highest, and may be accepted as represented by $C_{18}H_{36}O_2$ while Acetic ($C_2H_4O_2$) is one of the lowest. Of the latter Oleic is perhaps the most important, and it may be represented by the formula $C_{18}H_{31}O_2$.

Oils are salts of glycerine. 71

Chevreul was thus able to show that with most of these fatty acids, glycerine formed one, two or three salts, and that tallow was chemically a

CHEMISTRY.

mixture of tri-stearine, tri-palmine, with tri-oleine. The last is the less solid constituent already alluded to. The presence of a certain proportion of oleic acid in an oil gives origin to the well-known character of its being non-drying. Its less solid feature disqualifies it (in its unaltered state at least) for the candle trade, and we have here an easy way of determining one qualification of an oil intended for that industry, *viz.*, freedom from oleic acid. All this comes so naturally and simply to the student of modern chemistry, that the astonishment seems justified that the candle-makers sixty years ago did not at least see that it would be desirable to separate the less solid constituent of tallow from the mass. Doubtless the fear of the loss in weight deterred them from so doing. It may almost be said that the present age is pre-eminently that in the world's history when the words "waste material" have been discarded, and the expression "by-product" made to take their place. The waste from one industry is utilised by another. Upon Chevreul's discoveries of the nature of fatty oils being made known, two questions arose:

Freedom from oleic acid. 72

By-products. 73

(1) Which of the constituents of tallow is the most serviceable for the candle-maker?

(2) Is the presence of glycerine advantageous or otherwise?

The first was answered in favour of stearine, palmine being next in value, but it was equally easy to show that the pure stearic acid would be far superior to pure tri-stearine. This led to the isolation of the glycerine, a by-product which has since become of the greatest value, and to the separation of the oleic acid, a second by-product, highly valued in itself in soap-making, and for softening wool and dressing leather.

Uses of oleic acid. 74

Superiority of moulded to dipped candles. 75

PURIFICATION OF OILS.

PURIFICATION OF OILS.

Removal of glycerine. 76

REMOVAL OF GLYCERINE.—Space cannot be afforded to deal with this subject in the detail which its great importance deserves; we can but summarise the principal methods. It may be accomplished—

1st—By presenting to the fatty acid a more powerful base than glycerine. When this is done the acid leaves the glycerine and forms a soap with the new base. Common soap is a compound of a fatty acid with soda as a base, soft soap only differing in having potash. Both the soaps and the alkaline bases used in their formation are soluble in the above cases, but it might be possible to use a base which would form an insoluble soap. This would be precipitated, and thus removed from the soluble glycerine. Thus, if oil or fat were to be boiled with dry lime or with oxide of lead, the compouud formed with the fatty acids in displacement of glycerine would be insoluble.

By a base. 77

The base first used and still employed by the refiner of tallow was milk of lime, in the proportion of about 14 per cent. of lime to the weight of tallow. The lime soap thus formed is afterwards decomposed, and the fatty acids liberated. This is best effected by means of dilute sulphuric acid, gypsum being formed, and a mixture of stearic, palmitic, and oleic acids obtained. By a further stage the oleic acid is removed, and the mixture of stearic and palmitic acids washed and prepared for the candle-maker.

Mixed stearic and palmitic acids. 78

2nd—The fatty acids may be separated from the glycerine by an opposite process, *viz.*, by presenting a stronger acid to combine with the glycerine, or to decompose it; for this purpose sulphuric acid is generally employed. This method has been greatly improved and simplified by Dr. Bock of Copenhagen, and is perhaps the most scientific of all the operations at present in use.

By an acid. 79

3rd—The most recent process is, however, that in which advantage has been taken of the fact that if heated in a confined atmosphere of steam

By dissociation. 80

PURIFICATION OF OILS.

the fatty acids will separate from the glycerine. If heated, however, above a carefully-determined point, they are resolved to lower compounds, and at still higher temperatures, they are reduced to their ultimate elements—carbon, oxygen, and hydrogen. The dissociation by heat has, therefore, to be carefully regulated, and the danger of this process lies in the risk of explosion from the high pressure exerted in the boilers or heating chambers. If the boiling with lime or with sulphuric acid be combined with a high pressure, the quantity of the alkali or of the acid used may be greatly diminished and thus in a measure the principles of the first and third or the second and third processes may be combined.

Advantages of the Removal of Glycerine.
81

But whichever principle the manufacturer may choose to adopt, the glycerine is now entirely removed from the fat or oil employed in candle-making. The result of this has been that the modern candles are far less greasy and much more luminous. They have come to approach the more expensive wax and spermaceti candles which owe their superiority over the old-fashioned tallow candles to the fact that the ingredients of which they are made do not contain glycerine. One immensely important result of the modern chemical manufacture of candles is that it is now possible to produce the solid fatty matters necessary from a vastly larger number of crude materials than formerly. This has had the effect of cheapening candles, and thus the modern advances may be said to have proved that even tallow has its rivals as a candle material, both in point of quality and cheapness. Were the purest tallow only to be used, the third process of separating glycerine, namely, dissociation at a temperature of 570° with superheated steam at 600° at the same time injected into the boiling tallow would, in all probability, be mainly used. The employment of sulphuric acid allows of vegetable as well as of animal oils being utilised. Indeed, mixtures of all sorts of fats, such as kitchen stuffs, slaughter-house refuse, residues of tallow-melting, and waste materials from refining fish and other oils may be purified *direct* by the process of using sulphuric acid and steam together. To accomplish this it is not even necessary that closed vessels should be employed, so that the risk of explosion is entirely removed, and the appliances greatly cheapened. The process, as modified by Dr. Bock, has the advantage of being perfectly applicable to all vegetable as well as to animal oils, and the stearic acid formed is much more solid and even contains a proportion of altered oleic acid which is not only itself solid but seems to increase the solidity of the prepared stearic or mixed stearic and palmitic acids.

Advantage of using Sulphuric Acid.
82

USES.
83

USES OF OILS, FATS, &c.

It would perhaps serve no verygreat purpose to attempt a classification of the uses of oils. In the above remarks they have been grouped according to a Chemico-Industrial system, but among many other uses for oils the following may be specially mentioned and separately dealt with—

I.—As Articles of Diet.
II.—For Illumination.
III.—For Industrial Purposes, *e.g.*, paint, lubrication of machinery, preparation of oil-cloths, mordants in dyeing, &c., &c.
IV.—For Candle-making.
V.—For Soap-making.
VI.—For Medicinal purposes.

As already explained, the essential oils are excluded from considertion in the present article, a place having been assigned to them under **Perfumery**. To the Natives of India, Oils and Fats are chiefly of interest as ARTICLES of DIET, or as a means of ILLUMINATION. They but very

USES.

rarely employ oils in lubrication, and the oil-consuming industries have not progressed sufficiently to necessitate separate notices. House painting is practised by the upper classes only; the manufacture of oil-cloths is to the people of India not only unknown but in their simple lives unnecessary. The dyers, however, fully understand the advantages of certain oils in the production of some of their best tinctorial results. One of the most important Indian uses of oil, and one comparatively unknown in Europe, is the anointment of the person with mustard and a few other oils. The use of soap as a personal detergent cannot be said to be more than a luxury and indeed, to the mass of the people of India an unknown luxury. Crude soap is, however, manufactured, in nearly every town and village, to be used by the washermen and dyers and to a small extent superior soaps are also made (chiefly by the *North-West Soap Company* at Meerut), which are yearly more and more entering into competition with imported or foreign soaps. Candles are also largely employed, and though statistics of their manufacture cannot be obtained, crude tallow dip candles are to some extent manufactured and a large import trade exists in candles from Europe.

The most valuable phase of the oil industry of India is, however, an agricultural one, namely, ***the cultivation of oil seed crops for the export market.*** In the chapter below devoted to the Indian trade in oils and oil-seeds, the reader will discover the magnitude of this interest. Here it need only be stated that the following are the chief agricultural crops of this nature arranged alphabetically:—

1. **Arachis hypogœa**—Ground nut; Vol. I., 283.
2. **Brassica**—Rape and Mustard, Vol. I., 525, 532.
3. **Gossypium**—Various species of Cotton, see Vol IV., 33-37.
4. **Guizotia abyssynica**—Niger seed; Vol. IV., 186.
5. **Linum usitatissimum**—Linseed; Vol. V., 71.
6. **Papaver somniferum**—Opium seed; Vol. VI.
7. **Ricinus communis**—Castor oil; Vol. VI.
8. **Sesamum indicum**—Til or Gingelly; Vol. VI.

Each of these will be found dealt with in detail in its own alphabetical place in this work, but they may be here discussed collectively. According to the Agricultural Statistics of British India (published by the Government of India) the following were the surveyed areas under oil-seeds during the years named:—

	1884-85.	1885-86.	1886-87.	1887-88.	1888-89.
	Acres.	Acres.	Acres.	Acres.	Acres.
Madras	1,169,779	1,364,093	1,548,194	1,509,347	1,566,935
Bombay and Sind	1,976,867	2,013,527	1,971,978	1,813,996	1,796,284
Bengal	Not known.		Not known.		
N.-W. Provinces	447,343	677,450	732,738	695,019	574,835
Oudh	192,832	314,934	298,037	258,090	269,662
Panjab	1,061,518	759,013	631,688	821,170	891,564
Central Provinces	1,632,822	1,705,017	1,516,370	1,416,164	1,302,523
Lower Burma	30,013	28,251	28,071	25,331	36,705
Upper Burma	...	...	...	...	233,648
Assam	146,837	163,353	162,753	154,597	156,107
Berar	856,765	939,082	697,466	586,443	534,535
Ajmir-Merwara	...	...	6,745	18,356	17,871
Manpur (in Central India)	...	...	...	992	1,142
TOTALS	7,514,276	7,964,720	7,594,040	7,299,505	7,381,811

USES.

The above returns do not include Bengal and many minor parts of India, so that they by no means indicate the total acreage under Oil-seeds, In connection with Linseed (p. 60) for example, it will be found that the estimated area under that crop in Bengal is 1,500,000 acres. The productionof Rape and Mustard, of Til and other oil-seeds is also very considerable in the Lower Provinces. But another source of error in these returns lies in the fact that Cotton and Poppy seeds are obtained as by-products from other crops. These seeds, while returned in the foreign traffic of Oil-seeds, are not drawn from the surveyed area of Oil-seeds. The above returns, therefore, chiefly allude to Linseed, Rape, Sesamum, and Castor, and do not even embrace Cocoa-nut production. The figures in the table convey, however, a fairly accurate conception of the agricultural importance of these crops, which will be confirmed by the perusal of the remarks regarding the trade in Oil-seeds, given in the concluding pages of this article.

EDIBLE OILS.
84

I.—EDIBLE OILS.

The following may be accepted as the principal EDIBLE Oils, as also those which are employed to ANOINT the body:—

Aleurites moluccana.
Anacardium occidentale.
Ant-grease—White Ants or Termites; Vol. IV., 471.
Bassia butyracea.
B. latifolia.
B. longifolia.
Brassica campestris—Sarson or Colza is the chief oil used for anointing the person.
B. juncea—Indian mustard oil largely eaten.
Buchanania latifolia—Chironji Oil.
Butter.
Carthamus tinctorius.
Cocos nucifera.
Cucumis Melo.
Dugong oil, or the oil of the Sea Hog; Vol. III., 197.
Eruca sativa.
Garcinia Morella, Vol. III., 475.
Ghí, or Clarified Butter.
Gossypium herbaceum, and other species; Vol. IV., 34.
Guizotia abyssynica—Niger seed. Very extensively used as a substitute for Ghí; Vol. IV., 187.
Helianthus annuus—Sunflower-oil; Vol. IV., 210.
Impatiens racemosa, Vol. IV., 336.
I. Roylei, Vol. IV., 336.
I. sulcata, Vol. IV., 336.
I. Edgeworthii.
Juglans regia, Vol. IV., 550.
Lard.
Linum strictum—Cultivated in Afghánistán for its oil.
L. usitatissimum—Linseed oil.
Olea europœa—Olive Oil.
Papaver somniferum—Opium seed oil.
Perilla ocimoides.
Pistacia vera.
Prinsepia utilis.
Prunus armeniaca.
P. persica.
Sesamum indicum.

Theobroma Cacao.
Viburnum coriaceum.

EDIBLE & ILLUMINATING OILS.
85

II.—ILLUMINATION OILS.

The following are the chief oils specially mentioned by Indian writers as employed for illumination in lamps, *chirags*, &c., but these are only more liquid than the oils said to be suitable for candles.

Aleurites moluccana.
Amoora Rohituka.
Arachis hypogœa.
Argemone mexicana.
Bassia butyracea.
B. latifolia.
B. longifolia.
Brassica campestris.
Calophyllum inophyllum—Pinnay or Domba Oil; Vol. II., 30.
C. tomentosum.
C. Wightianum.
Cannabis sativa—Hemp-seed Oil; Vol. II., 121.
Carappa moluccensis, Vol. II., 141.
Carthamus tinctorius—Safflower; Vol. II., 194.
Celastrus paniculatus, Vol. II., 238.
Cerbera Odollam, Vol. II., 256.
Citrullus Colocynthis, Vol. II., 329.
C. vulgaris, Vol. II., 332.
Cnicus arvensis, Vol. II., 378.
Cocos nucifera—Largely used for lamps.
Eruca sativa, Vol. III., 267.
Garcinia Morella, Vol. III., 475.
Ghí—Used by the people of Lahoul; Vol. III., 491.
Gossypium herbaceum, and other species; Vol IV., 33-37.
Guizotia abyssynica—Important oil among the aboriginal tribes of the Central tableland of India, such as in Chutia Nagpur.
Helianthus annuus, Vol. IV., 210.
Hibiscus cannabinus, Vol. IV., 236.
Impatiens racemosa, Vol. IV., 336.
I. Edgeworthii, Vol. IV., 336.
Jatropha Curcas, Vol. IV., 546.
Linum usitatissimum, Vol. V., 71.
Litsæa sebifera, Vol. V., 83.
L. umbrosa, Vol. V., 84.
L. zeylanica, Vol. V., 85.
Manalú Oil of Kanara; Vol. V., 143.
Mesua ferrea, Vol. V., 237.
Papaver somniferum, Vol. VI., Pt. I., 21.
Persea gratissima, Vol. VI., Pt. I.
Petroleum, Vol. VI., Pt. I.
Pistacia vera, Vol. VI., Pt. I.
Pongamia glabra, Vol. VI., Pt. I.
Prinsepia utilis, Vol. VI., Pt. I.
Prunus armeniaca, Vol. VI., Pt. I.
P. communis, Vol. VI., Pt. I.
P. persica, Vol. VI., Pt. I.
Putranjiva Roxburghii, Vol. VI., Pt. I.
Ricinus communis—Extensively used for lamps in Bengal; Vol. VI., Pt. I.
Santalum album, Vol. VI., Pt. II.

ILLUMINATING OILS.

Sarcostigma Kleinii, Vol. VI., Pt. II.
Schleichera trijuga, Vol. VI., Pt. II.
Sesamum indicum—Gingelly or Til largely used as a lamp oil; Vol. VI., [Pt. II.
Tallow or **Murungana,** Vol. VI., Pt. II.
Theobroma Cacao, Vol. VI., Pt. II.
Vateria indica, Vol. VI., Pt. II.
Viburnum coriaceum, Vol. VI., Pt. II.
Xanthium strumarium, Vol. VI., Pt. II.

OILS FOR ARTS & INDUSTRIES. 86

III.—OILS USED IN THE ARTS AND INDUSTRIES.

(OTHER THAN CANDLE AND SOAP MAKING.)

It has already been explained that in India this is an unimportant class, The chief feature of these oils consists in their DRYING OR NON-DRYING property. As this subject has already been discussed, and brief lists of the better known examples of each kind have been given above (p. 450), nothing further need be said in this place. The more important are those employed for lubricating machinery (non-drying oils) and for paint (drying oils).

CANDLE-MAKING. 87

IV.—CANDLES & THE MATERIALS USED IN CANDLE-MAKING.

HISTORY.—Perhaps the simplest and earliest form of illumination, after the torch (λαμπάς, a torch, a light, or lamp), was the rush-light. In some parts of Scotland, this is used even at the present day, and in England, it was to be seen down to the close of the eighteenth century, and was generally known as the *cruisie*. It was made of stone, metal, or pottery, in the shape of an oblong shallow basin with a tapering spout or nozzle at one end, in which the wick rested, and a rounded, somewhat deeper, portion behind in which the oil was mainly contained. According to some writers this form is an adaptation from a still more primitive conception in which a bivalve shell served the purpose of the *cruisie*. Be that as it may, the rush-lamps of Britain were similar to the commoner sort of *chirágs* which are met with in India now and which have been in use for centuries. The Greek and Roman lamps of this type were often enclosed and had a small hole on the top to admit of being refilled with oil, and very often possessed in addition a handle at the opposite extremity to the nozzle. Sometimes instead of one nozzle, they were so constructed as to have 2, 3, 5 or more wicks, and in place of resting on a ledge or bracket were often suspended by means of a chain. In India may be seen in use almost every form hitherto discovered among the ancient ruins of the world. At Chittagong the pottery *chirág* is partly enclosed and has a tapering handle at the opposite end to the nozzle. In Mysore, completely closed earthen *chirágs* or lamps, with 3 or 5 wicks, are met with, and these are suspended by a chain which is also made of pottery. The sweetmeat-seller of the streets—the *halwai*—has, in every province of India, a brass lamp attached to the brim of his basket. This may be described as a 3-or 5-wicked *chirág*, borne upon an elegantly-curved pedestal. The *shama-dán* of the houses and shops of India consists of an upright stand on which is placed an open circular basin for oil, with 4, 5, 7, or 9 short spouts or nozzles in which wicks are deposited. But although the *cruisie* lamp of Britain may be said to survive in India to the present time, the wick is never so primitive as that which prevailed in Scotland less than half a century ago. Instead of the pith of a rush, the wick in the Indian *chirág* consists of a woven tape or thread of cotton, or simply of cotton twisted into a wick of a desired thickness.* In some respects these *chirágs* are thus improvements

* See the article Cotton (**Gossypium**) for ome account of the Indian consumption of cotton in the manufacture of wicks. (*Conf.* with Vol. IV., 61.)

USES.

on the *cruisie*, but they are far inferior in point of illuminating power to candles, and still more so to the modern improved oil lamps. It is, therefore, to be expected that, with the advances of civilisation, they should year by year steadily disappear in India. Already the consumption of candles has assumed a by no means inconsiderable proportion. The value of the imports of foreign candles were in 1874-75 R4,81,937, and ten years later, *viz.*, in 1884-85, R10,50,227, last year (1888-89) R7,79,759. During these same periods, the imports of kerosine and other mineral oils were valued (in 1874-75) at R27,17,112; (in 1884-85) at R1,15,82,217, and last year at R1,86,12,803. These figures powerfully demonstrate the rapidity with which the *chirág* and *shama-dán* are being displaced by the candle and the kerosine-oil lamp.

Candle-making.

From this point of view, it may not be considered out of place to attempt to lay before the reader, *1st*, a brief history of the growth and present position of the European candle industry; *2nd*, to show the extent to which effort has been made in India to meet its own wants in the matter of candles; and *3rd*, to very briefly discuss the merits of the Indian animal, vegetable, and mineral oils which may possibly come to be used, or are at present used, either in India or Europe for the purpose of being made into candles.

The use of wax candles, as a source of artificial light, dates in Europe from the middle ages, and there is every reason to believe that in Asiatic countries, such candles were also made long before the application of the much cheaper substance, tallow, was thought of. In a further page, it will be seen that the process of making tallow candles, as followed in India, is quite different from the European practice, and more closely resembles the method adopted in the wax-candle trade. There could for India be, perhaps, no safer or more effectual step towards improvement than to educate our Native candle manufacturers in the art which, for nearly half a century, has disappeared from Europe. It was a vast improvement on the clumsy method followed by the *batti-saz* (or tallow-candle manufacturers) of the Indian towns and villages of to-day. Although long since superseded in Europe by the advances made in machinery and the economies accomplished through the aid of chemistry, it had the advantage of being adaptable to a people not possessed of much capital. The allusion to Indian-made tallow candles has, however, been given in this place mainly from a supposition that the survival of the method originally pursued in the wax-candle trade may be accepted as an adaptation to tallow of a practice which may have existed in ancient times in Asia, as we know it did in Europe. At all events, wax-candles are used, during certain religious observances, by the Muhammadans of India, and have been so employed by the Chinese for many centuries. Wax-candles are also regularly used at Buddhist temples. It is thus probable that the discovery of the art of making candles from bees'-wax was contemporaneous in Europe and Asia. Being, however, expensive they were in Europe ultimately, but not until modern times, superseded by tallow-candles, and it seems probable that the use of tallow has only been introduced into India within the past few years.

Indian Tallow Candles. 88

An adaptation from the art of making wax-candles. 89

CLASSIFICATION OF CANDLES.

Classification of Candles. 90

Candles may be referred to two classes, according to the manner in which they are made, *viz.*, "dipped" and "moulded," but they may also be viewed with reference to the material or materials of which they are composed. Thus, for example, there are wax-candles, tallow-candles, paraffine candles, spermaceti-candles, composition-candles, stearine-candles, and palm-oil candles. A candle may be defined as a cylindrical

USES. rod of solid fatty matter enclosing a central wick. Its two component parts are, therefore, an oil or fatty matter and a wick. The modern improvements have been directed towards each of these components, and there is now nearly as much science displayed in the manufacture of the wick as in the purification of the oil or fat.

Materials for Candle-Making. 91

MATERIALS USED FOR.

The following are a few of the chief materials of which candles are made:—

Oil-Seeds. 92

I.—OILS FROM OIL-SEEDS.—The chief Indian fatty oils have already been enumerated and classified, and the further particulars regarding these will be elaborated by the succeeding pages, so that it is only necessary to fix their position here as the materials of which candles are or may be made. The cocoa-nut oil is perhaps, however, one that deserves special mention. See the history of its introduction by **Price & Co.**, Vol. II., 441.

Bees'-wax. 93

Wax-candles. 94

II.—BEES'-WAX AND WAX-CANDLES.—Bees'-wax is even to the present day a substance of the greatest importance in the candle trade, although the manufacture of candles from it entirely has greatly declined. England obtains her supplies chiefly from Corsica, but smaller amounts are also furnished by India, Ceylon, North America, and Brazil. The following facts regarding the Indian foreign trade in bees'-wax may be found of some interest in the present connection, since it has been found impossible to discover anything definite regarding the internal trade or the extent to which the Indian demand for wax-candles is met by an indigenous industry.

The exports of Indian wax were in 1875-76 valued at R6,21,890, in 1880-81 at R5,45,110, and in 1888-89 at R4,20,959. The trade would thus seem to be falling off, and perhaps a larger amount of wax is now being used up in India. The imports of foreign wax (excluding wax-candles) were valued in 1875-76 at R20,010, in 1880-81 at R1,45,467, in 1885-86 at R32,244, and in 1888-89 at R99,272; thus while the exports of Indian wax have shown a tendency to decline, the imports of foreign wax have greatly improved. The re-exports of foreign wax are unimportant; they were valued in 1875-76 at R180, in 1880-81 at R34,062, in 1885-86 at R5,411, and in 1888-89 at R11,978. The bulk of the imports of foreign wax came from the Straits Settlements and Hong-Kong, and was delivered mainly at Bombay.

Bees'-wax has a peculiarity which has determined the method by which alone it can be made into candles, namely, that it contracts to a very considerable extent on cooling. Were candles to be made by throwing the hot liquid wax into moulds, it would cool first of all, on the outer surface, in contact with the mould, and finally assume the form of a tube with the wick lying loose in the interior. To avoid this wax-candles are made by

European wax-candles. 95

suspending and stretching the wicks from a ring, held over the basin of melted wax. By means of a ladle the wax is poured through openings leading to the wicks, and as it trickles down it cools and adheres to the wicks. This is repeated, layer upon layer, until the desired thickness is obtained. While still hot, the candles are rolled on a table of hard wood or marble, kept wet; in this way they are consolidated and made to assume the desired form. Recently, this process has been greatly improved by **Messrs. Reiss & Co.**, wax heated to the plastic extent only being forced through a tube along which the wick has been arranged.

Other Insect waxes. 96

Before being used for candles the bees'-wax is bleached by being cut into thin ribbons and exposed in glass-houses to the sun until the whole of the yellow colour has been removed. It is more expeditiously bleached by means of sulphuric acid and potassium bichromate, the chromic acid liber-

USES.

ated removing the colour. Chlorine destroys the wax and hence cannot be used to bleach it, but cream of tartar or alum may be employed, and the French manufacturers actually use these reagents. English wax-candles and English refined wax are superior to the same articles made on the continent of Europe.

Other kinds of Wax. 97

III.—OTHER KINDS OF WAX.—In addition to bees'-wax there are several kinds of vegetable waxes and waxes produced on plants through the parasitic action of insects. The best known of the latter class is the so-called Indian white wax. This is produced by the action of **Ceroplastes ceriferos** on the young twigs of **Terminalia Arjuna** and several other trees. *Spons' Encyclopædia* recommends that an effort should be made to ascertain if it would pay to propagate this insect. A reference has been addressed on the subject, by Her Majesty's Secretary of State, to the Government of India, and the subject is now being investigated by the Indian Museum authorities at Calcutta. The promised forthcoming report may, therefore, be awaited with interest, but from what the writer has seen of the samples collected in connection with the enquiry it is to be feared India is not likely to take a very important place in the world's supply of these waxes. At present, however, little or nothing is known regarding the Indian vegetable waxes. For further information see the account below.

IV.—TALLOW AND TALLOW-CANDLES.—Tallow is simply beef or mutton fat, or a mixture of both. If it be intended to make moulded tallow-candles the finest mutton suet can alone be used, but for "dip" candles the refuse from the mould candles or the cheapest tallow (a mixture of all animal fats) may be employed. "Dips" or the candles made of tallow, by the process technically designated "dipping," were in use long before the invention of moulding was thought of.

Tallow Candles. 98

In 1875-76, India exported tallow to the value of R70,122, in 1880-81, R37,420, in 1885-86, R51,914, and in 1889-90, R71,675. Tallow was imported in 1875-76 to the value of R42,157, in 1880-81, R56,010, in 1885-86, R49,668, and in 1888-89, R60,622. As in the case of wax so with tallow, the exports are falling off and the imports improving. The trade is chiefly conducted with Bombay, and of the imports in 1885-86, R37,588 and in 1889-90, R44,890 seem to have come from Great Britain. Whether or not this improving trade in wax and in tallow can be taken as an indication of the growing importance of the (*batti-saz*) candle-maker's industry of India must be a matter of opinion, since the manufacture is mainly in the hands of single workmen scattered all over the country.

CANDLES MADE IN INDIA.

INDIAN CANDLES. 99

INDIAN-MADE DIP CANDLES.—The following account of the art of candle-making, as practised in India, expresses very nearly all that is known regarding the industry which from the consumption of candles, must, however, be extensive:—

"The process of candle-making as practised in Lucknow by the chandler or *Batti-saz* is very simple, and only the rudest kinds are made, resembling what are called in England 'dips.'

Native-made Dips. 100

"The Native chandler splits a long bamboo and makes a large hoop with it, and this he suspends from the ceiling, or else he has a rough round table. At intervals in the circumference he cuts notches or grooves, and from these suspends country cotton thread. He boils up fat (*charbi*) in a caldron, and when it is at boiling point he takes a huge spoon or ladle (*Karchhá*) with a hole in the bottom, and filling it so places it that the cooling liquid trickles down the suspended thraed. These candles are made without reference to weight, but solely with reference to length, which ranges from a span (*bálisht*) to an ell (*háth thar*).

INDIAN CANDLES. **Native-made Dips.** "Six seers of raw *charbi* cost about one rupee, and when boiled are reduced to $5\frac{1}{2}$ seers. The fuel consumed is 5 seers of wood, costing one anna. Two men will make up a *pachenel* (miscellaneous) lot of candles in a day with these materials, and one and a half *chittacks* of country thread to make wicks. The cost of thread will be about one anna. The workmen receive only 6 *paisa* each per diem, and one anna for miscellaneous expenses: total R1-6 per diem. Thus, for an expenditure of R1-6, the *Batti-sas* produces $5\frac{1}{2}$ seers of candle. These he sells wholesale at $2\frac{3}{4}$ seers for the rupee. This gives him 10 annas profit. The retailer also sells by weight at $2\frac{1}{2}$ seers per rupee.

Coloured Candles. **101** "Coloured candles are manufactured in the same way. The colour is added to the fat when it is boiling, and is estimated to add about one anna to the cost price per seer. Wholesale dealers sell coloured candles by weight, but retailers sell them by number." (*Hoey, Trade and Manufactures of Northern India.*) It is interesting to add that the Lahore Gazetteer (p. 100) alludes to a new industry started in that city in which tallow candles are actually moulded. These are described as clean and free from smoke "as compared with the oil *shama-dán* or *chirág*," and hence "it is no wonder they are coming into favour." **Balfour** states that candles are made at Vizagapatam, Goa, Malabar, Patna, Calcutta, and Berhampore. It has already been shown that the largest trade in both wax and tallow is conducted in Bombay, a fact that would perhaps point to a considerable candle industry as probably existing in the Western Presidency.

MOULDED CANDLES.—For an account of the various methods pursued in moulding candles, the reader is referred to the numerous technical works that exist on this subject, or to the writer's pamphlet on Candles, which appeared in the Selections from the Records of the Government of India for 1888-89. It need only be said, here, that recently a factory has been started in Lahore for the purpose of making candles by this process. It is not known to what extent the Bombay Steam Soap & Candle-works Company ever succeeded in making candles on the improved methods of Europe, but the Company having closed its works it may be assumed that competition with European articles was too keen.

It has already been remarked that perhaps as much science has been displayed in the preparation of the wick as in the purification of the fat. If saturated in borax it is found to be so weighted by the formation of an incandescent bead in the flame, that it turns over and thus protrudes and is rapidly oxidised. In that position it is consumed without necessitating the use of snuffers—the wick is trimmed automatically.

Spermaceti. **102** V.—OTHER FATS AND OILS EMPLOYED IN THE EUROPEAN CANDLE TRADE.—After tallow the most important of the animal oils, used in candle-making, is undoubtedly spermaceti. This is obtained chiefly from the oil bladders of two species of toothed whales, *viz.*, **Catadon macrocephalus** and **Physeter macrocephalus.** The oil derived from those sources is, however, too highly priced except, for the most expensive candles. It is in great demand for lubricating machinery. Common whale-oil, or that obtained from the whalebone whale (or toothless whale), is much inferior, and is

Whale-oil or Train-oil. **103** generally known in trade as *train-oil.* This is rarely employed in candle-making except in the form of a refuse oil, which is purified for inferior candles.

Palm oil. **104** The principal vegetable oil used for candles is that known as *palm oil.* This is obtained mainly from the fruits of **Elæis guineensis,** a palm found in great abundance in Ashantee and Dahomey. A considerable amount of the palm oil of commerce is now obtained, however, from America, being expressed from the nuts of **Elæis melanocca.** It seems probable that a certain

INDIAN CANDLES. **Annual consumption of Palm oil.** **105**

amount of cocoa-nut oil is also being used for this purpose, but definite information cannot be discovered. The odour of cocoa-nut oil seems a strong objection to it. About 25,000 tons of palm oil are annually used by the British candle manufacturers.

Mineral oil. **106**

Perhaps the most remarkable advance in the manufacture of candles has been the utilisation of several mineral oils. It would be of the greatest interest to treat of this subject, but space cannot be afforded. The principal oils of this nature employed for candle-making are paraffine and ozokerit or earth-wax. (*Conf.* with **Petroleum,** Vol. VI.)

Paraffine. **Ozokerit.** **107**

Earth-oil. **108**

V.—SOAP & MATERIALS USED IN SOAP-MAKING.

SOAP MATERIALS. **109**

To the Natives of India soap is not of much importance. Soap-substitutes (see the article **Detergents,** Vol. III., 84-92) may be said to take its place. A crude soap is, however, extensively made and used by the washermen and dyers, but the North-West Soap Company, Limited, at Meerut, seem to be yearly gaining ground. Their soaps are by many regarded as quite equal to the best imported goods. Last year the value of their manufactures was returned at R1,00,867. **Mr. Baden Powell** alludes to soap of good quality as made at Sialkot and Gujrat. **Lieutenant H. P. Hawkes** (***Report upon the Oils of South India, pp. 54-55***) gives particulars of soap manufacture at Tanjore, and gives preference to the Illupu oil (**Bassia longifolia**) over all others for soap-making. Pinnay and Margosa oils he considers about equal in value to Gingelly and Ground-nut, but all these inferior to ***Illupu.*** Soap manufacture is described in many of the Gazetteers, &c., as, for example, the following :—Bombay Administration Report, 1871-72, pp. 366-373; **Bombay Gazetteers, Vol. III.,** 76, 250; **Vol. IV.,** 134; **Vol. VI.,** 57; **Vol. VIII.,** 261; **Panjáb Settlement** Report of Jhang, p. 63; Gazetteers :—**Amritsar, p.** 41; **Sháhpur,** 75; **Ráwal Pindi,** 90; **Lahore,** 100; **Gurdáspur,** 63, &c.

The following are the chief oils used by the Natives of India or exported for the purpose of making soap :—

Arachis hypogœa—Largely used for this purpose; Vol. I., 282.
Bassia butyracea, Vol. I., 405.
B. latifolia, Vol. I., 407.
B. longifolia—Crushed seeds used as soap; Vol. I., 416.
Camellia theifera, Vol. II, 83.
Cannabis sativa—Hemp seed, used in Europe for soft soap; Vol. II., 121.
Cocos nucifera—Said to be of growing importance in European soap manufacture, especially the inferior qualities; Vol. II., 437-444.
Helianthus annuus—Sunflower oil; Vol. IV., 211.
Linum usitatissimum—Used for making soft-soap; Vol. V., 71.
Melia Azadirachta, Vol. V., 212.
Olea europœa, Vol V., 485.
Persea gratissima, Vol. VI., Pt. I.
Sesamum indicum—Very important in India, and its use in Europe has been recently greatly extended.
Tallow or Murungana, Vol. VI., Pt. II.
Theobroma Cacao, Vol. VI., Pt. II.
Vateria indica, Vol. VI., Pt. II.

The Report of the Conference held at London in connection with the Colonial & Indian Exhibition for 1886 on the subject of oils suitable for candles and soap, states that "Of all the oils shown it was thought that **Arachis hypogœa, Bassia latifolia,** and **Cocos nucifera** were the most hopeful for the soap trade. Considerable interest was, however, shown in the possibility of India being able to extend the cultivation of the olive. It was thought if this could be done the consequences would be serious to some parts of Europe, but invaluable to the numerous industries that were driven to seek substitutes for the already too highly priced olive oil." Since the date of that report, strenuous efforts, it is believed, have been made to extend olive cultivation in Kashmír. In Quetta it is a fairly plentiful tree, and might be there more largely grown.

OILS FOR MEDICINAL PURPOSES. 110

VI.—OILS EMPLOYED FOR MEDICINAL PURPOSES.

The following are the Oils chiefly used in India:—

Albizzia Lebbek used in leprosy.
Allium Cepa—Onion, expectorant, &c.
A. sativum—Garlic, stimulant.
Amomum subulatum—Aromatic stimulant.
Anacardium occidentale—External irritant, Leprosy, &c.
Aquilaria Agallocha.
Arachis hypogœa—Ground-nut.
Argemone mexicana—External use in ulcers, &c.
Atalantia monophylla—Rheumatism.
Bassia butyracea—Rheumatism, &c.
B. latifolia—Said to be useful in cepalalgia.
B. longifolia—Detergent.
Benincasa cerifera—Anthelmintic.
Cæsalpinia Bonducella—Useful in convulsions.
Calendula officinalis.
Calophyllum inophyllum—External use in rheumatism.
C. Wightianum—Leprosy and cutaneous affections, scabies and rheumatism.
Carthamus tinctorius—Ulcers, &c.
Cedrus Libani, *var.* **Deodara**—Oil obtained from the wood—ulcers, sores in cattle, &c.
Celastrus paniculatus—**Oleum nigrum**, is used in ***beri-beri.***
Croton oblongifolius—Purgative.
C. Tiglium—The Purging Croton, a drastic purgative.
Cuminum Cyminum—Stimulant.
Cynometra ramiflora—Leprosy, &c.
Dalbergia lanceolaria—Rheumatism.
D. Sissu.
Dipterocarpus alatus,
D. lævis,
D. tuberculatus,
D. turbinatus, } Leprosy.
Dugong—Substitute for Cod-liver oil.
Gynocardia odorata—Chalmugra—Rheumatism, phthisis, leprosy, &c.
Holarrhena antidysenterica—A thick red oil used medicinally.
Hydnocarpus Wightiana—Ulcers.
H. venenata—Leprosy.
Jatropha Curcas—Purgative, also external application.
Lard.
Linum usitatissimum—External.
Mallotus phillipinensis—Cathartic.
Melaleuca Leucadendron—Cajput—Stimulant and diaphoretic.
Melia Azadirachta—Nim or Margosa—Anthelmintic and externally to ulcers, rheumatism, headaches, &c.
Mesua ferrea—External for sores.
Moringa pterygosperma.
Myristica malabarica—Ulcers.
Nerium odorum—Cutaneous diseases.
Olea europea—Olive oil.

Papaver somniferum—Demulcent.
Petroleum.
Pistacia vera—Demulcent and restorative.
Pongamia glabra—A very valuable oil for skin diseases and possibly useful in leprosy.
Ricinus communis—Purgative.
Sarcostigma Kleinii—Rheumatism.
Schleichera trijuga—Skin diseases.
Semecarpus Anacardium—Rheumatism and leprosy.
Sesamum indicum—Demulcent, &c.
Strychnos Nux-vomica.
Vernonia anthelmentica.
Zanthium strumarium.

OIL-CAKES.

OIL-CAKES. III

References.—*Jour. Royal Agri. Soc., England, Vol. X. (1849), 479-494; XII. (1852), 522; XIX. (1858), 515; XXIV (1863), 589; XXV. (1864), 234-239; Vol. XII., Second Series (1876), 296; XIV. (1878), 468; XX., 361; Anderson's Agricultural Chemistry; Morton's Cyclopædia of Agriculture, Art. Oil-cake, Vol. II., 513 to 516; Encyclopædia Brit., Art. Oil-cake, &c., &c.*

The Journals of the Royal Agricultural Society of England have for many years past contained numerous papers which deal with the introduction of the various oil-cakes as articles of cattle and sheep food, or as manures. Much less, however, is said of the uses of these by-products as manures than as articles of food, and it seems probable that in modern agriculture this feature of the oil-cake traffic is less valuable relatively in Europe than in India. The reader will find information regarding the properties of the chief oil-cakes of India scattered throughout this work under the scientific names alphabetically of the various plants from which they are obtained, such, for example, as under **Arachis, Brassica, Cocos, Gossypium, Guizotia, Linum, Papaver, Ricinus,** and **Sesamum.** In European commerce the distinction into these forms of oil-cake is, however, rarely made, and the interest may be said to be concentrated chiefly in the various qualities of cake as articles of cattle diet in which purity or adulteration constitute the chief features of interest. The exact properties of the individual forms of oil-cake do not appear to have been made the subject of any very special investigation. Linseed-cake **(Linum)** would seem to have been first appreciated, and a standard of value became current in trade according to the purity of the sample from admixture with other oil-cakes or adulteration with extraneous and injurious materials. The properties of Rape-seed cake **(Brassica campestris)** would appear to have next been recognised, and a percentage of admixture of this with Linseed was accepted by some purchasers and condemned by others. Mustard **(Brassica alba** and **nigra** of Europe and **Brassica juncea** of India) and Castor-seed **(Ricinus)** were also early viewed as injurious admixtures, and the duty of the Agricultural Chemist, on being consulted by the farmer as to the value of a sample of cake, largely came to be the ascertainment of the presence of castor-seed or of other injurious materials. Shortly after the American War of Secession a strong interest was awakened in cotton-seed cake. Experiments were tried both with the decorticated and the entire seed, and results of such value attained that a position was gradually recognised for this new form of cake. Many other minor forms of cake have also been experimented with, such as ground-nut **(Arachis)** and hemp **(Cannabis)**, but these, instead of assuming a separate position, have apparently drifted into the category of cheap and not very objectionable admixtures. To this class also would seem to belong one or two other oil-seeds and cakes that are exported from India, such as *Til* **(Sesamum)**, Niger-seed **(Guizotia)**, safflower (**Carthamus**), cocoa-nut, &c., &c. It is, indeed, somewhat surprising that a seed so extensively grown in India as *Til*, and one in which the exports to Europe are by no means insignificant, should never, apparently, have attracted separate recognition (*Conf.* with p. 475). For many years past the exports in this oil-seed have been steadily increasing; the trade has expanded from 779,333 cwt. in 1870-71 to 1,775,559 cwt. in 1889-90; but in 1883-84 they were 2,843,382 cwt. The bulk of these exports go to France, but there is nothing to show to the contrary that the cake obtained on the expression of the oil is not being largely used in admixture with standard cakes. The chemical properties and physiological actions of sesamum cake do not appear to have been carefully determined, and it would, therefore, seem a subject well worthy the consideration of Agricultural authorities in Europe to extend their enquiry beyond the mere mechanical examination of prepared cakes to a determination of each of the separate products. Some of these may be far more injurious than is supposed, while others may not have attracted the recognition which they deserve from being used purely as admixtures in the articles sold under various trade names as forms of Linseed-cake.

The volumes of the Journal of the Royal Agricultural Society, cited above, are those which more especially deal with the subject of oil-cakes. These contain many observations of great practical value which it may not be out of place to review here very briefly, the effort being made to carry, as far as possible, the facts brought out by the various writers under the headings of the separate cakes dealt with. **Mr. J. Thomas Way**, then Consulting Chemist to the Society, furnished in 1849 one of the most complete articles which has as yet appeared; though in some features of the enquiry his experiments and chemical results have been since more thoroughly elaborated.

LINSEED CAKE. **112**

I.—LINSEED-CAKE.

Mr. Way says: "The consumers of linseed-cake in this country are almost unanimous in the belief that the different samples of this article are of varying value as food for stock. Where such an opinion as this is general, it would argue little wisdom to dispute its correctness: it only remains to investigate the cause. When linseed, ground into coarse powder and digested with a small quantity of water, with the aid of heat, is exposed to strong pressure, two products are obtained—the one, an oil of well-known characters, linseed-oil; the other, the cake which remains in the press. No other substance but oil (except a small quantity of water) is separated from the cake; and the two products, therefore, correctly represent the composition of the seed from which they are derived. Linseed is known to consist principally of mucilage, or gum, sugar, oil, and albuminous matter—the three former being substances devoid of nitrogen, the latter having the same composition as the flesh of animals, or the gluten of wheat. Now, as linseed-oil contains no nitrogen, it is obvious that the cake must be richer than the seed in albuminous principles in the exact proportion of the oil which it has lost by pressure. Given, then, the composition of any sample of seed, and the quantity of oil which is expressed from it, we have by the simplest calculation the composition of the cake." Proceeding with his enquiry into the supposed different feeding values of the linseed-cakes of various countries **Mr. Way** says: "There are, indeed, two distinct circumstances which might influence the value of cake, independently of its chemical characters—the flavour, which is more or less agreeable to stock, and the state of division—the fineness or otherwise to which the seed is ground. To this latter circumstance, in all probability, very great differences in the feeding properties of specimens of cake might with justice be referred." "Both these causes of variation—the mechanical condition of the cake and

LINSEED.

its flavour—lie, however, out of my province, which is confined to the chemical composition of the substance in question:—

1st—then, we must inquire, Is there any real difference in the chemical composition of various cakes; and, if so, does the distinction apply to those of different manufacture or origin?

2nd—Is the difference (supposing there to be any) sufficient to account for the observed effects in feeding? and

3rd—Is it to be attributed to adulteration, or to necessary variation in the composition of the seed?"

To answer these questions **Mr. Way** performed an extensive series of analyses. While commenting on the chemical composition of linseed he pointed out that the mucilage, sugar, and oil are the fat-forming constituents and the albuminous the muscle-forming. To variations in the proportions of these constituents he accordingly devoted his energies, examining both samples of commercial cakes and of linseed. "In order," he concluded, "that the reader may trace the differences, and that he may also observe the absence of any such distinction in the case of nitrogen, I shall here offer a table of the average composition of the specimens from different sources:—

Mean Composition of Linseed-cake from different Countries.

		Nitrogen.	Oil.	Water.	Ash.
French	8 specimens	4·72	9·06	7·60	7·80
American	7 "	4·74	11·41	7·60	6·35
English	9 "	4·57	13·52	8·60	7·27
German and Dutch	3 "	4·65	9·84	7·98	9·56
Russian	2 "	5·14	11·86	8·88	8·39
Italian	2 "	5·03	11·84	9·03	7·55
Sicilian	2 "	4·72	6·80	9·46	8·02

"From that statement it would appear very evident that the specimens of French, American, English, German, and Sicilian cakes are on *the average* practically alike in regard to nitrogen. Neither should we be inclined to believe that the Russian or Italian would have furnished an exception to this rule, had a sufficient number of samples been examined. On the other hand, there is an obvious difference between the quantity of oil in the cakes of different countries."

Mr. Way then summed up his opinions as follows: "We may conclude this subject by recapitulating the conclusions to which we have been led:—

"1. The samples of cake differ considerably both in the proportion of albuminous matter and of oil contained in them.

"2. That in respect to the former (the albuminous matter) there would appear to be no general distinction between home-made * or foreign cakes.

"3. That in the proportion of oil there is reason to believe a general distinction does exist, more especially between English and French cakes, and in favour of the former.

"4. That there is no reason to believe that linseed-cake is adulteratedat any time with sand or other earthy matter—adulteration by other matters being also rendered unlikely.

* I say "home-made." This term merely implies that the seed was *pressed* at home, there being no evidence to show that the seed in any of the English cakes described was other than foreign seed.

LINSEED.

"5. That different samples of seed, free from admixture of other seeds, or from impurity of any kind, present variations in the proportion of albuminous matter amply sufficient to account for those found in specimens of the cake.

"But whether the differences in chemical composition of the cake are such as to account for observed differences in feeding properties, it is left with the reader to decide."

There are two features in the above five conclusions that seem worthy of comment. The foot-note to No. 2 leaves it open to doubt but that a certain share of the merit of the English cake may have been due to the imports of Indian Linseed. **Mr. Way** does not, unfortunately, seem to have had a sample of Indian cake and seed to compare with the others he examined, but the exports from India to England of linseed at the time of his experiments were very considerable. The other point of interest is contained in his fourth conclusion. A perusal of the numerous papers that have appeared within the past few years in the Journals of the Royal Agricultural Society will suffice to show that at the present day adulteration is being practised to an alarming extent. Indeed, it might almost be said that interest in the subject is now entirely confined to the discovery of easy methods by which adulteration may be detected.

In a paper by **Mr. W. Wright**, on the ***Management of Breeding Cattle*** published in the Journal for 1858, it is stated that the total value of linseed-cake consumed in England during that year might have been estimated at £2,000,000. In 1878 **Mr. J. Algernon Clarke** published in the Journals of the Society a paper on ***Practical Agriculture.*** He there states that in 1878 the United Kingdom imported 190,225 tons of oil-cake, and 1,998,130 quarters of linseed, all but a small portion of which was undoubtedly used for feeding purposes, either as linseed or when made into cakes, and equivalent, at about 14 tons of cake per 100 quarters of seed, to about 280,000 tons of oil-cake. This, he estimates, amounted to about 1 ton of cake to every 100 acres of land cultivated (*Vol. XIV., 468*). "The East India qualities," says **Mr. Wright**, "in general are tolerably even, seldom vary to any appreciable extent, at the respective ports of shipment. They yield the greatest quantity of oil, and after pressure the cake is no doubt more nourishing and valuable than any other. But the bulk of the supplies is furnished by other countries more subject to variation of climate, and uniformity of quality in different seasons is not obtainable. Inferiority in quality is usually accompanied by adulteration, a variety of seeds being found mixed with the linseed, such as wild rape (**Brassica campestris**), wild mustard or charlock (**Sinapis arvensis**), seeds of **Lolium perenne,** of dodder (**Cuscuta epilium**), and of willow-weed and millet (**Panicum miliaceum**). The bulk of the linseed being imported from the Black Sea, the standard of quality and price is chiefly regulated by the importation from thence." **Mr. Wright** then deals with the chief cause of adulteration and the difficulty that exists in checking it. A deputation of Crushers waited on a firm of Greeks in whose hands a large share of the Black Sea import trade rested, in order to remonstrate with them as to the injury that was being done to the trade through adulteration. "The deputation were shown," says **Mr. Wright,** "a sample of pure linseed, and also one containing the admixture complained of, and being requested to state their opinion what price such linseed, on arrival here, would command in the market, were forced to admit that competition for seed amongst crushers, had much influence over the quality of the article supplied, and that the ready sale which could be found for an inferior quality removed all inducement from the seller to ship a genuine article."

Obstacles to sale of pure seed.

"Our brethren across the Atlantic," says **Mr. Wright,** "have studied

LINSEED.

the art of manufacturing linseed-cake. What they produce is faultless to the eye, and, if not always perfection to the taste, it is thin, tender, flaky, and finds a ready sale in this country at the highest market price of the day, the question of genuineness being left entirely for subsequent determination. The cake pleases the eye and is always in good demand; what more is required? Farmers have only themselves to blame in this respect. A considerable quantity of cake is now used for feeding sheep, and for this purpose cake of a tender kind such as this is preferred: *it is produced by the addition of bran in its manufacture.* Thus the farmer has to pay £10 or £11 per ton for cake which pleases the eye in this country, but omits to consider that for this he has to pay dearly for the admixture of bran, which costs £4 or £5 per ton, and in the United States, probably about half that price."

Mixture of Bran.

Many subsequent writers hold that the admixture of bran is, however, a commendable practice, as it checks (by supplying starch) the tendency of pure cake to cause indigestion. **Mr. Wright**, however, furnishes some additional information to what has already been given as to the composition of linseed-cake. The finest quality, he says, is made as follows:—300lb of ground linseed are mixed with 28lb of ground-nut cake, 28lb of rape-seed cake, and 28lb. of bran. The second qualities are made of a smaller admixture of linseed with a considerable part of the undermentioned articles; and the third qualities are made wholly of the last named of the following articles, *without any linseed at all*:—

Composition of the cake of commerce.

Foreign linseed-cake.	Castor-oil cake.
Dodder-seed cake.	Rice-husks.
Poppy-seed cake.	Cotton-seed cake.
African ground-nut cake.	Rape-seed cake.

"Sometimes the whole of the above are mixed and worked together; but a supply of each is not always on hand; and some or all, according to circumstances, are introduced to make up the desired quality."

The late **Professor A. Voelcker**, in a lecture delivered before the Royal Agricultural Society in 1863 dealt with the subject of the *Adulteration of Oil-cakes.* He there stated that "The great demand for all kinds of oil-cake, more especially the great demand for linseed-cake, has led to an amount of adulteration of which the practical farmer is hardly aware." "When we find one cake nearly as white as a poppy-cake, another very dark, and another like American barrel-cake possessing the nice colour of linseed itself, we have here good *primâ facie* evidence that the light cake is probably mixed with white-poppy cake, the dark-coloured with rape-cake, and that the third is in all probability the pure linseed-cake." Doubtless as a practical observation these opinions are correct for linseed-cake as actually met with in trade, but it may be pointed out that India possesses a linseed that is naturally as white as poppy seed, and *it* would accordingly afford an almost white-cake of pure linseed. To what extent the Indian white linseed finds its way to Europe is a subject of some doubt. From its being largely used as an article of human diet in India it would probably afford a considerably more wholesome cake than the ordinary red form. But to continue the review of **Professor Voelcker's** opinions:—"I shall first direct your attention to those admixtures which are in themselves injurious, and secondly, to those which become injurious during the time of keeping the cake. There are also matters added to oil-cakes which deteriorate the quality of the meat produced by imparting to it a bad flavour, so as to lower its market value. Let me first point out what are the characters of genuine or pure linseed-cake. A good pure linseed-cake ought to be made of nothing but pure linseed—practically speaking—pure, not abso-

Colour of.

LINSEED.

lutely pure; for such seed is not to be found in the market. The condition of linseed as imported varies in reality to an enormous extent. Many oil-cake dealers guarantee their cake as genuine, made from genuine seed as imported; but this is really no guarantee whatever." The Professor then exhibited samples of linseed as imported which varied from Bombay good quality with 4½ per cent. admixture to Rijeff common with 70 per cent adulteration. "Now," continued the Professor, "if the seeds that occur in linseed were all of an indifferent quality, that is to say, of a character not injurious to life, the injury would not be so great, but some of them are poisonous. From several samples of linseed I have separated the seeds and ascertained their botanical characters. In one particular sample I counted not less than 29 different kinds of weed-seeds, and among them the following which were more or less injurious: the common darnel, which is frequently present in considerable quantities in the inferior samples of Petersburg seed; corn-cockle, which often produces very serious effects on the animal system; wild radish, which occurs in some samples of Alexandrian seeds, and is very pungent; wild rape, which is not, properly speaking, a rape, but rather a mustard; charlock, or the common wild mustard. All these are seeds which it is positively known are injurious to the health of animals. But there are others which, as I stated at the beginning, impart a disagreeable taste to the meat of cattle fed upon inferior cakes. The Gold of Pleasure, or **Camelina sativa,** is such a seed, giving a disagreeable taste and also a deep yellow colour to the fat of animals. From the appearance of Camelina cake, you would think it ought to be nutritious; but it is an inferior description of cake, because it deteriorates the quality of the meat. Another seed of an injurious character is the purging flax." The Professor next explained to his hearers that a good cake should have a light colour and give an agreeable odour when dissolved in water. "When linseed-cake has been kept for a length of time, its mucilaginous properties more or less disappear. Mucilage is a substance that is very apt to spoil when kept in a damp place. If a cake does not become gelatinous on being mixed with water, it is not one of the best descriptions; but then the reverse does not follow as a matter of course. A cake may become gelatinous, and yet be inferior. *Wild mustard and rape-seed very commonly occur in inferior linseed;* when such cakes are mixed with water a more or less pungent smell of mustard is developed." "Among the common materials used in adulteration is bran, a material which possesses some feeding properties, and may with advantage be given with linseed-cake; but it is much more desirable to buy bran and pure oil-cake separately; for it is impossible to ascertain in what proportion they are mixed." "By mixing linseed with bran, and adding at the same time rape-seed, you make up for the deficiency of nitrogenous matter by the rape-seed. Thus a clever oil-cake mixer may readily produce such a mixed cake as will exhibit upon analysis the same amounts of oil, flesh-forming, and albuminous matter and the other constituents as are usually found in genuine pure linseed-cakes. The mere analysis, therefore, does not give any idea of the purity of the cake."

Meat flavoured by Cake.

Moulded Cakes poisonous.

Professor Voelcker concluded his lecture on oil-cakes by alluding to the poisonous nature of moulded cakes. The mixture of bran, he says, "is very apt to produce mouldiness. The cake is more porous and softer than pure linseed, and consequently far more liable to deterioration in a moist atmosphere than pure cake. In fact, it would seem that all mixed cakes are more liable to this degeneration than pure ones of whatever seed they may have been formed." **Professor Voelcker** thought that the poisonous property of mouldy cake was due to the growth of a fungus. "One sample of such cake," the Professor stated, "had, to his knowledge, killed not less

LINSEED.

than fourteen sheep, three horses, and a pony belonging to a gentleman who gave them only small quantities, the whole of the animals having been killed within two days." **Professor Voelcker**, in an essay on *The Influence of Chemical Discoveries on the Progress of English Agriculture* (Journal, Vol. XIV., New Series, 571), seems to take, however, less exception to rape-seed cake than might be inferred from the above passages. "Linseed and Rape-cake," he there says, "especially the former, are largely used for feeding and fattening purposes, and if pure and in good condition, no food is considered to equal linseed-cake for rapidly fattening sheep and oxen" In the annual issue of the Journal for 1884, **Professor Voelcker** publishes some useful facts regarding *Hard-pressed and Indigestible Linseed and Decorticated Cotton-cakes.* "Several cases," he says, "were brought under my notice in which the use of linseed-cake was suspected to have done injury to stock. A careful examination of the suspected cakes, however, showed that no ingredients prejudicial to health were present, nor that the cakes in question were mouldy or in a condition of incipient decomposition. The linseed-cakes complained of were, however, without exception, very hard-pressed—some as hard as a board, and on analysis proved to be greatly deficient in oil. Owing to the improved machinery which of late years has been introduced pretty generally in the American and in not a few English oil-mills, linseed is much more thoroughly crushed and harder pressed than in former years, and is thereby deprived more efficiently of its oil. Of all the constituents of food, ready-formed oil or fat unquestionably is the most valuable. Linseed-cake, comparatively speaking, poor in oil, is consequently less valuable than cake richer in that substance. In round numbers 1℔ of fat goes as far as 2½℔ of starch or sugar in fulfilling similar functions in the animal economy. Apart from the greater feeding and fattening value of linseed-cakes rich in oil, such cakes possess the further advantage of being softer and more readily broken up into small bits than hard-pressed cakes, which, as a rule, are deficient in oil." **Professor Voelcker** then furnished the following analyses of five samples of hard-pressed linseed:—

Hard-pressed Cakes.

COMPOSITION OF HARD-PRESSED LINSEED-CAKES.

	No. 1.	No. 2.	No. 3.	No. 4.	No. 5.
Moisture	12'02	11'03	12'40	11'01	11'10
Oil	6'76	7'93	7'37	6'33	6'83
* Albuminoids	31'75	27'81	28'31	32'94	25'87
Mucilage, sugar, and digestible fibre.	33'62	35'99	36'13	35'07	36'82
Indigestible woody fibre (cellulose).	10'37	7'16	9'39	8'90	10'73
† Mineral matter (ash)	5'48	10'08	6'40	5'75	8'65
	100'00	100'00	100'00	100'00	100'001
* Containing nitrogen	5'08	4'45	4'53	5'27	4'14
† Including sand	0'19	4'74	1'35	0'15	3'65

Commenting on these analyses **Professor Voelcker** says that sample No. 5 refers to an adulterated linseed-cake, which, moreover, was made from dirty, that is to say, badly screened seed; "but No. 4 cake I found," he adds, "was made entirely from unusually clean linseed. Unfortunately it was very hard-pressed, and poor in oil in consequence, and I have no

LINSEED.

doubt all the more indigestible because it was so rich in albuminoids." The person who submitted that cake for report informed **Professor Voelcker** that it had killed two cows, upon which the Professor wrote: "On analysis I did not find a trace of any metallic or mineral poisons in it, nor did the microscope reveal any injurious or questionable weed-seeds; the cake, indeed, was made, as just stated, from unusually clean linseed and nothing else, but, unfortunately, it was as hard as a board. I think it very likely that it caused the death of the cows, not because it contained any positively poisonous ingredients, but because it was poor in oil and rich in nitrogenous compounds, and was in a mechanical condition in which such cake, as usually given to stock, is unquestionably most indigestible."

"For comparison with the preceding analyses, I append the following three—one representing the composition of a pure linseed-cake of *average* quality, somewhat poor in oil; the second showing the composition of *superior* linseed-cake, and the third that of a pure linseed-cake of the *best* quality:—

COMPOSITION OF THREE PURE LINSEED-CAKES.

	No. 1.	No. 2.	No. 3.
Moisture	13·25	12·44	12·06
Oil	10·33	12·16	13·76
* Albuminous compounds	32·06	29·56	29·31
Mucilage, sugar, and digestible fibre	30·80	31·18	29·96
Indigestible woody fibre (cellulose)	8·26	8·67	10·13
† Mineral matters (ash)	5·30	5·99	4·78
	100·00	100·00	100·00
* Containing nitrogen	5·13	4·73	4·69
† Including sand	0·40	1·29	0·64

The reader should consult the article **Linum usitatissimum**—Linseed—Vol. V., 56 to 77, for further information.

RAPE & MUSTARD 113

II.—RAPE AND MUSTARD OIL-CAKES.

One of the earliest statements regarding Rape-seed-cake is that published by **Mr. J. Thomas Way** in 1849. In his paper on this subject he gives the following tables of the analysis of rape-cake and rape-seed:—

ANALYSIS OF RAPE-CAKE.

	Nitrogen.			Oil.	Water.	Ash.
	1st Analysis.	2nd Analysis.	Mean.			
Sample No. 1	5·29	5·17	5·23	11·63	7·06	5·70
Sample No. 2	5·62	5·54	5·58	10·62	6·62	10·41

RAPE & MUSTARD.

ANALYSIS OF RAPE-SEED.					
NITROGEN.			Oil.	Water.	Ash.
1st Analysis.	2nd Analysis.	Mean.			
4·17	4·26	4·21	37·84	6·44	3·31

"All that can be said from the above analysis of rape-cake is, that it is very much the same with that of linseed-cake. Till lately, however, the hot flavour of rape-cake has been an insurmountable objection to it as a substitute for the more costly article. Rape-cake has been more often employed as manure. The following analysis of its ash, made for me by Mr. Eggar, will not, therefore, be without interest :—

Valuable manure.

Composition in 100 parts of the ash of Rape-cake.

Sand and Silica	13·07
Phosphoric Acid	32·70
Carbonic Acid	2·15
Sulphuric Acid	1·62
Lime	8·62
Magnesia	14·75
Oxide of Iron	4·50
Potash	21·90
Soda	...
Chloride of Potassium	0·17
Chloride of Sodium	0·40
	100·00

"The ash of rape-cake is, therefore, in every respect the counter-part to that of linseed-cake. A ton of the cake will contain 128℔ of mineral matters, one-third of which is phosphoric acid, one-fifth potash, and one-seventh magnesia. Even although, as in the second specimen of cake analysed, the total quantity of ash should be greater, the above numbers will hold good, the excess being sand."

In Morton's *Cyclopædia of Agriculture* it is stated that "Rape-cake which has for a long time been used on the Continent, in the form of rape-dust, as an excellent manure, has lately been imported into this country in large quantities; and is now used both for feeding cattle and sheep, and for manuring the land, much more generally than was the case some years back." Then, again, after the remark that rape-oil is apt to turn rancid and the quantity of it left in the cake is liable to give a flavour objected to by cattle, the following occurs : "Sheep, however, as shown by Mr. Pusey, do not dislike rape-cake so much as cattle, and thrive upon it remarkably. Rape-cake being much cheaper than linseed has been for this reason recommended for the fattening of sheep."

Rape-dust.

Rape-cake recommended for Sheep.

From the above sketch of the chief opinions that have been advanced regarding rape and mustard cake mixed with linseed, it will be seen that both are objected to in cakes intended as cattle food, a prevalence of mustard, according to some writers, rendering the mixed cake almost, if not entirely, poison. In support of this statement it need only be necessary to quote one or two of Professor Voelcker's remarks. "Wild mustard and rape-seed very commonly occur in inferior linseed; when such cakes are mixed with water a more or less pungent smell of mustard is developed. I may observe, in passing, that rape-cake ought always to be

RAPE & MUSTARD.

tested in this manner; for rape-cake, especially that which is sold as Indian seed, very generally contains a large amount of mustard seed, and becomes so pungent that it is extremely injurious to cattle. I have here a sample of a cake sent to me for examination not long ago, which had killed three oxen. It is a rape-cake of the description just named, Indian rape-cake containing a good deal of wild mustard." But many writers speak almost favourably of rape-cake, and even **Professor Voelcker** himself, in one of his last papers (quoted above at the end of the remarks under Linseed-cake), links it with linseed as a useful feeding cake. The explanation of these somewhat conflicting opinions would seem to be a confirmation of the contention advanced by the writer under **Brassica** (Vol. I., pp. 520 to 534) that greater care is necessary than is apparently made by the exporters of rape and mustard-seed from India. Little or no true mustard-seed is grown in India. The mustard of Indian commerce is the seed of **Brassica juncea**, a perfectly distinct plant from the mustard of Europe. At the same time the properties of the various forms of seeds exported from India as rape, differ from each other as widely as the mustard of India from the mustard of Europe. It seems highly probable therefore that the evil effects observed by many writers from pure rape-cake and from linseed-cake which contains rape or mustard, are explicable by the theory that certain of these seeds are more injurious than others. This hypothesis would, therefore, seem to call for more careful chemical investigation into the nature of the separate seeds of **Brassica** obtained in India. The habit of growing rape and Indian mustard intermixed with linseed is so wide-spread in this country that it seems likely many years will have to elapse before the *rayats* can be induced to alter their time-honoured practices. The drying property of linseed is greatly lessened by the admixture, but to the people of India this is no serious disadvantage, so far as they are concerned, in the consumption of oil. For the export trade it is quite otherwise, and, since the bulk of the linseed is grown for the foreign markets, it would seem at least desirable to be able to recommend the avoidance at least of the cultivation of a form or forms of **Brassica** which not only lower the value of the linseed, but render the cake to a large extent a dangerous article of cattle food.

For further information the reader should consult the article **Brassica**, Vol. I., 520 to 534.

COTTON-SEED-CAKE. 11.

III.—COTTON-SEED-CAKE.

The reader will find a certain amount of information on this subject in the article **Gossypium** (*Vol. IV., 33-37*), but it seems desirable to supplement the opinions there advanced in the light of the present article, namely, with regard to specially prepared oil-cakes which are to a large extent combinations of more than one form of the by-products of Oil Mills. It seems unnecessary, however, to do more than give here the opinions of the late **Professor Voelcker**. In his lecture delivered in 1863 on this subject, the Professor said:—"Allow me now to direct your attention to cotton-cake. Cotton-cake of the best character is hardly ever now met with in the market; the horrible American war has cut off our supply. Cotton-seed has a hard shell, which, in some varieties, amounts to one-half the weight. The best cake is made from the kernels only; but it is not made in this country, because we have not the proper machinery for shelling seed." "In the whole seed cotton-cake there is sometimes such an excess of husk or indigestible cotton-fibre present, that the animal which is fed upon it has not the power to digest it: a mechanical stoppage takes place in the lower intestines inflammation of the whole intestinal canal

COTTON-SEED-CAKE.

ensues, and the animal dies. In these cases the death is frequently mistaken for a case of real poisoning; but there is nothing either poisonous or deleterious in cotton-seed; nevertheless, it acts injuriously, by causing a mechanical obstruction, and the result is the same as that sometimes produced by a strong irritant or metallic poison. There is, indeed, great danger in giving the whole cake made of the seed indiscriminately, that is to say, in too large a proportion; it ought always to be given in the form of meal, together with roots or other succulent food which have a tendency to keep the bowels open."

In his later paper published in 1884, **Professor Voelcker** deals with the degeneration of decorticated cotton-seed-cake, due to hard pressing, from the use of more perfect means of expressing the oil. The same result is attained as has been shown with hard-pressed linseed-cake, *viz.*, a cake poor in oil and unduly rich in albuminoids. He furnishes the following analyses of nine samples of cotton-cake taken at random from the analytical results of a large number of cotton-cakes examined:—

Composition of Decorticated Cotton-cakes of variable quality.

	No. 1.	No. 2.	No. 3.	No. 4.	No. 5.
Moisture	9·28	8·59	7·95	8·75	8·95
Oil	12·93	22·10	16·57	15·70	10·23
* Albuminoids	46·37	42·87	42·37	49·02	49·04
Mucilage, sugar, and digestible fibre	21·92	17·29	22·85	17·52	22·92
Indigestible woody fibre	3·47	3·31	3·10	3·51	3·21
Mineral matter (ash)	6·03	5·84	7·15	5·50	5·65
	100·00	100·00	100·00	100·00	100·00
* Containing nitrogen	7·42	6·86	6·78	7·85	7·04

Composition of Texas Decorticated Cotton-cake.

	No. 6.	No. 7.	No. 8.	No. 9.
Moisture	10·15	10·08	11·45	10·95
Oil	15·26	11·66	10·60	8·76
* Albuminoids	38·37	40·69	38·62	40·87
Mucilage, sugar, and digestible fibre	19·84	24·65	25·57	23·29
Indigestible woody fibre (cellulose)	10·83	7·03	8·26	10·83
Mineral matter (ash)	5·55	5·89	5·50	5·30
	100·00	100·00	100·00	100·00
* Containing nitrogen	6·14	6·51	6·22	6·54

"The preceding analyses show that the proportion of oil in decorticated cotton-cake varies from 8¾ to 22 per cent, and the percentage of albuminoids in round numbers from 38 to 49 per cent. A food containing nearly as much as half its weight of nitrogenous constituents and only 10 per cent. of oil, pressed into a hard cake, is not suitable for herbivorous animals. Such food requires to be ground into meal, and largely diluted with starchy or similar meals that are comparatively poor in nitrogenous compounds, in order to be a wholesome and safe food for ruminants."

COTTON-SEED-CAKE.

Speaking of cotton-cake **Mr. W. T. Carrington** in 1878 (*Pastoral Husbandry*) wrote : " A mixture of equal parts of linseed and decorticated cotton-cake is found a most suitable food with grass, the cotton-cake checking the purgative nature of the grass. A daily allowance of this mixture, commencing with 4℔ increasing to 6℔, as the beans approach ripeness, and costing from 2*s*. 6*d*. to 3*s*. 6*d*. per head per week, generally pays well. **Professor Voelcker**, in his essay on *The Influence of Chemical Discoveries on the Progress of English Agriculture*, says : " There are two varieties of cotton-cake—one is made in England from Egyptian cotton-seed, shell and kernel crushed together, and the other is principally imported from New Orleans, and made in America from decorticated seed." Decorticated cotton-cake is also made, he says, to a small extent in England from the kernels imported from America. The high percentage of nitrogenous matter and the low amount of oil is dealt with by the Professor in much the same spirit as has already been conveyed. He recommends that it should be more carefully broken up than is necessary with linseed-cake, and given also more sparingly, being mixed by about twice its weight of Indian-corn or barley-meal, or meal rich in starch and comparatively poor in nitrogenous compounds. Its superiority for sheep feeding, when grass is poor or scanty, is also alluded to, one-half to three-quarters of a pound per head of decorticated cotton-cake being the daily allowance recommended.

The reader should consult the article **Gossypium herbaceum,** Vol. IV., 33 to 37, for further information.

POPPY-CAKE. 115

IV.—POPPY-CAKE.

Of Poppy-cake, **Professor Voelcker** says : " It is a good cake when it can be had in a fresh condition. It is remarkably sweet to the taste, and is nutritious, but in England, there is not a sufficient demand for it, and it is frequently a drug in the market. If stored, it becomes mouldy and acrid to the taste, and is then more or less injurious to cattle."

Professor Anderson of Glasgow (*Agricultural Chemistry*) gives the following as the result of his analysis of poppy-cake :—

Water	6·56
Oil	11·04
Nitrogenous compounds	34·03
Respiratory compounds (mucilage, gum, &c.)	23·25
Fibre	11·33
Ash	13·79
	100·00

A very nearly similar result was obtained by **Soubeiran & Girardin** as follows :—

Water	11·0
Oil	14·2
* Organic matters	62·3
Ash	12·5
	100·
* Containing nitrogen equal to	7·00
Prot. compounds	43·75

In France and Belgium, where the poppy is largely grown for its oil, the cake is highly valued as a feeding substance, and is also extensively employed as a manure.

The reader should consult the article **Papaver somniferum,** Vol. VI., Pt. I., 19-21, for further information.

V.—EARTH OR GROUND-NUT CAKE.

EARTH OR GROUND-NUT CAKE. 116

Professor Voelcker says of this cake: "Earth-nut is a useful and very nutritious cake when made of the kernel; but if the husk is ground up with it, it partakes of the same disagreeable properties which characterise the whole seed cotton-cake. The earth-nut is an almond-like food; it is to some extent indigestible. The cake must not be confounded with nut-cake, which, as now sold in the market, is nothing more or less than the refuse kernel of the palm-nut. The American or earth-nut cake seldom passes as such into the hands of the consumer, because it is ground up with other materials and made into linseed-cake. There is another kind of nut-cake which is only fit for manure. It is made of a bean grown in the Cape de Verd Islands. Three or four beans are sufficient to produce a very powerful affection of the bowels, and in doses of an ounce this seed becomes a rank poison. Several cases of poisoning from the use of cakes which contained this bean have been brought under my notice at various times." The poisonous cake, alluded to by **Professor Voelcker**, would appear to be castor-oil, for further information regarding which the reader should consult the remarks below on that subject. Earth-nut cake is an altogether very different substance, and a large trade might easily be created in it from India, but at present it is chiefly bought up by the cake-makers and used for adulterating linseed-cake.

Professor Anderson furnished the following analyses of earth-nut cake:—

	Decorticated nut.	Entire seed.
Water	8·62	11·56
Oil	8·86	12·75
Nitrogenous compounds	44·00	26·71
Respiratory compounds (*i.e.*, mucilage, sugar, &c.)	19·34	45·69
Fibre	5·13	
Ash	14·c5	3·29

The reader should consult the article **Arachis hypogæa**, Vol. I., 282, for further information.

VI.—TIL OR SESAMUM-CAKE.

TIL OR SESAMUM-CAKE. 117

As has already been remarked, it is significant that while the exports from India of the seed which yields this cake have year after year for some time past increased steadily, no writer on the subject of oil-cakes has apparently dealt separately with the one under notice. That it is being employed by the cake-makers in the fabrication of what is known in the trade as certain qualities of linseed-cake, there would seem to be no doubt. **Soubeiran & Girardin** have published the following as the analysis of the cake:—

Water	11·0
Oil	13·0
* Organic matters	65·5
Ash	9·5
	100·0
* Containing nitrogen equal to	5·57
Prot. compounds	34·81
The ash contains soluble salts	0·57
Phosphates calculated as bone phosphates	3·20

TIL OR SESAMUM-CAKE.

As a manure, this cake would, therefore, appear to be far less valuable than rape, castor, or hemp, but as an article of cattle food it is probably very wholesome, since the relation of oil to nitrogenised compounds is less arbitrary than in most oil-cakes. The seed also is very small, contains no indigestible husk, and is largely eaten in India as an article of human food —facts which all tend to confirm the opinion that sesamum-cake is perhaps one of the most wholesome of all. **Professor Anderson** gives the following as the results of his analysis of the cake which, bye-the-bye, he includes in a *List of the Principal Varieties of Cattle Food*:—

Water	10·38
Oil	12·86
Nitrogenous compounds	31·93
Respiratory compounds (*i.e.*, mucilage, sugar, &c.)	21·92
Fibre	9·06
Ash	13·85
	100·00

The article Oil-cake in the *Encyclopædia Britannica* shows the chemical composition of this cake as—

Water	8·06
Oil	11·34
* Albuminous bodies	36·87
Mucilage, sugar, digestible fibre, &c.	25·05
Woody fibre	8·14
Mineral matter (ash)	10·54
	100·00
* Containing nitrogen	5·90

The reader should consult the article **Sesamum indicum**, Vol. VI., Pt. II., for further information.

CASTOR-CAKE 118

VII.—CASTOR-CAKE.

In India this is one of the most highly valued of all oil-cakes as a manure. In Europe it has acquired an evil repute as one of the most dangerous of substances when used in the fabrication of inferior qualities of linseed-cake. In commerce it appears to be sometimes designated Jatropha manure. In **Morton's** *Cyclopædia of Agriculture* the following passage occurs on this subject:—"This cake can only be used as a manure; for feeding purposes it is entirely unsuited, as the oil still remaining in the cake is exceedingly purgative and poisonous. This Jatropha manure was recently analysed by **Mr. Thomas J. Herapath** who found it to contain in 100 parts—

Valuable manure.

Water	10·24
Organic matters	81·88
Ash	7·88
	100·00
Percentage of nitrogen in fresh cake	4·20
Percentage of nitrogen in dry cake	4·68

CASTOR-CAKE.

The ash contained in 100 parts:—

Soluble salts	6·193
Carbonates	21·070
Insoluble phosphates	53·554
Silica, &c.	19·183
	100·000

Perhaps a more instructive table of the analysis of castor-cake is that published by the late **Professor Anderson** of Glasgow in his *Agricultural Chemistry* as follows:—

Water	12·31
Oil	24·32
Albuminous compounds	21·91
Mucilage, sugar, fibre, &c.	35·38
Ash	6·08
	100·00
Nitrogen	3·20
Silica	1·96
Phosphates	2·81
Phosphoric acid in combination with alkalis	0·64

The reader should consult the article **Ricinus communis,** Vol. VI., Pt. I., for further information.

VIII.—COCOA-NUT-CAKE.

COCOA-NUT-CAKE. 119

Professor Voelcker, in his Essay on *The Influence of Chemical Discoveries on the Progress of English Agriculture*, says: "Cocoa-nut-cake and palm-nut-kernel-cake and -meal are produced at Liverpool and other places in England, and are much appreciated for their fattening properties. These cakes contain from 14 to 15 per cent. of albuminous compounds, and variable proportions of oil, and are better adapted for fattening stock than for young growing animals or store stock."

The analysis of palm-kernel cake is given as—

Water	9·50
Oil	8·43
* Albuminous bodies	30·40
Mucilage, sugar, fibre, &c.	40·95
Mineral matter (ash)	10·72
	100·00
* Containing nitrogen	4·50

The reader should consult the article **Cocos nucifera,** Vol. II., 443.

IX.—NIGER SEED-CAKE.

NIGER SEED-CAKE. 120

Although the seeds from which this cake is obtained are largely exported from India, no information exists as to the extent the cake is actually used for feeding cattle. **Professor Anderson,** however, accepts niger seed-cake as one of *The Principal Varieties of Cattle Food,* and gives it the following composition:—

Water	6·23
Oil	6·58
Nitrogenous compounds	25·74
Respiratory compounds (*i.e.*, mucilage, sugar, &c.)	42·18
Fibre	11·15
Ash	8·12
	100·00

In India this cake is considered valuable for milch-cows.

The reader should consult the article **Guizotia abyssynica, Vol. IV.,** 186-188.

HEMP-CAKE. 121

X.—HEMP-CAKE.

Professor Voelcker gives the analysis of a sample of hemp-cake which he made in 1875 as follows:—

Composition of a sample of Hemp-cake.

Moisture	10·57
Oil	11·17
* Albuminous compounds (flesh-forming matter)	29·56
Mucilage, sugar, and digestible fibre	18·03
Woody fibre	24·20
Mineral matter (ash)	6·47
	100·00
* Containing nitrogen	4·73

He remarks: "Hemp-cake is not equal to linseed-cake in feeding quality; but in my judgment it is worth more as an article of food than whole-seed cotton-cake. About £8-8*s.* per ton would be a fair price for hemp-cake of good quality."

The reader should consult the article **Cannabis sativa, Vol. II., 121.**

TRADE IN OILS AND OIL-SEEDS. Prepared Oils. 122

TRADE IN OILS AND OIL-SEEDS.

Prepared Oils.—The enormous amount of oil-seeds exported from India is out of all proportion to the quantity of prepared oils. It has been repeatedly pointed out that mills for the preparation of oil in India would seem likely to prove highly remunerative. The following were the exports from India of prepared oils for the past few years ending March 1889:—

	1879-80.	1880-81.	1881-82.	1882-83.	1883-84.	1884-85.
Gallons	4,205,815	4,999,184	4,305,176	3,644,632	4,337,151	5,120,504
Rupees	56,94,532	58,11,394	46,82,274	41,62,768	49,59,768	53,40,411

	1885-86.	1886-87.	1887-88.	1888-89.	1889-90.
Gallons	3,617,299	4,039,491	4,162,099	3,841,732	4,339,399
Rupees	39,55,629	45,88,119	47,10,555	43,03,061	54,39,452

Of these exports nearly three-fourths are castor oil: of cocoa-nut oil, the only other of any importance, 1,525,910 gallons, valued at R16,53,379, were shipped to foreign ports, chiefly to the United Kingdom and France in 1888-89.

The above quotations include small quantities of animal, essential, and mineral oils in addition to the vegetable oils. The exports of animal oils do not seem to be of much importance, though the value has gradually increased from R3,478 in 1880-81 to R9,116 in 1888-89. The import trade is also improving, the figures in 1880-81 and 1884-85 having been R24,246 and R33,253 respectively, and in 1888-89 R41,013. The trade in essential oils is of more importance; but the exports have decreased in value from R3,40,452 in 1882-83 to R1,85,917 in 1884-85; they were in 1888-89 R2,67,797. The imports have improved from R29,611 in 1880-81 and R41,542 in 1884-85 to R46,994 in 1888-89. The export trade in mineral oils is not extensive; during 1884-85 it was valued at R27,165, and last year R15,343. The imports, on the other hand, are of great value. They consist almost exclusively of kerosine oil from the United States. This oil has increased from 9,692,269 gallons in 1880-81 to 26,299,091 gallons in 1884-85 and 38,285,559 gallons in 1888-89. The value of the kerosine oil imported into India has been returned at R1,11,00,507 in 1884-85 and R1,76,79,373 in 1888-89, or say, £1,700,000.

TRADE. Oil-seeds. 123

Oil-seeds.—In Mr. J. E. O'Conor's Trade and Navigation Returns, the oil-seeds are given under the heading "Seeds." A few of these may possibly be used as spices and condiments; or may be distilled for their essential oils. This is an unavoidable difficulty, but the amounts of doubtful or undefined seeds are comparatively small, so that the returns under "Seeds" may be accepted as meaning "Oil-seeds." In his review of the sea-borne foreign trade for 1884-85, Mr. O'Conor says of these seeds: "This trade has developed in recent years into one of the first importance, exceeding greatly the trade in wheat, rice, jute (both raw and manufactured combined), indigo, or tea, and being exceeded only by cotton and opium." The figures below exhibit the way in which this trade has augmented since 1879-80:—

	Cwt.	R
1879-80	7,091,469	4,68,58,929
1880-81	10,229,109	6,34 52,089
1881-82	10,466,098	6,05,40,987
1882-83	13,139,206	7,20,03,365
1883-84	17,355,588	10,08,37,583
1884-85	18,250,688	10,74,52,035
1885-86	17,280,147	9,94,83,498
1886-87	15,866,604	9,19,85,598
1887-88	16,060,400	9,38,50,242
1888-89	15,569,978	9,56,17,559
1889-90	15,794,742	10,62,75,533

The trade in the principal kinds of oil seeds was as follows in the last four years:—

	1885-86. Cwt.	1886-87. Cwt.	1887-88. Cwt.	1888-89. Cwt.	1889-90. Cwt.
Castor	670,537	610,893	764,293	585,769	894,631
Earthnut	655,670	945,895	1,261,637	827,997	1,394,191
Linseed	9,510,139	8,656,933	8,422,703	8,461,374	7,146,896
Poppy	695,097	612,654	456,308	730,455	482,893
Rape	3,721,840	2,659,649	2,081,300	3,061,586	3,904,150
Til or Jinjili	1,759,343	2,114,484	2,747,270	1,537,444	1,775,559

A material increase in the price of Rape-seed accounts for the larger value of a smaller quantity of seeds exported.

The trade in seeds is carried on mainly with England and the Continent. England takes the bulk of the linseed exported, but some also goes to the United States, and to the Continent is consigned the bulk of the other descriptions of seeds.

(*W. R. Clark.*)

OLAX, *Linn.; Gen. Pl., I., 547.* 124

Olax nana, *Wall.; Fl. Br. Ind., I., 576;* OLACINEÆ. 125

Vern.—*Merom met'*, SANTAL.

References.—*Brandis, For. Fl., 75; Rev. A. Campbell, Econ. Prod., Chutia Nagpur, No. 8800.*

Habitat.—A small undershrub found in the hot valleys of the North-West Himálaya from Nepál to the Panjáb, ascending to an altitude of 5,000 feet.

MEDICINE. Fruit. 126

Medicine.—The FRUIT is employed medicinally by the Santals (*Rev. A. Campbell*).

O. scandens, *Roxb.; Fl. Br. Ind., I., 575.* 127

Syn.—O. BADOR, *Ham.;* ROXBURGHIA BACCATA, *Koenig.*

Vern.—*Dheniani*, HIND.; *Koko-aru*, BENG.; *Rimmel*, KOL.; *Bodo-bodo-ria*, URIYA; *Hund*, SANTAL; *Harduli, urchirri*, MAR.; *Kurpodur, murki, malle, turka-vepa, bápanamushti, kotiki*, TEL.; *Lélu, chaung-lélu, lai-lu, joung-lai-lú*, BURM.

References.—*Roxb., Fl. Ind., Ed. C.B.C., 55; Voigt, Hort. Sub. Cal., 31; Dalz. & Gibs., Bomb. Fl., 27; Brandis, For. Fl., 75; Kurz, For. Fl. Burm., I., 233; Gamble, Man. Timb., 81; Elliot, Fl. Andhr., 23, 99, 113; For Ad. Rep., Ch. Nagpur, 1885, 29; Jour. As. Soc., Pt. II., No. II., 1867, 80, 81; Gazetteers:—Bombay, XV., 429; N.-W. P., IV., lxix.*

Habitat.—A large, woody, climbing shrub, with a trunk often as thick as a man's thigh; found in the tropical forests of the Western Himálaya from Kumáon to Assam, and extending into Central and South India. It is found also in Burma and the hotter parts of Ceylon, and is distributed to Java. In Burma it ascends to the pine forests, at an altitude of 3,500 feet. It is abundant on the wet ground near rivers and in ravines, and is most destructive to the trees, which it covers with its numerous branches and dark green foliage.

MEDICINE. Bark. 128 **Medicine.**—A preparation of the BARK is given for poverty of blood during fevers (*Revd. A. Campbell*).

FOOD. Fruit. 129 **Food.**—The FRUIT is employed in Hazaribagh for making *sherbet* (*Revd. A. Campbell*).

TIMBER. 130 **Structure of the Wood.**—Porous, yellowish-white, soft. Weight 38lb per cubic foot. It is not put to any industrial purpose.

131

OLDENLANDIA, *Linn.; Gen. Pl., II., 58.*

[*Ic., t. 822;* RUBIACEÆ.

132 **Oldenlandia corymbosa,** *Linn.; Fl. Br. Ind., III., 64; (?) Wight,*

Syn.—O. BIFLORA, *Lam.* (*and of Roxb. not of Linn.*); O. RAMOSA, *Roxb.*; O. HERBACEA & SCABRIDA, *DC.*; O. ALSINIFOLIA, *Don*; HEDYOTIS GRAMINICOLA, *Kurz*; H. ALSINÆFOLIA, *Br.*

There cannot be a doubt but that the allied species (**O. diffusa** and **O. Heynii**) are used indiscriminately with the one here dealt with. Indeed, it seems doubtful if any practical purpose is served by separating plants so nearly allied to each other as some of the so-called species in this troublesome genus (*Watt, Cal. Exhib. Cat.*).

Vern.—*Daman-papar,* HIND.; *Khetpapra,* BENG.; *Piriengo,* NEPAL; *Popalo, kasuri,* GOA; *Verri néla vému,* TEL.; *Wal-patpaadagam,* SING.; *Parpata,* SANS.

References.—*Roxb., Fl. Ind., Ed. C.B.C., 142; Dalz. & Gibs., Bomb. Fl., 116; Rev. A. Campbell, Rept. Econ. Pl., Chutia Nagpur, No. 8194; Sir W. Elliot, Fl. Andhr., 191; Rheede, Hort. Malab., X., t. 38; O'Shaughnessy, Beng. Dispens., 400; U. C. Dutt, Mat. Med. Hindus, 176, 305, 312; Murray, Pl. & Drugs, Sind, 195; Gazetteers:—Bombay, XV., 435; N.-W. P., I., 81; IV., lxxiii.; X., 311.*

Habitat.—An abundant annual weed, found throughout India from the Panjáb southward and eastward, ascending the hills to an altitude of 5,000 feet. It is distributed also to Ceylon, Malacca, and the Philippines and to Tropical Asia, Africa, and America.

DYE. Root. 133 **Dye.**—Revd. A. Campbell (*Econ. Pl., Chutia Nagpur*) states that "this plant is the source of the Madras red dye known in commerce as chay-root." This of course is not the case, as the commercial chay-root is obtained from **O. umbellata**, but no information seems available as to whether this species also may not yield a similar red dye, and Mr. Campbell's remark may therefore mean that it is so used.

MEDICINE. Plant. 134 **Medicine.**—This PLANT is considered by Native physicians to be a valuable medicine in the treatment of fevers which have their origin in derangements of the bile. The entire plant is used. A simple decoction is employed in bilious fever with irritability of the stomach, and it also enters into the com position of many compound decoctions which are administered as tonics and febrifuges (*U. C. Dutt*). It is given too in jaundice and supposed diseases of the liver. In the Konkan the JUICE of the LEAVES is applied locally for burning of the palms of the hands, and soles of the

Juice. 135 Leaves. 136

feet in fever, and this preparation, mixed with a little milk and sugar, is given internally for burning at the pit of the stomach (*Dymock*). MEDICINE.

SPECIAL OPINIONS.—§ "*Konticary, khetpapra, and goluncho.*—I have heard that Native physicians employ these in the form of decoction, with good effect, in chronic malarious fevers" (*Surgeon J. Ffrench Muller, M.D., I.M.S, Saidpore*). "A valuable febrifuge decoction, largely used in bilious fever" (*Assistant Surgeon S. C. Bhuttacharji, Chanda, Central Provinces*). "Is also an anthelmintic" (*Surgeon-Major P. N. Mukherji, Cuttack, Orissa*). "A decoction made of this herb along with coriander seed, &c., is very efficacious in checking long-standing fever" (*Assistant Surgeon R. Gupta, Bankipore*).

Oldenlandia umbellata, *Linn.; Fl. Br. Ind., III., 66.* 137

Commercial Name.—CHAY-ROOT, sometimes called INDIAN MADDER.

Syn.—O. PUBERULA, *G. Don*; HEDYOTIS UMBELLATA, *Lam.*; H. LINARIFOLIA & PUBERULA, *Wall.*

Vern.—*Chirval*, HIND.; *Surbuli*, BENG.; *Surbuli*, URIYA; *Kalhenyok*, LEPCHA; *Saya, emburel cheddi, imbural, saya-wer*, TAM.; *Cherivelu, chiri-veru*, TEL.; *Sayanmull*, SING.

References.—*Roxb., Fl. Ind., Ed. C.B.C., 141; Elliot, Fl. Andhr., 43; Lindley & Moore, Treasury of Botany, II., 808; Ainslie, Mat. Ind., II., 101; O'Shaughnessy, Beng. Dispens., 400; Dymock, Mat. Med. W. Ind., 2nd Ed., 404; Bidie, Cat. Raw Pr., Paris Exh., 113; Baden Powell, Pb. Pr., 446, 447; Drury, U. Pl., 240; Birdwood, Bomb. Pr., 299; McCann, Dyes & Tans, Beng., 43, 44, 45; Liotard, Dyes, 54; Smith, Dict. Econ. Pl., 106; Smith, Dic., 1006; Kew Off. Guide to Bot. Gardens & Arboretum, 79; Simmonds, Trop. Agri., 374; Gazetteers:—Mysore & Coorg, I., 436; III., 28; Ind. Forester, X., 547; Manual of Administration, Madras, I., 288, 360.*

Habitat.—A common biennial plant, met with from Orissa and Bengal southward to Ceylon, and in North Burma; occurs on sandy soils. It grows wild very abundantly in the Puri district, and is extensively cultivated in many parts of Madras, near the sea-coast. In its wild state it is a low widely spreading almost stemless plant, but under cultivation it assumes a more erect habit and grows six or eight inches high.

Dye.—The ROOT-BARK of this plant (commercially known as chay-root), with alum as a mordant, gives a beautiful red dye, which is fast, and was formerly much employed in Madras for dyeing the handkerchiefs for which that town was once so famous. Curiously, although the plant occurs abundantly in some parts of Bengal, it is not there employed as a dye, but a considerable export trade in the ROOTS exists between Puri in that Province and Madras. For dyeing purposes the roots of the wild plant are considered preferable (*Drury*). They are shorter and thicker than the cultivated state, and are said to yield one-fourth part more colouring matter. Roots of two years' growth are preferred. It is said that chay-root rapidly deteriorates when kept in a dark place, such as the hold of a ship, whence probably the want of success that has attended attempts at dyeing with it in Great Britain. DYE. Root-bark. 138 Roots. 139

Cultivation.—The Chay-root is much cultivated in sandy situations on the Coromandel coast, especially at Nellore and Masulipatam. From the Madras Revenue Proceedings for April 1885, No 463, the following interesting report may be reprinted:—"This cultivation is confined to the very sandy soils on the coast. In black soils, the root cannot penetrate sufficiently and consequently the plant will not flourish. Sandy soils, being in themselves very poor, require a great deal of manure. Preparations are made in May or June, and thirty cattle or sixty sheep should be picketed for at least ten nights on an acre of land. This mode answers CULTIVATION. 140

 Dyeing with Chay-root.

CULTIVATION. better actually, than manure from a distance. The soil requires to be made very loose, and for this purpose, it is ploughed nearly every day for a month. The seeds are sown in August, and before doing so, the ground is well moistened by means of *chattis* or watering pots, a well or pit in the sand being usually excavated close by. Immediately after the seeds are sown, the ground is again watered, and the process is repeated three times daily till the young shoots appear, after which the water need only be applied twice a day, night and morning. Cowdung should be mixed with the water once a day for the first fifteen days. Perpetual watering is the chief thing to be attended to, and this must be increased as the root of the plant grows longer, in order that the moisture may penetrate to an equal depth. Weeding has also to be constantly performed. The plants flower and seed three times during six months, at the end of which time they die down and the roots are taken up.

"For this purpose the soil is loosened with an iron spud and the roots carefully removed, so that they may not be broken. They are often a yard and upwards in length. The roots when taken out are dried in the sun for five days and then tied in large bundles of 2½ maunds.

"The expenses for an acre of land are as follows :—

	R.	a.	p.
Manuring	2	8	0
Ploughing	9	9	6
Weeding	7	0	0
Watering	6	0	0
Digging up roots	1	13	6
Schist	1	8	0
TOTAL .	28	7	0

"The returns are 1½, 2, or 2½ *putis* of roots according to the quality of the soil and degree of pains taken with the cultivation. Each *puti* is worth R25." The returns cannot, therefore, be considered large when compared with the constant care and attention required to secure a good crop."

No returns are forthcoming as to the outturn of the spontaneous chay-root, but in places where it is abundant the Revenue Department lease out the right to collect it. In Guntúr, Madras Presidency, the rent sells for over R7,000. The same land can be worked only every third year for spontaneous produce.

DYEING. 141

METHODS OF DYEING.—Only the bark of the root contains the dyeing substance, the woody portions being quite useless.

Both old and new cloths may be dyed with this substance, but in old cloths the resulting colour is not so good as in the new; the freshness of the colour depends on the texture of the fabric (*McCann*). The process of dyeing, as described by the *Balajis* of Ganjám, is as follows :—

"To dye a piece of cloth for a woman's garment 16 cubits long, make a mixture of 1½ seers gingelly oil, 8 tolas of alum, 3 seers of clear water, and 2 seers of well pulverised chay-root, and wash in this the cloth thrice on three successive days, drying it in the sun after each washing. On the fourth day, wash the cloth again in clear water, and dry in the shade. Then boil the cloth for two hours in a bath of one *viss* of pulverised chay-root and 20 seers water, and dry it in the sun; the next day wash the cloth in clear water and again dry in the sun. Then again, wash the cloth in water containing the alkaline ashes used by washermen (called *chakuli*, *kamum*, and *kharum*) and again dry in the sun : afterwards wash again in clear water and dry in the sun. Then boil the cloth for two hours in a bath of 4 seers of pulverised chay-root and 15 seers of water and dry in the sun. The next day wash the cloth in clear water and dry" (*McCann, Dyes and Tans, Bengal*). Various modifications of this process appear to be adopted by the dyers at different places, but the preceding is the one most generally pursued, and the others merely differ in unimportant items. It will be seen that the essential process is the boiling with pulverised chay-root and the use of alum as a mordant. The whole process is a very tedious one, sometimes lasting over a month, and during the rainy season all operations are suspended. For producing the brilliant red Madras handkerchiefs known as *Bandanas*, in which at one time a considerable export trade to the West Indies existed from Madras, the chay-root was employed; but since cheaper, though not so durable, substitutes have been discovered, this trade has disappeared, and the use of the chay-root is now almost entirely confined to the dyeing of the cotton clothes worn by the Natives of India as turbans, dresses, &c. No export trade exists in the dye-stuff.

MEDICINE. Leaves. 142

Medicine.—The LEAVES are considered by Native physicians to have expectorant qualities. A decoction in doses of about one ounce given twice daily is prescribed in colds, coughs, and asthma. The leaves dried and pounded are sometimes mixed with flour, and made into cakes which are eaten by those who suffer from consumption or asthma. The ROOT is said to be used as a specific in snake-bite and in various cutaneous disorders; but of its virtues in these cases no exact information is obtainable (*Ainslie*). Root. 143

OLEA, *Linn.; Gen. Pl., II., 679* 144

Olea cuspidata, *Wall.; Fl. Br. Ind., III., 611;* OLEACEÆ. 145

Syn.—O. FERRUGINEA, *Royle, Ill., 257, t. 65, f. 1.*

Considerable confusion seems to exist in some books as to whether this is merely a variety of the Wild Olive (**Olea europea** *var.* **sylvestris**) or an altogether different species. The remark on **O. cuspidata** in the *Flora of British India* seems to favour the latter idea. There it is noted that, "This was supposed a variety of the Wild Olive by **Dr. Stewart. Brandis** says it differs by the more lax inflorescence; the upper surface of the leaves glossy not dull grey, the lower surface reddish instead of white, smaller fruit, the absence of spines and a more distinctly marked heartwood. The cuspidation of the leaves, distinct in **Wallich's** specimen and relied on by **Boissier** as a diagnostic mark, proves variable in the Indian plant." It seems probable that both **Olea cuspidata** and **O. sylvestris**, whether varieties or different species, exist side by side, indigenous in North-West India, but as yet very few attempts have been made to separate the two by those who are interested in the subject. Specimens sent to Kew of Wild Olives found in Balúchistán by **Mr. J. H. Lace**, of the Indian Forest Department, have been identified as **Olea europea**, while in the same region **Olea cuspidata** appears also to occur. **Aitchison**, in his *Catalogue of Panjáb and Sind Plants*, regards **O. cuspidata** as a variety of the Wild Olive, but the differences between the two seem to the writer to be very constant and well marked and very nearly of the nature described in the *Flora of India* on the authority of **Brandis**. As it seems impossible, however, to separate the two in the economic uses made of them by the Natives of India, the information with regard to the Indian Wild Olive will be given under the head of **Olea cuspidata**, and merely the various attempts at cultivating introduced specimens dealt with under that of **Olea europea**.

Vern.—*Kau, kan,* HIND.; *Kao, kohú, ko, kau, kan,* PB.; *Shawan, trikh-shawan, káo, zaitún,* PUSHTU; *Khau, khan,* SIND; *Khwan, shwan,* BILUCH.; *Zeitun, sjetun,* ARAB.; *Zaitún,* PERS.

References.—*Brandis, For. Fl., 307, t. 38; Gamble, Man. Timb., 257, 258; Stewart, Pb. Pl., 139, 140; Aitchison, Cat. Pb. & Sind Pl., 186; Murray, Pl. & Drugs, Sind, 153; Baden Powell, Pb. Pr., 587; Atkinson, Him. Dist., 313; Watson, Report, 36; Balfour, Cyclop., III., 19; Kew Off. Guide to Bot. Gardens & Arboretum, 95; Gazetteers:—*

The Indian Olive.

Mysore & Coorg, I., 60; Hazára, 12; Dera Ismail Khan, 18;Ráwalpindi, 82; Shahpur, 69; Gurdáspur, 55; Simla, 10; Ind. Forester, IV., 345; V., 329; XIII., 59, 70; Agri.-Horti. Soc., Ind., Journ., XIII., 284; XIV., 86.

Habitat.—A moderate-sized deciduous tree of Sind, the Sulaiman and Salt Ranges, the North-West Himálaya and Kashmír, eastward as far as the Jumna and found up to an altitude of 6,000 feet. It is distributed also to Afghánistán and Balúchistán.

OIL. Fruit. 146 **Oil.**—In Afghánistán, a small amount of OIL is extracted from the FRUIT, and is said to be used medicinally (*Irvine; Bellew*). At Kohát, in the year 1851, an attempt on a large scale was made to extract oil from the wild olive of India. The oil was excellent, but the quantity obtained insufficient to repay a continuation of the experiment, as only one gallon of oil was got from a maund of dry fruit. It is doubtful, however, whether the method by which the oil was extracted was the proper one, at any rate, it appears to have been altogether opposed to the Italian process, as the fruit was dried in the sun and pressed in a common *kolhú* (oil press) instead of being collected into small mounds, the oil expressed by its own weight being first allowed to run off into receptacles, the fruit then pressed and the rest of the oil extracted, before fermentation set in. So much, however, is evident that oil can be made from the fruit of the Indian tree, and it seems probable that by grafting from the European species and by improved methods of extraction, the yield might eventually be improved.

MEDICINE. Oil. 147 Leaves & Bark. 148 **Medicine.**—An OIL is extracted from the fruit, which is used medicinally, as a rubefacient (*Irvine; Bellew*). The LEAVES and BARK are bitter and astringent; and are employed as an antiperiodic in fever and debility.

FOOD & FODDER. Fruit. 149 Leaves. 150 **Food & Fodder.**—The FRUIT is eaten by the natives of the countries where it abounds, although there is not much pericarp on it, neither is it pleasantly tasted (*Murray; Gamble*). The LEAVES are sometimes given as fodder for goats.

TIMBER. 151 **Structure of the Wood.**—The timber has a whitish sapwood and a large amount of heartwood which is hard and from light to olive brown in colour. Weight 68 to 82lb per cubic foot, averaging 73lb.

DOMESTIC USES. 152 **Domestic Uses.**—The timber polishes well and is much prized in turnery. It is used for the handles of agricultural implements, for cotton wheels, walking sticks, and combs. The crooked timbers are largely employed in boat-building on the Indus. It is also burned as fuel in regions where it is abundant. It might be utilised as a substitute for box-wood for engraving purposes, and in place of the wood of the European olive for inlaying work, since it is often prettily marked. The twigs are sometimes employed in making the twig bridges of the Western Himálaya, but for this purpose they are not considered so good as those of **Parrotia Jacquemontiana.**

153 **Olea dioica,** *Roxb.; Fl. Br. Ind., III., 612; Wight, Ill., t. 151.*

Syn.—O. WIGHTIANA & HEYNEANA, *Wall.*

Vern.—*Atta-jam,* BENG.; *Kala kiamoni,* NEPAL; *Kulumb,* C. P.; *Parjamb,* BOMB.; *Karambu,* MAR.; *Koli,* TAM.; *Parjamb, burra-nuge, mudla,* KAN.

References.—*Roxb., Fl. Ind., Ed. C.B.C., 36; Voigt, Hort. Sub. Cal., 547; Kurz, For. Fl. Burm., II., 157; Beddome, For. Man., 154; Gamble, Man. Timb., 257; Dalz. & Gibs., Bomb. Fl., 159; DC., Prodr., VIII., 286; Lisboa, U. Pl. Bomb., 98, 165; Drury, Useful Pl. of India, 318; Balfour, Cyclop., III., 19; Gazetteers:—Bombay, XV., 74, 437; XVII., 25; Settle. Rept., Chanda Dist., App., VI.*

Habitat.—A glabrous tree, 30 to 60 feet in height, found on the lower hills of Assam and Bengal, and throughout the Deccan Peninsula.

Medicine.—The BARK is said to be used as a febrifuge in the Central Provinces. MEDICINE. Bark. 154

Food.—The tree bears a FRUIT which is about the size of a Spanish olive, and is reported to be eaten in curries and to be also pickled. The fresh fruit is an article of food in famine times. FOOD. Fruit. 155

Structure of the Wood.—White, strong, and close-grained. TIMBER. 156

Domestic Uses.—The timber is extensively used by the inhabitants of Assam for various domestic purposes. DOMESTIC USES. 157

Olea europea, *Linn.* 158

THE OLIVE.

References.—*DC., Origin Cult. Pl., 279; Stewart, Pb. Pl., 139; O'Shaughnessy, Beng. Dispens., 432; Irvine, Mat. Med., 120; Baden Powell, Pb. Pr., 587; Christy, Com. Pl. & Drugs, V., 43; Royle, Prod. Res., 360; Balfour, Cyclop. III., 20; Smith, Dic., 297; Kew Off. Guide to Bot. Gardens & Arboretum, 95; Simmonds, Trop. Agri., 392; J. Harris Browne, The Cultivation of the Olive; Agri. Horti. Soc., Ind.:—Trans., I., 235, 401; III., 41; IV., 92; Journal, I., 401; II., 50, 297; III., 248; X., 116, Sel. lxxxix., clxxii.; Gazetteers:—Mysore & Coorg, I, 62; Bannu Dist., 23; Rāwalpindi, 15; Hazára, 13; Peshāwar, 13, 27; Ind. Forester, V., 329; XI., 55; Settle. Rept., Kohat, 28-29.*

Habitat.—This small shrub-like tree, most probably a native of Western Asia, is now introduced and cultivated in almost all the countries around the Mediterranean. It thrives well in dry climates like those of Syria and Assyria, and has been successfully acclimatised at the Cape of Good Hope, in some parts of America, and in Australia.

CULTIVATION IN INDIA.—Attempts have been made to cultivate this olive in various parts of India, but with small success. The tree grows well but is not productive. It flowers, but the fruits seldom set, and in cases where they have done so they have dropped off before becoming mature. Some trees of it exist in the Botanical Gardens at Calcutta which were introduced in 1800, "but up to the present time," says **Firminger** (1874), "they have never borne." CULTIVATION. 159

In 1842 several importations of European varieties of the Olive tree were made to the Bombay Botanic Gardens at Dapooree and Kaira. Although they grew well they did not flower. At Bangalore in the Government Gardens it attained a height of twenty to thirty feet, but did not flower freely and only fruited occasionally (*Agri.-Horti. Soc. Journals*).

In the Panjáb at Sialkote, an attempt was made to introduce the European olive from trees brought from Europe. They grew well, and appeared quite healthy, but though they flowered plentifully, the fruit would not mature. On attaining the size of a large pea it turned yellow and dropped off (*Indian Forester*). So far as the writer can ascertain, no experiments have been made by grafting the European olive on to the ordinary wild olive of India (**Olea cuspidata**), and the manuring of the trees seems in some cases to have been entirely neglected. Possibly if a good European variety were grafted on to the Indian species and the plants were heavily manured, which is always done in the cultivation of the tree both in Europe and in Australia, more successful results might be obtained. On several occasions the attempt has been made with more or less success to introduce the olive into Kashmír. It is believed the subject is presently occupying the attention of the Administration, and it would appear that that State (of Indian regions hitherto thought of) offers the best prospect of ultimate success.

O. glandulifera, *Wall.; Fl. Br. Ind., III., 612; Wight, Ic., t., 1238.* 160

Syn.—O. PANICULATA, *Roxb.* not of *R. Brown* (an Australian species).

Vern.—*Gair, galdu, garúr,* KUMAON; *Gúlili, raban, sira, phalsh,* PB.

References.—*Roxb., Fl. Ind., Ed. C.B.C., 35; Brandis, For. Fl., 309; Gamble, Man. Timb., 258; DC., Prodr., VIII., 285; Atkinson, Him. Dist., 313, 744; Ind. Forester, X., 34; Gazetteer, N.-W. Provinces, X., 313.*

Habitat.—A moderate-sized tree, found on the outer Himálaya from the Indus to Nepál, between 2,500 and 6,000 feet. It is distributed also to the Nilgiris and Anamalai Hills in South India.

MEDICINE. Bark. 161 Leaves. 162

Medicine.—The BARK and LEAVES of this species are astringent and used as an antiperiodic in fevers (*Atkinson*).

FOOD & FODDER. Leaves. 163

Food & Fodder.—The LEAVES are given as fodder principally to goats.

TIMBER. 164

Structure of the Wood.—Pale brown, moderately hard, close and compact, weight, on an average, 54 to 55lb per cubic foot. It is durable, capable of taking on a good polish, and is not liable to be eaten by insects.

DOMESTIC USES. 165

Domestic Uses.—The timber is employed in regions where the tree occurs, in house construction, carpentry and turnery, and for agricultural implements (*Brandis; Gamble*).

Olea robusta, *Kurz,* see **Ligustrum robustum,** *Bedd.;* OLEACEÆ; Vol. IV., 640.

Oleander, see **Nerium odorum,** *Soland;* APOCYNACEÆ; Vol. V., 348.

Olibanum, see **Boswellia,** *Roxb.;* BURSERACEÆ; Vol. I., 511.

Onager, see **Horses, Asses, &c.,** Vol. IV., 289.

Onion, see **Allium Cepa,** *Linn.;* LILIACEÆ; Vol. I., 169.

166

ONOBRYCHIS, *Gærtn.; Gen. Pl., I., 511.*

References.—*Fl. Br. Ind., II., 141; Aitchison, Bot. Afgh.! Del. Com., 57, 58.*

Several species of the leguminous genus—**Onobrychis**—exist in India and a far larger number on the sandy and stony soil of Afghánistán or Balúchistán. None of them are cultivated as fodder plants, like the Saintfoin of Europe (**Onobrychis sativa**), and that species apparently does not occur in India.

167

ONOSMA, *Linn.; Gen. Pl., II., 864.*

168

Onosma bracteatum, *Wall.; Fl. Br. Ind., IV., 178;* BORAGINEÆ.

Syn.—O. MACROCEPHALA, *Don.*

Vern.—*Gaozabán* (ox's tongue), BENG.; *Gul-i-gao-zabán* (the flowers), KASHMIR; *Gao-zabán,* TAM.

References.—*DC., Prod., X., 66; Pharm. Ind., 127; O'Shaughnessy, Beng. Dispens., 495; Moodeen Sheriff, Supp. Pharm. Ind., 133; Dymock, Mat. Med. W. Ind., 1st Ed., 478; 2nd Ed. 571; S. Arjun, Bomb. Drugs, 95; Irvine, Mat. Med. Patna, 32; Birdwood, Bomb. Pr., 48.*

Habitat.—A hispid, erect herb, found in the Western Himálaya, from Kashmír to Kumáon, ascending to altitudes of 11,500 feet.

MEDICINE. 169

Medicine.—Royle says this is the *Gao-zabán* of Indian medicine, and that it corresponds with the βουγλωσσος of the Greeks. Under the name of *Gao-zabán* the hakims, in the bazárs, sell parts of a plant, imported usually from Persia, composed of fragmentary leaves covered with tuberculated glands, each terminated by a hair. The flowers (*Gul-i-gao-zabán*) are purple, trumpet-shaped, and fully an inch and a half in length. (The *Flora of British India* says the flowers of **O. bracteatum** are only ½ inch long.) The perfect plant has as yet not been seen in the bales that come to India, so that the determination is still doubtful.

MEDICINE.

Probably several species of **Onosma,** as well as other Boraginaceous plants (see **Echium sp.,** Vol. III., 200), are sold under the collective name *Gao-zabán,* either as substitutes or as adulterants, although it would seem that the *Gao-zabán* of **Aitchison (Coccinia glauca)** agrees most closely with the description of what is usually sold in the bazárs under that name. In India *Gao-zabán* has long been much esteemed by the Native practitioners as tonic and alterative, and is much prescribed as a decoction in rheumatism, syphilis, and leprosy (*Arjun*). **O'Shaughnessy,** however, declares that it is a matter of little importance what the *Gao-zabán* of commerce really is, as it has no great medicinal value, being merely emollient, mucilaginous, and perhaps slightly diuretic.

Special Opinions.—§"The plant mentioned as being sent to England under the name of *gaw-zabán* was forwarded by me to **Dr. E. J. Waring,** and he speaks of this in the *Pharmacopœia of India,* p. 127. It was one of those plants which were raised by myself from the nuts of *gaw-zabán* which are always found with it in the bazár. The plant was about six or seven months old and did not flower. It was very small and stemless, with only five or six radical leaves. The largest leaves were about six inches long and six inches broad, and were covered with small dots, like white glands. When dry, the leaves are brittle and break easily. Some of the plants raised by me were more than two years old, but neither flowered nor thrived well. Some dry and pink flowers are sold separately in the bazár under the Persian name of *gule-gaw-zabán,* which means the flowers of *gaw-zabán,* but I am almost sure that they are not the produce of this plant. I have never found them in the bundles of *gaw-zabán.* Although *gaw-zabán* is not a very active medicine, yet it is a good refrigerant and demulcent, and proves itself useful in many of the affections in which medicines of such nature are indicated. Few medicines are equal to it for relieving the excessive thirst and restlessness in febrile excitement. It is also of great service in relieving functional palpitation of the heart, irritation of the stomach and bladder, and strangury. It is used in the form of an infusion prepared with either cold or hot water in the proportion of 1 to 20. Dose ℥ii to ℥iv frequently or *ad libitum*" (*Honorary Surgeon Moodeen Sheriff, Khan Bahadur, G. M. M. O., Triplicane, Madras*). "*Gao-zabán* is frequently used in combination with *Bunufsha* from a quarter to one tola in water as decoction in fevers. They bring on free perspiration" (*Meer Camer Ali, Native Doctor, Bhognipore Division, Etawah*). "The *gao-zabán,* as obtained from the bazár, is undoubtedly a Borage. The *gule-gao-zabán* flowers have all the characters of those of Boraginaceæ. They seem to be the flowers of an **Onosma,** the leaves of which are not used medicinally by the Natives. In fact, the leaves of one and the flowers of another species of **Onosma** seem to be used. *Cacalias* (*sic.*) have smooth leaves and can have no relation to the *Gao-Zabáns* of the bazárs" (*Assistant Surgeon S. C. Bhuttacharjee, Chanda, Central Provinces*).

Onosma echioides, *Linn.; Fl. Br. Ind., IV., 178.* 170

Syn.—O. hispida, *Wall.;* O. arenarium, *Waldst.*

Vern.—*Ratanjot,* Hind.; *Newar maharangi,* Nepal; *Ratanjot, maharanga, lál-jari, koame,* Pb.

References.—*Stewart, Pb. Pl., 366; Aitchison, Cat. Pb. & Sind Pl., 95; DC., Prodr., X., 61, 66; Boiss., Fl. Orient, IV., 181; Murray, Pl. & Drugs, Sind, 172; Baden Powell, Pb. Pr., 366; Atkinson, Him. Dist., 744; Agri.-Horti. Soc., Ind., Journ., XIV., 52.*

Habitat.—A hispid, biennial plant, frequent throughout the Western Himálaya, from Kashmír to Kumáon. It is widely distributed from Siberia and Cabul to France.

DYE. Root. 171 **Dye.**—The ROOT is used in the Panjáb Himálaya and Trans-Indus region as a red dye for wool (*Stewart*), being applied with *ghi* and the acid of apricots. It is employed by the Lambas of Lahoul, Spiti, and Kanawar to stain their images. It is also utilised as a substitute for alkanet (**Anchusa tinctoria**) and imparts a rich red colour to medicinal oils
MEDICINE. and fats, *e.g.*, Macassar oil (*Stewart*). In Nepál these roots brought from
Root. the Tibetan border are boiled with oil and used as a dye for the hair.
172 **Medicine.**—The bruised ROOT is used as an application to various
Leaves. forms of cutaneous eruptions. The LEAVES are said to possess alterative
173 properties, and the FLOWERS are prescribed as a stimulant and cardiac tonic
Flowers. in rheumatism and diseases of the heart (*Stewart*).
174

175 **Onosma Emodi,** *Wall.; Fl. Br. Ind., IV., 179.*

Syn.—O. VESTITUM, *Wall.*; MAHARANGA EMODI, *DC.*

References.—*DC., Prodr., X., 71; Murray, Pl. & Drugs, Sind, 172.*

Habitat.—A hispid, sub-erect plant which is found in the Alpine Himálaya at altitudes between 10,000 and 13,000 feet from Garhwal to Nepál and Sikkim.

DYE. Roots. 176 **Dye.**—Murray, in *Plants and Drugs of Sind*, says that the ROOTS of this plant also "make an excellent dye for wool and silk."

177 **O. Hookeri,** *Clarke; Fl. Br. Ind., IV., 178.*

References.—*H.f. & T.; Herb. Ind. Or.; Hooker, Him. Journ., II., 84.*

Habitat.—A hirsute plant with perennial root-stock, found in alpine Sikkim at altitudes between 12,000 and 14,000 feet.

DYE. Plant. 178 **Dye.**—The *Flora of British India* states that this PLANT affords the best Lepcha red dye. Further information with regard to it does not seem to exist in works on Indian products.

Onyx, see **Carnelian,** Vol. II., 174.

Opal, see **Carnelian,** Vol. II., 175.

Ophelia, see **Swertia,** *Linn.*; GENTIANACEÆ; Vol. VI., Pt. II.

Ophiocephalus, see **Fish,** Vol. III., 388.

179 OPHIORRHIZA, *Linn.; Gen. Pl., II., 63.*

180 **Ophiorrhiza Mungos,** *Linn.; Fl. Br. Ind., III., 77;* RUBIACEÆ.

THE MONGOOSE-PLANT.

Vern.—*Sarahati,* HIND.; *Gandhanákuli,* BENG.; *Kiri-purandán,* TAM.; *Chettu,* TEL.; *Kajo-ular,* JAVANESE; *Dal-kattiya, wal, eka, weriya, mendi,* SING.; *Sárpákshi,* SANS.

References.—*Roxb., Fl. Ind., Ed. C.B.C., 235; Elliot, Fl. Andhr., 62; Ainslie, Mat. Ind., II., 198; O'Shaughnessy, Beng. Dispens., 400; Pharm. Ind., 42; U. C. Dutt, Mat. Med. Hind., 317; Irvine, Mat. Med., Patna, 62; Agri.-Horti. Soc., Ind., Journal (Old Series), VI., 14.*

Habitat.—A suffrutescent herb, met with in the Khásia mountains and Assam, up to an altitude of 2,000 feet. It is found also in Burma, Tenasserim, the Andaman and Nicobar Islands, the mountains of the Western Peninsula and Ceylon. It is distributed to Sumatra and Java.

MEDICINE. Plant. 181 **Medicine.**—All the parts of this PLANT, bu more especially the ROOT, are intensely bitter, and may be used as stomachic tonics. In Ceylon, it
Leaves, Roots & Bark. has a high reputation as a remedy for snake bite. The LEAVES, ROOTS,
182 and BARK are made into a decoction and given in doses of half an ounce; but Roxburgh entirely discredits the idea of its having in this respect any medicinal virtues. Among the Sinhalese, it has got the name of *naga walli*, from the fact that it is supposed by them to be a specific against the bites of the ribbon snake. Its name, **O. Mungos** (the Ichneumon plant),

MEDICINE.

is derived from the fact that it is supposed to furnish the Mongoose with an antidote when bitten in its conflicts with poisonous snakes. **Kœmpfer**, in his *Æmenitates Exoticæ*, gave a particularly high character to the plant, which, he says, he used successfully as a decoction in putrid fevers and various malignant diseases, and in several cases as an antidote against the bites of mad dogs both in men and in other animals (*Agri.-Horti. Soc. Ind., Jour.*) **Horsfield**, in his account of *Java Medicinal Plants*, says this herb has been confounded with **Ophioxylon serpentinum** both by **Murray** and by **Burmann**, but that they are very distinct botanically, and the **Ophiorrhiza Mungos** is altogether inert, while **Ophioxylon serpentinum** is a valuable medicinal plant. The last mentioned plant by many writers also gets the name of Ichneumon plant (*Jones, Asiat. Researches, IV.*), and is said to be a valuable febrifuge (*vide* **Rauwolfia**, *VI., Pt. I*). In India, **Ophiorrhiza Mungos** is not used internally, but the roots are sold as a charm against snake-bites (*Irvine, Mat. Med., Patna*). 183

Ophioxylon serpentinum, *Linn.*, see **Rauwolfia serpentina**, *Benth.*; [APOCYNACEÆ; Vol. VI., Pt. I.

Opium, see **Papaver somniferum**, *Linn.*; PAPAVERACEÆ; Vol. VI., Pt. I.; [also **Narcotics**, Vol. V., 331.

OPLISMENUS, *Beauv.*; *Gen. Pl., III., 1104.* 184

[*India, 13*; GRAMINEÆ.

Oplismenus Burmanni, *Retz*; *Duthie, Fodder Grasses of Northern* 185

Syn.—O. BROMOIDES, *Boj.*; PANICUM BURMANNI, *Linn.*; P. HIRTELLUM, *Burm.*

Vern.—*Chusa* (Pilibhit), N.-W. P.; *Chimakal gadi, utaniya, wataniya* (Chanda) *ghor-chubba* (Seoni), *verwa* (Balaghat), C. P.

References.—*Roxb., Fl. Ind., Ed. C.B.C., 99*; *Rumph. Amb., VI., 14.*

Habitat.—A common grass, generally growing under the shade of large trees on the plains or at low elevations on the hills, which extends in its distribution from Oudh and Banda to Saharanpur and the Jhelum valley.

FODDER. 186

Fodder.—Duthie in his *Fodder Grasses of Northern India* says that cattle eat this grass with relish, while it is young, and that **Symonds** says it makes good hay.

O. colonus, *Kunth.*, see **Panicum colonum**, *Linn.*; Vol. VI., Pt. I.

O. compositus, *R. & S.*; *Duthie, Fodder Grasses of N. India, 13.*

Syn.—PANICUM COMPOSITUM, *Linn.*; P. SYLVATICUM, *Lam.*

Reference.—*Roxb., Fl. Ind., Ed. C.B.C., 99.*

Habitat.—This grass is found in the Himálaya at Simla, Kumáon, Massúri, and at Dehra Dun at the foot of the Massúri hills.

FODDER. 187

Fodder.—Roxburgh, who describes it under the name of **Panicum lanceolatum**, says it grows under the shade of trees, and that cattle are not fond of it.

O. frumentaceus, *Roxb.*; see **Panicum frumentaceum**, *Roxb.*, Vol. VI, [Pt. I.

Opobalsamum, see **Balsamodendron Opobalsamum** *Kunth.* [BURSERACEÆ; Vol. I., 369.

OPOPANAX, *Koch.*; *Gen. Pl., I., 922.* 188

Opopanax Chironium, *Koch.*; UMBELLIFERÆ. 189

Vern.—(The gum resin)—*Juvashur*, HIND.; *Jawe-shi*, BENG.; *Juwashur* BOMB.; *Juwashur*, ARAB.; *Gawshir, gaushur, juvashur*, PERS.

References.—*Ainslie, Mat. Ind., I., 281*; *O'Shaughnessy, Beng. Dispens., 361*; *S. Arjun, Bomb. Drugs, 67*; *Murray, Pl. & Drugs, Sind, 198*; *Irvine, Mat. Med. Patna, 44*; *Year-Book Pharm., 1876, 625*; *Bird-*

wood, Bomb. Pr., 42; Smith, Dic., 299; Treasury of Bot., II., 817; Kew Off. Guide to the Mus. of Ec. Bot., 76.

Habitat.—This plant, which somewhat resembles a parsnip in its general appearance, is a native of Southern Europe and Asia Minor.

GUM-RESIN. Root. 190 **Gum-resin.**—The ROOT yields a milky juice which hardens and becomes a GUM-RESIN possessed of properties somewhat similar to those of **Gum Ammoniacum** and known by the name of **Opopanax.**

MEDICINE. 191 **Medicine.—Opopaanx** is said to be much used by the Indian hakims who consider it an excellent stimulant and antispasmodic and prescribe it internally in doses of grs. 2-10, for uterine affections, flatulence, colic, and convulsions (*Irvine*). **Ainslie, Arjun,** and **Dymock** say they have never seen this gum-resin in India, and that *Javashir* or *Gawshir* is the name for Persian Galbanum. **O'Shaughnessy,** however, remarks that **Opopanax** is found in all the bazárs of Bengal, and is even exported from India to Europe.

192 **OPUNTIA,** *Mill.; Gen. Pl., I., 851.*

Several species of **Opuntia** are found in India, but only one **(O. Dillenii)** is described in the *Flora of British India.* It is naturalised in this country, and has overrun immense tracts of the hotter regions, such as in Madras, Bengal, the North-West Provinces, and the Panjáb.

[CACTEÆ.

193 **Opuntia Dillenii,** *Haw.; Fl. Br. Ind., II., 657; Wight, Ill., 114;*

THE PRICKLY PEAR.

Syn.—CACTUS INDICUS, *Roxb.*

Vern.—*Nág-phaná,* HIND.; *Nág-phana, pheni-mama,* BENG.; *Samar,* MAR.; *Zhorhatheylo,* GUZ.; *Nága-dali,* TAM.; *Nága-dali,* TEL.; *Nag-phansi, chappal, chappal-send,* DEC.; *Naga-mulla,* MALAY.; *Sha soung lit wa, ka la saw,* BURM.; *Kodu-gaha,* SING.

References.—*Roxb., Fl Ind., Ed. C.B.C., 395; Voigt, Hort. Sub. Cal., 62; Brandis, For. Fl., 245; Gamble, Man. Timb., 208; Stewart, Pb. Pl., 101; DC., Prods., III., 472; Origin Cult. Pl., 274; Mason, Burma & Its People, 420, 748; Royle, Ill., I., 9, 223; Ainslie, Mat. Ind., II., 217; Dymock, Warden & Hooper, Pharmacog. Ind., Pt. III., 99; Bidie, Cat. Raw Pr., Paris Exh., 119; Lisboa, U. Pl. Bomb., 160, 199; Liotard, Paper-making Mat., 11; Smith, Dic., 288; Kew Reports, 1877, 37; Kew Bulletin, 1888, 165-175; 1889, 26; Ind. Agr., Oct. 9th, 1886, Oct. 19th, 1889; Ind. Forester, IX., 625; Gazetteers:—Bombay, V., 126; Mysore and Coorg, I., 68; N.-W. P., IV., lxxii.; Panjáb, Karnál, 16; Rohtak, 14; Sialkot, 11; Rájputána, 27; Agri.-Horti. Soc., Ind. (Transactions), VI. (App.), 1-84; VII., 68; (Journals), VI., 38; VIII., 223; X., 8.*

Habitat.—A ramous bush, indigenous in America, but naturalised all over India, from Bengal and Madras to the Panjáb, and found in the Himálaya up to an altitude of 5,000 feet. **Roxburgh,** who described it under the name of **Cactus indicus,** believed it to be a native of India. As a matter of fact, however, it is most probable that it was introduced by the Portuguese. At all events, nothing was known of the various species of **Opuntia** by the Greeks and Romans; they were first described by the Spanish and Portuguese writers of the sixteenth and seventeenth centuries, as curiosities then introduced from the New World. Being introduced, they spread rapidly throughout the Mediterranean countries, and were most probably brought thence to India, where the large areas left uncultivated, favoured their peculiar habit of growth, in arid regions, in which no other plant could grow. When the cochineal insect was brought to India in 1795, this species of **Opuntia** was then so prevalent in India as to lead the writers of that date to speak of it as the indigenous species on which fortunately the cochineal insect was found to thrive as well as on the species of **Opuntia** which had been brought with the insect. (*Conf.* with remarks in Vol. II., 398.)

Fibre.—A coarse kind of fibre can be obtained from **Opuntia Dillenii** which has been suggested might prove useful as a paper material. **Dr. Bidie** of Madras, speaking of this fibre, wrote:—"The prickly pear, which so abounds and has become such a nuisance in the country, that large sums are expended annually in cutting it down and burying it, might prove valuable as a paper material." **Liotard**, in his "Materials in India Suitable for the Manufacture of Paper," describes it as a coarse fibre easily cleared of extraneous matter which he is inclined to think would pulp well. The samples of the fibre shown at the Colonial and Indian Exhibition were, however, by the paper-makers who examined them pronounced quite worthless as compared with other equally plentiful materials. FIBRE. 194

Medicine.—The FRUIT is considered a refrigerant, the LEAVES, mashed up and applied as a poultice are said to allay heat and inflammation (*Ainslie*). In the Deccan the baked fruit is given in whooping cough (*Lisboa*). **Dymock** (*Pharmacog. Ind.*) states that a SYRUP of the fruit appears to increase the secretion of bile when given in tea-spoonful doses three or four times a day and to control spasmodic cough and expectoration. In Dacca the MILKY JUICE is given as a purgative in doses of ten drops mixed with a little sugar (*Taylor, Topography of Dacca*). The ripe fruit when eaten has the power of dyeing the urine red (*Miller*). MEDICINE. Fruit. 195 Leaves. 196 Syrup. 197 Milky Juice. 198

SPECIAL OPINIONS.—§ "Said to be useful in gonorrhœa. The hot-leaf applied to boils hastens suppuration; the leaf made into a pulp is applied to the eyes in cases of opthalmia" (*Surgeon-Major D. R. Thomson, M.D., C.I.E., Madras*). "The leaves, roasted and mashed, form a very good poultice, especially in scorbutic ulcers" (*Native Surgeon T. R. Moodallier, Chingleput, Madras Presidency*). "Juice used as a purgative, also used as a demulcent in gonorrhœa" (*W. Forsyth, F.R.C.S., Edin., Civil Medical Officer, Dinajpore, North Bengal, & C. M. Mukerji, M.B., Civil Medical Officer, Dinajpore*). "The pulp of the leaf is used as a poultice in guinea-worm" (*Surgeon-Major J. North, Bangalore*).

Food & Fodder.—The FRUIT is pear-shaped, but stands erect on the margins of the thick fleshy leaf-like joints of the stem. It is covered with thin sharp spines and tufts of minute spinules. It contains a somewhat refrigerant and highly refreshing pulp, which in times of scarcity is used as a food by poor people. Superior qualities are in Europe and America specially cultivated on account of their fruit—the Prickly Pear. In Mexico, the LEAVES also of a species of **Opuntia** are cooked in hot ashes, and, the outer skin with the thorns being removed, the slimy, sweet, succulent interior is eaten. FOOD & FODDER. Fruit. 199 Leaves. 200

In Spain experiments have been made, with a view to extracting ALCOHOL from the fruits of the prickly pear, and proposals have been made in India also to try and utilise for this purpose some of the tracts of **Opuntia** that exist in Mysore. It is said that in Spain the industry might be made a very profitable one, and in India, where the prickly pear grows and flourishes without any attention, on the sandy wastes of Madras, Sind, and Balúchistán, very little care would, no doubt, produce a good class of fruit. The cost of cultivation would be very small indeed, and large tracts of land on which the prickly pear would grow, might be had for almost nothing, so that, if the industry can be *proved* a profitable one in Spain, it ought, in India, to be tenfold more so. The jointed, juicy columnar stems, and the leaves, deprived of their spines, are said to form an excellent supplementary food for cattle. Alcohol. 201

Domestic Uses.—The prickly pear is much used in India as a hedge-plant about fields, and for fences around the homesteads. For these purposes it is invaluable, being both impenetrable and uninflammable. Formerly, when villages were so often subjected to sudden predatory attacks in the rough DOMESTIC USES. 202

ORCHIS latifolia.

The Marsh Orchid.

DOMESTIC USES.

anarchical times that preceded the British rule in India, this plant was much valued as a source of defence around isolated villages, and it is said that Tipú Sahib strengthened the defences of Seringapatam by surrounding the fortifications with deep plantations of prickly pear. Practically, however, as **Sir George Birdwood** points out, the unrestrained growth of this plant around Indian villages is very conducive to the generally insanitary state that prevails in them, chiefly because the close growth of the flat jointed stems branching out in every direction makes it difficult from the nature of the plant to keep the ground underneath free from decaying vegetation. If this be kept clear, the prickly pear is as harmless to the health of man as any other hedge. Another complaint made about these hedges is that they harbour reptiles, but this also is dependent on the want of clearing away of undergrowth and decaying vegetable matter.
203 The prickly pear is, however, very apt to spread not only from its growth by cuttings, but also from the fact that the birds which eat the fruits cannot digest the hard seeds and these are cast out on the ground and germinate; but it will be found that in its growth it has a distinct preference for waste arid soils that will grow nothing else, and that it does not flourish so freely on rich well-cultivated land. Owing to the immense tracts of country covered with prickly pear in India, various attempts have been made to eradicate them or to utilise them for industrial purposes. In the way of eradication it was proposed in the Madras Presidency some years ago to utilise the masses of prickly pear that grow there by sowing tree seeds among them, in the hope that the masses of **Opuntia** might, in the first place, act as nurses to the seedlings, protecting them from the sun and browsing animals, and afterwards the trees growing up would shade the prickly pears to such an extent as ultimately to destroy them. The results of some of those experiments in Trichinopoly were very successful, the trees grew up, shaded the **Opuntia**, and ultimately destroyed it; but in other districts no such favourable effects were produced, and in dry rainless regions, where the **Opuntia** is most abundant, the experiment necessarily fails.

For an account of the utilisation of **Opuntia Dillenii** as a food-plant for the Cochineal insect, see **Coccus Cacti**, Vol. II., 398-405.

Orange flower or **Neroli**, see **Citrus**, *Linn.;* RUTACEÆ; Vol. II., 333.

Orchil, see **Lichen**, Vol. IV., 636.

204 **ORCHIS**, *Linn.; Gen. Pl., III., 620.*

Most of the species of this large and important genus of the ORCHIDEÆ have tubers which, when properly prepared, are capable of yielding *Salep* (*q. v. Vol. VI., Pt. I.*). The salep of Indian commerce, obtained from this genus, is principally derived from the following species:—**O. latifolia, O. laxiflora,** and **O. mascula**—the first mentioned is a native of India, but the supply of the tubers is mainly drawn from across the Himálaya. The species of Orchids are, however, far less important sources of Indian *Salep* than are other Orchids—natives of the Himálaya. (*Conf.* with **Eulophia**, Vol. III., 290.)

205 **Orchis latifolia,** *Linn.; Fl. Br. Ind., VI., 127 (proof copy furnished by Sir J. D. Hooker);* ORCHIDEÆ.

THE MARSH ORCHID.

Syn.—O. LATIFOLIA β INDICA, *Lindl.;* O. HATAGIREA, *Don., Prod.*

Vern.—*Salap* or *salab* (the tubers), HIND., PERS., AFGHAN.

References.—*Boiss., Fl. Orient., V., 71; Aitch., Afgh. Del. Com., 113; Ind. Forester, XIV., 371; Gazetteers, N.-W. Prov., X., 318; Agri.-Horti. Soc., Ind., Journ., II., Sel., 436; XIV., 56.*

Habitat.—An orchid which grows usually in damp meadows and marshes, and is widely distributed from Europe over Temperate Asia. In

India it occurs in the Western Temperate Himálaya, from Nepál to Kashmír and Western Tibet at altitudes between 8,000 and 12,000 feet. It was found by **Aitchison** in Khorassan in moist meadow land at an altitude of 3,000 feet.

Medicine.—The TUBERS of this plant, along with those of allied species, are collected in considerable quantities in Persia and exported, by way of Herát, to India. MEDICINE. Tubers. 206

Orchis laxiflora, *Lam.; Boiss., Fl. Orient., V., 71.* 207

Vern.—*Salap, salab* (the tubers), AFGHAN.

References.—*Aitch., Afghan. Del. Com., 114; Ind. Forester, XIV., 371.*

Habitat.—This plant has a similarly wide distribution with the preceding, being a native of the greater part of Central and Southern Europe, Turkey, the Caucasus, and Asia Minor, and reaching Persia, Afghánistán, Tibet, and even the mountainous frontier of India. It is, however, not described by **Sir J. D. Hooker** as met with in India proper.

Medicine.—"The TUBERS of this are collected with those of **O. latifolia** and, mixed with them, are exported as *salab* or *salap*" (*Aitchison*). MEDICINE. Tubers. 208

O. mascula, *Linn.; Boiss., Fl. Orient., V., 68.* 209

THE SALEP ORCHID.

Vern.—*Salep misrie*, BENG.; *Salum*, GUJ.

References.—*Fleming, Med. Pl. & Drugs, as in As. Res., Vol. XI., 189; Irvine, Mat. Med. Patna, 97; Med. Top., Ajmere, 151; Butler, Top. & Stats., Oudh & Sultanpur, 44; Agri.-Horti. Soc., Ind., Journ., II., Part I., 226; Part II., Sel., 216, 435; Pro., 441, 498; IV., Sel., 205; Pro, xxviii.*

Habitat.—Like the two preceding species **O. mascula** has a wide distribution throughout Central and Southern Europe and Asia Minor and Persia. It apparently does not occur in India.

Medicine.—The TUBERS are one of the largest sources of European salep, considerable quantities of it, prepared in Macedonia and Greece, being imported into India through Bombay. MEDICINE. Tubers. 210

O. sp. (?) 211

Food.—**Stewart**, in his *Panjáb Plants*, mentions that **Bellew** describes an **Orchis** the thick fleshy LEAVES of which in Afghánistán were cooked with *ghí* and eaten by the Natives, but says he doubts if this could have been an **Orchis**, as he (**Stewart**) has never heard of the leaves of one of these being eaten. FOOD. Leaves. 212

ORIGANUM, *Linn.; Gen. Pl., II., 1185.* 213

Origanum Marjorana, *Linn.; Fl. Br. Ind., IV., 648;* LABIATÆ. 214

THE SWEET MARJORAM, *Eng.*; MARJOLAINE, *Fr.*; MAJORAN, *Germ.*; Αμαρακον, *Gr.*

Syn.—O. WALLICHIANUM, *Benth.*; MARJORANA HORTENSIS, *Mœnch.*

Vern.—*Murrú*, BENG.; *Ban tulsí*, KUMAON; *Murwo*, SIND; *Marrú*, TAM.; *Murwa*, DEC.; *Mizunjúsh, mardakusch*, ARAB.

References.—*Voigt, Hort. Sub. Cal., 456; Ainslie, Mat. Ind., I., 213; Dymock, Mat. Med. W. Ind., 2nd Ed., 618; Irvine, Mat. Med. Patna, 71; Year-Book Pharm., 1880, 82; Atkinson, Econ. Prod., N.-W. P., Part IV., Foods, 41; Birdwood, Bomb. Pr., 65; Kew Off. Guide to Bot. Gardens & Arboretum, 105; Gazetteer, Mysore & Coorg, I., 64; Agri.-Horti. Soc. Ind., Journ., X., 89; (New Series), IV., 28; V., 34, 42.*

Habitat.—This herb, a native of Southern Europe, North Africa, and Asia Minor, is cultivated in gardens throughout India, especially in the South, for its seeds. Almost naturalised in Kumáon.

Oil.—An essential OIL is distilled from the SEEDS which is used as a perfume and for hot fomentations in acute diarrhœa. OIL. Seeds. 215

MEDICINE. Seeds. 216 **Medicine.**—The SEEDS are considered astringent and are used as a remedy for colic, for which purpose also the LEAVES (Leaves. 217) are eaten along with those of **Gynandropsis pentaphylla** (Vol. IV., 192). **Irvine** says the DRIED PLANT (Dried Plant. 218) is used in perfumes and as a stimulant.

SPECIAL OPINION.—§ "Cultivated near Bombay under the Portuguese name of *Mazarona*. The Bombay plant has been identified in London as **O. Marjorana**" (*W. Dymock, Bombay*).

FOOD. Leaves. 219 **Food.**—The LEAVES are used as a seasoning herb.

220 **Origanum vulgare,** *Linn.; Fl. Br. Ind., IV., 648.*

This also is called MARJORAM by Europeans in India.

Syn.—O. NORMALE, *Don;* O. LAXIFLORA, *Royle.*

Vern.—*Sáthra,* HIND.; *Mirzanjosh,* PB.; *Mridu-maruvamu,* TEL.; *Buklut-ul-gezal, sutur,* ARAB.; *Oushneh, mizangosh,* PERS.

References.—*Voigt, Hort. Sub. Cal., 456; Stewart, Pb. Pl., 171; Elliot, Fl. Andhr., 117; S. Arjun, Bomb. Drugs, 103; Year-Book Pharm., 1874, 628; 1879, 467; Birdwood, Bomb. Pr., 65, 225; Agri.-Horti. Soc., Ind., Journ., V., 42; Journ. Linn. Soc., X., 74; Gaz., N.-W. Prov., X., 315; C. P., 118.*

Habitat.—An herb which occurs plentifully on the Temperate Himálaya, from Kashmír to Sikkim, at altitudes between 7,000 and 12,000 feet, and is distributed to Europe, North Africa, and North and West Asia. On some parts of the Himálaya, as for example, in the Simla District, it is one of the most abundant of herbs, miles of country having a plant or two at least every few feet apart.

MEDICINE. Oil. 221 **Medicine.**—It yields a volatile OIL, used as an aromatic, stimulant and tonic in colic, diarrhœa, and hysteria, and which is applied locally also in chronic rheumatism and toothache, and dropped into the ear in earache. It is said to stimulate the growth of hair and also to act as an emmenagogue.

FOOD. Plant. 222 **Food**—The PLANT is used in the Panjáb as a pot-herb and, according to **Aitchison**, it is in Lahoul eaten as a vegetable.

223 **ORMOSIA,** *Jacks; Gen. Pl., I., 556.*

224 **Ormosia glauca,** *Wall.; Fl. Br. Ind., II., 253;* LEGUMINOSÆ.

Vern.—*Chuklein,* LEPCHA.

Reference.—*Gamble, Man. Timb., App., xvii.*

Habitat.—A tall tree of the Central Himálaya, found abundantly, for example, in Nepál.

TIMBER. 225 **Structure of the Wood.**—Greyish white, moderately hard.

DOMESTIC. Seeds. 226 **Domestic.**—**Gamble** remarks that the bright scarlet SEEDS of this tree are employed by the Lepchas as a bait to catch jungle fowl.

227 **O. travancoria,** *Bedd.; Fl. Br. Ind., II., 253.*

References.—*Gamble, Man. Timb., 116; Bedd., Fl. Sylv., t. 45; Ind. Forester, III., 22, 23.*

Habitat.—A tall tree, with finely grey-downy branchlets, met with in the Tinnevelly and Travancore hills.

TIMBER. 228 **Structure of the Wood.**—As yet it is scarcely known except to Natives, but is said to be very valuable and used for many domestic purposes (*Ind. For.*).

229 **OROBANCHE,** *Linn.; Gen. Pl., II., 984.*

A genus of parasitic plants which grow on the roots of many plants, chiefly field crops, such as the mustard, tobacco, poppy, &c. A great many species are described, but it is probable that the greater number of these are only varieties, due to the different foster-plants on which they grow. A good few special Afghán species have been described by **Aitchison** (*Afgh. Del. Com.*), one of which, parasitical on two species of **Artemisia**, and another parasitical on a Chenopodiaceous plant, are said by him to be eaten before the flowers expand. **Mr. J. F. Duthie** describes a species of **Orobanche** which is used in Afghánistán as a dye for silk. He says:—"A decoction of this plant is made in water, carbonate of soda is added, which makes a dye that gives a pale yellow colour to cotton and silk fabrics" The most important member, however, of this genus is **O. indica**, a parasite on mustard and tobacco plants, which often causes great damage to these crops. **O. cernua**, *Lœff.*, is frequently found on the poppy plants of the Patna District (*conf.* with **Papaver somniferum**, Vol. VI., Pt. I.); **O. nicotianæ**, *Wight, Ill., 179, t. 158*, appears to be the peculiar form met with on the tobacco crops of the Deccan.

Orobanche indica, *Ham.; Fl. Br. Ind., IV., 326;* OROBANCHACEÆ. 230

Syn.—PHELIPÆA INDICA, *G. Don;* (?) P. ÆGYPTIACA, *Boiss.*

Vern.—*Sarsán banda, bhatua ghás,* HIND.; *Turi sim,* SANTAL; *Khargaingi,* PUSHTU; *Vakumba,* GUJ.; *Bambaku,* DECCAN; *Bodu,* TEL.

References.—*Roxb., Fl. Ind., Ed. C.B.C., 467; Revd. A. Campbell, Econ. Prod. Chutia Nagpur, No. 7851; Aitchison, Afgh. Del. Com., 94; Trop. Agri., III., 615.*

Habitat.—A ramous, hairy herb, found throughout the plains of India, especially in mustard and tobacco crops, and distributed also to Central and Western Asia. It appears as a yellow-stemmed, succulent, asparagus-like shoot growing perpendicularly from the root-stock of the tobacco plant, sometimes singly, sometimes in clusters of five or six together. Although a noxious parasitic weed, it is provided with rootlets which descend into the earth alongside of its foster plant. Sometimes it grows in such quantity that when in blossom, it gives the fields a general purple or blue colour. As to its effects on the crops which it infests, the general opinion seems to be that it does no material injury to strong healthy plants, but injures greatly, even if it does not kill, weakly crops on poor soils. Some few years ago it gave considerable anxiety to the tobacco cultivators of South India, and an extensive official correspondence ensued, with the result that it seemed to be agreed the only effectual cure was to cut it off as it appeared, and thus prevent flowering.

FODDER. 231

Fodder.—It is sometimes used as cattle fodder, and there is every reason to believe that, although cattle may at first be averse to it, they in time become greedily fond of it, and yield more and better milk than when otherwise foddered. (*Conf.* with the article **Pests.**)

OROXYLUM, *Vent.; Gen. Pl., II., 1040.* [*1337;* BIGNONIACEÆ. 232

Oroxylum indicum, *Vent.; Fl. Br. Ind., IV., 378; Wight, Ic., t.* 233

Syn.—CALOSANTHES INDICA, *Bl.;* BIGNONIA INDICA, *Linn.;* B. PENTANDRA, *Lour.;* SPATHODEA INDICA, *Pers.*

Vern.—*Ullu, arlú, kharkath, pharkath, sauna, assar sauna, shyona,* HIND.; *Sona, sanpatti, násoná,* BENG.; *Pomponia, phunphuna,* URIYA; *Arengi banu, arengebaung, somepatta,* KOL.; *Bana hatak,* SANTAL; *Soizong,* RAJBANSHI; *Kering,* ASSAM; *Cherpong,* MECHI; *Totilla, karamkanda,* NEPAL; *Dhatte,* GOND; *Mulin, miringa, sori, tátpalang, tatmorang,* PB.; *Tattunúa,* C. P.; *Tantun, tetu, ulu, karkath, sauna-assar, phalphura,* BOMB.; *Tetu,* MAR.; *Tetu,* GUJ.; *Pana, vanga, achi, vanga-maram,* TAM.; *Pamania, pampana, dundillam, dondlup,* TEL.; *Teta,* KAN.; *Baladah,* AND.; *Kyoungyabeng, kyoung-sha,* BURM.; *Totilla,* SING.; *Syonáka, munduka-purna,* SANS.

References.—*Roxb., Fl. Ind., Ed. C.B.C., 495; Kurz, For. Fl. Burm., II., 238; Gamble, Man. Timb., 275; Stewart, Pb. Pl., 148; Mason, Burma & Its People, 411, 794; U. C. Dutt, Mat. Med. Hind., 320; Dymock, Mat. Med. W. Ind., 2nd Ed., 547; S. Arjun, Bomb. Drugs, 91; Baden Powell, Pb. Pr., 372; Lisboa, U. Pl. Bomb., 104; Bomb. Gaz., XIII., 27; XV., 64; Ind. Forester, III., 204.*

Habitat.—A glabrous tree, found throughout India, up to altitudes of 3,000 feet, and distributed to Burma, Ceylon, Malaya, Malacca, and Cochin China.

DYE & TAN. Bark. 234 Fruits. 235

Dye & Tan.—The BARK and FRUITS are used as a mordant in dyeing and tanning.

MEDICINE. Root-bark. 236

Medicine.—The ROOT-BARK is much used in medicine by the Natives of India. It is supposed to have astringent and tonic properties, and is employed in diarrhœa and dysentery. It is an ingredient of the compound decoction called *dasamula*, which is used in remittent and puerperal fevers, inflammatory diseases of the chest, affections of the brain, and other diseases, which are supposed to be caused by derangements of all the humours. The JUICE expressed from the root-bark, after it has been enclosed within some leaves and a layer of clay and roasted, is mixed with the gum of **Bombax malabaricum** (*mocharasa*), and given in dysentery and diarrhœa. In otorrhœa, the use of an oil has been recommended in Sanskrit medicine, prepared by boiling a PASTE made of the root-bark of **Oroxylum indicum** with sesamum oil (*U. C. Dutt*). In Bombay, the root-bark, ground up and mixed with turmeric, is said to form an efficient application to the sore-backs of horses and bullocks (*Gamble*; *Dymock*). Among the Gonds, **Dr. Evers** says, the BARK is used in decoction, as an application to rheumatic swellings. He states that he has made a trial of the powder and infusion of the bark, and has found it most powerfully diaphoretic, and slightly anodyne. He used it several times with marked success in acute rheumatism (*Ind. Med. Gazette*).

Juice. 237 Paste. 238 Bark. 239

SPECIAL OPINIONS.—§ "A good diaphoretic when administered internally as a powder or infusion—*vide Ind. Med. Gazette* for March 1875, page 66" (*Civil Surgeon B. Evers, M.B., Wardha*). "This is one of the ten roots of the Hindu Materia Medica" (*Assistant Surgeon S. Arjun, L.M., Gorgaum, Bombay*).

TIMBER. 240

Structure of the Wood.—Yellowish-white, soft, and devoid of heartwood. Weight 30lb per cubic foot. It is said to be so soft as to be unfit for use.

DOMESTIC. Seeds. 241

Domestic.—The large thin membraneous SEEDS are employed in lining hats, and are placed between two layers of wicker work to make umbrellas

ORPIMENT.

242 **Orpiment,** see also **Arsenic, I.,** 322.

YELLOW ARSENIC, YELLOW SULPHIDE OF ARSENIC, *Eng.*; AUREUM PIGMENTUM, *Lat.*

Vern.—*Hartál*, HIND.; *Horitál*, BENG.; *Hartál wilayiti, hartal warki* (the lamellar form), PB.; *Varkhihartála*, BOMB.; *Haritála*, MAR.; *Artál* GUJ.; *Táram, ari-táram, ponnari-tárakam, tálakam*, TAM.; *Haridalam, tálakamu*, TEL.; *Cháliyam*, MALAY.; *Hsae-dan*, BURM.; *Haritála, vansapatri haritala*, (the lamellar form), *Pinda hari-tala*, (the lump form), SANS.; *Arsániqún, Zarni-khe-asfar*, ARAB.; *Zarni-khe-zard*, PERS.

References.—*Pharm. Ind.*, 341; *U. S. Dispens.*, 15th Ed., 1716; *Ainslie, Mat. Ind.*, I, 499; *Irvine, Mat. Med. Patna*, 37; *Moodeen Sheriff, Supp. Pharm. Ind.*, 57; *U. C. Dutt, Mat. Med. Hindus*, 41; *Sakharam Arjun, Cat. Bomb. Drugs*, 171; *K. L. De, Indig. Drugs, Ind.*, 17; *Manual, Geology of India, Pt. III.* (*Ball, Econ. Geol.*), 162; *Baden Powell, Pb. Pr.*, 63, 102; *Buck's Report on Dyes of N.-W. P.*, 46; *McCann, Dyes & Tans., Beng.*, 1.

SOURCES. 243

Sources.—Orpiment, which derives its name from a corruption of the Latin *Aureum pigmentum*, occurs in two forms in the bazárs of India. It is met with either as smooth, shining, gold coloured scales, *vansapatri haritála*, or as yellow opaque masses, *pinda haritála*, and is found in small

SOURCES.

quantities in Kumáon in the North-Western Provinces, whence the best quality is derived. Inferior qualities come from Rangoon, or are brought into India overland from Swát and Kashgar. The trade seems to be entirely in the hands of Natives.

DYE. 244

Dye.—The "lump" orpiment is used for dyeing a yellow colour, and in sizing some country-made papers, which are, by its use, preserved from the ravages of insects.

MEDICINE. 245

Medicine.—The "foliaceous" orpiment is used in Hindu medicine as an alterative and febrifuge. "For internal use," says **Dutt**, "it is purified by being successively boiled in *kánjika*, the juice of the fruit of **Benincasa cerifera**, sesamum oil, and a decoction of the three myrobalans, for three hours in each fluid. The dose of orpiment is from two to three grains." Orpiment is said by Hindu physicians to cure fevers and skin diseases. It is employed in fevers in combination with mercury and aconite. In skin diseases it is used in various combinations as an external application as well as internally. As a depilatory it was in extensive use among the ancient Hindus.

Among the Afghans it is employed in the treatment of syphilis and skin diseases.

Orris Oil, see **Iris florentina**, *Linn.*; IRIDEÆ; Vol. IV., 497.

ORTHANTHERA, *Wight; Gen. Pl., II., 778.* 246

Orthanthera viminea, *Wight; Fl. Br. Ind., IV., 64*; ASCLEPIADEÆ. 247

Syn.—APOCYNEA VIMINEA, *Wall.*; LEPTADENIA VIMINEA, *Bth., & Hook.* In the *Genera Plantarum* the genus **Orthanthera** has been reduced to **Leptadenia**, but **Sir J. D. Hooker**, in the *Flora of British India*, takes a different view of the subject, and says "the long sepals and salver-shaped corolla are such strong generic characters that I do not follow the *Genera Plantarum* in uniting this genus with **Leptadenia**."

Vern.—*Mahur*, HIND.; *Máhúr*, N.-W. P.; *Chapkiya*, KUMAON; *Matti, mowá, lánebar, khip, chapkia*, PB.; *Kip, khip*, SIND.

References.—*Gamble, Man. Timb., 266; Stewart, Pb. Pl., 146; Murray, Pl. & Drugs., Sind 162; Atkinson, Econ. Prod., N.-W. P., Pt. V., 44, 80; Him. Dist., 313, 794; Royle, Ill. Him. Bot., 274; Liotard, Paper-making Mat., 36; Gazetteers:—N.-W. P., IV., lxxiv.; Panjáb, Gurgáon, 17; Muzaffargarh, 22; Dera Ismail Khan, 13; Rájputána, 30; Ind. Forester, IV., 233; XII., App. 2, 17.*

Habitat.—A glabrous shrub of North-West India, extending along the base of the Himálaya up to an altitude of 3,000 feet from the Panjáb and Sind to Oudh.

FIBRE. Plant. 248

Fibre.—This PLANT yields a fibre of which ropes are made. Near Delhi the plants are steeped for four or five days, the fibre extracted, and a strong and durable rope made of it. The fibre is remarkable for its tenacity and length (*Royle*). In Sind the unsteeped STALKS are made into ropes for Persian wheels, a purpose for which they are said to be admirably adapted, as they do not rot readily from moisture.

Stalks. 249

FOOD. Flower buds. 250

Food.—The FLOWER-BUDS are eaten by Natives as a vegetable.

ORTOLAN.

Ortolan. 251

GREENHEADED BUNTING, *Eng.*; ORTALON, ORTOLAN, *Fr.*; GARTENAMMER, *Germ.*; ORTOLANO, *It.*

Vern.—*Bergherie, bageyra, jam-johara*, HIND.

References.—*Ainslie, Mat. Ind., I., 286; Hooker, Him. Journ., I., 98; Balfour, Cycl. Ind., III., 52; Encycl. Brit., XVIII., 53.*

252 **Habitat.**—The Ortolan (**Emberiza hortulana,** *Linn.*) is a native of most European countries, the British Islands alone excepted (in which it but rarely occurs), and of Western Asia. The true Ortolan of Europe is but occasionally seen in India; but the social lark (**Calandrella brachydactyla**) and the black-bellied Finch-lark (**Pyrrhulanda grisea**) which are abundantly met with, especially in North-West India, are popularly known as Ortolans, and "at certain seasons," says **Ainslie,** "are much sought after by the Europeans in India, who consider them a great delicacy."

Hooker, in his *Himálayan Journals*, mentions having seen in Sikkim abundant flocks of a kind of lark called by Europeans "Ortolan," but not the real ortolan, although they were very fat and formed an excellent table substitute for that bird.

(*G. Watt.*)

253 **ORYZA,** *Linn.; Gen. Pl., III., 1116-1117.*

Bentham & Hooker, in their *Genera Plantarum,* state that botanical writers have described some twenty species as belonging to the genus **Oryza,** but they add that scarcely five of these can be well distinguished, and that even the five more easily recognised forms are very generally viewed as but varieties of one species, *viz.*, **O. sativa,** *Linn.* The chief forms of that species are inhabitants of the East Indies, though some are indigenous to Australia also, and most of them have been widely cultivated from very ancient times, throughout the warmer regions of both hemispheres. In fact, they are now almost naturalised in America and Africa. The Brazilian plant recognised under the name of **O. subulata,** *Nees,* is, according to **Bentham & Hooker,** probably only an abnormal state of **O. sativa** with the glumes elongate and acuminate after the fashion of those in **Poa** or **Festuca.**

In the *Genera Plantarum* **O. coarctata,** *Griff.*, is said to possess the strongest claim to an independent position, and the plant, there alluded to, is stated to be a native of the Himálaya. The writer has found, however, the greatest possible difficulty in determining that species. **Griffith** speaks of **O. coarctata** as a name which he affirms was given by **Hamilton,** but upon unsatisfactory grounds. To the plant, under which that synonym is quoted, **Griffith,** in fact, assigns the name of **O. triticoides,** *Griff. Notul. Ad. Pl. As., III., 8-10* (**O. coarctata,** *Ham. Icon., t. CXLII.*), and he speaks of it as a species which occurs in brackish water. His figure, as also his description, denote the plant as having very pronouncedly toothed leaves with the inflorescence composed of smaller spikes adpressed to the common axis, in such a manner as to simulate **Triticum** or even **Lolium.** The spicules are also sub-secund and turned towards the outside. **Griffith's** plate shows the spikelets lanceolate, acuminate, awnless, and in his description he speaks of the inner glumes (*palea*) as being nine-nerved and beakless, except the middle denticulation which is prolonged to a small point. He remarks also that it is prominently keeled, especially on the upper half. **Griffith's** plant is thus very unlikely to be the **O. coarctata** of the *Genera Plantarum,* but is probably a species more nearly allied to **O. sativa** than are **O. granulata,** *Nees,* or **O. officinalis,** *Wall.* Indeed, the writer can see no sufficient grounds on which **O. triticoides,** *Griff.*, could be separated from **O. coarctata,** *Roxb.* Both these authors allude to **Dr. Buchanan-Hamilton** as the discoverer of the plant, and **Roxburgh** adds that it is a native of the delta of the Ganges. In fact, it seems highly probable that the plant under consideration is nothing more than one of the forms of the so-called wild state of **O. sativa. Griffith's** figure on this supposition would have been made from an imperfectly expanded panicle. The writer has come to recognise several forms of wild **O. sativa**—one, for example, which is probably **O. rufipogon,** *Griff.* (possibly the origin of the red coloured rices of India), and another with greatly elongated spikelets of a pale greenish-white colour, which somewhat resembles **Griffith's** figure of **O. triticoides.** Although both **Roxburgh** and **Griffith** seem to have viewed the wild forms of **O. sativa,** *Linn.*, as of necessity awned, the writer has seen several samples quite awnless and others with only a short beak instead of an awn.

WILD RICES.

Hamilton was strongly of opinion that the cultivated rices of India were derived from more than one species, not only from the fact that some were awned, and others awnless, but also because of their diversified peculiarities under cultivation. He specially cited the fact that two kinds were frequently sown mixed on the same field, the one producing its grain in two months or so, and the other long after the first had been reaped; these two plants, he maintained, had shown no tendency after countless centuries of cultivation to lose their respective peculiarities, and hence presumably they were different species. The chief difficulty in accepting **O. triticoides** as one of the forms of **O. sativa** is the fact that **Griffith** speaks of the larger of the inner glumes as nine-nerved. The writer has seen no authentic specimen of **O. triticoides**, but though he has examined several hundred samples of wild and cultivated rices, he has not as yet met with any specimen in which the large glume possesses more than five nerves. **Dr. George King**, Director of the Botanic Gardens, Calcutta (to whom this doubtful point had been referred), has most obligingly furnished interesting notes, and forwarded, for inspection, a selection of the specimens of **Oryza** from the Herbarium. Unfortunately these specimens do not embrace any of **Griffith's** obscure plants, so that the author can offer but the merest conjecture as to the species figured and described by that distinguished botanist. One point alone seems certain, *viz.*, that the salt-marsh-loving plant described by **Griffith** cannot be the Himálayan species alluded to by **Bentham & Hooker**, under the name **O. coarctata**, *Griff.*

I.—WILD RICES.

I. WILD RICES. 254

It may not be out of place to give here a few brief notes on the botanical characters of the numerous specimens of wild and cultivated forms of **Oryza**, which the writer has cursorily examined in the hope of being able to establish a possible scientific classification of the cultivated rices. Two peculiarities, which have apparently been but indifferently alluded to by previous writers, seem to be of some importance, *viz.*, the configuration of the receptacular extremity of the pedicel of the spikelets, and the superficial structure of the outer wall of the inner glumes. In all the forms of **Oryza** examined, the receptacle (a term which will doubtless be readily understood, though not employed by agrostologists, as popularly expressing the club-shaped apex of the pedicel,) has been observed to consist of two facets, one on a slightly higher level than the other. The margin of these facets is fringed by a scale-like fold which in some of the wild forms is so greatly developed as to appear as an actual pair of scales outside the outer glumes. The upper facet is that on which the flower is in reality inserted, but the larger one of the inner glumes is generally so developed as to force its lower extremity with the subtended outer glume into the depression of the lower facet. The degree to which this distortion of the receptacle exists has been observed to be coincident with other characters and to help greatly not only in the separation of the species of **Oryza** but in the isolation of the wild and cultivated forms of **O. sativa**. Thus, for example. the outer glumes in **Oryza**, though they appear in all the species to take their origin at a common node (the point of dehiscence from the receptacle) and to be connivent around the base of the inner glumes, are often so obliquely distorted in their conformation to the receptacle as to appear as if the one took its origin at a point in some cases as much as an $\frac{1}{8}$th of an inch above and within the other. This separation of the receptacular facets is met with both in the wild and cultivated rices, so that it appears safe to relegate the forms according to this peculiarity, as derivable from each other or at least from a common ancestor, the more so since this association is borne out by other common characters such, for example, as the habitat of the forms (marsh-loving or dry land rices), structural peculiarities of the veins of the leaves and sheaths, the length and colour of the inner glumes, of the awn, of the keel and the hairiness, as also the quality of the grain. But to return to the second peculiarity upon which the author

ORYZA granulata. The wild Rices of India

WILD RICES. has been induced to lay some stress, *viz.*, the superficial structure of the inner glumes, it may be stated that under a fairly powerful lens the glumes of all the forms of **Oryza sativa** are seen to be built up of squarish granulations arranged with the utmost accuracy in parallel lines from the base to the apex. In **O. granulata**, *Nees*, on the other hand, the surface is rendered woolly-looking, through the presence of irregular roundish granulations which possess none of the uniformity in shape or arrangement of those of **O. sativa**. In **O. officinalis**, *Wall.*, the surface peculiarities approach nearer to those of **O. sativa** than to **O. granulata**, but the granulations are larger and the surface is at the same time almost tomentose, a fact which gives these glumes a papery appearance instead of the semi-translucent aspect of the glumes of **O. sativa**.

There are thus (excluding **O. triticoides**, *Griff.*=**O. coarctata**, *Ham. & Roxb.*) three easily recognisable forms of **Oryza** in India. These may be briefly dealt with in alphabetical order :—

255 **Oryza granulata**, *Nees.*

Habitat.—A species found on dry soils at altitudes up to about 3,000 feet. Specimens of it have been collected by **Wight, Griffith, Simons, Hooker, Beddome, Brandis, Clarke,** and **Kurz** from Sikkim, Assam, Burma, Bengal (Parisnath and Rájmahál Hills), Malabar and Courtallum. Indeed, it seems probable this may possibly be the Himálayan plant alluded to in the *Genera Plantarum.*

Botanical Diagnosis.—A perennial species, with an almost woody root-stock and thin round firmly-made branching stem, which, for an **Oryza**, is profusely clothed with short thin leaves, resembling somewhat those of many species of **Panicum**; ligule deeply fimbriate. Inflorescence simple, spicate; pedicels short dilated upwards, but folds of the receptacle very nearly on the same plane though their membranous scale-like margins are fairly distinct. Outer glumes very small, the one placed against the larger inner glume considerably the smaller though broader at the base, both glumes also spread away horizontally in a conspicuous manner not observed in any other form of **Oryza**. Inner glumes 0·225 inches long, obtuse at both extremities, awnless but ending in a disk-like process; surface glabrous though covered with irregularly-shaped and scattered roundish granulations.

Uses. 256 **Uses.**—**Mr. C. B. Clarke**, in a letter to the author, calls this **O. coarctata**, *Griff.* He lays great stress on the characters of the root-stock and of the leaves, and from its frequenting rocky woods he denominates it "wood rice." "You will see," he writes, "that the species is altogether remote from any cultivated rice. They tell me, however, that the flavour of the grain is so good that it is hand-collected by children." This plant does not appear to have ever been cultivated, and, though at one time disposed to view hybridization from it on certain forms of **O sativa** as the possible origin of some of the hill dry-crop rices, the writer has for the present abandoned that idea. The characters of **O granulata** are so striking that they would readily appear in any hybrid state. The most careful study has failed to discover a cultivated rice, possessed of the peculiar granular structure of the inner glumes, a peculiarity which is probably far more exclusive than the nature of the stem and root-stock on which **Mr. Clarke** lays so much stress. The writer has had the pleasure to have had sent him the Madras Herbarium sets of **Oryza**, and amongst these he found a sheet collected by **General** (then **Captain**) **Beddome** on which a note occurs to the effect that it is the wild rice of Malabar. In the same collection there are some good sheets of **Leersia hexandra** which are said to have been also collected in Malabar, growing in paddy fields and also in the Annamallay forests up to 3,000 feet. In foliage and granular structure of the surface of the glumes these specimens closely resemble **O. granulata**, although they are perfectly distinct in the absence of the outer glumes and in the Eragrostis-like inflorescence. As a weed of rice-fields this plant has, however, an incidental interest to the subject here dealt with, and it seems likely that **Leersia** or **Hygroryza** may be the "wild rice" of many authors as, for example, the *hama* of Kashmir (*Conf.* with p. 623). WILD RICES.

Oryza officinalis, *Wall.*

257

Habitat.—A tall sparsely branched species with very broad multi-nerved leaves and profuse (almost umbellately) branched panicles. This plant has been collected by **Hooker & Thomson**, and by **Jenkins, Kurz, & Duthie**. The area of its distribution may be given (from the above record of its collection) as from Sikkim and the Khásia Hills to Assam (Gowhatty) and Burma (Pegu and Arracan). The most northern record of its existence appears to be that of the specimens recently collected by **Mr. J. F. Duthie** from Chánda in the Central Provinces.

Botanical Diagnosis.—A perennial plant, with a sub-woody root-stock (much as in **O granulata**); stem tall, scarcely if at all branched save from the creeping root-stock. Leaves very large, 8 to 18 inches long, broadest near the middle where they are often as much as 1 to 1½ inches in diameter: sheaths very long (often 6 to 8 inches) completely embracing the sub-woody stem, never inflated with air chambers as in aquatic forms of **Oryza**; ligule irregularly jagged. Inflorescence sub-umbellately branched, profuse, the spikes borne on long naked peduncles, each having a tuft of longish hairs at its point of origin. Spikelets on very short pedicels, the receptacular apex of which is only very slightly distorted into two almost horizontal ears. Outer glumes narrow, lanceolate, entire, connate, and spreading from the same plane; inner glumes smaller though broader than those of **O. granulata**, more flattened and covered all over with long spreading hairs which become more prominent on the keel, surface composed of parallel bands of squarish corrugations, larger glume shortly beaked with a pair of short tail-like glands on either side, keel almost winged, especially on the upper half of its length.

Although this species is well marked and possesses many very striking features, it must be admitted to afford the transition between **O. granulata** and **O. sativa**. Many of the cultivated forms of the latter species are seen to possess certain of the characters of this plant. Thus its remarkable inflorescence, forming umbellate divisions, borne on long naked peduncles, is frequently met with in certain cultivated rices, which are otherwise unmistakably forms of **O. sativa.** In the same way, hairy glumes are a by no means infrequent character of certain alpine dry-crop rices, and these also possess the hard sub-woody root-stock of this species. Indeed, it seems highly probable that if hybridization has taken any part in the production of the forms of cultivated rices, it is to this species that attention should more specially be directed, in any attempt to ascertain the origin and character of such forms It is by no means unusual to read, in popular works, of Himálayan rices grown at great altitudes and on perfectly dry soils. Indeed, at one time, some trouble and expense were incurred to procure seed of a rice from Nepál that might be grown in the most northern tracts of Europe. To the rice sought was assigned by some writers the name of **O. nepalensis,** by others of **O. mutica**. **Dr. Buchanan-Hamilton** alludes, however, to a reputed alpine rice in Nepál, which he afterwards found to be a form of rye, and therefore not rice at all. Whether or not this same delusion passed into modern literature it would be difficult to tell, but the author's experience of Himálayan botany and agriculture leads him to affirm that no alpine hill rice can be absolutely grown on dry soil, though the amount and duration of actual flooding or humidity necessary in one form is much greater than in another. He has seen no rice grown much above 7,500 feet in altitude, and no hill rice that could not be regarded as a form of **O. sativa,** though, as already remarked, some of the hill rices manifest peculiarities that may with more extended study be traceable to hybridization. Dry Alpine Rice.

ORYZA sativa.

The wild Rices of India

WILD RICES. 258

Oryza sativa, *Linn.*; *Roxb.*, *Fl. Ind.*, *Ed. C.B.C.*, *306.*

RICE, *Eng.*; RISO, *It.*; RIZ, *Fr.*; REISS, *Germ.*; RYST, *Dutch*; ARROZ, *Port.*; ARROZ, *Sp.*

Vern.—*Dhán, chával*, HIND.; *Chál, chánwol, dhán*, BENG.; *Uri, úri horo* (wild rice), SANTAL; *Dhán*, or when husked *chául* (long-stemmed rice=*rábaná*), URIYA; *Deodhán*, C. P.; *Dhán, sáthi* (in Partabgarh wild rice is called *pasáhi* or *passarí*, also *tinní*), OUDH; *Dhán, munji* (a small-grained form), *sáthi* (60 days' rice), *pusai* (a wild bearded rice), *lehi* (a wild rice with small yellow grain, also bearded), *jarhan* (late rice), *phasai* (wild rice in Fatehpur and Rampur), [*Dhán* in ear, *chánwal*, husked, *chila*, the chaff, *bhát*, cooked], N.-W. P.; *Dein, táni*,* KASHMIR; *Dhán, tái, shálian, munji* (in Jhang), *shalí* (in Hazara), *shol* (in Peshawar), PB.; *Garri, sál* (Mount Abu), RAJ.; *Sari* [*sugdási*, one of the best white qualities], *chánwar* (husked rice), *chuno* (the husk), SIND; *Chánval*, DECCAN; *Dángar* (Broach, &c.), *bhatta* (Belgaum), BOMB.; *Tandula, bhát*, MAR.; *Chokha*, GUZ.; *Arishi, nellú*, TAM.; *Bhatta, nellu*, MYSORE; *Biyam, errajilama vadlu* (*nevari dhanyamu*, wild rice), *vudlu, urlú*, TEL.; *Akki*, KAN.; *Ari*, MALAY.; *Saba, san, chán*, BURM.; *Hál, úru-wí*, SING.; *Dhánya, vrihi* (*nivára*, wild), SANS.; *Arruz*, ARAB.; *Biranj*, PERS.; *Motsj, ko*, JAPANESE; *Lua*, COCHIN-CHINESE; *Arús, rúz, rús*, EGYPT.

References.—*Roxb., Fl. Ind., Ed. C.B.C., 306-308; Voigt, Hort. Sub. Cal., 711; Thwaites, En. Ceyl. Pl., 357; Trimen, Sys. Cat. Cey. Pl., 106; Dalz. & Gibs., Bomb. Fl., Supp., 98; Stewart, Pb. Pl., 257; Aitchison, Cat. Pb. & Sind Pl., 157; Kuram Valley Rept., Pt. I., 9, 25, 105; DC., Orig. Cult. Pl., 385; Rev. A. Campbell, Rept. Econ. Pl., Chutia Nagpur, No. 7821; Graham, Cat. Bomb. Pl., 235; Mason, Burma & Its People, 475, 818; Sir W. Elliot, Fl. Andhr., 52, 134; Rheede, Hort. Mal., V., 196, 201; Burmann, Fl. Ind., 85; Thesaurus, Zey., 108; Linnean Soc.:—Trans., XXIX., 170; Journ., XIX., 56; Hooker, Himálayan Journals:—I., 155; II., 105; Griffith, Journal of Travels in India, &c., several scattered passages; Posthum. Papers (Notulæ ad Plantas Asiaticus, Pt. III., 5-10, also Icones Pl. Asiatic, 139, 142, 144; Pharm. Ind., 254; U. S. Dispens., 15th Ed., 1716; Ainslie, Mat. Ind., I., 338; O'Shaughnessy, Beng. Dispens., 635; Moodeen Sheriff, Supp. Pharm. Ind., 191; U. C. Dutt, Mat. Med. Hindus, 267, 296; K. L. De, Indig. Drugs Ind., 81; Murray, Pl. & Drugs, Sind, 8; Bidie, Cat. Raw Pr., Paris Exh., 18, 70-71; Bent. & Trim., Med. Pl., 291; Birdwood, Bomb. Prod., 110; Baden Powell, Pb. Pr., 231; Drury, U. Pl. Ind., 321; Duthie, Fodder Grasses of N. Ind., 20; Useful Pl. Bomb. (XXV., Bomb. Gaz.), 186; Forbes Watson, Indian Prod., 40-41; Econ. Prod. N.-W. Prov., Pt. IV. (Cultivated Food-Grains), 17; Royle, Ill. Him. Bot., 415, 419; Church, Food-Grains, Ind., 8, 17, 66-76; Indian Fibres & Fibrous Substances, Cross, Bevan, King & Watt, 46-47; Folkard, Plant-Lore, 513; Kew Bulletin:—1888, 284-291; 1889, 13; Simmonds, Waste Products, 227; Tropical Agriculture, 313-335; Maiden, Native Plants of Australia, 49; Mueller, Sel. Extra Tropical Plants, 268; Shortt, Man. Ind. Agri., 246-250; Ayeen Akbary, Gladwin's Trans.,:—I., 75; II., 499, 521; Ain-i-Akbari, Blochmann's Trans. I., 57; Milburn, Oriental Commerce:—(1813), II., 236; (1825), 297; Buchanan-Hamilton, Journey through Mysore & Canara, &c., numerous passages; Account of Kingdom of Nepal, 73, 88, 222-226, 274, 284; Glossary of Anglo-Indian Terms, Yule & Burnell, 495, 577; Treasury of Botany, II., 826; Smith, Dict. Econ. Pl., 352; Paxton, Bot. Dict., 409; Spons' Encycl., 219, 1826; Balfour, Cyclop. Ind., 53-54, 412-419.*

Habitat.—Cultivated throughout India, but met with also in a truly wild condition wherever marshy land occurs in tropical areas. Its chief wild habitat is, however, from Madras and Orissa to Bengal, Chittagong, Arracan, and possibly even to Cochin-China. From the belt of moist tropical lands thus briefly indicated it extends northwards to the Nilghiri hills, to the North-West Provinces, and, according to some writers, even

* *Hama* (? *sama*) is said to be wild rice and *kre* acclimatised rice.

WILD RICES.

to the Panjáb. **Dr. King** found it truly wild in one locality of Central India, **Mr. J. F. Duthie**, at Chánda and Wardha in the Central Provinces and at Mount Abu in Rájputana. It is frequent on the margins of tanks in Chutia Nagpur and in Assam. **Thomson** is said to have collected it in the Panjáb, but **Stewart** expressly says that he had carefully looked for it, but to no purpose. Mention is not made of it in Sind, but **Dr. Dymock** writes to the author that a wild rice occurs in the swamps near Bombay; in Balúchistán, Afghánistán, and Persia it nowhere occurs in a wild state, and is in these countries even only very occasionally found under cultivation.

DeCandolle very probably places undue stress on the wild habitat where **Roxburgh** first found it, *viz.*, the Telegu country. It is no doubt wild there, but abundant evidence exists that it is prevalent throughout the southern and eastern tracts of India, indicated above, and that it extends from the swampy expanses (there of no uncommon occurrence) far to the interior of the country wherever sufficient water and the requisite temperature are met with.

Botanical Diagnosis.—Roxburgh was so satisfied as to the wild stock being the parent of the cultivated rices that he gave a botanical description of it and contented himself by specializing some of the features of the chief cultivated forms known to him. Following his example it may be affirmed that the majority of the wild states of **O. sativa** appear to be annuals which grow in dense tufts and produce coarse much branched radical stems, the erect portions of which root at the deflected nodes. They form spongy sheaths (through the presence of air-chambers in their texture) which are intended to support the ascending stems as these in their growth have to keep pace with the rise of the level of water, since rice is killed when completely submerged. The lower sheaths are often devoid of a lamina; ligule large, membranous, bifurcated into sickle-shaped arms on either side of the stem, each arm having seven pairs of erect rigid hairs. Inflorescence a panicle of spikes on short peduncles which have an ochreate-like hairy scale (frequently as in **O. officinalis** a distinct tuft of hairs) at the point of origin of the spikes; pedicels short with the receptacular apex large and often greatly distorted, the upper and lower facts sometimes fully ⅛th of an inch apart; the scale-like ears of these facets are sometimes very large, and, in exceptional cases, one or both are even produced into what might be called additional outer glumes. Outer glumes large, very often tri-dentate, midrib prominent; inner glumes variously shaped, but in the wild states considerably elongated, being, as a rule, o·325 inches in length, curved, pointed at both extremities, and in the great majority, the larger one is produced into a long awn which is frequently distinctly articulated and possessed at its base of two glandular processes which correspond to the extremities of the lateral nerves: surface more or less hairy, especially on the keel and nerves, structure manifesting the regular rows of shining granulations described above.

While these characters convey some of the main features of the wild states of **O. sativa**, it cannot be said that they serve absolutely to isolate wild from cultivated forms. In the writer's opinion they separate completely, however, the other two species, and the progression from the wild to the cultivated forms seems to be in the reduction of the size and alteration of the shape of the receptacle in the loss of the awn, and in the shortening but widening of the combined outline of the inner glumes—a character consequent doubtless on the formation of a shorter, though thicker and more compact grain. It would be beside the scope of the present work to deal further with the botanical value of the characters based, in these remarks, on the facets of the receptacle and their corresponding scale-like rims; in some specimens, as already remarked, these even assume the form of additional glumes. Suffice it to say that of the specimens examined by the writer, the plants collected by **Mr. J. F. Duthie** in marshes of Mount Abu manifest these structural peculiarities to the greatest extent, but only certain flowers do so, hence, if at all of value structurally, they would appear to be so as rudiments of what are now functionless organs. Of a similar cha-

ORYZA sativa.

Wild varieties of O. sativa.

WILD RICES.

racter also is the fact that while the vast majority of forms of **O. sativa** possess only one grain, certain forms have two or even three grains.

It will naturally be understood, from the admission already made of imperfect knowledge, that the task of separating the wild and cultivated forms into varieties or races is by no means easy. While not desiring it to be understood that this task has been accomplished, the author has thrown the wild specimens he has examined into groups which correspond to the chief forms of cultivated rices. These may be briefly indicated :—

259 **Var. 1, rufipogon.**—This seems to correspond to the plant of that name described by **Griffith.** The specimens examined by the writer belong to the Saharanpur Herbarium. Leaves, very large, much resembling those of **O. officinalis,** *Wall.* Inflorescence, copiously branched, forming an erect head, in general appearance somewhat like some of the forms of **Sorghum vulgare.** Branches of panicle, with a distinct tuft of hairs at their point of origin. Receptacle, thick, short, with prominent ear-like rims. Outer glumes, connate, entire; inner glumes, short; combined outline, stunted or rounded, with a distinctly articulated caducous awn, or awnless. The specimens of this plant were collected by **Mr. J. F. Duthie,** near Aligarh in the North-West Provinces. They are probably the source of most if not of all the red coloured rices of India. From their general appearance they would seem to require much less water than do the other wild states of this species. By cultivation they might readily be supposed to have given origin to many of the awnless white-grained *chotan áman* rices.

Griffith's specimens of **O. rufipogon** were collected at Hubbegunge and Nubbegunge in Eastern Bengal, growing on the margins of *jhíls.*

260 **Var. 2, coarctata.**—As already remarked, the writer has seen no authentic specimen of **Hamilton's coarctata,** but he suspects the plant described by both **Roxburgh** and **Griffith** to be the stock from which the deep water and brackish-loving cultivated rices have been derived, and that it may even be the source of some of the coarser forms of *borán áman.* The characters of the plant have already been fully detailed, but the cultivated forms referable to this stock are awned or awnless rices with spicate panicles of long pale-coloured grains. The points of branching of the inflorescence are not fringed with hairs.

261 **Var. 3, bengalensis.**—The writer proposes to place in this section all the remaining forms of wild rice which are found in Bengal and other regions, frequenting the margins of tanks and the deep *jhíls* or marshes in the interfluvial tracts. These are very generally known to the people of India as *urí* or *jhara* rices. They may be said to be characterised by having shorter but stouter stems with larger shining inflated sheaths than in either of the above forms. The inflorescence forms open sparsely flowered compound spikes with woolly scales at the points of branching. The receptacle is often only slightly distorted; outer glumes connate, inner glumes pale green or white, elongated, the upper portion generally empty; keel and nerves hairy, otherwise glabrous; awn very long, never naturally caducous.

This is possibly the source of some of the *áus* and intermediate *áus* and *áman* rices of Bengal and of other parts of India, and possibly of many of the qualities of *borán áman* which luxuriate in deep inundations. It is a large coarse grower, with much branched and distorted stems. When young, it is exceedingly like some of the forms of *áman*; but as the water rises, it grows with marvellous rapidity, keeping above the surface and spreading all over the marsh. At this stage the floating leaves are small and unlike those of the domesticated rices. The grain is not only exceedingly like that of cultivated rice, but it is regularly collected and eaten. *Urí* becomes in some districts the cultivators' greatest enemy, as it spreads from the marsh over the neighbouring and inundated fields of *áman,* and from its hardier nature often almost exterminates the cultivated plant. A remarkable peculiarity, and one which largely accounts for the difficulty experienced in eradicating this weed, is that the moment the grain is ripe it falls from the ear into the water and is thus self-sown. The fishermen collect the grain, however, by binding the *urí* into tufts and in this way preserve the ripe grain until harvest, when in their palmyra-palm canoes they float about, carefully cutting off the ears of rice or shaking the grain

WILD RICES.

into their primitive barges, and they thus often reap a good harvest from the self-sown *urí* crop. (See mode of exterminating *jhara* rice as practised in Dacca, p. 544.)

262 **Var. 4, abuensis.**—This is probably the most temperate form of wild **Oryza sativa** which has as yet been collected. It is apparently a much smaller plant than the prevalent wild form (that dealt with under No. 3) and its most striking features are the great distortion of the receptacle, the large size of the inner glumes and grain, the generally dark colour and hairiness all over the surface of the inner glumes, and the short articulated or caducous awn. It seems very probable that this imperfectly awned plant is the source of many of the best qualities of awnless *chotan âman* or *rowa* rices of Bengal and of the superior qualities of Upper India, of Madras and of the hills—those which are grown on high lands and furnished with but a limited supply of direct water. If this conjecture proves correct the plant will doubtless be found in many other localities of Upper India even ascending the Himálaya wherever sufficient water occurs to afford a wild habitat for rice.

The names suggested for the above four forms of wild **O. sativa** should be viewed as purely provisional. There could doubtless be a larger series of forms collected, and consequently a better classification established, with more extended study. The writer has been forced to adopt the above arrangement more from the exigencies of the extensive collection of cultivated rices which he has examined than from a satisfactory study of wild forms. But even the classification here suggested by no means provides a place for all the distinctive forms of cultivated rice. Thus, for example, it leaves out of all consideration the *boro* and the *raida* crops, not to mention many other special rices such as the scented rices, the glutinous rice, &c., &c., all of which possess, agriculturally, so well marked properties as to call for some possible solution of their origin. There would, indeed, appear to be few subjects of Indian economic botany which have remained in greater obscurity than the study of the wild forms of **Oryza.**

The suggestion made by **Dr. Buchanan-Hamilton** that all the forms of cultivated rice could not have come from a common ancestor, the writer most heartily concurs in, though he thinks there is no necessity to believe the independent wild stocks as botanically anything more than varieties of a common species. But, if this position be admissible, they are varieties that have preserved their individual characters even when carried to the most diverse conditions of climate and soil. If species, we are confronted with a problem which demonstrates the defective nature of botanical characters and terminology, since, according to all accepted notions of species, they are practically inseparable.

II.—CULTIVATED RICES.

II.—CULTIVATED RICES. 263

Roxburgh adopted, what is perhaps the most convenient classification of the agricultural rices, *viz.*, a system based on their peculiarities of cultivation—the *early* and the *late* rices. He does not give botanical characters for the sixteen forms which he specialises, but there are certain facts regarding these plants worthy of consideration, since the accuracy of all **Roxburgh's** work lends a value to his most casual statements. The late rices, he remarks, are the "great crop." Of these, he mentions eight forms, all of which are *awnless* and afford when cleaned *white grains.* Of his early rices, on the other hand, four are awned and yield red or coloured grains; one is awned but yields a white grain, while three are awnless and afford white grains. Of the early rices six have coloured husks, while two are white or pale; of the late rices, on the other hand, four have coloured and four white husks.

The general conclusion to be drawn from an analysis of **Roxburgh's** cultivated rices agrees with the writer's own observations, *viz.*, that the progression in value is from the awned to the awnless forms, and from the

ORYZA sativa.

Cultivated forms of O. sativa.

CULTIVATED RICES.

coloured to the colourless. Under each of his two sections Roxburgh mentions one form as more cultivated than all the others put together, and both these are awnless and produce white grains. On the other hand, it must be admitted that the most prevalent form of wild rice (**var. bengalensis** above) has a white, not a coloured, husk and is found in fairly deep water, though not so deep, and never in brackish water, as is the case with the plants here regarded as corresponding to **O. coarctata,** *Ham.* These facts would seem to suggest that the better class rices, found on fairly dry lands, had been derived from the coloured stocks (**var. rufipogon** and **var. abuensis,** above) either retaining their colour or losing it and becoming beardless, according to the degree of cultivation and the nature of soil and climatic conditions under which they had been developed. Further, that most, if not all, the inundation-grown white and bearded rices with long inner glumes, hairy along the keel and nerves, otherwise glabrous, have been produced from the prevalent Bengal wild form. In other words, the red rices have probably come from **O. rufipogon** and the blackish coloured forms from **O. abuensis,** but both these forms must have yielded to cultivation and produced also the higher class beardless white rices, the former giving origin to the profusely panicled forms with short, roundish, white grains, and the latter the sparsely panicled state with large, very hairy glumes and long white grains. The isolation of the better classes, by the above hypothesis, would leave the long semi-glabrous glumed states of inferior, mostly awned rices, as having come from the more tropical swamp types of the wild **Oryza sativa.**

The above remarks, which throw the cultivated rices of India into four chief sections, from structural peculiarities, have been offered purely as a conjectural explanation of certain well-known kinds of rice. One of the best sources of confirming or correcting these theories would be a careful study of the conditions under which the wild plants are found and their seasons of flowering and fruiting. It must, in fact, be admitted that we have to fall back on **Dr. Buchanan-Hamilton's** idea that the chief differences between the thousands of forms of cultivated rices, hinge on their properties and peculiarities under cultivation. These peculiarities the Indian cultivator, through the time-honoured practices of his ancestors, is able to recognise far more accurately than botanical science has as yet been able to explain. He determines the suitability or otherwise of this form and that to its contemplated environment with a degree of confidence quite inexplicable. If these properties can be shown to be hereditary from the wild plants, from which the slight structural peculiarities would seem to indicate their origin, a safe basis of reasoning would be secured. It would then become possible for practical suggestions to be offered as to the desirability of substituting a better stock to this cultivator and that than they presently possess. To blindly urge on the *rayat* the advantages likely to follow on growing Carolina or other superior rices, without the knowledge of their suitability to the proposed new environment, is to court the distrust of a community which, in the matter of rice culture, knows far more than Western science has as yet been able to account for.

General Properties and Uses of Oryza.

DYE. Husk. 264

Dye.—Most writers allude to the fact of a dye being obtained from rice HUSK. Samples were sent from Lahore to **Mr. Thomas Wardle,** and in his Report on the Dyes of India the following remark occurs: "This substance contains a small amount of a pale yellowish-brown colouring matter, and when boiled in water the infusion may, by the use of various processes, be made to dye light shades."

General Properties and Uses of Rice. (*G. Watt.*)

ORYZA sativa.

FIBRE. Straw. 265 Stems. 266 Root. 267

Fibre.—Attention has been frequently directed to the subject of rice STRAW, and more especially the lower portion of the STEMS and ROOTS (left generally to enrich the soil) as a paper material. Little progress, it is believed, has been made in the direction of the utilisation of this material, nor, indeed, would it appear likely that, for many years at least, rice straw will come to be used for that purpose. The straw is of too great value to the cultivators to be offered for sale, and the roots are too troublesome to collect, even were it not the case that their retention in the soil is agriculturally of importance. The following passage may here be given as showing the use to which straw generally is put in Europe:—

"The use of 'Straw,' from the CEREALS Wheat, Oats and Rye—has of late years greatly extended both in this country and throughout the continent of Europe, as well as in the United States of America, either alone or as an admixture with rags and other material, for all classes of paper, as these countries, equally with England, suffer from a deficient supply of *Raw Material;* but in England, owing to the increased consumption for agricultural and feeding purposes, and influenced also by the scarcity and high prices lately ruling for 'Esparto' in many districts, 'Straw' has become very difficult to obtain, and considerable quantities have in consequence been imported from Holland and Belgium, both raw and as bleached pulp" (*Routledge, Bamboo as Paper-making Material, p. 25*). Straw pulp (bleached) realises £26 to £27 per ton, but is not "likely to be introduced to any considerable extent" (*p. 38*).

MEDICINE. Grain. 268

Medicine.—U. C. Dutt (*Hindu Materia Medica*) alludes to the opinions held by Sanskrit medical authors, regarding the use of rice, as an article of diet for the sick and convalescent. *Yavága* is the Sanskrit name of a dish made of powdered rice boiled with water. Sometimes this dish is flavoured with ginger, long pepper, and other such substances. "*Lájá* is paddy fried in a sand bath. The husks open out and the rice swells into a light spongy body. It is considered a light article of diet suited to invalids and dyspeptics." "*Bhrishta tandala* (in the vernacular *Muri*) is rice fried in a sand bath. This is also a light preparation of rice and is given to sick persons as a substitute for boiled rice. It is also much used by the poorer classes for tiffin and early breakfast. *Prithuká* (var. *Chura*)—to prepare this, paddy is moistened and lightly fried. It is then flattened and husked. This preparation of rice is given with curdled milk (*dadhi*) in dysentery. It is well washed and softened in water or boiled before use. *Páyasa* is a preparation of rice with nine parts of milk. *Tandulámbu* is water in which unboiled rice has been steeped. This sort of rice is sometimes prescribed as a vehicle for some powders and confections." **Baden Powell** says rice is "occasionally used in diseases of the urinary organs and catarrh; also externally as an application to burns and scalds." In a special report from Nepál, *Hakwa* is said to be rice slightly fermented and subsequently dried. "It is considered a light and nutritious article of diet in illness and is used also as a vehicle for medicine." **Ainslie** remarks: "In a medical point of view, rice may be said to be of a less aperient quality than any other grain, and is, therefo e, invariably ordered as the safest and best food in all dysenteric complaints; for which purpose, in the form of gruel, it is excellent. The *Vytians* are very particular as to the kind of rice they prescribe, supposing the rices of different crops to have very different effects." Rice is, however, regarded as less nutritive than wheat. Being entirely free from laxative properties it constitutes a light and digestible article of food, especially in convalescence from diarrhœa or dysentery. The small amount of nitrogenous matter and potash which rice contains, perhaps accounts for the fact that it has been supposed to occasion the scurvy in workhouses in England when given in place of potatoes. In the *Pharma-*

ORYZA sativa.

The supposed connection of rice with cholera.

MEDICINE.

copœia of India rice-water is recommended as an excellent demulcent, refrigerant drink in febrile and inflammatory diseases, dysuria and other affections requiring this class of remedies. It is rendered more palatable by being acidulated with lime juice, and sweetened with sugar. This decoction is also recommended as an enema in affections of the bowels. **Dr. Waring** commends the use of rice poultice as a substitute for that of linseed-meal. Rice starch is used for similar purposes to other starches.

There appears to be no ground for the statement that a diet of rice tends to injure the eyes more than for the belief that it causes cholera. The consumption of new rice is generally however held to be unwholesome. But perhaps, as being of historic interest, the following account of its supposed connection with the production of cholera may be here given:—

"The first great epidemic outbreak of cholera originated in the town of Jessore in the year 1817, and spreading up the valley of the Ganges, attacked and decimated the army (one of the largest ever assembled in India) of the **Marquis of Hastings**, then engaged in operations against **Scindia** in Central India From there the epidemic extended in a northwesterly direction over the greater part of the civilized world. Cholera had been known in India before then but chiefly as an endemic disease, one or two records only existing of epidemic outbreaks between 1503 and 1756, and even these apparently did not spread beyond narrow limits and certainly did not extend outside India." The medical authorities of India were, however, in 1817 called upon for some explanation of the new form of the disease, and various opinions were offered. That given by **Dr. Tytler**, the then Civil Surgeon of Jessore, is in conformity with a very generally accepted view held by the Natives of the present day. **Mr. J. Westland** (*Report on the District of Jessore*) gives the following account of **Dr. Tytler's** opinions and the action taken: "**Dr. Tytler** records some interesting information regarding a special cause, to which he attributes a somewhat exaggerated importance, calling it "a great Truth which has under the favour of the Almighty been disclosed at this station, where the disease first broke out." The heat and humidity of the season had not only brought to early maturity the autumn crop of rice, but had imparted to it an unusual richness of flavour. The supply of new rice was abundant and cheap, and it was eagerly sought after; even before it became fully ripe, it was "devoured with avidity by Natives of all descriptions." To this extensive use of immature rice **Dr. Tytler** ascribes the outbreak, and from his observations it is clear that it was a predisposing cause. He mentions this as an opinion generally received and openly declared by the Natives. He states that in many places attacked by cholera, though nothing had been done to cut the jungle and drain the pools of stagnant water, the mere prohibition by the Magistrate of the sale and use of new rice had been effectual in causing an immediate decrease in the disease; and he mentions one instance coming under his personal observance. "On 2nd September, the use of new rice was absolutely forbidden in the jail, and on that day cholera disappeared from the jail, one case occurred after that, namely, a case of a female prisoner who, having surreptitiously obtained and eaten a small quantity of new rice, was attacked by cholera a short time after."

SPECIAL OPINIONS.—§ "Rice water makes a good substitute for barley water" (*Surgeon-Major A. S. G. Jayakar, Muskat*). "A wild rice called '*Devobháta*' is found in swampy spots near Bombay; it is carefully collected for medicinal use" (*W. Dymock, Bombay*). "*Soarú*=boiled rice" (*V. Ummegudien, Mettapollian, Madras*). "Rice is used as poultice" (*Surgeon-Major Robb, Civil Surgeon, Ahmedabad*). "A decoction of rice

MEDICINE.

called *pitcht* or *conjí* is found to be useful in checking atonic diarrhœa. It has also been used to make starch bandages. It is employed as a vehicle for enemas of opium and gallic acid owing to its cheapness" (*Assistant Surgeon Bhagwan Das (2nd), Civil Hospital, Rawal Pindi, Panjáb*). "Rice is used as an astringent drink in cholera and dysentery" (*Civil Surgeon G. C. Ross, Delhi, Panjáb*).

FOOD & FODDER. Grain. 269 Chaff. 270 Straw. 271

Food & Fodder.—That the grain of rice is one of the chief articles of human food in India, need scarcely be stated here. The section of the present article under the heading RICE deals almost exclusively with that subject. In many parts of India (as, for example, Manipur) rice is the chief article of food given to horses and cattle, and throughout India the straw of the better qualities is invariably collected, cut up into small pieces, and given along with several flavouring liquid preparations, oil-cake or grain, designated the *Currie*. The chaff and waste broken fragments obtained in winnowing and husking constitute also important articles of human and cattle food. These subjects will, however, be found so sufficiently dealt with below, in the numerous passages which have been extracted from standard works under provincial chapters, as to require no further consideration in this place. It will also be learned from these quotations that in some parts of India certain rices are grown for a short period as green fodder for cattle or are ploughed in green as manure; the half cropping thus accomplished is regarded as beneficial to the soil.

Husking. 272

HUSKING.—The subjects of thrashing and husking paddy will also be found dealt with in the provincial chapters. The latter is a troublesome process which it took the American pioneer cultivators some time to accomplish. In India a large part of the rice sold in the shops, and which is exported to Europe as an article of human food, has been husked by being first half-boiled, then dried in the sun, and finally husked by the ordinary pestle and mortar. This "par-boiled" rice, as it is called in the trade, is an important article, but one which the higher caste Hindus are forbidden to eat since it may have been boiled by low caste persons. Husking without boiling is a tedious process, the more so since it is done entirely by hand labour. This is, however, said to be facilitated by exposing the grain to the sun before treatment in the mortar. In the brief account which has been given of Yarkand rice (under the Panjáb) a description of a mechanical contrivance for husking rice will be found in which water is the motive power. In the plains of India, rice is frequently husked by the same appliance as is used in pounding bricks. A pestle suspended from the end of a beam is made to fall with considerable force on the grain. A woman, standing at the further end of the heavy beam, alternately rests and removes her weight from its extremity, and thus allows the pestle end to fall upon the rice. A second person attends to the grain, and as the pestle is raised a fresh handful is swept underneath. In this contrivance, the pestle, as it has been here called, works very generally on a flat stone. In using the ordinary pestle and mortar it is a common practice for the stroke to be given alternately from the right and left shoulders. The mortar may be made of stone, but it is commonly an hour-glass-shaped log of wood about two feet in height with a central cavity excavated for half its depth and only very slightly greater in diameter than the pestle which is usually about 3 to 4 feet long and about 3 inches in thickness. The mortar is as a rule firmly fixed in the ground inside the enclosure, in front of the peasant's hut, and the operation of husking, which may be almost daily witnessed, requires not only considerable agility but great exertion.

Preparations. 273

FOOD PREPARATIONS AND BEVERAGES.—Throughout India strong opinions are held in favour of or against the rices of certain districts as arti-

ORYZA sativa.

Food Preparations of Rice.

FOOD & FODDER.

cles of food. In connection with the account of the rices of the North-West Provinces mention has been made of Philibit rice as one of the most highly prized. Under Thana in Bombay a passage is given regarding scented rices, but such rices are not uncommon in other parts of India. Orissa, for example, is famous for its expensive perfumed rices. The rices of Burma are disliked by the people of Bengal and Madras. **Ainslie**, alluding to this subject, says: "The chief distinction, with regard to appearance and taste, betwixt the Bengal and coast rice would seem to be, that the former is whiter, boils drier, and is more delicate, in flavour; it is commonly, on those accounts, preferred by the people of rank, to eat with curry; and the Patna isdeemed the best. But the Natives of both coasts do not like the rice of the higher provinces; they call it dry and insipid, and say it is apt to bring on constipation."

Spirits.
274 The numerous preparations of rice made in India will be found alluded to in the provincial chapters, also under **Medicine**, and as a volume might be written on the subject, space cannot be afforded for more details than will be found scattered throughout this article. The Indian use of rice in brewing and distillation is both universal and extensive. Spirits from palm-juice and from rice (*sura*) are alluded to by **John Huyghen van Linschoten** as having been largely consumed in Southern and Western India 300 years ago. *Sura* is also frequently mentioned in the Institutes of **Manu**, a work which, according to most Sanskrit scholars, was written about 2,000 years ago. In an appeal to the King of Portugal **Linschoten** deplored the extent to which the Portuguese soldiers in India were acquiring from the Natives the pernicious practice of using spirits in place of the wine of their own country, and declared that if steps were not taken to check the growth of this acquired form of intoxication the army would be completely. demoralised. A kind of beer (*Pachwai*) made from rice is in almost general use throughout India.

Beer.
275 The following extract from the Gazetteer of Bengal may be given as an example of the passages which occur on this subject, not only in connection with the accounts of each district of the Lower Provinces, but it may be said of every district of India :—

"In Nadiya the liquid preparations made from rice are the following :—

"(1) *Amání* or water in which boiled rice has been steeped till the liquor becomes sour; used by the women of the lower classes as a cooling drink in the hot weather; it is not sold.

"(2) *Pachwái*, fermented rice liquor drunk by the low-caste husbandmen, and the Bunás, or aboriginal labourers; sold at about a penny the quart bottle.

"(3) *Dhenomad*, distilled rice liquor, is extensively used throughout the district, and sold at the rate of from 1*s*. 3*d*. to 2*s*. per quart bottle according to its strength and quality" (***Statistical Account of Bengal, Nadiya, by Sir W. W. Hunter, page 68***).

The reader is referred for further particulars on the subject of rice beer and spirits to the following articles, "**Malt-Liquors**," Vol. V. 124-140, "**Narcotics**," Vol. V. 330, 332-334, and "**Spirits**," Vol. VI., Pt II. In the remarks under **Narcotics**, particulars will be found of the methods of distillation generally pursued, and it will also be seen that the people of India have not only known and practised the art of distillation from time immemorial, but that their apparatus is so simple that no fiscal regulations, however stringent, could, in the present state of Indian civilisation, completely control, still less prohibit, the rural production and consumption of spirits. In a foot-note to **Linschoten's** remarks regarding spirits distilled from palm-juice and rice, the editors of the English revision (Burnell,

FOOD & FODDER.

Tiele & Yule, 1885) say: "It is quite a mistake to suppose that the Natives of India have learnt the vice of drunkenness from Europeans. Passages in the Maha Bhárata, &c., show that drunkenness was common, and this may be gathered also from the law-books, which forbid the use of wine (*e.g.*, Mánava-dharma-Castra, XI., 146 ffg.). In the Abhilashitárthacintámani (a Sanskrit Manual for kings and princelings, of 1128 A.D.), though most indecent and foolish amusements are allowed, drinking spirits, &c., is prohibited; but several different kinds of spirits are referred to."

The European uses of Rice in distillation will be dealt with in the concluding chapter of this article, INDIAN TRADE IN RICE.

Rice Flour. 276

RICE FLOUR.—Throughout India a certain amount of rice is reduced to flour and eaten in the form of cakes. Rice flour accordingly appears in all the trade returns; thus, for example, Burma regularly exports large quantities of that article, on an average say 50,000 tons. **Dr. Shortt** states of Madras that in preparing rice flour the grain "is either steeped in water and then pounded in a mortar and the flour subsequently dried in the sun or the grain itself is dried in the sun and then ground in a mill. Rice cakes under the name of 'Oppers' are in general use among most of the lower orders of Europeans, East Indians, and Native Christians. The rice flour is mixed with water and allowed to ferment over night. This process is frequently assisted by the use of toddy or the cocoanut tree sap. The next morning, it is baked into cakes. Frequently cocoanut milk is added to improve the flavour of the cakes. Different kinds of sweet cakes are prepared from rice flour." The substance sometimes known in trade as rice "flour" is the dust and wastage obtained at the power mills. This is chiefly exported for the purpose of feeding pigs and cattle (*Conf.* with p. 653 the para. on RICE MILLS).

CHEMISTRY. 277

Chemistry of Rice.—In his *Food Grains of India*, **Professor Church** gives the following passages which may be accepted as conveying the chief facts on this subject:—"The analyses which have been made of a large number of samples of 'cleaned' rice, give figures which are wonderfully accordant, considering the great differences in the appearance of the specimens and the very diverse conditions under which they have been grown. The fibre and adventitious earth are sometimes rather high from imperfect cleaning of the grain, but the nitrogenous constituents or albumenoids oscillate within narrow limits—probably nine samples out of ten will be found to contain not less than 7 per cent., and not more than 8.

COMPOSITION OF RICE.

	In 100 parts.	In 1 lb.	
		oz.	grs.
Water .	12·8	2	21
Albumenoids .	7·3	1	74
Starch .	78·3	12	231
Oil .	·6	0	42
Fibre .	·4	0	28
Ash .	·6	0	42

"The nutrient-ratio is 1 : 10·8 and the nutrient-value 86½. One hundred parts of rice contain no more than ·065 part of potash and 284 parts of phosphoric acid."

Professor Church continues:—"Two pounds of cleaned rice weigh 5 pounds after boiling. The liquor is either thrown away or is drunk as a

ORYZA sativa.

History of Rice.

CHEMISTRY.

beverage after the addition of a little common salt, or is given to stall-fed milch cows. Where rice constitutes the almost entire food of the population, the throwing away of the water in which it has been boiled involves the loss of some of the mineral matter in which rice is notoriously deficient, and is to be deprecated; no more water should be used in cooking this grain than can be absorbed by it."

The following brief note, supplementary to the above, has obligingly been furnished by **Dr. Warden**, Professor of Chemistry, Calcutta, on this subject. "Rice has been examined by **Lethebyn**, **Payen**, and others. **Payen** gives the percentage composition of dried rice, as, nitrogenous matter 7·55, carbohydrates 90·75, fat ·8, and mineral matter ·9. In chemical composition rice closely resembles the potato; one hundred parts of dried potato according to **Lethebyn's** analysis contains, nitrogenous matter 8·4, carbohydrates 88, fat ·8, and saline matter 2 8 parts."

DOMESTIC & SACRED. 278

Domestic and Sacred.—The sacred uses of rice will be found sufficiently indicated in the chapter which deals with its HISTORY. The reader is, therefore, referred to that part of this article, but there are certain minor domestic uses that may be referred to in this place. In Nepál and elsewhere baskets are made of rice straw; and plaited or twisted into ropes, it is generally used by the peasants in this country for the same purposes as straw in Europe. In the rural parts of the country necklaces and other personal articles of adornment are sometimes made of rice stems and necklaces of unhusked rice. In Kashmír and elsewhere sandals are also made of the straw.

Baskets. 279

Personal Ornaments. 280

Sandals. 281

RICE.

HISTORY.

HISTORY. China 2800 B. C. 282

Writers on this subject are agreed that the earliest mention of rice cultivation is connected with China. According to **Stanislas Julien**, a ceremony was established in that country about 2800 B.C., by the Emperor Chin-nung, in which the sowing of five kinds of grain is the chief observance. The reigning Emperor himself has to sow the rice, but he may delegate the sowing of the other four kinds to the princes of his family. A ceremony traceable to so great antiquity, if there be no possible doubt as to rice having always been, as it is at the present day, the grain assigned to the Emperor, when viewed in the light of the adaptability of large portions of China for rice cultivation, naturally led **DeCandolle** to presume that the plant may have been a native of that country. He does not, however, restrict its wild habitat to China, but admits that it has been found both in India and Australia, under such conditions, as to allow of little doubt that it is a native of these countries as well. **DeCandolle**, in fact, simply affirms that rice cultivation in India, though subsequent to that of China, in point of date of first record, has been a valued crop since classic periods. One of its Sanskrit names *Dhánya*, means "the supporter or nourisher of mankind." "By the Hindus it is regarded as the emblem of wealth or fortune. On a Thursday, in the month of *Pausha* (December to January), after the new paddy has been reaped, a rattan-made grain measure, called *rek* (in Bengali), is filled with new paddy, pieces of gold, silver and copper coins, and some shells called *cauries*, and these are worshipped as the representative of the goddess of fortune. This apparatus is preserved in a clean earthen pot and brought out for worship on one Thursday in each of the following Hindu months, namely *Chaitra*, *Sráwana*, and *Kártika*. Such is the form of the domestic goddess of wealth of an agricultural people living chiefly on rice."

Early Sanskrit Names.

"The three principal classes of Rice are *Sáli*, or that reaped in the

HISTORY.

Used in Hindu Ceremonials.

cold season, *Vríhi,* or that ripening in the rainy season, and *Shashtika*, or that grown in the hot weather in low lands." The above passage has been taken from **Dutt's** *Materia Medica of the Hindus,* a work compiled from Sanskrit medical authors. But it may be said that rice enters at the present day even more extensively into the ordinances of Hinduism, than has been denoted by **Dutt's** allusion to the goddess of wealth. It consequently was very probably similarly employed in ancient times. Neither the plant nor its grain can, strictly speaking, be said, however, to be held sacred, though certain rices are used as votive offerings at many religious ceremonies. It is forbidden to eat rice without having first washed. Young females desiring husbands offer dressed rice to the gods. It is used at the observances after birth of a male child and at the consecration of a Brahmanic disciple. The Brahmans, when performing the marriage rites, after having recited prayers, consecrate the union by throwing rice flour coloured with saffron on the newly-married couple. The Sanskrit word *syála* denotes the custom of the bride's brother scattering fried grains at the marriage ceremony. And later in life women, who desire male children, present offerings of rice and saffron at the temples. In Cuttack a form of rice known as *biáli* is prohibited from use at religious ceremonies, since a tradition prevails that it is less pure than *sarad* rice, because of its having been produced by the sage Viswámitra and not by Brahmá, the author of the universe.

In the *Ordinances of Manu* (translated by **Burnell & Hopkins**) the word "food" is given as synonymous with "rice." Thus in Lecture V., 144, it is enjoined that having eaten "(rice) food" the mouth should be rinsed. More frequent mention is, however, made of "hermits' (wild) rice," and in these latter passages there would appear to be no doubt the allusion is intended for **Oryza** and not food in its generic sense, which might have originally embraced other grains besides rice. Thus the twice-born are directed to offer the five great sacrifices, "with hermits' (wild) rice," with various pure (substances), or with herbs, roots, and fruits. Wild rice in Sanskrit works is spoken of as *nivára,* and, according to lexicographers, various Sanskrit names are given to the cultivated rices such as *dhánya, vríhi, syáli, jíva-sadhana* (this is to say means of subsistence or rice), and *tanonu, shashtika* (a quick-growing rice), and *makushthaka* (a peculiar form of kidney-bean or rice). Most of these names are traceable to roots which denote life, existence, subsistence. Thus *dhá* means to support, conceive; it gave origin to *dhátri* a founder, a creator; to *dháman,* a dwelling-place, home; and ultimately to *dhána,* grain, *i.e.,* rice. So in like manner *jíva* comes from *jív,* to live. It would thus seem that in their origin, the Sanskrit names for rice were associated with the most primitive conceptions of the human mind, and hence it is just possible they may have matured into specific significance at comparatively later periods. At all events, we find even *vríhi* (which many writers accept as the most direct Sanskrit name for the grain) associated also with other food materials. Thus, for example, we meet with *vríhi-kancana* (a synonym for *masúra*) as denoting the pulse **Lens esculenta,** with *vríhi-rájika,* the millet **Setaria italica,** and *vríhi-kanga,* the millet **Panicum miliaceum.** So again, special forms of rice are specified as *vríhi-bheda, vríhi-sreshtha,* the latter a synonym for *sáli-dhánya.* The word *vríhi* would appear to have been derived from the root *vrí,* to choose or select. As a possible historic fact, which would have a bearing on the cultivation of rice in Upper India, it may be said that the word *vríhi,* while it does not occur in the *Rig-veda,* is met with in the *Atharva-veda,* and is thus traceable to a period probably at least 1000 B. C.

Wild Rice.

The most general vernacular name for wild rice is, however, *urí,*

HISTORY.

and the plant occurs plentifully in Madras, Orissa, Bengal, Cachar, Burma, Cochin-China, and Ceylon. It has also been found at numerous isolated localities throughout the more northern tracts of India wherever a large supply of water exists. Some doubt, however, may be entertained regarding the remarks of popular writers, on the more northern tracts, since rice often persists self-sown in swampy lands under such conditions as to be readily mistaken for the indigenous plant. In fact, it may be said that the undoubted wild habitat is within the region where the word *urí* occurs in greatly diversified tongues as its most general name.

Botanical Evidence.

The genus **Oryza** and, indeed, the tribe of grasses to which it belongs, may be said to be tropical, South Asiatic, Malayan, Polynesian, Australasian, and American. We possess in this fact a reason why the statements of writers, who would extend the Indian wild habitat of **Oryza** into the upper drier tracts of the Peninsula, and even across the great mountain chain on its north into Central Asia, should be accepted with caution. There are several allied grasses which often occur in rice-fields and natural swamps, which the writer has found to be sometimes designated "wild rice." These yield edible grains which undoubtedly resemble rice, though botanically they are quite distinct. The most fruitful cause of misleading statements as to the occurrence of wild rice proceeds, however, from the fact already mentioned, *viz.*, that the aquatic inferior forms of **Oryza sativa** manifest great facility in becoming naturalized wherever favourable circumstances are offered. Of this nature very probably are the wild rices of certain authors on Panjáb and Kashmír Agriculture. The writer has, however, seen samples of rice, stated to be quite wild, which were collected by **Dr. G. King** in Central India, and by **Mr. J. F. Duthie** in Alígarh in the North-West Provinces and on Mount Abú in Rájputana. At Azimghar, wild rice is reported to bear the name of *tenni*. **Roxburgh** gives the name *urí*, as used by the Telegu people to denote the cultivated sorts of rice collectively, *urlu*, the grain in husk, and *brium*, the grain after being husked, that is to say, "Rice." The wild plant, he informs us, is called by the Telegus *newari*. In passing it may be remarked that **Roxburgh** gives *unú*, *dhánya*, and *vríhi* as the Sanskrit names for rice, but the first mentioned does not appear in any of the modern dictionaries and is probably a Coromandel supposed Sanskrit word.

Moodeen Sheriff says the plant is known to the Tamil people as bearing the Sanskrit name *tandalam*. In dictionaries *tanonu* occurs. In Mahratta rice is known as *tándála*, and wild rice, as already remarked, is at Azimghar called *tenni*. **Sir Walter Elliot** (an author who carries great weight in matters connected with the Telegu country) gives *vori*, *mattakarulu*, and *erra jilama vadlu* as its Telegu names. He speaks of wild rice under the name of *nevari dhányamu* (a name which, like many others used by the educated people of India, is directly traceable to the Sanskrit). The word *arísi* in Tamil denotes 'husked rice,' in Malayal the field crop is *arí*, and in Kanarese, *akki*. The field crop in Tamil is *shálí*, a word used almost throughout India (from Assam to the Panjáb and Kashmír) to denote a special class of rices.

Vernacular Names.

In Burma, rice is known as *sán* or *sa-ba*. In China, *tau* is the general term, when husked it is *mí*, and glutinous rice is *no*. In the tract of country between China and India, inhabited by warlike, antagonistic tribes who until recently had little or no dealings with each other and mostly speak monosyllabic languages, rice bears many undoubted aboriginal names. Thus, for example, in the Garo Hills it is generally designated *mí* (a word which may be identical with the Chinese). In the Naga Hills the two chief crops are known as *kezi* and *thedi* or *chedi*. In the Khásia Hills the collective name for rice is *u-kyba* (*u* "male," or "the"

HISTORY.

and *kyba*, "rice"). Rumphius tells us that the Malay name for the field crop is *padí* (a word anglicised in modern books into "Paddy"), that the grain is *bras* and when cooked *nassi*. In Malabar, he says, rice is known as *neli*, in the Celebes as *pare*, in Java as *bras*, and in Amboyna as *hala*.

Throughout the area of the South Asiatic supposed wild habitat (from Madras to Cochin-China), an extensive series of undoubted indigenous names occur, not only for each form of the cultivated rice, but for every process of its cultivation and manipulation, until it reaches the stage of being cooked, when each special preparation has similarly assigned to it a separate name. It would expand the present article too much to attempt to give numerous examples of this redundancy. Rumphius cites its occurrence in the various languages of the Malayan countries, and in Bengal, every district almost has a complete vocabulary of such names and terms. Thus opening Sir W. W. Hunter's *Statistical Account of Bengal* by chance, in Vol. VII., page 71, the names in Maldah are found and at 238 in Rangpur. These may be exhibited by way of illustration:

Names for Parts of the Plant & Preparations.

Parts indicated by names.	Maldah.	Rangpur.
The seed	*Bihan.*	*Bij* or *bichhán.*
Plants a foot high and ready for transplanting	*Phúl.*	*Neochá bichhán.*
Plants throwing out ears	*Gambhar.*	*Kanch thor.*
When the ears have appeared	*Phúlan.*	*Phulán.*
When the substance or as it is called the milk has formed in the grain	*Dudhi-khotan.*	*Dudh bhara.*
When the grain is ripening	*Dhán.*	*Kálá paká.*
When ready for reaping	*Pakká Dhán.*	*Purá paká.*
Unhusked rice	*Dhán.*	*Dhán.*
Husked rice	*Chául.*	*Chául.*
Husk	*Jus.*	*Jus.*
Husked after being boiled	......	*Ushná chául.*
Husked after being simply soaked in water	*Aroá.*	......
Husked after being ripened in the sun	......	*Atáp chául.*
Fragments of rice broken in husking		*Khud.*
Boiled rice	*Bhát.*	*Bhát.*
Boiled, then parched in flat pieces	*Chirá.*	*Churá.*
Rice soaked boiled, dried and husked, afterwards blown out by cooking in hot sand	*Murí.*	*Murí.*
Rice husked in heated sand, the husk coming off naturally as the grain expands	*Khái.*	*Khái* or *lái.*
Cakes from rice flour	*Pishtak.*	*Pithá.*
Spirits from rice	*Dhánimad.*	*Denomad.*
Rice beer	*Pachwái.*	*Pachwái.*
Liquor made of rice boiled with milk, sugar, ghí, and spices	*Paramanna.*	......
Rice mixed in water and left over nightfall till it becomes sour	......	*Pantha bhát.*

The above by no means represents all the terms in use, since these are extended to every stage of agricultural operations and culinary processes. Among the aboriginal tribes of Chittagong, three chief terms are in use:—germinating rice, *gej*, when forming the ear, *thor*, and when fully ripe, *pakna*. In Burma rice in husk is *saba*, husked rice, *tsan*, and when cooked, *tamin*. Wild rice is known as *nat-saba* or *saba-yaing*, the

ORYZA sativa.

History of Rice.

HISTORY.

former meaning rice given by the *nats* or spirits, and the latter literally wild as opposed to cultivated.

In Northern India the names are mostly derivative, and many of them of undoubted Sanskrit origin. Rice may be wild in Kashmír, but until the writer has examined a sample of the *hama* (alluded to by **Mr. Lawrence** in the account of Kashmír below), he feels it to be safer to assume that *hama* is only an ancient or more completely acclimatised stock than *kre*, if indeed, as already suggested, it does not prove to be a species of **Leersia.** No authentic mention occurs of wild rice in connection with the Panjáb; indeed, **Stewart** expressly says that he had failed to find it; the doubtful case mentioned by **Aitchison** and quoted by **DeCandolle** may, therefore, be set aside. Nor has wild rice been recorded in Sind, Balúchistán, Afghánistán, Persia, Arabia. **Aitchison** alludes to rice cultivation in Afghánistán up to altitudes of 7,500 feet, but **Boissier**, *Flora Orientalis*, does not even refer to it as cultivated, and, indeed, he describes only two grasses of the Tribe ORYZEÆ. **DeCandolle** writes: "'According to **Aristobulus**,' says **Strabo**, 'rice grows in Bactriana, Babylonia, Susida;' and he adds 'we may also say in Lower Syria.' Further on he notes that the Indians use it for food and extract a spirit from it. These assertions, doubtful perhaps for Bactriana, show that this cultivation was firmly established, at least, from the time of **Alexander** (400 B.C.), in the Euphrates valley, and from the beginning of our era in the hot and irrigated districts of Syria. The Old Testament does not mention rice, but a careful and judicious writer, **Reynier**, has remarked several passages in the Talmud which relate to its cultivation. These facts lead us to suppose that the Indians employed rice after the Chinese, and that it spread still later towards the Euphrates—*earlier, however, than the Aryan invasion into India*. A thousand years elapsed between the existence of this cultivation in Babylonia and its transportation into Syria, whence its introduction into Egypt after an interval of probably two or three centuries. There is no trace of rice among the grains or paintings of Egypt."

The above passage has been here reproduced to show the opinions held by so eminent a botanist as **DeCandolle**, on the origin of rice cultivation and its distribution to Central Asia, Persia, Egypt, and ultimately Europe. To return, however, to India it may be said frequent mention of it occurs in the more modern Sanskrit works. Thus, for example, it is very often alluded to in the *Ordinances of Manu*—a fact which perhaps might be accepted as agreeing with the opinion that the work was written about 100 to 500 A.D., the author being a Panjábi Brahman who wrote for the Deccan and South India. In other words, that great work was written in the very region where the *urí* or *nevári* was a prized luxury attainable only by the rich or by religious devotees.

Commenting on the identity of the Persian and Tamil names, *shâli*, **Crawford** wrote (*Agri.-Horti. Soc., Ind. Jour., Vol. I.* (*New Series*), *Selections, 13*):—"This leads to the belief that the grain was most probably introduced into Persia from Southern India in the course of that maritime trade which is known to have been carried on for ages between the ports on the western coast of India, where the Tamil is the vernacular tongue and those on the Persian Gulf. Had this cereal reached Persia from Northern India, its name, as in the case of wheat, would have been traceable to the Sanskrit or one of its derivations."

Yule & Burnell, in drawing attention to the resemblance of the Tamil word *ariśi* to the Arabic and Greek names for rice, say:—

"There is a strong temptation to derive the Greek *ὄρυζα*, which is the source of our word, through It. *riso*, Fr. *riz*, &c., from the Tamil *ariśi*,

HISTORY.

rice deprived of husk, ascribed to a root *ari*, to separate. It is quite possible that Southern India was the original seat of rice cultivation. **Roxburgh** (*Flora Indica, II., 200*) says that a wild rice, known as *newári* by the Telinga people, grows abundantly about the lakes of the Northern Circars, and he considers this to be the original plant. It is possible that the Arabic *al-ruzz* (*arruzz*) from which the Spaniards directly take their word *arroz*, may have been taken also directly from the Dravidian term. But it is hardly possible that *ὄρυζα* can have had that origin. The knowledge of rice apparently came to Greece from the expedition of **Alexander**, and the mention of *ὄρυζα* by **Theophrastus**, which appears to be the oldest, probably dates almost from the life-time of **Alexander** (*d.* B.C. 323). **Aristobulus**, whose accurate account is quoted by **Strabo**, was a companion of **Alexander's** expedition, but seems to have written later than **Theophrastus**. The term was probably acquired on the **Oxus**, or in the Panjáb. And though no Sanskrit word for rice is nearer *ὀρύζα* than *vríhi*, the very common exchange of aspirate and sibilant might easily give a form like *vrísi* or *brísi* (Comp. *hindú, sindú*, &c.) in the dialects west of India. Though no such exact form seems to have been produced from old Persian, we have further indications of it in the Pushtu, which **Raverty** writes, *sing.* 'a grain of rice,' *w'rijza'h*,* *pl.*, 'rice' *w'rijzey*,* the former close to **Oryza**. The same writer gives in Barakai (one of the uncultivated languages of the Kabul country, spoken by a 'Tajik' tribe settled in Logar, south of Kabul, and also at Kanigoram in the Waziri country) the word for rice as *w'rizza*, a very close approximation again to **Oryza**. The same word is, indeed, given by **Leech**, in an earlier vocabulary, largely coincident with the former, as *rizza*. The modern Persian word for husked rice is *birinj* and Armenian *brinz*. A nasal form, deviating further from the hypothetical *brísi* or *vrísi*, but still probably the same in origin, is found among other languages of the Hindú Kush tribes, *e.g.*, Burishki (Khajuna of **Leitner**), *bron;* Shina (of Gilgit), *briún*; Khowar of the Chitral valley (Arniyah of **Leitner**), *grinj*."

The area of wild rice, corresponding to that in which *urí* or *arí* and other such names are used (and that too in the most diversified aboriginal languages) leads naturally to the supposition that in India rice cultivation may have spread from there all over the rest of the peninsula and ultimately across the Himálaya. This diffusion of the knowledge of so valuable a crop seems, as **DeCandolle** suggests, to have taken place even prior to the Aryan invasion, and it may be added that the aboriginal names for the wild plant would in that case most probably have accompanied its cultivation until they gradually became absorbed in the more cultured tongues, some of them being even fused into the earliest Sanskrit conceptions. Considerable light would be thrown on this subject by a careful search in Sanskrit literature with the view of establishing the relative ages of the Sanskrit terms, as, for example, when they came to be unmistakably assigned to this grain. Rice is not likely, however, to have been known to the Sanskrit authors much before the date of the later *Vedas*, and, as already remarked, the names used in Upper India for the plant and the grain are mostly derivative, many of them of Sanskrit origin.

The redundancy in names that exists within the area of wild habitat (and which is by no means so comprehensive outside that area) seemed to support the general conclusion. In order, however, to obtain some information on this subject, drawn from Persian and Arabic literature, the writer

* *jz* is **Raverty's** mode of expressing the sound of French *j*, generally rendered in India *zh*.

ORYZA sativa.

History of Rice.

HISTORY.

addressed a letter to **Mr. C. J. Lyall** (Secretary to the Government of India in the Home Department), a well known Arabic scholar. **Mr. Lyall's** reply is so full of interest, in linking together the various theories and opinions which have been briefly reviewed above, that the liberty may here be taken of quoting some of its more instructive passages:—

"There can be no doubt that the Arabic names for rice are not derived from the Tamil *ariśi*. Rice cannot be considered a crop of the Arabian peninsula—that is, of the people to whom the Arabic language properly belongs. It is indeed now grown in the lowlands along the western shores of the Persian Gulf, and possibly in Yemen, but is of recent introduction there. The great tract in which the Arabs first became acquainted with it is beyond the boundaries of Arabia, *viz.*, Babylonia and the tract between the Upper Euphrates and Tigris. The Arabian names for rice are ***aruzz, uruzz, urz, uruz, aruz, áruz, ruzz***, and ***runz***, the last a form used by a tribe called 'Abd-al-kais, settled in 'Omán and Bahrain. The last form points to the double ***zz*** in the other forms being a substitute for ***nz***. The Aramaic form in use in Babylonia was ***uruzzā*** or ***aruzzā***, evidently the original of the Greek ὄρυζα. It was here, and not in Bactriana, that the Greeks first became acquainted with the grain and the name. The Aramaic name was evidently borrowed from the Persian. I am not aware of any example in old Persian literature of the original of the modern form ***birinj***; but the Arabo-Aramaic forms clearly point to ***virinzi*** or ***virinza*** as this original. This is confirmed by the Armenian ***brinz***, the Pushtu ***wrizha***, and the Khowar *grinj* (*g* initial pointing to an older ***w***).

"Now ***virinzi*** is exactly the equivalent we should expect of the Sanskrit ***vríhi***. *H* in Sanskrit regularly appears as *z* in old Persian (*sahasra*=*hazanra*; *hima*=*zima*; *hridaya*=*zereda*, *&c.*). But Persian words containing *z* corresponding to Sanskrit *h* are not loan-words from Sanskrit, but *sister-words.* They point to the time when the two branches of the Aryan race dwelt together and respectively developed their phonetic peculiarities from a pre-existing original tongue common to both. It is certain, therefore, from a comparison of ***virinzi*** and ***vríhi***, that the Persians did not borrow the cultivation of rice from India, but that that cultivation existed in the tract where the two races dwelt together, before the Indo-Aryans descended to the plains of the Panjáb or the Perso-Aryans occupied Airyana or Erán. If this tract was Bactriana, which **Strabo** mentions as a great rice country, the name may have originated there. It seems perfectly clear that it cannot have originated within India, or be connected with any Tamil name now current. ***Vríhi, virinzi***, is the parent of the Semitic names and also of the Greek ὄρυζα, which was borrowed from the Aramaic of Babylonia; through ὄρυζα it is also the parent of all the European names except the Spanish, which is taken direct from Arabic."

In a subsequent letter **Mr. Lyall** wrote to the Editor:—"I recently came across an interesting article on rice by **Victor Hehn*** in which he mentions a much older word than ὄρυζα for rice, *viz.*, ὀρίνδης ἄρτος, used by **Sophocles.** Now while ὄρυζα is, as I have shown, derived from the Aramaic form of the word, ὀρίνδη is evidently Persian, and equivalent to the hypothetical ***virinzi*** or ***virinza***, of which we have no literary example in that language. In **Sophocles'** time, I need hardly observe, the Greeks had much direct intercourse with, and knowledge of, the Persians."

It will thus be seen that botanical and philological evidence do not entirely agree. If the old Persian word, from which the Arabic, Greek, and European names are taken, is not derivable from the Sanskrit, but is

* Culturpflanzen und Hausthiere, Art. "Reis."

HISTORY.

cognate with it, rice cultivation might be supposed to have proceeded from the original home of the Aryan race. If this be supposed to have been anywhere in Central Asia, it is practically impossible to believe that the wild plant could ever have been found there. But there is no evidence which goes to show that its cultivation in Southern Asia was not so ancient as to have allowed of its diffusion into the Aryan home at a period prior to the division of that great branch of the human family. Indeed, the chief objection to this hypothesis is the fact that so valuable a plant is not pointedly alluded to in the earliest Vedas. But even that difficulty is not serious, since a pastoral people, like the early Aryan invaders, may not have appreciated its importance until they became localised and took to agricultural pursuits. In concluding this brief sketch of the main ideas regarding the early history of rice culture, it may be repeated that the chief wild habitat of the plant is, roughly speaking, from Southern India to Cochin China. That belt of land has often vast marshy expanses with low intervening mountains, possesses a tropical climate, as also strong periodic seasons of rainfall; inundations suitable for the growth of rice are therefore of annual recurrence. Its cultivation appears to have spread from thence eastward to China, perhaps 3000 years before the Christian era, and at perhaps a slightly more recent date, westward and northward, throughout India to Persia and Arabia and ultimately to Egypt and Europe. An enlightened people like the Chinese might be supposed to have more readily realised the importance of rice cultivation than would the aboriginal tribes on the hills and swamps of Lower India, where a sufficiently abundant supply for their wants could be gathered from the wild plant. This assumption is practically justified by the fact that the aboriginal tribes of Australia have continued satisfied to the present day with their wild rice (which they designated *kineyuh*), and have never thought of its cultivation, nor indeed of the cultivation of any plant.

CULTIVATION OF RICE.

CULTIVATION. 283

Some few years ago, it was currently stated that Rice was the staple food of the people of India. This is true of Bengal, and perhaps of Madras and Burma, but taking the people of India as a whole, millets and pulses collectively might, with a greater approximation to accuracy, be given as the chief food materials. In many parts of India rice is, indeed, as much a luxury as wheat is in others, and even in rice-producing countries the poor have often to eke out subsistence by greatly supplementing rice diet with other coarser and less expensive articles. In the Returns of Agricultural Statistics of British India for 1888-89* (exclusive of Bengal) it is shown that the total land still available for cultivation amounts to 89,814,481 acres[1] with 45,524,640 acres under forests,[2] 117,580,990 acres not available for cultivation,[3] and 134,653,065 acres cropped[4] with an additional area of yield of 14,158,424 cropped more than once[5]. Of this cultivated (surveyed) area 26,810,806 acres were found to be under rice[6]. As pointed out, however, these figures do not include Bengal, the richest agricultural province in India, but of which unfortunately no actual survey has as yet been made. This is the more to be regretted, since Bengal is India's chief rice-field. Some idea of the vast importance of Bengal may be derived from the following passage taken from the Administration Report for 1882-83 (p. 12):—"The dis-

* Since this was in type the Editor has received the returns for 1889-90. The following are the figures now published:—[1] 93,654,706; [2] 47,230,763; [3] 117,434,999; [4] 136,168,899; [5] 13,921,673; and [6] 27,866,447.

ORYZA sativa.

India's Rice Area.

CULTIVATION

tricts of the whole of Bengal proper, or the great alluvial and deltaic plain between the Himaláya and the Bay of Bengal, and the province of Orissa, or the alluvial territory between the hills and the sea, connecting Bengal with Madras—a level area of nearly one hundred thousand square miles unbroken by a single hill, rich in black mould, and of boundless reproductive fertility, subject to recurrent inundation and enjoying natural facilities such as no other country in the world possesses for internal commerce and irrigation,—constitute the great rice-producing area of India, which is ordinarily much more than self-supporting. The surplus produce of this area finds its way, generally speaking, to three directions, from which the rice-trading operations of Bengal are conducted. First, of course, is Calcutta; the imports into Calcutta have to find food for the metropolis and also for foreign exportation. In the second place come Behar and the North-West Provinces, where the demand for rice is always in excess of the local supply. Lastly, there is Chittagong, which is the centre of a considerable export trade by sea." The region indicated by the above passage by no means includes the whole of the province ruled over by His Honour the Lieutenant-Governor of Bengal. It consequently does not embrace all the region of unsurveyed rice-fields which exists in the Lower Provinces. The Government of Bengal, however, in its reply to question three of the Famine Commission of 1878, furnished a table of estimated agricultural statistics which gives further interesting particulars. It showed the total area of the districts as 92,196,240 acres (excluding Chittagong Hill Tracts). Of these, 54,645,468 acres were returned as cultivated and 48,634,497 as under food crops. In some of the northern districts and in Behar, rice cultivation is supplemented more or less by *rabi* or dry-weather crops, chiefly wheat, barley, and pulses, and by *bhadoi* or intermediate crops consisting mainly of Indian-corn and millets. In the lower districts, where rice only is grown, the early and the late rice crops correspond to the *bhadoi* and *rabi* of the drier tracts. Thus, for example, in Eastern and Southern Bengal, and in Orissa, where the rainfall is heavy and inundations frequent, millets and spring crops become rare or almost disappear, and, as a rule, only two kinds of rice are sown. These are the early and the late (the *áus* and *áman* of Bengal or the *beali* and *sarad* of Orissa), the latter being by far the more important of the two. Two crops of rice are thus by no means infrequent, and the actual area under rice is thereby greatly increased. According to the calculations worked out in connection with the Famine Commissioner's Report, the area yielding rice in Bengal was determined as 37,500,000 acres. It seems probable, however, that the estimates framed in connection with the Famine Commission erred on the safe side by understating the actual area and yield. Thus, for example, **Mr. A. P. MacDonnell**, in his enquiry into the question of ***Food-grain Supply in Behar and Bengal,*** showed that for 16 districts for which fairly trustworthy statistics were available, the total cultivated area was 25,989,447 acres, of which 23,279,225 were under food crops, the analysis of the food crop area being 12,824,190 acres under rice, 5,125,622 under *bhadoi* crops and 5,042,328 under *rabi*. These 16 districts were, however, in Behar and the northern drier tracts where comparatively less rice is grown than in the districts which produce early and late rice. Since the date of the enquiry instituted by the Famine Commission it is also probable that the area of rice culture may have considerably expanded. Fortunately, we are now in a position to form a more definite idea of certain Bengal crops. Through the exertions of the Director of Land Records and Agriculture, "Forecasts" of the outturn of important crops are prepared, and the District Officers who furnish the data for these forecasts are becoming able to gra-

Two Chief Crops.

CULTIVATION

dually eliminate the errors and difficulties that have rendered it hitherto impossible to furnish trustworthy statistics for Bengal. Through the courtesy of **Mr. M. Finucane** (the Director of the Department) the author has been furnished with advance copies of these forecasts, as also with valuable notes and suggestions bearing on the subject of rice culture in Bengal. Taking the Forecast for 1889 of "Winter Rice," the returns compiled show the total area of the sub-divisions comprising Bengal to have been 97,817,600 acres, the proportion cultivated to have been 55,407,360 acres, and the share of that cultivated area held by winter, *i.e*, *áman*, rice to have been 33,294,720 acres. The crop which followed, *viz*., the *bhadoi* of 1890, may be here discussed. The forecast showed the total area under *áus* and other forms of rice, jute, sesamum, hemp, indigo, certain pulses and maize, &c., to have been 14,593,280 acres. This practically amounts to the statement that a little less than 26½ per cent. of the cultivated area of Bengal is twice cropped. **Mr. Finucane** is of opinion that the proportion of *bhadoi* which actually yields rice is perhaps about 25 per cent. of the area of winter rice. Accepting that figure as approximately correct we thus learn that Bengal, during a period of 12 months (for which returns of both winter and *bhadoi* crops exist), had 41,618,560 acres devoted to rice. A comparison of these figures with those given above will reveal the fact that the Famine Commission's Report was framed on more trustworthy data than even the Commissioners themselves gave assurance to. The want of a definite survey has compelled statisticians to publish returns connected with Bengal as the merest generalisations. The writer has, however, found, in connection with articles of Bengal commerce, that these so-called generalisations agree very closely with commercial statistics which are compiled from sources altogether independent of Government. He has thus been induced, from time to time, to place a much greater degree of dependence on Bengal statistics, than the officials responsible for their preparation, would, from their guarded language, authorise. In the case of rice it is evident that the present returns show a relative expansion in each particular which is explicable by the lapse of years since the date of **Mr. A. P. MacDonnell's** exertions to furnish the Famine Commissioners with details regarding the Food-grain Supply of Behar and Bengal.

Area of Rice in Bengal. 284

41,618,560 acres.

Accepting, therefore, the above Bengal rice area as correct, the total for all India last year would have been 68,429,366 acres. This will be seen to be about 10 million acres less than the total area returned in the Agricultural Statistics as having been, during the same period, devoted to "Other Food Grains." But the millets and pulses which are grown in Bengal would, to allow of actual comparison, have to be added to the area of "Other Food Grains," shown above, before a comparison could be drawn regarding the importance of rice *versus* millets & pulses to the people of India, and an even larger and more important correction would have to be made for the rice and wheat which is exported to foreign countries. The export trade in cereals and pulses is of necessity drawn from the surplus over local demand, but a careful examination of the trade shows a large cultivation of cereals exclusively intended for the foreign markets and thereby manifests an increased importance to the people of India in a mixed diet of millets and pulses. Even, however, after all deductions have been made, rice occupies the foremost place amongst the food crops of India, as it takes up nearly three times the area devoted to wheat and twice that which yields millets if taken as distinct from pulses. But this conclusion after all is only what might have been expected, since by extending the enquiry to embrace China, Australia, America, and the West Indies, rice is found to be the most valuable (single) cereal of

Total Indian Area 68,429,366 acres. 285

Relative Value of Rice to other Crops. 286

ORYZA sativa.

The Importance of Rice

CULTIVATION

all, for considerably more than half the human family depend upon it as their staple article of diet.

The following table exhibits the relative importance of rice, wheat, and other food-grains n the surveyed areas of India :—

Pages where further details occur.	PROVINCES.	Acres of rice.	Acres of wheat.	Acres of Other Food-grains (=barley, oats, millets, & pulses).
I. 523	Bengal	No statistics.	No statistics.	No statistics.
II. 572	Madras.	6,285,806	20,360	13,927,269
III. 588	Bombay (including Sind) .	2,239,198	2,312,930	17,823,577
IV. 602	N.-W. Provinces	4,338,923	3,479,279	17,052,358
V. 602	Oudh	2,439,228	1,489,921	6,433,384
VI. 614	Panjáb	690,565	7,371,977	11,787,458
VII. 625	Central Provinces . . .	3,785,566	3,531,941	4,811,519
VIII. 627	{ Upper Burma	1,605.936	9,185	1,186,486
	{ Lower Burma	4,067,606	*Nil.*	49,790
IX. 632	Assam	1,262,791	12	61,665
	Berar	19,840	942,029	2,878,260
	Coorg	74,499	*Nil.*	1,611
	Ajmír-Merwara	758	9,548	163,294
	Pargana of Manpur (Central India)	90	2,831	2,254
	TOTAL .	26,810,806	19,170,013	76,178,925

It will thus be seen that, after Bengal, of the other provinces, Madras, the North-West Provinces, Lower Burma, and the Central Provinces (in the order named) are the chief rice-producing regions, but it may safely be conjectured that Bengal possesses very nearly twice as much rice lands as all the rest of India taken collectively.

The Report of the Famine Commission (with the special publications which it originated) possesses by far the most valuable data regarding the food-supply of the people of India. That publication has already been placed under liberal contribution in the above remarks, but a further passage may be here given regarding the relative consumption of rice as compared with other food substances. The total population of India (dealt with in the Report) is given as 181,350,000, the total acreage, under food crops, as 166,250,000, the outturn of food as 51,530,000 tons, and the requirements of India as 47,165,000 tons, thus leaving a surplus for export, or for the more luxurious consumption of the rich, of 5,165,000 tons. Another part of the report, dealing with a slightly different population (*viz.*, 191 millions), shows 67 millions as eating rice. Commenting on the subject of the *staple food of the people* of India the Commissioners say :—"In the Panjáb, the North-West Provinces and Oudh, in Behar and the northern part of the Central Provinces, and in Guzerát, the poorer classes live on the millets grown in the rains, and on barley and gram ; the richer classes eat principally wheat and rice. In Bengal proper and Orissa, and the eastern portion of Central India, rice is the principal food, the coarse, early rice being mainly taken by the poor, the finer, late rice by the rich. In the South or Mahratta-speaking part of the Central Provinces, in Berar, in the Bombay-Deccan, and the northern part of Madras, the two large millets (*jowár* and *bájra*) form the principal food, the Brahmins generally living on imported rice and wheat. In Mysore the ordinary food is the small millet, *rágí*. In the southern part of Madras and western districts of Bombay rice is chiefly consumed, though there is also a

CULTIVATION

good deal of millet grown and eaten. All classes mix pulses with their food, the nitrogenous matter which is found in the pulses supplying an ingredient of which little exists in the cereal grains, and which is necessary for the proper nutrition of persons who rarely eat meat."

In the following pages a somewhat extensive series of quotations will be found grouped under the provinces to which they refer. The writer has found the available material on rice cultivation too imperfect and often almost so contradictory as to not allow of his writing a separate provincial article. The preferable course, as it seemed to him, has been followed, *viz.*, of allowing district writers to speak for themselves, even though this has involved a certain amount of repetition.

I.—CULTIVATION IN BENGAL.

BENGAL. 287

CULTIVATION Bengal, *pp. 523—572.* **Madras,** *pp. 572—581.* **Mysore,** *pp. 581—585.* **Coorg,** *pp. 585—588.* **Bombay,** *pp. 588—600.* **Sind,** *pp. 600—601.* **N.-W. P. & Oudh,** *pp. 601—614.* **Panjab,** *pp. 614—622.* **Kashmir,** *pp. 622—624.* **C. P.,** *pp. 625—626.* **Burma,** *pp. 627—632.* **Assam,** *pp. 632—637*

References.—*Food-grain Supply of Behar & Bengal, by A. P. MacDonnell; Cultivation of Rice in Bengal, by C. B. Clarke, Kew Bulletin, Dec. 1888, pp. 284-291; Report on Agriculture of Lohardaga, by B. C. Basu; Report on Agriculture of Dacca, by A. C. Sen; Agri.-Horticultural Society of India: Trans. II., 88, 231-244; IV., 79-82, 100, 101, 113, 120-124; V., 36, 37; VI., 239, 246, Pro. 8; VIII., 107, 108, 160, 161, 470; II., Sel. 500, 503; IV., 104; Sel., 143; VI., Pro. 71; VIII., 81-88, 178-181, Pro. 29, 166; XII., 109, Pro. (1862), 8, 9 (1864), 47; XIV., 273-283. New Series, II., 24-90, 102-136, 177-190, Sel., 8-12; III., Sel., 5, 6, Pro. 5; IV., Pro., 6, 27; Grierson, Behar Peasant Life, 215-223; Westland, Report Jessore District; Suggestions on Cultivation & Introduction of Useful Plants in India by H. H. Spry, 133; Buchanan-Hamilton, Statistical Account of Dinagepore; Reports of the Director of Land Records and Agriculture; General Administration Reports of the Provinces; Forecasts of Crops in Bengal; Statistical Account by Sir W. W. Hunter: Vols. I. (24-Parganas), 134-138, 324-325; II. (Nadiya), 33, 64, 68-69; (Jessor) 184, 242, 247; III. (Midnapur), 38, 79-83; (Hugli), 204, 329-331, 340; IV. (Burdwan), 69-70; (Bankura), 245-246; (Birbhum), 345-346, 350, 354; V. (Dacca), 25, 82-83, 90-91; (Bakargánj), 171, 202-205; (Faridpur), 276, 296-305; (Maimansingh), 390, 419-421; VI. (Chittagong Hill Tracts), 71, 74; (Chittagong), 160-161; VII. (Maldah), 70-72, 74, 92; (Rungpur), 234-240; VIII. (Rajshahi), 58-59; (Bogra), 208-209; IX. (Murshidabad), 99-104; (Pabna), 301-302; X. (Darjiling), 92-95; (Jalpaiguri), 271-273; (Kuch Behar), 279-382; XI. (Patna), 109-111; (Saran), 274-276; XII. (Gayá), 82-84; (Sháhábád), 168, 230-233; XIII. (Tirhut), 28, 81-82; (Champaran), 228, 260-261, 291; XIV. (Bhágalpur), 116-118; (Santal Parganas), 270, 335-337; XV. (Monghyr), 83, 90-91; (Purniah), 281-286, 293; XVI. (Hazáribágh), 99-101; (Lohárdagá), 337, 338-340; XVII. (Singbhum), 79-80; (Bonai State, Chutia Nagpur), 176-177; (Jashpur State), 208; (Manbhum), 310-312; XVIII. (Cuttack), 58, 99-102; (Balasor), 263, 289-291; XIX. (Puri), 93-95; Orissa Gazetteer: Vol. I., 146-147 (rice consecrated into holy food before the idol at the temple of Jagannath at Puri); Prices of rice during and since the 14th century, 327; Vol. II., Kandh betrothal ceremony—rice carried by the bridegroom's father, 82, adopted into Hinduism as Rákshasa marriage (a ceremony derived from the aboriginal tribes); name of child determined by manner grains of rice fall to the bottom of water, 81; Appendix, 14-15; 36, 44-46, 50 (soils suited to rice); 51 (rice blights); the poor who live on a daily mess of rice only, most liable to elephantiasis, 65, also cutaneous diseases, 65; the Kol swears by touching cooked rice, 77; rice cultivation, 80, 96; rice cultivation in Cuttack, 130-133; area under rice, and yield, 135.*

Area under, and Traffic in, Rice.

Area. 288 *Conf. with p. 521.*

According to the review of the figures of rice area in Bengal given above, there were last year 33,294,720 acres under winter or *áman* rice and approximately 8,323,880 acres under *Bhadoi* (the *áus, boro,* and other minor crops of the grain). These figures thus show a total area producing rice of close on 42 million acres.

One of the earliest allusions to the Bengal traffic (on the part of Europeans) is that given by **John Huyghen van Linschoten,** whose journal was

ORYZA sativa.

Bengal and Calcutta.

CULTIVATION in Bengal. **Area.**

published in 1596. "The country," he says, "is most plentiful of necessary victuals, specially rice, for that there is more of it [in that country] than in all the East, for they do yearly lade divers shippes [therewith], which come thether from all places, and there is never any want thereof, and all other things in like sort, and so good and cheape that it were incredible to declare." This Bengal traffic which has certainly been of vast importance for more than 300 years may be examined according to four separate sets of records—(1) Internal or carried by road, rail, and river; (2) Trans-frontier; (3) Coastwise (that is, dealings between one Bengal port and another or between Bengal and other Indian ports); and (4) Foreign Trade.

Trade: Internal. 289

(1) INTERNAL TRADE OF BENGAL.—While fairly accurate statistics exist for the rail-borne and chief river traffic from district to district within Bengal and from and to other provinces, little or no information exists regarding the road traffic of the province as a whole, though a good deal is known of the supplies carried along the roads into the chief cities. A large margin must, however, be allowed for this defect since there can be no doubt but that the seaports drain supplies by small country-boats down minor rivers and by carts along the roads, and the amounts thus carried escape registration. But accepting the published figures of railway, river, and road trade as approximately conveying an idea of the Bengal Internal Trade in rice, the following balance sheet may be given of the chief transactions dividing rice as is customary into "Unhusked" and "Husked":—

Balance sheet of the Rice Trade of Bengal treating the City of Calcutta as a foreign country with which the Province has transactions.

Kinds of Rice.	Gross Imports.	Gross Exports.
	Mds.	Mds.
Unhusked (=Paddy)	11,64,469	14,11,407
Husked (=Rice)	9,35,116	1,53,22,290
TOTAL .	20,99,585	1,67,33,697
Deduct Imports .		20,99,585
Net Exports from the Province to Calcutta and to other Provinces		1,46,34,112

Of the Provinces which supplied Bengal with paddy and rice along internal routes, Assam, in the year under consideration (1888-89), was the most important The returns from which the above balance sheet has been compiled show Assam to have supplied Bengal with 7,34,169 maunds paddy and 5,37,793 maunds rice. Of the Provinces to which Bengal exported, the most important are the North-West & Oudh, which took 8,29,390 maunds of rice, and Assam which got 3,35,324 maunds. But of course by far the most important outlet for Bengal paddy and rice is to the capital. Calcutta with its large population is not only a great consuming centre for rice, but being the chief emporium in the foreign export trade, it naturally drains by rail, road, and river large supplies from the province of Bengal and smaller quantities from beyond the province. The returns here dealt with, show Calcutta to have obtained from Bengal 13,95,627 maunds paddy and 1,39,62,982 maunds rice, with smaller quantities from Assam (1,066 maunds paddy and 53,324 maunds rice), from the North-West Provinces & Oudh (2,843 maunds rice), and from the Panjáb (84 maunds rice). Calcutta exported also small quantities to Bengal and other provinces, but these have been provided for in the above balance sheet, which shows the net exports from the province. A critical examina-

CULTIVATION in Bengal.

tion of the rail and river trade shows Bengal Province to have had in 1888-89 a net export, by internal routes to other provinces, of 71,858 maunds and Calcutta a net import from all sources by these routes of 1,45,70,024 maunds.

Trade.

The chief river marts in the Calcutta trade are Backergunge & Sahebgunge, 16,73,363 maunds; Midnapore, 13,59,473; Jhalokati, 6,48,105; Dinagepore, 4,39,661; Hooghly, 3,36,049; Barrisál, 3,03,763; and some 16 other marts each with between two and a half and one lakh of maunds traffic. The chief marts tapped by inland steamers are Midnapore, Kutwa, Kulna, Ghattal, and Santipore. The road traffic is almost exclusively from the 24-Parganahs. Burdwan is the largest exporting district which employs the railway to convey its surplus to Calcutta. In the year under notice, it exported 5,31,611 maunds, was followed by Bhulpore with 3,63,303, and other stations with smaller amounts. Along the course of the Eastern Bengal State Railway, Bogra furnished the largest supplies.

Husked and Unhusked Rice.

It may, perhaps, be here pointed out that an error is involved in totalising the weights of husked and unhusked grain, since a loss in weight would necessarily occur on the paddy being husked—a loss of one maund reduced to 25 seers. It was, however, deemed a more instructive course to allow of this error than to reduce all weights to that of husked grain, since in India it seems probable the interchange of unhusked rice is largely to meet the supply of seed. At all events, the traffic in paddy (=unhusked) is very much smaller than in rice.

Land Trade. 290

(2) TRANS-FRONTIER LAND TRADE OF BENGAL.—The following may be given as the transactions across the land frontier of Bengal in paddy and rice:—

Kinds of Rice.	Gross Imports.	Gross Exports.
	Mds.	Mds.
Unhusked (=Paddy)	6,50,703	22,835
Husked (=Rice)	3,88,278	24,691
TOTAL .	10,38,981	47,526
Deduct Exports .	47,526	
Net Imports .	9,91,455	

It will thus be seen that the province obtained a net import in its favour of 9,91,455 maunds. The chief countries with which these transactions were made were, as is usually the case, Nepál, Sikkim, and Bhután, but the imports came entirely from Nepál.

Coastwise Traffic. 291

(3) THE COASTWISE TRAFFIC OF BENGAL.—The experience gained in connection with all other commercial articles is no less true of rice, namely, that it is extremely difficult to give a tabular statement for a required period of the intra-portal transactions within the province and the inter-provincial exchanges, by sea along the coast of India. The imports by one port are to a large extent the exports from another, but the period of record closes necessarily with a large proportion of the exports at sea and therefore not included as imports. The final destination of a cargo may also, and frequently is, altered by telegraphic instructions, so that what has been recorded as an export destined for one port, actually appears in the receipts of another. Allowing for these errors the following balance

ORYZA sativa.

Bengal Trade

CULTIVATION in Bengal.

Trade.

sheet may be given of the Bengal coastwise transactions in paddy and rice for the year 1888-89:—

Kinds of Rice.	Gross Imports.	Gross Exports.
	Mds.	Mds.
Unhusked (= Paddy)	25,716	3,36,249
Husked (= Rice)	5,83,805	57,18,333
TOTAL .	6,09,521	60,54,582
Deduct Imports .		6,09,521
Net Exports .		54,45,061
(or in cwt. .		3,889,329)

Of the imports of rice in husk 22,871 maunds came from British ports within the province, and of husked rice the major portion was received also from British ports within the province, namely, 5,49,710 maunds, Burma having supplied only 34,059 maunds. The bulk of these imports were derived from Balasore and Chittagong and were taken by the port town of Calcutta. Of the exports of rice in husk Madras took 2,98,418 maunds, and of the husked rice Bombay imported 34,17,891 maunds, Madras having stood as the next important recipient of Bengal husked rice, *viz.*, 13,56,086 maunds, while the greater part of the balance was shown as intra-portal transactions within the province. But analysing these figures more fully by removing from consideration the purely intra-provincial transactions, it is seen that Bengal had a net export of husked and unhusked rice of 54,45,514 maunds, and that the share of that trade taken by the port town (Calcutta) amounted to 40,85,732 maunds or 2,918,380 cwts.

Foreign. 292

(4) FOREIGN TRADE OF BENGAL IN RICE.—Under this heading is recorded the Bengal transactions with foreign countries. Following the system pursued above a balance sheet of this trade may be given:—

Kinds of Rice.	Gross Imports.	Gross Exports.
	Mds.	Mds.
Unhusked (= Paddy)	*Nil.*	56,677
Husked (= Rice)	75	89,83,808
TOTAL .	75	90,40,485
Chur or flattened rice .		5,583
TOTAL .		90,46,068
Deduct Imports .		75
Net Exports .		90,45,993
(or in cwt. .		6,461,423)

The share taken in these foreign transactions by Calcutta amounted to 81,37,232 maunds, or 5,812,309 cwt.

Total Trade. 293

Now, if we add together the net foreign exports by sea, the net coastwise exports to other provinces, and the net exports to other provinces by rail, road, and river less the net trans-frontier imports, we shall obtain an approximate conception of the amount of rice which left Bengal during the year 1888-89:—

	Mds.
Net foreign exports	90,45,993
Net coastwise exports	54,45,514
Net rail and river exports	71,858
	1,45,63,365
Less trans-frontier net imports	9,91,455
Grand total of net exports from Bengal .	1,35,71,910

in Rice. (G. Watt.)

ORYZA sativa.

CULTIVATION in Bengal.

Trade.

But the rail and river-borne trade into Calcutta must have mainly afforded the supply from which both the people of the city and its suburbs (say, 684,658 souls) obtained their rice as also met the sales in which the foreign and coastwise transactions were made, that is to say, that portion of the Bengal foreign and coastwise trade which actually took place from Calcutta. Making, therefore, the correction for the other than Calcutta internal, coastwise, and foreign rice trade of Bengal, the following may be given as the actual share taken by Calcutta:—

Calcutta Trade. 294

Routes.		Imports.	Exports.
		Mds.	Mds.
Internal	(a)	13,96,693	4,19,127
	(b)	1,40,19,233	4,26,775
Coastwise	(a)	25,697	35,138
	(b)	5,74,994	46,51,284
Foreign	(a)	*Nil.*	52,932
	(b)	47	80,78,891
	(c)	*Nil.*	5,471
TOTAL		1,60,16,664	1,36,69,618
Deduct exports		1,36,69,618	The balance here shown as a net import is the amount available for local consumption or about ⅞ths of a seer per head of population daily.
Net Imports		23,47,045	

(a) = Unhusked rice or paddy. (b) = Husked rice.
(c) = Chúr, parboiled, and flattened rice.

Local Consumption. 295

It would seem a safe statement to affirm that the reserve stocks in a city like Calcutta are of necessity very nearly constant; they are, if anything, a probably slightly increasing total year by year. But if this be so, the net import may be taken as very nearly the actual consumption. It has already been pointed out that an error is admitted in the acceptance of paddy as of equal value with husked rice. This error can at once be corrected, however, throughout the tables given. But in any attempt at estimating the Calcutta consumption of rice it is of more consequence to endeavour to obtain returns for the imports by road since, from a radius of, say, 50 miles around so important a rice-consuming centre as Calcutta, local producers must to a large extent convey their surplus stocks by carts and thus escape the registration that is preserved on the railways and larger rivers. The Bengal Government in its annual report of 1888-89 on the *River-borne Traffic of Lower Bengal and Inland Trade of Calcutta* fortunately are able to furnish much additional information regarding Calcutta, which cannot be supplied for the whole of the province. This chiefly amplifies the imports shown as obtained by Calcutta along the roads. The exports from the city, given by Government, are, to within a few maunds, the same as has been shown in the above table (if the same system be pursued in totalizing weights of paddy and rice together), but the imports are very considerably enhanced, leaving in consequence an increased net import as available for local consumption. The table of Calcutta Trade in rice, compiled by the Government from actual records of transactions, may be here usefully republished, since it shows the traffic for three years and exhibits the various routes by which the grain is usually carried.

ORYZA sativa.

Bengal and Calcutta.

CULTIVATION in Bengal.

Trade.

The final conclusion shown agrees so nearly with the Calcutta Trade, which in this paper has been worked downwards from the total provincial traffic and estimated production, that it may be accepted as confirming the accuracy of both systems of exhibiting the Bengal and more especially the Calcutta rice trade :—

Specification of Routes.		Imports. 1886-87.	Imports. 1887-88.	Imports. 1888-89.	Exports. 1886-87.	Exports. 1887-88.	Exports. 1888-89.
		Mds.	Mds.	Mds.	Mds.	Mds.	Mds.
By boat	Rice	93,31,007	1,22,10,563	1,14,74,755	1,96,342	2,59,981	2,70,646
	Paddy	9,49,737	12,12,192	11,63,888	1,79,556	1,01,153	4,15,858
By Inland Steamer	Rice	1,20,919	1,04,899	2,56,264	32,170	32,082	18,140
	Paddy	6,495	435	3,133	...	11	138
By E. I. Railway	Rice	25,66,811	19,10,278	15,03,132	21,063	49,477	1,27,078
	Paddy	26,465	21,881	2,26,284	...	1	3,158
By E. B. S. Railway	Rice	7,85,630	9,03,176	7,85,082	9,506	8,957	10,911
	Paddy	28,611	4,357	3,388	...	...	54
By road	Rice	6,84,472	7,90,517	6,35,646	1,84,831	1,32,603	1,43,557
	Paddy	95,400	65,642	47,061	1,22,245	1,06,054	1,51,378
By sea	Rice	4,15,079	4,16,788	5,59,084	1,07,00,494	1,44,83,517	1,23,76,560
	Paddy	1,14,901	40,603	24,983	1,34,618	1,14,045	85,625
Total	Rice	1,39,03,918	1,63,36,221	1,52,13,963	1,11,44,406	1,49,66,617	1,29,46,892
	Paddy	12,21,609	13,45,110	14,68,737	4,36,419	3,21,264	6,56,211
Grand total in rice after converting paddy into rice, at the rate of 25 seers of rice to a maund of paddy		1,46,67,424	1,71,76,915	1,61,31,923	1,14,17,168	1,51,67,407	1,33,57,024

CULTIVATION in Bengal.

Trade.

"The quantity of rice imported into Calcutta during the past year was 11,22,258 maunds, or 6·86 per cent. less than in 1887-88, but 13,10,045 maunds, or 9·42 per cent., in excess of the figures for 1886-87. The boat traffic, which amounted to 75·42 per cent. of the total trade, was 7,35,808 maunds, or 6·02 per cent. less than in the previous year. The traffic carried by rail showed a decrease of 18·59 per cent., namely, 4,07,146 maunds, or 21·31 per cent. on the East Indian line, and 1,18,094 maunds, or 13·07 per cent., on the Eastern Bengal line. The quantity carried by road routes exhibited a decrease of 1,54,871 maunds, or 19·59 per cent. on the trade of 1887-88, but in the supplies imported by inland steamers there was an increase of 1,51,365 maunds, or 144·29 per cent., and in the sea-borne imports of 1,42,296 maunds, or 34·14 per cent."

The Collector of Customs, Calcutta, makes the following remarks on the condition of this trade during the past official year:—

Consuming countries. 296

"Ceylon is the principal consumer of Bengal rice, and the shipments to it were large and above the average, though less than in the previous year. The population of Ceylon must have increased of late years with the extension of tea-growing and the consequent influx of coolies who require Bengal rice.

"The United Kingdom is the second largest consumer of our rice, and the exports to that country were large. The consumption of rice in Europe is now said to have reached the very large quantity of one million tons. The partial failure of the continental potato crop may have done something to stimulate the consumption of rice, but the enormous quantity that has come to be absorbed must indicate its growing popularity as a staple article of food. There were no great fluctuations in prices during the year, and those concerned in the trade—shippers, millers, and speculators—are all said to have done well.

"The heaviest decline was to Mauritius, but the shipments of the previous year had been very large ones. Less was shipped also to South America, the West Indies, and Arabia. To Germany there was a considerable decline, with an advance to France of 46,698 cwt. To the eastern coast of Africa there was a decline, which, however, was more than made up by larger shipments from other ports.

"The exports of paddy declined in the past year in comparison with 1887-88 by 30 per cent. Ceylon is the only country that has taken this grain in quantity, and the shipments to that island advanced by 39 per cent., the decline shown being due to a falling off to 'other countries.'"

The reader should compare the remarks given in this chapter on the Bengal and Calcutta Trade with the allied subjects briefly reviewed below under the sectional headings of—YIELD PER ACRE, TOTAL PRODUCTION, CONSUMPTION PER HEAD OF POPULATION, AND DISPOSAL OF AVAILABLE SURPLUS.

Varieties & Races of Bengal Rice.

Varieties & Races. 297

Some few years ago the writer was directed by the Government of Bengal to take over temporary charge of the Economic Museum, Calcutta, with the view to reorganising and naming its collections. About the same time he was also, by the Government of India, placed in charge of the Central Office of the Calcutta International Exhibition. In these two capacities he had the opportunity of examining a very extensive series of specimens of rice, numbering over 4,000. These were the rices of Bengal alone, and he therefore was induced to affirm that if a similarly exhaustive collection had been made in all the other provinces of India, the total number of named cultivated rices might have been found to be little short of 10,000 It should not, however, be understood from this statement that that very

ORYZA sativa.

Rice Crops.

CULTIVATION in Bengal.

large number were all forms which could be separately recognised by the uninitiated. They came from the various districts of Bengal and were readily distinguished by the local cultivators, with whom they bore distinctive names and were supposed to possess properties that rendered the one suitable for one class of soil, nay even for a particular field in the holding, while quite unsuited for another. An attempt to classify these forms of rice by any European standard would very naturally have rejected many as being identical, but experience has shown that such a reduction on the part of any person, save perhaps a *rayat*, would very probably lead to serious consequences. **Mr. C. B. Clarke**, an experienced and accurate botanist, remarking on the marvellous intuitive knowledge which the hereditary cultivators possess, in recognising the forms of rice, while dealing with the *urí* (wild rice), says: "I do not know how, in the young state, the cultivator tells the *urí* from the *áman*. I cannot." One might be prepared to suppose that there was something in the texture, colour, or even habit of the growing wild rice which enabled the cultivator to recognise it, but it is far more surprising to find him pick up a handful of dry grain and affirm that it would be found suitable to a particular method of cultivation, while he rejects an almost precisely similar grain as unsuitable.

Races or Forms of Rice. 298

Speaking generally, rices may be grouped by their colour, size, or shape, or according as they are awned or awnless. The colour of the husk or enclosing glumes gives, however, no positive indication of the colour or shape of the contained grain, so that a classification as to colour and shape, &c., would have to first take into account the peculiarities of the glumes and then of the grain. But when every effort had been spent, a hundred forms would suffice to break down completely every theory which had been established, so that attention would naturally have to be turned to the seasons of sowing, methods of cultivation, and peculiarities of climate and soil as affording more trustworthy data for classification; the peculiarities of the grain itself might then come to aid in the elaboration of the classification. The following brief account of the chief rice crops of Bengal was drawn up, by the writer, for the Calcutta International Exhibition, and as it affords the main ideas it may be here reproduced:—

"As far as the plains are concerned, rice crops may be referred to two or three primary groups, according to the method of cultivation, season of the year when cultivated, and length of the period required for ripening. These groups receive various names in different districts and provinces, but correspond to each other pretty constantly, being earlier or later, as the result of special peculiarities in climate and soil or season of rainfall. The average condition in Bengal may be expressed as follows:—

Aus. 299

1st.—The Áus or Bhadoi Crop.

"The forms thrown into this group are the early or autumn rices. They are sown from April to May, on comparatively high sandy lands not inundated during the rains. The seed is generally sown broadcast, and the field is carefully kept free from weeds during May and the first half of June. The crop is harvested from July to August, or even not till September.

"The forms of *áus* are the least valuable of all the rices; about one-sixth of Bengal rice belongs to this group.

Aman. 300

2nd.—Áman Crop—Winter Rices.

"This crop comprises the late or cold-season rices. Owing to their ripening on inundated fields, they are sometimes called the floating rices. They are referred to two important sub-groups:—

Chotan aman. 301

"(a) *Chotan áman.*—The early and better sorts of *áman* are of

CULTIVATION in Bengal.

this nature. They are generally sown on seed-beds, and when about nine inches in height they are transplanted into the fields. There are many local kinds of *chotan áman*, of which *ropa* or *rowa* or *roya* and *shal* are most favourably mentioned. The cultivation of the rices of this sub-group extends from May to October. They do not require deep water, and are often transplanted into the same field upon which a crop of broadcast *áus* is already well established, the crop of the latter being reaped from July to August, and the *áman* continuing to mature till October or November. The better classes of *chotan áman* are, however, grown by themselves.

(*b*) *Boran áman*.—These are much coarser forms of *áman*, and grow habitually in deep water. They are sown for the most part broadcast in *bhíls* or low-lying lands: they are only occasionally transplanted. The crop ripens in December or January. **Boran aman. 302**

The weeding required for the *áus* crop is generally sufficient for a combined *áus* and *áman* crop. As the height of the water rises over the inundated fields the *áman* crop is often observed to grow with marvellous rapidity, as much as nine inches having been recorded in 24 hours at the beginning of the rains. When submerged through a sudden flooding for more than three days, the crop is completely destroyed. This is the chief danger to the *áman* rices.

The *áman* is the principal crop of rice in the plains, after the harvest of which the land generally remains undisturbed until the end of February, when preparations for the new *áus* crop commence. Sometimes, however, winter crops of pulses and oil-seeds are taken off the higher *áman* lands.

3rd.—*The Boro Crop.*

The hot season rices come under this heading. They are transplanted from the seed-bed or sown broadcast from December to February and harvested in April to May. The forms of this group yield an abundant crop of very coarse and hard rice, chiefly consumed by the poorer classes. They are quick-growing rices, one kind of which is known as the *shatia* or 60-days' rice, because in that period, from sowing to harvest, it yields its crop. By some writers there is an *áus shátia* rice as well as a *boro*. Only a very limited amount of *boro* rice is cultivated; they are suitable for *churs* or low-lying lands, and may be grown in 10 or more feet of water notwithstanding strong currents. They are, however, of much value to the poor, since the coming of this crop in the hot season tends to lower the then high rates of other classes of rice. A peculiar kind of *boro* rice is known as *raida* or *bhasha-naranga*. This is sown along with the ordinary *boro* rice in December. The young stems are shorn when the *boro* crop is removed, but this does not seem to injure the *raida*. It continues to grow, and yields its crop in September or October, having been thus 10 to 11 months on the field. **Boro. 303**

Five Crops of Rice a Year.—A proprietor of an estate with fairly mixed soil according to this system might have three, if not four or even five, harvests of rice every twelve months, thus:— **304**

(1) *Áus* harvest, from July to August.
(2) *Chotan áman*, from October to November.
(3) *Boran áman*, from December to January.
(4) *Boro*, from April to May.
(5) *Raida*, from September to October.

ORYZA sativa.

Classification of

CULTIVATION in Bengal.

Two harvests are all but universal in Bengal, with an occasional third but smaller one; two crops are frequently taken off the same field. Of these groups of rices, the *áus, boro,* and *raida* cannot be used at religious ceremonies as offerings to the Hindu gods; but these, together with the *boran áman*, are the rices eaten by the million, the finer classes of *áman* being, from their high price, restricted to the rich. A remarkable fact which may be here noted is that the *áus, áman*, or *boro* rices of one district are often so different from those of another, that if interchanged the one will not grow on the fields on which the other has flourished for centuries. Here the European farmer is confronted with a problem scarcely known to his scientific agriculture; but although it is difficult to follow his reasonings, the rice cultivator of India will detect the one from the other with a perfectly marvellous degree of certainty.

Scented rice. 305 *Conf. with pp. 570, 595.*

Scented and Other Special Rices.—Some forms of rice are scented, while the majority have no smell whatever. Scented rices are common, for example, in Orissa, Thana, Behar, &c., and are much prized by certain classes of people. Of the scented rices, *benáphuli, kamini, bans-mati,* and *rándhuni-pagla* (or cook-maddening) are considered the best. The better class Natives eat the long thin white *chotan áman* rices, chiefly cultivated upon higher lands; while the short, thick, more or less reddish rices—the so-called Patna rices—are those eaten by the mass of the people of India. The Muhammadans prefer an absorbent rice, such as that from Pilibhit. In Burma, amongst many high class rices, a grain is grown which, while largely used for industrial purposes, is regarded as unwholesome as an article of food. One of the most curious peculiarities and one recently brought to light regarding rice is that, while the great mass of rices contain only one grain within the husk, two or even three grains are regularly present in certain rices."

Rowa 306

A FURTHER CLASSIFICATION OF CULTIVATED RICES.—**Mr. C. B. Clarke**, in his paper on the *Cultivation of Rice in Bengal* (published in the *Kew Bulletin* of December 1888), divides the rices into three crops: (1) *Rowa*, (2) *A'man*, and (3) *A'us (owsh)*. **Mr. Clarke's** main object appears to have been to impress on his readers the fact that there is actually a *rowa* crop, a fact which, for some reason unknown to the writer, **Mr. Clarke** appears to think, is not generally credited in India. The hard-and-fast separation of *rowa* from *áman* which **Mr. Clarke** insists on can hardly, however, be said to exist. It most certainly does so in Eastern and Central Bengal, the region of which **Mr. Clarke** more particularly wrote, but the transition from the higher lands on which the better class *rowa* is grown, to the deeper submerged tracts on which the *áman* (of which he speaks) is found to luxuriate, is a gradual one and whole districts might be said to belong to the middle region which could neither be classed in **Mr. Clarke's** *rowa* nor in his *áman*, while other districts are, almost (as far as these two crops are concerned), either entirely *rowa* of good quality or entirely *áman* of a very inferior kind. Thus, for example, the *rowa* rice of Dacca is scarcely as good as the *áman* of many other districts such, for example, as Burdwan and Hooghly. **Mr. Sen** very properly remarks (*Report on System of Agriculture of Dacca, p. 28*), that fully one-third of the whole produce of Dacca belongs to the *áus* and *boro* classes of rice, and that even the *áman* paddy, especially the long-stemmed variety, is a coarse and inferior grain. In his note on the rices of Bengal, however, **Mr. Clarke** omits from consideration the *boro* and *raida* crops and gives only the most general statements regarding the forms he specialises. His urgency to have the *rowa* separately established serves, however, a useful purpose, since, in the writer's opinion, the superior class of *áman* rice, grown on high lands, is botanically distinct from

CULTIVATION in Bengal.

the long-stemmed *áman* of the *jhíls*. **Mr. Clarke's** words on this subject may be here quoted:—

"I should warn you that many points in the above description have been controverted; and in particular it has been asserted, not only by Calcutta English newspapers, but by Government officers, that there exists no such rice as I have described as *rowa*." As far as the writer can discover, however, every recent book and report which treats of Bengal rices clearly establishes the existence of the *rowa* crop. What many authors differ in, is the propriety of the *chotan* and *boran áman* crops being viewed as sections of one crop or as entirely separate crops. And there is much to be said in favour of the latter opinion, for they are not only grown under quite different conditions, but, as has already been stated, they are probably derived from different stocks. The seasons of sowing and harvesting, however, are practically identical and in a great many parts of Bengal the *áman* crop consists chiefly of *chotan áman*, while in others it is mainly if not entirely *boran áman*. The term *rowa*, moreover, does not, strictly speaking, denote a form of rice, but is rather the process of transplanting from a seed bed. But transplanting is not confined to *chotan áman* rice; *áus* is sometimes once, and *boro* crops are frequently twice, transplanted. Moreover, broadcastings of what **Mr. Clarke** alludes to as *rowa* are by no means unusual.

Degree of Rainfall and Season suitable for Rice Culture.

Season of Rice Culture. 307

Mr. Clarke's remarks regarding the water requirements of the *rowa* crop are instructive and may be here quoted:—"The *rowa* crop depends very little on the early setting in of the rains (the Government printed reports I am aware imply a contradiction of this statement), but almost entirely on the rain holding on in the autumn. The Bengalí expects rain at least once a week during October, and enough to prevent his rice drying up before quite the end of the month; if he gets this he has a full crop (and I say whether he dibbled early or as late as 15th August to 10th September). If the rain holds on steadily (*i.e*, sufficient in quantity and at sufficiently short intervals to prevent the *rowa* field getting dry) to the beginning of November (as I have known it do nearly half the seasons I have witnessed) the crop is a bumper. If the rain stops by the 10th to 15th October, so that the *rowa* fields then dry up, there may be a three-quarter crop. If the rain stops before the end of September, as it did in Orissa in 1869, the crop (at least all the August-September dibbled) fails. Also, if the rain stops by the middle of October, so that the rice dries up, a late down-pour in November or December (which I have seen twice) does harm rather than good." **Mr. A. P. MacDonnell**, however, expresses in perhaps more precise language the necessities of the rice crops of Bengal for water.

Amount of Rain necessary. 308

He says:—"My instructions suggested the possibility that the varied experience of 1873 might afford some clue to the establishment of a fixed connection between an ascertained deficiency in the rainfall at any one place and a consequent failure of the crops there. I regret I have failed to establish any definite connection of the kind." The facts contained in his work will make it quite clear, he continues, "that the character of a harvest depends, within certain wide limits, much more on the seasonable distribution of the rainfall than on its absolute quantity. Although a well marked deficiency in the rainfall will certainly entail a deficient crop-yield, yet the magnitude of the deficiency will depend on the distribution of the rain which fell. For the *bhadoi* and late rice harvest it would seem that the distribution most favourable to agriculture—the husbandman's ideal year—is when premonitory showers, falling in May or early June, facilitate that spade husbandry which, to secure a really good crop, *must* precede

ORYZA sativa.

Rotation with Other Crops.

CULTIVATION in Bengal.

ploughing operations. The rain in the end of June and in July should be heavy: then should come an interval of comparatively fair weather, in which weeding operations may be successfully prosecuted. The September rains must be heavy, shading off into fine weather with October showers. On the sufficiency of the September rains, more than of any other month, depends the character of the winter rice crop—the chief harvest in these provinces. Finally, periodic showers from December to February (inclusive) are essential to a good *rabi* harvest."

Amount of rain necessary.

In the Famine Commission's Report, much valuable information has been tabulated, as to the degree of rainfall necessary for the rice and other crops of Bengal, and the deficiencies or rather seasons of deficiency that are calculated to produce famine. The following passage may be here given as of special interest in this connection:—"Of the rice crops there are in Bengal three main varieties—(1) the early rice, which is also known as the summer or autumn rice; (2) the main crop or winter rice; and (3) the late or spring rice. The first is sown on the highest, the last on the lowest lands; a year of excessive rainfall is as good for the former as it is bad for the latter.

The question of famine or no famine depends solely on the main crop, which is sown between April and June, and reaped between November and January according to the district concerned—its soil, climate, rainfall, &c.

There are three critical periods in the life of this crop during which rain is required, *viz.*:—

(1) in May and June for sowing, when light rains are required;

(2) in July and August for weeding, transplanting, and sub-soiling, when heavy rain is required; and

(3) in September and October for maturing the growth and filling out the ear, when moderate rain is necessary.

The last of these stages is the most critical, and it is to the failure of the rains in September and October that all famines in Bengal have been due." "Excessive rain has never been known to cause famine in Bengal, even indirectly by causing floods. It damages the rice crop and *rabi* crop only when it falls at two stages of their growth, *viz.*:—(1) the earlier—*i.e.*, before the young plants are high and strong enough to bear it, and (2) the later—when the grain is formed in the ear and the plant is beaten down by the rain. Exceeding good and promising crops have often been damaged by heavy rain, especially at the latter stage of their growth, but never to such an extent as to cause any widespread distress. The *bhadoi* and early rice crops are sown in April and May, and reaped in August and September. They require only a moderate rainfall at regular intervals. The *rabi* crops and late rice are sown in October and November, and reaped in April and May. They are more independent of rainfall than either of the other two classes of crops."

Rotation: Manure. 309

Rotation of Crops and Manure.

In the passages quoted below from local reports (which are intended to convey a general idea of the rice cultivation of Bengal as a whole) incidental allusions will be found to the subject of Rotation of Crops and Manures. The following passage, from the Famine Commission's Report, is believed to be as fully applicable to Bengal now as it was some 16 years ago when first penned:—"The manuring and rotation of crops as practised in Bengal does not affect in the slightest degree the question of famine. These matters vary so widely in different districts and in different parts of the country; the use of manure is so uncommon—save for the most valuable of the non-food crops, such as indigo, sugar-cane, jute, &c.,—and the rotation of crops so seldom practised and when practised is based upon such indefinite principles, that no precise report on the subject can be furnished. Cow-dung, which in England and other countries is given back to the soil as manure, is, in Bengal, used for fuel throughout the whole length aud breadth of the land, except in forest and jungle tracts, on which cattle graze, where it is wasted. The only other available manures are—(1) The products of each household dust-heap, comprising ashes, rice-water, vegetable refuse, &c. The amount of manure thus collected barely suffices for one or two fields near the homestead on which some valuable garden or non-food crop is grown. (2) For sugar-cane and other valuable and garden crops, oil-cake is used. The use of bones and other animal manures is prevented by caste prejudice. As far as the great staple rice-crop of the province is concerned, all that seems necessary to secure a bumper-crop is timely and abundant rain. The use of European artificial manures is of course beyond the means of the people, even if their caste prejudices would allow it. A large portion of the Deltaic tracts is well manured by the silt deposit of the annual floods of the Ganges and other large rivers." In another part of their report the Commissioners allude to the subject of manure in the following sentences:— "The importance of manuring is known much more widely than its use is practised. Every cultivator has his manure-heap on which the sweepings of the house and of the cattle-shed are thrown; but the cattle dung is almost universally collected and dried for use as fuel, except during the rainy months, and the droppings of the cattle that are not stall-fed, but turned out to graze on waste lands, are lost. It is roughly reckoned that each cultivating family with its cattle produces enough manure for an acre of land yearly. Hence, where the population is dense, and as in Bengal and the North-West Provinces each family holds about four acres, the land gets more manure than in sparsely populated tracts like the Central Provinces and the Deccan, where the holdings are three or four times that extent." On the subject of the degree of manuring of rice lands through silt deposits from the rivers, **Mr. C. B. Clarke** has much to say that deserves careful consideration. Before proceeding to give one or two passages from **Mr. Clarke's** paper, however, it may be as well to state that, as in the words of the Famine Commissioners, it is only a "portion" of the Deltaic tract which modern Indian writers maintain is so fertilized. In this respect the Gangetic basin is quite dissimilar from the Nile Valley, for, instead of the rivers regularly overflowing their banks and throwing down extensive deposits, the inundation of the major portion of Bengal to form the annual rice-fields is purely and simply due to rainfall. **Mr. Clarke** described the "anastomosing rivers of Central and Eastern Bengal (as in all alluvial deltas) as usually the most elevated grounds, and on these very commonly the villages stand. As you walk in the cold season from one river to the next you insensibly descend from the river-bank till you come to the 'bheel' which at this season may be shallow, or mud, or wholly dried up; and you insensibly ascend similarly from the *bheel* till you come to the bank of the next river. Now this *bheel* (or *jhíl*) will begin to swell in spring—sometimes as early as April (when the April showers are strong), sometimes not till June if these showers are light, and May, dry. As each band around the *jhíl* gets softened it has to be tilled and planted with rice before it is submerged more than a very few inches. The *bheel* will swell till it covers the whole country except the narrow belt by the rivers occupied by the villages; and the getting in of the *áman*" (the rice which **Mr. Clarke** assigns to this submerged tract) "may thus extend for three months." "In the *rowa* fields, as I describe them, the rice gets only rain-water, and the water drains from the fields into the rivers. **Liebig** says that the reason why rice can be grown in Bengal every year in

CULTIVATION in Bengal.

Rotation: Manure.

River Silt. 310

Liebig's Theory. 311

ORYZA sativa.

Characteristics of the Rice Area.

CULTIVATION in Bengal.

Liebig's Theory.

the same fields without manure is because the rivers annually replenish the soil by a layer of r'ch silt. My critics see the rice round Calcutta town and along the two principal railways in Bengal is all *aman*; and they think that rice could not grow for ever without the aid of the silt. I called their attention to **Roxburgh's** *Flora Indica*, II., p. 202, where he maintains that much of the rice land in Asia receives no help except what the air and rains yield (*i.e.*, as fully explained, neither inundation nor any material quantity of manure); yet that those fields have for probably thousands of years continued to yield annually a large crop of rice, thirty to sixty fold. **Roxburgh** barely ventures to suggest that a recuperation of the soil takes place between January and June. On the West side of the Great Bengal Plain, from Rajmahal and Burdwan to Orissa, the hills rise in many places gradually, and we see a great quantity of terraced rice; it is clear that these fields cannot be inundated from the rivers. As we proceed further into the Western hills we find V-shaped valleys terraced in narrow curved platforms; the rain-water is led down gradually from one to another; there is on the outer edge of each platform a little bank, usually not more than 6 to 12 inches high, and the rice is therefore never more than a few inches deep in water; this terraced rice is a sub-variety of *rowa* in my classification, and it gets nothing but air and rain-water; it is the kind of rice land **Roxburgh** refers to, his account being drawn up in the Circars immediately south of Orissa. Not rarely, in this kind of rice cultivation, a bank is drawn across the upper part of the valley and a tank formed immediately above the rice platforms; from this reserve the rice can be watered after the rain stops and the crop almost ensured. Now in all these cases it is manifest that the rice-fields can get no manuring by silt from the rivers.

Fields drain into the Rivers.
312

"In the North and East of Bengal the hills rise very suddenly from the plains; there is little terraced rice, but there is a broad belt of land in which there is enough clay in the soil for *rowa* that extends from Mymensingh to Chittagong. In the most valuable rice land, as on the right bank of the Brahmaputra, throughout the Zilla of Mymensingh, the water runs from the rice-fields to the Brahmaputra, and the fields are never deep in water. The land in the thoroughly inundated districts, as south-west of Dacca and Faridpore, is too sandy to grow *rowa* rice, and it is much less valuable than the belt of *rowa* land near the base of the hills. **Roxburgh** says in the place above cited that "the best rice lands are those that are overflowed annually by the inundations of large rivers." But he is speaking of the Circars; he probably never saw the fine *rowa* land of Mymensingh, Comilla, and Noakhali. The gross produce of swamp rice may be about as large as that of *rowa*, though raised at a greater cost of labour; but the value of the crop per acre is very much less.

"I am prepared to go further and to doubt whether even swamp rice gets much silt. Where the water from a muddy rapid river gets through its bank (which is the highest part of the country) and spreads out over the lower country beyond, it loses its velocity very fast, and therefore drops all its silt in the "bank," *i.e.*, near the river it has left. If this were not so the *bheels* would soon all fill up, and the "banks" (they may be a quarter of a mile wide) would not be the highest parts of the country. In fact, in swamp rice, the water is perfectly still and black, evidently all or very nearly all rain-water. Where the land is well silted, as in the case of large sandbanks, it is impossible to grow any rice but a little *owsh* (*áus*), and usually not that till the land has been made more tenacious by a few crops of indigo or some other leguminous crop. I have no doubt of the most important fact that a large class of rice-fields in Bengal

CULTIVATION in Bengal.

annually cropped without manure or silting, maintain a perennial fertility; though, like **Roxburgh**, I am not prepared with an explanation. **John Scott**, who was in 1870 Curator of the Calcutta Botanic Gardens, was of opinion that "in these rice-fields the recuperative power of the soil is sufficient, under the sun and rain of Bengal, to go on growing the present crops of rice indefinitely" (*Kew Bulletin*). It will be found below in the remarks regarding some parts of Bengal as, for example, Jessore and the Sundarbans, that the soil is believed to be perennially fertile and produces an abundant harvest year after year, not only without the aid of manure, but in many cases without any cultivation whatsoever. But in these cases it is admitted the annual inundations fertilize the soil so that all that is necessary is to transplant the seedlings from the nursery or simply to plaster the seed in the plastic mud left by the receding high water. In concluding the present paragraph it may be said that the system of manuring rice, and especially the nursery beds in Bombay, should be read in continuation of the above remarks, as also the information furnished regarding the system of manuring which is practised in the Panjáb.

Fertility of Rice Fields. 313 *Conf. with pp. 590-593 and pp. 615-619.*

Yield per Acre; Total Production; Consumption per head of Population; and Disposal of Available Surplus.

Yield and Production. 314

The information collected by **Mr. A. P. MacDonnell** afforded perhaps the chief data on which the Famine Commissioners framed their opinions on the subject of yield per acre and total production of food-grains in Bengal. Important though these subjects are, they do not appear to have been dealt with, by any subsequent writer, in the same detail, and hence though the figures given by the Famine Commissioners are, for 1874 they may be here reproduced:—"Although it is not worth while in the existing defect of knowledge to attempt to make any definite estimate of the probable food produce per acre for Bengal, it may be as well to indicate broadly the limits within which this estimate must lie. The population being 60 millions and the acreage of food-crops about 48⅓ millions, the consumption (at six maunds per head per annum, a slightly larger quantity being assumed because of the less nutritious quality of rice)—

	Maunds.
would be	36,00,00,000
Seed grain at ·50lb an acre	3,00,00,000
Wastage at 5 per cent.	2,50,00,000
Cattle food at half maund an acre	2,70,00,000
TOTAL	44,20,00,000
Add export of rice average say 600,000 tons	1,68,00,000
Add an equal amount for surplus retained to meet deficient years, &c.	1,68,00,000
	47,56,00 000
Or say	17,000,000 tons.

10 mds. per acre.

This quantity if produced by 48⅓ million acres gives an average produce of almost 10 maunds of food-grains per acre. If, as is probable, the area is under-estimated, the average produce must be less. This conclusion, vague as it is, shows that for the whole province it is impossible to assume a higher rate of production than at the outside 10 maunds per acre, although no doubt in the more densely-populated districts the outturn is considerably higher."

The Famine Commissioners in another part of their report return to the question of yield per acre of food-grains and of rice. Their remarks may be still further quoted:—"As regards produce per acre, so various

ORYZA sativa.

Yield per Acre.

CULTIVATION in Bengal.

are the soils, so different is the rainfall, so numerous are the modes of cultivation, the crops and their outturn in the various districts of Bengal, that it is quite impossible to fix any absolute standard figure representing the average outturn of an acre of land under any given food crop. As a rule, the cultivators themselves do not know it with any tolerable degree of accuracy. From such experiments and inquiries as have been made in Bengal from time to time, and after analysing all the statistics available on the subject, **Captain Ottley, R.E.**, came to the conclusion that as regards rice—the great staple crop of Bengal—"the average outturn for a number of years of all classes of land will be about 15 maunds per acre. This may be accepted as the nearest possible approximation to the truth, but not as a safe basis for calculation." **Mr. MacDonnell's** extreme caution in advancing any one figure of yield as applicable to the whole of Bengal cannot be too strongly commended. He wrote of his own calculations :—"No other character, however, than approximations to the truth is claimed for these estimates." "This question of average rates of produce has been one of the most perplexing with which I have had to deal. One maund, more or less, per acre may alter the complexion of a conclusion." It will thus be seen that it is impossible to advance any one figure as being likely to express an average yield for the whole province. The writer has, however, tabulated all the returns to which he has had access and struck the mean with the result that it would appear 10 maunds of *áman* and 8 maunds of *boro* and *áus* would be fair average yields per acre. These figures are, however, so very much below those determined by test harvests in Bombay, the Central Provinces, and Assam that they may be regarded as below rather than above the mark. **Sir W. W. Hunter** (*Indian Empire, p. 383*) gives, without any reservation, the average yield as 15 maunds of clean rice. One of the most instructive series of papers on the subject of Bengal rice will be found in the *Journal of the Agri-Horticultural Society of India* (*Vol. I., New Series, 1867*). The Secretary issued to all district officers throughout the province a series of questions on the yield per *bíghá*, the classes of rice grown, cost of cultivation, &c., &c. The replies are full of interest and should be carefully considered by any one desirous of learning something of this most important but very indifferently investigated crop. In doing so the caution is necessary that the *bíghá* is by no means a constant quantity. It would appear that the average of all the figures there given would justify 27 to 33 maunds of paddy or say 22 maunds of clean rice being the nearest approximation to the yield per acre. But so great are the extremes of the returns, not necessarily due to inaccuracies, but as a consequence of the variability of yield of different classes of rice or the productiveness or otherwise of the soils, that no average could be given for the whole of Bengal. It seems highly probable, however, that a yield of 22 maunds is more than double the actual yield per acre over the total area. If, therefore, we accept 10 maunds for *áman* and 8 maunds for *áus* and *boro* over the average of these crops given by the Director of Land Records and Agriculture, the total production in Bengal would be 39,95,38,240 maunds or, say, 14,269,223 tons. On working through **Mr. A. P. MacDonnell's** *Food-grain-supply* it becomes apparent that ¾ths of a seer is the average daily consumption of rice in Behar, and perhaps ⅔rds of a seer would be correct for Bengal proper, including Orissa. If the total population (say, 66 millions) were admitted to consume these amounts daily without intermission, they would require annually for their sustenance 42,32,92,325 maunds (or 15,117,583 tons). It is thus obvious that either the estimates of yield per acre, adopted in the above calculation, are considerably under the mark, or that the daily consumption recorded is what the people could

15 mds. per acre.

Concentrated Enquiry made in 1867.
315

22 mds. per acre.

Consumption per head.
316

CULTIVATION in Bengal.

Consumption per head.

consume of rice, but do not all necessarily do so every day. This latter conclusion is more than likely the correct explanation, since the rich always do eat large amounts of other food materials and the poor have to eke out subsistence by greatly supplementing their diet with other less expensive articles than rice. The calculations given in the opening sentences of this chapter (as taken from the Famine Commission's Report) were accordingly framed to embrace not rice only but all food-grains and pulses, whereas this review is intended to try and approach to some definite conception of the Bengal actual consumption of rice. In the above estimates of yield and production no provision has been made for seed. If we assume the average amount of seed necessary for an acre of rice land to be 30lb, 42 million acres would require 3,15,00,000 maunds to be annually reserved for that purpose. And we have still further to allow for rice used as cattle food, wastage, exports, and reserve stock. If a figure be worked out for each of these items and the total be deducted from production, there would remain a very much smaller quantity than determined above (from an assumed daily consumption per head) as the amount used up by the people of Bengal. From the remarks already given of the Bengal Trade in rice, it has been shown that in the year 1888-89, the net export of husked and unhusked rice collectively was 1,35,71,910 maunds. The Famine Commissioners allowed 6,88,00,000 maunds of food-grains as representing the wastage at 5 per cent., the amount used up as cattle food and the surplus retained to meet deficient years. The available information regarding rice does not allow of figures being given to represent the share of that amount which should be debited against rice production, but from general convictions obtained while reading up the extensive literature on rice, the writer is disposed to think that 40 million maunds would be a liberal allowance. But the figures here discussed must be accepted more as denoting the lines on which it might be desirable to treat the subject, than as absolute facts. Each and every one of the statements made can be corrected without the general principles enunciated being materially disarranged. We thus see that the following items have to be debited against the Bengal rice production, the balance being the amount available for consumption :—

Rice Production. 317

	Maunds.
Estimated production from 42 million acres	39,95,38,240
Amount for seed 3,15,00,000 maunds . . . . . }	
Exports (net). 1,35,71,910 maunds . . . . . }	8,50,71,910
Loss, Reserves, &c., 4,00,00,000 maunds . . . . }	
Amount of rice which in 1888-89 was available for Local Consumption in Bengal	31,44,66,330

Daily Consumption. 318

Assuming the above calculation to be correct the 66 millions of people in Bengal would each have consumed a little more than ½ of a seer (or say 1lb) of rice daily. It will thus be seen that this final conclusion of daily consumption, derived from the estimated total production less net exports, &c., and worked out to population, agrees very closely with the statement made by **Mr. MacDonnell** that in Behar the daily consumption might be put down at ¾ths of a seer and for the rest of Bengal at ⅔rd seer.

Surplus. 319

Mr. A. P. MacDonnell, in his valuable work on *Food-grains in Behar and Bengal*, deals in an able manner with the question of available surplus (which meets the export trade and pays rent) and also with the reserve stocks, held up to meet the requirements between harvest or carried forward against the possibility of future deficient supply. **Mr. MacDonnell's** work, it will be recollected, was written in connection with the Bengal Famine of 1874, but his opinions on village economy are no less true of normal than of abnormal seasons. He explains his elaborate and district to district

ORYZA sativa.

Available Surplus and

CULTIVATION in Bengal.

Surplus.

investigations in the following introductory remarks:— "I submit that in rural Behar and Bengal a quantity of food-grain, equal to the absolute wants of the people during the intervals between harvests, is always in the district. The people must live, and all experience teaches the lesson that they, or, what for my purpose is the same thing, their *mahajuns*, keep in stock a provision at least sufficient to carry them on from one harvest to another. The inter-harvest periods being short, people will not run the risk of impairing, by exportation, the sufficiency of this provision, unless simultaneous importation, altering possibly the composition of the provision, leaves its absolute magnitude unaltered.

"Now, necessary and usual as is the retention of a portion of each crop for subsistence till the next crop comes in, the disposal of another portion to pay rent is, as things go in Behar and Bengal, as necessary and imperative. Everywhere a very considerable portion of the cultivator's rent is paid from the sale-proceeds of food-grain; and the grain on which the rent is thus financed for is, I submit, over and above the provision for subsistence made from each crop, and referred to in the last paragaph. In the case of those cultivators who lock up in their own store-rooms the provision necessary to carry them over the interval before next harvest, this is obviously so; it is less obvious, but not less true, in the case of the needy classes who are in the *mahajun's* hands. For it is manifest that if a *rayat* of this latter description be only partially indebted, his partial independence, like the greater independence of his well-to-do neighbour, will show itself in the reservation of some provision for the immediate future; the grain, therefore, on which he finances for rent will be over and above that provision. If he be wholly in the *mahajun's* hands, the food-supply necessary for his subsistence, and for rent payment alike, goes to the *mahajun*. But as the latter must, and does, support the *rayat* under pain of losing his principal and interest together, it comes to the same thing in the end, as if the *rayat* had, with the *mahajun's* sanction, retained the provision for subsistence and made over the rent-grain.

"In fine, it is, I submit, a proposition generally true that the minimum food-supply, necessary to support from harvest to harvest a district, which is a surplus-producing district, may be, in considerations dealing with the disposal of such surplus, looked on as a fixed quantity, and, with the wants it supplies, eliminated from the argument.

"With a view to tracing out the manner in which this surplus is disposed of, it next became requisite to determine the quantity of food-grain which in each district is thrown on the market in financing for rent. To the solution of the question it was necessary to know, first, the actual amount of rent liquidated from the sale-proceeds of food grain; secondly, the average prices at which food-grain is sold to realize this amount. On both points local enquiry was necessary, and these local enquiries were, at my request, conducted by the various district officers and the results communicated to me. The results are interesting and valuable, not only as far as the purpose in hand is concerned, but also as suggestive of trains of enquiry which will doubtless be followed up.

"Conjoined with the question of the proportion of the rent liquidated by the sale-proceeds of food-grain was the determination of the gross rental of each district. In those districts which enjoy the advantages of the Road Cess Act this was an easy matter; but in those districts into which the Act has not been, or was being, introduced there was some slight difficulty. I think, however, the difficulty has been surmounted, and that each section, in which the information was of use, contains a close approximation to the aggregate rents collected by zemindars in the particular district. I may note, in passing, that the rental of every district is very

CULTIVATION in Bengal.

Surplus.

disproportionately large compared with the amount which, in the shape of land revenue, such district contributes to the imperial exchequer.

"The determination of the share of rent liquidated by the sale-proceeds of food-grain, the ascertainment of the actual amount of this share in cash, and the knowledge of the average rates at which food-grain is sold, rendered the calculation of the quantity so sold a matter of no difficulty. The point thus reached, or the knowledge thus acquired, enabled me to make a tripartite division of the local food-grain supply into (*a*) the supply necessary for absolute wants, (*b*) the supply necessary to liquidate rent claims, (*c*) the residue, if any. This brought me to the consideration of the questions of district trade and stocks in hand, questions of high importance, but regarding which there is a complete want of precise information. This want will, it is hoped, be supplied when our present system of inter-district trade registration shall have surmounted the difficulties incidental to all new arrangements. At present, however, the information to be gathered from this source is interesting, more for the promise of improvement it holds out, than for its intrinsic worth; more for the light it throws on the nature of the various commodities in which each district trades, than for the precision with which it gauges the magnitude of such trade.

"It was, however, necessary for me to estimate the extent to which food-grain is kept in stock in each district, and this necessarily involved the question of the district's trade in that commodity. Impelled by this necessity on the one hand, and having in ascertained facts but a slender and not often significant basis for argument on the other, I was obliged to have recourse to speculation to supplement the points in which actual experience was defective. For the speculation I venture to make, I claim no further value than that it affords some clue to an opinion on matters confessedly obscure and intricate. I venture to suggest that in a district which produces a surplus, the quantity of grain sold to pay rent charges forms the grain fund as it were, from which exportations are in the first instance made. The terms of the argument, if I be permitted to dignify the speculation by such a name, presuppose a surplus, portion of which is sold to defray rent, and the sequel will show that every district with which I shall deal, except one, produces such a surplus. The grain, therefore, sold to pay rent is over and above the quantity required for consumption, and over and above a quantity which exists in addition and is held in reserve. In proportion to the sufficiency of this quantity held in reserve, the grain which, in liquidation of rent charges, passes into the grain-dealer's hands is in the home market a drug, and in ordinary years continues to be so. There being no market for it at home, it being superfluous as a reserve, it must be exported.

"But seeing that a residue still exists, in addition both to the provision for subsistence and to this grain on which the rent is financed for; seeing that this residue, or a large portion of it, comes sooner or later into the grain-dealer's hands, whether in satisfaction of debts, or in financing for other wants, it is not fanciful to say that it is from the rent-grain exportation is first made, and not from the residue. How can a distinction be drawn? To this I answer, that the certainty (as I shall demonstrate) with which at fixed seasons grain is thrown on the market to meet rent charges, and the fact that at these seasons, immediately following each harvest, prices are cheaper than at other times, induce traders to buy in and export. This fact then indicates a distinction. The grain sold to pay rent is sold at stated times and at cheap prices; the grain sold for other necessities not so pressing is sold at various times and at dearer prices. Traders from outside (who carry on a large business this way),

ORYZA sativa.

Rice Cultivation

CULTIVATION in Bengal.

Surplus.

or home traders who supply foreign demand, will certainly export first the grain which affords the largest margin to cover profit and expenses. Afterwards, if the demand be greater than they can supply from this stock, they will export other grain, which, bought up by them at more unfavourable times and dearer prices, affords less of a profit.

"There are special circumstances in connection with the rice cultivation and export trade of North Behar which support the speculation that the grain sold to pay rents is all exported, but those special circumstances will come more properly under the districts in which they occur. Here I shall sum up by stating that in every district of which I have treated, the grain sold cheapest is the grain sold to pay rents; and that as this will give the largest profit in foreign markets, it is the grain which is probably first exported; *i.e.*, the minimum exportation of ordinary years, if the district reserves permit of its being all exported.

"Then arose the question—Is the residue of the year's surplus the maximum quantity reserved? If not, what is the limit of accumulation of reserves, and how is this limit maintained?

"Suppose that a surplus-producing district this year holds 50,000 tons of grain in reserve, and that next year's harvests are average harvests—Will these 50,000 tons which have not been drawn on this year be added to the reserve the ensuing year furnishes afresh? And, if so, where is this accumulation to stop? These are the most abstruse, as they are among the most important questions of district economy.

"I cannot undertake to summarise the modes of treating them in each case that came up for discussion. Here I shall only say that, if in the case supposed the ensuing year be an average year, there is no doubt the previous year's reserves partly swell the export grain fund; such of them as do not swell the export grain fund are utilized by freer consumption which cheap prices, consequent on the abundant supply, permit; much of them is wasted, grain being a very perishable commodity. In my estimates of production I contemplate a good average year; short years on the one hand, and bumper years on the other, being excluded from the calculation of the average. But short years are of the more frequent recurrence, and those often recurring short seasons are great solvents of surplus accumulations, whose *raison d'être* is the knowledge, begotten of experience, that short years will come."

METHODS OF CULTIVATION, SOWING, REAPING, &c.

Provincial methods.

320 The chief peculiarities of the Rices of Bengal may be discovered from the following quotations from important works which treat of the main divisions of the province.

Eastern and Central Bengal.

I.—Eastern and Central Bengal.

CULTIVATION OF RICE IN THE DACCA DISTRICT.

Dacca.

321 **Mr. A. C. Sen**, in his recent Report on the Agriculture of Dacca (pages 28-33), furnishes some useful information regarding rice. He divides them into three crops, sub-dividing each of these again into two sections. His classification may be thus exhibited:—

I. Aman . . .
- (*a*) Long stemmed—
  - 1. *Rayenda.*
  - 2. *Baoa.*
  - 3. *Khama.*
  - 4. Ordinary form.
- (*b*) Transplanted or *Rowa*—
  - 1. *Shal.*
  - 2. Ordinary *Rowa.*

CULTIVATION: Eastern & Central Bengal.

Dacca.

II. Aus . . . (a) Ordinary—
1. *Bheslan.*
2. *Boaila.*
3. *Shaita.*
4. *Surajmani.*
(b) Lepi—
1. *Shaita.*

III. Boro . . . (a) Ordinary.
(b) Lepi.

Mr. Sen's remarks so fully exemplify the rice culture of a large portion of Eastern Bengal that his account may be quoted freely :—

Long-stemmed paddy.

Long Stemmed. 322

"This variety of paddy is extensively grown in the Dacca district. The low lands, the sides of jheels, and low plains on which from 5 to 15 feet of water accumulates during the rains are selected for this crop. It has a remarkable power of growth, frequently shooting up to the extent of 12 inches in the course of 24 hours as the inundation rises, and in the case of some varieties, such as *rayenda* and *baoa*, attaining a length of from 10 to 20 feet. The greatest dangers to which it can be exposed during the season of inundation are a high and sudden rise of rivers, by which it is overtopped, and a strong current of water, by which whole fields are sometimes uprooted and carried away to long distances.

"*Soil.*—The soil most suited to this variety of paddy is the stiff clay deposited on the bottom and edges of jheels.

"*Mode of Cultivation.*—It has been stated above that the plants sometimes attain a length of even 20 feet. During the harvest, however, the whole plant is not cut off, but only the ears with about one-and-a-half feet of the straw are removed. The lower portion of the straw remains, and is locally known as *nara*. It is too coarse to be used as fodder even by the cultivators, who are far from being over particular as regards the quality of the food given to their cattle. It is sometimes used as fuel, but in most places it is gathered in heaps in the field and set fire to. The land is immediately after ploughed. This is generally done in December. It is then ploughed once or twice again and left exposed to the sun and rain till *Chaitra* (15th March to 15th April) when the large clods, of which the field is probably very full, are patiently broken down by the *intamugar*, and a somewhat rough tilth obtained. Advantage is now taken of a shower of rain to give the field one or two ploughings and harrowings more when the field is ready for sowing. This is generally done at the beginning of *Baisakh* (15th April to 15th May). About 15 seers of paddy is broadcasted over a *bigha* of land.

"In the moist low-lying places where water begins to accumulate early in *Baisakh*, the sowing cannot be so long delayed. The varieties of the long-stemmed paddy grown in such places are the *rayenda* and *baoa*, and these are sown at the end of *Magh* or early in *Falgoon* (15th February to 15th March). The harvest time is the same as that for other varieties of *áman* paddy, namely, *Aghan-Pous* (15th November to 15th January), so that these varieties of paddy remain in the field for ten months in the year.

"The seeds germinate in four or five days, when the field is rolled twice with the ladder. After the plants have attained a height of 4 to 5 inches, the soil is loosened by the rake, which also serves to thin the crop. After this the only operation to be done till the harvest time comes is a weeding. In some places even this weeding is done away with. The crop is ready for the harvest in *Aghan*. The yield per *bigha* varies between 3 to 12 maunds. The average will be about four and-a-half maunds."

Aman and Aus mixed.

Mixed. 323

"In many places in the district it is customary to sow *áman* and *áus* together in the same field. The advantage of this system is that if one of the crops fail, the ryot can rely on the other. In an unusually favourable year, even two full crops may this way be obtained from the same field in a single season, but generally a half crop of paddy, and a 12-anna crop of *áman* are got this way.

"*Tillage.*—The straw of the previous year's crop is collected in heaps, and burnt, and the field is ploughed. If the ground is sufficiently dry, the plough is followed by the ladder, otherwise a ploughing only is given. This is generally done in *Magh* (15th January to 15th February). After an interval of two to ten days the field is cross-ploughed, and the ladder is used twice. After three or four ploughings more have been given, the land becomes ready for sowing, which is generally done in

ORYZA sativa.

Rice Cultivation

CULTIVATION: Eastern & Central Bengal.

Dacca.

Chaitra. Twelve seers of *áus* are mixed with 6 seers of *áman* and sown broadcast over a *bigha* of land. The sowing is preceded by a ploughing, and followed by a ploughing and two rollings with the ladder. The seeds germinate in two to three days when the field is once ploughed, and twice rolled with the ladder. This process is technically termed *ubhani*. After about five or six days, the ladder is again used with the object of crushing the clods, or more properly shaking them a little, so that the seedlings below them may have an opportunity of coming out. This operation is termed *batar*. The interval between the *ubhani* and *batar* is longer when the field is dry, for in such a field the shoots are longer in appearing when the seedlings come above the clods, and the field looks green, the ladder is used once more, and this operation is termed *jaoai*, and it is done with the object of thinning the crops, as well as by somewhat checking the upward growth, of making the plants stronger at the bottom The *jaoai* is followed by a harrowing with the rake. After about six days the field is weeded. A second weeding is sometimes given in about a fortnight. The *áus* is reaped in *Ashar* (15th June to 15th July). During the rains, when the *jhara* (wild rice) paddy makes its appearance, this is pushed below the water with the aid of a piece of bamboo, so as to make it rot there, for otherwise the whole field would be choked up. This process is termed *dagao*.

"*N.B.*—When the same variety of paddy is sown successively more than two years on the same field, it degenerates into a wild paddy, which has the peculiarity of shedding the grain on the slightest touch, and hence it is known as *jhara*. The *jhara* paddy can hardly be distinguished from the wild paddy, growing in the jheels of the Madhupur jungle.

"*Diseases.*—There are several kinds of insects that do considerable injury to the paddy crop. One of them, known as *panari*, eats away the green leaves of the plant. Another black insect attacks the crop when the ears are just forming, and hundreds of fields are sometimes lost this way. The pest generally shows itself when no rain falls for days together in *Kartik*. Storms in the month of *Kartik* also prove injurious to the crop.

"*Harvest.*—The áman is harvested as usual in *Aghan*."

324

Transplanted Áman.

"The transplanted paddy is grown in the district in two different classes of land, namely, in the upper reaches of the valleys of the Madhupur jungle and in the comparatively high and old *dearahs* of the Brahmaputra and its branches. The paddy grown in the Madhupur valleys is a special variety, and is known under the name of *shaldan*; the transplanted paddy of other places goes by the general name of *rowa* or transplanted. The mode of cultivation followed there does not materially differ from that prevalent in West Burdwan; only it is somewhat simpler.

"*Nursery.*—Seedlings are prepared in a nursery, for which a plot of suitable land is selected, either close to the *rayat's* homestead or in a corner of the field to be afterwards transplanted. It is ploughed four or five times in *Baisakh*, and in *Jeith*, as soon as a little rain-water has collected on it, the *lepichanga* is past over it several times, so as to have the ground regularly plastered. While the preparation of the nursery land is going on, the necessary quantity of seed (6 seers for every bigha of land to be transplanted with) is weighed and soaked in an earthen pot for 24 hours. It is then drained and kept in a corner of the house covered with mats, leaves, &c., and weighted. The seed begins to germinate in two to three days, when it is sown broadcast so thickly that the grains somewhat overlap one another. The seed is not covered with the earth. When the seedlings are from a foot to 18 inches high, they are fit for transplantation.

"*Tillage.*—In case of the jungle valleys, the first thing to attend to is to repair the embankments thrown across the valleys for collecting water. The field is prepared by ploughing it in the mud two or three times. The seedlings are then transplanted about half a cubit apart either way, putting in three to four plants in the same place. The transplantation is generally done about the 15th of *Sravan*. The paddy is generally harvested in *Aghan*. It is better to do this as early as possible, for otherwise great damage is sometimes done by boars, monkeys, and other wild animals. In the case of the *dearah* land, about two ploughings are given in the dry field. This is done as soon as the previous crop, generally *khesari*, is off the ground. On such lands two rain crops are sometimes grown in the same field in the same season. As early as possible the field is sown with *jute* or *áus paddy*, generally of the *shaita* variety. The jute or the paddy is harvested early in *Sravan*, and then the field is immediately ploughed two to three times and transplanted with the *áman* paddy. The *áman* under such treatment seldom yields a full crop."

CULTIVATION: Eastern & Central Bengal.

Dacca.

325

Aus Paddy.

"Of the several varieties of *áus* paddy under cultivation here, the *boaila* and *shaita* stand sandy soil best, and the *shaita* has the additional advantage of taking only sixty days (whence its name, *shat* or sixty) to ripen from the time of putting in the seed. This class of paddy is grown (1) on the high grounds of the Madhupur jungle, where sufficient water cannot be collected for the cultivation of the *shal* paddy, and (2) on the comparatively high and sandy *dearah* lands. The *áus* paddy cannot be grown on lands on which more than two feet of water accumulates during the early part of the rains, for it grows to a length of three to three-and-a-half feet only, and does not rise with the rise of water, as is the case with the long-stemmed paddy. The land that grows *áus* paddy is also considered the best for the jute crops, and therefore, with the increase of the cultivation of jute, the area under *áus* paddy is rapidly diminishing. This is somewhat to be regretted, for this variety of paddy not only supplies the *rayat* with a food-grain, but provides his cattle with an excellent fodder.

"*Tillage.* The land on which *áus* paddy grows is light and easily workable. It generally bears two crops in the year—the *áus* paddy or jute during the rains, and one of the pulses or mustard during the cold weather. As soon as the *rubbi* crop is off the field, it is ploughed and harrowed repeatedly to get the land ready for the *áus* paddy as rapidly as possible. The preparation must be hastened, especially on the *chur* land, for here a late crop is sure to be lost by the rise of the rivers. On such lands the cultivator is sometimes obliged to make a green fodder of his *áus* crop, and in some years the rivers rise so quickly that even the fodder is lost. The sowing time, therefore, is different in different localities. In the *churs* of Meghna, it is sometimes sown so early as the end of *Magh*. In the highlands of North Manikgunge, again, the sowing is often delayed till the beginning of *Baisakh*.

"*After-Treatment.*—As soon as the plants have come out, the field is rolled with the ladder. After about a week, when the plants have grown to a height of 5 or 6 inches, the field is harrowed with the rake preliminary to the first weeding. The tines of the harrow will root out a good many weeds and somewhat thin out the crop. The most troublesome and laborious operation required to be done in an *áus* field is the weeding. In doing this the *rayats* mutually help each other, and hired labour is also often resorted to in order not to lose time.

"The harvest time extends from the end of *Ashar* to the beginning of *Bhadra*."

Boro Paddy.

326

"This class of paddy is of far greater importance here than in West Bengal. The places in the Dacca district, where it is most extensively grown, are—(1) the sides of the *jhils* and streams of the Madhupur jungle; (2) the *churs* and edges of the Meghna, and its numerous branches and creeks subjected to strong tides; (3) and in some of the *churs* of the Padma.

"*Soil.*—The soil best suited to the *boro* paddy is a mixture of clay and decayed vegetable matter or humous soil, and this is the nature of the soil on which it is grown within the Madhupur area. On the *dearahs* of the Meghna, which is the greatest *boro* tract in the district, the soil is a fine sandy loam, rich in organc matter, and always kept moist by being flooded at every tide. This class of paddy is generally transplanted, but there are one or two varieties of it that are sown broadcast and known as *lepidhan*.

"As soon as the rains are over, a plot of ground, from which the inundation water has just receded, is chosen for the nursery. The *kalmi* creeper and other aquatic grasses are removed, and the place is worked into a soft mud by treading on it. If a piece of sufficiently soft ground cannot be had, the land chosen for the nursery is ploughed three to five times, and then plastered by passing the *lepichanga* over it two or three times. When the nursery ground is being thus prepared, the seed is put in an earthen pot and soaked for 24 hours: it is then drained and kept under cover till the germination begins. The seed is now sown broadcast on the nursery so thickly that the grains touch one another and even overlap. The seed is not covered, and has therefore to be carefully watched to see that no injury is done to it by birds and other animals. The plant comes out in five or six days. Thirty seers of paddy sown on one-fourth *bigha* of land gives seedlings sufficient for transplanting two *bighas*. If the natural moisture in the nursery proves insufficient, as sometimes happens, the ground should be artificially watered. The seedlings are fit for transplantation as soon as they have grown to a height of about 9 inches, but seedlings intended for the *dearahs* of the Meghna subjected to strong tides are not removed till they have attained a length of from 14 to 18 inches. The usual time for putting in seed on the nursery is the first week of *Kartik*, and that for transplanting is *Pous*. No tillage is

ORYZA sativa.

Rice Cultivation

CULTIVATION: Eastern & Central Bengal.

Dacca.

generally needed for the *boro* land. All that is usually done is to transplant the seedlings on the soft mud left by the inundation water, or that is always to be seen on the edges of tidal rivers. Now and then, however, *boro* paddy is grown on somewhat stiffer ground, and, when this is the case, the land has to be prepared by three to five ploughings. Except on the edges of tidal rivers, the *boro* paddy requires to be artificially watered, and this is done by the canoe-shaped vessel known as the *duni* (in Burdwan as *donga*). At Mirpur, the cultivators water their *boro* paddy at every full or new moon. The fields are all situated on the banks of the Toorag, which here is subjected to tides, but the water does not rise high enough to reach the fields.

"*Injuries, &c.*—The crop is sometimes injured by hail, floods, drought, and wild boars.

"The harvest time is *Baisakh.* The yield per *bigha* is 5 to 12 maunds, and thus, of all varieties of paddy grown in the district, the *boro* gives the greatest outturn. It is also the least expensive to grow, and has the additional advantage of not requiring any working bullocks for its cultivation. The *boro* land therefore fetches the greatest rent. At Mirpur, the rent varies between R2 and R4-8 per *bigha.* On the *dearahs* of the Meghna nursery land is very scarce. The land generally is low, and is not left dry by the inundation water sufficiently early to grow seedlings on it. The soil is also too sandy for a good nursery bed. When, therefore, a suitable plot of nursery land is found in any locality, it is turned into an immense nursery ground by the people of the neighbourhood, and even by men living at a long distance. The *chur* No. 64, or *Bahar chur,* about 2 miles by ¾ths, is, this way, almost wholly occupied by nursery grounds. When I was passing by this *chur* one afternoon, at the end of December 1888, I saw some 150 boats going away from the place laden with seedlings."

Lepi. 327

Broadcasted Boro Paddy or Lepi.

"In some of the islands of the Padma large areas of land are sometimes to be seen almost on the same level with water at low tide, and covered with mud so soft and deep that one standing on it is put in danger of being buried alive. The *rayats* have discovered a method of cultivation to grow paddy on it known as the *lepi.* No ploughing, harrowing, or anything of the kind is needed, nor is it possible on such lands. All that is necessary is simply to sow the seed broadcast, and plaster (*lepa*) the mud over it. This is, however, not an easy task owing to the unstable nature of the ground. The man sowing the paddy has to support himself on a plantain tree, a piece of bamboo, or the like. The land is flooded at every tide, but that does not injure the seed on account of the mud over it being plastered. Before sowing the *lepi* seed receives the same treatment as the seeds of *boro* and *rowa* paddy.

"Is *lepi* a variety of *boro* or a result of cultivation? In some books the *lepi* has been taken for a variety of the *boro* paddy. This, I think, is a mistake, for not only can the variety of *boro,* usually grown this way, be grown by transplantation like other varieties, but a variety of *áus* paddy, known as *shaita,* can also be, and is often, grown as *lepi.* Near Kaliakur, again, when the land intended for the long-stemmed paddy is too wet to sow the seed on it in the usual way, the *lepi* system is often resorted to by the cultivators. The *boro* is sown as *lepi* in *Aghan,* and the *shaita* in *Pous.* They are both, however, reaped at the same time, namely, in *Baisakh.*

"*Yield* per *bigha* of different varieties of paddy—

(1) *Aman*	(a) *shaldan*	3 to 10 maunds.
	(b) ordinary transplanted	3 to 7 „
	(c) long-stemmed paddy	3 to 10 „
(2) *Aus*		4 to 6 „
(3) *Boro*	(a) ordinary	5 to 12 „
	(b) *lepi*	4 to 6 „ "

As still further exemplifying the character and value of the rice crop in Eastern and Central Bengal, the following passages may be given from a few of the Gazetteers. To reprint all the admirable articles that occur in each of the 24 volumes of **Sir W. W. Hunter's** Statistical Account of Bengal, would necessitate the allotment of a special volume of this work to rice alone. The numerous references given above will enable the reader, however, to discover rapidly the more important accounts of Bengal rice, and, by the careful study of these and other such passages, a fairly accurate conception can be obtained of the minor variations, both in the character of the rice grown and the systems of culture pursued. For example, while the *áman* rice is the chief crop of Bengal, in some districts, it might be said that *áus* was of co-equal importance, just as

CULTIVATION: Eastern & Central Bengal.

in the northern districts and in Behar, *áus* rice is practically unknown. In a like manner transplanted *áman* (*rowa* rice) is plentiful in some districts, unknown in others. *Boro*, in still another tract, becomes the chief crop, while, in others, it is never cultivated, though, in some cases, land suitable for it may be plentiful. The passages here quoted, as exemplifying the character of Bengal rice, are believed to fully demonstrate this variability, while they make known the more interesting facts of methods of culture, yield, cost, &c., &c.

Cultivation of Rice in the Farídpur District.

Farídpur. 328

"Rice forms the staple product of Farídpur; namely, *áman*, or winter rice; *áus*, or autumn rice; *boro*, or rice grown in deep water; and *ráidá* rice. Of these four, the first two are more generally cultivated, and form the chief staple food of the District, but the latter are almost always consumed by the peasant who raises them. The *áus* and *áman* rices are sown broadcast, the seeds being generally intermixed on lands neither very high, nor yet too low. Where the *áus* rice is separately cultivated, however, it is generally planted in pretty high and dry ground in rotation to a crop of sugar-cane. The *áman* rice grows luxuriantly in rather low grounds where the rain-water collects. Some of the best qualities of *áman* rice are transplanted from nurseries into carefully prepared land, which has received repeated ploughings early in the rains, until the whole field is worked into knee-deep mud. All the superior kinds of rice are derived from the *roá* or transplanted crop. The husbandman generally keeps this rice for sale, and uses the coarser varieties for his own consumption. In lands where the *áus* and *áman* are sown intermixed, the former can easily be reaped if the water rises slowly, but in the event of a sudden or rapid rise, would be destroyed. The *áman* rice is of two races, the *baran* and *chhotná*, each comprising several forms. Of these races, the former is regarded by the Hindus as sacred, but the latter is not. A Sanskrit verse (*sloka*) is quoted from the *Sástras*, containing a precept to avoid the use of *chhotná* rice, on account of its ripening before the setting in of the cold weather. As already mentioned, *áman* and *áus* rice are frequently sown together in the same fields during the early rains in March and April, but never later than the 10th of May, in places that are annually flooded. They grow rapidly with the rise of the water on the inundated lands, the stem sometimes reaching fifteen feet or upwards in length, according to the depth of water in places where it grows. The *áus* crop being reaped first in June, July or August, the pruning which the *áman* thereby necessarily undergoes, instead of doing any injury to the crop, rather improves it, as the shoots become more numerous and stronger after this cropping. It should be stated here that, although the fields are very carefully weeded both before and during the rains, it is impossible to rid them of a species of wild grass which ripens almost at the same time with the *áus*; hence the *áus* paddy is seldom free from the seeds of this grass. Indeed, this is so generally the case, that the presence of these wild seeds is regarded as a sure indication of *áus* rice. The *áman* rice is generally reaped in November or December, but there is an early species, *áswini*, which is harvested in the Hindu month of that name, corresponding to the English September. In the same way, one early variety of *áus* ripens and is cut at the end of May. The other two kinds of rice, *boro* and *roá* (transplanted *áman*) are cultivated altogether on a different system. The *boro* is planted in the low beds of marshes and swamps or on the borders of shallow receding rivers, such as the lowest parts of *chars*, *koals*, and *lep chars*, as they are locally called. The seed is sown in nurseries in October or November, transplanted in January or February, and reaped in May or June. The *roá*, or transplanted *áman*, is grown on comparatively high lands, which are seldom or never submerged during the rains. The plants are raised in nurseries in May or June, transplanted in June or July, and reaped in November or December. The finer varieties of rice are obtained from the *roá* crop, and the next quality from the ordinary *áman*. Superior rice is seldom obtained from the *boro* crop, and never from the *áus*. Besides the defect of the *áus* rice already pointed out, namely, its admixture with grass seeds, there are others which render it a very inferior and undesirable article of food. The grain is coarse, never wholly free from a layer of reddish or brownish colouring matter when husked; has a tendency to clot together when boiling; and has scarcely any taste. Ordinary *áman*, although generally also coarse grained, is free from these objections, and more agreeable to the palate. *Boro* rice is generally coarse and heavy, and less sweet than *áman*, but superior in every respect to *áus*. All the rice sown in the high lands and in shallow water, when ripe, is cut close to the ground, so as to leave as little stubble as possible, and to save all the straw for the cattle. In deep water, however, only the ears are cut off; the

ORYZA sativa.

Rice Cultivation

CULTIVATION: Eastern & Central Bengal.

stems remain in the fields till the water subsides, when it is either burnt for manure or collected for household fuel, or for thatching purposes." (Vol. V, 296-305.)

Jessore.

329 CULTIVATION OF RICE IN THE JESSORE DISTRICT.

Rice cultivation may be said to be referable to two widely different sections, the one corresponding to the upp and more cultivated tracts, and the other, to the lower swamps of the S derbans. The former may be taken as representative of the fertile cis Gangetic region which extends from Calcutta, through Jessore, Nudd a, Bírbhúm, and Murshidábád, to the Rájmahál Hills, while the latter is aracteristic of the lower Gangetic delta—a recently reclaimed, and in some respects temporarily cultivated, tract, which produces,almost without the aid of the plough, an abundant crop of exceptionally long-stemmed rice. Two extremely interesting accounts of the Jessore rice cultivation occur in Mr. J. Westland's *Report on the District of Jessore*; the one, by Mr. W. G. Deare, gives particulars mainly of the upper section of the district, and the other, by Mr. Westland himself, deals with the great problem of the utilisation of the Sunderband swamps. The latter is unfortunately too exhaustive to be quoted in its entirety in this work, though it may be remarked that it would richly repay perusal.

The following are the main points brought out by Mr. Deare in his account of rice cultivation in Jessore proper (*App. Rep. List. of Jessore by J. Westland*):—

330 *For Áman Paddy.*

"Operations begin about the 20th February, sometimes the beginning of March if there has been much rain in February, by firing the stubble of the previous year's cultivation. This process goes on during all March, and, as a field is cleared, the ashes are ploughed into the soil. The stubble is fired after a hot dry day about 4 o'clock in the afternoon, and it is a remarkable sight, as the evening closes, to see vast *bheels* on fire. The smoke from these *bheels* sometimes becomes disagreeable, as dense clouds of it are borne along by the evening breeze, enveloping the landscape as it were in a thick fog. After the spring showers fairly set in, the firing process is discontinued and ploughing is pushed on rapidly.

331 *For Aus.*

The preparation of the land begins about the 20th December or first week in January, as the higher lands on which this description of paddy is grown becomes sufficiently dry to admit of the plough being used. But as the winter crops are also raised on such lands, ploughing is delayed till March; by the 20th March ploughs are in full work.

332 *For Boro and Ráidá.*

No ploughing is necessary. The paddy is sown in *bheels* and swamps after the inundations subside, about the middle of December. Preparations begin by removing the rank weeds, called *kalmi,* that have grown so luxuriantly during the rains, and subsided in thick layers over the ground as the water receded. The paddy is then sown in about 10 inches to a foot of water, being transplanted from nurseries.

Boro mixed with *ráidá* in proportion of nine of the former to one of the latter, is sown on a plot of soft earth immediately after the inundation has receded, about the close of October. This plot may be considered the nursery in which the paddy grows till fit to be transplanted to *bhíls,* from which the deposit of *kalmi* weeds has been removed as described above.

Sowing and Planting.—Planting out is unknown in this part of the country, except in the cultivation of *boro* paddy. The paddy is sown in a nursery about the close of October; by the middle of December the water in the *bheels* is sufficiently low to enable the cultivator to clear away the accumulation of weeds. The seedlings are then transplanted to the *bheels,* and sown in about 6 to 10 inches of water and allowed to grow for about a fortnight or three weeks, when they are again transplanted to deeper water and allowed to mature.

Harvest.—The reaping of *áman* paddy begins about the first week in December, and is continued throughout the month to the 15th of January. The *áus* harvest begins about the 15th July, and is continued to the close of August. The *deega* crop is cut about the first week in December. *Boro* paddy is cut in April, and

CULTIVATION: Eastern & Central Bengal. Jessore.

with it the tops of the *ráidá* are taken off. The *ráidá* matures about the middle of October, when it is cut.

"*Yield.*—During the past cold season, I personally cut, weighed, and tested the yield of paddy in various parts of the sub-division of Naral and on the best lands, that is, lands that had suffered least from the effects of the cyclone, I found that a *bigah* yielded as much as 12 maunds. From inquiries I have made in various places I find the average amount of raw produce from a *bigah* of *áman* paddy may be fairly set down at 13 maunds, value R11. Of course the value fluctuates with reference to the supply and prevailing market rates. But the above, so far as I can ascetain—and I have made a careful inquiry,—would be what a *rayat* would expect as a fair average return. The average yield of *áus* would be 8½ maunds per *bigah*, value R6. *Boro* and *ráidá* may be classed with *áman* paddy, but the value of *boro* appears to be greater, averaging about a rupee a maund higher. This may be accounted for. During the *áman* harvest paddy is so abundant generally that prices invariably fall, whereas *boro* is cut while paddy is at its highest value. There is also more labour expended on the *boro* crop.

"*Soils suitable for Rice.*—There are two descriptions of land in this part of the district, designated by the rayat *boro jami* (land) and *ashari jami*. The lowest portions of *bheels*, in short swamps, are placed under the head *boro jami*. No ploughing is ever needed, and the rent paid for such land is R1-10 to 1-12 per *bigah*, while for the higher dry lands classed under *ashari jami*, on which are sown the *áman*, *áus*, and winter crops, the rent paid is R1-4 to 1-8 per *bigah*. It appears a higher rent is paid for *boro jami* land in consideration of less labour being expended in preparation and the saving in cost of ploughing.

"*Method of Cultivation.*—The mode of cultivation is certainly primitive. The implements of agriculture are of the most defective and imperfect form; hoeing is entirely dispensed with as too laborious an operation. The soil is scratched with a rudely constructed plough, the handle of which communicates but little power of directing it, and the share scarcely penetrates the ground to a depth of 3 inches. The business of the harrow is performed by an instrument like a ladder on which the husbandman stands. Bullocks are used for ploughs and harrows, but they are small meagre specimens. In the soft alluvial soil of Lower Bengal instruments and oxen, such as they are, seem to answer all purposes of agriculture. Manure is never used. The soil seems to be so very fertile that it will bear crop after crop sown without intermission. It must, however, be taken into account that the annual inundation has doubtless a fertilizing influence. The weeds that grow and decay on the land ploughed into the soil, together with the ashes of burnt stubble, tend to enrich the ground. Irrigation is not necessary, and is seldom employed. In the cultivation of the *boro* crop it is sometimes resorted to, when a drought prevails in March and April. The water evaporates rapidly under a vertical sun, and it becomes necessary to supply the loss by conducting water along narrow cuts leading from the deeper parts of the *bheel*. The tabular statement will show the number of times the land is ploughed for each crop.

"*Cost of Cultivation.*—It is difficult to form a correct estimate of the cost of cultivation on account of the various customs that prevail. The *gátá*, or mutual help system, is observed in some villages. Five or ten rayats, each the owner of a plough and a pair of bullocks, form a *gátá* or club to help one another in ploughing their lands; no expense for ploughing is necessary beyond the first cost of instruments and bullocks. But in estimating the expenses and profits of a rayat, we must take the cultivation of a whole year. Take, for instance, a *bigah* of ordinary *ashari jami* land on which the rayat has raised an *áman* and a winter crop, the following table will show the cost and profit:—

COST OF CULTIVATION.	R	a.	p.	VALUE OF CROP.	R	a.	p.
Ploughing . .	2	0	0	Paddy . .	11	0	0
Weeding . .	1	0	0	Mustard . .	6	0	0
Seed . .	1	8	0	Peas . .	3	0	0
Reaping . .	2	0	0				
Watching . .	1	0	0	TOTAL	20	0	0
Litigation . .	0	8	0	Deduct cost	9	8	0
Rent . .	1	8	0				
				Profit	10	8	0
TOTAL .	9	8	0				

ORYZA sativa.

Rice Cultivation

CULTIVATION: Eastern & Central Bengal. Sundarbans.

333

The following are the chief facts brought out by **Mr. Westland** regarding—

Cultivation of, and Trade in, Rice in the Sundarbans.

"The second great trade connected with Jessore is the rice trade, and the subjects connected with it may be shortly stated thus. The south of the district, and especially the Sundarbans, form a great rice-producing tract. From the Jessore Sundarbans and from the Backerganj Sundarbans through those of Jessore, there is a continual flow of rice to the westward towards Calcutta. Rice also goes northward, spreading itself all over the sub-divisions of the sudder and of Jhenida, which do not, as a rule, produce sufficient rice for local consumption.

"The fitness of the Sundarbans to serve as a grand rice-supplying tract was pointed out by **Mr. Henckell** so long ago as 1784 and 1785.

"*Clearing Forest.*—The clearing of Sundarban forest is a most arduous undertaking. The trees intertwine with each other to such an extent, that each supports and upholds the other, and some of the trees are of an immense size, one sort, the *jín* tree, of which a good specimen is seen at Morrellganj, spreading and sending down new stems till it covers, perhaps, an acre of ground. Trees like these cannot be cut down and removed in bulk, they must be taken piece-meal, and the tree must be cut up into little pieces before an attempt is made to cut it down. But the trees are not the only difficulty, for there is a low and almost impenetrable brushwood which covers the whole surface. This brushwood has simply to be hacked away bit by bit by any one who attempts to penetrate into the forest.

"*Dangers of Sundarban Cultivation.*—And there is no small danger from wild beasts while all this is going on. Alligators one is not likely to come across, except on the immediate banks of rivers; but tigers are not unfrequent, and occasionally break out upon the defenceless forest-clearers if the latter approach their lair too closely. A great number of these accidents one never hears anything about, but the occasions on which one does hear of such depredations through their occurring near inhabited places, are very frequent.

"Sometimes a tiger takes possession of a tract of land and commits such fearful havoc that he is left in peace in his domain. I am not writing of things which may occur, but of things which have occurred. The depredations of some unusually fierce tiger, or of more than one such tiger, have often caused the retirement of some advanced colony of clearers, who have, through their fear, been compelled to abandon land which only the labour of years has reclaimed from jungle.

"Suppose, however, that the Sundarban cultivator has got over these difficulties, and the equally formidable, though less prominent, difficulties entailed by a residence far from the haunts of men, his dangers have not yet passed. Unless the greatest care be taken of the land so cleared, it will spring back into jungle and become as bad as ever. So great is the evil fertility of the soil, that reclaimed land neglected for a single year will present to the next year's cultivator a forest of reed (*nal*). He may cut it and burn it down, but it will spring up again as thick as ever; and it takes about three eradications to expel this reed when once it has grown. The soil, too, must be cultivated for ten or twelve years before it loses this tendency to at once cover itself with reed jungle.

"*Reclaiming Rayats.*—The first and heaviest part of the clearing of any plot of land is usually done at the expense of the proprietor, the person who has settled with Government for the land; and when the clearing has proceeded to a certain point, he settles rayats upon the lands thus partially cleared and *they* bring it into cultivation. These rayats call themselves "*abád-kari*" or reclaiming rayats, and esteem themselves to have a sort of right of occupancy in their lands. When these rayats thus begin, they occasionally themselves extend their lands by additional clearings, but it may, I believe, be stated, as a general rule, that the greater part of the actual clearing work is done at the expense of the capitalist, and not of the rayats.

"*Migratory Cultivators.*—When a sufficient number of people are gathered together, they tend of course to form a settlement, and to remain permanently where they are. But the furthest advanced parts of the cultivation and some also of those which are not new or remote from old lands, are carried on upon a different principle. A large number of rayats, who live and cultivate lands north of the Sundarbans—that is, near the line of rivers which crosses the district from Kochua, through Baghahat and Khulna, to the Kabadak—have also lands in the Sundarbans, held under different landholders.

"*Seasons of Cultivation.*—The cultivating seasons in the Sundarbans are later than those farther north, and the plan which is followed by these double cultivators

CULTIVATION: Eastern & Central Bengal.

Sundarbans.

is this. The months of *Cheit, Bysack,* and *Jeth,* are spent in cultivation at home. The rayat then having prepared his home cultivation embarks his ploughs, and his oxen, and his food, in a boat and takes them away bodily to his "*abád,*" or Sundarban cultivation. *Assar, Sraban,* and *Bhadro,* are spent in ploughing and sowing and preparing the crops there, the rayat building for himself, with materials he has partly brought with him, a little shed under which he lives. The water gets high in *Sraban* and *Bhadro,* but that is little impediment to cultivation. Many of the lands under rice cultivation are below high-water mark, but the planting is easy, for rice sown on higher lands is transplanted into these low lands when it is strong enough to bear the water.

"The rayat now again comes home, and these outposts of cultivation are absolutely abandoned—large extents of cultivated rice fields and not a symptom of human habitation. By the end of *Agrayan,* the rayat has cut and stored his home-cultivated rice, and he then goes to the Sundarbans, re-erects his hut which has probably been destroyed during his absence (or lives in the open) and reaps his Sundarban rice. At that season of the year (*Pous* and *Magh*) reapers or "*dawals*" crowd to the district and they are extensively employed all over the rice fields of the Sundarbans. When the rice is cut and prepared for sale, the *byapáris* are sure to come round and buy it up, and the zemindar will also send his agents round to collect the rents from the rayats. The rayat has sold his grain, and paid his rent, and the rest of the money he can bring back with him to his home.

"*Settlements of Cultivators.*—While a great many cultivators in the more remote parts of the Sundarbans follow this method, there are in the nearer parts large settlements of rayats who dwell permanently near the land they have under cultivation. But it must be remembered that these tracts are after all sparsely inhabited, and that many of the rayats who dwell in them, besides having a holding near their own houses, have another eight or ten miles away, which they visit only occasionally when they have work to do. The great fertility of the land renders it easy for rayats to hold large areas under cultivation, and thus, what with resident large cultivating rayats and non-resident rayats, we do not find in the Sundarban tracts a population at all equal to what the amount of cultivation would lead us to expect.

"*Absence of Villages.*—There is another thing to be noticed with reference to the dwellers in these regions, namely, that they do not tend, as in other places, to group themselves into villages. Probably this is one result of their having holdings so large that it is most convenient to live near them. But, whatever the cause, many of the village names on the maps represent no sites of villages as we usually understand a village, but represent great seas of waving paddy with homesteads dotted over them, where families live apparently in perfect seclusion. This description, however, hardly applies to older settled tracts, such as pergunnah Hogla.

"*Embanking.*—I have neglected to note another feature in the reclamation and cultivation of these Sundarban lands, namely, the embanking of water inlets. It is a characteristic of deltaic formation that the banks of the rivers are higher than the lands further removed from them, and the whole of the Sundarbans may be looked on as an aggregation of basins, where the height of the sides prevents the water coming in to overflow the interior. Many of these basins are so formed, that, left to themselves, they would remain under flood, as they communicate with the surrounding channels by *khals* which penetrate the bank; and a great part of reclamation work consists in keeping out the water, and thus bringing under cultivation the marsh land inside.

"*Extent of Rice Fields.*—It is difficult to give an idea of the wealth of rice fields that one sees in passing during harvest time along the rivers which intersect the Sundarban reclamations. In other parts of the country, one's view is always restricted by trees or by villages, but in these Sundarbans it is different. You look over one vast plain, stretching for miles upon each side, laden with golden grain; a homestead is dotted about here and there, and the course of the rivers is traced by the fringes of low brushwood that grow upon their banks; but, with these exceptions, one sees in many places one unbroken sea of waving *dhán,* up to the point where the distant forest bounds the horizon. Of course this is not always the view; one cannot reclaim a whole estate in one day. In places where reclamation has only more recently begun, a fringe of half a mile broad on either side of the river contains all that has as yet been done by the extending colony.

"*Injury done by Cyclones.*—These colonies sometimes suffer most severely from cyclones. Their houses and their fields are only a foot or two above high-water mark, and when the cyclone wave pours up the great streams of the Passar and the Haringhatta, and from them spreads all over the country, the inundation works cruel havoc among these low-lying isolated villages. The grain in their fields is spoiled; their houses are torn away, and all their stores are lost; their bullocks

ORYZA sativa.

Rice Cultivation

CULTIVATION: Eastern & Central Bengal.

Sundarbans.

are carried away, and many of them drowned; and they themselves reduced to the extremest shifts to save their own lives. The cyclone of 16th May 1869 destroyed 250 lives near Morrellganj alone, and the loss it caused to property was something immense. One almost wonders how, in some of those storms, the whole country is not at once swept bare, for there is no shelter from the storm and little obstruction to the swelling waters. Liability to cyclones must put a practical limit to the extension of cultivation: for the nearer one gets to the sea, the greater the danger; and the more the forest is cleared away, the smaller the barrier placed between the cultivator and the devouring wave.

"*Harvest.*—In the Sundarbans the rice crop is reaped about the first fortnight of January, the soil easily retaining up till that time all the moisture necessary for the growth of the grain. The method of reaping, too, is different from that which prevails over the rest of the district, for, as the straw is of absolutely no value in the Sundarbans, the crop is reaped by only cutting off the heads, and the straw is subsequently burnt down.

"*Sale of Grain.*—I have now to show how the grain finds its way to market, and here I have first to observe that rayats cultivate in two ways—either under advances from the merchant, or without such advances. Many rayats in the Sundarbans are well enough off to cultivate with their own capital, but several also receive advances from merchants, who for this purpose send their men all over the country about *Bhadro* (August-September), and then again send their people after harvest to collect in ships the grain which has been thus pledged to them. Zemindars also make advances in some cases, but the zemindars of these lands, that is, the large zemindars, are mostly absentees, and receive back their advances in money, so that the matter does not influence the distribution of trade. The small *taluqdars* are different, and usually take a close interest in their rayats.

"A great quantity of rice, however, is cultivated without any sort of advances, and the rayats dispose of it themselves, either taking it to the *hât* themselves, or delivering it on the spot to a trader, or *byapâri*, who comes to purchase it. The latter method is probably the more frequent one in the case of very remote clearings, but in those which are situated within reach of the *hât*, the rayat takes his grain to sell it there. There is a line of *hâts* situated in the north of the Sundarbans to which grain in this way is brought—Chandkhali, Paikgachha, Surkhali, Gauramba, Rampal (or Parikhali), and Morrellganj; and from long distances the grain is brought up by rayats in their boats to these *hâts*.

"Of these *hâts*, the chief is Chandkhali, and Monday is the *hât* day, convenience of trade causing that only one day in each week, instead of two, should be set aside as *hât* day. If one were to see Chandkhali on an ordinary day, one would see a few sleepy huts on the river-bank and pass it by as some insignificant village. The huts are many of them shops, and they are situated round a square, but there are no purchasers to be seen, and the square is deserted. On Sunday, however, ships come from all directions, but chiefly from Calcutta, and anchor along the banks of the river and of the *khal*, waiting for the *hât*. On Monday boats pour in from all directions laden with grain, and others come with more purchasers. People who trade in eatables bring their tobacco and turmeric, to meet the demands of the thousand rayats who have brought their grain to market and will take away with them a week's stores. The river—a large enough one—and the *khal*, become alive with native crafts and boats, pushing in among each other and literally covering the face of the water. Sales are going on rapidly amid all the hubbub, and the *byapâri* and *mahajans* (traders and merchants) are filling their ships with the grain which the rayats have brought alongside and sold to them.

"*System of Traffic.*—The greater part of the traffic thus goes on on the water, but on land, too, it is a busy sight. On water or on land, there is probably a representative from nearly every house for miles around; they have come to sell their grain and to buy their stores; numberless hawkers have come to offer these stores for sale; oil, turmeric, tobacco, vegetables, and all the other luxuries of a rayat's life.

"By the evening the business is all done, the rayats turn their boats homewards; the hawkers go off to the next *hât*, or go to procure more supplies; and with the first favourable tide, the ships weigh anchor and take their cargoes away to Calcutta, and to a smaller extent up the river. By Tuesday morning the place is deserted for another week.

"At this Chandkhali *hât* alone 3,000 or 4,000 rupees worth of rice on an average changes hands every *hât* day, and during the busiest season the amount probably reaches twice that quantity; and about 1,500 boats are brought up by people attending the *hât*, boats being almost the only means of travelling here. And the rice alone does not measure the amount of trade at this *hât*; for, as we shall afterwards see, the traffic in firewood equals the rice trade in value, and much surpasses it in bulk.

O. 333

CULTIVATION: Eastern & Central Bengal. Sundarbans.

"Chandkhali is after all only one out of many *háts*, and besides the trade that is done in the *háts*, there is an immense traffic carried on, less conspicuously, by traders stationed all over the Sundarbans. Some of these have large ships, and with them visit the clearings and fill their ships close to where the grain grows. Others stationed at some village, buy up grain when they can get it, and ship it of themselves or sell it to larger traders. And everywhere there will be found a class of traders called "*fariass*," who insert themselves between the more petty sellers and the regular trader or *byapári*, buying up in very small quantities, and when a certain bulk has been accumulated, waiting for the *byapári* to come to buy, or taking the grain to him to sell it.

"In these ways, then, the rice passes from the hand of the cultivator into that of the trader (*byapári*) or the merchant (*mahajan*). The trader is a man who has a capital, perhaps, of R300 or R400, he sometimes exports his purchased rice himself, taking it to the merchant in Calcutta or elsewhere, who will buy it, and so give him money to use for a second similar transaction; or he will sell it on the spot to the larger exporting merchants, men who have large firms in Calcutta, and have agencies in the producing districts.

"*Export Routes.*—The principal export from the Sundarbans is to Calcutta, and there is a general westward motion of the grain through them, the produce of the Bákarganj Sundarbans passing through the Jessore rivers. The routes adopted for this traffic are nearly the same that they were a hundred years ago."

Cultivation of Rice in the Bákarganj District.

Bakarganj. 334

"Rice is the only cereal grown to any extent in the District, and is divided into three crops, the *áman*, or winter rice; *áus*, or autumn crop; and *boro*, or spring rice. The *áman* yields the finest grain, and is the staple crop of the District. It is sown on the setting in of the rains in April or May, transplanted from the beginning of June to the middle of August, and reaped in November and December. This crop requires to be carefully protected, and in a low-lying district like Bákarganj, covered with a complicated network of rivers and watercourses, its cultivation is attended with some risk, as the crop will not grow unless the ears of corn can keep well above the water. *Aman* rice may be divided into two sorts, *viz.*, coarse (*motá*) and fine (*chikan*).

The *áus* crop is sown in spring and the early part of the hot weather, and reaped in August. In many parts of the District it is transplanted like the *áman* rice, but in the northern portion it is simply sown broadcast. The third rice crop, the *boro*, although not equal in importance to the *áman* or *áus* crops, is cultivated, to a considerable extent, on the alluvial river accretions, and on other low-lying grounds. It is generally sown broadcast in December, and is reaped in April, or May, but is sometimes transplanted. It yields an abundant crop of a very coarse and hard rice, chiefly consumed by the poorer classes, who value it because it comes in at a season of the year when no other rice is ready. It is a quick-growing grain, and one variety of it gets the name of *shátiá*, the period from seed-time to harvest being only sixty days." (*Vol. V., 202.*)

Cultivation of Rice in the Nadiya District.

Nadiya. 335

In Nadiyá the staple crop of the District is rice, which is divided into the following four varieties, namely:—(1) *Aus*, or autumn rice, sown in May (*Baisákh*) and reaped in August and September (*Bhádra*); (2) *Aman*, or winter rice, planted in the months of June or July (*Ashár*), and harvested in November (*Agraháyan*); (3) *Boro*, or spring rice, planted in January or February (*Mágh*), and reaped in March or April (*Chaitra*); (4) *Jali*, sown in April or May (*Baisákh*), and reaped in October or November (*Kártik*). *Aus* rice is sown on comparatively high land, after it has been ploughed up and moistened by the early showers of rain in the end of May, and is not transplanted. *Aman* is sown in low, moist land, and transplanted a month later in low lands, which are then covered with shallow water. *Boro* rice, after being sown, is also transplanted to low, marshy land (*Statistical Account of Bengal, Nadiya, by W. W. Hunter, page 64*).

Cultivation of Rice in the Maldah District.

aldah. 336

The staple crop here, as elsewhere in Bengal, is rice, of which the following are the four chief varieties:—(1) *Boro*, sown in November and December, and reaped in April or May. It is grown on low-lying and marshy lands, and requires to be transplanted two or three times before coming to perfection. The grain is coarse and chiefly used by the cultivators themselves.

(2) *Bhadai*, sown in April and May, and reaped in August and September. It

ORYZA sativa.

Rice Cultivation

CULTIVATION: Eastern & Central Bengal.

is sown broadcast on high lands and on the banks of rivers. It requires no irrigation, nor is it transplanted, but it must be weeded when about five inches high. This crop which is identical with the *áus* of Eastern Bengal, is largely grown in all parts of the District. The grain is coarse, and consumed by the poorer classes.

Maldah.

(3) *Aman*, sown in June and July, and reaped in November and December. It is sown in low-lying lands which go under water during the rains, and does not require transplanting. It is extensively cultivated throughout the district, and, together with the *haimantik*, forms the main harvest on which depends the food-supply of the year. It may probably be identified not so much with the *áman* as with the *Kártik sáil* of Eastern Bengal.

(4) *Haimantik.*—This crop requires transplanting, and more resembles the ordinary *áman*. It is sown in June or July, transplanted in July or August, and harvested in November and December." (*VII., 70-72.*)

Rangpur. 337

Cultivation of Rice in the Rangpur District.

"Rice forms the staple crop of the Rangpur District. Two principal crops are sown and reaped during the year, namely, the *áus* or *bahi* or autumn rice, and the *áman* or *haimantik* or cold-weather rice. These two great genera are both divided into different species, and these again are sub-divided into very many varieties.

Aus or Autumn Rice is divided into three species,—*Káinán áus dhán, áus dhán*, and *jáli áus dhán.*

(*a*) The first-named species grows best on high-lying lands. The seed is sown broadcast on high rich lands, from which crops of sugar-cane, tobacco, and mustard have been obtained. This species of rice is sown in *Chaitra* and *Baisákh* (April and May), and reaped in *Bhadrá* and *Aswin* (August and September).

(*b*) The second species, *áus dhán*, grows best on ordinary land, neither too high nor too low. It is usually sown broadcast in *Phálgun* (February-March), on lands from which a crop of winter rice has been obtained, and reaped in *Ashár* and *Srában* (June—August).

(*c*) The third species of *áus* rice, *jáli áus dhán*, requires a low moist soil, and is generally sown in *Mágh* and *Phálgun* (January—March), in the beds of rivers and marshes, and reaped in *Ashár* and *Srában* (June—August). A crop of *áman* or winter rice is often sown in the same field with *jáli áus*, and at the same time. The *áman* rice springs up after the removal of the *áus* crop, being brought forward by the rains, and a second crop is also obtained at the time of the usual winter harvest.

Aman Rice forms the great winter rice crop of the District. It is divided into two species, *ropá* or *royá dhán*, which is transplanted; and *buná, boná* or *bhuiyá*, which is sown broadcast.

(*a*) The *ropá* or transplanted *áman* rice is sown in the first instance upon high land. When the seedlings are about a foot high, after the early rains have moistened the soil, they are gradually transplanted to marshy lands covered by about ten inches of water. In the eastern part of the District, between the Tístá, Dharlá, and Brahmaputra rivers, a variation is often introduced into the cultivation, and the rice is transplanted twice. First, when the shoots are about a foot high, they are transplanted into high dry land, which is well manured, and weeded. When about two feet high, they are retransplanted to wet, marshy soil. This practice is said to render the plants more hardy, and to save seed, the shoots from a single grain being often divided into nine or ten plants. This doubly transplanted rice is called *gáchhi dhán. Ropá dhán* is sown in the months of *Chaitra, Baisákh*, and *Jaishthá* (March—June) transplanted in *Srában* and *Bhadrá* (July—September), and reaped in *Agrahayán* and *Paush* (November—January). In cases where the plant is transplanted twice, the first transplantation takes place a little earlier; the second transplanting goes on in *Aswin* and *Kartik* (August—October). The peasantry enumerate no less than a hundred and seventy varieties of *ropá* rice.

The second or broadcast species of *áman* rice is sown in the beds of marshes and rivers in the months of *Phálgun* and *Chaitra* (February—April), and reaped in *Agrahayán* and *Paush* (November—January). This rice is frequently sown in the same field with the *áus* rice mentioned above. The growth of the plant keeps pace with the rising of the water in the marshes, during the rainy season, the stem sometimes growing to a length of twelve feet. This species of rice is not very extensively cultivated." (*VII., 234-237.*)

Cultivation of Rice in the Lower Bengal Generally.

Bardwan. 338

Perhaps one of the most instructive accounts of Lower Bengal Rice Culture is that given by the Director of Land Records and Agriculture in his first Annual Report. It, however, refers to the *Bardwan District*, more

CULTIVATION: Eastern & Central Bengal. Bardwan.

especially what might be called the interfluvial Hooghly-Damudar tract extending to Midnapur :—

"Paddy is by far the most important crop of the division. There are very large areas, especially in the western part of it, that grow nothing but paddy.

Classes.—All the different varieties of paddy cultivated in Bengal may be grouped under three primary classes distinguished from one another by marked characteristics—(1) the *áus* or early, (2) the *áman* or late, and (3) the *boro.* (1) The *áus* is a coarse variety difficult to digest and eaten by the poorer classes alone. It is grown on high lands, and requires much less water than the other two classes. When sown broadcast, as is the general practice, it is a good deal more troublesome to grow than the *áman*. It also yields a smaller outturn and fetches a lower price. But it supplies the rayat with a food-grain and his cattle with fodder at a time of the year when both are very scarce. It is also off the field early enough to permit of preparation of the land for the *rabbi* crops, the winter vegetables, including potatoes or sugarcane. (2) The *áman* class includes most of the varieties of paddy, and is grown over a larger area than any other crop. It is cultivated on low lands with a clay soil, and requires much more water than the *áus.* The finest varieties of paddy belong to this class. (3) The *boro* is a coarse paddy, some of the varieties being the coarsest known. It is less nutritious than the other varieties. It is grown on soft mud on the sides of rivers, canals, or lakes. Edges of rivers subjected to strong tides are of all places the most suited to growing this class of paddy.

Varieties.—It would serve no useful purpose to enumerate the more than one thousand different varieties of paddy grown in this division. Almost every considerable village has a variety of its own, and every year sees the extinction of some of the old varieties and the appearance of some not known before. Paddy is perhaps the best instance known of the variations which plants have undergone under cultivation. Originally an aquatic grass, the one characteristic which it has most persistently retained amidst all the changes brought about by differences in climate, soil, and mode of cultivation is the need of a large quantity of water for its proper growth. According to the popular saying, "*dhán pán netya snan*"—paddy and betel should have a bath every day. It is the belief of the rayats that give the paddy but this one thing needful, and it will grow in any soil and under any climate. Indeed, the facility with which it adapts itself to the different classes of soil from the stiffest clay to the lightest of sands, and from the peaty to the saline, is simply wonderful. Compared with the advantages of a proper supply of water, all other questions in its cultivation, namely, the quality of the seed used, the nature of the soil on which it is grown, the manures applied, and the mode of cultivation adopted, are things of very minor importance."

The Director then proceeds to give a detailed account of each of the three chief crops named above. Unfortunately space cannot be afforded to reproduce the report in its entirety, but the following passages convey the chief points that seem necessary to amplify the information already given :—

1.—Aus. 339

Rotation.—The high lands on which *áus* paddy is grown generally produce two crops in the year. As soon as the paddy has been gathered, the field is prepared for one of the *rabbi* crops ; in this part of the country one of the pulses or oilseeds.

In *dearah* lands the paddy is sometimes followed by wheat or barley. Potatoes and *áus* paddy form a rotation in some places.

On such lands *áus* paddy is the only rain crop grown excepting in jute-growing districts, where the following rotation is generally adopted :—

First year . { 1. *A'us* paddy.
2. One of the pulses or oilseeds, or the two mixed together.

Second year. . { 3. Jute.
4. One of the pulses or oilseeds, or the two mixed together.

Sugarcane is grown on *áus* land, but it is not a rotation crop. It takes a full year to mature, and can be grown only at long intervals, say every third or fourth year the crop which precedes and follows it being paddy.

A'us paddy is one of the best cleaning crops and is often grown as such. If an orchard is to be made or plantains are to be grown, it is on the standing *áus* paddy that the trees are generally planted. When this plan is not adopted, it becomes sometimes extremely difficult to put down the weeds, specially the *ulu* (**Saccharum cylindricum**).

ORYZA sativa.

Rice Cultivation

CULTIVATION: Eastern & Central Bengal.

Bardwan.

Harvest.—The harvest time for *áus* paddy extends from the last week of *Bhádra* to the second week of *Aswin* (last three weeks of September). The *áus* paddy is harvested while yet slightly green, for, if allowed to ripen fully, it will shed some of the grain, and *áus* straw being brittle, will get broken. *A'us* paddy again, excepting what the ryots keep for seed, is intended for immediate consumption. The reaping implement used is the sickle. *A'us* paddy is cut close to the ground and laid in the field in parallel lines for nearly a week. It is afterwards made into sheaves and taken to the threshing-floor, or put in heaps of some 100 to 150 sheaves each, the tops and sides of which are carefully smoothed to let rain water escape easily without penetrating into them.

Yield.—The outturn per bigah varies from 4 to 8 maunds of grain and 3 to 7 *pons* of straw.

Heavy rains are injurious to *áus* paddy.

Diseases.—A kind of black insect, resembling the common mosquito, sometimes eats away the cellular tissue of the leaves.

Transplanted A'us.—Instead of broadcasting the seeds, *áus* is sometimes transplanted. In this case the cultivation is much simpler and cheaper. For the preparation of the field fewer ploughings suffice. Ordinarily six ploughings are given: four when the land is dry and two in the mud.

1.—Áman Paddy.

340 *Soil.*—As stated before, *áman* paddy is grown on low-lying lands, generally clayey. It requires such a large quantity of water that high lands, unless situated very close to tanks, canals, or any other reservoirs of water, are not suited to growing this class of paddy. Some of the best varieties require a clay soil and about one-and-a-half feet of water almost from the time of planting to the harvest time.

The method of cultivation differs according to the comparative height of the land to be sown with *áman* paddy, that is, according as it is situated just below the *áus* land or situated lower down remaining under water for the greater part of the year. In the first case the land is generally loam. The paddy is either sown broadcast or transplanted. In the other case the soil is almost invariably clay, and transplantation is the general rule.

First Case where Áman Paddy is sown broadcast.—After the previous crop of paddy has been harvested, the first shower of rain is taken advantage of in giving the land one or two ploughings. This is generally done in *Magh* (15th January to 15th February). If the field be unclean and full of weeds, one more ploughing is given in *Falgún* or *Chaitra* (March or April), otherwise the field is left untouched till the beginning of the rainy season. At the end of *Baisákh* or the beginning of *Jaistya* (middle of May) after giving one or two ploughings, the seed is sown broadcast at the rate of 10 seers per bigah. It is ploughed in and the sowing is finished by passing the ladder over the field once or twice. Sometimes the sowing is not preceded by any ploughing, but after the seeds have been broadcasted the field is ploughed twice: once along and once across.

The only other operation before reaping is weeding. Generally one weeding is sufficient, but sometimes two to three weedings will be needed. Thinning the plants takes place along with the first weeding. This is, however, not the case everywhere. In some places the after treatment of broadcasted *áman* paddy is very much like that of broadcasted *áus*. When the plants are about seven or eight inches high, the field is twice lightly ploughed with the *langla* or smaller plough at intervals of about a week. After the second ploughing the ladder is passed over the ground once. These ploughings are followed by weedings. Two weedings are generally given. Transplantation from one part of the field to another is sometimes necessary.

Second Case where Áman Paddy is transplanted.—The low-lying clay lands receive very little tillage. Some cultivators are of opinion that these lands ought not to be ploughed in summer, for, by ploughing, grasses will be destroyed, and the success of the paddy is dependent on the growth of these grasses. In *Baisákh* (15th April to 15th May), after a heavy shower of rain, the land may be once ploughed, while the soil contains still a large amount of water. This ploughing of the wet land instead of destroying the grasses will encourage their growth. At the end of *Jaistya* or in *Ashár*, when the land has been quite saturated with moisture, the embankments of the field should be repaired and water collected in it. By two or three ploughings the grasses should be mixed up with the mud, and the seedlings of paddy planted in. There is no doubt some truth in the statement that grasses serve the purpose of green manuring, but that this is a one-sided view of the matter is evident from the fact of paddy grown on lands which are not early ploughed, being often subjected to what the cultivators call the disease of *kadamara*.

in Bengal. (G. Watt.)

ORYZA sativa.

CULTIVATION: Eastern & Central Bengal.

Bardwan.

Practically such lands receive no treatment whatever till the first or second week of *Ashár* (latter half of June) when the clay, softened by frequent heavy rains, is made into mud by one or two ploughings. Five or six days after this, the ladder is used once, and then more ploughings are given, when the field is ready for the transplantation of seedlings.

The only other operation gone through on the field till the harvest time comes is to weed the field two or three times.

Being a crop of the rainy season, *áman* rice does not generally require artificial irrigation. In *Kártik* (October 15th to November 15th), just before the plants begin to blossom, artificial irrigation is sometimes necessary.

Manures.—In some places where the cultivators are more intelligent, *áman* lands almost invariably receive some dung, generally at the rate of 20 baskets per bigah. Some well-to-do ryots in this division sow broadcast on the mud about a maund of oil-cake per bigha just before transplanting the seedlings—a practice which, so far as I know, only prevails in the Burdwan Division, and which might with advantage be encouraged and introduced elsewhere.

Diseases.—Injury is sometimes done to *áman* paddy by insects. (1) One called *Shanki poka* eats away the tender leaves of the young plant. This insect disappears as soon as the heavy rains set in. (2) At this stage of its growth the plant is also sometimes attacked by a mosquito-like insect, the same little creature that feeds on the *áus* paddy. (3) A black fly also occasionally attacks paddy when the ears are being formed. It sucks in the milky juice of the soft grains. These flies sometimes come in immense numbers: 50 to 100 flies may be counted on a single ear. (4) *A'man* paddy often suffers from *kadamara.* In a year of excessive rainfall, and when the field has not been properly ploughed, mosses likewise do injury to this crop.

Harvest.—The processes of harvesting and threshing of *áman* paddy are very much the same as of *áus* paddy. *A'man* paddy after being harvested may either be threshed as soon as the plants have dried by a few days keeping, or stacked.

Cost of Cultivation:—

	R	a.	p.
20 bags of dung	1	4	0
To spread the same	0	6	0
Four ploughings at 6 annas each	1	8	0
Planting 2 men	0	8	0
*Weeding 2 men	0	8	0
Reaping	0	12	0
Binding and threshing	1	4	0
Nursery	0	10	0
Rent	2	0	0
Total	8	12	0

	R
Yield.—10 maunds of paddy grain	10
10 pons of straw	2
Total	12

In clay soils somewhat more of grain and a little less of straw are obtained than what is given above.

1.—*Boro Paddy.*

341

This class of rice crop is not extensively grown in the Burdwan Division. Its cultivation is confined to the south-eastern part of the division, the southern extremity of the Hooghly and the Midnapore Districts, and even in these places the cultivation of *boro* paddy assumes importance only in years in which the *ámun* paddy has been wholly or partially destroyed by floods. The following is an account of the way the cultivation of *boro* is carried on in the Ghatal sub-division.

Two different crops of *boro* may be obtained in a year. It may be cultivated either as a winter or a rain crop.

When it is grown as a winter crop two different methods are adopted, and according as the one or the other of these methods is followed, the *boro* is termed the *kalpina boro* or the *chatá boro.*

* Sometimes more than R2 is spent on this head.

ORYZA sativa.

Rice Cultivation

CULTIVATION: Eastern & Central Bengal.

Bardwan.

Tillage Operation.—There is not much tillage required in the cultivation of this class of paddy.

(1) *Kalpina boro.*—When *boro* is planted in low-lying fields in which water left by the overflow of rivers is still standing in the beginning of *Agrahayán*, it is termed *kalpina boro.* If the ground is in the form of soft mud, which is generally the case with such lands, no ploughing is needed. The only thing to be done for the preparation of the ground is to remove the aquatic grasses that generally grow in such places, and bury them in the soil to serve as green manuring. If the ground is not quite soft, one or two ploughings are given.

(2) *Chatá boro.*—When boro is grown on dry lands with the aid of artificial irrigation, it is termed *chatá boro.* The embankments on the lower sides of the field are repaired and water is collected. The land is then ploughed twice as in the cultivation of *ámun* paddy, and the field is ready for transplantation.

Nursery.—The seeds are first sown in a nursery. For this purpose a plot of ground is generally selected containing soft silt brought down by the flood. The water standing on it is removed and the ground is levelled by passing over it several times the stalk of a palm leaf. If such ground cannot be had, a plot of land is flooded artificially and is ploughed several times, and then the water is removed if there be any standing on it. The seeds are sown on the soft mud, but not when water is standing on it. The grain requires to be carefully prepared before sowing in the nursery. It is a peculiarity of this class of paddy that only newly-threshed grain will germinate properly. For three days and nights the grain is alternately dried in the sun and exposed to the night dews. It is then put in a bag, which is kept under water all the night and dried all the day. This process is repeated for three days and nights. If the seeds have all germinated by this time, they are immediately sown. Otherwise they are filled into a bag and covered with blankets. After a day or two the seeds are taken out and broadcasted in the nursery at the rate of 50 seers per bigah. The seedlings obtained from 50 seers of grains will be sufficient for 6 to 10 bigahs of land. When the seedlings have grown about 2 inches, the nursery ground requires to be watered at intervals of a week. For *kalpina boro* the plants in the nursery are allowed to grow from 10 to 15 inches before they can be transplanted, for in this case the transplantation takes place on lands containing much water. Smaller plants would do for *chatá boro.*

Transplantation.—As in case of *áman* paddy.

After Treatment.—After the field has been transplanted, it is alternately irrigated and dried till the harvest time comes.

Variety.	Time of sowing seeds in the nursery.	Time of transplantation.	Time of harvesting.
(1) Winter .	*Ashin* and *Kártik* .	*Agrahyán* . .	*Chaitra, Baisakh.*
(2) Rain .	*Jaistya* . . .	*Ashár* and *Srávan* .	*Bhádra* and *Aswin.*

Yield.—The outturn is 7 to 8 maunds of grain. The winter variety gives a better outturn.

Murshidabad.

342

CULTIVATION OF RICE IN THE MURSHIDÁBÁD DISTRICT.

"Rice forms the staple crop in Murshidábád district, as elsewhere in Bengal. The rice crop is divided into four great classes, known as *áus*, *ámun*, *boro*, and *jáli.* The *áus* crop, which is sometimes also called *bhadai*, from the name of the month in which it is reaped, is sown in April and May and harvested in August and September. It is a coarse kind of rice, and is chiefly retained in the district as the food of the lower classes. It is usually grown on dry land, and never in the marshes. Convenience of irrigation is the circumstance that mainly governs the selection of land for its cultivation. Provided that water can be readily obtained, the dry and moist nature of the soil is of secondary importance. Fields which border on rivers or *khâls* are most frequently chosen. It is sown broadcast, and not transplanted. There is one variety of the *áus* crop the cultivation of which differs considerably from that which has been just described. It is distinguished from the common *bhadai* by the name of *kartiki*, and is also known as *jhanti*. It is sown in July and reaped in October. It grows for the most part on moist lands, and is sometimes transplanted.

The *ámun* or *haimantik* is the principal crop of the District, and constitutes the bulk of the rice that is consumed by the well-to-do classes, and exported to foreign markets. It is sown in July and August, occasionally as late as September, and reaped in December and January. It generally undergoes one transplantation, but sometimes it is allowed to grow up as it is sown broadcast. Well-watered or marshy lands

CULTIVATION.

are best suited to its cultivation though it can be grown on high lands. The *ámán* rice is sub-divided into an immense number of subordinate varieties, which differ from each other in the fineness of the grain, flavour, fragrance and other particulars.

The *boro* is a coarse kind of marsh rice, sown in January or February, and reaped in April, May, or June. It grows on swampy lands, the sides of tanks, or the beds of dried-up watercourses. It is transplanted, sometimes more than once.

The *jáli* rice is not much cultivated. It is sown in spring and reaped during the rainy season. It grows on low river banks, which remain moist even during the hot months owing to sub-soil percolation (*Statistical Account of Bengal, Murshidábád, by W. W. Hunter, IX., pages 1-102*).

II.—Chutia Nagpur.

Chutia Nagpur.

CULTIVATION OF RICE IN THE LOHÁRDAGÁ DISTRICT.

Lohardaga. 343

Mr. B. C. Basu, in his Report on the Agriculture of Lohárdagá, gives much useful information, regarding the rices of that district, which is more or less applicable to the whole of the mountainous tract of Chutia Nagpur. He divides the rices into lowland and upland, the former practically corresponding to *áman*, and the latter to *áus* rices. The former **Mr. Basu** sub-divides into broadcasted rice (*buná dhán*) and transplanted rice (*ropá dhán*). The account given of the Palámu sub-division is sufficiently like that in sudder sub-division, that the description of the latter may be accepted as applicable to the whole of Chutia Nagpur. **Mr. Basu** writes :—

I.—Low-land Rice.

344

Importance.—Low-land rice (*don dhán*) is by far the most important crop of the sub-division. Roughly speaking it occupies one-third of the total cultivated area, and supplies about two-thirds of the total cereal food produce of the country.

Varieties.—The varieties of low-land rice are extremely numerous, and need not be mentioned by name. But they all fall under one or other of the three following classes, *viz.* :—

(1) *Guruhan, jorhan* or *barká* rice, *i.e.*, as its vernacular names mean, rice of heavy growth. *Jorhan* rice, as it is oftener called, is grown only in the *gárhá don* lands. It is reaped in *Aughrán* (November-December), and is known as winter rice. The *jorhan* group includes a very large number of varieties, and comprises all the finer kinds of rice.

(2) *Lahuhan* rice, or literally rice of lighter growth. In Bengal this class of rice is known as *chhotná, i.e.*, small. It is reaped in *Kártik* (October-November) and known as autumn rice. Like the *jorhan* the *lahuhan* division includes a large number of varieties, all of which are of a coarse character. *Lahuhan* rice is grown on *chaunrá don, i.e.*, on the higher-lying terraces, and from this cause is extremely liable to drought.

(3) *Tewán* rice, sown in *Mágh* (January-February), and reaped in *Baisákh* (April-May), and thus corresponds to the *boro* rice of Bengal. There are no distinctive *tewán* varieties. The two varieties commonly grown as *tewán* are *kánáo* or *jenjné* and *álsangá gorá*, the first of which belongs to the *lahuhan* division, and the second is a variety of upland rice.

The *don* lands are about equally divided between *guruhan* and *lahuhan* rice, but the importance of the former is greater than that of the latter, inasmuch as *guruhan* rice, which is grown on *gárhá don*, is more productive and much less subject to drought than *lahuhan* rice, which is confined to the *chaunrá don*. *Guruhan* rice is the stand-by of the ráyats in a year of drought, like the last, and confers on the country a large measure of its immunity from famine.

Tewán rice is grown to a very limited extent in the country. It is grown on land which remains perpetually wet owing to the water of some rivulet or spring flowing over it. Such land is met with in many villages, but the cultivation of *tewán* is not in favour with the mass of the ráyats, as it requires to be watched day and night against the village cattle, who roam about freely in those months during which the *tewán* occupies the field.

The majority of the varieties of paddy grown in the country are coarse, but there are several very fine ones grown by zemindars and well-to-do ráyats. These include the *kanakchur, lakshibhog, nakhchirni, kalamdáni, &c.*

ORYZA sativa.

Rice Cultivation

CULTIVATION: Chutia Nagpur.

Lohardaga. 345

Methods of cultivation.—There are two methods of growing low-land rice, *viz.*, (1) that in which the seed is sown broadcast in *Jeyt-Asár* (June); and (2) that in which a nursery is made in *Jeyt-Asár*, and the seedlings transplanted therefrom to the fields in *Srávan* (July-August). Both *guruhan* and *lahuhan* rice can be grown in either way, but *tewán* is invariably sown broadcast. The two methods are entirely different from each other, and will be separately treated. Rice which is grown by sowing the seed broadcast will be, for shortness' sake, spoken of as *broadcasted rice*, and rice which is transplanted as *transplanted rice.*

Every variety of low-land rice (excepting *tewán*) can be grown in either way, but the finer varieties of *jorhan* rice are usually transplanted for this reason, that if grown by sowing broadcast, the grains become larger and coarser.

Coarse paddy is always steeped once or twice in boiling water before being husked, but the finer varieties do not receive any such treatment, and are eaten *ároá* or unsteeped. Steeped rice is known as *ushná* and is easier of digestion.

Soils.—Rice does best on alluvial soil or *pánkuá* as it is called in the Five Parganás, where alone in the sub-division this class of soils is met with. Next to alluvial soil is black *nágrá*, a strong clay soil, and next to the latter is *khirsí* or clay loam.

A.—Broadcasted Rice (*Buná dhán*). *Cultivation.*—If the soil remains naturally moist, the first ploughing is given directly after the winter-rice harvest is over; otherwise immediately after the first fall of rain in *Mágh.* During this month and *Fálgun* two or three ploughings are altogether given, and the soil left exposed to the action of heat and cold. In *Cheyt* the field is manured with cowdung, as much being applied as the rayat's stock of manure can afford. A good dressing will be 4 tons per acre, but few rayats can afford to make such liberal use of their manures. The usual time for applying manure is immediately after the *Sarhul* festival, which takes place at different periods during *Fálgun*, *Cheyt*, and *Bysákh*, but in most places in *Cheyt.* The manure is first distributed over the field in small heaps, and then spread out by the *kodáli* or spade, or buried in by ploughing.

In *Bysákh* or *Jeyt*, as soon as rain has fallen, the field is ploughed once more; the larger clods, if any, are broken by the mallet (or the *dhelphunrá*); the harrow (*mehr* or *chowk*) is then passed over the land. The time for sowing extends from late *Cheyt* to early *Asár*, according as rain begins early or late. *Cheyt*-sown rice comes up stronger and gives a better outturn than late-sown rice. After the land has been prepared, the seed is sown broadcast at the rate of about one *pukka* maund per acre, and buried in by a light ploughing. The harrow then passes once more over the field, which is thus left till *Sravan.*

There are two ways of sowing paddy, according as the soil is in the form of dust or is moist. In the first case, the sowing is called *dhuri-buná* or dust-sowing; in the latter, it is called *rash-buná* or moist-sowing. The first method of sowing (*dhuri-buná*) has been described above; in the second case (*rash-buná*), the only difference is that the soil being moist, the clods have not to be broken, and the last ploughing after sowing is not necessary, the harrow only passing over the ground after the seed has been sown. The last ploughing is essential to *dhuri-buná* in order to bury the seed deep enough. In this case the seed will remain in the ground for a fortnight or even longer, *i.e.*, as long as rain has not fallen, and germinates directly after the soil has been moistened by the first shower of rain. Of the two ways of sowing, the *dhuri-buná* or dust-sowing is preferred, as it gives a better outturn than the other.

Rice fields, whether sown or transplanted, are ploughed three or four times. The first ploughing is known as *chirni* (*lit.* cutting open), the second as *dobar*, the third as *uthão*, and the fourth as *purão* or finish.

Repairing árs.—The *árs* or embankments enclosing the rice-terraces have to be repaired every year with a view to make them as far as possible watertight. The *árs* of fields intended for broadcasted rice are repaired usually in *Mágh*, that is, at the time they are ploughed for the first time. There is however, no fixed time, excepting that the repairs must be done before the sowing time, and as the soil gets very hard in *Cheyt* and *Bysákh*, the most convenient period for doing the repairs is after the soil has been moistened by the winter showers. The repairs consist in earthing up any gaps that may have been made during the previous rains, and in adding earth to the top and sides of *árs*, and then making them strong enough to resist a sudden rush of water into the terrace. The repairing of *árs* is an item of heavy labour, and requires at least about twelve men for one day per acre.

Regulating water.—From the time the paddy is sown till the end of the rainy season, the amount of water in the terrace has to be constantly regulated. When too much rain has fallen, the rayat makes a narrow opening on the lower side of the

CULTIVATION: Chutia Nagpur. Lohardaga.

terrace, and allows the surplus water to flow off. But in case there has been not rain for a day or two, the opening is earthed up, and sufficient water is kept in the terrace. When the rainy season is about to close, that is, about the end of *Bhádra* (middle of September), the *árs* are made as watertight as possible, the rayat's endeavour being to keep in as much water in the terrace as he possibly can.

Weeding.—After the paddy has been sown broadcast, nothing is done till *Srávan*, beyond regulating water in the terrace in the manner described above. In *Srávan* the plants have become 8 to 10 inches high, and are found more or less infested with grasses. In order to destroy the latter, the field is once ploughed and harrowed; this operation is known as *bidháli* or *bidhána*. The rice plants, together with the grasses, get partly buried in the mud; the latter are largely destroyed, but the former come up again in about a fortnight. About a month later on, that is, in *Bhádra*, those grasses which have escaped being destroyed are carefully handpicked and serve as excellent fodder for cattle: sometimes, two weedings are given. After weeding nothing is done till harvest time. The method of harvesting rice will be described later on after all the different ways of growing it have been described. In the western parganás rice is weeded only by well-to-do and intelligent rayats, as the operation is equally troublesome and expensive. In the police circle of Silí, adjoining the district of Mánbhum, the weeding is followed by an operation called *gáchi-kátá*. It consists in filling up gaps with the plants removed from thickly-grown spots. This insures an equal distribution of the plants over the field, and thus encourages them to tiller, that is, to throw up fresh shoots. The *gáchi-kátá* operation is common in Bengal, but is not so much essential in Chutia Nagpur, where the soil being much poorer, the rice plants do not tiller so freely as they do in the fertile soils of Bengal; and consequently they do not suffer so much from being close to one another.

Lewá-buná.—This is a method of growing sown rice adopted in various emergencies. It is followed only to a limited extent, and being cheaper and less troublesome is practised largely by the poorer rayats. It is followed under the following circumstances :—(1) when the rayat wants to save labour; (2) when there has been a heavy shower of rain, and land rendered thus unfit for sowing in the ordinary way; (3) in case there are not sufficient seedlings to plant the fields intended for transplanted rice; and (4) if through drought in the early part of the season, broadcasted rice fails. As a rule, *lewá-buná* is resorted to only in the last instance. The yield is much inferior in ordinary years, and as the roots of the plants do not go deep enough, *lewá-buná* rice is very precarious in years of deficient rainfall.

On *gárhá* or low-lying terraces almost any variety of *guruhan* or winter rice can be grown in the *lewá-buná* way. On *chaunrá* or high-lying *don*, two varieties are generally grown, *viz.*, *karhání* and *kándo* or *jenjne*. Both of these are of very light growth, mature quickly, and require a considerable amount of rain for their growth. *Karhání* rice is highly prized by the aboriginal tribes, as it gives a rice-beer much stronger in quality than any other variety of rice.

If the field is intended for *lewá*, it is ploughed twice in *Mágh* and left as such. In *Asár*, after a heavy shower of rain, the land is worked up into mud by the plough and the harrow, which go over it three or four times. The mud is then allowed to settle down for a day or two; the supernatant water which comes above the mud is then drawn away, and the seed, which has been previously made to germinate, is sown broadcast. The germination is effected by soaking the cleaned seed in water for 24 hours, and then putting it in baskets, which are placed one over another with a view to cause partial fermentation; the top-most basket is covered over with a stone slab. The seed is kept in the baskets for another 24 hours. The sowing takes place on the third day, when the sprouts have just appeared: the germinated seed is known as *ánkowá*. When the plants have struck root, the field is treated in the same way as for broadcasted rice. Weeding is, however, seldom given, the yield of *lewá-buná* rice being too small to meet the expense of the operation. *Lewá-buná* rice is seldom manured for the same reason.

If *lewá-buná* is resorted to in any other circumstances, the land is ploughed, harrowed, and worked up into puddle in the same way as for transplanted rice. The seed is made to germinate and sown in the way described above.

B.—TRANSPLANTED RICE (ROPA DHAN). *Seedbeds.*—The seedlings are grown on carefully prepared nurseries, and then transplanted in the rice-fields. There are two kinds of nurseries or *birás*—*matihání birá* or up-land nursery, and *lehbirá* or low-land nursery. The *matihání birá* gives stronger plants than the *lehbirá*, which, like *lewá-buná* rice, is more precarious. The latter, however, requires less seed, and comes up quicker than *matihání*. The seedlings of all the finer varieties of rice are grown in *lehbirá*. Each of the two kinds of nurseries is described below. 346

The cultivation of land intended for up-land rice nurseries and the method of sowing the paddy seed are the same as for broadcasted rice. The only differences are that the number of ploughings in the case of the nursery is larger, being six or

ORYZA sativa.

Rice Cnltivation

CULTIVATION: Chutia Nagpur.

Lohardaga.

seven and often more; the application of manure more liberal, and the seed sown very much thicker. *Matiháni birá* is manured with ashes and cowdung, which has been previously burnt with a view to destroy the seeds of grasses and other weeds. The quantity applied is usually about double that given to fields to be sown with rice. The quantity of seed sown in *matiháni birá* per acre is seven to eight *pukka* maunds.

Matiháni birá is made in *bári* land or in portions of the *dihári* lands, *i.e.*, those adjoining the homesteads. Upland nursery lands are known as *cheerá*, and besides yielding the seedlings are annually cropped with various *rabi* crops, such as mustard, brinjals, wheat, and barley.

The time for sowing the paddy seed in up-land nurseries is from late *Jeyt* to end of *Asár*.

The cultivation of *lehbira* is similar to that of *lewá-buná*, or mud-sown rice. The land is occasionally broken open by one or two ploughings in *Mágh*. *Lehbirá* is manured with raw cowdung a few days before the seed is sown. The germinated seed is sown on the puddle made by repeated ploughings and harrowings. The rate per acre is considerably less than that for *matiháni*, and is about 6 maunds per acre of nursery. No weeding is given to *lehbirá*. The seed is sown in *Asár* and early *Srában*, and the seedlings get ready for planting out in from twenty to thirty days.

One acre of a rice nursery will afford sufficient seedlings for planting about 6 to 8 acres of paddy lands. It is calculated that one *kát* of land requires 40 *pailás* (= one *kát*) of seed if sown broadcast, 45 *pailás* of seed if the seedlings are grown in up-land nursery, and 35 *pailás* if they are grown in low-land nursery.

Cultivation.—In *Asár*, when the soil has been softened down by the rain, the land is broken by the plough, the embankments or *árs* are repaired, and water kept in the terrace for a few days with a view to cause grasses to rot. The water is then let off, a second ploughing is given, and water again shut in for a few days. The third ploughing (which is the last but one) is followed by a harrowing with the *chowk*, and a day or two after, the land, if uneven, is levelled by the *kárhá* or the *mher* (harrow). On the day of the planting, the land is once more ploughed, and the *chowk* goes over it round and round, so that the soil and water get mixed up into a uniform puddle, on which the seedlings are planted in the manner described below. The number of ploughings is usually four, the names being the same as given under broadcasted rice.

Manure.—Transplanted rice-fields seldom or never receive any manure, but the nurseries have to be plentifully manured. The rayats' stock is too limited to spare manure for transplanted rice, although it is admitted that the application of cowdung will be of considerable benefit to it. I have found a rayat here and there who, being well off, will only grow transplanted rice; this he manures liberally and gets in return better crops than his neighbours. The reasons why transplanted rice is not extensively grown, and why most rayats prefer to divide their rice-fields between broadcasted and transplanted rice, have been detailed in a preceding paragraph.

Transplanting.—The seedlings are first uprooted from the nursery and made into bundles, which are then carried to the field. Twelve women (the lighter field work being usually done by Kols and low caste Hindu women) are calculated to uproot seedlings enough in one day for one acre of land, and twenty-four women to transplant the same. The bundles of seedlings are thrown into the fields at adequate intervals; each woman takes a bundle, and with her thumb and forefinger makes a hole into the mud into which she thrusts a few plants. The seedlings are planted at intervals of 6 to 8 inches. The reason for close planting is that rice does not tiller so well in the poor soils of Chutiá Nágpur as they do in Bengal. The *gáchis* or clusters of rice plants do not become half so thick, I believe, as they do in Bengal, showing how poor the soil of the country must be in comparison with that of the Bengal plains. The time for transplanting the seedlings extends over *Srávan* and the first week of *Bhádra*. Late transplanted rice is extremely liable to various insect pests.

Regulating water.—After the seedlings have been transplanted, and as long as they have not struck root in the ground, the rayat takes care that the terrace is not overflooded with water. After the plants have come up and appear to be growing, the quantity of water in the terrace has to be regulated in the same way as for broadcasted rice. No further treatment is necessary till the harvest time.

As a rule rice-fields are not irrigated. During the rainy months the rice-terraces remain more or less flooded, and the cultivator has only to regulate the quantity of water by alternately opening up and closing some part of the embankments. At the disappearance of the rain he tries to make *árs* as water-tight as he can, and in this way water can be left in the terrace for periods varying with the position and character of the soils. In ordinary years there are always a few showers in *Hastá* or *Háttiá* (first fortnight of *Aswin*) and in *Chitrá* (last fortnight of the same month), so that *lahuhan* or autumn rice, which is grown on *chaunrá don* and more liable to drought, does not suffer from want of moisture in the soil. The *gárhá* low lands, from their low position and the clayey character of their soils, are more or less retentive of moisture, and although they considerably suffer from drought never turn out

CULTIVATION: Chutia Nagpur.

a total failure. In the event of the drought intervening in *Aswin* and *Kártic*, as was the case last year, the loss falls very heavily on *lahuhan* rice. In many villages I found last year that the rayats did not take the trouble of reaping their autumn rice, which was merely straw and husks, and was eaten down by their cattle. In the majority of villages the outturn of autumn rice barely reached 4 annas.

Lohardaga.

Autumn rice flowers in late *Bhádra* and the grains fill up in *Aswin*. Water is essentially necessary during this period, or otherwise the ears do not push up from inside the blades, and the grains remain empty, and become what the people call *piläi*. For the same reasons a few showers in *Aswin* and *Kártic* would considerably benefit winter or *guruhan* rice.

In years of normal rainfall the necessity for irrigating paddy is not much felt by the rayats. This partly counts for the almost utter absence of provisions for irrigation in the country. It is, however, freely admitted by the people that the yield of rice can be largely increased by irrigation even in a year of normal rainfall. The subject of rice irrigation has been treated in detail in the section on irrigation, to which I beg accordingly to refer.

Rice Harvest. Period.—The rice harvest extends over the two months of *Kártic* and *Aghrán*, that is, from the middle of October to the middle of December. *Lahuhan* or autumn rice is harvested in *Kártic*, and *guruhan* or winter rice in *Aghrán*.

Threshing Yard.—Before the harvest begins every rayat prepares his *kharyán* or threshing yard by scraping grasses off a convenient plot of ground, and then cleaning and making it even with a plastering of cowdung. If the rice-fields are near at hand, the *kharyán* is made in *bári* land, or if they lie at a distance, on an adjoining plot of fallow *tánr*. In the latter case he builds a temporary straw hut, from which he can watch his rice-heaps. Rocky places, when they are found in any village, are invariably turned into *kharyán*, as these are very easily kept clean, and do not require to be changed every year.

Harvesting.—The work of harvesting is done both by men and women in Chutiá Nágpur, but in the Five Parganás women are seldom employed for this purpose. Sixteen men and twenty women will harvest one acre of rice in one day, the actual number depending on the growth of rice. On the plateau of Chutiá Nágpur the rice plant never attains a height of more than 3 feet, and the straw is not long enough to be made into *bicháli* or bundles, as is usually done with the heavier varieties of rice in Bengal. The stalks are accordingly cut at some distance (6 to 8 inches) from the ground, the stubble being subsequently fed down by cattle. In the Five Parganás lying below the Chutiá Nágpur plateau the soil is more fertile, the rice plants become of longer and heavier growth, and the straw long enough to be made into bundles. The heavier varieties of rice (*guruhan*) are, when the rayat so wishes it, cut at the base with a view to keep the straw intact and convert it into *bicháli*. Most portion of the *guruhan* and the whole of *lahuhan* rice is, however, cut some way up the stalk, as in Chutiá Nágpur. The rice plants, as they are reaped by the sickle, are left on the ground in small bundles, which are subsequently collected into larger bundles, and these latter into loads, which are then carried to the *kharyán* and arranged into hollow circular heaps called *chakras* with the ears on the inside. There they remain until the time for threshing comes.

Threshing.—The method of threshing rice is the same as in Bengal. From five to seven bullocks are tied in a line, and are driven round and round over the rice stalks which have been spread out in a circle. The bullock at the centre is not tied to any post as is done in Bengal and Behar. Occasionally two or three lines of bullocks are employed in order to expedite the work. Four men and ten bullocks (five in a line) may thresh one acre of rice in six hours from morning till mid-day. Two men drive the two lines of bullocks, and the other two sift the straw with a pitchfork, or *ákáin* as it is called. Towards the end, the straw is well sifted by the hand, and the grains allowed to drop below on the ground; the straw is then gradually removed, layer after layer, and the paddy, together with small broken bits of straw, left on the ground of the *kharyán*. The uncleaned paddy is then taken in a *sup* or winnowing basket, which the rayat holds as high as his head, and shakes it to and fro to let down the mixed grain and chaff. The paddy grains, being heavy, fall at his feet, while the lighter chaff and dust are blown away to a distance by the breeze, caused by the shaking. The cleaned paddy is then measured out and either stored or disposed of in various ways. *Gorá-cátá* rice—rice reaped at the base of the stalks in order to convert the latter into *bicháli*—is threshed in the Five Parganás by beating the bundles over a flat wooden plank, usually the flat surface of a *sagar* wheel. This process takes more time, but the straw is much more valuable.

Watching.—During the harvest and threshing time the cultivator has to be perpetually on the watch to see that the paddy is not stolen away by dishonest labourers, in the *kharyán* he builds himself a rude triangular straw hut in which he would sit up and sleep at night so long as the threshing is not finished. As a further precautionary measure, it is a regular custom with the rayat to strew ashes in several lines

ORYZA sativa.

Rice Cultivation

CULTIVATION: **Chutia Nagpur.** **Lohardaga.**

over the surface of the conical heaps of paddy, and place a dungball at the top so that the least tampering with them will bring down the ball and upset the ash lines. This is especially done to prevent frauds on the part of servants in charge of the *kharyán*.

Harvest Wages.—Harvest wages are paid in various shapes, and are considerably greater than ordinary day wages. Money wages are not usually paid In some places the reaper gets one load out of every 21 loads of paddy he reaps. In this way he can earn from five to ten seers of paddy in a day. In other places he gets a bundle of paddy stalks every day he reaps. This may contain five or six seers of paddy. The usual daily wages is three seers of paddy and one *poá* of cleaned rice.

Yield.—The yield of paddy ranges from 20 maunds at the best to 8 maunds per acre. Ten maunds by weight will, I believe, represent the average yield per acre. In the Five Parganás, the rice lands are more productive, and the average there is greater by a couple of maunds or so.

The yield of straw is on the average about two maunds and a half. It should be remembered that the paddy grown in the district is of very light growth, and the bulk of straw is left in the field as stubble.

Storing.—When the paddy has been threshed and cleaned, the first thing the rayat does is to satisfy the *sáhu* or grain-lender from whom he may have borrowed in the previous months; he then delivers over the customary quantities to the *lohár* or ironsmith and to various other village functionaries. Out of the remainder he keeps apart a certain portion for family consumption for one or two months, and the surplus is then stored for future use. There are two ways of storing paddy and rice in vogue, namely, in *morás* and in *delís* or large bamboo baskets. The *morá* is made of loose straw, and is of spherical shape. The paddy is placed inside the straw, which is bound round and round with straw and *chop* or bark ropes. The straw remains about a couple of inches thick over the grain, and is made tight aud hard by being bound up with ropes. A *morá* may contain 20 maunds of paddy; often much less. Paddy kept in *morás* enjoys almost perfect immunity from weevils, and keeps on the whole better than when it is stored in bamboo baskets. These latter are of a roughly cylindrical shape with a concave bottom. They are first plastered on the inside with mixed earth and cowdung in order to make them air and water-tight. After being filled up with paddy the tops are covered up with straw, which is subsequently plastered over with the dung mixture. *Delis* are not in common use by the rayats for storing grain, and are not so safe against weevils as *morás*.

Cost of Cultivation.—It is rather difficult to estimate the cost of cultivation with accuracy. Agriculture is not carried on in India on the same business principles as in England; the ráyat never keeps account of his income and outlay. In the estimates of cost of cultivation given in this chapter, the wages of a man have been taken at 4 pice per diem; the daily wages of temporary field servants vary from 5 to 9 pice, but as they are generally engaged for the year, their rate of wages does not exceed 4 pice per diem. The hire of a plough has been taken at 6 pice per diem, the price at which it is actually let out in certain parts of the country. Neither manure nor straw have been valued, as they are seldom bought and sold, excepting in the immediate neighbourhood of Ránchí, Lohárdagá, and such important places.

Cost of Cultivation of Paddy sown broadcast.

	R	a.	p.
4 ploughings with harrowings (3 ploughs will plough one acre in one day 3 × 4 = 12 ploughs, at 1 anna 6 pies)	1	2	0
Carrying and spreading manure (4 men for one day)	0	4	0
One maund of seed paddy	1	0	0
One ploughing and harrowing at *bidhána* (3 ploughs at 1 anna 6 pies)	0	4	6
One weeding (20 women at 1 anna per diem)	1	4	0
Repairing embankments in Mágh (16 men at one anna per diem)	1	0	0
Reaping (16 women at 1 anna per diem)	1	0	0
Carrying the paddy loads to the threshing yard	0	2	0
Threshing (4 men and 10 bullocks will thresh produce of one acre in one day at 1 anna for man, 6 pies for bullocks)	0	9	0
Cleaning and winnowing (3 men for one day)	0	3	0
Rent of first class *don*	3	0	0
Ábwábs (at 2 annas per rupee not unusual)	0	6	0
TOTAL	10	2	6
Yield per acre of first class *don*, 12 maunds of paddy at 1 rupee per maund	12	0	0

in Bengal. (G. Watt.)

ORYZA sativa.

CULTIVATION: Chutia Nagpur. Lohardaga.

Cost of Cultivation per acre of transplanted paddy.

	R	a.	p.
Ploughing ⅛th of an acre for up-land nursery 5 times with harrowing = ⅝ths of an acre once, will take 2½ ploughs one day at 1 anna 6 pies	0	3	3
Spreading manure (2 men for one day)	0	2	0
Seed paddy, 45 seers, at R1 per maund	1	2	0
Pulling out seedlings (8 women for one day)	0	8	0
Carrying the bundles of seedlings to the field (4 men for one day)	0	4	0
Ploughing one acre of field 4 times = 12 ploughs at 1 anna 6 pies	1	2	0
Harrowing (2 men and 4 bullocks)	0	3	0
Repairing embankments	1	0	0
Planting (one man to distribute the bundle and 20 women to plant)	1	5	0
Reaping and threshing yard	1	2	0
Threshing and winnowing	0	12	0
Rent and ábwábs of one acre of first class *don*	3	6	0
TOTAL	11	1	3
Yield of one acre of first class *don*-12 maunds at one maund for the rupee	12	0	0

Pests.—Low-land paddy is subject to various insect and fungoid pests. The chief among them are—

(1) *Bhenku* is a fungoid disease appearing in particular years, causing the blade to whiten and dry up. The people are ignorant of its cause and remedy. The amount of loss caused by *bhenku* is never great in Chutiá Nágpur Proper. At Jheria in Mánbhum, I was told by many people that *bhenku* causes great damage to the paddy crop in some years, reducing its yield by as much as 50 per cent. There is a belief current among the people that if too much rain falls in *Adrá* (last half of May) and cools down the soil, *bhenku* is likely to be more abundant and injurious. This belief may be well-founded. The study of the insect and fungoid pests of farm crops, and above all, of paddy is very important.

(2) *Bonki*, an insect pest which causes considerable damage to rice that has been transplanted late. The insects are minute, thin, worm-like caterpillars, which eat into the blades and ear-stalks. The affected plants become white in colour. The ráyat's remedy consists in spells and charms, one of them being to scatter bamboo shoots cut into small bits, accompanied by certain spells or *mantras*. The time when the young caterpillars appear is the latter part of *Bhádro* and the early part of *Aswin*. About this time the leaves of paddy which has been sown or transplanted early in the season have become too stiff to be eaten by the young worms, but late transplanted paddy affords young and delicate leaves on which they greedily feed themselves. This explains why paddy is infested with *bonki* only when it has been planted late.

(3) *Chátrá* or *rátá* is, I believe, a fungoid disease characterized by the plants dying away in patches.

(4) *Gándhá or makhi.*—This is a small winged insect appearing in swarms at blossoming time. The fly emits a most offensive odour, which is said to dry up the milky juice inside the young paddy grains. It cuts into the ear, which falls down and dies. The pest is prevalent in a year of excessive rain.

Besides the above, paddy is infested with numerous insects, which on the aggregate must cause a deal of injury to it. Mr. Sen has pointed out in his report on Burdwan that the main cause of the appearance of insect pests is that the ráyats close up their rice fields and allow the water to become stagnant. This he considers to be a serious mistake. Whatever this may be in Burdwan, it is not so in Chutiá Nágpur. Here the ráyat, by shutting up water at the end of the rainy season, tries to save the paddy from drought, and although it runs the risk of being infested with insects, the ráyat accepts this evil as the lesser of the two.

347

C.—TEWAN OR SUMMER PADDY.—*Tewán* paddy, as I have said before, is grown on land which remains moist during the summer months. Its cultivation resembles in every respect that of *lewá-buná* or mud-sown rice. The usual time for sowing *tewán* is the month of *Mágh*. It is reaped in *Bysák* and *Jeyt*, and occupies the field for about three months. The field should be constantly watched against cattle, which at the time are freely pastured all over the village lands. From this cause the cultivation of *tewán* paddy is not in general favour, and many fields, which

ORYZA sativa.

Rice Cultivation

CULTIVATION: Chutia Nagpur. Lohardaga. 348

are naturally suited to it, are left fallow in the summer months. *Tewán* land fetches much higher rents, as they afford two crops of paddy in the year.

2.—*Up-land Paddy* (*Gorá dhán*).

Extent of Cultivation.—In Chutiá Nágpur up-land paddy ranks next to low-land paddy in order of importance. It occupies, I believe, about one-third of the total cultivated area of the uplands. *Gorá*, the two millets—*gondli* and *máruá*, and maize—are the stand-by of the ráyats during the three months (*Bhádra*, *Aswin*, and *Kártic*) previous to the great winter paddy harvest.

Rotation.—*Gorá* takes part in two distinct rotations, both of which have been treated in detail in the section on Rotation of Crops. Briefly speaking, it follows either *máruá*, or *urid*, according to one rotation, and is grown on fallow land according to the other. On good fertile land it may be taken also after *gondli*.

Varieties.—There are several varieties of *gorá* paddy, distinguished mainly by colour of the grains. All the varieties are, however, coarse in character. It need hardly be said that *gorá* is only another name for *áus* paddy.

Soil.—*Gorá* paddy does well on a light sandy loam; heavy clay soils are unsuited to its growth, as clay interferes with the proper diffusion of the roots.

Cultivation.—*Gorá* is always sown broadcast, and never transplanted as is frequently done in Bengal. In Chutiá Nágpur Proper it is never manured for fear of grasses which cowdung invariably gives rise to; in the Five Parganás, however, land intended for *gorá* is manured in the same way as *máruá*, which latter is very scantily grown in that part of the country.

Gorá land requires to be thoroughly worked up by repeated ploughings and exposure to heat and cold.

"*Báro másh, tero chás*"
"*Tábe karo gorá ásh*;"

that is, "plough 13 times in 12 months, then hope for gora paddy." The grasses must be thoroughly burnt up by heat by the time the land is fit for sowing in *Jeyt* or *Asár*.

On land which has remained fallow for one or more years according to the fallow system of rotation, the first ploughing is given in the month of *Bhádra*. It is followed by one or two ploughings more in the same month, and the land is left as such till the month of *Mágh*. During the intervening months the ráyat is too busy with his harvest to be able to afford time for the ploughing of *gorá* land. On lands which have borne a crop of *máruá* or *urid*, the first ploughing is given in *Mágh*, after the soil has been softened down by a shower.

From *Mágh* till sowing time the land is ploughed and cross-ploughed at intervals of a fortnight or so. No other treatment is necessary before sowing, which takes place in late *Jeyt* and early *Asár*. The sowing is immediately preceded by a ploughing, and may be either *dhulibuná*, *i.e.*, in dry soil, or *rashbuná*, *i.e.*, in the moist soil according as the soil at the time of sowing is dry, or has been moistened by a shower. The seed is sown broadcast at the rate of about one maund per acre.

On the third day after sowing the land is ploughed once lightly and harrowed with the *chowk* with a view to bury the seed. I cannot account for the fact why the ploughing should be given on the third day, and not immediately after sowing, as is usually done with many other crops. Any grasses that may have sprung up, are carefully hand-picked, when the crop is about a foot high. Nothing else is done till harvest time.

Diseases.—There are two insect pests to which up-land paddy is subject. These are—

(1) *Bhuská*—An ant-like insect which eats into the roots of *gorá* dhan, and thereby causes some amount of injury.

(2) *Nánuá.*—A spiny caterpillar, about an inch long, which causes considerable damage to *gora* and to *máruá* in August and September by eating away the leaves. A few specimens of this caterpillar I sent to Mr. E. C. Cotes of the Indian Museum through the Agricultural Department for identification. *Nánuá* is said to be more abundant in the jungly villages; the jungles probably affording the proper habitat for the moths. It is said that the *nánuá* was unknown some 10 or 15 years ago. Its ravages are not limited to *gorá* and *máruá*; *kher* or thatching grass has been largely destroyed by this insect in many parts of Chutiá Nágpur Proper.

Harvest.—*Gorá* paddy is harvested in the last fortnight of *Bhádro* and the first fortnight of *Aswin*. It is reaped and threshed in the same manner as low land paddy. Any detailed description is therefore unnecessary.

CULTIVATION: Chutia Nagpur.

Lohardaga.

Produce and Cost of Cultivation.—The produce of grain usually varies from 5 to 10 maunds by weight of unhusked paddy per acre: on the average 8 maunds valued at R8. The cost of cultivation per acre is as follows:—

	R	a.	p.
Ten ploughings at 4 annas 6 pie per ploughing . .	2	13	0
One maund of seed	1	0	0
One weeding by 20 women at one anna	1	4	0
Harvesting, threshing, &c.	1	8	0
Rent of one acre for six months	0	6	0
TOTAL .	6	15	0
Yield of paddy 8 maunds valued at R1 per maund . .	8	0	0

Gorá paddy is invariably steeped before husking in the same way as the varieties of low-land paddy."

It is perhaps unnecessary, after the very exhaustive account given above, to say anything further regarding Chutiá Nágpur. The following brief passages represent the main features of the region extending towards the Santal country (Rájmahál Hills, &c.) on the one hand, and Orissa on the other.

CULTIVATION OF RICE IN THE MÁNBHÚM DISTRICT.

Manbhum. 319

"The three principal crops of rice grown in Mánbhum are *gorá dhán, nuán,* and *halmuntik* or *áman.*

"*Gorá dhán* is sown broadcast early in May on table-lands and on the tops of ridges, and is reaped at the beginning of August.

"While engaged on famine relief operations in Bánkurá District during 1874, Mr. Macaulay observed that in the Fiscal Division of Maheswará, bordering on Mánbhúm, the growth of *áman* or winter rice was confined to the trough-like depressions which lie between the undulating ridges of the surface, and that the outturn even in these favoured spots was extremely precarious. 'The only security against almost annual failures of the rice crop in Maheswará lies,' writes Mr. Macaulay, 'in the substitution for *áman* rice of some crop requiring less moisture, and capable of successful cultivation on comparatively sandy soil. Such a crop is to be found in the south and south-west of Bánkurá, in Mánbhúm, and in the north-western parts of Midnapur. It is called *tetká* and *cháli* rice, terms signifying grains which resemble one another in all their conditions so closely, that they may be considered as practically identical. The cultivation of this rice is gradually moving northwards, as its early outturn, powers of endurance, and rapid growth on the poorest soil are becoming more and more known. I will enumerate a few of the characteristics of this grain:—Sowing begins about the 1st May, and reaping about the 1st August. The return therefore is rapid, and the harvest early. It may be grown broadcast, or it may be transplanted. In the former case, the ordinary moisture of high grounds in the rains is sufficient; and if the ground is low, the water must be drained off, as excess of moisture destroys the plant. In the latter case, the clay should be kept at a pasty consistency, but nothing more. The rice grows well on high and sandy soil; in fact, I have seen clearings in the *sál*-jungle sown with it as a first crop. It will, however, amply repay cultivation on richer grounds, provided no water is allowed to stand in the fields. The cost of cultivation is only R2 to R2-8 per *bighá*, and the outturn on poor lands is 5 *maunds*.

"Even on these, 8 or 9 *maunds* can be secured by careful preparation of the ground; while on richer soil, as much as 12 and 15 *maunds* have been produced. The grain of *tetká cháli* rice is somewhat smaller than that of *áman*. In other respects the difference is slight. Its value in the market is generally a little lower. I have known the paddy to sell for thirty-six *sers* for the rupee when *áman* paddy was selling at thirty-four, and this is the proportion generally maintained.'

"The *Nuán* or *Aus Rice,* which forms the autumn crop, is sown as soon as possible after the first good fall of rain generally in April or May. It is cultivated on the middle and higher levels of the terraced slopes described above. This rice is generally sown broadcast on the fields, but is occasionally transplanted. In either case, the crop is reaped at the end of September or beginning of October.

"*Aman,* or *Winter Rice,* is cultivated on the lowest levels of the terraced slopes, and on moist land lying beneath the embankments of tanks.

ORYZA sativa.

Rice Cultivation

CULTIVATION: Chutia Nagpur.

"It is sown in a nursery after the first showers of rain at the end of May and beginning of June, and is subsequently planted out in the fields. Harvest takes place in December and January.

"The *áman* rice is the most important crop of the district. *Aus* rice, as I have stated above, is not very extensively grown, and the produce of the *gorá dhán* is inferior to that of either of the low-land crops." (*Vol. XVII., 310.*)

Singbhum. 350

CULTIVATION OF RICE IN THE SINGBHÚM DISTRICT.

"The character of the rice cultivation of Singbhúm is determined, as in Hazáribágh and Lohárdagá, by the physical conformation of the surface of the soil. Nothing resembling the great level rice plains of Bengal Proper is to be met with throughout the district. Everywhere the face of the country is undulating and broken up by alternate ridges and depressions, which for the most part radiate from small central plateaux and form the channels of small streams. The ordinary kinds of rice can only be grown in the bottom of these depressions and on the lower levels of the slopes; but a variety called *gorá dhán*, which is peculiar to the Chutiá Nágpur Division, is cultivated on the tops of the ridges themselves. The system of constructing embankments across the upper ends of the hollows, and thus storing water at a high level, which reaches the crop partly by artificial drainage and partly by natural percolation, has not been so fully developed in Singbhum as in Hazáribágh and Lohárdagá. It is, however, largely resorted to, and the kinds of land are classified by their position with reference to these embankments. Thus, there are three classes of land:—

"(1 *Berá*, land of the best quality, which commands a supply of water throughout the year;

"(2) *Bád*, land of an inferior quality, but so situated as to be within the reach of artificial irrigation.

"(3) *Gorá*, land lying on the crest of a ridge above the level of the reservoir, and therefore entirely dependent on the natural rainfall. Homestead land and pasturage grounds are not assessed, and have no distinctive names. Three crops of rice are raised in Singbhúm District, *gorá, bád*, and *berá*. taking their names from the quality of land on which each is grown. *Gorá*, early or high land rice, is sown broadcast on high land, just after the opening showers of rain in May, and is reaped in September. *Bád*, or autumnal rice, is sown on second class lands in June, and reaped in the end of October and November. This crop is both sown broadcast and transplanted. *Berá*, winter rice, is grown on low-lying land of the best quality, situated below a reservoir, and commanding water throughout the year. It is sown in July in a nursery, and subsequently transplanted. The crop is reaped in December. Except in villages occupied by the Hindu caste of Kurmís, the general style of cultivation in Singbhúm is primitive, and the land undergoes scarcely any systematic preparation for the crop.

"Within the last few years, however, the Kols have made a considerable advance in the methods of cultivation, and now get three crops in the year where formerly they had only one."

III.—Orissa.

Orissa. 351

It is perhaps unnecessary to deal with the rice cultivation of this province in great detail. Large tracts of Orissa much resemble the rice lands which have been dealt with above under Burdwan, while other portions resemble the swamp rice lands of the Sundarbans. The reader is referred to Sir W. W. Hunter's *Orissa*, a work which will be found to deal very fully with many features of the rice traffic. The systems of tenure which prevail will be found to be fully discussed, as also the soils suitable for rice, and the yield and price of the grain. The following passages may, however, be specially selected as exhibiting some of the chief features of Orissa rice culture.

Purí. 352

CULTIVATION OF RICE IN THE PURÍ DISTRICT.

"No well-tested statistics have been obtained, but the following remarks are taken from a number of returns which have from time to time been officially submitted to, and accepted by, Government. Purí is strictly a rice-growing district. Of rice crops, the following are the most important:—(1) The Biáli, which is sown on high but moist land, in June or July, and reaped in October or November. Its principal varieties are Sáthiká, the Kuliá, and the Aswíná. (2) The Sárad, which is sown on middling high land in July, and is reaped in December. A hundred varieties are included under the generic name of Sárad. Of these, the ten following are the most important, Khaira, Kalásur, Bánkoi, Matará, Rangiasina, Nripatibhog, Gopálbhog, Básubati,

CULTIVATION: Orissa. Puri.

Bandiri, Narsinhbhog. (3) The Dálua rice, sown on low wet lands in November or December and reaped in April. Its most important varieties are the Piá and Kasundá.

"The Sárad, or winter rice crop, is generally transplanted by the more diligent husbandmen; and by this process a much larger return is obtained. If not transplanted the following are the operations required for the winter crop:—(1) Ploughing, March or April; (2) sowing, May or June; (3) weeding, after first rainfall; (4) harrowing and ploughing, July or August; (5) second weeding, August; (6) laying (arranging the crop for convenience of cutting) December; (7) reaping, December or January; (8) threshing, January or February; (9) scalding (a little water is poured upon the paddy, which is then placed over a fire until the water is evaporated for the purpose of loosening the husks); (10) cleaning (husking)" (*Orissa by W. W. Hunter, Vol. II. app., p. 14*).

Balasore. 353

CULTIVATION OF RICE IN THE BALASORE DISTRICT.

"Rice is the staple crop of Balasore as of all other districts in Orissa. It is divided into five great genera, and forty-nine principal varieties. (1) The *Dálna* rice, sown on low lands in December or January, and reaped in March or April, grown chiefly in the Fiscal Divisions of Báyang and Káyámá. It is a coarse, red, unwholesome grain. Its principal varieties are the *Dálua, Lakshminárayanpriyá, Bámanbáha, Antarakhá,* and *Sarishphul.* (2) *Sáthiyá* rice, sown on high lands in May or June, and reaped in July or August; common throughout the district. Its principal varieties are, *Dudhsará,* a fine white, and *Kalásuri,* a coarse red, grain. (3) *Niyáli* rice, sown on high lands in May or June, and reaped in August and September; common throughout the district" (*Orissa by W. W. Hunter, Vol II. app. page 44*).

Cuttack. 354

CULTIVATION OF RICE IN THE CUTTACK DISTRICT.

"The staple crop, in common with the other Districts of Orissa, is rice. The following is a list of the principal rice crops, with their varieties:—(1) *Biáli,* grown on high land, the banks of rivers and on the outskirts of villages. It is sown broadcast in May and reaped in September, the soil in which it grows being called *Dofasli,* or 'two crop' land. There are two distinct species of *Biáli* rice, the *Sáthiyá* and the *Bara,* each subdivided into many varieties. The *Sáthiyá* derives its names from the time it takes to come to maturity which is believed to be exactly sixty days. A tradition relates that the *Biáli* rice was not made by Brahmá, the author of the Universe, but invented by the sage Viswámitra. It is accordingly considered less pure and prohibited in religious ceremonies. The higher classes seldom use it, as it is a coarse grain difficult to digest, and apt to bring on diarrhœa in stomachs unaccustomed to it. The chief subdivision of the *Sáthiyá* species is the *Hárvasáthiyá,* and the principal varieties of the *Bara* species are the *Bakri, Inkri, Madiya, Chauli,* and *Jiraisáli.* (2) *Sárad* rice is of a better quality and includes two great species, the *Laghu* and *Guru.* The former is sown in May on comparatively high land, and is reaped in November. Its thirteen most important varieties are, the *Chhota-champá, Motra Rangiasina, Niyáli, Háruá, Lanká, Bodla-champá, Sara, Nardá, Mánt, Bangri-panchi, Palásphul,* and *Bhutmúndi.* The *Guru* species of *Sárad* grows on low lands, being sown in May and reaped in November or December. Some varieties are sown broadcast on low marshy ground; others are carefully reared in nursery fields, and transplanted stalk by stalk to higher and dryer soils. Ground covered with a foot of water gives a good crop, but the coarser sorts will grow in six feet, although all must be sown to begin with in solid land. For some, a soil having an admixture of sand (*Dorasá*), while for others a soil not sandy (*Mátál*), is best adapted. Some require to be sown early in the season, and others late. Certain varieties will not grow unless the land is thoroughly weeded, while others flourish in spite of every thing. In order to provide against the uncertainties of the season, the husbandmen sow both the species which require flooding or which will not suffer from it, and those that will flourish with but a moderate rainfall." (*Orissa by W. W. Hunter, II., 130-131 app.*)

Behar. 355

IV.—Behar.

Mr. George A. Grierson, in *Bihar Peasant Life,* gives an interesting account of rice culture, but chiefly in the sense of supplying an accurate vocabulary of the terms and sayings connected with that crop. The following list of terms may be found useful:—

Dhán = rice whether the crop or threshed.

Cháur = husked rice, but there are two kinds of husked rice, *usina usna* or *josanda,* rice eaten by the poorer classes which is parboiled before being husked, and *arwa,* eaten by the rich which is not parboiled before being husked.

CULTIVATION: Behar.

Bhát = plain boiled rice.
Khichri = rice boiled with pulses.
Golhath = rice boiled to a mash.
Panihata = a dish prepared by adding water to the rice left from last night's supper.
Dhanhar = land that has borne a rice crop.
Khilmar = land recently reclaimed for rice culture.
Birár or *benga* = a nursery for rice seedlings.
Bihan or *mori* = seedling rice.
A'nti or *ántiya* = bundles of rice ready for transplanting.
Khâru = twice transplanted Boro rice.

According to **Mr. Grierson** there are two chief crops of rice in Behar with a subordinate third crop. These are:—

356 *1.—Broadcast Sown Rice.*

This is known by various names, such as *báwag* or *báog*; in Gaya *bogera*, and in Patna, *bogha*. Of the broadcast rice, the following special kinds are enumerated:—(1) *sáthi*, this is a red rice which ripens in sixty days. It is sown in the month of May-June and reaped in July-August; (2) *sokna* sown with the first fall of rain in May-June and cut in September: this crop is called *bhadaiya*; (3) a very extensive series which are sown in February-March and reaped in November-December. **Mr. Grierson** then deals with the special forms of broadcast sown rice which are peculiar to certain districts of Behar. As he enumerates some 50 kinds, space cannot be here afforded to deal with them in this work. In Champárun and Tirhut there is a white-bearded rice known as ***kherha.*** In Shahabad a small black and white grain which, like the *sáthi*, ripens in sixty days. "In Patna on the first fall of rain, which generally takes place in the asterism of *Rohni*, in the month of May-June, the sowing is commenced. Paddy which is sown broadcast is divided into two classes —a red, which is considered superior, and is called *lalgondiya*, and a black, which is considered inferior, and is called *kára bogha*. The former kind includes *karhanni* as the principal. The ear is black."

357 *2.—Transplanted Rice.*

"This is generally sown with the first rains in May-June. It is transplanted in July-August (= month of *Sáwan*). In Patna the custom is to commence transplanting on the 5th of *Sáwan*, after holding a festival, called *nakpancho*, or the 5th of the asterism (*nakhat*). The regular harvest is held in November-December. Before this, however, some is cut for the ceremony of *bisun pirit* at which the Bráhmans are feasted on the new grain."

Scented Rices. *Conf. with pp. 532, 595.*

358 **Mr. Grierson** then mentions some seventy forms of transplanted rice as known in the districts of Behar and concludes his account of them in the following words:—"Of all the above rices, the most esteemed is *samjíra*. It is a fine kind, and when cooked its fragrance fills the house. The next best is *basmati* or *basmatiya* which is not quite so fine as the first. The *selha* may be considered as the third best."

359 *3.—Boro Rice.*

This, **Mr. Grierson** says, is a poor kind of rice which is sown in September-October or even in November, in the mud of the banks of streams and lakes. It is transplanted several times in December to February.

The most instructive account of the Rices of Behar is, perhaps, that contained in **Mr. A. P. MacDonnell's** Food Grain-supply of Behar and Bengal. The main facts brought out by **Mr. MacDonnell** have also been reviewed

CULTIVATION: Behar.

in the Famine Commission's Report. Thus, for example, it is stated that "The six districts of the Behar Division (excluding Patna) are those to which **Mr. MacDonnell** devoted most of his time and strength, and his conclusion is that the cultivated area is there approximately 11,040,000 acres supporting a population of 11,530,000 people, or about an acre per head; that the average outturn of food produced is about 11 maunds, or two-fifths of a ton per acre; the people require for food about a quarter of a ton per head per annum, and thus about one-seventh of a ton remains as an annual surplus to cover seed-grain, wastage, cattle-food, and export." The proportion of this food-supply drawn from the rice crop is shown by the table given by the Commissioners for the sixteen chief districts investigated by **Mr. MacDonnell.** It is explained that, unlike Lower Bengal, Behar possesses only one main rice crop—the winter rice—and that like the North-West Provinces it has a *rabi* crop (wheat, barley, pulses) and also a *bhadoi* or intermediate harvest chiefly of Indian-corn and millets. In Behar the cultivated area relative to the uncultivated is very much higher than in the rest of Bengal. For example, in the Saran District, out of a total of 1,698,000 acres, 1,566,000 are under the plough. Throughout this rich and fertile country the system of agriculture pursued is very nearly identical in most of the districts and differs chiefly from that pursued in the North-West Provinces in the greater importance of late or winter rice. "The high lands are sown with millets and pulses in the rains, or with early rice, and the same land is generally sown again with *rabi* in the autumn, the dense population in these tracts creating a large supply of manure. The winter rice is grown year after year in the same fields. Unlike Behar, in the eastern and southern divisions of Bengal, and in Orissa, where the rainfall is much heavier and inundations are frequent, the millets and spring crops become rarer and almost disappear, and as a rule only two kinds of rice are sown, the early and the late (*áus* and *áman* in Bengal or *beali* and *sarad* in Orissa), the latter being by far the more important of the two" (*Fam. Com. Rep., I., 78*). The *Rabi* and *Bhadoi* crops of Behar correspond to the late and early rice crops of the rest of Bengal.

The following brief passages taken from the Gazetteers convey the chief facts regarding Behar rice cultivation:—

Cultivation of Rice in the Patna District.

Patna. 360

"Rice, which forms the staple crop of the district, consists of two great divisions: the *kartiká* or early rice, and the *aghani* or winter rice, each of which is again subdivided into many varieties. A small part of the district is cultivated with *boro* or spring rice, which may be regarded as a third division.

"The *aghani* or winter rice forms the chief part of the rice grown in Patna. It is sown broadcast after the commencement of the rains in June and July, on lands that have previously been ploughed three or sometimes four times. After a month, or a month and a half, when the young plants are about a foot high, they are generally transplanted. Each plant is pulled out from the land, which is soft with standing water, and planted again generally by women, in another field, which has been first ploughed and smoothed. The young plants are placed in rows from two to three inches apart. Rice that has not been transplanted gives a comparatively poor crop, and all the finer sorts of rice are therefore treated in this manner. Should no rain fall in September or October, and if the water generally procured from the artificial watercourses is exhausted, the plants will wither, and become only fit for fodder. But if seasonable showers fall, the rice comes to maturity in November or December, and is then reaped. A second crop of *khesári*, sometimes mixed with linseed, is often sown in October among the rice when it is in flower (*gábh*). At that time there is generally a little water on the fields. The seed is sown broadcast, and if it is seen to germinate, the water is run off so as not to injure the young crop. When a second crop has been sown, the paddy is reaped below the ears, and the stalks are left standing, to be afterwards used as fodder." (*Hunter's Statistical Account of Bengal, Vol. XI., 109.*)

CULTIVATION: Behar. Saran. 361

CULTIVATION OF RICE IN THE SÁRAN DISTRICT.

"Rice is perhaps sown over a larger area in Sáran than any other crop; but it is not so important a staple as in the neighbouring district of Tirhut, from which large quantities are imported *viá* Sohánsí and Rewá *Ghats*. According to the *pargané* returns, rice is sown on 366,000 acres. It consists of two great crops—the *bhadaí* rice or autumn rice, and the *aghani* or winter rice, the latter being by far the larger crop.

The *bhadaí* is generally sown on high ground. The field is ploughed over several times when the early rains set in, and the seed sown broadcast in June. As soon as the young plants are a few inches high, the ground is weeded. The crop is reaped in September. The following are among the chief varieties—(1) *sathi*, 60 days' rice; (2) *sarha*; (3) *katki* or *munga*; (4) *kárhání*.

Aghani rice is sown on low ground. In June, after rain has fallen, a nursery is selected, ploughed three or four times, and the seed sown in it. When the young plant has come up, another field is prepared for transplantation. The rainy season has now thoroughly set in, and the whole soil is reduced to the consistency of batter. In this the young rice plants are put, in bunches, at a distance of about nine inches from one another. This variety is much more extensively cultivated than the autumn rice, but it is liable to be destroyed, if rain fails at the time of transplanting, or in September. It is sometimes drowned by a too sudden rise of water; but the *kalunji* variety is said to be able to keep pace with the inundation, if it does not rise too fast. It is harvested either in December or January."

MADRAS. 362

II.—CULTIVATION IN MADRAS.

References.—*Returns of Agric. Stat. of Br. Ind.* (*1888 89*), *4-9; Madras Man. of Admin. I., 285; II., 83, 105; Famine Commission Report, App. III., 170-174; Benson, Manual Saidapet Expr. Farm, 43; Nicholson, Coimbatore Manual, 214-216; Boswell, Nellore Manual, 388-391; Agri.-Hort. Soc., Ind., Transactions, VII., 86; Jour. (old series) XIV., Pro.* (*1866*), *58.*

Area. 363 Conf. with *p. 522.*

Area under, and Traffic in, Rice.

The *Agricultural Statistics of British India* (published by the Government of India) show that the Madras Presidency had in 1888-89 6,285,806 acres under this crop. It seems probable that 12 maunds per acre might be accepted as a safe average yield, hence the total acreage may be supposed to have produced 7,54,29,672 maunds (say, 2,693,916 tons). This statement is, however, the merest guess, since the yield is variously stated at from 6 to 40 maunds, and 15 is probably the average of all fairly good crops (excluding the extremes from consideration).

Rice is raised in the largest quantities in the alluvial and highly-irrigated districts of Tanjore, Godavery, and Kistna on the east coast, and in Malabar and Kanara on the west coast, where the rainfall is abundant. Of all cereals it is the one most grown in this presidency, and it is remarkable that the people of Madras regard Bengal and Burma rice as not suited to them; these rices have in fact the reputation of producing indigestion and diarrhœa.

Trade. 364

The Madras trade in rice may be briefly reviewed for the year 1888-89. The returns of internal trade show that the presidency had a net export of 26,50,665 maunds (taking paddy and rice conjointly), but that large amount went chiefly to the port towns to feed the city populations and to meet the demands of foreign trade. The coastwise transactions (excluding those from port to port within the presidency) resulted in a net import of 11,68,852 maunds coming chiefly from Bengal (the gross imports from which were 12,03,197 maunds). The figures of foreign trade show a net export of 25,77,136 maunds or 91,326 tons, the gross exports having been 91,346 tons, valued at R62,43,149.

The total export trade in the year 1888-89 showed an increase of 12,562 tons in quantity and over 9¼ lakhs in value, chiefly owing to the brisk demand from London and Bombay and to the good harvest in the Godavery and Tanjore Districts. Of the total quantity exported, 70 per

CULTIVATION in Madras.

cent. went to Ceylon; 11 per cent. to Bombay; 8 per cent. to Goa; and 4 per cent. to the United Kingdom.

METHODS OF CULTIVATION, SOWING, REAPING, &c.

Provincial Methods. 365

Space cannot be afforded to do more than denote some of the chief peculiarities of rice cultivation in South India, and even these will be best discovered by the perusal of a few of the better chapters on this subject taken from District Manuals. In the Madras Manual of Administration, useful information is given as to the soils suitable for rice cultivation and the general agricultural principles pursued. The following passage from that work may, therefore, be here quoted as an introduction to a series of passages from the District Manuals. Of land irrigated from tanks it is stated—

"Almost the only crop grown on this description of land is paddy. Where an abundance of water is available throughout the year two crops are grown annually, but most of the land produces only one crop. The irrigation water is usually supplied on each occasion to a depth of from 1 to 6 inches, and the supply is given according to its abundance on every day, every other day, or every third day, during the growth of the crop. A depth of from 8 to 12 feet of water may be taken as the common consumption in the growth of a complete crop of paddy. In many parts of the Presidency the practice is to sow the paddy broadcast on a semi-liquid soil which is brought into this state by frequent waterings and stirrings. The practice is also common of transplanting young seedling paddy plants from nursery beds into the field. Beyond weeding and watering the paddy land receives no attention while under crop. Some varieties of paddy are only six months on the ground, others only four months. The crop is cut when dead ripe and is thrashed by striking the sheaves on a log of wood and by the treading of cattle. The agricultural practice in dealing with irrigated land is said to be capable of improvement in the direction of a more economical use of water; as long, however, as the ryot is unable to apply manure to his irrigated land, he is under the necessity of using a large quantity of irrigation water in order to secure silt in its place. There are many practical difficulties in the way of procuring manure; there is also a deficiency of capital to invest in it" (*M. M., I., 285*).

In the Famine Commission Report it is stated that in Madras and Mysore "leaves are commonly trodden into the mud of the rice fields to increase the fertility of the soil." This practice would seem in some measure to take the place of the *ráb* manure so commonly adopted in Bombay.

Mr. Benson (in the Saidapet Farm Manual) does not, however, speak favourably of rice cultivation in Madras. But his remarks are instructive and may be reproduced here:—

"Both the *car* and *sumba* varieties of rice have been extensively grown from time to time. It has been found that the amount of water usually taken for irrigating the crop is far in excess of its actual requirements, which is the more to be deprecated, for the crop is of a low type, and produces a poor food; it develops habits of idleness and unthrift in the people from the ease with which it may be grown, and consequently the adoption of perfunctory methods of cultivation; it is insanitary in its effects, causing, under the system of the native cultivator, the presence of large bodies of generally stagnant water—a fertile source of malarious fever; and, finally, it monopolises water which might be far more usefully employed for irrigating maize or wheat, either of which crops might be grown with a little of the water now used for paddy, whilst the food they would produce would be equal or greater in quantity and of far better quality What the effects would be of an outbreak of a disease on this crop, which might, as the potato disease did in Ireland, destroy the greatest food resource of the people, it is difficult to contemplate; and yet the extension of irrigation works goes on, and consequently that of paddy culture. Paddy is at present the bane of South Indian farming."

The following notes from district reports may be given in alphabetical order:—

S. Kanara. 366

CULTIVATION OF RICE IN THE SOUTH KANARA DISTRICT.

"The staple crop is rice. Cocoa-nut gardens are numerous along the coast, and areca plantations in the interior, gram, beans, hemp, *ragi*, sugar-cane, tobacco,

ORYZA sativa.

Rice Cultivation

CULTIVATION in Madras.

and cotton are grown, but not to any extent. The land is thus classified according to its capacity for irrigation—*byle*, or rich wet land; *majal*, or middling wet land; and *bettoo*, or land watered only by the rainfall. On *byle* land of the best quality, three rice crops can be raised in the year; on the best *majal* land, two crops; *bettoo* land produces only a single crop. The *zeneloo* or *carty* is the earliest rice crop of the season, on whatever description of land it may be grown. The seed is usually first sown in nurseries, which are highly manured, and the plants are afterwards transplanted. In about two months after transplantation, the crop comes into ear, and about twenty-one days more is ready for reaping. Experiments have been made to introduce Carolina rice, but have not been generally attended with success."

Chingleput. 367

CULTIVATION OF RICE IN THE CHINGLEPUT DISTRICT.

Rice cultivation manifests one or two features of a peculiar kind in this district which will be discovered from the following extract from **Mr. C. S. Crole's** Manual :—

"The cultivation of paddy on Nunjah lands is made in the following process:—There are two modes in which paddy is cultivated—*Séttukál* and *Puluthikál*. In the first, the land is irrigated by water from tanks and channels, ploughed four or five times both lengthways and breadthways, and thus reduced to a puddle. The soil is then mixed with leaves and plants, dung ashes, &c. This process is called manuring. The land thus prepared is again ploughed four or five times. The clods of earth are thereby broken up and the land turned into regular mire, also the leaves and the manure become rotten. A harrow is then drawn over the surface in order to smoothen it. Then seed corn is put into a pot and steeped with water, kept so for three days, it is taken out and sown broadcast in the fields thus prepared, while there is a little water in it. The next day the field is drained and is left for four or five days to get dry. Within that time the seeds sprout up to the height of 1 inch. It is then watered little by little. The sprouts grow up gradually. If the field is regularly weeded, the growth will be commensurate to the fertility of the soil. Three *merkáls* of seed of 8 measures each are sown on an acre of land and will generally produce a 100 *merkáls* of paddy in return if the growth flourishes.

There is another mode of cultivation in which the seed is sown in a small plot of ground rendered miry, and is left to grow for 40 or 50 days, and when it has shot up to about a foot high the young rice is transplanted into the field where it is to grow and which is ploughed and prepared for that purpose. The field is kept inundated for a considerable length of time and the weeds are regularly plucked out. This third kind of cultivation requires four *merkáls* of seed per acre. If too many seeds are sown in any portion of a field, the surplus plants are transplanted to other parts. This is common both to *Puluthikál* and *Séttukál* cultivation. Even this yields a good harvest. The grounds for this manner of cultivation by transplantation are that there is a particular season for the cultivation of every product; that it is impossible to plough all the fields and sow the seeds therein in one and the same season, and that sowing in nurseries taking place at the same time as the field that is regularly ploughed is sown, the transplanted cultivation gets ripe for harvest about the same time that the other fields do. This kind of cultivation by transplantation is very common in this district.

Pulithikál or dry seed cultivation is commenced when the *Nunjah* land is wet with rainfall. It is then ploughed lengthways and breadthways four or five times. Then it is manured with dung ashes and other rubbish and again ploughed four or five times when the seed is sown. The growth is produced by the moisture of the ground. It is again left for a month and a half or two to grow in the moisture caused by the rainfall and not by irrigation. After this period the fields are kept constantly irrigated, weeds plucked out, and equal distribution of plants attended to. There will be a good yield only if there be sufficient rain between the sowing season and the time of irrigation, otherwise the produce will be but meagre. The principal points of difference between dry-seed cultivation (*Puluthikál*) and sprouted-seed cultivation (*Séttukál*) are that the former does not require irrigation for two months after sowing, and that the same affords greater facilities for cattle than the latter in point of ploughing. In some cases before commencing both the above sorts of cultivation, sheep and cattle are folded in the fields to be cultivated. The fields are then ploughed. Indifferent cultivators and those who hold lands beyond their capabilities do not add to the fertility of the fields by using manure and leaves, either previous or subsequent to tilling. This kind of cultivation does not yield a good crop. If the soil is a very fertile one there may be a pretty good crop. If the cultivation be only seasonable, it does not matter much; even if there be a little deficiency in manuring there will still be a good crop. The chief descriptions of paddy cultivated in Nunjah

CULTIVATION in Madras.

lands in the districts are *Samba Kár* and *Mana Kattei*. Of these, *Samba* is of the best, its rice being white in colour; *Kár* ranks next. The third and last is *Manakat*. The rice of the two latter is a little red. Of one and each of these sorts there are different kinds with separate names."

Of the *Samba* kinds *Kodam Samba* is cultivated in the beginning of the cultivation year, *viz.*, about May or June. This is besides generally cultivated in villages irrigated by river channels and in those watered by tanks containing the remains of the previous year's supply of water. The cultivation yields a good crop too. In many places a second crop is raised about the month of December or January on lands on which a first crop of *Kodam Samba* has been already raised during the year. All the other kinds of *Samba* are generally cultivated from the month of July onwards. *Sirumany* or *Kalavan Samba* and *Chinna Samba* are raised abundantly in this district, and the other kinds are raised in certain parts of this district to a certain degree. *Vádam Samba* and *Jádam Samba* are raised in dry seed cultivation in great quantities. The vast extent to which they are raised is attributable to their not being so readily liable to wither even though rain fails. Whereas, if the rainfall is short or the lands get dry a little, the cultivation is seriously affected with regard to the other kinds. In places where there is not a large supply of water, three months' *Samba* and two months' *Samba* are cultivated. They rarely yield a good crop. In many places two crops of *Kár* are raised in a year. *Muta Kár* and *Perum Kár* are raised in this district in large quantities. The raising of the first crop generally commences about the month of May and June, and the second in December. The second crop is raised on the very same fields on which the first crop has been already raised. The various kinds of *Mosánam* are cultivated to a good extent from the month of July, the season for *Samba*. *Pum Púlai* is cultivated in the foreshores of tanks and pools just previous to the rainy season; and even if water after rainfall stands to a height of a half or three-fourths of a yard, it shoots up higher than the level of the water and yields a good outturn. Some of the species of *Kár* are raised to a slight degree in certain villages in dry seed cultivation.

Manakat, the 3rd class, is cultivated on a small scale about the beginning of the official year, *i.e.*, in April or May, in lands irrigated by wells, &c. In some villages where there is a sufficient supply of water, the above kind of paddy is cultivated by sowing seeds or by transplantation about the month of January or February in lands where a harvest of *Samba* has been reaped. This yields a tolerably good crop. *Pisini* is raised in dry seed cultivation on high level *Nunjah* lands. This grain is cultivated in dry (Punjah) lands which are not without moisture. No irrigation is required to raise this kind of paddy. Like Punjah crops moisture alone contributes to the growth of this corn. The outturn is not great (*Rámasvámi Ayyar Mirasidar in D. M., p. 36, et. seq.*).

CULTIVATION OF RICE IN THE COCHIN DISTRICT.

Cochin. 368

Rice forms the staple of cultivation, and some fifty separate races are locally distinguished. The best land supports three crops annually. Next to rice in importance, cocoa-nut engages the attention of the cultivators.

CULTIVATION OF RICE IN THE COIMBATORE DISTRICT.

Coimbatore. 369

"Rice is said to be of almost endless variety, but may be roughly divided into three months (*kár*) and five months (*samba*) paddy; the former is grown both as a first and second crop, the latter, save very occasionally, only as a second or single crop, and in the cold weather. The best known varieties of these two sorts are tabulated below. Dry land paddy is also added, but is rare.

Three months—*Kar*.	Four to six months—*Samba*.	Dry land.
Kuruvei.	*Samba.*	*Suruna-velli.*
Annathánam.	*Kalingaráyan samba.*	
Aruvathám-kodei.	*Kártigai samba* four months.	
	Molagi or *Semmolagi.*	

Either variety succeeds *ragi* or other so-called dry crop, or paddy. The treatment, season, and yield are different in each taluk owing to the variety of water-supply and soil.

ORYZA sativa.

Rice Cultivation

CULTIVATION in Madras.

Common features are as follows:—

First crop (*kár*). The ground is prepared in June and July in the usual way by swamping, cross-ploughing several times, green-manuring and smoothing with the beam. Seed is generally sown broadcast at the rate of from 24 to 30 Madras measures, or from 1 to 1½ bushels per acre; in the fourth week the land is weeded and the plants thinned out if necessary, and bare places planted up. There is no subsequent weeding; water is let on as often as possible, and harvest takes place in September, October. There are occasional exceptions to the rule, *e.g.*, in some lands of *Nerúr* (Karúr taluk) *samba* is grown as a first crop from July to January, being transplanted in the usual way; *kuruvei* as a second crop, transplanted from January to May: *Kuruvei* is, in other lands in this village, either sown broadcast or transplanted.

The quantity of seed sown varies greatly according to the ryots, and this point is not clearly made out. In Erode, Karúr and Udamalpet it is represented that sowing in nurseries takes more seed than sowing broadcast, whereas the reverse is said to be the case in other taluks. The estimates of seed everywhere seem excessive, *e.g.*, 30 to 36 measures (84 to 100lb) per acre transplanted, and from 24 to 60 measures (68 to 168lb) for broadcast sowing. Full enquiry, however, shows that seed is used extravagantly; the nurseries are sown so thickly that much must be wasted, and the seedlings are transplanted in bunches of from four to ten according to size and the quality of the ground. If the soil is rich and the plants good, four or five seedlings will form a bunch and be planted at from 6 to 8 inches apart. It might be ascertained by experiment whether the use of so much seed is needful or useful.

Second-crop paddy (by which is understood a crop following a first crop either of *ragi* or other dry grain or of paddy). This may be either *kuruvei* or *samba*. With good irrigation, as in Erode, the Cauvery channels of Karúr, and under the Amaravati in the better channels and in good seasons, *samba* is the rule as a second crop. This is usually transplanted from nurseries prepared in October. After the first crop is reaped in October, the land is again prepared by ploughing, green-manuring, treading and levelling, and the seedlings rapidly transplanted in November-December.

There is little difference between the second crop and the single crop paddy either in yield, kind or season; it may be briefly stated that, as a rule, except in Satyamangalam, the first crop is a short one to take advantage of the water and season before the chief crop is grown. The quantity of seed required is said somewhat to exceed that for the first crop.

In certain taluks there are differences both in crop and in practice. In Polláchi, where the south-west monsoon brings considerable rain, and in a few lands in Udamalpet, paddy is grown on dry land, as a first crop. It is sown broadcast in June and reaped in September-October. In Udamalpet the practice is similar to that in Pollàchi, but is rare. In Karúr, and occasionally elsewhere, samba and kuruvei paddy are grown in gardens as a cold weather transplanted crop from November to March; it is watered every day; a quantity equal to about one inch being let on after transplantation, for about ninety days out of four months, allowing for rain and a cessation of ten days before reaping; according to experience, the surface of the soil must always be covered with water. Hence about seven feet is the quantity raised and used, and about six inches falls as rain. The outturn is heavy, probably owing to a good physical condition of the soil, *viz.*, from 3,000 to 4,300lb. The yield and the quantity of water used are noticeable; the quality is also said to be superior to that of wet-land paddy. It is also to be noted that paddy so cultivated sends down short tap-roots which are wanting in wet-land paddy; this is readily explained by the difference in the condition of the sub-soil.

The chief pest is an insect called *navei-púchi*, whitish, and from half to one inch long; it bores into and feeds on the tender plant, which then produces only empty husks." (*in D. M. by F. A. Nicholson, pp. 214-216.*)

Ganjam. 370

CULTIVATION OF RICE IN THE GANJAM DISTRICT.

It has been stated:—" Agricultural operations commence in June, during which month the rain of the south-west monsoon usually begins to fall. In June the early dry grains and paddy seed (Rice) intended for transplanting are sown. Rice is sometimes sown broadcast, but is usually transplanted from specially-prepared seed beds. In July and September an ample and continued supply of water is essential to the growth of the young plants. The reaping of the rice or paddy crop commences soon after the 1st November and sometimes lasts until the 15th January, according as the season has been early or late. An early season betokens, as a rule, a favourable harvest. The dry grain crops (*i.e.*, those grown upon unirrigated land) and early paddy are reaped between the 1st September and the 15th October. The after crop of dry grains continues, however, to be reaped from the middle of February to the beginning of

CULTIVATION in Madras.

April. A second crop of rice in Ganjam is almost unknown; it occurs, however, in a tract of land not far from Ichapore, bordering upon the sea." (*M. M. of Adm., II., 78.*)

Godavery. 371

CULTIVATION OF RICE IN THE GODAVERY DISTRICT.

Rice is said to be the chief cereal. There are two important kinds (*a*) Rice transplanted (white paddy), of which five distinct sorts are recognised. These are sown in May and July, and reaped in November and January. But two other sorts are recognised, and these are sown in June and reaped in October. The white rices are raised on marshy land. The second great section (*b*) are the black rices: these are sown in June and harvested in October.

Kistna. 372

CULTIVATION OF RICE IN THE KISTNA DISTRICT.

In the Madras Manual of Administration it is stated there are three classes of rice crops grown in this district; these are *púnausa* (early crop), sown in May or June, and reaped in September; *pedda* (great or middle) sown from July to September, and cut between November and February; and *peira* (late crop), sown in November and December, and gathered in February and March. "Rice of all kinds is sown in *regar* or black soil. The area under rice in 1883-84 was 357,394 acres or 18·7 per cent. of the cultivated area. The price of the best rice per *garce* (9,974℔) was, in the same year, R370." (*M. M., p. 83.*)

Madura. 373

CULTIVATION OF RICE IN THE MADURA DISTRICT.

It is stated that the rices grown are of several varieties, which differ from one another considerably in respect of productiveness, rapidity of growth, weight of grain, colour, taste, and wholesomeness. A statement is given of some 30 kinds which shows their seasons of sowing and reaping and the number of days which each kind takes to reach its various stages of growth. The period which they occupy the soil may be said to be from August to March. The finest kinds are those known as *Samba* and *Milagi*: these forms germinate in five days, are transplanted in 25 days, form the ear in 105 days, and are harvested in 45 days more, thus taking in all six months. The progression of merit may be said to extend from the rapid to the slow growing forms. But the author of the Manual may be allowed to deal with the chief features of the Madura rices, as it would serve no useful purpose to give only an abstract of his opinions:—

"It is observable that *Pisánam* rice is not now grown in the Madura District, but at the beginning of the last century was extensively cultivated. One of the Jesuit missionaries of that time speaks of *Samba* and *Pisánam* as the two principal varieties grown in the Ramnad zemindari. This is the more curious, as *Pisánam* appears to be extensively grown in both Tanjore and Tinnevelly. I suppose it has been discovered that the soil and climate of Madura are better suited to *Milagi*.

Rice appears to be grown on almost every description of soil indifferently, the only essential being a constant supply of water; but the richer soils yield of course much larger crops than the poorer. The seed is sometimes sown broadcast in the rice-fields, but, as a rule, it is sown in nurseries specially prepared. In the former case the expense of cultivation is light, but the produce is not very remunerative; in the latter, the crops are generally good. Sometimes the young plants raised in a nursery are transplanted into a second nursery and afterwards retransplanted into the field, and where this is done the yield is much larger than is ordinarily obtained.

The times of ploughing, sowing, transplanting, &c., vary very considerably in different seasons according as the rains are early or late, plentiful or the reverse, and also according as lands lie near to or far from sources of irrigation and are rapidly or slowly supplied with water. But probably, as a rule, the ploughing is done in June or July after the early rains have thoroughly softened the ground and the seed is sown in the nurseries at the end of July or beginning of August. After about 30 days the seedlings are fit for transplanting, and in January the crop is ready for the sickle.

When retransplanting is resorted to, the young plants are suffered to remain in the second nursery about thirty-five days. Weeding is done about a month after sowing or transplanting as the case may be. During the whole time the plants are in the ground they must stand in about 2 inches of water, which should not be allowed

ORYZA sativa.

Rice Cultivation

CULTIVATION in Madras.

to stagnate; or they will become weakly and diseased and will not yield a good crop unless they be refreshed and stimulated by pretty constant showers.

Madura.

The best farmers in Madura like light falls of rain at intervals of about 30 days.

In lands well supplied with water as soon as the January crop is got in, preparations are made for a second crop of rice (*Mási Kódei*) which is raised in the same mode as the first and comes to maturity about May. Or, if the season be unpropitious during January and February, the second sowing will be in March or April or May, and the second reaping in July or August. The second crop will then be called the *Adi-Kódei*. Sometimes there will be three crops of rice raised in succession in a period of 13 or 14 months. Rapidly-growing and inferior kinds of rice are always grown for second and third crops.

Where the irrigational supplies are sufficient the second crop is either one of rice, plantains, sugar-cane or betel vines according as the nature and properties of the soil favour one or the other of those products; but where the supplies are scanty and insufficient for the highest kind of crops, *ragi, kambu, cholam, varagu* or rapeseed will be cultivated immediately after the *kalám* or principal rice crop has been reaped.

The manure used in rice cultivation consists of every kind of refuse procurable. No one can be more careful than the Madura ryot in utilising every thing (excepting always human excrement) which can assist him in raising his crops. The dung and urine of cattle, goats, horses, asses, sheep, and bats, ashes, lime, sweepings of houses, husks of paddy and other grains, the skins of fruits, decayed leaves of various kinds, bark, muck from the tanpit, milk-hedge, *varagu*, and other straw are some of the many substances commonly used as manure, and it appears that many of them are applied only in particular circumstances and on highly scientific principles. The practice of diluting manure with large quantities of water seems to be known to the ryot and various modes of making, altering, and correcting soils are well understood. The fluviatile deposits found in the beds of channels and tanks are largely made use of. The principle of raising different crops in rotation is also understood and to a limited extent acted upon, but the peculiarity of the mode of agriculture necessarily followed in a country where deep ploughing generally ruins the soil, precludes the ryot from going exactly the same road as the English farmer.

Rice is commonly boiled or eaten hot or cold in distinct grains with salt, *chatni*, curry, &c., or it is ground into flour from which are prepared cakes and sweetmeats. It is also used in many ways too numerous to describe." (*D. M., p. 102.*)

Cultivation of Rice in the Malabar District.

Malabar 374

Of this district it is stated in the Madras Manual of Administration that—

"Rice is sometimes sown broadcast, but is usually transplanted from nursery beds. The first or *kanny* crop is sown in April and May, and cut in August and September. The second or *makaram* (January) crop is sown in September and October, and reaped in January and February. These are the principal rice harvests, but there are intermediate crops in some places; and a third, known as *poonja*, is sown in February, and reaped in April or May. The greater portion of land, however, bears only one crop. Within the last twenty years, rice cultivation has considerably extended, but very little improvement has taken place in the quality of the rice, although experiments have been tried in the district with Carolina paddy." (*M. M., II., 105.*)

Cultivation of Rice in the Nellore District.

Nellore. 375

Mr. Boswell, in his District Manual, says there are some ten different kinds of rice grown on wet lands. He then gives the following instructive passages regarding the systems of cultivation pursued:—

"A few other descriptions are occasionally met with in various localities, but the extent to which they are cultivated is insignificant, and particulars are not, therefore, gone into. Wet cultivation is carried on under two distinct systems, termed *Veligada* and *Kudappa*. The former may be explained as the dry-ploughing, and the latter as the wet-ploughing system. *Veligada* cultivation is only adopted where the supply of irrigation will not suffice early in the season for the land to be ploughed wet under *Kudappa*, and where the soil may be loamy or light sandy and so favour that mode of cultivation. Heavy *regada* soils cannot well be cultivated under the system. The quality of the irrigation supply and the nature of the soil determine, therefore, the mode of cultivation to be pursued.

"Whilst a large series of forms of rice are cultivated under *Kudappa*, only *peda sannavari, vada sannavari*, and *kalinga sambava* are regularly put down under *veligada*; sometimes, but very seldom, *pishanam*, and *mosanam* are thus cultivated as well. With the first early rains the loamy and light sandy soils to be sown under *veligada* are at once ploughed and the operation is repeated at intervals, until the

CULTIVATION in Madras.

Nellore.

land may have been ploughed some five or six times. The seed is then either sown broadcast and ploughed in twice, or else it is drilled in with the *gorru* and covered up and levelled with the *guntaka*. Should rain not fall within the second or third day, the *guntaka* is used once more. The seed is only drilled, as a rule, in villages where the *gorru* may be kept up for dry cultivation. The seed sown under *veligada* is usually 2* *tums* or 64 *manikas* per *gorru*. For nearly two months paddy thus planted is dependent on rain. After one month the *gorru* or teeth leveller is used to loosen the soil, and about the end of the second month, but sometimes earlier if the weeds are very thick, water is let on, and the whole is once well weeded, or it is done if heavy rain sets in. Whilst the crop is being weeded, the plants are thinned in parts and transplanted in places where the crop may have partially failed. After weeding, the crop is regularly irrigated as far as may bepracticable, and a second weeding, if necessary, takes place during the third month. Weeding is a very much heavier business under *veligada* than *kudappa*. The period during which paddy can be put down under *veligada* is limited to August and September, whilst under *kudappa* it can be sown at any time that water may be available, for one of the several crops is sure to suit at any season of the year. As a rule, however, *kudappa* cultivation may be considered to commence late in June or during July, and continue till February; for after that it is rarely resorted to, save in second crop land, or in submerged land that could not have been sown before. The early *kudappa* cultivation towards the end of June or during July is generally under the river channels or river-supplied tanks. The crops most extensively put down under these sources are those known as *pishanam* and *sannavari*, which take six months to mature, and yield the white or superior description of rice. These two kinds are only cultivable till about the end of August, after which the common descriptions, which mature in a less period, are sown, but these details have already been rendered, in the statement showing the periods at which the cultivation operations of the several wet grains are carried on. The mode of cultivating each description of paddy under *kudappa* is the same, irrespective of the season at which put down. The land is first ploughed, with the water on it, from five to eight different times, either each time separately, or twice at the same time, with an interval of a few days between each operation according to local usage. The number of times it is ploughed will depend on the soil in some measure, as the loamy soils work up far more speedily than the heavy. The interval between the ploughings is requisite to allow time for the grass and weeds to rot and be killed, in some measure, by being ploughed into and worked up with the slush the paddy field comprises when fully ploughed. Upon the completion of the requisite ploughing, the banks are made up, and the water is partly let off and the beam-leveller is used to level the surface of the field and make it as smooth as practicable. The field is next flooded, and the seed, which will invariably have been previously steeped and closely packed in leaves, is sown broadcast over the field. The quantity of seed expended varies somewhat, but is generally about 2½† *tums* or 80 *manikas* per *gorru*. The water is left on the whole field the first night, but is completely drawn off in the morning. Subsequent irrigation is regulated with reference to the position of the soil, whether high or low-lying; and the nature of the soil, whether porous or retentive of moisture. In either of the former cases the field will be flooded on the third, fourth, or fifth day, and again let off; in the latter cases from the sixth to the ninth day, and at varying intervals afterwards, depending very much on the same conditions up to about the fifteenth or twenty-fifth day, where the crop is regularly irrigated. After thirty or forty days, the latter most frequently, the crop is weeded, and any thin parts are transplanted with plants taken from where the paddy may be overthick. Again, after another month, it is weeded a second time if requisite. In some villages it is usual for the teeth leveller or bushes heavily weighted, to be drawn over the young crop to submerge the weeds. This is done when the plant is a month old, and with a view of reducing the labour of weeding. After this treatment the paddy rises again, but the weeds and grasses are in some measure killed. What weeds may afterwards remain are removed by the hand at the close of the second month. Water is afterwards let on and off regularly according to the soil and season. Should an unfavourable wind prevail, and the plant be attacked by any grub, the usual course is to cut off the water altogether for a time and let the soil dry. No irrigation is necessary for the last twenty or thirty days before the paddy is harvested.

"From the foregoing details it will be observed that the system of cultivation throughout the district is that of sowing the paddy broadcast. Transplanting the seed prevails only in the extreme northern part of the district bordering on the Kistna, and in the Sriharikota villages formerly appertaining to the Chingleput

* Equivalent to about 23·87 Madras measures per acre.

† Equivalent to about 30 Madras measures or 40 seers per acre.

ORYZA sativa.

Rice Cultivation

CULTIVATION in Madras.

Nellore.

district. Where the irrigation is as regular as it now is under the anicut-supplied tanks and those fed from the river channels, the ryots may eventually realize the very great advantage to be derived from transplanting the paddy. The advantage which the transplanting system may be considered to have over the broadcast system is that the former admits of well-grown and vigorous young plants raised in highly manured beds being transplanted, which readily take root, and having the start, are able to vanquish the weeds which at once spring up beside them. Under the latter, or broadcast system, the weeds sprout with the young paddy, and, as the latter grows, it has to contend for existence against the weeds, and the growth of the paddy is consequently more or less checked. The ryots object to the transplanting system on the score of the labour involved; but, although the actual labour requisite at one time may be greater, it would not collectively be as great as that which must necessarily be expended in the several operations of weeding. The transplanting system may be regarded as particularly advantageous for the poorer descriptions of soil, inasmuch as a strong and healthy plant once transplanted in poor land is able to thrive better than seed sown in land deficient in strength, which affords the plant but slight chance of growing vigorously at first. Another and important advantage is, that the transplanted crop will mature with irrigation for a less period by some twenty or thirty days than the broadcast-sown crop. This latter question affects only the *kudappa* cultivation, not the *veligada*; under the latter, the period that water is required is less than under the transplanting system. (*D. M., pp. 388-391.*)

Tanjore.
376

Cultivation of Rice in the Tanjore District.

"Rice is the staple crop of the district, and is raised almost entirely by artificial irrigation. It is grown chiefly in the delta of the Cauvery, and to a much smaller extent in the upland portion of the district, under tanks fed by the local rainfall. The rice grown in Tanjore consists chiefly of two species, *viz.*, *kár* and *pishanam*, each including minor varieties. A few coarser sorts are sown broadcast, but this mode of cultivation is very limited, being carried on only in a few places beyond the delta, and there on rain-fed land. In all cases of irrigated cultivation, young plants are raised in seed-beds and transplanted. The *kár* is planted in June and reaped in October. The *pishanam* in July and August and reaped in January and February." (*M. M. of Adm., p. 127.*)

The former are the inferior, and the latter the superior, class rices of the district. *Kár* is referred to two kinds, *kadappu* and *kár* proper, and the *pasánam* is similarly spoken of as of two kinds, *samba* and *pasánam* proper. The following account of Tanjore rice cultivation is taken from the District Manual :—

"The dates of actual commencement and termination of agricultural operations vary in the different parts of the district. This is due to the first supply of freshes reaching the eastern portion of the delta later; not that a flood takes any number of days in travelling within the district, but because the early flood is almost wholly utilised westward, and the cultivators at the eastern end have to wait until the more copious and, therefore, uninterrupted series of freshes reach them. Thus, in the further west of the delta, the *kár* cultivation (planting) begins in May and the crop is harvested in September. The *pasánam* cultivation begins in September and the crop is harvested in January when (oftener a few days before) a third crop, a pulse called *vayal-payaru*, * is sown and picked in two months, *i.e.*, in March. This pulse requires no ploughing and grows chiefly with the aid of the moisture that remains in the fields and succeeds even if there be no rain afterwards. In the extreme east of the delta, on the other hand, where more than one crop is hardly ever cultivated, the cultivation begins in August and sometimes (as when freshes are later than usual) in September and the crop is harvested in March. Where only one crop is grown in the year it is more generally one of seven months' duration from planting to reaping. The general mean may be taken broadly as four months, from July to October, for the *kár* species, and as six months, from September to February, for the *pasánam*.

Wherever two crops are raised, the second is usually one of the quicker-growing varieties of *sembá*. Taking the delta as a whole, the proportion of *kadappu kár* species to the entire paddy crop may be taken to be one-sixth.

The *kár* species draws more water and yields a greater abundance of crop than the *pasánam*. It will bear almost any amount of moisture without being injured. In times of old when the rainfall in October and November used to be very heavy with, at the same time, high floods in the rivers, it was usual to reap the

* Means field pulse, the prefix being intended to distinguish it from the dry grains and pulses which are grown more generally on high level lands.

CULTIVATION in Madras.

Tanjore.

kár crop in the very midst of rain and with water standing in the fields 2 feet deep.

Throughout the delta the invariable mode for irrigated cultivation is transplanting. *Kuttálei* is the only variety of paddy sown broadcast, but it is grown chiefly in the upland parts of the district. In some villages in the central part of the delta, one of the varieties of *pasánam*, called *ottadan*, is planted along with *kadappu* proper; when the latter, which grows quicker, is reaped, the ends of the *ottadan* plant, then about half-grown, are cut along with it, and the latter then shoots up all the more luxuriantly for it and comes to maturity at the usual season of the *pasánam* harvest. In this case the seeds of the two varieties are mixed together (in the proportion of ¾ *kadappu* and ¼ *ottadan*) before they are put down in the nursery and the young plants of both, which are hardly distinguishable, are planted promiscuously.

Allusion has been made to the pulse (*Vayal-payaru*) grown as a third crop on paddy lands. More generally, however, a dry crop forms an auxiliary to only a single crop of paddy, being adopted as an alternative where, as often happens, the available supply of water is not sufficient for two paddy crops, or the *mirásidárs* cannot afford the requisite labour. The dry crop is either the pulse above mentioned, *gingelly* or *ragi*; the latter two require ploughing, but it is only *ragi* that is planted. This additional crop is raised either before or after the paddy season. In the first case it is sown or planted, as the case may be, with the aid of the summer shower or the first fresh when it happens to come earlier than usual, but cannot be relied on for beginning operations for the cultivation of a regular wet crop. In the second case, if the dry crop is *gingelly* or *ragi*, the field is ploughed when there is yet moisture in it. The crop is then sown or planted in February and reaped in May, being matured entirely by the summer showers (*D. M., p. 354, et seq.*).

Manure.—As a rule, in the case of irrigated *nansei* or rice cultivation it is only the fields on which two crops are raised that are manured every year, and even these not invariably, for it is not always that the required manure is available. Single-crop lands are generally manured once in five years; in some cases not at all, the latter being often the case when the holdings are very large, and also when, owing to the *mirásidár* not residing on the spot or other causes, the cultivation is neglected. In the more western and better irrigated parts of the delta the alluvia brought down by the river freshes and the stubble, which is ploughed into the soil, are generally believed to furnish the required manure, and ordinarily manure is considered essential only where the soil is poor or the rice lands depend for their irrigation on the drainage of those situated higher up, which brings them little or no alluvial matter. In the case of dry cultivation, as also of irrigated cultivation under tanks and wells in the upland parts of the district, manure is more generally employed, great pains being taken to procure it.

The mode of manuring generally adopted is by enfolding sheep and goats or cattle; the former being more commonly employed. Cattle manure preserved in houses and in cart and cattle stands, vegetable mould, ashes and other refuse of cook-rooms, night soil and silt scraped from the beds of dried-up tanks are also used, but all these, except perhaps vegetable mould, can be found in sufficient quantities for only small holdings. (*D. M., p. 348.*)

Rotation.—In the case of irrigated cultivation there is no rotation of crops, except that the land usually devoted to paddy is sometimes planted with betel vines or sugar-cane. Such instances, however, are rare." (*D. M., p. 351.*)

III.—CULTIVATION IN MYSORE.

MYSORE. 377 *Conf. with p. 522.*

References.—*Mysore & Coorg Gazetteers; Agricultural Statistics published by the Government of India.*

METHODS OF CULTIVATION, SOWING, REAPING, &c.

Provincial Methods. 378

Next to *rágí*, rice is the most important crop in Mysore, though the area under cultivation is gradually declining. In 1870-71 it covered 25 per cent. of the cropped area, while the statistics of 1874-75 show it to have fallen to 10 per cent. Rice is grown to a more or less extent in all the districts of the Mysore State. It is most extensively cultivated in Shemoga (161,000 acres), Hassan (64,783 acres), Mysore (59,891 acres), and Kolar (54,412 acres).

Three methods of cultivation are practised in Mysore, namely (1st) the *bara batta* or *punaji*, in which the seed is sown dry on the fields; (2nd) *mole, batta*, in which germinated seed is sown in fields reduced to a puddle; and (3rd), the *náti*, in which the seed is transplanted from a nursery after it has attained a foot in height.

ORYZA sativa.

Rice Cultivation

CULTIVATION in Mysore.

The extracts from the Gazetteer of Mysore and Coorg, by Mr. L. Rice, given below, fully describe the above-mentioned methods of rice cultivation:—

379 Dry Seed Cultivation.—In the *hain* crop the following is the management of the dry seed cultivation. During the months of *Phálguna, Chaitra,* and *Vaisákha,* that is, from February till May, plough twice a month, having, three days previous to the first ploughing in *Phálguna,* softened the soil by giving the field water. After the fourth ploughing the fields must be manured with dung, procured either from the city or cow-house. After the fifth ploughing the fields must be watered either by rain or from the canal; and three days after, the seed must be sown broadcast and then covered by the sixth ploughing. Any rain that happens to fall for the first 30 days after sowing the seed must be allowed to run off by a breach in the bank which surrounds the field, and should much rain fall at this season, the crop is considerably injured. Should there have been no rain for the first thirty days, the field must be kept constantly inundated till the crop be ripe; but if there have been occasional showers the inundation should not commence till the forty-fifth day. Weeding and loosening the soil about the roots of the young plants with the hands, and placing them at proper distances, where sown too close or too far apart, must be performed three times; first, on the 45th or 50th day; secondly, 20 days afterwards; and thirdly, 15 days after the second weeding. These periods refer to the crops that require seven months to ripen. In rice which ripens in five-and-a-half months, the field must be inundated on the 20th day; and the weedings are on the 20th, 30th, and 40th days.

380 Sprouted seed cultivation.—Mr. Rice gives the following description of the method followed in the sprouted seed cultivation in the *hain* crop. The ploughing season occupies the month of *Ashádha* (June-July). During the whole of this time the field is inundated and is ploughed four times; while at each ploughing it is turned over twice in two different directions which cross each other at right angles. This may be called double ploughing. About the first of *Sravána* the field is manured, immediately gets a fifth ploughing, and the mud is smoothed by the labourers' feet. All the water, except one inch in depth, must then be let off, and the prepared seed must be sown broadcast. As it sinks in the mud it requires no labour to cover it. For the first twenty-four days the field must once every other day have some water, and must afterwards, till ripe, be constantly kept inundated. The weedings are done on the 25th, 35th, and 50th days. In order to prepare the seed it must be put into a pot, and kept for three days covered with water. It is then mixed with an equal quantity of rotten cowdung, and laid on a heap in some part of the house, entirely sheltered from the wind. The heap is well covered with straw and mats; and at the end of three days the seed, having shot out sprouts about an inch in length, is found fit for sowing. This manner of cultivation is much more troublesome than that called dry seed; and the produce from the same extent of ground is in both nearly equal; but the sprouted seed cultivation gives time for a preceding crop of pulse on the same field, and saves a quarter of seed.

381 Cultivation by transplanting the seed.—Mr. Rice, says that two distinctions are made in this method, the one called *baravági* or by *dry-plants*; and the other called *nírági* or by *wet-plants*; and that for both kinds low land is required.

382 (*A.*) *Baravági* or *dry-plants.*—"Labour the ground at the same season, and in the same manner as for the dry seed crop. On the first of *Jéshta,* or in May, give the manure, sow the seed very thick, and cover it with the

CULTIVATION in Mysore.

plough. If no rain fall before the 8th day, then water the field, and again on the 22nd; but if there are any showers, these waterings are unnecessary. From the 45th till the 60th day, the plants continue fit to be removed. In order to be able to raise them for transplanting, the field must be inundated for five days before they are plucked.

The ground on which the dry seedlings are to be ripened is ploughed four times in the course of eight weeks, commencing about the 15th of *Jéshta*, but must all the while be inundated. The manure is given before the fourth ploughing. After this, the mud having been smoothed by the feet, the seedlings are transplanted into it, and from three to five plants are stuck together into the mud at about a span distance from the other little branches. The water is then let off for a day; afterwards the field, till the grain is ripe, is kept constantly inundated. The weedings are performed on the 20th, 35th, and 45th days after transplanting.

383 (*B.*) *Nirági* [or *wet-plants.*—In the month of *Phálguna* (February-March) plough the ground three times while it is dry. On the first of *Jéshta* inundate the field; and in the course of fifteen days plough it four times. After the fourth ploughing smooth the mud with the feet, sow the seed very thick, and sprinkle dung over it; then let off the water. On the 3rd, 6th, and 9th days water again; but the water must be let off, and not allowed to stagnate on the field. After the 12th day inundate until the seedlings be fit for transplantation, which will be after the 30th day from sowing. The cultivation of the field into which the seedlings are transplanted is exactly the same as that for dry seedlings. The plot on which the seedlings are raised produces no crop of pulse; but various kinds of these grains are sown in the fields that are to ripen the transplanted crop, and are cut down immediately before the ploughing for the rice commences. The produce of the transplanted crop is nearly equal to that of the dry seed cultivation; and on a good soil, properly cultivated, twenty times the seed sown is an average crop.

384 RICE CULTIVATION IN THE KÁR SEASON.—Mr. Rice states that the *kár* crops, according to the time of sowing, are divided into three kinds. "When the farm is properly stocked the seed is sown at the most favourable season, and the crop is then called the *kumba kár;* but if there be a want of hands or cattle, part of the seed is sown earlier, and part later, than the proper season; and then it produces from 30 to 50 per cent. less than the full crop. When sown too early the crop is called *tula kár;* when too late it is called *mésha kár*. The produce of the *hain* and *kumba kár* crops is nearly the same.*

"No *tula kár* dry seed is ever sown. The ploughing season for the *kumba kár* dry seed is in *Bhádrapada* (August), and the seed is sown about the end of *Márgasira* (December). In the *mésha kár* dry seed the ploughing commences on the first of *Chaitra* (March) and the seed is sown at the feast of *Chaitra Paurnami* in April.

The *tula kár* sprouted seed is sown on the 1st *Kártika* (October), the ploughing having commenced with the feast *Navarátri*, in September. The *kumba kár* sprouted seed is sown in *Pushya* about 1st of January. The ploughing season occupies a month. The ploughing for the *mésha kár* sprouted seed commences about the 15th of *Chaitra*. The seed is sown about the 16th of *Vaisákha* (May).

"The *kumba kár* transplanted rice is cultivated only as watered seedlings. The ground for the seedlings begins to be ploughed in the end of *Kártika* or middle of November, and the seed is sown on the 15th *Pushya* or end of December. The fields on which the crop is ripened are begun to be ploughed about the middle of *Márgasira* (1st December), the trans-

* *Kumba*, or *kumbha*, is the sign Aquarius; *Tula* is Libra; *Mesha* is Aries.

ORYZA sativa.

Rice Cultivation

CULTIVATION in Mysore.

planting takes place about the 15th of *Mágha* or end of January. The *tula kár* transplanted rice also is sown as *nírági*, about the 30th of *Asvija* or middle of October, and in a month afterwards is transplanted. The *mésha kár* transplanted rice is also sown as watered seedlings, about the 15th of *Vaisákha* (May), and about a month afterwards is transplanted. The regular *kár* crop of the transplanted cultivation does not interfere with a preceding crop of pulse; but this is lost when, from want of stock sufficient to cultivate it at the proper time, the early or late seasons are adopted. The various modes of cultivating rice give a great advantage to the former; as by dividing the labour over great part of the year fewer hands and less stock are required to cultivate the same extent of ground than if there was only one seed time, and one harvest.

385 REAPING AND PRESERVATION.—The manner of reaping and preserving all the kinds of rice is nearly the same. About a week before the corn is fit for reaping, the water is let off, that the ground may dry. The corn is cut down about four inches from the ground with a reaping-hook called *kudagalu* or *kudagu*. Without being bound up in sheaves it is put into small stacks, about 12 feet high; in which the stocks are placed outwards and the ears inwards. Here the corn remains a week, or if it rain, fourteen days. It is then spread out on a threshing floor, made smooth with clay, cowdung, and water; and is trodden out by driving bullocks over it. If there has been rain, the corn after having been threshed, must be dried in the sun; but in dry weather this trouble is unnecessary. It is then put up in heaps called *ráshi*, which contain about 60 *kandagas* or 334 bushels. The heaps are marked with clay and carefully covered with straw. A trench is then dug round it to keep off the water. For twenty or thirty days (formerly, till the division of the crop between the Government and the cultivator took place) the corn is allowed to remain in the heap. The grain is always preserved in the husk, or, as the English in India say, in *paddy*. There are in use here various ways of keeping paddy. Some preserve it in large earthen jars that are kept in the house. Some keep it in pits called *hagevu*. In a hard stony soil they dig a narrow shaft, 15 or 16 cubits deep. The sides of this are then dug away so as to form a cave with a roof about two cubits thick. The floor, sides, and roof are lined with straw; and the cave is then filled with paddy. These pits contain from 15 to 30 *kandagas*, or from 83½ to 167 Winchester bushels. When the paddy is wanted to be beaten out into rice, the whole pit must at once be emptied. Other people again build *kanajas*, or store-houses, which are strongly floored with plank to keep out the bandicoots or rats. In these store-houses there is no opening for air; but they have a row of doors one above another, for taking out the grain, as it is wanted. Another manner of preserving grain is in small cylindrical stores, which the potters make of clay, and which are called *wóde*. The mouth is covered with an inverted pot; and the paddy, as wanted, is drawn out from a small hole at the bottom. Finally, others preserve their paddy in a kind of bag made of straw, and called *múde*. Of these different means the *kanaja* and *wóde* are reckoned the best. Paddy will keep two years without alteration, and four years without being unfit for use. Longer than this does not answer, as the grain becomes both unwholesome and unpalatable. No person here attempts to preserve rice any length of time; for it is known by experience to be very perishable. All the kinds of paddy are found to preserve equally well. That intended for seed must be beaten off from the straw as soon as cut down, and dried for three days in the sun, after which it is usually kept in straw bags.

386 CONVERSION OF PADDY IN RICE.—"There are two manners of making paddy into rice, one by boiling it previously to beating; and the other

CULTIVATION in Mysore.

by beating alone. The boiling is also done in two ways. By the first is prepared the rice intended for rajas and other luxurious persons. A pot is filled with equal parts of water and paddy, which is allowed to soak all night, and in the morning is boiled for half an hour. The paddy is then spread out in the shade for fifteen days, and afterwards dried in the sun for two hours. It is then beaten to remove the husk. Each grain is broken by this operation into four or five pieces, from which it is called *aidu núgu akki*, or five-piece rice. When dressed, this kind of rice swells very much. It is always prepared in the families of the rajas, and is never made for sale. The operation is very liable to fail; and in that case the rice is totally lost.

"Rice prepared by boiling in the common manner is called *kudupal akki*, and is destined for the use of the sudras, or such low persons as are able to procure it. Five parts of paddy are put into a pot with one part of water, and boiled for about two hours, till it is observed that one or two of the grains have burst. It is then spread out in the sun for two hours; and this drying is repeated on the next day, after which the paddy is immediately beaten. Ten parts of paddy, by this operation, give five parts of rice, of which one part goes to the person who prepares it for his trouble. Ten seers of paddy are therefore equal in value to only four seers of rice.

The rice used by the Brahmans, and called *hasí akki*, is never boiled. On the day before it is to be beaten, the paddy must be exposed two hours to the sun. If it were beaten immediately after being dried, the grain would break, and there would be a considerable loss. Even with this precaution many of the grains break; and, when these are separated from the entire rice to render it saleable, the *hasí akki* sells dearer than the *kudupal akki*, in the proportion of nine to eight.

"The beating is performed chiefly by women. They sometimes, for this purpose, use the *yáta*, or a block of timber fastened to a wooden lever, which is supported on its centre. The woman raises the block by pressing with her foot on the far end of the lever, and by removing her foot allows the block to fall down on the grain. The more common way, however, of beating paddy, is by means of a wooden pestle, which is generally about four feet in length, and three inches in diameter, which is made of heavy timber, and shod with iron. The grain is put into a hole formed in a rock or stone. The pestle is first raised with the one hand and then with the other; which is very hard labour for the woman." (*I., 102-107.*) It is obvious from the above process, which is not confined to Mysore alone, but is common to the people of Northern India and elsewhere, that the grain must by constant hammering be broken into small pieces. In this state it resembles, what has been above described, the sort of rice called *aidu núgu akki*, which is eaten only by the rajas and the wealthier classes.

IV.—CULTIVATION IN COORG. COORG. 387

Area under, and Traffic in, Rice. Area. 388

The Agricultural Statistics published by the Government of India show that there were 74,499 acres under rice in Coorg during the year 1888-89. The statistics of trade are not, however, shown separately, so that it is not possible to discover the balance between imports and exports. *Conf. with p. 522.*

METHODS OF CULTIVATION, SOWING, REAPING, &c. Provincial Methods. 389

The following extracts from the *Coorg Gazetteer* by Mr. Lewis Rice give a concise account of the methods of rice cultivation followed in Coorg.

"*Rice.*—This is the staple product of Coorg. The numerous valleys throughout the land have, from ancient times, yielded an unfailing supply every year for home consumption and for exportation to the Malabar Coast. The rice-valleys are most extensive in South Coorg—in the neigh-

CULTIVATION in Coorg.

bourhood of Virajpet and in Kiggatnad—where some fields are of considerable breadth and several miles in length; but owing to the surrounding low deforested hills, which yield little fertilizing detrition, the soil is of a quality inferior to those fields of the narrower valleys near the Ghâts, where the ground is terraced at considerable pains, but every field large enough for the use of the plough.

"The lower and broader fields of a valley having a rivulet running through them, are called *bailu-gadde,* and those terraced up along the sides, and chiefly depending on the rainfall, are named *maki-gadde.*

"The rice cultivated throughout Coorg, and in general use, is the large-grained *dodda-batta,* which is also exported. A finer and more palatable kind is the small rice *sanna-batta,* and a red variety the *késari*; for parched rice the *kalame* is the kind used.

YIELD.—"Except in a few valleys in North Coorg, there is annually but one rice crop, but its return is so rich that the ryots may well be satisfied, and allow their wretched cattle rest and their fields to lie fallow, or to 'sun themselves,' as the natives say, for the remainder of the year. Whilst in the low country and in some places of North-Coorg, the average return of one crop is from ten- to twenty five- fold, that in most parts of Coorg proper is from forty- to sixty- fold, and in seasons of extraordinary fertility from eighty- to a hundred- fold.

AGRICULTURAL IMPLEMENTS.—"The agricultural implements are few and of the rudest kind. The plough, constructed by the ryot himself, consists of a sampige-wood ploughshare, with an iron point, a handle of paliwood, and a pole of sago palm-wood for the yoke, and is so light that the farmer carries it to the field on his shoulders. Its value hardly exceeds one rupee. The *Tawe,* which answers to our English harrow, is generally a simple board to which a split bamboo is fastened to connect it with the yoke. The driver standing on the board adds to the efficiency of the operation, be it for pulverizing dry ground, as in the *Múdu-shime* or eastern district, or smoothing and levelling the wet fields. A strong sickle and a *mamoti* or hoe complete the stock of farming implements.

SEED AND LABOUR.—"To cultivate 100 butties of land, which is equivalent to an area yielding 100 butties at 8 seers by measure of paddy or rice in the husk, a farmer requires either a pair of bullocks or a pair of buffaloes, one plough and two labourers. On Monday he does not plough with bullocks, but with buffaloes only, considering Monday as the day of the bullock's creation.

MANURE.—"Whatever of cattle manure and dry leaves has been collected during the year, is, in the dry season, carried by women to the fields in large baskets, and deposited in little heaps which are there burnt and the ashes subsequently strewn over the ground.

PLOUGHING.—"With the first shower in April and May the ploughing commences. On a propitious day before sunrise the house lamp—'*Táli-akki-balake*' (dish-rice-lamp)—which plays a conspicuous *role* on all festive occasions is lighted in the inner verandah; the house people assemble and invoke their ancestors and *Kaveri Amma* for a blessing; the young men make obeisance to their parents and elders, and then drive a pair of bullocks into the paddy-fields, where they turn the heads of the beasts towards the east. The landlord now offers cocoanuts and plantains, rice and milk to the presiding deity of his *Nád,* and lifting up his hands in adoration to the rising sun invokes a blessing. The oxen are yoked and three furrows ploughed, when the work is finished for that morning. Of the turned-up earth they take a clod home to the store-house or granary, praying Siva to grant them an hundred-fold increase.

This recognition of the source of material well-being is followed by

CULTIVATION in Coorg.

personal industry that should command success. From 6 to 10 in the morning the ploughing is continued, till all the fields are turned over two or three times. Then the borders are trimmed, the channels cleaned, and the little banks between the fields repaired to regulate the water. By the end of May, one part of the fields, which commands a permanent water-supply and has been well manured, is prepared for a nursery, by repeated ploughing and harrowing, whilst the whole field is submerged.

SOWING.—"For every 100 *butties* of land from 2 to 2½ *butties* of grain are required for seed. The seed paddy is heaped up on the north side of the house, watered for three days and then covered up with plantain leaves and stones till it begins to sprout. The nursery ground has meanwhile been again ploughed and harrowed and the water allowed to run off so that the grain when sown is just embedded in the soft mud. After 20 or 30 days the blades have attained a height of about one foot and the seedlings are ready for transplanting.

TRANSPLANTING.—"Pleasing as are young corn and clover fields in Europe, there is no vegetation there which surpasses in beauty the brilliant green of a rice nursery. The eye is irresistibly attracted to these bright spots and rests upon them with the utmost delight. Regulated by the monsoon rain the rice-planting takes place during July and August. The women covered with leaf umbrellas called *Goragas*, that rest on the head and protect the whole of the body, pull out the plants from the nursery and tie them in small bundles, which are collected in one spot.

"Meanwhile, the submerged fields are repeatedly ploughed and levelled with the *Tawe*, 'till the soil is soft as treacle, white as milk the foaming surface' when all the men of the house, placed in a line and standing almost knee deep in the muddy fields, begin the transplanting in which the women are not expected to join. The bundles are conveniently deposited over the fields; each man takes a handful of plants at a time into his left, and with the right hand presses with great rapidity 6 or 8 seedlings together into the mud keeping a regular distance of about 6 inches. Before the completion of the largest field a space of about 10 feet wide is left throughout the whole field. This is the Coorg's race-ground, and offers right good sport which greatly exhilarates their monotonous task. All the men engaged in the work—and fifteen are reckoned for 100 *butties* of land—may run, but four or five only obtain a prize. Wearing merely a pair of short drawers they are eager for the run, for which their powerful legs well qualify them. The signal is given and away they scramble and plunge and stagger in the deep mud, roars of laughter greeting the unfortunate wight who sinks in. Having reached the opposite bank they return the same way, and hard is their struggle as they near the winning-post. The first comer is awarded with a piece of cloth, the second with a bunch of plantains, the third with a jack-fruit, the fourth with a basket of oranges, and the fifth with parched rice. When all the fields are planted, a feast for the people is given by the landlord.

WATERING.—"As a protection against the evil eye, some half-burnt bamboos about 6 feet high are erected in a line throughout the middle of the fields. It is now the farmer's business to regulate the water-supply in each field and to fill up holes made by crabs in the embankments. Also the weeding is attended to and any failures are replanted. At the end of October when the ears of grain are fully out, huts on high posts are erected, one for every hundred *butties*, for the watchman who guards the crops against wild beasts, occasionally firing off a gun.

REAPING.—"In November or December the paddy gets ripe and the Feast of First Fruits or *Huttari* is celebrated, after which the paddy may be reaped. The water is drained off the fields, the paddy cut down with

ORYZA sativa.

Rice Cultivation

CULTIVATION in Coorg.

sickles close to the ground and spread out to dry; after five or six days it is bound up in sheaves, carried home and stacked in a heap, the ears turned inside.

THRESHING, WINNOWING.—"In January and February, chiefly on moonlight nights, the sheaves are taken down to the threshing floor, spread round a stone pillar fixed in the middle and trodden out by bullocks and buffaloes, when the paddy is winnowed, the best quality reserved for seed and the rest stored up in a granary, for home consumption or for sale, the price varying from R2 to R4 a *butty* of 80 seers. A threshing machine introduced by **Captain Mackenzie** excited the astonishment and admiration of the natives; but the hand labour of the two coolies for turning it appeared to them too severe and impracticable for large quantities of paddy. A winnowing machine would find greater favour." (*III., 29-32.*)

BOMBAY. 390

V.—CULTIVATION IN BOMBAY.

References.—*Travels in India by Jean-Baptiste Tavernier (Bull's Transl.), I., 49, 142; Hove, Journ. Tour in Bombay (1787), 13, 16, 36, 123, &c.; Agri.-Horti. Soc., Ind. Jour. (Old Series) II., Sel., 162; Crop Experiments, Bombay, 1884-85, 16; Report, Rev. & Agri. Dept., Bombay, 1884-85, 60-70; Bombay Gazetteers:—II. (Surat), 65; (Broach), 406; III. (Kaira), 47; (Panch Mahals), 233; IV. (Ahmedabad), 57; V. (Mahi Kántha), 371; VI. (Surat), 247; VIII. (Káthiáwár), 186; X. (Ratnágiri), 147; (Sávantvádi), 424; XI. (Kolaba), 95; XII. (Khandesh), 150; XIII. (Thana), 287; XIV. (Thana), 113; XV. (Kánara), 16, 30; XVI. (Nasik), 99; XVII. (Ahmadnagar), 267; XVIII. (Poona), 517; Pt. II. 36; XIX. (Satara), 162; XXI. (Belgaum), 248-250; XXII. (Dhárwár), 275-277; XXIII. (Bijapur), 321; and XXIV. (Kolhapur), 164-165; Sind Gazetteer:—9, 10, 20, 170, 215, 249, 268, 317, 427, 493, 534, 559, 571, 573, 591, 628, 631, 654, 670, 745, 759, 820, 851, 859-860.*

Area under, and Traffic in, Rice.

Area. 391 The Agricultural Statistics of British India show Bombay Presidency (including Sind) as having had in 1888-89 2,239,198 acres under rice.

Conf. with p. 522.

According to the crop experiments which were performed at various times under orders of Government, the average yield per acre may be put down at 20 maunds, so that the total production may possibly have been

Yield 20 mds. per acre. 392 4,47,83,960 maunds or, say, 399,757 tons. A greater importance may be attached to this figure of yield per acre than is claimed for the similar figures given in connection with other provinces. A very extensive series of actual experiments, with selected fields, was performed, under European supervision, and the figure here given is the average of all the returns. It seems possible, however, that it may have been slightly increased through some of the returns being for paddy, not cleaned rice. This would represent an error of 25 per cent.—in such cases—excess of the actual yield of clean grain.

Trade. 393 An inspection of the published figures of the trade in rice reveals the fact that by internal routes the Presidency obtained a net import from other Provinces of 5,71,067 maunds (taking paddy and rice conjointly). The coastwise transactions were of even still greater importance, since the Presidency obtained a net import (excluding intra-provincial exchanges) of 24,48,226 maunds. The sources of this supply were chiefly Bengal, Madras, and Burma. The figures of the foreign trade, however, show a net export of 8,46,391 maunds (or 30,228 tons), the gross export having been 30,436 tons, valued at R29,42,599. A small export trade also took place across the Sind frontier, the largest consignment being destined for Lus Bela. It may thus be said that the ultimate result of all these transactions was a net import of 21,72,902 maunds into the Presidency (including its port towns), so that that amount may be added to the estimated

CULTIVATION in Bombay.

production in order to discover the actual consumption of rice. This worked out to the total population of Bombay and Sind might be said to be a consumption of ¾ of a seer per head. The exceptionally high yield, here accepted as a fair average, chiefly accounts for this apparent greater consumption of rice in Bombay than in Bengal, Madras, and other Provinces. More careful crop experiments would seem likely to raise the Bengal and Madras yields, and perhaps slightly lower that of Bombay, but any such corrections as found necessary could be made in the calculations here given without seriously affecting the general contentions advanced. There is this much to be said for Bombay rice cultivation that, agriculturally, it is carried out on a more careful and scientific system than in Bengal. It is, so to speak, with the exception of certain tracts near the coast, of a semi-exotic character, and its cultivation is not, as in Bengal, a matter of casting the seed on imperfectly ploughed fields (or, indeed, on fields not ploughed at all), but it is grown under a high class system of agriculture in which manuring and rotation form integral parts. Where transplanting is not followed the seed is often drill-sown, a practice which nowhere exists in Bengal, and which may be taken as manifesting the high class cultivation followed in Bombay. A greater yield might be looked for, therefore, as a necessary consequence of such agriculture. An interesting feature of the Bombay rice cultivation is the fact that it is mainly, if not entirely, in the production of the high class rices which belong to the *rowa* section of *áman*. A special series of forms, of what might be called *boro* rices (grown in brackish water), are found in Bombay; these are extremely interesting. They are inferior red rices which are chiefly eaten by the poorer classes. The cultivator of the better qualities of rice, grows the crop purely for the purpose of selling the grain. As a rule, he cannot afford to eat it himself, and thus not being dependent on rice for his own sustenance, he can sell the produce in small instalments according to the condition of the market.

Yield. 394

Superior System. 395

Drill Sown. 396

The numerous crop experiments, performed by Government, have revealed many interesting facts, but, for a complete review of these, unfortunately, space cannot here be afforded. They are, however, published in a series of annual volumes which the reader should find no difficulty in procuring. One or two instances may, however, be specially cited. A field specially selected for test-reaping was located in Salsette (near Bombay). The seed sown per acre was 100lb, the yield of grain from the same 4,220lb and of straw 14,253lb, the total value of the produce having been put down at R73-12-0. The remark made against that experiment states that the crop was obtained from "unirrigated, *unrábed* rice, grown every year on a land reclaimed from the sea-side and sweetened by heavy sweepings-manure. Manured in June, July, and August with sweepings from the Bándra slaughter-house at 120 cart-loads per acre. Cultivation good, but the most persistent growth of weeds in the field stunted the crop. Season though marked by excessive rainfall, was not unfavourable to rice crops in the districts. The seed was sown in the highly-manured soil and the seedlings, where very thick, were removed and transplanted elsewhere. The yield was locally estimated at a 12 annas crop." The Commissioner of the Division commenting on the experiments performed with rice, specially alludes to the Salsette trial. "All the rice experiments of the Thana District," he says, "show a very high outturn, proving how remunerative rice cultivation can be made by good farming and liberal manuring." "The Salsette experiment," he adds, "should encourage further reclamation schemes. There is still much land available."

Heavy Manuring. 397

In many parts of Bombay the ground is pressed firmly against the young seedling rice plants. This is done by a mud-roller called the *kodata*. "This consists in rolling the standing crop with a broad bar

Rolled. 398

ORYZA sativa.

Rice Cultivation

CULTIVATION in Bombay.

of wood hollowed on the lower side in the direction of its length, before the stems have hardened, while the field is still full of water. Its objects are two-fold. It checks the natural tendency of the crop to excessive growth of straw at the expense of the grain-yield and thus corresponds to the transplantation of seedlings raised in a nursery, and it also keeps down the weeds which have grown up since the bullock-hoeing and hand-weeding. The *hodata* cannot be done unless the rain is sufficient to flood the field with impounded water."

A careful investigation into the question of the cost of cultivating an acre of rice land in Belgaum resulted in the opinion being formed that this might be put down at R25. Among the items of expenditure may be mentioned R4-8 for manure, cost of rice seed at 80fb per acre R2, hand weeding R2, and hoeing, ploughing, &c., R2-14. Commenting on the report on the above test case it is pointed out that rice is grown in Belgaum without *rab*. The seed is sown direct|by a 6-coulter drill and the seedlings are not transplanted. The average value of the produce of an acre of rice land, worked out from 37 of the experiments tabulated in the *Reports of Bombay Crop Experiments*, the writer has found to be R40-6-3.

Rotation Manuring 399

Rotation of Crops, Manuring, &c.

Throughout the Reports of Crop Experiments frequent mention is made of these subjects. The rotation which is generally pursued will be discovered from the special passages quoted below from district reports of rice cultivation. The subject of manuring, however, seems worthy of special consideration. As has already been remarked, rice, in Bombay, is cultivated under a more rational system than in Bengal, Madras, and Burma—the provinces that produce the largest amount of the grain. The fields are elaborately ploughed, carefully weeded, and the land richly manured. As might be expected a considerably higher acreage yield is thereby obtained than in the provinces where rice is essentially the distinctive feature of their agriculture. In such provinces the ease with which the grain is obtained has engendered a slovenly system of agriculture which has extended its baneful influence to every feature of peasant life. It is thus not a matter of congratulation that the rich alluvial *rowa* soils of Bengal yield good harvests, year after year, of rice of a quality scarce equalled by the painstaking cultivators of tracts less suited to the crop. This gift of nature to insalubrious regions can scarce be used as an argument that improvement is impossible. **Mr. C. B. Clarke**, however, in his paper on Bengal rice, seems to think improvement problematic. Concluding his remarks regarding the *rowa* rice he says :—" The *rowa* rice is the food of the upper classes. It is to be noted that the cultivator except at harvest (when the weather is set fine and he works easily) does not do a day's work on his *rowa* land in the year, except about a couple of days at dibbling time. Those who propose to teach the Bengalee cultivator a more elaborate system of rice farming must ponder this fact well." Again, while speaking of the dilatoriness of the Bengalee cultivator in preparing his seed bed, **Mr. Clarke** says :—" One reason for this is the deep-seated Bengalee principle never to do to-day what can possibly be put off till to-morrow, and to do everything incompletely." In fact, **Mr. Clarke** seems to think the improvement of Bengal rice cultivation lies more in the direction of fostering a greater spirit of promptness than in better and more thorough ploughing or manuring; accordingly he concludes :—" My view is that rice is the very last crop on which we should attempt to give the Bengalee instruction." But the better results obtained in Bombay and elsewhere, with a superior agriculture, in which a liberal system of manuring is one

Conf. with average yield for all India, p. 644, also the paras. on manuring, p. 642.

CULTIVATION in Bombay.

Manuring.

of the cultivator's axioms, seem unmistakably to demonstrate the possibility of improvement. Whether the Bengal cultivator desires to better his position in life by obtaining a higher return from his land is of course an altogether independent question, and one which will probably be finally solved by the advance of education and the concomitant increase of luxury demanding more from the soil than is at present taken from it.

But to return to the discussion of the system of manuring pursued in Bombay this may be said to be referable to two important sections:—

(*a*) The heavy manuring of land in which rice is drill-sown or broadcasted.

Rab. 400

(*b*) The *ráb* cultivation, or that in which a nursery (similar to the *rowa* of Bengal) is richly manured by burning on its surface a mixture of farmyard manure, leaves and boughs of trees, straw and pulverised soil.

Regarding the former very little need be here said, since it differs in no respect from heavy manuring for any other crop, and is alluded to freely in the special passages quoted below. *Ráb* cultivation is, however, while not confined to rice cultivation, so peculiarly the system by which the better class rices of Bombay are raised that it requires to be separately considered. The reader will find much valuable information on this subject in **Mr. Ozanne's** Report on Ráb Experiments in 1884-85 and numerous other papers published by the Government of Bombay. These have been briefly reviewed by **Professor R. Wallace** *India in 1887*) and an explanation of the system offered which may be here reproduced:—

"The word *Ráb* literally means cultivation, but in its restricted and common meaning it denotes the growing of the young plants—principally of rice, but also of *Rági* (**Eleusine Coracana**) and *Varai* (**Panicum miliaceum**)—on seed beds which, as a part of their preparation, having burnt on them dung or certain herbaceous substances to be hereafter described. *Ráb* may also signify the material burnt.

"The site selected for the seed-bed is usually on land more or less elevated, to overcome waterlogging and consequent destruction of the young plant. It is banked or 'bunded' to prevent surface washing, and to retain the necessary amount of moisture. The burning is done in the dry weather before the advent of the south-west monsoon.

"The materials used are:—

Materials used. 401

(1) Cowdung collected in the dry weather, and preserved in the form of cakes, or, in rare instances, smeared on to the ground while wet.

(2) The loppings from trees or brushwood cut green, when the leaves are about full grown. The *Ain*, **Terminalia tomentosa,** is the tree most appreciated by the natives for the purpose.

(3) Dead leaves.

(4) Grass, usually coarse and rank.

(5) Inferior straw, after it has been beaten down by the tread of bullocks on the threshing floor, or husks.

(6) Pit manure or dung collected during the wet season, along with ashes and village refuse.

(7) Earth in a pulverulent condition.

"When dung is plentiful it may be used alone, spread carefully over the surface in a thin layer between one and two inches thick, or the other materials may be used in place of dung; but the most serviceable common methods adopt a combination of substances,—(1) Dung broken into small pieces from the cake condition, and evenly distributed; (2) a layer of leaves or loppings, also carefully reduced to uniformity by chopping where necessary; (3) grass; (4) finely-divided straw, to close the openings between the stems of the coarse grass, and prevent (5) the earth, which forms the fifth and final layer, from falling through. The function

ORYZA sativa.

Rice Cultivation

CULTIVATION in Bombay.

Manuring.

Ráb.

of the earth, which is often mixed with an equal amount of pit manure, is to keep the materials together and exclude air, so that the burning may be prolonged. This is also induced by firing the layer on the lee side, so that it burns against the wind. To make the conditions necessary for smother burning more perfect, the surface is not disturbed before the *ráb* is deposited, at least in the case of heavy soils, that do not readily break down under the influence of the sun; where the land is light, it is sometimes ploughed and levelled before being *rábed.*

"*The limits of Ráb.*—Thana seems to be the stronghold of this system of cultivation—over 81 per cent. of the total area cropped requiring *ráb* in the preparation of the seed beds.

"From the Konkan—the low-lying land, principally under rice, extending along the western seaboard of India, north and south of Bombay, and in which Thana is situated—*ráb* cultivation stretches over the ridge of the Western Ghats, and may be seen at Poona. It is essentially confined to districts of very heavy rainfall; not only so but to districts noted for the intensity and continuity of the early monsoon.

"The effects of *Ráb* are:—

(1) The change of the dung and other organic or herbaceous substances by burning into a useful form as regards the ash for the support of the young plants. The nitrogen is given off into the air.

(2) The destruction of the roots and also the seeds of weeds lying near to the surface.

(3) The action of the heat on the surface soil tends to make it more open and to liberate potash and phosphoric acid from the soil substance, in the same way as clay burning acts at home.

"The heat is moderate, though, at the same time, considerable. On scraping down with a knife, through the surface of a *pucca* (good) seed nursery, I found immediately after the fire had passed that the soil was so hot half an inch down that I could not comfortably hold my finger on it.

"The great advantages of a seed-bed are that the young plants early acquire a habit of vigorous growth, and can be brought forward on well-tended, restricted patches of land, while the larger areas are being suitably cultivated in preparation for them. The best part of the season is thus secured for the crop to grow in. The finer and more valuable, and at the same time more tender, varieties of rice pass the uncertain or risky stage of their existence in a place where care can be bestowed upon them.

"As already stated in the case of rice cultivation in Madras, seed-beds are prepared and grown from in other parts of the country and the advantages reaped without the use of *ráb;* but everything seems to point to the impossibility of *ráb* being dispensed with in certain rainy regions.

"By referring to my remarks on the decay of dung when kept in an extremely wet condition, I believe a full explanation may be found of the importance, I should say almost the necessity, of burning the manure before it is mixed with the soil.

"We know that marsh gas and low forms of carbon, organic acids which injure crop plants, are produced by decomposition of humus in very wet soils where the supply of air is restricted. What the precise forms are that the injurious substances take I do not pretend to say, but from the evidence before me I can assert with confidence that there is a matter of importance tending in the direction indicated that deserves investigation.

"On page 19 of the Thana Report it is stated that 'the yield of the experimental plots manured with dung gave 50·9 per cent. of the standard (the cowdung *ráb* plot), while the unmanured and unrábed plots showed 60·3 and 73·5 per cent.'

"The extent of the investigations, the evidently careful and unbiassed

CULTIVATION in Bombay.

Manuring.

Rab.

manner in which they were carried out, and the results recorded, enable me to attach greater importance than usual to experiments extending over such a short space of time. They are further confirmatory of native experience and explanatory of native practice.

"The records of the valueless character of dung applied to soil under certain conditions in India must be looked upon as specially important in support of the theory which I have propounded, because the results were obtained independently, and before the theory, which had its origin in entirely different facts, had been thought of.

"It is interesting to notice, further, what is related as occurring in a district of heavy rainfall, but under circumstances which vary materially from the above:—'In the south of the Presidency (Bombay) the ante-monsoon storms are heavy, and allow of the sowing in May.'

"The moisture is not excessive, and in such tracts rice may be drilled or sown in seed-beds, *manured but not rábed.* The early and more tender stages of the growth of the plant are passed without the manure being deluged to its injury.

"In the case of Konkan, rain does not come till June, and, 'there is a continuous downpour in June and July.'" (*Wallace, pp. 208-212.*)

Transplanting. 402

Professor Wallace adds that *rabbing* might be viewed as a civilised adaptation of the system (pursued by aboriginal tribes) of burning the surface soil, along with the herbage and trees, hewn down on selected spots of primeval forests, technically known in India as *Jumming*. There can be no doubt of the beneficial effects of manuring rice-beds with ashes and of the advantages of effecting certain decompositions of the surface soil by the heat engendered, but it does not seem proved that the system could not advantageously be extended to other rice-producing regions. The limitation of the practice within Bombay to the tracts suitable for rice cultivation might only indicate that the crops grown beyond these limits did not require, or were not improved by, transplantation. The superior results obtained from transplanting rice are well known throughout India. Healthy seedlings are pushed forward in a manageable spot where care and attention can be given them, so that by the time the fields are ready they are in a forward condition and thus get a start, which allows of a good harvest being fully matured by the season at which the crop has to be reaped. By delaying sowing until the period of the year when the fields are ready, enough time is not left for the maturing of the grain and the ripening of the crop: It is this climatic shortness of the rice season that in the agriculture of, perhaps, some 4,000 years, has doubtless given birth to the countless series of agricultural forms not only suited to the peculiarities of soil, but to the duration of the time in which the rice can grow.

METHODS OF CULTIVATION, SOWING, REAPING, &c.

Provincial Methods. 403

The following passages may be here given as expressing the main features of the cultivation of rice in Bombay.

In the Bombay Statistical Atlas, **Mr. Ozanne** gives much interesting information regarding the Presidency as a whole; he marks out the area of rice cultivation, and indicates the portions of it where the *ráb* system is pursued. His remarks may, therefore, be fitly given as introducing the district accounts from Gazetteers and other such publications:—

"In the Konkan districts and in the moist areas with very heavy rainfall all along the eastern slopes of the Sahyádris, rice is grown continuously in low lands, generally embanked to

Chief Vernacular Names.

IN HUSK (PADDY).	CLEAN RICE.	RICE STRAW.
G. *Dángar bhát.*	*Chokha, tándul.*	*Parál.*
M. *Bhát sál.*	*Tándul.*	*Pendha.*
K. *Bhatta nellu.*	*Akki.*	*Bhattad hullu.*

ORYZA sativa.

Rice Cultivation

CULTIVATION in Bombay.

impound the rain-water. In these tracts irrigation is impossible, and the crop is solely dependent on the rainfall, which is entirely derived from the south-west monsoon. The rains burst suddenly and heavily and are very continuous but terminate early. These are the tracts where rice seed-beds are prepared by burning layers of cowdung or brushwood with subordinate layers of leaves, grass, rice, straw, and earth. The seed-beds are on higher land than the fields into which the seedlings are transplanted—a necessary precaution to prevent flooding when the seedling is in its early stages. This, the *ráb* cultivation of rice, has been declared by the Forest Commission, on adequate evidence from local inquiry and careful experimental cultivation, to be good farming, and, under the peculiar conditions of the country where it is practised, the best known method of cultivating rice. Other systems known and practised are risky and incapable of extension beyond their present small areas in *ráb* tracts.

Reasons for *Rabbing*.

"The peculiar circumstances referred to are:—

(1) the absence of heavy showers before June;
(2) the very heavy early rainfall;
(3) the heavy continuousness of the early rain;
(4) the early closure of the rains;
(5) the absence of rain from the north-east monsoon;
(6) the absence of facilities for water storage.

"The other limited systems of growing rice in *ráb* tracts are (*a*) the direct sowing of the seed in areas where the seedling is left to mature to harvest—a system only possible with the coarse variety of rice sown in lands on the sea coast, either reclaimed from the sea or liable to partial immersion from the sea or tidal creeks—and thus largely impregnated with salt; (*b*) the raising of the seedlings in land heavily manured, either artificially or naturally. This system is of very limited application and is only a means of preparing seedlings where *ráb* is not available or where *ráb* seedlings have been lost. It is risky and seldom remunerative, and its requirements limit its use to very insignificant areas.

"The *ráb* cultivation of rice ceases in South Surat at the river Pár, chiefly because rice ceases to be the most important staple, and the manure is required for sugar-cane and other more remunerative crops. Its cessation is not due to superiority of skill on the part of the cultivators, for *ráb* is fully utilised in Bassein and Máhím in North Thána, where the cultivation is as skilful and admirable as in any tract in the whole Presidency.

"*Ráb* again ceases on all sides where the peculiar conditions of *ráb* tracts are wanting. Thus, in the Karnátak outside the Mallád, and to some extent within its limits, water storage facilities exist. Here rice is drilled in terraced rice-fields, with the advantage either of heavy showers in May which enable the seed to be sown and to get a good start before the burst of the regular rains, or of irrigation water stored in tanks by which the fields are flooded and the seed sown before the rains begin. This is the mode of cultivation in the rice tracts of Dhárwár and Belgaum. The variety of rice is different from that grown in *ráb* tracts. It is inferior and has lost some of the characteristics of an aquatic plant. This is clear, because if the early rain is not propitious, *jowári*, a dry crop millet, *can* be grown instead of rice. Occasionally rice is seen on one terrace and *jowári* on the next, and even rice and *jowári* are sown in the same field. In Gujarát, where rice is grown in low-lying lands, embanked as in the Konkan, the seed is raised in seed-beds manured but not *rábed*. This cultivation is possible, owing to local peculiarities of rainfall, even when irrigation is not available, but a large proportion of the rice in Gujarát is grown either under tank and well irrigation or as a pure dry-crop. In the latter case it is drilled as in Dhárwár and Belgaum, but here (*i.e.*, in Broach

CULTIVATION in Bombay.

chiefly) it is a row crop in cotton fields. It is thus almost a pure dry-crop. This short account shows that rice, essentially an aquatic plant, is grown even as a dry crop. The extremes are met with in the row crop rice of Broach and in the salt marsh rice of the Konkan. In the latter case the seed is sown broadcast and left to grow where it falls. The seed is treated before sowing to cause artificial germination. Where the land is very salt it is not touched till it has been inundated by the rain. It is then stirred with the bullock hoe and the sprouted seed is sown on the surface of the water. It falls to the bottom and takes root. It will stand complete immersion for 10 days. Sometimes it is sown from boats. Where the land is less salt, it is carefully picked with the hand hoe in the hot season and left in the rough till the rain falls, when the sprouted seed is sown broadcast. (Manuring. Rab.)

"The artificial germination of the seed, which is not confined to salt rice, or even to rice, is caused in several ways.

"There are very many varieties of rice, but it is sufficient for present purposes to show only how the varieties are influenced by the conditions of cultivation and rainfall as has been done.

"One interesting method of cultivation most prevalent in Ratnágiri and Kánara must be noticed. It is practised with hot weather rice, generally a late crop after the reaping of the monsoon crop of rice. The field is kept flooded to kill weeds for a time. The water is drained off. Manure is applied and the seed sown in December. It is watered from time to time and ripens in March. The water is brought from a dammed-up stream or from a well or tank. This is called *váyangana* rice. Hot Weather Rice. 404

"Rice straw is not a nutritious fodder. The grain is easily separated from the ear by beating on a board or on sheet rock. The glumes are persistent, and clean rice is only separated by pounding with pestle and mortar. The cultivators usually sell the rice in husk, and it is so kept by the traders till a sale has been ensured. The pounding is confined chiefly to large centres, of which Surat and Kalyán may be considered the most important. The straw is largely used for fodder, because where rice is a staple, other fodder is scarce. It is also used as a thatching material and has the great property of keeping off white ants.

The following selection of district accounts of rice cultivation may be here given as more fully exemplifying the Bombay methods:—

Cultivation of Rice in the Thana District. Thana. 405

One of the earliest European travellers who mentions the rice of Bombay is, perhaps, **Jean Baptiste Tavernier**, who describes the rice-field, seen by him on a march south from Surat in the year 1645. He makes special mention of the musk-scented rices of Surat. His remarks may be here quoted:—"All the rice which grows in this country possesses a particular quality, causing it to be much esteemed. Its grain is half as small again as that of common rice, and, when it is cooked, snow is not whiter than it is, besides which, it smells like musk, and all the nobles of India eat no other. When you wish to make an acceptable present to any one in Persia you take him a sack of this rice." A little more than 100 years ago, Dr. **Hove** wrote in admiration of the rice cultivation in this district and made the remarkable statement that *the farmers had lately only introduced from Guzerat the system of transplanting from a seed-bed.* **Hove's** remarks on this subject are of so much interest that his passage on this subject may be quoted:—"The rice fields by their voluptuous growth pointed me out that the soil (Thana District) is superior to Bombay. The plantations are all laid out in the plains and divided in equal squares. Each of these plantations is banked up about two feet high, wherein the

Scented Rices. *Conf. with pp. 532, 570.* 406

ORYZA sativa.

Rice Cultivation

CULTIVATION in Bombay.

Thana.

water is preserved a long time, after the rains have ceased. The farmers here have lately introduced a method of cultivating their rice from the Guzerat which I have never observed on the coast of Guinea. It is the following:—In the beginning of the rains, they sow in a level field a certain quantity of rice very thick, from whence, when it is grown to the height of 5 or 6 inches, they transplant it into small tufts into the other fields as soon as the rains permit them. This mode, which is only attended with little more trouble than the former, saves to the farmer not only a considerable quantity of seeds, but likewise brings him a triplicate crop." Then again, he returns to this subject:—"The soil of this district (Island of Salsette) is remarkably rich where the land is already occupied. The rice stands prodigiously promising, and it is a pleasure seeing the planters, who are chiefly females, transplant, which they perform in point of square and line, with the utmost exactness. They are never hindered from their employment, may it rain ever so hard, for that purpose they have a cover made in the shape of an upper turtle shell, from the leaves of the **Ficus Religiosa**, and they appear to a stranger at first sight as some unknown animals."

In the Gazetteer rice is said to be grown over the entire district and to hold the first place among cereals; the acreage devoted to it is about 63·9 per cent. of the tilled area.

"The first step in rice cultivation is to manure the land in which the seed is to be sown. A cultivator in the opener parts is obliged to sow his rice in his field, but where he has upland, *varkas*, near, he sows it in a plot of sloping land close to his field. The nursery is manured in March or April, or even earlier, by burning on it a collection of cowdung and branches or grass covered with earth, to prevent the wind blowing the ashes away. At the same time the earthen mounds, *bándhs*, round the fields are repaired with clods dug out of the field with an iron bar, *pahár*. Early in June, when the rains begin, the seed is sown and the seed-bed ploughed very lightly and harrowed. If the first rainfall is so heavy as to make the soil very wet and muddy, the seed-bed is ploughed before the seed is sown. In this case no harrowing is required. The field in which the rice is to be planted is then made ready, and, after ploughing, is smoothed with a clumsy toothless rake, *alvat*. After eighteen or twenty days the seedlings are fit for planting. All are pulled up and planted in the field in small bunches, *chud*, about a foot apart. In August the field is thoroughly weeded. Through June, July, and the early part of August, the rice can hardly have too much rain, but in September and October, the husbandman likes to see smart showers with gleams of sun. Scanty rain leaves the ears unfilled, while too much rain beats the rice into the water and rots it. By the end of October the grain is ripe and is reaped with a sickle, *vila*, gathered into large sheaves, *bhára*, taken to the threshing-floor, *khale*, and piled in heaps, *udvas*. At the threshing-floor much of the grain is beaten out of the sheaf by striking it on the ground, what remains is trodden out by buffaloes tied to a pole, *kudmad*, in the centre of the threshing-floor. The empty grains are separated from the full grains by pouring them from a winnowing fan on a windy day. Sometimes, instead of having them trod by buffaloes, the hasbandman seizes the sheaves in his hands and dashes the ears against a block of wood to separate the grain from the straw. By this process the straw is not made unfit for house-thatching as it is when trodden by buffaloes, but much grain and labour are wasted. The grain is then carried to the land-holder's house, where the outer husk is taken off by passing it through a large grindstone, *júte*. Instead of *bhát*, the rice is now *tándul*, but it is still *vene tándul*, that is, fit only for grinding into meal. To make it *sadih tándul*, and fit to eat with curry, the rice has to be further cleaned by putting it into a hole in a board in the floor of the house and pounding it with a pestle, *musal*. The inner husk, *konda*, is thus got rid of. In Bhiwndi, Kalyán, Panel, and other towns and villages rice-cleaning employs a large amount of labour. Instead of in a hole in the floor, three or four men with heavy pestles pound the rice in a huge wooden mortar like a gigantic egg-cup, *ukhali*. After it is cleaned the rice is sent in great quantities to Bombay.

There are two great divisions of sweet rice, *halva*, which wants little water and ripens between August and October, and *garva*, which requires a great deal of water and does not ripen till November. Of early *halva* rice there are eight or ten kinds, but as they are generally eaten by the grower, they do not come much into the market, and are called by different names in different parts of the district. The four best known varieties of *halva* are: *kudai* with a red, purplish, or white husk, which

CULTIVATION in Bombay.

Thana.

is generally grown in uplands; *mal jamin*; *torna*, with a white husk, which is grown both in fields and uplands and ripens in the beginning of *A'shvin* (September-October); and *sálva* and *velchi*, both with red husks, which ripen in *A'shvin* (September-October). Between the early or *halva* and the late, or *garva* classes are four or five medium kinds which ripen before *Diváli* (October-November). Of these, three may be mentioned: *máhádi* with a yellow husk and reddish grain; *halva ghudya* with a yellow husk; and *patni halvi* with a white husk. Of late, or *garva*, rice there are more than a dozen kinds, and, as they come much into the market, their names vary little in dfferent parts of the district. The best known varieties are: *garva ghudya* with a yellow husk, *dodka, garvel, ámbemohor, dángi*, with a red husk, *bodke* very small and roundish, *garvi patni, támbesál* with a red husk and white grain, *ghosálvel*, and *kachora* with a purplish husk and white grain. The price of these different varieties change according to the season. But taking the price of *kudai* at sixteen *paylis* or 89 pounds the rupee (2*s.*), the relative rupee prices of the other kinds are: for *torna* 46½ pounds, for *sálva* 41 pounds, for *velchi* 42½ pounds, *máhádi* 46½ pounds, *patni halvi* 44½ pounds, *garva ghudya* 35¼ pounds, *dodka* 42½ pounds, *garvel* 39⅝ pounds, *ámbemohor* 35¼ pounds, *dángi* 42½ pounds, *bodke* 42½ pounds, *garvi patni* 42½ pounds, *támbesál* 39½ pounds, *ghosálvel* 42½ pounds, and *kachora* 70½ pounds.

The tillage of salt rice differs greatly from the tillage of sweet rice. The land is not ploughed, no wood ashes are used, the seed is sown broadcast on the mud or water and left to sink by its own weight, and the seedlings are never planted out. Salt rice ripens November along with the late sorts of sweet rice. It has to be carefully guarded from salt water and wants a great deal of rain. The straw is not used as fodder but burnt as ash manure. The grain is red and comes much into the markets, being greatly eaten by the poorer Kolis and Kunbis as it is cheap and strengthening. Salt rice is of two chief kinds: *munda*, about 46½ pounds the rupee or 2⅞*d.* (1 1/16 ans.) a *páyáli*; and *kusa*, about ⅛*d.* (1 pie) cheaper." (*Bomb. Gaz., XIII., pt. I.,* 287.)

CULTIVATION OF RICE IN THE RATNÁGIRI DISTRICT.

Ratnagiri. 407

Rice is said to hold in this district the fourth place, as it occupies about 14·08 per cent. of the whole area under tillage.

"There are three modes of growing it as a rainy season crop. The first and commonest by transplanting seedlings, the second by sowing sprouted seed, and the third by sowing dry seed broadcast. Dry weather rice crops, called *váingan*, are grown by watering fields which have yielded a rainy weather crop. The places chosen for a dry weather rice crop are generally hill-side terraces well supplied with water. Land tilled in this way often yields a large outturn, but as it is already exhausted by the rainy season crop, before the rice is sown it wants heavy manuring and careful ploughing. The *váingan* rice crop ripens about the end of March. Of fifty varieties of rice, about forty, ripening in September, are called early, *halva*; the rest, ripening towards the end of October, are called late, *mahan* or *garva*. These varieties of rice differ much in value, the late sorts being generally the best. Their prices in ordinary season vary from ⅞*d.* to 1¼*d.* a pound (R35 to R48 a *khandi*). Rice is the common food of the well-to-do, and is eaten by the poor on marriage and other special occasions. It is used in the manufacture of ink and by washermen in making starch. Rice spirits are sometimes distilled, but, from the cheapness of palm liquor, are in little demand. (*Bom. Gaz., X., 147.*)

CULTIVATION OF RICE IN THE BELGAUM DISTRICT.

Belgaum. 408

Rice is said to be chiefly grown in Khánápur, Belgaum, and Sampgaon. There are five modes of rice tillage: three regular modes, and two extra modes which are used only when the regular modes fail. These five modes are described in the Gazetteer as follows:—

"The first and best form of rice tillage is called *rop* (M.) *natihackhona* (K.), or planting, but many husbandmen shrink from it because of the cost and the heaviness of the labour. In Khánápur and Belgaum during April or early May, a small nursery or seed-bed, a plot to which water has easy access, is covered with leaves, wood, straw, and rubbish, and this covering is burnt in late May before the first rainfall. At the same time the fields into which the seedlings are to be planted are being got ready. The field-banks are mended, the water-ways cleared, stiff plants and stalks are cut out, and as much of the ground as possible is covered with grass, weeds and rubbish, and burnt. When the first rain falls the seed-bed is thrice ploughed and harrowed. When well soaked it is covered with a thick broadcast sowing of rice in husk. The ploughing of the fields into which the seedlings are to be planted is not begun until the bullocks sink in the mud to the knees, a dreadful toil both to man and bullocks. Every field is thrice ploughed, and after the third ploughing, to clear it of

ORYZA sativa.

Rice Cultivation

CULTIVATION in Bombay.

Belgaum.

roots, is harrowed with a long-toothed harrow. In a good season, that is, heavy rain with gleams of sun, after five weeks or early in June the seedlings are fifteen to eighteen inches high and fit for planting. When the seedlings are ready, if possible in a break of bright weather, cowdung-ashes, litter, and leaves decayed to dust in the manure-pit are brought from the village, spread equally over the field, and trodden deep into the mud. When the field is manured the surface is levelled by dragging over it a loaded board called *hendor* (M.) or *karadu hodiyona* (K.). A day or two later, still if possible in fine weather when the field is not deep in water, the seedlings are rooted by the hand out of the seed-bed and brought to the fields in baskets. A rake with short teeth, ten to twelve inches apart, is drawn over the smooth ground to mark the lines in which the seedlings are to be set. The workers, who are generally women, follow with baskets from which they take small handfuls of eight to ten plants, and ten to twelve inches apart and as far as possible opposite the middle of the interval of the next row, thrust them about a foot deep. Except so much as is wanted to flood the lower fields the water is kept in the field, and when each field has had its share the channel to it is blocked. Two weedings are given, but as the field has been so carefully cleaned, the weeds are seldom strong. In ordinary years planted rice is ready for cutting in November or December.

The second mode of growing rice is the *kivri* or *kurgi*, that is, the seed-drill plan. This system is adopted in the hope that enough rain will fall within a week after the seed has been sown to make the soil muddy. It saves much labour, but should the rain hold off for about a fortnight the ground becomes heated and the seed suffers from the dryness and is eaten by birds and lizards. At best the outturn is small.

The third method is adopted when the early rain is so heavy that the seed-drill cannot be worked. Furrows are made by the light plough and the seed is sown in the furrow. This furrow-sowing system never yields a good crop.

When one of the three regular modes fails, in the hope that the harvest may not be entirely lost, sprouted seed or *málaki* (K.) is sown. A sackcloth or matting bag is filled with grain, dipped in water, and laid in a warm close place. In three or four days the seeds sprout and are thrown thick and broadcast on the field.

The fifth mode of growing rice is to root out the sprouted rice seedlings where they have come too thickly and plant them into the bare fields. This is the rice-grower's forlorn hope. It is called *surdi* (K.) or the cold crop, perhaps because it does not ripen till the close of the cold weather.

Ripe rice is reaped, and thrashed either by striking the ears against a board, or by beating them with a stick. After winnowing, the grain is carried home and dried in the sun. The husks of as much as is wanted for immediate use are beaten off in a stone mortar, *ukhal* (M.) or *varalu* (K.) by a wooden pestle, *musal* (M.) or *vanaki* (K.), and the rest is stored in high cylindrical baskets called *kungi*, the opening in which within and without are closed by a coating of cowdung.

In parts of Khánápur near the Sahyádris two crops of rice are grown every year. The first crop is sown with a seed-drill about the end of June, or is sown sprouted in August. It ripens towards the end of October and is called the *Kártik* or October-November crop. The second crop is sown sprouted in November and December, and ripens towards the end of April. It is called the *Vaishákh*, that is, the March-April, or the *sugi* crop. The April crop is reckoned better than the October crop because it is not exposed to the cold weather winds." (*Bomb. Gaz., XXI., 248-250.*)

Dharwar.
409

Cultivation of Rice in the Dhárwár District.

It is stated that rice occupies about 6·42 per cent. of the tillage area of this district:—

"It is grown almost wholly in the woody waste which is locally called *malládu* or hill land. Rice wants much and constant moisture. When it depends on rainfall alone rice is always uncertain, but this element of chance rather fascinates the people. Most rice-land is independent of simple rainfall for its water supply. The low-lying lands are watered from ponds and much is also watered by drainage from neighbouring high grounds guided by water-courses or *kalvás*. Failing pond water, irrigation is supplied from wells or, more commonly, from holes fed by underground soakage from ponds. The rice soil is red towards the extreme west, and further east it is a light coloured clayey mould. This clayey soil, by the action of water, tillage, and weather, becomes stiff, compact, and very retentive of moisture. This kind of rice soil is poor, middling, or good according to its situation. In high and exposed sites it is poor and shallow, even with care and manure able to bear only one crop of poor rice; in middle situations, neither very high nor very low, it is middling, of some depth, and where there is moisture enough, yields two crops, one of rice and the other of pulse; in low lands or valleys it is of superior richness, of a rich dark brown, and yields excel-

CULTIVATION in Bombay.

Dharwar.

lent after-crops. Regular rice-fields are divided into level compartments a few feet to fifteen or twenty yards broad and varying in length according to the land-holder's pleasure or the position of the ground. The slope of the ground or hill-side is generally carried into a series of terraces each one or two feet higher than the one immediately below it, and the front of each is guarded or raised by a foot high embankment forming part of the descending step. The effect of a hill or rising ground terraced in rice plots is extremely pleasing. The three kinds of rice land require almost the same labour. After harvest the poor soil seldom holds moisture enough to allow of its being ploughed; middle class soil, even when not moist enough to yield an after-crop, is always damp enough to be ploughed, and the ploughing is a gain, as it makes the land more ready to receive the occasional dry-season showers. The upturned grass and stubble roots die and rot, and the stiff clods crumble in the heat and air. At the end of March manure is laid in heaps. In early April the clods are broken by the leveller or *korudu*, or if still very hard, by labourers with clubs. In fields which have not been ploughed after harvest nothing, except the laying of manure, can be done till the first rains of late April or May, when the field is ploughed and the clods are broken by the mallet. The manure is then scattered broadcast from a basket, the surface is turned by the heavy hoe or *kunti* and the leveller or *korudu* follows. Nothing further is done till rain enough falls to admit of sowing, for which a small seed-drill or *kurgi* is generally used. An acre of rice-land on an average wants three to five loads of manure. If more is laid on and the rains are abundant, the crop will gain greatly; but with light rain in highly-manured land the crop will grow too freely and will probably dry without coming to ear. From the 25th of May to the end of June, as soon as the village astrologer has fixed the lucky day, the seed-drill is decked with green leaves, the husbandman bows before it, and sowing begins. The drill is closely followed by the *balle-sal-kunti* or light hoe to cover the seed, and the *korudu* follows to level the surface. In about eight days the seed sprouts, and in eight days more weeding begins with the *yadi kunti* or grubber and is repeated generally once in ten or twelve days. In two months the seed-drill is used for weeding, as the crop is too high and the fields are too full of water for the grubber. The weeds are always left to rot where they grow, and this constant supply of vegetable matter is one chief cause of the peculiar richness of the soil. The surrounding ridges are repaired, the earth cut from the front is heaped on the ridges, the beds are filled with water, and the leveller is passed over the crop. This gives the soil a smooth and beaten surface into which the water does not readily sink, but remains in pools.

The rice harvest begins about the 15th of November in the drier land, but many hollows, where water lies deep, are seldom ready for reaping before the end of December. An unusually dry or wet season may hasten or delay the harvest a fortnight either way. When rice is reaped it is left to dry on the field. It is then tied in sheaves, built, ears outwards, in a stack, and left to season for a month. A pole is fixed in the field, and the ground for a few yards round the pole is beaten hard and cowdunged to prevent cracks. The floor is cleaned and swept, and the loosened sheaves are scattered over it, and six or eight muzzled bullocks packed side by side in a line are slowly driven over the sheaves round the pole. This goes on till all the grain has been trodden from the straw. The straw is then removed, and fresh sheaves are laid and trodden. Winnowing follows thrashing. Rice is winnowed by filling with grain a flat basket which is raised at arm's length and slowly emptied into the air with a slight and regular shake. The winnowing wind blows aside the dust and the leaves, and the clean heavy grain falls on the ground. When a heap has been collected the grain is carried to the village, the outer husk is removed by a wooden hand-mill or *tolulikalu* (K.), and, as before, is a second time winnowed. When the operations are over, the rice is stored in a large round basket or wattle-and-daub safe, raised a little from the ground on beams laid across large stones, and roofed with thatch. Every husbandman's house has one grain basket in which rice and almost all other grains are stored. The only grains which are generally stored in pits are Indian millet, wheat, gram, and cajan pea. Nine chief kinds of rice are grown in Dhárwár. Of these, two, *ámbemori* and *konksáli sanbhatta*, are of good quality; three, *bedarsáli*, *somsáli*, and *hakkalsáli*, are of medium quality; and four, *dodigan*, a large grained variety, *hempgam* or red, *kerekgan* or black, and *gensáli*, are of poor quality. All are sown at the same time, and are reaped one after the other at short intervals. In a fair proportion of rice-fields sugarcane is grown once every third year. Where the soil has good natural moisture sugarcane is grown without watering, and, where the water-supply is plentiful, with as much watering as may be necessary. The only cane which is grown without any irrigation, except a single flooding of the land when it is planted, is the small grass cane which is locally known as *nol-kabbu*. The cane which does not succeed without occasional watering

CULTIVATION in Bombay.

during the dry season is the large or garden cane locally known as *gabras dali.* Green crops of *mug, pávta, matki,* and gram are also grown after rice in hollows which hold their damp till late in the year. Except in red and light coloured soils, a second crop of cane is seldom grown without watering." (*Bomb. Gaz.*, (*XXII.*, *275-277.*)

SIND. 410 *Conf. with p. 522.*

VI.—CULTIVATION IN SIND.

The following passages, in amplification of what has already been said of Bombay and Sind conjointly, convey the main features of the rice cultivation in the province of Sind :—

"In the cultivation of rice the ground is ploughed once, so soon as it is sufficiently dry, and about the middle of April; if water be procurable from the *kacha* wells generally dug for this purpose, the seed is sown by means of a drill attached to the plough. When water is not readily obtainable the soil is enriched with manure to force the growth of the plants, and to allow of their being prepared for transplanting about the middle of June. The land is afterwards flooded to a depth sufficient to allow the heads of the plants only appearing a little above the water, and this condition is carried out during their growth. Rice crops are subject to injury from rats, blight, crabs, drought, or accidental over-flooding."

Shahbandar. 411

Cultivation of Rice in the Sháhbandar District.

"Rice is the chief staple, being 76 per cent. of the whole cultivation in the division, and next to it comes *bájri,* which is in the proportion of 13 per cent. There are two distinct methods of cultivating the rice plant in this division. The first, which is common to the rice-growing districts of the Bombay Presidency, consists in preparing, in the first instance, a nursery-bed, in which the seed, usually in the proportion of 130lb to the acre, is sown. Here, again, there are two different ways of preparing these nursery-beds, which are technically known as *bijárani* and *khamosh.* By the first, the ground is well manured and ploughed several times, the seed, being sown by means of a *nári,* or funnel, during the last ploughing. Being sufficiently moist of itself, the soil does not require any irrigation, the plants being usually ready for transplantation in forty days. They are then taken to other fields, previously ploughed over several times, but *not* manured, these, in some cases, being four or five miles distant from the nursery-beds, and here they are regularly planted out. By the second plan, the stubble is burnt, which, with manure, is mixed with the soil of the intended bed, but not ploughed into it. The seed is sown with the hand. These nursery-beds are irrigated from *kacha* wells, and the plants are generally ready for transplanting in about twenty-five days. After transplanting, the plants are generally watered so as to ensure their being covered for two-thirds of their height. Some of the finest rice lands so cultivated are situate in the Mirpur, Batoro taluka, on the Khorwáh canal, and here is produced a fine description of white rice known by the name of *sugdási,* other kinds are known as *ganja, motia, satria,* and *lári.* The average yield *per acre,* in good land, is about 7 maunds or 560lb of cleaned rice, and in inferior soils 4¼ maunds or 340lb. The average profit, after deducting expense of seed and cultivation, is R5 for the good, and R3 *per acre* for the inferior lands. The second method of cultivating the rice, which is practised in the southern portion of the Sháhbandar and Ghorabari talukas, where the lands lie low, is to sow the seed broadcast in a soil which is seldom previously ploughed up for its reception. No transplanting is carried out, but the land receives a slimy deposit from the inundation waters, and is partially flooded at high tides. Little or no labour is required in this kind of cultivation, as there are no canals to clear, water-courses to make, or land to plough. The high tides irrigate the crops sufficiently without the help of the cultivator, and such rice-lands as these are, in consequence, in great request. The returns are heavy also, the crop *per acre* often reaching as much as 14 maunds, or 1,120lb of cleaned rice, and the net pecuniary profit to R15 *per acre.*" (*Gaz., p. 759, 760.*)

Indus Delta. 412

Cultivation of Rice in the Indus Delta.

One of the most remarkable methods of growing rice is that pursued on the swampy tracts near the mouths of the Indus. The following passage, taken from Balfour's *Cyclopædia,* fully deal with this subject, and it need only be remarked in explanation that by *bhúl* is meant the swampy alluvial deposits formed by the river which are only slightly raised above the level of the sea :—

"Should the river, during the high season, have thrown up a *bhúll,* the zemindar, selecting it for cultivation, first surrounds it with a low wall of mud about three

CULTIVATION in Sind. Indus Delta.

feet in height. These *bhúlls*, being formed during the inundation, are often considerably removed from the river branches during the low season. When the river has receded to its cold weather level, and the *bhúll* is free of fresh water, he takes advantage of the first high spring-tide, opens the bund, and allows the whole to be covered with the salt water. This is generally done in December. The sea water remains on the land for about nine weeks, or till the middle of February, which is the proper time for sowing the seed. The salt water is now let out, and as the ground cannot, on account of the mud, be ploughed, buffaloes are driven over every part of the field, and a few seeds of the rice thrown into every foot-mark, the men employed in sowing being obliged to crawl along the surface on their bellies, with the basket of seed on their backs, for were they to assume an upright position, they would inevitably be bogged in the deep swamps. The holes containing the seed are not covered up, but people are placed on the bunds to drive away birds until the young grain has well sprung up. The land is not manured, the stagnant salt water remaining on it being sufficient to renovate the soil. The rice seed is steeped in water, and then in dung and earth for three or four days, and is not sown until it begins to sprout. The farmer has now safely got over his sowing, and as this rice is not, as in other cases, transplanted, his next anxiety is to get a supply of fresh water, and for this he watches for the freshes which usually come down the river about the middle and end of February, and if the river then reaches his *bhúll* he opens his bund and fills the enclosure with the fresh water. The sooner he gets this supply the better, for the young rice will not grow in salt water, and soon withers if left entirely dry. The welfare of the crop now depends entirely on the supply of fresh water. A very high inundation does not injure the *bhúll* cultivation, as here the water has free space to spread about. In fact the more fresh water the better. If, however, the river remains low in June, July, and August, and the South-West monsoon sets in heavily on the coast, the sea is frequently driven over the *bhúlls* and destroys the crops. It is, in fact, a continual struggle between the salt water and the fresh. When the river runs out strong and full the *bhúlls* prosper, and the sea is kept at a distance. On the other hand, the salt water obtains the supremacy when the river is low, and then the farmer suffers. Much *bhúll* crop is destroyed in the monsoons and during heavy gales. The rice is subject to attacks also of a small black sea crab called by Natives *kookaee*, and which, without any apparent object, cuts down the growing grain in large quantities, and often occasions much loss. If all goes well, the crop ripens well about the third week in September, and is reaped in the water by men, either in boats, or on large masses of straw rudely shaped like a boat, and which, being made very tight and close, will float for a considerable time. The rice is carried ashore to the high land, where it is dried and put through the usual harvest process of division, &c., and the *bhúll* is then, on the fall of the river again, ready for its annual inundation by sea water." (*III.*, 415.)

VII.—CULTIVATION IN THE N.-W. PROVINCES & OUDH.

N.-W. P. & OUDH. 413

References.—*Wright, Mem. Agric. of Cawnpur; Hoey, Trade & Manuf. N. Ind.*, 4, 80; *Duthie & Fuller, Field & Garden Crops*, 15-20; *Atkinson, Him. Dist. (X., N.-W. P. Gaz.)*, 685; *Agri.-Horti. Soc., Ind.:—Transactions, I.*, 19, 45; *II.*, 88; *III.*, 82-90, 117-224, 166-171, *Pro.*, 265; *IV.*, 62, 63, 100, 101, 113, 120-124; *V.*, 35-37, 60-62; *VI.*, 246; *VIII.*, 95, 161, 162, *Pro.*, 391; *Journ. (Old Series), VIII., Pro.*, 112; *Gazetteers:—(N.-W. P.), I. (Banda)*, 86, 90, 93, 115, 119; *Hamirpur*, 150, 156; *Jhansi*, 252; *Lalatpur*, 312, 315, 316, 340; *II., Shaharanpur*, 160, 167; *Aligarh*, 375, 384; *III., Bulandshahr*, 24; *Meerut*, 225, 226, 231, 232; *Muzaffarnagar*, 463, 464-465, 466, 483 *munji rice*; *IV., Eta*, 30; *Etáwa*, 248; *Mainpuri*, 525 (*wild rices*); *V., Budaun*, 28-29, 85; *Bijnor*, 267, 269, 299, 336; *Bareilly*, 541, 557-559 (*Pilibhit rice*); *VI., Cawnpore*, 28, 32, 134, 138, 154, 254; *Gorakhpur*, 321-324, 331 (*Carolina*), 334, 339, 368, 411, 414 (*Nepál rice*), 474, 477, 501, 503, 525, 533, 539; *Basti*, 558, 587, 589-590, 646, 694, 700, 704, 735, 745, 769, 775, 781, 784; *VII., Farukhabad*, 34; *Agra*, 448-449, 458; *VIII., Muttra*, 41; *Allahabad*, 30-31; *Fatehpur*, 18-19; *IX., Sháhjahanpur*, 45, 132; *Moradabad*, 47, 122; *Rampur*, 16, 33; *X., Himalayan Districts*, 685; (*Oudh*) *I., Introduction, IX., Bahraich*, 153 (*extent twice cropped*), 154, 155, 159; *Bara Banki*, 234, 243, 244; *Fyzabad*, 414-415, 419, 423; *Gonda*, 498, 522, 528; *II., Hardoi*, 17, 23; *Kheri*, 159, 160, 182, 186; *Lucknow*, 314-319, 329; *III., Partabgarh*, 79, 83 (*wild rice*); *Rae Bareli*, 183, 196; *Sitapur*, 355, 356, 358, 366; *Sultanpur*, 422, 424; *and Unao*, 523, 524.

ORYZA sativa.

Rice Cultivation

CULTIVATION in the N.-W P. & Oudh.

Area under, and Traffic in, Rice.

Area. 414 *Conf. with p. 522.* The area shown as under rice in these provinces (in the Agricultural Statistics of the Government of India for 1888-89) is as follows:—

North-West Provinces	4,338,923 acres.
Oudh	2,439,228 „
	6,778,151 acres.

If we accept 10 maunds an acre as an average yield, the total production would have been, for the year under reference, 6,77,81,550 maunds or, say, 2,420,768 tons.

Trade. 415 The returns of internal trade show these provinces to have a net export of 1,38,620 maunds. Of the gross exports (chiefly paddy), the Panjáb took 5,78,659 maunds, Rájputana and Central India, 2,81,845 maunds, while Bombay got 82,489 maunds. Little Exports to Foreign Countries. 416 Very little rice from these provinces would appear to be exported to foreign countries, as Bombay and Calcutta port towns are shown as drawing only 3,081 and 2,843 maunds, respectively. The transfrontier land trade with Nepál, Thibet, &c., last year showed a net import to the credit of the province 6,90,623 maunds. Deducting the net export by rail and adding the net transfrontier import to the estimated production, the balance would be the amount available for Consumption per head. 417 local consumption which expressed to population (44 millions) would be a little more than $\frac{3}{20}$ths of a seer per head. In Bareilly **Mr. Moens** estimated the production at 14·8 maunds. **Messrs. Duthie & Fuller** say that in the Yield. 418 Meerut, Rohilkhand, and Benares Divisions and in North Oudh broadcasted and unirrigated rice may be assumed to yield an average of 12 maunds per acre, while, in the drier districts towards the centre and south of the Provinces, 10 maunds is the highest average which can be safely taken. The outturn of transplanted and irrigated rice may be estimated at 16 maunds per acre, the produce being superior to that of broadcasted rice in quality as well as quantity, and hence it commands at least 50 per cent. higher prices in the market. These figures of outturn are, however, of *unhusked* rice and must be reduced by at least 25 per cent. to arrive at the yield of husked rice. From these opinions, as well as a large series of other statements and experiments, it would seem that the Bengal yield of 10 maunds of rice to the acre would be a safe average. But the extremely small amount this represents as the daily consumption per head of population must be admitted throws considerable doubt on the degree of dependence that can be placed on any factor as expressing the average for a province. In the case of Bombay the returns of yield show the people to be consuming more rice than in Bengal and the North-West Provinces, an idea which is probably open to grave doubt.

Rotation of Crops, Manuring, &c.

Rotation: Manuring. 419 "No particular rotation is followed; in damp localities it often alternates with sugar-cane, and in the western districts of the Provinces with gram, barley or peas. But it is commonly grown year after year in the same land, and, moreover, when broadcasted and cut early, is generally followed by a crop in the succeeding *rabi*, and the land is thus drained by two crops within the year."

"Rice is almost always sown alone, the peculiar conditions of its cultivation not suiting any other crop. Occasionally the greater millet (*Juár*) is sown mixed with it, but more as an insurance against an overlight rainfall than in the hope of gathering a double crop. The suitable soil is stiff clay which commonly forms the bed of the drainage depressions and basins, in which rice cultivation most frequently occurs. Rice can even

CULTIVATION in the N.-W. P. & Oudh.

Manuring.

be grown on *usar* or saline clay, provided that an ample supply of water be given, and evaporation from the soil be checked by never allowing the surface to become dry. Manure appears to be very little used for broadcasting rice. The nurseries in which transplanted rice is raised are generally heavily manured, but the application of manure to the fields in which the seedlings are transplanted is only reported from the districts of the Benares Division in the Gogra-Ganges Doáb, where cattle are said to be herded on rice fields, and earth impregnated with saltpetre is occasionally use as a top-dressing" (*Field and Garden Crops*).

Provincial Methods. 420

METHODS OF CULTIVATION, SOWING, REAPING, &c.

The account of rice cultivation in these provinces, as given in **Messrs. Duthie & Fuller's** *Field and Garden Crops*, is so comprehensive and so fully embraces the main peculiarities that it may be here freely reproduced :—

Forms of Rice grown.—"The varieties which rice has developed are more numerous and more strongly marked than those of any other crop. In the District of Bareilly about 47 distinct varieties are enumerated, and it is probable that in the provinces their number considerably exceeds 100. Their names, however, vary so greatly from district to district as to be of little or no assistance in identification, and hence no useful purpose would be served by giving a list of them here. Judged by their leading characteristics the varieties may be thrown into three classes—the *first*, including those with a tall habit of growth, with the ear protruded from the sheath, feathery and drooping, and with thin, usually yellow-husked grain; the *second*, including varieties with a shorter habit of growth and stouter stems, with the ear not so prominent and carried more erect than that of the preceding, and with thick yellow or red-husked grain; and the *third*, comprising the common varieties of paddy, with short, strong stems, ear partially enclosed in the sheath and grain-husk dark coloured or black."

"The varieties of the first class are the most highly prized, the commonest being those known as *naha, bánsmatti, bánsphal*, and *jhilma*. The *seondhi* and *sumhára* are the principal varieties of the second class, while *sathi* (so called from its growth covering 60 days) is far the most important of those included in the third class, and, if its area be alone regarded, the most important of all the varieties. *Munji* is a term of varying meaning, denoting in some places (*e.g.*, Muzaffarnagar) high class rice, and in others being merely a general term for rice sown broadcast and not transplanted. This leads to another and a much simpler method of classification, in which the varieties may be grouped according to the method of their cultivation, as (1) those transplanted from seed-beds, and (2) those sown broadcast. As a general rule, the finer varieties, falling under the first two classes above named, are raised in seed-beds and planted out, while the coarser kinds are sown in the field broadcast. It may be mentioned that a kind of rice (**Hygrorhiza aristata**, *Nees.*) is commonly found growing wild round the edges of the lakes and marshes, being known as *passari, passai* or *phasahi*, and a sub-variety as *tinni* (Partabgarh). The grain is eaten by the poorer classes, being often collected by sweeping the plant heads with a basket."

The cultivation reaches its maximum in the belt of districts underlying the Himálayas, and increases very largely as we go eastwards. This merely, of course, illustrates the fact that a plentiful supply of water is the first requisite for rice growing.

Sowing and Reaping.—There is greater latitude in the period for sowing and harvesting rice than in the case of any other crop, it being

ORYZA sativa.

Rice Cultivation

CULTIVATION in the N.-W. P. & Oudh.

sown in all months from January to July, and harvested in all months from May to November. The rice, however, which is sown before the commencement of the monsoon rains, bears but a very small proportion to the total, and the seasons in which the greater portion is grown are June to August for broadcasted, and June to November for transplanted rice. Taking first of all broadcasted rice, by far the greater portion is sown on the break of the monsoons, and is ready for cutting in from 2 to 2½ months, *i.e.*, in *bhádon* (August) or *kuár* (September), and hence it is often known as *bhadoi* and *kuári*. The rapidity of its growth is signified in the name of one of the commonest varieties, which is called *sathi*, or 60-days rice. But a certain amount of broadcasted rice is sown two months before the monsoon rain can be expected, and in this case there are two methods of cultivation. Either the rice germination is promoted and its growth stimulated by frequent and copious irrigation until the rains break, or taking advantage of a fall of rain in April and May, the ground is ploughed up and sown, but the seed is allowed to lie unirrigated, and the young plants should not come up before the advent of the rains induces germination. The method is a very risky one, since, if the seedlings come up before the rains commence, they are speedily dried up and the crop ruined. The principal object in early sowing is to be able to harvest early and get the rice crop off the ground in time to be followed by one in the *rabi*, and by having the seed in the ground by the time the rains commence, the first fall is utilised in bringing up the young plants instead of in merely preparing the ground for ploughing.

"Nearly the whole of the transplanted (or *jarhan*) rice is sown in seed-beds at the beginning of the rains, planted out after a fortnight or three weeks, and cut in *aghan* or November, whence it is also called *aghani*. A very small proportion, however, called *boron*, *jethi*, or hot weather rice, is sown in January, planted out in February, and cut in May. This is only practised in slimy soil, along the edges of tanks or beds of rivers, which are planted with rice as the water becomes shallow from evaporation. Great labour of an especially disagreeable kind is required, and this method of cultivation is, therefore, chiefly confined to the fisher and boatmen caste. The area under *boron* rice in 1880-81 in the 30 temporarily-settled districts of the North-West Provinces was only returned as a little over 5,000 acres."

Tillage.—"A great portion of the rice land in the Sub-Himálayan districts is prepared by being dug over by the mattock during the cold and hot weather months, when the soil has been softened by a fall of rain. Labour is cheap in these districts, and practice has produced dexterity, and in consequence an acre can be dug in this manner to a depth of six inches for about R2-8, while at the contract rates allowed in Doáb Districts it would cost at least R8 or R10. For land not dug in this way, the number of ploughings varies according as the crop is to be sown broadcast or planted out, being two or three in the first case, and from four to six in the second. The soil is pulverised and weeds collected by a rough harrow made by fixing a row of pegs in the ordinary log clod crusher. If the land lie at all saline the harrow is not used since by rendering the earth more compact it is said to facilitate evaporation, which brings of course the salt to the surface.

Sowing.—"For sowing, the soil must be thoroughly moist, but may be a miry slush, on the surface of which the seed is scattered and harrowed in. If the rice is sown broadcast 40 seers to the acre are held sufficient. If seedlings are to be raised in a nursery much thicker sowing is followed. It is a common practice, especially when the weather at sowing time is

CULTIVATION in the N.-W. P. & Oudh.

very wet, to give an artificial stimulus to germination by soaking the seed in water for a night, and then leaving it for a couple of days covered with damp grass. If the crop is to be transplanted, the nursery should be about $\frac{1}{18}$th the size of the field. The seedlings are taken up when about a foot high, and planted out in regular lines at distances of six inches, from two to six seedlings being planted together.

Irrigation.—"For rice which is grown in the hot weather months, frequent and copious irrigation is absolutely necessary whether the district be moist or dry. Rice sown at the commencement of the rains and cut in August or September under ordinary circumstances needs no watering, but the transplanted varieties, which are not ready for harvesting till November, need two or three waterings after the rains have ceased. Of the total area under rice in the 30 temporarily-settled North-West Provinces Districts, only 15 per cent. is returned as irrigated, and this may be presumed as the proportion which transplanted bears to broadcasted rice.

"The rain water is carefully economized by surrounding the field with a bank which prevents any great loss of water by surface drainage. Irrigation, if required at all, is required in such quantity that wells are almost, if not quite, useless for the purpose, and the crop can only afford the less costly water which can be derived from tanks, rivers, or canals. The effect of the Ganges canal on rice cultivation is seen very clearly in the Muzaffarnagar District, where transplanted and irrigated rice, which was formerly almost unknown, now occupies 50 per cent. of the total rice area.

Weeding.—"At least one weeding is, as a rule, given to broadcasted rice. Planted rice is reported in Cawnpore to be more frequently weeded than broadcasted, but in Allahabad it requires no weeding at all. The explanation of the discrepancy is to be looked for in the previous preparation of the field; if the weeds were thoroughly eradicated then, subsequent weedings might be rendered unnecessary.

Harvesting.—"The crop is cut with sickles in exactly the same manner as wheat or barley. The most common method of threshing is by beating out the grain with sticks, but it appears that in some localities the grain is trodden out by cattle, the ears having been previously separated from the straw, which is too succulent to break up into chaff as is the case with wheat or barley. The straw called (*pial*) is used for cattle fodder when all else fails, but is very innutritious, and possibly this may be the reason why the agricultural cattle of rice districts are the worst in the Provinces. The grain after being threshed out does not lose its husk, and in this condition is known as *dhán*. The husk is separated by pounding the grain either with a wooden pestle (*mansari*) in a mortar (*akhali*), or in the lever mill known as the *dhekoli*. The husking is sometimes facilitated by soaking the grain in warm water and allowing it to dry. Of course so rude a process destroys some portion of the produce, and of the 60 to 70lb of cleaned rice which can be obtained from 100lb of *dhán*, from 10 to 15 per cent. will be broken and crushed and of little value.

Diseases & Injuries.—"Rice has most to fear from the green fly called *ganduki* or *tanki*, and since the attacks of these insects do not commence until towards the end of August, it is the finer varieties which suffer most. Strong and healthy plants suffer much less than backward ones, and this furnishes another reason in favour of sowing being as early as possible" (*Duthie & Fuller's Field and Garden Crops, pp. 17-19*).

The following district reports on rice cultivation in these provinces may be here quoted in amplification of the facts already given:—

Bareilly. 421

CULTIVATION OF RICE IN THE BAREILLY DISTRICT.

"Rices are by far the principal crop of the autumn harvest, and in Bareilly proper occupy indeed more land than any other crop of either harvest. The so-called Pili

ORYZA sativa.

Rice Cultivation

CULTIVATION in the N.-W. P. & Oudh.

Bareilly.

bhit rices are grown not in this district, but the Tarái. There is, however, a large trade in such rices at Pilibhít, and hence the name. Rice cultivation is thus described by **Mr. Moens.**

"The seed is first steeped thoroughly for a day, then wrapped in straw or cloth for three days, and usually sown on the fourth; but if the field is not ready by that time, it is re-dried in the sun, and will remain for 15 or 20 days fit for sowing. The sowings are called according to the time and method of cultivation employed. (1) *Gaja*—These are the first sowings made in *Baisákh* (April-May). The field is filled with water, and thoroughly ploughed four or five times over with the water on it till the earth is converted into a fine mud (till it is *gauj*). The water is then let off, and the field allowed to become half dry (*ant*), *i.e.*, the surface is allowed to dry to a depth of three or four inches. It is then sown and thoroughly irrigated every third day till the rains. The crop is cut in *Sáwan* (July-August). The produce is heavy, but the cultivation is expensive and laborious, and only possible where water is close at hand. (2) *Bhijua*— If a *rabi khet* has been selected, two ploughings are given in the ordinary way, otherwise four or five. The field is then irrigated, and when the land is half dry the seed is sown in *Baisákh* or *Jeth* (April-May or June) and left.

"If the weather keeps hard and dry the seed germinates, but does not spring up till the first rains. If, however, rain falls shortly after sowing the seed springs up, the young shoots are parched and killed by the hot weather that follows, and the crop is lost. It succeeds best in years when the rains set in late. The crop is cut in *Bhádon* (August-September), and the field can then be thoroughly prepared for a *dosáhi rabi* crop. This method is chiefly prevalent to the north of the district, and is much encouraged by the zamindárs. Where rents are taken in kind, and water is easily obtainable, *anjana*, *sathi* and *seorhi* are the kinds chiefly sown thus. (3) *Kúndher**—This is very similar to *gaja*. Land is selected on the very edge of a jhil or pond, and thoroughly dug up with a *kasi* and divided in *kiyáris* (beds); water is then let in and the land ploughed three or four times. The seed is then sown and ploughed in. The sowing is in *Phálgun* (February-March), and the field kept constantly wet. The crop is ripe in *Asárh* (June-July). *Sathi* is usually selected for this kind of cultivation. The land is usually let for *kúndher* in bits or *párs* of about two *kacha* bighas each, at so much per *pár*: money rates are almost always paid. (4) *Ratiha* or *rasota*—These are the regular sowings in the ordinary *rat* or season, hence the name. They are either (*a*) *khandhar*, where the rain of flood-water is collected in the *kiyarís*, the ground ploughed, and the seed sown wet on the water and ploughed in; the water is let off when the seed sprouts, and for four days afterwards no water is given; after that any amount is beneficial, so long as the top of the shoot is not covered; or (*b*) *kukhana*, where the ground is ploughed and sown broadcast in the ordinary manner.

"The *ratiha* sowings are between the last ten days of *Jeth* (May-June) and the middle of *Sáwan*, not later; and the crop is ripe in *Kuar*. *Karttik*, or *Aghan* (September, October, November or December), according to the kind of rice, and time of sowing; four to seven ploughings are given. The land is very rarely manured, as the rice would then run to straw and be laid, and weeds would be encouraged: five sers per *kacha bigha*, or 86lb per acre, is the usual allowance of seed. *Ratiha* sowings are rarely irrigated artificially; the rainfall gives sufficient water. Well-irrigation is never used for rice. For a full crop water is required up to fifteen days before the commencement of harvest. The necessary amount is generally supplied by the natural rainfall. If *jharúa* grass springs up, the field is weeded once, otherwise not. Rice is sown, as a rule, in *mattiyár* soils, but *sáthi*, *banki*, *dharilla*, and even *sankharcha*, are also sown in *dúmat*. If possible, the sowings commence on a Wednesday, the cutting on a Sunday. At the first cutting the produce of one *kacha biswa* is given to the *kherapati*† or a *fakir*. The seed is either sown broadcast, which is the ordinary method, or in a nursery or panir,‡ and the young plants transplanted. No delay must take place in this work, so that the plants may be as short a time as possible above ground: a calm day is selected for the purpose. As soon as the transplanting is completed in a *kiyári* the water is let in to overflow the plants. The harvest time is regulated by the time of sowings, which is early or late according to the rainfall. Broadly speaking, the coarse rices are sown and cut early; the finer kinds are sown early and cut late."

* The word *kúndher* is elsewhere in Rohilkhund applied rather to a variety of rice than a method of sowing rice.

† The *kherapati* is the village god.

‡ The *birha* of Fatehpur and Allahabad, *bihnaur* of Benares, and *khet biyár* of Gorakhpur.

CULTIVATION in the N.-W. P. & Oudh.

Bareilly.

"The operation of husking (*chhatáo*) the rice is performed by men of the Banjára caste. According to the contract most in vogue, they retain the chaff and three-eighths of the grain, returning the remaining five-eighths to their employers. It is usually reckoned that in forty seers of the paddy or unhusked crop there are 27½ clean rice, 2½ broken rice (*kinki* or *khanda*), and 10 of husk (*chanus* or *ghut*). The last is the established perquisite of the ponies who accompany the Banjáras on their wanderings.

"To destroy a moth (*tirha*) by which the rice is injured, the plants are smoked with aniseed (*ajwain*) or mustard-oil carried along their tops on a lighted cowdung cake. Other enemies of the crop are the *bakúli*, a green caterpillar, rust (*agaya*), and the weeds or grasses known as *dhonda bhangra* (*Verbesina prostrata*), *bansi*, and *gargwa*. The seed of the *dhonda* is eaten by the cultivators, the *gargwa* by cattle, and the *bansi* by buffaloes. The average produce of unhusked rice, as ascertained by frequent experiment, amounts to about 1,218℔ per acre, of which 837℔ will be cleaned rice, 76℔ broken rice, and 305℔ husk. The straw which is used as fodder, will average from 1,300 to 1,400℔ per acre. The best rice is raised in the northern and eastern parganahs, in the southern only *sáthi* and the inferior kinds are grown. Land suited for *sáthi* rents at R2 to R3-3-0 per acre; for *anjana* and similar rices from R3-6-0 to R4-6-0 per acre, and for *jhilma úsbás*, and the superior rices, at R4-12-0 to R7 per acre. The crop is very variable, and in an average period of five years one failure, three second-rate harvests, and but one of the first class may be expected" (*North-West Provinces Gazetteer, V.*, 557).

Gorakhpur. 422

Cultivation of Rice in the Gorakhpur District.

"The principal growth is rice (*dhan*), for whose culture and irrigation the moist tarai soil and numerous streams of Bináyakpur and Tilpur afford exceptional facilities. More of it indeed is grown than of all the crops put together, and the rice fields often present an unbroken expanse of some miles in extent. In parganah Haveli also the crop occupies a large area, and it is met with everywhere in the district, though to a small extent only in the southern and eastern tracts. A species of rice called *boro* must be elsewhere described, as its cultivation and time of reaping differ from those of the ordinary *dhán*. *Dhán* may itself be divided broadly into two classes—*bhadui* and *aghani*.

The former is sown in *Jeth* (May-June) on land which has been left fallow since the autumn harvest of the former year. The ground is ploughed in *Pús* (December-January) or *Mágh* (January-February) in order that the sun may penetrate and warm without hardening it overmuch. The field is again ploughed before sowing in *Jeth*. It is considered advantageous if a shower or two have fallen before this; but whether it rains or not, the seed must be sown by the end of the month just named. Seed sown before rain falls is called *dhuria báwag* (*i.e.*, dusty or dry sowing). The soil best adapted to receive the crop is that lying low enough for the water to lodge, but not too low, as excessive flooding is injurious. If no rain falls before sowing, and unless the soil is very cold and moist, it is usual to irrigate the field directly after that process. It is for this purpose that the Thárús of Bináyakpur dam up the small streams, which they then divert by numerous channels (*kulas*) into their fields. As soon as the water has collected, naturally or artificially, to a depth of about three inches, the field is ploughed once more. This rather rough treatment is said not to injure the seeds, but to eradicate weeds which would otherwise choke the young crop. In *Asárh* (June-July) any grass or weeds which may have sprung up are weeded out by women and children, who receive as wages about 2¾ seers of rice a day. This process is called *nirai*. The amount of seed sown on the recognised *bigha* varies slightly in different parts of the districts, the highest being 28 seers in Bánsgáon, the lowest 22 seers in the Sadr Tahsíl. After sowing, the crop is generally dependent on the rains, and is ruined if they fail. As this kind of rice thrives most when the water around it is not too deep, its sower prefers a season of light and sustained to one of sudden and heavy rainfall. The field have strong *merhs* or banks of about two feet high to retain the water. The crop grows rapidly and is cut in *Kuár* (September-October), or sometimes at the end of *Bhádon* (August-September), from which latter month it probably derives its name. Its best varieties are—*Jhali, kapúrchini, gajesar, bendi* (white and black), *muttri, bánsphúl, parni* or *padni, dudha, sátha* or *sáthi, anjanawa, sina, kauria, gajbel*, and *bandela*.

The second kind of dhán, *aghani*, is sometimes distinguished from its synonym *jarhani*, but no perceptible difference between the two would appear to exist. There are indeed two varieties of *aghani*, but these are varieties rather of cultivation than species, and the term *jarhani* applies to both. *Jarhani* in fact merely denotes the winter (*jára*) as opposed to the *Bhádon* or *bhadui* crop. Of the two varieties, the first (*chhitua*) is generally sown or scattered (*chhitua*) over fields which have lain fallow

ORYZA sativa. **Rice Cultivation**

CULTIVATION in the N.-W. P. & Oudh.

Gorakhpur.

for some time and have been prepared, like those for the *bhadui dhán*, some months beforehand. Often, however, a field in which gram, *kirao* or linseed has been sown is selected for crop. The stalks of the former one, being dug into the ground and mixed with the soil about two months before the rice is sown, form a kind of manure. The seed is sown in *Asarh* about a maund to an acre and just as the crop has begun to rise from the ground, it is ploughed up again and dug into the earth. After a time it sprouts afresh with greater strength than before.

It is cut generally in *Kartik* (October-November). The second variety, *behan*, is so called from *behan* or *bihan*, a cutting or seedling. This crop needs two fields. The first, called *khet biyar*, is ploughed twice or thrice in *Mágh* (January-February) and has high walls. In *Asárh*, after the first good fall of rain, it is ploughed and the water made to mix well with the soil. A plank heavily weighted is then dragged over it, and when the earth has become quite soft and slushy the seed is sprinkled broadcast and the plank taken over once again. About 30 to 35 seers of seed are sown to the acre. After a month the plant is usually ready for transplantation to the second field, which has been carefully ploughed for some time previously. If the crop is good one, a *biswa's* growth in the *biyar* field is enough for planting a *bigha* in the new one. The plants, which are one or one-and-a-quarter feet in height, are stationed in their new home at distances of some two inches from one another. As it is necessary to complete this work quickly, a great number of hands are employed, the average being a dozen men or women to the authorised bigha. These persons, if hired labourers, get two *rasias* of rice and a quarter ser of *charban*, or, if they prefer it, two annas a day. A considerable quantity of water is needed for this crop, and the walls of the fields are usually high and strong, so as to keep in the rainfall. The harvest is most often in *Aghan* (November-December). For carrying the crop to the threshing floor the labourers get either two annas daily or one sheaf in 16, or if the harvest be poor and labour plentiful, one in 24 only. This kind of rice being cut very late, it is impossible usually to grow spring crop on the same land. The same fields are therefore used year after year for this crop alone. When it is cut, stalks of about ten inches high are left in the field; in the hot weather these are burnt and, as soon as any rain falls are dug into the ground, forming a valuable manure. Amongst the best kinds of *aghani dhán* are the following:—Finer (*Mihin*) grains, *phen, gauria, beharni* (white and black), *syám jira*, and *gurdhi*. Coarser (*mota*) grains, *harbelas, rájal sahdiya, karga, nainjot*, and *angetha*. The *aghani* rice is, as a rule, more valuable, and yields for the same area a larger outturn than the *bhadui*, but the latter of course leaves the land vacant for a spring crop. In Sidhua Jobna a class of rice called *sengar* is largely grown on lakes or ponds where the depth of water during the rains prevents the ordinary kinds of rice being grown. Its peculiarity is that it floats on the top of the water, and that the growth of the plant, whose roots are fixed in the soil below, keeps pace with the rise of the surface even when that rise is sudden. It is cut in November, very often from boats, if the rains have been late and the water has not subsided.

In a good season the yield of rice is very great, and rice itself is the staple food of the poorer classes throughout the district. The outturn per acre of this and other crops will be shown on a later page. The process of threshing the rice or rather of treading it out with bullocks, is the same as elsewhere, and known as *dauri*! But thoroughly to separate the grain from the husk, to turn the *dhán* into *chánwal*, another process is required. The rice is placed in a *dhenki* or wooden mortar and pounded with a pestle, which hinged on a fulcrum, falls by its own weight and is lifted by the pressure of a foot on its lighter or pedal end. Three sers of dhán yield two of *chánwal* and one of chaff (*bhusa*).

The husking is usually the work of hired labourers, who receive as wages one seer in twelve of the grain" (*North-West Provinces Gazetteer* (*Gorakhpur District*), *VI.*, *321-324*).

Budaun.
423

Cultivation of Rice in the Budaun District.

"The cultivation of rice differs according to the variety sown, the principal varieties being *sathi, jhabdi*, and *khonder*. Of these, the most common is the *sathi*, so called because its crop ripens about sixty (*sáth*) days after appearing above ground. The seed, which is steeped in water all night before sowing, is, after the beginning of the rains, sown broadcast in moist or marshy ground prepared for its reception by two or three ploughings: and the crop is weeded once or twice during its growth. The *jhabdi*, on the other hand, is sown at the beginning of the hot weather, and is slow in growth, being rarely ready for reaping until December. As the young crop has to brave the fierce heat of the summer winds, it must of course be sown in the neighbourhood of a swamp, or some other spot where irrigation is easy. The crop is watered regularly until the beginning of the rains and its field is, before the sowings, ploughed about half a dozen times. *Khonder* is a variety resembling *jhabdi*, the only difference between the culture of the two being that before receiving the former the field must be first prepared by irrigation" (*North-West Provinces Gazetteer, Vol. V.*, 28).

CULTIVATION in the N.-W. P. & Oudh.

CULTIVATION OF RICE IN OUDH PROPER.

Oudh. 424 *Conf. with p. 522.*

Rice is extensively cultivated. In the introduction to the *Oudh Gazetteer* the following passage occurs which is of interest in connection with rice:—

"The principal *kharif* staples are rice, Indian-corn, and the millets, and the choice of crop is determined by the lay and character of the soil. Rice grows best in low stiff land, where the water accumulates first and is most slowly absorbed, maize on a light soil raised slightly above the floods. The yield of the first is sometimes as much as twenty maunds per *bigha* or 2,600℔ per acre, but three-fifths of that is considered a fair outturn; the latter will occasionally yield four cobs to the stalk, but it is seldom that more than three are fertile, and the agriculturist is contented with two good heads. The yield is heavier than that of rice, 3,300℔ per acre being an outside, and 2,000℔ a fair average, crop per acre. The smaller millets are less productive, grow on inferior soils, and exact less trouble in cultivation. Among the inferior crops, which are cut during the rains, are *mendwa, kákun*, and *kodo*, diminutive grains which form the principal diet of the very poor. The finer kinds of rice, which, instead of being sown and reaped on the same land, are transplanted in August from nurseries near the village site, do not ripen till the end of November, and form the most valuable item of the *henwat* crop. The average yield is at least 20 per cent. greater than that of the early autumn varieties, and the grain is smaller, better flavoured, and commands a rather higher price. The taste of the native differs diametrically from that of the English market, and the consideration in which the different kinds of rice are held varies inversely with their size" (*Oudh Gazetteer, I., ix.*).

It will thus be seen that, according to the remarks above (if the *bíghá* be taken at ⅝ths of an acre), the yield of rice would be 12½ maunds per acre, or 3,03,90,350 maunds for the whole province.

CULTIVATION OF RICE IN THE NORTH-WEST HIMÁLAYA.

N.-W. Himalaya. 425

"This widely-distributed grain is, as may be supposed, the principal rain-crop in the lowlands, and is also largely cultivated in the hills up to 6,500 feet, where some of the most valuable varieties are raised in the deep, hot valleys. In the hills, the agricultural year commences about the middle of February, when the land has to be prepared for the rice crop, which is usually sown where *mandua* has been raised in the previous season. The manure from the cattle-sheds is spread over the ground which is then ploughed and freed from stones. The terrace walls are repaired and the roots of the *mandua* from the last crop are collected and burned. In *Baisakh* (March-April) or *Jeth* (April-May), the land is ploughed again and the seed is sown in the furrows, which are closed by a flat log of wood drawn along them. When the young plants have risen to some three or four inches in height, a large rake or harrow is drawn over the ground to remove the weeds and thin the plants. Where water is abundant, the better sorts of rice are sown in a highly-manured and irrigated nursery (*bihnora*) or seed-bed. This is first flooded with water and then ploughed until the soil becomes a semi-liquid mass. Manure is then added and the seed is sown on the top and covered over with leaves, especially those of the *chir*, which are said to decompose easily in water and form an excellent top-dressing manure. The young plants are transferred (*ropa*) from the nurseries by the women and children in June-July to the open field. The manure used is commonly the sweepings of the cattle-pens, which are collected in regular heaps on a place set apart for it in the field, usually that in which the cattle have been regularly penned (*khatta*), to economise the collection of their droppings. Leaves also are collected and allowed to rot in heaps on the field, and twigs and branches of trees are burned and the ashes made use of. The latter are usually taken from the village forests and cost nothing but the labour in gathering and stacking them. When the field is a small one, the earth is loosened and the weeds removed by a small iron sickle (*kutala*). In July-August the weeds are again removed, whilst the land is kept inundated with water, and by the end

ORYZA sativa.

Rice Cultivation

CULTIVATION in the N.-W. P. & Oudh.

N.-W. Himalaya.

of August the poorer high land varieties are ready, and by the end of September or beginning of October the finer sorts grown on the low lands. Rice is cut from the root and stored on the field in stacks (*kanyúra*) with the ears inwards. There it is left for four or five days to dry, and after that the grain is trodden out by cattle on a threshing-floor paved with slates (*khala*) or simply by men on mats (*moshta*). The stalks (*puwál*) are made up in bundles (*púla*) and stored round a pole or in the fork of a tree and afford food for cattle and bedding for the poor. The grain is taken home, and, after being dried on the roof of the house, is stored for use in boxes (*bhakár*) or in baskets plastered with mud or cowdung, called *korangas* or *dálas*. Unhusked rice is known as *dhán* in Kumaun; and before husking it is again dried in the sun and then pounded in a wooden or stone mortar called an *ukhal*. The pestle (*musal*) in use is tipped with iron, and the grain is pounded three different times before the clean rice or *chánwal* is produced. The chaff (*chila*) is used as fodder for cattle, and the husk (*pithi*) of the third pounding, by the poor. Winnowing is performed by a shovel-shaped basket (*supa*) which is held at such an angle to the wind as allows the chaff to fly off, or the grain is placed on the ground and the basket is used as a fan. One *náli* or about four pounds of rice-seed produces in irrigated land 35 *nális* of unhusked or one-half that amount of husked rice, and rice-seed in up-land unirrigated land about half as much. Dry up-land rice ripens from early September; common irrigated rice from early October and the better irrigated sorts from the middle of October. In Dehra Dún there are three principal varieties, the *chaitru*, *haltyu*, and *kyári* or transplanted. The first, which is also known as *chambu* or *anjana*, is sown in unirrigated land in March-April (*Chait*) and is cut in August-September. *Haltyu* is sown a month later in similar land and is cut in September; it is also known as *anjani* and *naka*. The *kyári* furnishes rice of the best quality; the seeds are sown in nurseries in April-May, and the young plants are transferred in the following two months to well-irrigated fields, where they are carefully weeded. The principal varieties are the *ramjawáin* and *básmati*, and these grow best in warm valleys and along the great rivers where there is much moisture. *Chánwal* cooked in water is called *bhát*, but the broken grains (*kanika*) when cooked are called *jaula*. *Khijri* is a mixture of rice with *urd* or *bájra* boiled together in water; and *khír* is rice boiled in milk. The commoner varieties are often made into bread, and in that case the grain is only husked once and the inner husk is left on to be ground into flour, called *baghar* in Garhwál" (*Atkinson's Him. Dis.*, *685*).

Nepal. **426**

Cultivation of Rice in Nepál and the Central Himálaya.

Some of the finest and most prized rices of the North-West Provinces (indeed of India), such as the Philibit, are known to be imported from Nepál. The writer has failed to find any modern account of Nepál rice, but the passage given by **Dr. Buchanan-Hamilton** is, like most of the labours of that great pioneer student of Indian economic questions, so full of interest, that no further apology seems necessary for its reproduction here, than the statement that comparatively few modern writers have the pleasure of being able to consult **Dr. Hamilton's** works:—

"In Nepal, rice is the great crop, and the ground fit for it is of two kinds, which differ in the manner, and in the time of their cultivation, so as to make two harvests of rice; but no one field, in one year, produces two crops of this grain.

Colonel Krikpatrick, indeed, mentions that some fields yield two crops of rice successively, the one coarse, and the other fine, besides affording, in the same year, a crop of wheat. This, however, I presume, does not allude to Nepal Proper, but to some of the warmer valleys in the dominion of Gorkha, as where he goes on, to describe the expense of cultivation, he mentions the ploughings, an operation which is not employed in the agriculture of the Newars.

The first kind of ground produces the crop called *gheya*, is the highest, and there is no necessity for its being absolutely level, as the fields are not inundated. From the 13th of March to the with of April, this ground is hoed, and, having been well manured with dung collected in the streets, it is hoed again. A week after this, the field is hoed two or three times, and is well pulverized with the mallet. About the 12th of May, after a shower of rain, the field is slightly hoed, and the mould is broken, and smoothed with the hand. Small drills, at a span's distance from each other, are then made by the finger, which is directed straight by a line. At every span-length in these drills are placed four or five seeds of the rice called *uya dhan*, which is the only kind cultivated in this manner. The seed is covered by the hand, and a very small quantity only is required. In about five days the young corn comes up in small tufts, just as if it had been transplanted. From the 13th of

CULTIVATION in the N.-W. P. & OUDH.

Nepal.

June to the 15th of August, when the corn is about a cubit high, the weeds are removed with the spud. About the latter period, slugs, worms, and insects, fill all the moister fields in Nepal, and in order to be rid of them, the farmers keep a great number of ducks which, at this season, they turn into the fields, to devour the vermin. The *gheya* crop ripens about the 1st of September, and by the middle of the month the harvest is finished. The ears only are cut off, and next day the grain is beat out, and generally dried in the streets. Very little of the crop is made into *hakuya*, a process that will be afterwards mentioned. After the *gheya* crop has been cut, the field is in general cultivated with radishes, mustard or some other crop, that is usually sown about the time.

By far the greater part of the rice ground, and that the lowest and the best, is of the kind which produces the crop of rice called *puya*. The kinds of rice which are cultivated in this crop are very numerous, and it would be tedious to mention their names, as I have no observations to make on any one in particular. The fields which produce this crop must be perfectly level, as they are inundated during the greater part of the process of cultivation. Therefore, as the plain is by no means even, it has been divided into terraces not above two feet wide. The numerous springs and rivulets that issue from the surrounding hills have been conducted with great pains to irrigate these terraces, and have been managed with considerable skill.

The cultivation of the *puya* crop commences between the 13th of May and 12th of June, during which the field is hoed two or three times, and manured with dung, if any can be procured. At any rate, it is always manured with the kind of earth called *koncha*, which I have already described. The banks that confine the water are then repaired; and about the 12th of June, when either by the rain or by the irrigation from aqueducts, the fields have been inundated, and the soil has been by the hoe reduced to mud, the seedlings which have been raised in plots sown very thick, are transplanted by the women. The men perform all the other parts of the labour. This is a time of festivity as well as of hard work; and the people are then allowed a great freedom of speech, to which they are encouraged by large quantities of intoxicating liquors, in a share of which even the women indulge. The transplanting ought to commence from the 12th to the 15th of June, and ought to be finished by the *Amavasya* of *Asharh*, but this is a moveable feast. On the *Krishna Chaturdasi*, which happens on the day preceding the *Amavasya*, the Maharani or Queen, with her slave girls (*ketis*) transplant a small plot within the palace, and it is reckoned an unlucky circumstance when this is not the last planted field in the valley. The fields are always kept under water and weeds are not troublesome. The few that spring up are removed by the spud. This crop begins to ripen about the 15th of October, and by the 1st of November the harvest is completed, after which a considerable portion of the land is cultivated for wheat or other winter crops.

Hakuya Rice. 427

The *puya* rice is cut down close by the ground. The finer kinds of rice are immediately thrashed, as is likewise all that which is intended for seed, but the greater part is made into what is called *hakuya*. This is done with a view of correcting its unwholesome quality, for all the grain produced in the valley of Nepal is thought by the natives to be of a pernicious nature. The manner of preparing *hakuya* is as follows: The corn, immediately after having been cut, is put into heaps, ten or twelve feet in diameter, and six or eight feet in height. These are covered with wet earth and allowed to heat for from eight to twelve days, and till they may be seen smoking like lime-kilns. After this the heaps are opened, and the grain is separated from the straw by beating it against a piece of ground made smooth for the purpose. Both grain and straw are then dried in the sun. The grain is called *hakuya* and the straw is the fuel commonly used by the poor for firewood is very dear. According to the accounts received by Colonel Crawford, this manner of preserving rice was discovered by accident. Many years ago one of the towns was besieged by an enemy that came so suddenly as not to allow the citizens time to gather in the crop, which had just then been cut. The citizens, rather than allow the enemy to benefit by their corn, determined to throw it into the water and cover it with earth. In this manner it remained about a week, when the enemy were compelled to retire. When the grain was taken up it was found to have begun to rot, but necessity having compelled the people to eat it, they found, to their astonishment, that it was much better and more salutary than the grain which had been prepared in the usual manner. It is only the Newars that eat this *hakuya*.

The crops of rice in Nepal appeared to me very poor when compared with those of Bengal, and, if my Brahman was rightly informed concerning the extent of a *rupini*, they are really so. The *rupini* produces four *muris* of paddy, or $9\frac{376}{1000}$ bushels, but near Calcutta the *bigha* (supposed to be of the same extent) of good

ORYZA sativa.

Rice Cultivation

CULTIVATION in the N.-W. P. & Oudh.

Nepal.

ground produces often 640 sers, or $19\frac{82}{100}$ bushels. The difference of price, however, in the two countries makes the value of the produce of Nepal the greater of the two. I have already stated that the value of four *muris* of paddy in Nepal is usually 13 M. 2 A. 2 D., or about $5\frac{3}{4}$ rupees. But near Calcutta in harvest the usual price of 640 sers of paddy, is R5-5-4. If no error has been made in estimating the extent of a *rupini*, the acre of good land in Nepal produces rather more than 28 bushels of paddy, or rice in the husk. Immediately after the *puya* crop has been cut, the ground is formed into beds by throwing the earth out of parallel trenches upon the intermediate spaces. On these about the middle of November is sown wheat, or sometimes a little barley. These ripen without further trouble, and are cut from the 12th of April to the 12th of May. The seed for a *rupini* is stated to be one *pati*, and the produce is stated to be two *muris*. This would make the seed about the fifth part of a bushel an acre, and the produce about fourteen bushels; but this seems to me greatly exaggerated. I have never seen more wretched crops, and most of the fields of wheat are quite choked with hemp **(Cannabis sativa)** which in Nepal is a troublesome and useless weed. The wheat and barley are mostly used for making fermented or distilled liquors." (*Buchanan-Hamilton, Account of Nepal, 222-226.*)

Dry Alpine Rice.
428
Conf. with p. 501.

Much has been written on the subject of an alpine form of rice (similar to the *gheya* crop described above), but which is said to be cultivated in Nepál without being irrigated to any greater extent than is the case with wheat or barley. **Balfour** apparently alludes to this under the name of **O. nepalensis**, and suggests that it is the ancestor of all the Indian hill rices. There is, however, nothing to show that any of the hill rices differ botanically from the better class transplanted rices of India, and, what is more remarkable, all those examined by the writer belong to the *áman* or flood-land rices; none of them apparently correspond to the *áus* or dry-land crops of the plains, even although some are grown on unirrigated lands. The terraced fields of the Himálaya are flooded by the water being skilfully carried, often for miles along the hill sides, to the highest field of a village plot of lands and then allowed to overflow from one field to another until the series of terraces stand from 3 inches to 1 foot under water. As soon as the grain has formed, this water is let off or dries up naturally and the final ripening of the grain thus takes place in fields which, if inspected then for the first and only occasion, might (as doubtless may in some cases have happened) give origin to the statement that some of the hill rices are grown on perfectly dry fields. In the more humid valleys, especially in the Central and Eastern Himálaya, the moisture of the atmosphere and consequent dampness of the soil, allows of rice being grown, however, on fields that are never actually flooded. Such rices, in fact, in that respect, are the representatives of he *áus* rices of the plains (though not botanically so), but that any sucht rice occurs in the alpine zones of the Himálaya does not appear even probable. It thus seems likely that the origin of the statement that an alpine dry land rice (suitable for cultivation in England) existed in the higher tracts of Nepál, is traceable to **Colonel Kirkpatrick** (*Nepaul, p. 282*). A century ago that distinguished officer wrote that there was found in the alpine valleys, inhabited by the Limbus, a kind of dry rice which was there known as *takmaro*. But **Dr. Buchanan-Hamilton**, who devoted considerable attention to the study of the products of Nepál, failed to discover any purely alpine rice and came accordingly to the opinion that *takmaro* was synonymous with *uya* — the hill name for rye. (*See Buchanan-Hamilton's Account of Nepal, pp. 88, 274, &c.*)

The writer has no personal acquaintance with a strictly speaking alpine rice, and is, therefore, strongly disposed to think some such mistake, as that indicated by **Dr. Buchanan-Hamilton**, may have been made. In the absence of direct evidence to the contrary, he, however, feels called upon to reproduce one or two passages from authors who appear to believe in the existence of a Himálayan dry alpine rice. **Mr. Liotard**, after dealing with the upland rices of the United States, which, he says, were originally obtained from Cochin China and Nepál, concludes with the following remarks:—

CULTIVATION in the N.-W. P. & Oudh.

Dry Alpine Rice.

"On the north of India we find the Bara rice of Peshawar and the Joomla rice of Nepal, both no doubt of the **Oryza mutica** species, growing in high altitudes, the latter especially amidst snow and frost, under such special circumstances as to have merited a name of its own among botanists who call it the **Oryza nepalensis.**"

Bara Rice. 429

The Bara rice is much prized by Natives in Upper India. It grows in the valley of Shaik Khan, opposite the fort of Bara, about nine miles south-west of Peshawar, and seems to be of some historical interest, for **Dr. Collis**, in sending a sample of it to **Dr. Bonavia**, the Honorary Secretary to the Agricultural and Horticultural Society of Oudh, wrote in 1863:—

"This rice, which it is said will only grow in a few fields near Shaik Khan, is so much prized by Natives that, before the conquest of Peshawar by the Sikhs, the Cabul Sirdars had agents to watch the fields in order that none of it might be removed, but that the entire crop might be sent to them. When Runjeet Singh began to threaten Peshawar in 1822, twenty seers of Bara rice was among the presents sent by Yar Mahomed Khan in hopes of propitiating him. Even now a great deal of it is sent to Cabul. Of course every zemindar, who cultivates rice in the neighbourhood of the fort or river of Bara, brings his crop into the market as Bara rice. I can, however, vouch for the genuineness of the specimen I now send you, which I got at the time that I was obliged to visit sick prisoners in the fort at Bara once or twice a week."

Dr. Bonavia says that the quantity sent by **Dr. Collis** was about two handfuls, that it all germinated in Oudh where he had it sown, and that it produced a very fine crop which yielded about a maund of paddy; that the crop was much admired by all the natives, who said it was very different from the kinds of rice they had been accustomed to sow. Some of it was exhibited in sheaf at an agricultural show: it received the first prize for rice in husk, and was much thought of by all natives who were conversant with the qualities of good rice. **Dr. Bonavia** intended sending a large packet of this rice seed to each Deputy Commissioner in Oudh, and he hoped that, if it were distributed in small quantities to persons who are likely to care for it, the Agricultural and Horticultural Society will be able to disseminate it throughout the Province in two or three years. The further results of this enterprise are not, however, traceable in any of the books and papers to which I have access.

Joomla Rice. 430

The Joomla rice is a native of the lofty mountains of Joomla in Nepal, north-west of the great valley; and as its cultivation is carried on at high elevations, and under circumstances of frost and snow, it was thought that it might become acclimatised in the north of Europe. Accordingly, such information, as was obtainable, was gathered by the Resident in Nepal, who also sent samples of the rice seed to the Agricultural and Horticultural Society of Calcutta. Part of this supply was forwarded to the Honourable Court of Directors in London, and a portion was sent to the Colonial Secretary at the suggestion of Lord Auckland, who was of opinion that Canada might, perhaps, answer for its cultivation.

The mountain rice is also abundantly grown at altitudes of 6,000 and 7,000 feet in the Sewalik hills and in the up-land valleys of North-West Himálaya. **Dr. Balfour** says that it has been reared successfully on the banks of the Thames at Windsor. This sowing was evidently made from the sample of the Nepal or Joomla rice seed sent to England, but there is no distinct information as to this."

Rice Cultivation

CULTIVATION in the N.-W. P. & Oudh.

Nepal.

There is, however, no greater reason for supposing the ordinary hill rices, grown, for example, at Simla, or the cold-season rices cultivated on any part of the plains of India, could not be produced on the banks of the Thames, than that wheat, barley or oats, brought from Scotland, could not be grown during the summer at Simla or the winter on the plains. Still less is there anything to show that the ordinary Himálayan rices differ in essential characters from those of the plains, except, perhaps, what is readily accounted for by their long cultivation in a colder climate. Both are essentially aquatic crops, and neither of them could withstand the smallest touch of frost still less be grown at such altitudes as to necessitate their existing through the winter snow to ripen in spring, nor could they mature in an autumn which was liable to showers of snow or biting frosts. The hill rices of the Himálaya are off the fields by October, and thus before the autumn influence is seen on the deciduous foliage of the neighbouring trees and hedgerows. It is, however, a by no means unusual occurrence to find wheat, barley, rice and maize growing side by side on the Himálaya, the last two plants having adapted themselves completely to the temperate climate without which neither wheat nor barley can be grown. But this is after all a very different matter from the further statement that rice has been found as an alpine or arctic crop. The writer has certainly never seen rice at higher altitudes than 7,500 feet, but in Sikkim, **Sir J. D. Hooker** tell us, he found it growing at 8,000 feet, in exceptionally warm moist localities.

It may here be pointed out, in concluding these brief remarks regarding Himálayan rices, that the mode of treating paddy before husking it, described above in the extract from **Dr. Hamilton's** work, appears to be peculiar to Nepál. The writer has come across no other passage either dealing with this Nepál practice or describing anything that could at all correspond to it in connection with the provinces of India. Could it be the case that the high value put on certain Nepál rices in India, is dependent upon this somewhat remarkable method of treating paddy? (For further information regarding the rices of temperate regions see the remarks below under Kangra, Kashmir, as also Yarhand.)

PANJAB 431

VIII.—CULTIVATION IN THE PANJAB.

References.—*Agri.-Horti. Soc.-Ind.—Journ. (Old Series), VI., Pro., 116; VII., 232; VIII., Pro., 37; XIII., Sel., 49, 50; Gazetteers:—Amritsar, 35, XIII., XV.; Ambala, 49, XIII., XIV.; Bannu, 138, 145, XI., Delhi, 113, 114, XII., XV., Dera Ghazi Khan, 82; XI. Dera Ismail Khan, 116, 126, XII.; Ferozepore, 65, 69, XI.; Gujrat, 78, XI.; Gujranwala, 53, XI.; Gurdaspur, 51, XI.; Hazara, 124, 129, 130, XI.; Hisar, 48, XI.; Hoshiarpur, 92-93, XI.; Jalandhar, 44, XI.; Jhang, 107, 118, XI.; Karnal, 69, 73, 74, 172, 176-178, 185, XI.; Kohat, 101, 105, XI.; Kangra, 58, 59-60; Lahore, 86, 87, 88, 89, XII.; Ludhiana, 134, 141, XII.; Montgomery, 103, 104, 109, XI.; Mooltan, 92, 95, 96, 98, XIII; Muzaffargarh, 91-92, 93, XI.; Peshawar, 144, 146, XIII; Rawalpindi, 81; Rohtak, 91, 93, 94; Shahpur, XI.; Sialkot, 65, 66, 68, XII.; and Simla, 39, 53, 57, 59.*

Area. 432 *Conf. with p. 522.*

Area under, and Traffic in, Rice.

The Agricultural Statistics of British India, published by the Imperial Government, show the Panjáb to have had 690,565 acres under rice during the year 1888-89. This is the smallest rice area among the important provinces of India.

Trade. 433

The traffic in rice recorded under Rail, Road, and River, shows a net import of 5,90,297 maunds. In fact, it would appear the Panjáb is largely dependent on other provinces for its rice-supply. The gross imports from the North-West Provinces and Oudh amounted (in the year under consideration) to 5,78,658 maunds or 96 per cent. of the total imports. Bengal

CULTIVATION in the Panjáb.

furnished the Panjáb with 49,465 maunds. The Panjáb exports of rice, 23,617 maunds clean rice to Rájputana and Central India, 7,325 maunds to Karachi, and 5,876 maunds paddy to the North-West Provinces and Oudh.

Consumption per head. 434

From the returns of yield analysed below it may be assumed that the average outturn is 11 maunds an acre, which would give the total production as 75,96,215 maunds or 271,293 tons. To this amount should, however, be added the net import shown above, and this total expressed to population (18,850,437) shows a daily consumption per head of $\frac{1}{25}$th of a seer.

Rotation: Manuring. 435

Rotation of Crops, Manuring, &c.

The rotation of crops generally adopted in the Panjáb will be discovered from the series of district reports quoted below. As to manuring it may be said in general terms that the seed-beds are highly manured and the fields into which transplanted rice is grown are manured to a limited extent also, but broadcast-sown rice is scarcely, if at all, manured. As in most other parts of India through which canal irrigation has been extended, the cultivation of rice may be said to have increased with the facilities for water, but the quality of the rice grown has rather fallen off than improved. A somewhat interesting form of manuring or rather of clearing the fields of deleterious weeds, chiefly algæ and other inferior forms of vegetable life, was made known by the writer in the *Selections from the Records of the Government of India for 1888-89*, pp. 67-81. He dealt with the subject under the botanical name of the plant used (**Adhatoda Vasica**) and the original article should be consulted for full particulars. The opinions, there recorded, of the numerous officers consulted, fill many pages. It may briefly be stated in the words of the original note that—

Basuti. 436

"During a brief botanical excursion to Suní in the Sutlej valley (north of Simla), the writer witnessed the cultivators scattering the leaves of this plant over recently flooded fields, which were being prepared for the rice crop. Concluding that the same ideas prevailed in the Sutlej valley as in Oudh, with regard to some special virtue over any other plant, which the leaves and twigs of this bush possessed as a green-manure, the cultivators were asked whether the manure they were adding was only given to rice land or to other crops as well. The question seemed to afford amusement, for the cultivators hastened to explain that the *básútí* was not given alone as a manure, but also as a medicine or poison to kill the aquatic weeds that otherwise would greatly injure the rice crop. They pointed to fields that had been treated in this way; and there could be no mistake these were clean or free from the green scum caused through floating **Lemnæ** (duckweeds) and submerged **Charæ**. It was further explained that, before flooding the field, it was carefully manured with farm-yard manure, then flooded. The water in a few days, it was explained, became green through the growth of the weeds. When all the weeds which the soil or water were likely to produce had sprung up, a cultivator, with a large apron in front of him, full of the cuttings of *básútí*, each about six inches in length, walked through and through the flooded field, sowing or scattering the *básútí* until at a distance the field looked as if a crop of some plant, not unlike tobacco, was being grown in water. After a time these twigs are supposed to impart an objectionable flavour to the water, which completely kills the aquatic weeds. The *básúti* is then gathered off the field; and the rice crop sown or transplanted into it, as the case may be. Sometimes the *básútí* is ploughed into the soft mud, and thus made to act as a manure. Instead of *básútí*, when a green manure alone is desired, the leaves of the *Toon* (**Cedrela Toona**) or the *Neem* (**Melia**

ORYZA sativa.

System of Clearing Rice

CULTIVATION in the Panjab.

Basuti.

Azadirachta) are utilised; but these, with the *básúti*, are the only plants employed as green-manure in the Sutlej valley. In all these cases it will be observed that plants which possess powerful properties are resorted to in place of numerous other weeds, which might more conveniently be used. This same idea seems to prevail in many other parts of India—as, for example, the almost universal opinion, that the leaves of **Calotropis gigantea**, the *ák* or *madár* (a plant with a most powerful milky sap), is a valuable manure for rice land, and a specific against the injurious growth of *reh* efflorescence. It is difficult to understand what particular merit that green-manure could have, over any other green-manure, in neutralising the *reh* salts, but, as remarked, the idea that it does possess some such property is very widespread in India. It is thus noteworthy that the green-manures in most general use in India are, like the *básúti*, as far removed chemically and botanically as they well could be from the crop intended to be cultivated, and are plants with powerful active principles.

It has occurred to the writer that it may be possible the same practice and with the same object as recorded in the Sutlej valley, may be followed in Oudh and in other parts of India, and that the merit attributed to the *básúti* as a green-manure is not, strictly speaking, deserved. It, at least, seems more natural to suppose that the strongly fœtidly-scented leaves would impart an injurious flavour to the water, sufficient, as believed in the Sutlej valley, to kill aquatic weeds, than that a few twigs of this plant could have a special merit as a manure. If this supposition prove correct, it may further be found that the habit of using the plant in the construction of wells (a use reported in the *Oudh Gazetteer*) may be connected with the knowledge that the green scum so common on every sheet of water in India will not be formed in the presence of a few twigs of this plant. This idea is, however, only thrown out as a suggestion, in the hope of directing attention to the subject; for it may be found that a plant with such strong properties should, in the construction of wells, rather be discouraged than encouraged—at least in the case of wells intended for drinking purposes. If it be the case that the *básúti* leaves have the power to destroy the injurious weeds found in submerged fields, it would seem desirable to make this fact known, since, as far as the writer is aware, this property is not understood outside the limits of the Sutlej, or perhaps of one or two neighbouring valleys in the North-West Himálaya. Indeed, it is quite possible that this extremely plentiful plant, when its properties are fully examined, may prove to be capable of further development. For example, a decoction of it might be found useful in destroying animal pests to other crops, such as sugarcane, tea, and coffee.

Useful to Destroy Pests. 437

In the meanwhile it seems desirable that enquiries should be made in different parts of India regarding the practice of applying the plant to the fields; and the reason why it is so used, and why, also, it is chosen to bind the sides of wells" (*Simla, the 22nd June 1887*).

"It may be added that since the above was first published, the writer has accidentally come across one or two passages in **Dr. J. E. T. Aitchison's** *Kuram Valley Flora*, which are of interest as pointing to the belief in the special merits of **Adhatoda** as a manure being held by the people of certain portions of Afghánistán. Speaking of the Kuram district, **Dr. Aitchison** says: 'The germination of rice-seed in the seed-nurseries is supposed to be hastened by shading with the young branches of **Adhatoda Vasica** and **Sophora mollis.**' Again, of Thal and Badishkhel he writes: 'The young leaves of **Adhatoda Vasica** are largely collected and mixed with the grain in the rice-nurseries, to hasten the process of germination by the heat generated during their decomposition.' In nearly every province of India certain cultivators put great faith in green-manure for rice;

CULTIVATION in the Panjab.

Basuti.

and in most districts the grain is frequently made to sprout before being sown. Some interesting information of this nature will be found in the *Mysore Gazetteer*, Vol. I., pages 52, 110, 112, and 113—the plants specially mentioned being **Calotropis gigantea, Dodonea viscosa, Mirabilis, Datura, Solanum,** and **Ocymum.**—[*Ed., Sel. Rec. Gov. of Ind.*]"

The above preliminary note was circulated by the Government of India to all local Governments, and, as already remarked, the replies received are too numerous to be reproduced here. One or two passages may, however, be given as showing the extensive manurial use to which the plant is put. The Government of Madras furnished a collective reply, dated April 1888, which embraces the chief facts brought out by the District Officers. In that letter occurs the following:—

Madras Reply.

"It is only in Nellore that the practice of applying the plant to rice-fields and using it to bind wells, is said to obtain; and the Collector has no doubt that the reason is, that it destroys aquatic weeds. He also states that natives place the leaves among clothes to keep off insects. In North Arcot it is reported that the belief exists, that in flooded lands the plant kills water-weeds. In some parts of this district and of Tanjore the twigs of the plant are used to bind wells; but whether in the belief that it will prevent the growth of scum is not stated. In Chingleput the leaves are used, with those of other plants, as manure, and especially for saline soils, the leaves being supposed to counteract the saline properties of the soil. In the Vriddháchalam taluk of the South Arcot district and in some parts of the Kistna districts the application of the leaves as manure is believed to be beneficial to crops blighted or diseased. In the latter district the leaves are also used for covering wet seeds about to be sown for a second crop in order to induce sprouting. The plant is also grown in hedges round betel gardens to protect the creeper from the hot winds. In parts of Coimbatore, Tanjore, and South Arcot the leaves are applied to fields as manure; but they are only regarded as one of several kinds more or less suitable for the purpose." It may be here added that **Dr. Ainslie**, in the beginning of the present century, alluded to the Madras system of manuring rice-fields. "The rice reared on marshy land, or rather, that rice which requires being flooded, is usually manured with leaves and branches of trees." The fields in Malabar, which are not flooded, are manured, he says, with ashes and cowdung like other dry grain fields (*conf.* with *Ráb* manuring in Bombay).

Panjab Reply.

From the Panjáb many of the replies received confirmed the use of the plant in rice cultivation. The following passages from these replies may be specially cited here:—

Kangra.—The leaves of the *Básúti* are applied to the fields simply as a manure, and are used for covering safflower, in order to make it yield a valuable red colour, and ripening mangoes and plantains, which are kept covered by these leaves for a few days. The leaves, when applied after being warmed a little, afford relief in boils, pain in joints, ophthalmia, &c., and are valuable in hydrophobia. They are given as a medicine to cattle in case of pain in the stomach. The washerman burns the plant and uses the ashes for washing clothes as *sajji* (impure carbonate of soda). The tanner extracts a yellow colour from the leaves to dye skin. The smell of the leaves is known to prevent bleeding from the nose. The charcoal produced by burning the stems is used in manufacturing gunpowder. The roots are medically employed for removing cough, bile, impurity of blood, asthma, fever, phthisis, vomiting, seminal weakness, costiveness, pain in the chest, enlargement of the spleen, gonorrhœa, syphilis, and unsteady eyesight (*Chanot*).

The Commissioner states that in addition to what is said above it is

CULTIVATION in the Panjab.

Basuti.

known to him that this plant is used by zemindars as a lining to support *kutcha* wells, and is frequently thus applied in the Rawal Pindi district.

Peshawar.—"The plant grows abundantly on the stony and hilly lands throughout the district, but it is never grown in the cultivated land. The black *Bahekar* is useful for phthisis, consumption, spleen, cough, wounds, and boils. The white *Bahekar* is used in colic wounds, also for *Zehrbad* (Anthrax) to horses. Its seeds are eaten in colic and its leaves are taken internally in the form of a decoction. In the treatment of anthrax its seeds are generally administered. Its leaves also afford relief in this disease when they are pounded up and given with salt. The leaves of black *Bahekar* are scattered over flooded fields (of irrigated land), which are being prepared for a rice crop, and the zemindars state that by this means the yield of the rice crop is much augmented. Stems of black *Bahekar* afford a good charcoal, which is used in the manufacture of gunpowder. The plant is also extensively used as fuel."

The antiseptic property of the plant in killing minute organisms while being harmless to higher forms of life, thus established by native opinion, has been abundantly shown in a recent paper by **Mr. Hooper** in the *Pharmaceutical Journal and Transactions* (April 7th, 1888). **Mr. Hooper** writes of the chemistry of the leaves: "A well defined alkaloid appears to be the most important constituent; it constitutes the bitter principle, and, to all intents and purposes, is the active principle. It occurs in white transparent crystals belonging to the square prismatic system, without any odour, but with a decidedly bitter taste. It is soluble in water with an alkaline reaction, and in ether, but more so in alcohol." It would be beside the present purpose to give here the whole of **Mr. Hooper's** analysis of the substance. The above passage shows that simple maceration in water suffices to extract the active principle, and **Mr. Hooper's** practical experiments with it may therefore be preferentially cited here. "A sample of pond water," he writes, "containing **Spirogyra** and numerous animalcules was mixed with a few drops of a strong infusion of **Adhatoda** leaves. The chlorophyll gradually disappeared from the weeds and the cells became broken up. The oxygen was given off with less frequency, and at length ceased, some insect pupæ rose to the surface of the water and there died. Numerous **Paramecia** remained active for some time, but eventually succumbed to the action of the poison. In 24 hours the beaker containing the water that had been thus treated showed only a brown mass lying at the bottom; while some water in a beaker by its side, without this treatment, contained the green aquatic weeds evolving oxygen, and the animalcules alive."

Again "An aqueous solution of the alcoholic extract of the leaves was tried upon flies, fleas, mosquitoes, centipedes and other insects, and in every case the application met with poisonous results. The solution appeared to kill them without previous intoxication. On the higher animals the leaves do not seem to have such an effect. A quantity of the alcoholic extract representing 15 grains of the leaf was given to a small dog, and the administration was not attended with any inconvenient symptoms. These experiments show that the reputed use of the leaves in destroying injurious weeds in submerged fields is founded upon very scientific principles." "The poisonous properties of the **Adhatoda** on the lower orders of animal life will, perhaps, find for it a use in destroying insect pests, and make it an important addition to the *Materia Agricolarum* in India."

Mr. Hooper thus concluded his remarks in almost the same words as used by the writer fully a year before the date of the article in the *Pharmaceutical Journal, viz.*, that a "decoction of it might be found useful in destroying animal pests to other crops, such as sugarcane, tea, and coffee."

CULTIVATION in the Panjab.

The chemical investigations thus briefly reviewed have placed the utility of **Adhatoda** beyond any possible doubt, and its extended use should, therefore, not only be strongly urged on the attention of rice cultivators of India wherever aquatic weeds prove troublesome, but the subject should commend itself to tea, coffee, and indigo planters as well worthy of their consideration. Not only so, but it is probable we have in **Adhatoda** an antiseptic at the door of every Indian peasant which, if not found useful in the treatment of sores on his own person, could, at all events, be extensively used in the cure of the troublesome maggot-infested wounds of his cattle (*conf.* with the account of **Adhatoda Vasica** in Vol. I., p. 109.)

METHODS OF CULTIVATION, SOWING, REAPING, &c.

Provincial Methods. 438

Rice is grown to a small extent in most of the districts of the Panjáb. It is most extensively grown in Ambala (123,474 acres), in Kargra (110,763 acres), in Gurdaspur (62,871 acres), in Karnal (58,569 acres), and Sialkot (55,280 acres). The other districts have from 30,000 to 466 acres under the crop. In the Gazetteers of the various districts of the Panjáb, a table is given in the Appendix which shows the acreage yield of the more important crops. Under rice, the variation in yield (doubtless a natural enough one, since there can be no doubt of high and low yields according to the nature of the soil and the system of cultivation pursued) ranges from 189℔ per acre in Gurgaon to 2,240℔ in Kohat. The mean of all the returns for the province would be 705℔, but, excluding districts of minor importance and inferior yield, the average (and apparently a fair one) would be 829℔ or, say, 11 maunds. From this calculation Kohat has been left out of consideration, as the return published is an abnormally high one, and, if correct, is applicable only to the Kohat acreage, *viz.*, 2,813.

The following passages may be here given from the various district reports which have appeared, in connection with the settlements, or in the Gazetteers :—

CULTIVATION OF RICE IN THE HOSHÍARPUR DISTRICT.

Hoshiarpur. 439

"The several kinds of rice grown in the district may be divided as follows :—

1st class—Básmati, chahora, begami.
2nd class—Jhona, ratru, sukhchain, munji, sáthi, kalona, kharsu.

"The total area under rice is 33,656 acres; of this more than half is grown in the Dasúah *tahsíl*, and the area under first class rice in that *tahsíl* is 4,085 acres. Unfortunately no classification of rice was made in Una; but the total area under first class rice in the district is probably over 5,000 acres. Rice is cultivated only in marshy land, or in land copiously irrigated by a canal or stream. In one large village in Garhshankar Moránwáli it is grown in well-irrigated land; but this is most unusual, and the reason here is that the land is *dabri*, well suited for rice cultivation, and the water only three or four feet from the surface, so that as much water as is necessary can be given to the crop with very little labour. First class rice requires a constant flow of water; for the second class it is sufficient if plenty of water is given : if it stagnates no harm is done. Heavy floods, if they top the plant and cover it for two or three days, destroy it, but the mere passing of a flood over a crop does it no harm. The land is prepared by three or four ploughings. *Munji* (the commonest rice) is sown in March or April, the other kinds in June or beginning of July. Rice may be sown either broadcast or after raising seedlings in small beds. The broadcast sowings are of two kinds :—

(1) *Watrán* when the moisture has sufficiently subsided to allow of ploughing and sowing;

Rice Cultivation

CULTIVATION in the Panjab.

Hoshiarpur.

(2) *Kadwán* or *kadu*, when the seed is steeped in water for two or three days, and then scattered broadcast in the mud. When sown by raising seedlings the process is called *láb*, and this, though more laborious, is more profitable. The best kinds are always sown by the *láb* method. *Sathi* or *kalona* are always sown *kadwán*, the other kinds may be sown in any way it pleases the cultivator. Some weeding is required for rice sown broadcast; that sown by *láb* requires none. The earliest sowings ripen in September, the later ones in October and November. *Sathi* is supposed to ripen in sixty days.

Sathi pake sathín dini = *Sathi* ripens the sixtieth day,
Je páni mile athín dini = If it gets water every eighth day.

"*Kharsu* is a very coarse rice grown in poor alluvial soils, where the river has deposited some soil, but not yet sufficient for the better crops. Grasshoppers (*toka*) are fond of the young shoots; and pigs, which abound in the high grass of the *Chhamb*, do much harm by uprooting the fields of rice. High winds also are considered bad when the plant is nearly ripe. The crop should be cut before the grains are quite ripe (*hargand*), otherwise much of the grain is lost. Threshing is dóne by the treading of oxen without the wooden frame (*phala*) used in ordinary threshing. The rice straw is of little use, except for bedding and litter; it contains no nourishment, and cattle will not eat it unless very hungry. The grains are husked by pounding them in a large wooden mortar (*ukhál*) with a pestle (*mohla* or *músal*). As to the outturn **Captain Montgomery** writes :—

"Experiments were made on 41·8 acres, the result being an average outturn of 378 seers (or about $9\frac{1}{2}$ maunds) an acre. I am unable to give the average outturn of the different classes, but most of the experiments were made in *munji*, *jhona*, *sathi*, *básmati*, and *chahora*. **Mr. Temple** considered that some of the best rice-growing villages produced 60 maunds an acre; this appears to me quite incorrect, even if *kacha* maunds were meant, a produce of 25 maunds an acre is an excessive average, though special plots may grow as much. I am inclined to think that ordinary rice-growing land will not produce more than 9 or 10 maunds, and the better *básmati* and *chahora* lands about 12 maunds" (*Panjáb Gazetteer, Hoshiarpur District, pp.92-93*).

Karnal.

440

Cultivation of Rice in the Karnal Districts.

"Rices are divided into two well-defined classes, the fine rices, varieties of **Oryza sativa**, the grains of which cook separate, and which are known to the people under the generic name of *ziri*; and the coarse rices, varieties of **Oryza glutinosa**, the grains of which agglutinate when boiled, and of which the principal sorts are *munji* and *sánthi*. This and the following paragraphs refer to the fine rices only. The *ziri* proper is a small rice with a short straw; the principal varieties are *ramáli* and *rámjamáni*; the latter of which has a particularly hard fine grain. *Sunkar* and *ansári* are coarser rices, chiefly grown where there is fear of too much water, in which case their long straw gives them an advantage. Rice grows only in stiff soil. It is usually grown in low-lying *dákar* so as to take advantage of the drainage water; but if the water-supply is sufficient, the best rice is grown on fine stiff soil on a slope where the water is perfectly under control. The seed-beds are ploughed four or five times and carefully prepared, manure is spread on them, and the seed sown broadcast and very thickly on the top of the manure.

More manure is then spread over the seeds and the whole is watered. Four days after they are again watered, and after the fifth or sixth day, they must be kept wet till they are ready to plant out. The rice-field is ploughed twice, and such manure given as can be spared. It is then flushed with some three inches of water, and a *sohágga* (toothed), if there are weeds, is driven about under water (*gór* or *gán dena*). If the weeds are obstinate, the plough must be used again under water. When the *sohágga* has worked up the mud into a fine pulp, *Jhinwars* and *Chamárs* take the seedlings (*pod*) in handfuls (*júti*) and plant them one by one in the water, pressing in the roots with their thumbs. An acre will take 500 to 600 *jútis* which will cost, if

CULTIVATION in the Panjab.

Karnal.

bought, R1-4. It will take ten men to plant it in a day, and they get 2½ to 3 seers of grain each day.

The field is weeded once at least. At first the whole field must be kept under water continuously; for each seedling throws out five to ten new shoots, which cannot make their way unless the ground is pulpy, and it is on the abundance of these shoots that the crop depends. The water must not be more than six inches deep, or the shoots will be drowned before they get to the air, and it must not be changed, as it would carry away all the strength of the manure and the soil. When the ears once begin to form, the ground must be kept well wetted, but not too slushy, or the plants will fall. If the crop is wholly under water for more than four days, it dies. The reaping must be done directly the grain is ripe, or it will fall out of the ears into the water. Thus hired labour is a necessity, and the payment is 5 to 6 seers of unhusked rice. If the water is deep, and the plants, as cut, have to be put on bedsteads to keep them out of the water, the reaping is slow: otherwise the same as other small cereals.

The rice is thrashed in the ordinary manner; but the grain has to be husked in the *okal*. Standing rice is called *dhán*, as is the unhusked grain, in contradistinction to husked *cháwal*. The husking is generally done by the women of the house. If done by a labourer, he returns 18 seers *cháwal* from every 30 seers of *dhán*, keeping about 2 seers of good rice and as much of broken bits which he will grind up and eat as bread. The rest is husk, which is useless. The straw (*puráli*) is very poor fodder, and is used largely for bedding for cattle, and for mixing with manure, or is even ploughed in fresh. But it is also given to cattle to eat. Rice suffers much from *khúd* or *kokli*, apparently aquatic larvæ or other animals that eat the young sprouts. Water birds, too, play terrible havoc with it when it is ripening. If the whole plant dries up, it is called *malain*; if the grain only, *patás* is what is the matter with it.

Coarse rice (*v. s.*) is of two kinds, *munji* and *santhi*. The peculiarity of the former is that it cannot be drowned out, the straw lengthening as the water deepens. It is therefore sown in spots liable to flooding. It will stand two feet deep of water; and if the ripe plant falls into the water, the grains do not fall out as they do with *ziri*. The peculiarity of *sánthi* is that it ripens within an extraordinarily short time (nominally 60 days, hence its name) from the sowing. It is sown all over the Nardak, and generally wherever there is no irrigation, as the rains will usually last long enough to ripen it. Huen Tsang noticed its quick growth with admiration when he visited the Nardak 1,500 years ago. *Sánthi* has a short straw, and does with but little water, it being sufficient if the soil is thoroughly moist after the shoots are once up. The young shoots are liable to be eaten, and if the water gets very hot they will sometimes rot, but the plant is wonderfully hardy, and when the stalks have once grown up, hardly anything hurts it. Both kinds are sown at once where they are to grow. After two or three ploughings cattle are sent into the water to walk about and stir up the mud, or the *gán* or toothed *sohágga* is used under water. The seed is sown broadcast on the *gádal* or fine mud. No manure is used nor is the crop irrigated. The *puráli* or straw is better fodder than that of *ziri*, but still not good. The coarse rice forms a staple food of the people, the fine rices being sold and seldom eaten by them" (*Panjáb Gazetteer, Karnal District, 176-177*).

Cultivation of Rice in the Kángra District.

Kangra. 441

The following passage from the *Gazetteer* of this district may be accepted as conveying the main features of the Panjáb hill rices, when taken in conjunction with that which follows regarding Kashmír and what has already been stated under Nepál:—

"Rice is the staple product of the Upper Kángra valleys,* where is combined the abundance of water with high temperature and a peculiar soil which favours its growth. It is grown also in the irrigated parts of Dehra and Núrpur, where the produce, though inferior to that of Kángra, is still of good quality. Coarser kinds of rice are also grown without irrigation in the more elevated portions of the district. The people recognise upwards of sixty varieties of rice. The most esteemed kinds are:—*begami, basmati, jhinwa, nakanda, kamádh,* and *rangari*. Each of these varieties has its special locality. Thus Rihlú is famous for its *begami* and Pálam for its *basmati*. These are the finest rices. Of the coarser kinds grown in the Kángra valley, the best known names are *kathíri* and *kolhena*; and of the inferior produce of unirrigated lands *rora, kalúna, dhákar,* &c. On land which can command irrigation, the rice is not sown till the beginning of June. In districts dependent upon rain, the

* In *talúkas* Pálam and Rihlú rice occupies 78 per cent. of the total acreage under autumn crops, and the percentage would be higher were certain hill lands, which belong to these *talúkas*, excluded.

CULTIVATION in the Panjab.

Kangra.

seed is thrown into the ground as early as April, and the later the season of sowing the less chance of the crop reaching maturity. The harvest time is during the month of October.

"There are three methods of cultivation. The first and simplest, called *bátar*, is where the seed is sown broadcast in its natural state; on unirrigated lands this is the universal method. In the second method the seed is first steeped in water and forced under warm grass to germinate, and then thrown into the soil, which has been previously flooded to receive it. This method prevails wherever water is abundant, and is called *mach* or *lunga*. Under the third system, called *úr*, the young rice about a month old is planted out by hand at stated intervals in a well-flooded field. This practice involves much labour and is seldom followed, except in heavy swampy ground where the plough cannot work. The yield, however, of transplanted rice is always greater than under either of the other methods. The growth of weeds in the rice fields is very rapid; but the people have a simple and most effectual mode of ridding themselves of them. About the month of July, the crop, weeds and all, is deliberately ploughed up. Immediately after the operation, the whole appears utterly destroyed; but the weeds alone suffer, being effectually extirpated by this radical process, while the rice springs up again more luxuriantly than ever. This practice is called *holdna*, and the crop is worthless which does not undergo it. Rice is always sown by itself and never mixed. The grain is separated from the husk by the use of the hand pestle and mortar; women are usually employed upon this labour, and, when working for hire, receive one-fourth of the clean rice as their wages. Rice has a very extensive range. In Kángra proper it is seen as high as 5,000 feet above the sea, and in Kúlu in the valley of the Bias it grows as high as 7,000 feet" (*Panjáb Gazetteer, Kangra District, 156*).

Kashmir.

442

Cultivation of Rice in Kashmír.

Through the kindness of Mr. W. R. Lawrence (Settlement Officer in Kashmír) the writer has been favoured with proof copies of the passages in the forthcoming Settlement Report which relate to rice. Unfortunately space cannot be afforded to deal with these fully, district by district, but the following extracts convey some of the more instructive ideas brought to light by Mr. Lawrence:—

Varieties grown.—There are numerous varieties grown in the various *tahsils* and these may be referred to two important sections, *White* and *Red* Rices. Some 53 forms are specially mentioned by the cultivators in Donsu, and of these 41 are white rices. In the Lal *tahsil* some 40 forms have been recorded, red rices being the most extensively grown, whereas in the Phak *tahsil* white rices are more frequent. The white rices command the best price, but they are regarded as more exhausting to the soil and have to be followed by a red crop. White rice requires a good soil, is more delicate than the red and gives a smaller outturn of grain. Red rice is accordingly the food of the poorer classes.

Speaking of the Donsu *tahsil* Mr. Lawrence says:—

"The best of all rices are the *Kanyun* and *Basmati*. These are delicate kinds and will not grow on elevated ground exposed to the cold; the *kanyun* produces a round white very soft grain, and the *basmati* gives a long white soft grain. *Basmati* always commands a higher price than other kinds of rice and both in the market and in the communication allowed by the State, fetches R1 per *kharwar* more than any other variety. It is described as '*Asubiol*,' *i.e.*, the seed germinates more quickly and ripens sooner than any other rice. Consequently it can be sown later. It gives a small crop and the variety cannot be grown above Lakripura. It requires careful husking, and whereas it costs 12 annas to husk a *kharwar* of *shali* of the ordinary kind, it costs R1 to husk the same quantity of *basmati*.

"I have pursued my enquiries regarding the relative merits of the *wattru* or broadcast sowing of rice and of the *nihali*, or system of transplanting rice, and again I find that, given the same quality of soil and the same condition of manure and irrigation, the *wattru* system gives the larger outturn.

"*Wattru* is certainly the prevalent system in the Donsu *tahsil*. It requires good soil, and can only be successfully followed when the fields are of a fair size and are what is known as *bad-wattru* fields, which take four *traks* of seed. In smaller fields such as are found on terraced land (*ach-karu*) or (*dulwushu*) which are known as *lar-wattru* and which take from 1 to 3 *traks* of seed, the *nihali* system is perhaps the

CULTIVATION in Kashmir.

more economical. The seed for *nihali* and *wattru* rice is sown at the same time, *viz.*, from 15th *Baisakh* to 15th *Jeth*, but the seedlings from the former are not planted out till 36 days after sowing. In Narkara rice is sown a month earlier than in the other villages, but it ripens at the same time. The result of this early sowing is a greater outturn of rice. (In the Shahabad Ilaka near Vernag the sowing time commences at the beginning of *Baisakh* and there the outturn is very large.) In the interval of the 36 days, the sun and air have benefited the soil, and in addition to this the shepherds have gathered their sheep on the dry land reserved for *nihali*. One thousand sheep will enrich 4 acres of land in a single night. They are not allowed to stand in one place as their droppings would make the land too powerful, but are kept moving about. For every 5 *traks* of land manured by the flocks of sheep, the shepherd receives 4 *traks* of *shali*. It is calculated that the presence of a flock of sheep will raise the produce from 25 to 40 or 50 *kharwars* per *kharwar* of land. The secret of this enormous benefit derived by the *nihali* land, is that by the time the sheep come on to the ground, their food and strength has improved from the spring grass, and in consequence their manure is of greater value than it was a month before when the *wattru* lands were flooded for rice.

"But, generally speaking, I find that the *nihali* system is regarded as a compromise and as a 'dernier resort' by the people. If water is plentiful and the soil fair, *wattru* is the system followed. If water is not forthcoming at the time of sowing, the land is reserved for *nihali*, and if the land is at a high elevation the *nihali* system will be adopted. In low lands and near the swamps *nihali* cannot be grown. The advantages of *nihali* are, that the system clears the ground of weeds (*kach*), and that it requires less labour. Two *khushabas* are ample for *nihali*, whereas four *khushabas* are necessary for *wattru* rice.

Wild Rice. 443

"*Wild Rice.*—I have already described the *hama* grass. It is exactly like the rice plant, but it has rather narrower leaves. It gives a grain which is eaten by the poor people, and is in great request for Hindu temples It comes into flower earlier than the rice, and there is a saying in Kashmiri '*Hama niyul drao,*' '*shethi duo hrao*' which means 'cut your rice two months after *hama* comes into flower.' Besides the *hama* which only experts can distinguish from the rice, the cultivator has great difficulty with *kre*. *Kre* is rice which has sprung up from seeds dropped at the preceding harvest. The outturn of *kre* is small, and if it is not eradicated during the first season, and is reproduced for a second year, it causes considerable trouble and loss. In order to fight against *kre*, the cultivator invariably sows a new variety of rice each year. It may be that he substitutes white rice for red, or one variety of red for another variety of red, but the change enables him to distinguish with ease between the objectionable *kre* and the real rice plant.

"Everywhere I find evidence to the effect that red rice is a hardier plant and generally a heavier cropper than the white rice. It is always the custom to sow red rice in the field nearest the water-course. The water comes in cold from the channel and white rice would fail. When the water has been lying for a time on the first field it becomes warm enough to pass on to the second field, where white rice may be raised.

"*Operations.*—Rice cultivation is hard work for man and beast, and in former days the Government recognizing the importance of the rice harvest left the cultivators to themselves during the cultivation of the rice fields. Rice in Kashmir is like the most elaborate garden cultivation. It must be watched night and day. A break in irrigation or delay in weeding are fatal. Hence it follows that the present system of *begar* is most prejudicial to rice cultivation, for men are taken away at critical times, and the fields dry up and the weeds choke the rice. From a revenue point of view, it would pay the State to exempt all rice-growing villages from *begar* from the 15th *Baisakh* (April) to the end of *Katak* (October). Such forced labour as might be required during those months should be either taken from dry villages, or contractors should be entertained who could either import labour from India or procure labourers from the city population.

"It is a very important fact to bear in mind that rice cultivation in Kashmir is essentially of the nature of *petite culture*. The best rice which I have yet seen has been in the Shahabad Ilaka, where the holdings are very small, or in villages like Ichgam, Ichkot and Ompura, where the holdings are respectively 2 *ghumaos*, 4 *kanals*; 2 *ghumaos*, 5 *kanals*; 2 *ghumaos*, 2 *kanals*. Each plant should be worked with the hand and not a weed should be seen.

"Considering their size, the little cattle of Kashmir do good work. When at least three ploughings have been accomplished, the cattle will sometimes be required after the first *khushaba* to plough up the young crop. This process is known as *haji*, and perhaps ten pairs of oxen with their tails tied up to the yoke will be turned into one field. The *haji* must be done quickly. Cattle are also used after the first

ORYZA sativa.

Rice Cultivation

CULTIVATION in Kashmir.

khushaba for the *gupan nind* which consists of driving the cattle up and down the fields and mixing the soil with the water.

Rotation.—Greater attention is paid to rotation of crops in Donsu than in Lal and Phak. Much of the land is described by the villagers as Rassami, *viz.*, land which bears a good crop one year and requires rest the next year. This rest is given not by a fallow but by a change of crop. A common rotation is three years of rice followed by four years of cotton. Cotton clears the soil and frees it from weeds. Another common rotation is rice followed by maize or mash or barley, or wheat. All agree that this rotation of maize, mash, or barley improves the succeeding rice harvest by 2 *kharwars* of *shali* per 4 acres of land, but opinions differ regarding wheat. There can, however, be little doubt that the exposure to sunshine and air during the time when the land is under dry crops must benefit the soil. I have asked some of the more prosperous of the cultivators, as an experiment, to plough up the rice fields directly the rice crop is harvested. The land is still soft from the *popasag* or final watering given to swell the ears of grain, the cattle are in good condition, and the only objection the people have to my proposal is, that they have no leisure to plough, and that ploughing before the spring will increase the difficulty of *kré*."

Manure.—Sheep manure is the best, next comes cow manure. Horse-dung is not used for rice fields. It must be remembered that cow-dung is the chief domestic fuel, both for cooking and for *kangars*. Little need be added here regarding manure. In the *Gamdu* fields, by which is meant fields near enough to the homestead to be well manured, well watered, and well looked after, the supply of manure is ample. To every acre about 100 baskets, or 16 or 17 *kharwars* of dung is given. It is either put on to the land when dry and ploughed in dry with a view to the harvest of the following year, or if strength is wanted for the current year, the manure is scattered by hand over the fields submerged by water. A frequent practice is to heap manure on to the first field which receives irrigation, and from this field liquid manure finds its way to the fields below. Fields outside the *gamdu* radius receive manure about once every third year, and also receive a dressing of turf clods (*chak*). Clods vary in quality; the best are those dug from the sides of water channels which are rich in silt.

Diseases.—The *rai* disease to which I have allued in previous reports occurs in the lower villages of Donsu, and it seems evident that the cause of the disease is extreme warmth at night. The same idea prevails here as in other *tahsils* that *rai* is caused by the summer lightning known as *haji bawan*. Higher up the rice crops of Donsu are exposed to a very serious drawback. The tahsil is situated in the phrase of the people on the left hand (*Khowar*) of the sun's course. Towards September when sunshine and warmth are much needed, the left-hand villages are shaded by the mountains, and the cold wind from the heights checks the formation of grain. A large quantity of rice never reaches maturity, and is unfit for anything save green food for horses. Such rice is known as *Handru*. Another drawback to which the left-hand villages are liable is that the snows on the left-hand mountains melt very rapidly.

Yarkand. 444

Cultivation of Rice in Yarkand.

Vern.—*Shál*, the plant; *Gurunj*, the grain.

The following interesting account may be here reprinted from the Yarkand Mission's Report as it deals mainly with a contrivance used to husk rice:—" Principally grown in Yarkand Division. Sown in April and cut in September and October. Returns six to eight-fold. The seed is husked on the banks of the canals by a pounder-mill worked by water. It is called *súcana*, and consists of a horizontal axle-beam, in the middle of the shaft of which are four immovable flanges or paddle-boards against which the water plays. The shaft is supported at each end by an upright socket post on each side of the mill-stream, and at one end of it are two long clappers which project six or eight inches on opposite sides. As the shaft revolves, by the water playing against the paddles, these clappers alternately, in turn, catch and release the handle of the pounder, which works on a fulcrum block, and is so adjusted that the clapper in revolving with the shaft shall catch and depress the short hand of the lever, and thus raise the pounder, which projects down at right angles from its other end, and let it fall again, as in course of revolution it releases the depressed lever head. The pounder is a round-ended bar of wood fixed at right angles into the head of the long hand of the lever, and plays upon the rice in a wooden mortar. The *sucana* may be built by any one on a suitable stream by permission of the District Governor, and on payment of a fee to Government of sixty *tanga* (about fourteen rupees), and this clears of all further taxation on its operations. The miller's charge is one *chárak* in fifteen of husked rice. The entire mechanism of the mill is in wood-work, and the several parts are very neatly put together. We saw

CULTIVATION in Yarkand.

several of these mills at work on the canals in the line of our march, and were impressed by the stride in civilisation in advance of what we had left behind us in Kashmir, where the laborious and by no means graceful operations of the pole pestle and mortar are a prominent feature in the scene peculiar to the banks of the Bidasta" (*Report of a Mission to Yarkand in 1873, 77*).

IX.—CULTIVATION IN THE CENTRAL PROVINCES.

CENTRAL PROVINCES. 445

References.—*Annual Reports of the Director of Land Records and Agriculture; Experimental Farm Reports; Gazetteers:—Balaghat, 18; Barpali, 28; Bastar, 31, 35; Betul, 50; Bhandara, 64, 66; Bilaspur, 113, 115, 116, 120.*

Area under, and Traffic in, Rice.

Area. 446 *Conf. with p. 522.*

The Agricultural Statistics of the Government of India show these provinces to have had 3,785,566 acres under rice in the year 1888-89. Taking, what appears to be a fair average, 12 maunds per acre as the yield, the total production from the above acreage would have been 4,54,26,792 maunds or, say, 1,622,385 tons.

Trade. 447

The trade of the Central Provinces in this staple (which is almost confined to husked or clean rice) may be here briefly reviewed. The gross imports amounted to 1,30,282 maunds and the exports to 9,62,044 maunds, thus showing a net export of 8,31,762 maunds. About 99 per cent. of the imports came from Bengal, while of the exports the largest quantities went to Rájputana and Central India (3,52,651 maunds); Berar (3,17,841 maunds), and Bombay (2,44,024 maunds), all these exports consisted entirely of cleaned rice. It is impossible to discover how much of the Central Provinces rice finds its way to foreign markets, but if any leaves India it must be drawn from the exports to Bombay and Calcutta. In the year under review, only 32,687 maunds were consigned to Bombay.

Foreign Exports. 448

By deducting the net export from the estimated production and expressing the balance to head of population (9,838,791 millions) the daily consumption in these provinces would be ½ seer.

Consumption per head. 449

METHODS OF CULTIVATION, SOWING, REAPING, &c.

Provincial Methods. 450

Rice is grown throughout these provinces wherever water is available. The chief districts are, however, Bhandara, Chanda, Raipur, Bilaspur, Sambalpur, and Jabalpur. By far the major portion of the rice of these provinces is broadcasted, *viz.*, 2,855,200 acres, with 2,825,404 acres of that unirrigated. The remainder 930,365 acres is under transplanted rice, of which 508,058 is unirrigated. The entire rice obtained in Raipur, Bilaspur, and Sambalpur is broadcasted on unirrigated land. Indeed, irrigation rice fields are almost confined to Bhandara, Chanda, and Balaghat (= Nagpur Division), which conjointly possess 422,279 acres of irrigated transplanted rice.

Yield. 451 **12 maunds per acre.**

The Director of Land Records and Agriculture in his Annual Report of 1890 gives the results of numerous crop experiments with rice conducted in nearly every district. The lowest recorded return was 160℔ unhusked rice and the highest 4,940℔ per acre. These were for transplanted rice, and they show an average of 12 maunds. For broadcasted the average was found to be 14 maunds. But the chief districts which grow transplanted rice, taken by themselves, showed an average yield of 19¾ maunds. An average yield of 12 maunds for these provinces would therefore seem, if anything, to err on the side of underestimating the total production.

The list of references given will enable the reader to discover special articles which deal with the rice cultivation of these provinces. Space cannot be here afforded for more than one quotation, but it may be said that the more extensive level tracts manifest the same peculiarities as in the North-West Provinces, and the hilly districts possess a system of rice cul-

ORYZA sativa.

Rice Cultivation

CULTIVATION in the Central Provinces.

Bhandara.

452

ture scarcely different from that which has been detailed above under Chutia Nagpur in Bengal

CULTIVATION OF RICE IN THE BHANDARA DISTRICT.

"There are so many different kinds of rice that it were tedious to enumerate them. It is commonly believed that the local dry measure termed *kooroo*, equal to 20℔, can be filled to overflowing with a single ear of each separate kind, and it is certain that there are some hundreds of varieties of rice in Bhandara. Their names are quite bewildering; and it may be that the same variety passes under different names in different parts of the district. The most highly appreciated is the *chinnoor*; it is generally eaten by the better class of landholders only, and is much thought of. Its selling price in the Bhandara bazar at the end of January 1867 was 8 seers 12 chittacks the rupee. Other varieties of the first rank are the *chuttree, radhabalum, ambehmohur, kaleekamood*; none of these are equal to *chinnoor* rice, though ranking near it. *Ramkeyl, pissoo, doodhramkeyl, telassee* form the class which Brahmins in our offices and landholders' agents and other well-to-do people use in general. Most people hold land in one or other of the Nagpur districts, and they grow a small supply of the first class rice which comes into service for feasts and on holidays. The artisan class, our servants, and the great body of cultivators are content with the rice known as *pandhnee, sareehanee, ekloomhee, dhoul*, &c., &c. These varieties now fetch in the country markets about 13 seers the rupee; and *kamkeyl* is selling at 11 seers. *Dhers*, day labourers, and the lower strata of society, are glad to get the *chigna, chippena, rarsurree, hathibog, ratta*, and such kinds. Their average price is now about 15½ seers the rupee. Besides these numerous varieties of the genus rice, all of which are raised by labour and that of a more constant and exhausting character than other cereals require, there is yet another sort which is generally believed to be raised by the gods and is therefore called *deodan*, the gods' rice. It springs up of itself in swamps and other marshy places and is collected by the lower classes, who may be seen sweeping, with a piece of stick, this self-sown rice into their baskets. The gathering season is rather early, as the constant supply of water quickens the growth. This *deodan* is bartered to dealers and other well-to-do classes for regularly grown rice. Sometimes a tithe of the *deodan* sweeping is made over by its collector to the village landholders. The poorer classes eat *deodan* as long as it lasts. Their superiors are glad to get hold of it, as it comes into use on the *Rissheepunchum* (in August) and *Shirawundewadsee*, celebrated 11 days after the *Rissheepunchum*. The peculiarity of these festivals are that nothing raised by animal labour can be eaten. In the times of the Mahratta dynasty *deodan* was specially distributed from the Government storehouse (*ambarkhana*). Now, those who can beg no god-grown rice must eat Indian corn (**Zea mays**) or vegetables which have been raised by man's labour alone. Failing these a day's fast is the only alternative. It would seem that, at some time, the marshes or ponds, in which *deodan* is now yearly found, was regularly sown. It may have been once or oftener, but the produce now picked up must have had some seed, though the way in which it now yearly produces itself is curious and may well be ascribed by a simple people to the direct intervention of a divine power. It is well known that the seeds of rice which are lost in the moist ground at reaping, themselves spring up and produce abundantly. An instance in my own knowledge may be given. A field of 18 acres, well situated, was one year not sown as the favourable season had been allowed to pass. At harvest some 60 *kandies* or more than one ton of husked rice was reaped. This was, perhaps, half what that field would ordinarily have produced had the usual seed and labour been given to it."

CULTIVATION in Burma. 453

X.—CULTIVATION IN BURMA.

References.—*Burma Gazetteers, I., 404, 424-426; Statistical and Historical Account of the Thayet District by Colonel H. A. Browne, 1874, p. 83; Agri.-Horti. Soc., Ind., Transactions, II., 72, 73; VI., 114, App. 127, Pro., 5; VIII., 246, 354; Journ. (Old Series), II., 252; III., 215-217, Pro., 71, Sel., 1, 2, 271, Pro., 285, V., Sel., 124-127, Pro., 46, 53, 64; VI., Sel., 8, 9, Pro., 71; V., 59-61, 226-228; VIII., Sel., 59; IX., Pro., 37; X., Pro., 115; XII., Pro. (1862), 39.*

Area under, and Traffic in, Rice.

Area. 454 *Conf. with p. 522.*

The Agricultural Statistics of British India, published by the Imperial Government, show that the province of Burma had 5,673,542 acres under rice during the year 1888-89. Of this area, 4,067,606 acres were under cultivation in Lower Burma and 1,605,936 in Upper Burma. The following figures show the principal rice-growing districts of the two sections of the province: Upper Burma, Ye-u (250,000 acres); Minbu (176,880 acres); Shwebo (172,858 acres); Katha (166,400 acres); Pakokku (142,025 acres); Kyaukse (136,355 acres); and Sagaing (115,000 acres). Lower Burma, Pegu (678,200 acres); Bassein (496,003 acres); Akyab (451,418 acres); Hanthawaddy (403,983 acres); Thongwa (394,194 acres); Henzada (297,119 acres); Tharrawaddy (290,661 acres); Amherst (286,872 acres); and Prome (250,210 acres). For the other districts, in Upper Burma, the statistics fluctuate between 84,000 and 1,838 and, in Lower Burma, between 134,201 and 1,249 acres.

Yield. 455 **15 maunds per acre.**

It is commonly stated that 1,600lb of paddy per acre is the average yield in Burma. That quantity, by deducting 25 per cent. as loss of weight in husking, would show the yield to be 15 maunds of clean rice. Expressing that yield to the acreage returned, the total production may be said to have been (for the year 1888-89) as follows:—

	Maunds of Rice.	Total Production.
Lower Bruma . .	6,10,14,090	} 8,51,03,130 maunds
Upper Burma . .	2,40,89,040	} or 3,039,397 tons.

Trade. 456

The transfrontier trade to and from Upper Burma is not published, but the available statistics show the land traffic to and from the lower province, the transactions with Upper Burma being viewed as between a foreign country. These may be briefly reviewed:—Imports of rice and paddy 13,964 cwt. and exports 2,738,723 cwt., thus showing a net export of 38,00,663 maunds. The average exports to Upper Burma during the past nine years have been 20,92,188 maunds. But viewing the transactions between Upper and Lower Burma only, the net export from the latter to the former province was 38,11,262 maunds. That amount has, therefore, to be added to the estimated production of the upper province. The figure thus obtained would have given to the population of Upper Burma a daily consumption of 1⅓ seers per head. But as the amount thus shown is, judging from the similar results obtained for other provinces of India, exceptionally high, it may be pointed out that there are four possible errors:—The population accepted is a mere estimate and may be considerably below the mark; there were in Burma during that period a large army of soldiers and camp followers which materially increased the rice-consuming population; no allowance has been made for the exports beyond the frontier of Upper Burma; and the figure of yield (15 maunds of rice to the acre) is that worked out for the lower province and may be too high for the mountainous tracts of Upper Burma.

Consumption per head. 457

ORYZA sativa.

Rice Cultivation

CULTIVATION in Burma.

Trade.

But in addition to its exports to Upper Burma, the lower province has an extensive trade by sea to the provinces of India and to foreign countries. Besides Upper Burma, it has also a small land traffic with Siam and Karennee. The following balance sheet of the rice traffic of Lower Burma may, therefore, be given for the official year ending 31st March 1889:—

Destination.	Exports.	Imports.	Net Exports.
	Maunds.	Maunds.	Maunds.
Foreign Countries . . .	1,99,52,090	310	1,99,51,780
Indian Ports	5,35,498	1,05,482	4,30,016
Land Traffic	38,34,212	20,749	38,13,463
TOTAL .	2,43,21,800	1,26,541	2,41,95,259

Foreign. 458

These figures have been purposely made to exclude from consideration the transactions to and from ports within the province, and thus to exhibit the net exports from Lower Burma. Owing to the disturbances in Upper Burma the exports to that province, during 1887 and 1888, were abnormally high, and the comparison with those of the succeeding year, therefore, showed an apparent, though not real, failing off, since the total exports were, during these years, adjusted by the temporary decline and again restoration of the foreign trade. A slight confusion is also occasioned through the fact that the financial year ends (31st March) in the middle of the rice season. An average of the transactions carried out during a period of years would, however, admit of the correction of this cause of confusion. To allow the above balance sheet to be compared with other published statements, it may, for example, be said that the average gross exports (under the three headings shown in the table) for the past nine official years ending 31st March 1890, have been 3,95,10,308 maunds or, say, 1,411,082 tons, and the average net export of any period of years would appear never to have exceeded 40 million maunds.

It need scarce be here added that paddy and rice have in these figures of Burma trade (as in those of the provinces of India) been taken conjointly. The error thereby admitted into the calculations is, perhaps, more serious in the case of Burma than in any of the provinces of India, since the coastwise exports are in nearly equal quantities of paddy and rice. But even this fact is greatly minimised by the immensely greater quantity of rice exported to foreign countries. If we accept the balance sheet as fairly correct, the net export deducted from the estimated production would leave the amount which in the year in question was available for local consumption. That quantity expressed to head of population (*viz.*, 3,736,771) would be about 1 seer per day.

Mr. W. T. Hill, Director of Land Records and Agriculture, has recently published the following note on the Burma traffic in rice:—"From the figures given on page 28 of **Mr. O'Conor's** Review of the Trade of India in 1889-90, it would appear that the exports of rice from Burma had of late years tended to fall off. There is, however, every reason to be satisfied with the prospect and, if the figures are carefully examined, it will appear that the exports have, on the whole. increased to a very satisfactory extent.

Figures for the financial year (*i.e.*, the year ending 31st March) are apt to be very misleading. That date falls in the middle of the busiest season, while the end of the calendar year is also the end of the rice season. The rice exported within any calendar year is, almost without exception, the

CULTIVATION in Burma.

Trade.

produce of the harvest which is gathered in January of that year. The following table gives the exports of rice since 1878 by the calendar year :—

YEAR.	EXPORTS OF RICE FROM LOWER BURMA		
	To Upper Burma.	To other countries.	TOTAL.
	Tons.	Tons.	Tons.
1878	53,323	789,464	842,787
1879	41,837	802,289	844,126
1880	5,677	849,841	855,518
1881	5,243	919,181	924,424
1882	39,764	1,039,866	1,079,630
1883	37,840	910,804	948,644
1884	87,127	763,801	850,928
1885	96,363	928,838	1,025,201
1886	77,044	953,934	1,030,978
1887	125,177	953,664	1,078,841
1888	150,668	776,398	927,066
1889	58,504	954,059	1,012,563
1890 (actuals for nine months).	37,000	1,068,000	1,105,000
1890 (estimate for year) . .	50,000	1,160,000	1,210,000

"Now it is not clear why, when the rice trade of Lower Burma is under discussion, the exports to Upper Burma should be excluded. From 1879 to 1883 the exports to Upper Burma were comparatively small, while in 1887 and 1888 they were very large indeed, and the result was that the exports to other countries were much lower than they would have been if production in Upper Burma had been normal. There have been four very remarkable harvests in the series of years for which figures are given. Those of January 1882 and 1890 were much above average, while in January 1884 and 1888 the crop was very short. If these facts are taken into consideration, it does not appear that, even if the exports to Upper Burma are set aside, the rice trade tends to fall off. The exports did not reach 900,000 till 1881. They never fell below that again except when the harvest partially failed in two years. From 1886 to 1889 (excluding the year when the harvest was short) the exports to other countries were about 950,000, a very substantial increase as compared with previous years if 1882, when the harvest was abnormal, be excluded. The very large increase of the present year is due partly to extension of cultivation and partly to the crop being a bumper one. The crop of January 1890 was probably no better than that of January 1882, but the exports to other countries will be 1,160,000 tons as compared with 1,040,000 tons in 1882, or an increase of 120,000 tons in eight years.

"But, as already mentioned, there is no reason why the exports to Upper Burma should be excluded; and, if the total exports from Lower Burma be compared, it will appear more clearly that the progress on the whole has been steady. In the five years 1881—85 the total exports only twice exceeded 1,000,000 tons; in the next five years they have only once fallen below 1,000,000 and that was in a very exceptional year, when the exports from Akyab alone fell from 160,000 tons to 90,000 tons owing to destruction by salt-water inundation.

"It does not appear that the trade is likely to be affected to any serious extent by the exports from Saigon, Siam, and Java to Europe and the Straits. The exports of rice from Saigon between the 1st January 1890 and the 13th October amounted to 440,971 tons as compared with 260,714 tons in the corresponding period in 1889, but in that year the crop in

ORYZA sativa.

Rice Cultivation

CULTIVATION in Burma.

Trade.

Saigon was a very short one. However, even with these large exports from Saigon, only 16,202 tons of rice went to Singapore and 33,893 tons to Europe. Again, between the 1st January 1890 and the 31st August there were only 8,466 tons of rice exported from Java to Europe and 2,872 tons to Singapore, while the exports of rice from Siam to Europe from the beginning of 1890 have, according to the latest available information, been 72,000 tons."

In drawing a comparison between the figures given by **Mr. Hall** and those used by the writer in the foregoing remarks, it should be borne in mind that the one statement deals with the financial, and the other with the calendar, year. The serious effect of this will be seen by the abnormally high exports of 1888 to Upper Burma and the coincident low foreign transactions as compared with the immediate reversion of these figures during the three months of 1889 which fell into the official year ending 31st March.

Mills. **459**

Mills and Milling.—There are now at work in Burma 45 mills for cleaning and husking rice. These employ during the working season 27,084 persons and retain permanently in their employment 4,922 persons. The milling trade of Burma is accordingly a very important one. (*Conf.* with Chapter on Trade, p. 653.)

Provincial Methods. **460**

METHODS OF CULTIVATION, SOWING, REAPING, &c.

"By far the most fertile land is found in the delta of the Irrawaddy, and it is here that the largest amount of rice and that of the best quality is grown. There are five methods of raising a crop practised in different parts of the country:—

1st—On the ordinary swamp land in low-lying plains where the rainfall is sufficient.

2nd—On level land from which the rain water runs off too quickly and irrigation has to be resorted to.

3rd—On land near the river bank which is submerged and cannot be planted till after the highest rise.

4th—In hill clearings.

5th—A hot-weather crop obtained by irrigation either by means of dams or by water-wheels.

"For the first three, ploughing commences about June, that is, as soon as the annual rains have softened the soil and rendered possible the use of the primitive plough which is, in truth, little else than a harrow with long teeth. The ploughman stands on the bar and the harrow is dragged about the water-sodden field till a smooth surface of mud has been obtained. In the meanwhile nurseries have been prepared on somewhat higher spots and the seed sown in them broadcast. By about July or August the fields of the first two classes and the plants in the nursery are ready and the young plants are dibbled in, two together at intervals. This portion of the agricultural labours is generally undertaken by the women and children, neighbours often aiding each other. In the case of the riparian lands the plants cannot be put out till about September, the harvest begins in November and is over by January. The crop is cut with a sickle and collected at the threshing floor in some places in carts, in others on sledges. This is simply a portion of the field where the soil has been made flat and hard and a stake driven in to the ground in the middle. The sheaves are taken from the stack and arranged in two circular lines round the stake, one on the other, with the ears pointing in one row inwards and in the other outwards, and cattle are driven round and round treading out the grain. It is then sifted, in some parts of the country, by a hand-winnowing machine (fans fastened to a spindle being made to revolve and blowing away the light grain and straw as the rice in the husk, poured in at the top, passes down between the revolving fans and a bell-shaped mouth at the end of the machine), but usually the method is more primitive; a tall tripod is raised and a platform put up between the legs at five feet or so from the ground, over this and reaching down to the ground is a mat which serves as a kind of shoot to guide the grain in its descent; hanging from the top of the tripod is a flat basket, a man stands on the platform and, receiving from one below the baskets of unwinnowed rice he fills the hanging basket and upsets it, the good grain falls on to the mat and thence on the ground in an accumulating heap and the light grain and chaff are blown away by the wind.

"The hot-weather crop is planted in January, February, and March, and is reaped about three months afterwards. The water necessary is usually obtained by throwing

in Burma. (G. Watt.)

ORYZA sativa.

CULTIVATION in Burma.

a dam across a stream, but at Meng-doon in the Thayet district, and nowhere else, a self-acting water-wheel is used which is thus described by **Colonel Horace A. Browne**—

"'It is simply a large paddle-wheel, constructed altogether, with the exception of the shaft, of bamboos. The shaft is made of some hard wood. The floats are pieces of coarse bamboo mat-work. Between each pair of floats are two bamboo baskets sloping outward at an angle of 45° from the centre of the periphery of the wheel. The frame-work on which the wheel rests consists, generally, of about eight moderate-sized jungle-wood posts, four of which are planted close to the water's edge, and the other four parallel to them further out in the stream. The shaft of the wheel rests on a cross-piece between the four posts, two on each side, which are lowest down the stream. The wheel is placed so that the floats are just under the surface of the water. The current causes the wheel to rotate. The mouths of the buckets are pointed down stream; they have a slanting upward direction as they emerge from the stream, so that they retain the greater part of the water with which they have been filled when below the surface, until the rotation of the wheel brings them to the top when they discharge their contents into troughs above the edge of the river bank.

"'The banks of the Ma-htoon, where a view can be obtained at once of several of these spider-like machines in ceaseless motion their shafts humming loudly, and the waters splashing and sparkling all over them, form a singularly, interesting spectacle.'*

"The toung-ya or hill-gardens are by no means peculiar to this country. They are known in Mysore, in the Central Provinces, where, according to **Captain Lewin**, they are called 'dhai-ya' and in Assam where they are called 'jhoom,' and this system of cultivation is probably common everywhere in India amongst hill tribes; it is the only one which they can follow but it is most wasteful, and is the worry of the lives of our forest officers who never seem to lose an opportunity of inveighing against it. Having selected a site on the side of a hill, and the more thickly covered with bamboos and forest the better, the cultivator and his family set to work in April and fell everything. After two months' drying the fallen trees and dried brushwood are set on fire and burn for several days, and some of the larger logs even for weeks, the ashes fertilizing the soil. An ingenious method of lightening labour is sometimes adopted. The cutting commences from the bottom and the lower trees are only slightly cut on the upper side, as the woodmen ascend the hill they cut deeper and deeper and at last cut the trees completely through. Thus cut they fall on those below them and by their weight knock them down and the felling thus continues to the bottom. It is only in suitable localities that this system can be adopted.

"Immediately after the first fall of rain, the surface is slightly broken up with a kind of hoe and the ashes mixed with it, and the seed, usually rice and cotton or sesamum and cotton, is sown broadcast, and from this time onwards the principal labour is in keeping down the weeds. The rice (which is not of the same kind as any of that grown in the plains) and sesamum are reaped in September or October, and the cotton balls are picked in December—April. After this the hill-clearing is abandoned.

"Where this system prevails the land is never measured, but the revenue is assessed on the families working. The number of these hill-clearings in 1868-69 was 51,352 and in 1877-78, 77,707" (*British Burma Gazetteer, Vol. I., p. 424—426*).

Husking Paddy.—"Cleaned rice can be bought in most bazars and rice-husking has in some parts of the country near large towns become a regular trade, but over by far the larger part of the country the rice for the family use is husked by the females of the family as it is wanted, the stock of unhusked grain being kept in granaries adjoining the house. The grain is first twice milled by hand. The mill consists of a solid cylinder of wood about two feet in diameter, the upper surface roughened by radiating lines a quarter of an inch deep being cut in it; on this works another cylinder, the lower surface similarly roughened, with an opening through it in the shape of an inverted truncated cone; to one side of this upper piece is loosely fastened the end of a long pole and by working this backwards and forwards the upper cylinder is made to revolve and to husk the grain which is passed in at the top and comes out between the two portions of the mill. The rice is then winnowed either by throwing it up into the air when the lighter husk is blown away or in a hand-worked winnowing machine, an European invention of late introduction into Burma.

"The inner pellicle has now to be removed and this is done by pounding, either by hand with a long, heavy, massive pestle in a large wooden mortar, or in a some-

* Statistical and Historical Account of the Thayet District. By Colonel H. A. Browne, Rangoon, 1874, p. 83.

 Rice Cultivation

CULTIVATION in Burma.

what similar mortar let into the ground; in the latter case the pestle consists of a short but thick bit of wood let into a long one near the end and at right angles; this long one is supported on two low uprights near the end furthest from the mortar in such a way that the end of the pestle rests on the mortar; one or more persons by standing on the further end depress it into a hollow made for it in the ground and raise the other end, and as they step off the pestle falls with great force on the rice which has been placed in the mortar; this alternate stepping on to and off the 'pestle-bar' continues until the work is finished" (*Burma Gazetteer, Vol. I., p. 404*).

Burmese Diet.—"There are ordinarily two meals a day, one at about eight in the morning and the other at about five in the evening. The staple article of food is plain boiled rice; with it is taken a kind of soup or thin curry of vegetables, chillies, and onions, a pinch or two of fish-paste, a little salt, a little oil and, if it can be afforded, fish or meat; often fried salt-fish is added to the meal. As a condiment is used a mixture of fish-paste, chillies, and onions fried together. The rice is placed in the flat lacquered dish, called *byat*, and the soup and condiment are placed in small cups and the whole family eat together" (*Burma Gazetteer, Vol. I., p. 404*).

ASSAM. 461

XI.—CULTIVATION IN ASSAM.

References.—*Annual Reports of the Department of Land Records and Agriculture, 1886-87 to 1888-89; Assam Gazetteers, Vols. I. & II.; Agri.-Horti. Soc., Ind., Transactions, III., 58; Jour. (Old Series), IX., 240-246; X., Sel., 1; XII., Pro. (1860), 8, 9, 39; (New Series) II., 91-102, Pro. (1870), 78, 79; III., Sel., 5.*

Area under, and Traffic in, Rice.

Area. 462 *conf. with p. 522.*

The Agricultural Statistics of British India, published by the Imperial Government, show that the Province of Assam had 1,262,791 acres under rice during the year 1888-89. If this be supposed to yield 13½ maunds clean rice per acre, the total production would have been 1,70,47,678 maunds or, say, 608,846 tons.

Yield. 463

Trade. 464

By rail, road, and river, Assam exported 13,26,352 maunds of paddy and rice combined and imported 3,52,542 maunds, thus showing a net export of 9,73,810 from the province. The exports of paddy alone were a little in excess of rice, namely, 7,35,235 maunds of the former and 5,91,117 of the latter. The neighbouring province of Bengal drew the major portion of these supplies—11,12,002 maunds in all, which consisted of 7,34,169 maunds of unhusked, and 3,77,833 maunds of husked, rice: 54,390 maunds were received by the town of Calcutta, a portion of which may, therefore, have gone to foreign markets.

The statistics of the river-borne trade (which are included in the above statement) display some features of interest. For this purpose the trade of the Brahmaputra and Surma valleys are considered separately. The Brahmaputra valley received most of its imports in cleaned rice (not paddy), from Eastern and Northern Bengal (*viz.*, 1,85,065 maunds and 1,12,976 maunds, respectively), while its exports amounted to only 41,793 maunds, mostly paddy, the major portion going to Northern Bengal. The Surma valley carried on its chief export transactions with Dacca (6,49,634 maunds of paddy and 5,22,318 rice), while its imports were nominal, *viz.*, 24,296 cleaned rice and no paddy. Thus the Surma valley is the chief exporting portion of Assam, while the imports go mainly into the Brahmaputra, apparently to supply the large demands of the tea gardens. The statistics of both the valleys show an import of 3,52,996 maunds and an export of 13,00,819 maunds (paddy and rice together). Of the imports, 57,254 maunds were conveyed by boat and 2,95,742 by steamer. Of the exports, 12,91,415 were carried by boat and 9,404 by steamer.

Consumption per head. 465

Assam has a very insignificant trade with the countries on its frontier; 1,994 maunds being imported from Naga, Mishmi, Lushai, and Tipperah Hills, while 6,619 maunds were exported, mainly, to Bhutan and Towang.

The above figures show a net export from Assam of 9,78,435 maunds, which, deducted from the estimated production, would leave a surplus for

in Assam. (*G. Watt.*) **ORYZA sativa.**

CULTIVATION in Assam.

the people of 1,60,69,243 maunds or expressed to population, would represent a daily consumption of a little more than ⅓rd seer per head.

Provincial Methods; also Forms Grown. 466

Varieties and Races of Rice grown in Assam.

The Annual Reports of the Department of Land Records and Agriculture published by **Mr. H. Z. Darrah** contain much interesting information on this subject. The attempt to classify the Assam rices became a necessity in connection with the numerous crop experiments which were conducted under the orders of Government. It was felt that the yield was so diversified, through the complexity of forms, that it became imperative to treat each race separately. And, doubtless, future more extended enquiry into the rices of India generally will render a similar course indispensable, at least in the provinces that produce two crops a year. An average provincial yield per acre, such as has been shown in the above provincial chapters, may serve a useful purpose in affording a starting point of comparison in agricultural and commercial returns, but the factor to be adopted as the average can only be approximated to truth when (as in Assam) the actual yield of each form has been first determined and then followed by a calculation worked out to cover the error (incident to all the figures given by the writer) due to the relative extent of cultivation of each race. While fully conscious of errors on every hand and of the imperfect nature of the available information regarding rice cultivation and trade in India, the writer has thought it the better course to follow a definite plan as he may thus suggest valuable corrections which in time may lead to the production of an article on this very important staple that could lay claim to greater accuracy than the present The results obtained by **Mr. Darrah,** in the yield of the various kinds of Assam rice, fully justify this admission of imperfection. So diversified are the returns, that without a knowledge of the actual acreage of each crop, it would be unsafe to hazard a definite statement as to the total production and available surplus. Following the plan adopted, however, with other provinces an average production of 13½ maunds on the total Assam area of rice has been assumed in order to show a figure equivalent to local consumption, after the balance has been struck of imports and exports.

Yield. 467

In his **Annual Report** for 1887-88, **Mr. Darrah** discussed the Assam rices under two sections which practically correspond to the chief Bengal crops, *viz.,* Early or Summer and Late or Winter Rices. These terms allude of course to the season of harvest. He then sub-divided each of these into the two great geographical sections of the province, *viz.,* Assam proper or the Brahmaputra valley and Sylhet and Cachar. In his subsequent report he adopted an elaboration of this classification according to the names of the chief recognisable races of the crop. This latter system may be here discussed while bringing together all the facts regarding each form which have appeared from time to time in **Mr. Darrah's** reports—

(A).—RICES OF THE BRAHMAPUTRA VALLEY.

1. *Sáli*—
   - (*a*) *Lahi dhán.*
   - (*b*) *Bor dhán.*
2. *Áhu*—
   - (*c*) Broadcasted.
   - (*d*) Transplanted.
3. *Bao*—
   - (*e*) Broadcasted.
   - (*f*) Transplanted.

(B).—RICES OF SYLHET OR SURMA VALLEY.

4. Low-land *Áus.*
5. High-land *Áus*—
   - (*g*) Transplanted.
   - (*h*) Broadcasted.
6. *Murál̄i.*
7. *Sáil.*
8. Low-land *Áman.*
9. High-land *Áman.*—
   - (*i*) Transplanted.

ORYZA sativa.

Rice Cultivation

CULTIVATION in Assam.

(B).—Rices of Sylhet, &c.—*contd.*
(*j*) Broadcasted.

10. *Kataria.*
11. *Sáil bura.*

(C).—Rices of Cachar.

12. *Dumai.*
13. *Murálí.*
14. *Asra.*
15. *Sáil.*

Thus it will be seen **Mr. Darrah** deals with the Assam rices under 15 main sections with, in some cases, sub-sections; but it will be discovered by the remarks that follow that these cannot exactly be regarded as 15 distinct races. Several are the exact equivalents of each other in the three areas and have been separately dealt with more from agricultural and commercial considerations than structural peculiarities.

(A.) 468 *1. Sáli* of the Brahmaputra Valley—

This, **Mr. Darrah** says, is the general term applied to all transplanted rice which is grown on land lower than that required for *áhu,* and higher than that needed for *bao.* "It is sown about May and June, transplanted in July and August, and reaped in December and January. Some varieties differ but slightly from transplanted *bao. Sáli* is always transplanted. *Lahi dhán* comprises the finer varieties of *sáli,* which are transplanted rather earlier than the other kinds and on comparatively higher lands. *Bor dhán* includes the larger grained varieties of *sáli,* transplanted later than *lahi dhán* and on comparatively lower lands." **Mr. Darrah** shows that the average yield, out of 85 experiments with the Brahmaputra *lahi* form of *sáli,* was 1,159℔ per acre of paddy. Out of 241 test harvests, he gives the average yield of *bor dhán* as 1,821℔. "The experiments in *sáli* of both kinds number up to date 2,439. They were made over 814·62 acres with a total crop of 1,247,947℔, and yield a result of 1,532℔ of threshed grain per acre, or, roughly, nearly 20 maunds." This, it will be observed, is the yield of clean paddy, but following the rule adopted with other provinces a deduction of 25 per cent. should be made to obtain the yield of rice or, say, 15 maunds.

469 *2. Áhk*—

This is the name given to "the very numerous kinds of rice which grow on high lands, require but little rain, are sown from March to June, and reaped from June to September. *Áhu* is sometimes transplanted, but only in the case of the later varieties. Usually, it is sown broadcast. Many varieties germinate well without a drop of rain."

The average yield of transplanted *áhu,* in the Brahmaputra valley, was found to be 1,380℔ of paddy out of 92 experiments, and of broadcast sowings out of 305 test cases it was found to be 1,300℔ paddy. **Mr. Darrah** states that the total number of experimental harvests to test the yield of *áhu* rice have "up to date been 1,105, the total area cut 381·59 acres, the total crop 504,190℔, and the average outturn 1,311℔ of threshed but unhusked grain per acre, or, roughly, about 17 maunds." This reduced to rice by a deduction of 25 per cent. would be, say, 13 maunds an acre.

470 *3. Bao dhán*—

This "comprises the varieties grown on the lowest lands which will support rice. It is sown in March and April, and cut in November and December. When transplanted, the operation is performed in July and August. This is said to correspond to the *áman* and *kataria* rices of Sylhet and to *asra* of Cachar."

In the Brahmaputra valley transplanted *bao* was found to give the average yield of 2,240℔ per acre and of broadcasted the average, in 152 experiments, was found to be 1,254℔. Discussing the final results of all the experimental harvests with *bao* rice Mr. **Darrah** says: "The average outturn per acre for 155 experiments, with a total crop of 68,647℔, over 54·25 acres, is 1,265℔, or, roughly, 16 maunds" (of thrashed but not husked grain or, say, 12 maunds of rice). "In these experiments, however, nearly all the crop cut had been sown broadcast." (*Conf.* with 8 and 9 *Áman* below.)

CULTIVATION in Assam.

(B.) 471

4 & 5. Aus of the Surma Valley—

The former (highland form) is sown in the higher parts of the district of Sylhet usually broadcast but occasionally transplanted. It is put down in March and April, and harvested in August and September. The latter (lowland *áus*) is grown in the lower parts of the district, but is never transplanted. It is sown in January and February, and cut in May and June.

These rices are often grouped along with *muráli* as early rices: "they are practically the same as the *áhu* of the Brahmaputra valley and the *dumai* and *muráli* of Cachar. Taken together the early rice experiments in Sylhet have, during the six years up to date, been 218 in number; 57·96 acres, bearing a crop of 65,491℔, have been cut, and the general average has been 1,130℔ or, roughly, over 14 maunds of paddy or, say, 10½ maunds of rice."

472

6. Muráli of Sylhet or the Surma Valley—

This is generally sown on lower land than in the case of *áus* of the highland form but on higher than the lowland. "It is put down in February and March, and cut in June and July." (*Conf.* with Nos. 4 and 5 above.)

473

7. Sáil—

This rice is always transplanted. "It is sown in April and May, transplanted July and August, and reaped November to January. It is grown on land almost as high as that which grows *áus*."

474

8 & 9. Áman—

As in the case of *áus* this may be referred to two sections owing to the height of the ground on which it is grown. "In the higher parts of the district it is transplanted like *sáil*, but is sown and gathered about a month earlier than that crop. In the lower parts of the district one variety of *áman* is always sown broadcast, as the early rise of the water allows no time for transplanting. This is the long-stemmed kind, which grows with a steady increasing flood, unless swamped by a sudden rise of water." These *áman* rices are, therefore, sown in March and April and harvested in October and November or a month earlier than *sáil*. The above description of long-stemmed *áman* correspond, very closely, with the *rayenda*, *baoa*, and *khama* rices of Eastern Bengal. (*Conf.* with pp. 542 *et. seq.*)

Mr. Darrah in dealing with the yield of *áman* rices says:—"Experiments in lowland and highland *áman* and *kataria* have up to date been grouped together under the generic name of *áman*. The corresponding crop in the Brahmaputra valley is *bao* and in Cachar *asra*. There have been 386 experiments during the last six years; an area of 104·71 acres, with a crop of 157,348℔ having been cut, and the net result is an average outturn of 1,502℔, or, roughly, nearly 19 maunds per acre, or, say, 14¼ maunds of clean rice." The *áman* rices of the Surma valley thus give about 3 maunds an acre better yield than the corresponding *bao* rices of the Brahmaputra. (*Conf.* with No. 3 above.)

475

10. Kataria—

This is said to be a form *ánam* which is sown in April and May, transplanted in May and June, and reaped in October and November.

CULTIVATION in Assam. 476 *11. Sáil bura—*

This is "grown on the lowest lands, and therefore mainly in Sunamganj and its neighbourhood. It is always transplanted and generally irrigated. It is sown in October and November, transplanted in December and January, and cut in April and May." This "crop stands alone, there being nothing like it in either the Brahmaputra valley or Cachar." It is transplanted when the main winter crop is being cut and reaped at the same time as the *boro* rices of Bengal to which group it probably belongs. "The lands used are those portions of the *haors*, or large natural depressions in Sylhet, which are left dry by the receding of the water in the cold weather. Embankments (locally called *ails*) are formed and the water from the lowest part of the *haor*, where it is never quite dry, ladled up to the land occupied by the crop." Out of 163 experiments, the average of 1,545℔ was obtained, or 19¾ maunds paddy equivalent to 14½ maunds rice.

(C.) 477 *12. Dumai* rice of Cachar—

Is said to be the Cachar rices which correspond with the forms of *áus* which are always sown broadcast on high lands in April and May, and reaped in July and August. It is never transplanted. In his report for 1888-89 **Mr. Darrah** says of this form of Cachar rice: "*dumai* and *muráli* have hitherto been included in one under the head *áus*, or early rice. During the last years, including 1888-89, 87 experiments in early rice were made, involving the cutting of a crop of 28,106℔, on an area of 23·13 acres. The result gives 1,215℔, or, roughly, 15 maunds, as the average outturn of an acre." That return was of clean but not husked paddy, but, making the usual reduction of 25 per cent. for husking, the yield would be 11¼ maunds of rice.

478 *13. Muráli* of Cachar—

This is sown in March and April, and is sometimes transplanted about May. The crop is reaped in June and July.

479 *14. Asra—*

This is sown in March and April in low-lying lands, and is reaped in December. It is never transplanted. "It belongs to the same class as *áman*, a variety never grown in Cachar, as it requires deeper water than *ásra*." In five experiments the average yield was found to be 1,574℔ per acre or 19¾ maunds of paddy, which reduced to rice would be 14¾ maunds per acre.

480 *15. Sáil* of Cachar—

It is practically the same as the *sáil* of the Brahmaputra valley and the *sáil* of Sylhet. Out of a total of 123 experiments, the average yield of this rice was found to be 1,585℔, or 19¾ maunds paddy equivalent to 14¾ maunds rice.

CONCLUDING NOTE ON THE CULTIVATED RICES OF INDIA.

Classification of Rices. 481 The most careful consideration of every remark made by **Mr. Darrah** regarding the above 15 forms of rice has failed to enable the writer to assort them under the four great sections which on certain structural peculiarities have been established in the chapter devoted to the botanical account of **Oryza** (*Conf. with pp. 503-507*). This is, however, not much to be wondered at, for with even good dried specimens before one it is not always possible to pick out an *áus* from an *áman* or from a *boro* rice. But it has to be admitted that the use of such terms is to be deprecated if the agricultural conceptions of these forms are, like the botanical, so meagre that it becomes possible to speak of a *boro* rice as a variety of *áman*. This can only be the case by supposing the terms interchangeable in the various provinces of India in which case it would be advisable to relegate them with the thousands of other local names and refrain from their use in any

CLASSIFICATION of Rices.

system of classification. This naturally leads one back to the belief that **Roxburgh** and other authors who classified rices by their seasons of growth laid down the principle upon which it is desirable the rices should be agriculturally grouped; at all events, until such time as botanical research can frame a classification by which the local forms of *áus, áman, boro, sáil, áhu, kataria*, &c., &c., can be arranged under new and arbitrary names not open to the ambiguity occasioned through the possibility of a so-called *áman* not being *áman*. There can be no doubt, however, that the forms of Assam rice, specified by **Mr. Darrah** are local manifestations which are recognised under the names given, and that as special adaptations to the environment in which they are found, they are of vital importance in agricultural statistics and rural economy, since they could scarce be interchangeable, give different yields to the acre, and possess in the markets special properties that recommend them to different purchasers. These remarks, therefore, have been made purely and simply to guard against the false impression that an *áman* rice from Assam or any other province belongs of necessity to the same class in Bengal. There is no other reason for wishing to establish the *áman, áus* and *boro* rices of Bengal as denoting the primary sections of the cultivated forms of **Oryza sativa** than the fact that Bengal is the chief rice-producing country. These three classes were recognised by Sanskrit writers under the names of *Sáli, Vríhi*, and *Shashtika*, and thus lay claim to being the classes formed by ancient writers. But if future study can establish these, on a combination of structural peculiarities and agricultural manifestations, then it would certainly be desirable that the rices of all the provinces of India should be spoken of as belonging to these or other recognisable groups.

Improvements in Rice cultivation, &c. 482

IMPROVEMENTS IN RICE CULTIVATION AND QUALITY OF STOCK.

Effects of Improved and Scientific Agriculture.

Throughout the District Gazetteers, the remark occurs, almost in every volume, that no improvement appears to have taken place within historic times in the methods of cultivating this very important article of food. In connection with the remarks about Bombay, however, it will be found that a distinguished botanical traveller (**Dr. Hove**) affirms that the people of Thana had only recently (1787) adopted the system of transplanting the seedlings from a specially prepared nursery. In most parts of India the history of this highly beneficial method is involved, however, in the same obscurity that characterises every feature of rice culture, so that it cannot be discovered from whence and how this system was diffused. The numerous district accounts which have been freely quoted in the provincial passages above, while they frequently contain much in common, will be found to each manifest some special peculiarity. These departures from the standard system are often doubtless direct adaptations to the requirements of climate or soil, but some, as in the case of the high manuring pursued in Bombay, are outcomes of a better system of agriculture such as that which characterises the Western Presidency. That great improvements could be effected, there would appear to be no doubt. The argument, in support of this statement has already been advanced that the superior yield in one province as compared to another, can easily be shown to be more a result of better cultivation than of any inherent adaptability of soil or climate. But if this needs further proof than afforded by present Indian agriculture the case of America may be cited.

Carolina Rice. 483

Experiments at Acclimatization of Carolina Rice.

According to all accounts, India gave towards the close of last century a mere handful of rice seed to America. Cultivation in the hands of Euro-

ORYZA sativa.

Carolina Rice.

IMPROVEMENTS in Cultivation.

pean planters resulted in the New World, not only becoming a rice-producing area, but in the finest and most profitable of all known rices being those developed by American enterprise. The result of the American success has not been a stimulation in India to improvement by selection and better culture, but in what it is feared must be characterised as a fruitless endeavour to bring back to the mother country the improved stock. The extent of these experiments, in the acclimatization in India of Carolina rice, have been fully made known, in the *Memorandum Regarding the Introduction of Carolina Rice into India*, published in 1879 by the Revenue and Agricultural Department of the Government of India. It would, perhaps, serve no very distinct purpose to republish the main facts there brought out, suffice it, therefore, to say that failure (with very few exceptions) was the universal verdict, but failure not so much to grow the rice and produce a large yield, but failure to induce the people to grow it or to cut the produce when grown. The following passage taken from **Mr. Charles Benson's** *Saidapet Experimental Farm Manual and Guide* may be given not only as showing the degree of success that has attended Carolina cultivation in Southern India, but as representative of all the other parts of India where its cultivation was found possible:—

"Many experiments have been made with this variety, both with seed imported direct, and with seed raised in this country; the results have been highly satisfactory, and have proved that the native system can be adapted to its growth, if only less water be used, deeper cultivation be adopted, and more attention be paid to drainage: good results can be obtained either by sowing the seed at once in the paddy bed, or by sowing it in a seed bed and afterwards transplanting the plants into the paddy bed; either procedure may be justified by the peculiar circumstances of the rayat: it has been found that remunerative crops of country paddy can be produced on a soil that altogether fails in producing a fair crop of Carolina paddy, and also that a soil which one year produces a good crop of this grain may, in the next year, not be able to produce even half a crop. Thus, soils which in 1870 produced as much as 2,000℔ per acre of Carolina paddy yielded in 1871 only half as much, though equally as well cultivated and manured: the explanation of this is simple; the soils which produced such good crops of Carolina paddy in 1870 were in a fresh state at the time the paddy was sown; they had not been under wet cultivation for several years, while those sown in the following year with this paddy had been under wet cultivation in the previous year, and had become more or less reduced, by the stagnation of the irrigating water in the sub-soil, to the condition of the more inferior descriptions of paddy land, or, in other words, to a sour, unhealthy, soapy condition, in which atmospheric agencies had been able to have no effect for good, in which plant food was either locked up and rendered inert, or formed into combinations more or less injurious to vegetable life, but whose surface to a depth of 2 or 3 inches was more or less healthy and capable of producing plants of a low type. It is easy to understand why, on a soil in this condition, Carolina paddy seldom does well, while country paddy may thrive. The former feeds deep down in the soil, it has long deep roots, varying in length, under average circumstances, from 4 to 6 inches, while few of the rootlets are horizontal. It is very different with the roots of country paddy; nearly the whole of its rootlets radiate over the surface of the ground at right angles to the stem of the plant. They are seldom more than 2 or 3 inches in length, and are so interwoven as to form a sort of network well suited for supporting the plant on the semi-liquid mud in which this variety delights to grow. Hence a soil in a bad physical state, if it only has a

IMPROVEMENTS in Cultivation.

healthy surface 2 or 3 inches in depth, may produce fair crops of country paddy, though it will entirely fail in producing a crop of Carolina paddy. On such a soil the Carolina paddy would start as well as the country paddy, and would continue to grow as well, until its roots began to penetrate below the aërated soil, when its growth would be checked, and it would gradually turn yellow and sicken.

"It was expected by many that the sample imported would deteriorate; this does not necessarily follow, for, although deterioration must take place if it be grown under very unfavourable circumstances, if properly cultivated it will not. In preparing rice from ordinary Carolina paddy, by means of the pestle and mortar, in the usual manner adopted by natives, the following results were obtained:—

Yield. 484

	Per cent.
Clean rice	67·24
Small rice	8·48
Tour	5·24
Husks, &c.	19·04
	100·00

Regarding the cultivation of the crop the following advice is given:—

"*Soil, &c.*—Carolina paddy has a very strong aversion to stagnant water; therefore avoid all land on which, from imperfect drainage, water is likely to collect and stagnate. Remember that water may stagnate in the soil as well as on its surface. Select good healthy paddy-land, resting on a moderately porous sub-soil and with a surface sufficiently high, that it will not become a swamp during the rains, while it is yet sufficiently low to admit of being watered at a moderate cost. In the endeavour to avoid a soil that is too retentive take care that you do not select one that parts too readily with its moisture, as a soil with this defect is as objectionable as one that is too retentive. Endeavour to select a soil that is neither too retentive nor too porous; in every *taluq* there are hundreds of acres of wet land that fulfil these conditions. These observations refer chiefly to the physical or mechanical condition of the soil. It is of much greater importance to secure a soil in a good physical condition than one in a high manural condition, as by the application of a suitable quality and quantity of manure the last-mentioned condition can so readily be secured. Land which is put under irrigation for the first time will nearly always produce good crops of Carolina paddy.

"*Preparation of the Soil.*—The soil may be ploughed and worked in the dry state for sowing, or it may be ploughed under water and puddled. I much prefer the former arrangement; the latter is sometimes attended with advantages, but the evils which result from the wretched state in which the soil is left by the process are frequently of a very serious kind. But, whatever plan you adopt, take care to give the soil a liberal application of good manure during the time it is being prepared for the reception of seed.

Manure. 485

"*Seed and Sowing.*—Choose a fresh sample of seed; it should be of a rich golden colour; the sample should be uniform both as regards colour and the size of the grain; if the seed is mixed with the seed of country paddy (which is the case almost invariably with all country-grown samples), and you cannot get a pure sample, you may, by using a fan and a bamboo sieve, remove the greater portion of the indigenous grain. Before you finally select the seed you should test its vitality. You may do so in the following way:—

"Take a shallow vessel—a soup-plate will answer admirably; place about 1 inch deep of good garden soil in the plate, and scatter over the

ORYZA sativa.

Carolina Rice.

IMPROVEMENTS in Cultivation.

soil 100 seeds; cover this with a piece of muslin, and over the muslin place about half an inch of soil. Keep the soil damp. After a few days the muslin with the upper soil may be lifted off and the condition of the seed ascertained. The number of seeds that have germinated will give the percentage of vital seeds. But you had better make duplicate experiments and take the average of the results obtained as the percentage of vital grain; ascertained in this way the results of the test are more trustworthy.

Planting. 486

"Having selected suitable seed, and having prepared your land for its reception, you may proceed with the sowing. You may sow the seed at once on the land, or you may sow it in seed-beds and afterwards transplant the seedlings into the field. If you adopt the former of these arrangements, you may sow, say, at the rate of fifteen measures per acre on either dry or puddled soil; if on dry soil, you may take the seed in its natural state and broadcast it or drill it in; but, if you prefer to sow on a puddled surface, you may sprout the seed before you sow it, but when this is done you can only sow broadcast. When transplanting is preferred, take care that the nursery-bed is *deeply* dug and thoroughly manured. When this is done the roots of the seedlings are small and compact, and the plants can be raised with ease and without suffering any injury in the operation. To produce seedlings for transplanting, sow eight measures or about 18℔ of seed in the nursery-beds, for each acre of land to be planted; less will probably suffice, but this quantity will provide against any loss. The seedlings may be lifted for transplanting, when three or four weeks old: they may be planted in the field in the puddled, or wet soil, at distances of from six to nine inches apart, a couple of seedlings being planted together. It may be good practice either to sow the seed at once in the field or to sow it first in the nursery, and afterwards to plant out the seedlings in the field. Thus, when the area to be sown is large, the season late, labour scarce, and seed plentiful and cheap, we would adopt the former plan, and the latter, when the seed is scarce and dear, the area to be cropped small, the weather favourable, and labour plentiful.

Irrigation. 487

"*Irrigating.*—In whatever way the land is sown or planted, it should be moderately watered on the same day; this enables the seed to meet with a good bed and fixes the roots of the seedlings. It is seldom necessary to flood the land except to kill weeds or insects, especially after the plants begin to shade the soil. The endeavour should be to keep the land constantly damp, and not alternately in a state of puddle, and dry and filled with large cracks, as is too commonly the case with wet land under ordinary paddy culture."

"*After-cultivations.*—These are of a very simple nature; it is only necessary to keep the crop free of weeds and to thin out a few of the plants if the crop grows too luxuriantly."

Throughout India, especially in Government Experimental Farms, small plots of land continue from time to time to be thrown under Carolina and other improved forms of American rice, but, as already stated, the results have not even yet passed beyond the experimental stage, and in no part of India can it be said that these superior rices have found their way into the hands of the ordinary cultivator such as has been the case with American forms of cotton. **Mr. Liotard** (in the Memorandum already alluded to) thus sums up the results of the experiments throughout India:—"No very definite conclusions can be drawn from the experiments summarised in the preceding chapter. In the Madras Presidency a fair amount of success has attended the trials; the failures are, as reported by the Superintendent of Government Farms, in many instances, to be attributed to ordinary causes under control in average seasons, and to mismanagement or to wilful neglect. In the Bombay Presidency the results

IMPROVEMENTS in Cultivation.

do not appear to have been decisive either way. In Bengal, in the North-Western Provinces and Oudh, in the Central Provinces, and in British Burma, there have been more failures than successes; and the general result would lead to the belief that, though Carolina rice can be successfully grown in these provinces if it is properly cultivated, yet the people do not like this rice, and it is liable to risk from which the indigenous variety is free. In the Panjáb the poorer and more numerous classes live on dry grains, and the wealthier prefer their own very superior rice grown in Kangra and thereabouts. In Assam, in Berar, and in Mysore, the experiments resulted in failure, and a continuance of them has been deprecated by local authorities. In the Andaman Islands the trials have succeeded satisfactorily, and the Carolina rice has displaced, or nearly displaced, the Bengal variety which had hitherto been grown there.

"On the whole, it would seem that the Carolina rice seed was too widely and too indiscriminately distributed. It might, perhaps, have been more satisfactory if one or two well-managed experiments had been made in each district and a trustworthy record kept of the results. This is a lesson learnt for future guidance."

Mr. C. B. Clarke (*Kew Bulletin, 1888, p. 289*) thus characterises the Government efforts towards improving the quality of rice grown in India:—"The Carolina rice has a large grain; it does not follow that the produce per acre would be larger, far less that it would be more valuable than that of Bengalee small-grained kinds. Government for 20 years past has been sending round this Carolina rice for trial in Bengal. A bag of seed rice is sent to each collector; the collector hands it over to some Bengalee gentleman supposed to take an interest in agriculture; the reports finally got in by the Bengal Secretariat are found, on the whole, unfavourable. In 1870, the Bengal Government called on the Calcutta Botanic Garden to try Carolina rice. The Curator, **Mr. John Scott**, grew it on ordinary rice land outside by Bengalee cultivators in their own way, merely supervising them to see that the rice was forward in the seed-bed, the land thoroughly cleaned, and the dibbling early. There was a very heavy crop, but no native dealer would purchase it, and it was finally bought by a European merchant for export to London. Government is still (up to three years ago) sending round bags of experimental Carolina rice to the collectors. The Bengalees distinguish shades of flavour in rice; they do not like American or Burmese rice; they do not like large-grained coarse rice; and they do not like newly harvested rice, as they say it disagrees with them extremely. The *rowa* rice in Mymensingh, harvested in December, is kept in raised, well-thatched granaries till the following August, when the Calcutta traders' large boats arrive; it reaches Calcutta just aged enough for the Calcutta Babus to eat. In the 1874 famine Government imported Burmese rice largely into Behar and distributed rations of it to those employed on the famine relief works. But these recipients largely sold their rations to traders, by whom the Burmese rice was exported back to Burma."

Burmese Rice Objected to in Bengal. 488

On the general subject of improvement of rice culture **Mr. C. B. Clarke** offers some salient remarks with which this chapter may be allowed to close:—"Next I come to proposals that have been made to teach the Bengalees how to grow rice better. A favourite proposal is to give them an English plough, which shall go deeper than the native cultivator and bring up fresh soil. I pass by the practical difficulty that in none of the terraced fields, and in none of the small fields, without a revolution in boundaries and customs, could such a plough be used. The plough is the most perfect implement yet devised for setting in creeping grasses, as couch; it cuts the creeping rhisome into convenient lengths, and by the turnover

European Ploughs. 489

OṘYZA sativa.

European Methods

IMPROVEMENTS in Cultivation.

buries the fragments deep. If in England a farmer is about to put in a large field of barley or beans and sees a patch of couch grass in the fallow, he misses that patch with the plough, and at some dry time in early summer he turns in a party of boys to tease the grass thoroughly out. Some bad (or about-to-quit) farmers will plough a field full of creeping grass to get a corn crop; they get an inferior corn crop and the field is then found (in vulgar parlance) to be "clean run out," or (as I should rather put it) to be so thoroughly foul that it will take two years at least and a heavy outlay to get it straight. Now in India we have, not one or two, but many creeping grasses to contend with; the safety of the Bengalee cultivator is that he has a hard pan, impervious to creeping grasses, which his cultivator travels upon but never has broken. He gets the creeping grasses well out of the top 4-6 inches of his soil, and has a full crop on his shallow tilth if the water is right. I may add that if a Bengal field was ploughed with an English plough just before dibbling, I doubt whether the rice would get a firm enough hold.

Creeping Grasses. 490

"I have never inspected a Government experimental farm. **Dr. Watt** tells me that on one of these he has known the skilled European agriculturalist plough deep with an English plough in April, as soon as the first rains had softened the ground a little. The field was then left till July or August when it was tilled in the native fashion and dibbled with rice. By this plan a much better crop of rice was obtained in a season when the rain ran short than in the adjoining native fields.

"I have no doubt this would be the result. The English skilled agriculturalist in this case would have been careful to have the land perfectly clean before he commenced on it with his English plough.

"I am not at all sanguine of success in applying the English plough in Bengalee hands to rice-growing in Bengal. I fancy it would be exceedingly difficult to induce a Bengalee cultivator to clean his land thoroughly before ploughing, or to undertake any extra labour to ensure a better crop in a bad season; he would say that if the rain should not last on the average time that would be the will of Providence.

"Moreover, I think there is something easier than deep ploughing which might be done to assist the crop when the rain stops too early.

Manure. 491

"*Manuring Rice Fields.*—A second favourite proposal is that the Bengalee cultivator should be taught to manure his rice. It has been urged* on the Lieutenant-Governor of Bengal (in a formal agricultural council), that the lowest classes in Bengal should be taught chemistry, but I will pass that by. The Bengalee cultivator has little manure and he applies what he has mainly to his cold weather crops. There is a considerable quantity of cowdung used for fuel. It might be possible to forbid by police *ukase* the burning of cowdung in Calcutta and its suburban municipalities; I do not think it would be remunerative to purchase extraneous manures. The effect of manure may be considered as similar to that of deep ploughing, and it must be recollected that it is quite possible to get corn too strong. The rice crop, when a full one, often suffers before harvest by getting laid to the mud and water when the country is drying up in November; this is especially the case with the *Aman*.

"I concluded my first (1868) paper on rice by saying that I did not think we had much to teach the Bengalees in rice-growing; and this statement did not, I fear, conduce to the popularity of that paper. I will venture here to mention a few points where I see the best chance of improvements being effected.

* Conf. with pp. 534-537; 557; 562; 581; 586; 589-594; 599; 603; 609; 615-619; 624, &c.

Conf. also with p. 644—average yield per acre.

IMPROVEMENTS in Cultivation.

"One most important point is that the *rowa* should be dibbled out as early as possible. This is in general not done; the rain comes and by the middle of July the cultivator tills his field, but the plants for dibbling are not ready; a month or six weeks later often he cultivates his field over again and dibbles it. One reason of this is the deep-seated Bengalee principle never to do to-day what can possibly be put off till to-morrow, and to do everything incompletely; he says, "it is not quite as it should be, Sahib, but it will act (*chullibai*)" The real difficulty in introducing any improvement in Bengal is to infuse some energy into the native character. In the majority of the seasons I have seen in Bengal I know no reason why the rice that was dibbled between 1st August and 7th September might not have been dibbled between 1st July and 7th August. The advantage of promptness would outweigh, I believe, all the deep ploughing and manuring that will ever be introduced in Bengal. If the rain holds on through October it nowise injures the forward rice; while, if the rain stops between 1st and 15th October, the forward rice gives nevertheless an excellent crop. Many of the simple savages on the frontiers of Bengal exhibit much more punctuality in business than the civilized Bengalees I cannot help suspecting that the magnificent rice crops I have seen (in the Naga Hills in the extreme north-east no less than in the Kolhan in the extreme south-west) are due to their promptness in dibbling; but I have never spent a whole season among these people.

Seed-beds. 492

"A difficulty is encountered, it is true, in dibbling rice in Bengal from the uncertainty of foreseeing the exact time when the land will be ready for dibbling, so that it is impossible to raise the seed-bed to fit; the rice must be a certain length for dibbling and cannot stand over long in the seed-bed, so that the native cultivator plants his seed-bed in fair average time; rather late than otherwise. It thus often happens that the field is wet enough for tilth before the seed-bed is ready for dibbling; and in one season I saw in Burdwan the water came so late that the seed-bed rice was seriously injured (and some dead) before it could be dibbled and a deficiency in the Burdwan crop ensued. But this difficulty could surely be met by some combination among the cultivators to have a series of seed-beds "to follow in succession."

Tanks. 493

"*Railways and Tanks.*—For the increase of the gross rice produce in Bengal (the real object of Government), the most important point is improved communications. There are still very large areas in Chota Nagpore, Assam, and the margin of the Soonderbun, which would be made to grow rice, if communications gave a good market. Also irrigation tanks could be largely increased in Chota Nagpore, where the long gradual slopes at the head of the valleys lend themselves to such. Communications in Bengal mean the enlightenment of the population; not merely by the penetration of European enterprise, for no well-educated Bengalee wishes to live in the wilderness, and the opening of a railway produces a line of good schools, where before, at great direct cost to Government in paying discontented teachers, no satisfactory schools could be kept.

Model Farms. 494

Whatever is to be taught the Bengalee cultivator must be taught by example. It is no use whatever to lecture him; it is absurd to expect him to adopt a new, outlandish, and troublesome process unless you show him clearly that it is remunerative. You must therefore have model farms in central accessible situations where this can be shown."

DISEASES & PESTS.

Diseases. 495

The subject of the injury done to the rice through insects, crabs, and other animals has been incidentally alluded to by many of the writers quoted above. The clearing of the fields of algæ and other injurious

Diseases.

aquatic weeds has also been fully dealt with in the paragraph on manuring practised in the Panjáb. But no mention has been made of the subject that for the past 300 years at least has figured so prominently in the writings of travellers (see **Rheede** and other early authors), *viz.*, the marvellously rapid way in which the rice-fields become oceans which swarm with fish. These afford in many parts of the country a valued addition to the diet of the peasantry, but as they do not appear to do serious injury to the crop, the reader is referred to the article **Fish** for further information. The insect and fungoid pests of rice have also been reviewed in the articles **Fungi & Fungoid Pests**, Vol. III., pp. 455-458; also "**Pests,**" **Insect**, Vol. VI., to which the reader is referred for details of the injury actually sustained by this crop.

TRADE.
496
Conf. with pp. 519-523.

INDIAN TRADE IN RICE.

In the foregoing pages an estimate has been made of the total area under rice and of the probable yield. These figures have been worked ou tfor the year 1888-89—the most recent year for which alone it was found possible to get all the required statistics. Foreign and other trade returns are, however, available for more recent years, and in the present chapter an effort will, therefore, be made to utilise these. After the elaborate treatment, however, which has been pursued with the internal trade of one year for each of the provinces separately, it seems only necessary to deal here with the foreign trade. Before proceeding to do so, however, it may be as well to bring together the main facts which have been brought out :—

**Conf. with p. 521.*

Total area under rice in 1888-89 has been given as 68,429,406 acres.*
Estimated production of rice—

Bengal (*Conf. with p. 538*)	14,269,223 tons.
Madrasl(*p. 572*)	2,693,916 ,,
Bombay and Sind (*p. 388*)	399,757 ,,
North-Western Provinces and Oudh (*p. 602*) . .	2,420,768 ,,
Panjáb (*p. 615*)	271,293 ,,
Central Provinces (*p. 625*)	1,622,385 ,,
Burma (*p. 627*)	3,039,397 ,,
Assam (*p. 632*)	608,846 ,,
TOTAL .	25,325,585 tons.

But though reference has been made to Mysore and Coorg, to Berar and the Nizam's dominions, to Rájputana and Central India, to Nepal, Kashmír, and other Native States, the areas under rice in these regions have not been dealt with. Nor can a figure be discovered that could at all be accepted as conveying any definite idea of the additional supply of rice from these unsurveyed areas which should be added to the above estimates for the chief provinces of India. If we, however, allow that these tracts collectively may have possessed in 1888-89 one million acres which yielded an average of 10 maunds an acre, the total produce from these areas might be put down at 357,142 tons. Thus, allowing for every possible error, the total annual production of rice in India does not certainly exceed 26 million tons, or an average over the entire cultivation of $10\frac{1}{2}$ maunds per acre. From the remarks which have already been made it will be seen that, judging from the returns of Bombay, the Central Provinces, Madras, Assam, and Burma, the writer believes that an average production of 10 maunds for Bengal is either considerably under the mark or, if correct, an abundant proof is obtained from that fact that far less is taken from the soil than it might be made to yield.

Total Production: 26 million tons.
497
Conf. with p. 589.

TRADE.

Transfrontier Traffic. 498

TRANSFRONTIER TRAFFIC.—From the following table, which analyses the returns of the land transfrontier trade, it will be seen that, for the past six years, India has obtained a net import of 33,87,833 maunds from countries beyond its frontier, the greater portion (about 95 per cent.) coming from Nepál.

Analysis of the Transfrontier Trade in Rice for the past six years, the returns of Upper Burma being taken separately for the first three years, since that province was, from 1887-88, no longer treated as Transfrontier and does not, therefore, appear in the returns from which the table is compiled for the years 1887 to 1890.

Years.		Imports in Cwt. Gross.	Imports in Cwt. Total.	Exports in Cwt. Gross.	Exports in Cwt. Total.	Net Import of Rice and Paddy in Cwt.
1884-85	Rice . .	534,588		105,990		
	Upper Burma .	*28*		*927,080*		
	Paddy . .	865,604		34,952		
	Upper Burma .	*Nil.*	1,400,134	*897,462*	140,942	1,259,192
1885-86	Rice . .	809,446		143,775		
	Upper Burma .	*12*		*685,048*		
	Paddy . .	1,056,716		30,901		
	Upper Burma .	*602*	1,866,152	*1,074,531*	174,676	1,691,486
1886-87	Rice . .	882,547		164,305		
	Upper Burma .	*35*		*1,376,418*		
	Paddy . .	1,153,333		35,841		
	Upper Burma .	*1,529*	2,035,880	*578,086*	200,146	1,835,734
1887-88	Rice . .	1,054,332		117,863		
	Paddy . .	1,443,695	2,498,027	45,995	163,858	2,334,169
1888-89	Rice . .	502,636		129,806		
	Paddy . .	821,134	1,323,770	42,514	172,320	1,151,450
1889-90	Rice . .	854,245		94,942		
	Paddy . .	1,240,210	2,094,455	37,970	132,912	1,961,543
Averages for six years	Rice . .	772,972		126,113		
	Paddy . .	1,095,948	1,869,738	38,028	164,142	1,705,595
						or 23,87,833 mds.

ORYZA sativa.

Indian Trade

TRADE. Foreign. 499

Foreign.—It has been admitted in the provincial notes that an error is occasioned through treating rice and paddy conjointly. That error is, however, of no more serious consequence in the returns that have already been dealt with than it will be seen in those of the foreign trade. While, therefore, to keep up the comparison with the provincial notes, the two have been combined (in the tables of this chapter), they have been shown separately as well.

The following table may be given of the total and of the average exports and imports of rice and paddy by sea to and from foreign countries for the past six years. The period of six years has been chosen so as to show an equal number of years before and after the date of the trade with Upper Burma being no longer treated as Transfrontier.

Years.		Imports in Cwt.		Exports in Cwt.		Net Export of Rice and Paddy in Cwt.
		Gross.	Total.	Gross.	Total.	
1884-85	Rice .	4,071		21,702,136		
	Paddy .	2,081	6,152	349,396	22,051,532	22,045,380
1885-86	Rice .	1,271		27,813,844		
	Paddy .	446	1,617	408,751	28,222,595	28,220,978
1886-87	Rice .	1,898		26,460,500		
	Paddy .	31	1,929	418,772	26,879,272	26,877,343
1887-88	Rice .	1,484		28,148,706		
	Paddy .	1,153	2,637	385,351	28,534,057	28,531,420
1888-89	Rice .	8,319		22,768,229		
	Paddy .	3,833	12,152	376,412	23,144,641	23,132,489
1889-90	Rice .	29,557		26,774,251		
	Paddy .	15,524	45,081	324,655	27,098,906	27,053,825
Averages for six years.	Rice .	7,766		25,612,944		
	Paddy .	3,844	11,594	377,222	25,988,500	25,976,906 or 3,63,67,668 mds.

If now we add the average net import by land and deduct the average net export by sea, the balance from the estimated production of 26 million tons, or 728 million maunds, would have approximately been the amount of rice consumed by the people of India during 1888-89, since the reserve stock would probably be a constant or slightly increasing quantity. The averages in preference to the actual external traffic have been taken, since (as already explained) the commercial year of rice transactions does not correspond with the financial, under which all Government statistics are compiled. The rice thus consumed in India, expressed to the population (253,890,000) would be about 0·3 of a seer or, say, ⅝ths of a ℔ per head daily. This is, however, very probably in excess of the actual consumption, since a considerable quantity of the grain is used up in feeding cattle and horses, and a still larger quantity lost through unavoidable wastage or through the perishable nature of the grain. A similar reduction of daily consumption would also have to be made for the increase of population. An average per head of population for all India is, moreover, a misleading figure, since there are large tracts where rice might almost be said to be unknown even as an agricultural crop or, at all events, tracts where the people certainly never eat any rice. On the other hand, even in rice-eating provinces diet is always largely supplemented by other articles of food, the proportion increasing with poverty and luxury, for the poor have to eke out existence with cheaper materials, and the rich, and especially the rich non-Hindu populations, consume a consider-

TRADE.

able amount of more expensive and even animal ingredients of diet. The Famine Commissioners, while dealing with a smaller area and less population than the foregoing chapters have aimed at embracing, stated that, for the population of 166 millions, the outturn of food-grains might be put at 51½ million tons, and the actual requirements of the people at 47 million tons, thus leaving an available surplus of 5 million tons for export. In the table given above it will be found the average net export of rice by sea to foreign countries for the past six years has been 25,976,906 cwt. or 1,298,845 tons, and the corresponding figures for wheat may be here shown, *viz.*, 17,333,500 cwt., 866,675 tons. But an important consideration must be here alluded to: Burma is the chief country concerned in India's export trade in rice. Out of a total cultivated area of 5,673,542 acres, it has been shown that 4,067,606 were under rice. That area has been estimated to have produced 8,51,03,130 maunds or, say, 3,039,397 tons of rice. Taking the average net export of rice from Burma for the past nine years (in preference to any one of the recently disturbed annual records of transactions), Burma would appear to have exported 3,95,10,308 maunds, or a little over than 46 per cent. of its total production. Burma, therefore, expressly cultivates rice as an article of trade, and it is accordingly slightly misleading to deduct the exports from India (proper) and Burma from an estimate of the production in these two countries conjointly. A relatively far larger amount of rice remains with the people of India (proper) than has been shown. In fact Burma might practically be viewed as taking even a larger share in the foreign trade than the published figures manifest, for by coastwise it, to some extent, restores to the provinces of India the amounts *they* give to foreign countries.

Available Surplus. **500**

The following table is given by **Mr. J. E. O'Conor**, in his *Review of the Trade of India* for 1889-90, in order to exhibit the share taken by each province in the foreign exports of rice:—

Exports of Rice (husked) in cwt. (000's omitted).

Year.	Burma.	Bengal.	Madras.	Bombay.	Sind	Total.	Value in R (000's omitted).
1880-81 .	16,730	6,717	2,363	927	32	26,769	89,717
1881-82 .	16,690	7,617	1,549	614	49	28,519	82,496
1882-83 .	21,249	7,838	1,319	552	71	31,029	84,401
1883-84 .	16,994	7,394	1,843	521	80	26,832	83,289
1884-85 .	13,507	6,035	1,403	677	80	21,702	71,228
1885-86 .	19,084	6,879	1,181	521	149	27,814	91,672
1886-87 .	18,216	5,902	1,564	639	139	26,460	87,648
1887-88 .	17,879	7,996	1,438	764	72	28,149	92,251
1888-89 .	14,205	6,417	1,538	589	19	22,768	78,453
1889-90 .	18,259	5,992	1,654	799	70	26,774	100,473
Average for 10 years.	17,491	6,878	1,585	660	76	26,681	86,162

Commenting on the above table **Mr. J. E. O'Conor** remarks:—"A glance at the quantities exported in the last ten years shows that this trade does not make much progress. It is in fact, as has often been remarked before, no longer practically an Indian monopoly but, rather, is subject to a very keen competition."

"From Siam and Cochin-China alone the exports of rice are not much less than half those of all India. Then there is the rice from Japan and Italy, and we must also reckon with the competition of other grains which compete with Indian rice for distillation."

ORYZA sativa.

Exports of Indian

TRADE.

"The countries to which Indian rice was exported last year are these—

Exports.

		Cwt. (000's omitted).
Europe—		
United Kingdom		5,177
Malta		953
France		207
Egypt		7,398
Germany		176
Other countries		66
	TOTAL	13,977
Asia—		
Ceylon		3,269
Straits		3,916
Arabia and Aden		1,191
Persia and other countries		346
	TOTAL	8,722
Africa—		
Mauritius, Réunion, and East Coast		2,014
Other countries		256
	TOTAL	2,270
America—		
West Indies		269
South America		1,426
Canada		53
	TOTAL	1,748
Australia		56

"Thus we have 52 per cent. for Europe and 32½ per cent. for Asia, leaving about 15 per cent. for the coolies of Mauritius and Réunion, the West Indies and South America, and for Australia. The European demand was slightly smaller last year than in either of the two preceding years when it was equal to 55 and 53 per cent. of the whole respectively. It is possible that the present year may again show a slightly smaller relative demand for Europe in consequence of the Italian Government having raised the duty on rough (cargo) rice closer to the level of the duty on cleaned rice, and thus interfered with an active re-export trade carried on by Italian millers with great profit from State bounties.

"It is practically almost impossible to ascertain the ultimate destination of a great part of our rice in Europe. It will be observed that, of the total quantity assigned to Europe in the foregoing table, say 14 million cwt., nearly 8½ million cwt., are placed against Egypt and Malta. In the first of these cases the rice is shipped to Port Said for orders to be received there by the portmaster as to port of destination; in the second case, the rice is stored for reshipment to other Mediterranean ports. The system obscures the effects which changes of fiscal legislation in Italy, Turkey and Austria, the States lying between Turkey and Russia, or in other countries, may have upon the direction and extent of our rice trade with Europe."

Commercial Forms, &c. 501

THE PURPOSES FOR WHICH RICE IS EXPORTED.

Commercial Forms, Prices, &c.

The purposes for which rice is exported are generally stated to be three-fold, for *food*, for *starch*, and for *distillation*. The relative amounts used for each of these purposes is, however, a subject regarding which the greatest possible diversity of opinion prevails. The following passages from **Mr. O'Conor's** Trade Reviews exhibit some of the chief opinions that have been advanced :—" Of the Burma rice, the greater part is used for distillation or for conversion into starch. It is a thick, coarse grain, which, when boiled, is repulsive in appearance to those who are not accustomed to it, nor is its flavour equal to that of Bengal rice of the better qualities. 'Table' rice, as it is called, exported from Bengal, is the kind of Indian rice which is chiefly used for food in Europe. We have no accurate knowledge of the ultimate destination of a large part of the rice shipped from India and Burma, nominally to the United Kingdom. From Burma especially it is the practice for ships loaded with rice to clear for channel ports, such as Falmouth, for orders, on receiving which they proceed to their destination, which, as often as not, is some port in the northern seas of Europe, this rice being very extensively employed in distillation in Holland and Germany. Much of the rice which is recorded as carried hence to Egypt (*i.e.*, Port Said), Malta, Italy, and Gibraltar, similarly goes thither only for orders as to its ultimate destination of which we remain ignorant. However, we know that the bulk of the rice sent to Europe is consumed in England, Germany, Holland, and Belgium, and France. A good deal of that which is imported into England is re-exported. Thus, in 1876, the imports of rice into England were about 6⅛ million cwt., and the exports thence were about 3⅓ million, about half of which was sent to the Continent (Russia, Austria, Germany, France, Belgium, Portugal, Italy, Turkey), little less than a third of the whole exports to Spanish West Indies, and smaller quantities to the United States and elsewhere." "The consumption of Europe is practically met by Indian rice, supplemented by the rice of Northern Italy (Piedmont and Lombardy), Spain, and Egypt, and to a smaller degree by the rice of Siam and Cochin-China, Japan and Java. Very little rice is received in Europe from the United States, and this country cannot be regarded as competing with India. Carolina rice is very much superior to any other quality of rice known in commerce, and it fetches more than double the price of the best Bengal rice, but its fineness and high price operate against its consumption except in the preparation of confectionery and puddings and such like luxuries; hence the consumption is limited." "Broadly stated, the chief use of *half of our exports of rice is for conversion into liquor* in Europe, of which some comes back to India for consumption as cheap gin and brandy. Some portion is made into starch, of which a good deal comes back to India as size in cotton goods, and some is eaten by Europeans. Of the remainder of our exports, the greater part is eaten by Asiatics in the immediate vicinity of India or in colonial possessions in Africa and America" (*An. Rev. of Trade, 1878-79, p. 24*).

TRADE: Purposes for which Rice is Exported.

Burma Rice used for Distillation or Starch. 502

Bengal Rice used for Food. 503

Messrs. Fraser & Co., in their Annual Report for 1882, allude to the trade in rice for distillation as follows :—" There was yet an outlet, one last resource, to which the eyes of all instinctively turned, and in which centred the most ambitious hopes of speculators, and this was the sudden great demand for distillation. It was even predicted that something like 100,000 tons would be absorbed by this imagined illimitable source of consumption. Yet how soon were these expectations to be dispelled. It was not until the middle of July when the use of rice for purposes of distillation in large quantities first engaged universal attention. The high and increasing price of maize prevented this cereal from being employed to any great extent with advantage, and distillers, anxious to find some suitable grain to take the place of this, their favourite article,

TRADE: Purposes for which Rice is Exported.

Distillation.

turned to rice as most likely to answer the purpose. Several cargoes were sold to Scotland and the Continent, thereby raising the current quotations, mostly for the lower varieties, about 3*d*. It was soon found, however, that the looked-for benefit, which was to spur on the market and assist speculators, would be very unimportant, and that absolutely nothing could save the gradual but persistent downfall of prices; added to which when the cargoes which had previously been bought by distillers came off coast, a few of these were also offered for sale, and disposed of to the highest bidders, the value of maize having by this time been considerably reduced, and preferred, even at a higher ratio, of its expensive substitute."

Mr. O'Conor in his Review for 1883-84 again returns to the question of the consumption of rice in distillation. Thus, of the exports of rice from India, he says:—"Of that portion which was sent to Europe it is not possible to say how much was intended for distillation, though it is known that there has been an enormous increase in the employment of rice for this purpose in recent years. Thus, the Commissioners of Inland Revenue in their report for the year 1883-84 observe, with reference to the production of spirit, that the most noteworthy circumstance of the year was the large increase in the use of the rice and corresponding displacement of grain. The consumption for this purpose in 1880 was only 13,352 cwt., in 1882 it was 142,101 cwt., and in 1883 it was as much as 229,292 cwt. In France, Germany, and Italy its use for the same purpose has become equally common. The British Consul at Genoa, writing this year, says that the distilleries of Genoa, Milan, and Naples worked double tides last year in anticipation of an increase in the spirit duty—which came into effect from July 1883—and that there were great importations of rice in consequence." The subject of the use of rice in distillation continues to be referred to from time to time, but no definite information as to the extent of the trade appears even as yet to have been determined. Thus, for example, **Mr. O'Conor** remarks in his report of 1888-89:—"The rice sent to Europe is more largely used for distillation and the manufacture of starch than for food."

Glutinous Rice. 504

COMMERCIAL FORMS.—The peculiar glutinous rice of Burma has been well known for many years. In an early part of this article reference has been made to **Rheede's** botanical account of it under the name of **Oryza glutinosa.** **Ainslie** speaks of it as "never used as bread, but commonly prepared as a sweetmeat" There would seem, however, to be little or nothing to justify its botanical separation from many of the other forms of **O. sativa,** but it is noteworthy in concluding these remarks that it is not only recognised commercially, but by some writers as botanically, distinct from the ordinary rices of Bengal which, from a trade point of view, are in Europe classed according to their merits as different qualities of "Patna" or "Table" rice. A form of Bengal rice said to be suitable for distillation is that recognised in the trade as *Cazla*. The bulk of the rice exported from Burma is returned as "Cargo rice" or as "Cleaned rice." The former is a mixture of one part paddy with five parts cleaned rice. But Burma rice is also referred to two trade classes "Ngatsaing" and "Ngayouk," the former generally fetching a slightly better price. It bore in 1880-81 R84 to R115 for 100 baskets and the latter R80 to R112, but these prices considerably declined until, in 1882-83, they stood at R57 to R84 and R57 to R82.

Patna. 505 **Cazla.** 506 **Cargo Rice.** 507 **Cleaned Rice.** 508 **Ngatsaing.** 509 **Ngayouk.** 510

Prices. 511

PRICES.—In recent reports the wholesale price is shown to have considerably risen, but so great are the almost daily fluctuations that it would serve no very useful purpose to quote here a long series of figures. A review of the fluctuations in price during the year 1888-89 will be found in the Report of the Trade and Navigation of Burma, pp. 19-22. During

TRADE: Purposes for which Rice is Exported.

Prices.

the early part of April arrivals of paddy were meagre and are said to have caused an advance of rates from R97 to R105 per hundred baskets. Later on supplies came in more freely and prices fell to R98 and R100 for *ngatsaing* and *ngayouk* qualities respectively. "White rice" about the same time sold for R241 to R250 per hundred baskets and "Cargo rice" at R190 to R194 per 100 baskets.

The Burma basket, as a measure of grain, was formerly equivalent to $1\frac{1}{8}$ bushels or expressed to liquids, equivalent to 9 gallons. Within recent years there is reason to believe that the keen competition of exporters has resulted in the size of the Burma basket having decreased, so that the figures of trade recorded under that standard are not of great value. Taking the bazár retail prices, therefore, as a more trustworthy means of discovering the fluctuations of the trade, prices of rice may be said to have been greatly adjusted with the opening out of railway communication. They have fallen in non-rice-producing provinces and correspondingly risen in the chief rice areas. The following table of average prices for the provinces of India from 1864 to 1889 may here be given:—

Quantity sold per Rupee in sers and decimals of a ser of 80 tolas.

	Average for five years ending					Actuals for			
	1865.	1870.	1875.	1880.	1885.	1886.	1887.	1888.	1889.
Burma—									
Tenasserim . .	23·07	22·83	19·36	16·06	17·27	15·88	16·00	12·86	12·18
Pegu (Deltaic) .	15·41	14·13	16·90	12·91	14·19	14·59	14·28	12·71	13·18
Pegu (Inland) .	13·05	13·29	20·76	13·34	14·77	13·37	13·37	11·70	13·03
Upper Burma .	Not shown.						10·25	10·58	12·18
Arakan . .	19·23	21·37	21·97	14·29	15·66	15·23	18·00	13·58	15·35
AVERAGE FOR THE PROVINCE .	17·69	17·90	19·74	14·15	15·49	14·77	14·38	12·28	13·18
Assam—									
Surma . . .	30·61	18·41	23·47	15·82	21·56	14·24	16·61	20·78	15·66
Brahmaputra .	18·00	17·00	16·51	12·78	15·61	15·22	16·64	17·18	14·83
AVERAGE FOR THE PROVINCE .	24·30	17·70	19·99	14·30	18·58	14·73	16·62	18·98	15·24
Bengal—									
Eastern . .	27·92	21·03	21·42	15·45	21·17	16·53	19·18	19·16	13·92
Deltaic . .	24·44	18·61	18·25	15·13	18·46	16·94	20·05	18·90	14·04
Central . .	27·85	22·44	21·15	18·44	21·61	21·06	23·03	20·43	15·31
Northern . .	28·51	23·73	23·26	19·82	20·23	20·40	22·20	20·53	15·11
Orissa . . .	30·42	22·25	28·11	18·87	25·05	19·77	22·09	20·99	17·08
Chota Nagpur .	26·22	21·05	23·40	23·50	24·42	25·18	23·77	18·00	17·76
Behar, South .	26·47	22·50	18·62	16·57	17·53	17·54	18·96	16·76	14·25
Behar, North .	21·74	19·70	19·96	16·88	18·60	18·29	19·78	17·82	13·95
AVERAGE FOR THE PROVINCE .	26·69	21·41	21·82	18·08	20·88	19·41	21·13	19·07	15·18
N.-W. Provinces—									
Eastern . .	16·04	13·84	15·14	14·91	15·77	15·13	15·05	14·25	12·11
Central . .	14·89	14·08	14·33	13·49	13·74	13·40	12·90	11·78	11·58
Western . .	13·02	11·26	13·42	12·33	12·91	12·75	12·29	11·50	10·80
Sub-Montane .	19·71	15·65	15·19	13·91	14·85	14·71	13·98	13·46	13·69
AVERAGE FOR THE PROVINCE .	15·91	13·71	14·52	13·66	14·32	13·99	13·55	12·77	12·04

ORYZA sativa.

TRADE.
Prices.

Prices of Rice in the Provinces.

	Average for five years ending					Actuals for			
	1865.	1870.	1875.	1880.	1885.	1886.	1887.	1888.	1889.
OUDH—									
Southern . .	17·45	13·57	16·71	14·94	16·08	16·59	14·53	14·09	13·64
Northern . .	20·87	16·31	17·22	15·12	17·02	17·20	16·24	14 84	13·83
AVERAGE FOR THE PROVINCE	19·16	14·94	16·96	15·03	16·55	16·89	15·38	14·46	13·73
RAJPUTANA—									
Eastern	Not	shown.	9·15	9·26	10·40	10·13	10·06	9·96	9·39
Western	Not	shown.	8·37	6·56	7·70	8·83	9·40	8·18	7·85
AVERAGE FOR THE PROVINCE .	...	...	8·76	7·91	9·05	9·48	9·73	9·07	8·62
Central India . .	Not	shown.	8 49	7·75	9·80	9·63	9·73	9·57	8·71
PUNJAB—									
Southern .	11·05	8·57	10·95	8·96	10·08	10·00	9·92	10·00	9·92
Central . .	14·55	10·69	12·02	11·04	11·55	11·67	12·01	11·11	10·54
Sub-Montane .	15·26	11·48	12·26	11·15	12·45	11·81	11·76	10·91	10·88
North-Western .	11·66	9·21	10·48	9·35	11·38	12·19	10·50	10·46	11·09
Western . .	10·30	8·16	8·95	8·06	10·32	10·33	9·81	9·86	9·38
AVERAGE FOR THE PROVINCE .	12·56	9·62	10·93	9·91	11·15	11·20	10·80	10·47	10·36
Sind and Baluchistan	15·12	12·73	14·44	11·22	14·78	14·26	13·87	11·48	9·60
BOMBAY—									
Konkan . .	10·47	9·45	12·60	9·97	12·79	11·81	11·49	12·03	10·78
Deccan . .	11·27	9·43	11·77	9·12	11·69	10·53	10·80	10·78	10·42
Khandesh . .	8·86	6·93	10·51	8·67	11·36	10·21	10·33	9·90	9·56
Guzerat . .	7·92	6·87	10·56	8·75	11·40	10·05	10·39	9·82	8·77
Kattywar . .	Not	shown.	9·51	8·52	9·95	9·08	9·01	8·30	7·97
AVERAGE FOR THE PROVINCE .	9·63	8·17	10·99	9·00	11·43	10·33	10·40	10·16	9·50
CENTRAL PROVINCES—									
Western . .	13·80	9·64	13·57	11·07	13·53	11·67	11·04	11·03	10·44
Central . .	19·42	13·55	16·84	14·79	16·94	13·69	12·74	13·56	12·43
Eastern . .	51·87	32·29	39·71	30·09	39·49	24·04	19·58	17·62	15·99
	28·36	18·49	23·37	18·65	26 68	16 46	14·45	14·07	12·95
Berar . . .	10·92	8·97	11·48	9·40	11·12	18·82	9·83	9·80	9·10
Nizam's Territories .	12·21	9·87	11·40	8·96	10·34	10·66	10·55	10·01	9·35
MADRAS—									
Malabar Coast .	13·15	11·83	13·86	10·84	14·36	14·05	15·06	14·60	12·44
South (Central) .	11·70	9·77	15 44	10·12	14·77	13·08	14·60	14·27	13·38
Central . .	11·30	9·85	13·87	9·95	12·93	12·45	14·33	14·08	13·12
East Coast, North .	17·56	15·91	19·39	12·60	15·38	12·99	13·79	14·06	12·21
East Coast, Central	14·46	13·08	18·13	12·29	15·77	14·38	15·07	15·07	13·39
East Coast, South .	14·09	11·95	16·10	10·96	15·72	15·37	14·81	14·94	13·38
Southern . .	11·58	9·83	13·52	10·57	14·75	14·56	13·46	13·62	12·78
AVERAGE FOR THE PROVINCE .	13·40	11·74	15·76	11·05	14·81	13·84	14·45	14·38	13·03
Mysore . . .	13·37	9·65	12·68	8·94	11·47	10·64	12·23	12·21	10·80
Coorg . .	Not	shown.	15·93	11·24	16·64	15·31	16·31	15·52	14·21
AVERAGE FOR ALL INDIA .	16·87	13·45	14·83	11·95	14·56	13·77	13·33	12·76	11·60

TRADE. Duty. 512

EXPORT DUTY AND INCIDENCE OF LAND REVENUE.—Rice and Opium are the chief, indeed the only, articles which bear an export duty before leaving India. In the case of the former this is estimated to be about 15 per cent. In the Administration Report of Burma for the year 1887-88, it is stated that the cultivators received on the average £7 per 100 baskets of paddy. "The average incidence of revenue on rice lands was 3*s*. 3*d*. the acre and the amount paid in revenue was about one-thirteenth of the value of the whole yield." The imposition of an export duty has within recent years been urged as one of the chief reasons of rice not becoming a more important article of export from India than it is at the present day. Thus, **Mr. J. E. O'Conor** wrote in the Annual Review of Trade for 1882-83 (*p. 78*): "For food our rice has to compete with European rice (that of Lombardy in particular), and with the rice of other Asiatic countries as well as with Madagascar rice and with rice produced in the Southern States of the Union. Thus, whereas some twenty years ago we did a fairly large business in rice with China, that trade has almost ceased to exist, Cochin-China and Siam as well as Japan having driven our rice out of the market. The course of prices in Europe too shows the extent to which competition with Indian rice is carried. On the whole, it is clear as anything can be that the argument of those who maintained the propriety of the export duty on rice, because India had a monopoly of the trade, is now, whatever it may at one time have been, wholly unsustainable. I do not myself believe that it ever was fairly sustainable. And all those who have given close attention to the facts of the question must have learnt with satisfaction from the last Financial Statement that the duty is to be retained only until the condition of the Indian exchequer will permit of its removal without inconvenience. The revenue from the duty is no doubt large, but the greater the sum the greater is the burden on one of the largest trades of the country and the staple industry of an important province."

The duty still continues to be levied, and during the past five years the amount collected has varied from R61,26,726 to R75,64,985 or, say, £500,000 per annum.

MILLS AND MILLING.

MILLS. 513

An important feature of the Indian rice trade is the position which power-mills have attained. There are now at work 45 rice-mills in Burma and 2 in Bengal. These employ 5,122 permanent and 27,617 temporary hands and have a yearly outturn of 18,661,974 cwt., valued at R6,36,46,062.

These mills in Burma had their origin in the very high cost of labour in that province, and the disinclination of the Burman agriculturist to undertake any work that he can avoid. In Bengal the rice crop is, as a rule, husked, whether it is required for local consumption or for export, by the members of the agriculturist's family in the *dhenki*, worked by hand or rather by the foot, before it is put on the market. In Burma the grower takes his unhusked rice to the market and leaves to the buyer the trouble of husking it. It soon became evident to exporters that, if a large trade was to be done it was impossible to export unhusked rice, and that husking mills were a necessity. These mills have now been in existence for many years, the oldest being more than thirty years old. They give employment to many thousands of Madras coolies who go across the Bay to the rice ports for the rice season, which may be said to last from December to May, earning high wages (from 12 annas to one rupee daily) during the season. The mills were for a long time greatly handicapped by the cost of fuel, imported coal (from England chiefly) being used. The cost was very great and the question of the disposal of the husks became serious, for the only method of dealing with them seemed

MILLS AND MILLING. to be to throw them into the rivers and creeks. Any one travelling in Burma in those days had constantly before his eyes great floating masses of rice husks, covering the streams for miles and so thick as to be a real impediment to navigation. When decomposition set in, the smell was very offensive. Some twelve years ago, however, it was discovered that a good way of disposing of the husk was to burn it in the mills, and now all the mill furnaces are fed with the husks. The cost of coal also is lower, and as Bengal coal is now being shipped to Burma in large quantities, the rice mills are working under favourable conditions. Every year the proportion of white rice (rice which has been entirely cleaned in the mills) grows larger, and the proportion of cargo rice (which is partly unhusked) grows smaller. Of the total quantity of rice now exported from Burma, about 70 per cent. is "white" rice.

The process of milling involves the production of a good deal of bran
Dust. 514 or dust. This stuff was formerly thrown away like the husk, but is now sold to Chinamen who use it for feeding pigs and cattle both in Burma and the Straits.

(*W. R. Clark.*)

515 **OSBECKIA**, *Linn.; Gen. Pl., I., 744.*

A genus of Melastomaceous plants, twenty-nine species of which are found in the Indian Peninsula. They are mostly herbs, sometimes shrubs, and are worth cultivating on account of the beauty of their flowers. Otherwise they are of little economic value.

516 **Osbeckia crinita**, *Benth; Fl. Br. Ind., II., 517;* MELASTOMACEÆ.

Syn.—O. STELLATA, *Don;* O. STELLATA, var. β., *DC.*

Vern.—*Number*, LEPCHA.

References.—*Don. Prodr., 221 partly; Gamble, Man. Timb., 199.*

Habitat.—A shrub of the East Himálaya and Khásia hills, found at altitudes between 4,000 and 8,000 feet: it is common about Darjíling.

TIMBER. 517 **Structure of the Wood.**—Light brown and moderately hard.

OSMANTHUS, *Lour.; Gen. Pl., II., 677.*

518 **Osmanthus fragrans**, *Lour.; Fl. Br. Ind., III., 606;* OLEACEÆ.

Syn.—OLEA FRAGRANS, *Thunb.;* O. ACUMINATA, *Wall.*

Vern.—*Shilling, silang*, KUMAON; *Tungrung*, LEPCHA.

References.—*Roxb., Fl. Ind., Ed. C.B.C., 35; Brandis, For. Fl., 309; Gamble, Man. Timb., 257; Lisboa, U. Pl. Bomb., 223; Kew Off. Guide to Bot. Gardens & Arboretum, 144.*

Habitat.—A small tree found in the Temperate Himálaya from Garhwál to Sikkim at altitudes between 4,000 and 7,000 feet. Sometimes wild, but more often cultivated for the sake of its sweet-scented flowers. It is distributed to China and Japan where it is widely cultivated.

DOMESTIC. Flowers. 519 **Domestic.**—The FLOWERS are used in China to flavour tea, in Kumáon to keep insects away from clothes.

520 **O. sp.**

Vern.—*Silingi*, NEPAL; *Chashing*, BHUTIA.

Reference.—*Gamble, Man. Timb., 257.*

Habitat.—A small tree with opposite coriaceous leaves, found at Tonglo near Darjíling at an altitude of 10,000 feet.

TIMBER. 521 **Structure of the Wood.**—White, hard, close-grained, seasons well, mottled on vertical section. Weight 53lb per cubic foot.

OSTODES, *Bl.; Gen. Pl., III., 299.*

522 **Ostodes paniculata**, *Bl.; Fl. Br. Ind., V., 400;* EUPHORBIACEÆ.

Vern.—*Bepari*, NEPAL; *Palok*, LEPCHA; *Walkakúna*, SING.

References.—*Kurz, For. Fl. Burm., II., 403; Gamble, Man. Timb., 365.*

Habitat.—A large evergreen tree, found in the forests of Sikkim at altitudes between 2,000 and 6,000 feet. It is common also in the forests of the Khásia hills and of Martaban, and is distributed to Java.

Gum.—It yields a gum, which is used as size in paper manufacture (*Gamble*). **GUM.** 523

Structure of the Wood.—White and soft. Weight 26lb per cubic foot. **TIMBER.** 524

OSYRIS, *Linn.; Gen. Pl., III., 227.*

Osyris arborea, *Wall.; Fl. Br. Ind., V., 232; Wight, Ic., t. 1853;* [SANTALACEÆ. 525

Syn.—O. WIGHTIANA, *Wall.*; O. NEPALENSIS, *Don.*

Vern.—*Bakardharra, bakarja,* KUMAON; *Popli,* BELGAUM; *Jhuri,* NEPAL.

References.—*DC., Prodr., XIV., 633; Brandis, For. Fl., 399; Gamble, Man. Timb., 320; Royle, Ill. Him. Bot., 322; Grah., Cat. Bomb. Pl., 177; Dalz. & Gibs., Bomb. Fl., 223; Watt, Sel. from Records Gov. Ind., Rev. & Agri. Dept., I., 59; Gazetteer, N.-W. Prov., X., 317; Ind. Forester, XI., 369; Agri.-Horti. Soc., Ind., Journ., IV., 260.*

Habitat.—A glabrous shrub or small tree occurring in the Sub-tropical Himálaya from Simla to Bhután, at altitudes up to 7,000 feet. It is found also on hills in the Deccan Peninsula, in the Central Provinces, and in Ceylon.

Medicine.—The infusion of the LEAVES has powerful emetic qualities. **MEDICINE.** Leaves. 526

Domestic.—Dr. **Royle** mentions that in Kumáon the LEAVES of **Osyris arborea** are used as a substitute for tea, and this is probably the Green Tea of Bischar (=Bisháhr) which **Moorcroft** (*Travels, I., 352*) describes as being imported into Ladak under the name of Maun or Bischar Tea, the produce of a shrub growing on a dry soil, especially about Jhagul between Rampur and Seran. The leaves are gathered from July to November, and, after infusion in hot water, are rubbed and dried in the sun. The first infusion is reddish in colour and is reckoned nauseating: the second, which is used, is yellowish green. Writing of the use of these leaves as a substitute for tea, **Dr. Watt**, in his Selections from the Records of the Government of India, Revenue and Agricultural Department, says:—"The leaves smell remarkably like tea, when specially prepared; but unfortunately the infusion has powerful emetic properties which require long usage to conquer. **Dr. Royle** suggested that experiments should be made in the cultivation of the plant, in order to discover if the emetic property could be removed by careful cultivation. The discovery of tea proper in Assam, and the greatly extended cultivation of that plant, have left the matter of **Osyris** tea in the position in which it was at the beginning of the present century, when it first attracted the attention of the public. **DOMESTIC.** Leaves. 527

The plant is very common around Simla. It is closely allied to sandalwood, but seems to possess no properties that would justify its cultivation, since tea can be produced quite as cheap. [Pt. II.

Otaheite apple, see **Spondias dulcis,** *Willd.*; ANACARDIACEÆ.; Vol. VI.,

OTTERS; *Blanford, Mammalia in Fauna of Br. Ind., Pt. I., 181.*

Otters. 528

The Otters, which belong to the Sub-family **LUTRINÆ,** constitute the last sub-division of the MUSTELIDÆ. They form a group, the different species of which are externally very difficult of discrimination, but are readily determined by the shape of their skulls.

529 **Lutra aureobrunnea,** *Hodgson; Blanford, Mammalia in Fauna of Br. [Ind., Pt. I., 186.;* MUSTELIDÆ.

THE HIMALAYAN OTTER.

No other naturalist except **Hodgson** has met with this species, but it is evident from his collections that, besides **L. vulgaris, L. ellioti,** and **L. leptonyx,** another otter is found in Nepál (*Blanford*). It is much smaller and the skin is coarser than in the other species.

Habitat.—Found by **Hodgson** in Nepál.

530 **L. ellioti,** *Anderson; Blanford, Mammalia in Fauna of Br. Ind., Pt. I., [185.*

THE SMOOTH INDIAN OTTER.

Syn.—L. MONTICOLA, *Hodgson;* L. TARAIYENSIS, *Blyth;* L. NAIR, *Cantor, Blyth, partim.*

Habitat.—This species also is found throughout India. It is distinguished from **L. vulgaris** by the smaller size of the animal, the greater comparative breadth of its skull, and its coarser shorter hair.

DOMESTIC. 531 **Domestic.**—Probably this species also is used by fishermen in a similar manner to the common otter (*q.v.*). Its fur is not so valuable.

532 **L. leptonyx,** *Horsfield; Blanford, Mammalia in Fauna of Br. Ind., [Pt. I., 187.*

THE CLAWLESS OTTER.

Syn.—L. INDIGITATA, *Hodgson;* AONYX LEPTONYX, *Cantor.*

The skull is much shorter than in other Indian forms. The claws are rudimentary, sometimes altogether wanting; the third and fourth toes on all feet considerably longer than the others.

Habitat.—This species inhabits the Himálaya, generally at low elevations; it is found also in Lower Bengal (being common near Calcutta) in Assam, Burma, and at low elevations on the Nilgiris in Madras Presidency.

DOMESTIC. 533 **Domestic.**—This animal, together with the other species, is kept tame by the fishermen in Lower Bengal.

534 **L. vulgaris,** *Erxleben; Blanford, Mammalia in Fauna of Br. Ind., [Pt. I., 182.*

THE COMMON OTTER.

Syn.—L. NAIR, *F. Cuv.; Jerdon partim;* L. INDICA, *Gray.*

The otter found commonly over India is usually known as LUTRA NAIR, but there seems to be no constant characters by which this can be distinguished from the LUTRA VULGARIS of Europe.

Vern.—*Ud, ud bilão, pani kutta,* HIND.; *Sag-i-al,* PB.; *Lad, pan-manjar, jal manjar, jal-manus,* MAR.; *Nirunai,* TAM.; *Niru-kuka,* TEL.; *Nirnai,* KAN.

References.—*Jerdon, Mam. Ind., 86; Sterndale, Mam. Ind., 150; Forbes Watson, Ind. Survey of India, 382; Encyc. Brit., XVIII., 69.*

Habitat.—The common otter is found throughout the Palæ-arctic region, extending into the North-West Himálaya. The Indian form commonly known as **L. nair** appears to occur over the whole of India and Ceylon.

DOMESTIC. 535 **Domestic.**—Otters are easily tamed when captured young. **L. vulgaris** is very commonly semi-domesticated by the fishermen in various parts of India, who employ the members of this and other species to drive fish into their nets. On the Indus they are found specially serviceable in porpoise catching. The fishing otters are kept on the prow of the boat and slipped like dogs when required.

Skin. 536 The SKIN of this species is the one most valued by furriers. It is rufous and beautifully soft. In India, **Baden Powell** says, it is much prized for making caps and for *poshtíns* or fur jackets.

OUGEINIA, *Benth.; Gen. Pl., I., 518.* 537

[*t. 391;* LEGUMINOSÆ.

Ougeinia dalbergioides, *Benth.; Fl. Br. Ind., II., 161; Wight, Ic.,*

Syn.—DALBERGIA OUGEINENSIS, *Roxb.*

Vern.—*Sándan, asainda, tinnas, timsa,* HIND.; *Tinis,* BENG.; *Bandhona,* URIYA; *Ruta,* KOL.; *Rot,* SANTAL; *Sandan-pipli,* NEPAL; *Tewsa,* BHIL; *Sér, shermana, tinsai,* GOND; *Rutok,* KURKU; *Shúnjan, pánan, tinsa, sáldan,* N.-W. P.; *Telus, sannan, sándan,* PB.; *Tunnia,* BANSWARA; *Tinsa, karimattal, kala phalas, tinnas,* C. P.; *Tiwas, tunus, tunnia, telas, sandan, timsa,* BOMB.; *Kala palas, tewas, tanach,* MAR.; *Dargu, tella motuku, nemmi chettu, manda motuku,* TEL.; *Kari mutal,* KAN.; *Tinisa sejanduna,* SANS.

References.—*Roxb., Fl. Ind., Ed. C.B.C., 532; Brandis, For. Fl., 146; Beddome, Fl. Sylv., t. 36; Gamble, Man. Timb., 119; Dalz. & Gibs., Bomb. Fl., 78; Stewart, Pb. Pl., 72; Rev. A. Campbell, Rept. Econ. Pl., Chutia Nagpur, No. 7513; Sir W. Elliot, Fl. Andhr., 111, 133, 177; U. C. Dutt, Mat. Med. Hindus, 321; Birdwood, Bomb. Prod., 328; Baden Powell, Pb. Pr., 577; Drury, U. Pl. Ind., 176; Atkinson, Him. Dist. (X., N.-W. P. Gaz.), 308; Useful Pl. Bomb. (XXV., Bomb. Gaz.), 58, 272, 278, 393; For. Adm. Rep., Chutia Nagpur, 1885, 6, 29; Man. Madras Adm., I., 313; Report on Gums & Resinous Substances of India, 24, 33, 35; Settlement Reports Central Provinces, Chindwara, 110; Seonee, 10; Baitool, 127; Gazetteers:—Bombay, VII., 35; XIII., 27; XV., 33; XVI., 18; Panjáb, Gurdaspur, 53; N.-W. P., I., 88; IV., lxx.; Ind. Forester, I., 275; III., 23; IV., 292, 322; VIII., 101, 114, 116, 129, 388, 412, 414, 416; X., 61, 222, 325; XI., 367; XII., 188 (xxii.); XIII., 120, 127; XIV., 147, 151; Balfour, Cyclop. Ind., I., 878.*

Habitat.—A moderate-sized deciduous tree, under certain circumstances gregarious, found chiefly in the intermediate zone of the sub-Himálayan tract from the Sutlej to the Tísta, ascending to 5,000 feet, but distributed also to Central India and the west coast.

Gum.—It yields an astringent red GUM, very similar to *Dragon's Blood.* GUM. 538

Medicine.—The BARK when incised furnishes a kino-like exudation, which is used in cases of dysentery and diarrhœa (*Lisboa*). According to Campbell (*Econ. Prod., Chutia Nagpur*), a decoction of the bark is given among the hill tribes, when the urine is high coloured. In the Central Provinces the bark is said to be used as a febrifuge. MEDICINE. Bark. 539

Fodder.—The LEAVES appear after the blossoms, and are in summer given as fodder to cattle, for which purpose the branches are lopped off. FODDER. Leaves. 540

Structure of the Wood.—Sapwood small, heartwood mottled, light-brown, sometimes reddish-brown, hard, close-grained, tough, and durable. It takes a beautiful polish. Weight 55 to 60℔ per cubic foot. It is a very valuable timber, specially those specimens that grow on the Godavery, but is comparatively rare in these parts. The tree, however, is a small one, and it is difficult to obtain planks over nine inches broad (*Brandis, Gamble, Balfour*). TIMBER. 541

Domestic.—The wood is used for agricultural implements, carriage poles, wheels, and furniture, also for building purposes (*Gamble*). The bark is astringent, and is employed to poison fish, for which purpose many trees are stripped of their bark (*Lisboa*). DOMESTIC. Wood. 542 Bark. 543

Ovis, see **Sheep**; Vol. VI., Pt. II.

OXALIS, *Linn.; Gen. Pl., I., 276.* 544

A genus of acid herbs, consisting of over 200 species, which are chiefly Tropical and Temperate South American and South African. Three species are indigenous to the Indian Peninsula; but practically only one of these is recognised as of economic value.

545 **Oxalis acetosella,** *Linn.; Fl. Br. Ind., I., 436;* GERANIACEÆ.

THE COMMON WOOD SORREL.

References.—*Stewart, Pb. Pl., 37; Agri.-Horti. Soc., Ind., Journ., XIV., 45; Gaz., N.-W. Prov., X., 307; Smith, Econ. Dict., 385; Kew Off. Guide to the Mus. of Ec. Bot., 16.*

Habitat.—An herb found all over the Temperate Himálaya from Kashmír to Sikkim, at altitudes between 8,000 and 12,000 feet, and distributed to Europe, North Asia, North Africa, and North America.

MEDICINE. 546 **Medicine.**—Although at one time this found a place in the London Pharmacopœia, yet in India no account appears to exist of any supposed medicinal virtues inherent in this species. In Europe, it was introduced into the Pharmacopœia as a refrigerant in fever, and as an antiscorbutic in scurvy, but has now fallen into disuse.

547 **O. corniculata,** *Linn.; Fl. Br. Ind., I., 436; Wight, Ic., t. 18.*

THE INDIAN SORREL.

Syn.—O. REPENS, *Thunb.;* O. PUSILLA, *Salisb.*

Vern.—*Anbóti, seh, a'mrulsák, amrul, chalmori,* HIND.; *A'mrulsák, umulbet, amrúl, omlóti, chuka-tripati,* BENG.; *Tandi chatomarak,* SANTAL; *Chengeri tenga,* ASSAM; *Amrul, amlika, ambuti,* N.-W. P.; *Amrúl, chalmori,* KUMAON; *Amlika, amrúl, surchi, trawuke, khatta mithá, chukha,* PB.; *Ambutí, bhui-sarpati,* BOMB.; *Bhin-sarpati, ambuti,* MAR.; *Anbóti-ki-bhaji, ambuti,* DEC.; *Puliyárai, puliyárai-kírai, paliakiri,* TAM.; *Puli-chintá, pulla-chanchali, pallachinta, anbóti-kura,* TEL.; *Pullam-purachi-sappu,* KAN.; *Poliyárala,* MALAY.; *Amlaloniká, changeri, amlika, ambashta, chúkrika, shúklika,* SANS.; *Hememdab, hemda, homadmad,* ARAB.

References.—*Roxb., Fl. Ind., Ed. C.B.C., 389; Voigt, Hort. Sub. Cal., 191; Stewart, Pb. Pl., 37; Sir W. Elliot, Fl. Andhr., 158, 159; O'Shaughnessy, Beng. Dispens., 257; Irvine, Mat. Med. Patna, 122; U. C. Dutt, Mat. Med. Hindus, 124, 290, 295; Dymock, Mat. Med. W. Ind., 2nd Ed., 121; Dymock, Warden & Hooper, Pharmacog. Ind., I., 246; Med. Top., Ajmere., 153; Birdwood, Bomb. Prod., 145; Baden Powell, Pb. Pr., 331; Drury, U. Pl. Ind., 3, 24; Atkinson, Him. Dist. X., N.-W. P. Gaz.), 307, 708, 744; Useful Pl. Bomb. XXV., Bomb. Gaz.), 148, 196; Gazetteers:—Bombay, XV., 428; N.-W. P., I., 79; IV., lxix.; Mysore and Coorg, I., 69; Agri.-Horti. Soc.:—Ind., Jour., XIV., 11; Journ. As. Soc., Pt. II., No. II., 1867, 81; Ind. Forester, III., 237.*

Habitat.—A very variable, caulescent weed, found abundantly throughout the warmer parts of India and Ceylon, ascending the Himálaya to 7,000 feet. It is found very frequently in cultivated places or near human dwellings. In distribution it is cosmopolitan.

MEDICINE. Leaves. 548 **Medicine.**—The LEAVES are considered cooling, refrigerant, stomachic, and antiscorbutic. Various preparations, in which this plant forms a principal ingredient, are much esteemed in the treatment of fevers, dysentery, and scurvy. A *ghrita* (*v.* **Ghí**, Vol. III., 495), prepared from the fresh juice of **Oxalis corniculata,** and the leaves of the same plant reduced to a paste, together with *ghí* and curdled milk, is recommended by Hindu physicians as a useful medicine in diarrhœa, dysentery, prolapsus of the rectum, tympanitis, piles, and difficult micturition. The fresh juice of the leaves is given to relieve the intoxication produced by **Datura** (*U. C. Dutt*).

The fresh leaves made into a curry are said to improve the appetite and digestion of dyspeptic patients (*Moodeen Sheriff*). Bruised with or without water, they are formed into a poultice and applied over inflamed parts, by which means, **Moodeen Sheriff** says, "great cold is produced, and pain and other inflammatory symptoms are relieved."

In the Konkan the plant is rubbed down with water, boiled, and the juice of white onions added. This mixture is applied to the head in bilious headaches. **Baden Powell** and also **Atkinson** mention its almost universal use in the North-West Provinces and the Panjáb as a specific applied externally for the removal of warts. **Baden Powell** at the same time mentions the use of the juice as an application for the removal of opacities of the cornea. MEDICINE.

SPECIAL OPINIONS.—§ "The leaves are useful in dysentery of children" (*Assistant Surgeon N. L. Ghose, Bankipore*). "Juice of the leaves is used as an antiscorbutic" (*Surgeon A. C. Mukherji, Noakhaly*). "Antiscorbutic, stomachic; leaf and whole plant are used fresh as a salad" (*Apothecary T. Ward, Madanapalle, Allahabad*). "The juice of the leaves is used in dysentery, also as dentifrice and to improve foul breath" (*Civil Surgeon J. H. Thornton, B.A., M.B., Monghyr*). "Expressed juice of the leaves made into a sherbet with a little sugar, often prescribed in dysentery to allay thirst" (*Civil Surgeon S. M. Shircore, Moorshedabad*). "Cooling and useful in biliousness" (*Assistant Surgeon S. C. Bhattacharji, Chanda, Central Provinces*).

Food.—In some parts of the country it is eaten both raw as a salad and cooked as a POT-HERB, particularly in times of drought when its refrigerant qualities are said to be very grateful to the consumers. In Madras, it is cultivated as a vegetable and sold to the natives especially Muhammadans, who are very fond of it (*Moodeen Sheriff*). At Poona, during the famine of 1877-78, the seeds were eagerly sought after and eaten by the poorer classes. FOOD. ot-herb. 549

Domestic.—The JUICE is useful in removing stains of iron mould. DOMESTIC. Juice. 550

OXEN, BUFFALOES & ALLIED SPECIES OF BOVINE ANIMALS. 551

The sub-family BOVINÆ, one of the great divisions of the tribe of Ruminants, is broken up by naturalists into three groups:—(1) the BISONTINE to which belongs the Yak (**Poephagus grunniens**) of Central Asia; (2), the TAURINE, sub-divided by Blyth into (α) *Zebus*, a genus represented by the humped domestic cattle of India, (β) *Taurus*, the humpless cattle with cylindrical horns, represented by the cattle of Europe, (γ) *Gavaeus*, humpless cattle with somewhat flattened horns, represented by the Gaur of India and South-Eastern Asia; (3), the great group of **Bovinæ** known as the BUBALINE, which is represented by the varieties of the domestic buffalo of India (**Bubalus bos**) and the wild buffalo (**Bubalus arni**).

Here, however, it has been thought expedient to follow the usual practice in this work, namely, to describe shortly the different species of BOVINÆ indigenous to India, in their alphabetical order, and to conclude with a general article on the subject of Oxen.

Bos indicus.

THE INDIAN ZEBU OR HUMPED OX. 552

References.—*Jerdon, Mam. Ind., 301; Murray, Vertebrate Zool. Sind, 55; Encycl. Brit., V., 244.*

Habitat.—The genus *Zebu*, containing the humped cattle of India, comprises not only the several varieties of domestic cattle, but also others that have run wild. In the really feral state, the Indian Zebu is probably extinct. On the seacoast, near Nellore in the Carnatic, there is a herd of cattle that have run wild for many years; but they are merely escapes from the domestic state, and although in several other places also there are small herds of apparently wild cattle, no specimens of the real wild Zebu seem now to exist. Humped cattle are found in the greatest perfection in India, but they extend eastwards to Japan, and westward to the Niger in Africa.

Characters.—Zebus differ from the European domestic cattle not only Characters. 553

INDIAN OXEN.

in the hump-like fleshy protuberance on the shoulders, but also in the number of their sacral vertebræ, and in the character of their voice which is described as "a hoarse guttural grunt or half cough instead of the ringing bellow of the European species, in which both lungs and throat play an important part" (*Wallace*).

DOMESTIC & SACRED.

554 **Domestic & Sacred.**—Cattle in India comprise the chief wealth of the agricultural classes, and perform most of the agricultural operations of the country. With them, the *rayat* ploughs his land, raises water for irrigation when wells form the source of the water-supply, and conveys his produce to market in carts or in panniers on their backs, where the roads are unsuitable for wheeled vehicles. They are thus valued not so much for their flesh-producing qualities, as for their strength and activity, which enables the Indian *rayat* to perform the work done by the agriculturists in England with horse labour.

All the varieties of the Indian Ox are held sacred by the Hindus, who consider it a sin to kill them, and pollution to partake of their flesh, although the milk and products from milk are largely consumed in most districts and by all classes. For further information as to the different breeds of cattle see the general article **Oxen**, p. 665 *et seq*.

555 **Bubalus arni.**

THE WILD BUFFALO.

Syn.—BOS ARNI, *Ken & Shaw;* BOS BUBALUS (wild var.), *Jerdon.*

Vern.—*Arna* (the male), *arni* (the female), *jungli bhyns,* HIND.; *Gera erumi,* GOND; *Mung,* BHAGALPORE.

References.—*Jerdon, Mam. Ind., 307; Sterndale, Mam. Ind., 490; Balfour, Cyclop., I., 487; Encycl. Brit., IV., 442.*

Habitat.—The wild buffalo is found in the swampy Terai at the foot of the hills from Bhútan to Oudh, and in the plains of Bengal as far west as Tirhut, but increasing in numbers eastwards on the Bramaputra and in the Bengal Sunderbuns. It also occurs on the table-lands of Central India as far south as the Godavery. It is distributed to Ceylon, Burma, and the Malayan Peninsula. These animals inhabit the margins rather than the interior of forests and never ascend the mountains, but adhere to the most swampy portions of the districts they frequent. (*Jerdon.*)

Characters.

556 **Characters.**—The wild buffalo is somewhat larger than the domesticated variety. It measures 10½ feet and upwards from the tip of the snout to the root of the tail, and the height at the shoulder is often as much as 6½ feet. The colour of the hair is usually black. It is scanty over the body, but tufts occur on the forehead, over the eyes, and on the knees.

557 The horns are of two varieties, one very long and nearly straight, the other much shorter and well curved. The measurement of a fine specimen now in the British Museum was 12 feet 2 inches round the outside of both horns and over the forehead. The horns of the second variety rarely exceed 3 feet each in length. The wild buffalo, unless it has been much hunted, is by no means shy in disposition and may be easily approached. They abound in the swampy jungles of the regions above mentioned. They emerge from the forests to feed on cultivated lands, chiefly at night or in the early morning. They are immensely powerful animals, and a wounded one will occasionally charge and overthrow even an elephant. They rut in autumn, the female gestates for ten months, and produces its young during the hot weather. They usually live in large herds.

Speaking of the wild buffalo **Hodgson** remarks that "the tame species is still most clearly referable to the wild buffalo after ages of domestication, and tame buffaloes, when driven to the forests for pasture, often have

The Buffalo. (*W. R. Clark.*) **OXEN.**

INDIAN BUFFALO.

intercourse with wild ones." **Jerdon**, however, says that wild buffaloes dislike the tame ones and will even retire from the spot where tame ones are feeding. Unlike the tame animals, which are usually lean and angular, the wild buffaloes are uniformly in high condition.

Bubalus bos.

THE BUFFALO.

Syn.—BUBALUS BUFFELUS, *Blum.* 558

Vern.—*Bhains* (male), *mhains* (female), HIND.; *Bhainsá* (male), *bhains, bhainsi* (female), BENG.; *Jhotá, bainsa* (male), *majh, mainh* (female), PB.; *Karbo, harbou*, MALAY.; *Ky-wai*, BURM.

References.—*Wallace, India in 1887, 116-126; Shortt, Manual of Indian Cattle & Sheep, 107-112; Gazetteer, Mysore & Coorg, 171; Buchanan-Hamilton, Journey from Madras, I., 3, 116, 206; II., 114, 381, 382, 488; III., 57, 210, 356; Jerdon, Mam. Ind., 307; En. Cyclop. Brit., IV., 442; Balfour, Cyclop. Ind., I., 502.*

Habitat.—Tame buffaloes are found all over the plains and lower wills of India. They are semi-aquatic in their habits, and, during the hot season, may be seen rolling about in muddy holes half submerged or entirely under water with the exception of their noses.

Characters.—The buffalo is distinguished by its large flat horns—in some curved, in some straight; but in all marked with rings indicative of age. These are often very long in many breeds, measuring as much as 6 feet in length and 18 inches in circumference at the base. The general colour of the animal is usually blackish, but in some forms they vary from brown to grey in colour. Albinoes are not infrequent. They are generally more delicate in constitution than the dark-coloured animals. **Characters.** 559

The hair is usually scanty, although long and wavy, and affords but little covering to the body. The following average measurements of the ordinary country buffalo are taken from **Dr. Shortt's** valuable Manual on Indian Cattle:—"Length up to about 10 feet from the tip of the snout to the root of the tail. Height at the shoulder from 3 to 6 feet" They are not usually so leggy as the cattle of India, and the larger varieties, although they may not stand so high as large bullocks, would weigh much more.

Although ungainly animals, they are much more intelligent and docile than the cattle of India, and may be trained with the utmost nicety for agricultural purposes. In the cart they are used without a nose-string, and are guided by the touch of a wand, or the voice of their driver.

Buffalo cows are much valued by natives for milking purposes, as they yield a much larger quantity and richer quality of milk than the ordinary Indian cow. A good buffalo will yield from 6 to 12 seers of milk per diem, and this milk, being richer in cream, yields a larger proportion of butter and *ghi* than cows' milk does. The cows are, therefore, kept for dairy purposes, while the males are castrated and used as beasts of burden and draught.

The female buffalo breeds once every two or three years only, and usually produces in all six calves. For two years after parturition, she continues to give a large quantity of milk, the third year it decreases in amount, and for about two months before calving she is entirely dry.

Buffaloes are very foul feeders, subsisting on the coarser grasses and the refuse left by oxen. In many parts of the country they are almost entirely stall-fed on stable litter which they devour greedily. Notwithstanding their coarse appearance and habits, they are very delicate animals and are subject to the same ailments that affect country oxen.

The principal breeds of buffaloes in India are five in number. Others have been described, but they seem to be mere local varieties of these types.

INDIAN BUFFALO.

560 The *Jafarbadi* or *Nadhiali* buffaloes are distinguished by the immense development of their frontal bones, and the fact that their horns are directed at first backwards and downwards, then upwards, so that they are quite useless as weapons for attack. The horns of the male are much broader than those of the female. These animals are very mild-tempered and tractable in disposition, and the cows are, perhaps, the best milkers in India.

561 The *Ramnad* breed is the best in Madras, and is found principally in a district to the south-east of Madura. The horns in this breed are of medium length, flat, curving in at the tips, and have a general backward direction.

They are thick, low-set, and very deep in the chest. The hair is light-coloured, and a fringe of white hair usually occurs on the upper side of the ear, a good development of which is held by the natives as indicative of superior milking powers. The females are good milkers, although they do not yield so much as those of the preceding breed.

562 The *Gujarat, Talabdu, or Ganjal* breed is much smaller than the Jafarbadi, and yields less milk but of a richer quality. The horns are short.

563 The *Nagpur* buffaloes are massive, low-set animals with black hair. Their horns are immensely long and sweep downwards, backwards, and then upwards. The cows of this breed are excellent milkers and can stand more knocking about when in milk than animals of the other breeds. The males grow to an immense size and are much used by the natives for heavy draughts (*Wallace*).

564 *Deccani* buffaloes have moderately long horns which curve backwards, downwards, and then upwards, almost in the form of a half circle. Brown is the commonest colour of the hair of these animals, but duns and chestnuts also are found. They are comparatively small in size, but are good milkers and are more hardy in constitution than the other breeds.

565 **Gavaeus frontalis.**

THE GAYAL.

Syn.—BOS FRONTALIS, *Lambert.*; B. GAVÆUS, *Colebrooke.*

Vern.—*Gayal, methun,* HIND.; *Gobay goru,* BENG.; *Beunrea-goru,* ASSAM; *Methua,* MANIPURI; *Nunec,* BURM.

References.—*Jerdon, Mam. Ind., 306; Sterndale, Mam. Ind., 486; Balfour, Cyclop. Ind., I., 1184.*

Habitat.—The hill tracts of Chittagong, extending northwards to the head of the Assam Valley, Manipur and the Mishmi hills, and southwards to Burma. "It probably extends as far north and east as the borders of China" (*Jerdon*).

Characters.

566 **Characters.**—Compared with the Gaur, this is a heavy, clumsy-looking animal, but it is similarly coloured, and has, like it, white legs. Unlike the Gaur, it has a small, but distinct dewlap. It is better adapted than the Gaur for mountainous regions. The Gayal has a much milder disposition than the Gaur, is easily domesticated, thrives well and breeds in captivity. Over the area in which it occurs, the wild animals are decoyed by tame Gayal, and the Kukis are great experts at driving the wild animal, along with tame herds, to their villages. In these regions, the herds of Gayal are the most valuable property of the people. The milk is rich, and the flesh good.

It has been urged that the Gayal should be used in crossing with the ordinary cow to give new blood and strengthen the stock, and with this object the Committee of the Calcutta Zoological Gardens have on several occasions sent Gayal to Calcutta, and cross-bred animals, born in the gardens in Calcutta, have turned out very well, the Gayal imparting a straight humpless back to the cross-bred calf.

Gavaeus gaurus. BISON. 567

This is the BISON or GAUR BISON of some Indian sportsmen (*Jerdon*).

Syn.—BOS GAURUS, *H. Smith, Blyth.*; BIBOS CAVIFRONS, *Hodgson, Elliot*; BOS GAUR, *Traill, Hardwicke*; B. ASUL, *Horsfield.*

Vern.—*Vana-gao, ban-gao*, BENG.; *Gaur, gauri-gai, jangli khulga*, HIND.; *Kar-kona*, KAN.; *Perú-maú*, GOND.; *Katu-yeni*, TAM.; *Gaoiya*, MAR.; *Pyoung*, BURM.

References.—*Jerdon, Mam. Ind.*, 301; *Sterndale, Mam. Ind.*, 481; *Balfour, Cyclop. Ind., I.*, 1185.

Habitat.—The Gaur occurs in almost all the large forests of India from near Cape Comorin to the foot of the Himálaya, but is commonest in those of the Western Ghát and in eastern India from Chittagong to Assam. It is distributed also to the countries to the east of the Bay of Bengal from Burma to the Malayan Peninsula. Although sometimes found on low levels, it prefers hilly ground, but does not ascend the mountains.

Characters. 568

Characters.—It is a large, fine animal, measuring often 9½ to 10 feet in length, and standing as high as 6 feet at the shoulder. The general colour is a dark chestnut-brown, the legs from the knee downwards are of a rufous white. The head is large, the skull massive, the forehead is concave and defined between the horns by a prominent arch of bone. The skin above the eyes is wrinkled, which gives the animal an ill-natured look. The horns are of a pale greenish colour with black tips, curving outwards, upwards, slightly backwards, and finally inwards. They are somewhat flattened at the base, smooth and polished, and in old animals generally broken at the tips; a fine head measured 6 feet 2 inches from tip to tip round the outer edge and across the forehead, but some specimens are said to reach as much as 6 feet 11 inches.

The cow has a slighter and smaller head, a slender neck and no hump; it is much smaller, and the legs are of a purer white.

"The flesh of the Gaur is excellent if not too old, and the marrow bones and tongue are delicacies always preserved by the successful sportsman" (*Jerdon*). The Gaur is usually driven by a line of beaters towards the sportsman who remains concealed behind a tree. A wounded Gaur will sometimes charge, and when he does so it is without the slightest wavering, but most usually he seeks safety in flight. The Gaur is usually extremely shy and retiring in its habits, and very quick of hearing, so that extreme care has to be taken in stalking it. Hitherto all attempts at rearing it have failed. It is said not to live in captivity beyond the third year.

G. sondaicus. 569

THE BANTENG OR BURMESE WILD OX.

Syn.—BOS SONDAICUS, *Müller*; BOS BANTENG, *Raffles.*

Vern.—*Tsoing*, BURM.; *Banteng*, JAVANESE & MALAY.

References.—*Jerdon, Mam. Ind.*, 307; *Sterndale, Mam. Ind.*, 488; *Balfour, Cyclop. Ind., I.*, 1185.

Habitat.—It is found wild in the forests of Pegu and Tenasserim. In distribution it extends as far northwards as the Chittagong Hills and southwards to the Malayan Peninsula, Sumatra, Borneo, and Java. In the island of Bali near Java it is domesticated.

Characters. 570

Characters.—In size, this animal is much the same as the two last species. It resembles the Gaur in appearance, but the young and the cow are bright chestnut in colour, and in the old males, the horns become connected at their base by a great horny thickening under the skin. This animal thrives in captivity, and specimens have been sent to the Zoological Gardens both of London and of Calcutta.

YAK. 571 **Poephagus grunniens.**

THE YAK OR GRUNTING OX.

Syn.—BOS POEPHAGUS; B. GRUNNIENS.

Vern.—*Brong dhong, soora-goy, yak, bubul,* THIBETAN; *Kotass*, TURKI; *Ban-chowr* (the wild), *chaori gao* (the domestic yak), HIND.

References.—*Jerdon, Mam. Ind., 300; Sterndale, Mam. Ind., 489; The Fauna of Br. Ind., Blandford, Mammalia; Hooker, Him. Journ., I., 212, 214; II., 160; Drew, Jummu & Kashmir, 288, 353; Kinloch, Big game in India; Ward, Sportsman's guide to Kashmir and Ladak, 75; Balfour, Cyclop., III., 1104-1105; Encycl. Brit., XXIV., 725. Consult also Cunningham, Ladakh and Schlagentweits' Travels in Thibet.*

Habitat.—The high lands of Tibet, Sikkim, and Ladakh, the valley of the Chang Chenmo, and the slopes of the Kara Koram range (*Kinloch*).

The Yak is indigenous to High Tibet and is found extensively domesticated all over these lofty regions. It is only, however, in Eastern Thibet that the wild yak now commonly occurs, but it is a visitor, so the writer is informed by **Lieut. H. Bower**, "of Chang Chenmo, Kizil Jilga at the head of the Karakash river and of Kotass Jilga, west of the Sujet Pass." Amongst all quadrupeds the yak is the one found at the greatest elevations: "It is met with," says **Hooker**, "up to elevations of 20,000 feet." At the same time the range of temperature and altitude in which it can live is limited, for it can scarcely exist in summer at 8,000 feet. Even in the high valleys of Kashmír at upwards of 8,000 feet, it rapidly degenerates. The heat and the insects are its greatest enemies, and it seems to care little for the luxuriant vegetation of these comparative lowlands, but prefers the coarse hard fodder it can find in the more elevated ranges (*Balfour*; *Kinloch*).

Characters. 572 **Characters.**—The prevailing colour of the domestic yak is black but red, dun, partly coloured, and white ones are also seen. Some of the domestic yaks are hornless; but the majority have long, nearly cylindrical horns, pointed at the ends, and with a characteristic curve at first directed outwards, then upwards, forwards and inwards, and lastly, a little backwards.

The wild yak, the progenitor of this domesticated form, is a much larger animal, so much so that the Tibetans say the liver of a wild yak forms a load for a tame one. It is very fierce, and, when brought to bay, falls on its hunters with horns and chest, and often inflicts serious and even mortal injuries. The horns of a full grown bull are said to be as much as three feet long. In colour the wild yak is black all over, with occasionally a white streak on the forehead (*Hooker*).

The *dhso* or *zobo*, a cross between the yak and Himálayan hill cow, is common in the North-West Himálaya, and is used by the natives there in place of the yak. It is a very fertile hybrid; indeed, the brothers **Schlagentweit** declare they were informed that there was never found any limit to the number of generations (*Schlagenweit*; *Hooker*; *Balfour*).

Domestic Uses. 573 **Domestic Uses.**—In the regions in which it occurs, the yak is the ordinary beast of burden of the people. It is extensively bred in Bussahir and other parts of Tibet. The female goes nine months in calf, and drops one every two years, usually about the month of April. It ordinarily bears ten or twelve altogether. The average value of a young yak is twenty to thirty rupees.

In Spiti the people plough with the tame yak, it carries their loads, it furnishes them with milk, and its hair is woven into ropes. Much of the wealth of the people of Eastern Tibet consists in their flocks of yak. To these mountaineers it is invaluable for its strength and hardiness. Flocks of them may often be met in these regions, bearing two bags of salt or rice weighing 260℔ or four to six planks of pine wood slung in pairs along either flank each animal being led by a rope attached to a wooden ring

YAK.

passed through the nasal cartilage. The yak is regularly milked by the Tibetans and yields a very rich milk. In autumn, when the calf is killed for food, the mother will give no milk unless the herdsman gives her the calf's foot to lick or lays a stuffed skin before her to fondle. Much of the ordinary food of the Tibetans consists of the rich curd of yak's milk, either fresh, or dried and powdered into a kind of meal.

The hair of the yak is twisted into ropes or woven into a coarse kind of cloth which the Tibetans use to cover their tents. The gauze shades for the eyes used by these people in crossing the snowy passes are also made from yak's hair. The bushy tail forms the chowry or fly flapper well known in the plains of India, and the long hair is much esteemed by native women to supplement their own.

The flesh of the young yak is said to be very rich, juicy, and of a most delicate flavour, that of the older animal is sliced and dried in the sun to form jerked meat (called in Tibet *schat-tcken*). **Hooker** says it is a very palatable food, as the scanty proportion of fat in it prevents its becoming rancid.

The horns of the wild yak are used by the grandees of Tibet as cups at marriages and other feasts. They are filled with spirits and handed round to the company.

OXEN. 574

Vern.—*Bail* (the bull or ox), *gai* (the cow), HIND.; "*Maweshi, mál, mal jal* (in Patna), *dhúr* (cattle generally), *goru, gáy goru* (in Patna), *dhúr dánger* (horned cattle exclusive of buffaloes), *sanrh* (a bull branded with sacrificial marks and let go), *dhákar* (when not so branded), *baradh* (in Patna and N.-W. Tirhut), *barad* (a bullock), *harathi* (plough-bullocks), *adári* (in Patna), *audar* (a bullock unbroken to work), *jataha, basaha* (a bullock with excrescences on its body purchased and led about by religious mendicants), *gau, gay* (a cow), *puhilöth, puhilaunth gay* (a cow that has had one calf), *puráhiya* (one that has a calf yearly), *pachar* (one that breeds when five years old), *bahila* (in South-West Shahabad), *thakra* (a barren cow), *bardáël, gabhin* (in South Bhaugulpur), *phurli* (a cow in calf), *dudhar, dudhari* (a good milker), *chonrhi, chonrh* (one that gives little milk), *leru* (an unweaned calf), *bachhwa* (male), *bachhi* (female), (when they are from 1½ to 3 years old), *dodan* (in Shahabad), *do-dant, du-dant* (a calf with two teeth), *udant* (in the East), *adant* (when it has not yet got its true teeth), *osar* (to the West), *kalor* (in N.-E. Tirhut), *gaur* (in Patna), *phetain* (a heifer ready for the bull)," BENG. and BEHARI, (*Grierson, Bihar Peasant Life*, 287). "*Bail* (an ox or bullock), *bail, badiyá, dhor dangar* (cattle generally), *goí, goín, juár, juára* (a pair of plough oxen), *lád-an* (a pack bullock), *haryá* (a plough bullock), *gariha* (a cart bullock), *náta* (a dwarf or stunted ox), *bhúnr-bhundá* (an ox without horns), *unandi, jatah, nadiyá* (a deformed ox led about by religious mendicants), *gae, gau, gaiya* (a cow), *bahila* (a barren cow), *dhén* (a cow in milk), *lain* (a cow just after calving), *purebha* (a cow that gives two calves within a year), N.-W. Prov. & OUDH;" (*Crooke, Glossary of Rural and Agricultural terms*). *Bail* (a bull or ox), *gau* (a cow), (in Rohtak) *bachra* (an ox under two years old), *bahra* (from two to four years) *baladh* (an ox, fit for work) *dhanda* (an old ox past work) (a cow bears the names corresponding to those of the ox till she is four years old when she becomes) *gae*, (in Montgomery) *vachhá* (an ox till 1 year old), *vachhí* (a cow), *wairká* (an ox till 2½ years old), *wairki* (a cow), *vauhr* (an ox till 4 years old) *dhanap* (a cow till she calves) *bail* or *sanh* (after 4 years old) *gau, gao* (a cow after calving) PANJÁB; *Mai* (an ox), TAM.; *Mwa* (an ox), *kywai* (an English ox), BURMESE; *Ukyurz* (an ox), TURK.; *Bakara* (an ox), ARAB.

References.—*Wallace, India in 1887, 1-115; Shortt, Manual of Indian Cattle and Sheep, 1-106; Gazetteer, Mysore & Coorg, I., 166-171; Manual and Guide, Saidapet Farm, Madras, 71-81; Buchanan Hamilton, Journey from Madras, I., 3, 116, 165, 205, 206, 417; II., 5-8, 114, 180, 327, 380, 382, 488, 509, 562; III., 56, 335; Ayeen Akbery (Gladwin's Transl.), I., 168-172, 190, 263; II., 65, 69, 71, 136; (Blochmann's Transl.), 21, 148, 149;*

OXEN. **Mysore Cattle.**

Voyage of John Huyghen Van Linschoten to the East Indies, I., 257, 300; Grierson, Bihar Peasant Life, 287-293; Crooke, A Rural & Agricultural Glossary for the N.-W. P. & Oudh (various passages); Mason, Burma & Its People, 173, 670; Gazetteers & Settlement Reports (many passages); Reports of the Rev. & Agric. Dept., Govt. of India (many passages); Selections from the Records of the Govt. of Ind. (Home Dept.), No. LXIX.; Sel. from Records of Govt. of Bengal (Papers relating to cattle disease) (various passages); Sel. from Records of Govt. of Pánjab (New Series, No. XX.), (Cattle diseases), (various passages); Gunn, Report on Cattle disease in the Central Provinces (various passages); Mills, Plain Hints on diseases of cattle in India, 16-24; Agri.-Horti. Soc., Ind., Transactions, I., 17-18, 39-40; II., 218, 219, 250-253, 257 (App.), 310; IV., 110; V. (Pro.), 5, 7, 8, 14-29, 34-35, 45, 50; VI. (Pro.), 95-97, 120-123; VII., 111-114 (Pro.), 53-56; Journals (Old Series), I., 320-329; XIII., Sel., 27-30; XIV., 89-101, 171-180 (Sel.), 105-128, 177-188; (New Series), I., 25; III. (Pro.), 23, 24; VII., 365; VIII., 238-243, 253-259 (and many other scattered passages throughout their publications); Quarterly Journal of Vet. Sc. in India (April 1890), 186; Encycl. Brit., V., 244; Balfour, Cyclop. Ind., III., 64.

BREEDS. 575

I.—The Breeds of Indian Cattle.

Mysore. 576

I. Mysore Cattle.—The cattle of this district are justly celebrated, both for their swiftness and for their spirit and powers of endurance. The form which is perhaps best known, and most valued in India, is known as the *Amrit Mahal* breeds, the chief centre of which is the Hunsur grazing farm of the Madras Government. Though probably altered in character from time to time, the breed existed from a very early period in the Mysore kingdom, and was in all likelihood produced "by crossing a number of the best varieties from the districts of Mysore. At any rate, under the native rulers, it seems to have attained its greatest perfection" (*Wallace*). Dr. Buchanan-Hamilton, in his *Journey through Mysore, Canara, and Malabar*, in the year 1800, speaks highly of the breed of Mysore cattle, which he describes as follows:—"The race of oxen of this country may be readily distinguished from the European species, by the same marks that distinguish all the cattle in India; namely, by a hump on the back between the shoulders, by a deep undulated dewlap, and by the remarkable declivity of the *os sacrum*. But the cattle of the South are distinguished from those of Bengal by the position of the horns. In those of Bengal the horns project forward, and form a considerable angle with the forehead, whereas in those of the South the horns are placed nearly in the same line with the *os frontis*. In this breed also, the prepuce is remarkably large; and vestiges of this organ are often visible in females; but this is not a constant mark.

"Two breeds are most prevalent. The one is a small, gentle, brown, or black animal; the females are kept in the villages for giving milk; the oxen are those chiefly employed in the plough. This breed seems to owe its degeneracy to the want of proper bulls. As each person in the village keeps only two or three cows for supplying his own family with milk, it is not an object with any one to have a proper bull, and as the males are not emasculated until three years old, and are not kept separate from the cows, these are impregnated without any attention to improvement, or even to prevent degeneracy. Wealthy farmers, however, who are anxious to improve their stock, send some cows to be kept in the folds of the large kind, and to breed from good bulls. The cows sprung from these always remain at the fold, and in the third generation lose all marks of their parent's degeneracy. The males are brought home for labour, especially in drawing water, by the *capily*; and about every village may be perceived all kinds of intermediate mongrels between the two breeds. The cattle of the other breed are very fierce to strangers, and no body can

BREEDS.

Mysore.

approach the herd with safety unless he be surrounded by *goalas*, to whom they are very tractable. At five years old, the oxen are sold and continue to labour for twelve years. Being very long in the body and capable of travelling far on little nourishment, the merchants purchase all the best for carriage. To break in one of them requires three months' labour, and many of them continue very unruly. The bulls and cows were so restless that, even with the assistance of the *goalas*, I could not get them measured; but the dimensions of a middle-sized ox were as follows:—From the nose to the root of the horn 21 inches. From the root of the horn to the highest part of the hump 30 inches. From the height of the hump to the projecting part of the *ossa ischia*, 45 inches. From the hump to the ground 46 inches. From the top of the hip-bone to the ground 51 inches.

"The cows of this breed are pure white, but the bulls have generally an admixture of black on the neck and hind-quarters.

"These cattle are entirely managed by *goalas*: and some of these people have a considerable property of this kind: but the greater part of these breeding flocks belong to the rich inhabitants of towns or villages, who hire *goalas* to take care of them, and, for the advantage of better bulls, send to the fold all their spare cows of the village breed. In procuring bulls of a good kind, some expense is incurred, for the price given for them is from 10 to 20 pagodas (£3-7-1 to £6-14-2), while from 8 to 15 pagodas is the price of an ox of this kind. Care is taken to emasculate all the young males that are not intended for breeding before they can injure the herd."

After the capture of Seringapatam by the British forces, the breeding establishment was, in 1800, intrusted to the Native Government, but the inducements which had led the Native princes to keep up its efficiency being absent, the whole race of cattle degenerated to such an extent that in 1813 the management was taken over by the British Government, and a Commissariat officer (Captain Harvey) was placed in charge of the establishment. Till 1860, it remained under Government management, when, on account of the large expenditure involved and the frequent complaints of inefficiency in the cattle supplied from the establishment, the herds were broken up, and cattle for Government purposes were bought in the open markets. For six years, this arrangement was continued; but the prices for the best cattle soon rose to such an enormous extent that in 1867 it was found necessary to again organise the herds. As many of the original breed as could be found were purchased, and in December 1867, the Amrit Mahal herd, with 5,935 head of cattle, was again re-established. In 1871 there were 9,800 head of all sizes, exclusive of 1,000 young male cattle in the training depôt. It was arranged also that a certain number of bulls should be handed over to the Native Government of Mysore, and these were stationed at various points in the country for the purpose of improving the breed of cattle used by the *rayats*. "In 1886 the numbers had increased to 12,457. Up till this time 400 steers were required annually by the Madras Government, but because horses are to replace bullocks in the 2nd line artillery waggons, in future only 200 steers will be required annually. It is intended to materially reduce the numbers of the herd" (*Wallace, India in 1887*).

"The Amrit Mahal cattle comprise three varieties called the Hallikar, Hagalvadi, and Chitaldroog, from the districts whence they were originally produced" (*Shortt*). They are distinguishable from all other Indian cattle by their well-shaped, symmetrical heads and light yet strong build. They seldom attain a great height, the bullocks being on an average 48 to 50 inches high; but in proportion to their height they are, for Indian

OXEN. **Variations in the Mysore Breed.**

BREEDS. Mysore.

cattle, remarkably deep and wide in the chest, long and broad on the back, well ribbed up and strong in shoulder and limb. They are active, fiery, and vigorous, and can walk quickly for long distances, a quality particularly valuable for the military purposes to which the breed is usually applied. The cows of the breed are white, but the males have generally an admixture of bluish grey over the fore and hind quarters. Marled or broken colours appear most frequently in inferior specimens, where cross breeding has been practised. Early castration of the bulls renders them lighter in colour. "The horns are straight and long, extending from 2 to 3 feet in length, tapering and sharp-pointed, inclining upwards, slightly convex on the outer and concave on the inner side, approaching one another at the tip" (*Shortt*).

They are extremely fierce in their dispositions, dislike strangers, and would attack them if untied, although at the same time they are very tractable to the natives who work them and are known to them. The cows of this breed are very inferior milkers and yield only about 1 to $1\frac{1}{2}$ seers of milk per diem, so that they are of no use for dairy purposes, as the calf cannot be deprived of any portion of the milk without being materially injured in its growth, and after bad years, when the pasture has been insufficient and the milk not up to its average quantity, it is found that when the calves grow up they are decidedly inferior in quality, and when they are put to work they have often to be rejected on account of deficient strength.

The following account of the management of the Amrit Mahal herd may be abstracted from *Shortt's Manual af Indian Cattle and Sheep* :—

In the cold season when herbage is abundant, the calves are generally weaned from their mothers at the age of five months; but those that are born later in the year cannot be separated till after the hot weather. Heifers begin to breed between three and a half to four years old and bring forth six or seven times. Twenty cows are allowed to one bull. The average number of births annually is 50 per cent. on the number of cows, and the proportion of males and females born is nearly equal. The calves are castrated in November, invariably between the ages of five and twelve months, as their growth is supposed to be promoted by early castration, and it is attended with this important advantage that it prevents the cows being impregnated by inferior bulls and so keeps the breed from degenerating. The steers are separated from the herds after four years of age, and transferred to the Public Cattle Department when turned five years old, perfectly trained and fit for work.

577 Although the principal, the Amrit Mahal is not the only breed of the Mysore type of cattle; indeed, there are numerous variations, most of which are, however, inferior in size or in working powers or in both. Thus the *Madesvaram Betta* is a variety of the Mysore type that is bred in the jungles and hills on the South-Eastern frontier of Mysore. The animals of this class are larger than those of the Amrit Mahal breed, but are loosely made and not well ribbed up. They are very heavy, slow animals, and not so useful for military as for agricultural and draught purposes.

578 The *Kankahalli Breed* comes from Kankanhalli in the south-east of Mysore, and although somewhat similar to the preceding, the oxen are much smaller, being inferior in size to the Amrit Mahal breed.

579 The *Village* cattle of Mysore are usually smaller in size, and irregular in colour and shape, but still strongly exhibit the Mysore type. They are poor milkers, but excellent cattle for agricultural and draught purposes.

Nellore. 580 **II. The Nellore or Ongole Breed.**—Another breed of cattle widely celebrated in India is the Nellore one. The animals of this class are valued not so much for draught purposes as for their milking qualities.

BREEDS.

Nellore.

There are two varieties, the *large* and the *small*—of which the *large* stands from 15 to over 17 hands in height; their powers of draught are great, but they are slow and inclined to be leggy, and the females are not such good milkers as those of the smaller breed. The horns of both sorts are short and stumpy, usually 4 to 6 inches in length and very rarely up to 12 inches. The colour of the breed is white with black points, and frequently a slight shading of grey in the male. In ancient times the Nellore cattle were remarkable for the size of their dewlaps, but this, in more modern days, although still large, has become much diminished in size. They are chiefly used for draught, in carts and with the plough, their height and size being against their use as pack bullocks generally. The cows, as already remarked, are excellent milkers; "some of them," says **Shortt**, "have been known to yield 18 quarts (of 24 oz.) of good rich milk in 24 hours, and they rear a calf at the same time."

"The price of a first class cow is about 200 rupees; as much even as 300 rupees have been paid for a prize cow; that of a pair of bullocks, 150 to 350 rupees; the ordinary bullocks fetch from 100 to 150 rupees the pair." **Wallace** remarks that this breed is said to be degenerating, and that from ancient descriptions and drawings it would appear to be so. The influence of this breed extends as far north as the Kistna district, the cattle of which, although reckoned as a distinct breed by some authorities, are regarded by **Shortt** as simply a variety of the Nellore type.

Gujarat. 581

III. The Gujarat Breed.—This is locally known as *Yalabda* or indigenous, and is also found in two varieties, the *large* and the *small*, the former of which is decidedly the finest in North-Western India. The colour of this breed is from white to bluish grey in the cows, while many of the bulls are decidedly grey. The horns are somewhat bowed, curving at first outwards as they leave the skull, then upwards, and lastly, inwards. They project upwards without inclination either forward or backward. The dewlap of the male is large, and well developed in proportion to the sheath. The ears are large and pendant. The trunk is compact and well knit, and with a good depth. The limbs are powerful and well knit, the feet black and possessed of unusual hardness and durability. They are much valued on account of their docility and good temper, which renders the cows particularly suitable for crossing with Mysore bulls to produce bullocks for battery and transport purposes.

This breed also is said by **Wallace** to be degenerating in quality on account of the breaking up of pasture land by new cultivators.

Sind. 582

IV. Sind Cattle.—Sind cattle are compact in form and well built, but smaller than any of the preceding. The oxen are slow and lazy, but powerful, animals. The cows are remarkably good milkers. As regards colour of hair, the white or grey colour predominates; but there are many spotted or brindled cattle of this breed, the spots being usually of a brown or light brown colour. The horns are short, thick, blunt pointed, and project outwards and a little upwards.

Kathiawar. 583

V. Kathiawar or Gir Cattle.—These are of a special type quite different from the ordinary cattle of India. They are bred in the Kathiawar district, and are somewhat small in size; but the bullocks are good powerful draught animals, while the cows are excellent milkers and supply most of Western India with milk cattle. They have long, pendant, somewhat bell-shaped, lop ears. The head is neat and well formed, the horns are short, crumpled, something like those of the buffalo, inclined backwards and in the male, thick, while in the female they have the same characters, but are much thinner. The common colour of the cow is light brown, while the bull has a darker brown colour, but broken colours and brindling are common.

BREEDS.

Trichinopoly. 584 **VI. Trichinopoly Breed.**—The cattle of this breed are small and useful, mostly with horns 10 to 15 inches long, where they are not suppressed by firing before they begin to shoot out. They are mostly white in colour with the eyes, muzzle, hoofs, and hoof-heads black. They are strong draught animals.

Aden Cattle. 585 **VII. Aden Cattle.**—These are more compact and symmetrical than Indian cattle usually, and are much valued for their milking properties. Numbers have been imported with the object of crossing with native cattle to improve their value for dairy purposes. The bullocks are small but powerful. They are very tame and docile and can be handled either by Europeans or Natives without difficulty. They are not so hardy as the ordinary Indian cattle when exposed at work to the sun. The Aden bulls much resemble the Brahminy bulls of Bengal, and have very large humps extending far forward on to the neck. The ears are small, and the horns short and thick. The cows are small and light of carcase, hornless, or with very short horns.

Hill Cattle. 586 **VIII. Hill Cattle.**—The cattle seen on the Himálaya are small in size, mostly black in colour; but a few are grey or dun with dark, mouse-coloured points. The hump is very small, in some almost wanting. These animals are very hardy, short on the legs, and active in their habits, admirably adapted to pick up a livelihood on the mountain sides.

Village Cattle. 587 **IX. Village Cattle.**—Many other so-called breeds of cattle exist in India; but they are mostly crosses amongst those above described, and can scarcely be ranked as distinct races. They are usually smaller in size, but appear to do the work required of them fairly well in proportion to their dimensions and strength. To attempt to cross them so as to produce larger and finer breeds, would probably be unwise until their treatment by the *rayats* has been changed, as these animals pick up a living where a larger and finer animal would undoubtedly starve. "The only method of improvement," says **Wallace**, "that will end successfully, is that of selection, combined with greater attention and better treatment."

Burmese Cattle. 588 **X. Burmese Cattle.**—An interesting account of these cattle was given by **Vet. Surgeon R. Forest, A.V.D.**, in the *Quarterly Journal of Veterinary Science in India*, April 1890, from which the following details have been abstracted:—The, cattle of Burma are of a purely indigenous character and form, a type peculiar to that province and the neighbouring Shan States. The Burmese ox stands from 46 to 50 inches high behind the hump. The hump of the bull is well developed, that of the castrated male less so. There is a marked disparity in the size of the cow as compared with the bull, the former being probably stunted by the effects of too early breeding and too long suckling of the calf. The colour of Burmese cattle varies, but not to a great extent. Red prevails very extensively, after which come the various shades of straw, and next brown. Broken colours are rare. The trunk of the Burmese ox is well ribbed up, and it is a fairly muscular and symmetrical animal, possessing great strength for its size. The head is small, markedly so in the cow; the horns are usually short and stunted principally on account of the custom which prevails of paring them down to give the animals a juvenile appearance. When not interfered with, the horns take an upward and forward direction and grow to a length of from 10 to 18 inches. The neck is short and powerful, but loses a good deal of its heaviness after castration. The dewlap is somewhat small. The Burmese do not castrate their cattle till they have attained the age of maturity, so that the young bulls of 2½ to 5 years of age, of which there are a number in each herd, are the sires of all the calves born in a village. Castration is usually performed by crushing or beating, seldom by cutting. No

BREEDS.

Burmese Cattle.

system of selection of sires is possible, as the cattle of each village are all kept together, and the Burman has strong prejudices against the early castration of bulls, which he considers prejudicial to the growth and development of the male stock, and their future usefulness for agricultural purposes.

As milk-producers, the cattle of Burma occupy a very low place. The Burman cultivator seldom or never thinks of milking his cows. They are merely utilised to suckle the calves which they produce. This they go on doing, from the birth of one calf till that of another; indeed, it is not unusual to see two calves, one of over a year old and the other only a few months, being supported by the same cow.

As a rule, the Burman is very careful of his work oxen, does not work them during the heat of the day, and keeps them under shelter, either beneath his own house or in small lean-to sheds in connection with it during the night. During the day the herd is driven out to graze, except when the crops are on the ground, at which season he cuts food for them. Waste-lands in connection with holdings are always utilised for grazing purposes; a small amount of straw is stacked every year for the use of the cattle in times of scarcity.

Further than that above alluded to, the Burmese cattle receive no hand feeding, but, except for a few months of the year, the grazing is abundant, and the animals do not usually suffer.

A large number of milch cows are now imported from Calcutta every year, which are sometimes crossed by country bulls. This cross is larger than the indigenous breed. They are better milkers and fetch a higher price for draught purposes.

II.—Attempts at improving the Breed of Cattle in India. 589

The subject of the alleged deterioration of cattle in India is one that has, since early in the present century, occupied the attention of the administrators and agriculturists of the country. Many suggestions at possible means of improvement of the breed, and actual attempts at doing so, have been made, but without marked success, and, up to the present day, the ordinary cattle in use by the Indian agriculturist seem to be pretty much what they were at the time when the question was first raised. In 1839 the Agri.-Horticultural Society of India instituted annual cattle shows in Bengal which they continued for five years, and offered prizes to successful competitors for good specimens of country-bred cattle. These shows were discontinued in 1844, on the report of the Cattle Committee who considered that "the attempt to improve cattle and sheep by money premiums and medals had not held out sufficient encouragement in the number of cattle brought forward at the shows to induce a continuance of the annual exhibitions." In 1864 a special committee appointed by the Society reported (1) that there could be no doubt that the breed of cattle in Bengal had deteriorated; (2) that the causes of this degeneration were (*a*) the poorness of the food given to them, (*b*) the want of proper pasturage ground, (*c*) over-work and being worked at too early an age, (*d*) being allowed to breed too young, (*e*) the scarcity of good bulls, (*f*) stinting of the calves of milk. The members of this committee suggested that good bulls should be established in each village like the parish bulls in England; that *zamindars* should be required to give yearly prizes for good cattle, each in his own estate; that the *zamindars* should be induced to allow a certain area in each village to be kept for pasturage, instead of all the land being cultivated, as was the tendency at the time; that fodder depôts should be established on each estate, whence an abundant supply of good

OXEN. **Attempts at Improving**

BREEDS.

food for stall feeding might be obtained; and that for milking purposes, cross breeding with European stock should be introduced, while for draught and agricultural purposes good bulls of the country stock should be used. They remarked that the great object was not to introduce crosses which would, for most purposes, not be found so useful as the ordinary country breed, but rather to bring to perfection the cattle, of each breed, by affording the calves a fair share of milk and good pasture.

No action was, however, taken on these suggestions, and in 1887 we again find a Special Committee of the Agri-Horticultural Society lamenting the deterioration of cattle in Bengal and the disappearance of pasture lands, while suggesting that Government should take steps to teach the *rayat* to stall-feed and grow fodder for his cattle, as they considered that the restoration of pasture lands was now impossible. They were of opinion that for the improvement of the milch cows of Bengal, crosses with Jersey, Guernsey, and Ayrshire bulls would be most suitable with a view to the greater production of milk and butter. For the breeding of draught cattle, they thought the best specimens of the existing breeds of the locality should be chosen as sires.

In various other parts of India attempts have been made to improve the breed of cattle, by offering prizes for the best animals, exhibited at agricultural shows, but these attempts have, on the whole, been but fitful, and the whole question of the improvement of the breed of cattle in India remains pretty much in the condition in which it was at the beginning of the century. The introduction of animals of the humpless type is looked upon generally with disfavour by Indian agriculturists, as they declare, and that truly, that their object is to procure not so much good milking or beef-producing animals as to get strong, hardy, draught cattle, for which purpose the more delicate humpless animals of Europe are unsuitable.

In the beginning of the present century the Court of Directors of the East India Company sent out some English bulls to the cattle farm at Hissar with a view to improving the breed there, but that attempt was necessarily a failure, as the cattle at Hissar are bred for draught purposes, while those sent out were mostly of the short-horn type.

A number of bulls of European and Australian and the best indigenous breeds have of late years been purchased by private individuals and District Boards for breeding purposes, and crosses have been made, but as proper attention is not usually paid by the ordinary agriculturist to the feeding and bringing up of those more delicate cross-bred animals, the attempts have not proved successful, and until crops are grown for stall-feeding, pasture lands set apart, and the cattle better housed, no substantial improvement in the ordinary breed of cattle in India can be expected.

An experiment on an extensive scale with a view to improve the breed of cattle in the Himálaya, by an admixture of European blood, was begun in 1880 by the Government of India. Twelve head of Brittany cattle, *viz.*, four bulls and eight cows, were then imported direct from Brittany. This breed was chosen for the experiment, as they are small in size, of active habit, accustomed to mountains, and good milkers; and were, therefore, thought most suitable for crossing with the Himálayan breed. In January 1887, a second batch of cattle, consisting of three bulls and four cows, and in January 1888, a third batch of eight bulls and five cows, was imported. In 1890, the result of this experiment was found to be as follows:—15 bulls and 17 cows in all were imported. These were distributed to various hill districts in the North-West Provinces and the Panjáb, where they were given in charge to trustworthy European planters and settlers. At the end of 1889, the numbers of pure bred Brittany cattle had mounted up to 15

BREEDS.

bulls and 22 cows, but 8 bulls and 14 cows, both of the original and of the Indian-bred stock, had died. Seventy-three half-bred calves had been born, and 57 cows had been covered by the Brittany bulls, but had not calved at the date of the report A number of other cows were covered, but no record of the result was kept by the owners.

The large mortality in the imported cattle was principally due to rinderpest. Opinions as to the success or otherwise of the experiment are very various. Some of the people, with whom the cattle were placed, reported that the experiment "had not been successful, as the Brittany cows, although reputed good milkers, had not shown themselves such in India, and the *zamindars* and villagers did not like the Brittany bull, owing to the absence of the hump, but preferred the indigenous breed, as they were useful for ploughing purposes." Others reported that the experiment "would certainly in time be productive of good results, and that even already they saw good effects in the improved milk-producing and beef-yielding capabilities of the cross-bred animals." On the whole, however, when the difficulties of acclimatisation are considered, the experiment seems to have been already fairly successful, and to be not destitute of promise for the future; but it seems to the writer that attention should be paid to the improvement of the cattle of India, more in accordance with the lines laid down by the Agri-Horticultural Society in the report by their Cattle Committee alluded to above, than by the introduction of expensive and delicate breeds from Europe and the colonies, although these may be useful in improving the milk and beef-producing qualities of the indigenous breeds.

III.—Cattle Diseases in India.

CATTLE DISEASES. 590

The principal epidemic cattle diseases prevailing in India are rinderpest, anthrax in its several forms, epiozootic aphtha or foot-and-mouth disease, and, less commonly, pleuro-pneumonia.

Rinderpest. 591

Rinderpest.—(*Bossonto*, Bengal; *Gootee*, *chuchack*, North-West Provinces; *Kalawah*, *wah*, *zahmat*, Panjab; *Pitchinow*, Bombay; *Peya*, Madras) is the most common and fatal of the four diseases above mentioned.

"The amount of injury and loss resulting annually from the unchecked ravages of this disease in India is something enormous" (*Wallace*). It appears to be present throughout the year at all seasons, and carries off thousands of cattle annually, and as no direct Government interference can be attempted to prevent the disease, when it breaks out, spreading to wider areas, its ravages are very extensive.

Anthrax. 592

Anthrax.—(*Golafula*, Bengal; *Gutherewan*, North-West Provinces; *Goli*, *suth*, Panjab; *Odro*, Bombay; *Thaloreenova*, Madras) is highly fatal to cattle, sheep, and horses in India. It is seldom curable and appears under the most varied forms, about which there are great varieties of opinion, and much remains to be learned. It is a blood-germ disease Young cattle that are thriving best are most apt to be attacked with it, and few if any of those attacked, recover. It is specially virulent and common in some parts of the Central Provinces and of the Panjáb, but occurs epidemically all over India.

Foot-and-Mouth Disease. 593

Foot-and-Mouth Disease.—(*Khoorat*, *khoratie*, Bengal; *Khurpakka*, North-West Provinces; *Mohona*, Panjab; *Khurpakka*, *khur*, Bombay; *Mupaung*, Madras) is probably more contagious than either of the two above-mentioned diseases, and is certainly more widespread, but as it is very mild, and few cattle succumb to it, the people pay little attention to the attacks. At times, especially if it comes on during the rainy season, the hoofs may fall off before the new ones are ready, and maggots then attack the foot; but if proper care is taken of the

CATTLE DISEASES.

animals affected, no permanent evil results from this disease, and a fatal issue is not frequent.

Pleuro-pneumonia. 594 PLEURO-PNEUMONIA—*Pheepree,* HIND.—although occasionally playing considerable havoc in India, is altogether more local in its action, and less disastrous in its effects than it is in England. When it does occur, it is usually in the colder parts of North-West India (*Wallace*).

This completes the list of the more important communicable cattle-diseases in India, but others, such as tuberculosis and cowpox, are also common and often very fatal in their effects.

" Direct Government interference to prevent their extension will prove efficacious only if the areas under which disease has broken out are placed under strict quarantine, and the bullock traffic on the highways through infected parts is entirely suspended" (*Wallace*)—two provisos almost impossible in the present state of the country. Much, however, might be done through local effort to suppress their ravages by disseminating a knowledge of the nature of these diseases and the means of checking them among the village officials and leading Natives generally.

To attempt to coerce the Native by introducing strict quarantine measures must defeat its own ends until he comprehends the objects of these measures and the value of their results.

Ox-gall, see **Fel,** Vol. III., 322.

Oxides of Iron, see **Iron,** Vol. IV., 540.

Oxide of Lead, see **Lead,** Vol. IV., 602.

Oxide of Manganese, see **Manganese,** Vol. V., 144.

OXYBAPHUS, *Vahl.; Gen. Pl., III., 4.*

595 **Oxybaphus himalaicus,** *Edge.; Fl. Br. Ind., IV., 708;* NYCTAGINEÆ.

Vern.—*Punac, bhans,* PB.

References.—*DC., Prodr., XIII, 2, 430; Stewart, Pb. Pl., 182.*

Habitat.—A scrambling scabrous plant found in the northern tracts of the Pánjab, between altitudes of 6,000 and 9,000 feet. It occurs on the Western Himálaya from Kulu to Garhwál.

FODDER. 596 **Fodder.**—It is collected by the inhabitants of the regions where it occurs for winter fodder.

OXYRIA, *Hill; Gen. Pl., III., 100.*

597 **Oxyria digyna,** *Hill; Fl. Br. Ind., V., 58;* POLYGONACEÆ.

MOUNTAIN SORREL.

Syn.—O. RENIFORMIS, *Hook.;* O. ELATIOR, *Br., Royle.*

Vern.—*Amlu,* PB.

References.—*DC., Prodr., XIV., l., 37; Boiss. Fl. Orient., IV., 1004; Royle, Ill. Him. Pl., 314; Stewart, Pb. Pl., 184; Gazetteer of Simla Dist., 12.*

Habitat.—A small plant, with an acid flavour, which occurs in the alpine Himálaya between altitudes of 10,000 to 14,000 feet. It is found in Western Tibet up to an altitude of 17,500 feet, and is distributed to the mountains of Europe, North Asia, and America, and to the Arctic regions.

MEDICINE. Plant. 598 **Medicine.**—In Kanawar this PLANT is eaten as a cooling medicine (*Cleghorn*).

FOOD. Leaves. 599 **Food.**—The LEAVES have a pleasant sorrel taste, and in Chamba are eaten raw also in *chatnís* (*Stewart*).

OXYSTELMA, *Br.; Gen. Pl., II., 749.*

Oxystelma esculentum, *Br.; Fl. Br. Ind., IV., 17;* ASCLEPIADEÆ. 600

Syn.—PERIPLOCA ESCULENTA, *Linn.*; ASCLEPIAS ROSEA, *Roxb.*; OXYSTELMA WALLICHII, *Wight.*

Vern.—*Dudlutta, dudhiálatá, dugdhicá, kyirin, dudhi,* HIND.; *Dudhlutta, khirai, dudhiálatá, kirui, dudhi,* BENG.; *Dudhia-latá,* URIYA; *Gharote, ganí,* PB.; *Garay-khiri, dhudhi,* SIND; *Dúdhiká,* BOMB.; *Dudhári,* MAR.; *Pála kúra, dudi-palla, ourupalay, chiru pála, piuna-pála, dúdi pala, néla pála, chiri pala, sépachettu,* TEL.; *Tiktadugdha, dughdika,* SANS.

References.—*Roxb., Fl. Ind., Ed. C.B.C., 254; Voigt, Hort. Sub. Cal., 541; Grah., Cat. Bomb. Pl., 121; Dalz. & Gibs., Bomb. Fl., 150; Stewart, Pb. Pl., 146; Sir W. Elliot, Fl. Andhr., 43, 47, 132, 142, 153, 167; Sir W. Jones, Treat. Pl. Ind., V., 104, No. 3; O'Shaughnessy, Beng. Dispens., 457; U. C. Dutt, Mat. Med. Hind., 297; S. Arjun, Bomb. Drugs, 208; Murray, Pl. & Drugs, Sind, 161; Gazetteers, Mysore and Coorg, I., 56; N.-W. P., IV., lxxiv.; Agri.-Horti., Soc. Ind., Journal (Old Series), IX., 158; X., 18.*

Habitat.—A slender, glabrous climber, met with throughout the plains and lower hills of India from the Panjáb to Ceylon. It is found also in Burma and Java.

Medicine.—A decoction of the PLANT is used as a gargle in aphthous ulcerations of the mouth and in sore-throat (*Murray; Arjun*). In Sind the MILK is collected from the plant, dried, and used with warm water for washing ulcers (*Murray*). In combination with turpentine it is sometimes prescribed for itch. Probably on account of the milky juice which it exudes, native practitioners ascribe galactogogue properties to this plant. It has a very bitter taste and is said to possess marked antiperiodic properties (*Arjun*). The fresh ROOTS are in Orissa held to be a specific for jaundice (*W. W. Hunter*).

MEDICINE. Plant. 601 Milk. 602 Roots. 603

Food & Fodder.—It produces a FRUIT which is said to be eaten in Sind. During the famine of 1877-78, at Poona the FOLLICLE, and in the Khandesh District, the LEAVES, of this plant were eaten by the poorer classes. Oxen browse on the leaves, and the ROOTS are sometimes dug up and given to them as fodder.

FOOD Fruit. 604 Follicle. 605 FODDER. Leaves. 606 Roots. 607

OXYTENANTHERA, *Munro; Gen. Pl., III., 1211.*

A genus of Bamboos, five species of which are found in the East Indies and a sixth species widely distributed in the forests of Tropical Africa.

Oxytenanthera albo-ciliata, *Munro.* 608

Syn.—GIGANTOCHLOA ALBO-CILIATA, *Kurz.*

Vern.—*Wa-pyoo-galay,* BURM.

Reference.—*Munro, Trans. Linn. Soc. (London), XXVI. (1870).*

Habitat.—A bamboo of Burma with stems 20 to 30 feet in height.

O. monostigma, *Beddome.* 609

Reference.—*Beddome, Flor. Sylvat., II., ccxxxiii.*

Habitat.—A bamboo met with in the Anamalai Hills of South India.

O. nigro-ciliata, *Munro.* 610

Syn.—BAMBUSA GRACILIS, *Wall.*; B. NIGRO-CILIATA, *Büse*; GIGANTOCHLOA NIGRO-CILIATA, *Kurz.*

Vern.—*Lengka,* MALAY.

References.—*Beddome, Fl. Sylv., II., ccxxxiii.; Munro, Trans. Linn. Soc. (London), XXVI. (1870); Kurz in Indian Forester, I., 345.*

Habitat.—An arboreous bamboo, 30 to 60 (according to Zollinger up to 130) feet high, met with in the forests of the Western Ghàts, Chittagong, Burma, and the Andaman Islands, and distributed also to the Malayan Archipelago.

611 **Oxytenanthera Stocksii,** *Munro.*

References.—*Beddome, Fl. Sylv., II., ccxxiii.; Munro, Trans. Linn. Soc. (London), XXVI., (1870); Gazetteer, Bombay, XXV., 138.*

Habitat.—A bushy bamboo, 20 to 30 feet high, said to exist in the Koncan Gháts. Neither **Beddome** nor **Lisboa** had themselves seen this plant, but both remark that its leaves are said to be exactly like those of **Dendrocalamus strictus.**

612 **O. Thwaitesii,** *Munro.*

Syn.—DENDROCALAMUS MONADELPHUS, *Thw.*

Vern.—*Watte,* ANAMALAIS.

References.—*Beddome, Fl. Sylv., II., ccxxxii.; Munro, Trans. Linn. Soc. (London), XXVI. (1870).*

Habitat.—A very common bamboo on the Anamalai Hills at altitudes between 3,500 and 6,000 feet; it is also found in many other localities on the outskirts of the moist forests of the Western Gháts, and is frequent in the central parts of Ceylon (*Beddome*).

For a description of the properties and uses of the various genera of Bamboos, the reader is referred to **Bamboo,** Vol. I., 370; **Dendrocalamus,** Vol. III., 71 to 80.

613 **OXYTROPIS,** *DC.; Gen. Pl., I., 507.*

Oxytropis microphylla, *DC.; Fl. Br. Ind., II., 139;* LEGUMINOSÆ.

Vern.—*Niargal,* LADAK; *Táksha,* SPITI.

References.—*Stewart, Pb. Pl. (omissions), 72; Gazetteer, N.-W. P., X., 308.*

Habitat.—A stemless herb, met with in the Western Himálaya between altitudes of 11,000 and 16,000 feet; found also in Sikkim at similar altitudes.

FODDER. 614 **Fodder.**—It is browsed by sheep and yaks in Ladak and Spiti (*Stewart*).

615 **OYSTER.**

The Oyster belongs to the genus **Ostrea,** a well known and widely diffused group of Molluscs, members of which occur in many parts of the seas around the Indian Peninsula. This genus belongs to the third order of the Lamelli-branch Molluscs, *Monomya,* so called because the valves of its shell are closed by a single large adductor muscle. Seventy different species of the genus **Ostrea** have been distinguished; but the most important form and the only one that need be dealt with here is the Edible Oyster—**Ostrea edulis.**

For a description of the Pearl Oyster (**Mellagrina margaritifera**), see **Pearls,** Vol. VI., Pt. I.

616 **Ostrea edulis.**

THE EDIBLE OYSTER,

HUITSE, *Fr.*; AUSTERN, *Germ.*; OSTRICHE, OSTRICHA, *Ital.*; OSTREA, *Lat.*

Vern.—*Sipi, kalu, kustura,* HIND.; *Alie,* TAM.; *Cavati,* SING.; *Tirim,* MAL.; *Badlan,* ARABIC.

References.—*Ainslie, Mat. Ind., I., 287; Forbes Watson, Indust. Surv. Ind., I., 368; Balfour, Cyclop. Ind., iii., 68; Encycl. Brit., xviii., 106.*

Habitat.—The edible oyster occurs in many parts of the seas around the Indian Peninsula, but the best known and most valued oyster-beds are found on the coast near Karachi, Bombay, and Madras.

FOOD. 617 **Food.**—The oyster is much valued as an article of food by the Europeans in India; but the trade in them is almost entirely local, although small quantities are now transported by rail from Karachi to various stations in Sind and the Panjáb, even as far as Simla, for the use of European residents in those places.